GW01337132

The BIRDS of AFRICA
Volume II

The BIRDS of AFRICA
Volume II

Edited by

EMIL K. URBAN

*Department of Biology, Augusta College,
Augusta, Georgia, USA*

C. HILARY FRY

*Department of Zoology, University of Aberdeen,
Aberdeen, UK*

STUART KEITH

*Department of Ornithology, American Museum of Natural History,
New York, New York, USA*

Colour Plates by Martin Woodcock

Line Drawings by Ian Willis
Acoustic References by Claude Chappuis

1986

ACADEMIC PRESS
Harcourt Brace Jovanovich
London · Orlando
San Diego · New York · Austin · Montreal · Sydney · Tokyo · Toronto

ACADEMIC PRESS INC. (LONDON) LTD.
24/28 Oval Road, London NW1 7DX

United States Edition published by
ACADEMIC PRESS, INC.
Orlando, Florida 32887

Copyright © 1986 by
ACADEMIC PRESS INC. (LONDON) LTD.

All rights reserved. No part of this book may be reproduced or transmitted in any form or by any means, electronic or mechanical, including photocopy, recording, or any information retrieval system without permission in writing from the publisher

Birds of Africa
   Vol. II
   1. Birds—Africa
   I. Urban, Emil K.   II. Fry, C. H.   III. Keith, Stuart
   598.296   QL692.A1

ISBN: 0-12-137302-9

Production services by Fisher Duncan Ltd, 10 Barley Mow Passage, London W4 4PH

Printed in Hong Kong by Imago Publishing Ltd

ACKNOWLEDGEMENTS

We are most grateful to our authors for their untiring efforts and co-operation in preparing the species accounts, for providing us with unpublished notes and records, and for their patience during the protracted preparation of this volume for publication. We thank them all for their forbearance with editorial demands during a succession of drafts.

We would like to acknowledge with thanks the contribution of referees, many of whom have gone to great lengths to ensure that the species accounts are as accurate and comprehensive as possible. Families have been refereed as follows: Phasianidae, R. K. Brooke, M. Ridley, the late B. W. H. Stronach and P. B. Taylor; Rallidae, P. B. Taylor; also A. Brosset, W. R. J. Dean, the late D. M. Skead and S. L. Olson; Gruidae, G. Archibald and P. Konrad; Heliornithidae, P. Ginn; Otididae, P. S. Jones, G. J. Morel, H. Schulz and W. R. Tarboton; Jacanidae W. R. Tarboton; Rostratulidae, M. de L. Brooke and D. J. Pearson; Dromadidae, C. J. Feare; Haematopodidae, J. Cooper; Recurvirostridae, D. J. Pearson and A. J. Tree; Burhinidae, J. M. Mendelsohn; Glareolidae (Egyptian Plover), J. R. Jehl, Jr (coursers), C. J. Vernon, (Glareolinae), C. Erard; Charadriidae (Charadriinae), G. L. Maclean, D. J. Pearson and A. J. Tree, (Vanellinae), C. J. Skead; Scolopacidae (Phalaropodinae), D. J. Pearson, (all others), A. J. Prater; Stercorariidae, J. C. Sinclair; Laridae, J. Cooper and P. J. Grant; Sternidae, P. R. Colston and C. J. Feare; Rynchopidae, M. J. Coe and D. J. Pearson; Alcidae, E. D. H. Johnson; Pteroclidae, D. H. Thomas; and Columbidae, A. Brosset, R. J. Dowsett, F. Dowsett-Lemaire, D. Goodwin and R. de Naurois. As well as writing species accounts or refereeing, G. Archibald, the late C. W. Benson, P. L. Britton, R. K. Brooke, A. Brosset, P. J. Grant, P. A. R. Hockey, P. Konrad, G. J. Morel, M.-Y. Morel, D. J. Pearson, A. J. Prater, P. B. Taylor, A. J. Tree and W. R. Tarboton have assisted in many additional ways and we are most thankful to them.

We are greatly indebted to the Trustees and staff of: American Museum of Natural History, Department of Ornithology; British Library of Wildlife Sounds; British Museum (Natural History), Sub-Department of Ornithology; Carnegie Museum of Natural History; Cornell Laboratory for Ornithology; Durban Museum; Field Museum of Natural History, Bird Division; Los Angeles County Museum, Bird Division; Musée Royal de l'Afrique Centrale, Tervuren; Museum of Comparative Zoology, Harvard; Museum Alexander Koenig, Bonn; National Museum of Zimbabwe, Bulawayo; National Museums of Kenya, Nairobi; Percy FitzPatrick Institute of African Ornithology; Philadelphia Academy of Natural Sciences; Smithsonian Institution, US National Museum of Natural History, Department of Ornithology; and Transvaal Museum, Bird Department, for repeatedly affording study facilities and for loans of numerous study skins. We especially thank P. J. K. Burton, P. A. Clancey, P. Colston, G. R. Cunningham-van Someren, the late B. G. Donnelly, J. Fitzpatrick, I. C. J. Galbraith, F. B. Gill, D. Goodwin, J. Gulledge, A. Harris, M. P. S. Irwin, H. D. Jackson, A. C. Kemp, R. Kettle, P. Lorber, M. Louette, J. Mendelsohn, J. P. Myers, S. L. Olson, K. C. Parkes, R. Paynter, D. Read, R. W. Schreiber, L. L. Short, D. W. Snow, W. R. Siegfried, D. Steadman, M. A. Traylor, M. Walters, G. E. Watson, and D. S. Wood.

Further, we are indebted to G. Archibald and K.-L. Schuchmann for organizing meetings respectively at Bharatpur, India (International Crane Workshop) and Bonn, W. Germany (Symposium on African Vertebrates), and to the organizers of the Vth and VIth Pan-African Ornithological Congresses at Lilongwe, Malaŵi, and Francistown, Botswana. We participated in all of these meetings, presenting papers stemming from study for *The Birds of Africa*, and were afforded facilities to discuss the enterprise.

In writing accounts of Palearctic birds in Africa we have drawn freely on the detailed information in Cramp and Simmons, *The Birds of the Western Palearctic*, and Glutz and Bauer, *Handbuch der Vögel Mitteleuropas*, and it is a pleasure to record our debt to the authors of these invaluable compilations. We are especially indebted to S. Cramp and M. Wilson for providing pre-publication information from Volume IV of the former work (now published).

We have drawn heavily upon the generosity of correspondents, who have readily responded to queries affecting all aspects of African ornithology. G. C. Backhurst, G. R. Cunningham-van Someren, G. Nikolaus, A. J. Prater, M. D. Rae, T. Salinger, R. W. Schreiber, K.-L. Schuchmann and G. E. Watson provided us with thousands of bird measurements. We have benefitted greatly from the kindness, in providing us with their books and papers in manuscript prior to publication, of W. R. J. Dean and M. A. Huntley (Angola), P. Colston and K. Curry-Lindahl (Mt Nimba), M. K. Rowan (Columbidae), J. M. Thiollay (Ivory Coast) and F. Dowsett-Lemaire (Columbidae). S. M. Goodman, R. W. Storer and Sherif Baha El Din have provided us with manuscripts, records, food data and measurements from Egypt, and R. W. Schreiber and K. L. Garrett gave us complete data on African specimens in the Los Angeles County Museum. J. D. R. Vernon has provided distributional summaries for Morocco, M. Smart for Tunisia, M. W. Pienkowski for northwest Africa, P. W. P. Browne and B. Lamarche for Mauritania, G. Nikolaus and M. D. Rae for Sudan, and P. Becker for Namibia. M. de L. Brooke, J. Hinshaw, P. C. Lack, H. Lapman, E. Little (and colleagues at Augusta College's Reese Library and Aberdeen University's Queen Mother Library, Inter-Library Loan Departments), M. W. Pienkowski, A. J. Prater, C. Sibley, G. E. Watson, M. Wilson and staff of Oxford University Edward Grey Institute have been most helpful in obtaining library materials. We are no less indebted for help given in many ways by J. S. Ash, D. R. Aspinwall, P. C. Beaubrun, D. Baird, P. Becker, P. Bergier,

R. K. Brooke, A. Brosset, L. Bortoli, N. Bryant, B. Campbell, J. Carver, M. A. Casado, R. Chancellor, C. Chappuis, D. R. Collins, H. Q. P. Crick, A. A. Crowe, J. E. W. Dixon, R. J. Douthwaite, R. J. Dowsett, F. Dowsett-Lemaire, E. K. Dunn, J. H. Elgood, C. C. H. Elliott, H. F. I. Elliott, C. Erard, H. Fanshawe, G. D. Field, D. Fisher, R. W. Furness, J. P. Gee, P. Géroudet, P. J. Ginn, H. H. Gray, L. Grimes, J. Gulledge, S. I. Guttman, D. B. Hanmer, P. Hogg, D. J. Holmes, J. H. Hosken, S. Howe, R. Hudson, J. L. Ingold, M. P. S. Irwin, R. A. C. Jensen, M. C. Jennings, R. Johns, E. D. H. Johnson, M. Kelsey, P. J. Knight, B. Lamarche, C. Levassor, A. Lewis, Mrs. W. Liell-Cock, P. Lorber, A. Morris, R. de Naurois, J. H. van Niekerk, B. Ochando, R. Osborne, J. R. Peek, E. H. Penry, R. and J. Plunkett, D. Pomeroy, A. Prigogine, R. Ralph, J. F. Reynolds, M. Ridley, D. V. Rockingham-Gill, M. K. rowan, A. Sala, D. Schmidl, J. T. R. Sharrock, W. R. Siegfried, C. J. Skead, G. S. Smith, t. Stevenson, P. Steyn, R. Stjernstedt, the late B. W. H. Stronach, C. D. Taylor, K. Thangavelu, M. Thévenot, D. H. Thomas, P. Thomsen, R. Thomsen, D. A. Turner, J. P. Vande weghe, L. Walkinshaw, A. J. S. Weaving, G. R. Welch, H. J. Welch and J. B. Wood.

We should like to thank the following for permission to redraw black and white line illustrations from their own published or personal references:

G. Archibald (*Anthropoides paradisea*, *A. virgo*, *Balearica regulorum*, *B. pavonina*, *Bugeranus carunculatus* and *Grus grus*); C. T. Astley Maberly in Bokmakierie (1967) (*Eupodotis ruficrista*), in Honeyguide (1979) (*Neotis denhami*); G. W. Begg (*Vanellus albiceps*); G. R. Cunningham van Someren (*Actophilornis africana*); H. Deetjen in Journal für Ornithologie (1969) (*Fulica cristata* and *F. atra*); M. P. L. Fogden in Ibis (1964) (*Larus leucopthalmus* and *L. cirrocephalus*); P. G. H. Frost in the Bulletin of the British Ornithologists' Club (1975) (downy young of various Phasianidae); C. H. Fry (*Podica senegalensis*); D. Goodwin in 'Pigeons of the World', Cornell University Press (1983) (*Columba livia*), in Avicultural Magazine (1956) (*Columba livia*, *C. guinea*, *Streptopelia roseogrisea* and *S. semitorquata*); C. J. O. Harrison in Sandgrouse (1983) (*Sterna albifrons* and *S. saundersi*); P. A. R. Hockey (*Haematopus bachmani* and *Charadrius marginatus*); J. B. D. Hopcraft (*Actophilornis africana*); J. Horsfall (*Chlamydotis undulata*); T. R. Howell in University of California Publications in Zoology (1979) (*Pluvianus aegyptius*); J. H. Van Niekerk in Bokmakierie (1983) (*Francolinus swainsonii*); G. L. Maclean (*Vanellus melanopterus*, *V. coronatus*, *Pterocles burchelli*, *P. namaqua*, *Burhinus oedicnemus*, *B. senegalensis*, *B. vermiculatus*, *B. capensis*, *Cursorius africanus*) and in The Living Bird (1968) (*Pterocles namaqua* and *P. bicinctus*); K. A. Muller in Emu (1975) (*Rostratula benghalensis*); D. Pearson (*Gallinago gallinago*, *G. nigripennis* and *G. stenura*); G. P. Schaller (*Ardeotis kori*); W. Tarboton (*Actophilornis africana*); J. Terres (*Rynchops flavirostris*); N. Tinbergen in Behaviour (1959) (*Larus argentatus*); N. Tinbergen & G. Broekhuysen in Ostrich (1925) (*Larus hartlaubii*); J. Trollope in the Avicultural Magazine (1970) (*Turnix sylvatica*); A. J. Weaving (*Actophilornis africana*); B. Wood in the Bulletin of the British Ornithologists' Club (1975) (*Streptopelia hypopyrrha*).

We are especially pleased to have worked with Martin Woodcock who prepared the colour and black and white plates, Ian Willis who drew the black and white illustrations, Claude Chappuis who prepared the list of acoustic references, and Lois Urban who prepared the indexes.

We wish to thank the staff of Academic Press (London) Ltd and Fisher Duncan Ltd, particularly Jane Duncan, Moira Fisher, Jennie Morley and Andrew Richford for helping us through an increasingly arduous task.

With Volume III already in press, each Editor has accumulated vast files of correspondence and papers. We hope that we have acknowledged all of our numerous sources of help and apologize to—and ask the forbearance of—anyone whose name has been inadvertently omitted.

Finally, we would like to thank our three wives—Kathie Fry, Sallyann Keith and Lois Urban—for their patience, tolerance and understanding during the long hours we have spent preparing this volume.

July 1985

Emil K. Urban
C. Hilary Fry
Stuart Keith

CONTENTS

ACKNOWLEDGEMENTS v

LIST OF PLATES ix

INTRODUCTION

    Illustrations xi
    Superspecies xi
    Nomenclature and Systematics xii
    Range and Status xiii
    Maps xiii
    Description and Field Characters xv
    Voice xv
    General Habits and Breeding Habits xv
    Reference System xv
    References xvi

ORDER GALLIFORMES
    Phasianidae, guineafowl, Congo Peacock, quail, partridges and francolins (T. M. Crowe, S. Keith and the late L. H. Brown) 1

ORDER GRUIFORMES
    Turnicidae, button-quail (P. B. Taylor) 76
    Rallidae, rails, flufftails, crakes, gallinules, moorhens and coots (S. Keith; *Crex egregia* co-authored by P. B. Taylor) 84
    Gruidae, cranes (E. K. Urban) 131
    Heliornithidae, finfoots (C. H. Fry) 145
    Otididae, bustards (N. J. Collar, P. D. Goriup and P. E. Osborne) 148

ORDER CHARADRIIFORMES
    Jacanidae, jacanas (C. H. Fry) 180
    Rostratulidae, painted-snipe (E. K. Urban) 185
    Dromadidae, Crab Plover (the late C. W. Benson and F. M. Benson) .. 188
    Haematopodidae, oystercatchers (P. A. R. Hockey) 190
    Recurvirostridae, stilts and avocets (E. K. Urban) 193
    Burhinidae, thick-knees, stone curlews and dikkops (G. L. Maclean and E. K. Urban) .. 198
    Glareolidae 206
        Cursoriinae 206
            Pluvianus, Egyptian Plover (T. R. Howell) 206
            Cursorius, coursers (G. L. Maclean and E. K. Urban) 209
        Glareolinae, pratincoles (A. Brosset) 218
    Charadriidae 224
        Charadriinae, plovers (E. K. Urban) 225
        Vanellinae, lapwings (S. Keith) 250
    Scolopacidae 283
        Calidridinae, Gallinagininae, Scolopacinae, Tringinae, Arenariinae, sandpipers and snipe (D. J. Pearson) 283
        Phalaropodinae, phalaropes (E. K. Urban) 328
    Stercorariidae, skuas (B. L. Furness and E. K. Urban) 332
    Laridae, gulls (P. L. Britton) 340
    Sternidae, terns (P. L. Britton) 374
    Rynchopidae, skimmers (C. H. Fry) 412
    Alcidae, auks (E. K. Urban) 415

ORDER PTEROCLIFORMES
    Pteroclidae, sandgrouse (G. L. Maclean and C. H. Fry) 422

ORDER COLUMBIFORMES
Columbidae, pigeons and doves (G. J. Morel, M.-Y. Morel and C. H. Fry) 442

BIBLIOGRAPHY
1. General and Regional References 498
2. References for Each Family 501
3. Acoustic References 534
    Section A: Discs and Cassettes 534
    Section B: Most Important Discs and Cassettes by Region 536
    Section C: Institutions with Sound Libraries 536
    Section D: Individual Recordists 536

ERRATA, VOLUME I 538

INDEXES
1. Scientific Names 539
2. English Names 547
3. French Names 550

LIST OF PLATES

Plate		Facing Page
1	Francolins, partridges and quail	32
2	Francolins	33
3	Francolins	48
4	Francolins	49
5	Congo Peacock and guineafowl	96
6	Cranes	97
7	Button-quail, crakes and rails	112
8	Rails and flufftails	113
9	Rails, gallinules, coots and finfoots	160
10	Bustards	161
11	Bustards	176
12	Jacanas, painted-snipes, thick-knees, woodcock and snipe	177
13	Oystercatchers, avocets, stilts, Crab Plover, curlews and godwits	240
14	Lapwings	241
15	Plovers	256
16	Plovers, dotterels, Ruff and turnstones	257
17	Sandpipers	304
18	Sandpipers and phalaropes	305
19	Egyptian Plover, coursers and pratincoles	320
20	Shorebirds *in flight*	321
21	Shorebirds *in flight*	352
22	Skuas, gulls and auks	353
23	Gulls	368
24	Gulls	369
25	Terns and skimmers	400
26	Terns	401
27	Gulls *in flight*	416
28	Terns *in flight*	417
29	Sandgrouse	448
30	Pigeons	449
31	Doves	464
32	Doves and green pigeons	465

INTRODUCTION

The Birds of Africa was the brainchild of Leslie Brown, and his death necessitated a wholly new approach to authorship. It has become a multi-author work, under the joint editorship of one of the original authors, Emil K. Urban, co-equally with two members of the former Advisory Editorial Panel, C. Hilary Fry and Stuart Keith. Species accounts have been written by the best authority available for each family, and refereed by equally competent authorities, many, like the authors themselves, presently living and working in Africa.

The numerous published reviews of Volume I have been in general laudatory. The Editors have benefitted from constructively critical comments of reviewers and correspondents; they have also been their own severest critics, and have sought to introduce such improvements as could be made without materially altering the format established for the series with the first volume. With a large team of authors, the Editors have been obliged to ensure uniformity of treatment, which requirement has inevitably resulted in a long and detailed set of instructions to authors. During the course of preparation of this volume, the instructions themselves have changed and evolved, so that some inconsistencies may have slipped through uncorrected. The main departures from the format of Volume I are in the layout of legends interpreting the colour plates, and in adoption of the superspecies principle; these, and several less obvious innovations, are detailed below. Innovations notwithstanding, the treatment of species is in much the same format as in Volume I (in which see pp. 24–26: *Content and Layout of the Text*).

The Birds of Africa was originally conceived as a four volume work. However, to accomodate the mass of information on African birds that is continually coming to light, both published and unpublished, the work has been expanded to six volumes, three non-passerine and three passerine. The present volume covers galliforms to columbiforms. The cut-off date for information included in this volume is November 1983, although many significant 1984 and 1985 references have been included.

Illustrations

All of the Plates in this volume are by Martin Woodcock. As before, he has painted from skins, but he has also been able to familiarize himself with numerous birds in the field, by means of a three-month sketching visit to East and southern Africa. A new departure, intended to facilitate the identification of individual bird illustrations in each Plate, is that their names appear on full-sized outlines on the facing page. Black-and-white illustrations in the text, mainly of behaviour, were drawn by Ian Willis from photographs and other original sources.

Superspecies

The concept of the superspecies is invaluable in zoology to both museum systematists and field naturalists. The term was coined by Ernst Mayr (1931) as a translation of the *Artenkreis* of Rensch (1929). Mayr later (1963) defined it as "a monophyletic group of entirely or essentially allopatric species that are too distinct to be included in a single species". The definition was further refined by Amadon (1966) as "a group of entirely or essentially allopatric taxa that were once races of a single species but which now have achieved species status".

The first works to apply the superspecies concept to a whole continent were Mayr and Short (1970) for North America and Hall and Moreau (1970) for Africa. The former had a more rigid definition of superspecies as "groups of taxa which have barely crossed the threshold of species status" (p. 100). The concept was more loosely used by Hall and Moreau, and some of their superspecies comprise species that clearly crossed this threshold some time ago and now exhibit very considerable differences.

A brief explanation of the superspecies concept and its application in Africa and in this work may be fitting. In this huge continent numerous kinds of birds have vast ranges, and they vary geographically in plumage, size or behaviour. Some variation is gradual or clinal, and is a balance between differing selection pressures at each geographical extreme (producing differentiation) and unimpeded gene flow in between (promoting uniformity and countering differentiation). Commonly, variation is abrupt, particularly when different populations of a species are allopatric (isolated from each other by considerable distances). In many other instances isolates have expanded and met again, producing a spatially coherent species with abrupt changes at quite well marked boundaries between contiguous (parapatric) subspecies.

If the ranges of two essentially parapatric, closely related populations are beginning to overlap and they remain reproductively discrete despite frequent encounters, systematists must treat them as separate species, for that is how they treat each other. But well differentiated populations may hybridize where they meet, although the hybrid zone seems to be narrow, stable and contained (Prigogine 1980). The question then arises, should they be treated as the same or different species? The difficulty is even more acute when populations which are moderately well differentiated, yet sufficiently alike to be certainly of immediate common descent, are either allopatric or strictly parapatric. For practical purposes, such as the treatment which they will be accorded in this work, a decision has to be made as to whether they have differentiated to specific level; and the decision has to be based on an untestable prediction — how the two populations would behave reproductively were they ever to meet naturally.

To that extent the recognition of species is arbitrary, and if the choice were solely between a subspecies or a full species the consequences for a work such as this would be profound. Had we decided, for instance, that Black and Grey Crowned Cranes were conspecific, *Balearica pavonina sensu lato* would have received little more than half the text allocation given to what we in fact regard as two allopatric species, *B. pavonina sensu stricto* and *B. regulorum*.

Such geographical replacement species, or allospecies, of which there may be two or more, together constitute a *superspecies*. The use of this taxon not only emphasizes close affinity between congeneric species, but also reduces the practical difficulty stemming from limited taxonomic choice. Ideally, every sort of bird would receive textual treatment in direct proportion with its degree of distinction, brief if it is an ill-defined subspecies, and lengthy if it is a distinctive monotypic species without close relatives, like the Egyptian Plover *Pluvianus aegyptius*. For reasons of space, that ideal cannot be achieved here; but recognizing superspecies has permitted some textual abbreviation as a result of cross-reference between allospecies. It also allows the reader to extrapolate data about a well known bird to its poorly known allospecies with some confidence.

Most birds considered to constitute a superspecies have their breeding ranges shown in the generic diagnoses, using one map of Africa per superspecies, and their allospecies are also listed in the text, immediately after the species heading. Since their ranges do not overlap they are shown on the superspecies map simply in outline. In a few cases two birds may have the main attributes of a superspecies but their ranges are beginning to overlap (without hybridization); with them the area of overlap (e.g. the doves *Streptopelia vinacea* and *S. capicola*) or of possible overlap (e.g. the bustards *Ardeotis arabs* and *A. kori*) are made plain. Where an African bird has a European or Asiatic allospecies which is a non-breeding visitor to Africa, the latter is identified as such in the text but its breeding range is not shown in the superspecies map. Allospecies never recorded in Africa are ignored (e.g. the Asiatic Little Pratincole *Glareola lactea*, an allospecies of the African Grey Pratincole *G. cinerea*).

Nomenclature and Systematics

Authors have been required to assess critically the generic affiliations of a species, superspecific affiliations, specific boundaries, and the validity of described races. This has led to a few systematic innovations, although we have favoured conventional treatments and as a rule taxonomy follows that of Snow (1978) for species and White (1965) for subspecies.

Family sequence is that of Voous (1973). Species sequence was decided by the Editors and authors, and is intended to express evolutionary progression from primitive to derived species wherever possible.

English Names

Many African birds are known by different English names in different parts of the continent. The Editors, in selecting a single preferred name for each species, have had to make some difficult decisions. (No more than a single alternative name is provided for any bird, and that only when it is widely used.) For a full account of the principles used in selecting names, see S. Keith and L. Short, 'The English names of African birds' (Proc. VI Pan-Afr. Orn. Congr., in press). The main points are as follows:

1. We take a world view of English names; it is our intention that every African bird shall have a name that clearly distinguishes it from every other bird in the world. In cases of synonymy we either change the name of our bird or add 'African' to it. For instance, the name Purple Gallinule has been given to both *Porphyrio martinica* of the New World and *P. porphyrio* of the Old World; for the latter species we decided to use another available name, Purple Swamphen. 'White-winged Dove' is used for both *Zenaida asiatica* of the New World and *Streptopelia reichenowi* of Africa; we distinguish our bird by calling it African White-winged Dove. In another move toward global conformity, all members of the genus *Vanellus* are here called Lapwings, following Johnsgard (1981). For African species which go by different names in other parts of the world, we use the name most appropriate to its world range; if the names seem equally valid, we use the African name; e.g. for *Fulica cristata* we prefer Red-knobbed Coot (African) to Crested Coot (European). In the case of Palearctic species we use the European name, e.g. Kentish Plover for *Charadrius alexandrinus*, rather than Snowy Plover (North America) or Red-capped Dotterel (Australia).

2. Certain species which are locally the sole representatives of their genus or family—particularly in Britain—have arrogated the genus or family name to themselves, without a modifier. To these we give modifiers, e.g. Coot ('The' Coot) becomes Eurasian Coot, 'The' Knot becomes Red Knot, 'The' Turnstone becomes Ruddy Turnstone.

3. We tend to be conservative and to favour old, traditional names even if not completely accurate or appropriate. Usage has an important role, and we coin new names only for the most difficult cases, and then only after searching for already-published alternatives.

4. For some purely African species we replace restrictive regional names with descriptive ones appropriate for the species as a whole. 'Uganda Spotted Woodpecker' is inappropriate for a bird that occurs from Nigeria to Kenya, and we adopt the name coined for it by Short (1982), Speckle-breasted Woodpecker. But we favour geographical names when they describe all or nearly all of a bird's range, as Djibouti Francolin for *Francolinus ochropectus*.

5. Changes in taxonomy sometimes dictate changes in English names. Upon systematic reappraisal of francolins, *Francolinus psilolaemus* emerged as a full species; its component races had previously been assigned to other species and consequently no English name was

available for it. We name it Moorland Francolin, after its habitat. The Green-backed Woodpecker (*Campethera permista*) and Little Spotted Woodpecker (*C. cailliautii*) are now treated conspecifically. Both English names are inappropriate for the enlarged species since western birds (*permista*) have unspotted green backs while eastern ones (*cailliautii*) have so many spots on the back that the overall impression is not green. After discussion with L. Short, we have adopted Green-backed Woodpecker as the lesser of two evils.

Range and Status

In this section each species is either identified as endemic to Africa or, if not endemic, its world distribution is outlined at the start. Range and status are described generally in greater detail than was practicable in Volume I. It owes to the spate of recent books and lengthy papers, published mostly since 1980, on the status of birds in Algeria, Angola, Cameroon, Chad, Djibouti, Ethiopia, The Gambia, Kenya with Uganda and Tanzania, Libya, Mali, Natal, Nigeria, the Serengeti, Somalia, Togo, Tunisia, Zanzibar and Pemba, as well as the 'South African Avifauna' and 'Southern Birds' series. In addition, data have been incorporated from atlases and checklists in preparation of the birds of Botswana, Egypt, Kenya and Uganda, Mauritania, Morocco, Niger, Senegambia, and Sudan.

The Republic of Upper Volta has changed its name to Burkina Faso; this and some other political name changes have been too recent for us always to incorporate. The name Senegambia is used interchangeably with Senegal and Gambia.

Maps

As previously, range maps are based on the point-plotted distributions in *An Atlas of Speciation in African Non-Passerine Birds* (Snow 1978). We have, however, mapped more conservatively than in Volume I, shading regions containing many known occurrences of a species and leaving unshaded apparent gaps greater than about 500 km (or, where vegetational or topographical considerations dictate even 'tighter' boundaries, about 200 km). Map and text are complementary and in all cases should be consulted together.

Different shadings are employed to indicate (a) dense breeding range, (b) sparse breeding range, (c) occurrence on migration only, (d) dense nonbreeding ('wintering') range and (e) sparse nonbreeding range. 'Dense' and 'sparse' are, of course, relative and rather arbitrary: there are obvious difficulties in the delineation of ranges of widespread but rare species, or those having only a few far-flung breeding records. Where only the sparse density of occurrence is shown, it should be assumed that there is insufficient information to enable more detailed mapping, rather than that a species is uniformly rare. Where complex or unusual distributions make these conventions difficult to apply, specific keys are provided for that map. Arrows, in Volume I indicating supposed migration directions and routes, have been omitted except in the few instances where there is good observational evidence for migration, when *box arrows* are employed. By contrast, *simple arrows* are used sparingly to draw attention to isolated ranges or occurrences (Fig. 1).

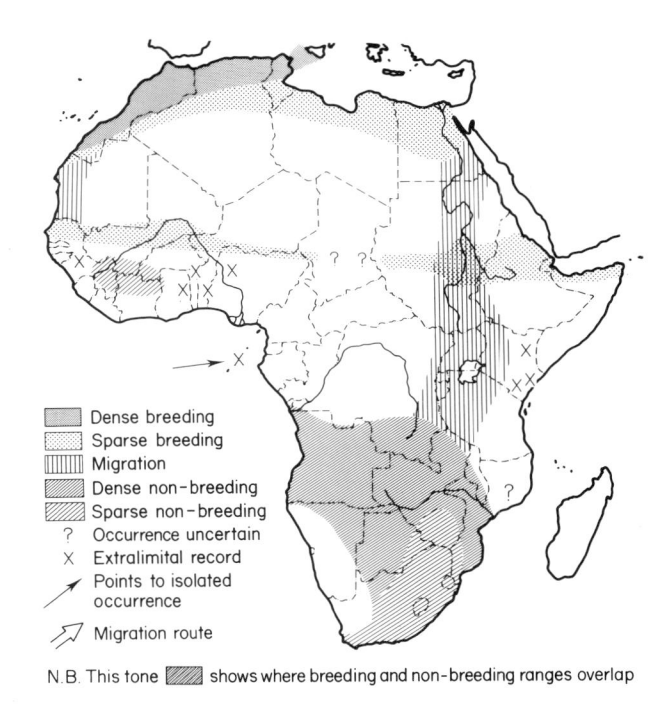

Fig. 1. Shading, symbols and arrows used on maps.

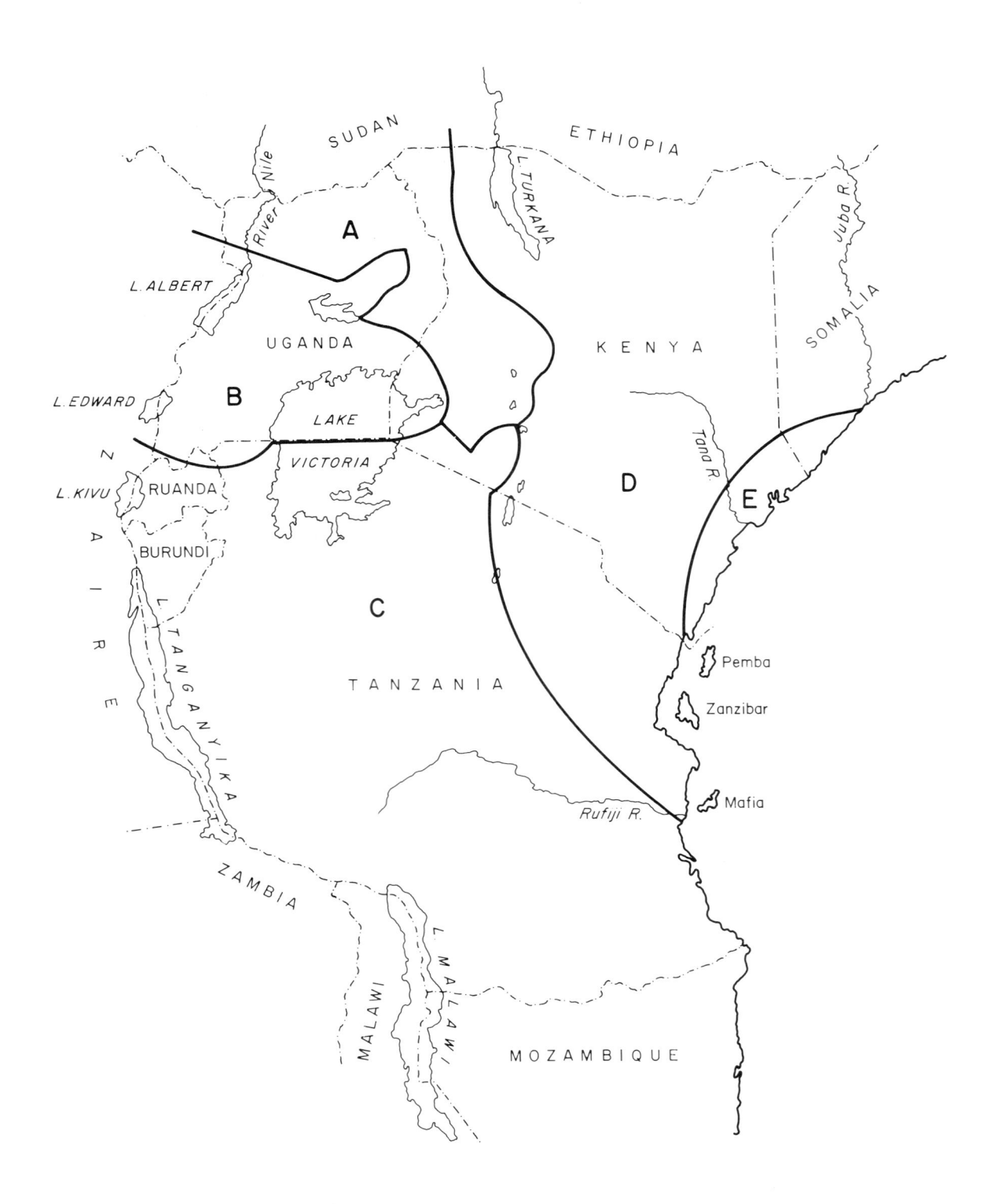

Fig. 2. Climatic zones of East Africa (Uganda, Kenya, Tanzania), referred to under specific Laying Dates (from Brown and Britton 1980 after J. F. Griffiths, 'Climatic zones of East Africa', *E. Afr. Agric. J.* 1958, 179–185).

Description and Field Characters

The 'thumbnail sketch' of appearance which in Volume I preceded the detailed description has been eliminated. On the other hand, more emphasis has been placed on Field Characters in the present volume.

In some instances where two or more species comprising a superspecies are very alike in plumage, descriptions of the second (and additional) allospecies are abbreviated and the reader is referred to the first, principal, allospecies.

Voice

The list of acoustic references which follows the bibliography at the end of the book (pp. 534–537) has been reorganized and greatly expanded; this has been the work of Claude Chappuis. Besides being a general reference list, its purpose is to provide researchers with sources for vocalizations in which they may be interested, and for the reader who would simply like to listen to African bird voices there is ample material in the section on discs and cassettes.

The material has been divided into four sections. Section A lists discs and cassettes, numbered 2–76, in roughly chronological sequence of publication; the numbers are those of Chappuis (1980). For each volume, only those containing vocalizations of species in that volume are listed, e.g. a record containing voices of just owls and nightjars will appear only in the reference list of Volume III, which covers those species. Section B is a short list of the most important discs and cassettes by region.

Section C lists those institutions which have large sound libraries; each of the six is given an identifying letter (A, B, C, F, N or S). Section D lists individual sound recordists who have recordings of particular species as referred to in the text.

At the beginning of the Voice section in each species account, if the bird has been tape-recorded we list sources for recordings as in Volume I; they are given a number or letter corresponding to that in the Reference List. The sequence of sources cited follows that in the Reference List; e.g. under Little Button-Quail, in the entry 'Tape-recorded (60, 73, C, F, 264)', 60 and 73 are a disc and a cassette, C and F are institutions, and 264 is an individual recordist. For certain well-known species like the Common Moorhen *Gallinula chloropus*, numerous recordings are available, of varying quality. Rather than trying to provide an exhaustive list of all recordings made of every species, we think it will be more useful to the reader to provide a short list of the best recordings, and as a general rule we have restricted the number cited in the Voice section to six. Claude Chappuis has undertaken the task of making these selections, taking into account not only quality of sound but variety of representative vocalizations.

General Habits and Breeding Habits

The opening phrase, categorizing how well known each bird is, has been discontinued; we leave it to readers to judge how much is known from the length and detail of these sections. Remarks like 'unknown' and 'no information' have been used more sparingly than hitherto. Where there is no information on a particular aspect of a bird's life history, the section concerned has usually been omitted. We would remind those with enthusiasm for discovery that there is barely a single species of African bird which is yet known adequately in all respects (see Volume I pp. 17–22: *Some Possibilities for Research*).

Readers are also reminded that Palearctic species which do not enter sub-saharan Africa are treated in less detail than Afrotropical ones, and that with such Palearctic birds we have concentrated on their habits in N Africa rather than in Europe. As before, for Palearctic migrants which winter but do not breed in Africa we have concentrated on their general habits in Africa and omitted their breeding habits.

Usually general habits are dealt with in 3–4 paragraphs, respectively treating habitat, nonbreeding (everyday) behaviour, foraging and migration.

For well known species, breeding habits are treated as follows. The first paragraph deals with mating system, dispersion, and courtship behaviour. Then—a departure from the format of Volume I—headed paragraphs give data about NEST, EGGS, LAYING DATES, INCUBATION, DEVELOPMENT AND CARE OF YOUNG and BREEDING SUCCESS/SURVIVAL. For less well known species the paragraphs are variously condensed. In Laying Dates, 'Region A' (to 'E') refers to East African regions defined and mapped by Brown and Britton (1980)—see Fig. 2.

Reference System

Text citations are detailed in two lists: (1) general and regional works, and (2) references for each family. As in Volume I, titles in (1) are not repeated in (2). Thus, the reader who fails to find a text citation in the appropriate family list should search the General and Regional References list.

References

Amadon, D. (1966). The superspecies concept. *Syst. Zool.* **15**, 245–249.

Brown, L. H. and Britton, P. L. (1980). 'The Breeding Seasons of East African Birds'. East African Natural History Society, Nairobi.

Chappuis, C. (1980). List of sound-recorded Ethiopian birds. *Malimbus* **2**, 1–15, 82–98.

Hall, B. P. and Moreau, R. E. (1970). 'An Atlas of Speciation in African Passerine Birds'. British Museum (Natural History), London.

Johnsgard, P. A. (1981). 'The Plovers, Sandpipers and Snipes of the World'. University of Nebraska Press, Lincoln, Nebraska.

Mayr, E. (1931). Birds collected during the Whitney South Sea Expedition. XII. Notes on *Halcyon chloris* and some of its subspecies. *Amer. Mus. Novit.* **469**, 1–10.

Mayr, E. (1963). 'Animal Species and Evolution'. Belknap Press (Harvard University), Cambridge, Mass.

Mayr, E. and Short, L. L. (1970). Species taxa of North American birds. A contribution to comparative systematics. *Publ. Nuttall Orn. Club* No. 9; Cambridge, Mass.

Prigogine, A. (1980). Étude de quelques contacts secondaires au Zaïre oriental. *Gerfaut* **70**, 305–384.

Rensch, B. (1929). 'Das Prinzip Geographischer Rassenkreise und das Problem der Artbildung'. Borntraeger Verlag, Berlin.

Short, L. L. (1982). 'Woodpeckers of the World'. Delaware Museum of Natural History, Greenville, Delaware.

Snow, D. W. (ed) (1978). 'An Atlas of Speciation in African Non-passerine Birds'. British Museum (Natural History), London.

Voous, K. H. (1973). List of recent Holarctic bird species. Non-passerines. *Ibis* **115**, 612–638.

White, C. M. N. (1965). 'A Revised Check List of African Non-passerine Birds'. Government Printer, Lusaka.

Order GALLIFORMES

A large cosmopolitan order of small to very large, mainly terrestrial birds. Earliest fossils date from mid-Eocene, i.e. c. 45 million years ago. The only taxonomic character which unambiguously indicates that the order is monophyletic is presence of a lateral foramen delimited by fused manubrial spines of sternum. There are 2 superfamilies: Megapodioidea (Australasian megapodes) and Phasianoidea (South American Cracidae, and the bulk of the world's galliforms, the Phasianidae). Nearest relatives appear to be the Anseriformes, perhaps via the screamers (Anhimidae) (Prager and Wilson 1981, Cracraft 1981, Dzerzhinsky 1982, C. G. Sibley, pers. comm.); but see Olson and Feduccia (1980) for another view. Only one family in Africa, the Phasianidae. This family may be divided into 6 subfamilies (T. M. Crowe et al. unpub., Mainardi 1963, Stock and Bunch 1982): Numidinae (guineafowl), Pavoninae (peafowl, Argus and peacock pheasants, Congo Peacock), Gallinae (junglefowl, Old World quail, francolins and partridges), Tetraoninae (grouse), Meleagridinae (turkeys), and Phasianinae (true pheasants and New World quail). Of these, the Numidinae, Pavoninae and Gallinae occur in Africa.

Family PHASIANIDAE: guineafowl, Congo Peacock, quail, partridges and francolins

A large and diverse family, including the smallest (quail), largest (turkeys), and most spectacular (peafowl) galliforms. 4 species, Junglefowl (*Gallus gallus*), Turkey (*Meleagris gallopavo*), Helmeted Guineafowl (*Numida meleagris*) and Common Quail (*Coturnix coturnix*) are domesticated animals. Many phasianids are popular gamebirds and/or are prized by aviculturalists. Represented in Africa by 10 genera.

Phasianids inhabit every African vegetation type from desert to dense forest, and from sea level to the upper limits of vegetation on high mountains. All are terrestrial, with rather short, stout, sometimes strongly spurred, tarsi. ♂♂ average larger than ♀♀. Heads usually small; bills short and robust, well suited for digging; necks short; bodies stout, rounded, with very large breast muscles; wings short, rounded, with little emargination of primaries; tails short, but long in Congo Peacock and Vulturine Guineafowl. Sedentary except for quail, which are migratory or at least nomadic. Prefer to escape by running; fly only reluctantly and when pressed, or to reach elevated roosts; flight direct, fast, with rapid wing-beats, usually low and not long sustained. Feed on a variety of animal and vegetable matter found on or immediately above or below ground level (although Crested and Vulturine Guineafowl *Guttera* and *Acryllium* spp. have been observed eating fruit in trees). All have well developed crops, powerful gizzards and extensive caeca. Syrinx relatively simple; voices loud, often raucous, but calls not elaborate. Most lay large clutches on the ground, and all produce precocial young, capable of limited flight within 2 weeks of hatching.

Subfamily NUMIDINAE: guineafowl

A well defined endemic African subfamily comprising 4 genera: *Agelastes* (2 spp.), *Guttera* (2), *Acryllium* (1) and *Numida* (1). Morphological (Cracraft 1981, T. M. Crowe et al., in prep), immunological (Mainardi 1963) and karyological (Stock and Bunch 1982) evidence suggests that guineafowl link Phasianidae and Cracidae. Large galliforms with bare heads and necks, often brightly coloured; most with feathers or a casque on crown. Heads relatively small in comparison with body. Bills short, stout, upper mandible arched. Necks relatively long, bodies stout, thickset. Wings short, rounded. Tails short (long in *Acryllium*). Tarsi short, stout, unspurred in most species, well adapted for running, walking and scratching. 3 forward and 1 strong hind toe. Almost exclusively ground foragers, using feet and bill to excavate food items. Plumage black, spotted and/or vermiculated with white. No obvious sexual dimorphism, although ♂♂ average larger than ♀♀, and ♂♂ of at least 1 species (Helmeted Guineafowl *Numida meleagris*) possess a cloacal protuberance in the breeding season. Voices raucous, often high-pitched; more musical and whistling in *Agelastes* spp. All appear sedentary. Inhabit nearly every type of African vegetation from dense tropical forest to subdesert steppe. The Helmeted Guineafowl has been widely domesticated and feral populations have become established in some areas, e.g. Cuba and W Cape Province (South Africa). Generally gregarious when not breeding; flocks fragment into monogamous pairs at onset of breeding season.

Genus *Agelastes* Bonaparte

Small guineafowl inhabiting dense primary forest. Head and neck bare, skin dull to bright pink; no tufts or plumes on crown. Body plumage black, or black with white collar, with varying amount of white vermiculation. 2 monotypic spp., both very little known. ♂♂ and some ♀♀ have 1 or 2 tarsal spurs. Spurs are bony outgrowths from tarsometatarsus itself (as in *Gallus*), and not from the hypotarsus as in most phasianids. Tarsal scales imbricated and in rows, as in francolins, and unlike other guineafowl genera. The most primitive members of the subfamily, nearest to an ancestral francolin-like stock (Ghigi 1936, Crowe 1978a), or perhaps from the same stock which gave rise to the junglefowl.

Plate 5
(Opp. p. 96)

Agelastes meleagrides Bonaparte. White-breasted Guineafowl. Pintade à poitrine blanche.

Agelastes meleagrides 'Temm.' Bonaparte, 1849 (1850). Proc. Zool. Soc. London, p. 145; no locality given = Ghana.

Forms a superspecies with *A. niger*.

Range and Status. Resident in primary forests of W Africa from Liberia to Ghana. Formerly occurred Sierra Leone and probably S Guinea. Not seen Ghana since 1963, may be extinct there. Very rare Liberia; formerly occurred Mt Nimba but rapidly became extinct when area opened up for iron ore mining. Status in Ivory Coast unclear; still some numbers at Tai National Park, but generally uncommon to rare. Severely threatened by hunting pressure and habitat destruction and will probably disappear except from a few protected areas. Possibly one of the most endangered birds in Africa.

Description. ADULT ♂: head and upper neck naked and red, pinkish on lower neck with scanty whitish filaments. A broad white collar on hindneck and breast. Rest of body plumage black, finely vermiculated white. Primaries dark brown, edged grey on outer webs; secondaries dull brown, vermiculated white on outer webs. Bill greenish brown; eye brown; legs greyish brown or greyish black; 1–2 spurs 5–8·5 mm. Sexes alike, ♂♂ average larger. SIZE: (7 ♂♂, 4 ♀♀) wing, ♂ 204–214, ♀ 197–205; tail, ♂ 148–157, ♀ 135–142; tarsus, ♂ 79–89, ♀ 76–82.
IMMATURE: head and upper neck covered with blackish downy feathers; no white collar, hindneck plain blackish brown. Belly pure white; rest of body plumage like adult, although some feathers tipped or tinged rufous; spurless.
DOWNY YOUNG: undescribed.

Field Characters. Larger than any forest francolin, but much smaller than other guineafowl. Bare bright reddish head and conspicuous white collar and breast contrasting sharply with rest of dark body plumage distinguish it from other forest guineafowl (*Guttera*) which have bright blue spotted plumage and crested heads with naked skin slaty grey.

Voice. No known recording. Contact call from feeding flock, a deep 'kok-kok'; 'cheeping' call given as flock reassembles on departure of predator. Also has a loud, ringing, melodious call (Bechinger 1964).

General Habits. Inhabits primary rain forest with thin undergrowth. Unlike Crested Guineafowl *Guttera pucherani*, unable to adapt to denser undergrowth of secondary forest; very rarely leaves protective forest cover.

Sometimes occurs in pairs or small groups, but typically occupies large territories in which groups of 15–20 birds constantly move about in search of food. Forages on ground, scratching with feet like domestic chicken (Pfeffer 1961). Ground below fruiting tree is favoured feeding place. When bird finds food, others move in quickly and try to displace it; much jostling and squabbling takes place, but no serious fighting (Bechinger 1964). Group scatters on approach of predator, reassembles using 'cheep' call to re-establish contact. Very shy and hard to approach, but easily trapped; also, hunters lure birds by imitating the call, then kill them with slingshots.

Food. Insects, small molluscs, berries and fallen seeds of forest trees; captives readily eat termites and rice.

Breeding Habits. Nest said to be on ground in dense undergrowth; clutch, *c.* 12 (Bechinger 1964). Eggs reddish buff with white pores; 45 × 35.

References
Bannerman, D. A. (1930).
Bechinger, F. (1964).
Collar, N. J. and Stuart, S. N. (1985).

Agelastes niger (Cassin). Black Guineafowl. Pintade noire. Plate 5

Phasidus niger Cassin, 1857. Proc. Acad. Nat. Sci. Philad., 8, p. 322; Cape Lopez, Gabon. (Opp. p. 96)

Forms a superspecies with *A. meleagrides*.

Range and Status. Resident in dense primary forests from Cameroon and Gabon south to *c.* 4°S in lower Congo basin, thence eastwards north of Congo R. to *c.* 28°E near western edge of western Rift Valley. Scarce; but probably less threatened by habitat destruction than White-breasted Guineafowl *A. meleagrides* because range much more extensive.

Description. ADULT ♂: head and neck largely naked, dull pink, with a short crest of downy black feathers from base of bill to occiput, and a few scattered black feathers on chin and throat. Nape and back of neck with scattered black feathers. Rest of plumage black, obscurely vermiculated buffy white on belly; wings and tail unmarked. Mid-belly browner, feather bases white. Bill greenish grey; eye brownish grey; legs greyish brown; 1–2 spurs 5–7·5 mm. Sexes alike, ♂♂ average larger. SIZE: wing, ♂ (n = 22) 200–218 (209), ♀ (n = 17) 193–208 (197); tail ♂ (n = 21) 150–165 (156), ♀ (n = 17) 144–152 (148); tarsus, ♂ (n = 21) 77–85 (82), ♀ (n = 16) 71–80 (74).

IMMATURE: resembles adult but spurless black feathers of upperparts with ochraceous tips; wing-coverts greyish brown, mottled with rufous and tipped buff; primaries paler and mottled rufous distally. Belly pure white; rufous vermiculations on breast and innermost secondaries.

DOWNY YOUNG: forehead, crown and hindneck blackish, with a narrow line of buff running back from each nostril to above eye, a median spot of same colour between eyes, and 2 conspicuous stripes of buff running from vertex down nape and hindneck, 1 on each side. Upper back and wings uniform dark rufous, lower back dark maroon, indistinctly barred with black. Throat and cheeks rich buff with rufous wash, this colour extending up as mottling onto lores and area behind eyes; chest rufous-buff. Remaining underparts whitish, with slight rufous wash, becoming darker on flanks, which are separated from middle of lower back by an ochraceous stripe on each side. Bill pale brown-grey, blackish on centre of maxilla; eye dark brown; legs pale brown-grey.

Field Characters. Small black guineafowl with bare pink head and neck, plain black plumage in adult, belly white in immatures. Lack of long plumes on pinkish head and unspotted plumage distinguish it from much larger Plumed Guineafowl *Guttera plumifera*.

Voice. Tape-recorded (217, 226). Short, musical, high-pitched 'kwee' repeated monotonously 2–3 times per s. Alarm call higher-pitched still, and given much more rapidly (5–6 per s).

General Habits. A scarce and/or elusive bird of dense primary forest. Travels in pairs or small parties of less than 10; shy, difficult to approach. Feeds on ground; presumed to roost in trees or bushes.

Food. Beetles, ants, termites, millipedes, hard seeds, green leaves, fruits, and small frogs.

Breeding Habits. In NE Zaïre breeds in almost any month, but mainly during the drier months Dec Feb. Eggs pale reddish brown, sometimes washed with yellow or violet; very deeply pitted; *c.* 42 × 34.

References
Chapin, J. P. (1932).
Lippens, L. and Wille, H. (1976).

Genus *Guttera* Wagler

Rather large guineafowl, inhabiting forest and dense scrub. Naked head and neck largely dull blue-grey (sometimes red on throat and/or around eye) with crest of long black feathers. Poorly to moderately developed blue-grey pointed gape wattles. Occiput with a small to well developed fold of skin. Collar spotted or with variable band of black feathers. Body plumage black, spotted white, with varying amount of blue sheen and no vermiculation. Tail rather long, cocked or erected in excitement or alarm. Tarsus unspurred, scales pentagonal and not in rows. Internally, characterized by presence of a hollowed-out hypocleidium of the furcula, through which a loop in the trachea passes. The hollow hypocleidium is better developed in *G. pucherani* (Chapin 1932), and may contribute to the relatively low pitch of its calls.

Crowe and Snow (1978) divide the genus into 3 spp., *G. plumifera*, *G. pucherani* and *G. edouardi*; but Ghigi (1936) and Crowe (1978a) regard latter pair as conspecific, based on intergradation in captivity and in the wild.

PHASIANIDAE

Plate 5
(Opp. p. 96)

Guttera plumifera (Cassin). Plumed Guineafowl. Pintade plumifère.

Numida plumifera Cassin, 1857. Proc. Acad. Nat. Sci. Philad., 8, p. 321; Cape Lopez, Gabon.

Forms a superspecies with *G. pucherani*.

Range and Status. Resident in dense forests from S Cameroon east of Sanaga R. to *c*. 4°S in lower Zaïre basin (Loango Coast), thence inland in forests of Congo basin east to western edge of western Rift Valley, furthest south in the forests west of Lake Tanganyika. Uncommon and probably not uniformly distributed within this range.

Description. *G. p. plumifera* (Cassin): Cameroon, Gabon and lower Congo basin east to *c*. 16°E, where intergrades with *G. p. schubotzi*. ADULT ♂: crown surmounted with a crest of long, straight bristly plumes. Rest of head and neck slate grey, largely bare, with moderately long gape wattles. Rest of body, most of the wings and tail black with white spots. Primaries brown; secondaries blackish brown with irregular light spots. Three outermost secondaries longitudinally edged with white. Bill bluish grey; occipital skin-fold poorly developed; eye brown; legs blue-grey. Sexes alike, ♂♂ average larger. SIZE: wing, ♂ (n = 15) 223–232 (228), ♀ (n = 11) 216–227 (222); tail, ♂ (n = 15) 127–135 (131), ♀ (n = 11) 118–128 (124); tarsus, ♂ (n = 14) 78–87 (83), ♀ (n = 11) 73–83 (77); height of crest at centre, (♂♀) 32.

IMMATURE: crest shorter; retains black downy feathers on head and neck, throat feathers black with white tips. Upperparts grey, barred blackish. Secondaries and their coverts grey, barred and vermiculated with black. Breast dusky with whitish spots and bars.

DOWNY YOUNG: forehead, crown and nape dark brown; crown-stripe rufous-buff; sides of head dark brown with irregular rufous-buffy streaks. Upperparts dominated by a dark brown stripe down mid-back, with 2 parallel stripes on either side, the inner buffy, the outer dark brown. Wing feathers mottled and irregularly barred with buff and rufous-buff, buff dominating distally. Throat buffy; remaining underparts rufous-buff. Bill light buff with broad blackish patch at base of culmen; eye grey; legs yellow.

G. p. schubotzi Reichenow: Zaïre from *c*. 17°E to eastern border. Patches of orange-yellow on skin of hindneck and in front of ear.

Field Characters. Might be confused in dim light of forest with rather larger Crested Guineafowl *G. pucherani*. Differs in long erect crest of black plumes, white (not blue) spots on black body, longer gape wattles, grey or blackish (not red) bare skin on throat. Much larger than Black Guineafowl *Agelastes niger*, and more gregarious.

Voice. Tape-recorded (226). A sharp cackle (Bates 1930); a nasal 'kak' different from Crested Guineafowl and a discordant chorus of 'ka-ka-ka-ka-ka' when alarmed (Chapin 1932). Wings in flight make whirring noise.

General Habits. Habitat primary forest, although occurs in very mature secondary growth; does not come to cultivated clearings. Shy and difficult to observe. More gregarious than *Agelastes* spp.; found in flocks of 20–40. When disturbed runs and scatters widely rather than flying; but takes refuge in thick foliage of tall trees if pursued, e.g. by dogs, remaining quiet and unseen until danger has passed. Traces on forest floor suggest that flocks forage together, scratching up large areas of leaf litter in search of food; little direct observation. Roosts at night in tall trees, flocks usually flying up together into same tree, using a different tree each night. Like other guineafowl descends to ground in early morning to forage.

Food. Seeds, fruits, leaves and invertebrates (including snails, slugs, millepedes, spiders, roaches, grasshoppers, crickets, hemipterans, beetles, termites and ants).

Breeding Habits. Probably monogamous like Crested Guineafowl.

NEST: simple scrape on ground among dry leaves.

EGGS: 1 clutch, (n = 10 eggs) 50·1–53·3 × 38·6–39·9; pale buff, with numerous pits darkened as if with dirt, becoming nest-stained, another clutch (n = 9 eggs) 47·5–49 × 37–38·5.

LAYING DATES: Zaïre Mar, Sept; probably no definite seasonality, although avoiding drier months.

References
Bannerman, D. A. (1930).
Chapin, J. P. (1932).

Guttera pucherani (Hartlaub). Crested Guineafowl. Pintade de Pucheran.

Numida pucherani Hartlaub, 1860. J. f. Orn., 8, p. 341; Zanzibar.

Forms a superspecies with *G. plumifera*.

Range and Status. Resident from Guinea-Bissau east through all forested areas to S Somalia (lower Juba R.), south to Natal, and in denser *Brachystegia* woodlands with underlying evergreen thickets in Angola and Zambia to c. 15°S. Generally much more common and widespread than supposed, since difficult to observe. Nowhere severely threatened, but range decreasing generally through destruction of forest and thicket habitat.

Description. *G. p. pucherani* (Hartlaub): E Africa from Juba R. south to Rufiji R. (Tanzania) and inland to eastern Rift Valley in Kenya. ADULT ♂: crown with short black crest of downy feathers, taller towards occiput. Face and upper neck naked, with small wattle at gape and large occipital fold. Narrow collar of black feathers on lower neck, broader on upper breast. Rest of body, most of wings and tail black, dotted all over with blue-white spots. Primaries brown, outer webs vermiculated bluish. Secondaries darker brown, outer webs of outer 5 whitish, forming conspicuous pale bar in wing; rest with lines of blue dots on inner webs and bright blue bars of confluent dots on outer. Tail feathers black, spotted blue-white. Bill greyish to yellowish; bare skin round eye and on throat red, on rest of head and neck greyish blue; eye red; legs dark olive brown to black. Sexes alike, ♂♂ average larger. SIZE: wing, ♂ (n = 42) 256–276 (268), ♀ (n = 22) 245–264 (257); tail, ♂ (n = 40) 130–160 (151), ♀ (n = 21) 136–154 (144); tarsus, ♂ (n = 38) 87–97 (93), ♀ (n = 20) 83–90 (86); height of crest in centre (♂♀) 26, in front 19. WEIGHT: ♂ 721–1573 (1149).

IMMATURE: differs from adult in being generally duller, barred finely with black and white above and below. Bare skin without bright adult colours, secondaries dark grey-brown, flecked with black and tipped with buff, innermost with blue barring at base. Tail feathers as adult but brownish distally.

DOWNY YOUNG: forehead, crown and nape dark brown; crown-stripe and tiny cap rufous-buff; sides of head dark brown with irregular rufous-buffy streaks. Upperparts dominated by a dark brown stripe down mid-back, with 2 parallel stripes on either side, the inner buffy, the outer dark brown. Wing feathers mottled and irregularly barred with buff and rufous-buff, buff dominating distally. Throat buffy, remaining underparts rufous-buff; legs yellow.

We follow Crowe (1978a) in considering *G. edouardi* and *G. pucherani* conspecific. Of the many races described, only 5 are recognized here; they differ in colour of bare skin and shape and extent of the feathered crest.

G. p. verreauxi (Elliot) (including former *seth-smithi*, *schoutedeni*, *pallasi*, *chapini*, and *kathleenae*): W Africa (except S Cameroon), Congo basin, Uganda and Kenya west of E Rift Valley, south to *Brachystegia* woodlands of Angola and W Zambia, east to c. 27°E. Front of crest longer than in *G. p. pucherani*; no red bare skin surrounding eye; red skin on throat less extensive, rest of bare facial skin and neck grey-blue. Eye brown. Broad plain black collar on lower neck. Blue in body spots more intense. Wing, ♂♀ (n = 159) 238–260. WEIGHT: 2 adult ♂♂ 910 and 1160.

G. p. sclateri Reichenow: occurs only in a small area of forest and forest/savanna mosaic in W Cameroon centred on the Sanaga R. Like *G. p. verreauxi*, but crest very short in front, higher at rear. Wing, ♂♀ (n = 12) 236–256.

G. p. barbata Ghigi: coastal forests from S Tanzania to Mozambique (16°S), west to Malawi. Intergrades with *pucherani* and *edouardi*. Crest short, no red on bare throat or around eye; bare skin of neck dull blue; eye red. Collar plain black without vermiculations. Body spots sometimes interspersed with chestnut blotches. Wing, ♂♀ (n = 28) 254–276.

G. p. edouardi (Hartlaub) (including the former *lividicollis* and *symonsi*): E Zambia east to Luangwa Valley, Zambezi Valley and from Mozambique (near Beira) to Natal. Like *barbata* but crest curly especially in front and skin on head and neck blue-grey, with a contrasting heavy white fleshy fold below ear and on hindneck. Broad black collar extensively blotched chestnut. Wing, ♂♀ (n = 34) 248–263.

Field Characters. Only guineafowl outside the Congo Basin in forest or dense thickets of secondary growth. When disturbed, often flies into trees, showing pale brown wing patches and bluish white spots on blackish plumage. If only head and neck visible, distinguished from Plumed Guineafowl *G. plumifera* by shorter, curly crest and red skin on throat or around eye in some subspecies. Crest distinguishes it from other guineafowl in same habitat.

Voice. Tape-recorded (9, 32, 39, 74, 75, F, 244, 264). Has a rhythmic call consisting of a series of 3–4 ringing, piping notes separated by a sharp 'kuk' or 'krek', the series usually ending with a reedy 'krraaa'. Several birds may call together; pairs possibly call antiphonally. Calling takes place during the day and also regularly at dusk or just after dark as birds are settling into roosts (Keith 1971). In forest, when moving along, utters a low 'chuk' contact call. Alarm call a rattling 'chuk-chuk-chukchukkerr' similar to that of Helmeted Guineafowl *Numida meleagris*, but much lower in pitch. Less noisy than Helmeted Guineafowl.

General Habits. Preferred habitat is forest edge, secondary forest and gallery forest; less of a forest bird than the Plumed Guineafowl, Black Guineafowl *Agelastes niger* or White-breasted Guineafowl *Agelastes meleagrides*. Occurs in dense stands of pure *Cryptosepalum* woodland in NW Zambia. While often difficult to observe, and not encountered until already alarmed, sometimes emerges into the open in full sunlight, in cultivation or along roadsides. Cocks tail like a bantam when excited or alarmed. Normally occurs in flocks of 10–30, rarely up to 50, outside the breeding season. Also seen in small groups, probably family parties, pairs, or singly. Flocks break up into smaller groups and pairs during breeding season. Presence in any area more likely to be detected by signs (scratchings and dropped feathers, especially near dust baths) than by direct observation. May remain undetected for years in well-known areas unless such signs found. Roosts in trees, normally in flocks; same general area may be used nightly. Descends from trees early morning (later in wet weather) and forages, maintaining close contact, through dense thicket or forest, sometimes in relatively open woodland with undershrubs. Normally feeds on ground but occasionally observed in trees feeding on fruit. Sometimes moves in close association with arboreal monkeys, feeding on food scraps falling from above (Hill 1974). Rests in dense cover in middle of day; sometimes emerges in open in full sunlight to dust or drink. Covers a large area daily. Seldom seen by day except with aid of dog, when flock flies noisily up to leafy treetops, at once becoming almost invisible in dense foliage and remaining quiet.

Food. Seeds and fruits (Rubiaceae, Amaranthaceae, Compositae, Malvaceae and Leguminosae), green leaves, bulbs, roots (e.g. of manioc) and invertebrates (beetles, grasshoppers, hemipterans, termites, millepedes, spiders and small snails). Vegetable matter makes up bulk of recorded crop contents, but crops may contain whole or broken snail shells, perhaps ingested for grinding other material as well as for food value.

Breeding Habits. Monogamous; pairs separate from main flock at onset of breeding season, remaining apart until young well grown. Cocking and spreading of tail, normally indicating alarm or excitement, may also be courtship display. ♂ courtship-feeds ♀. In captivity, ♂ spends much time in alert postures while ♀ feeds.

NEST: few found; a scrape on ground, among dead leaves in dense cover.

EGGS: normally 4–5, occasionally up to 7; clutches of 10–14 recorded South Africa almost certainly 2 ♀♀ laying in same nest. South African eggs deep buff to pinkish, with minute dark pores, very thick shelled; rounder, less pointed than those of Helmeted Guineafowl; in E Africa white, with many dark pores. SIZE: 50–55 × 37·8–43·5 (Clancey 1967); (n = 7, Zimbabwe) 51·3–54·6 × 40·6–42 (52·4 × 41·3).

LAYING DATES: Natal, Nov–Feb, peaking Nov–Dec, main rains; Zimbabwe, Nov–Jan; Zambia, Nov–Feb, peaking Dec, main rains; Mozambique, Nov; E Africa, Uganda, Regions A and B, Jan, Feb, May, Nov; Kenya, Regions D and E, Apr, June–Aug, Nov–Dec, during wetter months; W Africa where seasons clearly defined apparently prefers rains, but near equator in bimodal rainfall regimes breeding throughout the year.

INCUBATION: probably by ♀ only, since only ♀♀ obtained on nests.

DEVELOPMENT AND CARE OF YOUNG: both parents tend and actively guard young until well grown.

BREEDING SUCCESS/SURVIVAL: probably poor, since dry season flocks contain very few obvious immatures.

References
Clancey, P. A. (1967).
Wilson, K. J. (1965).

Genus *Acryllium* Gray

A monotypic genus confined to Somali/E African arid region. The largest and most brilliantly coloured of the guineafowl. Head small, crown bare and unornamented, only rudimentary cheek wattles, nape and sides of head to above ear covered by a band of short dense velvety chestnut feathers; neck longer and more slender than in other guineafowl. Body relatively slender and elongated; wings short, broad and rounded. Basic spotted plumage embellished with iridescent blues, lilac and white. Tarsi long with multiple spur-like bumps, longer in ♂♂ than ♀♀. Tarsal scales pentagonal. Intestinal caecum very long, presumably an adaptation to improve water conservation and allow microbial digestion of food high in cellulose.

Acryllium vulturinum (Hardwicke). **Vulturine Guineafowl. Pintade vulturine.**

Numida vulturina Hardwicke, 1834. Proc. Zool. Soc. London, p. 52; Tsavo, Kenya.

Plate 5
(Opp. p. 96)

Range and Status. Resident from extreme NE Uganda (Karamoja) north to S Ethiopia, Somalia, arid parts of N and E Kenya, south to Pangani R., Tanzania. Generally common to abundant; not threatened except by local hunting pressure.

Description. ADULT ♂: head and neck naked except for nape tuft of chestnut feathers. Feathers of lower neck, mantle, and chest form a hackle; each feather long, lanceolate, black with wide white shaft-stripes, edged bright cobalt blue, longest dotted with fine blue spots. Rest of upperparts black, spotted white intermeshed with white vermiculation. Tail long, trailing, central feathers much longer than outer, black, spotted and vermiculated with white. Lower breast and abdomen cobalt blue, becoming black centrally. Primaries dark brown, edged white on outer webs; outer webs of secondaries edged with lilac. Bill light grey; bare skin of face and neck dark blue-grey; eye red; legs black. Sexes alike, ♂♂ average larger. SIZE: wing, ♂ (n = 31), 290–311 (298), ♀ (n = 20) 285–296 (290); tail, ♂ (n = 23) 265–280 (272), ♀ (n = 20) 251–274 (260); tarsus, ♂ (n = 30), 100–117 (109), ♀ (n = 20) 94–103 (99). WEIGHT: 2 adult ♂♂ 1026, 1645, 2 adult ♀♀ 1135, 1523.

IMMATURE: much drabber than adult. Natal down partially retained on head; hackle apparent even at this stage. Greyish brown, barred and mottled with rufous-brown, buff and black; blue less intense than in adult. Primaries brownish black, barred with buff on outer webs and tips; secondaries darker, barred with buff and freckled black. Tarsi unspurred. Assumes adult plumage within 1 year.

DOWNY YOUNG: yellowish buff, mottled with dark brown, with centre of crown dark brown, extending down nape and along back; fine dark brown streaks on head and forehead; a dark curving streak below eye and 2 other curved cheek stripes. Bill and legs flesh.

Field Characters. A tall, long-legged, long-tailed, slender, brilliantly blue and black guineafowl, unlike any other galliform. If head and neck only seen in tall cover, chestnut nape feathers and lack of crown adornment distinguish it from Helmeted Guineafowl *Numida meleagris* and Crested Guineafowl *Guttera pucherani*.

Voice. Tape-recorded (36, 217, 244, 264, 296, 308). Easily distinguished from Helmeted Guineafowl. Most frequently heard call a metallic 'cheenk, cheenk, chickera, chik-chicker-chik-chik' alarm call given even by very young birds. Higher pitched than its counterpart call in Helmeted Guineafowl; often given when flying to roost. Low clinking contact calls in flocks, especially when foraging in thickets. Also a 2-noted call 'keuw-kwee' like that given by ♀ Helmeted Guineafowl, but delivered more slowly (call duration c. 1 s in Helmeted, 2 s in Vulturine). Normally rather silent.

General Habits. Inhabits semi-arid *Acacia/Commiphora* scrub, often with shrubs, e.g. *Acalypha*, but enters montane forest (Mt Marsabit, Kenya) and forages in tall riverine *Acacia* woodland. Sometimes enters very dense thickets when occurs with Crested Guineafowl. Becomes locally abundant and tame in national parks or reserves. Occurs in flocks of 20–30, occasionally more, for most of year, pairs separating in breeding season. Roosts in trees, generally selecting tallest *Acacia* in area, often in depression or watercourse, and surrounded by dense shrubs. Descends from roost at dawn, and forages actively away from roost until heat of day; rests in shade of bush until late afternoon, when forages again. Flocks usually frequent same area, but follow different routes daily. Ground near roosting trees extensively scratched up. When disturbed, flocks walk or run away, moving from cover of one bush to another, clustering in tight groups in shade. If hard pressed, fly for 50–100 m before settling and running again; does not take to trees unless harried. Although often foraging in open patches, usually has access to dense cover; reluctant to cross large stretches of open ground. Feeds mainly on ground, but occasionally climbs into bushes and low trees to feed on berries and fruits, e.g. of *Salvadora persica* and *Commiphora* spp.

Food. Seeds and leaves of grasses and herbs; berries and fruit (*Commiphora* spp. and *Ficus* spp.); green buds and shoots. Also insects, scorpions, spiders, small molluscs. Apparently entirely independent of fresh water; does not drink even where water readily available in dry season.

Breeding Habits. Little known in wild; studied in captivity (Grahame 1969). Rains stimulate breeding activity; flocks partially fragment into breeding pairs and small groups of unmated birds. ♂ posturing before ♀ rears up, bends head downwards, and partly spreads wings. ♀ may then crouch to encourage copulation, may take no notice or run away. If ♀ moves away, ♂ follows, sometimes attempting to place himself conspicuously before her. Displays little known; but mutual chasing seen at this time is probably ♂–♂ aggressiveness. In captivity ♂ escorted ♀ and spent much time in alert posture while ♀ fed. Courtship feeding very elaborate, ♂ running 3–4 m ahead of ♀ with food item in bill, then dropping it with head lowered and wings fanned.

NEST: simple scrape on ground in thick grass and/or bush cover.

EGGS: up to 13–15, laid on successive days; creamy white or pale brown, smooth and slightly pitted, broad oval, slightly pointed, hard shelled. SIZE: (n = 4, Somalia) 49·5 × 37·3; (n = 13, Kenya) 55·0 × 41·0; in captivity, 50 × 36.

LAYING DATES: Ethiopia, June; Somalia, June; N Kenya, Jan, Feb, June, Aug, Nov, Dec, peaking in June and Dec–Jan.

INCUBATION: in captivity by ♀ only. Period incubated by domestic chicken 23–25 days; another captivity record 27–28 days (Grahame 1969).

DEVELOPMENT AND CARE OF YOUNG: young leave nest almost immediately. At 10 days wing-feathers well developed, primary coverts tipped whitish, back brown. Blue feathers appear on breast at $c.$ 8 weeks, some down retained on head and neck until 16 weeks. Can fly at $c.$ 14 days. ♂ broods and feeds young in first few days; young thereafter follow both parents. Adults with small young less than 1 month old often join adult flocks, earlier than Helmeted Guineafowl, but scanty observations suggest course of breeding cycle generally similar.

BREEDING SUCCESS/SURVIVAL: preyed upon by all carnivores from lions to eagles.

Reference
Archer, G. and Godman, E. (1937).

Genus *Numida* Linnaeus

The most common and widespread guineafowl. Moderately large, head and neck bare, brightly coloured, adorned with gape wattles and a bony casque, and sometimes with cartilaginous bristles on cere. Plumage black with white spots (cryptic on bare black ash-covered ground after fires). Wings short, rounded; flight powerful, swift and direct when pressed. Tail short, depressed, not erected in display. Tarsi moderately long, stout and unspurred. Tarsal scales pentagonal, as in *Guttera* spp. and *Acryllium*. 2–3 species have been recognized, but intergradation in wild suggests that they are all conspecific (Crowe 1978a). Likewise over 30 subspecies have been ascribed to the single species, but only 9 warrant recognition.

Plate 5
(Opp. p. 96)

Numida meleagris **(Linnaeus). Helmeted Guineafowl. Pintade sauvage.**

Phasianus meleagris Linnaeus, 1758. Syst. Nat. (10th ed.), p. 158; Nile R., Nubia, Sudan.

Range and Status. Resident from Senegal to Ethiopia and Somalia, south to Namibia and South Africa. W African subspecies (*N. m. galeata*) domesticated throughout much of the world; introduced in SW Arabia and Madagascar. Occurs in all open-country vegetation from forest edge to subdesert steppe, from sea level to above 3000 m. Locally common to very abundant, especially in savannas mixed with cultivation. Limited locally by the availability of drinking water and elevated roosts. Affected by heavy hunting pressure and egg collecting, but rarely extirpated where habitat sufficient. Moroccan subspecies *N. m. sabyi* has recently become very rare; only 3 records in the 1970s (J. D. R. Vernon, pers. comm.).

Description. *N. m. meleagris* Linnaeus (including former *major, inermis, omoensis, macroceras, neumanni, toruensis, intermedia, uhehensis*): E Chad to Ethiopia south of 17°N, east to Rift Valley, south to borders of N Zaïre, Uganda, N Kenya; south of equator only in Uganda. ADULT ♂: head and neck bare, but lower hindneck covered by short, dense, black filoplumes, grading into broad greyish black collar finely barred with white. Crown with keratin-covered bony casque, red-brown at base grading to horn at apex. Rest of upperparts, including upperwing and tail-coverts, dark grey to black with white spots interspersed with a network of white vermiculations. Tail dull black, heavily spotted and vermiculated with white. Below, dark grey to black with white spots and much less vermiculation, spots larger than on upperparts, increasing in size posteriorly. Primaries and secondaries black with white spots; outer webs of secondaries with regular bars perpendicular to shaft, interspersed with dense white vermiculations. Base of upper mandible brownish horn, tip and lower mandible pale grey. Cere covered with dense cartilaginous bristles up to 6 mm long. Cheeks, area round eye, and most of neck grey-blue, hindneck black. Gape wattles round and blue; eye dark brown; legs dark grey to black. ♀ similar but smaller. SIZE: wing, ♂ (n = 198) 253–276 (269), ♀ (n = 113) 247–268 (259); tail, ♂ (n = 198) 164–176 (172), ♀ (n = 110) 160–179 (170); tarsus, ♂ (n = 198) 74–89 (84), ♀ (n = 103) 69–82 (79); casque (n = 311) 4–22 (12). WEIGHT: (n = 9) 1150–1600 (1300).

IMMATURE: natal down retained on head (even when almost in adult plumage); wattle and casque smaller. Subadult plumage, assumed at c. 5–6 weeks: upperparts drab grey-brown, barred with rufous-brown and freckled with black, each feather tipped with buff. Underparts as above except less rufous-brown barring.

DOWNY YOUNG: above mottled brown and grey with 2 longitudinal buffy white stripes margined with black straddling a much broader central brown stripe; below paler buff, throat whitish. Broad dark brown stripe on crown, with several light brown and buff longitudinal facial stripes. Underparts tawny buff grading to whitish on throat. Bill rufous, tip whitish; eye grey; legs rufous.

N. m. sabyi Hartert: NW Morocco, between Oum er Rbia and Sebou rivers. Like *meleagris* but facial skin pale blue-white, no cere bristles, wattles red, hindneck filoplumes very long and confined to mid-line, and collar violet-grey. Wing, ♂♀ (n = 4) 253–272 (267); casque, ♂♀ (n = 4) 14–18 (16).

N. m. galeata Pallas (including former *marchei*, *callewaerti*, *blancoui*): W Africa east to S Chad, south to central Zaïre and N Angola. Like *sabyi* except collar greyish and casque smaller. Smaller, wing, ♂♀ (n = 137) 240–268 (251); casque, ♂♀ (n = 137) 3–9 (6). Plumage of domesticated birds derived from this race white or a mixture of wild-type and white; tarsi shorter, stouter, orange and not grey or black; weight 30–50% greater than wild birds.

N. m. somaliensis Neumann: arid parts of NE Ethiopia and Somalia. Like *meleagris* but cere bristles very long (up to 24 mm), hindneck filoplumes very long and restricted to mid-line, and wattles somewhat pointed (not rounded) and blue with red tips. Wing, ♂♀ (n = 44) 250–279 (266); casque, ♂♀ (n = 44) 6–21 (13).

N. m. reichenowi Ogilvie-Grant (including former *ansorgei*): Kenya and central Tanzania to c. 13°S. Like *meleagris* but facial skin pale blue-white, wattle entirely red, much longer casque, hindneck filoplumes long and confined to mid-line, and much less vermiculation of wing, body and tail feathers, especially on outer webs of secondaries. Much larger, wing, ♂♀ (n = 121) 270–285 (282); casque, ♂♀ (n = 121) 21–42 (30).

N. m. mitrata Pallas: coastal and W Tanzania south to coastal Mozambique, west through Zambezi valley and Zimbabwe to S Angola and N Botswana. Like *reichenowi* but casque smaller (but still larger than that of *meleagris*), wattles pointed (not rounded) and blue with red tips, facial skin blue-grey. Wing, ♂♀ (n = 293) 263–287 (275); casque, ♂♀ (n = 293) 11–28 (19).

N. m. marungensis Schalow (including former *maxima*, *frommi*, *rikwae*, *bodalyae*): central African savannas and woodlands south of Zaïre basin south to W Angola and east in Zambezi basin to the Luangwa Valley in Zambia. Differs from *mitrata* in having casque yellow-ochre and much more squat and broad at base. Larger, wing, ♂♀ (n = 97) 270–302 (284); casque, ♂♀ (n = 97) 13–23 (19). WEIGHT: S E Zaïre, ♂ (n = 7) 1200–1777 (1613), ♀ (n = 13) 1280–1822 (1612).

N. m. damarensis Roberts (including former *papillosa*): arid parts of Botswana and Namibia south to c. 26°S. Like *mitrata* but cere covered more with cartilaginous papilli; casque taller but more withered basally and strongly backward-curving; mantle, body and wing spots and vermiculations larger and denser. Wing, ♂♀ (n = 102) 265–282 (273); casque, ♂♀ (n = 102) 17–27 (22). WEIGHT: 1 ♂ 1350, 2 ♀♀ 1150, 1600.

N. m. coronata Gurney (including former *transvaalensis* and *limpopoensis*): moister eastern parts of South Africa south of 20°S. Introduced in W Cape Province. Like *N. m. mitrata* but casque much better developed and mantle more streaked than barred with white. Wing, ♂♀ (n = 136) 263–283 (271); casque, ♂♀ (n = 136) 19–31 (25). WEIGHT: ♂ 1145–1816 (1475), ♀ 1135–1823 (1492); ♂ lighter than ♀ during breeding season (Siegfried 1966).

Field Characters. Large, gregarious, dark guineafowl with bluish bare head and neck, topped with a horn-like casque. White spots not obvious except at close range. Distinguished from other guineafowl by having casque instead of head feathers.

Voice. Tape-recorded (2, 9, 21, 36, 38, 50, 58). Commonest call a raucous nasal staccato alarm 'kek, kek, kek, kek, kaaaaaa, ka, ka, ka, ka, kaaaaa, ka, ka' is usually given by flock when threatened by predators or at roosting time. Much less frequently heard is soft metallic contact note 'cheenk, cheenk, cheenk' given as flock progresses through thick grass. ♀ has 2-note 'buck-wheat' call (to which ♂ responds with single 'cheeng'), which functions in pair-bond maintenance (Elbin 1979). The staccato and buck-wheat calls are far carrying. Adults with keets (chicks) are extremely aggressive and give a growling 'grrrrr, grrrrrrr, grrrrrrrr' before attacking intruder. Keets utter a soft 'peep', and occasionally a series of more melodious twittering notes. Adult-type calls are given by 4-month-old immatures (Maier 1982). No apparent geographical variation in vocalizations (Benson 1948).

General Habits. Inhabits all types of open country from forest edge to semi-desert, commonest in well wooded moist grass savannas with 600–1200 mm rainfall. Especially high concentrations in areas cultivated with wheat and maize. Critical habitat features are drinking water, cover and roosting sites. Flock home range variable with habitat; smaller (0·8–1·8 km²) in primary woodland, larger (7·6–21·2 km²) in secondary woodland (Ayeni 1983). Roosts gregariously on adjacent trees (or even telephone poles), occasionally flying to roosting area from some distance. Sometimes roosts may be traditional with up to 30 cm of droppings accumulating below. Descends from roost at first light, and at once commences foraging, moving away from roost area, often in single file toward a drinking hole; dominant ♂♂ tending to lead. Spends much of early morning and late afternoon foraging, preening or dusting. Remains relatively inactive in shade during hottest parts of day. In late afternoon flocks walk and feed their way to general vicinity of roosting area. Gregarious except during breeding season, when flocks of 15–40 fragment into singletons, pairs and small groups. Several flocks may converge on drinking and good feeding sites, or dusting or roosting areas. Flocks have complex social structure and may remain stable in individual membership over several years (Crowe 1978b). 'Pecking-order' dominance hierarchy evident among ♂♂ but not ♀♀, although ♀♀ join ♂♂ in inter-flock aggressive encounters.

In non-breeding season, relatively few aggressive interactions between flock members; restricted to pecks or short chases by ♂♂. When disturbed, flock compresses, individuals give staccato alarm call and run off to nearest dense cover where they 'sit tight' and remain silent. Flocks often drive off aerial and terrestrial predators and, where population dense, sometimes defend territories (Crowe 1978b and in prep.) Isolated individuals prefer to run on ground to escape predators, but sometimes take to trees. In aggressive encounters ♂♂ arch wings above back and compress them against sides, looking slim from in front but bigger and higher from the side.

Food. Quantitatively analysed in E and S Africa (Skead 1962, Angus and Wilson 1964, Grafton 1971, Mentis *et al*. 1975, Swank 1977). Omnivorous. Food year-round by volume, non-agricultural seeds (39%), maize kernels (17%), sedge tubers (20%), invertebrates (mainly insects) (12%) (Mentis *et al*. 1975). Invertebrate prey preferred when abundant, especially grasshoppers and termites; one crop contained more than 5000 harvester termites *Hodotermes mossambicus* (Steyn and Tredgold 1967). Often strips seeds from grass heads (e.g. *Panicum*, *Eragrostis*, sorghum and millet). Damage to maize restricted to fallen or discarded cobs. Occasionally takes newly sown grain. During dry season eats mainly bulbs, roots, and especially tubers of *Cyperus* spp.

Breeding Habits. At onset of breeding season ♂–♂ chases and fights increase in frequency, duration and intensity. Chases may last 5–10 min and involve several ♂♂. Fights terminated only when defeated bird crouches (similar to sexually receptive ♀) or runs from scene. ♀♀ give 'buck-wheat' call more frequently, attracting ♂♂ who respond antiphonally with 'cheenk' call. Monogamous, but pairs not stable during early phase of breeding. Later they stabilize and ♂ vigorously defends ♀ from other ♂♂ and predators. Courtship display of ♂ indistinguishable from arched-wing lateral display given in ♂–♂ aggressive encounters. ♂ courtship-feeds mate (Stokes and Williams 1971), who spends much time feeding or preening. Copulation occurs on ground (not in trees as reported by Skead, 1962) during cooler hours of day or at dusk. Pairs have no real territory, but do not range as widely as non-breeding flocks.

NEST: ♀ selects nest-site, usually in tall grass, at base of grass tussock, or under bush, well hidden in dense cover. Often at the interface of open grassland and woodland or dense bush. A scrape in earth, lined with grass stems and feathers; *c*. 25–32 cm across, up to 8 cm deep.

EGGS: 6–12, laid on successive days; larger clutches of 20–50 involve 2 or more ♀♀. Broad, slightly pointed ovals, shells very thick and hard (thus possibly resistant to grass fires). Colour variable, usually yellowish to pale brown with darker speckling, sometimes almost white. SIZE: (Morocco) 44–55 × 36–40 (49 × 38); (n = 9, W Africa) (47·2 × 38·5); (n = 14, Ethiopia) 51·0–53·5 × 40·0–42·5; (n = 4, Uganda) 55–58 × 40–41; (n = 6, Kenya) 50–58 × 37–41; (n = 16, Somalia) (50 × 38·8); (n = 7, Malaŵi) 46·4–53·3 × 38·2–41·4 (51·4 × 40·1); (n = 31, Zimbabwe) 49·3–54·0 × 39·5–42·2 (51·8 × 41·2); (n = 42, South Africa) 51·2–57·6 × 38·5–41·9 (53·6 × 40·8); (n = 172, South Africa) 44–56 × 35·5–42·5 (52·3 × 39·6); (n = 3, Namibia) 50·2–52·5 × 39·4–39·8. WEIGHT: *c*. 39.

LAYING DATES: South Africa, Cape Province, Sept–Dec, especially Oct–Nov; Transvaal, Natal, Oct–Mar, peaking Dec–Jan, in height of main rains; Namibia, Jan–Feb, May; Zambia and Zimbabwe, Oct–Apr, occasionally May, peak Nov–Feb during early main rains; Malaŵi, Dec–Apr, peaking in Dec–Feb; East Africa, Uganda, Regions A and B, Mar–Aug, Oct, Dec; Tanzania, Region C, in May–July and Dec, during and after rains; Ethiopia, any rainy month, mainly July–Sept; Central Ethiopia, June; Somalia, May–July; Nigeria, July–Aug; rest of W Africa east to Sudan, chiefly May–July, in main rains, sometimes later; Morocco, Mar–May (many records depend on broods of uncertain age). Almost all breeding occurs in or just after rainy seasons.

INCUBATION: begins when clutch complete. Period 24–27 days, by ♀ only, who is silent and sits very tight. Joint incubation by captive ♂ and ♀ recorded (Ayeni 1983). ♀ plucks and adds stems to nest material while incubating, but forages very little (Elbin 1979). ♂ not always faithful, consorts with other ♀♀, playing no further role until after hatching, when he returns and does *c*. 80% of brooding of keets during their first 2 weeks (Elbin 1979).

DEVELOPMENT AND CARE OF YOUNG: eggs hatch almost simultaneously. Hatching weight, 24–34 (30·1). Yolk reserve permits survival without food for several days, but keets can feed within 24 hours. Weight increases to 250 g at 35 days; 550, 70 days; 700, 105 days; 1100, 140 days; reach full adult weight (*c*. 1300 g) at *c*. 210 days. ♂♂ grow more quickly than ♀♀, are heavier than ♀♀ at 6 months; but ♀♀ thereafter overtake ♂♂ and usually remain heavier, especially during breeding season. Primaries and tail-feathers visible at hatching, contour feathers emerge 1–2 weeks later. At 21 days partly feathered, at 35 days in juvenile plumage, traces of down remaining to 85–100 days on head and throat. By 42 days completely covered with drab greyish brown juvenile plumage. Casque appears at 35–42 days. Moult into adult plumage begins at 28–35 days with first primary, then tail feathers at 35–42 days. Drab juvenile body plumage moulted between 42–84 days; plumage then still paler and less spotted than full adult. Can fly short distances at 14 days, strongly at 30 days. Primaries and secondaries grow rapidly and harden at 90 days. Wing-tips uneven at 56 days, but completely rounded by 84 days. Young cannot be sexed reliably by plumage or weight, but young ♀♀ give characteristic 'buckwheat' call at 60 days (Siegfried 1966), and can lay at 28–32 weeks (Ayeni 1983).

Both parents aggressively defend keets from predators and adult guineafowl; attack usually preceded by full frontal display with wings arched and spread and growling threat call. Pairs with young amalgamate to reform flocks when keets 1–3 months old.

BREEDING SUCCESS/SURVIVAL: many broods fail altogether if ♂ not in attendance; most are rapidly reduced despite efforts of both parents, especially if keets hatch during cold/wet periods.

References
Ayeni, J. S. O. (1983).
Crowe, T. M. (1978a).
Crowe, T. M. (1978b).
Elbin, S. B. (1979).
Maier, V. (1982).
Siegfried, W. R. (1966).
Skead, C. J. (1962).

Subfamily PAVONINAE: Congo Peacock

A small subfamily comprising some of the most spectacularly feathered galliform genera: *Pavo*, peacocks (2 spp.); *Polyplectron*, peacock pheasants (6); *Rheinartia*, Crested Argus (1); *Argus*, Great Argus (1); and *Afropavo*, Congo Peacock (1). All species are sexually dimorphic and have ocellated upperparts (faintly adumbrated in *Afropavo*), elaborate displays involving long to very long tails and/or tail-coverts, divergent tail-moult (except *Afropavo* for which tail-moult has yet to be described) and small clutches (2–4). Links between some or all of above genera are supported by electrophoretic (Gysels and Rabaey 1962), karyological (De Boer and van Bocxstaele 1981, Stock and Bunch 1982), plumage microstructure (Durrer and Villiger 1975), myological (Hulselmans 1963), and/or osteological (Lowe 1938, T. M. Crowe *et al.*, unpub.) evidence; although A. D. Stock (pers. comm.) suggests that the karyotype of *Polyplectron* spp. is similar to that of pheasants *Phasianus* spp. Karyological (Stock and Bunch 1982), myological (Hudson *et al.* 1966), and immunological (Mainardi 1963) researches also indicate closer links between most pavonines and guineafowl Numidinae than with other phasianids.

Genus *Afropavo* Chapin

An endemic, monotypic genus; affinities with Oriental Region peacocks suggested by electrophoretic (Gysels and Rabaey 1962), karyological (De Boer and van Bocxstaele 1981, Stock and Bunch 1982), myological (Hulselmans 1963) and osteological (Lowe 1938, T. M. Crowe *et al.*, unpub.) evidence. On the other hand, Ghigi (1949) suggests relationship with francolins *Francolinus* spp., Verheyen (1956) with guineafowl Numidinae, and Taibel (1961) with Cracidae (*Penelope*).

The discovery of *Afropavo* (Chapin 1938) was one of most sensational ornithological events of the 20th century, not merely because such a large and conspicuous bird had eluded discovery in an area reasonably well surveyed ornithologically, but also because of the implications of its probable Asiatic affinities. *Afropavo* is a very large galliform with iridescent green, violet and rufous plumage, long tail and sharply spurred (length *c*. 35 mm) tarsi. Smaller size and lack of long train with striking ocelli distinguish it from *Pavo*.

Afropavo congensis Chapin. Congo Peacock. Paon du Congo.

Afropavo congensis Chapin, 1936. Rev. Zool. Bot. Afr., 29, p. 1–6; Sankuru district, central Zaïre.

Plate 5
(Opp. p. 96)

Range and Status. Endemic, resident; confined to equatorial forest of central and E Zaïre, from longitude 20°E to 28°E and from latitude 2°N to 5°S. Generally uncommon and distribution may be limited by hunting and its well known susceptibility to disease and parasitism (Verheyen 1965, Lovel 1976). Snow (1978) suggests that mutually exclusive, complementary distributions of *Afropavo* and Black Guineafowl *Agelastes niger* may be due to competition. Greatest concentrations in comparatively dry forests of N Kasai. Totally protected by law within Salonga National Park (3·6 million ha) in centre of range. Secretive and apparently requiring largely unpopulated forested areas, so major threat to survival is forest destruction. However, small clutch size and ease with which hunters can call it up by imitating nasal alarm call of local antelope make local extermination by hunting a significant threat (Verheyen 1965). High regard in which it is held in tribal culture may counterbalance this, however.

Afropavo congensis

Description. ADULT ♂: head covered by short, black, downy feathers; crown surmounted by tuft of white bristly feathers c. 90 mm long, with crest of c. 14 black feathers immediately behind, longest measuring c. 35 mm. Throat only sparsely feathered with black down; base of neck and breast black with iridescent greens and violet. Remaining upperparts dark bronze-green. Tail black glossed with violet-blue and green. Underparts black with greenish gloss. Wings brown-black, innermost secondaries tipped green, some wing-coverts tipped glossy violet. Bill blue-grey; eye brown; naked skin between ear and eye blue; throat skin red; legs blue-grey, spurs whitish. ADULT ♀: crown bristles shorter and brown (not white). Head down brown-red and more dense, chin brown-white. Upperparts glossy iridescent green, with incomplete fulvous ocelli. Tail chestnut, barred and spotted black with broad green tip. Lower foreneck, breast and flanks rufous, mottled and barred with brown; rest of underparts black. Wings largely rufous, barred and flecked with black, more distinctly on secondaries than primaries. Smaller overall. SIZE: wing, ♂ (n = 32) 306–335 (318), ♀ (n = 14) 270–295 (286); tail, ♂ (n = 30) 206–240 (221), ♀ (n = 14) 169–205 (190); tarsus, ♂ (n = 33) 86–109 (99), ♀ (n = 15) 75–93 (86); 1 spur; ♂ 18–38 (28), ♀ 8–15 (11). WEIGHT: captive adult (sex?) 1010.
   IMMATURE: ♂, upperparts duller, back tinged brown, remaining feathers with narrow violet-blue borders. Underparts dull black. White bristles on crown shorter, tarsi without spurs. Adult plumage assumed in 2nd year. ♀, upperparts also duller; first plumage mottled cinnamon and black. Iridescent green feathers appear at 1st moult, last feathers to be replaced are primaries and secondaries. Sexual dimorphism in plumage apparent at 2 months.
   DOWNY YOUNG: crown, nape and hindneck black; sides of head tawny yellow, with narrow black line along eyelid and above eye; small black patch behind and touching eye, another on ear-coverts; back dark uniform black-brown, lighter on rump (lacking buffy longitudinal stripes of other African phasianid chicks); wings and tail deep cinnamon; underparts and sides bright creamy yellow, tinged with tawny rufous on all but chin. Bill dusky above, light fleshy buff below; legs light fleshy buff.

Field Characters. Much larger than any other forest galliform. Iridescent green, violet and rufous plumage markedly distinct from relatively dull dark plumage of forest guineafowl Numidinae and francolins *Francolinus* spp.

Voice. Not tape-recorded. Most frequently heard call a duet with 20–30 repetitions. ♂ utters high-pitched 'gowe', ♀ replies with low 'gowah'. Duets usually preceded by loud 'rro-ho-ho-o-a', increasing in volume. In another form of duet, ♂ calls 'ko-ko-wa' and ♀ responds with higher pitched 'hi-ho-hi-ho'. ♂ gives snore-like call when guarding nest area while ♀ incubates, attracts chicks with high-pitched call. Both sexes have guttural clucking calls as in other peafowl, and immatures twitter almost continuously.

General Habits. Inhabits primary rain forest below 1200 m with extensive undergrowth on well drained soil. Avoids seasonally flooded areas. Optimum temperature in captivity 20°C, relative humidity 90%. Often associates with tree *Celtis adolfi-fridericii*, fruits of which provide food. Usually in pairs or small parties, feeding on ground in dense understorey. In captivity avoids bright sunlight. Active in morning and late afternoon/evening, otherwise hides in undergrowth. Frequently very active well into night, especially when moon is full, calling often from 1900 h until dawn. Strong flier and jumper; can leap up to branches 1·5 m above ground without flapping wings. Calling is contagious, several birds answering initiators, with chorus lasting c. 10 min. Perches on high branches of trees at night (to roost), and when disturbed by dogs easily shot. When excited, both sexes cock and spread tails; ♂ also erects crest.

Food. Seeds and fruits of trees and plants of forest understorey (e.g. *Celtis adolfi-fridericii*, *Sapium ellipticum*, *Ficus* spp., *Strombosia grandifolia*, *Xymalos monospora* and *Caesalpinia decapetala*), ants (e.g. *Tallothyreus tarsatus*), termites and their larvae, other ground-living invertebrates and aquatic insects (e.g. *Orectopyrus* spp.) (Prigogine 1971).

Breeding Habits. Mating system monogamous. ♂♂ very aggressive, attacking one another and occasionally ♀♀. ♂ displays to other ♂♂ and ♀♀ standing erect with spread tail and wings like peacocks *Pavo* and turkeys *Meleagris*. Also displays laterally like other phasianids, giving ♀♀ side view. In captivity courtship display sometimes occurs in trees, ♂ and ♀ perched face to face, bowing head deeply, with tail spread out at angle of 45° (van Bemmel 1961). ♂ leads as pairs progress through vegetation and often defers to ♀ when both converge on same food item.

NEST: 1·5–3 m up tree in fork of major branch; no nest made; eggs laid in hollow or flat spot.

EGGS: 2–4, laid at 1–3 day intervals. Rufous-brown or cream, unspotted. SIZE: (n = 8) 56–63·3 × 43–46.9. WEIGHT: (n = 4, freshly laid) 67–71 (69.5).

LAYING DATES: Jan–May, Sept–Oct, but possibly at any time of year according to local conditions.

INCUBATION: by ♀ only; begins on day on which 2nd or 3rd egg is laid. Period: 25–28 days. Once incubation begins ♀ sits very tight, hiding head under wing, leaving only at dawn or dusk for short feed and drink. Does not leave nest at all during last week and becomes very agitated as hatching nears. ♂ remains near nest during incubation, regularly posturing with head and neck extended so that body is orientated horizontally. Then flaps wings vigorously 2–3 times, almost taking off, causing leaf litter to be blown away. When wandering around nesting area and not in this posture, ♂ often emits low snore-like call.

DEVELOPMENT AND CARE OF YOUNG: after hatching, chicks remain in tree, brooded by ♀ for 1–2 days. Thereafter they fly/tumble down to ground, and ♂ immediately attracts them with high-pitched call and broods them during night while ♀ roosts on nest-site. On 2nd and subsequent nights, ♀ broods them on low branch (c. 0·7 m). Both parents care for chicks, feeding them from their bills or immobilizing insect prey for them. By 6th day chicks fly distances of 5 m and more; totally independent at 1 month, but still sleep with ♀. ♂ continues to defend brood even at this age. ♀♀ fertile at 12 months, ♂♂ only at 18 months.

BREEDING SUCCESS/SURVIVAL: in captivity, only 22% of eggs hatched (Verheyen 1965); in wild, 2–3 young survive per brood.

References
Chapin, J. P. (1938).
Collar, N. J. and Stuart, S. N. (1985).
Delacour, J. (1977).
Lovel, T. W. I. (1976).
Verheyen, W. N. (1965).

Subfamily GALLINAE: quails, partridges and francolins

A poorly defined, possibly polyphyletic assemblage. Chromosome morphology (Stock and Bunch 1982) suggests that true pheasants (Phasianinae), New World quail (Odontophorini), grouse (Tetraoninae) and turkeys (Meleagridinae) may belong to a separate assemblage, while Old World quails and partridges (Perdicini), junglefowl (Gallini), peafowl (Pavoninae) and guineafowl (Numidinae) form its sister group. Moreover, myological (Hudson et al. 1966) and immunological (Mainardi 1963) evidence suggests that *Gallus* and *Coturnix* are far removed from other phasianids. However, osteological (T. M. Crowe et al., unpub.), myological (Hudson et al. 1966) and electrophoretic (Sibley and Ahlquist 1972) data suggest that partridges and francolins appear to grade with pheasants and quails. Therefore, phylogenetic relationships among these phasianids remain unclear and further study of a wide range of taxonomic characters is needed before their relationships can be determined with any degree of certainty.

Peters (1934) recognizes 51 genera distinguishable mainly by number and relative lengths of tail and wing feathers. 5 genera occur in Africa: *Coturnix* (3 spp.), *Alectoris* (1), *Ammoperdix* (1), *Ptilopachus* (1) and *Francolinus* (36). The Ring-necked Pheasant *Phasianus colchicus* has recently been introduced for shooting purposes at several localities in N Morocco, but has not yet become established in the wild.

Genus *Coturnix* Bonnaterre

An Old World genus of 8–10 spp. including blue quails (*Excalfactoria*). Small to very small, sexually dimorphic; legs short without spurs, wings short, rounded, tail short with 10–12 feathers (8 in *Coturnix chinensis*). Uppertail-coverts longer than tail-feathers; first primary longer than the 10th and subequal with 3rd, 2nd slightly longer than others. Reluctant to fly, but flight swift and direct, permitting long-sustained sea and desert crossings. Voices high-pitched, distinctive. Common Quail *C. coturnix* and Harlequin Quail *C. delegorguei* heavily exploited for food while migrating, and Common Quail has been domesticated.

3 spp. in Africa, none endemic; *coturnix* and *chinensis* also in Europe and/or Asia, *delegorguei* also in Madagascar and S Arabia.

PHASIANIDAE

Plate 1 *Coturnix coturnix* (Linnaeus). Common Quail. Caille des blés.

(Opp. p. 32)

Tetrao coturnix Linnaeus, 1758. Syst. Nat. (10th ed.), 1, p. 161; Europe, Asia, Africa (restricted type locality Sweden).

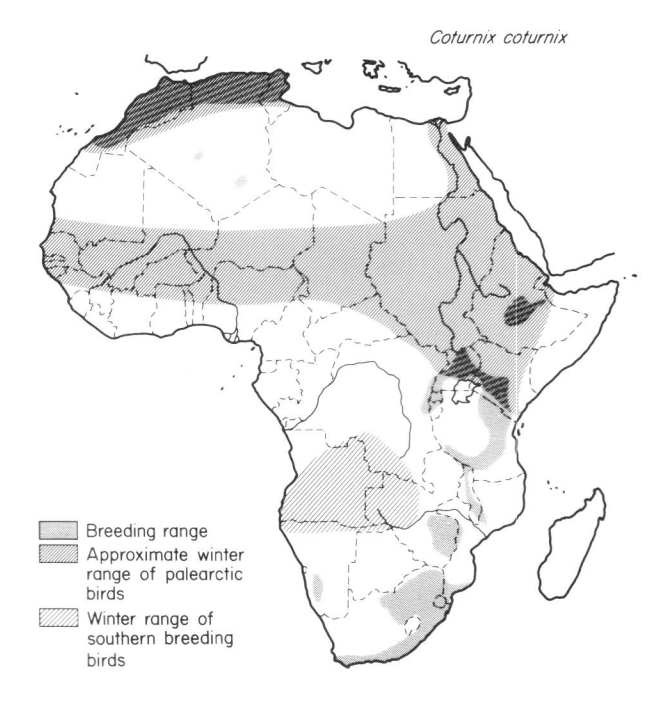

Range and Status. Africa, Malagasy Region, Hawaiian Islands (introduced), Atlantic islands, Eurasia east to central Siberia and south to India.

Nomadic resident, intra-African migrant and Palearctic visitor. Breeds in much of Morocco, especially on coastal plains and in mountains, in favourable years as far south as the Seguiet el Hamra; N Algeria south to subdesert at S edge of mountains, also in oasis of Reggane and in Hoggar Mts; Tunisia (throughout except S desert area), and occasionally Egypt. Breeding distribution south of Sahara patchy with large gaps; confined to montane grasslands in Ethiopia, E Zaïre, E Africa and south to central Zimbabwe and central Mozambique; another population both in mountains and at lower elevations in S Mozambique, South Africa and S Namibia. Palearctic migrants winter mainly north of equator from Senegal to Sudan and Ethiopia, some also in Morocco, Algeria, Tunisia and Egypt. Southern populations of *C. c. coturnix* winter in S Zaïre, Angola, W Zambia and N Namibia. Despite heavy exploitation by netting on migration (e.g. through Egypt and Sinai several million taken annually at beginning of century, 200,000 in 1968: Woldhek 1980), European *C. c. coturnix* still locally and seasonally abundant to very abundant. African *C. c. erlangeri* less exploited, locally abundant.

Coturnix coturnix

- Breeding range
- Approximate winter range of palearctic birds
- Winter range of southern breeding birds

Description. *C. c. erlangeri* Zedlitz: montane grasslands from E Zimbabwe and W Mozambique (including Mt Gorongosa) to Malaŵi, extreme NE Zambia (Nyika Plateau), Tanzania, E Zaïre, Uganda, Kenya, Sudan and Ethiopia. ADULT ♂: forehead, crown, nape dull dark brown edged rufous-buff. Narrow buff streaks on each feather of crown and wider streak above eye. Lores and sides of face buff, with dark rufous streak from gape to ear-coverts. Lower neck rufous. Mantle, back, upper scapulars and upperwing ochraceous brown, mottled with blackish brown and barred rufous-buff, each feather with lance-shaped, buff-white shaft-streak; lower back and uppertail-coverts dark brown with rufous-buff barring and streaking; tail brown with narrow rufous bars, each feather edged with buff. Throat buff with broad median blackish streak and lateral bar, forming an anchor-shaped mark (lacking in some individuals). Breast rufous to paler chestnut, palest on undertail-coverts, flanks broadly streaked black and buff. Belly dull grey. Primaries and secondaries dark grey, mottled and tipped buff, and irregularly barred rufous-buff, more pronounced on secondaries. Bill blackish brown; eye brown to red-brown; legs yellowish pink or pale brown. ADULT ♀: generally less brightly coloured. Throat lacks 'anchor' bar, and breast has some dark streaking or spotting; bill brownish black. Pale and rufous morphs occur: in rufous morph, throat, sides of breast and flanks chestnut rather than buff. SIZE: wing, ♂ 98–114 (104), ♀ 100–112 (106); tail, ♂♀ 31–40 (36); tarsus, ♂♀ 22–27 (24). WEIGHT: ♂ (n = 144) 75–111 (90), ♀ (n = 157) 81–122 (103).

IMMATURE: like ♀ but paler, flanks less heavily barred, upper breast spotted.

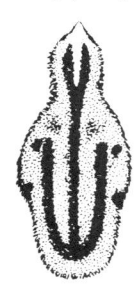

DOWNY YOUNG: above rufous-brown, with dark streak over forehead, 2 blackish streaks over crown; sides of head spotted darker. Upperparts rufous, each feather with a broad black central and 2 lateral blackish streaks. Base of wing and forewing blotched blackish, sides and underside faintly streaked darker. Wings black, blotched with rufous-buff. Underparts buffy. Eye dark brown, bill and legs pinkish flesh.

C. c. coturnix Linnaeus: breeds N Africa, Cape Province, Natal, Zululand, Lesotho, Swaziland, S Mozambique, Transvaal highveld and occasionally in central Namibia. Palearctic birds winter in equatorial Africa. Southern birds may overwinter in their breeding range or migrate to N Namibia, Angola, S Zaïre or W Zambia. Similar to *erlangeri* but slightly larger and generally paler due to less blackish brown and rufous on upperparts. ♂ in breeding plumage has much more distinct anchor-shaped mark on throat bordered below with white. SIZE: wing, ♂ 107–117 (112), ♀ 105–114 (111). WEIGHT (southern Africa): ♂ (n = 185) 76–115 (92), ♀ (n = 175) 80–132 (101).

Subspecific treatment follows Clancey (1976) and Bourquin (1980). Southern African birds, formerly separated as *C. c. africana* Temminck and Schlegel, differ too little from the nominate race to warrant recognition. Migratory birds breeding in South Africa and adjacent countries are considered to belong to the nominate race; the relatively sedentary populations of E and central Africa are placed in *erlangeri*, a name formerly used only for Ethiopian populations.

Field Characters. Small, dumpy and short-winged, slightly larger than Harlequin Quail *Coturnix delegorguei* and considerably larger than Blue Quail *C. chinensis*; in flight appears paler and more heavily streaked. Underparts buffy with black and white flank streaks. ♂♂ of other African quails have black and white throat pattern and blackish underparts, ♀♀ have darkly barred or spotted underparts. Outer webs of primaries dusky brown mottled or barred with buff. Voice distinct. Much larger than button-quail *Turnix* spp. and in flight wing appears more uniform (button-quail have pale

downy young

inner wing panel contrasting with darker primaries). Sometimes confused with juveniles of sympatric partridges *Alectoris* and *Ammoperdix* spp., and francolins *Francolinus* spp.

Voice. Tape-recorded (10, 20, 33, 74). Best known, most heard call is ♂ advertisement call, a high-pitched, tri-syllabic 'wheet whit-it, wheet whit-it' repeated 3–10 times, often rendered 'wet-my-lips'. Higher pitched than similar call of Harlequin Quail. ♂ also utters low growling or grinding 'mau-wau', and, when feeding ♀, 'grueee-grueeegrueee'. ♀ utters low melodious calls, variously rendered 'huee-huee', or 'bru-bru'; when inviting copulation with ♂, a soft 'rururururu'. When flushed both utter a trilling or spluttering 'whreeee' or 'scree-scree'. When uneasy, a soft 'trulili', and, in anger, a purring 'gurrgurrgurr', resembling purring cat. Most calls inaudible at distance except ♂'s 'wet-my-lips', audible at up to 400 m. Calling more frequent at dawn and dusk.

General Habits. Inhabits semi-arid, perennial grassland and herbage 0·4–0·5 m high with patches of short grass. May occur in other vegetation types, especially when migrating. Particularly attracted to agricultural lands (growing crops, fallow fields, pastures) and grasslands regenerating after being burnt. Low rainfall/high temperature regimes and areas with heavy bush/tree cover are avoided. Generally found at higher altitudes than Harlequin Quail in tropical Africa (1200–3000 m) but in southern Africa at 0–1800 m.

Normally remains within cover, perhaps to avoid aerial predators; very seldom seen unless flushed or crossing open space. Presence best indicated by calling, although difficult to locate due to ventriloquial nature of advertisement call. When flushed, flies with whirring wings very fast in straight line for 80–150 m, rarely further, and pitches suddenly in long grass or other suitable cover. Often flush as pairs, flying parallel for some distance and then crossing over. Roosts in coveys on ground at night.

Palearctic birds mainly or wholly migratory in winter, entering Africa in very large numbers. Southward-moving autumn migrants concentrate near sea coasts, and are then heavily exploited (especially in Egypt and Italy). Migratory movements in and around Mediterranean basin are highly complex and crossings are made along broad fronts. Birds fly few m above waves, and divert around, and perhaps follow, obstacles, e.g. moving ships, sometimes resulting in longer flight in same direction as ship. Autumn passage south across Mediterranean mainly late Aug–Oct, peaking mid-Sept, most migrants arriving at winter quarters in Oct to early Nov. Apparently cross Sahara in unbroken flight, not normally landing at oases. Western and eastern crossings may involve shorter traverse over stark open desert. Some birds breeding in N Africa cross Mediterranean northwards in mid-summer to southern Europe. Most migrants in sub-Saharan Africa winter from Senegal east to Ethiopia and north Kenya in semi-arid grasslands. In Senegal, reported at densities of 1–5/ha. Migrants leave austral wintering areas Feb–May, peaking in April, with some later stragglers. Birds ringed Italy recovered Senegal, Libya, Tunisia, Algeria, Morocco; ringed Holland recovered Morocco; ringed Tunisia (Cap Bon) recovered France, Belgium, Italy, Malta, Yugoslavia, Hungary, Albania, Greece, Bulgaria, Rumania, Morocco.

Movements within Africa of resident birds poorly understood; but in southern Africa more regularly migratory, less locally nomadic than in central Africa. Move into Natal–Cape Province in spring (Sept–Oct), arriving at breeding quarters in Oct. After breeding, depart north again to N Namibia, Angola, S Zaïre and W Zambia. Movement analogous to European race, but in opposite direction, and in smaller numbers. Some birds overwinter within normal breeding range if conditions (temperature/food) favourable. In central tropical Africa and Kenya–Ethiopia movements much more irregular, essentially nomadic. May appear in numbers following rains in some areas, possibly breed, and depart again, not reappearing in similar numbers for years.

Food. Reviewed in detail by Mentis (1978) and Bourquin (1980). Opportunistic feeders, taking mainly seeds of grasses, weeds, and some fallen grain, supplemented by worms and small ground-living animals and their larvae, including spiders, flies, beetles, hemipterans, ants, small grasshoppers, termites and small molluscs. Not a pest of grain crops, gleaning only fallen grain from ground.

Breeding Habits. In wild state poorly known; studied in detail in captivity (Orcutt and Orcutt 1976). Mating system probably monogamous, but may vary and include bigamy or successive polygamy. ♂♂ arrive on breeding grounds before ♀♀, and advertise presence with 'wet-my-lips' call. In captivity ♂ courts only own ♀ and ♀ responds to calls of only 1 ♂; individual voices recognizable. ♂ courtship-feeds ♀ who apparently selects nest-site, then responds to ♂ advertising call. ♂ then runs at ♀ and circles round her (Circle Display) with neck stretched, feathers ruffled and wing nearest ♀ dropped. Inviting copulation, ♀ crouches, spreads wings and utters inviting call. ♂ responds by approaching with both wings lowered, breast-feathers ruffled. He mounts, gripping feathers of ♀'s nape for 2–3 s during act, which may be rapidly repeated up to 10 times. Copulation frequent before egg-laying commences, thereafter reduced but continues until clutch complete. Established pairs loaf at nest together, ♂ not straying far from ♀.

NEST: in dense vegetation. ♀ makes small scrape and adds bits of grass to it, working on it daily at laying time, usually plucking nest material when sitting. Continues to pluck and add vegetation during incubation, and repeated egg-turning results in rim of even height around eggs. External diameter 14 cm, internal 7–9 cm, depth 2·5–4 cm.

EGGS: 2–14, usually 5–7, laid on consecutive days; 1 clutch of 14 probably involved 2 ♀♀; southern African mean (79 *C. c. coturnix* clutches) 7·2. Pointed ovals; smooth, glossy; colour variable: pale cream, yellowish or dark buff, speckled with deep olive and/or reddish brown with some blackish blotching; sometimes dark blotches large and extensive. SIZE: (n = 271, South Africa, *C. c. coturnix*) 25·9–33·1 × 20–25·1 (30·3 × 23·2); (n = 300, Palearctic, *C. c. coturnix*) 25–34 × 20–25 (30 × 23). WEIGHT: 8–9.

LAYING DATES: Morocco, Mar–June (mainly Apr–May); Algeria, from Mar (desert oasis) to July (coastal mts); Tunisia, Mar–May; South Africa: W Cape, Sept–Dec, after winter rains, Karoo Aug, E Cape, Nov–Jan (rains), Natal, Dec, Transvaal and Orange Free State, Mar–Apr; Namibia, Jan, May; Zimbabwe, June, Dec; Zambia, (birds in breeding condition Jan); Malâwi, Jan, Oct–Nov; East Africa, in any month of year, season prolonged on high plateaux east and west of Rift Valley, with peaks both in dry and wet seasons, most in driest months; Ethiopia, possibly July.

INCUBATION: begins when clutch complete. By ♀ only, who sits tight, and if disturbed sleeks feathers and creeps away among grass. Leaves nest several times daily for 7–20 min intervals. ♂ sits near incubating ♀ and may join her when off feeding. Period: 16·5–17 days (Orcutt and Orcutt 1976); 17–20 days (Cramp and Simmons 1980). Captive ♀♀ may be aggressive to ♂♂ near end of period. Eggs hatch synchronously; chicks perhaps hear one another in eggs assisting synchronous hatch.

DEVELOPMENT AND CARE OF YOUNG: chicks weigh 5·5 at hatching. Can flutter from ground at 11 days, fully fledged at 19 days. Sexually mature and able to breed in year after hatching, perhaps earlier; in N Africa, young hatched early breed in same season. Chicks can leave nest within 15 h of hatching, but may be brooded in nest for 1 night by ♀ who, at hatching, constantly moves about, uttering low peeping and trilling calls. ♂ does not approach nest at hatching, but remains nearby. ♀ broods young, usually in cover, and may chase ♂ from them if he approaches too close. Chicks twitter softly when brooded, utter shrill alarm while separated. ♀ constantly maintains vocal communication with chicks, uttering low trills, hoarse peeping call, and alarm call, a sharp 'pweet'; also a low 'kukkuk-kuk' warning call which causes them to disperse, crouch or freeze. One or both parents may feign injury if family party is pressed by predator. ♀ feeds chicks at first, but by 4th day they range away from her, and by 11th day range freely, maintaining contact by calling. Remain in family parties 30–50 days, possibly migrating in small family groups.

BREEDING SUCCESS/SURVIVAL: in Palearctic *c.* 50% of eggs produce chicks which survive to migrate south. Chicks especially susceptible to death by hypothermia when wet, e.g. sudden rain storm. Hawks and herons are predators in South Africa, with herons swallowing chicks alive. In flight very vulnerable to birds of prey (e.g. falcons and especially accipiters).

References
Benson, C. W. and Irwin, M. P. S. (1966).
Bourquin, O. (1980).
Clancey, P. A. (1967, 1976).
Cramp, S. and Simmons, K. E. L. (1980).
Orcutt, F. S. and Orcutt, A. B. (1976).

Plate 1
(Opp. p. 32)

Coturnix chinensis Linnaeus. Blue Quail. Caille bleue.

Coturnix chinensis Linnaeus, 1766. Syst. Nat. (12th ed.), p. 277; China and Philippines.

Range and Status. Africa, India, S China, SE Asia, Philippines and Australasia.

Nomadic resident in Sierra Leone, Mali (1 record), and from E Ivory Coast to Nigeria (local and uncommon breeder, north to Jos and Bauchi Plateaux), Cameroon (common in dry season in woodland of forest/savanna mosaic), S Chad, extreme S Sudan (uncommon) and W Ethiopia (uncommon). Occurs outside forest in Gabon and Zaïre, becoming widespread in Angola (central plateau), Zambia and Malâwi; widely distributed in E Africa (except arid NE Kenya and dry interior Tanzania), including Zanzibar and Pemba, but scarce and local. Range more restricted in Zimbabwe (mainly Mashonaland Plateau) and central Mozambique; sparsely distributed from E Transvaal and Zululand south along coastal areas to E Cape. Local and uncommon in most regions, but no evidence of threat to population.

Description. *C. c. adansonii* Verreaux: only subspecies in Africa. ADULT ♂: top of head, face, sides and back of neck, mantle, back and rump blue-black; centre of crown streaked black. Malar area white, bordered behind by black 'anchor' bars. Scapulars and tertials dark chestnut-maroon mottled and streaked with slate blue. Uppertail-coverts chestnut with blue streaking. Tail slaty black. Throat black separated from breast by broad white band, conspicuously edged black. Underparts, except sides and flanks, dark slate grey; sides and flanks chestnut, streaked blue-grey. Primaries and secondaries brown. Upper mandible black with blue edges, lower mandible dark blue; eye dark red, lighter at outer edge; legs bright yellow. ADULT ♀: pale streak on centre of crown; upperparts dark brown, faintly streaked with buff and mottled and vermiculated with rufous and black. Sides of face, round eye and throat pale rufous-brown. Underparts light rufous-brown, barred black from neck and breast downwards, bars heavier toward belly. Upperwing-coverts with fine dark bars and pale shaft-streaks. Bill horn-colour; eye reddish brown; legs duller yellow than ♂. SIZE: wing, ♂ (n = 9) 78–82 (80·0), ♀ (n = 5) 80–84·5 (81·6); tail, ♂♀ 26–32 (29·6); tarsus ♂♀ 18–20 (19·0). WEIGHT: (*C. c. adansonii*) 1 ♂ 43, 1 ♀ 44; (*C. c. chinensis*) 43–57.

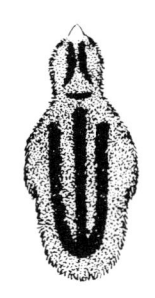

IMMATURE: like adult ♀ but has pale shaft-streaks on rump, upperwing and uppertail-coverts.

DOWNY YOUNG: (*C. c. chinensis*) generally brown above with buff crown streak, buff supercilium and wing-tips. Below light brown, chin and throat buff.

Field Characters. A tiny quail, length 12–13 cm; ♂ distinguished by dark bluish black body with contrasting white breast-patch, and in flight by conspicuous chestnut wing-patch. ♀ and immature are darker brown than ♀♀ of other quails, especially underparts which also have pronounced dark barring, and lack white eyebrow or throat-patch. Habitat also helps, most likely to be found in lush grass near swamps, in vleis and dambos. Whistling call distinctive. Similarly dark Böhm's Flufftail *Sarothrura boehmi*, occurring in same habitat, has longer body and neck, long dangling legs, slower wing-beats (without pauses) and white streaks on upperparts.

Voice. Tape-recorded (C). ♂ utters trisyllabic piping whistle 'kee kew yew', 1st note loudest. A squeaky triple whistle 'tir-tir-tir', or shrill 'swi' given when flushed. Calls rather infrequently, unlike other quails.

General Habits. Inhabits tall grassland up to 1800 m, often along edges of swamps or waterlogged areas; also in patches of this habitat within forest, and around edges of swampy forest. Avoids dense woodland, forest and arid areas. Very hard to locate; does not attract attention to presence by frequent calling; usually seen only when flushed, and then with difficulty. Flies rapidly, then pitches; flight swift and direct like other quails; not usually long sustained (20–40 m). Sometimes runs and freezes when chased by dog rather than flying. Emerges into open to feed just after dawn, but remains wary. Not gregarious; found singly, in pairs, or in small groups or family parties.

Possibly less nomadic than other quails. Near Lagos, Nigeria, arrives during peak of rains in late Mar, breeds and then departs when chicks well grown (Elgood *et al.* 1973). In extreme south of range (Natal, E Cape Province) may be summer-breeding resident, moving north again in winter. Most often seen in dry season when burning of long grass elsewhere forces it to concentrate in areas of moist grassland near swamps. Apart from such local movements, not known to migrate in main tropical range.

Food. Seeds of grass and weeds, green plant material, and some small insects and their larvae, including termites.

Breeding Habits. Mating system monogamous in captivity; ♂ courtship-feeds mate. Said to be territorial (Bannerman 1930), but little evidence. If pair flushed, only fly short distance and call to one another until they reunite.

NEST: small scrape on ground below grass tussock, sometimes lined with grass stems; occasionally near cultivation.

EGGS: 3–9. Broad ovals; without gloss, rough in texture and thick-shelled; uniform olive-brown with minute reddish brown and purple dots. SIZE: (n = 16, South Africa) 21·0–28·4 × 17–21 (25·5 × 19·8).

LAYING DATES: South Africa, Dec–Apr (summer rains); Zimbabwe, Jan–Apr, during rains; Zambia, Jan–Feb, height of rains; Malaŵi, Feb, Mar, May; East Africa: Kenya-Uganda, Regions A, B and D mainly May–July, late in rains and in drier mid-year break; N Zaïre, Nov, S Zaïre, Jan–Feb, at end of rainy season; W Africa, Apr–July, early rains; 'Cameroon', Nov (dry season or end of rains), W Cameroon, May–June, S Cameroon, gonads active in May. Records suggest breeding mainly in rainy season.

INCUBATION: by ♀ alone, although ♂♂ have been collected with well developed brood-patch, suggesting they may also incubate. Period: *c.* 16 days.

DEVELOPMENT AND CARE OF YOUNG: weight at hatching (*C. c. chinensis*) 2·9–3·5 (Bernstein 1973). Both parents care for chick; family parties remain together until young can fly well.

BREEDING SUCCESS/SURVIVAL: markedly affected by weather. Newly hatched unbrooded chicks survive only few minutes at ambient temperatures below 20°C; most sensitive during 1st 2–3 weeks.

References
Ali, S. and Ripley, S. D. (1969).
Clancey, P. A. (1967).

PHASIANIDAE

Plate 1
(Opp. p. 32)

Coturnix delegorguei Delegorgue. Harlequin Quail. Caille arlequine.

Coturnix delegorguei Delegorgue, 1847. Voy. Afr. Austr., 2, p. 615; Oury, upper Limpopo River, Transvaal.

Range and Status. Africa, Madagascar and S Arabia. Nomadic resident and intra-African migrant. Found throughout most of eastern and southern Africa from N Ethiopia and central Sudan to N Namibia and E Cape Province. Absent from deserts of SW and NE and from lowland forest. Distribution and movements poorly understood in rest of Africa, where few records except in Nigeria; west of Nigeria only a few scattered records from Senegal, Mali and Ivory Coast. Occurs on Bioko, São Tomé, Zanzibar and Pemba. Locally common to very abundant. Heavily trapped for food in some areas but no evidence of general population decline.

Coturnix delegorguei

☐ Main breeding range
☐ Sparse or irregular (status uncertain)

Description. *C. d. delegorguei* Delegorgue: African mainland, Bioko, Pemba and Zanzibar. ADULT ♂: crown and nape blackish brown, each feather edged with pale brown, those on mid-crown with buff shaft-streaks forming distinct medial streak; wide cream superciliary stripe extends down side of head and neck. Narrow supra-loral stripe and ear-coverts brown; lores white, extending backwards to form streak below and behind eye. Moustachial stripe black, extending backwards to black 'anchor' mark below ear-coverts. Upperparts variable, dull grey-brown to blackish brown, sometimes with fawn tinge mottled with grey-brown, usually darker on rump and uppertail-coverts, each feather with central cream shaft-streak and cross-bars; cream streaks and bars margined with black. Tail dark brown, each feather with whitish shaft-streak and cross-bars. Throat white, with median black streak spreading laterally to form conspicuous black 'anchor' mark, separated from upper breast by white band. Upper breast black, becoming deep chestnut on lower breast and belly, each feather centrally streaked black, widest on flanks; undertail-coverts chestnut. Primaries and secondaries grey-brown above, below pearl grey, unbarred; outer web of outermost primary edged creamy buff; underwing-coverts buff to white. Bill blackish grey; eye red-brown or chestnut; legs brownish white to pale pink. ADULT ♀: somewhat paler above and below. Throat white to pale buff, lacking conspicuous black 'anchor' bar. Breast and upper belly pale chestnut, darker on sides with scale-like white markings and some black spotting on breast. Wings unbarred, grey-brown above, grey below. Bill brown, base yellow. SIZE: wing, ♂ (n = 35) 91–100 (96), ♀ (n = 30) 93–105 (100); tail, ♂♀ 28–33 (31·7); tarsus, ♂♀ 24–27 (25). WEIGHT: ♂ (n = 11) 65–81 (72·4), ♀ (n = 16) 73–94 (78·5).
IMMATURE: like adult ♀ except crown and nape light brown, with pale crown-streak. Underparts paler, rather grey, lower throat and breast spotted and barred dark brown, each feather with terminal white shaft.
DOWNY YOUNG: crown, face and nape ochraceous buff; crown-streak olive-brown. Broad dark brown stripe down mid-back flanked on either side by narrower stripes of black and buff. Wings buffy yellow, darker basally. Underparts buffy yellow. Eye brown; legs pale pinkish.
C. d. histrionica Hartlaub: São Tomé. Darker generally in both sexes than *delegorguei*. Size similar, wing ♂ 91–100, ♀ 93–98.

downy young

Field Characters. ♂ slightly smaller and much darker on underparts than Common Quail *C. coturnix*. Upperparts duller, less streaked, but with more prominent narrow cross-barring. In flight Common Quail appears browner and more streaked. Primaries uniform brown and not mottled with buff. Good view of ventral surface of ♀ shows much darker black spotting on upper breast. White flank streaks of ♀ narrower and duller than in Common Quail, and undertail-coverts are rich orange-buff (paler in Common Quail). Sympatric with Common Quail in tropical Africa, but Harlequin usually predominates in lowlands, Common Quail in highland grasslands. Downy young much darker brown than Common Quail; immatures barely distinguishable. Larger and fatter than Little Button-Quail *Turnix sylvatica*, showing much shorter neck in flight. Upperwings of Little Button-Quail much paler (mottled at close range) contrasting with darker flight-feathers.

Voice. Tape-recorded (10, 74, C). ♂ advertisement call, 2–3 note 'whit-whit-whit' 'whit-whit-whit' repeated 5–6 times at short intervals; notes evenly spaced (whereas 3-note call of Common Quail has pause between 1st and 2nd notes). Resembles that of Common Quail, but more metallic, less liquid. ♀ answers with almost inaudible 'quick-ic' or 'queet-ic' at much lower volume. In alarm when flushed, a squeaky, trilling 'kreeee'. When disturbed, small chicks make high-pitched peeping calls similar to those of Common Quail.

General Habits. Inhabits open, rank grassland less than 0·3 m high with scattered bush cover, often of species with large seeds providing abundance of food, e.g. *Setaria* and *Brachiaria* spp., or *Sorghum purpureo-sericeum*. Sometimes frequents mixed cultivation and dense thornbush if grass cover available. On São Tomé, also in coconut plantations and in vicinity of airport. In Chad, does not occur north of 500 mm rainfall isohyet (Salvan 1968). In southern Africa more likely to frequent higher plateaux (occasionally to 2000 m) than in tropical Sudan, Ethiopia and Kenya, where normally in lowlands below 1200 m.

More gregarious than Common Quail, found in coveys of 6–20 during non-breeding season. Tends to concentrate locally, often in densities exceeding 10/ha. 200 may be on the wing at same time during exceptional concentrations (Jackson and Sclater 1938). Not often seen during day unless flushed, but sometimes feeds at edges of roads and tracks, affording reasonable view to observer approaching quietly. Flushes easily but reluctant to fly far; gets up with whirr, flies short distance, then drops down and runs. Escapes easily unless tracked with dog. Attracted to house lights during migration at night, when many stunned by crashing into buildings. Roosts in coveys on ground at night.

Nomadic rather than regularly migratory in East and central Africa. Movements seem generally linked with rainfall; may appear in enormous numbers at restricted localities, breed and then not reappear for years. However, probably some regular movement southwards occurs in Nile Valley and in W Kenya around Lake Victoria at Kavirondo Gulf. In southern Africa, birds head south to breed in rains, returning north again in dry season; but not to same areas annually. In Zambia, influx of birds noted during wet season but movements erratic and numbers vary greatly from year to year; some local movements not correlated with rainfall, and some birds present throughout dry season (P. B. Taylor, pers. comm.). In Chad and W Sudan, moves north to breed in wet season (June–July), returning south in dry season. Numbers increase in N Nigeria (Kano–L. Chad) during rains, but pattern of movements elsewhere in Nigeria confusing.

Along Kavirondo Gulf, tribesmen trap quail extensively. Mode of trapping is to erect tall pole with c. 12 small wicker cages, one below the other, suspended from it. Each basket contains ♂ or ♀ (♂♂ predominate) Harlequin. Their combined calling attracts other ♂♂ and ♀♀ who become ensnared in nooses laid out within maze of sticks (or planted *Euphorbia tirucalli*) at base of pole. Mazes are visited frequently by owners to prevent quail from strangling themselves. Large numbers must be taken, since as many as 80 quail caught in 1 pole/maze in single morning, and there may be up to 2 pole/mazes per km along northern shores of Kavirondo Gulf. Once gulf is rounded, quail can apparently spread over wider areas of taller grass and are less easily caught. Decoy quails fed on grain and termites quickly become very tame and easily handled. Most are eaten by owners in due course; a few retained each season to attract next year's migrants.

Food. Predominantly insects, including grasshoppers, beetles, bugs, caterpillars, ants and termites. Also seeds, especially of *Eleusine*, millet, green shoots and leaves, and some small ground molluscs.

Breeding Habits. Mating probably temporary and monogamous for breeding season. Possibly polygynous (van Someren 1956) and 2 ♀♀ may lay in same nest. Throughout range breeds in large compact groups, almost colonies, with similar country nearby completely unused. ♂♂ are very pugnacious at breeding grounds, fighting while jumping into air and facing one another; bills primary weapons. One may seize another and shake opponent violently. ♂ may even attack mate. ♂ display like Junglefowl *Gallus gallus*, i.e. laterally to ♀ with the wing closest to her fanned, occasionally courtship feeding ♀, who responds with faint 'twe, twe, twe' calls as she collects offering. Copulation then takes place as in Junglefowl, i.e. receptive ♀ crouches, ♂ stands on her back grasping her nape feathers with his bill, ♀ lifts and spreads her tail allowing cloacal 'kissing'.

NEST: usually in grassland with well spaced tufts, allowing sitting ♀ several escape routes. Small, well concealed scrape, sometimes with grass bower which helps conceal nest and shade contents. Sometimes lined with leaves or grass. Rarely found, even where breeding population very dense.

EGGS: 4–8, occasionally up to 22 (probably laid by 2 or more ♀♀ in same nest); in captivity, single ♀ may lay more than 20; mean (9 clutches) 4·8. Broad ovals; some gloss but generally distinguishable from those of Common Quail by relatively rougher shell; dull creamy buff, covered, mainly at broad end, with fine purplish brown freckles or blotching; pattern highly variable, but all eggs of each ♀ similar. SIZE: (n = 39, *C. d. delegorguei*) 27·4–30·9 × 21·7–23·5 (29·2 × 22·4).

LAYING DATES: E South Africa, Oct–Mar, peaking in Dec–Jan; S Mozambique, 'from about Oct' (Clancey 1971); Namibia, July–Sept; Zimbabwe, Oct–June, peaking Jan–Feb during rains; Zambia, Oct, Feb–Apr (rains); Malawî, Jan, Oct; Kenya and N Tanzania, May–June and Nov–Dec, in both rainy seasons but also Feb after unusual rains; Ethiopia, Apr–June (rains); Sudan, July (rains), Zaïre, Aug, Sept. São Tomé, probably all year (de Naurois 1981). Breeding everywhere associated with heavy rains, irregular and local, and can be successful even on waterlogged heavy clay soil.

INCUBATION: by ♀ alone. ♀ sits very tight; may move eggs to new nest if disturbed. Period: 14–18 days.

DEVELOPMENT AND CARE OF YOUNG: 5-day-old chicks can fly short distances. At 1 month mean weight of 7 chicks was 21 g (Dudley 1971). ♂ may stay with brood until chicks fly. One or both sexes may feign injury to distract potential predators, while chicks scatter and squat.

References
Clancey, P. A. (1967).
Trollope, J. (1966)
van Someren, V. G. L. (1956).

PHASIANIDAE

Genus *Ammoperdix* Gould

Small, sexually dimorphic, partridge-like birds, larger than quails *Coturnix* spp. but smaller than partridges *Alectoris*. Paler than partridges, flanks with horizontal, not vertical, black bars, outer tail-feathers red, conspicuous in flight. Body thickset, tarsi short, unspurred, although stance very upright. Tail with 12 feathers, somewhat rounded, but feathers subequal. 1st primary longer than 10th, subequal with 6th, and not much shorter than 3rd which is the longest. Run swiftly on ground, and seldom willingly take wing. Found in arid areas from NW India to NE Africa. Occur in small groups, locally common, but habits very poorly known. 2 spp., 1 in Africa and Asia, 1 in Asia.

Plate 1
(Opp. p. 32)

Ammoperdix heyi (Temminck). Sand Partridge. Perdrix de Hey.

Perdix heyi Temminck, 1825. Pl. col. livr. 55, pl. 328, 329; Desert of Akaba, Arabia.

Range and Status. NE Africa to Jordan, Israel, W and S Arabia.

Resident in NE Egypt between Gulf of Suez and lower Nile, extending south through E Egypt to NE Sudan. Status obscure, but probably still common or frequent in most of range; occurrence in NE Ethiopia not recently confirmed.

Description. *A. h. nicolli* Hartert: higher country of NE Egypt between Gulf of Suez and lower Nile. ADULT ♂: forehead, crown, cheeks, neck, upper back and upperwing-coverts and scapulars pale pink-buff, feather tips tinged grey when in fresh plumage; ear-coverts white, conspicuous. Uppertail-coverts and lower back sandy-buff. Central tail-feathers (obscured by uppertail-coverts) vermiculated grey and cinnamon; outer ones bright rufous, undersides paler rufous. Chin and throat orange-buff; breast, belly and flanks pinkish sandy-buff, tinged greyish, with long flank feathers pinkish grey, streaked with black and chestnut. Undertail-coverts pink-buff. Flight-feathers brown, mottled buff, with some heavier darker bars on outer primaries; undersides pale grey. Underwing-coverts and axillaries pale pinkish. Bill dull orange or yellow; eye brown to reddish brown; legs dull yellow. ADULT ♀: generally sandy-coloured; differs from ♂ in having forehead, crown and neck vermiculated grey and buff, with faint pale streak above eye. Lacks conspicuous white ear-coverts of ♂. Upperparts pale pinkish buff, more or less heavily vermiculated grey, some dark brown spots near tips of feathers. Chin and throat whitish, rest of underparts sandy-buff, flank feathers vermiculated brown, no conspicuous black flank-streaks. Wing-feathers more heavily marked brown than in ♂, tail similar. Eye brown; bill and legs duller yellow than in ♂. SIZE: wing, ♂ (n = 4) 129–139 (131), ♀ (n = 6) 123–132 (128); tail (*A. h. heyi*), ♂♀ (n = 21) 56–65 (60); tarsus, ♂ (n = 5) 33–35 (33·6), ♀ (n = 6) 33–34 (33·4).

IMMATURE: both sexes unmarked sandy-buff, primaries shorter and more pointed than in adult.

DOWNY YOUNG: sandy-buff above with some rufous mottling on upper wing and lower back; some blackish mottling on head, and a blackish eye-stripe; underparts buffy. Bill blackish; eye brown; legs pale buff.

A. h. cholmleyi Ogilvie-Grant: E Egypt and NE Sudan between Nile and Red Sea. Darker, more rufous-brown, rather than sandy. Chin-patch in ♂♂ chestnut, not orange-buff.

Field Characters. Smaller and paler than any sympatric francolin *Francolinus* sp., and barely overlaps with any (only in south of range); also not in range of Barbary Partridge *Alectoris barbara*. Larger than Common Quail *Coturnix coturnix*, much paler, and plumage overall much plainer, without heavy mottling and streaking. Face pattern and flank stripes of ♂ distinctive, and, in flight, both sexes show conspicuous rufous corners to pale tail.

Voice. Not tape-recorded. A loud, yelping 'quay', 'teu', or 'quake' rendered as 'kew-kew-kew', accentuated by echoing off rocks. In alarm 'wit-wit'. In flight, wings make sibilant or whirring sounds.

General Habits. Inhabits open, hilly and rocky desert country below 600 m, especially in and near wadi beds with some vegetation, seldom in flat open sandy or stony desert; most abundant in vegetated valleys, with shrubs and some grass. Normally in pairs or small groups, occasionally in flocks of up to 15; most groups 2–5. Normally silent, except in breeding season or at dusk and dawn. Allows close approach in vehicle, very reluctant to fly, always preferring to escape by running swiftly among vegetation or rocks. May squat and sit tight, then very difficult to see until it moves. Despite statement that it drinks frequently (Etchécopar and Hüe 1967), can live for years in absolutely waterless desert

(Arabia). Seeks shade during heat of day. Numbers probably much greater than often supposed, in Arabia (Hadhramaut) locally forming over 10% of total bird numbers and probably 15–20% of avian biomass (L. H. Brown, unpub. data).

Food. Seeds, berries, corms, bulbs, green leaves and insects. Commonly eats berries of *Salvadora persica* and *Commiphora* spp. and seeds of grasses and acacias.

Breeding Habits. Probably monogamous, since usually seen in pairs in and outside of breeding season.

NEST: shallow scrape, often shaded by bush or rock; sometimes lined with grass stems.

EGGS: 5–7. Rounded at broad end but rather pointed at narrow end, hard-shelled; uniform pale yellow-buff.

SIZE: (n = 50, *A. h. heyi*) 33·5–41·0 × 24·2–27·5 (37·0 × 26·7). WEIGHT: *c.* 14.

LAYING DATES: Sinai, Apr; no records African mainland.

DEVELOPMENT AND CARE OF YOUNG: young probably remain with parents until mature, and on basis of large coveys occasionally seen, several families may join together in non-breeding season (L. H. Brown, unpub. data).

References
Cramp, S. and Simmons, K. E. L. (1980).
Etchécopar, R. D. and Hüe, F. (1967).

Genus *Alectoris* Kaup

Stout-bodied, medium-sized galliforms, with strongly patterned heads and throats, plain upperparts, vertically barred flanks, and rufous outer tail-feathers. Sexes similar, ♂♂ slightly larger. Wings short, 1st primary longer than 10th, subequal with 6th, 3rd marginally the longest. Tail with 14 feathers, short, slightly rounded, outer feathers 10–25% shorter than inner. Legs short, strong; ♂♂ of some species have stout blunt spurs. Voices loud, cackling or crowing, not squealing or whistling.

7 spp. in Europe, W Asia and Africa; 1 sp., Barbary Partridge *A. barbara*, in Africa.

In 1964, 6 Chukar *Alectoris chukar* were released on Robben Island (33°40S 17°42E) near Cape Town, South Africa (Siegfried 1971), and a further 12 birds on Rooipoort (28°45S 24°05E), a game farm near Kimberley, in 1974. Latter introduction was unsuccessful (all birds killed by raptors or disappeared within 7 days). Robben Island birds have persisted, reaching a population of *c.* 300 as of December 1983 (A. Berutti, pers. comm.), but live under effectively artificial conditions (no major natural predators and preferred habitat is dense stands of Manitoka *Myoporum serratum*, an alien shrub).

Alectoris barbara (Bonnaterre). Barbary Partridge. Perdrix gambra.

Perdix barbara Bonnaterre, 1792. Tabl. Encyc. Meth., Orn., pt. 1, p. 208, pl. 94, f. 2; Morocco (ex Edwards, pl. 70).

Plate 1
(Opp. p. 32)

Range and Status. N Africa, Sardinia, S Spain (introduced), Canary Islands (introduced).

Resident, found almost throughout Morocco (absent only from some desert areas along Algerian border and in extreme south), N Algeria (common to abundant along coast and in mountains, extending south to edge of desert), Tunisia (throughout except extreme south), and most of coastal Libya, with break in centre isolating population in Cyrenaica. Separate population in Ajjer Mts (SE Algeria) and adjacent Libya (Ghat), and 8 birds recently observed in Hoggar Mts 70 km north of Tamanrasset (Mahler 1981). Former population on coast of NW Egypt probably extinct due to heavy hunting pressure (Sherif Baha el Din, pers. comm.). Elsewhere has declined locally due to hunting pressure and loss of habitat to agriculture, but still common or even abundant in remote areas and where protected; no race threatened at present.

Description *A. b. barbara* (Bonnaterre): NE Morocco, N Algeria, N Tunisia. ADULT ♂: centre of forehead, crown and nape dark chestnut. Cheeks, throat and broad streak above eye to upper hindneck ash grey; ear-coverts rufous-buff. Mantle, back, rump and uppertail-coverts dark grey, tinged olive on mantle and back, slaty on uppertail-coverts. Upperwing-coverts, scapulars and tertials grey-brown, tinged slaty, becoming cinnamon on greater and primary coverts. Tail mainly rufous-chestnut, with central feathers and bases of outer webs of others dark slate. Throat grey, separated from upper breast by chestnut gorget with white spots; upper breast paler grey; centre of breast pinkish cinnamon, rest of underside pinkish buff, tinged grey at sides of vent, flanks with vertical bars of rufous-cinnamon, black and white. Primaries dark brown-black, terminally edged buff on outer webs; secondaries dark olive-grey, tinged buff, freckled dusky on edges of outer webs. Underwing-coverts and axillaries pale grey, washed and mottled buff. Bill crimson to orange, yellower at tip; eye light brown; bare skin round eye pink or red; legs bright red with black claws. ADULT ♀: like ♂ but lacks spurs. SIZE: wing, ♂ 162–171 (166), ♀ 149–162 (156); tail, ♂ 89–106 (95·1), ♀ 75–91 (83·6); tarsus, ♂ 43–48 (45·1), ♀ 40–45 (43·1). WEIGHT: c. 500.

IMMATURE: plumage generally much paler, sandy; lacks grey on throat and upper breast, and chestnut gorget; fewer and less clear flank bars. Bill and legs yellow. Assumes adult plumage within 1 year.

DOWNY YOUNG: pale sandy-buff, with dark rufous streak along crown to hindneck, blackish streaks above and through eye; back streaked and barred blackish. Bill yellow; eye dark brown; legs pale yellow or pinkish buff.

A. b. spatzi (Reichenow): desert plains of SW and E Morocco, Algeria and Tunisia south of Atlas Mts, and in Tassili-n-Ajjer. Generally paler, crown and gorget rufous-brown, flanks less heavily barred. Slightly smaller: wing, ♂ 159–166 (162), ♀♀ 154–163 (158).

A. b. barbara (Reichenow): Cyrenaica. Very distinct; crown dark cinnamon, sides of head and chin blue-grey, gorget cinnamon spotted blue-grey, upperparts greyer, tinged pinkish rufous. Wing, ♂ (n = 15) 162–171 (166), ♀ (n = 8) 149–162 (156).

A. b. koenigi (Reichenow): NW Morocco. Upperparts less olive-brown than *barbara*, sides of head, throat and upper breast darker grey, breast deeper cinnamon; size similar.

Field Characters. The only partridge living in N and NW Africa except very local Double-spurred Francolin *Francolinus bicalcaratus*, which is duller, heavily streaked black, with chestnut crown, frequents different habitat and has totally different, raucous calls.

Voice. Tape-recorded (60, 62, 73, 246). Most common call 'kutchuk, kutchuk' or a rapidly repeated 'kakelik-kakelik' followed by a slower 'tchuk-tchouck-tchoukor-tchoukor'. ♂ advertisement call long-drawn, harsh 'krrraik', somewhat resembling call of Barn Owl *Tyto alba*. Alarm call when flushed a rapid, screaming 'chukachew-chew-chew-chew', or high-pitched squeal 'kooeea-kreeea'.

General Habits. Inhabits wide range of vegetation types, from thorny coastal maquis, sand dunes and dry rough grass to xerophytic vegetation on top of Atlas Mts at 3300 m. In absence of congeners Red-legged Partridge *Alectoris rufa* and Rock Partridge *A. graeca*, occupies habitat of both (Beaubrun and Thévenot in press). Occurs in broom and other scrub, subdesert steppe with stands of *Euphorbia* and Sumach (*Rhus*), sandy hills with sparse bush, rocky and stony hillsides with or without growth of juniper and other trees and shrubs; in semi-arid areas partial to dry wadis with low vegetation. Also occupies croplands and groves of palm, citrus, olive and *Eucalyptus*; in Morocco, commonest in habitat which includes mixture of woodland, clearings and agriculture, while absent from cedar and other dense woods (Beaubrun and Thévenot in press). Can survive without free drinking water.

Occurs in pairs or small groups, not in large flocks, dispersing in pairs in breeding season. Chiefly sedentary, but birds from highest parts of Atlas Mts forced to lower altitudes by heavy snow in winter, and, in arid parts of range, may extend further south into desert in wet years. Where hunted is shy and difficult to approach, runs rather than flies to escape, especially in thick vegetation. When forced to fly, flight is direct and swift; rises with shrieking alarm call, travelling several hundred m before settling again. Feeds mainly early morning and evening, resting in vegetation in heat of day, also partly at night in areas where much disturbed by day. Roosts on ground, usually under rocks or bushes.

Food. Seeds, fruits and leaves, supplemented by insects, especially ants. Succulent leaves of *Salsola*, *Lycium*, *Asparagus*, and fruits of *Euphorbia* form up to 33% of stomach contents, providing most moisture. Feeds on fallen grain of crops, but not an agricultural pest. Food of young often mainly ants.

Breeding Habits. Monogamous; pairs form in spring. ♂ utters advertisement call for several weeks before pairing. When calling, ♂ stands upright, neck, legs and toes fully stretched, body supported on tips of toes; collar spots and flank bars displayed, tail spread. Attacks other birds from side, with crown erect, tail spread, and wings drooped. Attacked bird runs away with crown feathers erected, neck collar displayed, flank stripes breaking into separate bars.

♂ displaying to ♀ in courtship-feeding runs round in semi-circle, breaking off to mock feed, or runs towards ♀ with little jumps. Also walks in semi-circle, head and neck level with tail, following ♀ if she flees, then resuming semicircular displays. Before copulation ♂ moves away and stops, ♀ approaches and crouches, facing away, crest erected and tail raised. ♂ straightens into upright posture, runs and jumps towards ♀, grasps her crown feathers, and stands on her back. Copulation lasts 2–3 s. After copulating, both run briefly in upright posture.

NEST: shallow scrape on ground, usually under bush or other vegetation, sometimes lined with bits of vegetation.

EGGS: 6–20, probably laid on consecutive days; clutches of 16–20 recorded in captivity may be laid by 2 ♀♀ in same nest; mean (25 clutches) 11·3. Smooth ovals; slightly glossy; pale yellow-buff, finely specked red-brown. SIZE: (n = 70, *A. b. barbara*) 36·8–44·5 × 27·8–31·8 (40·5 × 30·4). WEIGHT: c. 19.

LAYING DATES: Morocco, from end Feb (El Ayoun) and mid Mar (Atlantic coast south of Rabat) to mid June in high mountains. Similar breeding season likely in other countries, with most clutches laid Mar–May, earliest in lowlands, latest in mountains. In semi-arid areas does not breed at all in very dry years.

INCUBATION: begins with last egg; as in closely related Red-legged Partridge; probably by ♀ only if single clutch, ♂ may incubate 2nd clutch. Period: unknown, probably 24–25 days, as in congeners.

DEVELOPMENT AND CARE OF YOUNG: young hatch synchronously; no details known of behaviour of young or parents, but in other members of genus young leave nest almost at once, are immediately able to feed themselves, capable of flight at 10–12 days, and reach full size at 50–60 days. Breed in 1st year; both parents attend single broods; 1 parent only, if broods derived from 2 clutches. 1 pair with 12 young recorded (Heim de Balsac and Mayaud 1962). When adult broods young in early stages, makes scrape each time, but brooding soon reduced. Family parties remain together until autumn, when several broods merge to form larger groups, but no very large flocks recorded anywhere.

References
Beaubrun, P. and Thévenot, M. (in press)
Cramp, S. and Simmons, K. E. L. (1980).

Genus *Ptilopachus* Swainson

A monotypic genus endemic to N tropical Africa. Rather small and partridge-like, with short neck, stout body, short, unspurred tarsi, and bantam-like cocked tail of 14 feathers. 1st primary slightly shorter than 10th, 5th slightly the longest. Voice flute-like, high-pitched, pleasingly melodious. Largely confined to rocky country.

Ptilopachus petrosus (Gmelin). Stone Partridge. Poule de rocher.

Plate 1
(Opp. p. 32)

Tetrao petrosus Gmelin, 1798. Syst. Nat. 1, pt. 2, p. 758; Gambia.

Range and Status. Resident in rocky hills and small outcrops from Senegambia east to N Ethiopia, south to N Uganda and N Kenya (to Uaso Nyiro River). In W Africa mainly between 7–8°N and 17°N. Locally distributed, but often common to abundant. To some extent hunted for food, but no evidence of any need for conservation action.

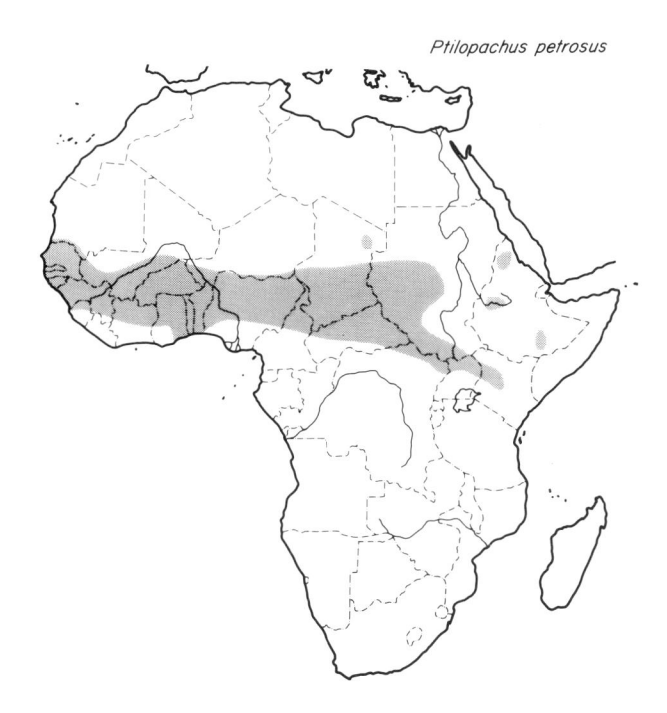

Description. *P.p. petrosus* (Gmelin) (including *brehmi, florentiae, saturatior, emini* and *butleri*): Senegal to Kenya. ADULT ♂: crown, mantle and upperwing-coverts greyish brown, feathers with chestnut shaft streaks (black on crown) and whitish edges. Rest of upperparts dull greyish brown, heavily vermiculated with white and buff. Chin whitish, each feather centrally blotched with dark brown; throat and lower neck greyish brown, feathers with black shafts, chestnut shaft streaks, and whitish edges. Centre of lower breast and belly uniform buff. Flanks greyish brown, feathers mottled brown and white, broadly streaked chestnut. Primaries and secondaries brown; primaries lightly, secondaries heavily vermiculated chestnut; tail dark brown, appearing black. Bill base red, tip dusky yellow; eye brown; naked skin around eye red; legs dark red. ADULT ♀: like ♂ but pale patch on belly creamy white. SIZE: wing, ♂ (n = 12) 115–128 (123), ♀ (n = 13) 116–126 (122); tail ♂♀ 63–76; tarsus, ♂♀ 29–33. WEIGHT: ♂ (n = 2) 190.

IMMATURE: like adult but more clearly marked; with distinct bars on back, underparts, rump, tail and innermost secondaries.

DOWNY YOUNG: forehead, centre of crown, and back dark blackish chestnut; above eye, face and underparts dark brown, speckled black.

P. p. major Neumann; NW Ethiopia. Larger, paler, with broader chestnut streaks on flanks; wing, ♂♀ 130–133.

A number of clinal subspecies, based on single characters, have been described, e.g. 6 recognized by Peters (1934) and 3 by White (1965). We recognize only 2 subspecies. In general, specimens from more arid areas tend to be paler and more rufous.

Field Characters. A small, plain, dark bird with comparatively long tail usually cocked like a bantam, restricted to rocky localities. Crested Francolin *Francolinus sephaena* also has dark tail often cocked but is otherwise very different, having erectile crest, prominent white supercilium, heavy white streaking on upperparts, different voice and habitat.

Voice. Tape-recorded (C, 215, 217, 235, 267, 277). High-pitched, monotonous, far-carrying, flute-like 'ouit-ouit-ouit'. Also described as liquid, piping 'rrr-weet, rrr-weet, rrr-weet', or 'weet-weet-weet', varied, often uttered in duet or chorus. Also rendered 'we've been-weetin-weetin-weetin', or 'will-we-weet, will-we-weet'. Higher pitched and more whistling than voice of any francolin.

General Habits. Inhabits dense growth among boulders at base of rocky hills from 600–1500 m, or, in some areas, flat-topped laterite hills. Also occurs in steep-sided, wooded, dry watercourses, and broken, eroded, woody country, and sometimes in nearby long grass and cultivation; in Mali even in open sandy Sahel zone in low, shady spots. Very agile climber, moving rapidly up cliff faces without flying. Usually seen in pairs or coveys of 3–4, occasionally 15–20. Makes for rocky hills and dense cover if surprised away from them. Very vocal, usually betrays presence at dawn and dusk by loud calling (at almost any time of day in rainy season) from rocky areas; secretive away from them. Runs when disturbed, very reluctant to fly; but when pressed, e.g. by dog, can fly fast and direct, rising with high-pitched cries, and usually pitching after short distance. Active in early morning and evening, usually resting in midday heat in shade of trees or dense bush. Can exist for long periods, probably permanently in some localities, without any regular water source. Roosts among rocks at night. Sedentary; often confined to single rocky hill.

Food. Seeds of grass and herbs, green leaves, fruits and buds; supplemented by insects.

Breeding Habits. Monogamous in captivity. Malzy (1962) suggests polygamous mating system, but little supporting evidence; Ogilvie-Grant (1896) described a single ♂ displaying to several ♀♀. May be stimulated to breed, at least in arid areas, by onset of rains, when small flocks typical of dry season split up into pairs. Each pair takes possession of small rocky area with a group of prominent boulders. In amongst these boulders, adults display, running after one another with tails cocked, calling continuously, in duet between pairs, and sometimes with many pairs close together. 1 pair's calling often stimulates response from others, resulting in concerted bouts of calling followed by relative silence. ♂♂ courtship-feed mates. In courtship, ♂ displays by hopping in front of ♀ with neck feathers ruffled, tail fanned and wings fanned and trailing on ground.

NEST: well concealed simple scrape at base of rock, tree or tuft of grass, sometimes lined with grass.

EGGS: 4–6, dull pale stone colour or ochre-yellow, glossless, slightly pointed at narrow end. SIZE: 31·8–36·6 × 24·5–25·3 (33·3 × 24·8). WEIGHT: c. 11.

LAYING DATES: Senegambia, Dec–July; Mali, Nov–Dec in N, Feb–May in S; Niger, June–July; Nigeria, July, Jan, Feb, very young chicks Apr; Central African Republic, all months; Sudan (Darfur), Jan–July, Dec (dry season); N Uganda, Dec, dry season; Ethiopia, Aug–Sept; Eritrea, Sept. Season possibly varies according to annual rainfall, in rains in drier parts of range, dry season in wetter areas.

DEVELOPMENT AND CARE OF YOUNG: family parties remain together until young are as big as adults, probably only separating in following breeding season.

Genus *Francolinus* Stephens

Large genus (41 spp.: 36 African and 5 Asian); static taxa perhaps derived from a more quail-like ancestor which colonized Asia in the early to mid Pliocene (26–17 million years ago) (Crowe and Crowe, 1985). Small (c. 200 g) to medium-sized (1100 g) galliforms; generally resemble quails (*Coturnix*) or partridges (*Perdix*, *Ammoperdix* and *Alectoris*); but are usually more strongly marked, have longer, more powerful, somewhat hooked bills and more upright stance. In plumage, most species sexually monomorphic and cryptically coloured above. Stout-bodied, short-legged; ♂♂ often average larger than ♀♀ and most species have strongly spurred tarsi. Tails short, with 14 feathers. 1st primary longer than 10th, 4th to 6th the longest. Plumage of downy chicks remarkably uniform within the genus, despite the broad range of habitats occupied, and generally resembles that of pheasants *Phasianus* spp. (Frost 1975). Many species have raucous grating or cackling voices, others more musical and whistling.

Within Africa, francolins are characteristic birds of most vegetation types, found especially in savannas and thornbush, but also in dense tropical forest, and in afro-alpine vegetation; in rainfall from 200 mm/year to over 2000 mm/year, and from sea level to 4000 m. Although common and conspicuous birds, extensively hunted for sport and food, few francolins have been studied in detail. For many, even ecological preferences are still poorly understood. Few accurate quantitative data on food, and very little or nothing known about breeding habits. Nests of several have never been found, and only a few have been studied in captivity. Thus the group deserves more detailed study.

The systematics of this genus have been studied in detail by Hall (1963) and Crowe and Crowe (1985). Since francolins are extremely sedentary, geographical barriers which have promoted only subspeciation of guineafowl (Numidinae) have allowed speciation of francolins. The number of species of African francolins recognized in various checklists remains fairly constant. However, many subspecies have been described: Peters (1934) recognizes 107, Hall (1963) 95 and White (1965) 89; a large number of these are either extremely local forms represented by only a few specimens in museum collections, or were collected along segments of relatively smooth clines and are distinguishable only by 1 or 2 quantitative characters, e.g. 'darker overall, wing somewhat longer'. Elevating such variants to the same level as qualitatively distinct taxa (e.g. subspecies of Latham's Forest Francolin *Francolinus lathami* and the Plumed Guineafowl *Guttera plumifera*), and those delineated by suites of characters which vary concordantly and step-clinally (e.g. subspecies of Helmeted Guineafowl *Numida meleagris*), obfuscates rather than elucidates patterns of geographical variation and the processes which may have brought them about. Therefore we used subspecies recognized by Hall (1963) as a starting point, but only those which are defined by several characters and which are represented in museum collections by adequate series of specimens are recognized below.

Hall (1963) divided the African francolins, except 2 small forest spp. (Latham's Forest Francolin and Nahan's Forest Francolin *F. nahani*), into 7 groups: (1) Red-tailed Group—Coqui *F. coqui*, White-throated *F. albogularis* and Schlegel's *F. schlegelii*. (2) Red-winged Group—Shelley's *F. shelleyi*, Moorland *F. psilolaemus*, Orange River *F. levaillantoides*, Red-wing *F. levaillantii*, Grey-wing *F. africanus* and Finsch's *F. finschi*. (3) Striated Group—

Crested *F. sephaena* and Ring-necked *F. streptophorus*. (4) Vermiculated Group—Double-spurred *F. bicalcaratus*, Clapperton's *F. clappertoni*, Heuglin's *F. icterorhynchus*, Harwood's *F. harwoodi*, Hildebrandt's *F. hildebrandti*, Natal *F. natalensis*, Red-billed *F. adspersus*, Cape *F. capensis* and Hartlaub's *F. hartlaubi*. (5) Scaly Group—Scaly *F. squamatus*, Ahanta *F. ahantensis* and Grey-striped *F. griseostriatus*. (6) Montane Group—Erckel's *F. erckelii*, Djibouti *F. ochropectus*, Chestnut-naped *F. castaneicollis*, Jackson's *F. jacksoni*, Handsome *F. nobilis*, Cameroon Mountain *F. camerunensis* and Swierstra's *F. swierstrai*. (7) Bare-throated Group (formerly *Pternistis*)—Red-necked *F. afer*, Yellow-necked *F. leucoscepus*, Grey-breasted *F. rufopictus* and Swainson's *F. swainsonii*.

Hall's classification has been modified slightly by Crowe and Crowe (1985) who depart from her within-group species sequences and suggest that the Ring-necked Francolin be shifted from the Striated Group to the Red-winged Group and that the Cameroon Mountain and Swierstra's Francolins form a 2nd superspecies within the Montane Group. Their results also confirm her speculation that Latham's Forest Francolin has affinities with members of the Red-tailed Group, and that Nahan's Forest Francolin has affinities with members of the Scaly Group.

Francolinus coqui superspecies

1 *F. coqui*
2 *F. albogularis*
3 *F. schlegelii*

Francolinus levaillantii superspecies

1 *F. levaillantii*
2 *F. streptophorus*
3 *F. finschi*
4 *F. africanus*

Francolinus shelleyi superspecies

1 *F. psilolaemus*
2 *F. shelleyi*
3 *F. levaillantoides*

PHASIANIDAE

Francolinus squamatus superspecies

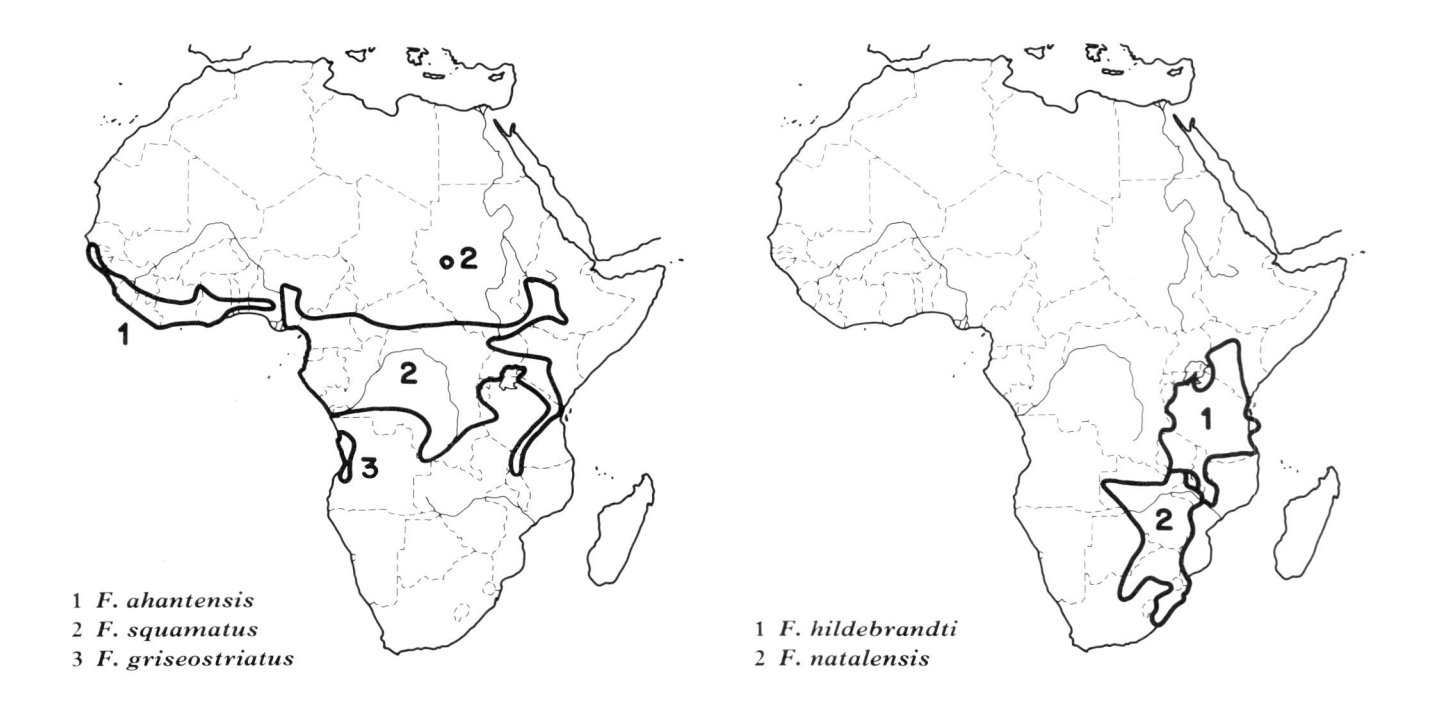

1 *F. ahantensis*
2 *F. squamatus*
3 *F. griseostriatus*

Francolinus natalensis superspecies

1 *F. hildebrandti*
2 *F. natalensis*

Francolinus bicalcaratus superspecies

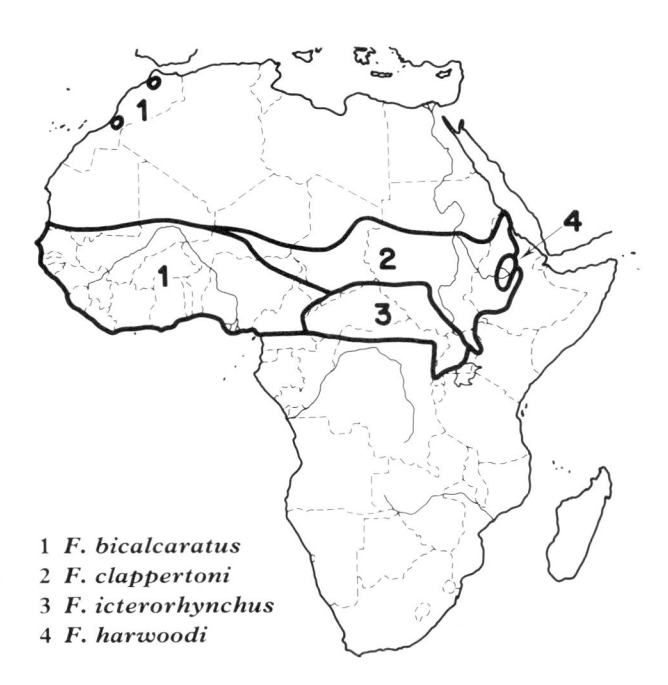

1 *F. bicalcaratus*
2 *F. clappertoni*
3 *F. icterorhynchus*
4 *F. harwoodi*

Francolinus **camerunensis superspecies**

1 *F. camerunensis*
2 *F. swierstrai*

Francolinus **jacksoni superspecies**

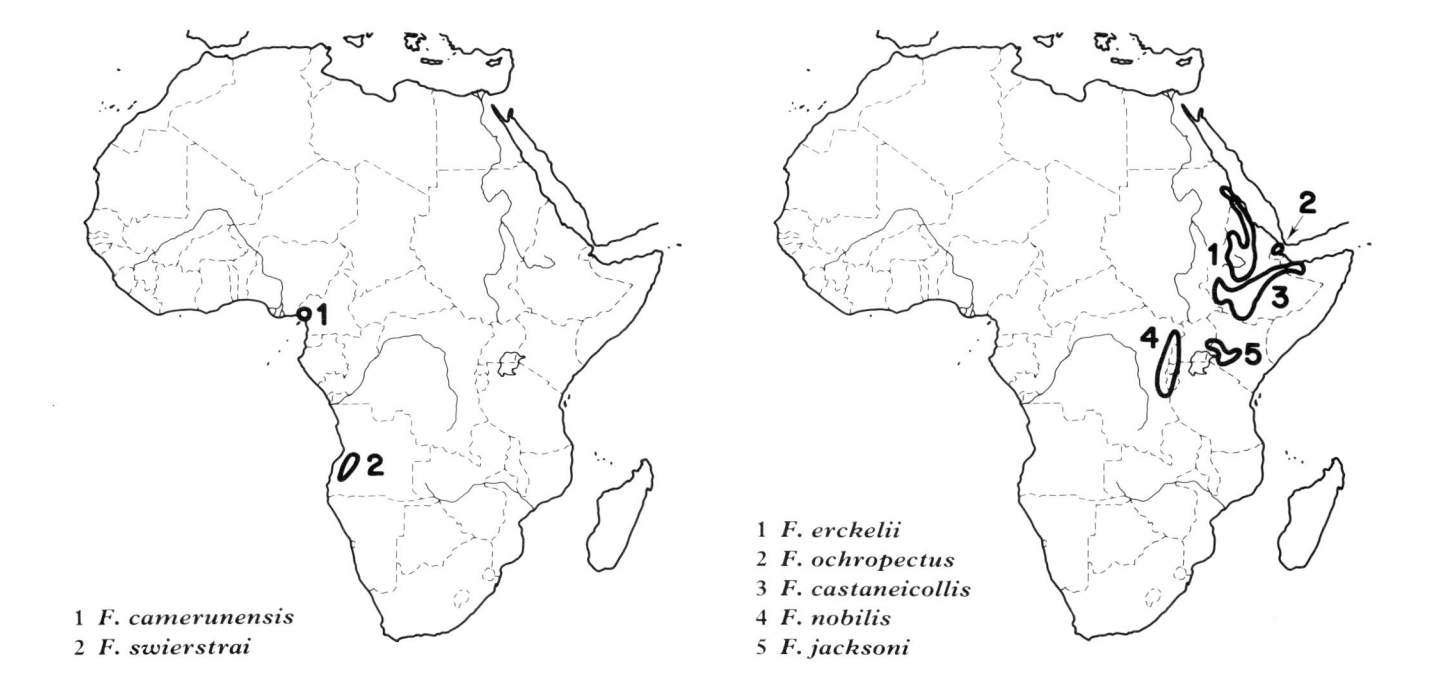

1 *F. erckelii*
2 *F. ochropectus*
3 *F. castaneicollis*
4 *F. nobilis*
5 *F. jacksoni*

Francolinus **afer superspecies**

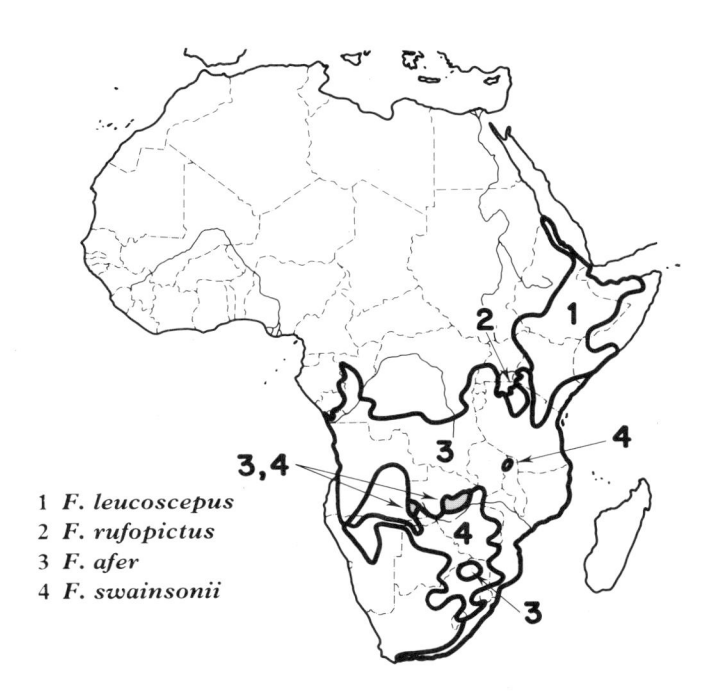

1 *F. leucoscepus*
2 *F. rufopictus*
3 *F. afer*
4 *F. swainsonii*

Plate 3
(Opp. p. 48)

Francolinus lathami Hartlaub. Latham's Forest Francolin. Francolin de Latham.

Francolinus lathami Hartlaub, 1854. Journ. f. Orn., 2, p. 210; Sierra Leone.

Range and Status. Endemic resident in lowland forest belt from Sierra Leone to Zaïre (south to Mayombe, Lusambo; east to Itombwe and Ituri Forest), Angola (Cabinda, Conde) and W Uganda (Bwamba, Budongo, Bugoma and Lugalambo Forests). Small outlying populations in S Sudan (Zande district) and S–central Uganda (Mabira and Kifu Forests). Mainly in lowlands, but up to 1400 m in Uganda. Uncommon to locally common. Threatened by forest destruction and hunting, since relatively easily trapped with snares baited with termites and shot when coming to areas baited with scattered grain.

Francolinus lathami

Description. *F. l. lathami* Hartlaub: Sierra Leone to *c.* 18°E in Zaïre. ADULT ♂: crown and nape glossy dark olive-brown; forehead and ear-coverts pale grey; eye-stripes black, bordered with white, extend backward joining at hindneck; lores black. Upper mantle black, feathers irregularly barred white and with a terminal heart-shaped white blotch. Lower mantle, scapulars, wing-coverts and back dark greyish brown (paler towards tail), heavily washed and mottled rufous-chestnut and faintly barred rufous-buff, each feather with a narrow white shaft streak. Rump and uppertail-coverts dark olive-brown vermiculated buff and black. Tail dark brownish grey. Chin and throat black. Rest of underparts mainly black, but breast feathers irregularly barred white with terminal heart-shaped white blotch, flanks mottled brown and chestnut with some white streaks, undertail-coverts barred white; lower belly whitish buff. Primaries blackish brown, 2nd to 5th outer webs edged white basally; secondaries dark brown, outer webs edged rufous, innermost vermiculated chestnut. Bill black; eye brown, eyelid pale greenish; legs yellow. ADULT ♀: much browner. Forehead brown, crown and nape lighter brown, face patch pinkish brown, ear-coverts brownish grey; black areas on hindneck, upper mantle and underparts replaced with brown; chestnut areas of back and wings replaced with black blotching and narrow rufous barring; uppertail-coverts and rump dull brown. SIZE: (24 ♂♂, 39 ♀♀) wing, ♂ 128–143 (136), ♀ 127–143 (134); tail, ♂ 67–78 (73·3), ♀ 66–78 (72·5); tarsus, ♂ 37–49 (41·4), ♀ 37–51 (40·6); 1 spur; ♂ 1·3–12 (8·3), ♀ 0·5–5 (2·8). WEIGHT: ♂ (n = 1) 254, ♀ (n = 1) 284.

IMMATURE ♂: like ♀ but crown mottled black and brown, chin and throat white, sides of head and ear-coverts brownish. Upperparts reddish brown, heavily mottled black, scapulars with ochraceous shaft streaks. Breast and abdomen brown with white cross-shaped markings outlined with black. Flanks lighter brown with whitish shaft streaks and faint vermiculations. IMMATURE ♀: feathers of crown and nape with broad black tips. Upperparts like ♀ but more rufous, especially in wing-coverts. Chin and throat white, breast brown with white streaks, rest of underparts white with brown streaks except flank feathers which are rufous barred black. Secondaries bright rufous.

DOWNY YOUNG: crown and back dark chestnut brown; face buff with chestnut eye-stripe from behind eye to nape; underparts dark buff. Bill horn, lower mandible dusky; eye grey-brown; legs pale yellow.

Possibly most closely resembles ancestral francolin, with nearest relative one of the Red-tailed Group (Crowe and Crowe 1985).

F. l. schubotzi Reichenow: Zaïre east of *c.* 20°E, Uganda, Sudan. ♂ has paler grey patch on face, back and wings darker chestnut with some dark blotching, black extending further down belly, lower flanks with very little brown; ♀ has rufous face-patch, lower neck and upper breast rich brown with little spotting, white of underparts, including spots, buff rather than white. Slightly larger: wing, ♂ (n = 11) 132–145 (140), ♀ (n = 22) 132–145 (138). WEIGHT: ♂ (n = 1) 254.

Field Characters. Small size and black and white barred underparts preclude confusion with any francolin except Nahan's Forest Francolin *F. nahani*. Nahan's has black and white streaked and mottled underparts, upperparts without chestnut markings, bare red skin around eye instead of grey or rufous face- and neck-patch, throat mainly white, and red legs.

Voice. Tape-recorded (48, 217, 226). Melodious dove-like 'coo' repeated 3 times, or 'kwee, coo, coo' repeated again and again in strings of 8. Also described are a prolonged series of rather uniform high-pitched whistles (Chapin 1932); a flute-like call (Lippens and Wille 1976); and a low clucking call while moving through forest (Bannerman 1951).

General Habits. Inhabits lowland primary forest and, occasionally, dense secondary forest; in Sudan, gallery forest. Found singly, in pairs or small coveys. Extremely shy and difficult to observe even where fairly common, though sometimes emerges onto forest tracks, especially after rain. Very hard to flush; sits tight until nearly stepped on, or else runs off through undergrowth. Flies fast, but only for short distances. Feeds by scratching among leaf litter, possibly even at night when sometimes heard calling. Calls from perches in trees and on ground.

Food. 90% arthropods, especially termites *Basidentitermes* spp. and ants *Psalidomyrmex* spp., also snails, beetles and other insects and their larvae. 10% vegetable matter, mainly fruits, especially of oil palm *Elaeis guineensis*, also seeds and green leaves.

Breeding Habits. Probably monogamous since found in pairs during breeding season.

NEST: no real nest, eggs laid on dry leaves between projecting buttresses of forest trees, such as *Piptadeniastrum africanum*.

EGGS: 2, rarely 3. Elongate-ovate; very hard and thick-shelled; uniform dark buff or light brown, sometimes rusty. SIZE: 36–42·5 × 25–28.

LAYING DATES: W Cameroon, Feb, Dec, S Cameroon, Dec (during dry season); Zaïre, Dec–Apr; Uganda, Aug (birds in breeding condition May–June).

BREEDING SUCCESS/SURVIVAL: hatching success 80%, much higher than average for forest birds (33%), possibly due to placement of nest between tree buttresses (Brosset 1974). Principal predators viverrids and snakes.

References
Bannerman, D. A. (1930, 1951).
Chapin, J. P. (1932).
Thiollay, J.-M. (1971, 1973).

Red-tailed Group

3 small francolins (*F. coqui*, *F. albogularis*, *F. schlegelii*) with ochre on sides of head and varying amounts of rufous on tail. Sexes similar in size but dissimilar in plumage (less so in Schlegel's Francolin *F. schlegelii*), ♂♂ with more ochre on head and neck, ♀♀ with stripes on face and neck. All species have: short yellowish legs with 1 spur, black bills (yellowish at base); feathered ceres; quail-like upperparts; black and white barred underparts; musical trumpet-like calls; roost on the ground in open-country habitat. Often sympatric with members of other groups, suggesting lack of competition.

Francolinus coqui (Smith). Coqui Francolin. Francolin coqui.

Plate 2
(Opp. p. 33)

Perdix coqui A. Smith, 1836. Rep. Exped. Centr. Afr., p. 55; near Kurrichane, Transvaal.

Forms a superspecies with *F. albogularis* and *F. schlegelii*.

Range and Status. Endemic resident; the most widespread African francolin. Range in W Africa apparently disjunct, with populations in S Mauritania, Mali (rare), W Niger and N Nigeria (mainly Kano to Lake Chad; absent in NW). A small population in S Ethiopia. Main range from Uganda (Ankole) and central Kenya (Eldama Ravine, Kitui, Sokoke Forest) south through E Rwanda, E Burundi, Tanzania, SE Zaïre (roughly SE of line from Kasaji to Marungu), Zambia (except Luangwa and middle Zambezi Valleys), Malaŵi, highlands of N Mozambique, Zimbabwe (except SE lowlands and E highlands), Angola (Huila and Malanje east to Lunda and N Moxico), N and E Botswana, Namibia (to Waterberg), South Africa (to N Orange Free State, S Natal) and S Mozambique (Sul do Save). An apparent gap separates this from population in highlands of Angola and savannas of S Zaïre and Congo. Uncommon and local in some areas, common in others. May be limited locally by disease, e.g. fowl pox (A. W. Potterill, pers. comm.). Sensitive to changes in grass cover; disappears locally with burning and overgrazing.

Description. *F. c. coqui* (Smith) (including *campbelli*, *vernayi*, *hoeschianus*, *angolensis*, *kasaicus* and *ruandae*): southern Africa north to Zaïre, central and NW Tanzania and Uganda. ADULT ♂: forehead, supercilium, sides of neck ochre, extending as half-collar towards hindneck, becoming paler on chin and throat. Crown, nape, hindneck, indistinct eye-stripe and ear-coverts rusty, crown and nape blotched with grey. Lower hindneck barred black and white; rest of upperparts and wing-coverts dark greyish brown, paler on rump and tail, feathers with creamy shaft streaks, rufous-buffy barring and grey vermiculations; tail dark reddish brown barred black. Breast, belly and flanks barred black and white, black bars more widely spaced on belly, a few rufous blotches on flanks.

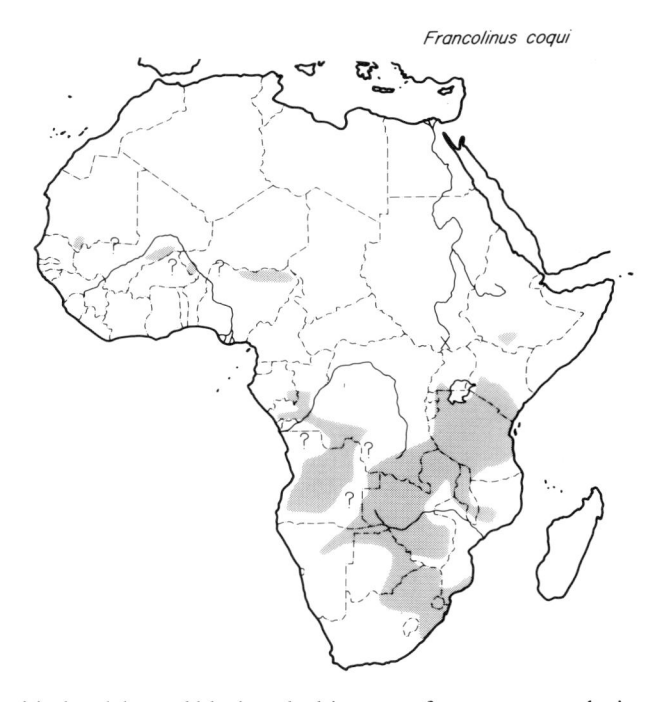

Undertail-coverts pale reddish brown barred finely black. Primaries uniform grey-brown; secondaries similar, but with faint rufous barring on inner webs. Bill black with yellow base; eye brown; legs yellow. ADULT ♀: like ♂ but narrow black border to crown, black and white stripes above and below eye, latter extending as necklace around throat. Breast light vinaceous-brown, not barred. SIZE: (221 ♂♂, 139 ♀♀) wing, ♂ 123–147 (134), ♀ 118–147 (131); tail, ♂ 65–95 (75·2), ♀ 63–91 (75·4); tarsus, ♂ 31–45 (37·8), ♀ 31–43 (37·1); 1 spur; ♂ 4–11 (6·9), ♀ 0·5–2 (0·9). WEIGHT: ♂ (n = 2) 278–289, ♀ (n = 4) 218–259.

IMMATURE: like ♀ but paler and more mottled than streaked with rufous-buff and brown above, and more buffy below with faint black and white barring.

DOWNY YOUNG: forehead and crown rufous-brown; sides of head white, superciliary, eye and moustachial stripes dark brown. In between dark superciliary and crown-stripes is additional buff superciliary stripe. Broad rufous-brown stripe bordered with black down mid-back, with narrower buff stripes on either side. Underparts buffy white, breast washed vinaceous. Wings rufous mottled with brown.

Local and clinal variation considerable, especially within nominate subspecies. Birds from Botswana and Namibia paler, with tails averaging c. 13% longer; those from more mesic areas (e.g. Zaïre and Zululand) more richly pigmented. Birds from Uganda and NW Tanzania ('*ruandae*') have broader black barring on underparts.

F. c. hubbardi Ogilvie-Grant: W and S Kenya from Rift Valley and W Highlands to Mwanza Province of Tanzania, also east to Kitui. Belly buff, unbarred; crown more heavily blotched grey; primaries washed rufous. Larger: wing, ♂ (n = 26) 135–152 (142), ♀ (n = 19) 128–140 (133).

F. c. maharao Sclater (including *thikae*): Ethiopia, Kenya and N Tanzania in highlands east of Rift Valley from Murang'a to Arusha. Black ventral barring narrower; crown stripes on either side. Underparts buffy white, breast washed rufous. Smaller: wing, ♂ (n = 15) 127–137 (131), ♀ (n = 5) 123–129 (126).

F. c. spinetorum Bates: W Africa. Belly buff (not barred black and white), and crown and primaries heavily washed rufous.

Field Characters. Small, sexually dimorphic francolin with black and white barred underparts, black bill and yellow legs. ♂ has black and white barred breast and appears yellow-headed in the field; ♀ has buffy throat, black and white face-stripes and necklace (suggestive of Red-winged Group) and vinaceous-brown wash on breast. Allopatric with Schlegel's Francolin *F. schlegelii*, but overlaps with similar-sized White-throated Francolin *F. albogularis* in Zaïre and Angola. ♂ White-throated easily told by lack of barring on underparts, but ♀ quite similar to ♀ Coqui; barring on underparts much finer, breast with much less vinous wash, face less ochre and with different pattern, with single dark stripe back from eye, no black necklace. Chestnut wing-patch shows in flight. Tendency to walk in slow, stooped manner in open and to 'freeze' when alarmed also distinctive.

Voice. Tape-recorded (34, 74, C, F). 2 calls commonly heard: advertisement call, given mainly in early morning, midday and late afternoon, series of 7–10 trumpet-like notes 'ter, ink, ink, terra, terra, terra, terra, terra, terra', 2nd or 3rd being loudest, latter ones falling away. Birds from E and W Africa, especially from Nigeria, give more condensed (i.e. more rapidly given) version of this call, approaching that of Schlegel's Francolin; 2nd call heard throughout day, 2 high-pitched, squeaky notes 'co-qui co-qui co-qui' (first note being accented) repeated in series. Warning call a harsh, subdued 'churr-churr'; soft contact notes given by members of moving covey.

General Habits. Inhabits grassland, savanna or well grassed woodland, especially miombo woodland, up to 2200 m. In drier country may be found in sand dunes with good bush cover. Avoids hilly and stony biotopes. Sometimes enters edges of cultivation but does no harm to crops. May be independent of water. Usually in pairs or small coveys; far more often heard than seen. Extremely difficult to flush, but once airborne flies well, pitching into grass after short distance. Unlike other francolins, often 'freezes' rather than running when surprised in open. Takes to trees rarely and only when extremely harassed, e.g. by dogs (Clancey 1967). Roosts on ground, often several birds close together. Moults during Feb–Apr in SE Zaïre.

Food. Grass and other seeds, fallen grain, small leaves, beetles, ants and other insects and their larvae. Also gleans ticks off grass stalks (J. R. Peek, pers. comm.).

Breeding Habits. Monogamous and probably territorial since pairs remain in same general area during breeding season (W. R. Tarboton, pers. comm.), and ♂♂ very pugnacious. Advertisement call heard much more frequently at onset of breeding season, rarely when hens are nesting.

NEST: slight hollow or scrape 10–12 cm across, lined with grass or leaves, placed in good cover at base of tuft of grass.

EGGS: 2–8, usually 4–5. Smooth ovals; colour variable—pale brown, cream, or white to pink with slight gloss. SIZE: (n = 32, *F. c. coqui*) 31·3–34·6 × 25·5–28·6 (32·7 × 27·4); (*F. c. hubbardi*) 42 × 31; (*F. c. spinetorum*) 34 × 27·3.

LAYING DATES: South Africa: Natal, Oct–Apr peaking Nov–Feb, Transvaal, Jan–Feb, Apr–May, July, Nov; Zimbabwe, Aug–June, peaking in Dec–Feb during rains; Angola, Aug, juveniles Feb–Apr, Aug, Nov; Zambia, every month except June, mainly Nov–Feb; Malaŵi, Jan–Feb, Apr, Sept–Nov; East Africa: Tanzania, Apr–Jun, Aug, Dec; Kenya, Jan, Apr–July, Oct–Dec; Uganda, May, Oct–Nov; Region B, Apr–May, Oct–Nov; Region C, Feb–May, Nov–Dec; Region D, Feb, May–July, Sept, Dec (i.e. during rains and early dry season); Ethiopia, possibly May–June; Zaïre, late Aug–Mar, during rains; Nigeria, July.

DEVELOPMENT AND CARE OF YOUNG: young remain with adults for at least several months forming family parties of 3–6.

BREEDING SUCCESS/SURVIVAL: a large proportion of the young fall prey to raptors, wild cats and snakes. Raptors are the primary predators of adults.

References
Clancey, P. A. (1967).
Meyer, H. F. (1971b).

Francolinus albogularis **Hartlaub. White-throated Francolin. Francolin à gorge blanche.**

Francolinus albogularis Hartlaub, 1854. Journ. f. Orn., 2, p. 210; Gambia.

Forms a superspecies with *F. coqui* and *F. schlegelii*.

Range and Status. Endemic resident, locally distributed in W African savanna from Senegambia (fairly common in Tambacounda region), Guinea and SW Mali (rare, mainly in Mandingo Mts) to central Nigeria (locally common) and N Cameroon (Benué Plain). Relict populations in SE Zaïre (Marungu to Upemba National Park), Angola (E Moxico) and Zambia (NW Balovale District, not seen in recent years). Uncommon to rare in much of range.

Description. *F. a. albogularis* Hartlaub: Senegambia. ADULT ♂: forehead and crown grey with variable amount of rufous blotching; lores and supercilium creamy white; ear-coverts grey; under eye and sides of neck ochraceous-buff. Upperparts brownish grey, heavily washed chestnut on wing-coverts, feathers with a cream shaft streak and faint rufous-buffy barring. Uppertail-coverts grey, faintly barred buff; tail dark brownish grey barred rufous buff. Chin and throat white. Rest of underparts buff, a few chestnut streaks on breast and flanks. Undertail-coverts rufous-buff, barred black. Primaries mainly rufous with brown tips and some brown barring; shafts pale. Bill black, base yellow; eye brown; legs orange-yellow. ADULT ♀: like ♂ but creamy streaking on back narrower and breast and flanks finely barred with black and white. SIZE: (4 ♂♂, 4 ♀♀) wing, ♂ 129–141 (135), ♀ 122–131 (127); tail, ♂ 68–69 (68·5), ♀ 60–70 (64·5); tarsus, ♂ 35–42 (39·8), ♀ 37–39 (38·1); 1 spur; ♂ 4–13 (9·1), ♀ 0·3–2 (0·9). WEIGHT: (n = 8, unsexed) 263–284 (276).

IMMATURE: like ♀ but ventral barring sometimes much more extensive; incomplete black necklace in ♂.

DOWNY YOUNG: crown dark brown; sides of head buff with dark brown eye stripe. Broad dark brown stripe on mid-back flanked by narrower stripes of buff. Wings and rest of upperparts chocolate brown mottled buff. Chin and throat white, rest of underparts buff washed rufous.

F. a. buckleyi Ogilvie-Grant: E Ivory Coast to Cameroon. Like *albogularis* but supercilium buff (not white), ventral black and white barring of ♀ forms faint necklace (as in Coqui Francolin *F. coqui*) and extends onto belly; slightly smaller: wing, ♂ (n = 16) 120–141 (126), ♀ (n = 10) 117–133 (123).

F. a. dewittei Chapin (including *meinertzhageni*): Zaïre, Angola and Zambia. Breast of ♂ dark chestnut streaked paler; rest of underparts reddish buff. Underparts of ♀ completely barred. WEIGHT: (SE Zaïre) ♂ (n = 1) 278, ♀ (n = 2) 284, 267.

Field Characters. A small francolin with white throat and chestnut wings conspicuous in flight. ♂ with unbarred chestnut and buff underparts easily told from ♂♂ of other 2 red-tailed francolins (Coqui *F. coqui* and Schlegel's *F. schlegelii*) which have yellowish throats, barred underparts; ♀ Coqui has black and white stripes on face and neck, vinous breast, broader bars on underparts and no chestnut in wings; ♂ and ♀ Schlegel's have yellowish throats, more rufous on upperparts, white underparts finely barred black.

Voice. Tape-recorded (268). High-pitched trumpet-like call 'ter-ink-inkity-ink' qualitatively similar to that of Coqui Francolin, but delivered much more rapidly; also bisyllabic 'ter-ink, ter-ink' similar to the 'co-qui' call of Coqui Francolin, but also faster.

General Habits. Inhabits open savanna, especially along tracks and on burned areas, open, rolling, hilly country with light scrub, bare areas recently burned, and bush growth in disused fields. In Nigeria confined to 'ironstone' country (C. H. Fry, pers. comm). Occurs in pairs or small coveys. Shy, prefers to escape by slinking off through grass; often sits tight, and hard to flush; flight low, rapid and direct with neck extended. Moult in SE Zaïre Apr–June.

Food. Grasshoppers, termites, beetles and other insects, grass seed and green plant material.

Breeding Habits. Probably monogamous, since occurs in pairs during breeding season.

NEST: scrape on ground lined with grass and leaves.

EGGS: 4–7, usually 6. Rounded; buff to pale brown with slight gloss, and fine pitting and brown speckling. SIZE: (n = 4) 31·9–32·8 × 26·3–26·4.

LAYING DATES: Senegambia, Sept–Oct; Nigeria, June, also 'dry season', Yankari (Elgood 1982); S Zaïre, Oct–Dec, early rains.

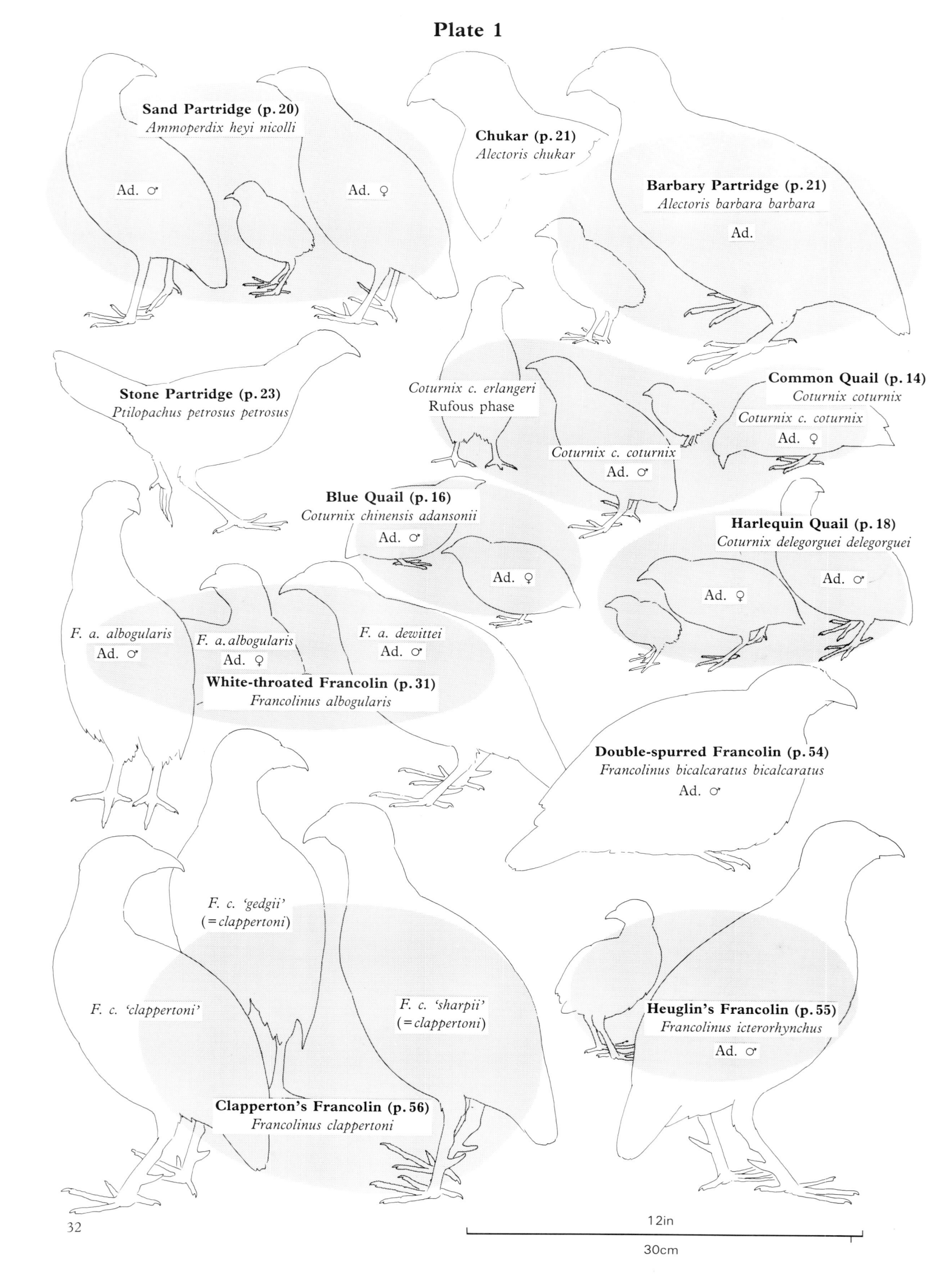

Plate 1

Sand Partridge (p. 20)
Ammoperdix heyi nicolli
Ad. ♂   Ad. ♀

Chukar (p. 21)
Alectoris chukar

Barbary Partridge (p. 21)
Alectoris barbara barbara
Ad.

Stone Partridge (p. 23)
Ptilopachus petrosus petrosus

Coturnix c. erlangeri
Rufous phase

Common Quail (p. 14)
Coturnix coturnix
Coturnix c. coturnix
Ad. ♀

Coturnix c. coturnix
Ad. ♂

Blue Quail (p. 16)
Coturnix chinensis adansonii
Ad. ♂
Ad. ♀

Harlequin Quail (p. 18)
Coturnix delegorguei delegorguei
Ad. ♀   Ad. ♂

F. a. albogularis
Ad. ♂

F. a. albogularis
Ad. ♀

F. a. dewittei
Ad. ♂

White-throated Francolin (p. 31)
Francolinus albogularis

Double-spurred Francolin (p. 54)
Francolinus bicalcaratus bicalcaratus
Ad. ♂

F. c. 'gedgii'
(= *clappertoni*)

F. c. 'clappertoni'

F. c. 'sharpii'
(= *clappertoni*)

Heuglin's Francolin (p. 55)
Francolinus icterorhynchus
Ad. ♂

Clapperton's Francolin (p. 56)
Francolinus clappertoni

12in
30cm

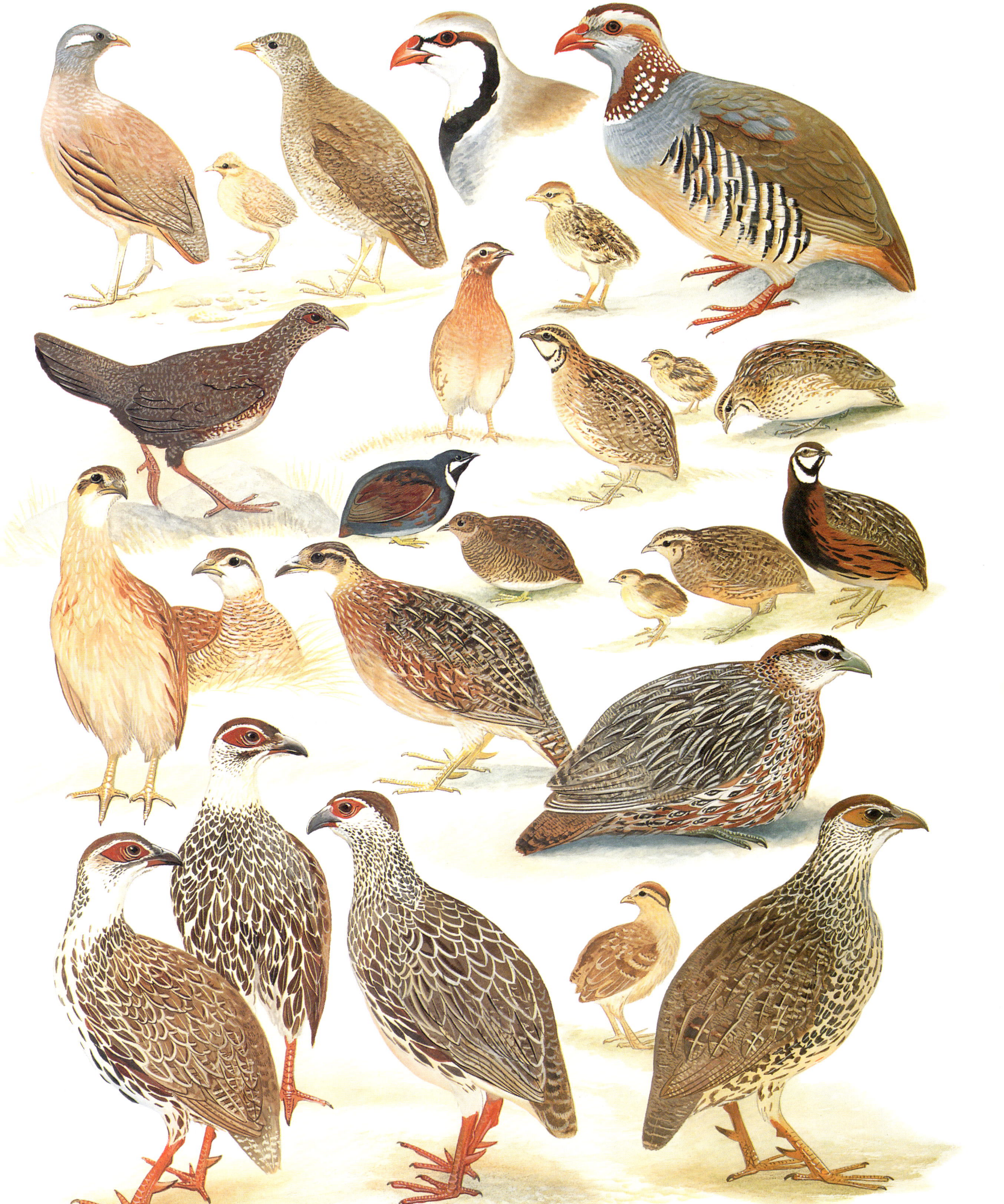
MWW 1979

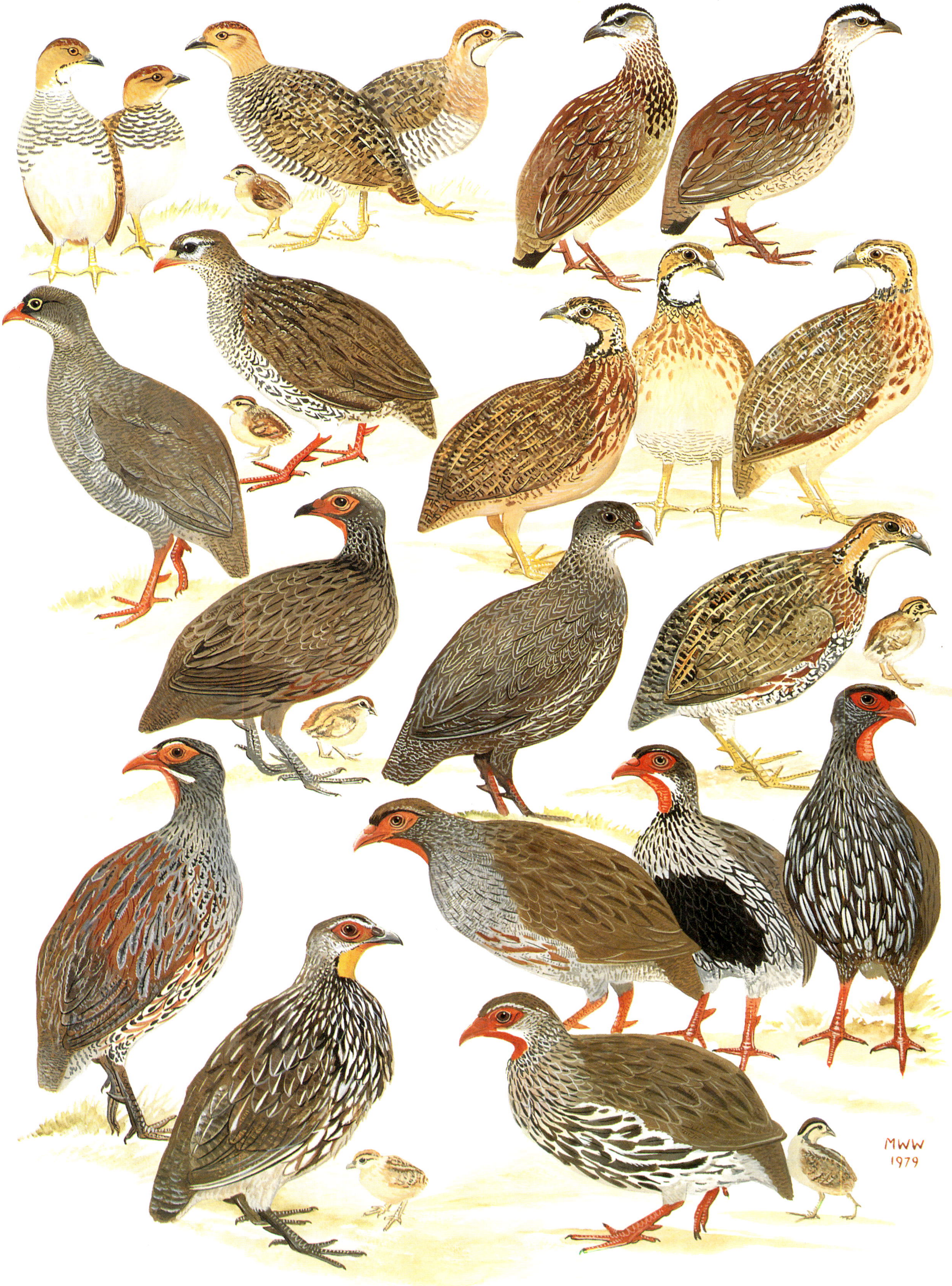
MWW
1979

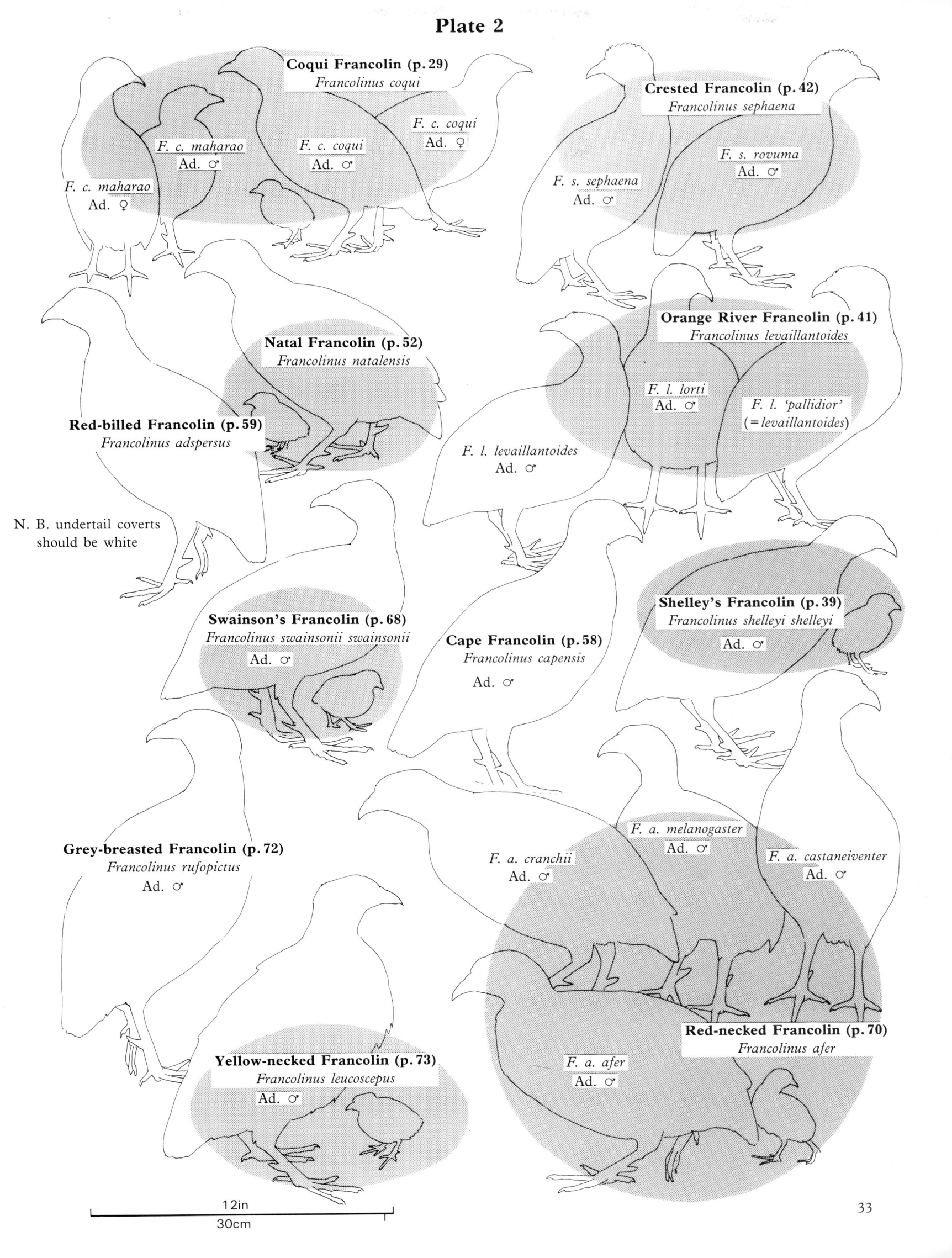

Plate 2

Coqui Francolin (p. 29)
Francolinus coqui

F. c. maharao Ad. ♂
F. c. coqui Ad. ♂
F. c. coqui Ad. ♀
F. c. maharao Ad. ♀

Crested Francolin (p. 42)
Francolinus sephaena

F. s. sephaena Ad. ♂
F. s. rovuma Ad. ♂

Natal Francolin (p. 52)
Francolinus natalensis

Red-billed Francolin (p. 59)
Francolinus adspersus

N. B. undertail coverts should be white

Orange River Francolin (p. 41)
Francolinus levaillantoides

F. l. lorti Ad. ♂
F. l. 'pallidior' (= *levaillantoides*)
F. l. levaillantoides Ad. ♂

Swainson's Francolin (p. 68)
Francolinus swainsonii swainsonii Ad. ♂

Cape Francolin (p. 58)
Francolinus capensis Ad. ♂

Shelley's Francolin (p. 39)
Francolinus shelleyi shelleyi Ad. ♂

Grey-breasted Francolin (p. 72)
Francolinus rufopictus Ad. ♂

Yellow-necked Francolin (p. 73)
Francolinus leucoscepus Ad. ♂

Red-necked Francolin (p. 70)
Francolinus afer

F. a. cranchii Ad. ♂
F. a. melanogaster Ad. ♂
F. a. castaneiventer Ad. ♂
F. a. afer Ad. ♂

12in / 30cm

Plate 4
(Opp. p. 49)

Francolinus schlegelii Heuglin. Schlegel's Francolin. Francolin de Schlegel.

Francolinus schlegelii Heuglin, 1863. Journ. f. Orn., 11, p. 275; Bongo River (= Bussere River), Bahr-el-Ghazal, Sudan.

Forms a superspecies with *F. coqui* and *F. albogularis*.

Range and Status. Endemic resident from W-central Cameroon (Adamawa Plateau), through N Central African Republic and S Chad (south of 10°N) to SW Sudan (Bahr el Ghazal). Generally uncommon to rare and local.

Description. ADULT ♂: forehead and crown greyish brown with some rufous mottling, grading to ochraceous on nape. Lores and ear-coverts brownish grey; supercilium and remainder of sides of head and neck ochraceous. Upperparts mainly rufous-chestnut, some grey mottling on back and blackish blotching on scapulars and tertials, feathers with creamy shaft streak and very faint black and buff barring; feathers on rump and tail-coverts vermiculated black. Tail rufous faintly barred black. Chin and throat ochraceous-buff; rest of underparts white, finely barred black, some chestnut blotching on flanks. Primaries brown, secondaries and wing-coverts rufous. Bill black, base yellow; eye brown; legs yellow. ADULT ♀: like ♂ but back browner and with more black blotching, creamy shaft streaks narrower, black barring on belly irregular. SIZE: (10 ♂♂, 8 ♀♀) wing, ♂ 121–133 (128), ♀ 118–126 (123); tail, ♂ 59–71 (66.7), ♀ 63–71 (65.1); tarsus, ♂ 34–41 (37.2), ♀ 31–37 (34.4); 1 spur; ♂ 4–12 (8.1), ♀ 0.3–1.5 (0.7).

IMMATURE: like ♀ but mantle and scapulars with rufous-buff barring.

DOWNY YOUNG: undescribed.

Formerly considered a subspecies of Coqui Francolin *F. coqui*.

Field Characters. A small francolin with yellowish face, neck and throat, rufous upperparts and wings, and white underparts narrowly barred black. Allopatric with Coqui Francolin; similar-sized White-throated Francolin *F. albogularis* has white throat, chestnut on wings but very little on upperparts which are mainly grey-brown; ♂ with unbarred buff and chestnut underparts, ♀ has underparts barred but duller and buffier. Schlegel's exhibits little sexual dimorphism, with both sexes barred black and white below as in Coqui Francolin ♂.

Voice. Tape-recorded (215, 217). Trumpet-like advertisement call 'ter, ink, terrrra' qualitatively similar to that of Coqui Francolin but delivered much more rapidly (even faster than that of White-throated); also somewhat lower in pitch.

Francolinus schlegelii

General Habits. Inhabits well grassed wooded savannas in close association with 'Ka' tree *Isoberinia doka*. Shy bird, rarely observed near human habitation. Occurs in pairs or small coveys, often mixing with other francolins (*F. icterorhynchus?*). Sleeps on ground in pairs huddled close together, facing in opposite directions, often at foot of tree. Flight slow, short and silent.

Food. Grass seed, leaves of *Isoberlinia doka*, grain, caterpillars.

Breeding Habits. Probably monogamous since usually seen in pairs during breeding season.
NEST: hollow on ground often lined with leaves.
EGGS: 2–5 (record of 10 probably erroneous or from 2 ♀♀). Smooth; cream coloured. SIZE: 33–38 × 24–28.
LAYING DATES: Sudan, Sept–Nov.

References
Cave, F. O. (1949).
Bannerman, D. A. (1951).

Red-winged Group

7 closely related spp. in 2 superspecies (*streptophorus/africanus/levaillantii/finschi* and *shelleyi/psilolaemus/levaillantoides*). Clearly distinguished from other francolins by distinctive, musical call type, contrasting with more raucous calls of most other francolins. Medium-sized (320–520 g) francolins with quail-like upperparts (feathers rufous-brown with lattice-work of rufous-buffy barring and buff shaft streaks); not sexually dimorphic in plumage or size (wing length); relatively short, yellow tarsi with 1 spur; feathered ceres; black bill (base of lower mandible usually yellowish). All species have more or less open habitats and roost on ground. In areas of sympatry, species segregate by altitude.

Francolinus streptophorus Ogilvie-Grant. Ring-necked Francolin. Francolin à collier.

Plate 4
(Opp. p. 49)

Francolinus streptophorus Ogilvie-Grant, 1891. Ibis, Ser. 6, vol 3, p. 126; Mangiki, Mt Elgon, Kenya.

Forms a superspecies with *F. africanus*, *F. levaillantii* and *F. finschi*.

Range and Status. Endemic resident. Small population in highlands of Cameroon (Foumban area) at 1050–1200 m. In Uganda occurs from Kidepo Nat. Park west to Nile and south to Katonga Valley; also at foot of Mt Moroto (S. Keith, pers. obs.). Found in W Kenya from S slopes of Mt Elgon to Samia Hills and Nyando Valley, and in NW Tanzania in Kibondo and Kasulu Districts. Patchily distributed and generally uncommon, but locally abundant in Gulu and Choa Districts of N Uganda, and fairly common in NW Tanzania.

Description. ADULT ♂: forehead and crown dark greyish brown; supercilium white, extending backwards to nape. Sides of head, nape and neck rufous-chestnut. Upper mantle barred black and white, extending as broad band across breast. Rest of upperparts grey-brown, feathers with narrow buff shaft streaks. Tail grey-brown. Chin and throat white. Underparts rufous-buff heavily blotched with dark brown. Undertail-coverts grey, densely vermiculated and irregularly barred with buff, each feather with a faint buffy shaft streak. Flight-feathers uniform brownish grey. Bill black, base of lower mandible yellowish; eye brown; legs pale yellow. ADULT ♀: like ♂ but crown dark brown, feathers edged lighter, and upperparts brown, feathers barred rufous-buff and with broader buff shaft-streaks. SIZE: (16 ♂♂, 9 ♀♀) wing, ♂ 141–167 (150), ♀ 139–160 (152); tail, ♂ 67–80 (73·9), ♀ 74–83 (77·7); tarsus, ♂ 34–46 (40·6), ♀ 36–50 (42·5); 1 spur; ♂ 1–3 (1·9), ♀ 1–2·5 (1·7). WEIGHT: ♂ (n = 2) 364, 406.
IMMATURE: undescribed.
DOWNY YOUNG: undescribed.

Field Characters. Most distinctive character is black and white barred breast and 'ring-neck'; other sympatric francolins have these areas streaked. Combination of black bill, chestnut face and nape, and yellow legs also distinctive, and grey upperparts plainer, less heavily marked than other species. Heuglin's Francolin *F. icterorhynchus* also has yellow legs but has yellow bill, grey face-patch, mottled back. Much larger Clapperton's Francolin *F. clappertoni* has red bill, face and legs, mottled upperparts and blotched underparts. For differences from Crested Francolin *F. sephaena*, see that species.

Voice. Tape-recorded (217). Advertisement call distinctive: 2 soft dove-like 'coos', 1st note with a lower pitch, followed by piping trill; also a very noisy flight call.

General Habits. In Cameroon occurs in wild uninhabited country covered with thin grass among rocks. In E Africa occurs on stony hill-sides with sparse cover and in wooded grassland from 600 to 1800 m. Also visits cultivation. Shy, skulking bird, but can be seen by roadside in early morning; spends hotter part of day in shade of bush. Runs into cover when alarmed, but also flies very fast, even for a francolin. Usually in pairs or small coveys.

Food. Insects, seeds.

Breeding Habits. Probably monogamous since usually in pairs during breeding season. Possibly territorial, ♂ giving advertisement call in early morning from a prominence, e.g. termite mound.
NEST: slight depression with little or no lining, often close to a rock.
EGGS: 4–5. Greyish buff with dark speckled pores.
LAYING DATES: Uganda, Apr, early in rains; W Kenya, Dec–Mar, in dry season.

References
Bates, G. L. (1930).
Britton, P. L. *et al.* (1980).
Jackson, F. J. and Sclater, W. L. (1938).

PHASIANIDAE

Plate 4
(Opp. p. 49)

Francolinus africanus Stephens. Grey-wing Francolin. Francolin à ailes grises.

Francolinus africanus Stephens, 1819. Shaw's Gen. Zool., II, pt. 2, p. 323; Cape Province.

Forms a superspecies with *F. streptophorus*, *F. levaillantii* and *F. finschi*.

Francolinus africanus

Range and Status. Endemic resident from W, SW, E and NE Cape Province to SW and E Orange Free State, Lesotho, SE Transvaal and highlands of Natal. Local and uncommon: in Natal Drakensberg occurs at densities of 1 bird per 30–100 ha (Mentis and Bigalke, 1985). Numbers apparently declining (Clancey 1967) and conservation measures needed. Management strategy for populations in Natal suggested by Mentis and Bigalke (1973).

Description ADULT ♂: forehead to nape dark brown, feathers edged rufous. Supercilium and side of head rufous; eye and moustachial stripes black, extensively flecked white, extending downwards along sides of neck and throat, broadening to meet on upper breast. Mantle and wing-coverts dark grey-brown, feathers barred rufous-buff and with buff shaft streak. Uppertail-coverts similar but lack shaft streak; tail grey-brown, barred buff. Chin white, throat white flecked with black. Upper breast mottled and irregularly barred black and white, lower breast tawny-buff, heavily blotched chestnut. Belly and flanks dark buff finely and irregularly barred black and white, flanks blotched with chestnut. Undertail-coverts grey-brown barred with buff. Primaries brownish grey, faintly barred and washed rufous-chestnut. Bill brownish black; eye brown; legs yellowish brown. Sexes alike. SIZE: (27 ♂♂, 20 ♀♀) wing, ♂ 144–169 (157), ♀ 142–163 (153); tail, ♂ 74–92 (84·5), ♀ 73–90 (81·1); tarsus, ♂ 38–48 (42·7), ♀ 37–47 (41·7); 1 spur; ♂ 1–7 (4·2), ♀ 0·5–3 (1·1). WEIGHT: ♂ (n = 6) 411–501, ♀ (n = 4) 385–410.

IMMATURE: like ♀ but duller, throat white.

DOWNY YOUNG: forehead, crown and nape rufous-brown, crown-stripe darker; sides of head buffy with dark eye and moustachial stripes, and an indistinct stripe in between. Broad dark brown stripe down mid-back flanked by narrower buff stripes. Rest of upperparts mottled and irregularly barred black, brown and buff. Underparts buffy, washed with rufous on breast.

Field Characters. Differs from most other red-winged francolins in having small amount of chestnut on wings, grey throat mottled with black, greatly reduced black and white gorget, and relatively narrow black and white barring on belly and flanks. Call also distinctive.

Voice. Tape-recorded (74, C, F, 296). Musical, whistling, high-pitched 'whee-hee-hee, wee-pe-ew' repeated in series with 2–5 s intervals. Also, soft whistling contact call and loud squeal when flushed.

General Habits. Inhabits low scrub and open grassy patches on hillsides and hilltops between c. 1800 and 2750 m; occurs at sea level in SW Cape Province. Avoids areas recently burnt and those with accumulated grass and leaf litter and relatively long grass. Occurs in coveys of 5–25, usually not more than 8. Sedentary, coveys seen year after year on same ground. Feeds mainly in first and last few hours of daylight; bulk of food intake late in day. Digs for bulbs and roots with bill. Remains inactive in cover during middle of day.

Call often given from prominence, mainly in early morning and late afternoon.

Food. Volumetric analysis of crop contents of Natal birds showed: bulbs and roots (especially Iridaceae, Amaryllidaceae, Cyperaceae) 70–75%, insect and other invertebrates 20–25% (mainly grasshoppers, ants and beetles in summer), and small amounts of seeds, fruits and other vegetable matter (Mentis 1973, Mentis and Bigalke 1973, 1981a).

Breeding Habits. Monogamous; probably territorial during breeding season, possibly all year, since remains in same area even after chicks hatch.

NEST: scrape, sometimes lined with feathers or grass, under tuft of grass.

EGGS: 3–8, up to 15 (2 ♀♀?), mean c. 5. Yellowish brown, sometimes speckled with brown and slate. SIZE: (n = 53) 36–41·8 × 28·7–32·5 (39 × 29·9).

LAYING DATES: Cape Province, July–Dec, Natal, Aug–Mar, peaking in Nov–Dec.

DEVELOPMENT AND CARE OF YOUNG: young hatched early may breed in 1st summer.

BREEDING SUCCESS/SURVIVAL: an approximately 1:1 old to independent young ratio indicates an annual mortality rate of c. 50%.

References
Mentis, M. T. and Bigalke, R. C. (1973, 1979, 1980, 1981a,b, 1985).

Francolinus levaillantii (Valenciennes). Red-wing Francolin. Francolin de Levaillant.

Plate 4
(Opp. p. 49)

Perdix levaillantii Valenciennes, 1825. Dict. Sci. Nat., 38, p. 441; Cape of Good Hope.

Forms a superspecies with *F. streptophorus*, *F. africanus* and *F. finschi*.

Range and Status. Endemic resident, distribution extremely patchy. In Kenya occurs at 1800–3000 m in W highlands from Mt Elgon to Trans-Nzoia and Mau; in Uganda from S Bunyoro and Mengo through Ankole (very common) to NW Tanzania (Bukoba), Rwanda and Burundi, and west into E Zaïre from highlands east of Lake Edward to Itombwe Mts. Occurs in S Tanzania from Rukwa to Selous Game Reserve and south to Kitulo Plateau, and in N Malâwi and NE Zambia on Nyika Plateau. Another population in SE Zaïre (Katanga Province), chiefly in Upemba National Park area. Occurs in NW Zambia in Balovale, Kabompo, Kalabo, Mongu and Mankoya Districts, and in highlands of W-central Angola in Huila and Huambo. In South Africa fairly widespread, from highveld of central and S Transvaal through E Orange Free State and W Natal to E Cape Province, and along coastal mountains to Swellendam; also W Swaziland and lowlands of Lesotho. Locally common but generally uncommon. In Natal Drakensberg occurs at density of 1 bird/25–50 ha (Mentis and Bigalke 1985). Disappears if habitat clean-burned, overgrazed, or allowed to go fallow.

Description. *F. l. levaillantii* (Valenciennes) (including *crawshayi*): South Africa, N Malaŵi and NE Zambia. ADULT ♂: forehead to nape brownish black, feathers edged rufous. Sides of head and supercilium ochraceous tawny; crown and moustachial stripes black, heavily flecked white, latter continuing down side of neck to join black and white gorget which extends around back of neck and upper breast. Stripe through eye and onto side of neck reddish ochre. Upperparts dark grey-brown, feathers mottled black, barred rufous-buff and with a creamy shaft streak. Tail dark grey-brown, finely barred rufous buffy. Chin and throat white, rimmed below with ochraceous-tawny above a black and white gorget. Remaining underparts buff; lower breast washed rufous and heavily blotched chestnut; belly indistinctly barred brownish black and heavily blotched chestnut, especially on flanks; undertail-coverts finely barred black. Primaries and outer secondaries rufous-chestnut, tips grey-brown. Bill blackish horn, base of lower mandible yellow; eye brown; legs yellowish brown. Sexes alike. SIZE: (23 ♂♂, 17 ♀♀) wing, ♂ 149–171 (162), ♀ 140–168 (158); tail ♂ 71–91 (80·9), ♀ 71–88 (77·4); tarsus, ♂ 44–53 (47·5), ♀ 42–49 (46·0); 1 spur; ♂ 1–8 (4·2), ♀ 0·5–3 (1·2). WEIGHT: (Upemba, Zaïre) ♂ 495, ♀ about to lay 515; (Uganda) ♀ (n = 1) 462; unsexed (n = 22, Natal) mean 487.

IMMATURE: like ♀ but paler and black and white gorget less distinct. WEIGHT: ♂ (n = 1) 495 (Zaïre).

DOWNY YOUNG: forehead and crown dark brown, crown-stripe darker. Sides of head buffy with dark brown eye and moustachial stripes, and an indistinct one in between. Wide dark brown stripe down mid-back edged darker, flanked by slightly narrower buff stripes. Rest of upperparts mottled and barred with brown and buff. Underparts buffy, darker on breast. Upper mandible pale horn, lower darker; legs pale yellow.

Remarkably little geographical variation in this species in spite of widely separated populations.

F. l. kikuyuensis Ogilvie-Grant: Angola and Zambia (except Nyika), to Zaïre and E Africa. ADULT ♂: like *levaillantii* but black and white gorget much reduced, breast chestnut with pale streaks, rest of underparts buff vermiculated and barred broadly black. WEIGHT: ♂♀ 500–560.

Field Characters. A richly coloured francolin with broad neck-stripe of reddish ochre, rich buff underparts and much red in the wings. Distinguished from all other red-winged francolins by ring of ochre separating throat from black and white gorget, and broad collar of black and white barring on hindneck, lower throat and upper breast. Voice also distinctive.

Voice. Tape-recorded (4, 74, 75, C, F). Advertisement call musical, high-pitched whistle 'whee-hee-hee-hee-whee-hee', much more like Grey-wing Francolin *F. africanus* than Orange River Francolin *F. levaillantoides*, for which it is often mistaken in field. 2-part call, 'whee-hee-hee' followed by 'heep', possibly antiphonal. ♀♀ give a clucking 'chook, chook, chook' when leading chicks. Alarm call, 'kourrr', also typical Red-winged Group squeal when flushed.

General Habits. Inhabits high altitude grassland (generally in moister situations than Orange River Francolin), especially on steeper slopes in sheltered valleys, reedy spots; on coast in stony grassland. Also found in rank grass along watercourses, patches of scrub and woodland, grassy clearings in second growth (Kivu), and on cultivation. Usually in pairs or coveys of up to 10. Not shy in Zaïre, where allows approach within 8–10 m (Verheyen 1953). Calls most often in early morning and at dusk. Although a good runner, sits tight and often does not flush until almost stepped on; when flushed usually flies considerable distance. When foraging, scratches on ground among dead leaves and other vegetation, and digs for roots with bill. Roosts in groups on ground. Complete moult takes place SE Zaïre Mar–Apr. In southern Africa moults Mar–May.

Food. In quantitative analysis (Natal) bulbs and corms (mainly Amaryllidaceae and Iridaceae) comprised 70–75% of crop volume, even higher during winter months (Mentis 1973). Also takes some vegetable matter, but usually less than 5% of crop volume. Rest of diet animal food, mainly ants, spiders, grasshoppers, millepedes and beetles, with largest quantities in summer months.

Breeding Habits. Presumably monogamous since often in pairs during breeding season.

NEST: shallow scrape lined with grass and/or rootlets, hidden at base of tuft of grass or sedge.

EGGS: (southern Africa) 3–12, mean 5. Oval; brownish yellow to olive-brown freckled slaty. SIZE: (n = 41, *F. l. levaillantii*) 37·3–44 × 29–34 (40·1 × 32·1).

LAYING DATES: E Cape Province, Mar–July; Natal Aug–Feb; Transvaal Dec; SW Angola, Apr; Zambia, Jan, Mar, July–Sept, Dec; Malawi, Jan, Mar, Nov–Dec, may breed only during rains to avoid grass fires; Zaïre, Shaba Province, July–Dec, during dry season; East Africa (Kenya and Uganda): Region A, Mar–Apr, Region B, Jan–Feb, Aug, Oct–Nov, in rains or early dry season.

DEVELOPMENT AND CARE OF YOUNG: young birds accompany parents in family parties until fledged and for some time thereafter.

References
Mentis, M. T. and Bigalke, R. C. (1973, 1979, 1980, 1981a,b, 1985).

Plate 4
(Opp. p. 49)

Francolinus finschi Bocage. Finsch's Francolin. Francolin de Finsch.

Francolinus finschi Bocage, 1881. Orn. Angola, pt. 2, p. 406; Caconda, Benguella, Angola.

Forms a superspecies with *F. psilolaemus* and *F. levaillantoides*.

Range and Status. Endemic resident in 3 areas: (1) on both sides of lower Congo River, around Brazzaville (Congo) and Kinshasa (Zaïre); (2) S Zaïre in Gungu District, Kwango Province; and (3) Angola from Cuanza Norte (Salazar) and S Milanje to N Huila (Caconda, Mt Moco, Cagandala National Park). Sparsely distributed and uncommon to rare in Angola; not uncommon around Gungu, Zaïre.

Description. ADULT ♂: forehead to nape and hindneck brownish grey, feathers edged lighter. Sides of head and broad stripe down neck rufous. Upperparts dark grey-brown, feathers vermiculated and irregularly barred rufous-buff and with buff shaft streak. Uppertail-coverts brownish grey, densely vermiculated and indistinctly barred buff. Tail brownish grey barred and vermiculated buff. Chin and throat white. Upper breast brownish grey, mottled and indistinctly barred buff. Lower breast and belly buff, heavily blotched with chestnut. Flanks greyish, irregularly barred with rufous-buff and blotched chestnut. Undertail-coverts grey with dense buff vermiculations and indistinct barring. Primaries and outer secondaries rufous-chestnut, tipped greyish brown. Bill black, base of lower mandible yellow; eye brown; legs pale yellow. Sexes alike. SIZE: (4 ♂♂, 9 ♀♀) wing, ♂ 162–170 (167), ♀ 158–174 (166); tail, ♂ 77–97 (88·1), ♀ 80–99 (84·7); tarsus, ♂ 43–54 (48·9), ♀ 40–53 (45·0); 1 spur; ♂ 8–10 (9·4), ♀ 1–10 (4·4).

IMMATURE: undescribed.
DOWNY YOUNG: undescribed.

Field Characters. The most distinctive red-winged francolin: only one with grey breast and without black and white stripes on face and neck or black and white breast and collar; underparts relatively uniform. Sympatric Red-wing Francolin *F. levaillantii* has extensive black and white breast and collar, chestnut blotches on belly contrasting more with buff of remaining underparts.

Voice. Not tape-recorded. Call a loud 'wit-u-wit' usually heard at dusk.

General Habits. Inhabits grassland near gallery forest (Gungu), wooded savanna, *Brachystegia* woodland, bare slopes above timberline (Mt Moco). Recorded foraging on ground among burnt grass and leaves (Hall 1963); often in pairs.

Food. Beetles, insect larvae and seeds.

Breeding Habits. Presumably monogamous since often found in pairs.
　NEST: placed in vegetation on the ground.
　EGGS: *c*. 5, light brown.
　LAYING DATES: Zaïre, Jan, Mar, July; Angola, June–July.

References
Hall, B. P. (1960, 1963).
Lippens, L. and Wille, H. (1976).

Francolinus shelleyi Ogilvie-Grant. Shelley's Francolin. Francolin de Shelley.　　Plate 2

Francolinus shelleyi Ogilvie-Grant, 1890. Ibis, p. 348; Hartley Hills, Umfuli River, Mashonaland, Zimbabwe.　　(Opp. p. 33)

Forms a superspecies with *F. psilolaemus* and *F. levaillantoides*.

Range and Status. Endemic resident. Distribution very patchy. Widespread in Kenya and N Tanzania from north of Mt Kenya to Loita Plains, Chyulu Hills, Crater Highlands and North Pare Mts. Only scattered records from elsewhere in Tanzania and 1 record from Uganda (Mulema Hill, Ankole). Occurs throughout Malaŵi and north through parts of N Zambia to extreme E Zaïre east of L. Tanganyika. In Zambia absent from Luapula Province and Luangwa Valley, but found from copper belt north and west to Mwinilunga and extreme S Zaïre; south through most of Zimbabwe (absent from SE) to W border of Mozambique. Occurs again in S Mozambique (probably most of Sul do Save) and NE South Africa (NE Transvaal and all Natal except SW). Generally rather uncommon but locally common. Disappears from overgrazed or frequently burned areas.

Description. *F. s. shelleyi* Ogilvie-Grant (including *uluensis*, *canidorsalis* and *sequestris*): S Uganda and SW Kenya south through Tanzania, S Zambia, S Malaŵi, Zimbabwe and Swaziland to E Transvaal and N Natal. ADULT ♂: forehead, crown and nape dark brownish grey, feathers edged buff or pale grey. Sides of head and supercilium rufous-buff; eye-stripe and moustachial stripe black flecked white, continuing as necklace round lower throat and upper breast. Remaining upperparts dark grey-brown, feathers barred rufous-buff and with a creamy shaft streak. Uppertail-coverts and tail pale greyish brown finely barred buff. Chin and throat white, upper breast strongly barred black and white. Lower breast and flanks broadly streaked chestnut. Belly irregularly and broadly barred black and white. Undertail-coverts finely barred black and white. Primaries and outer secondaries rufous-chestnut, tips brownish grey. Bill black, lower mandible yellow at base; eye brown; legs dull yellow. Sexes alike. SIZE: (76 ♂♂, 57 ♀♀) wing, ♂ 152–177 (162), ♀ 145–172 (157); tail, ♂ 76–97 (86·3), ♀ 74–93 (82·9); tarsus, ♂ 37–47 (41·8), ♀ 35–47 (40·5); 1 spur; ♂ 4–12 (7·9), ♀ 0·2–4 (1·2). WEIGHT: ♂ (n = 4) 420–480, ♀ (n = 3) 392–460.
　IMMATURE: like ♀ but paler, black and white barring on upper breast irregular, outer webs of primaries marked with buff. Rufous-buff on upperparts reduced or absent.
　DOWNY YOUNG: crown, nape and broad stripe down mid-back rufous-brown edged darker, flanked by stripes of rufous-buff. Sides of head buffy white, eye-, crown- and moustachial stripes dark brown; also some indistinct striping in between eye and moustachial stripes. Underparts buffy.

F. s. whytei Neumann: SE Zaïre, N Zambia and N Malaŵi. Belly less heavily marked with black and white, black and white barring on upper breast often extremely reduced, and throat buffy (not white).

Field Characters. Among Red-winged Group overlaps mainly with Red-wing Francolin *F. levaillantii* which differs in having underparts washed with yellow-buff (not barred), much more red in wing, broad neck-stripe and narrower ring around throat rich ochre, hindneck barred black and white, upperparts with much broader pale shaft streaks. In East Africa approaches Moorland Francolin *F. psilolaemus* which occurs at higher elevations and has much more red in wings. In South Africa approaches Grey-wing Francolin *F. africanus* which has throat white spotted black and fine barring on neck and rest of underparts, and Orange River Francolin *F. levaillantoides* which is paler above and has buffy underparts with chestnut streaks and very little barring. Call also distinctive.

Voice. Tape-recorded (21, 74, B, F). Advertisement call: musical 4–note 'I'll-drink-yer-beer' usually repeated 6–7 times, with longest pause between notes 2 and 3. By 5th repetition, another bird often answers. When flushed gives shrill alarm note.

General Habits. Occurs in open montane and other grassland, wooded savanna, thornveld, and open areas in *Brachystegia*, mopane and other woodland; often in stony terrain or among rocky outcrops. Locally enters cultivation. Found from coastal areas (Mozambique, Natal) to highlands, up to 2200 m in Zimbabwe (Inyanga Highlands), 2450 m in Malaŵi (Nyika Plateau) and 3000 m in E Africa. In Kenya, grassland and wooded gorges; replaced at higher altitudes by Moorland Francolin *F. psilolaemus*. In Zaïre and Tanzania found in wooded savanna in thick grass. Largely independent of water. Occurs singly, in pairs or small coveys. In digging for food makes characteristic cone-shaped hole 3–5 cm deep and 2–3 cm wide at top. Very difficult to flush, preferring to escape on foot. If flushed, flies only short distance and then sits tightly. Calls mainly just after dawn and at dusk, especially latter.

Food. Roots, bulbs, fallen grain, small molluscs, termites, grasshoppers, beetles and locusts.

Breeding Habits. Probably monogamous since commonly in pairs during breeding season.

NEST: among rocks, grass or herbaceous cover; well concealed shallow depression lined with grass or roots.

EGGS: (southern Africa) 4–7, normally 4–5; (Malaŵi) 3–4. Regular ovals; creamy buff to pinkish white, sometimes with fine brown spots. SIZE: (n = 4, Kenya) 40 × 33; (n = 11, Zimbabwe) 36·9–43·6 × 29·3–35·4 (40 × 31·5); (n = 30, South Africa) 35·2–40·4 × 28·7–32·3 (38 × 31·1).

LAYING DATES: South Africa: Natal and Transvaal, Aug–Jan; Zambia, Feb–May, Sept–Oct, peak in Sept; Mozambique, Jan; Zimbabwe, Jan–Apr, June, Aug–Dec, peaks in Mar–Apr and Sept–Oct; Malaŵi, May, Aug–Nov; East Africa: Region C, May–June; Region D, Mar, July (a dry season breeder).

INCUBATION: probably by ♀ alone. Period: *c.* 22 days.

BREEDING SUCCESS/SURVIVAL: low among chicks due to predation by raptors and small carnivores.

References
Clancey, P. A. (1967).
Meyer, H. F. (1971a).

Plate 3
(Opp. p. 48)

Francolinus psilolaemus G. R. Gray. Moorland Francolin. Francolin montagnard.

Francolinus psilolaemus G. R. Gray, 1867. List Birds Brit. Mus., Gallinae, p. 50; Shoa, Ethiopia.

Forms a superspecies with *F. shelleyi* and *F. levaillantoides*.

Range and Status. Endemic resident, confined to montane areas of central and SE Ethiopia (Shoa, Arussi Plateau, south to near Kenya border), Uganda and Kenya (Mt Elgon, Mau-Narok, Aberdares, Mt Kenya). Generally uncommon to rare in Kenya; locally frequent to common in Ethiopia.

Description. *F. p. psilolaemus* Gray (including *ellenbecki*): Ethiopia. ADULT ♂: forehead and crown dark brown, feathers broadly edged, barred and streaked rufous-buff, giving mottled appearance. Broad rufous-buff and narrow black stripes over eye, latter extends down to hindneck; eye-stripe and moustachial stripe black, freckled rufous-buff; areas in between stripes rufous-buff. Hindneck mottled black and pale buff. Rest of upperparts blackish brown (tinged rufous on upper back) feathers narrowly barred rufous-buff and with broad cream shaft streak; uppertail-coverts similar but have narrow shaft streak. Tail dark brown barred rufous buff. Chin and throat white heavily freckled dark brown. Upper breast rufous-buff, feathers spotted terminally brownish black. Lower breast, belly and flanks buff, finely barred black and heavily blotched chestnut. Undertail-coverts buff barred broadly black. Primaries and outer secondaries chestnut, tips grey-brown, inner secondaries with much less chestnut. Bill blackish brown, base of lower mandible yellowish; eye brown; legs pale yellow. Sexes alike. Birds from southern part of range tend to be larger and more richly pigmented on head and back, and spots on underparts darker. SIZE: (17♂♂, 11 ♀♀) wing, ♂ 150–176 (164), ♀ 151–172 (159); tail, ♂ 76–101 (85·8), ♀ 81–96 (87·4); tarsus, ♂ 38–51 (44·8), ♀ 40–48 (44·4); 1 spur; ♂ 4–10 (8·3), ♀ 0·3–3·5 (0·9). WEIGHT: ♂ (n = 2) 510, 530; non-breeding ♀ 370, ♀ with large ovaries 510.

IMMATURE: undescribed.
DOWNY YOUNG: undescribed.

F. p. elgonensis Ogilvie-Grant (including *theresae*): East Africa. Chestnut collar on hindneck, feathers with black terminal spots; much less brown blotching on throat; breast rufous-buff spotted black; rest of underparts greyish buff with narrow dark barring, lower breast, belly and flanks blotched rufous. Also average larger: wing, ♂ (n = 15) 168–180 (174), ♀ (n = 12) 152–173 (166).

Field Characters. In Ethiopia sympatric with Orange River Francolin *F. levaillantoides* which has greyer upperparts, unspotted white throat, whitish underparts barred black, dark chestnut spots on breast, much less red in wings. In East Africa isolated by habitat from Shelley's Francolin *F. shelleyi*, but overlaps with Red-wing Francolin *F. levaillantii* which differs in having hindneck barred black and white, broad black and white necklace on lower throat separated from throat by yellow-buff ring, unspotted chestnut breast, underparts without pale rufous blotches.

Voice. Not tape-recorded. Call almost identical to that of local form of Shelley's Francolin *F. shelleyi*; gives typical 'red-wing' squeal when flushed.

General Habits. Inhabits montane heath (*Erica*) moorland and grassland from 1800 to 4000 m. Found in pairs or small coveys (probably family parties). Calling contagious, another bird echoes advertisement call after few repetitions.

Food. Not described, but probably a mixed diet dominated by bulbs and roots.

Breeding Habits. Probably monogamous since usually in pairs during breeding season. Notes on 1 nest found at Mau Narok, Kenya (undated EANHS nest record card from P.H.B. Sessions): nest in rough grass contained 2 chicks and 3 eggs. Months when young birds seen with adults: (a) very small chicks, Jan–Mar, June–Aug; (b) half-grown chicks, Mar, May–Aug; (c) nearly full-grown chicks, June, Oct–Dec. Laying dates apparently all months except Sept–Oct, but most in 1st half of year. Young generally remain with family group for *c.* 6 months after hatching. Usually 4–5 young seen with adults when small, but covey diminishes with time and predators.

References
Jackson, F. J. and Sclater, W. L. (1938).
Sessions, P. H. B. (1967)

Francolinus levaillantoides (Smith). Orange River Francolin. Francolin d'Archer.   Plate 2

Perdix levaillantoides A. Smith, 1836. Rep. Exped. Centr. Afr., p. 55; country towards sources of Orange River.   (Opp. p. 33)

Forms a superspecies with *F. shelleyi* and *F. psilolaemus*.

Range and Status. Endemic resident disjunctly distributed in NE and SW Africa. Range in NE very poorly known; 3 apparently separate populations in N Ethiopia (Eritrea, N highlands), from extreme N Somalia to Rift Valley in Ethiopia and south in SE highlands to near Kenya border, and from extreme SE Sudan to NE Uganda (Kidepo Valley National Park, Mt Moroto). In SW occurs in Angola from Lobito and Mossamedes through Huila north to Gambos, and broadly through Namibia from Kaokoveld and Ovamboland to Damaraland and southeast through much of southern Botswana (north in E to Makgadigadi area (Nata, Kanyu)) to Transvaal (highveld), Orange Free State, lowlands of Lesotho and N Cape Province. Frequent to common in Ethiopia, locally common in southern Africa. Not threatened generally, but no recent records from Lesotho.

Description. *F. l. levaillantoides* (Smith) (including *pallidior* and *kalaharica*): Lesotho, South Africa, Botswana and Namibia north to Otjiwarongo (Damaraland). ADULT ♂: forehead to nape blackish brown, feathers edged rufous-buff; top of head bordered narrowly in front and broadly to rear by ochre stripe which continues down onto sides of neck; a 2nd ochre stripe from eye over ear-coverts and down sides of neck, separated from 1st by stripe of black and white mottling, and from white throat by another black and white stripe which continues as necklace across foreneck and upper breast; supercilium white flecked dark brown. Upperparts dark greyish brown, paler on rump, feathers barred rufous-buff and with a buff shaft streak, mantle and back blotched with chestnut; wing-coverts similar but paler and greyer. Tail grey-brown, barred and vermiculated buff. Chin and throat white. Upper breast and lower neck barred finely black and white, linking with eye- and moustachial stripes to form a small gorget. Lower breast and belly rufous-buff, breast heavily blotched chestnut and flanks and lower belly indistinctly barred black and white. Undertail-coverts brownish grey, feathers barred and streaked rufous-buff. Primaries and secondaries rufous-chestnut, tips grey-brown. Bill brownish black, yellowish at base; eye brown; legs dull yellow. Sexes alike. SIZE: (52 ♂♂, 34 ♀♀) wing, ♂ 142–173 (163), ♀ 146–175 (160); tail, ♂ 75–93 (87·2), ♀ 66–97 (86·8); tarsus, ♂ 39–48 (43·1), ♀ 37–48 (41·2); 1 spur; ♂ 3–12 (6·2), ♀ 1–8 (2·3). WEIGHT: ♂ (n = 6) 370–538, ♀ (n = 6) 379–450.

IMMATURE: like ♀ but duller, facial striping and gorget poorly defined, and underparts irregularly barred with black and buff.

DOWNY YOUNG: undescribed.

Geographical variation considerable, but some described subspecies are poorly defined. We admit only the most distinctive subspecies here.

F. l. jugularis Buttikofer (including *cunenensis*): N Namibia (Swakop River, W Damaraland) to Angola. Paler, breast broadly mottled black and white, chestnut spotting on underparts much reduced. Smaller: wing, ♂ (n = 9) 148–162 (156).

F. l. gutturalis (Rüppell): N Ethiopia. Black and white facial striping and gorget poorly defined; breast greyish buff heavily blotched with chestnut; flanks and belly buff heavily streaked black, flanks also blotched with chestnut. Size similar to *levaillantoides*: wing, ♂ (n = 11) 151–169 (161), ♀ (n = 5) 148–158 (154).

F. l. lorti Sharpe (including *archeri*): Sudan, Uganda, S Ethiopia, Somalia. Like *gutturalis* but much paler; throat white, underparts off-white without dark streaks, finely barred black; breast and flank spots darker chestnut. WEIGHT: ♂ (n = 3) 340, 369, 397; ♀ (n = 1) 400.

Field Characters. Distinguished from Grey-wing Francolin *F. africanus* and Shelley's Francolin *F. shelleyi* by lack of black and white patterning on underparts; Red-wing Francolin *F. levaillantii* differs in having broad black and white collar around hindneck, rufous stripe separating white throat from black and white necklace, more red in wings. Call also distinctive. For differences from Moorland Francolin *F. psilolaemus*, see that species.

Voice. Tape-recorded (74, F). Call, 'ki-bi-til-ee', repeated in bouts of 5–9; similar to the 'I'll-drink-yer-beer' call of Shelley's Francolin *F. shelleyi*, but faster, with shorter intervals between notes.

General Habits. Inhabits open grassland, wooded and bushy grassland, dry woodland, grassy mountain slopes strewn with boulders. Presence most often detected by its calling; calls mainly in early morning and late afternoon. Usually in pairs or small coveys of up to a dozen birds. Very difficult to locate, even with good hunting dogs. When discovered and pursued, may hide in Anteater *Orycteropus afer* or mongoose burrows.

Food. Bulbs, corms of *Moraea*, seeds, berries, fallen grain, insects (bugs, termites, beetles, grasshoppers).

Breeding Habits. Probably monogamous since usually in pairs during breeding season. Territorial; reacts strongly to playback of conspecific call.

NEST: scrape under tuft of grass.

EGGS: 5–8. Pale pink to yellowish brown, sometimes speckled brown. SIZE: (n = 17, *F. l. levaillantoides*) 34·4–40·8 × 28–32 (37·2 × 29·4); (n = 6, *F. l. lorti*) 40·9–41·8 × 30·6–31·8 (41·2 × 31·2).

LAYING DATES: Transvaal, Feb–May, Sept–Oct; Namibia, June; W Angola Aug; Ethiopia, Feb, Apr, Aug.

References
Clancey, P. A. (1967).
Parker, S. A. (1963).

Plate 2 *Francolinus sephaena* (Smith). Crested Francolin. Francolin huppé.
(Opp. p. 33)

Perdix sephaena A. Smith, 1836. Rep. Exped. Centr. Afr., p. 35; Marico River, W Transvaal.

Links Latham's Forest Francolin (*F. lathami*) and the Red-tailed and Red-winged Groups (which are comprised of small to medium-sized francolins which live in relatively open grassland or savannas, roost on the ground, and have reddish brown, quail-like upperparts, black bills, yellow legs, comparatively musical calls and feathered ceres) with the remaining species (most of which are relatively large, live in thickets or forest, roost in trees, and have grey-brown, unpatterned, scaly or vermiculated upperparts, black or red legs, raucous calls and cartilaginous ceres). The Crested Francolin is smallish and has a black bill and reddish brown, quail-like upperparts like Latham's and most of the Red-winged and Red-tailed francolins. However, it has a rudimentary cartilaginous cere and lives in thickets and woodland, roosts in trees, and has red legs and raucous calls like most members of the remaining groups.

Range and Status. Endemic resident in E and southern Africa; extends from SE Sudan, central Ethiopia and NW Somalia, south through Kenya, Uganda, Tanzania, Mozambique, S Zambia, Zimbabwe, to S Angola, N Botswana, N Namibia, E Swaziland and NE South Africa. Locally common to abundant and nowhere threatened.

Description. *F. s. sephaena* (Smith): from *c.* 20°S in SE Zimbabwe, SE Botswana and S Mozambique south to E Swaziland and NE South Africa. ADULT ♂: forehead very dark brown; crown dark brown, feathers edged grey; crown-, eye- and moustachial stripes black, latter flecked with white; supercilium white. Ear-coverts brownish grey; sides of head white freckled with chestnut. Hindneck, mantle, scapulars and back dull chestnut brown, feathers with a creamy shaft streak. Rump and uppertail-coverts dark buff densely vermiculated and indistinctly barred dark grey. Tail reddish brown, distal third black. Chin and throat white. Upper breast buffy, each feather with a dark triangular-shaped, chestnut-brown, terminal blotch. Lower breast and belly buff indistinctly barred black. Undertail-coverts rufous-buff indistinctly barred black. Primaries brownish grey; secondaries similar but vermiculated buff. Bill black; eye brown; legs dull red. ADULT ♀: somewhat smaller; more cryptically coloured above, densely vermiculated and barred with buff and creamy shaft streaks narrow. SIZE: (44 ♂♂, 55 ♀♀) wing, ♂ 137–170 (154), ♀ 135–166 (145); tail, ♂ 86–112 (98·5), ♀ 85–113 (95·7); tarsus, ♂ 42–51 (45·8), ♀ 39–52 (42·7); 1 spur; ♂ 5–18 (12·8), ♀ 0·5–4 (1·3). WEIGHT: ♂ (n = 17) 320–417 (368), ♀ (n = 9) 280–350 (315).
IMMATURE: like ♀ but paler with dorsal shaft-streaking broader and more clearly defined.
DOWNY YOUNG: forehead and face white. Black eye-stripe extends from behind eye only. Crown, nape and broad stripe down centre of back dark rufous-brown, margined with black and flanked by broad buffy stripes. Underparts buffy-white, darker on breast. Wings mottled brown and buffy.

Varies considerably, both geographically and within populations. We follow Hall (1963) and recognize only 5 subspecies.
F. s. zambesiae Mackworth-Praed: W–central Mozambique, central and SW Malaŵi, S Zambia, N Zimbabwe, N Botswana, NW Namibia, and S Angola. Like *sephaena* but crown and nape paler, reddish brown on back much richer, and underparts less marked with dark brown and vermiculated with grey.
F. s. granti Hartlaub: N and W Ethiopia, S Sudan, Uganda, Kenya (except coast), NE Zaïre and N–central Tanzania. Like *sephaena* but smaller and dark brown blotching on upper breast much reduced. Wing, ♂ (n = 161) 124–161 (144), ♀ (n = 95) 128–147 (136). WEIGHT: ♂ (n = 20) 265–335 (288), ♀ (n = 9) 220–270 (240).
F. s. rovuma Gray: coastal Kenya and Tanzania, S Malaŵi and N Mozambique from just south of equator to *c.* 20°S latitude. Like *granti* but dark brown blotching on upper breast extends throughout underparts in form of shaft streaks.
F. s. spilogaster Salvadori: E Ethiopia, Somalia and NE Kenya. Like *rovuma* but larger and dark brown ventral shaft-streaking much finer. Wing, ♂ (n = 28) 137–161 (151), ♀ (n = 24) 130–148 (143).

Field Characters. Diagnostic features are: black bill, red legs, white supercilium and reddish brown back with white striations. When running often cocks tail like bantam and raises crown-feathers. In flight, black tail conspicuous. Loud, high-pitched, rattling call unlike that of other francolins. The Ring-necked Francolin *F. streptophorus* in few instances of sympatry segregates by habitat, has yellow (not red) legs, upper breast barred with black and white (not blotched with dark brown), and has continuous ring of chestnut feathers around nape and neck.

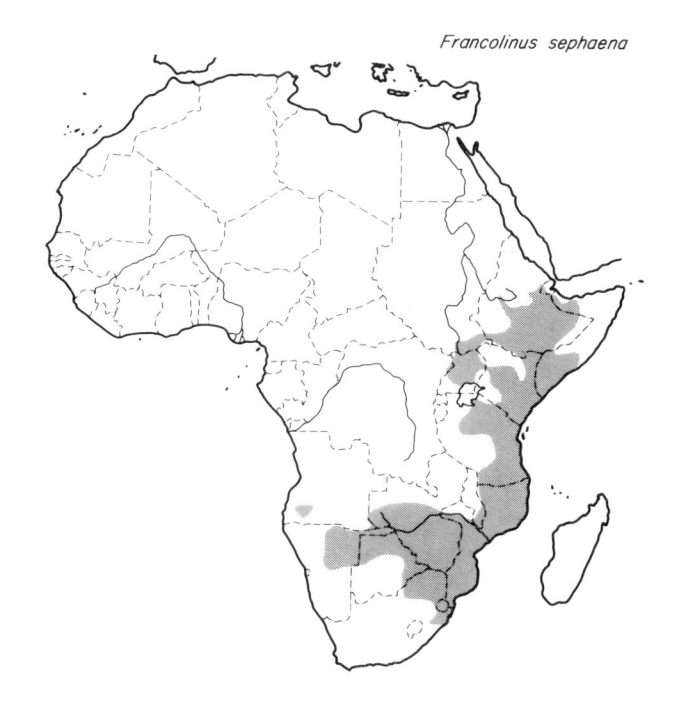

Voice. Tape-recorded (7, 10, 21, 50, 61, 72, 74). ♂ advertisement call a high-pitched squealing cackle, often 2 or more birds simultaneously, 'kerra-kreek' repeated rapidly 7–9 times, 2nd syllable accented and higher-pitched. Possible antiphonal calling 'kee' 'kek-kerra', 1st note given by 1 bird (♀?), 2nd being response of another bird (♂?).

General Habits. Inhabits thickets and dense bush along rivers with sparse ground cover or *Acacia*/*Commiphora* woodland with sparse grass cover, also forest edge and overgrown cultivation. In Somalia occurs from sea level to 2200 m, in tamarisk shrub in dry river beds and thickets of *Salvadora persica* and *Euphorbia* beside rivers and pans, also juniper forest (Mt Wagar); only found near water, its presence being an indicator of nearby water (Archer and Godman 1937), though elsewhere in its range it can be independent of water. Often on same ground as Coqui, Natal, Red-necked and Swainson's Francolins (*F. coqui, natalensis, afer* and *swainsonii*). Commonly found in pairs or small coveys of up to 7 birds; noisy; calls at dawn and after rain, first caller answered by chorus from all sides; sometimes calls in heat of day, when usually lying up in bush; also calls at night (J. H. van Niekerk, pers. comm.). Searches through elephant droppings for undigested matter and dung-inhabiting insects and their larvae. Normal gait slow and measured, with arched back and neck drawn in; when alarmed stands erect with neck stretched out, then runs off at great speed if cover insufficient. Flies very fast but for short distance only; escapes mammalian predators by flying up into bush or tree. Usually elusive and difficult to flush, but can become tame when not molested; noted feeding in bush camps and around edges of certain towns in Somalia (Archer and Godman 1937). Roosts in trees at night.

Food. Quantitative analysis of 167 crop contents from E Africa: insects (mainly termites) and their larvae made up $c.$ 32% of diet by volume, sedge bulbs $c.$ 16%, seeds, mainly of *Commelina* sp. and grasses, 27%, remainder miscellaneous green and other plant material. Elsewhere also berries, small molluscs, probably dung beetles and their larvae.

Breeding Habits. Probably monogamous since often in pairs during breeding season. Territorial during breeding season (W. R. Tarboton, pers. comm.), ♂♂ responding vigorously to playback of tape recordings of advertisement call.

NEST: well concealed shallow depression lined with grass beneath bush or shrub.

EGGS: Somalia 5–6; southern Africa 4–9, normally 6. Oval; very hard and thick-shelled; white to cream or pinkish, speckled with pale brown. SIZE: (n = 43, South Africa) 37·1–43·2 × 28·8–32·4 (39·3 × 30·4); (n = 21, Zimbabwe) 37·4–42·7 × 30–31·9 (39·3 × 30·9); Mozambique av. 37 × 29; Somalia 41 × 30·5.

LAYING DATES: South Africa, Oct–Mar, May; Namibia, Sept; Zimbabwe, Oct–Mar, May, no obvious peak; Botswana, Apr; Mozambique, June–July; Zambia, Dec–Feb; Malaŵi, Nov; East Africa: Uganda, Region A, Jan, Mar, Region B, June–July; Tanzania, Region C, Dec–Apr, Sept; Kenya, Regions D and E, May–July, Dec; Ethiopia, Mar–May, possibly Jan–June, Oct; Somalia, mainly May–June, but also in Oct, Nov and Feb.

INCUBATION: period $c.$ 19 days.

References
Clancey, P. A. (1967).
Swank, W. G. (1977).

Scaly Group

A superspecies comprising 3 medium-sized ($c.$ 500 g) francolins (*squamatus*, *ahantensis* and *griseostriatus*). Generally greyish brown above, feathers irregularly barred, streaked and vermiculated lighter; underparts greyish brown or creamy buff, feathers streaked darker and/or lighter, giving a scaly appearance. ♂♂ tend to be larger than ♀♀, but no marked plumage dimorphism. Ceres cartilaginous, bills and legs red or orange-red, usually with two spurs. Calls nasal and raucous; roost in trees. Largely confined to tropical forests within 15° of equator.

Plate 3
(Opp. p. 48)

***Francolinus squamatus* Cassin. Scaly Francolin. Francolin écaillé.**

Francolinus squamatus Cassin, 1857. Proc. Acad. Sci. Philad., 8, p. 321; Cape Lopez, Gabon.

Forms a superspecies with *F. ahantensis* and *F. griseostriatus*.

Range and Status. Endemic resident, almost entirely confined to belt of equatorial Africa between 10°N and 10°S. Distribution rather patchy; found from S-central Nigeria south through Gabon to extreme W Zaïre, and from Cameroon east in belt along northern edge of main forest block to Uganda, SW Ethiopia, central Kenya and highlands of N and E Tanzania and N Malaŵi. Also in highlands of E Zaïre, Rwanda, Burundi and W Tanzania, and in disjunct populations in Cabinda, central and S Zaïre, and Jebel Marra (W-central Sudan). Locally common to abundant, e.g. in vicinity of Owerri (Nigeria) outnumbers Double-spurred Francolin *F. bicalcaratus* by $c.$ 20 to 1. Forest destruction and trapping may lead to local extinction.

Description. ADULT ♂: forehead and crown uniform grey-brown, supercilium light grey; sides of head pale grey-brown, feathers tipped darker. Mantle to mid-back dark brownish grey, feathers irregularly margined with buff and densely vermiculated with black; remaining upperparts greyish brown densely vermiculated and irregularly barred with buff. Throat buff; breast grey-brown, underparts greyish buff, densely vermiculated darker, especially on feather margins, producing scaly effect. Undertail-coverts dàrk grey, densely vermiculated with black and edged with buff. Flight-feathers uniform grey, inner secondaries spotted with black. Upper mandible dark brown, lower mandible orange; eye dark brown; naked skin above ear greyish yellow; legs orange. ADULT ♀: like ♂ but averages smaller. SIZE: (47 ♂♂, 60 ♀♀) wing, ♂ 159–184 (175), ♀ 147–182 (163); tail, ♂ 81–115 (97·4), ♀ 73–110 (89·2); tarsus, ♂ 48–60 (53·9), ♀ 41–59 (49·1); spurs, 2 in ♂ (lower longer), 7–15 (11·5), 1 in ♀, 0·4–1·5 (0·8). WEIGHT: (E Africa) ♂ (n = 14) 432–565 (510), ♀ (n = 17) 377–515 (440); (SE Zaïre) ♀ (n = 7) 377–492 (437), ♂ (n = 1) 491; (Uganda) ♂ (n = 3) 519, 552, 553, ♀ (n = 1) 398.

IMMATURE: like ♀ but more rufous overall and with arrow-shaped black markings on upperparts. Underparts barred black and white.

DOWNY YOUNG: crown, hindneck and broad mid-dorsal stripe dark rufous-brown. Face and sides of head buffy white, with single dark brown eye-stripe. Rest of upperparts dark brown, washed with rufous and mottled with buff. 2 buff stripes parallel dark rufous-brown stripe on mid-back. Underparts buffy washed with rufous. Upper mandible brown, lower mandible flesh; eye grey; legs pale yellow or pale vermilion.

Exhibits considerable individual variation and smooth clinal geographical variation, with western birds being much more mottled and blotched with buff and rufous dorsally, and paler overall ventrally than eastern ones. Extremes in variation illustrated in Plate 3 with western 'squamatus' on the right and eastern 'usambarae' on the left. We do not recognize these clinal extremes as valid subspecies.

Field Characters. A widespread species, overlapping range and habitat with many others. A rather dull, dark bird of dense undergrowth. Rarely found in open habitat; characterized by scaly underparts and orange-red legs, also 'two-tone' bill. For differences from Cameroon Mountain Francolin *F. camerunensis*, Handsome *F. nobilis* and Jackson's Francolins *F. jacksoni*, see those species.

Voice. Tape-recorded (32, 217, 226, 264, 308). Advertisement call a high-pitched, nasal 'ke-rak' repeated 4–12 times, increasing in volume. Gathering call 'quarek quarek'; brood call 'chu-ri chu-ri'; alarm call 'kerak kak kak'; contact call low clucking and chuckling (audible for only a few m). Chick alarm call high-pitched cheeps, almost a trill.

General Habits. Inhabits evergreen forest with dense undergrowth *c.* 3 m high from 800 to 3000 m. Also frequents forest clearings and dense understorey of plantations, especially abandoned ones, and other secondary regrowth; and will persist in quite small remnants of forest/bush when remainder has been cleared. Usually found in pairs. Heavy fliers, often seek refuge in trees. Secretive, skulking birds, prefer to squat then run if threatened. Start calling before dawn, resume again at dusk until as late as 20.30 h (later on moonlit nights); often from termite mound or other prominence. Roosts at night in trees among secondary growth; sits well within leaf cover and very difficult to observe. Once settled in, begins a *c.* 20-min dusk chorus of advertisement calls. Apparently very sedentary, since found in same area over long periods.

Food. Fruits, seeds, snails, millepedes; termites, ants, other small insects. Also partial to cultivated plants, e.g. cassava, sweet potato, peanuts and rice. Vegetable material predominated in 8 crops collected in Zaïre.

Breeding Habits. Probably monogamous, since usually found in pairs. ♂♂ not as aggressive as other francolins, e.g. Hildebrandt's *F. hildebrandti*.

NEST: shallow scrape in ground under tuft of grass or bush, often lined with grass and/or feathers. 1 found in S Cameroon on top of a termite mound.

EGGS: 3–8, usually 6. Ovals, with very hard shells; buff to pinkish buff, pitted white. SIZE: (n = 5, Uganda) 41·4–42·7 × 32·7–34 (41·9 × 33·6).

LAYING DATES: Malaŵi, Aug; E Africa, every month except Feb, no obvious seasonality; Ethiopia, possibly Oct–Dec; Zaïre, Jan, Feb, May, June; S Cameroon, Oct–Dec (dry season), W Cameroon, Oct–Mar; Gabon, June–Aug.

INCUBATION: probably by ♀ alone; sits very tight, making no sound if flushed off nest.

BREEDING SUCCESS/SURVIVAL: low for chicks, which are taken by mongooses and genets. Adults trapped by humans.

References
Bates, G. L. (1930).
Bannerman, D. A. (1930).
van Someren, V. G. L. (1916).

Francolinus ahantensis Temminck. Ahanta Francolin. Francolin d'Ahanta. Plate 3

Francolinus ahantensis Temminck, 1854. Bijdv. Dierk. I, p. 49; Ahanta, Gold Coast. (Opp. p. 48)

Forms a superspecies with *F. squamatus* and *F. griseostriatus*.

Range and Status. Endemic resident in 3 apparently disjunct populations in W Africa: (i) coastal areas of S Senegambia, N Guinea–Bissau (rare and local); (ii) S Guinea, Sierre Leone, W Liberia; (iii) NE Ivory Coast (common) and Ghana through central Togo and central Benin to SW Nigeria, where common.

Description. ADULT ♂: forehead, lores and supercilium whitish, each feather with a small dark brown terminal blotch. Crown, nape and ear-coverts dark greyish brown; lower face from under eye to sides of neck whitish blotched with dark grey-brown. Sides of neck, hindneck and mantle black, feathers with rufous-buff shaft streaks and white marginal streaks (giving scaly appearance). Back, scapulars and wing-

46 PHASIANIDAE

coverts browner, feathers vermiculated and irregularly barred with buff. Rump and tail rich brown with dark vermiculations. Chin and throat white; underparts grey-brown heavily streaked with white. Primaries brown-grey with outer margins slightly mottled with brown. Secondaries greyish brown, vermiculated and mottled with black and rufous. Bill orange-red, base black; skin above ear-coverts pale orange; eye brown; legs orange. ADULT ♀: like ♂ but secondaries darker. SIZE: (13 ♂♂, 14 ♀♀) wing, ♂ 161–181 (172), ♀ 154–172 (164); tail, ♂ 90–109 (98), ♀ 80–109 (92·5); tarsus, ♂ 53–59 (56·4), ♀ 46–53 (50·2); spurs, 2 in ♂ (lower longer), 6–17 (12·5), 1 in ♀ 0·3–1·2 (0·6).

IMMATURE: like adult but with arrow-shaped black markings on mantle, scapulars and innermost secondaries; underparts ashy streaked with white.

DOWNY YOUNG: crown, nape and broad stripe down mid-back dark rufous-brown. Sides of head buffy-rufous with rufous-brown eye-stripe extending back from eye. Remainder of upperparts rufous-brown, mottled lighter, except for 2 narrow buffy stripes which run parallel to central rufous-brown stripe. Underparts buffy with rufous wash. Primaries dark brown, inner webs reddish, outer webs brown. Tail reddish brown, vermiculated darker. Bill brownish horn, lower mandible paler; eye brown; legs pinkish flesh.

Western populations, recognized by White (1965) as *hopkinsoni*, are paler and have reddish brown wing-coverts, but geographical variation largely clinal and does not warrant recognition of subspecies.

Field Characters. A large, dark francolin streaked white above and below, with orange bill and legs. Double-spurred Francolin *F. bicalcaratus* is paler and richer brown above, has buffy underparts with dark markings, prominent pale eye-stripe, and olive-green bill and legs. Tiny Latham's Forest Francolin *F. lathami* has white cheeks, black throat, and black and white barred underparts. Call also distinctive.

Voice. Tape-recorded (217, 253). Loud, high-pitched 'kee-kee-keree' repeated several times; even more squealing than that of Scaly Francolin *F. squamatus*. Also a less raucous 'kok-kee-keroo'. Sometimes 2nd bird (? ♀) gives different call, 'ker-weerk', simultaneously. Antiphonal calling with ♂ giving a clear rippling note, and ♀ replying with a vibrant undertone.

General Habits. Inhabits forest edge and clearings, secondary growth, dense cover by water and overgrown cultivation; in Nigeria, tangled scrub between gallery forest and farmed savanna. Shy, retiring and difficult to flush; on flushing often takes to perching in trees. Usually roosts in trees at night, but will remain on ground if disturbed at dusk. Associates in pairs or small coveys. Noisy at dawn, but may call at any time of day, often on moonlit nights.

Food. Seeds, small beans, cassava and large fruits; insects, including termites.

Breeding Habits. Very poorly known.
NEST: scrape lined with leaves in thick cover.
EGGS: 4–6, up to 12. Cream to pinkish buff. SIZE: av. 42 × 33.
LAYING DATES: Senegambia, Jan, Sept; Ghana, late Dec.

References
Bannerman, D. A. (1930).
Collier, F. S. (1935).
Colston, P. and Curry-Lindahl, K. (in press).
Holman, F. C. (1947).

Plate 3
(Opp. p. 48)

Francolinus griseostriatus Ogilvie-Grant. Grey-striped Francolin. Francolin à bandes grises.

Francolinus griseostriatus Ogilvie-Grant, 1890. Ibis, p. 340, pl. 10; Quanza River, N Angola.

Forms a superspecies with *F. squamatus* and *F. ahantensis*.

Range and Status. Endemic resident in escarpment zone of W Angola. 2 isolated populations, 1 in S Cuanza, W Milanje (Ndalo Tando, Pungo Andongo, Dondo, Cuanza Basin and Cuanza Gorge), 1 on escarpment in S Benguela District and extreme NE Huila District (Chingoroi and Caxito). Status unknown, most records based on specimens collected before 1910; not recorded since 1954, may be threatened by forest destruction.

Description. ADULT ♂: forehead rufous-buff, crown and nape grey-brown; sides of head pale greyish brown. Sides of neck, hindneck, mantle, back and scapulars chestnut with pale buff streaks and a few black markings. Rump and uppertail-coverts greyish brown with a few pale vermiculations and indistinct buff and black barring. Tail dark rufous-brown with indistinct black bars. Upperwing-coverts brown edged paler; primaries grey-brown with some tawny mottling on outer webs; secondaries mottled and barred brown and tawny. Chin and upper

Francolinus griseostriatus

IMMATURE: ground colour of upperparts a much richer cinnamon, each feather with blackish, not chestnut centre and with black triangular-shaped markings above and below. Breast less chestnut, belly whiter.

DOWNY YOUNG: undescribed.

Field Characters. Combination of rufous chestnut-streaked plumage, pale throat, unmarked face, red-orange bill and legs and forest habitat should distinguish it from other francolins within its range.

Voice. Not tape-recorded. Described as like that of the Scaly Francolin *F. squamatus*, i.e. a high-pitched, rasping 'kerak' (J. P. Chapin, unpub.).

General Habits. Inhabits dense undergrowth of gallery and secondary forest or dense thickets between 800 and 1200 m. Ventures into adjacent grass flats and former cotton fields in morning and late afternoon to feed. When disturbed, flies back to forest edge. Roosts in trees at night.

Food. Insects and other small arthropods, green shoots, seeds.

throat off-white, lower throat and breast streaked chestnut and greyish buff; rest of underparts rich buff, feathers with narrow rufous-brown shaft streaks. Upper mandible black with red base, lower orange-red; eye brown; legs orange-red. Sexes alike. SIZE: (5 ♂♂, 8 ♀♀) wing, ♂ 139–161 (153), ♀ 144–153 (148); tail, ♂ 87–102 (95), ♀ 85–96 (91); tarsus, ♂ 43–52 (48), ♀ 40–45 (42·4); 1 spur; ♂ 6–15 (11), ♀ 0·4–1·1 (0·6).

Breeding Habits. Unknown.

References
Collar, N. C. and Stuart, S. N. (1985).
Hall, B. P. (1963).

Francolinus nahani Dubois. Nahan's Francolin. Francolin de Nahan. Plate 3

Francolinus nahani Dubois, 1905. Ann. Mus. Congo, Zool. (4), 1, p. 17, pl. 10; Popoie, Aruwimi River, Zaïre. (Opp. p. 48)

Range and Status. Endemic resident in NE Zaïre (between Aruwimi, Nepoko and Semliki Rivers) and W and S-central Uganda (Budongo, Bugoma and Mabira Forests). Locally distributed; much rarer than Latham's Forest Francolin *F. lathami* in Zaïre. Few records for Uganda, but possibly not uncommon locally since 8 specimens collected in Bugoma over short period (Jackson and Sclater 1938). No information on status, but hunting may be a threat, since relatively easily shot along forest paths baited with grain. The population in the Semliki Valley is within the Virunga National Park, and would be best protected if the Virunga National Park were extended to include the E Ituri Forest.

Description. ADULT ♂: forehead to nape and ear-coverts blackish brown; posterior half of supercilium freckled black and white. Sides of head and upper back black, each feather edged white, giving a distinct mottled appearance. Back and scapulars black, extensively washed, vermiculated and irregularly barred rufous-buff. Rump and uppertail-coverts brown with irregular dusky barring. Tail dark grey-brown, vermiculated darker. Chin white; throat black, feathers edged white. Upper breast feathers black with broad white margins, giving distinct scaly appearance. Lower breast, belly and

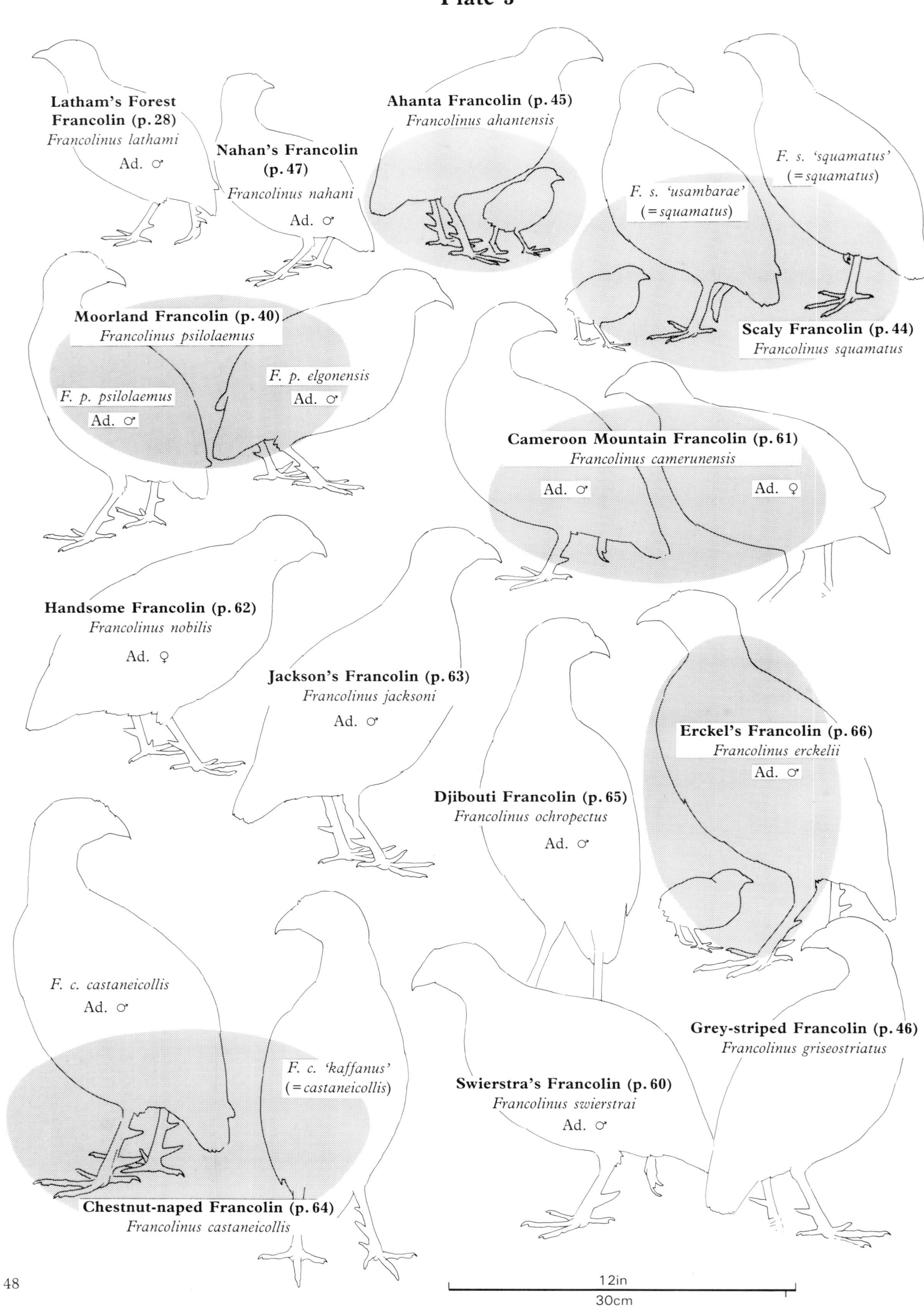

Plate 3

Latham's Forest Francolin (p. 28)
Francolinus lathami
Ad. ♂

Nahan's Francolin (p. 47)
Francolinus nahani
Ad. ♂

Ahanta Francolin (p. 45)
Francolinus ahantensis

Scaly Francolin (p. 44)
Francolinus squamatus
F. s. 'usambarae' (= *squamatus*)
F. s. 'squamatus' (= *squamatus*)

Moorland Francolin (p. 40)
Francolinus psilolaemus
F. p. psilolaemus Ad. ♂
F. p. elgonensis Ad. ♂

Cameroon Mountain Francolin (p. 61)
Francolinus camerunensis
Ad. ♂   Ad. ♀

Handsome Francolin (p. 62)
Francolinus nobilis
Ad. ♀

Jackson's Francolin (p. 63)
Francolinus jacksoni
Ad. ♂

Djibouti Francolin (p. 65)
Francolinus ochropectus
Ad. ♂

Erckel's Francolin (p. 66)
Francolinus erckelii
Ad. ♂

Chestnut-naped Francolin (p. 64)
Francolinus castaneicollis
F. c. castaneicollis Ad. ♂
F. c. 'kaffanus' (= *castaneicollis*)

Swierstra's Francolin (p. 60)
Francolinus swierstrai
Ad. ♂

Grey-striped Francolin (p. 46)
Francolinus griseostriatus

12in
30cm

48

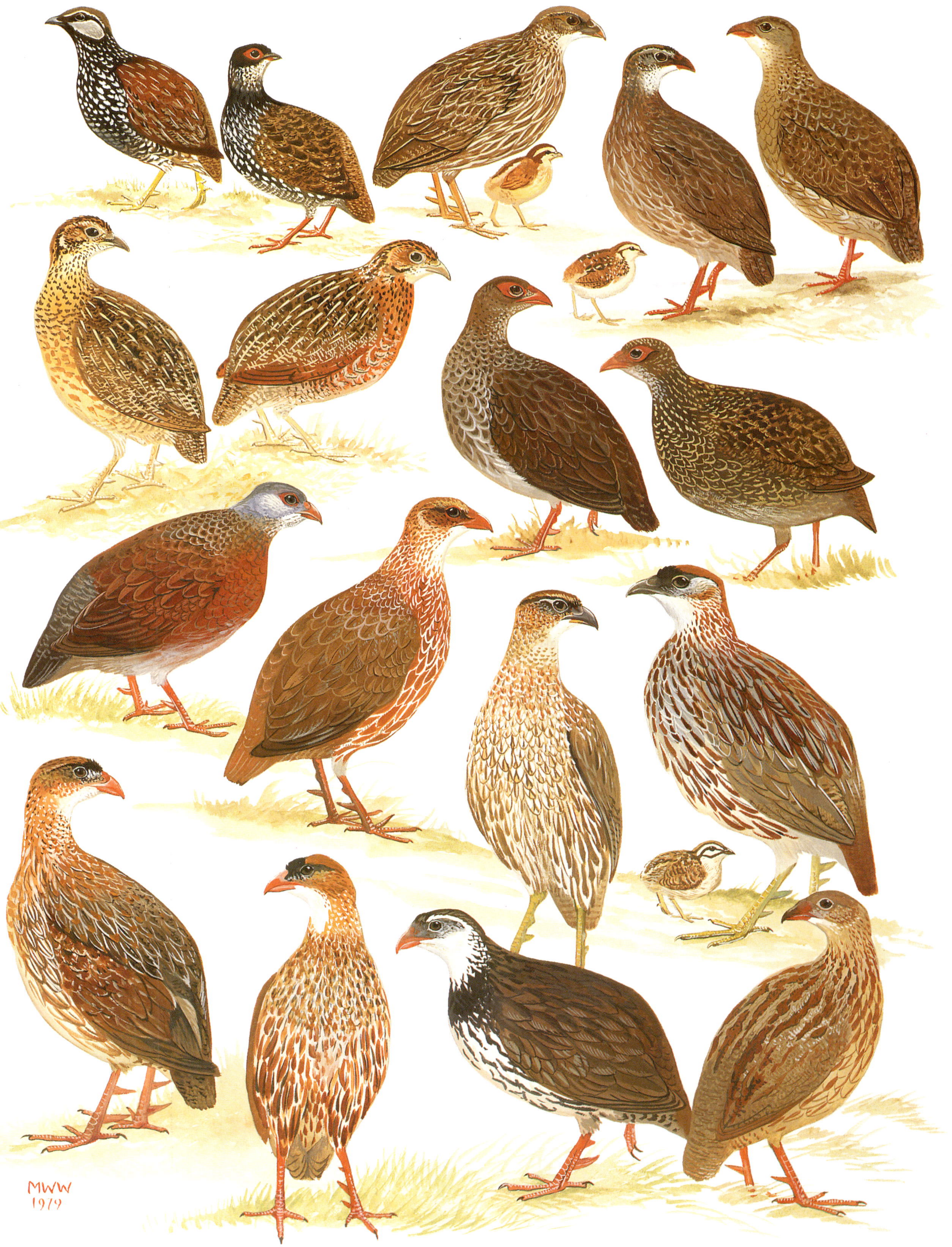
MWW
1979

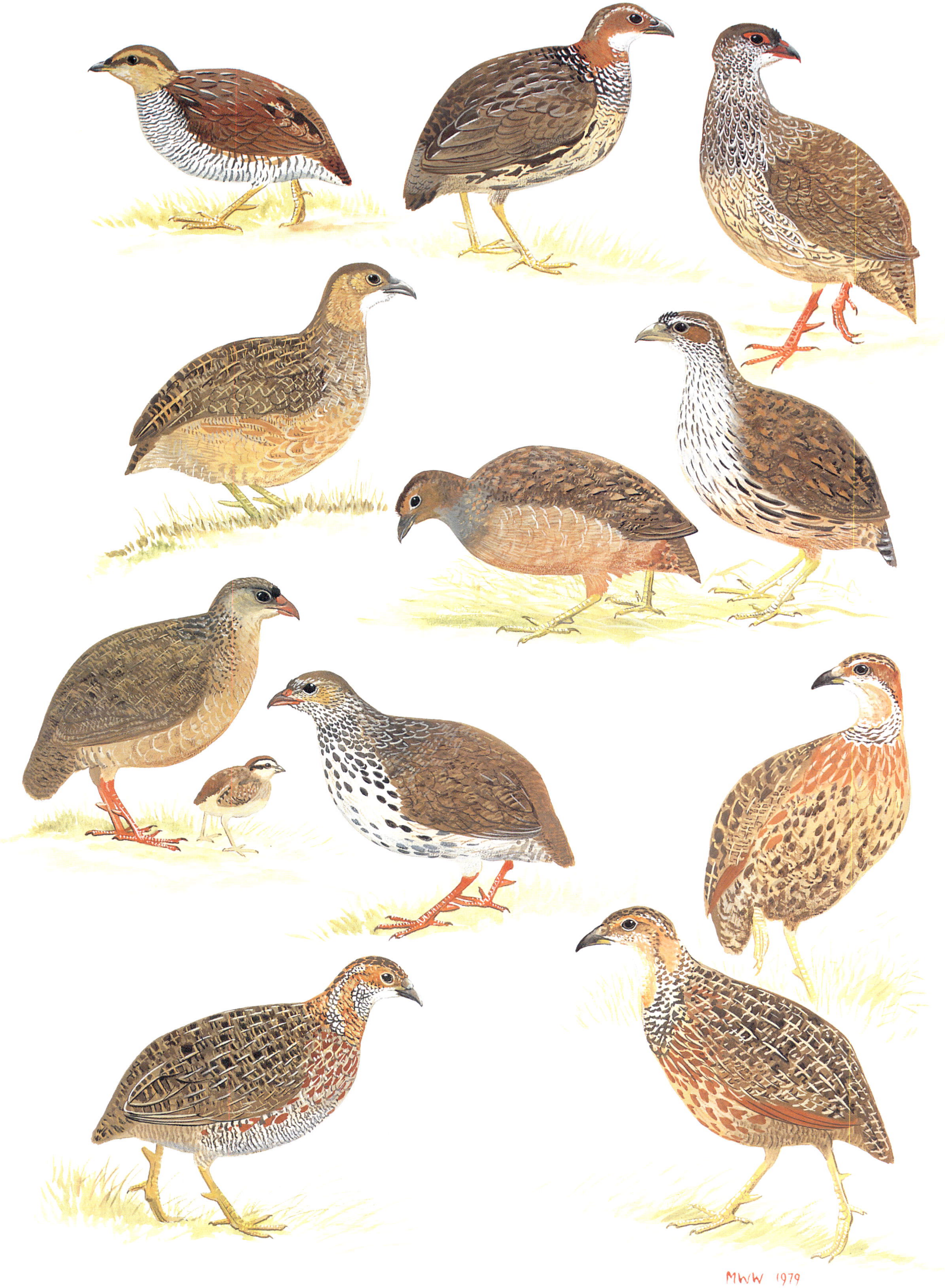
MWW 1979

Plate 4

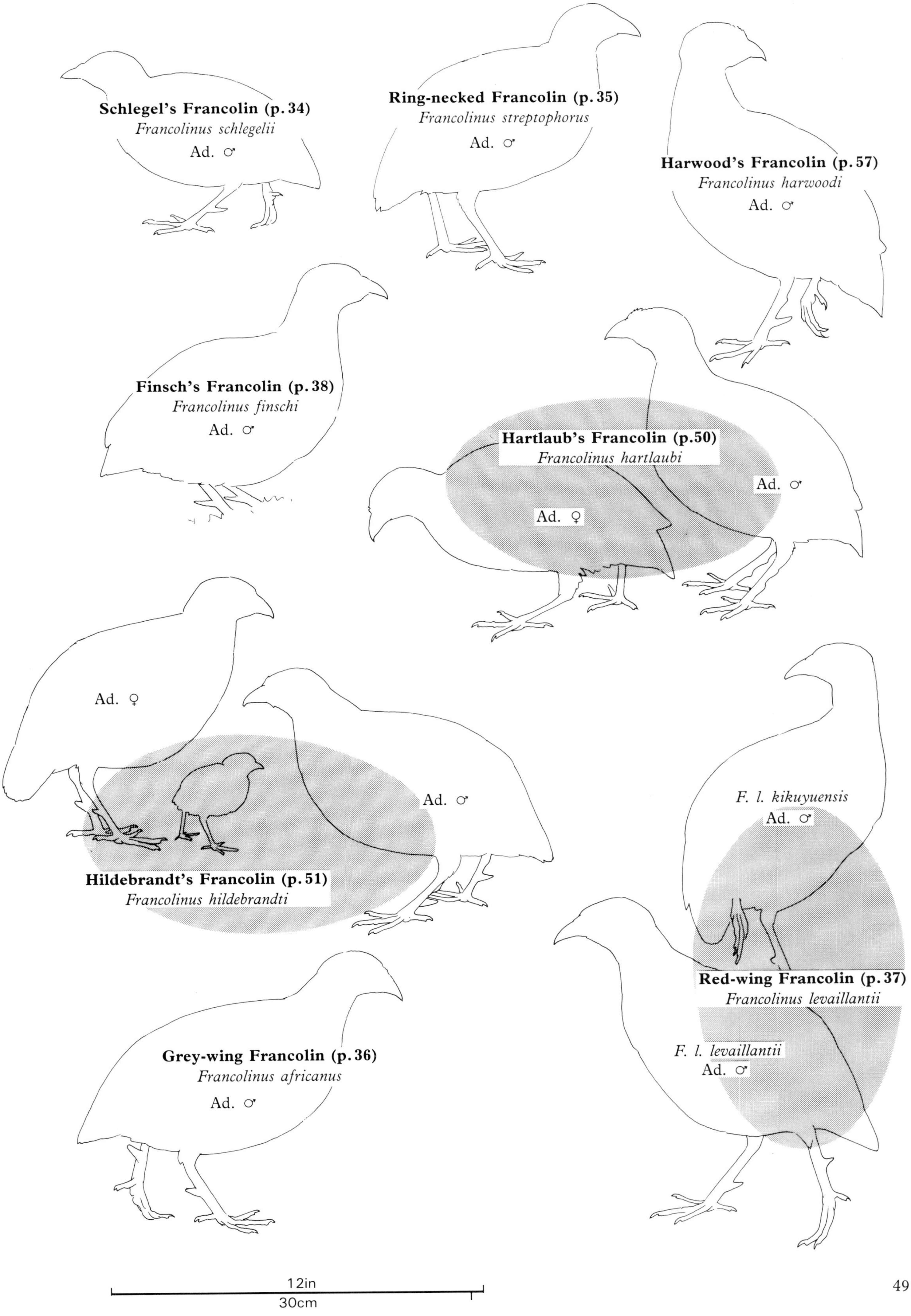

Schlegel's Francolin (p. 34)
Francolinus schlegelii
Ad. ♂

Ring-necked Francolin (p. 35)
Francolinus streptophorus
Ad. ♂

Harwood's Francolin (p. 57)
Francolinus harwoodi
Ad. ♂

Finsch's Francolin (p. 38)
Francolinus finschi
Ad. ♂

Hartlaub's Francolin (p. 50)
Francolinus hartlaubi
Ad. ♀
Ad. ♂

Ad. ♀

Hildebrandt's Francolin (p. 51)
Francolinus hildebrandti
Ad. ♂

F. l. kikuyuensis
Ad. ♂

Red-wing Francolin (p. 37)
Francolinus levaillantii

F. l. levaillantii
Ad. ♂

Grey-wing Francolin (p. 36)
Francolinus africanus
Ad. ♂

12in / 30cm

flanks black, mottled and streaked white. Undertail-coverts black. Primaries uniform greyish brown, paler on outer webs; secondaries black, irregularly barred and vermiculated rufous-buff. Wing-coverts rufous-brown, each feather with a central buff spot. Base of bill red, rest brownish black; eye dark brown, naked skin around eye red; legs red, without spurs, claws blackish grey. Sexes alike. SIZE: (9 ♂♂, 13 ♀♀) ♂ 129–141 (135), ♀ 128–142 (135); tail, ♂ 70–85 (79·5), ♀ 71–90 (79·0); tarsus, ♂ 35–41 (37·1), ♀ 34–41 (36·7). WEIGHT: ♂ (n = 2) 308, 312; ♀ (n = 3) 234–260.

IMMATURE: darker, legs grey.
DOWNY YOUNG: undescribed.

Field Characters. A small dark francolin with belly mottled and streaked (not barred) black and white, bare red skin around eye and red legs. Can be mistaken only for Latham's Forest Francolin which has underparts irregularly barred black and white, lacks red skin on face, has chestnut mantle and wings in ♂, grey or rufous neck-patch, black throat and yellow legs.

Voice. Unknown.

General Habits. Found only in dense primary forest up to 1400 m. Very shy, usually seen in pairs, often with Crested Guineafowl *Guttera* spp. Feeds by scratching on ground in leaf litter.

Food. Insects, small molluscs, green shoots, seeds and bulbs.

Breeding Habits. Probably monogamous since usually found in pairs.

NEST: only 1 found, in hollow of a tree trunk *c.* 1 m above ground.

EGGS: only 1 clutch of 4 eggs found. Smooth, glossy, pyriform, not thick-shelled; buff to pinkish with pale brown and purple speckling. SIZE: *c.* 36 × 26.

LAYING DATES: probably throughout year. 2 ♀♀ collected Uganda Apr had ovaries much enlarged.

Vermiculated Group

Hall's (1963) largest but least homogeneous group. 9 spp. (*hartlaubi, hildebrandti, natalensis, bicalcaratus, icterorhynchus, clappertoni, harwoodi, capensis* and *adspersus*) comprising 2 superspecies (*hildebrandti/natalensis* and *bicalcaratus/icterorhynchus/clappertoni/harwoodi*) and 3 more distantly related spp. Upperparts of all are greyish brown or grey, usually heavily vermiculated and/or streaked with buff, and underparts of many with buff U and/or V streaking. Although ♂♂ are significantly larger than ♀♀, only 2 members, Hartlaub's Francolin *F. hartlaubi* and Hildebrandt's Francolin *F. hildebrandti*, exhibit strong sexual dimorphism in plumage. Most have reddish or orange bills, reddish or orange legs with 2 spurs, cartilaginous ceres, grey or dark brown primaries with buff or white vermiculations and/or streaking, and raucous, grating calls. The Natal Francolin *F. natalensis* hybridizes with Swainson's Francolin *F. swainsonii* (one of the Bare-throated Group) in W Zimbabwe. All inhabit grass lands, scrub, cultivation in woodland, and/or acacia steppe, and roost in trees.

Plate 4
(Opp. p. 49)

Francolinus hartlaubi Bocage. Hartlaub's Francolin. Francolin de Hartlaub.

Francolinus hartlaubi Bocage, 1869. Journ. Acad. Real Sci., Lisboa, 2, p. 350; Huila, S Angola.

Range and Status. Endemic resident in hilly country from SW Angola in Benguela (Catengue) and central and SW Huila (including Iona National Park) to northern half of Namibia (Kaokoveld and Waterberg south to Rehoboth area). Generally local and uncommon but locally fairly common.

Description. ADULT ♂: forehead freckled very dark brown and white; crown and nape greyish brown, feathers margined lighter. Upper superciliary white with some brown freckling; lower black. Sides of head, chin and throat white, each feather with a dark brown shaft streak; ear-coverts rufous. Bulk of remaining upperparts very dark greyish brown, feathers heavily streaked, vermiculated and irregularly barred with rufous, grading to buff on uppertail-coverts. Tail blackish brown irregularly barred white. Chin, throat, breast and upper belly white, each feather with a dark brown shaft streak. Remaining underparts buffy, heavily washed with rufous. Undertail-coverts off-white, broadly barred with black and rufous-buff. Primaries grey-brown with lighter vermiculations on outer webs. Secondaries grey-brown, outer webs vermiculated with buff. Upper mandible horn, lower yellowish; eye brown; legs yellow. ADULT ♀: averages smaller;

upperparts like ♂ except rufous blotching and streaking extend to forehead and sides of head. Throat rufous-buff. Upper breast grey, mottled with rufous; rest of underparts rufous-buff, undertail-coverts darker and blotched black terminally. SIZE: (27 ♂♂, 19 ♀♀) wing, ♂ 131–172 (143), ♀ 131–148 (135); tail, ♂ 82–94 (88·9), ♀ 77–94 (85·3); tarsus, ♂ 34–57 (37·8), ♀ 32–40 (34·8); spurs, 1–2 in ♂ (usually 2, lower longer) 2–4 (3·2), 1 in ♀, 0·5–2 (0·8).

IMMATURE: forehead buffish; crown brown tinged rufous; nape dark greyish brown vermiculated with light buff; hindneck and sides vermiculated black and white, each feather blackish with some white distally and crossed by from 1 to 4 irregular white bars; mantle, scapulars and wing-coverts buffish grey vermiculated with black, feathers medially scored by a white shaft streak which flares apically; rump and uppertail-coverts buffy grey vermiculated with blackish. Face with eye region and supercilium pale buffish, becoming paler still and tinged greyish behind eye. Breast barred black and white grading to buffish grey on abdomen, each feather with whitish shaft streaks and spots and fine darker vermiculations. Underwing-coverts and axillaries buffish grey vermiculated darker. Primaries greyish brown, outer web peppered and inner with ill-defined buff barring; secondaries buffish grey with black vermiculations and broad black cross-bars. Tail greyish buff, coarsely vermiculated and irregularly barred darker (Clancey 1967).

DOWNY YOUNG: undescribed.

Geographical variation clinal; birds from centre of range tend to be darker than those to north and south.

Field Characters. A medium-sized, sexually dimorphic francolin. ♂ distinguished by black streaking below, ♀ uniform rufous-buff below. Bill relatively longer than Red-billed Francolin *F. adspersus*, which may occur nearby, and which has underparts finely barred black and white. Habitat, i.e. kopjies, also diagnostic.

Voice. Tape-recorded (255a). Territorial advertisement call is a duet between dominant ♂ and ♀; mixture of loud, grating cackles and high-pitched squeaks 'kor-rack, keerya, keerya, kew', 1st note at least given by ♀, 2nd and 3rd notes sometimes repeated several times. Heard throughout the year in early morning (more frequent and longer renditions during breeding season). ♂ alarm call high-pitched, squeaky 'kerr-ak', usually given when separated from ♀ during territorial dispute with other ♂. ♀ alarm call high-pitched 'keer'. Flight call a rapid (3–5/s) chattering 'krak' given only by ♂. Contact call low intensity 'cheeerrrr' given only by ♂ when close to or preening ♀.

General Habits. Inhabits dry ridges, hills and kopjies with scrub and bush. Usually 1 group of 3–4 per kopjie, probably a dominant pair and previous season's offspring. Seeks shade under ledges during heat of the day. May feed for up to 2 h in detritus beneath cliffs where there is shade. Roosts among rocks or on precipices. Particularly evident in early mornings when calls from the highest point of the kopjie. Occurs close by to Red-billed and Orange River Francolin *F. levaillantoides*, but strict habitat segregation: Red-billed occurs in bush along watercourses, Hartlaub's in kopjies, and Orange River in intervening grassland.

Food. Berries, bulbs, seeds, small snails, termites and other insects.

Breeding Habits. Monogamous and territorial, 1 pair per kopjie.

NEST: only 1 described; in hollow on cliff ledge.

EGGS: probably 4–8 or more. Uniform creamy and oval. SIZE: (n = 2, Namibia) 43·5 × 29·8, 42 × 29.

LAYING DATES: Namibia, May, June, Nov.

DEVELOPMENT AND CARE OF YOUNG: young and parents remain in family parties well after breeding season.

References
Clancey, P. A. (1967).
Kinahan, J. (1975).
Macdonald, J. D. (1957).

Francolinus hildebrandti Cabanis. Hildebrandt's Francolin. Francolin de Hildebrandt.

Plate 4
(Opp. p. 49)

Francolinus (Scleroptera) hildebrandti Cabanis, 1878. Journ. f. Orn. 26, p. 206, 243, pl. 4, f. 2; Voi, Teita district, Kenya Colony.

Forms a superspecies with *F. natalensis*.

Range and Status. Endemic resident from Kenya (Marsabit, Barsaloi, Malawa River, Voi) through most of Tanzania except NW (north to Biharamulo), extreme SE Zaïre (Musosa), NE Zambia south to c. 12°S, and S Malaŵi (Mt Mulanje). Introduced on Bwejuu Island near Mafia. Generally very locally distributed and uncommon.

Description. ADULT ♂: forehead and supercilium very dark brown, feathers freckled with white; crown uniform dark greyish brown, distal margins of each feather much lighter. Ear-coverts rufous-grey; sides of head, neck and throat white, each feather broadly streaked with dark greyish brown. Mantle black with broad creamy U-shaped streaking. Remaining upperparts dark greyish brown, heavily vermiculated buff and mottled rufous. Tail as upperparts but without rufous mottling. Chin and throat white, mottled black. Remaining underparts white, heavily blotched black, especially on breast; undertail-coverts barred broadly and irregularly with buff and grey. Primaries and secondaries uniform dark grey. Upper mandible black with red base, lower mandible red; eye brown; legs red. ADULT ♀: upperparts like ♂ but with faint buff barring. Sides of head and throat pale buff. Rest of underparts rufous-buff, feathers margined lighter, undertail-coverts barred brown and buff. Averages smaller. SIZE: (30 ♂♂, 29 ♀♀) wing, ♂ 158–189 (174), ♀ 151–179 (162); tail, ♂ 92–118 (105), ♀ 86–110 (97·2); tarsus, ♂ 44–62 (50·3), ♀ 41–56 (45·8);

52 PHASIANIDAE

2 spurs; ♂ (lower longer) 4–14 (10·4), ♀ 2–13 (8·4). WEIGHT: (East Africa) ♂ (n = 2) 600, 645; ♀ (n = 2) 430, 480.

IMMATURE: like ♀ but upperparts more distinctly barred black and rufous-buff; buff on underparts more extensive, with dark blotching reduced to spots and streaks.

DOWNY YOUNG: very distinctive. Crown rufous-brown with darker crown-stripe. Sides of head buffy with dark brown stripe which extends from front of eye back to nape (eye-stripe much reduced in most other francolin chicks). Remaining upperparts dominated by broad rufous-brown stripe down mid-back, flanked by buffy stripes. Rest of upperparts mottled with brown and buff. Underparts buffy. Legs dull brown.

Geographical variation clinal and not well described by subspecies. ♀♀ from southern end of range ('*johnstoni*') lack white streaking on mantle and are smaller.

Field Characters. ♀ with rufous-buff underparts very distinctive. ♂ with red legs, red lower mandible, black-spotted underparts and rather plain brown lower back, wings and tail. Unlike any francolin in its range or habitat, but on Kenya/Uganda border comes close to range of similar-looking Clapperton's Francolin *F. clappertoni*, which has pale eye-stripe, red eye-patch and streaked upperparts.

Voice. Tape-recorded (C, 217, 264, 296). High-pitched cackle 'kek-kekek-kek-kerak' given mainly at dawn and dusk, several birds usually calling simultaneously. Also single note 'kek' repeated 1 to many times at variable rate, faster when alarmed, and a low grating 'chuk-a-chuk' contact call. More like voice of Natal Francolin *F. natalensis* than other members of vermiculated group.

General Habits. Occurs from 2000 to 2500 m in dense scrub, thickets and bushed grassland on rocky hillsides; at higher altitudes, bracken–briar, and reaches lower heath zone in Arusha National Park. The classic place to see it in Kenya is Hell's Gate Gorge, near Naivasha, where it lives on the rocky, bushy slopes at the foot of the cliff where Lammergeiers *Gypaetus barbatus* often nest. Usually found singly, in pairs or in small coveys. Extremely wary bird; sits very tight even for a francolin. Much more often heard than seen. Perches and roosts in trees.

Francolinus hildebrandti

Food. Seeds, bulbs, tubers, insects and their larvae.

Breeding Habits. Probably monogamous, since normally found in pairs throughout the year. ♂♂ very pugnacious during early breeding season.

NEST: well concealed small depression lined with grass and leaves.

EGGS: 4–8; creamy white to pale brown. SIZE: (n = 4, Malawi) 37·8–41·5 × 30·5–33·2 (39·8 × 31·9).

LAYING DATES: Zambia, July (birds in breeding condition, Oct); Malawi, Apr–Nov, peak in June–July; East Africa, Kenya and N Tanzania, Regions C and D, Jan, Mar, May–Aug, Nov–Dec, no clear correlation with rainfall.

Reference
Jackson, F. J. and Sclater, W. L. (1938).

Plate 2
(Opp. p. 33)

Francolinus natalensis Smith. Natal Francolin. Francolin du Natal.

Francolinus natalensis A. Smith, 1834. S. Afr. Quart. Journ. 2nd Ser., p. 48; Durban, Natal.

Forms a superspecies with *F. hildebrandti*.

Range and Status. Endemic resident from *c.* 12°S in central Zambia (immediately south of range of Hildebrandt's Francolin *F. hildebrandti*) through extreme W Mozambique, Zimbabwe, extreme E Botswana, and Swaziland, to South Africa in Transvaal, W Orange Free State, Natal, and extreme N and E Cape Province. Locally common to abundant, but in South Africa distribution and numbers reduced due to land development.

Description. ADULT ♂: forehead, supercilium and sides of head very dark brown, feathers edged buff; crown dark grey-brown, feathers edged lighter; eye-stripe black. Ear-coverts greyish brown. Hindneck and mantle black, each feather broadly and irregularly barred and vermiculated with white. Remaining upperparts dark greyish brown (rump paler), feathers heavily vermiculated with buff and with black shaft streaks. Tail light brown, vermiculated darker. Chin and throat white, heavily spotted with black. Remaining underparts white with irregular broad black U-shaped streaking,

Francolinus natalensis

giving scaled appearance. Primaries and secondaries greyish brown, vermiculated buff on outer webs. Bill orange, base dull greenish; eye dark brown; legs orange to orange-scarlet. ADULT ♀: like ♂ but averages smaller. SIZE: (38 ♂♂, 28 ♀♀) wing, ♂ 150–186 (168), ♀ 149–167 (156); tail, ♂ 92–120 (101), ♀ 88–102 (95·2); tarsus, ♂ 42–57 (49·3), ♀ 40–52 (45·2); spurs, 1–2 in ♂ (usually 1), 4–21 (13·2), 1 in ♀ 0·4–2 (0·8). WEIGHT: ♂ (n = 4) 415–650, ♀ (n = 3) 370–400.

IMMATURE: like ♀ but paler generally and upperparts greyish brown, densely vermiculated and indistinctly barred with rufous-buff. Underparts buff with faint black barring, each feather with a prominent white shaft streak. Bill greenish, legs flesh coloured.

DOWNY YOUNG: crown rufous-brown, crown-stripe darker. Sides of head buffy white with dark brown eye-stripe extending behind eye. Broad rufous-brown stripe down mid-back flanked by buffy stripes. Rest of upperparts mottled with brown and buff. Underparts creamy buff with rufous wash on breast. Legs pale yellow.

Geographical variation clinal and characters do not vary concordantly; therefore we do not recognize subspecies. Birds from northern and southern ends of range ('*neavei*' and '*natalensis*' respectively) tend to be more richly pigmented with brown above, those in between being paler. Birds from northern two-thirds of range tend to be paler below than those in south. Hybridizes with Swainson's Francolin *F. swainsonii* in W Zimbabwe (Irwin 1971).

Field Characters. Orange bill and legs, plain brown upperparts and rather pale scaled underparts distinguish it from other francolins in most of its range. Lacks bare red face and throat of Red-necked *F. afer* and Swainson's *F. swainsonii* Francolins, while Red-billed Francolin *F. adspersus* is closely barred above and below. At northern edge of range meets Hildebrandt's Francolin from ♂ of which it must be distinguished with great care. Dark markings on underparts much narrower, edges of feathers pale, giving scaly effect, whereas Hildebrandt's, especially in southern populations '*johnstoni*', has large dark blobs on underparts.

Black and white mottling on hindneck of Hildebrandt's is more extensive. Crested Francolin *F. sephaena*, which also inhabits thick vegetation along rivers, is smaller and has white throat and supercilium, erectile crest, and often runs with tail cocked like a bantam.

Voice. Tape-recorded (21, 74, 75, F). Commonest call a 4-noted rasping 'ker-kik-kik-kik' (last 3 notes emphasized); faintly reminiscent of advertisement call of Crested Francolin. Also high-pitched, rasping 'krr-ik-krr' repeated 4–5 times. Another common call (possibly of alarm) is single 'kik' often repeated.

General Habits. Utilizes broad range of vegetation types from sea level to 1800 m. In Zimbabwe, dense thickets (especially on rocky hill-sides) and woodlands with good ground cover, dry riparian forest, and even in montane forest and bush. In South Africa, acacia scrub on rocky ground and in bush along rivers (where found alongside Crested Francolin). Also ventures into cultivation with sufficient cover. Found in pairs or small coveys. Most activities confined to early morning and late afternoon. Dustbathes in open at edge of thickets. Retires to bush during heat of day. Roosts in trees and thickets 3–4 m above ground. Slower on wing than Swainson's Francolin, preferring to dive into cover to avoid raptors (A. C. Kemp, pers. comm.). A bird collected in South Africa in April had nearly completed moult.

Food. Berries, cowpeas, small bulbs, roots, seeds, fallen grain, molluscs, beetles, termites, grasshoppers, and caterpillars. Also pecks in rhinoceros and elephant droppings, presumably taking undigested material, insects and their larvae.

Breeding Habits. Presumably monogamous since usually found in pairs during breeding season.

NEST: well concealed scrape lined with grass under bush or in dense grass.

EGGS: 2–8, sometimes up to 10 (2 ♀♀?), normally 5. Oval; buff to creamy white or yellowish white with no gloss. SIZE: (n = 22, South Africa) 39·1–46·8 × 31·7–36·6 (42 × 34·6); (n = 14, Zimbabwe) 36·1–44·4 × 28·4–34·3 (40·1 × 31·4).

LAYING DATES: South Africa, irregular, but most records Jan–Feb, Apr–July; Zimbabwe all months except Oct, peaking Mar–May (41 out of 76 clutches); Zambia, Mar–May.

INCUBATION: in captivity by ♀ alone. Period: 20 days.

DEVELOPMENT AND CARE OF YOUNG: family parties of up to 10 seen well after end of breeding season.

BREEDING SUCCESS/SURVIVAL: 1 record of predation by Spotted Eagle Owl *Bubo africanus* (Chenaux-Repond 1983).

References
Clancey, P. A. (1967).
Harrap, K. S. (1964).
Irwin, M. P. S. (1971).

54 PHASIANIDAE

Plate 1
(Opp. p. 32)

Francolinus bicalcaratus (Linnaeus). Double-spurred Francolin.
Francolin à double éperon.

Tetrao bicalcaratus Linnaeus, 1766. Syst. Nat., (12th ed.), p. 277; Senegal.

Forms a superspecies with *F. clappertoni*, *F. icterorhynchus* and *F. harwoodi*.

Francolinus bicalcaratus

Range and Status. Endemic resident. Common to very abundant throughout W African savannas between c. 5° and 18°N latitude from Senegal and Guinea to SW Niger, SW Chad and Cameroon. Commonest francolin in W Africa outside of forest. Status of disjunct subspecies *F. b. ayesha* in NW Morocco obscure, certainly at least rare due to hunting and habitat destruction (Meinertzhagen 1940, Cramp and Simmons 1980). Small numbers recorded at Essaouira in 1972, Forest of Mamora (1977) and probably Sous (1971). In 1979 Heinze and Krott (1979) searched intensively for this subspecies in the Mamora Forest and at Sidi Bettache, but heard it calling only at the latter locality. In Ivory Coast numbers of *F. b. bicalcaratus* outside of conserved areas have been reduced by hunting.

Description. *F. b. bicalcaratus* (Linnaeus) (including *thornei*, *adamauae*, and *ogilvie-granti*): W Africa. ADULT ♂: forehead and crown-stripe black; crown brown grading to rufous on nape; supercilium white; eye-stripe black extending barely behind eye. Sides of head white, feathers with broad dark brown shaft streaks. Neck rufous, feathers heavily blotched with black and edged white. Rest of upperparts dark greyish brown, heavily vermiculated with buff, feathers with broad U-shaped cream streaks. Uppertail-coverts pale grey, heavily vermiculated darker. Tail grey, mottled and barred brown. Chin and throat buff; rest of underparts chestnut-brown, feathers with inner black and outer creamy white shaft streaks, black one expanding terminally to form a teardrop with an irregular creamy buff 'window' centrally. Undertail-coverts grey, feathers broadly edged buff. Flight-feathers greyish brown irregularly vermiculated and streaked with buff, especially on inner webs. Upper mandible black, tip dark olive-green, lower mandible olive-green; eye dark brown; legs olive-green. ADULT ♀: like ♂ but averages smaller. SIZE: (40 ♂♂, 34 ♀♀) wing, ♂ 157–185 (170), ♀ 142–179 (159); tail, ♂ 71–84 (78·1), ♀ 68–84 (74·6); tarsus, ♂ 44–65 (56·4), ♀ 43–61 (52·1); spurs, 2 in ♂ (lower longer), 6–15 (11·4), 1 in ♀ 0·6–2·2 (0·9). WEIGHT: (Senegambia) ♂ (n = 5) av. 507, ♀ (n = 3) av. 381; unsexed, Ghana (n = 3) 400–505.

IMMATURE: generally duller; U-patterning on upperparts less distinct, supplemented with indistinct buff barring and streaking. Belly buff blotched with black; flanks barred black and white. Flight-feathers more distinctly barred with buff. Culmen dark horn, base of cutting edges and lower mandible grey-horn; eye brown; legs yellow with only 1 spur.

DOWNY YOUNG: crown rufous-brown, crown-stripe dark brown. Forehead and sides of head buffy with dark brown eye-stripe. Broad dark rufous-brown stripe down mid-back, flanked by 2 buffy stripes. Rest of upperparts mottled with rufous and buff. Underparts creamy buff. Wings greyish buff finely speckled darker. Legs yellow.

Several W African subspecies have been described. These do not warrant recognition since local variation is considerable and primary difference among suggested subspecies is based on darkness of plumage which varies clinally. In general, birds collected in wetter portion of range (S and E) are darkest.

F. b. ayesha Hartert: Morocco. Paler above, chestnut on underparts much reduced. Larger; wing, ♂ (n = 3) 176–182 (179), ♀ (n = 3) 156–176 (167).

Field Characters. Greenish legs and bill, chestnut underparts (appear dark brown at distance, but streaky and mottled close up) and lack of bare patch of skin around eye distinguish it from Clapperton's Francolin *F. clappertoni* and Heuglin's Francolin *F. icterorhynchus*. Ahanta Francolin *F. ahantensis* is darker with orange bill and legs. In Morocco, chestnut crown and black forehead and chestnut underparts distinguish it from Barbary Partridge *Alectoris barbara* which has crown and forehead chestnut, black and white barred flanks, red bill and legs.

Voice. Tape-recorded (37, 60, 217, 235, 267). ♂ advertising call, which varies considerably among individuals, is a grating, nasal 'ke-rak', repeated c. 2/s. Also rendered 'kok-ker kor-ker', 'kokoye-kokoye', and 'bebbrek-ek-kek kek-kek koak koak' ('koak koak' more often than other notes). Also an irregularly repeated 'ee-tek' (1st syllable quiet and musical), a grating 'quare-quare' like guineafowl, a low whistle given to mate and/or young, and a croaking call as it flushes.

General Habits. In W Africa inhabits broad range of vegetation types outside of forest and very thick bush, ranging from relatively moist savanna derived from agriculture, through *Isoberlinia* woodland, into arid savanna; most often found in cultivations (including oil-palm, cassava and cocoa plantations), in and around maize, rice, guinea-corn and other cereal fields, in and around groundnut, cotton, capsicum, okra and almost any dry zone farmland, in shrubby pasture, in thick cover by streams and rivers, dissected countryside (lower elevations); seldom far from human habitation and farming. In Morocco, moister Mediterranean vegetation in coastal lowlands; mainly interface between forest or woodland and open areas: clearings, wetlands, *Cistus* heaths, bushy wadis, palm groves and active and former cultivations. May seek refuge in forest edge when surrounding savanna is burned, emerging into open only to feed. Rarely enters dense, tall grassland and other relatively moist vegetation types during rainy season, presumably to avoid wetting plumage. Usually occurs in coveys of 2–12 (especially *F. b. ayesha*), but in W Africa, occasionally flocks of up to 40. Dominant ♂♂ may lead as groups progress through relatively open vegetation. Calling, feeding and other activities mainly in early morning and late afternoon; seeks shade of bush in heat of day. However, will feed in relatively open areas throughout day in rainy season. Calls usually from top of termite mound or other prominence. Prefers to escape by running. When flushed, flies strongly at height of 4–5 m for usually less than 200 m. Drinks in late afternoon, then proceeds to roost trees. Remains on roost later on wet mornings, presumably to avoid wet ground-level vegetation.

Food. Information available only from W Africa. Opportunistic feeder; predominantly (*c.* 80%) vegetable material including: fruits, roots, green leaves, seeds, millet, corn, rice, peanuts. Animal material (*c.* 20%) includes: small molluscs, frogs, termites, caterpillars, ants, beetles, grasshoppers and other insects and their larvae. In Senegambia varies with habitat: near permanent water mainly seeds of grasses, especially *Panicum*, *Echinochloa* and *Dactyloctenium*; in more arid habitat other seeds and insects (mainly ants and termites) taken (G. and M.-Y. Morel, pers. comm.).

Breeding Habits. Probably monogamous since occurs in pairs prior to and during breeding season.

NEST: shallow, circular scrape or natural hollow in ground, diam. *c.* 15 cm, sometimes lined with grass, twigs or feathers; placed among tussocks of grass, under bush or in dense undergrowth of abandoned cultivations.

EGGS: 5–7, normally 6. Oval to short pyriform; smooth, thick, slightly glossy, minutely pitted; uniform sandy or yellowish buff, sometimes spotted darker. SIZE: (n = 13) 39–45 × 32–36 (42 × 33). WEIGHT: 27.

LAYING DATES: Morocco, May–June; Senegambia, Jan–May, Aug–Dec in habitat with permanent water, in waterless habitat only during rainy season (Aug–Oct); Sierra Leone, mid Oct–mid Feb; Ivory Coast, Aug–Feb; Mali, Aug–Oct in Sahel, all year in south; Niger, Sept; Benin, Oct, Feb; Nigeria, through dry season, mainly Nov–Feb, overall Sept–Mar; S Cameroon, Oct–Dec (dry season).

INCUBATION: sitting bird may sit tight, running off eggs at last possible moment and taking wing immediately; or may leave eggs when intruder still at some distance, run through undergrowth and take wing as much as 50 m from nest.

BREEDING SUCCESS/SURVIVAL: mean of 4 chicks reared from mean clutch size of 5·5 (Thiollay 1970).

References
Bannerman, D. A. (1930).
Beaubrun, P. and Thévenot, M. (in press).
Cramp, S. and Simmons, K. E. L. (1980).
Walls, E. S. (1933).

Francolinus icterorhynchus Heuglin. Heuglin's Francolin. Francolin à bec jaune.

Plate 1
(Opp. p. 32)

Francolinus icterorhynchus Heuglin, 1863. Journ. f. Orn., 11, p. 275; Bongo, Bahr-el-Ghazal, Sudan.

Forms a superspecies with *F. bicalcaratus*, *F. clappertoni* and *F. harwoodi*.

Range and Status. Endemic resident, common to abundant in Central African Republic, N Zaïre, S Sudan and Uganda east to N Karamoja, Teso and Mengo.

Description. ADULT ♂: forehead black, grading to dark rufous brown on crown; eye- and moustachial stripes blackish brown; supercilium white freckled with brown. Sides of head and neck white, feathers with blackish brown shaft streaks. Nape, collar and mantle blackish brown, feathers edged buff forming U-shaped streaks. Rest of upperparts greyish brown, heavily vermiculated and indistinctly barred with buff. Chin and throat white. Breast and belly buff, feathers streaked with dark brown. Undertail-coverts broadly and irregularly barred dark brown and buffy white. Primaries and secondaries dark brown, densely vermiculated and indistinctly barred buff. Upper mandible dusky yellow, lower orange-yellow, tips of both black; eye dark brown, lids and bare skin behind eye dusky-yellow; legs orange-yellow. ADULT ♀: like ♂ but averages smaller. SIZE: (33 ♂♂, 19 ♀♀) wing, ♂ 158–181 (169), ♀ 145–171 (157); tail, ♂ 75–93 (85·2), ♀ 73–85 (77·0); tarsus, ♂ 54–65 (56·8), ♀ 44–54 (48·4); spurs, 1–2 in ♂ (usually 2, lower longer), 3–16 (10·2), 1 in ♀, 0·5–1·8 (0·8). WEIGHT: ♂ (n = 7) 504–588 (571), ♀ (n = 3) 420–462.

IMMATURE: like ♀ but barring on upperparts much more apparent.

DOWNY YOUNG: crown rufous-brown, border darker. Sides of head buffy with dark brown eye-stripe. Broad dark brown stripe down mid-back, flanked by 2 buff stripes. Rest of upperparts mottled brown and buff. Underparts buffy with faint rufous wash. Legs pale yellow.

Birds from Uganda, recognized as *F. i. dybowskii* by White (1965), tend to be somewhat darker, but geographical variation clinal and not deserving of subspecific status.

Field Characters. Yellow bare skin behind eye and orange-yellow bill and legs distinguish it from Double-spurred Francolin *F. bicalcaratus* (which has no bare skin around eye and greenish bill and legs) and Clapperton's *F. clappertoni* (which has red bare skin around eye, a black bill and red legs).

Voice. Tape-recorded (C). Hoarse, slow 'kerak-kerak-kek' given from tree, termite mound or other prominence.

General Habits. Inhabits open grassland, scrubby grassland or lightly wooded savanna, 500–1400 m, also cultivation. Usually found singly, or in pairs, or small coveys of up to 5. Difficult to see in natural habitat, more often located by its calling. Prefers to escape by running into cover, although sometimes perches in trees, especially if grass is wet. Flight heavy, with much beating of wings, usually for short distance. ♂ calls in early morning (but not before dawn) and late afternoon or after rain; ♀ may fly up into tree from which ♂ is calling. Nearby ♂♂ call in response to one another.

Food. Seeds, berries, grass, millet, beetles, hemipterans, millepedes, termites and ants.

Breeding Habits. Probably monogamous since found in pairs throughout much of year.

NEST: scrape in ground under bush or in dense cover.
EGGS: 6–8; pale greyish buff. SIZE: 41·5–43·5 × 31·5–35·0.
LAYING DATES: East Africa, Uganda, Regions A and B, Feb, Apr–July, Oct, no obvious correlation with rainfall; Zaïre, Sept–Nov, during latter part of rains.

References
Chapin, J. P. (1932).
Jackson, F. J. and Sclater, W. L. (1938).
Lippens, L. and Wille, H. (1976).

Plate 1
(Opp. p. 32)

Francolinus clappertoni **Children. Clapperton's Francolin. Francolin de Clapperton.**

Francolinus clappertoni Children, 1826. Denham and Clapperton's Travels, 2, app. 21, p. 198; (no locality = Bornu).

Forms a superspecies with *F. bicalcaratus*, *F. icterorhynchus* and *F. harwoodi*.

Range and Status. Endemic resident in semi-arid northern savannas and grasslands from extreme E Mali (Azzawakh), central Niger, extreme NE Nigeria (where it replaces Double-spurred Francolin *F. bicalcaratus*), Chad (throughout except extreme S and deserts), central and S Sudan (overlapping in SW with Heuglin's Francolin *F. icterorhynchus*), N Uganda and Ethiopia (W Highlands and Rift Valley). Generally patchily distributed and locally common to abundant.

Description. ADULT ♂: forehead dark brown grading to rufous brown through crown and nape; supercilium white; ear-coverts rufous brown; sides of head and neck white, feathers with dark brown shaft streaks; moustachial stripe dark brown. Mantle, back and wing-coverts pale greyish brown, feathers edged with buff forming U-shaped streaks. Uppertail-coverts and tail grey indistinctly barred buff. Chin and throat white. Rest of underparts creamy white, each feather with a prominent black, pear-shaped blotch and a fine buff shaft streak. Primaries greyish brown broadly edged buff. Secondaries brownish grey barred with buff. Bill black, base red; eye brown; naked skin around eye red; legs red. ADULT ♀: like ♂ but smaller. SIZE: (24 ♂♂, 18 ♀♀) wing, ♂ 170–193 (180), ♀ 150–178 (166); tail, ♂ 77–96 (87·7), ♀ 70–91 (81·7); tarsus, ♂ 57–72 (63·6), ♀ 48–63 (56·3); spurs, 1–2 in ♂ (usually 2, lower longer), 2–19 (13·4), 1 in ♀, 0·6–2·1 (0·9). WEIGHT: (Toro, Uganda) ♂ (n = 12) (604), ♀ (n = 10) (463).

Francolinus clappertoni

IMMATURE: like ♀ but less distinctly patterned.
DOWNY YOUNG: undescribed.

Geographical variation complex, but we prefer not to recognize subspecies. Little variation in size. Extent of black moustachial stripe varies geographically. Birds from N Ethiopia, NE Sudan and Eritrea ('*sharpii*', '*konigseggi*' and '*nigrosquamatus*') lack a stripe. Amount of brown blobbing below also varies, with birds collected west of *c*. 30°E being much less blobbed than those to east.

Field Characters. Distinguished from Heuglin's Francolin *F. icterorhynchus* and Double-spurred Francolin *F. bicalcaratus* by red eye-patch and base to bill, creamy white underparts spotted and streaked black, and red legs. Similar to ♂ of Hildebrandt's Francolin *F. hildebrandti*, whose range it comes very close to in E Uganda; latter has similar bill, legs and underparts, but lacks red face patch, and lower back, rump, wings and tail appear rather uniform brown in field. For distinctions from Harwood's Francolin *F. harwoodi* see that species.

Voice. Tape-recorded (217). Loud, grating 'kerak' repeated 4–6 times, like that of Double-spurred and Heuglin's Francolins and faintly reminiscent of call of Yellow-necked Francolin *F. leucoscepus*. Also a single note 'kek' repeated in bursts of 4–5, sometimes preceding the 'kerak'.

General Habits. Chiefly inhabits semi-arid, sandy grassland (e.g. *Hyparrhenia* spp.) with bushes and trees (e.g. *Acacia*, *Terminalia* and *Combretum* spp.) from sea level to 2300 m; also cultivation and rocky hillsides. Usually found in pairs and small coveys. Roosts in tall trees at night, and occasionally during day. Usually calls from termite mound, low branches or other prominences.

Food. Insects, seeds, berries, small molluscs.

Breeding Habits. In captivity mating system monogamous, and ♂♂ courtship-feed their mates. In wild normally occurs singly or in pairs, even outside the breeding season.

NEST: well concealed scrape on ground.
EGGS: clutch size not known. Very thick-shelled with distinct pores, dirty white or yellowish brown. SIZE: *c*. 43 × 33.
LAYING DATES: Mali, Aug–Sept; Chad, July–Sept, ♀ with egg in oviduct Feb (Friedmann 1962); Nigeria, Feb–Mar; Sudan (Darfur), Aug–Sept, SE (birds in breeding condition Mar, Nov); Ethiopia, Feb, Apr–Dec.
BREEDING SUCCESS/SURVIVAL: no more than 4 young observed with parent (Salvan 1978) but not known if due to poor survival rate or clutch size.

Francolinus harwoodi Blundell and Lovat. Harwood's Francolin. Francolin de Harwood.

Plate 4
(Opp. p. 49)

Francolinus harwoodi Blundell and Lovat, 1899. Bull. Br. Orn. Cl., 10, p. 22; Aheafeg (= Ahaia Fej and Haiafegg), Shoa, Ethiopia.

Forms a superspecies with *F. bicalcaratus*, *F. clappertoni* and *F. icterorhynchus*.

Range and Status. Endemic resident in highlands of central Ethiopia in and near gorges of Blue Nile and tributaries, e.g. Aheafeg (10°13′N 39°18′E), Jemmu Valley (9°58′N 38°55′E), near Bichana (10°26′N 38°16′E), Kalo Ford (9°54′N 37°57′E) and Muger River (9°28′N 38°36′E). Sight records from Gibe Gorge (8°15′N 37°35′E) and 39 km NE of Dembidollo. Locally common but range very restricted.

Description. ADULT ♂: forehead and supercilium black; crown dark brown. Ear-coverts grey; sides of head, chin and throat white streaked brown. Remaining upperparts pale greyish brown, densely vermiculated and indistinctly barred buff. Feathers of lower neck and breast strongly streaked black and buff in U-pattern, giving scaly appearance; belly largely buff with some black U-streaking. Undertail-coverts paler and streaked similarly, but in a V-pattern. Primaries greyish brown with dense tawny vermiculations and indistinct transverse barring. Upper mandible black with base and tip red, lower mandible red; eye brown; bare skin around eye red; legs red. ADULT ♀: like ♂ but slightly paler and browner below and belly less streaked, i.e. buff more extensive. Marks on lower belly V-shaped rather than U-shaped. SIZE: wing, ♂ (n = 5) 177–187 (181), 2 imm ♂ 180, 185, ♀ (n = 2) 165, 162, 1 imm. ♀ 165; tail, ♂ (n = 5) 75–86 (81·6), 1 imm ♂ 83, ♀ (n = 2) 72, 69, 1 imm ♀ 73; tarsus, ♂ (n = 5) 53–58 (56), 2 imm ♂ 53, 57; ♀ (n

PHASIANIDAE

= 2) 46, 47, 1 imm ♀ 46·5; spurs, 2 in ♂ (lower longer), 9–16 (12·8), 1 in ♀, 0·4–1·3 (0·7). WEIGHT: ♂ (n = 1) 545, ♀ (n = 1) 446, 2 imm ♀ 438, 414.

IMMATURE: upperparts of ♀ like adult ♀ but less distinctly barred, closer to adult ♂; underparts intermediate between adult ♂ and ♀.

DOWNY YOUNG: undescribed.

Field Characters. The only other francolin likely to ocurr in its proven range is the very different Erckel's Francolin *F. erckelii* which is much larger and has black face and yellow legs. Isolated by habitat from Moorland Francolin *F. psilolaemus* which lives in grassland above gorges and, in any case, is a member of the Red-winged Group, with striped back and yellow legs. Could be confused with Clapperton's Francolin *F. clappertoni* which has similar call and occurs in area of sight records of Harwood's. Both have red eye-patch and red legs, but Clapperton's has pure white throat and much heavier black blotching on underparts, belly white with black blotches rather than buff with V- or U-shaped marks, red only on base of lower mandible. Upperparts of Harwood's barred, but this not clearly discernible in the field (Ash 1978).

Voice. Not tape-recorded. A rasping 'koree', somewhat like that of Clapperton's Francolin.

General Habits. Inhabits dense and extensive *Typha* beds with scattered trees along shallow streams. Will move out into open sorghum fields to forage, but escapes by flying into cover of *Typha* beds. Also uses them for shade during heat of day. If sight records from Gibe Gorge and Dembidollo prove correct, habitat also includes open *Combretum/Terminalia* woodland in high, dense *Hyparrhenia* grassland, and areas of mixed shrubs and cultivated patches on plateaus. Roosts in trees, although not above level of reeds, also in *Typha* beds. Not particularly shy.

Food. Tubers (? *Dioscorea* sp.), grass seed (*Echinochloa* spp.), other seeds and berries (Commelinaceae; *Amaranthus* spp.), sorghum, unidentified berry-like fruits and termites.

Breeding Habits. Nest and eggs undescribed. Brood of 3 young *c.* 5 weeks old 20–21 Feb, thus 1st egg probably laid 2nd week Dec.

Reference
Ash, J. S. (1978).

Francolinus capensis (Gmelin). Cape Francolin. Francolin criard.

(Opp. p. 33)

Tetrao capensis Gmelin, 1789. Syst. Nat. (1st ed), pt. 2, p. 795; Cape of Good Hope.

Range and Status. Endemic resident in S and W Cape Province from lower Orange River Valley south through Little Namaqualand to Cape Peninsula, thence east in fynbos vegetation to eastern Cape (Uitenhage). Records from northern Cape probably erroneous. Introduced onto Robben Island near Cape Town, but present status unknown. Locally common and not threatened within normal range.

Description. ADULT ♂: forehead, crown, sides of head and neck dark greyish brown, feathers edged buff. Ear-coverts uniform brownish grey. Bulk of remaining upperparts dark brownish grey, each feather with 3–5 narrow buff U-shaped streaks. Uppertail-coverts grey with buff V-shaped markings. Tail blackish brown, feathers finely and irregularly barred with buff. Chin white; throat white, blotched with greyish black. Breast and belly dark grey, each feather densely vermiculated with white, with a white shaft streak and edged with a white U-shaped streak; lower abdomen greyish brown, each feather flecked with off-white. Undertail-coverts dark brownish grey with white V-shaped streaking. Primaries greyish brown with fine lighter vermiculations on outer webs. Upper mandible dark brown, base orange-red; lower mandible dull orange; eye brown; legs orange-red. ADULT ♀: like ♂ but averages smaller, and lower mandible and tarsi dull orange. SIZE: (12 ♂♂, 20 ♀♀) wing, ♂ 203–219 (212), ♀ 185–213 (196); tail, ♂ 108–128 (119), ♀ 98–125 (111); tarsus, ♂ 62–72 (67·8), ♀ 56–76 (61·5); spurs, 1–2 in ♂ (usually 2), 4–10 (11·5), 1 in ♀, 0·5–3 (0·8). WEIGHT: ♂ (n = 6) 600–915, ♀ (n = 4) 435–659.

IMMATURE: like ♀ but less distinctly patterned; browner above, greyer below. Wings greyish brown, transversely barred lighter.

DOWNY YOUNG: undescribed.

Field Characters. Large dark francolin with white streaked underparts, red and black bill, orange-red legs. Only other francolin in its range and habitat, the Greywing *F. africanus*, is much smaller, paler, barred black and white below, and has red in wings and yellowish legs. Cape also shows blackish tail in flight.

Voice. Tape-recorded (74, 75, C). Loud crowing 'kak-keek, kak-keek, kak-keeeeeek', with 2nd syllable accented. Also a spluttering flight call.

General Habits. Occurs in pairs or small coveys in areas of scrubby heath, especially coastal fynbos vegetation, and in sheltered scrub along streams and rivers. Has also taken to stands of introduced Australian acacias in vicinity of Cape Town, and is commonly found beside the road in Cape of Good Hope Nature Reserve. Not shy; comes onto farmhouse lawns if not hunted, and dustbathes regularly in open. Calls mainly in early morning and late afternoon. Prefers to escape by running, flying only with great reluctance. Roosts in trees at night if they are available.

Food. Bulbs, corms, fallen grain, shoots, seeds, berries, small molluscs, insects (especially termites and ants). Also aril-like funicle of introduced *Acacia cyclops*.

Breeding Habits. Presumably monogamous since usually in pairs during breeding season.
NEST: well concealed scrape lined with grass under bush.
EGGS: 6–8, sometimes up to 14 (2 ♀♀ ?), mean (25 clutches) 7·4. Brownish cream to pale pink or purplish pink. SIZE: (n = 14) 45·0–57·6 × 36·0–40·7 (48·3 × 38·3).
LAYING DATES: July–Feb, peaking in Sept–Oct, i.e. late in winter rains or early in dry summer.
BREEDING SUCCESS/SURVIVAL: mean number of chicks from 45 broods 4·2.

Reference
Clancey, P. A. (1967).

Francolinus adspersus Waterhouse. Red-billed Francolin. Francolin à bec rouge.

Plate 2
(Opp. p. 33)

Francolinus adspersus Waterhouse, 1838. Alexander's Exped. Int. Afr., 2, ap., p. 267; Great Fish River, Great Namaqualand.

Range and Status. Endemic resident in arid savannas of S Angola (S Huila, especially in Cunene floodplain; also Bicuari and Iona National Parks), Namibia (widespread wherever sufficient cover, from Cunene River to Orange River), floodplains of SW Zambia, W Zimbabwe, and N and E Botswana. Common to abundant.

Description. ADULT ♂: forehead and eye-stripe black, crown dark grey-brown very finely vermiculated with buff; ear-coverts dark grey, sides of head and throat dark grey-brown, feathers edged buff. Hindneck and mantle barred finely with black and white. Remainder of upperparts dark brownish grey, densely and uniformly vermiculated with buff. Tail greyish brown, sometimes with rufous wash and black vermiculations. Chin and throat white very finely barred with black. Remaining underparts uniformly barred finely with black and white. Primaries and secondaries greyish brown with buff vermiculations. Bill orange-red; eye brown, eye-ring yellow; legs orange-red. ADULT ♀: like ♂ but averages smaller. SIZE: (106 ♂♂, 111 ♀♀): wing, ♂ 157–194 (177), ♀ 150–178 (163); tail, ♂ 93–119 (103), ♀ 83–104 (94·8); tarsus, ♂ 48–60 (54·2), ♀ 44–53 (46·7); spurs, 1–2 in ♂ (usually 1), 4–24 (16·5), 1 in ♀, 0·5–1·9 (0·8). WEIGHT: ♂ (n = 17) 340–635 (461), ♀ (n = 26) 340–549 (394).
IMMATURE: decidedly different from adult; much browner overall and faint quail-like pattern of buffy barring and streaking on both upper- and underparts. Upper mandible dark brown, lower whitish horn; eye grey-brown; legs dirty yellow.
DOWNY YOUNG: crown dark rufous-brown, crown-stripe darker. Sides of head buffy with dark brown eye-stripe extending from behind eye only. Broad dark brown stripe down mid-back flanked by 2 buff stripes. Rest of upperparts mottled with buff and brown. Underparts pale yellowish buff, darker on breast. Bill brown; eye brown; legs pale yellow.

Field Characters. Large francolin with a feathered throat, red bill, yellow eye-ring, and finely barred black and white plumage. Swainson's *F. swainsonii* and Red-necked *F. afer* Francolins have red bare skin on throat and around eye. Cape Francolin *F. capensis* is dark but has scaled and streaked plumage, red and black bill; Natal Francolin *F. natalensis* has paler underparts with dark scaling.

Voice. Tape-recorded (35, 38, 74, F, 262). Commonest call a very noisy, highly variable, loud, relatively low-pitched, hoarse cackle 'ka-waark' repeated up to 10 times, sometimes with trailing 'krr', or 'ka-wak-wak-wak, ka-krr-krr-krr-krrr', notes 2–4 loudest.

General Habits. Inhabits floodplains, *Baikaiea*, acacia and mixed woodlands, low scrub, thickets interspersed with open ground, and edges of acacia woodland on Kalahari sand; usually not far from water, often along watercourses. Frequently found on same ground as Swainson's Francolin *F. swainsonii*. In non-breeding season associates in groups of 5–10 (up to 20) birds when feeding or dustbathing in open. Reluctant flier, preferring to escape on foot, but occasionally seeks refuge in trees. Active mainly in early morning and late afternoon, retiring to shade of bush in heat of day; but extends active period when overcast and cool. Dustbathes in open, but near cover. Calls mainly in early morning and late afternoon. Sedentary, but moves to higher ground during floods. Roosts in trees at night.

Food. In Zambia, red arils of seeds of *Guibourtia coleosperma* tree. In Namibia fruits of Devil Thorn *Tribulus terrestris*. Also shoots, greenery, berries, bulbs, beetles, termites, grasshoppers, small molluscs, bugs.

Breeding Habits. Presumably monogamous, since usually in pairs during breeding season. ♂♂ break away from coveys and defend territories with approach of breeding season and call while perched on prominence (e.g. termite mound or low branch of tree). Calling heard throughout day, but mainly in early morning and late afternoon, and appears to attract ♀♀ (T. M. Crowe, unpub.). Courtship display of ♂ lateral with wings fanned, similar to Waltzing Display of Junglefowl *Gallus gallus* (B. Donnelly, pers. comm.).

NEST: hollow scraped out under bush.

EGGS: 4–10, mean (6 clutches) 6·7. Oval; thick-shelled; creamy to brownish yellow. SIZE: (n = 65, southern Africa) 38·9–46·5 × 31·9–35·2 (42·2 × 33·4); (n = 9, Zambia) 39·9–44 × 32·8–35·6 (41·4 × 33·6)

LAYING DATES: N Namibia/Botswana, Jan, Mar, Apr–Aug, peaking in Apr–July; central and S Namibia, almost any month, mainly Dec–Apr; Zambia, May; Zimbabwe, Jan–Mar, May, July Aug; breeds in late rains or early dry season.

INCUBATION: period in captivity 22 days.

DEVELOPMENT AND CARE OF YOUNG: in captivity immatures moult into adult plumage at c. 3 months and spurs apparent on ♂♂ at c. 5 months.

Reference
Clancey, P. A. (1967).

Montane Group

7 species divided between 2 superspecies, (a) *swierstrai/camerunensis*, and (b) *nobilis/jacksoni/castaneicollis/ochropectus/erckelii*. *F. swierstrai* and *camerunensis* are medium sized and strongly sexually dimorphic in plumage but not in size; barred and vermiculated upperparts of ♀♀ suggest affinities with Scaly Group. Members of the *nobilis* superspecies are much larger (900–1100 vs c. 550 g); ♂♂ much larger than ♀♀ with strongly developed second spur but differing little in plumage. All members of Montane Group have cartilaginous ceres, most have red or orange-red bills and legs and streaked upper- and underparts (lighter below).

Plate 3
(Opp. p. 48)

***Francolinus swierstrai* Roberts. Swierstra's Francolin. Francolin de Swierstra.**

Francolinus swierstrai Roberts, 1929. Ann. Trans. Mus. p. 72; Mombolo, SW Cuanza Sul district, Angola.

Forms a superspecies with *F. camerunensis*.

Range and Status. Endemic resident in highlands of W Angola, locally distributed in Bailundu highlands and Mombolo Plateau, and isolated populations on Chela escarpment and Tundavala in Huila District and at Cariango in Cuanza Sul District. Restricted to a few relict patches of montane forest a few km² in extent, e.g. on Mt Moco and Mt Soque. Habitat reduced by forest destruction and population likely to be very small. Some conservation action being taken. Listed in Appendix 2 of the 1973 Convention on International Trade in Endangered Species of Wild Fauna and Flora of which Angola became a signatory in 1977.

Description. ADULT ♂: forehead black; crown dark brown, grading to black on nape; collar from sides of neck round hindneck black and white. Broad supercilium continuing over ear-coverts white; lores and area under eye black, separated by grey line in front of eye. Mantle, back, scapulars and wing-coverts grey-brown, feathers with lighter centres and black edges separated by narrow strip of slightly rusty brown. Rump and tail dark greyish brown, rump with a few black vermiculations. Chin and throat white. Broad band across breast mainly black with some white streaks; rest of underparts streaked black and white. Flight-feathers and underwing dark grey-brown. Bill red; eye brown; legs red. ADULT ♀: head and underparts like ♂. Back and scapulars slightly reddish brown blotched with fuscous, each feather pale brown lightly

Francolinus swierstrai

vermiculated with fuscous on outer web and with large subterminal fuscous spot and 1 or 2 bars on inner web, shaft streak pale buff. Rump, uppertail-coverts and rectrices pale brown vermiculated with dusky. Wing-coverts reddish brown, vermiculated with dusky and with faint pale shaft streaks and a few dark blotches on median coverts. Outer webs of primaries with rufous vermiculations. SIZE: (7 ♂♂, 1 ♀) wing, ♂ 160–181 (171), ♀ 167; tail, ♂ 90–105 (98), ♀ no data (tail missing in single specimen); tarsus, ♂ 49–58 (53), ♀ 43; spurs, 1–2 in ♂ (usually 2, lower longer), 7–16 (11·5), 1 in ♀, 0·3–1·0 (0·4).

IMMATURE: like ♀ but throat and supercilium pale buff, upperparts streaked and barred with rufous buff, flanks and belly barred black and white.

DOWNY YOUNG: undescribed.

Field Characters. Rather large francolin with black and white underparts; nothing else like it within its habitat. Red-necked Francolin *F. afer* has red face and throat, and lives at lower altitudes in different habitat.

Voice. Not tape-recorded. A shrill, harsh cry, not unlike that of Jackson's Francolin *F. jacksoni* (Hall 1960).

General Habits. Inhabits undergrowth in montane evergreen forest and forest edges; also on rocky and grassy slopes of mountainsides and in tall grass savannas on mountain tops and gullies. A bird flushed from bracken-covered bank of stream perched in tree (Hall 1960). Keeps to dense undergrowth; when disturbed, runs or flies into tree. Feeds among fallen leaves in undergrowth.

Food. Insects, grasses and seeds of Leguminosae spp.

Breeding Habits. Almost unknown. Half-grown young collected Sept; adult ♂ collected Sept was observed displaying and had enlarged testes; ♀ with enlarged ovaries Mar; evidence from Aug specimens suggests breeding May–July.

References
Collar, N. C. and Stuart, S. N. (1985).
Hall, B. P. (1963).
Hall, B. P. and Moreau, R. E (1962).
Traylor, M. A. (1960b).

Francolinus camerunensis Alexander. Cameroon Mountain Francolin. Francolin du Mont Cameroun.

Plate 3
(Opp. p. 48)

Francolinus camerunensis Alexander, 1909. Bull. Br. Orn. Club, 25, p. 12; Cameroon Mountain at 2300 m.

Forms a superspecies with *F. swierstrai*.

Range and Status. Endemic resident in montane forest on SE slopes of Cameroon Mountain between 850 and 2100 m; collected in 1909 at Buea and Musake. Total range estimated to be less than 200 km². Locally common, but may be adversely affected by forest destruction (by man and/or volcanic activity) and excessive hunting.

Description. ADULT ♂: forehead, crown and sides of head uniform very dark brown. Neck and remaining upperparts uniform dark brownish grey. Throat dark brownish grey, feathers edged buffy grey. Underparts dark grey, darker under tail, feathers with darker centres. Flight feathers uniform very dark brown. Bill and patch round eye red; eye brown; legs red. ADULT ♀: like ♂ but crown and throat mottled with rufous buff. Upperparts barred with off-white, uppertail-coverts densely vermiculated with off-white. Underparts streaked with black and white; undertail-coverts weakly barred with off-white. SIZE: (3 ♂♂, 7 ♀♀) wing, ♂ 167–175 (171), ♀ 157–169 (164); tail, ♂ 79–85 (82), ♀ 74–87 (81); tarsus, ♂ 52–68 (59), ♀ 56–61 (59);

Francolinus camerunensis

spurs, 1–2 in ♂ (usually 2, lower longer) 4–8 (6), 1 in ♀, 0·3–1·1 (0·5).

IMMATURE: like ♀ above. Underparts barred black and whitish, flank feathers with black and white tips. Patch around eye feathered. Bill, legs dusky red.

DOWNY YOUNG: undescribed.

Field Characters. Same size as duller, grey-brown Scaly Francolin *F. squamatus*, but distinguished by red face, blacker plumage, largely unmarked in ♂, ♀ with pale edges to feathers producing barred and scaly appearance.

Voice. Tape-recorded (217). High-pitched, musical, triple whistle.

General Habits. Preferred habitat dense undergrowth in primary and secondary forest, apparently avoids montane grassland. Very shy; prefers to escape by running, but if flushed by dogs will take to trees. Usually occurs in pairs and small parties; dustbathes in open in sunshine.

Food. Berries, grass seed and insects.

Breeding Habits: lays Oct–Dec (dry season breeder).

References
Bannerman, D. A. (1930).
Collar, N. C. and Stuart, S. N. (1985).
Serle, W. (1965, 1981).

Plate 3
(Opp. p. 48)

Francolinus nobilis Reichenow. Handsome Francolin. Francolin noble.

Francolinus nobilis Reichenow, 1908. Orn. Monatsber., 16, p. 81; Virunga Volcanoes, Zaïre.

Forms a superspecies with *F. erckelii*, *F. ochropectus*, *F. castaneicollis* and *F. jacksoni*.

Range and Status. Endemic resident from highlands west of L. Albert, Ruwenzori Mts and Impenetrable Forest (Uganda) through highlands of E Zaïre, Rwanda and Burundi to Itombwe Mts (S Kivu) and Mt Kabobo. Common to locally abundant, although trapped with snares by local people.

Description. ADULT ♂: head and neck dark brown, centres of feathers darker; ear-coverts pale grey, face grey, feathers with fine black shaft streaks. Back deep rufous, feathers edged grey; scapulars, lesser and greater upperwing-coverts still more rufous, scapulars almost entirely deep rufous or maroon. Rump, uppertail-coverts and tail dark brownish grey. Throat greyish buff, streaked dark grey. Breast and flanks rufous, feathers edged grey or greyish buff; abdomen grey, feathers tipped buffy; undertail-coverts blackish brown edged light brown. Underwing-coverts dark grey or brown. Primaries and outer secondaries grey brown, inner secondaries deep rufous. Bill and bare skin round eye bright red; bare skin above ear dull orange; eye brown; legs bright red. ADULT ♀: as ♂ but smaller and somewhat duller. SIZE: (11 ♂♂, 13 ♀♀) wing, ♂ 191–210 (198), ♀ 172–186 (178); tail, ♂ 99–116 (106), ♀ 97–102 (100); tarsus, ♂ 62–66 (64·6), ♀ 53–57 (55·1); spurs, 1–2 in ♂ (usually 2, lower longer), 4–22 (12), 1 in ♀ 0·6–1·7 (0·9). WEIGHT: ♂ (n = 2) 862, 895; ♀ (n = 3) 600, 648, 670.

IMMATURE: like adult but upperparts barred dark grey and rufous buff, and underparts paler.

DOWNY YOUNG: undescribed.

Birds from Ruwenzori Mts ('*chapini*') have narrower grey edging on belly feathers, but this character also varies within populations.

Field Characters. A large francolin with much rufous in plumage, especially wings and underparts, with red bill and legs, bare red skin round eye. Smaller Scaly Francolin *F. squamatus* has no rufous in plumage or red face. Within most of range it is the only francolin present.

Voice. Tape-recorded (32). Loud crowing, 'chuk-a-rik' repeated 4–5 times or squealing 'cock-rack' repeated 6–8 times; very noisy, especially morning and evening.

General Habits. Inhabits dense undergrowth from lower edge of montane forest up through bamboo zone to Afro-alpine heath zone at *c.* 3700 m. Usually shy and hard to see; escapes by running away through thick undergrowth. Reluctant to fly, and does so for only short distance, quickly dropping back into cover. However, pairs and small groups found along edges of roads in very early morning and late afternoon (Prigogine 1971; R. W. Smart, pers. comm.). Roosts in low trees or bushes; very vocal when going to roost.

Food. Seeds.

Breeding Habits.

LAYING DATES: E Zaïre (S Kivu), late Apr/May–Aug/Sept.

DEVELOPMENT AND CARE OF YOUNG: small groups observed are presumably family parties, as in other francolins.

References
Chapin, J. P. (1932).
Prigogine, A. (1971).

Francolinus jacksoni Ogilvie-Grant. Jackson's Francolin. Francolin de Jackson.

Plate 3
(Opp. p. 48)

Francolinus jacksoni Ogilvie-Grant, 1891. Ibis, p. 123; Kikuyu, Kenya.

Forms a superspecies with *F. erckelii*, *F. ochropectus*, *F. castaneicollis* and *F. nobilis*.

Range and Status. Endemic resident in the montane forest and Afro-alpine zone of Kenya from *c.* 2200 to 3700 m; from Mt Kenya and Aberdares to Mau Plateau and Cherangani Hills; also 2 records from Mt Elgon, 1 from Kenya and 1 from Uganda. Locally common to abundant, occurring at high densities in national parks. Not threatened, but some loss of habitat at lower fringes of forest in recent years.

Description. Forehead to nape greyish brown, feathers washed with rufous and edged with buff; lores dull rufous; ear-coverts pale grey. Sides of head below eye and neck off-white, feathers with broad chestnut shaft streaks. Mantle chestnut, feathers edged white; back dark olive-brown vermiculated black and washed with rufous; rump and tail rufous-brown; upperwing-coverts brown. Chin and upper throat white, lower throat streaked with chestnut. Lower neck, breast and belly rich chestnut, feathers edged white, with some black vermiculations. Sides, flanks, lower belly and undertail-coverts bright chestnut, feather edges broadly vermiculated black and grey. Flight feathers grey-brown. Bill dark coral-red; eye brown, lids red; legs bright red in front, darker behind. ADULT ♀: like ♂ but smaller overall. SIZE: (13 ♂♂, 8 ♀♀) wing, ♂ 203–234 (218), ♀ 195–217 (200); tail, ♂ 116–152 (129), ♀ 112–121 (116): tarsus, ♂ 67–97 (74·7), ♀ 59–79 (66·0); spurs, 1–2 in ♂ (usually 1), 7–24 (12), 1 in ♀, 1–2 (1·2). WEIGHT: ♂ (n = 1) 1064; 1 very large ♂ 1130 (Lynn-Allen 1951).

IMMATURE: generally duller above; scapulars, inner secondaries, tail-coverts, tail and outer webs of flight feathers barred darker brown; below, breast chestnut, otherwise barred black and white.

DOWNY YOUNG: undescribed.

Birds from Mt Kenya ('*pollenorum*') are somewhat darker than those from the western part of the species' range.

Field Characters. A very large francolin with white-streaked chestnut underparts and red bill and legs. Smaller Scaly Francolin *F. squamatus* is a dull grey-brown bird with scaly underparts; sympatric Moorland Francolin *F. psilolaemus* (one of Red-winged group) has striped face, buffy underparts and rufous flight feathers.

Voice. Tape-recorded (264). High-pitched, extremely loud series of cackles, more reminiscent of Scaly or Hildebrandt's Francolin *F. hildebrandti* than e.g. Yellow-necked Francolin *F. leucoscepus*. Described as like sharpening scythe with whetstone. Utters low clucking calls when feeding, and at close quarters in dense bush, probably to maintain contact.

General Habits. Occurs in all types of forest including *Juniperus*, *Podocarpus*, bamboo *Arundinaria alpina*, *Hagenia*, *Hypericum* and in moorland *Erica* and *Stoebe* thickets. Normally in dense shrubby growth rather than within tall forest, especially numerous in giant heath on Aberdare Mts, and in areas where bamboo has died over large areas, to be replaced by scrub, where perhaps attracted to abundant fallen seed. Usually remains within thickets where more often heard than seen; but often emerges into glades of short-cropped Kikuyu grass and even onto roadsides in national parks where conspicuous and nowadays often tame. In Aberdare Park approaches humans closely at picnic places to obtain scraps and visits mountain lodges with arc lights in early morning to pick up dead insects. In past, frequented routes followed by porters, collecting fallen grain or meal. Roosts in low trees, tall heath, or other dense scrub, and often takes to trees or bush when disturbed, e.g. by dogs. Like Chestnut-naped Francolin *F. castaneicollis* feeds in middle of day, not only in early morning and evening. In dry weather, active very soon after dawn, descending to ground from roost, and foraging before sunrise. Emerges into open ground to dry plumage in wet weather and to dust bathe in dry weather; in natural conditions dustbathes in earth thrown up by mole-rats; now also on roadside banks. Noisy in morning and again before evening when going to roost.

Food. Grass shoots and bulbous roots, berries in season, small snails, insects. Also feeds on fallen seed of dying bamboo (which seeds and dies over patches at intervals), achenes of some Compositae.

Breeding Habits. Monogamous, possibly throughout the year. Territorial; pairs evenly distributed, but spread out; small patch of heath and surrounding grassland may be occupied year after year by only 1 pair.

NEST: not described, but 1 found at edge of clump of bamboo.

EGGS: 3 or more. Glossy; pale brown. SIZE: $46 \cdot 5 \times 36$.

LAYING DATES: Aberdares–Mt Kenya, Dec–Jan, 1 record Aug, in driest season of year; on Mau highlands, more frequent, Jan, Feb, Aug, Oct, Dec; in high wet mountain ranges probably can only breed successfully in dry months.

DEVELOPMENT AND CARE OF YOUNG: up to 7 chicks/brood; young remain with adults in small family parties for about 8 months.

References
Hall, B. P. and Moreau, R. E. (1962).
Jackson, F. J. and Sclater, W. L. (1938).

Plate 3
(Opp. p. 48)

Francolinus castaneicollis **Salvadori. Chestnut-naped Francolin. Francolin à cou roux.**

Francolinus castaneicollis Salvadori, 1888. Ann. Mus. Civ. Genova, 26, p. 542; Lake Ciar-Ciar (= Chercher), Shoa. (Note by authors: Chercher not in fact in Shoa Province, but in Hararghe).

Forms a superspecies with *F. erckelii*, *F. ochropectus*, *F. jacksoni* and *F. nobilis*.

Range and Status. Endemic resident in highland forests of NW Somalia, Ethiopia south and east of Rift Valley, south to Kenya border, and west of Rift in upper Omo basin, Jimma and Kaffa, and recently discovered in extreme N Kenya near Moyale. Generally common to abundant; locally very abundant in Bale and Arussi Mts, Ethiopia; less common west of Rift and in Somalia. Not threatened.

Description. *F. c. castaneicollis* Salvadori (including *bottegi, ogoensis, kaffanus, gofanus*): NE Ethiopia and Somalia from Harar south to Arussi, Bale, and Sidamo, west of Rift to Kullo, Alghe, Arero, Burji and near Lake Stephanie. ADULT ♂: forehead black grading to rufous on crown and nape; lores black, feathers with white shaft streaks; ear-coverts and sides of face paler rufous freckled with buff. Mantle rufous streaked with black and buff; back and upperwing-coverts dark brown, feathers broadly edged with white and chestnut. Rest of upperparts brownish grey, vermiculated darker and indistinctly barred with buff. Tail dark brown, barred darker brown. Throat white. Upper breast chestnut, feathers centrally streaked black and white, edged with bright chestnut; flanks with fewer, broader white streaks, centred black; lower belly pale buff, undertail-coverts grey barred black. Flight-feathers uniform brownish grey, secondaries with some dark vermiculation. Underwing-coverts reddish brown, streaked blackish. Bill red; eye dark brown; legs coral red. ADULT ♀: smaller. SIZE: (31 ♂♂, 30 ♀♀) wing, ♂ 191–226 (210), ♀ 169–203 (186); tail, ♂ 123–143 (133), ♀ 99–128 (114); tarsus, ♂ 59–74 (67), ♀ 48–67 (56); spurs, 2 in ♂, of equal length, 9–20 (14·5), 1 in ♀, 1–2 (1·2). WEIGHT: ♂ (n = 6) 915–1200, ♀ (n = 4) 550–650.

IMMATURE: duller, rump and tail barred and vermiculated black and buff; bill sepia, base dull red.

DOWNY YOUNG: undescribed.

Specimens collected in Somalia tend to be paler than those from NE Ethiopia and those from Jimma and Kaffa provinces more richly pigmented with chestnut (a bird from Kaffa Province is illustrated in Plate 3); but variation clinal and does not warrant taxonomic recognition.

F. c. atrifrons Conover: Mega area, S Ethiopia; as nominate subspecies but upperparts lacking chestnut, belly buffy, not strongly marked with chestnut and black. Smaller, wing, ♂ (n = 2) 191–192, ♀ (n = 3) 162–176.

Field Characters. Large, same size as Erckel's Francolin, *F. erckelii*, but more brightly coloured, with chestnut head and neck, bright red bill and legs. Much larger than Harwood's Francolin *F. harwoodi*; much less heavily streaked below, and has no bare red skin round eye.

Voice. Tape-recorded (226). Varied, often noisy; a harsh 'kek kek kek kerak', often in duet or in chorus.

General Habits. Prefers forest, forest glades and undergrowth, occurring in Bale and Arussi Mts from c. 2500 m to highest plateaux at over 4000 m, in tall *Erica* scrub, and in Arussi in short burned *Erica* and grassland. In Somalia occurs in arid juniper forest, and at much lower elevations, 1200–2250 m. In Kaffa and Jimma areas, inhabits edges of broad-leaved forests, not inside densest growth. Optimum habitat appears to be *Hagenia–Hypericum* forests at 3100–3500 m with dense wet undergrowth of *Kniphofia* and giant *Lobelia* spp. Very tame and confiding in Bale Mts, often observed in open and can be approached to within a few m, even though to some extent persecuted. When pursued, retreats from open grassland into dense undergrowth of e.g. *Kniphofia* (Red-hot Poker lilies) where it remains, refusing to emerge into open again, or to fly, but moving through narrow passages between growth. Is then very easily trapped with string nooses, but nevertheless tameness suggests little persecution. In higher moorlands with *Erica* scrub less likely to be seen in open short grassland, and here flies more readily, perhaps because more open scrub cover of burned heath less safe refuge.

Roosts in trees at night, and on wet days may remain in trees for some time rather than descend to soaking undergrowth. In *Erica* scrub roosts close to, perhaps on, ground. Unlike most other francolins, does not only feed early morning and evening, but frequently also in middle of day or late morning, right out in open; this behaviour may be forced upon it by frequently soaking wet undergrowth. Small coveys, presumably family parties, frequently feed together in compact groups, walking slowly about picking up seeds or other food. Does not dig much, although legs are strong enough. Seldom feeds in croplands, but in some high cultivated areas (e.g. Arussi Mts up to 3100 m) may glean fallen grain in barley fields. Most food picked up off ground; does not reach up to strip heads of grasses or Compositae. Calls generally from within dense cover, more in morning and evening, but at any time of day on fine days.

Food. Seeds and some insects, including termites.

Breeding Habits. Probably monogamous and territorial.

NEST: scrape, under bush or other cover.

EGGS: 5–6. Smooth, much rounder than those of Yellow-necked Francolin *F. leucoscepus*; cream coloured. SIZE: (n = 3, *F. c. castaneicollis*) 46·3–48·0 × 36·0–37·5 (46·8 × 36·8); (n = 11, *F. c. castaneicollis*) mean 46·5 × 38. WEIGHT: *c.* 33.

LAYING DATES: Ethiopia, Jan–Mar, Oct, Nov, Dec, season extended, but perhaps preferring drier months; Somalia, May (wet) and Dec (dry).

DEVELOPMENT AND CARE OF YOUNG: parents remain with young until full grown; family parties number 4–8.

BREEDING SUCCESS/SURVIVAL: size of coveys (5–8) observed in Bale Mts suggest breeding success may often be good.

Francolinus ochropectus Dorst and Jouanin. Djibouti Francolin. Francolin des Somalis.

Plate 3
(Opp. p. 48)

Francolinus ochropectus Dorst and Jouanin, 1952. Ois. et Rev. Franç. Orn. 22, pp. 71–74; Plateau du Day, near Tadjoura, Djibouti.

Forms a superspecies with *F. erckelii*, *F. castaneicollis*, *F. jacksoni* and *F. nobilis*. Though intermediate both geographically and morphologically between *F. erckelii* and *F. castaneicollis*, it is sufficiently distinct from either to be retained as a valid, though weakly marked species.

Range and Status. Endemic resident, restricted to Forêt du Day (*c.* 1400 ha of relict juniper forest) and adjacent wadis, *c.* 25 km west of Tadjoura, Djibouti; altitudinal range 700–1780 m. Widespread within this area; population recently estimated at *c.* 5000 (J. Blot, pers. comm.). Principal threat to survival is habitat destruction—the forest was halved in size between 1977 and 1983; 7 tons of firewood are shipped out daily, and overgrazing and trampling by domestic animals are a major problem. Further threats are posed by land clearance for agriculture and disturbance of the forest by local people, tourists and the French army. The Djibouti Francolin is listed in Appendix II of CITES, to which Djibouti is not a signatory. There is an urgent need to stop further deforestation, and a captive breeding programme would be advisable (Welch and Welch 1984).

Description ADULT ♂: forehead black, feathers with white shaft streaks; crown dark rufous-chestnut, hindcrown grey; crown-stripe rufous-chestnut; white eye-stripe extends from eye to grey ear-coverts. Further below eye rufous-chestnut, feathers with buffy margins. Upperparts grey, streaked with rufous-buff. Tail with faint rufous hue. Throat white. Underparts white, feathers with U-shaped black streaking towards proximal end and narrow grey-black shaft streak which expands to form a terminal tear drop blotch of buff. Teardrops appear denser on upper breast. White streaking predominates on lower breast, belly and undertail-coverts. Flight-feathers greyish brown. Upper mandible black, lower black with some yellow; eyes brown; legs yellow. ADULT ♀: like ♂ but smaller, more rufous in tail, upperparts somewhat vermiculated. SIZE: (3 ♂♂, 1 ♀) wing, ♂ 200–209, ♀ 176; tail, ♂ 119–135, ♀ 110; tarsus, ♂ 60–65, ♀ 47·2; spurs, 2 in ♂ (upper longer), 17–22, 1 in ♀, 0·6.
  IMMATURE: like ♀ but more barred than streaked with buff and grey.
  DOWNY YOUNG: undescribed.

Field Characters. The only francolin so far reported within its range. A large bird, appearing very dark in the field.

Voice. Not tape-recorded. Call of ♂ a loud 'Erk-ka, ka, ka, k-k-k-kkk', the 'Erk' being the dominant sound, the rest of the notes becoming quieter and faster, ending in a chuckle (Welch and Welch 1984). Birds feeding in groups give soft, low conversational clucking.

General Habits. Breeds in inaccessible wadis with dense, lush vegetation including palms and ferns. Ascends to plateaus after breeding and lives in primary and secondary forest, where principal tree is *Juniperus procera*, 5–8 m high; main understorey spp. *Buxus hildebrandti* and *Clutia abyssinica*; in more open areas, *Acacia etbaica*. Where parasitic fig tree *Ficus* sp. is widespread, birds frequently feed on figs at forest edge. Requires dense over for roosting.
  Most active from dawn to 0800 h. Elusive during rest of day, spending much time roosting at height of up to 4 m from ground, where it sits very tight. More often heard than seen. Scratches on ground for seeds and also for termites in areas disturbed by warthogs.

Food. Figs, berries, grass and other seeds; termites.

Breeding Habits. Lays Dec–Feb; family party of 9 seen in Mar. Major predator the genet *Genetta genetta*.

References
Dorst, J. and Jouanin, C. (1954).
Welch, G. R. and Welch, H. J. (1984).

Plate 3 (Opp. p. 48) *Francolinus erckelii* (Rüppell). **Erckel's Francolin. Francolin d'Erckel.**

Perdix erckelii Rüppell, 1835. Neue Wirbelth., Vog., p. 12, pl. 6; Taranta Mts, Ethiopia.

Forms a superspecies with *F. ochropectus*, *F. castaneicollis*, *F. jacksoni* and *F. nobilis*.

Range and Status. Endemic resident fairly continuously distributed in highlands above 2000 m in central and N Ethiopia and Eritrea north and west of Rift Valley; separate population in hills of Red Sea Province, Sudan. In Ethiopia, frequent, locally common, but may be much less common than formerly, owing to widespread forest destruction over much of N Ethiopia.

Description. ADULT ♂: forehead and crown-stripe black; face below eye and sides of neck chestnut, feathers with narrow white shaft streaks; ear-coverts pale grey. Crown and nape chestnut. Mantle chestnut streaked with white; remaining upperparts greyish olive-brown, feathers with chestnut margins. Tail reddish brown, barred darker. Throat white, breast grey, feathers with chestnut shaft streaks; belly and flanks creamy white, feathers with chestnut shaft streaks. Flight feathers greyish brown. Bill black; eye dark brown; legs dull yellow. ADULT ♀: much smaller. SIZE: (13 ♂♂, 11 ♀♀) wing, ♂ 200–227 (216), ♀ 167–194 (185); tail, ♂ 120–142 (130), ♀ 98–131 (112); tarsus, ♂ 61–75 (66·2), ♀ 52–65 (55·7); spurs, 2 in ♂ (upper longer), 11–22 (15·6), 1 in ♀, 0·4–1·5 (0·7). WEIGHT: ♂ (n = 3) 1050, 1150 and 1590, ♀ (n = 1) 1136.
  IMMATURE: paler grey above, with buff quail-like streaks and bars on mantle and back, and dark bars on outer webs of flight feathers. Wing-coverts, rump, and tail similarly barred and streaked.

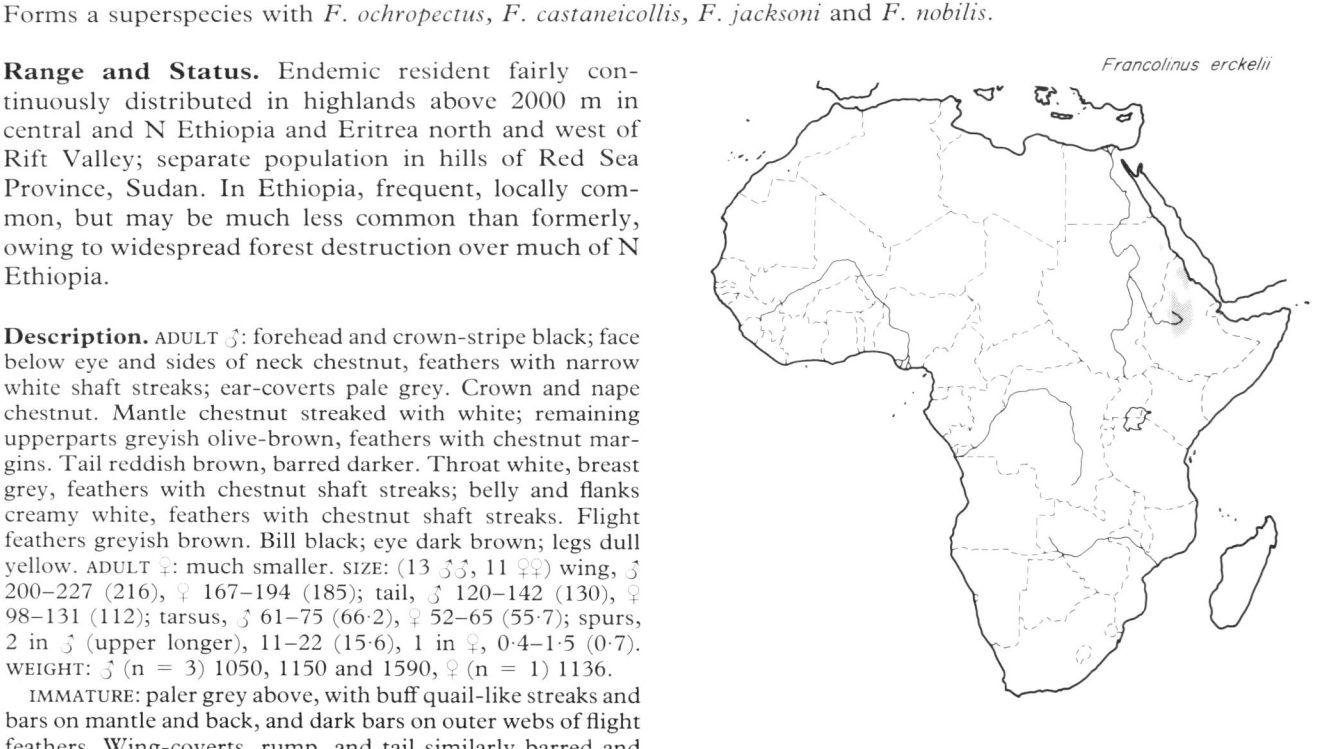

DOWNY YOUNG: dark brown stripe from forehead down centre of crown to nape; sides of head buff with black moustachial stripe, and another through eye down side of neck. Body above chocolate and black, with 2 lateral buff stripes; below plain brownish white.

Specimens from an isolated population in Sudan (recognized as *F. e. pentoni* by White (1965)) are somewhat paler overall.

Field Characters. A very large francolin; in most of range only one to occur in montane habitat between 2000 and 3500 m. At lower fringe of habitat may overlap with Clapperton's Francolin *F. clappertoni*, from which easily distinguished by dark face lacking eye-stripe and bare red skin round eye, yellow legs, much larger size, and call. May also overlap with Harwood's Francolin *F. harwoodi*, which is much smaller, has bare red skin round eye and red legs. More uniformly olive-brown above than Chestnut-naped *F. castaneicollis* or Djibouti *F. ochropectus* Francolins.

Voice. Tape-recorded (C). Advertisement call a long series of croaking, cackling notes, first loud and emphatic, then falling in pitch and volume towards end: 'errrk-erkk-erk-erk-rkkuk-kuk-ku', also a rasping 'kri-kri-kri-kri-wa-wa-wa-wa'. ♀ courtship feeding call 6 soft notes of *c.* 0·5 kHz and 0·08 s in duration. Somewhat similar structurally to those given by Chukar *Alectoris chukar* and Grey Francolin *F. pondicerianus*.

General Habits. Generally remains within patches of scrub cover, composed largely of forest remnants between 2000 and 3500 m, e.g. *Carissa, Maytenus, Rosa abyssinica, Rumex* with long coarse *Hyparrhenia* grass. May originally have preferred forest, now forced to adapt to scrub cover; but, where forest available (e.g. Menagasha near Addis Ababa, and Ankober), does not normally enter tall timber, but remains in thickets near edge. In Red Sea hills, Sudan, frequents woods or other dense cover along stream-beds, coming to much lower altitude than in Ethiopia, but still in hills. In high mountains occurs into giant heath *Erica arborea* zone, but does not frequent open short montane grasslands at same altitude (Semien Mts).

Very shy, far more often heard than seen, runs away from human observer uphill, and, if observer catches up with it, flies noisily back downhill. Normally rather silent, calling towards dusk and dawn, usually from a prominence. Feeds normally around and in scrub and forest edges, often entering cultivation to glean fallen grain, usually very early in morning. By day remains almost entirely within dense cover, seldom seen. On steep cliffs flies easily from ledge to ledge, plucking seedheads of grasses quite inaccessible to any other animal (including Ibex *Capra ibex*). Leans right out over precipice and, if it falls off, merely flutters back or to next ledge down. Can work way up or down near vertical precipices by alternately walking along grassy ledges and flying few m to next. Tends to drink in late afternoon, then proceeds to nocturnal roosting tree. Pairs seem entirely sedentary, usually feeding together and found in same general area from day to day.

Food. Fallen grain, grass seeds and shoots, berries and seeds of shrubs and herbs, notably *Rumex*; some insects.

Breeding Habits. Observations on captive birds suggest mating system monogamous. Paired individuals remain constantly in each other's company; ♂♂ courtship-feed their mates.

NEST: on ground, in scrape.

EGGS: 4–10. Dirty white to pale brown; very hard-shelled. SIZE: (n = 7) 44·2–48·0 × 36·0–37·5 (46·2 × 36·5). WEIGHT: *c.* 33.

LAYING DATES: highland Ethiopia, May, Sept–Nov (in rainy months); N Sudan, Apr, May (rains).

DEVELOPMENT AND CARE OF YOUNG: both parents accompany brood, and all remain together, forming small family parties, probably until subsequent breeding season.

BREEDING SUCCESS/SURVIVAL: baboons prey on eggs and young. Adults vulnerable to raptors.

Reference
Moltoni, E. and Ruscone, G. G. (1942).

Bare-throated Group

4 spp. (*swainsonii, afer, rufopictus* and *leucoscepus*) in a single superspecies. Large francolins (500–800 g) with cartilaginous ceres, red or yellow bare skin on throat and around eye, reddish to blackish (never yellow) bills, red legs (black in *swainsonii*) with a pair of spurs on ♂♂, and relatively uniform dark brownish upperparts and streaked underparts with varying amounts of vermiculation. Although ♂♂ tend to be much larger than ♀♀, there is little sexual dimorphism in plumage (♀♀ tend to have somewhat more strongly vermiculated upperparts). Inhabit lowland grasslands with some tree cover and bushy vegetation along watercourses. All roost in trees and have loud, grating advertisement calls.

PHASIANIDAE

Plate 2
(Opp. p. 33)

Francolinus swainsonii (Smith). Swainson's Francolin. Francolin de Swainson.

Perdix swainsonii A Smith, 1836. Rep. Exped. Expl. Cent. Afr., p. 54; W Transvaal near Kurrichane.

Forms a superspecies with *F. afer*, *F. leucoscepus* and *F. rufopictus*.

Range and Status. Endemic, resident in savannas and bushveld of southern Africa from S and S–central Zambia to Luangwa Valley, and NE to Lundazi; Malaŵi (near Mzimba); Mozambique (Tete District); south to Namibia, Botswana, Transvaal, Orange Free State, and Natal near Durban. Generally common to abundant. Benefits from moderate human activity, adapting well to cultivation, especially maize; has increased in abundance and extended range in Natal and Zimbabwe in recent years; has displaced Red-necked Francolin *F. afer* in farming areas around Harare, Zimbabwe.

Francolinus swainsonii

Description. *F. s. swainsonii* (Smith) (including *damarensis*, *gilli* and *chobiensis*): N Natal, Transvaal, S Botswana, N Namibia and SW Angola). ADULT ♂: upperparts greyish brown (including upperwing-coverts), finely vermiculated dark brown, feathers edged greyish, with narrow black shaft streaks. Tail greyish brown, freckled and vermiculated darker. Face bare on lores and round eyes, otherwise pale greyish brown. Hindneck and sides of neck paler, broadly streaked dark brown. Lower throat greyish brown, feathers edged grey, with drop-shaped dark brown shaft streaks, these becoming larger on lower breast and flanks, washed dull chestnut. Belly and vent buffy-grey; undertail-coverts grey-brown, streaked darker. Flight feathers greyish brown, vermiculated darker on lower half; axillaries and underwing-coverts pale brown, streaked darker. Upper mandible blackish, lower mandible dull red; throat and face red; legs black. ADULT ♀: like ♂ but smaller, generally rather darker and more heavily marked; above vermiculated dark brown, incompletely barred dark brown on mantle, rump, upperwing-coverts and scapulars. SIZE: (58 ♂♂, 26 ♀♀) wing, ♂ 172–208 (191), ♀ 158–190 (174); tail, ♂ 74–100 (90·1), ♀ 71–92 (81·7); tarsus, ♂ 50–68 (60·5), ♀ 47–58 (52·1); spurs, 1–2 in ♂ (lower longer), 5–23 (17·1), 1 in ♀, 0·2–1·1 (0·6). WEIGHT: ♂ (n = 41) 487–875 (732), ♀ (n = 63) 365–650 (510).

IMMATURE: like adult but generally paler and duller, less chestnut, throat covered with downy buffy white feathers, underparts faintly barred black and white, legs yellowish brown, bill dark with yellow base.
DOWNY YOUNG: crown dark rich brown margined with black. Face pale buffy with dark eye-stripe. Crown-patch extends caudally, narrowing at nape, widening again as it proceeds along mid-back. This back streak flanked by 2 buffy stripes. Underparts buffy.

2 rather poorly defined subspecies, *F. s. swainsonii* and *lundazi*. Populations in N Namibia, Botswana and W Zimbabwe are locally variable, and show features intermediate between *swainsonii* and *lundazi*. In general, birds from Botswana and Namibia tend toward *swainsonii* and those from W Zimbabwe toward *lundazi*. Hybrids between *F. s. swainsonii* and Red-necked Francolin *F. afer* known from Harare, Zimbabwe, and hybrids with Natal Francolin *F. natalensis* from Bulawayo, Zimbabwe.

F. s. lundazi (White): S Mozambique and N and W Zimbabwe; like *swainsonii* but smaller, paler, more buffy brown, less spotted below. Wing, ♂ 187–197, ♀ 164–175.

Field Characters. A large, dull brown francolin, distinguished from all sympatric species except Red-necked Francolin *F. afer* by red face and throat; from Red-necked by black legs, black upper mandible and lack of prominent stripes on underparts. Immatures more difficult to distinguish; but generally paler above then those of Red-necked Francolin, with which it is often seen.

Voice. Tape-recorded (74, B, F, 305). ♂ advertisement call a hoarse, rather deep-toned, rasping croak 'kwaaark, ker-dowaaark, dowaaark'; and phrase 'kowaaark, kwarrk, kwarrk, kwaarrk, krrk, krrr' descending in pitch and diminishing in volume towards end. During breeding season ♀ sometimes responds to this call with a 'kwee ke-ke-kwe', which sounds like the call of a human baby. Advertisement call can be distinguished with experience from that of Red-necked Francolin, which is often higher pitched, otherwise rather similar. Adults give clucking notes as they feed and a 'qua-qua-qua-qua-quak' flight call. Immature birds with parents give mewing call which elicits approach of adults who respond with clucking call (Cardwell 1971).

General Habits. Common large francolin of most of warmer parts of southern Africa. In South Africa optimum habitat is dense grassland with nearby water and cultivation (especially maize and wheat); also found in thornbush with sparse to good cover, bushveld, and along open vleis and dambos. In Zimbabwe in thickets along edges of any woodland and in riparian forest. Usually most abundant in the flatlands along and near watercourses, or in rank long grass near vleis. More dependent on water than any other southern African francolin except Red-necked, coming to drink in early morning and/or late afternoon. Usually found in pairs or small coveys, usually 2–3 birds, up to 8 (probably adults with recently reared broods). May often consort with other francolins, e.g. Red-necked, Red-billed *F. adspersus* and Natal *F. natalensis*. Roosts in low trees and bushes; feeds and calls in early morning (up to *c.* 1100 h) and late afternoon (from *c.* 1500 h), remaining quiescent in heat of day. On wet mornings occasionally perches in bushes and trees basking in sunshine, presumably to dry out plumage. Wary, keeping to cover, and, if disturbed, running rather than flying; if forced, flies short distance holding head below level of body, then lands again, and at once runs. A very fast and manoeuvrable flier capable of out-flying raptorial predators (A. C. Kemp, pers. comm.). If caught by hawk may 'lie dead' until let loose and then flies away minus some feathers. More inclined to feed in cultivation and take grain than other francolins, and may then occasionally be crop pest, concentrating on sprouting grain.

Food. Seeds, berries, grass leaves, roots, bulbs, tubers, supplemented by spiders, insects and larvae (including locusts, ticks, millepedes, beetles, grasshoppers and termites); some small molluscs. If available, prefers maize (especially newly sprouting plants), sunflowers, peanuts, seeds of weeds, bulbs of watergrass, leaves of lucerne and clover. Also feeds in rhino and buffalo dung pats. Quantitative study by crop weight in Transvaal showed 30% agricultural products (mainly maize, wheat, beans), 25% indigenous seed (mainly of pioneer grasses such as *Urochloa*, *Eleusine*, *Panicum* and *Digitaria* spp. and *Juncus* and *Concorus* spp.), 14% roots and corms (mainly of *Cyperus* spp.), 7% arthropods and 2% green leaves. In summer (Dec–Feb) arthropods made up to 20% of crop weight. Agricultural products predominated in winter-spring (May–Sept) (Kruger 1981).

Breeding Habits. Monogamous, territorial; ♂♂ in aggressive display raise crown-feathers, and spurs are used as weapons in fighting as in Junglefowl *Gallus gallus* (J. R. Peek, pers. comm.). Courtship display similar to Red-necked Francolin, i.e. lateral display with drooping wings (J. R. Peek and B. Donnelly, pers. comm., van Niekerk 1983). ♂♂ attract ♀♀ by standing upright with bill pointing skyward and giving advertisement call. Bare throat skin is moderately inflated during call and appears to be a richer red during breeding season. ♀♀ respond by assuming same upright posture and giving the 'human baby' call (**A**). After 1–2 minutes of calling ♂ chases after ♀, adopting a low-intensity lateral courtship display, i.e. body somewhat crouched, bill pointed downward, tail depressed and mantle feathers moderately erected (**B**). If ♀ does not run away, ♂ may then give high-intensity lateral courtship display, i.e. head upright, bill pointed forward, wings fanned, almost dragging on the ground (**C**). Calling frequency increases with approach of laying season, dropping off after eggs hatch.

NEST: scrape on ground lined with grass, leaves and/or feathers, well hidden in dense grass, shrub or bush cover.

EGGS: normally 4–8, 1 clutch of 12 recorded; mean 5·5. Oval; rather granulated and rough with numerous white pore marks; cream to pinkish buff. SIZE: (n = 69, *F. s. lundazi*) 41·4–48·2 × 32·4–37·2 (43·9 × 35·6); (n = 42, *F. s. swainsonii*) 39·0–47·6 × 32·0–38·3 (44·1 × 35·9).

LAYING DATES: South Africa, usually Dec–May, peaking in Feb–Mar, occasionally in every month, may breed twice in 1 year (Mentis 1970); Swaziland, late Mar; Botswana, Feb, Mar, May; Namibia, Apr, May; Zimbabwe, Nov–Aug, peaking in Feb–Mar; Mozambique, Jan, Apr, May, July, Aug, Dec; Zambia, May–July, Dec.

INCUBATION: probably by ♀ only. Period: *c.* 21 days.

BREEDING SUCCESS/SURVIVAL: relatively high, mean brood size of chicks in Transvaal *c.* 6 (Kruger 1981). Major predators monitor lizards, mongooses, snakes, baboons and Ground Hornbill *Bucorvus leadbeateri* (Peek 1972).

References
Clancey, P. A. (1967).
Kruger, F. J. (1981).
Peek, J. R. (1972).
Van Niekerk, J. H. (1983).

Plate 2

(Opp. p. 33)

Francolinus afer (P. L. S. Müller). Red-necked Francolin; Red-necked Spurfowl. Francolin à gorge rouge.

Tetrao afer P. L. S. Müller, 1776. Natursyst., Suppl, p. 129; Benguella, Angola.

Forms a superspecies with *F. leucoscepus*, *F. rufopictus* and *F. swainsonii*.

Range and Status. Endemic, resident; extending from Uganda, NE coastal Kenya and S Zaïre southwards through central Africa to N Namibia in west, and, in east, to E Cape Province. Introduced onto Ascension Island. Common to abundant. Less hunted than other bare-throated francolins because of preference for dense vegetation, but has decreased near Harare, Zimbabwe, perhaps as a result of competition with Swainson's Francolin *F. swainsonii* which has expanded its range into areas modified by agriculture. Numbers in Zaïre reduced due to poaching, especially along roads (Verschuren and Mankarika 1982).

Description. *F. a. castaneiventer* (Gunning and Roberts) (including *notatus* and *lehmanni*): S and E Cape Province north to Limpopo River in E South Africa. ADULT ♂: crown dark grey, forehead and crown-stripe black; sides of head, nape and sides of neck dark grey, feathers edged white (especially birds from N Transvaal and Swaziland); ear-coverts silvery grey. Mantle, scapulars, upperwing- and uppertail-coverts greyish olive-brown, each feather with blackish shaft streak, narrower on uppertail- and upperwing-coverts. Tail dull brown, indistinctly vermiculated darker. Lower throat and upper breast dark olive-grey, each feather with a black shaft streak flanked by streaks of white. Rest of breast, belly and flanks blackish brown with white streaks; birds from E Cape have black and white abdominal feathers edged chestnut. Lower belly to undertail-coverts and thighs greyish brown, feathers streaked and vermiculated dusky; axillaries and underwing-coverts greyish brown. Primaries and secondaries greyish olive-brown, paler on outer webs of primaries, vermiculated on inner webs of secondaries, forming somewhat paler wing-patch seen in flight, tertials with broad black shaft streaks. Bill red; bare skin of face and throat red or red-orange; eye brown; legs bright red. ADULT ♀: like ♂ but smaller. SIZE: (13 ♂♂, 8 ♀♀) wing, ♂ 171–210 (192), ♀ 170–188 (182); tail, ♂ 90–110 (102), ♀ 90–105 (96); tarsus, ♂ 56–64 (60), ♀ 51–64 (56); spurs, 1–2 in ♂ (lower longer), 5–19 (13·6), 1 in ♀, 0·3–1·4 (0·8). WEIGHT: (South Africa) ♀ (n = 1) 465, unsexed (n = 14) 444–765 (586); (SE Zaïre) ♂ (n = 3) 480–585, ♀ (n = 1) 370.

IMMATURE: generally duller, browner; throat still retains some white feathers; flight feathers mottled and barred with white. Bill dark horn; legs yellow.

DOWNY YOUNG: crown brown, crown-stripe black, hindneck blackish. Face buff, eye-stripe black, extending to encircle neck below in some chicks. Broad dark brown stripe down centre of back vermiculated and edged blackish, flanked by ochre-buff stripes; sides brown; throat buff; underside yellow-buff with some black blotches; wings patched brown and buff. Legs pinkish yellow.

2 well marked subspecies groups, named Black-and-white and Vermiculated/rufous-striped Groups (Hall 1963), intergrade along 2 broad fronts. One front extends from central Tanzania through central Malaŵi into Luangwa Valley, other

from NW Zimbabwe through N and central Angola. Members of much more variable Black-and-white Group are characterized by strongly marked black and white underparts and (except *F. a. afer*) comparatively large size (wing length usually exceeding 180 mm). Members of Vermiculated/rufous-striped Group are characterized by vermiculated underparts with rufous streaking, and smaller overall size (wing length usually less than 180 mm).

Black-and-white Group

F. a. afer (P. L. S. Müller): W Angola to extreme NW Namibia, intergrades with *F. a. cranchii* in highlands of W Angola. Like *castaneiventer* but moustachial stripe and supercilium white; upper breast pale brown, rest of underparts black, feathers edged white. Much smaller, wing, ♂ (n = 14) 157–181 (168), ♀ (n = 9) 153–175 (159).

F. a. swynnertoni (Sclater): interior of Mozambique south of Zambezi west to SE Zimbabwe at Headlands and Marandellas. Occurs up to 2200 m. Forehead, supercilium, face, sides of neck white; upperparts paler and greyer than in *castaneiventer*; breast and flank feathers white, finely vermiculated with grey and broadly streaked with black. Lower breast and belly black. Size similar to *castaneiventer*. Hybrids between this subspecies and Swainson's Francolin collected in wild near Harare, Zimbabwe. WEIGHT: ♂ (n = 1) 801.

F. a. melanogaster (Neumann) (including *loangwae*): Mozambique north of Zambezi, and E Tanzania north to Pangani, west to Korogwe, Mahenge, Songea; intergrades with *cranchii* near Songea and in Luangwa Valley in E Zambia. Like *swynnertoni* except supercilium and rest of face black.

F. a. leucoparaeus (Fischer and Reichenow): coastal Kenya from Tana River to Tanzanian border; like *melanogaster* but supercilium black and white, and face white with little black mottling. WEIGHT: ♂ (n = 3) 800–850, ♀ (n = 2) 525, 610.

Vermiculated/rufous-striped Group

F. a. cranchii (Leach) (including *intercedens*): W Congo through central Zaïre to Uganda north of Lake Victoria, and W Kenya and NW Tanzania round Lake Victoria; south to Angola, east through Zambia west of Luangwa Valley, N Malaŵi, S and SW Tanzania. Forehead and supercilium black or dark brown; face black mottled whitish; below streaked roufous, with greyish vermiculations. Smaller than black-and-white-bellied subspecies except *afer*. Wing, ♂ (n = 179) 157–198 (179), ♀ (n = 81) 149–179 (163). WEIGHT: (Kenya) ♂ (n = 2) 765, 765, ♀ (n = 1) 650; (Zaïre) ♂ (n = 3) 480–583, ♀ (n = 3) 370–412.

F. a. harterti (Reichenow): from north end of Lake Tanganyika in Ruzizi Valley to Bujumbura, Uvira, Baraka, and Kasulu in Tanzania. Like *cranchii* but smaller and chestnut streaking below darker, more maroon colour. Wing, ♂ (n = 20) 156–179 (171), ♀ (n = 10) 147–177 (160).

Field Characters. A large francolin, in southern Africa likely to be confused only with Swainson's *F. swainsonii*; distinguished by wholly red bill and bright red legs. Darker than either Grey-breasted *F. rufopictus* or Yellow-necked Francolin *F. leucoscepus* in Kenya-Tanzania, and, where ranges overlap, prefers moister habitats, e.g. thickets, scrub, and forest edges, rather than open steppe country. In flight, lacks conspicuous pale rufous patch in wing of Yellow-necked Francolin. May occur in same habitat as Scaly Francolin *F. squamatus*, from which distinguished by red not black bill, bare red throat; also larger, much darker, and has heavily streaked underparts.

Voice. Tape-recorded (20, 75, F, 305). Similar to voice of Swainson's, Grey-breasted, Yellow-necked and Red-billed *F. adspersus* Francolins. Generally higher-pitched, more squealing cackle, 'ko-waaark' repeated 4–8 times with accent on 2nd syllable. Said by Chapin (1932) to be almost identical with calls of Heuglin's Francolin *F. icterorhynchus*, a hoarse, croaking, slowly repeated 'k-rack-k-k, k-rack-k-k'.

General Habits. Habitat varies geographically. In southern Africa, wooded gorges and fringes of evergreen forest. In Zambia, *F. a. cranchii* in almost any habitat, especially rank grass and *Brachystegia* and Mopane woodlands; often alongside Swainson's *F. swainsonii* and other francolins. In Zimbabwe, thickets and areas with herbaceous cover, usually in moist situations and in evergreen growth. In Zaïre and Uganda in grassy plains with thickets. In Tanzania, bushed and wooded grassland; and coastal Kenya, in mixed long grass and forest patches. Where commonest large francolin (in most of tropical central Africa south of equator) it may inhabit several vegetation types; but, where it occurs alongside Swainson's or Yellow-necked Francolin, generally prefers denser cover in moister areas, especially along stream beds. Adaptable to limited human occupation and frequents cultivation freely, especially when near dense cover into which it can easily escape. Disappears from heavily inhabited cultivated country.

Roosts in trees or bushes inside dense cover. At dawn, or soon after, descends to ground, then forages until heat of day, and again in evening. Digs for tubers and bulbs and may occasionally damage crops by stripping pods or fallen or laid heads; but generally only gleans fallen grain. Calls mainly in early morning (even before dawn) and late afternoon. Generally hard to observe, remaining mainly inside dense cover, seldom seen in open for more than few moments; but in some areas (e.g. Uganda) more inclined to emerge into short cropped grasslands, and in other areas often seen on roads, where it may resort to dust. Hard to flush, preferring to escape by running in cover; but when pressed flies short distance, 50–100 m, alighting again in any available dense cover. Takes refuge from ground predators, e.g. dogs, in dense leafy trees, remaining silent. Calls mainly from ground, but also termite mound or other prominence. Normally occurs in pairs, at most small parties (probably adults with grown broods), never in large flocks; may associate with other species, especially Swainson's Francolin in southern Africa, in mixed groups.

Food. In Zaïre mainly small tubers, probably of sedges (*Cyperus* spp.). In southern Africa vegetable matter predominates, especially tubers and bulbs, shoots, berries, roots, seeds, fallen grain or legume crops, supplemented by insects and larvae, termites (even in flight), and some land molluscs. In Zimbabwe, during wetter/warmer months focuses on invertebrates, fresh grass shoots and seeds. Gleans ticks off grass stalks. In relatively cooler and drier months shifts to fallen seeds and underground bulbs and roots (Peek 1972). Also feeds in rhino and buffalo dung pats.

Breeding Habits. Monogamous, territorial, ♂ defending nesting area and aggressive to other ♂♂. Pairs also tend to remain in same area outside breeding season. Early in breeding season pair remains mainly in dense cover, frequently dustbathing. Courtship said to be similar to that of Junglefowl *Gallus gallus* (Peek 1972); ♂ gives lateral display with wings lowered to ground, though display somewhat more frontal than that of Junglefowl. During courtship, ♂ calls with all plumage sleeked, neck stretched up, flank-feathers fluffed out and whitish cheek-patches clearly displayed. Approaches apparently disinterested ♀ with high-stepping gait, neck upstretched. ♂ courtship-feeds ♀ (in captivity, with mealworms). Calling frequency drops from 20–25 bouts per day during courtship to 1–3 during laying period.

NEST: on ground, in small scrape made by ♀, in long grass usually at foot of bush or tree, usually partly lined with some weed or grass stems, few feathers.

EGGS: 3–9, laid on alternate days. Oval, with very thick shells; pinkish buff to uniform light brown, heavily pitted white. SIZE: (n = 11, *F. a. cranchii*) 40·3–46 × 32·7–34·9 (43·4 × 33·9); (n = 2, *F. a. leucoparaeus*) 34·5–40 × 29·5–30·5; (n = 24, *F. a. castaneiventer*) 40·4–46·5 × 30·2–37·2 (45·1 × 35); (n = 13, *F. a. swynnertoni*) 43·7–47·5 × 33·9–36·4 (45·5 × 35·2).

LAYING DATES: Natal, E Cape, Apr–Aug (cool dry season); Zimbabwe, Jan–July, Nov–Dec, wet and dry seasons, peaks in Apr and Dec; Angola, Mar–June, possibly until Aug; Zambia, every month except Aug–Oct, peaking in Feb–May late in or just after rains; Malaŵi, Jan–Aug, peaking in Feb–Jun; East Africa: Uganda, Region B, virtually every month with peak in Dec–Jan; Tanzania, Region C, Feb–Apr; July–Aug; Kenya, Regions D and E, May–July, Dec; N Zaïre, Dec–Apr, peaking in Mar, at end of rains or beginning of dry season. Most laying occurs late in rains, with chicks hatching when there is dense cover early in following dry season.

INCUBATION: begins with completion of clutch, probably by ♀ only. Period: in captivity 23 days.

DEVELOPMENT AND CARE OF YOUNG: eggs hatch together, and development rapid at first, slower later. Within 2 days first pin-feathers burst sheaths; at 4 days wing-feathers appearing; 6 days, bare patch behind eye visible; 8 days, flight feathers reach end of body and contour feathers sprouting on back and shoulders. Can fly at 10 days, primaries then project beyond tail. At 13 days feathers on neck, sides of body and rump opening; 18 days, bald patch behind eye and feathers on thighs developing; bill dark, tipped pale. 23 days, head downy, rest of body feathered. 39 days, showing sibling aggression, bare red facial skin now showing. At 45 days resemble large quail, below light grey spotted black, above mottled sandy brown; bill black, bare facial skin pink. 106 days, earth brown above, naked throat patch now showing, upper breast striped. 130 days, almost as big as adult ♀, face orange-red, small red throat-patch, culmen dark, sides of bill reddish; neck speckled, underside striped and spotted. Almost fully grown at 3–4 months; probably breeds at 1 year old. Both parents accompany broods, remaining with them until they are well grown, almost as big as adults. These family groups form small coveys of 4–6, but do not regularly join to form larger flocks. Families remain together perhaps to near beginning of next breeding season.

References
Clancey, P. A. (1967).
Peek, J. R. (1972).
Roles, D. G. (1973).

Plate 2
(Opp. p. 33)

Francolinus rufopictus (Reichenow). **Grey-breasted Francolin; Grey-breasted Spurfowl. Francolin à poitrine grise.**

Pternistis rufopictus Reichenow, 1887. Journ. f. Orn., 35, p. 52; Wembere Steppe, Tanzania.

Forms a superspecies with *F. afer*, *F. leucoscepus* and *F. swainsonii*.

Range and Status. Endemic, resident in plains and savannas of NW Tanzania from Serengeti and Crater Highlands to Wembere steppes and near Mwanza. Generally common, locally only frequent; has suffered in recent years due to habitat destruction resulting from spread of dense cultivation and pastoralism eastwards. Hybridizes with Yellow-necked Francolin *F. leucoscepus* in SE Serengeti National Park (Turner 1977, Britton 1980, B. Stronach, pers. comm.).

Description. ADULT ♂: forehead very dark brown, grading lighter on crown and nape, crown-stripe very dark brown, grading to white anteriorly. Upperparts grey, feathers with greyish black shaft streaks, and 3 narrower streaks of buff, greyish black and chestnut. Tail grey, barred and vermiculated irregularly with black. Moustachial streak white; breast grey streaked black; lower breast, belly and underside grey, feathers with central greyish black shaft streak and three narrower streaks of buff. Primaries and secondaries grey, irregularly barred and vermiculated with cinnamon to tips.

Francolinus leucoscepus 73

Francolinus rufopictus

Bill orange; bare skin round eye and throat orange yellow to coral pink; eye brown; legs brownish black. ADULT ♀: like ♂ but smaller. SIZE: (18 ♂♂, 24 ♀♀) wing, ♂ 193–222 (213), ♀ 180–199 (190); tail, ♂ 90–102 (95), ♀ 81–94 (86); tarsus, ♂ 62–72 (68), ♀ 54–63 (58), spurs, 2 in ♂ (lower longer), 10–20 (16·8), 1 in ♀, 0·2–1·3 (0·7). WEIGHT: ♂ 779–964 (848), ♀ 439–666 (588).

IMMATURE: like adult but upperparts grey-black with white central shaft-streak and barring and grey-buffy margins. Underparts broadly barred black and white.

DOWNY YOUNG: crown dark rufous-brown margined with darker brown. Sides of head buffy with 2 dark brown eye-stripes, 1 above, 1 below eye. Nape and central back stripe dark brown with 2 parallel stripes of buffy and dark brown; underparts buffy.

Field Characters. Large, very like Yellow-necked Francolin, and with similar habits; differing in orange (not black) bill and pink bare skin on throat and around eye. Distinguished from Red-necked Francolin *F. afer* by pink or orange bare skin, black legs, generally paler, duller colour.

Voice. Tape-recorded (C). Similar to that of Yellow-necked Francolin, loud, grating, 'ka-waaaark, ka-waaark, ka-waarrrk' or 'koarrrk-koarrrk-karrkkrrk-krrk-krrr' descending towards end. Alarm call when flushed a higher-pitched cackle.

General Habits. Inhabits grassland/acacia woodland ecotone (e.g. *Acacia tortilis*, *drepanolobium*, *xanthophloea* and *kirkii*), thickets of thornbush or other dense vegetation along watercourses. Lives in areas of higher (monomodal) rainfall (500–700 mm) than Yellow-necked Francolin. Ventures out into grassland only in early morning and late afternoon. Found singly, in pairs, or in small groups; may gather in considerable numbers in favoured feeding localities.

Food. Principally sedge *Cyperus* tubers, obtained by digging. Also seeds of grasses and weeds, some insects, including grasshoppers and termites. Feeds in cultivation where this not too dense, taking fallen grain and legume seeds, but also weed seeds.

Breeding Habits. Monogamous and territorial; calls morning and evening year-round from elevated points such as termite mounds and stumps. Calling intensifies with onset of rains, suggesting pre-breeding advertisement activity. Courtship displays similar to those of Yellow-necked Francolin.

NEST: scrape in long grass, lined with few bits of grass and feathers.

EGGS: 4–5. Rather rough to the touch; buff or pale brown, with chalky white pore spots. SIZE: (n = 3) 42·7–44·5 × 35·1–35·2 (43·4 × 35·2). WEIGHT: *c.* 22.

LAYING DATES: Feb–Apr, June, July, late in main rains, and in subsequent dry season.

DEVELOPMENT AND CARE OF YOUNG: broods of young accompanied by both parents; family parties form small coveys of up to 5–7, usually less. Immatures stay with parents until onset of next breeding season.

Reference
Schmidl, D. (1982).

Francolinus leucoscepus G. R. Gray. Yellow-necked Francolin; Yellow-necked Spurfowl. Plate 2
Francolin à cou jaune.
(Opp. p. 33)

Francolinus leucoscepus G. R. Gray, 1867. List Bds. Brit. Mus., p. 48; Ethiopia.

Forms a superspecies with *F. afer*, *F. rufopictus* and *F. swainsonii*.

Range and Status. Endemic, resident in Somalia, Ethiopia (including Eritrea), SE Sudan, NE Uganda, central Kenya and drier parts of N central Tanzania. Locally common, e.g. along eastern shores of L. Turkana, but generally only frequent. Adapts to moderate human population density. Becomes more common in cultivated country, but disappears in vicinity of very dense human populations. Has disappeared in last 30 years from large areas of Kenya where formerly abundant, disappearance hastened recently by availability of nylon for snares. Can easily be trapped at narrow entrances to cultivation through brushwood fences, and now thrives best in overgrazed pastoral areas. Not threatened at present.

Description. ADULT ♂: crown dull grey-brown (Eritrean populations with white shaft streaks), crown-stripe white flecked with brown; ear-coverts pale grey. Sides of neck brownish black, feathers edged white. Upperwing-coverts brown, centrally chestnut; rest of upperparts dark brown, feathers with conspicuous buff shaft streaks. Tail brown vermiculated with buff. Chest, breast, and most of underside dark brown, feathers broadly streaked with pale buff (broadening to form triangular patch at tip) and washed chestnut on flanks. Primaries, primary coverts and secondaries dark brown, outer primaries broadly edged buff, and large buff patch formed by plain buff inner webs; secondaries brown, vermiculated buff on outer webs. Bill black, base red; bare skin of throat yellow, around eye more orange; eye brown; legs dark brown to black. ADULT ♀: like ♂ but smaller. SIZE: (54 ♂♂, 38 ♀♀) wing, ♂ 184–216 (200), ♀ 170–216 (187); tail, ♂ 85–110 (102), ♀ 80–95 (91·2); tarsus, ♂ 61–67 (64·4), ♀ 53–69 (57·8); spurs, 2 in ♂, 8–22 (16·5), 1 in ♀, 2–8 (3·6). WEIGHT: ♂ (n = 173) 615–896 (753), ♀ (n = 223) 400–615 (545).

IMMATURE: like adult but less clearly marked, upperparts grey-buffy vermiculated with black, and feathers with white shaft streak and ill-defined black bars. Underparts grey, feathers with broad white central shaft streak. Throat and bare facial skin paler yellow, legs brown.

DOWNY YOUNG: forehead dark brown; crown light brown with dark brown crown-stripe; single eye-stripe extending from posterior eye margin to nape; nape and broad stripe down central back dark brown with 2 parallel stripes of buffy and dark brown on either side. Underparts buffy.

Several subspecies have been described, but are regarded as clinal variants, and therefore do not warrant taxonomic recognition.

Field Characters. A large francolin with yellow bare skin on throat and around eye. Within range usually the commonest francolin. Paler than Red-necked *F. afer*, and has blackish brown legs and blackish bill. Distinguished from Grey-breasted *F. rufopictus* in Tanzania by yellow bare skin and blackish bill; but hybrids on border of range may show varied characters. In flight, buff patch in wing conspicuous.

Voice. Tape-recorded (5, C, 217, 253, 264, 308). Commonest call is ♂ advertisement call, deep grating, repeated bisyllabic 'ko-warrrk, ko-warrk, ko-weeark', very much like advertisement call of Swainson's Francolin *F. swainsonii*; uttered from an eminence, or on ground within cover. Also, longer series of calls 'ko-weerrrrk-kweeerrrrk-kwerrrk-kwarr-karr-karr' falling away in pitch and volume towards end. Voice generally duller, rather lower-pitched than that of Red-necked Francolin *F. afer*, with which it overlaps in coastal Kenya. Sometimes a 'kerak' intermediate between its typical call and that of Clapperton's Francolin *F. clappertoni*.

General Habits. Preferred natural habitat light *Acacia/Commiphora* bushland with annual and some perennial grasses and rainfall of 200–400 mm/year (e.g. near Uaso Nyiro R. and Isiolo, N Kenya, or Selengai, S Kenya). More adaptable than often stated, though not occurring in very tall grass savanna with *Hyparrhenia-Combretum* and *Terminalia* associations. Found around slopes of Mt Kenya in cultivation and near forest edges up to 2400 m, and in rainfall up to 1500 mm/year or more. Adapts well to, and becomes commoner in, lightly populated woodland with some cultivation, interspersed with bush; but as human population, and probably hunting pressure, increase, it disappears, perhaps mainly through reduction of suitable dense cover for unmolested breeding. May be more abundant on metamorphic soils with abundant grit than on volcanic soils in similar rainfall; but most abundant on some seasonally waterlogged volcanic soils in W Kenya, Ethiopia, where sedges common.

Normally found in pairs, singly, or in small family parties of adults and young, up to 5–6. Larger numbers occur in favoured feeding areas to which many birds resort, often from some distance.

Roosts in low trees or bushes. Calls from termite mound or low tree branch at dawn, either from roost, or soon after descending to ground; and may walk or fly several hundred m to favoured feeding ground, e.g. patch of cultivation. Feeds for 1–1·5 h, then retreats to cover near nocturnal roost until late afternoon. On cloudy mornings feeding time extended by up to 2 h. On wet mornings, when grass soaking, often emerges into open spaces, e.g. roads, presumably to dry plumage; and in soft-soil areas lacking grit may come to roadsides to obtain gravel. In heat of day rests in shade of bushes, usually concealed among ground vegetation, or by trees with low branches, and may dustbathe. Seldom drinks, and can subsist in areas totally devoid of free water, obtaining water requirements from food; concentration along watercourses more due to vegetation than available water. Often follows elephants or rhinoceroses, at once scratching among fresh droppings probably to obtain undigested material; and where not persecuted may similarly approach motor vehicles, presumably mistaking them for large mammals! If disturbed, prefers to escape by running into and through cover; but flies readily, sometimes seeking refuge in tree. Flight swift, direct, often more than 100 m. Feeds again in late afternoon and evening, from *c*. 2–3 h before dark, usually in same areas as in morning. Well before dusk, returns towards roosting area and usually calls again before roosting. Roosts in same general area nightly, though not necessarily in same tree. Pairs normally frequent same areas for most of year. Entirely sedentary; and same area may attract different birds year after year, if habitat not destroyed.

Forages in cultivation by walking slowly about, picking up fallen grain, and, in semi-arid uncultivated areas, largely by scratching up soft ground with powerful legs, exposing favourite food of small sedge tubers. Catches emergent alate termites as they leave nest-hole, sometimes by running about, making short jumps into air and seizing them on wing. Formerly fed on locust hoppers, but was then killed in large numbers by control measures using bran poisoned with arsenic (L. H. Brown, unpub.). Occasionally climbs on leaning sorghum stems, or picks ripe grain from heads of *Eleusine* millets; but generally not a crop pest. Rarely, digs up recently sown seed and can do damage if germination long-delayed. Persecuted by African cultivators to point of extinction in many areas; not because a crop pest, but for food.

Food. Quantitative analysis of crop contents of birds collected in semi-arid natural bushland habitat, S Kenya and in Tanzania (Swank 1977, Stronach 1966): mainly tubers of sedges especially *Cyperus rotundus* (52·3% year-round by volume); fruits and seeds of herbs and grasses, notably *Commelina* sp., *Urochloa* sp., *Oxygonum* sp., 12·7%; other vegetable matter, including different parts of *c*. 28 spp. of plants, mainly seeds, 16·2% insects 18·8% by volume, of which 17·5% termites. Main dry-season food sedge tubers, but following rain, proportion of sedge tubers in food may fall to 7–18% by volume. More grass and herb seeds, and especially more insects then taken. Immediately after rain, and before grasses and herbs have seeds, insects become especially important. Elsewhere, may feed mainly on fallen grain or legume crops in cultivation at and after harvest.

Breeding Habits. Monogamous; ♂ and ♀ live together in territory for breeding season. Onset of rains probably stimulates breeding activity. Calling by advertising ♂♂ loud and continuous especially on fine mornings after night rainstorms. ♂♂ call from termite mounds, stumps, and fence posts, or form ground, occasionally from roost-trees. ♂ courtship-feeds ♀, and displays to her in open areas by rearing into upright position, arching neck slightly downward, lowering and partly spreading wings, exposing buff patches on feathers; runs round ♀ to face her repeatedly. She may crouch and copulation then may occur, ♂ mounting with flapping wings, holding feathers of back of ♀'s head.

NEST: simple scrape in ground, unlined or with a few grass stems and feathers.

EGGS: normally 3–8, usually 5; 1 nest with 17 from 2 ♀♀. Slightly pointed ovals, shells thick and very hard; cream or pale pinkish buff, speckled darker, with white pore marks. SIZE: (n = 17) 43·0–47·3 × 32·0–37·8 (45·2 × 35·5); WEIGHT: *c*. 31.

LAYING DATES: Ethiopia, Jan–June; Eritrea, Jan, Apr; Somalia, chiefly Apr–June, also Nov; Kenya, N Tanzania, every month but Feb, peaking in May–July. Breeds generally late in rains, with hatching in subsequent dry season; if choice of breeding season available, prefers hatching in cooler dry season. In some arid areas, if rainfall insufficient only a few pairs and no breeding at all.

INCUBATION: period 18–20 days.

DEVELOPMENT AND CARE OF YOUNG: primaries apparent at *c*. 2 weeks when chicks begin to fly short distances. At 4–6 weeks young have full set of juvenile primaries; outer juvenile primaries are pointed. From week 6 to 24 primaries moulted continuously until only outermost left, which is shed by week 28 when the birds reach adult plumage. Spurs appear at *c*. week 18 but are not sharp until 10–12 months, and continue to grow into 2nd year. Young leave nest within 24 h, remain with both parents and are defended, especially by ♂, until almost same size as adult ♀. In areas with 2 rainy seasons, family parties or coveys may break up after 6 months when fresh rains occur; but in monomodal rainfall areas may remain together longer (7–8 months), separating only at onset of following breeding season.

BREEDING SUCCESS/SURVIVAL: many broods fail (mean brood size 4·3 at 10–14 weeks) and survival to near adult size probably averages not more than 2 chicks/pair/year. Young can probably breed within 1 year, perhaps even earlier in bimodal rainfall regimes.

References

Archer, G. and Godman, E. M. (1937).
Jackson, F. J. and Sclater, W. L. (1938).

Order GRUIFORMES

A heterogeneous assemblage of small and large wading and terrestrial birds. Diagnostic characters anatomical; include several skeletal, circulatory, muscular, and digestive features such as lack of caeca and crop (for summary of principal anatomical features, see Sibley and Ahlquist 1972). Anterior toes very incompletely or usually not at all webbed; hind toe if present usually elevated. Of ancient age; polyphyletic; divided into 6–10 suborders with some possibly better treated as separate orders; 12 families (Wetmore 1960, Storer 1971, Feduccia 1980, Cracraft 1981). In Africa 2 suborders and 5 families: Turnicidae, Rallidae, Gruidae, and Heliornithidae in Suborder Grues and Otididae in Otides. Relationship to other orders unclear but probably closest to Charadriiformes and Galliformes.

Suborder GRUES

Small to large wading and terrestrial birds; bill variable in shape; nostrils impervious or pervious; lores feathered or with bristles; adult down feathers on both pterylae and apteria; tail short or reduced, consisting of only a few soft feathers; wing rounded; tertiaries sometimes as long as primaries; oil gland present; tarsal scales transverse. 8 families world-wide, 4 in Africa.

Family TURNICIDAE: button-quail

Very small, quail-like ground birds. Body short and rounded, wings short and broad, with 10 primaries, 8–10 secondaries; eutaxic. Flight normally rather poor and not sustained. Tail short, bill rather slender, shorter than head; nares impervious. Legs short but robust; hind toe lacking. Oil gland feathered; feathers with aftershaft; crop absent. ♀ usually larger than ♂. ♀ has enlarged trachea and inflatable bulb in oesophagus to amplify and project voice. Plumage cryptic; ♀ more strongly patterned and brighter than ♂; juvenile plumage often different from that of adult; adult plumage attained within 1 year.

Secretive, occurring in grassland, savanna and scrub and also denser vegetation. Run rapidly; difficult to flush and do not fly far; crouch and rely on cryptic colouration for concealment. Non-migratory except for some populations of Asiatic *T. tanki*, but many species nomadic. Eat seeds, grain and green shoots, small invertebrates. Drink continuously without raising head. Solitary or in family groups; where nomadic, occur in flocks. Behaviour little known except for studies in captivity. Thought to be successively polyandrous, but monogamous pair-bonds also likely at times. ♀ takes leading role in territorial and courtship activities. Breeding season prolonged; can nest opportunistically in any month. Nest on ground, a scrape lined with vegetation; sometimes with dome and even short approach runway; both sexes build. Eggs oval, smooth, glossy, usually well marked; clutch 2–7 in *Turnix*, 2 in *Ortyxelos*; egg laid daily until clutch complete. ♀ may lay for 2 or more ♂♂. Incubation 12–15 days, usually starting when clutch complete; by ♂ but ♀ may participate in early stages; hatching synchronous. Young cared for by ♂; often independent before fully grown and soon capable of flight. Age of 1st breeding may be as little as 4 months.

15 spp. in 2 genera; monotypic *Ortyxelos* endemic to Africa, *Turnix* mainly Indo-Australian, with 2 spp. in Africa (1 endemic).

Genus *Ortyxelos* Vieillot

Wings longer and broader than in *Turnix*, with conspicuous pattern of black and white, flight wavering and erratic. Plumage sandy brown above. Tail longer than *Turnix*, rectrices stiffer. Call, a low whistle.

Ortyxelos meiffrenii (Vieillot). Quail Plover; Lark Quail. Turnix de Meiffren.

Turnix meiffrenii Vieillot, 1819. N. Dict. Hist. Nat., 35, p. 49; Senegal.

Plate 7
(Opp. p. 112)

Range and Status. Resident and intra-African migrant, commonest in sahel belt between 12°N and 17°N, in arid and semi-arid scrubland and grassland, with discontinuous population in E Africa and S Ethiopia. Uncommon and local Mauritania; widespread and fairly common Senegal; locally common Mali; uncommon Niger, extreme N Nigeria from Sokoto to L. Chad, N Cameroon (Waza Nat. Park); widespread central Chad; locally common Sudan (Darfur, Kordofan); uncommon and local Ethiopia, Uganda (Karamoja) and NW to SE Kenya. Records from coastal Gambia, Ghana, Ivory Coast and, recently, coastal Kenya (G. A. Allport, pers. comm.) suggest irregular seasonal occurrences in wetter coastal grasslands (breeding?). Apparently absent from likely habitat between White Nile and Ethiopian highlands and from arid NE Africa. Range in Tsavo East Nat. Park, Kenya, has recently expanded southward (Lack 1975).

Description. ADULT ♂: forehead, crown, nape and hindneck rufous-brown, edges of feathers cream with black inner border. Lores, face and broad supercilium cream, tinged golden buff. Streak behind eye to side of neck rufous-brown; ear-coverts washed pale rufous-brown. Chin and throat white, throat feathers tipped pale golden buff. Mantle, scapulars and back rufous-brown, paler on upper mantle; feathers broadly fringed cream with black inner border. Pattern variable, some mantle and back feathers having cream spots, bordered black, instead of cream fringe, or variable cream spots adjacent to shaft, or black transverse bars or blotches. Rump and uppertail-coverts paler than back, uniform pale rufous-brown, with narrow faint buff tips. Tail pale rufous-brown, outermost 2 rectrices with buff to white fringe and outermost rectrix with dusky brown submarginal mark on outer web. Other rectrices tipped cream, with 2–3 transverse cream bars on distal half varyingly bordered black. Breast golden buff, feathers with rufous-brown tips to outer webs and white tips to inner webs giving a spotted effect. Sides of breast darker, more rufous-brown, with black-bordered cream spots; lower breast and belly cream, shading to white on flanks and undertail-coverts. Primaries blackish, outermost with outer web, tip, central half of inner web and most of shaft white, P6–9 with central patch, tip and narrow edge of outer web buff, darkening to rufous-buff at tip, with dark subterminal markings, P1–5 with broad white tip and rufous-buff spot at base of inner web. Secondaries blackish with broad white tips and narrow white edges. Tertials rufous-brown with tips and transverse bars cream, bordered black. Marginal upperwing-coverts rufous-brown, tipped cream; median and lesser upperwing-coverts cream with rufous-brown centres; greater upperwing-coverts white, tertial coverts with rufous-brown markings. Axillaries white; underwing-coverts cream. Bill yellowish horn, greenish grey or pale green, with bluish brown culmen; eye pale to rich brown; legs and feet whitish flesh to flesh or creamy yellow. ADULT ♀: as adult ♂ but breast deeper rufous-brown and outer 3 rectrices shaped differently, narrowing towards the tips, with dusky brown submarginal marks and broad white fringes. SIZE (4 ♂♂, 4♀♀): wing, ♂ 72–76 (73·0), ♀ 76–80 (77·8); tail, ♂ 29–32 (30·3), ♀ 33–36 (34·8); bill (culmen to base of feathers), ♂ 8, ♀ 7·5–8 (7·9); tarsus, ♂ 17–19 (18·0), ♀ 18–20 (19·0). WEIGHT: (2 ♂♂) 15·7, 19·5.

IMMATURE: upperparts considerably paler, more sandy or browner and less rufous than adults; feathers broadly fringed white and more vermiculated; rufous at sides of breast paler; wing markings less regular.

DOWNY YOUNG: not described.

Field Characters. Both sexes readily distinguished from button-quail *Turnix* spp. and all other ground-living grassland and scrubland species by combination of wing-pattern and flight action: blackish flight feathers with white tips, white on outermost primary and buff patch on outer primaries, white greater coverts; flight fluttering and erratic, reminiscent of a bush-lark *Mirafra* or even a large butterfly, an effect enhanced by short tail. Wings proportionately longer and broader than in button-quail, which have low whirring flight. May rise to some height and fly further than button-quail, dropping to ground like a bush-lark. On the ground, distinguished from quail *Coturnix* spp. by rufous-brown vermiculated upperparts, white on underparts and greater upperwing-coverts; from button-quail by broad buff supercilium and white greater coverts; from both by superficial resemblance to miniature courser *Cursorius* in pose and gait.

Voice. Not tape-recorded. Call a very soft low whistle, uttered on the ground; silent when flushed.

General Habits. Inhabits arid or semi-arid grassland, bushed grassland, thin scrub and acacia savanna; also quite dense bushland, sometimes with a few trees, and wetter coastal grassland. In Sudan never far from 'heskanit' grass *Cenchrus catharticus* (Lynes 1925); in Chad prefers *Aristida* grassland, especially *A. papposa* (common association with that grass is indicated by its local arabic name, translated 'father of *A. papposa*': Newby 1980). Can survive in arid areas without access to water. Altitudinal range 0–2000 m. Outside breeding season occurs singly or in pairs. On ground moves rapidly through grass in crouch, but also runs fast over open ground like miniature courser; stands, courser-like, in open and watches intruder. If disturbed, crouches and relies on cryptic colouration for concealment; flies up only when almost trodden on. Rises suddenly and silently, with no wing noise; flies in jerky undulations, often rising high and flying for some distance. Runs for short distance after dropping to earth, then stands and watches intruder before crouching. If approached quietly, may be observed in the open.

Mainly resident in eastern part of range (Sudan, Ethiopia, Kenya); probably also resident Senegambia, Chad, Mali and Niger. Also makes seasonal movements—only present (breeding) coastal Gambia and Ghana in cool dry season; vagrant Ivory Coast; in Nigeria present only in extreme north, mostly in dry season, suggesting movements beyond northern border during rains (Elgood 1982). Reportedly present in some parts of range only in wet season (Snow 1978).

Food. Grass-seeds; also termites (possibly to supplement moisture intake) and other insects.

Breeding Habits. No evidence of polyandry; solitary nester.

NEST: small scrape in bare ground near base of plants, lined with a few dry leaves and stalks; often with a rim of small pebbles.

EGGS: 2; oblate-oval with slight to medium gloss. Ground colour stone or cream, with inky purple shell markings and many surface blotches and spots of black, brown and grey. SIZE: (n = 3) 17·5–18 × 14·7–15 (17·7 × 14·9); 16·4–18 × 14·4–15 (Lynes 1925).

LAYING DATES: Senegambia, Jan, Mar, Sept–Dec; Ghana, 'winter' (Lamm and Hopwood 1958); Sudan, Jan (birds in breeding condition Nov–Feb); Ethiopia (probably breeds Mar); Kenya (birds in breeding condition Dec and Jan). Lays in cool dry season, both inland and at the coast.

INCUBATION: by ♂.

References
Lack, P. C. (1975).
Lynes, H. (1925).

Genus *Turnix* Bonnaterre

Plumage cryptic on upperparts, mostly grey, brown or black, often with pronounced bars or spots; brighter below. Tail small, feathers soft and short. Voice of ♀ is far-carrying and ventriloquial; utter booming, purring or droning calls, often rapidly repeated, or a low moaning. Calling sometimes accompanied by stretching of neck and lowering of head slowly (*T. suscitator*) or bird may remain motionless when calling (*T. sylvatica*).

Plate 7
(Opp. p. 112)

Turnix sylvatica **(Desfontaines). Little Button-Quail; Kurrichane Button-Quail. Turnix d'Afrique.**

Tetrao sylvaticus Desfontaines, 1787. Mem. Acad. Roy. Sci. Paris, p. 500; near Algiers.

Range and Status. SW Europe, Africa, E Iran, India east through S Asia to SE China, Philippines, Sulus, Java and Bali.

Scarce to locally abundant resident and intra-African migrant. Range much reduced N Africa; Morocco, 6 records since 1963, from W coast (Oualidia, Casablanca, Azzemour) and N coast (mouth of R. Moulouya) (J. D. R. Vernon, pers. comm.); Algeria, formerly bred along coast, only recent (1976) record from mouth of Oued Zour (Ledant *et al.* 1981); Tunisia, formerly bred along N coast to Cap Bon, present status uncertain, only recent record (1972) 10 km north of Sousse (Thomsen and Jacobsen 1979). Occurs almost throughout sub-saharan Africa except in forests and deserts, overlapping with Black-rumped Button-Quail *T. hottentotta* in moister areas. Status in W Africa uncertain; apparently widespread. In Nigeria common all year at Lagos and Tivland, not uncommon elsewhere (Elgood 1982, Gee and Heigham 1977); no definite record Cameroon but doubtless occurs (Louette 1981); elsewhere local and uncommon but probably overlooked. Present in Ouadi

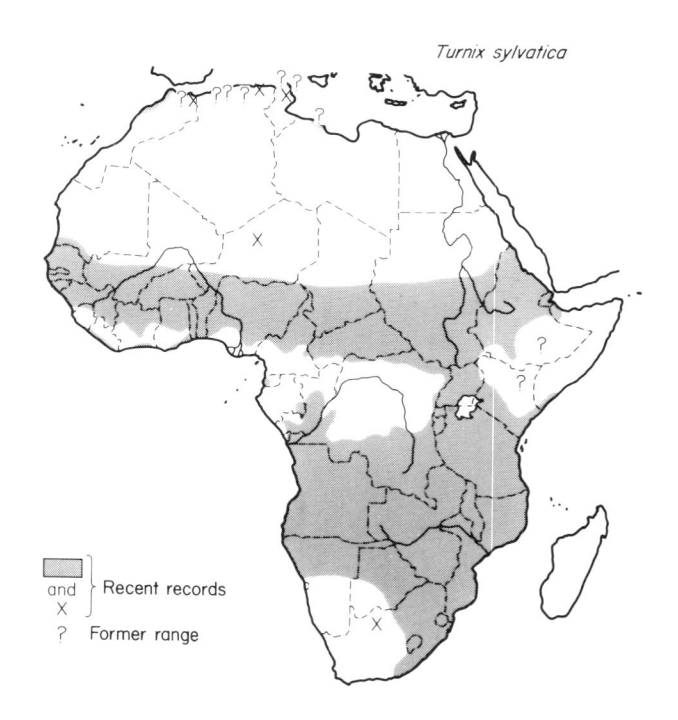

Achim Faunal Reserve, Chad (Newby 1980); moderate numbers Darfur and Kordofan, Sudan; frequent Ethiopia; widespread and locally common E Africa (especially coastal Kenya) but local Uganda; very common Zanzibar; uncommon Pemba. Present throughout Angola, including Cabinda; common to abundant Zambia, Zimbabwe and Malawi; status elsewhere in southern Africa not well-defined, but widespread; not uncommon on grassy flats throughout Kalahari. Generally much more numerous than Black-rumped Button-Quail.

Description. *T. s. lepurana* (Smith): Africa south of Sahara. ADULT ♂: forehead, crown and nape feathers blackish brown, sometimes with dull rufous-brown subterminal spots and always with dull rufous-brown to pale buff fringes; forehead often with pale buff streaks; central crown streak pale buff. Supercilium, lores, face, ear-coverts and sides of neck pale buff, feathers tipped rufous-buff to brownish black, darkest on malar region and sides of neck. Chin and throat pale buff to white. Hindneck cinnamon to rufous-brown, feathers having buff edging with blackish brown inner border, and variable transverse blackish brown barring (often absent). Mantle, scapulars, tertials, back, rump, uppertail-coverts and tail closely barred blackish brown and rufous-brown to cinnamon, each feather with broad white to grey or buff edging with black inner border; tertials also have broad pale creamy buff lateral markings. Sides of upper breast rufous-brown to cinnamon, feathers with creamy buff lateral spots and variable narrow blackish brown bars and streaks. Sides of lower breast and anterior flanks creamy buff, each feather with central black spot which is usually heart-shaped and has rufous-brown central area. Centre of breast plain orange-rufous. Lower flanks plain creamy buff with orange-rufous or golden buff wash. Belly pale creamy buff to white; undertail-coverts similar but tinged golden buff to orange-rufous. Alula and remiges dark sepia, remiges with paler inner webs and alula feathers tipped grey; outer webs of outer primaries edged rufous-buff to creamy buff, most prominent on P9–10. Inner primaries and all secondaries with outer webs narrowly edged buff and tips pale grey to buff. Outer webs of inner secondaries finely barred and blotched cream to pale rufous-buff. Marginal and lesser upperwing-coverts sepia, fringed cream to rufous-buff or grey. Median and greater upperwing-coverts rufous-brown to cinnamon, outer webs broadly fringed or almost completely creamy buff, this colour also extending over most of inner webs in some birds; each feather with blackish brown subterminal bar or spot, usually a blackish brown central spot, and variable blackish brown bars on inner web, barring more extensive on innermost coverts. Underwing-coverts golden buff or cream to pale grey fringed golden buff. Underside of remiges pale grey. Leading edge of wing creamy buff. Axillaries cream to pale grey, fringed golden buff. Bill grey to blue-grey, culmen and tip often darker; eye pale straw to creamy white; bare skin round eye pale blue; legs and feet pale flesh to pinkish white. ADULT ♀: like ♂ but more richly coloured: forehead to nape with brighter rufous-chestnut fringes, bold creamy buff flecks on forehead and prominent creamy buff central crown stripe; supercilium, lores and face to sides of neck pale cream, usually with bold brownish black feather-tips giving a well speckled effect; feathers of hindneck and upper mantle rufous-chestnut with pale rufous-buff to creamy buff or grey edges and variable dark bars (often absent); pale fringes of feathers of upperparts may be predominantly cream or grey: very variable, but pattern usually more pronounced than in ♂; centre of breast, flanks and undertail-coverts deeper orange-rufous. Dark spots on breast, flanks and upperwing-coverts larger and more prominent than in ♂. SIZE: wing, ♂ (n = 20) 73–80 (76·7), ♀ (n = 20) 82–92 (86·5); tail, ♂ (n = 19) 25–33 (28·9), ♀ (n = 17) 32–40 (35·2); bill (culmen to base of feathers), ♂ (n = 8) 10–11·5 (10·8), ♀ (n = 8) 11–12·5 (11·8); tarsus, ♂ (n = 19) 17–20 (18·4), ♀ (n = 20) 19–22·5 (20·6). WEIGHT: ♂ (n = 10) 32·2–43·5 (36·0), ♀ (n = 12) 39·0–53·7 (51·0), unsexed (n = 20) 31·6–79·0 (44·9).

IMMATURE: plumage acquired by post-juvenile partial to complete moult soon after fledging. Upperparts like ♂ but duller; pale areas of head may be more grey, less buffy, than adult; tertials with lateral whitish spots. Underparts duller and paler; blackish brown spots extend across breast; centre of breast washed pale orange-rufous. Upperwing-coverts duller than adult, dull cinnamon with dull dark brown barring and whitish spots. Remiges like adult but outer webs of all secondaries marked like those of inner secondaries of adult, and faint markings also extend to inner webs.
JUVENILE: forehead to nape as immature but central crown-stripe poorly marked. Feathers of mantle and back, scapulars, tertials and upperwing-coverts with whitish lateral spots, least evident on centre of mantle and back. Underparts white or creamy buff; sides of upper breast rufous-brown, feathers with two lateral white spots with dark inner border; rest of breast as immature; dark spots often V-shaped. Eye black or brown.
DOWNY YOUNG: head and upperparts rufous-brown, darkest on crown and centre of back; pale buff stripes down centre of crown, from lores over eye, and down each side of back. Rufous-brown areas of upperparts edged blackish brown. Entire underparts buff or creamy buff.
T. s. sylvatica (Desfontaines): N Africa. Larger; upperparts more strongly marked; sides of breast deeper-toned, more cinnamon; spots on breast larger. SIZE: wing, ♂ (n = 12) 83–92 (88·0), ♀ (n = 17) 91–101 (97·3).
Clancey (1978) named the richer-coloured southern and eastern populations of southern Africa *T. s. alleni*; they are darker and colder above than *lepurana*, feathers chestnut or reddish brown, barred blackish and edged grey or buff, and more richly chestnut on the breast (Irwin 1981). Because plumage varies considerably and intermediate plumages have been recorded from Zimbabwe (Irwin 1981), we prefer to retain these birds in *lepurana*.

Field Characters. For general distinctions from quail *Coturnix* spp. see Black-rumped Button-Quail. Most likely to be confused with Common Quail *C. coturnix* and Harlequin Quail *C. delegorguei*, which occur in similar habitat and are similar colour in flight; separable in flight by small size, very pale upperwing-coverts with dark spots contrasting with mid-brown upperparts; slower wing-beats interspersed with short glides, flight usually brief and low. On ground, told from quail by bold black spots at sides of orange-rufous breast, pale upperwing-coverts, lack of face pattern; also by pale blue orbital skin, longer thinner bill, longer neck and less bulky body. For differences from Black-rumped Button-Quail, see that species.

Voice. Tape-recorded (60, 73, C, F, 264). Most commonly heard call is advertising call of ♀, uttered by night and day: a deep, resonant, hollow, ventriloquial 'hoo', usually lasting *c*. 1 s and repeated every 1–2 s for 30 s or more. Call sometimes a double 'oo-up', or notes may be run together to produce an almost continuous sound, sometimes wavering. When calling, ♀ adopts upright posture (**A**), neck stretched and arched, throat swelling with each note, bill pointing downwards. Before calling, ♀ repeatedly contracts and expands chest, then draws head back between shoulders so that bill often rests on upper breast, and delivers call with bill closed, belly drawn in and chest alternately contracting and expanding. Pumping stage of call accompanied by gasping sound during which sides of neck are inflated. Other calls of captive ♀♀ during breeding season are long-drawn, monosyllabic rattle 'terrrr' and a growl (both to attract ♂?). Calls of ♂: a sharp 'tuc-tuc-tuc' in response to or initiating ♀'s advertising call; a kestrel-like high-pitched 'kee-kee-kee-kee'; a long drawn 'triii' given by separated ♂ in answer to ♀'s advertising call; a rattling call like ♀'s; a soft 'ick' when with ♀; a soft clucking call to young. When together, both sexes of pair utter low-pitched 'cree-cree-cree' and soft peeping.

General Habits. In N Africa principally inhabits dwarf palm or palmetto *Chamerops humilis* scrub, or found among asphodels (*Liliaceae*) and in other bushy coastal vegetation. Elsewhere inhabits dry to moist grassland, bushed grassland or tree savanna; grass often rank but may be short and tussocky with sparse cover. Also frequents cultivation, usually in fallow or neglected areas or clearings. Occurs on recently burnt ground where some cover remains, and in Zambia in long grass at edges of *Brachystegia* woodland and young conifer plantations. In cultivation and open grassland often prefers vegetation on or around termite mounds. Occurs on sandy soils in W and E Africa, but also on dark clay soils in Zambia, e.g. at dry edges of swamps. Altitudinal range sea level to at least 2000 m. Normally occurs in drier habitat than Black-rumped Button-Quail but some overlap may occur.

Occurs singly or in pairs; in non-breeding season in small groups or scattered flocks. Shy, but not as difficult to observe as Black-rumped Button-Quail; emerges from cover to feed in open short grass or at edges of roads. Can be approached to within 5 m by careful stalking; on close approach bird crouches and relies on cryptic colouration for concealment (a bird thus approached by car to within 1 m moved head in nervous jerks before finally running into cover, pausing on the way and stretching neck). Moves with slow creeping walk when feeding and runs swiftly through cover to escape predator; runs over open ground with rolling gait. Reluctant to fly and sits close before rising, but quite easily flushed by dog. Normally flies low for a short distance with rapid and noisy wing-beats, lands in cover and runs; more difficult to flush a 2nd time. In cover may be repeatedly flushed by dog or may freeze and even allow itself to be caught. Does not call in flight. On landing sometimes assumes upright stance with outstretched wings.

Pairs dustbathe and sunbathe together; when dustbathing, birds scuffle hurriedly with fluffed-out feathers, rubbing sides of head and body on ground; when sunbathing they remain side by side and either fully open both wings or lie on one side and open one wing. At night captive pair roosts together or with other pairs. In the wild, most obviously active in early morning and late afternoon.

Believed sedentary N Africa; wet-season breeding visitor to drier areas of N tropics. In Nigeria bird hit lighted window at night, Ibadan, June, suggesting migration (Elgood *et al.* 1973), but resident Lagos; wet-season visitor to Ouadi Achim Faunal Reserve, Chad (Newby 1980). Arrives Sudan from May on and numbers appear July with Harlequin Quail; resident Ethiopia but in W Eritrea absent in dry weather; at Uelle (NE Zaïre) only found in dry season (Jan–May), possibly a seasonal visitor from Sudan. Resident E Africa, but small numbers regularly attracted to lights at night, Ngulia (Kenya), suggest some movements, as do occasional influxes, e.g. at Kisumu (W Kenya) July; visitor in variable numbers to Tsavo E National Park (Kenya) in rains, remaining to breed after rains. Resident Malaŵi, but probably moves locally in Zambia (apparently absent Zambezi District Feb–Apr and occurs erratically Ndola, often absent during rains and influxes noted with Harlequin Quail Nov). Resident Zimbabwe with local movements. Namibian and Botswanan birds move east in non-breeding season (May–Dec) to Transvaal, Zimbabwe and Zambia, where found alongside presumed resident birds.

Food. Small seeds, especially of grasses, and many insects (including ants); chicks entirely insectivorous for first 10 days of life.

Breeding Habits. Solitary nester. Nature of pair-bond disputed; captive birds show successive polyandry, ♀ laying clutches for 2–3 ♂♂ and re-mating with ♂♂ which have previously reared broods; ♂♂ nest up to 7 times each per year (Wintle 1975). Normal polyandrous state may give way to monogamy when local conditions demand ♀'s help in rearing young (Trollope 1970). Captive birds formed apparently monogamous sustained pair-bond (Hoesch 1959, 1960, Trollope 1970), renewed when brood independent; in experiments ♀ always chose 1 of 2 ♂♂, often chasing the other ♂ off, and paired ♀ also attacked newly-introduced ♂, but in no cases were ♀♀ allowed to seek a new mate when ♂ incubating.

Sex roles reversed. ♀ attracts ♂ with advertising call, then chases ♂ and attempts to fight; if in breeding condition ♂ stands ground, and pair soon feeds and roosts together. When birds brought together, up to 5 days may elapse before pairing preliminaries observed; birds then move along together with rocking motion. Important actions of ♂ during pair-formation include adoption of horizontal position with stiff movements and staring eyes, and performance of scrape-ceremony: bird rests on breast, hind-body raised and tail vertical; pecks at soil and makes soft calls; bird may scrape out several hollows in soil, then 1 or both of pair toss material over shoulder towards scrape. Scrape-ceremony also performed by ♀. ♀ also displays by puffing out breast feathers and adopting incubating position (but display not directed at and does not affect ♂). ♀ courtship-feeds ♂ using 'Titbitting' display (see below). In breeding season both sexes allopreen, particularly ♀ near egg-laying when preening followed by pecking, lunging and attempts to mount ♂. ♂ more active in copulation, driving ♀ along, pecking at her head until she crouches, when ♂ mounts and grips ♀ with wings while copulating. Aggressive in breeding season; when birds re-introduced after being kept apart they adopt rigid pose with horizontal body and make rocking motion (submissive posture to reduce aggression?). ♀♀ kept together fight vigorously; paired ♂ attempts to expel any other ♂. ♀ pecks head and nape of other ♂ introduced to established pair; he responds by lying flat on ground to inhibit ♀. In experiments, 3 ♂♂ kept with ♀ all attacked strange ♂ introduced at any time (Wintle 1975). After 4 weeks of visual but not vocal separation, mates recognize each other again: in greeting, ♂ adopts submissive posture, crouching with breast touching ground, and ♀ allopreens his head.

NEST: in grassland, on ground under grass tussock or other herbage; a shallow scrape lined with grass, seed-heads and twigs; usually well concealed. In captivity both sexes search for site but ♀ makes final selection and builds nest (Wintle 1975), or both sexes build (Hoesch 1960). Nest dispersion, Namibia, 10 nests in *c.* 10 ha.

EGGS: 2–7; E Africa mean (5 clutches) 2–5 (3·6); Malawi (21 clutches) 3–4 (3·4); laid at daily intervals. Ovate-pyriform; smooth and glossy; whitish, cream or pale buff to grey or pale brown, heavily speckled and blotched with black, brown, grey, purple or yellow: very variable in colour and pattern. SIZE: (n = 14, E Africa) 20·2–23·0 × 15·7–19·0 (21·7 × 17·6) and 1 egg 31·0 × 23·0; (n = 60, South Africa) 20·3–26·2 × 16·9–20·0 (23·4 × 18·6). Lays again if clutch destroyed before incubation begun.

LAYING DATES: Nigeria, Jan, Apr, July–Aug; Sudan, June–Sept; Ethiopia; Apr–May, Sept; S Somalia, May; S Zaïre, July; East Africa, Region A, May, Aug–Sept, Dec; Region B, Sept; Region C, Jan–Mar, Sept; Region D, Jan, May–June, Nov–Dec; Region E, Apr–May, Oct, Dec; Zanzibar, Mar–May; Malawi, Jan–Feb, Apr–Nov; Zambia, Feb–May, July–Dec; Zimbabwe, all months, peaks Jan–Mar and Oct; Mozambique, Oct, Dec; Namibia, all months; South Africa, all months, peak Feb–Mar.

INCUBATION: information from captive birds only. Starts with last egg; by ♂ only; sitting bird leaves nest 8 or more times per day to feed and sunbathe for 5–15 min periods; hatching synchronous. Period: 12–15 days.

DEVELOPMENT AND CARE OF YOUNG: information from captive birds only. Precocial and nidifugous. Chicks leave nest with ♂ 4 h after hatching; ♂ uses 'Titbitting' display, offering food in bill while keeping head stiffly poised, with soft calls and shivering wings; chicks depend entirely on ♂ for food for 4 days and for feeding and brooding for first 10 days; after 10 days brooded only at night. After being fed chicks poke at ♂'s flanks until he settles down to brood with breast and belly feathers fluffed out. Chicks freeze when alarmed. ♂ leads brood to water and food. Chicks follow any moving bird and other ♂♂ are aggressive to them; not defended by parent. Can fly at 10 days; start to perform rocking-motion behaviour from 12th day; by 13th day daily maintenance behaviour fully adult-like; independent at 18–20 days. Fully grown at 35 days with full immature plumage; moult to adult non-breeding plumage completed before 7–8 weeks old; can breed from 15 weeks. Breeding cycle completed in *c.* 53 days.

BREEDING SUCCESS/SURVIVAL: information from Wintle (1975). In captivity, 2 chicks usually reared from clutches of 3 eggs, but broods of 3 or 4 may all survive. ♀♀ kill chicks, and older broods of chicks can kill younger ones. 36 days after brood hatches, ♂ will again consort with ♀ if latter has no mate; chases away fledged young and is on eggs again after 7–12 days. 2 ♂♂ each nested 7 times in one year; ♀ lays clutches for 2–3 ♂♂, re-mating with ♂♂ which have previously reared broods. ♀ attracts new ♂ within a day of current mate beginning incubation; ♀ therefore theoretically capable of breeding with 5 ♂♂ continuously throughout the year; potential population increase thus enormous, and kept in check presumably by high mortality rate, lack of suitable breeding conditions at some times of year, and lack of surplus ♂♂ in population: of 20 chicks raised to adulthood, 8 or 9 were ♀♀. Maximum age attained in captivity, over 9 years. Predation of adults by Wahlberg's Eagle *Aquila wahlbergi* recorded (Steyn 1980).

References
Cramp, S. and Simmons, K. E. L. (1980).
Hoesch, W. (1959, 1960).
Wintle, C. C. (1975).

82 TURNICIDAE

Plate 7
(Opp. p. 112)

Turnix hottentotta (Temminck). Black-rumped Button-Quail; Hottentot Button-Quail. Turnix nain.

Turnix hottentottus Temminck, 1815. Pig. et Gall., 3, p. 636; Cape of Good Hope.

Range and Status. Endemic; locally common resident and local intra-African migrant, in grassland and savanna, mainly in moist woodland belts; absent from arid areas. Status in western part of range uncertain; probably rare to uncommon and local Senegambia, Sierra Leone, Ivory Coast, Ghana, Nigeria, Cameroon, Gabon, Congo and Angola. Locally common Zaïre (especially E), Uganda and W Kenya (Trans Nzoia); only 1 record Tanzania (Songea). Widespread and local Zambia (mainly wet season visitor) and Malaŵi (possibly partly migratory). Uncommon and local S Mozambique. Sparse and localized Zimbabwe; occurs in rains, but locally common near Harare. In South Africa, resident in winter rainfall area of Cape Province (*T. h. hottentotta*), local and uncommon E Cape Province, Natal, Zululand, Swaziland and Transvaal (*T. h. nana*). Possibly commoner than records suggest, since easily overlooked; may be locally numerous in suitable habitat but is generally much less numerous than Little Button-Quail *T. sylvatica*.

Turnix hottentotta

Description. *T. h. nana* (Sundevall): Senegambia and Kenya to South Africa (E Cape Province). ADULT ♂: forehead and crown blackish, feathers with white edges and dull rufous-brown tips; forehead sometimes pale orange-rufous. Lores, face and broad supercilium pale orange-rufous to golden buff, feathers varyingly tipped blackish. Malar region and sides of neck more patterned, feathers with white bases, rufous centres and blackish tips. Chin and throat white to golden buff (pale orange-rufous in northern populations 'luciana'). Nape and hindneck dull dark sepia, feathers with white or buff spots at tips and rufous-brown and blackish barring. Mantle and back darker, feathers blackish with grey-brown tip, white submarginal bar and irregular rufous-brown barring (feather edges more orange-rufous in northern populations). Scapulars broadly edged white or golden buff, with more distinct rufous and blackish barring. Rump and uppertail-coverts blackish, feathers tipped golden buff or rufous-brown and varyingly barred distally with white, rufous-brown or golden buff. Feathers of sides of rump broadly tipped deep golden buff. Tail blackish, feathers fringed and striped golden buff. Breast deep golden buff to orange-rufous (deeper in northern populations), feathers on sides of breast and upper flanks tipped blackish with black-bordered subterminal white bar. Variable amount of barring on upper breast (centre usually unbarred or indistinctly barred). Lower breast, belly and lower flanks white, tinged golden buff to orange-rufous; undertail-coverts golden buff to orange-rufous. Axillaries grey-brown, fringed golden buff. Primaries dark sepia with paler inner webs; edge of outer web of P10 (sometimes 8–10) creamy buff to orange-rufous, broadest on P10. Tips of P1–6 (sometimes 1–10) buff. Secondaries paler than primaries, distally edged and tipped buff, inners with distal buff spots. Tertials sepia with white spots, and barred blackish, dark brown and rufous-buff. Alula and primary coverts dark sepia, tipped buff to golden buff; leading edge of wing whitish buff. Marginal upperwing-coverts dull brown, broadly fringed rufous-buff. Lesser and median upperwing-coverts rufous to orange-rufous with variable white tips and subterminal blackish brown barring. Underwing-coverts buff to grey, washed golden buff to orange-rufous; underside of remiges dark silvery grey. Bill dark brown with yellowish grey edge and lower mandible, and dark tip. Eye pale blue to silvery white. Legs and feet pale flesh to dusky white. ADULT ♀: like ♂ but forehead, supercilium, face and chin to breast orange-rufous, deepest on breast and palest on chin and throat. Upperwing-coverts like ♂ but median and greater coverts often orange-rufous with golden buff subterminal spots and blackish bars. Barring on breast confined to sides; on flanks less extensive than in ♂. SIZE: wing, ♂ (n = 7) 71–77 (73·9), ♀ (n = 8) 77–86 (81·1); tail, ♂ (n = 6) 21–27 (24·7), ♀ (n = 7) 28–32 (29·7); bill (culmen to base of feathers), ♂ (n = 7) 9–11 (10·0), ♀ (n = 6) 10–12 (11·2); tarsus, ♂ (n = 7) 19–21 (19·8), ♀ (n = 8) 19–21 (20·4). WEIGHT: 2 ♂♂ 40·2, 40·0; 2 ♀♀ 57·5, 62·4.

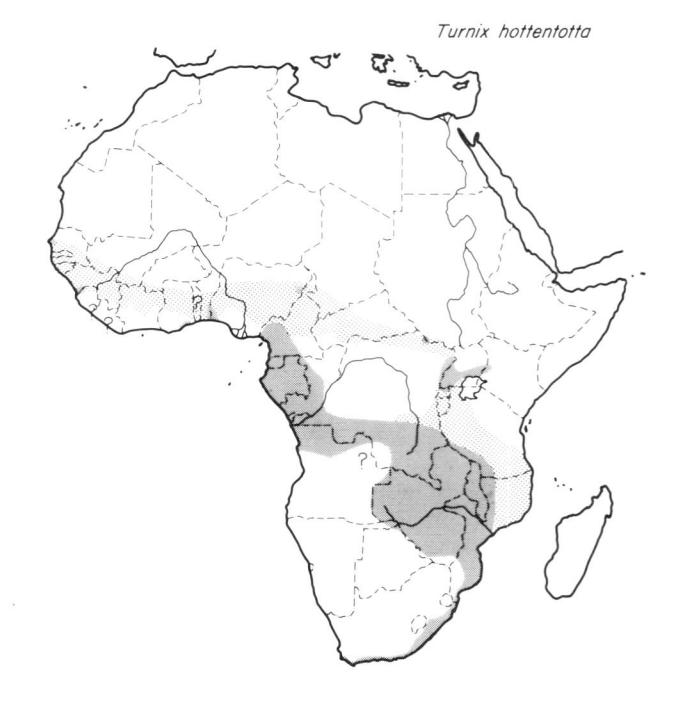

IMMATURE: similar to adult ♂ but with much less rufous in plumage and more heavily marked below. Face and supercilium washed pale orange-rufous, and spotted white; scapulars edged white or buff; no golden buff at sides of rump. In ♂, chin to belly white, heavily barred dark brown on breast and barred blackish and white on flanks; sometimes tinged golden buff on upper breast. In ♀, breast washed orange-rufous and barring less extensive in centre of breast. Upperwing-coverts with prominent white or buffish white spots. Eye pale brown. Legs and feet dark flesh.

DOWNY YOUNG: not described.

T. h. hottentotta (Temminck): South Africa, from SW Cape Province to Port Elizabeth. Paler above; conspicuously spotted black on breast, flanks and sides of belly; underparts paler rufous; rump paler; eye yellow.

We follow White (1965) and others in including *T. h. luciana* Stoneham and *T. h. insolata* Ripley and Heinrich in *nana*; although southern populations are somewhat paler than northern ones, variation is considerable and subdivision is not clear.

Field Characters. In flight, distinguished from quail *Coturnix* spp. by small size, less bulky body with noticeable neck, rounded head without pale central crown streak. Flight slower than that of quail, usually brief and low. Does not flush unless almost trodden on. Rarely flushes a 2nd time (runs after landing); no call when flushed. Harlequin Quail *C. delegorguei* and Common Quail *C. coturnix* are larger and paler; ♀ and immature Blue Quail *C. chinensis* are same size, dark above, and occupy similar habitat, but have typical quail shape and flight and almost unpatterned upperparts; Black-rumped Button-Quail has more patterned upperparts (pale feather edges and bars), including upperwing-coverts. Böhm's Flufftail *Sarothrura boehmi* differs in flight by slower wing-beats, longer body and neck, long dangling legs, red head (♂) and blackish plumage with white streaks or bars. Darker in flight, with paler patterning, than Little Button-Quail which has paler, browner upperparts, no black on rump, noticeably pale (but black-spotted) upperwing-coverts which form a distinctive pale upperwing panel, and paler remiges.

On ground, separable from quail by colour and pattern of underparts, face pattern, less bulky body, longer neck and longer thinner bill. Race *nana* distinguished from Little Button-Quail by barred breast and flanks; race *hottentotta* has spotted breast and flanks, as does Little Button-Quail, but latter's spots are more heart-shaped on the flanks; both races distinguished by darker upperparts, lack of pale upperwing-coverts, rufous on face of ♀, and lack of central crown streak.

Voice. Unconfirmed tape-recording (F) from Zimbabwe is a series of low-pitched resonant 'hoo' notes, uttered at a rate of *c.* 1·6 per s with hardly any pause between notes; pitch similar to that of longer 'hoo' notes of Little Button-Quail. Call closely resembles that attributed by Masterson (1969) to Black-rumped Button-Quail.

General Habits. Inhabits open grassland, savanna, plains; grass usually short and fairly open, often moist or inundated (e.g. moist dambos). Breeding habitat near Harare, Zimbabwe, on dark clay soil with 25–50 cm high grass growing in tufts with clear ground in between and small plant *Thesium brevibarbatum* common; surface flat so that pools do not form even when ground sodden (Masterson 1973). At Ndola, Zambia, immigrants in similar habitat, also at edges of thickets in grassland and in short dense grass. Normally avoids marshy areas and standing water, but has bred in marshy area of irrigated sugar cane (Masterson 1973). In East Africa inhabits grass on sandy soil but not in Zambia or Zimbabwe. Attracted to short-grazed grass and ground disturbed by cattle. Altitudinal range 0–1800 m. Little Button-Quail prefers rank grassland and cultivation in drier areas, but species' habitats may overlap slightly.

Normally solitary or in pairs; sometimes in loose aggregations. Very difficult to observe on ground; skulking and less likely to emerge from cover than Little Button-Quail. When undisturbed, gait is slow creeping walk; normal method of escape is fast run through more open grass; seeks cover in denser taller herbage and inside thickets when pursued. Unwilling to fly, even when pursued by dog, and can elude dog by running. Sits close before rising; when flushed flies low for short distance, lands and runs; very difficult to flush a 2nd time. Flight quite fast and whirring; does not call in flight.

Migratory status not clear. Resident in Kenya, Uganda and Cape Province; apparently vagrant Nigeria; occurrence erratic Cameroon. In Zambia partly a rains breeding visitor and most records Oct–Apr; influxes noted Ndola Nov (birds remaining less than 1 month) and Kafue Flats Jan. At Harare, Zimbabwe, occurs in rains only (breeds). Birds picked up at lighted buildings at Livingstone, Zambia, Jan and Zomba Plateau, Malaŵi, May, suggest movements; possibly partially migratory Malaŵi. Probably itinerant over much of range, arriving to breed during or at end of rains and remaining until habitat becomes unsuitable, e.g. has remained in Zambia until Apr and June in suitable habitat.

Food. Seeds and insects.

Breeding Habits. Solitary nester in open grassland; possibly polyandrous.

NEST: shallow scrape 50–75 mm diam., 20 mm deep, sparsely lined with grass blades and stems; usually placed at base of grass tuft with growing blades drawn down to form loosely made canopy; also placed below sheaf of fallen grass without any canopy.

EGGS: 2–6; Zimbabwe mean (8 clutches) 3·0, E Africa (6 clutches) 3·0. Sharply pyriform, pale greenish or greenish white (*nana*) or yellowish grey (*hottentotta*), very thickly speckled dark brown and yellowish brown or dark grey and olive-brown. SIZE: (n = 4, South Africa) 21·5–24·5 × 17·3–20; (n = 9, Zimbabwe) 21·3–24·0 × 17·7–19·2 (23·0 × 18·5); (n = 3, Kenya) 20·5–21·0 × 17·0–17·5 (20·7 × 17·3).

LAYING DATES: Nigeria, Dec–Jan; E Africa, Uganda, July, Oct; Kenya, May–July; Region A, May–July, Oct (peak June–July); Region B, Oct; Zambia, Jan–Feb, Oct, Dec, (gonads active Mar, Nov); Malaŵi, Apr; Zimbabwe, Jan–Feb, Sept–Dec; South Africa, Jan, Oct–Dec. Normally lays during or at end of rains.

INCUBATION: by ♂ only. Period: 12–14 days.

References
Horsbrugh, B. (1912).
Masterson, A. N. B. (1973).

Family RALLIDAE: rails, flufftails, crakes, gallinules, moorhens and coots

A large and fairly homogeneous family of world-wide distribution. Small to medium-sized (14–51 cm); plumage usually dull and cryptic, brown or grey, some bluish or reddish; bare parts may be brightly coloured. Sexes usually similar. Some species have flightless moult. Wings rounded, short to fairly long, normally 10–11 primaries; young of many species have claw at tip of alula. Tail short; bill long and slender to short and thick; legs fairly long; toes usually long and slender, hallux present and functional. Body laterally compressed ('thin as a rail'), enabling birds to slip away through vegetation without moving it. Schizognathous; holorhinal; no occipital foramina or supraorbital furrows. 14–15 cervical vertebrae, 2 carotid arteries, caeca long, aftershaft present in most species. Nostrils pervious, sometimes perforate; olfactory process well developed, providing good sense of smell. Nidifugous; downy young of nearly all species black.

Typically associated with marshes, both salt and fresh, but also found on open water, in forest and dense brush, even in dry fields. Since the more primitive forms are found in forest, adaptation to aquatic habitats may be more recent development (Olson 1973). Most species non-specialized and omnivorous, able to adapt to new habitats and food sources. Though flight over short distances appears weak and fluttering, many rails are capable of sustained long-distance flight, being long-distance migrants and successful colonizers of oceanic islands. Some species flightless (none in Africa). Generally shy and difficult to see, inclined to skulk in thick vegetation and slink away on foot rather than flush. Flick tail while walking. Good swimmers and agile climbers. Very vocal, with wide repertoire; some species duet.

132 spp. world-wide; 26 in Africa.

Subfamily HIMANTORNITHINAE: Nkulengu Rail

We follow Olson (1973) in giving *Himantornis* subfamilial rank. Downy young unique in Rallidae, patterned dark brown and buff like galliform or anseriform bird. The most primitive rallid, with no close relatives, closest to the stock that gave rise to both Psophiidae and Rallidae, providing a link between the 2 families (Olson 1973). Large, with long, slender legs and short, decurved bill. Skull much more like that of trumpeters Psophiidae than that of any rail, and several features of post-cranial skeleton also similar to those of *Psophia*.

Subfamily endemic to Africa; 1 sp.

Genus *Himantornis* Hartlaub

Endemic to Africa. Monotypic. The most primitive rail, in many characters like Neotropical trumpeters (Psophiidae). 8 rectrices; wing eutaxic; no aftershafts; no tuft on oil-gland. Stout; long red legs with rather short toes; plumage dull brown; downy young not uniformly black. Largest African rail.

Plate 8
(Opp. p. 113)

Himantornis haematopus Hartlaub. Nkulengu Rail. Râle à pieds rouges.

Himantornis haematopus 'Temm.' Hartlaub, 1855. J. f. Orn. 3, p. 357; Dabocrom, Gold Coast.

Range and Status. Endemic resident. Ranges from Sierra Leone (Bintumane Peak), coastal Liberia and Mt Nimba east to Ghana; a single old record from Nigeria (Degema); and from Cameroon through Gabon to Mayombe Forest (mouth of R. Congo) and across N Zaïre to eastern border at Semliki R. (1 record also from Bwamba, Uganda) and S Kivu (Kamituga), and south to Sankuru R. Frequent to common.

Description: ADULT ♂: plumage highly variable. Forehead, crown and nape grey-brown, brown or dark brown, forehead usually paler; pale stripe over eye (sometimes indistinct). Sides of head and neck brown, sometimes washed with buff; chin and throat whitish. Upperparts, including scapulars and wing coverts, very variable; feathers may be almost uniform brown with narrow pale tips, or have dark brown or blackish centres, a subterminal band of brown to rich chestnut, and whitish or pale grey tips of variable width. Feathers of

Himantornis haematopus

underparts vary from plain brown with indistinct pale tips to blackish with broad grey tips. Primaries and secondaries brown, axillaries and underwing-coverts brown with narrow pale tips. Bill black, base of mandible light greenish or bluish grey. Bare skin on lores and narrow ring around eye black. Eye reddish brown to reddish orange; legs and feet red. Sexes alike. SIZE: (6 ♂♂, 6 ♀♀) wing, ♂ 211–235 (220), ♀ 200–222 (212); tail, ♂ 72–94 (85), ♀ 74–85 (80); bill, ♂ 36–41 (38), ♀ 33–40 (38); tarsus, ♂ 78–84 (81), ♀ 70–78 (75). WEIGHT: 1 ♀, 390.

IMMATURE: similar to adult but feathers of upperparts with dark centres and tawny margins; below, light grey-brown; chin, throat and belly whitish. Tip of bill pale horn, eye dark brown, legs and feet dull light red.

DOWNY YOUNG: broad stripe from forehead over head and back to tail blackish brown; broad stripe over eye, sides of head, chin, throat and underparts creamy white; lores, eye-ring and ear-patch black; band across breast brown; sides of body dull light brown. Eye dark brown; upper mandible blackish, lower mandible grey, feet dull pink.

2 other races have been described, *whitesidei* Sharpe and *petiti* (Oustalet). However, there is great individual variation within the ranges of all 3 'races', and characters of one 'race' may occur in birds of a different 'race'. Until more extensive plumage studies have been made correlating age, sex, season and locality, it seems best to treat the species as monotypic.

Field Characters. Easily distinguished from any forest rail by large size, short, deep bill, mottled plumage and long red legs. In plumage looks more like francolin than rail. Sympatric Scaly Francolin *Francolinus squamatus* has dark, mottled plumage and red legs, but legs short and has different gait and behaviour.

Voice. Tape-recorded (48, 53, 303). Song antiphonal, a series of phrases of 6 notes, 'ko-KAW-zi-KAW-hu-HOOO', repeated in quick succession for series of phrases of 6 notes, 'ko-KAW-zi-KAW-hu-HOOO', repeated in quick succession for several minutes. Each phrase lasts *c.* 1·5 s. Notes loud, raucous and far-carrying. Sometimes single or the hoot of an owl. At dusk and during the night song is delivered from perch up to 20 m high in tree; during the day, from on or near the ground. Sings every month of the year, mainly at night, especially during full moon, and most often just before dawn.

General Habits. Inhabits rank vegetation along streams in lowland rain forest; occasionally found in mangroves. Its short toes are adapted for walking on hard ground; often found away from streams. Posture more upright than other rails. Feeds on the ground, picking for food items among dead leaves and sticks; feeds by day as well as at dawn and dusk. Roosts in bush or low tree. Shy and seldom seen; mainly caught in traps.

Food. Frogs, snails, millepedes, ants, beetles, hard seeds.

Breeding Habits. Probably monogamous and territorial. Song frequent during dry season in Gabon, when birds not breeding, suggesting song may play role in maintenance of pair bond (A. Brosset, pers. comm.).

NEST: only 1 found: large, thick structure sturdily built of coarse twigs and leaves, diameter 35 cm, depth 35 cm, placed 1·2 m up in bush in primary forest, on plateau far from water (A. Brosset, pers. comm.)

EGGS: 1 clutch of 3. Wide oval, not equally pointed at both ends, without gloss; creamy white with small spots and blotches of red-brown, latter most numerous at small end. SIZE: 49–50·5 × 38·1–38·9.

LAYING DATES: Cameroon, Sept; Gabon (nest), Feb; Zaïre, probably Feb, Mar, Sept (♀ with egg in oviduct, Feb).

References
Chapin, J. P. (1939).
Chappuis, C. (1975).

Subfamily RALLINAE

We agree with Olson (1973) that all rails other than *Himantornis* are so alike morphologically as not to warrant further subfamilial separation; hence we place them in a single subfamily, Rallinae.

Genus *Canirallus* Bonaparte

A primitive genus sharing 2 skeletal characters with *Himantornis*: a much expanded procoracoid process and very slender, square-shafted tarsi with wide articulations. Bill with high, flat-ridged culmen and large, deep nasal fossa; tenuous nasal bar; underparts and longish, fluffy tail chestnut. Downy young dark brown, not black. Closely allied to *Rallicula* of New Guinea.

2 spp., 1 in Africa, 1 in Madagascar.

Plate 8
(Opp. p. 113)

Canirallus oculeus (Hartlaub). Grey-throated Rail. Râle à gorge grise.

Gallinula oculea 'Temm.' Hartlaub, 1855. J. f. Orn., 3, p. 357; Rio Boutry, Gold Coast.

Range and Status. Endemic resident in lowland rain forest, in Liberia, Ivory Coast and Ghana; in Nigeria (only 2 old records, may possibly be extinct there: Elgood 1982); and from Mt Cameroon to Gabon, mainly along coast, and across N Zaïre to Ituri district and S Kivu (Bikili, Kibimbi). 1 record from Bwamba, Uganda (J. T. Weekes, pers. comm. to J. P. Chapin, 1952). Appears to be uncommon, but that may be due partly to its secretive habits.

Description. ADULT ♂: forehead, forecrown, lores, cheeks, chin and throat grey; crown and nape dark brown. Sides of neck, foreneck, breast and upper belly reddish chestnut. Belly, lower flanks, thighs and undertail-coverts brown with russet wash and pale bars. Primaries blackish brown with large white spots; secondaries similar but with outer web olive. Tertials and wing-coverts olive with russet wash and pale spots or bars. Axillaries and underwing-coverts blackish with large white spots. Bill black, sides and base of lower mandible green. Eye red, eye-rim yellow-green, legs and feet brown. Sexes alike. SIZE: (8 ♂♂, 5 ♀♀) wing, ♂ 167–179 (173), ♀ 167–180 (173); tail, ♂ 59–70 (64), ♀ 58–70 (65); bill, ♂ 37–41 (39), ♀ 37–38 (37); tarsus, ♂ 49–55 (52), ♀ 50–57 (52).
IMMATURE: forehead, face and throat brown instead of grey; upperparts with russet wash; underparts dark reddish brown.
DOWNY YOUNG: blackish brown.
(We agree with Chapin (1939) that *C. o. batesi* Sharpe differs too little from the nominate race to warrant recognition.)

Field Characters. Quite unlike any rail in its lowland forest habitat. Nkulengu Rail *Himantornis haematopus* is larger and heavier with deep bill and no red in plumage. When running away in front of observer it would show mottled grey and brown, rather than plain olive upperparts, and red legs. If Grey-throated Rail were to flush (unlikely), white-spotted dark wings would be conspicuous (wings of Nkulengu Rail unspotted).

Voice. No known recording. Voice in wild state not known. Captive birds gave a loud snore, a soft coo and a short 'chunk'.

General Habits. Confined to lowland rain forest, including secondary forest, where it lives in ravines and along forest streams bordered with rank undergrowth. Extremely shy and very rarely seen; most specimens caught in traps.

Food. Skinks, snails, slugs, small crabs, millepedes and insects including beetles, larvae and caterpillars, and ants.

Breeding Habits. 1 nest situated in swampy bottom of ravine among *Raphia* palms and undergrowth of *Canna*-like plants; consisted of broad grass leaves and was placed on a stump. Another nest very probably of this species (Bates 1927) placed among roots of uprooted tree on stream bank. Both nests contained 2 eggs: glossy, long ovals equally pointed at both ends; creamy buff with mauve-brown and lavender spots. SIZE: 43–44 × 30·5–33 (43·4 × 32).
LAYING DATES: Cameroon, Feb, Apr, July; (Zaïre, breeding condition, Sept, Nov, Dec). Breeds during rainy season.

References
Bates, G. L. (1927).
Chapin, J. P. (1939).

Genus *Sarothrura* Heine

A distinctive genus confined to Africa and Madagascar. Strongly sexually dimorphic (rare in Rallidae), ♂♂ with much chestnut in plumage, especially on head and neck; ♀♀ dark with pale spots or bars. Wings short and rounded; tail fluffy, decomposed, bill short and slender. Eggs white, unmarked. Moaning voices distinctive. ♂♂ very similar to forest-dwelling, chestnut-plumaged *Rallicula* of New Guinea, which is also sexually dimorphic, though less strongly than *Sarothrura*. Olson (1973) believes the 2 genera closely related, with *S. pulchra* the closest link, being most similar in plumage and having 'primitive' characters of forest habitat and slender tarsus. If so, occupation of grassland and marshes by other *Sarothrura* spp. (except. *S. elegans*) may be secondary adaptation. *S. ayresii*, with white patch on wing, may be link with *Coturnicops* of E Asia, N and S America.

9 spp.; 7 in Africa, 2 in Madagascar.

Sarothrura pulchra (J. E. Gray). White-spotted Flufftail; White-spotted Crake. Râle perlé.

Plate 8 (Opp. p. 113)

Crex pulchra J. E. Gray, 1829. Griffith's Anim. Kingd., 8 (Aves, 3), p. 410, col. pl.; no locality (= Sierra Leone, *ex.* Latham, Gen. Hist. Bds., 9, p. 379, no. 16).

Range and Status. Endemic resident, from Gambia and S Senegal (Casamance R.), east to Nigeria (north as far as Zaria and Kano), Central African Republic (Haut Kemo), Zaïre (absent from SE), S Sudan (Benengai) and W Kenya (Kakamega, Kaimosi, Yala River, Nandi), and south to N Angola (north of 12°S), extreme NW Zambia (Salujinga) and extreme NW Tanzania (Bukoba). Questionable sight record from Mali (marsh by R. Niger, on Guinea border: Lamarche 1980). Common to locally abundant.

Description. *S. p. centralis* Neumann: Congo and Central African Republic through Zaïre to S Sudan, Uganda, W Kenya, NW Tanzania, NW Zambia and N Angola. ADULT ♂: entire head, neck, mantle and breast bright chestnut red, paler on chin and throat, darkest on top of head. Back, rump, upper belly, flanks and wing-coverts black with conspicuous white spots. Lower belly and thighs olive-brown with smaller whitish spots or bars. Uppertail-coverts, tail and undertail-coverts chestnut-red, tips of tail feathers sometimes darker. Flight feathers blackish brown with small white spots; axillaries and underwing-coverts blackish brown with narrow white bars. Bill blackish, eyes medium brown, legs and feet brownish grey. ADULT ♀: foreparts as ♂. Back, rump, uppertail-coverts, upper belly, flanks and wing-coverts black, narrowly barred reddish buff. Tail chestnut barred black; lower belly and thighs olive-brown barred buff. Flight feathers blackish brown, a few small buff bars on inner secondaries and traces of same on other feathers. Bill dark grey, eye light brown or brownish grey, legs and feet dark grey. SIZE: wing, ♂ (n = 232) 76–88 (81·1), ♀ (n = 78) 77–89 (79·3); bill, ♂ (n = 30) 15–19 (16·6), ♀ (n = 18) 15–18 (16·7); tarsus, ♂ (n = 30) 28–32 (30), ♀ (n = 18) 28–32 (30). WEIGHT: ♂ (n = 9) 39–49 (45·3), ♀ (n = 2) 41–43 (42).

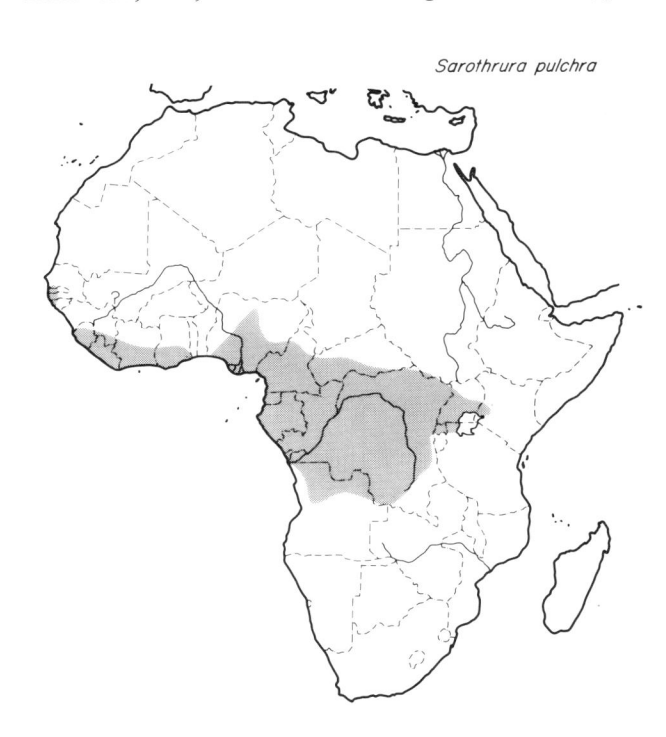

IMMATURE: ♂, similar to adult but duller. Nape, hindneck, mantle and rump brown with rufous wash; back and wing-coverts dull black, spots off-white or washed rufous; centre of breast, belly and flanks largely dull brown, a few flank feathers black with white spots. ♀, as adult but duller.

DOWNY YOUNG: blackish, thighs and belly browner.

S. p. pulchra (J. E. Gray): S Senegal to Niger Basin south of 6°30'N: central Nigeria (Jagindi, Bauchi Plateau); N Cameroon east of 11°E and north of 4°30'N. Black barring on upperparts of ♀ narrower than in *centralis*, buff bars somewhat broader and more rufous. Birds from N Cameroon somewhat larger, but extensive overlap prevents recognition of *S. p. tibatiensis* Bannerman.

S. p. zenkeri Neumann: extreme SE Nigeria, coastal Cameroon and N Gabon. ♂, flanks and abdomen tinged chestnut, white-spotted feathers tipped brown instead of black, resulting in duller appearance. ♀, dark bars somewhat blacker than *centralis*, much blacker than *pulchra*; width of bars about as *centralis*. More richly coloured than other races. Wing (71–92) smaller than *pulchra* (78–92) or *centralis* (76–89).

S. p. batesi Bannerman: S Cameroon (except coast) south of 4°30'N. ♂ like *pulchra* and *centralis*. ♀ like *zenkeri* but black barring more intense, chestnut barring above paler, chestnut of head somewhat paler. Similar to *zenkeri* in size (wing 70–83).

Field Characters. Distinguished by spotted plumage and forest habitat from all *Sarothrura* spp. except Buff-spotted Flufftail *S. elegans*. ♂ distinguished from Buff-spotted Flufftail ♂ by white spots on blacker upperparts, red mantle and lack of black bars on tail. ♀ easily told from ♀ Buff-spotted Flufftail by extensive red on foreparts, buff-barred black upperparts.

Voice. Tape-recorded (32, 53, C, 235, 264, 296). Song, unlike that of any other *Sarothrura*, a series of short, rapidly repeated, high-pitched notes arranged in groups, typically 6–10 per group, with breaks of 1·5–2·0 s between groups; reminiscent of tinkerbirds *Pogoniulus*, e.g. Golden-rumped Tinkerbird *P. bilineatus*, but with more ringing and insistent quality. Song broadly divided into 2 types: (a) long (up to several minutes) and monotonous; notes do not vary in pitch, speed or intensity; (b) shorter, usually less than a minute; begins rapidly and gradually winds down, each group of notes being slower and lower in pitch than the previous one. In both types falsetto notes not infrequent. In type (a), second bird sometimes joins in with lower-pitched, less ringing 'ker-ker-ker ....'. In type (b), second bird often joins in with similar song. Neither duet is antiphonal. Song of ♀ described by Serle (Keith 1973) as being similar to ♂ but softer, higher-pitched and faster, almost a trill. Call, a shrill, rapid 'kik-kik-kik ....'.

General Habits. Habitat chiefly lowland rain forest, usually near water (swamps, streams and pools inside forest; banks of rivers), but also found on forest floor up to 400 m from water. Also follows rivers and streams out into drier zones in gallery forest, dense thickets, shrubby growth and rank herbage. Exceptionally in papyrus and more open vegetation by lakes. Not in montane forest, though ascends to *c.* 1600 m in some areas. Shy and skulking; can sometimes be seen in open if observer remains motionless, but ducks back into cover at slightest movement. Normally remains in thick cover, creeping about like mouse without moving vegetation. Tail cocked and flicked up while walking. Seen bathing in drainage ditch, also crossing brook by jumping from stone to stone. Birds in captivity built roosting platform of palm fibre (Yealland 1952). Sedentary; no evidence of any movements.

Food. Chiefly insects, including caterpillars, ants and beetles, spiders and snails. Small frogs, earthworms, and some vegetable matter, including seeds, also taken.

Breeding Habits. Monogamous; solitary nester.
NEST: few described; placed in forest on damp ground, by small pool, or on low, rotten tree root in shallow water in swamp. Small, oval mound 10 cm high × 23 cm long, concealed by covering of dead leaves like those nearby; or flimsy, domed structure of wet, dead leaves. Interior lined with dry plant stems, dry leaves, decayed fibres or grass. Entrance, a slit at side.
EGGS: 2. Oval, white with some gloss. SIZE: (n = 6) 21·5–22·1 × 30·0–30·9 (21·8 × 30·2).
LAYING DATES: Sierra Leone, (immature Oct); Liberia, (♀ in breeding condition, Sept); Nigeria, (♀♀ with egg in oviduct Sept, ♀ 'just finished laying', Dec, immature Oct); Central African Republic, (immature Apr); Zaïre, Apr, May, Nov, (immature May), 2 immatures Dec; Uganda, (♀ with egg in oviduct Apr, immatures May, July). Breeds during the rains, probably seasonally in southern areas, where rainy season marked, less so closer to equator where season less definite.
INCUBATION: carried out by both sexes.

References
Keith, S. *et al.* (1970).
Pye-Smith, G. (1950).
Ripley, S. D. and Heinrich, G. H. (1966).

Plate 8
(Opp. p. 113)

Sarothrura elegans (A. Smith). Buff-spotted Flufftail; Buff-spotted Crake. Râle ponctué.

Gallinula elegans A. Smith, 1839. Ill. Zool. S Afr., Aves, pl. 22; near Durban, Natal.

Range and Status. Endemic resident, from SE Nigeria, Bioko (Fernando Po) and Cameroon east to S Sudan and S Ethiopia and south to N Angola, Zambia, and through forested parts of eastern Africa (including Zanzibar and Pemba) to E Cape Province, South Africa. Vagrant Strand (Cape): *Bokmakierie* 36, 66. 2 records from Liberia (Mt Nimba and Firestone Plantation) and 1 from Somalia (Wagar Mts). Frequent to locally common.

Description. *S. e. elegans* (A. Smith): Ethiopia, Somalia and E Kenya to South Africa. ADULT ♂: head, neck and breast light reddish chestnut. Upperparts from mantle to uppertail-coverts, scapulars, tertials and wing-coverts sooty black closely spotted buff. Tail barred black and pale chestnut. Rest of underparts black closely spotted white, spots becoming buff on lower flanks and undertail-coverts. Flight feathers brownish black, outer webs with small buffy spots and bars. Axillaries and underwing-coverts blackish narrowly barred white. Bill dark grey, base of lower mandible paler. Eye brown; legs and feet brownish grey. ADULT ♀: upperparts from top of head to rump and wing-coverts warm brown with buff spots, edged black; spots on head tiny, giving barred effect. Face and sides of neck finely barred black and buff; chin and throat creamy white. Tail reddish brown with indistinct black bars. Breast olive-buff with black scalloping; rest of underparts white barred black and washed with olive-buff. Rest of wing as ♂. Bare parts as ♂ but bill more horn-coloured. SIZE: wing, ♂ (n = 41) 83–94 (88·7), ♀ (n = 17) 85–93 (89·6); bill, ♂ (n = 15) 15·5–17·5 (16·3), ♀ (n = 6) 15–17 (15·3); tarsus, ♂ (n = 16) 22–25 (23·8), ♀ (n = 6) 23–25 (23·9). WEIGHT: ♂ (n = 3) 32–45 (40·3), ♀ (n = 3) 49·5–51 (50·2), 1 imm. ♂ (Kenya) 32, 1 imm. ♀ (Uganda) 35.

IMMATURE: uniform sepia-brown, paler on belly (shading to white in centre).

DOWNY YOUNG: black; bill black without white markings, legs and feet black.

S. e. reichenovi (Sharpe): W Africa to Uganda and Angola. ♂ generally darker above, spots larger and coarser.

Field Characters. Distinguished by spotted plumage and wooded habitat from all flufftails except White-spotted Flufftail *S. pulchra*. ♂ distinguished from ♂ White-spotted Flufftail by buff spots on upperparts, lack of red on mantle and black bars on tail. ♀ very different from ♀ White-spotted Flufftail, mainly plain brown with buff spots, lacking red except in tail.

Voice. Tape-recorded (17, 21, 32, 53, 74, 75). The bird's remarkable song has been object of much superstition and speculation. Believed to be the wail of a banshee, the noise made by a chameleon in the agonies of giving birth, and the sound of a chameleon mourning for his mother, whom he has killed in an argument over some mushrooms; has also been attributed to skink, large land snail, climbing mammal, tree snake, 'crowing crested cobras', and Puff Adder *Bitis arietans*. Consists of series of hollow, moaning notes like tuning fork, 'oooooooooooo ...', starting softly and increasing in intensity, ending abruptly. Note longer than any other flufftail, 3–4 s, given at rate of 6–8/min; sometimes immediately followed by higher-pitched 'eeeeeeee' (from second bird?). Song may die away into series of soft 'jug-jug-jug' notes, probably annoyance call. Conversational duet between ♂ and ♀ (often preceding song), low, rather hoarse moan not unlike Chestnut-headed Flufftail *S. lugens*, and irregularly spaced, hollow, bubbling notes. 'kookookook ..... kookarook ... kook-kook ....', etc. Contact calls, 'moo' and 'mair' (Ash 1978). Other calls include growling, humming, whining, mewing and hissing. ♀ gives low, crooning call to chicks. Sings mainly at night, often for several hours, sometimes all night; also by day, especially in wet weather. Usually sings from bush or tree, typically 1–3 m above ground, once at 8 m (Gillard 1976); also from ground.

General Habits. Tolerates wide variety of habitats but generally associated with forest or thick bush. Broad altitudinal range, from lowland rain forest to bamboo forest at 2600 m (Kenya) and Juniper/*Podocarpus* forest at 3200 m (Ethiopia). Generally favours forest edge, clearings, second growth and more open types of forest, but can also occur in dense forest. Also found away from tall trees in bushy ravines, brush-covered hillsides, dense thickets and scrub, and invades old, overgrown cultivated lands, banana plantations and gardens as long as there is plenty of ground cover. Sometimes enters grassland where congeners not present, but usually not far from forest or bush. Can occur along streams but not tied to water like White-spotted Flufftail: may be 1–2 km from nearest water.

Shy and skulking. Very hard to see even when singing, since it remains motionless and well concealed in dense vegetation, or may move away on approach of observer and sing from post farther ahead. Ventriloquial quality of song adds to difficulty of spotting bird. Mainly terrestrial but also arboreal, especially when singing; will fly from bush to bush and walks with agility on branches. In evening sometimes sings from branch in sun (Pakenham 1943). Forages on ground, picking at moss and other vegetation. When alarmed darts like mouse into undergrowth (Astley Maberly 1935b); only flies if pressed; flight rather clumsy.

Movements: no evidence of regular migration. Habitat not seasonal, and birds present throughout the year South Africa. Some local vagrancy indicated by birds captured in or near houses in wrong habitat (grassland and savanna), flying against lighted window at night, or even on city street. Status of birds far from normal habitat L. Turkana Apr (1 captured in hut, others seen in sparse, short grass) uncertain (Keith 1973).

Food. Chiefly insects, including termites, a small roach and 'hard-shelled black insects' (Pakenham 1943); also snails, spiders, seeds and a little vegetable matter.

Breeding Habits. Monogamous; solitary nester; territorial. ♂ bold in defence of nest (Cottrell 1949): rushed towards approaching humans in distinctly menacing manner, fluffing out feathers like broody hen, uttering low growls and hisses. ♂ and ♀ will attack people's legs if chicks approached (Oatley and Pinnell 1968). ♂ sitting on eggs gave snake-like hissing and plaintive 'peep-peep' (Porter 1970).

NEST: on ground, well hidden in grass, ferns or tangled vegetation; made of grass and sometimes leaves; lined with dry grass and rootlets; usually with dome or roof and entrance at side, but open nests recorded.

EGGS: 3–5; mean (7 clutches), 3·86. Large oval; glossy, white, sometimes stained. SIZE: 25–30·5 × 19–22.

LAYING DATES: Nigeria, (downy young June, imm. Dec); Cameroon, Sept, Oct; Zaïre, Sept; Uganda, (imm. Dec); Ethiopia, (♂ with testes slightly enlarged Apr); Kenya, (♂♂ breeding condition Feb, Apr, May); Mozambique, Sept; Zimbabwe, Apr; South Africa, Sept–Mar.

INCUBATION: by both sexes. ♀ with 2 tiny chicks seen foraging on ground in bushy undergrowth. Survival possibly aided by snake-like hiss given in defence of nest.

References
Astley Maberly, C. T. (1935b).
Cottrell, C. B. (1949).
Pakenham, R. H. W. (1943).

Plate 8
(Opp. p. 113)

Sarothrura rufa (Vieillot). Red-chested Flufftail. Râle à camail.

Rallus rufus Vieillot, 1819. Nouv. Dict. Hist. Nat. 28, p. 564; Africa.

Range and Status. Endemic resident. Isolated population in Ethiopia (Shoa, Kaffa and Wollega districts); scattered records from Sierra Leone, Togo and Nigeria; and more continuous range from Cameroon and Gabon through Central African Republic (Bangui) and N and E Zaïre to Burundi, central Uganda, W and central Kenya, mountains of NE Tanzania, and Zanzibar and Pemba; south through W Tanzania and S Zaïre to Angola, most of Zambia (except south) and most of Malaŵi. Apparently absent from Zambezi and lower Luangwa Valleys. Present throughout most of higher ground in Zimbabwe and adjacent highlands of Mozambique; and from central Transvaal and N Zululand (and probably S Mozambique) south through Natal and along coast to Cape. Also Botswana (Okavango area) and Namibia (Omanbonde). Recent records from coastal Kenya near Mombasa unsubstantiated. Recent range extensions proven from tapes of voice (Erard and Vielliard 1977, Ash 1978) suggest the species is often overlooked by observers unfamiliar with its calls, and its range may be more extensive than shown on the map. The most widespread member of the genus, found at elevations up to 2700 m; frequent to common.

Description. *S. r. rufa* (Vieillot): central Kenya, S Zaïre and Angola to South Africa. ADULT ♂: entire head, neck, mantle and breast light reddish chestnut, darker (sometimes washed black) on top of head and hindneck. Lores and cheeks somewhat darker than paler face; chin and throat paler chestnut, chin almost buffy. Rest of upperparts, including tail, scapulars, tertials and wing-coverts black with short white streaks, becoming spots on tail. Rest of underparts blackish brown closely streaked white, palest in centre of belly. Primaries and secondaries blackish brown; axillaries and underwing-coverts brownish black narrowly barred white. Bill, upper mandible blackish, lower mandible blue-grey. Eye dark brown; legs and feet dark grey-brown. ADULT ♀: entire upperparts from top of head to tail black dotted with spots or broken bars of buff. Sides of head and neck buffy, finely scaled and barred black. Chin, throat and centre of belly off-white, rest of underparts buff with black bars. Scapulars, tertials and wing-coverts as upperparts but barred, not spotted with buff. Rest of wing as ♂. SIZE: wing, ♂ (n = 63) 71–80 (75·8), ♀ (n = 26) 71–81 (76·8); bill, ♂ (n = 63) 13–16 (14·4), ♀ (n = 26) 12·5–15 (13·8); tarsus, ♂ (n = 62) 19·5–23 (21·7), ♀ (n = 25) 19·5–23 (20·2). WEIGHT: 4 ♂♂ 30–42 (33·8).

IMMATURE: dull black above, dull grey-black below; chin, throat and centre of belly whitish.

DOWNY YOUNG: brownish black. Bill black, tip white and base deep pink. Eye black, legs purplish black (Liversidge 1968).

S. r. elizabethae Van Someren: Central African Republic, NE Zaïre, Uganda, W Kenya, Ethiopia. Upperparts of ♂ with longer white streaks, no spots; upperparts of ♀ tend to be barred rather than spotted buff.

S. r. bonapartei (Bonaparte): Sierra Leone to Gabon. Small (wing 66–73). Upperparts of ♂ as *elizabethae*; upperparts of ♀ tend to be streaked rather than spotted or barred.

Field Characters. ♂ distinguished from ♂ Chestnut-headed Flufftail *S. lugens* and ♂ Böhm's Flufftail *S. boehmi* (found in similar or adjacent habitats) by chestnut extending to mantle and lower breast; from Böhm's Flufftail by blacker upperparts with shorter white streaks, and spots on tail. ♀ told by buff, not white markings on upperparts. Both sexes distinguished from Striped Flufftail *S. affinis* by lack of chestnut in tail.

Voice. Tape-recorded (11, 20, 21, 53, 74, 75). Song a series of clear notes, each 0·6–0·8 s long, with intervals of 0·4–0·7 s between notes, lasting up to a minute or more but typically shorter. Notes are on one pitch but increase in intensity, giving the impression of rising in pitch ... 'wooah'. They are shorter and higher-pitched than any African *Sarothrura* species except White-spotted Flufftail *S. pulchra*. Infrequently, a second bird answers antiphonally with a higher-pitched 'wah' which immediately follows the 'wooah'. Sometimes the singing bird 'warms up' with some low grunts, then incorporates them into the song, 'g'wooah' (similar action noted for Böhm's Flufftail). In some songs the 'wooah' note is immediately followed by a grunt which can be quite loud, possibly made by a second bird ... 'wooah-boo' or 'haw-boo'.

Common call, a series of loud notes rising in pitch, 'kei-kei-kei ...' or 'dueh-dueh-dueh ...'. As in song, a grunt may be incorporated into the notes. Other calls include a deep grunting 'boo' or 'wuk' which may be answered by a second bird on a higher pitch; a low growl, and a plaintive chirp. Warning note, a high-pitched hum. ♀ calls chicks with 'dueh' note; chicks answer 'cheep'.

Calls day and night. Rain stimulates calling; calling decreases after nesting has begun and during dry season.

General Habits. Inhabits permanent swamps and marshes in reed beds, rushes, tall, dense, tussocky grass and other tall, rank vegetation, including papyrus; also found beside rivers and around ponds and dams, always in thick vegetation. During rains in Zambia, moves out of reedbeds and swamps to occupy taller, denser parts of seasonally inundated grassland (P. B. Taylor, pers. comm.). In Zambia occurs in same dambos as Böhm's Flufftail, occupying lusher, taller vegetation in centre (Böhm's is in shorter grass at edge). Generally at low or medium elevation, once at 2700 m (Molo, Kenya).

Feeds mainly at dawn and dusk. Runs through vegetation at great speed; chicks can run as fast as adults. At dusk seen to creep about in tall vegetation tangles *c.* 1·5 m above ground (Pakenham 1943).

Sedentary; individuals on Pemba spend weeks or months in same few m^2 of territory (Pakenham 1943). 2 birds away from usual habitat in South Africa were probably strays, not migrants.

Food. Snails, small ants, termites, minute flies and water bugs; small seeds, including grass seeds.

Breeding Habits. Monogamous; territorial. On Pemba, breeding pairs in continuous habitat spaced at distances of 50–100 m; lone pair in isolated locality had smaller territory.

NEST: shallow grass cup lined with narrow, reed-like leaves, well hidden in clump of grass in marsh; diam. outside 13 cm, inside 8–9·5 cm, depth of cup 3–4 cm, edge of cup 21 cm above ground; with or without slight dome.

EGGS: 3–5, laid at 1–2 day intervals; mean clutch size (12 clutches) 3·75. Oval; white. SIZE: (n = 11, South Africa) 19·2–21·5 × 25·8–28·6 (20·5 × 27·3).

LAYING DATES: Sierra Leone, May, July; Nigeria, Dec; Cameroon, Mar, May, July; Zaïre, Jan, Feb, May, July, Aug; Angola (breeding condition, May); East Africa: Uganda, Oct; Tanzania (breeding condition Mar, Apr, July); Pemba, Jan–Apr; Region D, Feb, Mar; Malaŵi, Feb–Apr; Zambia, Jan–Mar (breeding condition, Oct); Mozambique, May, Nov; Zimbabwe, Jan–Mar, May, Dec; South Africa: Transvaal, Jan; Natal, May, Dec; Cape, Jan, Nov, Dec. Breeds during rains in southern Africa; breeding season indeterminate in equatorial regions.

INCUBATION: by both sexes. Period: 14 days from day last egg laid. At Cape nest, ♂ incubated in morning, eggs left uncovered in afternoon, ♀ incubated late afternoon and evening (Broekhuysen *et al.* 1964).

DEVELOPMENT AND CARE OF YOUNG: hand-reared ♂ developed as follows (Liversidge 1968): at 3 weeks legs became less black, white on bill reduced to tip, chick developed 'jump-run'—jumped up with a twist as if to kick off from ground, and dashed off with rapid initial rush. At 4 weeks bill and feathers on tail elongated, black feathers appearing on wing, head covered with dark quills, chin paler, and hint of pale medium ventral line. At 6 weeks feathers of mantle developed white streaks, chin and median stripe to vent white, rest of plumage sooty black; bill black with pink nares and base. At just under 7 weeks, first buff neck feathers appeared, bird gave deep double call note.

References
Broekhuysen, G. J. *et al.* (1964).
Keith, S. *et al.* (1970).
Liversidge, R. (1968).
Pakenham, R. H. W. (1943).

Plate 8
(Opp. p. 113)

Sarothrura lugens (Böhm). **Chestnut-headed Flufftail. Râle à tête rousse.**

Crex lugeus (sic) Böhm, 1884. J. f. Orn. 32, p. 176; Ugalla country, interior of Tanganyika Territory.

Range and Status. Endemic resident, known in scattered localities from Cameroon (Obala, Ngaounyanga) to NE Zaïre (Faradje), central Angola (Chitau) and W Tanzania (Ugalla), and from NE Zambia and SE Zaïre to Zimbabwe (Inyanga). Generally uncommon; locally common NE Zambia.

Field Characters. ♂ distinguished from ♂ Red-chested Flufftail *S. rufa* and ♂ Böhm's Flufftail *S. boehmi* (found in similar habitat) by lack of red on sides of neck, throat and breast; from *rufa* by lack of red on mantle; from *boehmi* by blacker upperparts, spots on tail. ♀ told from ♀ *rufa* by white, not buff marks on upperparts; from ♀ *boehmi* by chestnut wash on head.

Voice. Tape-recorded (53, 204, 253). Song, a series of moaning notes lasting up to 1 min. Speed and pitch variable: lower notes last *c.* 1 s, higher ones *c.* 0·5 s. Sometimes notes increase in intensity toward middle of song, and at the end may die away into short 'exhaustion' notes, 'hoo-boo'. Call, a series of rapid, loud, far-carrying notes, 'koh-koh-koh-koh....', 3/s for 30–45 s. Notes loudest in middle of series, dying away at the end to a grunt. 2 birds of a pair may call at once, asynchronously. Response to playback, an irregular, low 'annoyance' call like that of Böhm's Flufftail.

General Habits. Inhabits grassy clearings, such as patches of savanna in lowland forest and dambos in miombo woodland, also grassy marshes in savanna and marshes beside lakes. Prefers tall, lush grass and dense vegetation 0·7–1·5 m high.

Food. Insects, small black ants and seeds.

Description. *S. l. lugens* (Böhm): Cameroon to Zaïre and W Tanzania. ADULT ♂: whole head down to hindneck and malar region rich chestnut. Chin and throat off-white. Sides of neck, upperparts from mantle to uppertail-coverts, scapulars, tertials and most wing coverts black with fine white streaks. Tail similar but streaks become spots. Underparts similar to upperparts but white streaks much broader, producing overall paler effect, especially on centre of belly. Narrow white line around leading edge of bend of wing. Primaries, secondaries and underwing-coverts dark greyish brown. Bill dusky brown, underside of lower mandible whitish. Eye dark greyish brown; legs and feet very dark brown. ADULT ♀: similar to ♂ but chestnut of head paler and more buffy, and heavily streaked black. Overall tone of body and wings browner, streaks on mantle somewhat broader, wings with spots or bars rather than streaks. Pale area on belly more extensive. SIZE: wing, ♂ (n = 31) 75–82 (78·2), ♀ (n = 15) 76–82 (79·1); bill, ♂ (n = 29) 13–15 (14·4), ♀ (n = 13) 13–15 (14·1); tarsus, ♂ (n = 29) 19–22·5 (20·6), ♀ (n = 13) 20–22 (20·9).
 IMMATURE: uniform blackish with dingy white chin.
 DOWNY YOUNG: black; bill black, base and tip white.
 S. l. lynesi (Grant and Mackworth-Praed): Angola, Zambia and Zimbabwe. Like *lugens* but smaller: wing, ♂♀, 70–74 vs. 75–82.

Breeding Habits. Nests and eggs have never been found. Laying dates (derived from young birds or collected ♀♀ containing eggs): Cameroon, Apr, July, Sept; Zaïre, Mar–Apr; Zambia, Mar, Dec.

References
Keith, S. *et al.* (1970).
Roux, F. and Benson, C. W. (1969b).

Sarothrura boehmi Reichenow. Böhm's Flufftail; Streaky-breasted Pygmy Crake. Râle de Böhm.

Plate 8
(Opp. p. 113)

Sarothrura boehmi Reichenow, 1900. Vög. Afr. 1, p. 272 (in key), p. 290; Likulwe R., Katanga, Belgian Congo.

Range and Status. Endemic resident and intra-African migrant. From Cameroon, Zaïre and Kenya to E Angola, S Tanzania (Mufindi), Zambia and Zimbabwe. Unsubstantiated sight record Mali, November (Lamarche 1980). Vagrant taken at sea *c*. 150 km off coast of Guinea (10°N × 15°W). Usually uncommon but may be locally numerous (estimated 100 + calling ♂♂ on large grass plain, Zambia; P. B. Taylor, pers. comm.)

Field Characters. ♂ distinguished from ♂♂ of Red-chested Flufftail *S. rufa* and Chestnut-headed Flufftail *S. lugens* (which occur in similar or adjacent habitats) by amount of chestnut in plumage, which extends to hindneck and upper breast in this species, to mantle and lower breast in Red-chested, while in Chestnut-headed it is confined to top of head and hindneck. Upperparts have brownish wash but appear black in flight, as in the other 2, but head and neck of ♂ appear somewhat paler, more orange chestnut than Red-chested. ♀ told from ♀ Red-chested by white, not buff markings on upperparts, though these not easy to see in flight; very like ♀ Chestnut-headed but lacks chestnut wash on head. ♀ and imm. not unlike dark brown ♀ and imm. of Blue Quail *Coturnix chinensis*, which occurs in same habitat, but distinguished in flight by longer body and wings, long legs, slower wing action without pauses, white markings on upperparts.

Voice. Tape-recorded (53, C, F, 253, 258, 296). Song of 2 types: (a) a rather deep, hollow 'hooo' repeated at 2 s intervals, and (b) a higher-pitched 'er'. In (b), each note lasts 0·3–0·4 s, with intervals of 0·6–0·7 s between notes, and may be repeated up to 25 times. Note (b) is preceded by a low grunt, audible only at close range, probably connected with the mechanism of sound production, not a separate note of the song. The combination may be written 'g'wer'. In reaction to playback, song trailed off into annoyance call, a low 'cuk-cuk-cuk....'. In another reaction to playback, ♂ continued calling while second bird, presumed ♀ of pair, joined with agitated 'cuk' call. Incubating birds hiss when disturbed. Call of chicks, 'peep, peep'.

General Habits. Breeding habitat similar to that of Striped Crake *Aenigmatolimnas marginalis*, with which it occurs – areas of short grass temporarily inundated during the rains, such as edges of rivers, dambos and swamps, and open, grassy flats. Altitudinal range, 500–1890 m. Retreats toward equatorial regions during dry season, when breeding habitat liable to be burnt over. All records from southern part of range (Zambia, Malaŵi, Zimbabwe) are Dec–Mar, i.e. during rains. Migrants taken at night around buildings, S Tanzania, May (Anon. 1983; N. E. Baker, pers. comm.). Bird collected at sea off Guinea was presumably migrant blown off course. Relatively long wing suggests migratory tendency. Flight strong and direct.

Description. ADULT ♂: head, neck and upper breast light chestnut, somewhat darker on top of head and hindneck. Chin and throat white. Amount of chestnut on lower throat and breast variable. Upperparts, including tail, scapulars, tertials and wing-coverts sooty black with fine white streaks. Underparts white, with broad black streaks except on centre of belly. Primaries, secondaries, axillaries and underwing-coverts dark grey-brown. Bill brown, lower mandible paler. Eye brown, legs and feet greenish brown. ADULT ♀: entire upperparts from top of head to tail, scapulars, tertials and wing-coverts dull sooty black, scalloped and barred with white. Chin, throat and underparts white, barred black on breast and flanks. SIZE: (23♂♂ 13♀♀) wing, ♂ 82–88 (84·8), ♀ 81–88 (84·7); bill, ♂ 13–14·5 (13·8), ♀ 13–14·5 (13·7); tarsus, ♂ 18·5–21·5 (19·6), ♀ 18–21·5 (19·5).

IMMATURE: dull sooty black; chin, throat and centre of belly white.

DOWNY YOUNG: black. Bill black with pale tip.

Food. Small seeds, especially grass seeds, and small insects.

Breeding Habits. Monogamous; solitary nester in grassland.

NEST: small pad or shallow saucer of grass *c.* 2·5 cm thick, placed 2·5–7·5 cm above wet ground in grass tuft, part of which is pulled down to form roof over nest.

EGGS: 2–5; mean (33 clutches Zimbabwe), 3·8. Oval; pale cream to white, sometimes with few tiny brown spots round larger end. SIZE: (n = 31) 26·3–28·1 × 18·5–20·3 (27·1 × 19·2).

LAYING DATES: Kenya (birds in breeding condition, May; imm., Sept, judged to be from egg laid July); Malaŵi, Zambia, Zimbabwe, Jan–Mar. Laying takes place during the rains.

INCUBATION: carried out by both sexes.

DEVELOPMENT AND CARE OF YOUNG: day-old chicks described by Neuby Varty (1953) as 'an inch tall and black, mostly legs and feet, and very active . . . . always on the move': He believed they never returned to the nest after hatching and he did not see parents near chicks after they left the nest.

BREEDING SUCCESS/SURVIVAL: some evidence of desertion and loss of eggs following discovery or disturbance by humans. Eggs believed often eaten by swamp rat *Otomys* sp., and it may be advantageous for birds to breed early in rains before rat population has recovered from reduction by burning of grass in dry season. 1 nest destroyed when ox stepped on it.

References
Keith, S. *et al.* (1970).
Neuby Varty, B. V. (1953).

Plate 8
(Opp. p. 113)

Sarothrura affinis (Smith). Striped Flufftail; Chestnut-tailed Crake. Râle à queue rousse.

Crex affinis Smith, 1828. South African Commercial Advertiser, Vol. 3, p. 144; Cape Province.

Range and Status. Endemic resident; discontinuously distributed in widely separated areas of montane grassland from S Sudan (Imatong Mts), Kenya, extreme N Tanzania (Ndassekera), extreme S Tanzania (Matengo Highlands), Malaŵi, extreme NE Zambia (Nyika Plateau), E Zimbabwe, and South Africa from Transvaal to Cape. Uncommon.

Description. *S. a. affinis* (Smith): South Africa. ADULT ♂: top and sides of head light chestnut, becoming buffy on chin and throat. Upperparts, including scapulars, tertials and wing-coverts black with long whitish streaks. Tail light chestnut; underparts streaked black and white. Rest of wing dark grey-brown; outer web of outer primary white; a few pale streaks on axillaries and underwing-coverts. Bill black, base of lower mandible flesh. Eye brown; legs and feet flesh brown. ADULT ♀: upperparts from top of head to uppertail-coverts, scapulars, tertials and wing-coverts dark brown, scalloped and barred whitish. Tail chestnut with black bars. Sides of head and neck buff, finely speckled black. Chin and throat white; underparts whitish to buff with black spots and scales, becoming bars on flanks. Undertail-coverts rusty with black bars. Rest of wing as ♂, but small buff marks on outer webs of flight feathers. SIZE: (7 ♂♂, 6 ♀♀) wing, ♂ 70–76 (72·9), ♀ 68–76 (71·7); bill, ♂ 13–14 (13·5), ♀ 12–13·5 (13); tarsus, ♂ 15·5–17·5 (16·5), ♀ 15–16·5 (15·7). WEIGHT: 1 ♂ 28·8; 1 ♂ *S. a. antonii*, *c.* 30.

IMMATURE: not known; an imm. *S. a. antonii* in partial adult plumage is mainly plain blackish with some adult feathers on upperparts and some dull chestnut on head and tail.

S. a. antonii Madarasz and Neumann: Zimbabwe to S Sudan. Chestnut on neck of ♂ extending onto upper breast; larger (wing 76–85). Populations from Kenya and Sudan may be different subspecies but more material needed (Keith *et al.* 1970).

Field Characters. The only flufftail in its habitat for much of its range, but in South Africa enters marshes inhabited by Red-chested Flufftail *S. rufa*. ♂ easily distinguished by unbarred chestnut tail (tail of ♂ Red-chested black with white spots). ♀ similar to ♀ Red-chested but tail barred chestnut and black.

Voice. Tape-recorded (53, 253). Song, series of up to 30 moaning notes *c.* 1 s long, given at intervals of *c.* 1·5 s; more forceful and higher-pitched than moaning song of Chestnut-headed Flufftail *S. lugens*: about same pitch as song of Buff-spotted Flufftail *S. elegans* but notes much shorter. Notes are all on one pitch but intensify in middle of song, as in other flufftails, giving impression they are rising in pitch. Rattling, tinny note lasting 2–3 s believed to come from ♀ (Benson and Holliday 1964). Sings both during day and at night.

General Habits. Typically inhabits dry uplands with long or short grass and sometimes admixture of bracken and *Protea*; in South Africa often near forest edge, also in crops (lucerne, millet). Tends to avoid swampy situations, but in Kenya found in open moorland bogs, 3350–3360 m, and in South Africa enters marshy vleis (seasonally?).

Shy and skulking; sometimes so reluctant to fly that it can be stepped on or picked up by hand. Flight feeble; wing-beats rapid. No evidence of long-distance migration, but may move locally to damper situations when grasslands become too dry or are burnt. Apparently resident in breeding areas throughout year.

Food. Small insects, including small beetles, seeds and some vegetable matter, also possibly termites and snails.

Breeding Habits. Monogamous; solitary nester.
NEST: bowl of rootlets built into grass tuft, shielded by canopy of loose grass.
EGGS: 4–5; mean (4 clutches) 4·3 (2 clutches of 1 egg presumed incomplete). Oval, rounded or pointed; smooth, sometimes shiny, white or off-white. SIZE: (n = 6, South Africa) 25–26·7 × 17·7–20·4 (26·15 × 19·3); (n = 5, Zimbabwe) 23·8–25·5 × 19·5–20·0.
LAYING DATES: Sudan, (♂♀ breeding condition, May); Kenya, May; (♀♀ breeding condition Feb); Tanzania, Jan; Zambia, (♂ breeding condition Jan); Zimbabwe, Jan; South Africa, Dec, Feb.
INCUBATION: by both sexes.

Reference
Benson, C. W. and Holliday, C. S. (1964).

Sarothrura ayresii (Gurney). White-winged Flufftail; White-winged Crake. Râle à miroir. Plate 8
Coturnicops ayresii Gurney, 1877. Ibis, p. 352, pl. 7; Potchefstroom, Transvaal. (Opp. p. 113)

Range and Status. Endemic resident and intra-African migrant. Discrete population in Ethiopia (Addis Ababa area: Akaki, Gafersa, Sululta Plain; and Kaffa Province: Charada). Not seen since 1948 despite extensive recent searching. Recent sight records from Zambia (Chingola) and Zimbabwe (Harare); status there not known. Recently rediscovered South Africa, where previously known for certain only from 4 old specimens from Potchefstroom, Bloemfontein and King William's Town. In 1975 bird found dead near Heidelberg, Transvaal (Wolff and Milstein 1976); and in 1982 populations discovered in Franklin Marsh, E Griqualand, Natal (35 birds counted, Oct 1982–Jan 1983) and Belfast, Transvaal (5 or more, 9 Dec 1982–9 Jan 1983: *Bokmakierie* 36, 66).

South African birds threatened by draining of vleis, building of dams and grazing of habitat by cattle.

Description. ADULT ♂: top of head and nape blackish with chestnut bases to feathers. Hindneck, sides of neck and upper mantle dark chestnut, becoming lighter chestnut on lower throat and sides of breast. Lores and cheeks blackish; sides of head chestnut with fine black scaling. Chin and throat whitish, washed with pale chestnut at sides. Rest of upperparts, scapulars, tertials and wing-coverts black with fine white streaks. Tail barred black and chestnut. Centre of breast and belly white, flanks and undertail-coverts streaked black and white. Primaries dark grey-brown, secondaries white. Axillaries, underwing-coverts and narrow line round bend of wing white. Bill purplish brown, eye brown, legs and feet purplish flesh. ADULT ♀: similar to ♂ but blacker, less rufous on head and neck. Breast rufous brown scaled with dark brown; flanks black spotted with white. Inner secondaries as back, outer ones white. SIZE: (13 ♂♂, 12♀♀) wing, ♂ 73–79 (76), ♀ 75–80 (76·6); bill, ♂ 12–13 (12·3), ♀ 12–13·5 (12·5); tarsus, ♂ 17–19·5 (18·3), ♀ 16–20 (18·5). WEIGHT: 1 ♀ 14, but specimen already partly dry.
IMMATURE: brownish black, throat and centre of belly white, breast feathers barred white and dusky.

Field Characters. Both sexes distinguished from all other flufftails by white patch in wings, conspicuous in flight, and from all marsh and grassland spp. except ♀ Striped Flufftail *S. affinis* by tail barred red and black. At rest ♀ told from ♀ Striped Flufftail by blacker, white-marked upperparts, rufous on hindneck and breast. Appears smaller in flight than Böhm's Flufftail *S. boehmi*.

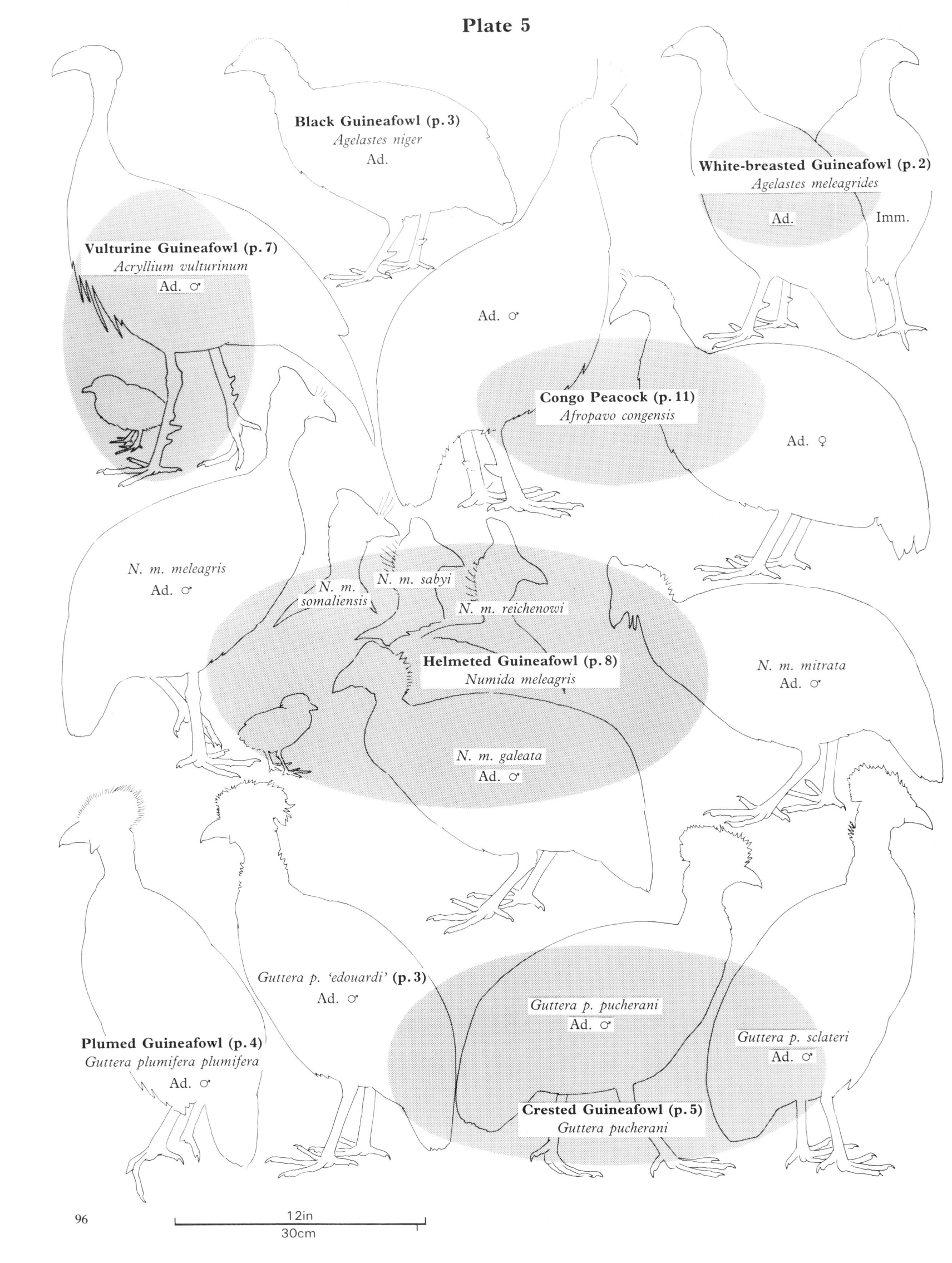

Plate 5

Black Guineafowl (p. 3)
Agelastes niger
Ad.

White-breasted Guineafowl (p. 2)
Agelastes meleagrides
Ad.   Imm.

Vulturine Guineafowl (p. 7)
Acryllium vulturinum
Ad. ♂

Ad. ♂

Congo Peacock (p. 11)
Afropavo congensis

Ad. ♀

N. m. meleagris
Ad. ♂

N. m. somaliensis   *N. m. sabyi*
N. m. reichenowi

Helmeted Guineafowl (p. 8)
Numida meleagris

N. m. mitrata
Ad. ♂

N. m. galeata
Ad. ♂

Guttera p. 'edouardi' (p. 3)
Ad. ♂

Guttera p. pucherani
Ad. ♂

Guttera p. sclateri
Ad. ♂

Plumed Guineafowl (p. 4)
Guttera plumifera plumifera
Ad. ♂

Crested Guineafowl (p. 5)
Guttera pucherani

12in
30cm

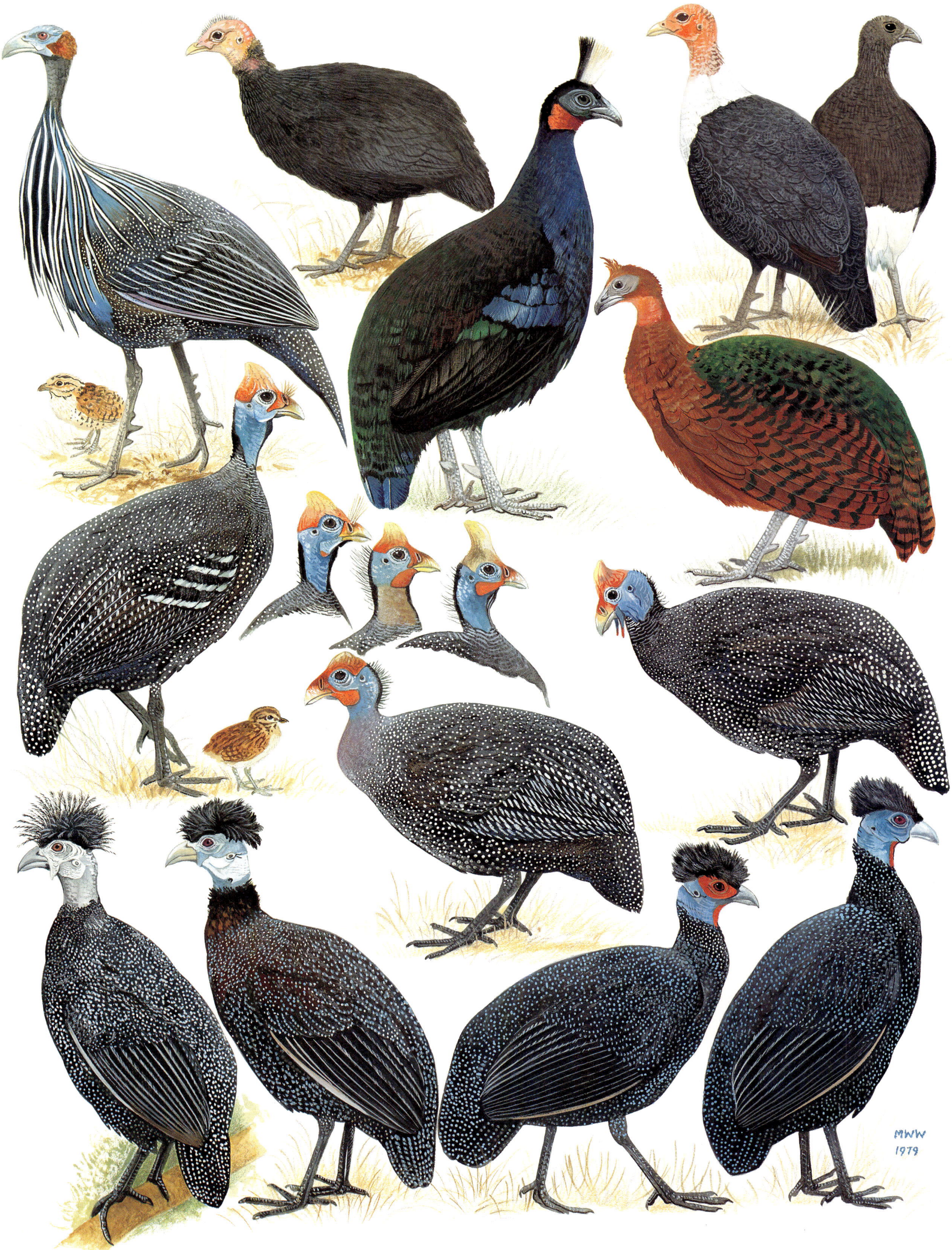
MWW
1979

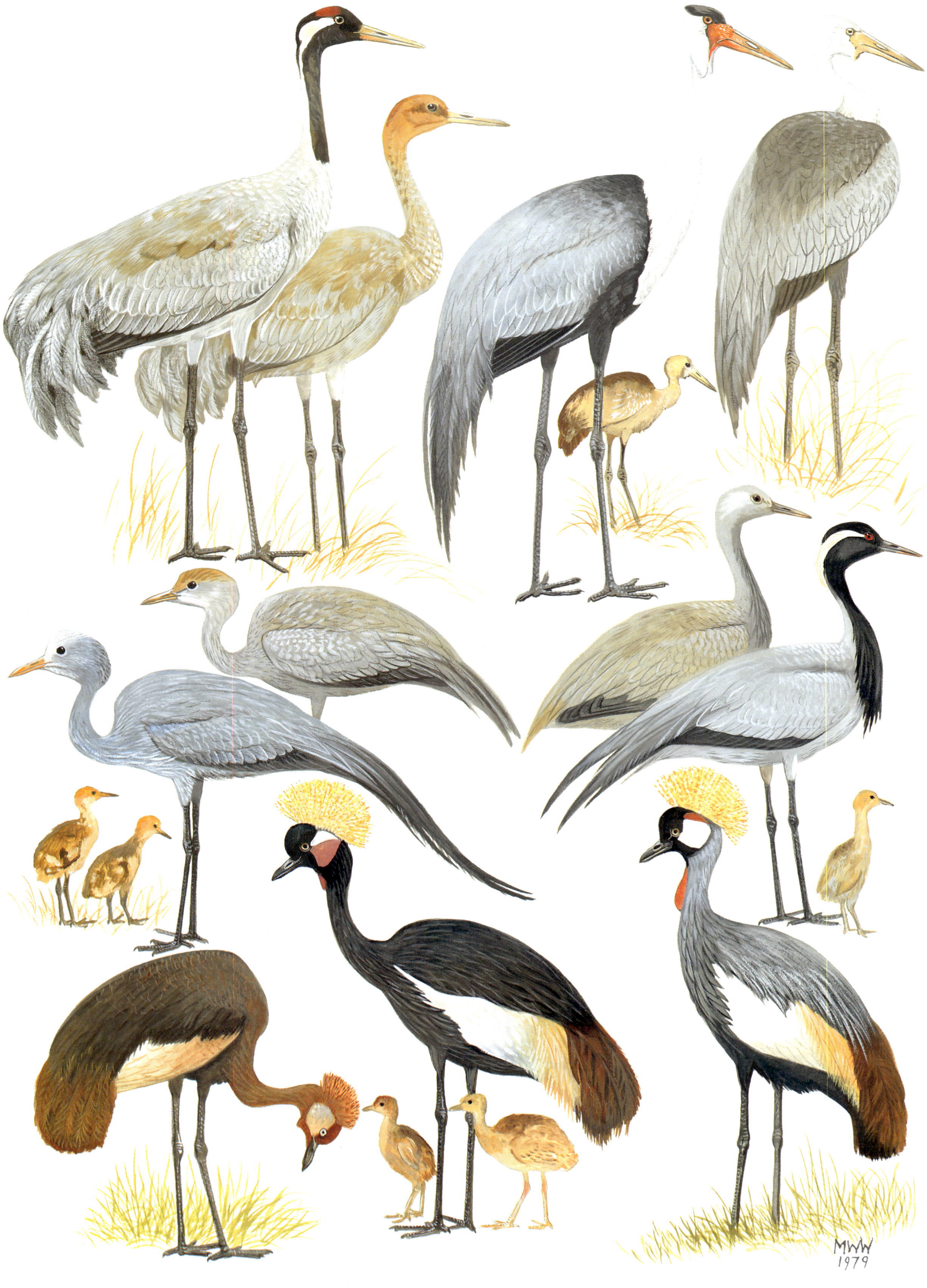
MWW
1979

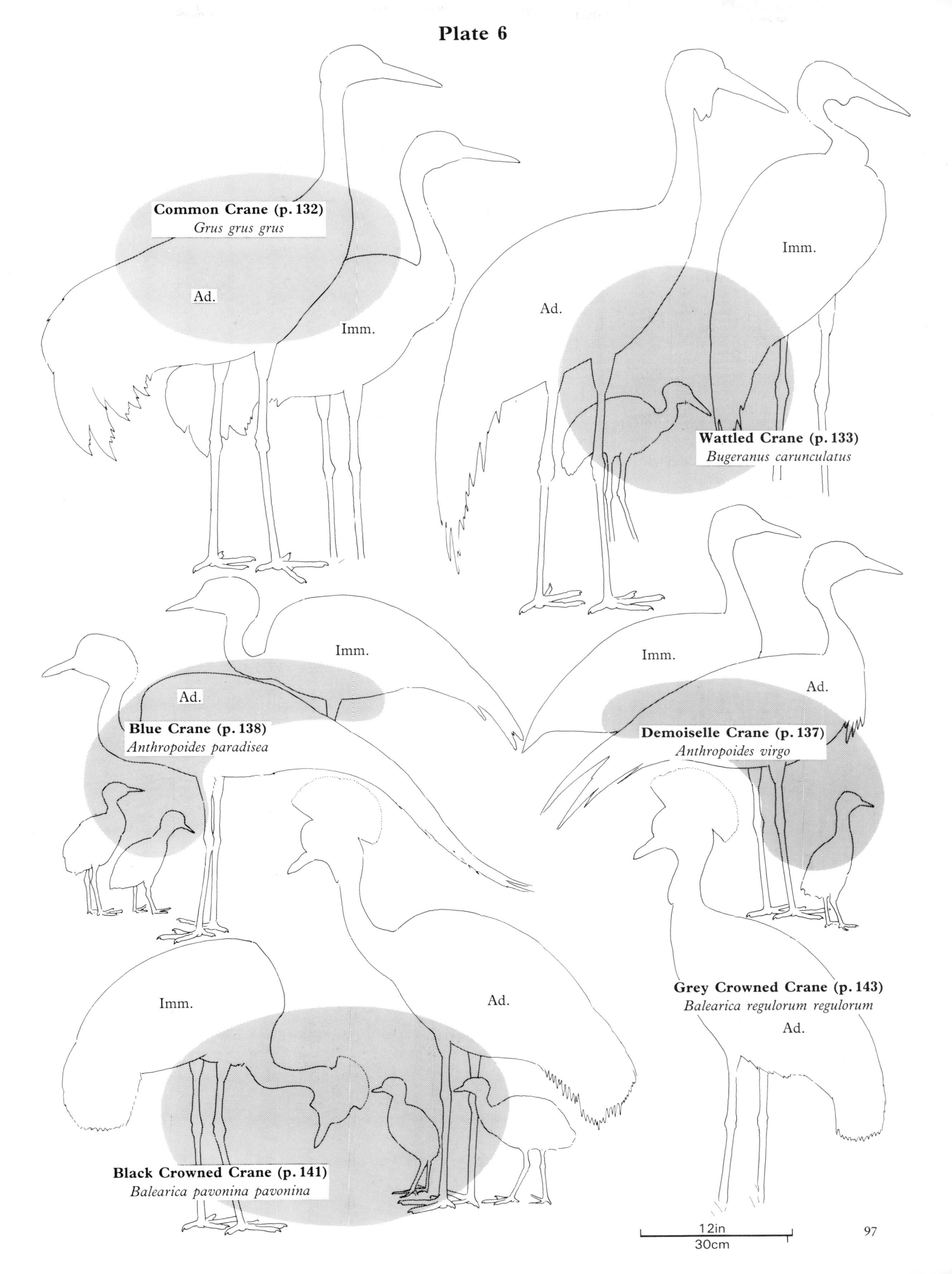

Plate 6

Common Crane (p. 132)
Grus grus grus

Ad.

Imm.

Wattled Crane (p. 133)
Bugeranus carunculatus

Ad.

Imm.

Blue Crane (p. 138)
Anthropoides paradisea

Ad.

Imm.

Demoiselle Crane (p. 137)
Anthropoides virgo

Imm.

Ad.

Grey Crowned Crane (p. 143)
Balearica regulorum regulorum

Ad.

Imm.

Ad.

Black Crowned Crane (p. 141)
Balearica pavonina pavonina

12in
30cm

97

Voice. A very deep 'oooh, oooh' (J. C. Sinclair, pers. comm.).

General Habits. Breeding habitats Ethiopia: (a) 1-acre marsh with rushes and marsh orchids, ankle-deep water; (b) flat, grassy plain partly inundated during rainy season. Other habitats: pan-like marsh (Zambia); wet vleis with knee-high grass (especially *Leersia hexandra* and *Hemarthria altissima*) or taller (1 m) rank grass and clumps of *Cyperus* (Zimbabwe); vleis with short sedges and water grass, long grass by dam, and beds of reeds (*Phragmites australis*) and bulrushes (*Typha*) (South Africa). Though occurring in similar habitat to Red-chested Flufftail *S. rufa* in Franklin Marsh, the 2 spp. apparently occur in pockets, with little mixing (Mendelsohn *et al.* 1983). When flushed from reed bed, birds scramble to the reed tops, take off at considerable speed on whirring wings and fly some distance before crashing back into the reeds.

Arrives Ethiopian breeding grounds June/July, vacates them in dry season; Gafersa dry in Apr and Sululta Plain dry in May with cattle feeding on it. Distance travelled not known; possibly to permanent marshes in SW Ethiopia since Kaffa bird taken May. Strong-flying bird seen Zimbabwe Feb by Hopkinson and Masterson (1977) suggests at least local movement. Evidently some local movement in South Africa; birds absent Franklin Marsh Nov, when water high, present Oct and Dec when water lower.

Food. Only water insects recorded.

Breeding Habits. No nest or eggs yet found. Ethiopian birds in breeding condition, July, and young bird unable to fly taken Sept 22 suggest breeding in August. South Africa (Franklin Marsh), 2 partly grown birds with stunted wings, unable to fly far, seen 6 Jan (Mendelsohn *et al.* 1983).

References
Guichard, K. M. (1948).
Hopkinson, G. and Masterson, A. N. B. (1977).
Mendelsohn, J. M. *et al.* (1983).
Wolff, S. W. and Milstein, P. le S. (1976).

Genus *Crex* Bechstein

Medium-sized rails more often found in dry grassland than other African species. Bill short and stout, nasal bar tenuous, wings broad and rounded. Underparts grey (*egregia*) or rufous (*crex*); *crex* with rufous wing-patch. African Crake *C. egregia* linked with neotropical *Porzana albicollis* by Benson and Winterbottom (1968), due to similarity in plumage, and placed in same superspecies, but Olson (1973) notes that *P. albicollis* has broad, flat nasal bar of typical *Porzana* and considers them unrelated. Olson, with some hesitation, maintains *egregia* in monotypic genus *Crecopsis*, but we prefer to link it with *Crex*.

2 spp., *egregia* endemic to Africa, Corncrake *C. crex* breeding in Palearctic and wintering in Africa.

Plate 7
(Opp. p. 112)

Crex egregia (Peters). African Crake. Râle des prés.

Ortygometra (*Crex*) *egregia* Peters, 1854. Monatsb. K. Akad. Berlin, p. 134; Tete, Zambezi.

Range and Status. Endemic resident and intra-African migrant. Widespread south of Sahara and locally common except in rain forest and desert. Occurs from Senegambia (uncommon resident; scarce visitor July–Dec) to Nigeria (widespread and locally common except Sahel zone), Cameroon, Zaïre, S Sudan and Darfur (where probably scarce breeder) and Ethiopia; south through Angola (generally distributed except SW) to N Namibia, N Botswana, and through Burundi (abundant on plain of lower Ruzizi R. and shores of Lake Tanganyika) and East Africa (widespread, locally common) to Malaŵi (common below 1500 m in rains, Dec–Apr), Zambia (locally common Nov–May), Mozambique (locally not uncommon), Zimbabwe, and South Africa to E Cape Province (Ferndale). Vagrant São Tomé, Bioko (Fernando Po) and coastal Namibia (Luderitz).

Description. ADULT ♂: forehead, crown and nape black, feathers narrowly edged olive-brown. Narrow white line above lores. Lores, sides of head and neck, lower throat and breast grey, sides of breast washed olive-brown. Chin and upper throat white. Upperparts, including tail, wing-coverts and tertials black, the feathers broadly edged olive-brown, producing mottled appearance. Belly, flanks and undertail-coverts barred black and white. Primaries and secondaries dark brown; axillaries and underwing-coverts blackish narrowly barred white. Bill pink, reddish or purplish with distal portion grey, dusky culmen, cutting edge blue-grey or blue-violet. Eye red or orange, bare skin round eye salmon pink, legs and feet pale brown or grey, sometimes with reddish or pinkish tinge. Sexes alike. SIZE: (8 ♂♂, 12 ♀♀) wing, ♂ 118–133 (123), ♀ 117–131 (124); tail, ♂ 36–48 (41·6), ♀ 37–42 (39·7); bill, ♂ 23–27 (24·8), ♀ 21–26 (23·4); tarsus, ♂ 38–43 (40·8), ♀ 39–43 (40·7). WEIGHT: 1 ♂ 137, 4 unsexed 110–117 (113·5).

IMMATURE: resembles adult but top of head and hindneck

Crex egregia

Intra-African migrant

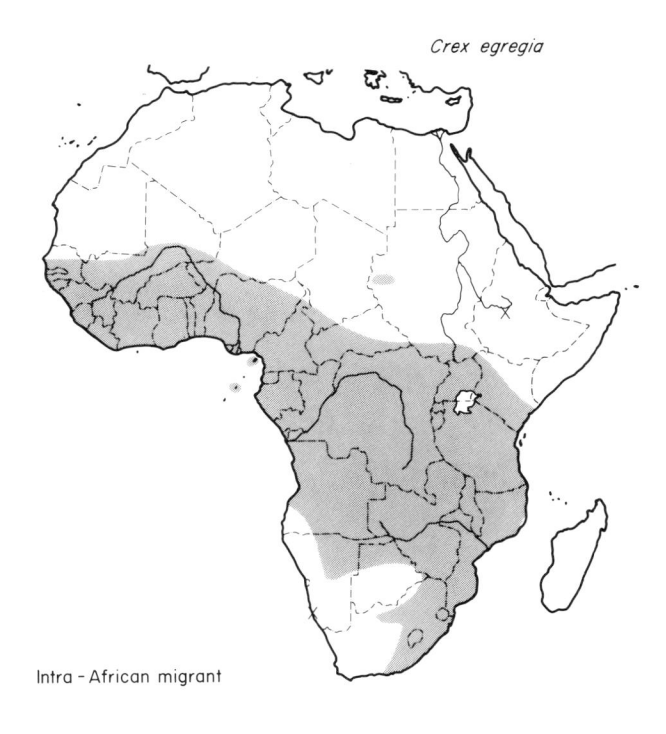

brown, face, neck and breast dull brown; some have grey on ear coverts and behind eye. Barring on underparts less distinct, brown and off-white, not black and white.

DOWNY YOUNG: black; eye, legs and feet blackish brown.

Field Characters. Slightly smaller than Corncrake *Crex crex*. Similarly striped upperparts appear generally darker, feathers edged darker olive-brown; feather edges of Corncrake often pale, giving pale streaked appearance. From side view dark barring on flanks and lack of rufous on underparts apparent. In flight lacks rufous wing-patch of Corncrake and wings appear shorter, more rounded, wing-beats deeper; tail shorter, more rounded, appearing all dark. Bill appears longer and heavier than that of Corncrake, gonydeal angle shallow; bill of Corncrake often looks short and conical, with sharp gonydeal angle. Forehead flatter and crown less rounded than Corncrake; this, combined with bill shape, gives the head a longer, flatter appearance. Other sympatric crakes are considerably smaller and have white streaks or spots on upperparts. Young African Crake not fully grown is closer in size to Striped Crake *Aenigmatolimnas marginalis* or Spotted Crake *Porzana porzana* but upperparts duller, darker, less patterned, lacking any white, and undertail-coverts barred.

Voice. Tape-recorded (F, 258). Common call, indicating both aggression and alarm, a single hard 'kip', 'kup', 'tsack' or 'chack'; loud, sharp, and quite low-pitched, usually repeated several times and occasionally run together as a rapid series: 'kip kip kir-ip-ip kip kip kip . . . '. Frequently given by bird before or after it flies from long grass; also given when flushed, and on approach of intruder. Neck stretched up during call. Also a short, harsh, grating churr quite similar to that of Corncrake, at rate of 2–3 notes/s. Alarm call of young, loud, chittering 'chi-chi-chi . . .'.

General Habits. Inhabits freshwater swamps, reedy marshes, rank vegetation by rivers and streams, seasonally inundated grasslands and grassy edges of dams, marshes, dambos, and rice fields; also far from water in dry grassland, airfields, tall grass savanna, grassy areas in dry woodland, weed patches, edges of cultivation, maize and cotton fields; sometimes close to human habitation.

Normally singly or in pairs, but up to 6 together on migration. Less shy and easier to flush than other crakes, and newly arrived migrants often seen in open. When flushed usually flies less than 100 m, occasionally further; crouches immediately on landing. When feeding in open allows quiet approach of man and dog to within 5–10 m; bird first stretches body and neck up for good look, then crouches. May make vertical jump followed by short run before pausing and watching intruder. If observer remains motionless, bird resumes feeding. Can be approached in car to within 1 m—bird remained crouched, alternately fluffing and flattening plumage. Active all day, mainly at dawn and dusk, also after rain and during light rain. While foraging emerges onto grassy, muddy or laterite tracks, picking up seeds and insects and turning over dead grass. Probes into bases of grass tussocks; on open ground makes short, rapid rushes to catch running insect; also digs in ground. Roosts in depression in grass tussock.

Moves away from equator both N and S to breed during rainy seasons, returning after breeding, though also breeds in equatorial region. Grassland breeding habitat frequently burnt during dry season, forcing emigration. In Nigeria moves north to breed in rainy season (June–Sept) returning south in dry season, and similar movements noted Senegambia, Ivory Coast and Cameroon. Large wintering population at Belinga Airfield, Gabon, Dec–Feb, during 'off-season' of northern birds, while none at all there Feb–Nov. Birds were thin and gonads not enlarged (Brosset 1968). Present and probably breeding W-central Sudan (Darfur) Aug. Present all year in equatorial regions (e.g., Zaïre, Burundi) but commoner at certain times of year, suggesting influx of migrants. Non-breeding birds present near Mombasa, Kenya, May–Dec, presumably from southern African breeding population. Present most southern African countries only during rains (South Africa, mainly Oct–Feb; Mozambique, Oct–Mar; Botswana, Feb, Apr; Zimbabwe arrives at beginning of rains; Malaŵi, Dec–Apr). South African records from other months may be birds that failed to migrate rather than resident population. Present Zambia late Nov–May: arrives seasonally inundated Luangwa Valley Dec; arrives Itawa (near Ndola) early Dec, leaving late Apr–mid-May; earliest date 2 Dec, latest 14 May; influx noted on short-grass wet plains, Balovale, Dec–Jan. Arrives to breed S Tanzania (Iringa) Jan.

Food. Snails, earthworms, insects and their larvae (ants, beetles, grasshoppers), seeds and vegetable matter.

Breeding Habits. Monogamous. Territorial, on wintering grounds as well as when breeding. Size of winter territories, Kenya (n = 14) 2·4–6·7 ha (3·7 ha). In fighting at winter territorial boundary, 2 birds on grass track made flying jumps at each other, with vigorous pecking. Pair-formation, with increased calling and courtship behaviour, observed on winter territory just before departure. In courtship chase, both birds fluff out body feathers; ♀, in the lead, retracts neck but ♂'s pose is upright. ♀ crouches and ♂ immediately mounts and copulates for 2–3 s. Both then stand side by side in normal pose before moving off to feed. Similar chases occur on breeding grounds (Zambia). In Zambia, 8 pairs bred in area of 41·6 ha (density of 1 pair/5·2 ha).

NEST: shallow cup of dry grasses, placed either in scrape or depression on dry ground, hidden under tussock of grass or small bush, or just above ground in thick grass or other herbage near or over standing water; nest occasionally floating on water. Diam. 215 × 225 outside, 110 × 115 inside; external depth 40, depth of cup 20.

EGGS: 3–8; mean clutch size (n = 44) 5·4. Ovate; polished, glossy, dull white, light cream or pinkish cream with large and small blotches and spots of red-brown and lilac, mainly at broad end. SIZE: (n = 27, Nigeria) 37·0 × 24·0 and 34·8 × 26·0–32·0 × 23·8 and 32·3 × 22·5 (34·7 × 24·6); (n = 68, southern Africa) 32·8–36·0 × 24·2–27·0 (34·3 × 25·2). WEIGHT: 9·6.

LAYING DATES: Sierra Leone, July; Nigeria, July–Sept in north, June–Sept in centre, Apr–Nov in south; Zaïre, Jan (imm. May); Angola, Apr (breeding condition Jan); W Kenya, May–June (breeding condition, July); central Kenya highlands, June; N Tanzania, June; SW Tanzania, Mar (breeding condition, Jan); Malaŵi, Jan–Mar; Zambia, Dec–Mar; Zimbabwe, Dec–Mar; South Africa (Natal/Zululand), Oct–Feb. Breeds during rains.

INCUBATION: by both sexes. Period, *c.* 14 days.

DEVELOPMENT AND CARE OF YOUNG: young tended by both parents; when disturbed escape by running into dense cover.

BREEDING SUCCESS/SURVIVAL: adults killed by Serval, *Felis serval* (Aspinwall 1978), house cats (C. H. Fry, pers. comm.) and Black-headed Heron *Ardea melanocephala* (R. McVicker, pers. comm.).

References
Benson, C. W. (1964).
Shuel, R. (1938).
Taylor, P. B. (1985).

Plate 7
(Opp. p. 112)

Crex crex (Linnaeus). Corncrake. Râle des genets.

Rallus crex Linnaeus, 1758. Syst. Nat. (10th ed.), 1, p. 153; Europe, restricted type locality Sweden, *ex.* Fn. Svec.

Range and Status. Breeds Palearctic from NW Europe to central Siberia; winters mainly in Africa.

Palearctic migrant. Occurs chiefly on passage N Africa but occasionally winters Morocco, Algeria, Tunisia, Egypt (Beni Hassan, Feb: Short and Horne 1981). In W Africa noted in Mali, Ghana, Nigeria, but probably more widespread since regular Morocco both spring and autumn and there are Oct and Apr records from Mauritania. Single winter records from Cameroon (Dec) and Congo (Jan). More widespread in Zaïre, in grasslands bordering main forest block; 2 records from Angola. Some winter Sudan but in Ethiopia only a passage migrant; in Kenya, Uganda, and NW Tanzania only scattered winter records. Main winter range from central Tanzania and Mozambique through Zambia, Malaŵi and Zimbabwe to N Botswana and South Africa to eastern Cape Province. Vagrant W Cape. Locally common within this range but must be more abundant than records suggest since bulk of world population winters in Africa. However, less common now in Morocco than during 1950s (J. D. R. Vernon, pers. comm.), perhaps reflecting decline in W European breeders.

Description: ADULT ♂ (breeding) (occasionally seen in Africa): eye-stripe greyer, cheeks, sides of neck, foreneck and breast generally grey rather than brown. ADULT ♂ (non-breeding): feathers of forehead, crown, hindneck and upperparts, including tail, tertials and scapulars, blackish with greyish buff edges. Lores and ear-coverts buff; broad streak over and behind eye brownish grey. Sides of neck, foreneck and breast tawny with slight wash of grey; chin and throat white tinged buffy. Rest of underparts buffy white with tawny bars on flanks, thighs and undertail-coverts. Primaries and secondaries brown with rufous tinge; narrow leading edge of creamy white around bend of wing. Wing-coverts rich tawny; some pale bars on median and greater coverts. Axillaries and underwing-coverts ferruginous. Bill dark brown, pale horn or brownish flesh; eye light brown or hazel; eye-ring brick red, legs and feet pale greyish to pink or yellowish flesh. ADULT ♀: very similar to ♂ in both plumages but tends to be warmer brown, less grey. SIZE: wing, ♂ (n = 15) 139–150 (144), ♀ (n = 6) 130–145 (135); tail, ♂ (n = 38) 41–52 (47), ♀ (n = 24) 40–49 (44); bill, ♂ (n = 37) 20–25 (21), ♀ (n = 25) 19–23 (21); tarsus, ♂ (n = 36) 37–43 (40), ♀ (n = 23) 35–40 (38). WEIGHT: monthly measurements of European birds, Apr–Oct, ♂ (n = 36) 135–210 (155–180); ♀ (n = 16) 119–197 (138–158); heaviest Sept–Oct. 3 ♂♂ from Tanzania (Dar-es-Salaam) Apr, 180, 183, 194; 1 ♂ Nairobi Nov, 140; 1 ♀ Nairobi Dec, 129.

IMMATURE: as adult but feathers of upperparts with brown rather than greyish edges, sides of head, neck and breast tawny buff without grey, underparts less distinctly barred.

Field Characters. Same size as African Crake *Crex egregia* and with similarly streaked upperparts, but in flight shows tawny wing patch, lacking in African Crake. Breast buffy (not grey) and flank bars tawny, not blackish. For additional details, see under African Crake.

Voice. Tape-recorded (53, 62, 73). Territorial call on Palearctic breeding grounds, a rasping, mechanical double note, 'raak-raak' or 'crake-crake'. Alarm call when surprised, a loud 'tsuck'.

General Habits. On migration and in winter found in grassland and savanna and in drier grassy areas bordering marshes and rivers, from the coast up to 3000 m; also in rank grass around sewage ponds. Generally avoids wet places, but occasionally in damp and even flooded grass.

Generally solitary, but occasionally forms groups on migration. More secretive than African Crake; rarely in open, though occasionally feeds on open tracks in grassland and at edges of dirt roads. When surprised in open makes short rapid rush, then pauses to watch intruder. When flushed usually flies only short distance; sometimes lands in bush or thicket. Most active at dawn and dusk, also during drizzle. In Zambia, winter territories occupied continuously by single birds for 3–4 months (P. B. Taylor, pers. comm.); average size of 4 territories, 6·7 ha.

Passes through Morocco Aug–Oct (mainly Sept) and Feb–May (mainly Mar–Apr). Noted in spring in oases of S Algeria. Rare or irregular Tunisia (mainly spring) and only 2 records Libya (Apr and Sept). Passes through Egypt Aug–Oct (common Sept) and Mar–Apr (less common than in autumn). Eastern birds enter Africa across Red Sea and Gulf of Aden. Occurs Sudan Sept–Apr and Ethiopia Nov–May. Bulk of Kenya records Oct–Dec and Mar–Apr, extreme late date June 2, Nairobi. In Tanzania recorded only Feb–Apr. Most E African records from highlands, but Apr records from Dar-es-Salaam and May record Mombasa may indicate spring passage up coast. Extreme dates Zambia 21 Oct and 9 Apr. Ringing recoveries W Germany (Heligoland) and Sweden to Zaïre, Nov–Dec; France to Angola, Mar; Britain to Congo, (date?).

Food. Insects (beetles, flies, grasshoppers, dragonflies, ants) and other small invertebrates (snails, slugs, worms, spiders). Some plant material.

Reference
Dahm, A. G. (1969).

Genus *Rougetius* Bonaparte

Medium-sized rail of uncertain affinities. Bill medium-long, upperparts uniform brown without patterning typical of most *Rallus*, underparts cinnamon-rufous. Tenuous nasal bar makes affinity with gallinules or *Amaurornis* unlikely.

Monotypic genus endemic to Ethiopia.

Rougetius rougetii (Guérin-Méneville). Rouget's Rail. Râle de Rouget.

Plate 8 (Opp. p. 113)

Rallus rougetii Guérin-Méneville, 1843. Rev. Zool. p. 322; Ethiopia.

Range and Status. Endemic resident, Ethiopia; widespread in highlands 2000–4100 m, from Eritrea to Kaffa (Maji Plateau), Gamu-Gofa and Sidamo Regions and Bale Mts. Common to locally abundant.

Description. ADULT ♂: top and sides of head, back, rump, tail and entire wing dark olive-brown. Hind neck and mantle similar but contrastingly paler. Pale buffy line from upper mandible above lores to eye. Chin whitish, shading to pale cinnamon on throat. Sides of neck, breast, belly and upper flanks cinnamon-rufous. Lower flanks and vent dark olive-brown. Undertail-coverts white. Bill red; eye reddish chestnut; legs and feet dark red. Sexes alike. SIZE: (8 ♂♂, 8 ♀♀) wing, ♂ 124–136 (131); ♀ 124–137 (131); tail, ♂ 45–54 (49), ♀ 41–51 (46); bill, ♂ 30–35 (33), ♀ 28–34 (31); tarsus, ♂ 46–57 (51), ♀ 44–51 (47).

IMMATURE: somewhat paler than adult. Bill brownish, eye light brown.

DOWNY YOUNG: unknown.

Rougetius rougetii

Field Characters. Combination of unstreaked upperparts and cinnamon-rufous underparts easily distinguishes this species from all other African rails.

Voice. No known tape-recording. Advertising call a loud, ringing, repeated 'Wreeeee-creeuw, Wreeeee-creeuw...'; given mainly in morning and evening but also at other times of day and on moonlit nights. Several birds may call together. Alarm call, a shrill, piercing 'dideet' or 'di-dii'.

General Habits. Lives in marshy situations in montane grasslands and moorlands. Found in lush grass, reeds and bushes along streams and around ponds, in open marshy meadows, in small pockets of tussocky marsh grass and lobelias in wet hollows, and in *Alchemilla* bogs. Also occurs on dry ground, among heaths or *Alchemilla*.

Though sometimes shy and wary, and always retaining some typical ralline caution, it also becomes accustomed to man, and has been seen at midday feeding on bare mud by small pond within 50 m of busy highway. Forages in open meadows and also in shallow water, hopping from stone to stone. Seen to dive into water at waterfall like dipper (*Cinclus*), and reappear in calmer water (von Erlanger 1905). Flicks tail while walking, showing conspicuous white undertail-coverts.

Food. Seeds and aquatic insects, especially water beetles; crustaceans and small snails.

Breeding Habits. Monogamous; solitary nester. Pairs already formed Apr, Arussi Province, and closely spaced—12 pairs seen in 1 km (Dorst and Roux 1973).

NEST: pad of dead rushes on ground or in rushes over water.

EGGS: 4–5, laid at 1-day intervals; mean (3 clutches) 4·7. White, creamy or ivory with faint buff tint, covered, especially at large end, with fine speckles and irregular spots of red-brown, lilac and grey. SIZE: (n = 7) 42·4–46·8 × 30·6–31·6 (44·1 × 31·1).

LAYING DATES: Mar–Oct.

INCUBATION: by ♀, probably also by ♂.

DEVELOPMENT AND CARE OF YOUNG: occurs in family parties of 3–6 outside breeding season, so young apparently remain with parents until fully grown.

BREEDING SUCCESS/SURVIVAL: 2 broods found by J. S. Ash (pers. comm.) each had 1 chick only; 1 was 10 days old, the other two-thirds grown.

References
Brown, L. H. (1966).
von Erlanger, C. (1905).
Harrison, C. J. O. and Parker, S. A. (1967).

Genus *Rallus* Linnaeus

Bill long and slender, wings short and rounded; plumage of upperparts patterned in all except *caerulescens*. Body specialized, very slim, adapted to semi-aquatic life in reedy marshes. Skull, sternum and pelvis narrow, legs slender.

9 spp.; 2 in Africa (1 endemic), in same superspecies.

Rallus aquaticus Linnaeus. Water Rail. Râle d'eau.

Rallus aquaticus Linnaeus, 1758. Syst. Nat. (10th ed.), 1, p. 153; Europe, restricted type locality, Great Britain.

Forms a superspecies with *R. caerulescens*.

Range and Status. Breeds Palearctic, N Africa, to Iran, China and Japan; winters N Africa to SE Asia.

Resident and Palearctic migrant. Breeds N Morocco (many localities, locally common), N Algeria (mainly within 100 km of coast), Tunisia south to Sfax, Libya (Cyrenaica) and Egypt (Nile delta and Wadi El Natrun). More widespread in winter; very common in Morocco, and extends far into Sahara in Algeria (many localities, south to Daïet Tiour, El Goléa and Aïn Amenas) and Libya (Brak, Sebha). Also winters along Nile south to El Kab, near Luxor (Short and Horne 1981), and 3 records from western desert oases, Egypt.

Description. *R. a. aquaticus* Linnaeus (only race in Africa). ADULT ♂ (breeding): entire upperparts including forehead, crown, hindneck, tail, secondary coverts and tertials olive-brown with blackish feather centres. Lores dark grey or blackish; stripe over eye, sides of head and neck, chin, throat, breast and upper belly dark bluish slate. Flanks, axillaries and underwing-coverts barred black and white; flank feathers elongated and tipped buff. Lower belly and thighs slate with pale bars; undertail-coverts buff (shorter feathers) and white (longer feathers). Primaries, outer secondaries and primary coverts dark brown. Narrow white leading edge to inner wing and bend of wing. Bill mainly red; culmen and tip of lower mandible dark brown. Eye red, orange or orange-brown. Legs and feet flesh-brown. ADULT ♂ (non-breeding): as breeding but chin whitish. Sexes alike, except ♀ slightly smaller. SIZE: wing, ♂ (n = 126) 119–132 (125), ♀ (n = 124) 110–121 (116); tail, ♂ (n = 60) 47–59 (53), ♀ (n = 56) 45–55 (48·4); bill, ♂ (n = 55) 39–45 (41·4), ♀ (n = 58) 34–40 (37); tarsus, ♂ (n = 63) 39–46 (42·6), ♀ (n = 58) 36–41 (38·5). WEIGHT: monthly measurements Sept–Mar, ♂ (n = 170) 88–190 (114–144), ♀ (n = 145) 74–138 (98–107); heaviest Dec–Feb.

IMMATURE: upperparts, tail and wings as adult, somewhat darker on crown, hindneck and rump. Lores brown with pale line from base of culmen to eye. Sides of head and neck whitish mottled brown; chin and throat whitish; rest of underparts whitish to buff, deepest on breast and lower belly. Breast and flanks with variable amounts of dark barring and mottling.

DOWNY YOUNG: black with bluish gloss. Small red bare area on back of head; bill black, banded ivory.

Field Characters. Differs from closely related Kaffir Rail *R. caerulescens*, latter having plain upperparts and bright red legs, but their ranges do not overlap. Larger than the 3 crakes which occur in same habitat (Spotted Crake *P. porzana*, Little Crake *P. parva*, and Baillon's Crake *P. pusilla*), but size hard to judge in poor light or when birds half hidden in vegetation. Best distinguishing character at all ages is long, thin bill. Adults with slaty underparts told from Little and Baillon's Crakes by unbarred white and buff undertail-coverts, lack of white streaking on upperparts; from Spotted Crake by colour of underparts and lack of spots. Immature like immature Little Crake but undertail-coverts unbarred, no white streaks on upperparts.

Rallus aquaticus

Also winters in breeding range

Voice. Tape-recorded (62, 73). Commonest call, used for advertisement, display, alarm and warning, a duet, often antiphonal. ♂ gives low grunting or growling note, ♀ higher-pitched, shriller note sometimes lengthened into pig-like squeal. Timing, duration and form of call very variable. ♂ and ♀ notes often coincident, giving impression of single bird calling. Calls typically at dusk, but also any time of day or night. Breeding season song a shrill, piercing 'tyick', and (by ♀ only?) a trill, 'tyueeerrr', given separately or together, e.g. 'tyick-tyick-tyick tyueeerrr'. Other grunting, groaning and purring calls associated with display and tending of young. Call of young, a plaintive 'pee'.

General Habits. Inhabits reed patches, marshes, lakes, streams and banks of rivers and canals. Generally shy but in winter ventures some distance out into the open away from cover and may tolerate human presence. Moves in quick, erratic manner, tail held up or down, flicked in alarm. Flight weak and fluttering, legs trailing. Swims short distances. Forages on mud, on land or in shallow water, taking items on or below the water or from emergent vegetation. Washes food in water before swallowing it. Climbs vegetation to reach fruits. Has regularly-used paths between feeding areas but also flies from one to another. Generally solitary except in breeding season; small groups may sometimes form in autumn and winter but birds aggressive so usually well spaced out. Often has loosely defined winter territory from which all other rallids driven. 2 birds face each other with neck stretched up, standing high on toes, and give territorial call; then move forward and stab at each other with bills. Loser becomes submissive and moves off with head turned away, head and neck held close to ground. Winner calls, moving head to and fro, wings partly outstretched. Roosts at night except on migration, when rests by day, and during early part of breeding season, when calls at night.

Passes through Morocco Sept, Algeria Sept–Oct, Egypt Sept–Nov. Present coastal Libya Oct–Apr, and in Saharan localities Dec–Apr.

Flightless for 3 weeks during autumn moult.

Food. Small fish, shrimps, crayfish, frogs, worms, leeches, snails, spiders, insects and their larvae. Kills and eats small birds and mammals; also eats carrion. Some plant material eaten in winter.

Breeding Habits. Monogamous, pair-bond lasting during breeding season. ♂ displays to ♀ by bowing head, touching breast with bill, raising wings to show barred flanks, raising tail and alternately fanning and closing undertail-coverts. ♀ circles ♂ giving soft crooning notes; then they rub bills and allopreen. ♂ courtship-feeds ♀.

Territorial and aggressive; attacks intruder (sometimes ♂ and ♀ together) with charging-attack (body lowered, neck stretched forward), which may lead to fighting (attacking with bill). After driving off intruder, stands erect and gives territorial call. Breeding density Europe 0·25–2·0 pairs/ha; in high-density areas nests 20–50 m apart.

NEST: substantial cup of leaves and plant stems in thick vegetation on ground near or in water; diam. 13–16 cm, height 7 cm; surrounding vegetation often pulled down to form loose canopy over nest. Nests in water built up if water level rises.

EGGS: 5–16 (6–11) in Europe, laid at 1-day intervals; 2 broods usual. Blunt oval; smooth, glossy, creamy white with spots and blotches of red-brown, mainly at large end. SIZE: (n = 120) 32–40 × 24–27 (36 × 26). WEIGHT: 13.

LAYING DATES: Morocco, May–June; Algeria, May–June; Tunisia, June; Egypt, Apr–June.

INCUBATION: begins when clutch complete; carried out by both sexes but ♀ does more than ♂. Period: 19–22 days.

DEVELOPMENT AND CARE OF YOUNG: fledging period: 20–30 days. Hatching usually synchronous; in asynchronous hatchings ♂ tends first-hatched chicks while ♀ continues to incubate. Later, both sexes tend young. Young brooded on nest for first few days; food brought to brooding bird who passes it to young. Food-begging young raise wings, stretch up and jump at parent's bill. Young transported by adult in bill. Young start to feed themselves at day 5, start becoming independent at 20–30 days, long before flying stage (7–8 weeks). Young stay together in loose group. Adults and chicks freeze on approach of man; injury-feigning by adult also recorded. Incubating bird stays put and even attacks intruder.

References
Cramp, S. and Simmons, K. E. L. (1980).
Ledant, J.-P. *et al.* (1981).
Mayaud, N. (1982).

Plate 7
(Opp. p. 112)

***Rallus caerulescens* Gmelin. Kaffir Rail; African Water Rail. Râle bleuâtre.**

Rallus caerulescens Gmelin, 1789. Syst. Nat., 1, Pt. 2, p. 716; Cape of Good Hope.

Forms a superspecies with *R. aquaticus*.

Range and Status. Endemic resident. Rare W Africa: Sierra Leone (Ribi R. area); Cameroon (Ndop); São Tomé (1 record). Uncommon to rare in highlands of Ethiopia. Main range from E Zaïre (Kivu, Katanga), Rwanda, Burundi (common, lower Ruzizi R.), NW Uganda (Pakwatch) and W-central Kenya south through Tanzania, Zambia (widespread except Luangwa and Zambezi Valleys), Malaŵi (probably common below 1500 m), Mozambique (widespread and locally common in south, especially along coast), Zimbabwe and much of eastern South Africa to Cape; and east to interior highlands of Angola (widespread) and through N Botswana to N Namibia. From sea-level to 3000 m.

Description. ADULT ♂: crown and hindneck blackish; face, sides of neck, foreneck and breast slate. Throat pale grey becoming whitish on chin. Upperparts including wing-coverts dark vinous brown, rump and tail darker. Flanks black with narrow white bars; belly brownish grey, feathers tipped buff. Undertail-coverts white, median feathers barred black.

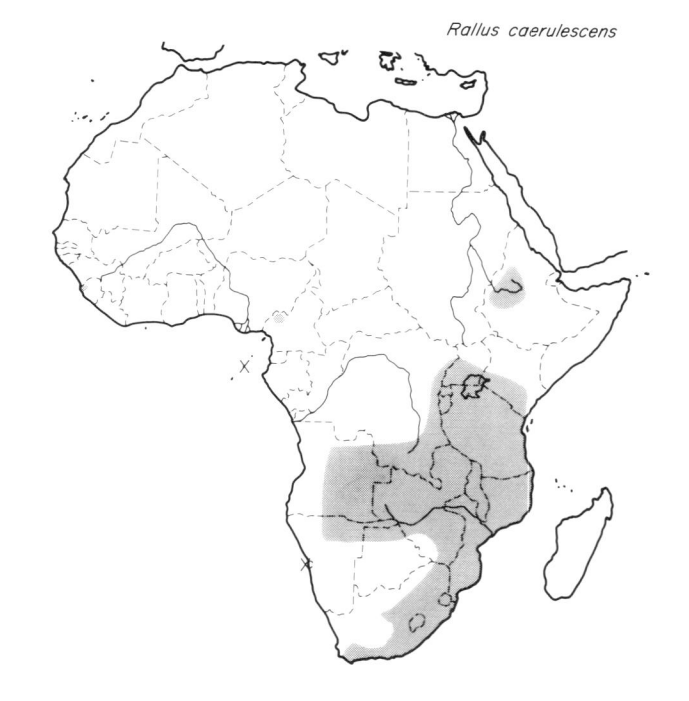

Primaries and secondaries dark brown; underwing-coverts and axillaries black barred white. Bill bright red, culmen brown; eye red or red-brown; legs and feet red. Sexes alike, except ♂♂ larger and heavier. SIZE: wing, ♂ (n = 54) 109–135 (122), ♀ (n = 38) 105–126 (115); tail, ♂ (n = 8) 43–48 (46), ♀ (n = 5) 38–44 (41); bill, ♂ (n = 55) 42–59 (52), ♀ (n = 38) 40–50 (46); tarsus, ♂ (n = 8) 40–46 (42), ♀ (n = 5) 33–41 (38). WEIGHT: ♂ 146–205, ♀ 120–170.

IMMATURE: upperparts and wings as adult but mantle darker. Chin and throat white, face, sides of neck, breast and flanks blackish brown, breast mottled. Centre of breast and belly buff. Flanks barred white in younger, rufous in older birds. Later still, upperparts become lighter, resembling adult; blackish brown of underparts pales to dull sooty brown; flank bars change from rufous to white; outer undertail-coverts tipped rufous buff, inner ones white with black subterminal bar. Bill, legs and feet dark brown at first, becoming redder with age.

DOWNY YOUNG: black; legs and feet slate. Bill pink, distal half black.

Field Characters. Range does not overlap that of Water Rail *Rallus aquaticus*, which is similar but has streaky upperparts. In Ethiopia overlaps with similar-sized Rouget's Rail *Rougetius rougetii*, which has cinnamon-rufous underparts. Adult distinguished from other rails by long red bill and red legs. Brown immature has duller bill and legs, but bill is still long and thin, not short and stubby as in most sympatric rallids.

Voice. Tape-recorded (74, 75, 264, 296). Common territorial 'song' is excited chatter given in chorus; 1 bird starts, and others in the area join in: begins suddenly as rapid trill, with notes run together, and gradually winds down into spaced, single notes, at the same time losing volume. Notes are high-pitched and shrill but have a low, 'pumping' undertone like that of Water Rail. Other birds may join in with series of low clucking or grunting notes. When fighting, a shrill 'ri-ri-ri'; in threat display, a deep, puffing 'krock'. Warning, a low growl; warning to chicks, squeaky 'zii-zii'; parents leading chicks give short snores.

General Habits. Inhabits reed beds and dense, rank growth in swamps and marshes and beside lakes, rivers and streams; in Cameroon also in paddy fields. During rains in Zambia, moves out from permanent reedbeds to adjacent seasonally inundated dense tall grass (P. B. Taylor, pers. comm.).

Active all day and especially at dawn and dusk, but not much at night. In South African summer (Dec–Jan) active mainly 0500–1200 h and 1500–2000 h; in winter (May–Aug) 0700–1800 h.

Forages on mud, also in shallow water, at edges of reedbeds, and on floating mats of dead vegetation. Probes deeply with bill (usually slightly open) into mud or grass tussocks, and in shallow water immerses head and neck. When feeding in open, walks with long, even, rapid strides, legs flexed. At other times moves in nervous, jerky manner, flicking tail more frequently when annoyed or uneasy. When alarmed, flattens plumage and stands upright, tail cocked. Swims well if undisturbed. Bathes in shallow water, then sunbathes with slightly open wings. Climbs well, up to 4 m above water in reeds *Phragmites*. Flies low with dangling legs. Flies to cover if disturbed in open; otherwise escapes by running through vegetation. Within dense cover not particularly afraid of human intruder, may remain within *c.* 3 m, but freezes on approach of predator, e.g. African Marsh Harrier *Circus ranivorus*.

Mainly resident, but some evidence of regular movement: in E Africa, both resident and wanderer; in Transvaal, considerable influx Apr–May (Schmitt 1976).

Remiges moulted simultaneously, rectrices usually earlier (sometimes at same time). Flightless for *c.* 3 weeks, remaining in dense cover. In Transvaal flightless Aug–Nov. Adults moult before incubation or after chicks independent.

Food. Crabs, worms, insects and other aquatic animals; also carrion.

Breeding Habits. Monogamous; territorial. One South African territory covered *c.* 150 m of river before hatching but only *c.* 50 m in 2 weeks after hatching. At territorial boundary 2 rivals face each other with elongated stance, then jump at each other and into air, attacking with bills. One bird driven away; or they fight for short period and then resume feeding. Territories abandoned after breeding season. Tolerates nesting Black Crakes *Amaurornis flavirostris*, Common Moorhens *Gallinula chloropus* and Yellow-billed Ducks *Anas undulata* in territory.

NEST: shallow cup made of aquatic plants (*Carex, Juncus, Typha, Rorippa*) placed among aquatic vegetation over water, well concealed; outer diam. 150–200 mm, depth 50–100 mm, depth of cup 20–50 mm, rim of cup 100–400 mm above water.

EGGS: 2–6; mean (16 clutches) 4·1. Creamy white, profusely spotted with light brown, red-brown, purple and grey, especially at blunt end. SIZE: (n = 29, South Africa) 36·5–40 × 25·8–28·7 (38·5 × 27·5).

LAYING DATES: Ethiopia, Aug; Zaïre, (breeds Upemba Nat. Park during dry season); Kenya, May, June; Malaŵi, Feb, Mar; Zambia, Jan (breeding condition May); Mozambique, Jan; Zimbabwe, Jan–Mar; South Africa: Transvaal, Jan, Feb, July, Sept–Dec; Natal, Jan, Aug–Oct, Dec.

INCUBATION: by both sexes. Period: *c.* 20 days. Single-brooded; relays if clutch fails. Incubating bird remained standing on nest and threatened human intruder (P. J. Whitehouse, *in* Schmitt 1976).

DEVELOPMENT AND CARE OF YOUNG: weight at hatching 9·5 g. At 7 days, greenish sheen visible on down of chicks. At 2–3 weeks feathers appear on underparts and back; eye olive. At 4–5 weeks flight feathers appear; lower mandible all black, base of upper mandible still pink. At 8–9 weeks flight feathers fully developed; bill

RALLIDAE

black except for pink nostril spot; eye becoming brown. Chicks reach adult weight at 3 months.

Adults build roosting platforms 200–500 mm above water level to be used by chicks after they leave the nest. Adults aggressive in defence of chicks: bend legs, lower head, stretch neck, fluff out feathers and attack hand of human holding chick. Young independent at 6–8 weeks.

BREEDING SUCCESS/SURVIVAL: known predators are African Marsh Harrier *Circus ranivorus* and Slender Mongoose *Herpestes sanguineus*. Fires in reed beds known to kill birds as well as destroying habitat.

Reference
Schmitt, M. B. (1976).

Genus *Porzana* Vieillot

A 'catchall' genus, possibly polyphyletic, very difficult to categorize because of differing taxonomic treatment; 21 spp. listed by Ripley (1977) but only 13 by Olson (1973), who assigns many of Ripley's *Porzana* spp. to other genera. The 3 African spp. (none endemic) are small, rather squat, short-necked, short-billed birds with mottled upperparts, greyish underparts and barred flanks. *P. parva* sexually dimorphic (♀ with buffy underparts).

Plate 7
(Opp. p. 112)

Porzana parva (Scopoli). Little Crake. Marouette poussin.

Rallus parvus Scopoli, 1769. Annus 1, Hist.-Nat., p. 108; probably from Carniola.

Range and Status. Africa, Europe and W Central Asia; winters Africa, Mediterranean basin, Middle East and W Pakistan.

Palearctic migrant and formerly (still?) local resident. Bred Algeria (Lakes Halloula and Fetzara) mid-19th century; recent breeding suspected Biskra and Djelfa (Cramp and Simmons 1980). Bred once Egypt (Nile Delta). Passage migrant in considerable numbers through N Africa, Morocco–Egypt; many recent records Algeria and almost common on coast of NW Libya (Tripoli). Winters Egypt and possibly elsewhere N Africa. Small wintering population Senegambia (Richard-Toll). Passage migrant Sudan (uncommon, chiefly autumn), and Ethiopia (rare), and single record Somalia (Hargeisa, Sept) presumed migrant. Only 4 other records south of Sahara: Nigeria (at least 3 near Kano, Dec; Wilkinson *et al.* 1982); Uganda (several near Butiaba, Dec); Kenya (3, Thika, Jan; P. B. Taylor, pers. comm.); and Zambia (1, Ndola, Mar; Taylor 1980). However, large numbers breeding western USSR, considerable migration through N Africa, and occurrence in deserts of Algeria (Beni Abbès, Timimoun, El Goléa, Ouargla) and Libya (Ghat) suggest wider subsaharan distribution and greater numbers than records indicate.

Description. ADULT ♂: overall colour of upperparts dull olive-brown from forehead to tail and upper side of wings, paler on scapulars, darker on flight feathers. Dark centres to feathers of mantle to tail, scapulars and tertials, blackest on back. Variable amount of white spots and streaks on mantle, back, scapulars and tertials. Face, sides of neck and underparts bluish-slate, chin slightly paler. Rear flanks and vent brownish slate narrowly barred white. Undertail-coverts barred black and white. Axillaries and underwing-coverts dark grey-brown. Bill green with red spot at base. Eye and narrow ring of bare skin around eye red. Legs and feet green. ADULT ♀: upperparts, wings and tail as ♂. Lores and supercilium pale grey. Chin, throat and foreneck white. Ear-coverts and sides of neck buff-brown. Breast, belly and flanks pinkish buff. Lower flanks, lower belly and undertail-coverts barred black and white. SIZE: wing, ♂ (n = 24) 99–111 (106), ♀ (n = 22)

99–109 (103); tail, ♂ (n = 21) 49–60 (51·4), ♀ (n = 15) 50–58 (52·1); bill, ♂ (n = 25) 17–20 (18·5), ♀ (n = 22) 16–19 (17·4); tarsus, ♂ (n = 25) 30–34 (31·9), ♀ (n = 20) 29–32 (30·8). WEIGHT: ♂ (n = 10) 30–72 (50), ♀ (n = 11) 36–65 (49).

IMMATURE: upperparts and wings as adult but more white spots on mantle, scapulars and wing-coverts. Stripe over eye, chin, throat and underparts creamy white. Narrow brown bars on breast; barring on rest of underparts as ♀.

DOWNY YOUNG: black, with oily green gloss on head and upperparts.

Field Characters. Slim shape produced by relatively long neck, wings and tail useful distinction from other crakes. Smaller than Spotted Crake *P. porzana* and Striped Crake *Aenigmatolimnas marginalis*, undertail-coverts barred black and white. Upperparts and wings, especially in flight, paler, duller and more uniform than Baillon's Crake *P. pusillus*. Pale scapular stripe evident at rest and in flight (Taylor 1980). Flight more powerful, less fluttering than Baillon's Crake. On underside, ♂ similar to ♂ and ♀ Baillon's Crake but barring fainter and restricted to lower flanks and lower belly; bill with red spot at base. Pinkish buff underparts of ♀ distinctive. Immature less heavily barred below than immature Baillon's Crake, but best told by other characters.

Voice. Tape-recorded (53, 62, 73). Advertising call (song) of ♂, loud 'kuck' or 'kweck' repeated at 1–2 s intervals for up to several min, then accelerating and descending in pitch, trailing away at end; later notes softer and more guttural. Last part of song sometimes omitted. Sings mainly at night, sometimes during day. Advertising call of ♀, fast, hard trill, often preceded by one or two sharp 'keck's, 'keck-krrrrrrrrr'. Also duets with ♂, answering with sharp 'kik'. ♂ on migration S Spain gave shortened, somewhat hesitant song (Chappuis 1975). Alarm call, sharp 'tyiuck'; contact calls include low 'gug-gug-gug'.

General Habits. Inhabits swamps and ponds overgrown with reeds, bulrushes (*Typha*), sedges and rank grass, also seasonally inundated grassland (Zambia). Prefers more aquatic habitats than Spotted Crake, and walks out from cover onto floating vegetation over deeper water. Swims and dives readily.

Forages while walking, swimming or wading; picks food from mud, water surface or vegetation; does not probe. Generally skulking, but can be seen in open and even becomes fairly bold when accustomed to human presence. Usually singly or in pairs, but numbers congregate on migration.

Passes through N Africa late Aug–Oct and Mar–May (sometimes June), reaching Senegambia and Somalia Sept, Sudan Oct. Most N African records in spring, suggesting autumn migrants may overfly region.

Food. Mainly insects and seeds of aquatic plants, also spiders, worms, snails and some aquatic vegetation.

Breeding Habits. Monogamous; solitary nester; territorial.

NEST: cup of leaves and stems in thick vegetation by or over water, often on platform. Size: diam. 11–20 cm outside, 10–16 cm inside, height 2–9 cm. Built in 3–7 days, probably by both sexes.

EGGS: 4–11, laid at 1-day intervals; Germany mean (32 clutches) 6·75. Short oval, smooth, glossy; yellow-buff with brown spots. SIZE: (n = 145) 28–34 × 19–23 (30 × 22). WEIGHT: 8.

LAYING DATES: Egypt, Apr.

INCUBATION: begins before clutch complete; by both sexes. Period: 15–17 days per egg, 21–23 days per clutch. Hatching asynchronous.

DEVELOPMENT AND CARE OF YOUNG: young brooded and fed by both parents, but after few days can feed themselves. Fledging period 45–50 days; independent before fledging.

References
Taylor, P. B. (1980).
Wilkinson, R. *et al.* (1982).

Porzana pusilla (Pallas). Baillon's Crake. Marouette de Baillon.

Rallus pusillus Pallas, 1776. Reise versch. Prov. Russ. Reichs, 3, p. 700; Dauria.

Plate 7
(Opp. p. 112)

Range and Status. Africa, Madagascar, Palearctic, Oriental and Australasian Regions.

Resident and Palearctic migrant, some local movement southern Africa. Breeds Morocco (Larache, probably several other localities). Formerly bred Algeria (L. Zana) and Egypt (Nile Delta); may still do so. Suspected breeding Tunisia (present in breeding season, L. Ichkeul). Passage migrant N Africa, and recorded in deserts of Morocco (R. Dra), Algeria (Beni Abbès, Laghouat, El Goléa) and Libya (Sebha, Jaghbub). 2 records Senegambia (Nov, Jan); uncommon to rare resident Ethiopia, and one record highlands of N Somalia near Ethiopian border. Widely distributed and locally common resident from E Zaïre, central Uganda and central Kenya to Angola, N Namibia, N Botswana and E South Africa to Cape. Also at mouth of Orange R., and 2

Porzana pusilla

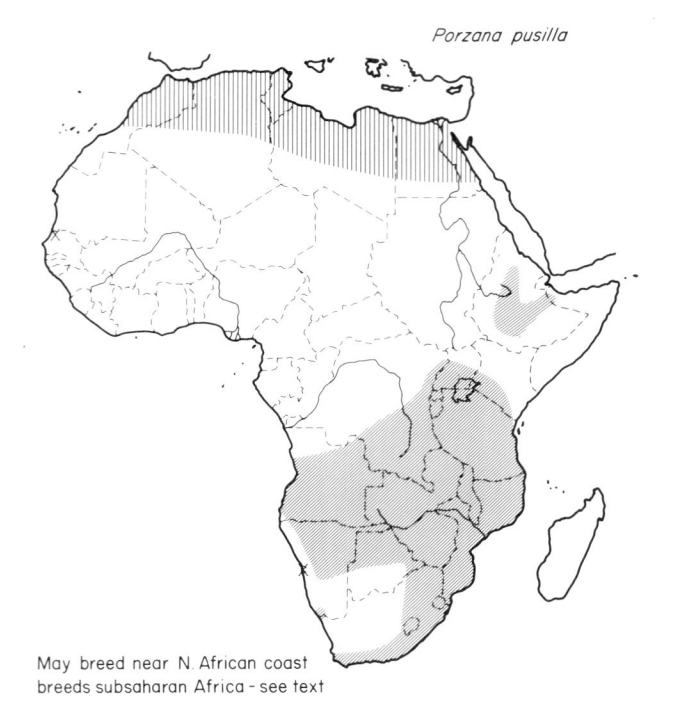

May breed near N. African coast
breeds subsaharan Africa - see text

records coastal Namibia (Swakopmund: P. Becker, pers. comm.). Winter range of Palearctic migrants not known; some probably join resident population southern Africa, others may winter W Africa; however, W Palearctic population small and declining, so numbers wintering Africa may not be great.

Description. *P. p. intermedia* (Hermann): Africa. ADULT ♂ (breeding): top of head and hindneck red-brown with indistinct black streaks. Face, sides of head and neck, and chin to upper belly bluish slate. Upperparts, including tail, scapulars, tertials and most wing-coverts red-brown with broad black and narrow white streaks, tail with little white. Lower belly barred grey and white; flanks and undertail-coverts barred black and white. Flight feathers and primary coverts dark greyish brown. Axillaries and underwing-coverts dark greybrown with indistinct white barring. Bill dark green or greenish grey, culmen darker. Eye red; legs and feet greenish grey or olive-green, sometimes yellowish. ADULT ♂ (non-breeding) and ADULT ♀ (breeding): same but chin pale grey. ADULT ♀ (non-breeding): chin whitish. SIZE: wing, ♂ (n = 19) 89–97 (92·9), ♀ (n = 11) 87–96 (91); tail, ♂ (n = 20), 39–48 (43·2), ♀ (n = 12) 40–46 (43·2); bill, ♂ (n = 21) 16–18 (17·2), ♀ (n = 12) 15–17 (16·2); tarsus, ♂ (n = 24) 26–30 (28·3), ♀ (n = 15) 25–29 (27). WEIGHT: 2 ♂♂ 42, 45, 2 ♀♀ 46, 51.

IMMATURE: upperside and wings as adult. Lores whitish; sides of head mottled brown and white. Underparts from chin to upper belly white, mottled and barred brown on breast and upper flanks. Rest of underparts as adult. Bill dark horn, base paler. Eye brown or olive; legs and feet flesh brown.

DOWNY YOUNG: black, with oily green gloss except on breast and belly.

P. p. pusilla Pallas (Egypt, 1 record): paler, more ashy below; well-marked brown eye-stripe.

Field Characters. Smallest of the crakes, with dumpy appearance, short wings and weak, fluttering flight. Upperparts and wings deeper, richer, more orange-brown than Little Crake *P. parva*, and more heavily patterned, with prominent black and white streaks. Upperwing-coverts patterned (unpatterned in Little Crake). Below, similar to ♂ Little Crake but barring on belly and flanks heavier and more extensive; no red spot on green bill. Immature more heavily barred below than immature Little Crake, but best told by shape, flight, and colour of upperparts.

Voice. Tape-recorded (62, 73). Advertising call (song) of ♂, a hard dry trill or rattle, somewhat like winding of watch, lasting 1–3 s, repeated at 1–2 s intervals. Pitch fairly even but may waver up or down during call. Sings mainly at night. Agitated ♂ gives gravelly, low, growling 'chorr-chorr-chorr-chachachachacha'. Alarm call, 'tyiuk', similar to Little Crake. Warning call to young, sharp, grating 'jup' or 'check'. Contact call of young, shrill 'pleep' or 'pee'. Frog-like bubbling also reported.

General Habits. Inhabits marshes, swamps and edges of lakes, ponds, dams and streams, in thick vegetation, especially reeds, rushes, sedges, tall dense grass, and reedmace (*Typha*); also seasonally inundated grasslands. Solitary except when breeding; swims and dives readily. Forages on mud at edge of ponds, along banks of streams, among matted beds of broken reeds and *Typha*, and while walking on floating vegetation, wading in shallow water, or swimming. Probes in mud as well as picking items from surface. Generally shy and skulking, but sometimes closely approaches observer standing quietly in open (P. B. Taylor, pers. comm.). Active dawn to mid-morning and mid-afternoon to dusk. When flushed, flies low for short distance.

Most N African migrants recorded Mar–Apr, suggesting overflying of region in autumn. Some local movement away from seasonally inundated grasslands, Zambia and Zimbabwe, probably to nearby permanent swamps.

Food. Insects, chiefly aquatic, including many beetles; also worms, snails, small crustaceans, seeds and some green plant material.

Breeding Habits. Monogamous; solitary nester; territorial. ♂ in courtship flight circles at height of 6–8 m giving advertising call. Allopreening noted. In defence of nest against human intruder gives distraction display, raising feathers, lowering head and drooping wings to show white outer web of outer primary; and intimidation display, with back feathers raised and wings fanned out from body, giving repeated grumbling call.

NEST: small cup of rushes, sedges, grasses, dead leaves and reed or other stems, placed in reeds or other thick vegetation by or over water, with surrounding vegetation pulled down over it to form roof. Diam. 9–10 cm outside, 7 cm inside, height 8–10 cm, depth of cup 3 cm; probably built by both sexes.

EGGS: 2–7 Africa, 4–11 Europe, laid at 1-day intervals; Africa mean (7 clutches) 4, Europe mean (18 clutches) 7·4. Short oval, smooth, with or without gloss;

olive, buff or ochre, thickly spotted brown, especially at large end. SIZE: (n = 4, Africa) 27·6–29·9 × 20·0–21·5 (29 × 20·75); (n = 95, Europe) 25–31 × 19–22 (29 × 21). WEIGHT: (n = 18, Europe) 5–9 (6).

LAYING DATES: Morocco, May, (♀ about to lay Apr); Egypt, Apr; Ethiopia, July; Kenya, (imm. Jan, probably from eggs laid Nov–Dec: P. B. Taylor, pers. comm.); Tanzania (♀ with large ova Apr); Malaŵi, June (♂ with large testes Mar); Zimbabwe, Jan–Mar; Botswana (birds breeding condition Jan); South Africa, Sept, Nov–Jan (juvenile Apr). Breeds during or just after rains.

INCUBATION: begins before clutch complete; by both sexes. Period: 14–16 days per egg, 17–20 days per clutch. Hatching asynchronous.

DEVELOPMENT AND CARE OF YOUNG: young brooded, tended and fed by both parents, but after few days can feed themselves. Fledging period c. 35 days. Independent before fledging.

References
Baur, S. (1980).
Benson, C. W. (1964).
Benson, C. W. and Pitman, C. R. S. (1966b).

Porzana porzana (Linnaeus). Spotted Crake. Marouette ponctuée.

Plate 7
(Opp. p. 112)

Rallus porzana Linnaeus, 1766. Syst. Nat. (12th ed.), 1, p. 262; Europe (= France, ex. Brissonian reference).

Range and Status. Breeds Palearctic east to central Siberia and Iran; winters Africa, S Europe, Middle East, S Asia.

Palearctic migrant. Winters N Africa (common Egypt, scarce Morocco, once Tunisia), Senegambia (common, delta of Senegal R.), and from Sudan through E Zaïre, Burundi (common, delta of Ruzizi R.) and E Africa (scarce) to Angola, N Namibia, N Botswana and N South Africa to Natal and Lesotho (once). Centre of abundance Zambia, Malaŵi, Zimbabwe and probably Mozambique. Passage migrant N Africa, Sudan and Ethiopia; in E Africa, much commoner on passage than in winter. Isolated records Mauritania (Port Etienne), Mali (Bamako), Nigeria (Lagos, Kano, Malamfatori) and Chad (Abéché), but probably more common W Africa than records indicate since regular migrant through N Africa.

Description. ADULT ♂ (breeding) (sometimes seen in Africa): as non-breeding but chin, throat and foreneck become clear grey without spots. ADULT ♂ (non-breeding): crown and wedge in centre of forehead streaked brown and black; nape and hindneck similar but with small white spots. Broad stripe from base of culmen over and behind eye slate grey. Lores and patch behind eye mixed brown and buff; pale line across lores sometimes present. Feathers at base of bill blackish. Chin and throat dull grey with indistinct pale spots. Mantle, back, rump, uppertail-coverts, scapulars and wing-coverts greenish olive-brown streaked black, covered with small white spots or streaks. Tertials similar but inner webs mainly dull yellowish buff. Tail olive-brown with black streaks, some feathers narrowly fringed with white spots. Breast dull olive-brown with greyish wash, spotted white. Belly off-white; flanks barred dark brown and white. Undertail-coverts cream or buff. Primaries and secondaries dark grey-brown, outer web of outer primary white. Narrow white line along leading edge of wing. Axillaries and underwing-coverts barred dark brown and white. Bill greenish or yellow, sometimes orange, with red base, darker at tip. Eye yellowish or reddish brown. Legs and feet olive-green to apple-green. ADULT ♀: similar to ♂ but in both plumages less grey on face, throat and breast, sides of head and neck somewhat more spotted. SIZE: wing, ♂ (n = 46) 117–128 (122), ♀ (n = 27) 111–123 (118); tail, ♂ (n = 45) 42–54 (47·5), ♀ (n = 26) 44–52 (47·3); bill, ♂ (n = 45) 18–22 (19·7), ♀ (n = 27) 17–20 (18·4); tarsus, ♂ (n = 45) 32–37 (34·1), ♂ (n = 27) 30–35 (33). WEIGHT: (European birds, Mar–Oct) unsexed (n = 142), 61–147, monthly means 71·2–96. 1 ♂ Zimbabwe, Feb, 84. Unsexed: 1 Sudan Oct, 57; 2 Kenya Apr, 82, 110; 1 Zambia Mar, 88.

IMMATURE: similar to ♀ but lacking any grey. Eye-stripe covered with tiny white spots. Face and sides of neck mottled brown and white. Chin and throat off-white; breast dull brown with scattered pale spots. Eye greenish.

DOWNY YOUNG: black, with greenish gloss on head, throat and upperparts.

Field Characters. Smaller than African Crake *Crex egregia* but larger and bulkier than Striped Crake *Aenigmatolimnas marginalis* and *Porzana* spp. Distinguished by spotted plumage and buff undertail-coverts; latter exposed by frequent cocking of tail. Closest in size to Striped Crake, which has striped upperparts and reddish cinnamon undertail-coverts. Lacks slaty underparts of ♂ Little Crake *P. parva* and ♂ and ♀ Baillon's Crake *P. pusilla*. ♀ Little Crake is mainly pinkish buff below, not brown, and lacks spots. Immatures of both are mainly white below with some dark barring. Yellow bill and white loral line conspicuous in good light. In flight, shows white leading edge to wing and pale trailing edge to inner wing; legs project less than Striped Crake, wings longer.

Voice. Tape-recorded (53, 62, 73). Advertising call or song of ♂, a short, sharp, whip-like note, 'whick-whick-whick', repeated at rate of about 1 per s, in series often lasting for several min. ♀ sometimes joins ♂ in duet, answering with softer version of ♂'s call. Common (contact?) call, a single 'click' or double 'trick-track'. Alarm call, a hard 'ä'. Generally silent in Africa, but advertising call occasionally heard in spring (Libya, Senegambia, Zaïre). Sharp 'tck' uttered by bird chasing away Pied Kingfisher *Ceryle rudis* (P. B. Taylor, pers. comm.).

General Habits. Inhabits dense vegetation in shallow standing water or on wet ground, in dams, swamps, sewage farms, dambos and seasonally flooded ground.

Usually solitary but may form pairs or small groups on migration. Not particularly shy in Africa; feeds in the open during the day, sometimes far from cover, often joining flocks of Wood Sandpipers *Tringa glareola* and other waders (P. B. Taylor, pers. comm.). Extensive preening in open also recorded. Tolerates quiet observer standing in full view; ignores cars (Parnell 1967). Forages by probing in mud or pecking at vegetation, walking on mud or wading in water, also swimming and walking on lily pads. Turns over floating vegetation; stretches up to pluck seeds from overhanging grass stems. Flies direct from cover to open areas to feed; if disturbed, flies or runs back to cover. Gait deliberate, with long strides and nodding head, or a crouching run, with neck stretched out and tail cocked. Dives readily and swims for short distances. Flies low when migrating, usually within 3 m of ground. Temporary territoriality on winter quarters suggested by numerous observations of intra-specific aggression; fight between 2 birds, probably of different sexes, observed Zimbabwe (Brooke 1974). Normally roosts at night, in thick vegetation, but during migration flies at night and roosts by day.

Movements: Morocco-Libya, regular and locally common in spring (Mar–May) but few autumn records, suggesting most southbound birds overfly region. Recorded Senegambia Sept, Chad Oct and Nigeria Nov. Arrives Egypt early Sept and reaches Sudan and Ethiopia Sept, but no E African records before late Nov. Earliest Zambia arrival date Dec 11, but by late Dec has reached Zimbabwe, Botswana and Namibia. Itinerant in winter quarters, seldom staying long in one spot (up to 3 weeks, Zambia). Peak of occurrences in S-central Africa Feb–Mar, but some still present Apr (late dates Zaïre 18 Apr, Zimbabwe 30 Apr). Most E African records Apr–May, and some present Sudan and Ethiopia until May. Northern birds must start spring migration earlier, since movement through Europe chiefly Mar–Apr.

Food. Insects, grass seeds, small fish 1–2 cm long. In Europe, wide variety of plant and animal material (see Cramp and Simmons 1980).

References
Brooke, R. K. (1974).
Parnell, G. W. (1967).

Genus *Aenigmatolimnas* Peters

Very close to *Porzana* but separated by much deeper bill, longer legs, long hind toe, nasal bar very broad, almost vertical, bony nostril smaller. Sexually dimorphic.
Endemic to Africa; monotypic.

Plate 7
(Opp. p. 112)

Aenigmatolimnas marginalis (Hartlaub). Striped Crake. Marouette rayée.

Porzana marginalis Hartlaub, 1857. Syst. Orn. West-Afr., p. 241; Gabon.

Range and Status. Africa; vagrant Aldabra (Indian Ocean).

Endemic resident and intra-African migrant. Vagrant Algeria (Biskra) and Libya (Wadi Turgat). Broadly but sparsely distributed from Ghana, Togo and Nigeria through Cameroon and Gabon to Zaïre and Kenya, south to Zambia, Zimbabwe, Namibia and South Africa. Uncommon, with apparent gaps in distribution and few breeding records.

Description. ADULT ♂: forehead, crown, nape and hindneck dark brown, shading to buffy brown on sides of head and neck. Chin and throat white. Mantle, scapulars, tertials and wing-coverts dark brown, the feathers edged laterally with white, producing striped effect. Rump and uppertail-coverts blackish brown, feathers with rusty edges. Tail similar but feather edges paler, more buffy. Breast pale orange-buff; belly off-white; flanks dark olive-brown, feathers edged whitish. Lower flanks and undertail-coverts reddish cinnamon. Primaries and secondaries dark olive-brown, outer web of outermost primary white. Underwing-coverts dark greyish brown, narrowly edged white. Bill apple green shading to bluish white on lower mandible, culmen blackish. Narrow rim of rich orange or yellow skin above eyes. Eye golden brown. Legs and feet jade green. ADULT ♀: similar to ♂ but forehead to hindneck dark grey, shading to lighter grey on sides of head and neck. Breast and flanks grey, feathers fringed white, producing slightly scalloped effect. Bare skin around eye pale green or yellow, bill greenish, eye sometimes dark brown. SIZE: (3 ♂♂, 6 ♀♀) wing, ♂ 104–109 (107), ♀ 104–109 (106): tail, ♂ 44–49 (46·5), ♀ 42–47 (43·5); bill, ♂ 17–19 (18), ♀ 15–19 (17·5); tarsus, ♂ 35–37 (36), ♀ 33–37 (35). WEIGHT: 1 ♀, 61.

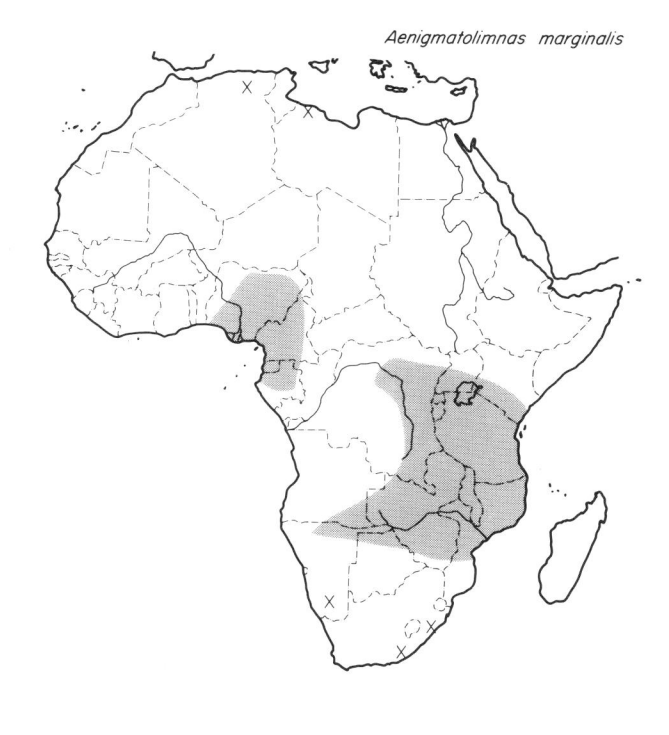

IMMATURE: browner above than adult, lacking streaks; forehead to nape paler, more rufous, sides of face and body tinged rufous, breast rufous. Flanks and thighs tinged grey-brown; chin, throat, belly and undertail-coverts off-white. Bill yellowish horn, legs blue-grey, eyes brown.

DOWNY YOUNG: black; cere creamy white.

Field Characters. Distinguished from Corncrake *Crex crex* and African Crake *C. egregia* by smaller size and white stripes on upperparts. Best told from all plumages of *Porzana* spp. by reddish cinnamon lower flanks and undertail-coverts, visible as bird flies away (Spotted Crake *P. porzana* has pale buff undertail-coverts); and in side view by deeper bill and lack of barring on underparts. Further distinguished from Spotted Crake by lack of spots, from Spotted Crake and Little Crake *P. parva* by lack of red base to bill, and from Baillon's Crake *P. pusilla* and ♂ Little Crake by lack of slaty underparts. Legs and toes longer than other crakes, project more in flight. Wings shorter than Spotted Crake and wing-beats shallower; pale leading edge to wing less prominent.

Voice. Tape-recorded (296). Call, taped in ditch where bird previously observed by R. Stjernstedt and D. Aspinwall, matches description of suspected song in Hopkinson and Masterson (1975): a long series of clucking or ticking notes run together rapidly, lasting *c*. 1 min. Calls at night. Grunts or growls given if afraid or protecting chicks. Family party gave repeated 'chup' or 'yup'.

General Habits. In breeding season inhabits seasonally inundated, often tussocky, grasslands which dry out and are often burned over during dry season. Prefers areas with muddy patches, shallow pools and finer grasses up to 1 m high, avoiding permanent marshes with heavier growth of sedges and reeds. Breeds away from larger streams liable to heavy flooding, preferring habitats on somewhat higher ground, such as edges of marshes, rice fields, shallow temporary pans, short-grass dambos, gravel pits and banks of drainage ditches. In Ghana, marshy savanna with small scattered thickets into which birds retreat when flushed. Habitat outside breeding season similar: inundated grassland and pool edges (Kenya, Mombasa); fine reeds near stream (Namibia); short sedge and grass clumps fringing shallow water (Zambia, Luangwa Valley).

More secretive than other crakes, hard to flush even with dog, seldom seen in open. Slinks away when disturbed, creeping through vegetation like rat, or 'freezes', when it may be caught by hand. Bird pursued by dog submerged in several cm of muddy water (P. B. Taylor, pers. comm.). Bobs tail vigorously while feeding. Though night-time migrant, probably mainly diurnal; captive bird bathed in early morning sun and roosted under clump of grass.

As breeding habitats dry out, retreats toward local streams and marshes, but also makes longer movements towards equatorial regions. Migrants (wintering?) present Nov–Mar on airfields, Gabon (Brosset 1968). Migrants have hit lighted windows at night Nigeria, Kenya and Zambia, and vagrants have reached Aldabra, Algeria, Libya and extreme southern Africa. Lack of geographical variation prevents distinguishing of migrants in tropical regions from local birds.

Food. Small snails, grasshoppers and other insects. Captive birds ate earthworms and a spider.

Breeding Habits. Preponderance of ♀♀ in skin collections suggests polygamy. Solitary nester in grassy marshes. 7 nests found in area of *c*. 24 ha Zimbabwe; closest nests 45 m apart.

NEST: typically placed in tuft of old dry grass, sometimes in sedge clump; normally 10–25 cm above water, occasionally floating; sometimes above damp ground, rarely on it. Shallow bowl or platform made of grass (*Setaria, Eragrostris, Sporobolus, Leersia, Aristida* and *Paspalum*), sometimes rushes and sedges. 1 nest measured '82 cm (*sic*) diameter' (Pitman 1965), presumably an error for 8·2 cm. Surrounding vegetation pulled down to form roof over nest, aiding concealment. Sometimes

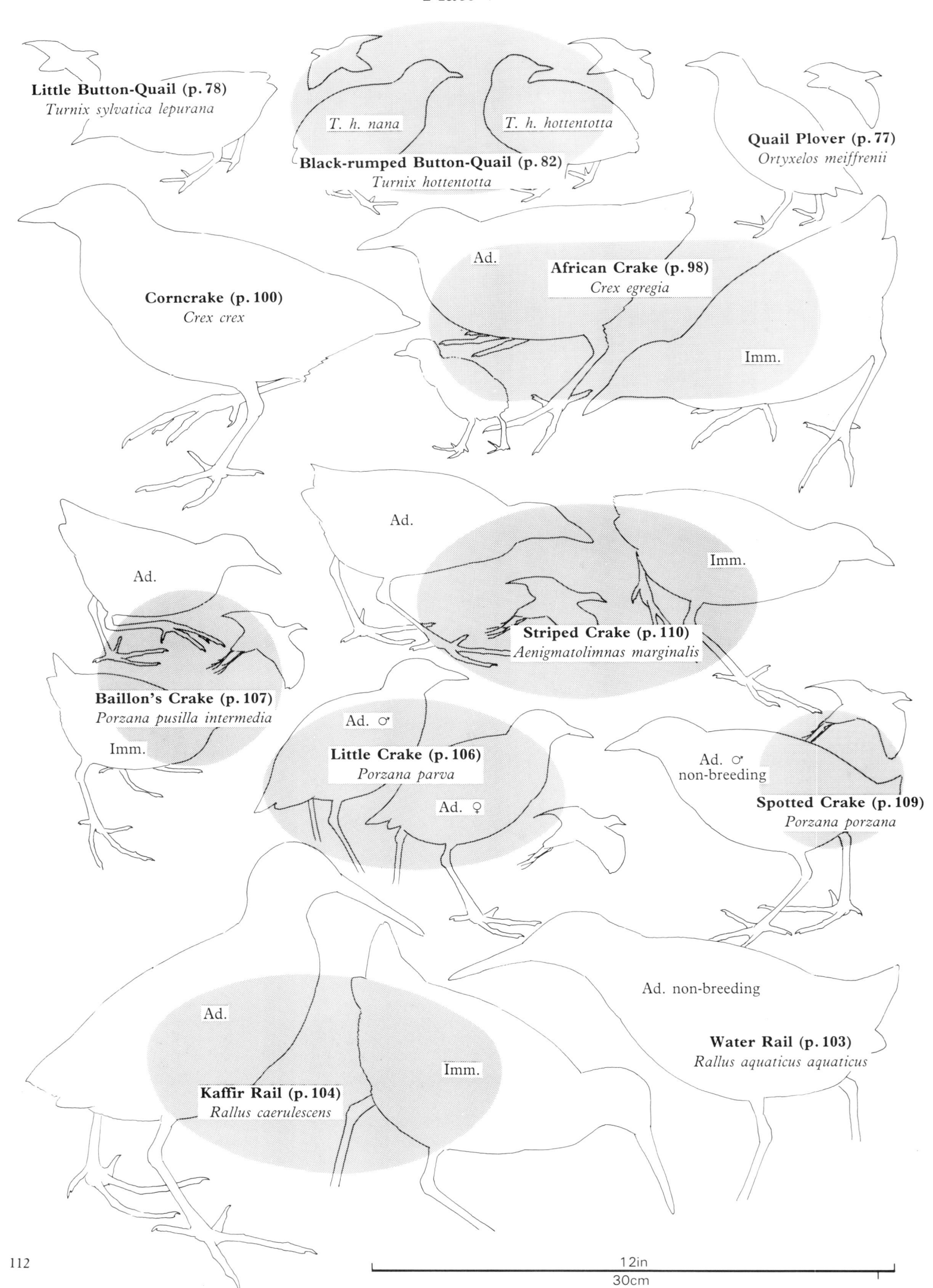

Plate 7

Little Button-Quail (p. 78)
Turnix sylvatica lepurana

T. h. nana   *T. h. hottentotta*

Black-rumped Button-Quail (p. 82)
Turnix hottentotta

Quail Plover (p. 77)
Ortyxelos meiffrenii

Ad.

African Crake (p. 98)
Crex egregia

Imm.

Corncrake (p. 100)
Crex crex

Ad.

Imm.

Striped Crake (p. 110)
Aenigmatolimnas marginalis

Ad.

Baillon's Crake (p. 107)
Porzana pusilla intermedia

Imm.

Ad. ♂

Little Crake (p. 106)
Porzana parva

Ad. ♀

Ad. ♂ non-breeding

Spotted Crake (p. 109)
Porzana porzana

Ad. non-breeding

Water Rail (p. 103)
Rallus aquaticus aquaticus

Ad.

Imm.

Kaffir Rail (p. 104)
Rallus caerulescens

112

12in
30cm

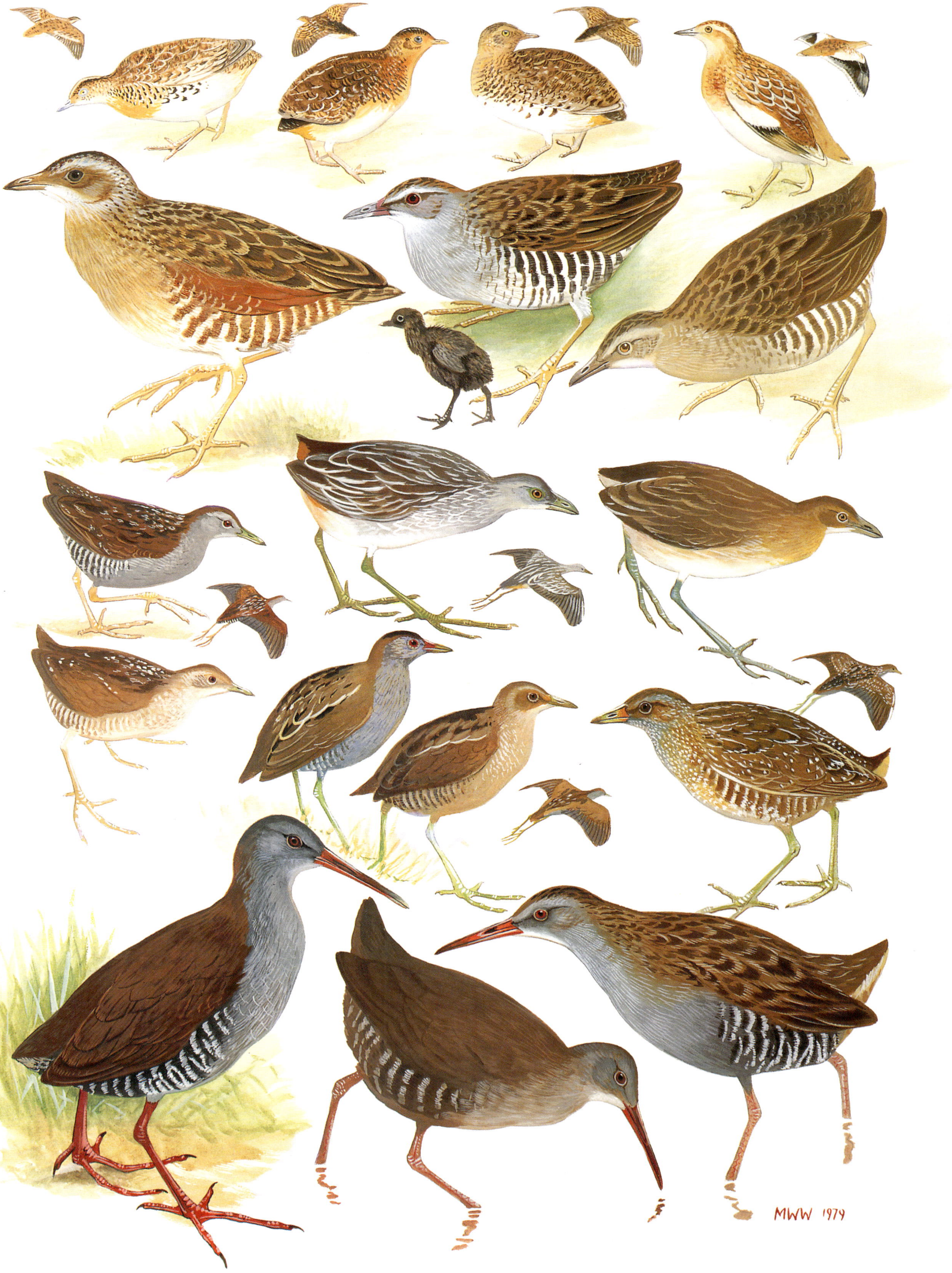
MWW 1979

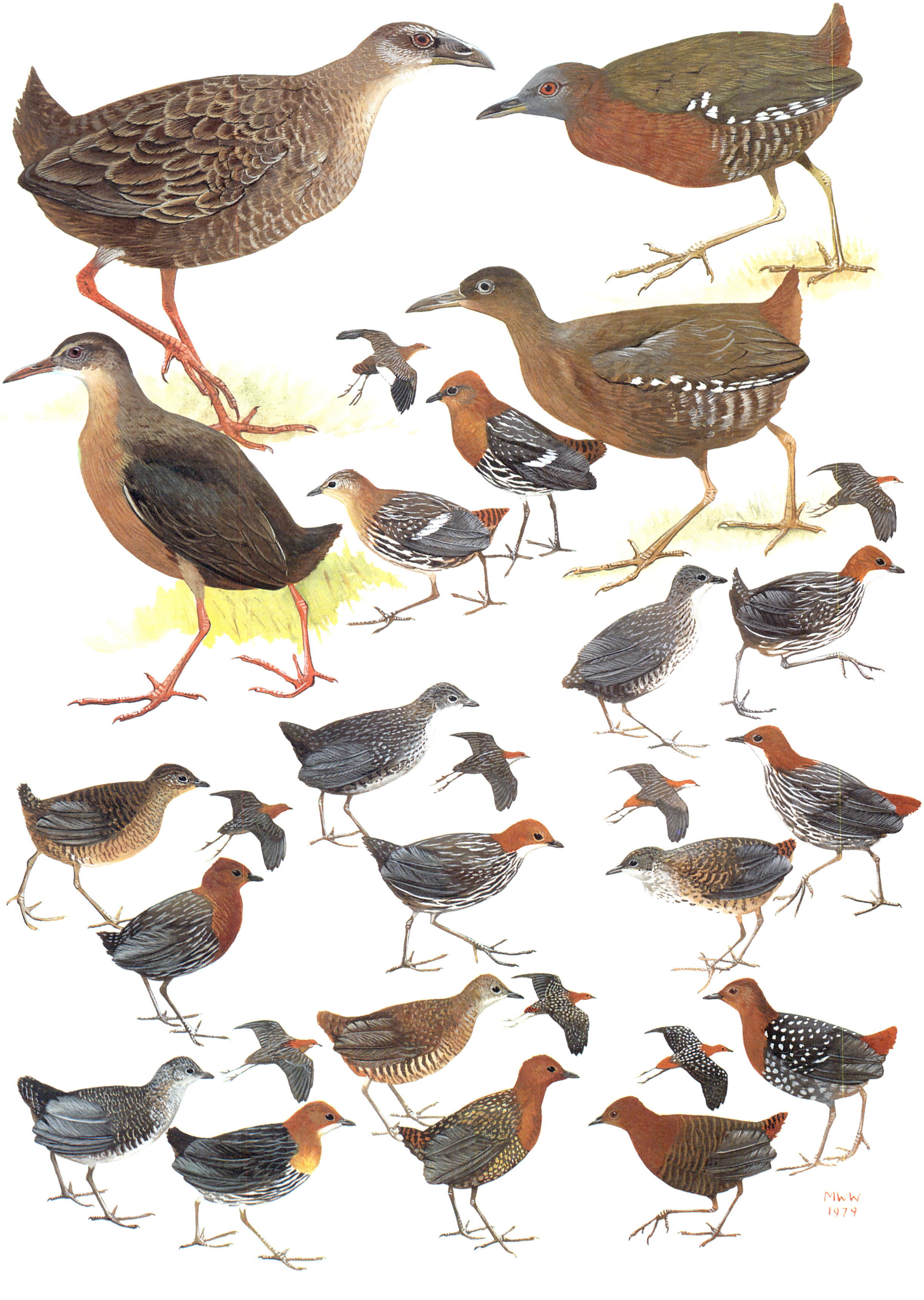
MWW
1979

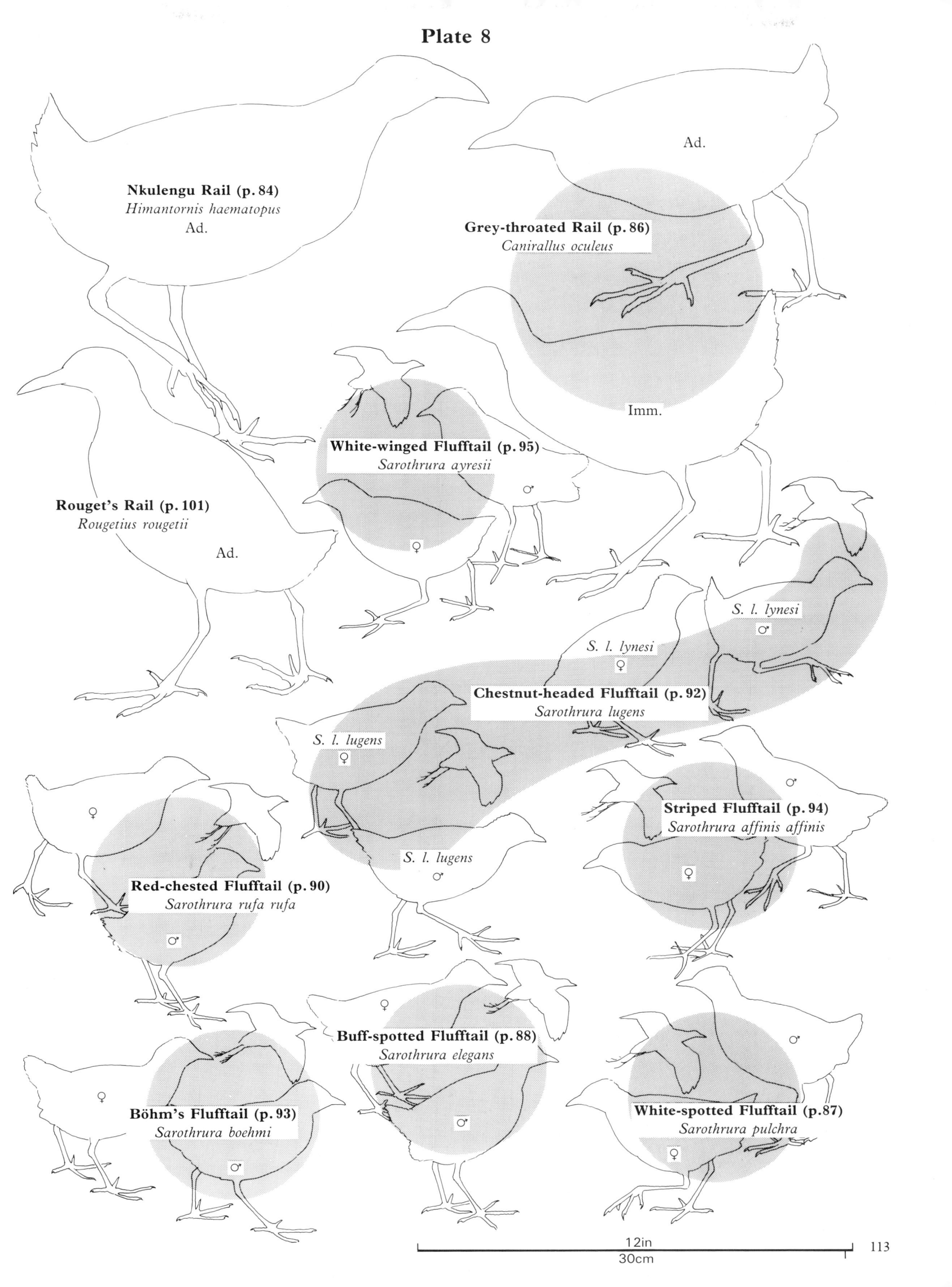

Plate 8

Nkulengu Rail (p. 84)
Himantornis haematopus
Ad.

Grey-throated Rail (p. 86)
Canirallus oculeus
Ad.
Imm.

White-winged Flufftail (p. 95)
Sarothrura ayresii
♀

Rouget's Rail (p. 101)
Rougetius rougetii
Ad.

S. l. lynesi ♀
S. l. lynesi ♂

Chestnut-headed Flufftail (p. 92)
Sarothrura lugens
S. l. lugens ♀
S. l. lugens ♂

Striped Flufftail (p. 94)
Sarothrura affinis affinis
♀

Red-chested Flufftail (p. 90)
Sarothrura rufa rufa
♀
♂

Böhm's Flufftail (p. 93)
Sarothrura boehmi
♀
♂

Buff-spotted Flufftail (p. 88)
Sarothrura elegans
♀
♂

White-spotted Flufftail (p. 87)
Sarothrura pulchra
♀
♂

12in / 30cm

113

114   RALLIDAE

builds dummy nests, partially constructing and then abandoning them.

EGGS: 4–5, laid at 1–2 day intervals; mean (17 clutches) 4·35. Oval, smooth, somewhat glossy. Ground colour pink, yellow, buff or cream, with variable overlay of red-brown spots and blotches, usually coalescing to form broad band around larger end. Some also have grey or violet marks. SIZE: (n = 24, Zimbabwe) 26–33 × 20–23·5; (n = 4, Zambia) 29–31 × 17–20 (n = 24, several other countries) 28·7–33 × 19·7–23 (29·6 × 21·2).

LAYING DATES: Ghana, June; Nigeria, Aug; Kenya, June (♀♀ breeding condition, May; ♀ carrying nest material, Nov; juvenile Sept); Tanzania (♂ breeding condition, June); Malaŵi, Jan–Feb; Zambia, Jan (birds breeding condition Dec); Zimbabwe, Dec–Mar; Namibia, Feb–Mar. Lays at beginning of rains.

INCUBATION: carried out by both sexes.

DEVELOPMENT AND CARE OF YOUNG: presumed family party of 4 Zimbabwe May, creeping among aquatic vegetation on pan by river: birds were 'not so agile as jacanas and had to flutter over the thin parts' (Brooke, *in* Pitman 1965).

BREEDING SUCCESS/SURVIVAL: in 4 out of 6 nests Zimbabwe, eggs broken, possibly by rat or mongoose, and possibly as result of prior human disturbance (Hopkinson and Masterson 1975). Predation recorded by African Marsh Owl *Asio capensis* and bullfrog *Pyxicephalus adspersus*.

References
Benson, C. W. (1964).
Hopkinson, G. and Masterson, A. N. B. (1975).
Pitman, C. R. S. (1965).

Genus *Amaurornis* Reichenbach

Bill short and stout; plumage, including that of immature, plain, unpatterned. Tarsus longer, more slender than *Porzana*. Skeleton of African Black Crake '*Limnocorax*' *flavirostris* almost identical to that of *Amaurornis phoenicurus* (Olson 1973), and *Limnocorax* is therefore here merged with *Amaurornis*.

6 spp. world-wide; 1 (endemic) in Africa.

Plate 9
(Opp. p. 160)

Amaurornis flavirostris (Swainson). Black Crake. Marouette noire.

Gallinula flavirostra Swainson, 1837. Bds. W. Afr. 2, p. 244, pl. 28; Senegal.

Range and Status. Africa; vagrant Madeira. Endemic; mainly resident, but locally migratory. The most widespread and numerous rail in Africa, found almost throughout, including forested areas, from northern edge of Sahel zone to Cape; sea level to 3000 m. Absent only from northern Africa and desert regions of NE and SW. Present on some offshore islands (Zanzibar, Pemba, Mafia), but absent from Gulf of Guinea Is. Common to abundant.

Description: ADULT ♂ (breeding): entire plumage black or nearly so. Mantle, breast, upper belly and wing-coverts dark slate. Primaries and secondaries brownish black. Bill greenish yellow; skin around eyes red; eye dark red; legs and feet bright pinkish red. ADULT ♂ (non-breeding): as breeding, but legs dull red. Sexes alike. SIZE: (10♂♂, 10 ♀♀) wing, ♂ 98–113 (106), ♀ 92–112 (101); tail, ♂ 39–49 (43), ♀ 37–48 (41); bill, ♂ 25–27 (26), ♀ 22–27 (24); tarsus ♂ 38–45 (42), ♀ 37–42 (39). WEIGHT: ♂ (n = 37) 78–118, ♀ (n = 31) 70–110. Weight of individual varies greatly within a year.

IMMATURE: chin and throat whitish, rest of plumage uniform olive-brown, becoming greyer with age.

DOWNY YOUNG: black. Bill pink with black band around centre; eye grey; legs and feet slate. Weight at hatching, 8–9 g.

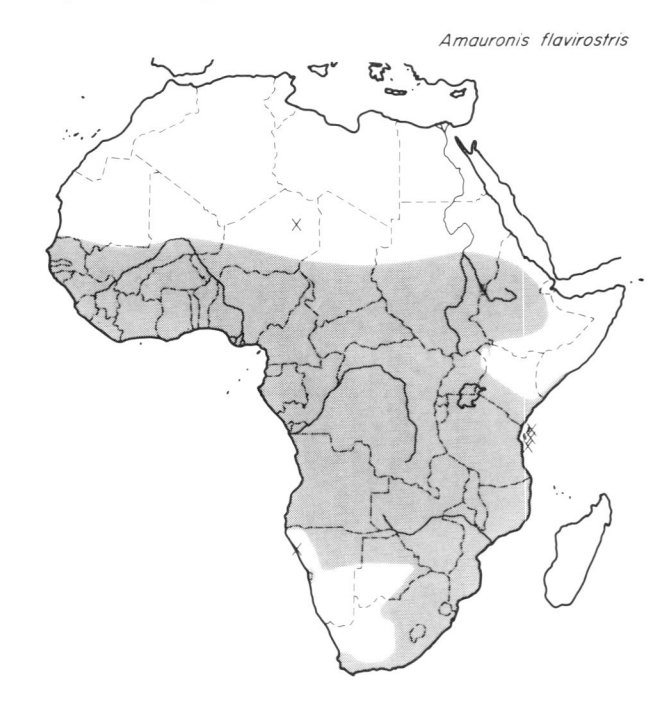

Field Characters. Adult, with black plumage and red legs, unlike any other African rallid; brown immature completely lacks any barring, spotting or streaking, and has brown undertail-coverts. Brown immature Kaffir Rail *Rallus caerulescens* is larger, with longer bill, and has barred flanks and white undertail-coverts.

Voice. Tape-recorded (10, 21, 38, 53, 74, 75). Advertising call or song takes the form of a duet. Bird 1 gives rather harsh chatter which increases in volume and often ends in crowing 'krraaa'. Bird 2 joins in with a softer, almost dove-like purring or crooning call, which may be a series of single notes or often a phrase 'coo-crrr-OOO'. First part of song is not antiphonal, as either bird may initiate it and point of entry of partner varies. Sometimes they continue to call seemingly independently, and may or may not end at same time. At other times, notes at end of song are precisely timed and antiphonal, thus:
Bird 1   'krrrok—krraaa —krrrok—krraaa'
Bird 2   'krrooo   krrooo'
Alarm call, a single rather sharp 'tyuk' or 'chipp', repeated every few s. Parent's call to young, a soft 'pu-pu'.

General Habits. Inhabits all swampy places as long as there is some cover: vegetation fringing open water, rank grass, sedges, reeds, papyrus, swampy thickets, bushes and other vegetation beside lakes, ponds, pans and rivers. Partial to ponds covered with water-lilies and other aquatic vegetation. In forested regions occurs in dense undergrowth in boggy clearings and along swampy forest streams, and in open country may occupy broad, grassy marshes. Very adaptable, and in drier zones will accept tiny streams with only thin cover.

Active by day but not by night. In Transvaal active in summer (Dec–Jan) 0500–1200 h and 1500–2000 h; in winter (May–Aug) active 0700–1800 h. Active after rain. Usually in pairs, but may gather in groups of up to 10. Bolder and easier to see than other crakes; in populated areas may become fairly tame. Often emerges from vegetative cover to feed along open shore of lake or river bank, in grass and cultivation near dams, even on dry ground some distance from water. Picks food from surface of water, from aquatic vegetation and from mud. Often perches on hippopotamuses, and has been seen feeding on back of one; perches on warthogs, apparently gleaning ectoparasites. Bobs head and jerks tail as it walks; treads slowly and deliberately over lily pads and other floating vegetation like jacana, and may spread wings to keep its balance. Swims well, and escapes danger by diving. If surprised on land flies up into bush or dives into reeds or other cover and runs off through vegetation; also 'freezes'. Normally flies low over water; flight weak, with dangling legs, or strong and direct. In evening climbs papyrus stems to several m above water, possibly to roost.

Chiefly sedentary; in N Ghana, N Nigeria and Sahel Zone of Sudan appears to be local migrant, appearing commonly with the rains in places where absent in dry season. In E Africa occupies temporary waters, and presumed migrants have been captured at night at Ngulia, Kenya.

In South Africa, flight feathers moulted annually Dec–Mar, making the birds flightless for up to 3 weeks. Tail feathers usually moulted at same time, sometimes earlier. 1 case of suspended moult reported; and 1 ♀ moulting its tail contained fully developed egg (Schmitt 1975). Body feathers moulted Sept–Apr, over a period of 5–8 weeks, with peaks in Oct and Mar, indicating 2 body moults a year. Flightless ♂ recorded Zaïre, Dec (Chapin 1939); in SE Zaïre there are 2 complete moults a year, in Apr–June and Oct–Nov (suggesting birds may breed twice a year, Verheyen 1953).

Food. Insects and their larvae, snails, frogs, crustaceans, small fish, worms, and seeds and other parts of aquatic plants; also heron eggs and carrion.

Breeding Habits. Monogamous; territorial. ♂♂ establish territories at beginning of breeding season, relinquish them when breeding finished. In Transvaal, territories were found to consist of *c.* 100 m of river frontage, with little depth (Schmitt 1975).

NEST: deep cup of reeds, rushes, sedges, creepers, grasses or other aquatic plants; in vegetation on or just above water, sometimes floating; also on ground in grass tussock near water. Nest in forest was placed on bed of fern stalks above mud (Bates 1930); nests on islets in Lake Victoria placed up to 2·5 m high in bushes, possibly as protection against sudden changes in water level (Pitman 1929). Outer diam. 100–300 mm, depth 80–175 mm, depth of cup 50–90 mm, rim of nest 200–500 mm above water level. ♂ may build extra nests in which no eggs are laid, possibly as roosting platforms. Immatures sometimes help build nest.

EGGS: 2–6, usually 3; once 2 ♀♀ laid 6 and 4 eggs in 1 nest (Pitman 1929). Elliptical oval or elongate ovate; ground colour white, stone, cream, buff or warm pinkish buff, marked with fine spots all over or with larger spots concentrated at blunt end; spots light brown, red-brown or chestnut over pale purple or slate. SIZE: (n = 27, Uganda) 29·5–33·4 × 22·0–24·4 (31·4 × 23·2); (n = 21, South Africa) 31·0–35·7 × 23·0–26·0 (33·8 × 24·2).

LAYING DATES: Senegambia, Dec–Mar, May–June, Aug–Sept; Liberia, Nov; Nigeria, Jan, June, Aug–Nov; Gabon, Mar; Zaïre almost all months, season not clearly defined; Sudan (Darfur), Oct; Ethiopia, Apr–Oct; Burundi, Mar; East Africa: Uganda, all months, with peaks in Apr and Nov; Kenya, Mar–Dec; Tanzania, Apr, Oct–Nov; Region A, Jan, May–July, Sept–Oct; Region C, Mar, June, Oct–Nov; Region D, all months except Nov; Region E, May; breeding peaks during or just after main rains. Zambia, Jan–Aug, Nov–Dec; Malaŵi, Mar–May, July–Sept, Dec; Mozambique, May–June; Zimbabwe, Jan–Feb, June, Aug–Oct, Dec; South Africa: Transvaal, Oct–Mar, peaks Nov and Jan;

'summer' breeding only, probably due to severity of winter climate; Natal, Aug–Mar (summer and winter); Cape Province, Dec.

INCUBATION: period: 13–19 days.

DEVELOPMENT AND CARE OF YOUNG: chick 1 week old is black with some greenish sheen; bill black with pink tip and pink patch around nostrils, eyes olive, legs and feet black. Wing claw brownish black, well developed and used in climbing. At 2–3 weeks body feathers present (down only on head and neck), and flight feathers are developing; pink tip of bill greatly reduced, legs becoming brownish. At 3–4 weeks flight feathers half grown, no pink tip to bill, pink patch around nostril smaller and paler, feet dark brown. At 4–6 weeks plumage olive-brown, flight feathers fully developed, nostril marks even smaller, eyes brown. At 8 weeks bill completely black. Between 2 and 4 months of age young undergo partial moult in which throat becomes whiter, breast grey; at same time bill becomes dark green, eyes and skin around eyes dark red. Able to breed within year of hatching; flight feathers first moulted c. 1 year after hatching.

Young tended by both adults, sometimes also by immature helpers, which roost around nest. Mutual preening by adult and helper noted (Brooke 1975). In Transvaal, chicks of 1st brood stay with parents for up to 6 weeks, chicks of 2nd brood for up to 8 weeks. 1st brood chicks may be deserted sooner to make away for 2nd brood. Pair with single young observed away from cover in banana plantation in Rwanda (Curry-Lindahl 1956).

BREEDING SUCCESS/SURVIVAL: eggs eaten by Monitor Lizard *Varanus niloticus*; Slender Mongoose *Herpestes sanguineus* and Water Rat *Otomys irroratus* suspected of eating eggs and chicks. Young preyed on by African Marsh Harrier *Circus ranivorus* and Purple Swamphen *Porphyrio porphyrio*.

References
Brooke, R. K. (1975).
Pitman, C. R. S. (1929).
Schmitt, M. B. (1975).

Genus *Porphyrio* Brisson

Medium to large rails with purple, blue and green plumage (except 1 sp.) and brightly coloured bill, frontal shield and legs. More specialized than *Gallinula*, adapted for walking on floating vegetation; toes long, legs modified in various ways, proportionately longer and more slender. Skull strong, bill short and heavy, nostrils small, jaw muscles well developed; ribs pneumatic.

5 spp. world-wide (including '*Notornis*'); 3 in Africa (1 endemic, 1 a vagrant only), of which 2 form a superspecies (*P. alleni*, *P. martinica*).

Plate 9
(Opp. p. 160)

Porphyrio alleni Thomson. Allen's Gallinule. Talève d'Allen.

Porphyrio Alleni Thomson, 1842. Ann. and Mag. Nat. Hist., 10, p. 204; Idda, Niger R.

Forms a superspecies with *Porphyrio martinica*.

Range and Status. Africa and Madagascar. Extralimital records from Britain, France, Denmark, W Germany, Spain, Italy, Sicily, Cyprus, Azores, Ascension, St Helena, Comoro Is., Rodriguez I., possibly Madeira.

Resident and intra-African migrant. Vagrant to N Africa (Morocco, Tunisia, Egypt, possibly Algeria). Widespread and locally common in most countries in subsaharan Africa, from coasts up to 1900 m, from Senegambia to Ethiopia and SW Somalia, south to Namibia, N Botswana and E South Africa to Port Elizabeth, with stragglers west to Rondevlei (Cape). Absent from most of Somalia, N Kenya and dry interior of South Africa. In general, common and widespread in rainy seasons, less common and localized in dry seasons. Common on Zanzibar, Pemba and Mafia Is. but vagrant to Banc d' Arguin, Bioko (Fernando Po) São Tomé and Pagalu (Annobon).

Description. ADULT ♂: forehead, crown, nape and sides of head black. Hindneck bright purplish blue; chin blackish, throat, sides of neck, breast and flanks blue-violet. Mantle, back, scapulars and tertials dull green with a bronzy wash. Rump, uppertail-coverts and tail dark green; belly, vent and thighs and central undertail-coverts blackish; lateral undertail-coverts white. Inner wing-coverts as back, outer ones blue-green. Primaries and secondaries blue-green on outer web, dull black on inner web. Axillaries blue-green; underwing-coverts blackish, those near edge of wing washed violet-blue. Bill dark red, frontal shield greenish blue; eye redbrown, red or yellow; legs and feet dark red. Sexes alike. SIZE: wing, ♂ (n = 10) 148–162 (156), ♀ (n = 13) 141–164 (152); tail, ♂ (n = 11) 60–68 (65·2), ♀ (n = 13) 61–73 (66); bill, ♂ (n = 11) 23–25 (24·4), ♀ (n = 13) 22–25 (23·3); tarsus, ♂ (n = 11) 49–56 (51·7), ♀ (n = 13) 46–54 (50·3). WEIGHT: ♂ (n = 4) 154–172 (165), 2 ♀♀ 112, 145; 1 juv. ♂ 140, 2 juv. ♀♀ 102, 126.

IMMATURE: crown and hindneck dark brown, becoming medium brown on sides of head and buff on cheeks, sides of neck and foreneck. Chin and throat creamy white. Upperparts, including scapulars, tertials and wing-coverts mottled dark brown and buff, darker on rump; outer wing-coverts with blue-green centres. Breast, flanks and thighs pale buff, belly white, undertail-coverts rich buff. Flight feathers and axillaries as adult; underwing-coverts blackish with white mottling. Bill brown with red base; frontal shield olive-brown; legs and feet red-brown.

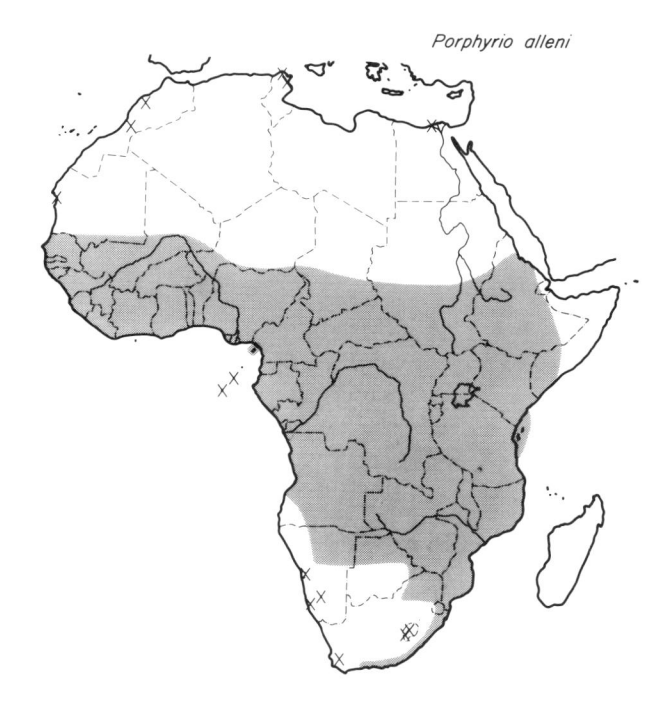

DOWNY YOUNG: black, browner below, tipped silvery around face and chin.

Field Characters. Like Purple Swamphen *P. porphyrio* but tiny (smaller than Common Moorhen *Gallinula chloropus*), with diminutive bill, blue-green frontal shield, and no light blue on face or throat. For distinctions from vagrant Purple Gallinule *P. martinica*, see that species.

Voice. Tape-recorded (53, 63, 73, 74). Variety of harsh, nasal calls, including dry 'keck' or 'kup', more drawn-out 'kerk', given singly or rapidly repeated; series of sharp notes ending with churring, 'kik-kik-kik-kik-kik-kyer-kyer-kyer-kyer-kiurr-kiurr-kurr-kurr'; and harsh, duck-like quacking. A high-pitched 'kli-kli-kli ...' given in flight; alarm, a sharp 'click'. Querulous anxiety note given if intruder close to nest when eggs are hatching.

General Habits. Inhabits marshes, reed beds, inundated grassland, papyrus swamps, rice fields, and sedges, rank grass and other thick vegetation beside lakes, ponds and rivers. Partial to ponds with water-lilies and other floating vegetation.

Generally shy and retiring, less so near human habitation. Most active from dawn to 0930 h and from 1745 h to dusk, but also feeds during middle of day and by moonlight. Forages while swimming like Common Moorhen, taking insects and plant material from water; also walks on water-lily leaves and other floating vegetation, picking up food items. Turns over floating lily pads and gleans food from under surfaces (Fry 1966); holds leaf down with feet after turning it over with bill. Partial to flower heads of water-lilies; breaks off head with bill, holds it with foot and tears off pieces with bill. Also holds food in toes and conveys it to mouth. Climbs waterside bushes and creepers to feed on fruits; climbs up to 2 m on *Typha* stems, presumably to feed. Sometimes makes feeding platforms of reeds *c.* 30 cm above water from which it can reach flowers and seed heads of reeds. Also feeds in short dry grass at edge of dam. Aggressive toward conspecifics when feeding, and sometimes parasitizes other species—seen robbing African Pygmy Goose *Nettapus auritus* of water-lily head (P. B. Taylor, pers. comm.). Density of non-breeding population on dam near Mombasa, Kenya, Sept–Dec, 25 birds/ha (P. B. Taylor, pers. comm.). Short flights fluttering and clumsy, but capable of strong long-distance flight. Jerks tail while walking; swims well and can dive. When running for cover, lowers head, raises tail, and runs rapidly with long strides. In evening sometimes climbs to high perch, possibly to roost.

A few remain resident but with onset of rains in N tropics most migrate northward (Nigeria, Cameroon, Chad) and most in S tropics southward. Mainly resident in Zaïre, but 3 in Kivu (June) believed to be migrating (Prigogine 1971)—one perched in tree top far from marsh. Seasonal influxes E Africa, e.g. Mombasa (non-breeding), SW Tanzania (breeding); wanders, e.g. to L. Turkana, and in July–Aug 1977 thousands appeared at L. Baringo in dry central Kenya (Britton *et al.* 1980). Resident Zambia, Malaŵi and Zimbabwe, but much more widespread during rains Dec–Apr. Wet season visitor to NE Botswana and N Transvaal, though resident on permanent waters of N Botswana and NE Namibia.

Flightless moult recorded Cameroon Dec, SE Zaïre in dry season, June–Aug and Oct–Nov. Rectrices dropped before remiges; remiges dropped simultaneously.

Food. Flowers and seeds of reeds and sedges, seeds, stems and leaves of grasses and other marsh plants, seed heads of water lilies; fruits of thorn bush *Drepanocarpus lunatus*; insects, spiders, worms, molluscs, crustaceans, fish eggs and small fish.

Breeding Habits. Monogamous; occurs in pairs year round. Territorial; nests well spaced even where birds are common.

NEST: rather loosely constructed of water plants, with deep or shallow cup; placed just above water, sometimes woven into surrounding vegetation, which may be pulled down over it to form dome; situated typically in

RALLIDAE

reeds, grasses or tangled vegetation at edge of water, but also in open swamps and rice fields.

EGGS: 3–8; mean (33 clutches) 4·4. Oval, occasionally oval-pyriform, smooth, with slight gloss; dirty white, pinkish cream, pale brown or light red-brown, covered with small, sharp, distinct spots and specks of red-brown over pale purple or ashy. SIZE: (n = 41, Nigeria) 31·8–39·2 × 23·6–27·5 (36·2 × 26·1); (n = 14, Malaŵi) 35·0–39·5 × 25·5–27·0 (36·8 × 26·3); (n = 24, Zimbabwe) 34·5–39·0 × 24·7–28·0 (36·5 × 26·2).

LAYING DATES: Nigeria May–Oct, especially at height of rains; Cameroon, Aug; Zaïre, SE, probably in second half of rainy season, beginning Jan–Feb (Verheyen 1953); Ethiopia, Apr, June, Sept–Oct; E Africa: Kenya, Sept–Oct, Zanzibar and Pemba, May–Aug, Region D, Apr, June, Aug–Sept; prefers dry months following long rains. Zambia, Dec–Apr; Malaŵi, Feb–Apr, June, Sept–Dec; Zimbabwe, Dec–Apr (mainly Jan–Mar), unseasonal records May, Sept; Namibia, Apr; Botswana (breeding condition, Dec, present NE Botswana only Feb–May).

INCUBATION: by both sexes; begins with 1st egg. Incubating birds sit tight on approach of intruder; when flushed may sneak off through vegetation but may also run away in full view across floating vegetation or even fly.

References
Brooke, R. K. (1968).
Hudson, R. (1974).
Serle, W. (1939).

Plate 9
(Opp. p. 160)

Porphyrio martinica (Linnaeus). Purple Gallinule; American Purple Gallinule. Talève pourprée.

Fulica martinica Linnaeus, 1776. Syst. Nat. (12th ed.) 1, p. 259; Martinique, West Indies.

Forms a superspecies with *P. alleni*.

Range and Status. Warmer parts of New World from S United States to N Argentina and S Brazil. Wanders widely; vagrants have reached E Canada, Greenland, Britain, Norway, Switzerland, Azores, Bermuda, Ascension, St Helena, Tristan da Cunha (frequent), Falkland Is., South Georgia and Africa.

Vagrant from New World. 1 record Liberia (90 km off coast at 5°05′ N × 10°22′ W); 21 records from SW Cape Province, South Africa, where found almost annually in recent years (most Cape Town; also Stellenbosch, Touws River, off Cape Columbine, Dassen I., and 100 km WNW of Cape Town).

Description. ADULT ♂: head, sides of neck, breast, upper belly and flanks deep purplish blue. Narrow line from sides of breast across upper mantle and hindneck light turquoise-blue. Upperparts from mantle to tail, scapulars, tertials and inner wing-coverts dull green, washed blue on mantle and becoming browner on rump and tail. Lower belly, vent and thighs dull black with purplish wash; undertail-coverts white. Primaries and secondaries dark grey-brown on inner webs, blue-green on outer webs. Outer wing-coverts light greenish blue. Axillaries blue-green; underwing-coverts dark grey-brown with some narrow white bars, those closest to edge of wing greenish blue. Bill red with yellow tip; frontal shield bright pale blue. Eye red; legs and feet yellow. Sexes alike. SIZE: wing, ♂ (n = 11) 179–191 (185), ♀ (n = 13) 172–184 (178); tail, ♂ (n = 9) 64–81 (70·3), ♀ (n = 8) 63–71 (67·7); bill, ♂ (n = 11) 27–31 (30), ♀ (n = 14) 25–30 (27·6); tarsus, ♂ (n = 12) 62–68 (65·4), ♀ (n = 14) 58–65 (61·8). WEIGHT: (a) normal: ♂ 203–269, ♀ 213–291; (b) 4 vagrants reaching South Africa, 119, 137, 150, 160.

IMMATURE: crown to hindneck warm brown, becoming lighter and buffier on sides of head and neck and foreneck. Chin and throat creamy white. Upperparts from mantle to tail, scapulars, tertials and inner wing-coverts dark brown, with bronze-green tinge on back, scapulars and wing-coverts. Breast and flanks buff-brown, belly and vent creamy white washed with buff, undertail-coverts white. Outer wing-coverts and flight feathers similar to adult but less blue. Axillaries and underwing-coverts dark grey-brown, former washed blue-green, latter broadly tipped white. Bill yellow-green with pale tip and brown-pink base; frontal shield brown or grey. Eye brown; legs and feet brown or yellowish.

Field Characters. Similar to Allen's Gallinule *P. alleni* but larger; head purple, not black, frontal shield brighter and paler, bill tipped yellow, legs yellow (not red) and undertail-coverts entirely white. Immature like that of Allen's Gallinule but upperparts and wing-coverts more or less uniform without pale edges to feathers, belly paler, undertail-coverts pure white (not buff).

Voice. Tape-recorded (63, 73). Typical call a sharp, high-pitched 'kyik' or 'kyek', sometimes with a booming undertone. Also common is a loud 'kur', often led up to by a series of 'cooks'—'cook, cook, cook, cu-KUR-cu, cu-KUR-cu', not unlike protest of domestic chicken. Other calls include a high-pitched squeal, a rapid 'ka-ka-ka-ka', low ticking, and a low, reedy buzz.

General Habits. Inhabits grassy marshes, overgrown swamps, ponds and lagoons (especially if overgrown with water lilies and other floating vegetation), rank vegetation beside rivers, even roadside ditches.

Usually found close to or in cover, but less shy than congeners; feeds in open by day on ponds, sometimes in fields near water, even in gardens if unmolested. Walks on lily pads and other floating vegetation in search of food, and turns over lily pads to glean underside. Jerks tail while walking; swims and dives easily; flies readily, long legs dangling. Easily climbs to top of marsh plants, and often seen there in evening; climbs trees to feed, and will also fly up into them.

Movements: of 20 dated records from South Africa, all but 2 fall in period 22 Apr–2 July, when birds migrating north from Argentina, Uruguay and S Brazil, whence African vagrants may originate (Silbernagl 1982). Favourable weather patterns and westerly winds were present just before the birds' arrival. 19 of the 21 vagrants immature. A further record near Cape Town 6 July 1983 (age?).

Food. Vegetable matter: pondweeds, sedges, willows, water-lilies, fruits, grass seeds, grain and rice; also aquatic insects, dragonflies, grasshoppers, spiders, frogs, and eggs and young of herons.

References
Lambert, K. (1969).
Silbernagl, H. P. (1982).

Porphyrio porphyrio (Linnaeus). Purple Swamphen; Purple Gallinule. Talève poule-sultane.

Plate 9
(Opp. p. 160)

Fulica Porphyrio Linnaeus, 1758. Syst. Nat. (10th ed.), 1, p. 152; Asia, America, = lands bordering the western Mediterranean Sea.

Range and Status. Africa and Mediterranean through Middle East and Oriental Region to Australasia and islands of SW Pacific; Madagascar.

Resident; distribution patchy. Very sparsely distributed N Africa: Morocco, breeds in marshes along lower Loukos R. and at mouth of R. Moulouya, and noted at Casablanca; Algeria, breeds L. Tonga (20 individuals, Mayaud 1983; abundant, Ledant *et al.* 1981) and L. Boughzoul and in marshes near Oran and Algiers, and recorded from oasis at Touggourt; Tunisia, breeds L. Kelbia and L. Ichkeul; Egypt, Nile delta and along Nile to Kom Ombo. In subsaharan W Africa occurs (breeding) only in Senegambia; from inland delta of Niger R. (Mali) to Chad (L. Fittri); and on Ghana–Togo coast. One record Sudan (Kosti: G. Richards and F. Lambert, *fide* P. B. Taylor); uncommon to rare in Ethiopian highlands. Ranges more continuously from highlands of E Zaïre, Rwanda and Burundi (common), Uganda (common around L. Victoria and on L. Kyoga and nearby lakes, rare elsewhere), Kenya (L. Turkana, central highlands and Mombasa) south through SE Zaïre and Mozambique to NE Botswana and E South Africa, and along coast to Cape; also west to NE Namibia and SW Angola.

Numbers have greatly declined N Africa this century, partly due to drainage and disturbance; however, numbers also fluctuate south of Sahara, and may vary considerably from one region to the next. Uncommon and very local Nigeria, with recent decline in numbers, and apparently uncommon L. Chad, but large numbers at L. Fittri, just 200 km to the east (Salvan 1968). Generally common to uncommon, but abundant in some localities.

Description. *P. p. madagascariensis* (Latham): Egypt and Africa south of Sahara. ADULT ♂: hind crown, nape, hindneck, sides of neck and upper mantle dull purple. Face, throat, chin and upper breast light blue. Lower mantle, back, scapulars and tertials bronze-green, becoming dark olive on rump and tail. Lower breast and flanks dull purple, belly dull black. Undertail-coverts white. Primaries and secondaries black, outer webs dark purple-blue; wing-coverts dark purple-blue.

Axillaries and underwing-coverts black, some of latter washed purple-blue. Bill and frontal shield crimson; eye crimson; legs and feet dark pink. Sexes alike. SIZE: wing, ♂ (n = 15) 229–268 (251), ♀ (n = 14) 226–258 (243); tail, ♂ (n = 4) 85–88 (86·5), ♀ (n = 5) 81–94 (87·2); bill, ♂ (n = 15) 38–43 (40·5), ♀ (n = 14) 37–41 (38·5); tarsus, ♂ (n = 15) 81–99 (92·1), ♀ (n = 14) 84–91 (86·8). WEIGHT: ♂ (n = 11) 528–687 (636), ♀ (n = 8) 480–737 (556).

IMMATURE: similar to adult but crown to upper mantle dark grey, only tinged with purple. Face grey, washed with light blue; chin and throat white. Underparts dull violet-grey, palest on belly. Bill and frontal shield dark grey; bill becomes dull red before adult plumage attained.

DOWNY YOUNG: black, with pale filoplumes on head, mantle and wings. Bill white, some purple-red at base; eye dark grey; legs and feet pink.

P. p. porphyrio (Linnaeus): NW Africa. Bronze-green of upperparts and wings replaced by purple. Somewhat larger: wing, ♂ 250–275 (265), ♀ 245–264 (259).

Field Characters. Huge size, purple plumage, massive red bill, red shield and red legs distinguish this species from all other African rails. Vagrant Purple Gallinule *P. martinica* is considerably smaller, with red and yellow bill, blue-white shield and yellow legs. Allen's Gallinule *P. alleni* is much smaller, with greenish blue frontal shield.

Voice. Tape-recorded (34, 53, 63, 73, 74, 76). Has a wide vocabulary; significance of many calls not known. 'Song' (possibly territorial), "long (8–15 s) unbroken series of powerful plaintive nasal rattles, without preamble and reaching crescendo, with almost human tone; rendered 'quinquinkrrkrrquinquinquinkrrkrr . . .'" (Cramp and Simmons 1980). Other calls include trumpeting alarm, 'gooweh', deep, grunting 'unk-unk-unk', short 'ank' or 'aak', short clucking 'cuk' (contact?), reedy groan or wail, a dry, screechy, heron-like 'wraaah', and a low hoot given every few s. 2 'interacting adults' recorded by T. Harris produced a short 'tok', a high, liquid 'wik', and a low, short reedy cackle. Flying bird gave a loud 'oork' (T. Harris, pers. comm.). Calls at night in chorus as well as during day. Normal call of young, a sparrow-like chirp. Adult leading juvenile to cover when disturbed gives quiet 'hon'; juvenile answers with low grunts.

General Habits. Inhabits extensive permanent swamps and perennial shallow lakes with dense growth of reeds, papyrus, sedges, *Typha* and other aquatic vegetation, especially with water-lilies; also edges of large dams and sewage farms, and in Egypt, islands in Nile. Occurs from sea level to *c*. 2500 m (once 3000 m, Mau Narok, Kenya).

Solitary, in pairs, or groups of 12 or more. Feeds mainly in early morning and late evening. Pulls down tall stems of reeds and other plants with bill, shears off section and eats it, grasping it in 1 foot and tearing off portions with bill. Pulls up younger stems with bill, splaying legs for support. Eats only pith of older stems, discarding outer tissue; younger leaves eaten whole. Climbs stems of reeds and other aquatic plants to strip off flower and seed heads. Nibbles low shoots, pecks at objects on ground, turns over stones and floating matted vegetation for invertebrates and digs in ground for roots. Picks insects from water surface; reaches under water and pulls up roots, young shoots and tubers with bill. Once observed climbing into tree in heronry to eat eggs of Cattle Egret *Bubulcus ibis* and Yellow-billed Egret *Egretta intermedia* (Strijbos 1955). Food sometimes carried to water and washed before being eaten. Bite very strong (C. H. Fry, pers. comm.)! Builds platforms for feeding and roosting.

Bathes while standing in shallow water, ducking head and flapping wings, then leaves water to preen and oil plumage. Oil applied to feathers of body, mantle and scapulars with bill, and to less accessible parts of body with wing-tips, which are rubbed over oil gland. Preening often followed by sun-bathing; bird stands with partly spread wings extended out from body and pointed downwards.

Generally rather shy, but bolder in early morning and late evening, walking out over water-lily pads or climbing into view at top of tall plant stems to sunbathe. Flight heavy and unwieldy, legs usually dangling but extended in lengthy flight. When startled in open may also 'paddle' over surface of water in hurry to escape, flapping wings. Runs fast through tangled vegetation; rarely swims but can swim strongly. Occasionally climbs trees. Flicks tail frequently, in variety of situations, typically indicating uneasiness, as a fear response, and to warn of predator. Tail also flicked in aggressive display (see Breeding Habits) and as signal for young to follow.

Mainly sedentary, but sometimes moves locally after breeding in NW Africa and Egypt, where formerly wandered to Suez Canal in winter (Meinertzhagen 1930). In Gambia, only a non-breeding dry season visitor; in E Africa disperses to temporary ponds in seasons of unusually heavy rains (Brown and Britton 1980); some local movement also noted in normal seasons (P. B. Taylor, pers. comm.). Has wandered to Pemba Is. and scattered localities in desert.

Remiges and rectrices moulted simultaneously. In South Africa (Transvaal) moult is in Oct–Dec, while breeding; a ♀ with oviducal egg had remiges and rectrices two-thirds grown (Fagan *et al.* 1976). Body feathers moulted twice a year, Sept–Dec and Apr–May.

Food. Omnivorous, but takes mainly aquatic plant material: roots, stems, leaves, flowers and seeds; especially partial to tubers of water-lilies. Also insects, snails, worms, leeches, crabs, fish eggs, small fish, frog spawn, frogs, birds' eggs and young; some carrion.

Breeding Habits. Monogamous. Aggressively territorial, with many boundary squabbles, but in restricted habitats pairs sometimes associate in loose colonies. Territory defended mainly by ♂, also by ♀ and helpers. Nest dispersion dependent on density of vegetation and size of population; in Nigeria, 11 nests with eggs found within 3.2 ha (Serle 1939).

In aggressive encounters commonly uses hunch-backed posture; in more intense form, neck arched and head thrust toward opponent. Runs at opponent if it fails to retreat and pecks it hard on head and neck; but if opponent turns head away in appeasement gesture it may be allopreened instead, and sexual behaviour may follow. In defence of mate or young adopts gaping-threat posture; with plumage sleeked, holds neck straight out and gapes at adversary (often a different species), frequently chasing it as well. In territorial-threat posture, bird stands upright and flicks tail to display white undertail-coverts, while giving grating calls, 'kree' or 'kree-ik'. Fighting may follow if intruder does not retreat, involving pecking or grappling with feet. During or after aggressive behaviour, displacement activity such as violent pecking at food objects and bill-wiping may take place. Swanning-display used in response to disturbances near nest or chicks—bird fluffs out plumage and raises partly spread wings over rump.

Courtship behaviour usually starts with allopreening, infrequently accompanied by courtship-feeding. ♀ induces ♂ to allopreen her by turning head away and fluffing neck feathers. This ritualized activity functions to reduce aggression. Mutual preening at nest is concerned with plumage care. Courting ♂ may approach ♀ with vegetation in bill, bow to her while calling quietly, raise and flap wings, stand up straight, continue bowing, then bring wings forward, quivering them and calling loudly. Allopreening may lead to marking-time display in which birds walk on the spot without closing the toes (toes normally closed when lifted for walking); this may lead up to copulation. ♀ solicits copulation by adopting solicitation posture; ♂ steps or jumps onto her back and copulates, clinging with feet and flapping wings to aid balance. Copulation lasts 2–3 s, though ♂ may attempt to copulate several times while remaining on ♀'s back. Other birds present may preen ♀ before and during copulation. After copulation ♂ walks away and ♂ and ♀ preen themselves.

NEST: a large, bulky, loosely constructed structure with shallow cup and 1–2 access ramps. Built of reeds, rushes, coarse grass and other aquatic plants, sometimes with lining of grass blades, papyrus heads or ferns; placed in dense vegetation in or over water; surrounding stems often bent over it to form roof, aiding concealment. May be built up from water surface or attached to tall reed stems with underside clear of surface; occasionally up to 2 m above water. Built by both sexes, ♂ usually bringing material while ♀ builds; helpers bring material and sometimes help build. Nest also added to by incubating bird with material within reach.

EGGS: 2–6, laid at daily intervals; mean (33 clutches) 2·9. Long or broad oval; surface smooth or slightly rough, with some gloss; ground colour variable, creamy white, yellow-stone, pale green, deep buff, pink or pale red-brown, blotched and spotted with purple, red-brown and purple-brown over violet and grey. SIZE: (n = 30, Nigeria) 49·2–57·4 × 34·6–38·6 (53·9 × 37·4); (n = 20, South Africa) 53·0–59·9 × 35·2–40·0 (55·6 × 37·6). WEIGHT: 42.

LAYING DATES: Morocco, Mar–May; Algeria, Mar–May; Egypt, Mar–Apr; Senegambia, Nov–Feb; Ghana, July–Aug; Nigeria, Jan–Feb, July–Oct; Zaïre, 2 seasons, Mar–Apr and Aug–Oct (Verheyen 1953); Ethiopia, Mar, Oct–Dec; E Africa: Uganda, Feb–Apr, Aug–Oct; Kenya, Oct–Nov, Jan–Feb; Tanzania Feb; Region B, Mar–Apr, June, Region C, Apr, Region D, May–July, Nov–Dec; breeds during or late in rainy season. Malaŵi, Feb–Mar, June–Aug, Dec; Zambia, Dec–Jan, Mar–May, July; Zimbabwe, Jan, May–July, Sept–Oct; South Africa: Transvaal, Sept–Mar; Natal, Oct–Feb, Apr, Aug; Cape, Aug–Jan.

INCUBATION: begins with last or penultimate egg; by both parents, mainly ♀, and helpers. Period: 23–25 days. Eggs hatch synchronously. Changeovers at nest gradual, with little ceremony; arriving bird sometimes presents piece of nest material to sitting bird and then briefly allopreens.

DEVELOPMENT AND CARE OF YOUNG: at 25 days chick is mainly black with no sheen, much grey on face and flanks. At 44 days grey, not black, darker above than below, with whitish throat and undertail-coverts; legs grey-pink; wing-feathers not developed; size about that of African Jacana *Actophilornis africanus*. At 2 months still grey, darker on head; throat and undertail-coverts white, blue-green wing feathers partly grown.

Young tended and fed by parents and helpers. Brooded on nest for 1st few days, not receiving much food, then fed on tender plant shoots which are broken up by adults. Walks and swims soon after hatching. Starts feeding itself at c. 14 days, but parents continue to feed it at 25–40 days. Still following parents around at 44–60 days, but probably independent soon afterwards.

BREEDING SUCCESS/SURVIVAL: chicks caught in open preyed on by Black Kite *Milvus migrans* and other predators.

References
Fagan, M. J. et al. (1976).
Hamling, H. H. (1949).
Holyoak, D. T. (1970).
Marshal, H. W. (1958).

Genus *Gallinula* Brisson

Medium to medium-large, rather heavy-set rails; plumage mainly dark, with row of white stripes or spots on flanks. Bill short and stout; frontal shield brightly coloured. Wings long. Legs fairly stout. Toes with narrow lateral membrane but not lobed except in '*Porphyriops*', which helps bridge gap between coots and gallinules (Olson 1973).

8 spp. world-wide (including '*Tribonyx*', '*Porphyriornis*' and '*Porphyriops*'); 2 in Africa (1 endemic).

Plate 9
(Opp. p. 160)

Gallinula chloropus (Linnaeus). Common Moorhen; Common Gallinule. Gallinule poule d'eau.

Fulica Chloropus Linnaeus, 1758. Syst. Nat. (10th ed.), 1, p. 152; Europe; restricted type locality England, *ex.* reference to Albin.

Range and Status. Almost world-wide in tropical and temperate zones, including many oceanic islands, except Australasia, where replaced by allospecies. Withdraws from parts of N temperate zone in winter.

Resident (with some local movements) and Palearctic migrant. Widespread N Africa; breeds Morocco south to edge of desert, northern half Algeria, including small areas of permanent water in Sahara (Dupuy 1969), Tunisia (rather scarce but south at least to Tatahouine; *c.* 20 pairs L. Kelbia), Libya (Fezzan: colony of 50 pairs Sebha, possibly resident Brak; may also breed coastal Tripoli), N Chad (Tibesti) and Egypt (abundant resident along Nile; also Wadi Natrun and Faiyum). Winters throughout breeding range (numbers increased by migrants from Europe), also Hoggar and elsewhere S Algeria, coastal Libya and inland at Hon, and Egypt at Dakhla and Kharga oases.

Palearctic birds also winter Senegambia (R. Senegal to Ziguinchor and Kedougou), central Mali, extreme N Nigeria (Kazaure), central Chad (L. Chad to Abéché) and N Sudan (south to Khartoum).

Subsaharan population resident, patchily distributed. Absent from forested regions; in W Africa occurs from Mali (where overlaps with wintering *chloropus*; 2500 counted L. Fati, Jan), Ivory Coast through Ghana (uncommon) and Nigeria (not uncommon, mainly in N) to Cameroon (rare). In S and E Zaïre locally abundant on lakes in E highlands, uncommon elsewhere; widespread and frequent in highlands of Ethiopia. Widespread and common in parts of E Africa, up to 3000 m, chiefly E Uganda, W and central Kenya, N Tanzania, Zanzibar and Pemba. Increasing on dams in Malaŵi below 1500 m but absent from L. Malaŵi and other large lakes. Uncommon Zambia; common on coast of Angola but scarce in interior. Locally common on permanent waters Zimbabwe and N Botswana; widely distributed Namibia and South Africa, even in dry interior; completely absent only from Namib and Kalahari deserts. Vagrant São Thomé; formerly bred Pagalu (Annobon), now possibly extinct.

Description. *G. c. meridionalis* (Brehm): subsaharan Africa. ADULT ♂: head and neck dull black, grading into dark slate on mantle. Rest of upperparts, scapulars and tertials dark olive-brown, tail brownish black. Underparts dark brownish slate, feathers of belly and vent variably tipped white. A line of broad white streaks along flanks; undertail-coverts white with black centre. Primaries and secondaries brownish black; narrow line around leading edge of bend of wing white; wing-coverts as mantle but variably washed with slate; axillaries and underwing-coverts brownish black narrowly tipped white. Bill bright red with yellow tip; frontal shield bright red. Eye red to red-brown; legs and feet yellow-green except for orange-red 'garter' at top of tibia. ADULT ♀: as ♂ but frontal shield smaller. SIZE: (12 ♂♂, 5 ♀♀) wing, ♂ 156–169 (163), ♀ 145–170 (161); tail, ♂ 62–77 (70·4); ♀ 63–73 (70·4); bill, ♂ 25–29 (27.1), ♀ 25–27 (26.2); tarsus, ♂ 44–51 (46.8), ♀ 45–48 (46.4). WEIGHT: ♂♀ (n = 117, South Africa) 173–335 (245).

IMMATURE: crown to upper mantle grey-brown, rest of upperparts and upperside of wings warmer brown, lacking olive tone of adult. Sides of head and neck buffy, variably speckled brown. Chin, throat, lower breast, belly and vent creamy white; foreneck, upper breast and flanks grey-brown, last with a few broad buff streaks. Undertail-coverts and rest of wing as adult, but more white on underwing-coverts. Bill dark olive-grey with yellow-green tip; eye brown. Legs and feet olive-grey to olive-yellow, garter yellow. WEIGHT: (n = 15, South Africa) 136–283 (203).

DOWNY YOUNG: black, upperparts somewhat glossy. Bill orange with yellow tip. Bare skin on head rose-red, above eye grey-blue, on throat yellow, on wing pink. Eye grey-brown.

G. c. chloropus (Linnaeus): N Africa, some wintering south to *c.* 17°N: wing-coverts same colour as back, without slate wash; larger (wing, ♂ 178–194, ♀ 169–184).

Field Characters. A familiar waterbird, often tame, adult likely to be confused only with Lesser Moorhen *G. angulata* (*q.v.*). Plumage dark grey and brown, lacking purple of gallinules, *Porphyrio* spp. Immature like small version of immature coots *Fulica* spp. but has buffy line on flanks and white undertail-coverts, grey to green, not white, bill and shield (coots of all ages have black undertail-coverts, no pale flank line). Immature Allen's Gallinule *Porphyrio alleni* and Purple Gallinule *Porphyrio martinica* are buffy to whitish below, without grey, and have some greenish in wings.

Voice. Tape-recorded (11, 38, 53, 62, 73, 74). Repertoire extensive; most familiar calls are advertising call, a single loud 'quarrrk' or 'krrraak'; and softer, less explosive 'krr-tuk' or 'krr-kuk', thought to indicate mild alarm. Anger or alarm indicated by loud 'kik-kik',

'kik-kuk' or 'kikker', sometimes rapid 'kik-kik-kik-kik'. Fighting birds emit short, sharp, clicking chatter, a shrill fast 'dee-da-da', and many sharp 'kik's. Flight call during nocturnal migration 'kek-kek-kek'. A low, musical toot, the 'murmur call', used in various displays, and by parents calling young. Call of young a hoarse, mewing 'phew'. Calls mainly during day, but 'quarrrk' call also given at night.

General Habits. Inhabits all fresh waters with some fringing vegetation: ponds, lakes, dams, streams, riverbanks, canals, marshes, rice fields. Occurs on lakes with water-lilies and other floating vegetation as well as on open water; on larger lakes remains fairly close to shore. Inhabits brackish waters on Namibian coast (P. Becker, pers. comm.).

Occurs singly, in pairs or family parties; outside breeding season forms loose feeding flocks. Forages while swimming, picking food from water surface or dipping head underwater, occasionally up-ending or diving. Gleans food from floating vegetation and emergent plants, and also feeds on land (meadows, cultivation) close to water. Seen to run out of cover and snatch dragonfly out of mist net (P. Becker, pers. comm.). Afrotropical birds (*meridionalis*) rather shy and wary, less tame than European birds. Swims well and dives readily; climbs nimbly, on reeds, swamp vegetation, and in trees and bushes where perches easily. Flight fluttering and somewhat weak-looking over short distances but strong long-distance flier. Roosts singly, in pairs or in small groups, on ground in reed bed or other dense vegetation, sometimes on branch of tree.

European migrants arrive N Africa (Morocco–Egypt) Sept–Dec, many remaining for winter; return passage Mar–May. Birds ringed Denmark recovered Morocco Nov, Algeria Feb; bird ringed Holland recovered Morocco Jan. Some continue down Mauritanian coast (Port Etienne, Banc d'Arguin) or down Nile to Sudan; others cross Sahara on broad front (noted in oases in Algeria, Libya, Niger and Chad; northbound in Algeria and Libya). Some dispersal of N African breeding birds in winter in Tunisia and Algeria, probably elsewhere. Breeds at and migrates through Boughzoul (Algeria) but not there in winter.

Afrotropical race *meridionalis* mainly sedentary; some local movements noted but no regular migration. Wanders to temporary pools in E Africa, and influxes noted on permanent waters. Mainly sedentary Zimbabwe but some local dispersion in rains. Vagrant to NE Chad (Fada) (Friedmann 1962); all other specimens from Chad are of nominate race.

Remiges and rectrices moulted simultaneously, at peak of breeding. Flightless birds noted Pemba Sept, Namibia Jan, Transvaal Sept–Mar, W Cape Jan–Apr. Primaries grow 5·0–5·3 mm per day, moult completed in 20–27 days (Fagan *et al*. 1976). Adult with half-grown wing-feathers seen with 2-week-old chick; moulting ♀ laid and incubated eggs and at same time tended young of previous brood. In Transvaal, body feathers moulted twice a year Aug–May; moult lasts more than 6 weeks.

Food. Wide variety of animal and vegetable matter: seeds and fruits of aquatic plants, water-lily leaves, filamentous algae, insects including dragonflies, worms, slugs, snails, crustaceans, tadpoles, birds' eggs and sometimes carrion.

Breeding Habits. Monogamous, though polyandry recorded. Resident birds may remain paired for several years, but where territory seasonal pair-bond may only last for breeding season. Territorial; pairs breed solitarily, nests usually well spaced. Territories defended vigorously during breeding season. After breeding some birds gradually reduce size of territory, which becomes mainly neutral, but still defend 'core' area; others become non-territorial. In Britain, breeding territories comprise 80–220 m of waterway and winter 'core' areas only 30–50 m (Wood 1974). Territories defended by aggressive displays: (1) charging—swimming or running at intruder with head and neck horizontal and frontal shield displayed; (2) splattering—a more intense form of charging in which attacker flaps wings while running at adversary on land or splattering over surface of water; (3) mutual retreat (usually from territorial boundary)—2 birds simultaneously lower heads, arch wings, fan white undertail-coverts and slowly swim away from each other; (4) fighting (not common; usually confined to breeding season, during takeover attempt by invader)—birds face off with necks and bodies stretched up, then kick at each other, trying to claw opponent's breast, often flapping wings and rising into air, or try to force opponent underwater; (5) swanning and churning (used in defence of nest or young against other species, including man): (i) swanning—like mutual retreat but undertail-coverts only raised, not fanned; (ii) churning—follows swanning, and involves slapping water with feet.

Heterosexual displays and behaviour include: (1) meeting and passing (performed throughout year)—a greeting ceremony in which members of pair lower heads and flick and sometimes fan tails; (2) bowing and nibbling—first bird (either sex but usually ♀) bows head, partner then nibbles feathers on back of its head. Performed on land or on display platform (loosely built structure of reeds 2–5 cm deep × 20–25 cm diam.) on water, throughout year in paired birds, and during courtship and possibly pair-formation; (3) courtship chasing (performed before egg-laying)—♂ chases ♀ around territory; both run with head and neck stretched forwards, body horizontal; chase may be followed by nibbling or copulation; (4) arching and coition—♀ stands with head and neck arched downwards, bill pointing at toes, then squats, keeping neck arched; ♂ then mounts her, flapping wings to keep his balance. Coition often followed by post-copulatory display in which ♀ turns head to one side while ♂ bows to her; after this displacement feeding and preening may also take place.

♀♀ compete for mates; heaviest win most agonistic encounters, select small ♂♂ with large fat reserves. ♀♀ paired to fat ♂♂ initiate more clutches in a season, since fat ♂♂ can incubate for longer than thinner ♂♂ (Petrie 1983).

NEST: saucer-shaped structure, neat or loosely built, of reeds, rushes and sedges; placed in tall reeds, sedges, grass tussock or other dense cover, on water or up to 1 m above it; sometimes in shrub overhanging water; other sites recorded are fork of tree a few cm above seasonal pool or at edge of vlei, and dead tree stump in small dam, nest being well hidden in creepers. In Europe, nests floating in more than 20 cm of water usually have ramp (Wood 1974). Built by both sexes, sometimes helped by immatures of earlier brood. ♂ usually brings material while ♀ builds. Nest may be added to during incubation.

EGGS: 3–9, laid at 1-day intervals; mean (134 clutches, *meridionalis*) 5·7. 2 ♀♀ sometimes lay in same nest. Broad ovals, smooth and somewhat glossy; creamy white, pale buff, pinkish buff or pale brown, with specks, spots and blotches of rust red or deep red-brown over lead blue, grey or pale purple, mainly at blunt end. SIZE: (n = 25, Ghana) 37–42 × 26·5–31: (n = 37, Nigeria) 39·6–45·6 × 28·1–30·7 (42·3 × 29·4); (n = 100; South Africa) 38·0–46·3 × 28·3–35·0 (42·3 × 30·6). WEIGHT: (n = 88, Britain) 21·–28·5 (24·9).

If nest raided, lays again up to 5 times. In Palearctic 2 broods fairly common, 3 unusual (only from experienced pairs). In South Africa (W Cape), in situation where food supply always available, 2 pairs bred continuously all year (Siegfried and Frost 1975). Time between hatching of successive clutches 33–65 (42) days; eggs laid c. 14 days after hatching of previous clutch.

LAYING DATES: Morocco, Apr–June, Aug; Algeria, May–June; Tunisia, May; Libya, May–June; Egypt (imm. Oct); Ghana, June; Nigeria, July–Aug; East Africa: Uganda, Feb–May, Kenya, June, July, Tanzania, Apr, June–Aug, Pemba, June–Aug, Dec. Region B, Mar, May–July, Nov, peaking June, late in main rains; Region C, Apr–May, Oct, in wet months; Region D, every month except Feb, peaking June–Sept during cool, dry season; Region E, June. Malaŵi, Mar, May–Sept; Zambia, Feb–Mar, July, Nov–Dec; Zimbabwe, all months; 155 records show concentration Jan–Apr (54%) and June–Aug (32%) (Irwin 1981). Namibia Oct–Apr, probably Sept; South Africa: Natal, Jan–Mar, June, Oct–Nov; Transvaal, Aug–Feb, June; W Cape, all months, peaking Sept–Oct.

INCUBATION: of 1st clutch begins with last egg (hatching synchronous); in 2nd and replacement clutches begins when clutch about half completed (hatching asynchronous). By both sexes, though by ♂ more than ♀; sometimes by immature helpers. Period: 21–22 days. In South Africa eggs incubated continuously throughout 24-h period, by ♂ 72% of time, by ♀ 28% (Siegfried and Frost 1975). ♀ does not incubate at night. Changeovers occur from 1–2 h after sunrise until 2 h before sunset; they are initiated by arriving bird, but soft calls made by incubating bird may summon partner to nest. ♂ incubates overnight for 14·5 h. First shift of ♀ in morning lasts 2 h 50 min, subsequent shifts av. 50 min; av. shift of ♂ by day 67 min. During day, eggs turned every 10–12 min, during night every 22·5 min (Siegfried and Frost 1975).

DEVELOPMENT AND CARE OF YOUNG: weight at hatching c. 25 g, at 10 days 35, 20 days 90, 30 days 140, 45 days 230, slowly increasing to c. 275 g at 70 days. Plumage: days 8–18, down becomes thicker on head, blue above eye starts to fade; days 19–30, body-feathers grow, tail-feathers begin growing, down confined to wings and tail; bill and shield become pinkish, legs and feet grey-black; days 31–45, primaries, secondaries and wing-coverts growing; bill, shield and legs become dark olive-green; days 46–65, wing-feathers continue to grow; able to fly at 60–65 days, sometimes earlier. No further plumage changes after day 65.

Young tended and fed more by ♂ than ♀; larger broods may be divided between the 2 parents and fed separately. Remain in nest for 1–2 days; by 3rd day can swim easily, and thereafter brooded in rapidly built brood nest similar to original nest. Several brood nests may be built, according to size of brood. Chicks brooded for 1st week, then less frequently, and by end of 2nd week only in cold or wet weather. Fed by parents for 1st week or so, then start feeding themselves. By 8th day partially independent and able to escape when disturbed by diving and swimming up to 3 m under water; parents may carry young to safety on their backs. At 21–25 days they find most of their own food, though parents may continue feeding them until day 42–45, and they remain in parents' territory for another 6 weeks before dispersing. Immatures from earlier broods may help to brood, tend and feed chicks. Age at 1st breeding, 1 year.

BREEDING SUCCESS/SURVIVAL: in South Africa, from 40 clutches 1 pair raised 33 broods averaging 4·0 young to full independence and flying, and a 2nd pair raised 32 broods averaging 3·5 young from 37 clutches (Siegfried and Frost 1975). Young bird recorded killed by turtles (Read 1982).

References
Siegfried, W. R. and Frost, P. G. H. (1975).
Wood, N. A. (1974).

Gallinula angulata Sundevall. Lesser Moorhen. Gallinule africaine.

Plate 9
(Opp. p. 160)

Gallinula angulata Sundevall, 1850 (1851). Ofev. K. Vet.-Akad. Förh., 7, p. 100; Lower Caffraria, i.e. Natal, type from the Umlazi River.

Range and Status. Endemic resident and intra-African migrant, widespread and locally common south of the Sahara from Senegambia (mainly in wet season) to Sudan (locally north to Wad Medani, also Darfur in rains) and Ethiopia (rare), south to NE Namibia (Windhoek), central Botswana and E South Africa (rarely west along coast to Oudtshoorn and Swellendam). In Ivory Coast, outnumbers Common Moorhen *Gallinula chloropus* in Korhogo marshes; in Nigeria frequent except SE; numerous S Cameroon, uncommon Chad. Widespread but not generally common Zaïre except E and SE; in Uganda only at Awoja and Entebbe. Locally abundant in rains in Kenya, Zambia and Zimbabwe; one taken 16 km off coast near Inhambane, Mozambique.

Description. ADULT ♂: head grey-black; upper mantle and underparts medium grey. Rest of upperparts, including scapulars, tertials and most wing-coverts, dark olive-brown; tail blackish brown. A line of broad white streaks on flanks; undertail-coverts black in centre, white at sides. Primaries and secondaries dark brownish grey, outer web of outer primary white. Narrow line along leading edge and bend of wing mixed white and grey. Axillaries and underwing-coverts grey narrowly tipped white. Bill yellow; culmen and frontal shield red. Eye red; legs and feet green or yellow-green, sometimes with pinkish tinge. ADULT ♀: similar to ♂ but upperparts lighter and browner. Face largely light grey, black confined to area around base of bill; throat silvery grey, rest of underparts paler grey than ♂, especially belly. Frontal shield smaller, less bright, and orange next to feathers. SIZE: (9 ♂♂, 3 ♀♀), wing, ♂ 125–145 (137), ♀ 130–135 (132); tail, ♂ 52–62 (56·3), ♀ 55–59 (57); bill, ♂ 19–21·5 (20·3), ♀ 19; tarsus, ♂ 35–39 (36·6), ♀ 36–37 (36·7). WEIGHT: ♂ (n = 3) 150–164 (158), 2 ♀♀ 99·3, 137.

IMMATURE: top of head, upperparts and wings (except flight feathers) olive-brown, greyer on mantle. Sides of head and neck and lower throat buffy brown; chin, throat, lower breast and belly creamy white, shading to light grey on lower belly and flanks. Undertail-coverts as adult. Scapulars and tertials edged pale buff, remainder of wing as adult. Bill brownish yellow, base of culmen dusky; legs and feet greyish green to dull yellow-green. WEIGHT: (n = 1) 130.

DOWNY YOUNG: black. Bill black with white tip and pink base. Frontal shield and base of culmen pale red-brown, pale purple next to forehead. Eye dark brown; legs and feet bluish grey.

Field Characters. Smaller than Common Moorhen, but better character is bright yellow bill (red with yellow tip in Common). Legs as Common Moorhen but lacking red garter. Frontal shield red, pointed (rounded in Common). Plumage of ♂♂ very similar; Lesser lacks white markings on belly and underwing, but this hard to see in the field. ♀ Lesser distinguished by pale grey face, throat and underparts. Immature has dusky yellow bill, pointed frontal shield, browner head and neck, brownish wash on breast, and paler underparts, without dark flanks of immature Common. The 2 spp. are separable by habitat in many areas, Lesser occupying temporary waters while Common remains in more permanent ones, though they can also occur together.

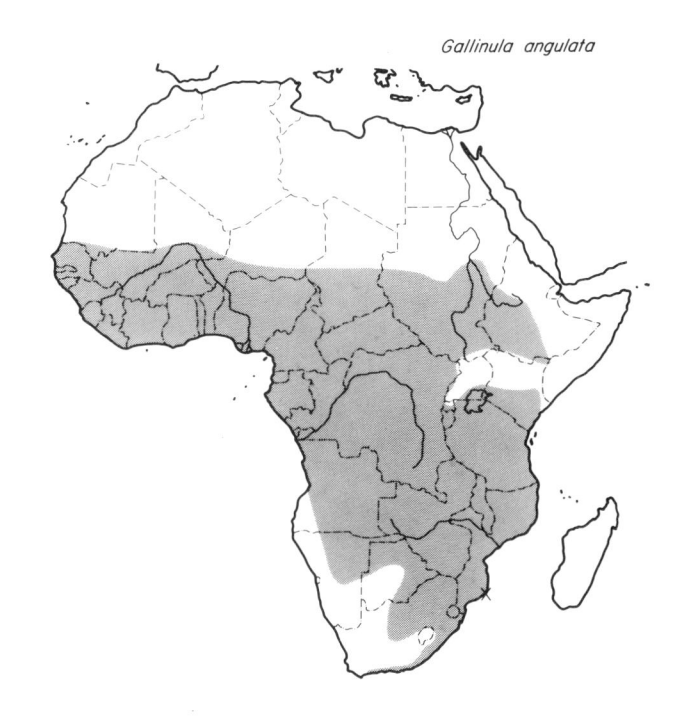

Voice. Tape-recorded (53). Calls resemble those of Common Moorhen but are somewhat quieter. A sharp alarm call, 'tik' or 'tek', made by bird near nest. Also gives subdued chuckling notes and a soft 'pyup'.

General Habits. Occupies permanent and temporary waters: papyrus swamps, reedbeds, marshes with rushes and open water, ponds with water-lilies, rank vegetation along forest streams and beside ponds, dams and rivers; shallow pans with emergent grass, rice fields, flooded farmlands, sewage ponds and seasonally inundated grasslands.

Much shyer than Common Moorhen and usually remains within cover, but sometimes tolerates proximity to man and may even become fairly tame; sometimes comes close to houses and footpaths in E Africa (Nairobi, Dar-es-Salaam). Swims less readily than Common Moorhen and less inclined to fly; flight as Common Moorhen. Sometimes walks on lily pads; jerks tail when walking or swimming. Forages on or at edge of water, on floating vegetation, or on open mud. Not aggressive while feeding, tolerates own and other species; though dominant over Spotted Crake *Porzana porzana*, feeds close to it (P. B. Taylor, pers. comm.).

A rains migrant. In W Africa some birds resident in wet southern latitudes, but in dry northern ones numbers increase during rains (when it breeds); later, it disappears from some areas as ponds dry out. In Chad resident except in Sahel zone, where purely a rains migrant. In Cameroon a possible nocturnal migrant with much fat was found in Nov, while in Gabon 4 birds were taken around buildings between 15 and 20 Nov, probably migrants (Brosset 1968). Mainly resident Zaïre, but birds in Kivu May–June were probably migrating (Prigogine 1971). Occurs Kenya and NE Tanzania most months, commonest Apr–July (post-breeding influx from southern Africa?). At Ngulia (Kenya), attracted to lights at night, Dec–Jan. In southern Africa largely a rainy season visitor. Of 124 specimens from SE Zaïre to South Africa, 88% were taken Dec–Apr (38% in January alone) (Benson and Irwin 1965). In Zimbabwe and Botswana breeds in semi-arid areas where pans and other temporary waters disappear during the dry season. In Zambia largely absent June–Nov, though some present all months except Sept where suitable habitat remains. Numbers build up at end of rains, May–June; over 1000 recorded Kafue Flats (Taylor 1979).

Food. Insects, beetles, molluscs; vegetable matter, including seeds and flowers of reeds; in captivity, termites.

Breeding Habits. Monogamous, territorial. Nests usually well separated, but in Nigeria 4 pairs nested in radius of 20 m (Serle 1939).

NEST: smaller and more compact than that of Common Moorhen; a grass cup 12–15 cm diam., placed on or a few cm above water in patch of grass; surrounding grass stems often bent down over nest and bound together to form roof.

EGGS: 3–9; mean (55 clutches) 5·0. Oval; slightly glossy; pale cream, pale buff or yellow-white, with some small spots and a few large angular blotches of light brown, red-brown or dark brown, with pale lilac or purple undermarkings. SIZE: (n = 82, Nigeria) 31·0–37·2 × 23·0–27·0 (34·1 × 24·8).

LAYING DATES: Senegambia (♂ breeding condition, Aug); Nigeria July–Sept; Chad Aug; Sudan Aug; Zaïre (imm. Aug); on high plateau of SE, Jan–Mar (second half of rains), at lower altitudes possibly Apr–May (Verheyen, 1953). Angola, Jan, Mar; East Africa: Kenya, June; Region C, Dec–Mar; Region D, Mar, July; Region E, June. Zambia, Jan–Mar; Malawi, Jan–Mar; Zimbabwe, Dec–Apr, peaking Jan–Feb; Namibia, Feb–Mar, probably Sept; Botswana, Feb; South Africa, Dec–Jan.

INCUBATION: begins before clutch complete; by ♀ and probably also by ♂. On approach of intruder bird usually slips away but sometimes sits tight. One flushed while eggs were hatching was back on nest in 10 min.

References
Benson, C. W. and Irwin, M. P. S. (1965).
Serle, W. (1939).
Taylor, P. B. (1979).

Genus *Fulica* Linnaeus

Large rallids of open water with plumage entirely or mainly black. Bill stout, somewhat compressed, with shearing edges; frontal shield large. Wing rather short and rounded; humerus longer than femur in most species, an adaptation favouring both high wing loading and diving. Pelvis narrower and more elongate than in *Gallinula*, tarsus more compressed, cnemial crest of tibia better developed, all adaptations, for diving. Toes lobed except in 1 species.

9 spp. world-wide, 2 in Africa, forming a superspecies (*F. atra*, *F. cristata*).

***Fulica atra* superspecies**

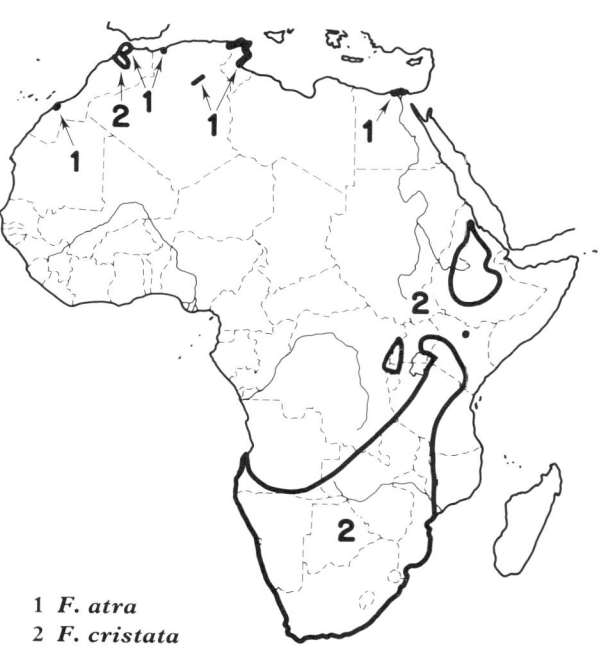

1 *F. atra*
2 *F. cristata*

Fulica atra Linnaeus. Eurasian Coot. Foulque macroule.

Fulica atra Linnaeus, 1758. Syst. Nat. (10th ed.), 1, p. 152; Europe, restricted to type locality Sweden, *ex* ref. to Fn. Svec.

Plate 9
(Opp. p. 160)

Forms a superspecies with *F. cristata*.

Range and Status. N Africa, Eurasia, Australasia.

Resident and Palearctic migrant. Locally common breeder N Africa, chiefly on larger lakes and marshes close to coasts, Morocco–Tunisia, and in Nile delta. Also breeds in Middle Atlas (Morocco), in central Algeria at El Goléa (small permanent colony?), in Tunisia south to Lac Affial near Feriana, Sidi Mansour, Gabes and Bou Grara, and occasionally on temporary waters (e.g. El Aïun, S. Morocco, 1955). Many European birds winter N Africa, especially Morocco (100,000) and Tunisia (311,620, 1973). More widespread in winter than summer, occurring coastal Libya, south into desert oases Morocco-Algeria, and along Nile at least to Khartoum (common to very abundant Egypt, uncommon Sudan, chiefly in north). In Khartoum Province occurs mainly in autumn and spring, suggesting birds winter further south (Macleay 1960). Some cross Sahara to winter in Senegambia (several hundred), Mali (usually in small numbers, central Niger delta to Gao; several thousand east of Timbuktu, Jan 1980), Nigeria (up to 500 in extreme north, once at Ibadan), Chad (Mao, Ouinanga Kebir) and W Sudan (El Fasher, Buram). Unconfirmed sight records Ethiopia, unacceptable record Tanzania.

Description. *F. a. atra* Linnaeus: only subspecies in Africa. ADULT ♂: head and neck black. Upperparts, including tail, scapulars, tertials and wing coverts dark slate grey. Underparts duller, brownish slate, undertail-coverts black. Primaries and secondaries dark brownish grey, secondaries tipped white, producing pale trailing edge to wing in flight. Narrow white line along leading edge of wing. Axillaries and underwing-coverts dark grey. Bill and frontal shield white, eye red. Tibia orange, tarsus mainly yellow, feet grey. Sexes alike. SIZE: wing, ♂ (n = 21) 211–229 (219), ♀ (n = 23) 197–213 (205); tail, ♂ (n = 18) 50–61 (55.4), ♀ (n = 22) 49–60 (53.4); bill, ♂ (n = 30) 28–32 (29), ♀ (n = 30) 26–29 (27.6); tarsus, ♂ (n = 37) 59–65 (61.7), ♀ (n = 38) 54–60 (56.8); WEIGHT: all months, range and means ♂ (n = 181) 676–1200 (835–953); ♀ (n = 190) 595–1150 (715–802).

IMMATURE: top of head and hindneck black. Sides of head dark grey, variably flecked with white, throat and foreneck mainly white with dark flecks. Upperparts and upper surface of wings dark brown, tinged slate on mantle and wing-coverts. Underparts mainly brownish grey, becoming paler in centre and mainly white on breast. Rest of plumage as adult except for white tips to some underwing-coverts. Bill grey with pink tinge, eye grey or brown, legs and feet grey.

DOWNY YOUNG: crown black; forehead, nape, sides of head, chin and throat orange to yellow. Otherwise grey-brown, down of mantle and wings tipped yellow. Bill red with white and black tip, shield red. Bare skin red on crown, blue above eye. Eye brown.

Field Characters. Not easy to tell from Red-knobbed Coot *F. cristata*, and sight records should be treated with caution. Lack of red knobs, character usually cited, not a good mark since knobs very small in non-breeding season and hard to see even at close range. For distinctions, see under Red-knobbed Coot.

Voice. Tape-recorded (62, 73, 76). Common (contact) call, a single, short 'kowp', 'koop' or 'kup'; sometimes a sharper 'kick' or 'kewk' or a softer 'cou'. 2 notes sometimes combined, e.g. 'kick-kowp'. Other calls include combat call (♂, a sharp, high-pitched, explosive 'pssi' or 'pyee'; ♀, a very short, croaking, coughed 'ai'); alarm call (♂, sharp, metallic variant of combat call, ♀, a rapid sequence of short notes, 'ai, oeu....'); other contact calls (♂, a mechanical 'p', 'dp', or 'ta'; ♀, a high, short, falsetto 'oeu'); and courtship call (♂, rapid sequence of low, hissing 'phsi' calls, ♀, soft, resonant 'oeu'). Parents call young with soft 'kt, kt' (Alley and Boyd 1950); call of young, a rasping 'creer', becoming shriller in alarm.

General Habits. Inhabits shallow lakes, large open marshes, fresh- and salt-water lagoons; on Nile, largest numbers at barrages, but also on slow-flowing stretches of river. Exploits temporary pools and seasonally inundated marshes for breeding. On wintering lakes keeps to deeper, more open areas than Red-knobbed Coot (Wood 1975).

Gregarious, forming large flocks outside breeding season, often associating with ducks. Aggressive, especially while breeding. Within flocks, shield-showing display together with rapid head movements used as individual spacing mechanism; charging attack (swimming towards opponent at high speed) used to establish dominance hierarchy. Flock forms tight pack on approach of gull or bird of prey and splashes up water. Forages by grazing on aquatic vegetation on surface of water, and by up-ending and diving for underwater vegetation and invertebrates; also feeds on land. Steals food from subordinate birds and ducks. Generally roosts at night but sometimes remains active. Flocks roost on open water or in vegetation.

Resident population partly nomadic according to changes in water level; breeds opportunistically in temporarily flooded areas. Migrants reach Morocco–Libya Sept, but main Egyptian wintering population does not arrive until Oct. Crosses Sahara on broad front, reaching Senegambia and Nigeria Nov; migrants pass SW Libya (Ghat) Oct–Nov, E Central Libya (Serir) Nov, Niger (Bilma) Nov, and a very early arrival NE Chad (Ennedi) 2 Sept. Fewer spring records, but present Nigeria and Sudan until Mar, late date NE Chad (Fada) 17 Apr. Present N Africa until Apr, Libya until early May (late date, group of 11, 17 June).

Over 40 birds ringed England, Belgium and Spain recovered Morocco; ringed France, Switzerland, Poland and Yugoslavia recovered Algeria; ringed Germany, Hungary and Poland recovered Tunisia. Chick ringed in Tunisia recovered Italy (Rome) 1·5 years later (Mayaud 1982).

Flightless during wing moult, chiefly July–Aug.

Food. Chiefly leaves, stems and seeds of aquatic plants, also grasses, algae, molluscs and insects; a few worms, frogs, small fish and other animals.

Breeding Habits. Monogamous; solitary nester in strongly defended territory. Mean distances between nests, 45–265 m; territory size 0·1–0·5 ha. For wide variety of displays used in territorial defence, pair-formation and maintenance of pair-bond, see Cramp and Simmons (1980).

NEST: bulky structure of plant material, sometimes strengthened with twigs of tamarisk and other bushes, lined with finer material; placed in shallow water, usually hidden in vegetation but sometimes in open. Diam., 25–55 cm, height above water 20–30 cm; built by both sexes, and added to during incubation and if water rises.

EGGS: 6–10, laid at 1-day intervals; mean (1131 clutches) 7·5. Oval, smooth, slightly glossy; pale buff spotted dark brown. SIZE: (n = 485) 44–61 × 33–40 (53 × 36). WEIGHT: (n = 283) 38.

LAYING DATES: Morocco, Mar–June; Algeria, Apr–May; Tunisia, Mar–June; Egypt (carrying nest material, Mar).

INCUBATION: begins 2nd–4th egg; by both sexes. Period: 21–24 days. Young hatch over period of several days.

DEVELOPMENT AND CARE OF YOUNG: fledging period variable; minimum 8 weeks, usually longer. At 8 weeks, mean wing length $c.$ 62% of final length, weight $c.$ 76% of final weight. 95% of final body weight and wing length attained at $c.$ 11–12 weeks; wing reaches maximum length at 12–13 weeks; weight continues to increase until 16th week. Body weight and wing length increase at different rates. At 5–6 weeks, weight 50% of final value, wing length only 30%; thereafter, wing length increases more rapidly (Visser 1974).

Young brooded on nest for 3–4 days while ♂ brings food; later fed and brooded by both parents. Brood division may take place, as in Red-knobbed Coot, each parent caring for part of brood. Parents with young less than 2 weeks old will tolerate or even adopt strange young resembling their own; thereafter drive strange chicks off, even if same age as their own (Alley and Boyd 1950). Young will follow and beg from any adult, even while being attacked by it, for first 8–11 days, thereafter learn to avoid adults in attitude of attack. At 3 weeks they can recognize their own parents; at 4–5 weeks they start finding their own food; at 9 weeks display territoriality, driving strangers from their parents' territory.

BREEDING SUCCESS/SURVIVAL: (England: 34–35% of eggs laid hatch, 21–23% of chicks fledge).

References
Alley, R. and Boyd, H. (1950).
Cramp, S. and Simmons, K. E. L. (1980).
Visser, J. (1974).

Plate 9
(Opp. p. 160)

Fulica cristata Gmelin. Red-knobbed Coot; Crested Coot. Foulque à crête.

Fulica cristata Gmelin, 1789. Syst. Nat., 1, pt. 2, p. 704; Madagascar.

Forms a superspecies with *F. atra*.

Range and Status. Africa, Madagascar, S Spain.

Resident, also nomadic. Distribution disjunct: Morocco, eastern and southern Africa. Locally common Morocco; 2100 on winter count, Jan 1972, of which 2000 on one lake (Dayet Annoceur); this now drained, but still winters other Middle Atlas lakes (J. D. R. Vernon, pers. comm.). Bred L. Halloula, Algeria, last century, may still do so (Cramp and Simmons 1980). Common to abundant Ethiopia, mainly in highlands and Rift Valley. In Zaïre mainly confined to lakes of E highlands, but numerous near Lusinga Station, Upemba N.P. (Verschuren 1978). In Burundi occurs in small numbers in delta of Ruzizi R. (Gaugris 1979). Avoids L. Victoria and other large E African lakes, though common L. Turkana and L. Naivasha. Common to abundant E Africa up to 3000 m, chiefly on

DOWNY YOUNG: dark grey, paler below, with orange-yellow collar, white tips to down of mantle and back. Bill red with white subterminal band; bare skin over eye grey-blue. Eye brown.

Field Characters. Not easy to tell from Eurasian Coot *F. atra*. At close range red knobs at top of shield distinguish it, but during non-breeding season they are small and very hard to see. Best character, good at distance, is shape of feathering between bill and shield (Deetjen 1969): in Red-knobbed Coot (**A**) it is blunt and rounded, ending at base of bill; in Eurasian Coot (**B**) it projects forward as pointed wedge between bill and shield. In flight, wing entirely black (wing of Eurasian Coot has narrow white leading and trailing edges). Immatures not safely distinguishable.

A   **B**

Voice. Tape-recorded (20, 34, 62, 66, 73, 74). Wide vocabulary; significance of many calls uncertain. Common (contact?) calls, sharp, shrill 'kik' or 'krik'; lower, reedy 'kek' or 'kert'; shrill, trilled 'krrt' intermediate in tone between first two; double clucking note, 'pickup'; and 'koop' or 'kup' with reedy overtone, like similar note of Eurasian Coot but deeper. Other calls: hollow 'hoo', sometimes repeated at regular intervals; deep, nasal, frog-like croak, and a drawn-out, muted 'ker'. Commonest call Morocco, Mar, rhythmic sequence of 3 notes, 'hu-hoo .... hoo'; 3rd note possibly made by 2nd bird. Alarm call, metallic, ringing 'cro-oo-k' or snorting 'tcholf'; group predator alarm call, high-pitched, nasal 'hue-hue-hue' (Wood 1975). Bird approaching nest gives low 'kiow, kiow', and birds passing each other or engaged in deep feeding for submerged vegetation give 'hinny' (van Someren 1956), which may be same as "strange, humming, breathy 'Vvvvv'" of McLachlan and Liversidge (1978). Call of young, plaintive 'cur-li'.

General Habits. Habitat chiefly open fresh water: ponds, lakes, lagoons, dams, permanent and temporary vleis, flooded plains; also swamps with reeds and papyrus, and sewage ponds. Small numbers sometimes on rivers and tidal lagoons, but prefers still rather than moving water. After breeding, pairs form flocks. Some birds in winter flocks appear paired. Non-breeders flock during breeding season. Normally not shy, but water-churning display (splashing water by slapping feet on surface) normally only used in defence of nests, ex-

highland lakes and dams. Irregularly distributed Zambia, usually in small parties, but up to 1200 counted on Kafue Flats. Uncommon in Angola, mainly confined to coastal plain of Mossamedes and Benguela; also recorded Pundo Andongo, Humpata, Iona National Park (W. R. J. Dean and M. A. Huntley, pers. comm.). In Malaŵi, occurs on open waters to *c*. 2500 m, but not on L. Malaŵi. Very local and uncommon Mozambique, possibly only non-breeding visitor. Common Zimbabwe, Namibia and Botswana in suitable habitat. Locally abundant South Africa: over 25,000 at times at Barberspan; over 30,000 counted at De Hoop Vlei, Bredasdorp (D. Skead, pers. comm.).

Though thriving E and S Africa, decreasing Palearctic. Nearly extinct S Spain. Formerly bred Tunisia and more widely Algeria. Reasons for decline not clear.

Description. ADULT ♂ (breeding): head and neck black. Upperparts and inner upper surface of wing dark slate, becoming dark brown on uppertail-coverts, rump, tail and outer wing. Underparts dark brownish grey, tinged slate on sides of breast and flanks. Undertail-coverts black. Axillaries and underwing-coverts dark grey. Bill and frontal shield white, sometimes tinged blue; 2 large red knobs at top of shield. Eye red, legs and feet mainly dark green. ADULT ♂ (non-breeding): sides of bill somewhat grey, knobs greatly reduced in size and dull red. Eye red-brown, legs and feet grey. Sexes alike; birds cannot be sexed even in the hand (Dean and Skead 1978). SIZE: wing, ♂ (n = 15) 219–239 (227), ♀ (n = 13) 208–224 (217); tail, ♂ (n = 15) 56–66 (59·4), ♀ (n = 17) 54–65 (58·7); bill, ♂ (n = 16) 30–36 (32.9), ♀ (n = 20) 29–33 (31); tarsus, ♂ (n = 14) 68–75 (71·8), ♀ (n = 19) 61–70 (66). WEIGHT: unsexed (n = 4016) 363–1236 (737) (Dean and Skead 1979).

IMMATURE: dark brown above, becoming flecked with white on sides of head and neck. Lores, chin and throat white, rest of underparts pale grey. Wings as adult. Bill grey, eye brown, legs and feet grey. WEIGHT: (n = 741) 243–1078 (579).

hibited toward human intruder by birds in winter in Morocco (Wood 1975). In group defence against Marsh Harrier *Circus aeruginosus*, grazing birds fled into water, formed tight pack, and stretched necks forward and upward, giving group alarm call (Wood 1975). Separated individual escaped by repeated diving and rejoining flock.

Feeds in flocks; forages by grazing (bill has specially adapted shearing edge). Foraging methods similar to those of Eurasian Coot, with emphasis on aquatic modes, such as diving down and pulling up underwater vegetation. On Kafue Flats (Zambia) grazes in areas where vegetation trampled by Lechwe *Kobus lechwe* (Douthwaite 1978). South of Sahara largely aquatic feeder, but also grazes on grassy shores of lakes and dams, especially when food scarce. Overgrazes green crops planted close to dams in Natal, where considered nuisance (D. Skead, pers. comm.). Feeds on floating algae while standing or swimming by repeated opening and closing of vertical bill partly immersed in water (P. B. Taylor, pers. comm.).

Generally sedentary, but also nomadic and opportunistic. Occupies temporarily flooded dams, vleis and floodplains, retreating as they dry out. Numbers fluctuate on permanent waters, indicating movement. At Barberspan (South Africa), factors influencing fluctuations are rainfall, water level, and availability of favourite food sago pondweed *Potamogeton pectinatus* (Skead and Dean 1977). Ringing recoveries show mean distance travelled from Barberspan 270 km, 70% recovered within 300 km, longest distances travelled Jan–Apr when population drops during rainy season (Skead 1981). Long distance recoveries from S Zambia, Namibia, Botswana, S Mozambique and (1072 km) near Cape Town. Birds ringed Rondevlei disperse up to 400 km (Winterbottom 1966). Birds ringed Tanzania (Ngorongoro Crater) recovered Kenya (Eldoret, Naivasha). Recovery rate of 16,500 birds ringed Barberspan 1955–1978 c. 1%. Present Lochinvar National Park, Zambia, Feb–Oct, peak abundance Mar–June, max. 900, May (Douthwaite 1978).

Body and tail feathers moulted all year. Flightless moult Kafue Flats June–July and Barberspan at any time of year (peaks Apr and Oct–Nov). Flightless period lengthy, 49–59 days, which may be adaptive and energetically economical, result of absence of pressure to moult quickly (Dean and Skead 1979). During flightless period prefers permanent waters with plenty of food. Tends to stay out on open water among full-winged birds rather than closer to shore where food more accessible and vegetation cover available, possibly to avoid predation.

Food. Chiefly aquatic vegetation, including algae; also grass, seeds, fruit, Arthropoda, water snails (Gastropoda), and crustaceans; occasionally carrion, e.g. ducks washed up on shore (D. Skead, pers. comm.). In L. Naivasha, Kenya, fed chiefly on unrooted submerged or floating plants; diet composed of seeds and fruit (7·3%), other plant material (90·1%), Arthropoda (2·5%), birds 0·4%; waterweed *Najas pectinata* occurred in all stomachs, mean proportion 81·6%; very little overlap with diet of ducks there (Watson *et al.* 1970). On Kafue Flats feeds on *Panicum repens* (leaves), *Aeschynomene fluitans* (stems, flowers, fruits, chlorophyllous aerial roots), *Polygonum limbatum* (stems, leaves and fruit), and *Najas pectinata* (stems and leaves) (Douthwaite 1978). In South Africa eats insects and aquatic plants, including *Marsilia* and especially sago pondweed *Potamogeton pectinatus*, but will also come ashore readily to pick up almost any type of food, including campers' scraps (Shewell 1959).

Breeding Habits. Monogamous; solitary nester. Strongly territorial, using charging attack (swimming toward rival at high speed) and more intense splattering attack (charging by running over surface of water, flapping wings). Paired birds allopreen.

NEST: bulky platform of reeds or other aquatic plant material, usually with ramp at one side; in shallow water or floating, in open or in vegetation, with no attempt at concealment. Cup lined with finer materials. Built by both sexes, occasionally with help of immature; rearranged and added to during incubation, and built higher if water rises. Builds many 'false nests', used as resting platforms.

EGGS: 3–11, laid at 1-day intervals; mean clutch sizes, South Africa: SW Cape (n = 90) 6·0; E Cape (n = 11) 4·6; Zimbabwe (n = 4) 5·0. 2 ♀♀ known to have laid in 1 nest. Long oval, smooth, with slight gloss; pale grey speckled dark brown. SIZE: (n = 56, South Africa and Zimbabwe) 47·0–59·5 × 35·0–39·5 (52·5 × 36·1). WEIGHT (calculated): 41. Size variation within clutch less than that between clutches.

LAYING DATES: Morocco, Feb–Sept, possibly double brooded (Mayaud 1982); Ethiopia, Apr–July, Sept–Dec; E Africa: Region A, Jan, Apr–July; Region B, Jan, Mar, Apr, July, Sept; Region C, May; Region D, all months; Zambia, Apr–July; Malaŵi, June, July; Zimbabwe, Jan–Sept; South Africa: Transvaal (Barberspan), all months, (peaks Feb–Mar and July); Natal, Mar–Apr, June; E Cape, June–Sept, Nov–Dec; SW Cape, Jan, Apr, July–Dec (peak Aug–Oct).

INCUBATION: by both sexes, with frequent nest-reliefs. Bird taking over turns eggs with bill. Period: 18–25 days. Incubating birds often sit very tight.

DEVELOPMENT AND CARE OF YOUNG: chicks hatch at 1-day intervals, leave nest within a day; can dive soon after hatching. Downy plumage retained at least until chicks size of Little Grebe *Tachybaptus ruficollis*. Down and long hairs remain attached to tips of growing feathers, giving chick long-haired appearance. Hairs later bleach, giving grizzled look. Breast feathers appear first, wing-feathers last.

Both parents feed and raise young, normally alone, but sometimes with help of immatures from earlier brood, particularly when food and nesting habitat abnormally abundant. Broods often divided between parents, unequally, particularly when young small (Dean 1980), and remain separate at least until fledged. Little intra-family hostility in broods. At Barberspan, helpers seen only with divided broods. With addition of helpers, each chick may have 1 or more birds tending it.

In E Africa, death rate among chicks very high (van Someren 1956). In South Africa, 35·5% of 510 eggs survived to hatching, 42·1% of 126 nests produced fledged young. Causes of loss of 56 nests were: predators 20, wave action 18, rise in water level 18. African Fish Eagle *Haliaeetus vocifer* is an important predator of adults and young (W. R. J. Dean, pers. comm.); other recorded predators: Tawny Eagle *Aquila rapax* and Grey Heron *Ardea cinerea*.

References
Dean, W. R. J. and Skead, D. M. (1979).
Watson, R. M. *et al.* (1970).
Winterbottom, J. M. (1966).
Wood, N. A. (1975).

Family GRUIDAE: cranes

Large terrestrial and wading birds, often gregarious, with elongated, streamlined body; long neck, wings and legs; short, broad tail. Plumage usually combination of grey, black and white; inner secondaries and tertials dense and elongated, often curving over tail. Bill straight, longer than head, laterally compressed, sharply pointed; lateral grooves on each side of lower mandibles; nostrils half way along upper mandibles in nasal grooves. 19–20 cervical vertebrae; trachea convoluted or straight. 10–11 primaries; 19–25 secondaries; 12 rectrices. Tibia partly bare; toes short, connected at base by membrane; hind toe small and elevated; claw of inner toe elongated. Sexes similar although ♂ often larger; voice of ♀ usually higher pitched. Monogamous, probably pair for life. Perform remarkable dancing displays. In sexual display give series of unison calls, either screaming, trumpeting, or honking and synchronous, or mellow and asynchronous. Stance upright when resting or alert; crouch in dense cover in breeding area. Flight slow, direct; in 'V' or line formation; sometimes use thermals to soar. Young precocial, with 2 down plumages.

In overall shape resemble large herons but flight silhouette different—head, neck and legs held straight out and declined slightly from horizontal, giving rather 'humped' appearance; head and neck never retracted as in herons, and wing-beat less regular, upstroke faster than downstroke. Similar to storks but bill shorter and utter loud, trumpeting cries in flight.

Cosmopolitan except S America, 15 spp. in 2 subfamilies: Gruinae (3 genera, 13 spp.) and Balearicinae (1 genus, 2 spp.); 6 spp. in Africa.

Subfamily GRUINAE

Trachea convoluted in varying degrees between clavicle and sternum; piercing, broken call; 11 primaries; plumage compact, without bushy crown; external condyle at distal end of tibiotarsus with indentation; hallux usually vestigial and short; egg usually slightly pointed at one end. 3 genera, *Grus* (10 spp.), *Bugeranus* (1 sp.) and *Anthropoides* (2 spp.) although current biochemical evidence indicates that *Bugeranus* and *Anthropoides* may be congeneric with *Grus* (J. L. Ingold, S. I. Guttman and D. R. Osborne, pers. comm.). All 3 genera represented in Africa, by 4 spp. All African species have ♂ and ♀ plumages and (except in *A. paradisea*) breeding and non-breeding plumages alike.

Genus *Grus*

The largest cranes. Adults with red, orange or black bare patch on head or head and neck; sometimes patch has a few hair-like feathers. Young with head and neck fully feathered. 11 primaries, the 11th vestigial and 3rd usually longest. Inner secondaries elongated, considerably exceeding primaries, often drooping over tail, giving effect of bushy tail. Trachea convoluted; voice loud, clanging and trumpeting. Unison call synchronous, of indeterminate duration; calls and postures sexually distinct; sexes usually stand side-by-side; ♀ gives 2–3 calls for each ♂ call; usually ♀ starts display and ♂ ends it. Never roost in trees and never stamp feet to flush prey.

132 GRUIDAE

Plate 6
(Opp. p. 97)

Grus grus (Linnaeus). Common Crane. Grue cendrée.

Ardea Grus Linnaeus, 1758. Syst. Nat. (10th ed.), 1, p. 141; Sweden.

Range and Status. Breeds N Europe to E Siberia, south to N Germany, Ukraine, Mongolia and N China; winters N and NE Africa and S Europe, east to N India and China.

Palearctic winter visitor NW Africa and eastern Egypt south to central Sudan and Ethiopia. Locally abundant to common Morocco (maximum numbers wintering in 1960, 430; in 1981, l'Oued Massa, at least 200 Jan, 76 Dec; Merdja Zerga, 32 Jan; Sebkha Bou Areg, 31 Jan; l'Oued Grou, 10 mid-Dec), Algeria, Tunisia (thousands north of Tunis), Egypt (many thousands on passage, Kharga), Sudan and Ethiopia with main wintering areas known in Sudan along Nile System, particularly between Blue and White Niles (great concentrations as recently as the 1960s, Wadi Medani-Gezira; present numbers unknown but reduction in numbers suspected: G. Nikolaus, pers. comm.), and along Bahr-el Arab (500–600 c. 10°N, 25°E), and in Ethiopia, chiefly Koka Reservoir (1000+, 8°N, 39'E). Uncommon to rare coastal NW Libya and NW Egypt; vagrant Libyan desert and Niger.

Grus grus

Description. *G. g. grus* (Linnaeus): only subspecies in Africa. ADULT ♂: forehead black, crown red, both naked except for a few black hair-like feathers. Nape black. Ear-coverts and upper half of hindneck and side of neck white, upper half of foreneck black; lower half of neck slate grey, continuous with underparts which are entirely grey. Lores black, naked except for a few black hair-like feathers; rest of face black, continuous with black chin and throat. Mantle to tail slate grey; uppertail-coverts edged black. Primaries, greater primary coverts and alula black; secondaries grey with black tips; inner secondaries with dark tips, much elongated and drooping over tail giving bushy effect. Rest of upperwing- and underwing-coverts slate grey. Bill olive to greenish grey; eye yellow, reddish orange, red or red-brown; legs and feet black. Sexes alike except that ♀ slightly smaller. SIZE: wing, ♂ (n = 8) 561–629 (593), ♀ (n = 8) 522–582 (559); tail ♂ (n = 8) 202–225 (210), ♀ (n = 8) 189–215 (200); bill ♂ (n = 11) 100–119 (109), ♀ (n = 14) 97–108 (102); tarsus ♂ (n = 11) 240–275 (258), ♀ (n = 16) 211–256 (238). WEIGHT: ♂♀ (n = ?) 3950–7000.

IMMATURE: overall browner, less grey. Head and neck brown without red crown or white stripe. Inner secondaries not elongated; underparts much lighter.

Field Characters. A large bird, distinguished from storks and herons by uniform grey body, black and white head pattern. Most easily confused with smaller Demoiselle Crane *Anthropoides virgo*, which has white tuft of feathers reaching part way down black neck, and similar flight pattern—grey wing contrasting with black flight feathers. Demoiselle is slimmer, however, with shorter bill, much less bushy 'tail', black of neck extending down onto breast, and different voice.

Voice. Tape-recorded (62, 73, 76). Gives: (a) low purr call when feeding; (b) flight intention call, a low abrupt note; (c) flight call, a loud, broken, high-pitched call in flight; (d) alarm call, a low-pitched, abrupt call, given in a series resembling machine-gun fire; (e) guard call, a loud call beginning at low intensity with a series of broken parts, changing to loud continuous structure, then ending abruptly, with ♂ noticeably higher-pitched than ♀, when fleeing or attacking; and (f) unison call which ♀ usually begins with long, loud, high-pitched scream followed by 3 short, abrupt calls; ♂ then emits a long series of loud, low-pitched, continuous calls that carry much further than ♀ calls. For every ♂ call, ♀ gives 2–3 short calls. When emitting unison call, both ♂ and ♀ at first hold neck slightly forward with bill pointing up, fold wings tightly against body and elevate tertials (see **A**). Then, ♀ bends neck slightly backward, lifts wings c. 20–30° above back (**B**); ♂ does the same but also droops primaries toward ground (**C**). This synchronized duet functions as threat to other cranes and as sexual display, also given during migration and on wintering grounds.

General Habits. In open wet and dry areas including grasslands, farmland, mudflats, sandbanks and along rivers and lakes; avoids wooded areas. In non-breeding season gregarious, in flocks of a few to 1000 or more. Walks slowly and gracefully, regularly stretching neck upwards to observe surroundings. Forages in fields, digging in earth with long bill for food. Bathes for several minutes, then preens for long time. Roosts on ground in large numbers, returning to roost at twilight. Sometimes roosts in water. Same roost used nightly and sometimes annually. Once every other year undergoes complete moult, losing all flight feathers at same time and being flightless for c. 6 weeks; moult usually completed before leaving for wintering grounds. On wintering grounds, performs 'dancing' behaviour singly, in pairs, or in groups. Dancing ♂ or ♀ opens

A **B** **C**

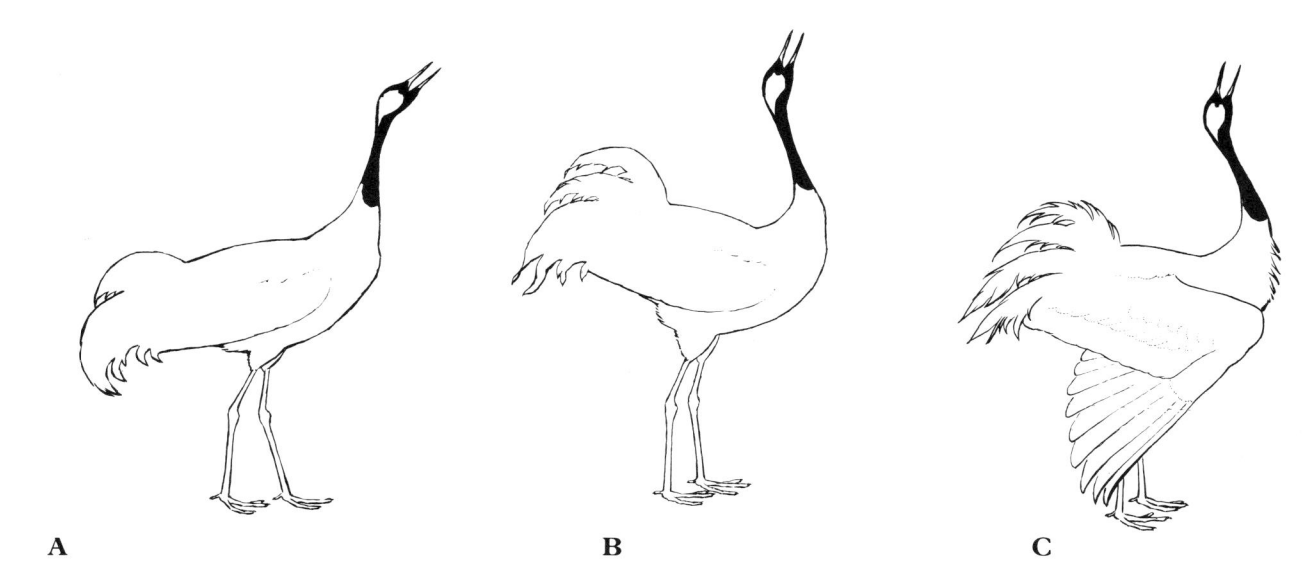

wings, makes figure of 8 or circle with fast steps, then retraces steps, stops, bows several times, jumps up *c*. 1 m to right or left, lands, picks up an object, throws it in air, catches it, stands erect very quickly, shakes plumage, then stops.

Wing-beats slow but powerful. Migrates in flocks of 10–50, sometimes up to 400, during day or night in long lines or in V-formation, at elevations of up to 6600 m for 14–21 h; mean speed 67 km/h over sea, 44 km/h over land (Alerstam 1975). Readily crosses wide stretches of water such as Mediterranean. Reaches Morocco, Algeria, Tunisia and probably western Libya via Gibraltar and southern Italy and Sicily, and E Libya and Egypt via Cyprus. Majority follow E Mediterranean shoreline to Egypt, Sudan and Ethiopia. Some also cross Saudi Arabia and Red Sea. Birds crossing at Gibraltar are mainly from Scandinavia; some move back and forth between Morocco and Spain during the winter. Other birds wintering in Africa are largely from E Europe and Russia. Birds migrating along E Mediterranean reach Nile mainly south of Cairo, then proceed along Nile to winter quarters (630, 12 Feb–12 Apr, junction of Blue and White Niles: Mathiasson 1963). Some do not move farther south but remain in Egypt as far north as Alexandria. Southbound migration starts early Sept, main bulk mid-Oct; northbound largely Mar–early Apr. Winter Sudan (Gezira and Port Sudan) largely Nov–late Feb.

Food. Plants and animals, including grass roots and shoots, cereal grain, berries, leaves of crops, peas, potatoes, olives, insects, snails, earthworms, frogs, lizards, snakes, eggs and young of birds, and small mammals. Damages grain crops in wintering areas but amount of damage not known.

References
Archibald, G. (1976).
Cramp, S. and Simmons, K. E. L. (1980).
Walkinshaw, L. H. (1973).

Genus *Bugeranus* Golger

With pair of pendent feathered wattles hanging down either side of chin. Unison call synchronous; of limited duration; calls and postures sexually distinct; usually both sexes stand side by side, begin with neck in a coiled posture; high-pitched voice with unusual quality; usually one ♀ call per ♂ call. 1 sp., endemic to Africa.

Very similar to *Grus* and included with it by some authors (e.g. Snow 1978). Current unpublished biochemical evidence supports this (J. L. Ingold, S. I. Guttman and D. R. Osborne, pers. comm.). Siberian Crane *G. leucogeranus* sometimes placed in this genus (Archibald 1976).

Bugeranus carunculatus (Gmelin). Wattled Crane. Grue caronculée.

Plate 6
(Opp. p. 97)

Ardea carunculata Gmelin, 1789. Syst. Nat., 1, pt. 2, p. 643; Cape of Good Hope.

Range and Status. Endemic resident, in 2 areas: Ethiopia, and central and southern Africa. In Ethiopia mainly central and southern Western Highlands and Southeastern Highlands, frequent, usually only 1–3 individuals seen at a time, but sometimes in flocks (e.g. 1 flock of 60, Tefki Marsh); population size unknown and no information since 1975. In central and southern Africa locally abundant to rare, but declining, endangered or recently extinct in some areas. Status 1979: Zambia—Kafue Flats 600–3000+ in peak seasons, Bangweulu several hundred, Busanga 150–300+, Liuwa 500+; Botswana—Okavango 1000+, Makgadi-

134 GRUIDAE

kgadi 2000 + (non-breeding); South Africa—Swaziland and Cape Province extinct, Transvaal 25–30 pairs, Natal 40 pairs, Orange Free State 2 pairs. Status 1982: Zimbabwe—mostly in east on Mashonaland plateau, flocks of 52 and 72 (Sept), 76 (Oct), 54 (July) and 42 (Aug), Rainham Dams; Malaŵi uncommon to rare (*c.* 100); Zaïre—frequent (several hundred); Angola—frequent (no more than 500); Namibia—uncommon (population unknown but sparse); Mozambique uncommon (less than 250); and Tanzania uncommon to frequent (*c.* few hundred). Populations declining due to loss of habitat, human interference and development projects. Major concentrations in Kafue Flats and Bangweulu threatened because of proposed hydroelectric schemes and in Okavango because of wetland reclamation. Total population, 1983, not more than 6000 and possibly as low as 4000 individuals.

Description. ADULT ♂ (breeding): forehead naked with red warty skin; crown slate; nape, neck all around, upper mantle and upper breast white. Rest of mantle, back, breast, belly and tail black. Lores and cheeks in front of eye naked with red warty skin; ear-coverts, throat and chin white; 2 white feathered pendent wattles with bare red anterior edge hanging from chin. Primaries and secondaries black; elongated inner secondaries slate grey with darker tips extending beyond tail; rest of upper and underwings slate grey. Bill reddish brown; eye orange-yellow, dark orange or red; legs and feet black. Sexes alike. SIZE: wing, ♂ (n = 7) 613–717 (669·7), ♀ (n = 7) 619–687 (634·1); tail, ♂ (n = 7) 233–270 (257), ♀ (n = 7) 227–295 (261·3); bill, ♂♀ (n = 21) 124–188 (166·9); tarsus, ♂ (n = 7) 298–342 (321·6), ♀ (n = 7) 232–330 (309·8).
IMMATURE: similar to adult with black, white and grey plumage except plumage more tawny, not so contrasting black and white. Crown less slate, more whitish. Face mainly feathered; little trace of warts. Wattles less prominent than adult's.
DOWNY YOUNG: buff to dark yellow above and below, darker above, wattles small, pale buff; white marks below eye; bill horn colour; eye brown; legs and feet blue-black.

Field Characters. Easily distinguished from sympatric cranes by huge size and white neck, but at distance beware superficial resemblance to Woolly-necked Stork *Ciconia episcopus*, which also has dark cap, white neck and dark wings and body. At closer range, Wattled Crane easily identified by grey wings, red face and throat wattles.

Bugeranus carunculatus

Voice. Tape-recorded (232, 296, C). Unlike other cranes, silent much of time. Gives: (a) low purr call, repeated when feeding, encountering other cranes, and tending nests and chicks; (b) flight intention call, a low abrupt note; (c) guard call, a loud, high-pitched scream when alarmed or showing agonistic behaviour; and (d) unison call. ♀ initiates unison call by coiling neck to shoulder level (see **A**); then shooting neck upward with head held *c.* 30° in front of the vertical (**B**), she emits shrill, high-pitched calls followed by short series of loud abrupt brief notes; does not move wings during this display. A split second after she starts calling ♂ emits shrill, high-pitched, long, partly broken call, followed by several short calls and ending with another long call. He also coils and extends neck like ♀ (**C** + **D**). With final call, he elevates humeri *c.* 20° above back (**E**). Usually

A

B

one ♀ call given for every ♂ call. The unison call given at onset of breeding; lasts 5–7 s; and rarely used in threat. Chicks give 'peeee ... peeee' calls until c. 10 months.

General Habits. Inhabits very large open areas including wet grasslands, open marshes and river edges, usually above 2000 m or below 1000 m in non-breeding season. Requires shallow water with sedges. Rarely enters deep water but occasionally wades deep enough to cover back. Of 769 individuals counted in Zambia, 86·5% were feeding in shallow water, 2·8% in deep water and 10·6% in uplands (Konrad 1981). Feeds for much of the day, occasionally at night, mainly by digging with large bill in soft or loosely packed soils. When feeding, sometimes immerses head and neck in water; also occasionally pecks for food. Shy; when approached within 100–200 m tends to walk or run away; rarely flies; flocks of 10–60 (Ethiopia) may react to intruder by running single file to another area. Flight slow, 130–131 beats/min (Cooper 1969).

Gregarious, in flocks of 5 to 300; occasionally solitary. Sometimes joins other cranes, ducks, geese, herons and large game mammals. In Ethiopia at sunset as many as 40 fly long distances to roost in 1 locality; some fly close to ground, others descend several hundred m in air (Cheeseman and Sclater 1935). At roost may dance (see Breeding Habits), but usually silently. Departs from roost around sunrise.

Makes local movements but they are poorly known. In Ethiopia present in highland marshes (Tefki) at end of long rains (Aug–Sept) but disappears as marshes dry up. In Botswana (Makgadigadi) numbers increase considerably during wet season (Jan–May); their origin unknown but probably at least some from Zambia (Kafue Flats) where cranes depart during periods of high floods (Jan–Mar) and return as water subsides (Apr–May) (Konrad 1981). In Zimbabwe moves west in rains to unknown areas but possibly to Botswana and Zambia (Morris, in press). In Mozambique, nomadic in low areas (e.g. Gorongosa and Banhine grasslands); probably moves to lower elevations in rains (West 1976). Also moves from higher to lower ground in rains in Malaŵi and probably in Ethiopia.

Occupies many of same areas in southern Africa as Grey Crowned Crane *Balearica regulorum* but avoids competing for food by digging for rhizomes rather than eating seeds on ground or in low-growing vegetation. Also avoids competing by nesting in austral winter (dry season) at higher elevations (over 2200 m) while Grey Crowned Crane nests in austral summer (wet season) usually below 1500 m.

Flight feathers moulted simultaneously, causing temporary flightlessness; at this time is found in small groups or by itself.

Food. Chiefly sedge tubers and rhizomes, including *Cyperus esculentus*, *C. rotundus* and *C. usitatus* (some rhizomes 3–5 mm diam.) and *Eleocharis dulcis* (rhizomes up to 8 mm diam.) (Kafue, Zambia). Also water-lily rhizomes (*Nymphaea* spp.). Occasionally grain, seeds, insects (Orthoptera, Coleoptera), snails and small fish, frogs and reptiles.

Breeding Habits. Monogamous; each pair defends a territory c. 1 km² (Konrad 1981), 18–40 ha (Tarboton 1984). Home range also large; in South Africa (Transvaal) 6 pairs with ranges of 130–180 ha (av. 155: Tarboton 1984). Pairs may not breed annually; in Transvaal nests on average every 14 months. In courtship bird sometimes spreads wings, runs a little way, then jumps several m straight up with wings spread, head held up and legs bent and dangling. Upon landing, may touch back with its head, bow, turn around and jump again. Mate may behave similarly. Sometimes both toss grass into air and call in unison (see Voice).

NEST: a circular platform several cm above water and usually completely surrounded by it; diam. (n = 4 nests) 76–183 cm. Usually in marshes with water up to 60 cm deep, knee- to shoulder-high grasses, sedges and bulrushes. Nests also on islets, rocks or mounds. Some nests used every year; abandoned nest of Spur-winged Goose *Plectropterus gambensis* also used once. Both

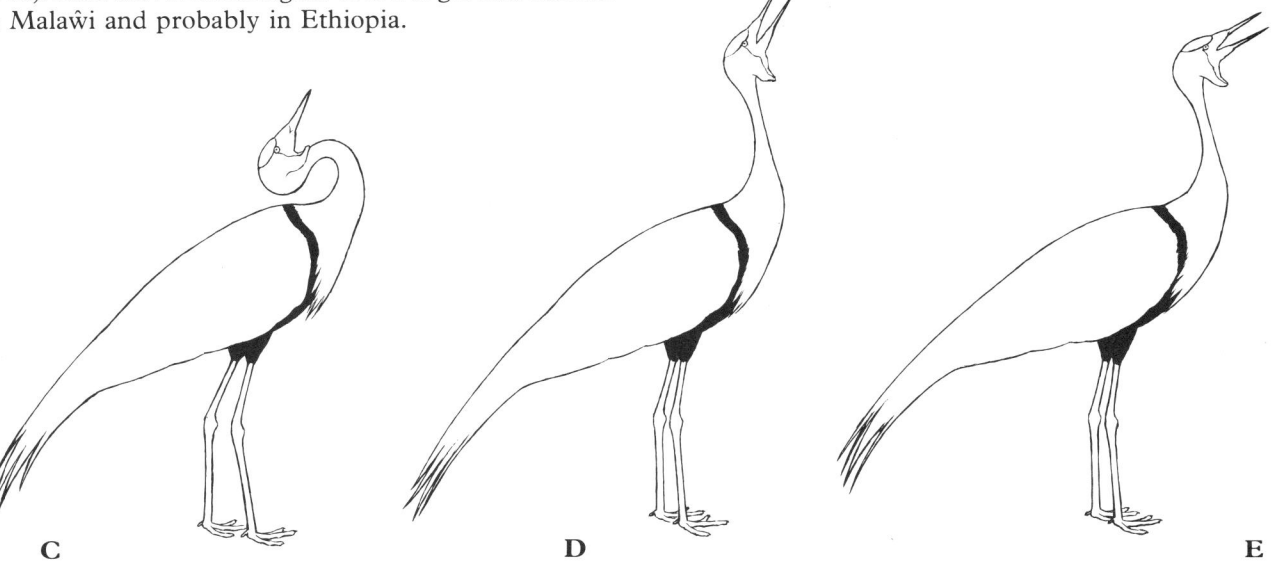

C　　　　　　　　　　　　D　　　　　　　　　　　　E

sexes construct nest with their bills, pulling, throwing and piling vegetation from immediate vicinity; area up to 1·5 m around nest denuded. In Zimbabwe nests constructed of *Scirpus muricinux* and *Echinochloa* spp., the latter a grass 97–115 cm long and 2·5–5 cm thick (Cooper 1969).

EGGS: 1–2; Zimbabwe mean (n = 21) 1·52; South Africa (Transvaal) (n = 17) 1·29; southern Africa (n = 95) 1·6; overall (n = 133) 1·55; captive birds (n = 41) 1·77; laid at 18-h interval. Oval or pyriform, deep cream, olive-buff or reddish fawn with brown, buff or reddish spots; spots merging at large end into almost solid dark background mass. SIZE: (n = 48) 91–117 × 60–72 (102 × 65). WEIGHT: 199–265; captive birds (n = 21) 199–258 (av. 234–243).

LAYING DATES: Ethiopia, Apr–Aug; Tanzania, late Dec–early Jan (estimate); Angola, Apr, June–July; Zambia, Jan–Dec; Malaŵi, May–July, Oct, Dec; Namibia, May; Zimbabwe, Jan–Dec with peak in dry season Apr–Sept; Botswana, Aug–Sept; Mozambique, June; South Africa, Jan–Dec. Southern populations tend to nest when floodwaters recede and sedges become available. Numbers breeding vary depending on amount of flooding, e.g. Kafue Flats (Zambia) 40% of pairs nest in normal years while only 3% did so in limited flooding years (Douthwaite 1974). Ethiopian populations nest mainly at height of long rains but also in dry season before onset of long rains.

INCUBATION: begins with 1st egg; period: *c.* 40 days (Collar and Stuart 1985), 32–33 days (Walkinshaw 1973), 38–40 days (West 1963), captive birds (n = 7) 32–35 (35) days (Conway and Hamer 1977). Both parents incubate throughout night and 95% of day; ♀ incubates 64% of time (Walkinshaw 1965). Several changeovers occur daily with both occasionally leaving nest unattended for *c.* 1 h (2 individuals changed on av. 4 times/day with ♀ incubating 238 min/day and ♂ 157 min/day: Walkinshaw 1965). In 1st week, only head of incubating bird is visible and mate stays nearby; thereafter, 1 parent forages several hundred m away, but both are present at hatching. At changeover both sometimes dance as in courtship, with relieved bird preening and resting close to nest, for up to 1 h; rarely do unison call at changeover (recorded once out of 8 changeovers: Walkinshaw 1965). If intruder approaches, bird at nest creeps away in crouched attitude for considerable distance before showing itself. Both distract intruder from nest and later from young by moving away with wings spread and legs bent, and by calling, walking about and occasionally dancing.

DEVELOPMENT AND CARE OF YOUNG: young leave nest a very short time after hatching; wander around nest area until *c.* 3 weeks old. One parent broods young at nest at night until *c.* 3 weeks old while other parent sleeps nearby. Both feed small young during day. At 5 days young 15 cm tall and covered with dark yellow to buff down; at 15 days, 30 cm tall and covered with dark brown down. At 40 days 60 cm tall, wattles prominent with most contour feathers present; at 55 days, 75 cm tall. At 67 days 90–100 cm tall, inner secondaries form slate grey 'tail'; at 93–99 days 100–110 cm tall, begin to develop black, white and grey colours of adult; begin to fly at 14–18 weeks; and at 21 weeks fly strongly. Remain with parents until 10–21 months old; then form flocks with other non-breeding birds and feed and roost with them. May become sexually mature at *c.* 4 years.

BREEDING SUCCESS/SURVIVAL: lowest reproductive rate of cranes (Konrad 1981); not more than 1 young/pair raised. Only 1 out of 118 pairs had 2 fledged young, 117 had 1 young each (Benson and Pitman 1964). Of 6 young, each with different parents, 3 survived to flying stage (West 1963). Of 552 individuals in Zambia (Kafue Flats, Bangweulu, Busanga, Liuwa) and Botswana (Okavango and Makgadigadi) in 1978–79, 64·3% (range 50·0–90·3) were breeding pairs of which 13% reared 0·13 young/pair (range 0·11–0·25) (Konrad 1981). In South Africa (Belfast population, Transvaal) 0.61 young are reared/pair/year (Day 1980). Breeding losses occur owing to human intruders robbing nests, capturing young incapable of flight and disturbing nesting pairs; also caused by *Varanus* lizards and flooding of nest.

References
Collar, N. J. and Stuart, S. N. (1985).
Douthwaite, R. J. (1974).
Konrad, P. M. (1981).
Tarboton, W. (1984).
Walkinshaw, L. H. (1973).

Genus *Anthropoides* Vieillot

Medium-sized cranes with short bill and neck; head entirely feathered and with ornamental plumes; long pointed feathers on crop; inner secondaries elongated without 'bushy' effect. Trachea convoluted, more than *Bugeranus*, less than *Grus*; voice loud. Unison call synchronous; of determinate length; calls and postures sexually distinct; sexes usually stand side-by-side; 1 ♀ call per ♂ call; both call with wings tightly closed throughout, not in various postures of other cranes. 2 spp. 1 endemic to Africa, the other a Palearctic winter visitor with a relict resident population in NW Africa. The 2 are quite closely related but insufficiently so to treat as a superspecies. Current biochemical evidence indicates that this genus may not be distinct from *Grus* (J. L. Ingold, S. I. Guttman and D. R. Osborne, pers. comm.).

Anthropoides virgo (Linnaeus). Demoiselle Crane. Grue demoiselle.

Ardea virgo Linnaeus, 1758. Syst. Nat. (10th ed.), 1, p. 141; 'In Oriente' = India.

Plate 6
(Opp. p. 97)

Range and Status. Breeds NW Africa, and from SE Europe through central Asia to N Mongolia. Winters Africa, India, Burma and China.

Resident and Palearctic visitor. Relict resident population now very rare: 6 apparently breeding cranes near Fez and 1 bird on nest in Middle Atlas, Morocco. No recent breeding Algeria (no records since early 20th century) nor Tunisia (no records since 1940s). NW African population migratory (e.g. rare, Tunisian coast in autumn) but winter range unknown; not known to cross W Sahara.

Palearctic visitor, rare to locally very abundant, NE Africa south mainly to *c*. 16°–9°N between NE Nigeria and W Ethiopia. Common passage migrant Egypt, but few winter there. Main wintering concentrations NE Nigeria (up to 900, Dikwa), central and S Chad (several hundred, Bahr Salamat drainage), central Sudan (several thousand, mainly south of Khartoum and north of Sobat River), and central W Ethiopia (locally common). In Nile System in far greater numbers than Common Crane *Grus grus*.

Description. ADULT ♂: forehead black, crown ashy grey, front half of hindneck black, hind half ashy grey, continuous with ashy to slate grey mantle to tail. White superciliary stripe ending with white plumes which droop over hindneck. Rest of face, chin, throat, side- and foreneck black. Lower foreneck and upper breast with long black lanceolate plumes. Rest of underparts slate grey. Primaries dark grey, secondaries slate grey tipped darker grey, alula and greater coverts dark grey; inner secondaries slate grey, drooping over tail; rest of upper- and underwing slate grey. Bill short, olive with reddish tip; eye red, crimson or red-brown; legs and feet black. Sexes alike except that ♂ slightly larger. SIZE: wing ♂ (n = 19) 466–516 (495), ♀ (n = 14) 440–490 (467); tail ♂ (n = 20) 157–190 (173), ♀ (n = 17) 141–176 (160); bill ♂ (n = 19) 62–71 (66·3), ♀ (n = 15) 58–69 (64·2); tarsus ♂ (n = 20) 175–205 (188), ♀ (n = 17) 160–190 (176). WEIGHT: (summer, USSR) ♂, 5100, 6400, 6950, ♀ 4900, 5250, 5700.

IMMATURE: black on head and neck with many ashy grey feathers; hindneck brown to slate grey; white plumes extending from superciliary stripe greyer, fewer; fewer black lanceolate feathers extending from foreneck and upper breast. Inner secondaries only slightly elongated. Rest of plumage similar to adult.

DOWNY YOUNG: 2 down plumages; 1st, greyish buff, tipped dark brown on back, and with pink or tan cast on head and neck. Bill pink; eye dark brown; legs and feet pink. 2nd down, all grey, underparts somewhat lighter. Weight at hatching 45 g.

Field Characters. Grey, with black foreneck and breast, white plumes behind eye, and long inner secondaries forming long 'tail', but slender, not bushy as in Common Crane. Distinctly smaller, more delicately built than Common Crane or Wattled Crane *Bugeranus carunculatus*; about same size as Black Crowned Crane *Balearica pavonina* but lacks bushy crown; wings grey and black. Can be hard to distinguish from Common Crane at distance; for distinction, see that species.

Voice. Tape-recorded (62, 73). Gives: (a) low purr call when feeding, meeting other cranes and attending nest and chicks; (b) flight intention call, a low short, raspy, abrupt call when preparing to fly; (c) flight call, a loud, broken, rather low-pitched raspy call; (d) alarm call, a low-pitched, broken, abrupt raspy burst of short calls somewhat resembling sound of a brief burst of machine-gun fire; (e) guard call, a loud, broken raspy call, of lower frequency in ♂ than ♀, given when preparing to flee or attack; and (f) unison call. ♀ initiates unison call by extending head and neck 45° beyond the vertical; then gives a series of brief, continous calls as she does so (**A**). ♂ gives several guard-like calls with neck held vertically and bill *c*. 45° above horizontal (**B**). Both give unison call with wings tightly closed; calls last 5–7 s. Chicks make peeping calls until *c*. 10 months old.

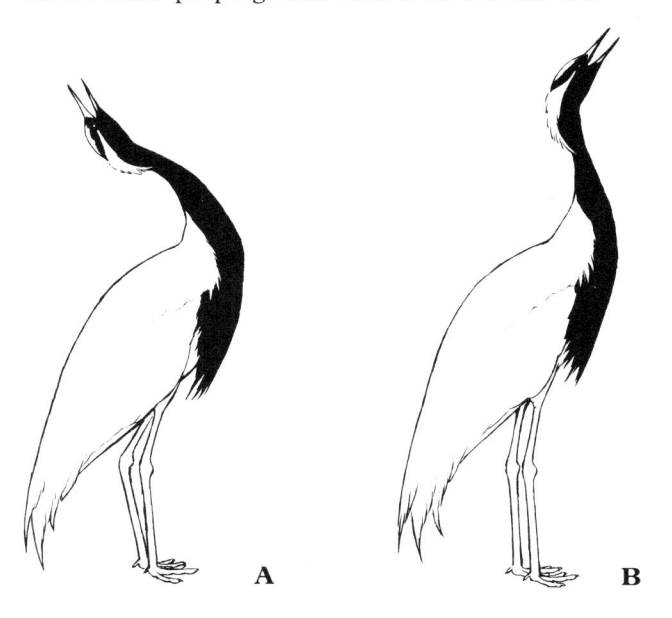

General Habits. Inhabits short grass *Acacia* savannas, sometimes 2–3 km from water; also marshes, cultivated fields, margins of freshwater lakes and rivers; usually found below 2000 m. Roosts at night in marshes, along shorelines and on sandbanks in rivers. Gregarious; in pairs or flocks up to several thousand; occasionally flock with Common Crane. Usually feeds during daylight; occasionally at night. Several pairs in flock may perform courtship dance (see Breeding Habits) on wintering grounds.

Migrates in V-formation, not in lines, up to 770 in a flock. Sometimes soars high on thermals. Migrates 330–1330 m above ground at 60–65 km/h. Most birds wintering in Africa breed probably in central Asia; one juvenile ringed USSR (Kherson, Ukraine) recovered Sudan (Dangola, Dec). Details of autumn entry and spring exit almost unknown. Probably migrates vast distances without alighting. Migrates along Nile, and along and probably across Red Sea (flocks of 500 pairs over Jiddah on Red Sea coast of Saudi Arabia: Trott 1947; but no records from Red Sea coast of Africa). In Sudan southbound birds may stop overnight; earliest arrivals, mid-Sept, most Oct; northbound birds, usually migrating at night, depart late Feb–Mar (1982–83, autumn, earliest to arrive Port Sudan, 6 Sept, 30 individuals; also 8 Sept, 250; 10 Sept, 110; and 12 Sept, 50; latest, Wadi Medani-Gezira, 20 Oct, 55; spring, earliest to depart Wadi Medani-Gezira, 3 Mar, 35; also Khartoum, 10 Mar, 100–500; latest Suakin, 22 Mar, 55: G. Nikolaus, pers. comm.).

Lose flight feathers 1 or 2 at a time; never completely flightless.

Food. Mainly seeds, especially of cereals and wild grasses; also worms, lizards, beetles and other large insects.

Breeding Habits. Monogamous; bond maintained for several years or for life. Defends territory in open country usually less than 1–2 km from water; nests usually 3–4 km apart; in Russia breeding territory is av. 10 km² in extent. Courtship includes circling dance performed early in day; both members of pair bow, toss twigs, jump up and down with wings half-spread and neck stretched out, spread throat feathers, and call. Sometimes other birds, with ear-tufts erected, black throat and breast plumes raised, and tail fanned, form rings up to 3 deep around dancing pair. At other times groups of courting birds will all run in one direction. As ♂ dances around ♀, she may initiate copulation by extending her wings and arching forward.

NEST: a shallow cup on dry ground; sometimes lined with grass; probably built by both sexes.

EGGS: 1–2, rarely 3; laid every 24–48 h. Oval to long oval, smooth, not glossy; olive grey to dark olive brown with a few rufous spots at large end. SIZE: (n = ?, Russia) 74–91 × 48–57 (83 × 53).

LAYING DATES: Morocco, Mar–June.

INCUBATION: begins with 1st egg; by both parents but mainly ♀; ♂ mainly acts as a sentinel. Period: 27–29 days. Bird on nest looks around with head held high. With danger, either flies off nest or first walks some distance from nest; usually does not hide but runs about. When eggs nearly ready to hatch, ♂ diverts danger by dancing; ♀ may feign injury, trailing one or both wings and keeping head and body low.

DEVELOPMENT AND CARE OF YOUNG: young leaves nest when dry, shortly after hatching, sometimes after 2 days. Returns to nest at night when both parents, but mainly ♀, brood it. Young fed by adults when very small, picks up food independently from 3–4 days on. Shows no special begging behaviour; instead parent makes it take food by repeatedly touching its bill tip. Legs and feet turn bluish grey within a few days of hatching. Has 2nd downy coat, largely grey to brown on back, greyish buff below. Flight feathers appear at 3 weeks; most contour feathers at one month; flies at 60–70 days; leaves parents at *c.* 10 months.

References
Archibald, G. (1976).
Cramp, S. and Simmons, K. E. L. (1980).
Walkinshaw, L. H. (1973).

Plate 6
(Opp. p. 97)

Anthropoides paradisea (Lichtenstein). Blue Crane; Stanley Crane. Grue de paradis.

Ardea paradisea Lichtenstein, 1793. Cat. Rerum Rariss., Hamburg, p. 28; inner South Africa.

Range and Status. Endemic resident, South Africa, Namibia, Botswana, Swaziland and Lesotho. Main concentrations in upland country in central and eastern South Africa: Cape Province, common to abundant with large numbers in Eastern Cape; Natal, common to abundant, except coastal districts where absent; Transvaal, common, mainly in west; Orange Free State, common to abundant. Elsewhere in southern Africa, uncommon to vagrant: Lesotho, in lowlands; Swaziland, mainly in west; Namibia, occurs mainly north of 20°S and south and east of Etosha Pan, rare to vagrant south of 20°S; Botswana, rare Makgadigadi; Zimbabwe, vagrant to rare Mashonaland with persistent reports of sightings. Locally migratory but details poorly documented. Total population not known but probably numbers several thousands; not endangered.

Anthropoides paradisea

and wings closed (see **A**); ♂ with neck vertical, bill 45° above horizontal, wings slightly open, tertials and primaries drooped to legs (**B**). Chick gives peeping calls until *c.* 10 months old.

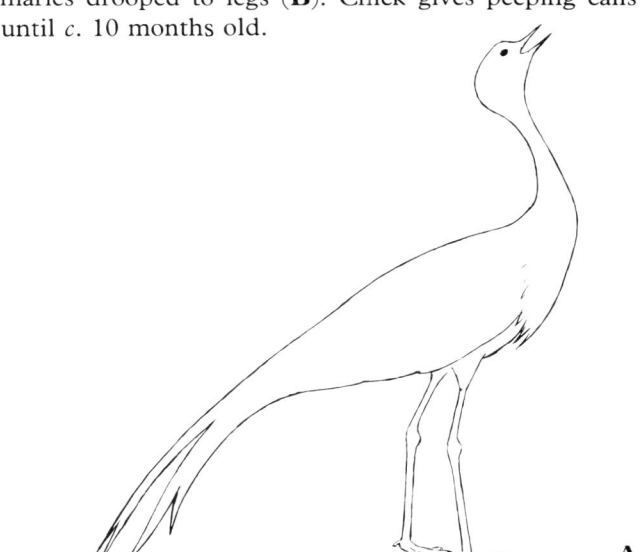

Description. ADULT ♂ (breeding): forehead to nape white with some grey. Foreneck dark grey with some ashy wash; rest of neck grey; lower foreneck and upper breast with long, pointed slaty grey plumes. Lores and upper cheeks slaty white; ear-coverts dark slaty grey. Plumage on crown and ear loose and long. Chin white with some grey, throat dark slaty grey. Rest of body slaty grey. Primaries and tips of secondaries black, inner secondaries much elongated, with black tips touching ground when walking, inconspicuous in flight; rest of wing slaty grey. Eye dark brown; bill pinkish yellow; legs and feet black. ADULT ♂ (non-breeding): ear-coverts and throat with less extensive dark slaty grey; plumes on foreneck and innermost secondaries not so long; otherwise like breeding ♂. Sexes alike except that ♀ smaller. SIZE: ♂♀ (n = 10) wing 514–590 (552·6); tail 202–265 (237·5); bill 81–98 (88·8); tarsus 205–252 (235·2).

IMMATURE: lighter blue-grey, top of head and sometimes neck light chestnut to tawny; inner secondaries not elongated.

DOWNY YOUNG: head and neck buffy yellow, back grey and buff; throat and breast white; rest of underparts grey and buff. Bill pale bluish grey; eye dark brown; legs and feet bluish grey. Weight at hatching, mean (n = 5) 97–109 (103).

Field Characters. Easily distinguished from other cranes, also herons and storks, by uniform grey-blue plumage with long black ornamental secondaries curving to ground in adult. Bill short, head looks rather swollen.

Voice. Tape-recorded (21, 42, 74). Gives: (a) low purr call when feeding, meeting others, and attending nest and chicks; (b) flight intention call, a low, short, raspy abrupt call when preparing to fly; (c) flight call, a loud, low call; (d) alarm call, a low-pitched, broken, abrupt, raspy, short call; often given in a burst, when resembles sound of machine-gun fire; (e) guard call, a series of loud, broken, raspy calls of higher frequency in ♂ than ♀ when preparing to flee or attack; and (f) unison call. Last is initiated by either ♂ or ♀, with a broken call similar to guard call; then both emit series of 5–10 short continuous calls; duet lasts *c.* 5 s. ♀ calls with neck 20° beyond vertical

General Habits. Inhabits open grasslands, especially upland country, usually near water; also along river edges and shallow lakes and reservoirs. Tends to be more in grasslands, less in marshes, than Grey Crowned Crane *Balearica regulorum* or Wattled Crane *Bugeranus carunculatus*. Gregarious; in pairs, family groups or, in non-breeding season, in flocks of 10–300. Walks or flies to feeding grounds, sometimes far from water. Feeds by stripping seed heads from grass, picking up grain and animals, and sometimes by digging. Sometimes roosts in water, occasionally with herons and storks. Individuals within a large flock may spread out over 0·5–3 km². Excellent flier; soars to great heights, where sometimes calls.

Locally migratory but details poorly known. Tends to spread to lower levels but not coastal areas in non-breeding season; returns to uplands in spring (Sept–Oct) to nest, stopping along way to perform breeding displays (see Breeding Habits). Sometimes non-breeding immatures remain in flocks at lower elevations in breeding season. In Transvaal (Graskop) appears early Sept, departs mid-Apr, mainly to west. In Natal appears early Sept, departs northwestward in mid-Mar. Large flocks of non-breeding birds occur in Cape Province, Natal, Transvaal and Orange Free State.

Moults primaries and secondaries simultaneously, once in about 2 years, becoming flightless. During this period escapes predators by running fast to nearest water and swimming to islands, wetland thickets and other inaccessible areas.

Food. Grain (especially maize and wheat), seeds of sedges and grasses, roots and tubers; also insects (especially locusts and grasshoppers), crabs, worms, fish, frogs, reptiles and mammals.

Breeding Habits. Monogamous, probably pairs for life. Calls when defending territory. Nests in territories 400–2000 m apart in short grass marsh with water 1–2 m deep as well as in dry pastures and hill slopes near water; from 900–2000 m in elevation. Courtship in wild poorly recorded; in captivity begins when one or both mates perform dance, running in a circle. Each may stop, give unison call (see Voice), throw grass into air, and jump high with wings flapping, even kicking at bits of material as they fall to ground. Each may repeat the dance and jump. Sometimes 1 may run in front of other for 150 m, stop and call. ♀ invites copulation by crouching low, spreading wings, holding up tail, and bending and pointing neck forward. ♂ then crouches on her back, flaps his wings, and sometimes brings wing-tips in contact with ground for support. After copulation, both give guard call and sometimes unison call.

NEST: slight scrape on ground, lined with small stones, or built with reeds and/or stones; av. diam. (n = 13) 44·5 × 51·5 cm (Walkinshaw 1963); sometimes no nest constructed. In captivity ♂, then ♀, inspects an area, then ♂ selects nest-site. Once site selected, both perform unison call up to 13 times in 10 min. Both sexes build nest; nest sometimes constructed within few m of old nests year after year.

EGGS: 1–3, usually 2, mean clutch (n = 61) 1·9; laid in mornings c. 1–3 days apart. Elongated, buff-brown with dark brown streaks and blotches, darker than most crane eggs. SIZE: (n = 112) 81–101 × 52–66 (92 × 60). WEIGHT: newly laid (n = 16) 168–202 (185), at hatching 142–183 (156).

LAYING DATES: Namibia, Dec–Jan; South Africa: Transvaal, Oct–Jan, Natal, Sept–Feb, mainly Nov–Dec, Orange Free State, Oct–Feb, Cape Province, Oct–Jan, occasionally to Mar. Breeds in austral spring and summer with rains.

INCUBATION: begins with 1st egg. Period: 30 days. Both parents incubate with ♀ spending more time on nest than ♂ (2 days observation, on average ♂ 66 min at a time, ♀ 94 min: Walkinshaw 1965). During day eggs incubated 88% of time (2 days' observation) with changeover up to 9–10 times; either bird incubates at night. Both may give unison call during changeover. During nesting period, each parent feeds for av. 81 min before returning to the nest (18 observations: Walkinshaw 1965). During incubation and brooding of young at nest, ♂ mostly guards nest-site, attacking intruders; ♀ tends to be secretive, either walking away from nest or remaining low and motionless on it with head extended in front of nest. Once young leave nest-site, ♀ as aggressive as ♂. When intruder present, either calls and attracts attention to itself by dancing. Sometimes 1 or both will move with wings opened toward intruder; either may also jump, flap wings, kick or peck at it. Also performs injury-feigning display.

DEVELOPMENT AND CARE OF YOUNG: young usually hatch in morning; usually 8–24 h apart; dry within 2–4 h. Remain usually at nest-site for c. 12 h; then led to higher ground by parents; do not remain in marshy area like young crowned crane. Fed by both parents but mainly by ♀ as early as second day. Parent may touch bill of young several times before young accepts food. After c. 10 days, parents show young location of food; and no longer feed them. By c. 15th day young feeds without parents' aid. Height at 3 weeks 35 cm, 4 weeks 55 cm, 5 weeks 70 cm. Leaves parents after 10 months.

BREEDING SUCCESS/SURVIVAL: hatching success, 90% (25 of 28 eggs hatched: Walkinshaw 1963). Adults sometimes killed with dressed seed; more than 200 birds killed in one incident; also shot at, mainly to chase birds away (Van Ee 1981).

References
Van Ee, C. A. (1966, 1981).
Walkinshaw, L. H. (1963, 1973).

Subfamily BALEARICINAE

Medium-sized cranes, with crown-like erect tuft of almost vaneless, straw-coloured feathers with spiralled shafts; bill short; nostrils oval; inner secondaries not elongated or modified. Large knob at base of culmen covered with black velvet-like feathers. Plumage loose. 10 primaries; flight feathers not moulted simultaneously. Hallux elongated and prehensile; no indentation of external condyle at distal end of tibiotarsus. Trachea straight; honking guard call lower pitched than in most cranes. Unison call asynchronous, of indeterminate length; sexes may or may not stand together when performing the unison call; neither sex moves wings when calling; either sex may begin or end it; postures associated with it not sexually distinct. Eggs oval, not pointed at 1 end. Sometimes roost in trees; also stamp feet to flush prey.

Subfamily endemic to Africa, with a single genus of 2 spp. These spp. are often treated as one (called *B. pavonina* by Peters 1934 and Snow 1978; and *B. regulorum* by White 1965), but we place them in a superspecies because: they have (1) different unison calls, (2) different coloured cheek patterns and necks and (3) different size throat wattles; and (4) recent electrophoretic data show differences of specific importance (Ingold *et al.*, in press).

***Balearica pavonina* superspecies**

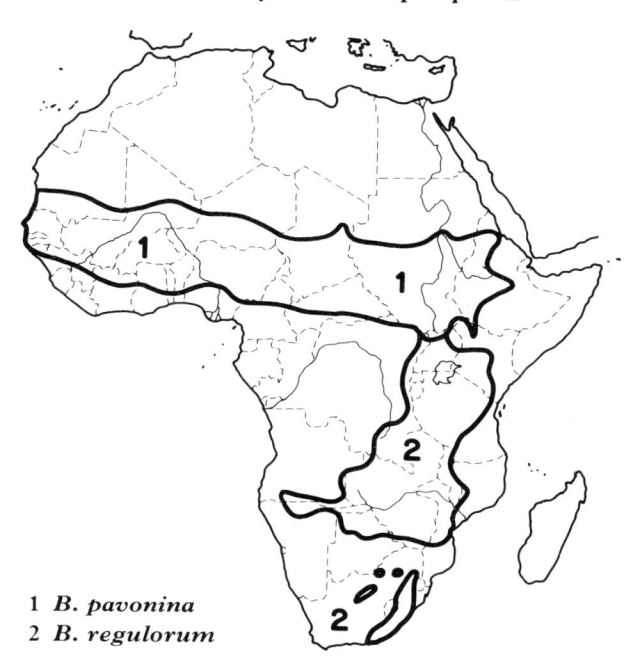

1 *B. pavonina*
2 *B. regulorum*

Genus *Balearica* Brisson

Balearica pavonina (Linnaeus). Black Crowned Crane. Grue couronnée.

Ardea pavonina Linnaeus, 1758. Syst. Nat. (10th ed.), p. 141; Cape Verde, Senegal.

Forms a superspecies with *B. regulorum*.

Plate 6
(Opp. p. 97)

Range and Status. Endemic resident with local seasonal movements; widespread, rare to locally abundant from Senegambia to central Ethiopia, N Uganda and NW Kenya. Known to occur as far south as Difule on Uganda–Sudan border, NW corner of Murchison Falls National Park and N Lake Turkana (Keith 1968, Jackson and Sclater 1938, and Owre 1966). Total population not known but as recently as the early 1970s numbered many thousands (e.g. 7000–10,000 Waza National Park, Cameroon; several thousand Malakal, Sudan). In W Africa concentrations exist in 2 main areas, Senegambia and the Chad basin, with intervening area sparsely populated. Numbers in W Africa reduced during 1970s; total population Senegambia and Mauritania, 2500–3500; Mali, Upper Volta, Ivory Coast, Ghana and Niger combined, only few thousand at most; NE Nigeria, low hundreds; now probably extinct in most of former Nigerian range. Status, 1980s in E and central Chad and eastward (*ceciliae*) unknown but this race apparently not threatened (Fry, in press). The subspecies *pavonina* however threatened in parts of its range, or on verge of extinction, e.g. Nigeria.

Description. *B. p. pavonina*: W Africa from Senegambia to Chad. ADULT ♂ (breeding): forehead and forecrown velvety black; bush of stiff feathers on hind crown; nape and neck slate grey. Mantle slate grey; rest of upperparts including tail black. Underparts slaty grey. Long lanceolate, dark slate grey feathers all around neck and on upper mantle, breast and belly. Side of head from crown to lores and posterior part of ear-coverts with black velvety feathers. Cheeks and area behind eye naked; skin mainly white with lower part pink. Chin velvety black; middle throat naked with single 2 cm long rosy pink wattle; rest of throat black. Primaries black, secondaries chestnut-maroon, inner secondaries broad and long, innermost wing-coverts straw-coloured; underwing-coverts and axillaries white. Eye white to pale blue; bill black; legs and feet black. ADULT ♂ (non-breeding): like breeding ♂ but lanceolate feathers less developed. Sexes alike except ♀'s crest slightly smaller. SIZE: ♂♀ (n = 6) wing 506–585 (547·5); tail 233–275 (244·3); bill 53–64 (56·5); tarsus 190–203 (196). WEIGHT: adults (n = 13) av. 3629.

IMMATURE: generally blackish grey with upperparts edged with rufous, underparts with sandy buff. Head and neck rufous; crest small, chestnut; white and some buff on wings. Lores, side of face and cheeks yellowish white.

DOWNY YOUNG: pale buff; head and back tawny; bare face and orbital area slaty. Bill slate grey; eye dark brown; legs and feet flesh-coloured.

B. p. ceciliae Mitchell: S Sudan, W Ethiopia, N Uganda and NW Kenya. Darker, neck black; cheeks and area behind eyes extensively red, not pink with only upper quarter white. Smaller, wing ♂♀ (n = 17) 470–567 (496·7).

Field Characters. This large slate grey bird with its tricoloured wings, velvety black crown, pink and white cheek-patch, pink throat wattle and unique golden crest cannot be confused with any African bird except its close relative, the Grey Crowned Crane *Balearica regulorum*. The latter has grey neck, larger throat wattle, and double call note 'o-wan' or 4-syllable 'ya-oou-goo-lung'. Colour of cheek patch varies between races of each species. Where their ranges come close, in northern East Africa and S Sudan–S Ethiopia, race *ceciliae* of Black Crowned Crane has cheeks mainly red, with only upper quarter white, while race *gibbericeps* of Grey Crowned Crane has cheeks mainly white, with red patch above and sometimes below the white. Both species show conspicuous white and straw patch on wing in flight.

Voice. Tape-recorded (3, 25, 40, 54, B, C). Gives: (a) low purr call when feeding, meeting other individuals, and attending nest and chicks; (b) alarm, flight, threat and guard calls: in each instance a loud single 'honk' or 'ka-wonk'; when giving guard call, each bird keeps its neck erect and holds its bill horizontally (see **A**); and (c) unison call in which both members of a pair produce a series of monosyllabic honk-like calls. Very occasionally the bird ends its unison call by producing a high-pitched booming at which time the bird inflates its gular sac, keeps its mandibles closed and rotates its head from side to side.

General Habits. Inhabits dry and wet open areas including marshes, damp fields and open margins of lakes and rivers; rarely associated with open water. In pairs, family groups or, in non-breeding season, large flocks of sometimes several hundred or more. Roosts in trees; leaves at sunrise (sometimes leaves before dawn) for feeding grounds; departs singly, in pairs or small groups; feeding grounds may be several km away, although sometimes within walking distance. Produces honk-like calls, especially when leaving or returning to roost. Tends not to call on ground. Feeds morning and afternoon in dry open areas; rests at midday. Picks up food, rarely digs for it. Sometimes 'stamps' feet, presumably to disturb insects. Flies only 100–150 m above ground; beats wings 120 times/min (Walkinshaw 1966). Dances at all seasons, but particularly before breeding; mostly in pairs, sometimes in large groups.

Makes local seasonal movements. In S Sudan begins to flock along Nile in Nov, reaching peak in late Feb–Mar. In Ethiopia (L. Tana) present Aug–Sept, absent Feb–June and at Gambela forms flocks of 250 in Mar. In Nigeria subject to local movements with seasonal changes of water level. In Chad, gathers in concentrations after breeding, then moves south (Newby 1979).

Flight feathers not moulted simultaneously.

Food. Plants and animals including grass and sedge seeds, millet, corn, rice, molluscs, crustaceans, millepedes, insects (grasshoppers and flies), fish, amphibians and reptiles.

Breeding Habits. Monogamous, probably pairs for life. Arrives at breeding area in flocks but in a few days separates into pairs; nests singly, in territories 0·5–1·0 km² in extent. When defending territory against conspecifics, ♂ moves to within few cm of intruder; both birds, with neck and head arched in curve and bill pointing to ground, stand motionless for 10–30 min, then they depart. Both sexes defend the territory. Early in breeding season members of a pair usually remain close together, walking 1–3 m apart and roosting together. When starting to dance, both bob whole body up and down in unison, leap toward one another with wings flapping, land, circle each other, then leap until 30 m apart, turn, and leap back toward each other. ♀ crouches during copulation; holds neck and head forward in straight line.

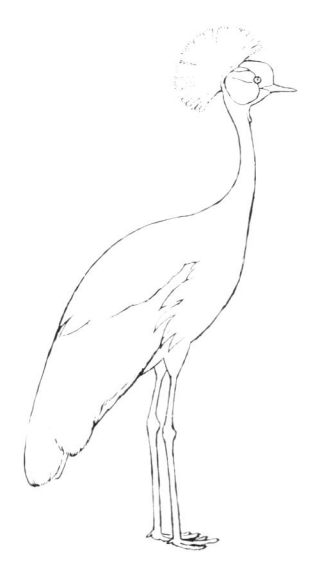

A

NEST: round, poorly constructed platform of reeds and grasses placed in short grass marsh in several cm of water; sometimes on dry land. Both sexes build nest by trampling vegetation and pulling, throwing and placing grass, sedges, bulrushes and other vegetation on it. Diam. at base (n = 8 nests) 69–109 × 71–140 (88–106). Occasionally more than one nest built.

EGGS: 2–3, oval, green or pale blue and small brown blotches. SIZE: (n = 36) 70–87 × 52–61 (78 × 57). WEIGHT: (n = 15) 122–168 (140).

LAYING DATES: Mauritania, Oct; Senegambia, Sept–Jan; Mali, Dec; Chad, Aug; Nigeria, July–Sept; Cameroon, 'in rains' (June?); Sudan, Sept–Nov; and Ethiopia, Aug–Sept. Breeding usually mid- to late rains, sometimes in dry season.

INCUBATION: begins with 1st egg. Period: 28–31 days. Both parents incubate; 1, probably ♀, does so all night while other roosts in tree 1–1·5 km away. During day 1 parent remains at nest, usually sitting on but occasionally standing over eggs, while other feeds, sometimes c. 1–2 km from nest, at other times as close as 100 m.

DEVELOPMENT AND CARE OF YOUNG: young hatch within a day of each other; leave nest with parents by 2nd day. Young remain near nest for several days, sometimes returning to nest at night. Parents defend them with distraction display, running about with drooping wings. 1st flight at 4 months; lores, sides of face and cheeks turn pink and white at 10–12 months; attain adult voice about 1 year; breed at 4 years.

BREEDING SUCCESS/SURVIVAL: eggs destroyed by Pied Crow *Corvus albus* (Nigeria).

References
Fry, C. H. (in press).
Urban, E. K. (1981).
Walkinshaw, L. H. (1964, 1966, 1973).

Balearica regulorum (Bennett). Grey Crowned Crane. Grue royal.

Anthropoides regulorum Bennett, 1833 (1834). Proc. Zool. Soc. London, p. 118; (South Africa).

Plate 6
(Opp. p. 97)

Forms a superspecies with *B. pavonina*.

Range and Status. Endemic resident, with local movements; from S Uganda and central and SW Kenya south through E and SE Tanzania, Zambia, Malaŵi, Zimbabwe and W Mozambique to eastern South Africa west to eastern Cape; also S Angola, N Namibia, N Botswana, Lesotho and Swaziland; locally common to abundant. Uncommon in Uganda in areas with less than 70–80 cm of rain per year; in southern Africa usually below 2300 m. In Kenya (Kisii district) 0·44 individuals/km², Uganda (large areas in south) 1/km² and Zambia (Kafue Flats) 1·0/km². Occurs as far north in Kenya as Maralal (130 km south of L. Turkana, Owre 1966) and in Uganda at L. Albert (Sclater 1924). Total population not recorded but probably numbers many thousands.

Description. *B. r. regulorum*: S Angola, S Zambia, S Malaŵi, Zimbabwe and S Mozambique southwards. ADULT ♂ (breeding): forehead, crown and nape velvety black; bush of stiff feathers on hindcrown. Neck light grey. Long, lanceolate, grey feathers all around neck; most at base of neck. Upper mantle slate grey with elongated grey feathers; rest of upperparts from mantle to tail and including scapulars dark slate grey to black; some scapulars and feathers on back also slightly elongated. Side of head from crown to lores and posterior ear-coverts with black velvety feathers. Cheeks and area behind eye naked, mostly white, outlined with black, and with small crescent-shaped area of carmine red at upper edge. Chin and upper part of throat with velvety black feathers; rest of throat with large red pendent wattle. Underparts grey, darker posteriorly, breast with elongated grey lanceolate feathers; smaller slate grey lanceolate feathers with grey shafts on lower breast and upper abdomen. Primaries black glossed green; outermost 1–2 secondaries black, next 2–3 with black inner webs, rest chestnut with black bases. Greater coverts white with buff tips; innermost wing-coverts golden, with white bases. Underwing-coverts white. Bill black, eye pale blue, legs and feet black. ADULT ♂ (non-breeding): like breeding ♂ but lanceolate feathers less developed. Sexes alike except that ♀'s crest is slightly smaller. SIZE: ♂♀ (n = 22) wing 523–642 (565); tail 212–256 (239); bill 57–68 (62); tarsus 183–234 (207); width of black on forecrown 37–48 (41). WEIGHT: ♀♀ 3575, 3970.

IMMATURE: generally grey, upperparts broadly edged with rufous, underparts sandy buff. Head and neck rufous, small chestnut crest. Lores, side of face and cheeks yellowish white; eye pale brown.

DOWNY YOUNG: crown and nape brown, neck fawn. Body fawn, darker above. Face with buff down. Upper mandible black, lower horn; eye dark brown; legs and feet slate to pink.

B. r. gibbericeps Reichenow: Uganda and Kenya south to northern Zambia, N Malaŵi and N Mozambique. Slightly larger than *B. r. regulorum*; bright red patch above and sometimes below white face; width of black on forecrown narrower, 18–32 (24).

Field Characters. Extremely similar to Black Crowned Crane *B. pavonina*; for differences see that species.

Voice. Tape-recorded (5, 20, 36, 64, 74, C, F). Gives: (a) low purr call when feeding, meeting other individuals and attending nest and chicks; (b) alarm, flight, threat and guard calls, usually a loud 4-syllable 'ya-oou-goo-lung', sometimes 'o-wan'; and (c) unison call in which both members of a pair first produce several guard calls, then give a prolonged low pitched booming sequence that may last more than a minute.

General Habits. Inhabits dry and wet open areas including grasslands, open riverine woodland, and shallow flooded plains; in Uganda prefers freshly ploughed fields to grasslands and short to long grass. Found in pairs or groups of up to 20, sometimes up to 200. Roosts along rivers and marshes; also trees, preferring open trees with good view. Leaves roost at dawn, or up to 1 h later on wet or misty mornings; returns just before sunset. Often walks from nest to feeding areas; may feed well away from nest and water; least active at midday when usually rests near water. Feeds by rapidly pecking at food, sometimes uproots plants; strips seedheads from grass. 'Stamps' feet to disturb insects. Sometimes walks among cattle like Cattle Egret *Bubulcus ibis*. May group with large mammals such as Impala *Aepyceros melampus*, Burchell's Zebra *Equus burchelli*, Puku *Kobus vardoni* and baboons *Papio* spp. Flight heavy and laboured; flaps wings 120 times/min; speed 37–56 km/h. Pants visibly in hot weather.

Display within flocks not infrequent, with pairs preening mutually (especially face and neck), calling, dancing and head bobbing (see Breeding Habits); display by 1 pair stimulates others in flock to display with as many as 60 birds displaying at same time (P. Konrad, pers. comm.). Displays rarely last more than a few minutes and are interspersed with feeding.

Probably only moves locally, but movements poorly documented. In Uganda flocks vary in number seasonally, being largest late in dry season and in early wet season, and smallest when the cranes are nesting (Pomeroy 1980a). In Tanzania (Serengeti Plains) scarce during dry season (Sept–Oct) (Frame 1982). In Zimbabwe most sightings July–Sept, fewest Nov–Jan, suggesting some seasonal movements that have not been documented (Boulton *et al.* 1982).

Flight feathers not moulted simultaneously.

Food. Prefers seed-heads of sedges *Cyperus* spp. and grasses *Cynodon* spp.; also eats grain, insects including grasshoppers, crickets, locusts, cutworms, and army worms, crabs *Potamon* spp., frogs and lizards.

Breeding Habits. Monogamous, probably pair for life; maintain territories 1–1·5 km apart usually at altitudes of 900–1500 m but sometimes as high as 3000 m. Both mates dance; either begins by bobbing head up and down several times; bows, then with wings spread, jumps 2–2·5 m up in air, lands, picks up and tosses objects, then jumps again. Single bird dances around mate, or both dance opposite each other. Sometimes performs unison call when dancing. When first giving unison call, gives guard calls, holding its head up with bill horizontal (see **A**), then gives boom; as it booms, it lowers its neck, holds its head at shoulder level with bill pointing up at *c.* 45° (**B**), inflates its gular sac, keeps its mandibles closed and rotates its head from side to side. Displays performed when dancing variable.

A   B

NEST: bulky pile of grasses, sedges and sometimes small twigs in wet grasslands or shallow marshes with tall grasses and sedges; prefers to nest in marshes with water a few cm to 1 m deep and with vegetation 1 m above water. Both sexes construct nest by stamping down vegetation for 5–20 m all around; then with bill, they drag mashed vegetation into heap in middle of circle; diam. (n = 6 nests) 76–86 × 51–52 (78 × 51) with centre of cup 8–18 (12) cm above water. Very rarely nests in trees: once (Zimbabwe) on top of Leadwood tree *Combretum inberbe c.* 20 m from edge of a dam in open area of grass and thorn scrub; nest 6 m above ground; made of pieces of climbing convolvulus *Ipomoea verbascoides*, 36 × 71 cm in diam. (Steyn and Ellman-Brown 1974). Very rarely may use abandoned nests of other species including Wattled Crane *Bugeranus carunculatus* and Secretary Bird *Sagittarius serpentarius*.

EGGS: 1–4; Kenya and Uganda mean (n = 41 clutches) av. 2·56, varying with altitude: generally clutches below 1500 m, av. 2·17, above 1500 m, av. 2·72 (Pomeroy 1980a); Zimbabwe, Zambia and Malaŵi (28 clutches) 2·33; southern Africa (34 clutches) 2·44; South Africa (Natal) (16 clutches) 2·93; overall (119 clutches) 2·52. Oval, pale blue without spots. SIZE: (n = 45) 76–93 × 50–62 (85 × 57); WEIGHT (at laying): (n = 13) av. 180 g.

LAYING DATES: E Africa: Kenya, Jan–Feb, May–July, Sept; Uganda, Jan–Dec with most Nov–Feb and May–July; Tanzania, Jan–May, Dec; Region A, Mar,

May–Oct; Region B, Jan–Dec; Region C, Jan–Mar, May, Nov; Region D, Jan–Dec; Zaïre, Mar, May–June, Nov; Zambia, Dec–May; Malawi, Jan–June; Zimbabwe, Jan–Mar, May, Nov–Dec with most Dec–Feb; South Africa, Jan–May, Sept–Dec. Breeds mainly in rains but also in dry season.

INCUBATION: begins with 1st egg or, in captivity, after all eggs laid (G. Archibald, pers. comm.). Period: 26–31 days (28 in captivity: Carthew 1966). Eggs incubated for 91% of the time; by ♂ 56%, 44% by ♀ (2 days observation). Bird at nest inconspicuous unless raises head above vegetation. If neither parent on nest, at least 1 usually remains nearby although occasionally nest unattended for a few h. Av. duration of 12 absences by 1 parent away from nest 129 min (Walkinshaw 1965). Changeovers up to 6 times/day; occasionally calls at nest-relief. Defends nest or young with broken wing display, bobbing head up and down, spreading wings to show white wing patches, or running and jumping around intruder with wings spread. May approach intruder within few m; puffs neck feathers and, flapping wings, jumps, kicks and stabs at it with bill. Both parents may approach intruder side by side with wings spread.

DEVELOPMENT AND CARE OF YOUNG: young leave nest within few h after hatching; parents may cover eggshells with reeds, possibly to make nest less obvious to predators (Steyn and Tredgold 1977). Young feed during daylight in surrounding marshes but not open fields. At 1 month coverts and flight feathers appear. At 2 months body parts grey with buff tips, crest feathers and white wing-coverts appear, primaries and tail-feathers black glossed green, inner secondaries black with remainder chestnut and black, and legs horn colour. At 100 days young 51% of weight, 91% of tarsal length and 82% of wing-length of adult, and capable of flight. At 4 months pink wattle appears; eye pale brown, legs black. At 6 months crown and nape velvety black, crest with spiral buff feathers with black tips, upperparts dark grey, lesser coverts white. At 12 months face bare, patterned pink and white; both mandibles black; wattle red; upperparts grey, golden plumes of inner secondaries visible. At 20–24 months face pattern and eye like adult; neck with grey plumes (Pomeroy 1980b).

BREEDING SUCCESS/SURVIVAL: Uganda, of 36 eggs, 20 hatched, 12 young reared (or c. 1 young/nest: Pomeroy 1980a); Tanzania (Serengeti and Ngorongoro), of 9 families, an av. of 1·3 (0–3) young survived to 3 months (Frame 1982); Zambia (Liuwa) 2–3 chicks often reared (Konrad 1981).

References
Frame, G. W. (1982).
Pomeroy, D. E. (1980a, b).
Walkinshaw, L. H. (1964, 1973).

Family HELIORNITHIDAE: finfoots

Elusive, solitary, grebe-like waterbirds of rallid affinity (Beddard 1890; Sibley and Ahlquist 1972 and references therein), keeping near overhanging vegetation on quiet backwaters. Plumage thick; body long, trunk broad and flat, neck very thin and head disproportionately small; legs set well back, tarsus quite strong but short, toes long, lobed and strongly clawed. Swim silently, graduated tail resting on surface and head moving back and forth in time with leg strokes; submerge to neck when alarmed; seldom dive. Skitter over surface, climb up woody growth, run fast on land, fly strongly; eat small animals.

2 genera, 1 neotropical, 1 paleotropical. Plumages mainly olive brown, with white stripe(s) behind eye. Bright beaks and brilliant black-and-yellow-banded, green, or red legs. 1st digit clawed (possibly used by juvenile for climbing). Build substantial flat nest of loose grasses in tangled vegetation by water. American species (and others?) has very short incubation period and ♂ has shallow brood pocket in skin below wing and carries 1 blind and almost naked chick on each side when swimming and flying (del Toro 1971). 3 genera recognized by Brooke (1984).

Genus *Podica* Lesson

1 African and 1 Asiatic sp. Much larger than neotropical *Heliornis* and relatively large-billed (particularly the Asiatic sp.). Sexually dimorphic, ♂ with black or grey and ♀ with white throat. Tail long, steeply graduated, stiff, with 16–18 rectrices, the outer one tiny. Small feathered carpal knob. Legs red or green; middle toe-nail pectinate in some old birds. In contrast with *Heliornis*, evidently only ♀ incubates and newly hatched young are downy with open eyes.

146 HELIORNITHIDAE

Plate 9
(Opp. p. 160)

Podica senegalensis (Vieillot). African Finfoot; Peters' Finfoot. Grébifoulque du Sénégal.

Heliopais senegalensis Vieillot, 1817. Nouv. Dict. Hist. Nat., 14, p. 277; Senegal.

Range and Status. Endemic, sedentary. Widespread in subsaharan Africa from sea level to *c*. 1800 m, except in arid regions; from Senegambia (north to Senegal R. delta) to Cameroon (north to Lake Chad) and whole of Congo basin east to Uganda (Entebbe) and NW Tanzania and south to Angola (where very few records, south to 15°S); Zaïre, Zambia, Malaŵi, Mozambique, Zimbabwe and South Africa (Transvaal, NE Natal and thence in streams near coast west to Knysna). Separate populations in low-lying E Kenya with NE Tanzania, and in Ethiopia. Seems uncommon or rare, but elusive, and in fact probably common on all perennial streams and rivers having lush, well wooded banks; 6 pairs on 13 km of river at Ruiru, Kenya (E. J. Carver, pers. comm.). Abundant E Rwanda (Akagera depression).

Description. *P. s. senegalensis* Vieillot: W Africa and Congo basin (except for range of *camerunensis*) to Uganda and NW Tanzania, intergrading in SE Zaïre with *petersii*; also Ethiopia. ADULT ♂ (breeding): forehead, crown, nape and hindneck black with green gloss; mantle dark brown, each feather with 2–4 mm diam. whitish spot behind 2 mm glossy greenish black tip; scapulars similar but brown-fringed and with additional (concealed) whitish spots; back, rump and uppertail-coverts dark rufescent brown, with a few small white spots. Tail stiff and blackish, each feather narrowly buff-tipped, with thick amber shaft and webs concave above. Narrow whitish line runs from eye down side of neck, also (obscurely) above lores. Lores, ear-coverts, chin, throat and foreneck dove grey, flecked with white on throat and foreneck. Upper breast finely barred dark brown and white, lower breast streaked with brown, belly creamy white; sides of breast dark brown with large round white or creamy spots; flanks and undertail-coverts rufescent brown, heavily barred white. Wings dark brown, outermost primary with white marks and all coverts with creamy spots; leading edge between shoulder and wrist largely white; underwing similar but coverts heavily white-barred. Small feathered bony knob near carpal joint (**A**). 1st digit with sharp straight claw 4 mm long (**A**). Bill coral red; eye red-brown; legs and feet red-orange, nails yellow; toes broadly lobed (**B**) Λ-shaped in section (**C**). ADULT ♂ (non-breeding): lores, chin, throat and foreneck white (see **D**). ADULT ♀ (see **E**): differs from ♂ in having forehead to hindneck brown; lores white with grey triangle in front of eye; chin, throat and foreneck white; ear-coverts and broad stripe down side of neck grey; bordered by narrow white stripe above and behind. Bill red, upper mandible and tip dusky; eye pale red-brown. SIZE: (*P. s. camerunensis*, 8 ♂♂, 8 ♀♀; *senegalensis* similar) wing, ♂ 189–206 (195), ♀ 152–178 (167); tail, ♂ 133–152 (139), ♀ 109–134 (122); bill to feathers, ♂ 39–45 (42·1), ♀ 35–40 (37·2), to skull, ♂ 45–50 (48·2), ♀ 41–45 (43·5); tarsus, ♂ 39–44 (41·2), ♀ 32–40 (36·0). WEIGHT: (n = 4) ♂ 627, ♀ 338, ♂♀ 550 and *c*. 879 ('1 lb 15 oz').

IMMATURE: like adult ♀ but all upperparts including mantle, back and scapulars warm brown (not black), immaculate or with few tiny white spots; breast and flanks tawny buff, flanks obscurely spotted; wing claw 18 mm long.

DOWNY YOUNG: down thick; forehead rufous, crown rich brown, rest of upperparts brown; lores black with small white mark above; cheeks, chin, throat and belly white, breast fawn. Maxilla shiny black with small egg tooth, mandible dark horn, mouth pale pink; eye rich brown and fully open; legs orange-yellow.

P. s. camerunensis Sjöstedt: S Cameroon, Gabon and N Zaïre east to 25°E. ADULT ♂: upperparts with blue gloss, unspotted or with few tiny white spots. Foreneck silky grey; no white line down side of neck. Breast and belly glossy black, flanks rufescent brown; obscure whitish bars on flanks, belly

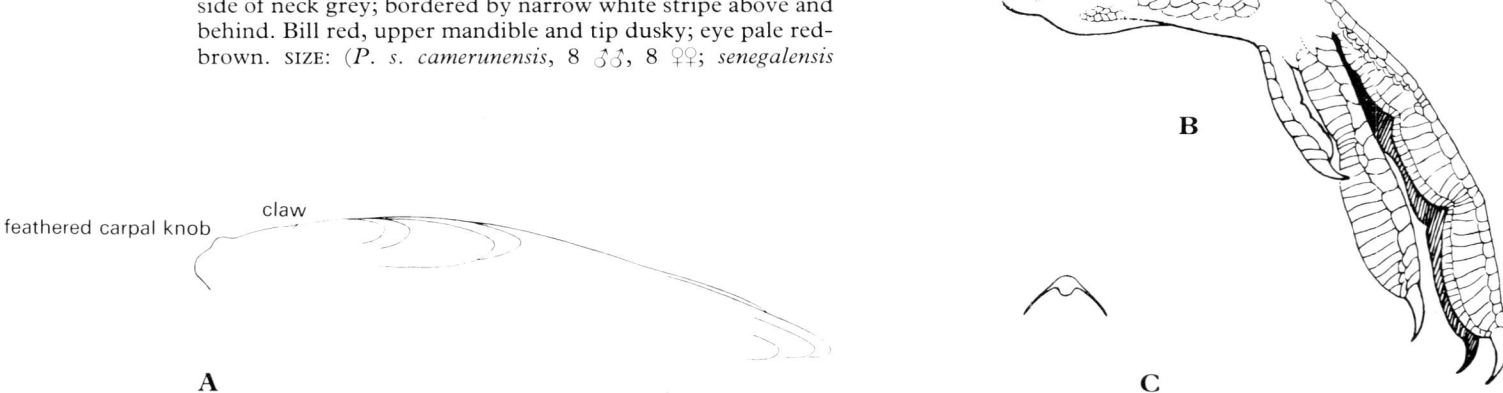

D   E

and undertail-coverts (some have white streaks on breast, large white spots on flanks, and undertail-coverts banded white). SIZE: see above (*camerunensis* and *senegalensis* may be dark and pale morphs respectively of one population; both occur together, with intermediates, in S Cameroon: Louette 1981).

P. s. petersii Hartlaub: SE Zaïre to South Africa. Like *senegalensis* but larger (wing, ♂ 220–252, ♀ 184–215).

P. s. somereni Chapin: Kenya, NE Tanzania. Like *camerunensis* but larger (wing, ♂ 220–235, ♀ 192); ♂ obscurely white-spotted above, and centre of breast and belly white.

Field Characters. A solitary, silent waterbird usually encountered swimming along overgrown bank of quiet, shaded river. Length *c*. 50 cm. Shaped like Darter *Anhinga melanogaster*, with flat body, small head, very thin neck and long black tail carried flat on surface, but smaller and has shorter neck. Further resembles ♂ Darter in having narrow white stripe down side of neck, but distinguished by white stripe being behind, not under, eye; by throat being grey (♂) or white (♀ and young); and by shorter, red bill (dark in young), orange-red legs and (except *camerunensis*) small white spots on dark brown mantle. Pumps head and neck rhythmically back and far forward with every leg-stroke, in more exaggerated fashion than common Moorhen *Gallinula chloropus* or any other waterbird.

Voice. Tape-recorded (53, 72, 217, 295, F). Usually silent, but has a reiterated bull-like roaring or booming, either subdued or very loud (the Upper Zambezi name 'Mumbooma' is doubtless onomatopoeic), sometimes alternating with fluty 'pay-pay', barked staccato 'kwark' (♀ responding to calling chicks), or explosive 'p-r-r-r'. ♀ chatters when chased by ♂. In hand, utters fearsome low growl passing into weak squawk. Chicks have shrill duckling-like 'pay, pay-pay, pay'.

General Habits. Inhabits perennial streams with thick growth of *Syzygium guineense* and secluded reaches of thickly wooded rivers, keeping to shade and hugging banks with overhanging green foliage; avoids stagnant and fast-flowing 'white' water. Also in creeks and mangroves, or edge of dense papyrus beds far from shore, and sometimes on lakes and dams by reeds or bare rocks; rarely found any distance from shoreline vegetation.

Most active in early morning and evening. Elusive and secretive, but will make close approach to motionless observer. Swims smoothly and quietly, with head moving back and forth as it paddles, tail fanned on surface and awash distally. Forages by swimming slowly along lush bank, picking invertebrates off vegetation and from surfaces of soil and water, often going under overhanging greenery or undercut bank or tangle of fallen branches. Also feeds by walking along bank, methodically exploring at every step. When afloat picks insects off bare rock; sometimes dives effortlessly, surfacing immediately; also rushes to make stab at bank. Immobilizes larger prey (snake) by beating it vigorously against low-hanging branch. Recorded jumping clear of water to pluck leaf (or insect and leaf) from tree, gulping it down with violent movement of head; also 'tests', then discards, floating vegetable matter, and seen carrying long leaf (probably for nest). Roosts singly in woody vegetation low over water, and by day often rests on belly on broad limb or sits half upright to preen. Once reported 'anting' (Whateley 1982). If alarmed, either sinks body up to neck or (uncommonly) dives without a splash, or skitters along surface like coot (*Fulica*) with flailing wings and pattering or dangling feet and trailing tail, eventually sinking back onto water or rising in full flight. Flight strong and duck-like, on long wings; drops back onto water cleanly. Alarmed bird also 'freezes' with head down and body low in water, or clambers out onto rock or bank where waddles duck-like or runs surprisingly fast. Climbs vegetation with alacrity, up to nest or 3–4 m up tree, half opening wings and possibly using clawed digit (Percy 1963). Traverses dense or swampy herbage and reeds easily and speedily. Flies more readily from ground or perch than from water.

Highly localized and not migratory; but some movement must occur since it readily colonizes new dams.

Food. Mainly small arthropods: adult and larval insects (dragonflies, mayflies, grasshoppers, beetles, mantis egg-cases), spiders, millipedes, crabs, shrimps and prawns. Also recorded—numerous small snails, grit, frogs, small fish, and a 50 cm snake *Philothamnus*. Small pieces of grass found in one stomach.

Breeding Habits. Monogamous; evidently territorial; pair restricts itself to few hundred m stretch of waterway, sometimes few km distant from next pair. In courtship one bird (♂?) repeatedly swims from cover to open water, raising and opening left and right wings alternately; other bird makes clapping noise probably by snapping bill shut, and emerges from cover to escort mate back into cover each time (Vernon 1983).

NEST: solitary; untidy, loosely constructed round or oval platform of coarse grass bents, dry sedge and bulrush leaves and a few very thin twigs, sited on flood detritus caught in fallen branches or on overhanging limb 1–2·5 m above surface, also up to 4 m above water on horizontal branch of tree and (once) on ground in rushes near water's edge. Platform 30 (28–38) cm across and 12–14 cm deep, almost flat or with cup c. 16 cm across and up to 8 cm deep. Top of nest is of finer, more pliable material than base; clutch laid on lining of soft green mossy material or a few tree leaves. Up to 170 dry leaves added, evidently after hatching.

EGGS: 2 (n = 7 clutches) or 3 (n = 2–3 clutches), interval not known. Short-oval or almost even-ended; glossy; drab brown, cream or pale buffy green ground, blotched, smeared and streaked with red- or purplish brown on underlying bluish grey speckles, mainly at broad pole. SIZE: (n = 11, mainly Zimbabwe) 52–58 × 38–42 (54·7 × 40·6); clutch of 3, South Africa (Transkei), 59 × 40, 58 × 40 and 55 × 39.

LAYING DATES: breeds to coincide with high water. Kenya Apr–May; Zambia Oct and Jan–Feb; Zimbabwe Sept–Dec and Apr. Also Liberia, small chick Dec, large chicks Nov; Nigeria, very small chick Oct; Zaïre, oviducal egg May, ripe gonads Apr–July.

INCUBATION: evidently by ♀ only; sits tight, with bill resting on foreneck or back; ♂ often rests nearby. When disturbed, ♂ noisily splashes away over water, but ♀ can be approached very close before slipping quietly off nest. She flops into water then rests and preens on land, and soon returns by walking up branch from water or by short flight. Pauses by nest to preen and dry herself before resuming incubation. Period: unknown, at least 12 days.

DEVELOPMENT AND CARE OF YOUNG: both downy young hatch on same day and soon sit up strongly and alertly. Very close brooded by ♀ for at least 2 days; in 1 case, when disturbed on 2nd day, young left nest by dropping 2·5 m onto water and swam away with parents. Whether normally spend longer in nest is unknown. On water chicks keep mainly with ♀. One-third and two-thirds grown chicks (163, 320 and 357 g, Liberia, also a full grown 'chick' in Cameroon) have sleeked trunk plumage but still retain thick down on head and neck (centre of crown feathered); bill black, eye brown-grey, legs yellow.

References
Ginn, P. J. (1976).
Jubb, R. A. (1982).
Percy, W. (1963).
Skead, C. J. (1962).
Pitman, C. R. S. (1962).

Suborder OTIDES

Medium to large terrestrial birds; nares pervious; tarsal scales hexagonal; no oil gland; body covered with dense friable powder-down; part of body down pink from porphyrin compounds. Egg-white proteins completely different from Grues (Sibley and Ahlquist 1972) although DNA-DNA hybridization studies show close relationship with some of its families (see Family Otididae) (C. Sibley and J. Ahlquist, pers. comm.). 1 family in the Old World.

Family OTIDIDAE: bustards

Medium-sized to very large terrestrial birds, including the heaviest of all flying birds. Bill short; neck fairly long and often strikingly slender; body short; tail short with 16–20 rectrices. Legs fairly long; tarsus longer than middle toe and claw; 3 toes. Wings broad and rounded; secondaries almost as long as primaries. Contour feathers with aftershaft. No preen gland; birds clean plumage by dustbathing. Sexes generally alike, but in larger species ♂ larger than ♀; plumage usually cryptic above, white, buff or black below. Inhabit semi-desert, grassland, savanna, open patches in scrub, light woodland and cultivated areas. Migrant in N-central and W Africa; elsewhere movements unclear, but most species move locally, perhaps in response to grass fires; seldom fly more than 200 m above ground. Walk with head moving back and forth, trot with head and neck held forward. Forage in steady walk through grass, stopping to pick up items as encountered, and nimbly running after low-flying insects. Feed on arthropods (mainly grasshoppers and beetles), flowers, shoots and seeds, and small vertebrates. Young chiefly insectivorous but take more plant food as age progresses.

Mating systems largely unelucidated, but many species apparently have dispersed leks, successful ♂♂ fertilizing several ♀♀ without any true bonds forming. Black-bellied *Eupodotis* spp. in southern Africa dwell in relatively dense cover, live solitarily and have distinct aerial displays, whereas white-bellied spp. occupy more open country, live in groups and do not have aerial displays. White-bellied *Eupodotis* spp. and the larger bustards display on the ground, some with remarkable self-advertisement postures involving neck-inflation and feather-erection. Some smaller species have penetrating monotonous calls.

Nest on ground, with or without scrape: rarely strands of surrounding vegetation are found under clutch, perhaps caused by trampling, but bustards are the only family of birds which never use bill in nest-building (C. J. O. Harrison, pers. comm.). Clutch size 2–6 in *Tetrax* (and *Sypheotides*), 2–4 in *Otis* and *Chlamydotis*, 1–2 in rest. During drought, clutch reduced or breeding suspended. Incubation by ♀ only, period 20–28 days except in *Ardeotis kori*. Young precocial and nidifugous (though incapable of following ♀ for some hours after hatching), cared for and fed, bill-to-bill, by ♀ and in some species possibly also by ♂. Fledge at 4–6 weeks and first breed at 1–6 years. Roost on ground in open, approaching site cautiously, sitting abruptly. Birds most active in early morning and from late afternoon often well into dusk, generally inactive in midday period.

Chiefly African, but also in Eurasia and Australasia. 21–24 spp. in 4–11 genera have been recognized; we recognize 22 spp. in 7–8 genera, with 18 in Africa (15 endemic). A homogeneous and ancient family; some evidence (e.g. hexagonal tarsal scales, egg-white proteins, Mallophaga) suggests lack of any relationship to Gruiformes. However, DNA–DNA hybridization studies show close relationship to Gruidae (cranes), Aramidae (limpkins), Psophiidae (trumpeters) and Heliornithidae (finfoots) but not Rallidae (rails) (C. Sibley and J. Ahlquist, pers. comm.).

Genus *Tetrax* T. Forster

The smallest bustard, with short bill and full tuft of elongated plumes on nape and hindneck. Legs relatively short; wing 4 times length of tarsus. Sexes dissimilar when breeding. 1 sp., N Africa, Europe, Asia.

Tetrax tetrax (Linnaeus). Little Bustard. Outarde canepetière.

Otis tetrax Linnaeus, 1758. Syst. Nat. (10th ed.), 1, p. 154; France.

Plate 10
(Opp. p. 161)

Range and Status. NW Africa and S Europe east to W China; northern populations winter south to N Africa, Turkey, Syria and Iran, occasionally Sinai (Egypt).

Uncommon resident Morocco (south of Tangier; formerly near Larache; south of Meknes; forest of Marmora; around Khenifra and Mazagan; no recent proof of breeding), and Algeria (Macta). Uncommon winter visitor, probably from SW Europe to Morocco (Larache, where 100 recorded Jan 1977 south of Tangier) and Algeria (L. Télamine); and rarely to Tunisia (Cap Bon, Feriana) and N Libya (Tripoli, Tobruk); formerly Egypt (north fringe of Nile delta). Formerly abundant and widespread NW Africa as both resident and migrant; now massively reduced in range and numbers for reasons unknown, and without special protection likely to become extinct in Africa in a few decades.

Description. *T. t. tetrax* (Linnaeus): NW Africa. ADULT ♂ (breeding): forehead, front and sides of crown dark brown flecked buff; centre of crown light orange-brown minutely vermiculated dark brown, with some dark and buff flecking. Ear-coverts, chin, throat and upper foreneck ash-blue. Rest of neck and breast black, with contrasting white collar running from upper hindneck to central foreneck (there forming a V), and broader white band across upper breast (bordered black below). Mantle, scapulars, back, and rump buffy light orange-brown with fine black vermiculations; scapulars more lightly vermiculated; uppertail-coverts ground colour whitish with some black irregular bars. Tail white mottled brown with 3–4 evenly spaced bars, central feathers suffused buffy. Sides of breast buffy light orange-brown with fine black vermiculations; rest of underparts white. Outer 4 primaries brownish with white bases (7th primary short and curiously emarginated: see Voice); inner primaries white with brown subterminal bar; secondaries white. Greater primary coverts brownish with white tips and bases; rest of wing-coverts white, suffused buffy light orange-brown and with light vermiculations; tertials similar but more coarsely vermiculated. Underwing white with grey on tips of primaries. Bill slaty brown; eye yellowish brown to brown; legs and feet dull greyish yellow. ADULT ♂ (non-breeding): similar to breeding ♂, but ash-blue, black and white of head and neck replaced by buff with black streaks on upper breast and bars on lower breast. ADULT ♀: like non-breeding ♂ but neck, breast and upperparts more heavily marked with complex patterns of streaks and bars, with bars and heart-shaped spots extending sparsely down flanks. Wing as in ♂, but 7th primary normal, secondaries and greater coverts with brown barring, median and lesser coverts more coarsely vermiculated. SIZE: wing, ♂ (n = 3) 240–247 (243), ♀ (n = 8) 236–252 (242); tail, ♂ (n = 3) 95–103 (98), ♀ (n = 6) 87–112 (98); bill, ♂ (n = 3) 21–25 (23·2), ♀ (n = 8) 20·0–24·5 (23·0); tarsus, ♂ (n = 3) 64–68 (66), ♀ (n = 8) 58–69 (62). WEIGHT: ♂ (France, summer) 940–975, ♀ (USSR, May) 740–910.

IMMATURE: as adult ♀, but buff mottling in primaries; immature ♂ has emarginated 7th primary.

DOWNY YOUNG: yellow- and whitish buff ground colour with complex pattern of bold black dots and lines.

T. t. orientalis (Hartert): Egypt. More greyish tinge to upperparts and, especially in ♂♂, wing-coverts; ground colour of back whiter (not buffish) in ♀♀. Wings longer, ♂ (n = 10) 245–257 (251) and tarsi shorter, 59–65 (62).

Field Characters. A very small, stocky bustard. Breeding ♂ has bold white gorget on thick black neck, white ring across black breast. Non-breeding ♂, ♀ and immature have short bill, uniform mottled brown head, neck and upperparts. In flight, rather short, rectangular wings with fast, shallow, duck-like wing-beats (producing whistling sound in ♂), black and white pattern different from Houbara *Chlamydotis undulata*. Larger Houbara has longer, thinner neck with black stripe, longer tail.

Voice. Tape-recorded (62, 73). Generally silent although utters low cackling 'ogh' when flushed. ♂ display call a short, sharp 'prrt', every 5–20 s; often preceded by brief accelerando foot-stamping and followed by brief flutter of wings, which produces a 'sisisi' (due to emarginated 7th primary); wings of ♂ also make continous 'sisisi...' in flight. Breeding ♀ makes low chuckling when flushed; 'youp' to call young; shrill squealing when attempting to distract intruder from young. Young give faint 'bui' or 'kri-i-i-i' when content, 'wehg' for contact, also a plaintive 'churr'.

General Habits. Inhabits flat or undulating open short grassland, steppe or pastures; very sensitive to modification of habitat and affected by disturbance from people. Generally gregarious; on wintering grounds forms flocks not segregated by age or sex. Forages on leguminous crops such as alfalfa, peas and beans. Flies powerfully with rapid wing-beats and no glides.

Winter migrants from Europe arrive Oct–Nov, depart Mar–May. Formerly, post-breeding movement noted from plains south of Meknes to Middle Atlas plateau, June.

Food. In Europe, plants and invertebrates including especially young shoots, leaves, flowers, grasses, beetles and grasshoppers; rarely small vertebrates.

Breeding Habits. ♂♂ use dispersed lek system, holding clustered display territories and copulating opportunistically (no pair-bond formed), ♀♀ nesting solitarily, although sometimes inside ♂'s territory (Schulz, in press). Territory size varies with density, generally 4–6 ha, but as little as 1 ha. At dawn and dusk, ♂ gives territorial display in which briefly but strongly stamps feet, calls and immediately flutters wings. In stronger light, performs a courtship display similar to territorial display, but foot-stamping weaker and bird also leaps 60–70 cm in air with fluttering wings. Display sites chosen by acoustic quality of ground, often enhanced by dried faeces on hollow natural object (e.g. anthill) to aid resonance of foot-stamping (Schulz, in press, H. Schulz, pers. comm.). During and just after display call, neck plumes partly raised. ♂♂ fly up to pursue passing ♀♀ and copulation usually occurs. Courtship consists of ♂, in very upright stance with erected neck-feathers, running behind ♀, often stopping, foot-stamping and giving display call with sideways movements of body and head (H. Schulz, pers. comm.). Copulation extremely rapid with ♀ squatting flat to receive ♂; difficult to observe in wild.

NEST: shallow, unlined scrape, *c.* 15–18 cm diam.; in grasslands or cultivated areas, often well away from place where ♀ fertilized.

EGGS: 2–6 laid at *c.* 2 day intervals. Elliptical; glossy olive green-brown, often streaked darker. SIZE: (n = 50) 48–57 × 35–41 (52 × 38); (n = 28) 35·5–41·5 × 49·0–55·5. WEIGHT: *c.* 41.

LAYING DATES: Morocco, Feb, Apr–July.

INCUBATION: probably begins before last egg; carried out by ♀ alone. Period: 20–22 days.

DEVELOPMENT AND CARE OF YOUNG: young cared for and fed only by ♀; follow her around keeping in contact with peeping calls; fledge at 25–30 days; full size at 50–55 days; remain with ♀ into 1st autumn.

References
Cramp. S. and Simmonds, K. E. L. (1980).
Schulz, H. (in press).

Genus *Neotis* Sharpe

Medium-sized to large bustards, close to *Ardeotis*, but with no crests and higher degree of colour patterning. Largest sp. (*denhami*) shows closest resemblance to *Ardeotis*, smallest sp. (*nuba*) to *Eupodotis*. Endemic; 4 closely allied spp.

Neotis denhami (Children). **Denham's Bustard; Stanley Bustard. Outarde de Denham.** Plate 10

Otis Denhami Children, 1826. In Denham and Clapperton's Travels, app. p. 199; Lake Chad. (Opp. p. 161)

Range and Status. Endemic resident and intra-African migrant; occurs W Africa from S Mauritania to around Lake Chad, and east through Central African Republic to central and S Sudan, N Uganda and N and W Ethiopia; south through W Kenya, Uganda, Rwanda, Burundi, W Tanzania, N Malaŵi, Zambia, NE (rarely W) and S Zaïre, S Congo, W Angola, and W Zimbabwe (c. 10 records, including Wankie National Park) to N and central Botswana (3 records, from Nxai Pan, Kwando R. and near Chukudu R.); central Mozambique, South Africa, Lesotho and Swaziland. Common locally, but persecuted by hunters in past decade in northern parts of range where almost extinct (Darfur, Sudan); 100–150 birds Kenya (T. Stevenson, pers. comm.); hunting probably responsible for disappearance from Malaŵi, where perhaps now only on Nyika Plateau; now uncommon South Africa (main breeding range Natal interior); probably less than 200 breeding birds Transvaal and 100–200 in E Cape Province.

Description. *N. d. denhami* (Children): W Africa to Sudan, Ethiopia and N Zaïre. ADULT ♂: forehead and crown black with broad whitish crown-stripe extending to blackish vestigial nuchal crest. Supercilium white, rest of face whitish, greyer on ear-coverts, with grey bare skin along moustachial line to below eye. Chin and throat whitish (rarely, black: '*N. d. burchelli*' Heuglin). Hindneck and sides of neck unmarked rufous; foreneck grey. Upperparts including tertials and scapulars finely vermiculated buff and dark brown, uppertail-coverts with greyish tinge. Tail dark brown with 2–3 broad creamy white bands, distal half of central feathers as back. Breast grey, paler towards belly; sides of breast rufous vermiculated with brown; rest of undersides whitish. Primaries and secondaries dark brown; inner primaries with extensive white on inner shafts and whitish tips, secondaries with white mottling and tips. Greater primary coverts greyish brown with white tips, median primary coverts whitish with grey-brown mottling, greater coverts white with dark brown mottling, median coverts dark greyish brown with white tips, lesser coverts as back. Outer areas of underwing greyish brown, central part of primaries white, central part of secondaries greyish, lesser underwing-coverts whitish with dark mottling. Axillaries white. Bill dark brown above, paler below; eyes brown; legs and feet pale yellow. ADULT ♀: as ♂, but smaller, with throat and upper chest transversely vermiculated with buff and blackish brown. SIZE: wing, ♂ (n = 4) 577–643 (604), ♀ (n = 3) 507–520 (515); tail, ♂ (n = 7) 293–346 (315), ♀ (n = 3) 249–279 (263); bill, ♂ (n = 9) 70–91 (84·4), ♀ (n = 3) 70·5–78·0 (74·2); tarsus, ♂ (n = 7) 160–178 (171), ♀ (n = 3) 134–147 (142). WEIGHT (kg): ♂ 9·0, 10·0, ♀ 3·0.
IMMATURE: like adult but outer primaries more pointed.
DOWNY YOUNG: buff; head and neck striped black, crown and mantle mottled black.

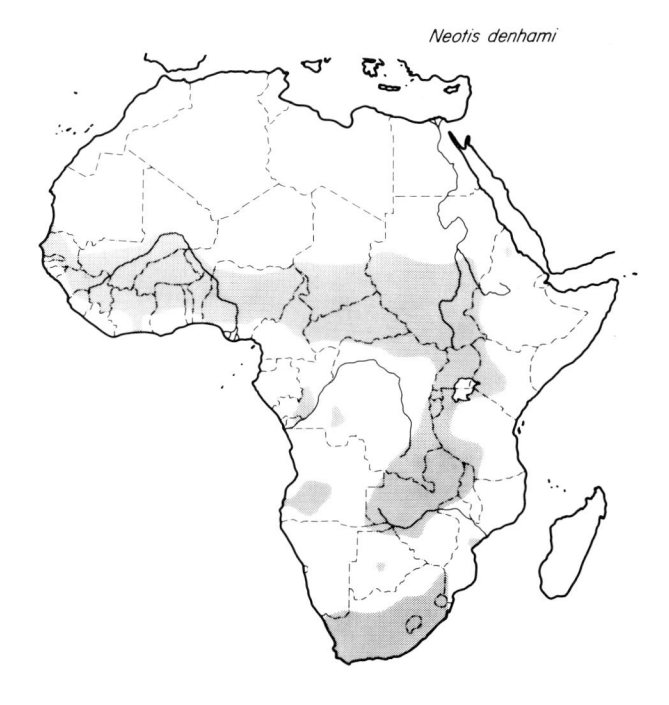

N. d. jacksoni Bannerman: E Africa, S Zaïre, Zambia, Angola, Botswana (non-breeding records from W Zaïre and S Congo probably refer to this race). Slightly smaller: wing, ♂ (n = 6) 528–610 (584); darker rufous on hindneck than in nominate *denhami*.

N. d. stanleyi (Gray): South Africa. Rufous on hindneck darker and richer, back paler than *jacksoni*; breeding ♂ with all-white feathers down front of neck.

Field Characters. A large bustard, distinguished by conspicuous broad white bars and panels on black closed wing, and from all except Ludwig's *N. ludwigii* by extensive rufous down hindneck. Lacks crest of *Ardeotis* spp. and often holds neck at slight forward angle. Larger than Nubian Bustard *N. nuba* with much darker upperparts, black crown, whitish chin and throat and (in flight) black tail-bars. For differences from Ludwig's Bustard, see that species.

Voice. Tape-recorded (72, 74, F, 233). Generally silent. Sometimes produces a guttural barking 'kaa-kaa', 'kia kia kia', and a resonant booming when displaying. A ♂ disturbed by observer gave a soft resonant 'ummm' and a ♀ near nest made a loud hissing (W. R. Tarboton, pers. comm.). A wounded bird gave a loud hoarse cry, and a tame bird repeated a usually quiet, throaty 'choerrie'.

General Habits. Inhabits grasslands, with or without tree cover, e.g. high plateau downland (Malaŵi, Kenya); also found in coastal macchia (South Africa) and usually in quite thick shrubland, light woodland or farmland (Nigeria); also dried marshland and arid scrub plains; sometimes cotton fields and other cropland; likes burnt grass. Highly site-faithful, often returning to same area a year later. In non-breeding season usually solitary, but sometimes in small flocks and loose concentrations (up to c. 20) on migration and at fires. When defending food source (active termitary), spreads and trails wings, raises and fans tail (**A**). Pecks in animal droppings (e.g. zebra *Equus* spp. and Wildebeest *Connochaetes taurinus*) for dung-beetles, and wades thigh-deep in water, apparently for frogs on grass stems above surface (Howells and Fynn 1979). Performs post-preening display by leaping up 7 m into air with beating wings, dropping back down with wings snapping open at last moment; also adopts a 'strutting' posture (**B**) with tail raised or lowered, sometimes preceded by a brief vertical jump. Flies little and usually only short distance except during migration when flies high, c. 150–450 m above ground.

Throughout range, nominate *denhami* moves north usually May–June, but occasionally as late as Aug, returning south in Sept–Oct and sometimes as late as Dec, in response to rains. Some remain in Sahel throughout year (Lamarche 1980). In Sudan often migrates in company of Arabian Bustards *Ardeotis arabs*. In Senegambia, non-breeding visitor mid-July to mid-Nov; in N Ghana common only in dry season. Central and E African *jacksoni* subject to less defined movements, but central populations may shift north-south in some months, since species occurs in SE Zaïre, May–Oct, while records for Botswana and Zimbabwe fall in period Aug–May (mostly Dec–Apr). E central Africa montane populations may move to lower altitudes, June–Aug. In South Africa, some winter movement off cold, high-lying areas, in some cases to eastern coastal regions.

Food. Beetles, grasshoppers, bugs, caterpillars and flies, ants, termites, and millepedes; in Mali, evidently feeds largely on nymphal Desert Locust *Schistocerca* (Malzy 1962). Also skinks, colubrid snakes, eggs and nestlings of ground-nesting birds, rodents; and flowers, leaves, shoots, berries, stems and roots.

Breeding Habits. Mating system unclear. Although strong circumstantial evidence of monogamy (close and extended association of ♂ with nesting ♀ witnessed in all cases in Malaŵi, Nyika Plateau: Wilson 1972), observations in South Africa suggest open-country dispersed lek with territorial ♂♂ displaying at least 700 m apart, in response to each other and to any ♀, with no ♂ participation in nesting (Tarboton, in press). ♂ self-advertising display includes billowing out white breast-feathers and bending back neck so that head close to back; bird then stands or walks about slowly (sometimes with bouncing motion) for up to 1 h (Tarboton, in press).

NEST: shallow scrape in ground, usually between tufts of grass (or in shade of bough) on or near crest of hillside.

EGGS: 1–2. Light brown, varyingly blotched darker. SIZE: (n = 17, *N. d. stanleyi*) 72–79 × 51–57 (76·0 × 55·1). WEIGHT: (estimated) 126.

LAYING DATES: Mali, July–Oct; Chad, June–Aug (during rains); Nigeria, May; N Zaïre, Dec–Feb; Central African Republic, Jan; E Africa: Region A, Jan–Mar; Region C, Mar, July; Zambia, Nov–Feb, July–Aug; Malaŵi, Oct–Jan, Mar, Aug; South Africa, Oct–Dec. Breeding apparently opportunistic at least in north of range.

INCUBATION: probably by ♀ only, although ♂ recorded sitting at nest with ♀ and may have been incubating.

A

B

References
Howells, W. W. and Fynn, K. J. (1979).
Tarboton, W. R. (in press).
Wilson, V. J. (1972).

Neotis ludwigii (Rüppell). Ludwig's Bustard. Outarde de Ludwig.

Plate 10
(Opp. p. 161)

Otis ludwigii Rüppell, 1837. Mus. Senckenb. 2, p. 223; Graaf-Reinet.

Range and Status. Endemic resident and partial migrant SW Angola, Namibia, lowlands of Lesotho and South Africa (dry interior of Cape Province south of Orange R., with small populations in high foothills of Drakensberg). Frequent to common. In South Africa formerly in E and NE; decline in recent years due probably to hunting (Brooke, 1984).

Field Characters. A fairly large bustard, resembling Denham's Bustard *N. denhami*, including rufous hindneck, but smaller, lacking extensive white on wing, tail less prominently barred. Head and foreneck dark brown (grey with black crown in Denham's), though at distance this pattern sometimes indistinct, bird having dark-hooded appearance.

Voice. Tape-recorded (74). Gives a deep resonant 'klump' or 'wup' at intervals of $c.$ 5 s in self-advertising display.

General Habits. Inhabits open grassy plains, rolling uplands with rocky outcrops and termitaria, light thornbush and broken veld, desert-edge and coastal desert; also farmland. Sometimes found alongside Denham's Bustard. In pairs or small groups of up to 6; occasionally in flocks of up to 20.

Subject to local movements that are poorly known; arrives to breed Little Namaqualand June–July.

Food. Grasshoppers, beetles, small reptiles and mammals, and vegetable matter including seeds.

Description. ADULT ♂: head dull brown, lores, chin and throat similar with whitish flecks; hindneck greyish white, foreneck and breast as head. Upper mantle dull orange; rest of upperparts brown with fine buff vermiculations; tail similar but vermiculations coarser, with up to 4 broad dark bars, ground colour whitish on proximal half. Sides of upper breast as upperparts, belly whitish. Outer primaries dark brown with pale bases, inner primaries whitish in centre. Secondaries dark brown with whitish spots half-way along shafts and whitish tips. Greater primary coverts dark brown with whitish spots, median primary coverts whitish, median coverts dark brown with whitish mottling; lesser coverts and scapulars as upperparts but more coarsely vermiculated, tertials similar. Underwing greyish brown with patch on middle of inner primaries and whitish mottling on secondaries. Axillaries whitish. Bill dark brown; eyes light reddish brown; legs and feet dirty yellowish green. ADULT ♀: as ♂, but smaller, face and throat off-white mottled dark brown and buff. SIZE: wing, ♂ (n = 6) 495–561 (536), ♀ (n = 5) 433–470 (452), AD (n = 4) 450–485 (464); tail, ♂ (n = 4) 235–263 (255), ♀ (n = 3) 205–256 (231), AD (n = 4) 218–235 (226), AD (n = 7) 210–260; bill, ♂ (n = 4) 53–64 (58), ♀ (n = 3) 43·5–63·7 (52·2), AD (n = 4) 43·5–50·8 (48·2), AD (n = 7) 46–60; tarsus, ♂ (n = 4) 133–149 (137), ♀ (n = 3) 114–122 (117), AD (n = 4) 117–126 (121), AD (n = 7) 110–140. WEIGHT: (kg) ♂ 3·1–7·3.

IMMATURE: undescribed.

DOWNY YOUNG: reddish streaked blackish.

Breeding Habits. Self-advertising display similar to Denham's Bustard, plumage being fluffed, tail raised over back, neck greatly inflated.

NEST: scrape on bare ground often among stones on crest of low ridge or slope of hill.

EGGS: 2, occasionally 1. Oval; light olive-brown, streaked and clouded brown and slate. SIZE: (n = 10) 68·5–79·0 × 51·2–56·4 (73·8 × 54·2). WEIGHT: (estimated) 118.

LAYING DATES: South Africa July–Aug, Oct–Mar; Namibia Dec–Jan.

Reference
Brooke, R. K. (1984).

154 OTIDIDAE

Plate 10
(Opp. p. 161)

Neotis nuba (Cretzschmar). Nubian Bustard. Outarde nubienne.

Otis nuba Cretzschmar, 1826. In Rüppell, Atlas Vög., p. 1; Shendi.

Range and Status. Endemic resident Mauritania (Tidjikia; Tichit; north of Nouakchott; Banc d'Arguin) (Lamarche 1980; J. P. Gee, P. J. Knight and N. Montfort, pers. comm.), Mali (common in Sahel and Sahara), Niger, Chad (north of 13°N) and N–central Sudan. Frequent but much less so in arid regions; vagrant N Nigeria (near Gadau, 1959). Threatened by hunting throughout range.

Description. *N. n. nuba* (Cretzschmar): Sudan. ADULT ♂: forehead and crown tawny-buff lightly marked with black. Nape greyish white, neck pale grey. Broad black supercilium extends to black vestigial nuchal crest; rest of face whitish but chin, throat and moustachial region black. Lower hindneck and upper mantle unmarked tawny-buff. Lower mantle and back tawny-buff lightly vermiculated black; rump and uppertail-coverts similar but washed with pale grey. Tail as rump but with large white area at base. Upper breast grey becoming tawny-buff on lower breast and lightly vermiculated dark brown. Rest of underparts whitish. Primaries blackish brown with large white patch on proximal half of central ones. Secondaries dark brown mottled white at base and with white tips. Greater primary coverts white with brown mottling and brown tips; alula dark brown. Greater coverts tawny-buff with pale tips and brownish flecking; lesser coverts, scapulars and tertials tawny-buff lightly vermiculated black. Underwing whitish with grey-brown trailing edge (more extensive on leading primaries). Axillaries white. Bill pale yellow with blackish culmen; eyes brown; legs and feet pale yellow or whitish. ADULT ♀: as ♂, but smaller and black patch restricted to centre of chin. SIZE: wing, ♂ 453, 474, ♀ (n = 4) 361–418 (395); tail, ♂ (n = 6) 249–266 (260), ♀ (n = 4) 210–236 (223); bill, ♂ (n = 8) 48·0–55·5 (50·8), ♀ (n = 4) 49·0–53·0 (50·6); tarsus, ♂ (n = 6) 117–130 (123), ♀ (n = 4) 95–107 (102). WEIGHT: (kg) ♂ at least 5·4.

IMMATURE: similar to adult, but black parts on head browner, and black on throat reduced to a stripe.

DOWNY YOUNG: not described.

N. n. agaze Vaurie: Mauritania to Chad. Paler and less vermiculated above; blue-grey below tawny-buff breast-band.

Field Characters. A fairly large, very pale bustard, distinguished in flight from Arabian Bustard *Ardeotis arabs* by extensive white on underwing. At rest strikingly similar to much smaller White-bellied Bustard *Eupodotis senegalensis* but face less white, more extensive black on chin (extending to bill), crown buff with broad black supercilium (not black and grey with white supercilium), upperparts with narrow black vermiculations. In flight resembles Denham's Bustard *N. denhami* or Houbara *Chlamydotis undulata* but tail unbarred and with white patches at base. For further differences from Denham's Bustard, see that species.

Voice. Not tape-recorded. Call, a shrill 'magur'. A low 'wurk' when approached (J. P. Gee, pers. comm.).

General Habits. Inhabits arid and semi-arid scrub and savanna on desert fringes.

Sedentary (Chad), but probably at least locally nomadic.

Food. Mainly large insects such as locusts; also ants, grass seeds and *Acacia* gum. Stomachs of 2 birds, Aïr (Niger), June, contained leaves and fruits of *Salvadora*, other leaves and shoots, various seeds, bugs, tenebrionid beetles, ants, and other small hymenopterans, a large elaterid beetle and white stones (Fairon 1975).

Breeding Habits.
NEST: on bare sand; once between forks of fallen branch.

EGGS: 2. Elliptical; greyish green, lightly blotched reddish brown and mauve. SIZE: (S Aïr, Niger) 70·1 × 47·0. WEIGHT: (estimated) 84.

LAYING DATES: Mali, July–Oct; Niger, Aug; Chad, presumed around July–Aug.

Neotis heuglinii (Hartlaub). Heuglin's Bustard. Outarde de Heuglin.

Otis heuglinii Hartlaub, 1859. Ibis, p. 344; between Zeila and Harar.

Plate 10
(Opp. p. 161)

Range and Status. Endemic resident, lowland N, E and S Ethiopia, Djibouti, NW and E Somalia and N Kenya (regularly east to Marsabit and south to *c.* 0°32′N at Garba Tula). Uncommon to frequent.

Description. ADULT ♂: forehead, crown (except white patch on posterior part), vestigial nuchal crest, face, chin and throat blackish brown, bordered dull white in down-curving line from nape across ear-coverts and uppersides of neck, meeting in V on upper foreneck; whitish merges into pale grey on rest of neck, upper breast and upper mantle. Rest of upperparts blotched and vermiculated buff and dark brown, tail similar but paler with brown subterminal band and whitish on extreme tips. Breast feathers filamentous, greyish with chestnut distal halves and black tips. Belly whitish. Primaries, greater primary and median coverts dark brown with whitish tips; central primaries with whitish on inner webs. Secondaries slightly paler with white tips, and whitish on edge of outer web, buff mottling at bases. Lesser primary coverts whitish. Greater coverts dark brown with white tips and buff-grey mottling towards bases. Scapulars and tertials dark brown, coarsely vermiculated buff. Lesser upperwing-coverts blotched and vermiculated buff and dark brown. Axillaries whitish. Inner area of underwing (middle of central primaries and secondaries) whitish, rest of underwing greyish brown; lesser underwing-coverts pale brown with whitish mottling. Bill slaty above, pale below; eyes brown; legs and feet yellowish white. ADULT ♀: as ♂, but smaller; black-brown on face and throat replaced by pale buff supercilium, greyish brown ear-coverts, whitish chin and buff throat; tail paler. SIZE: (3 ♂♂, 4 ♀♀) wing, ♂ 489–505 (495), ♀ 405–432 (423); tail, ♂ 185–190 (188), ♀ 160–180 (171); bill, ♂ 72–79 (76), ♀ 62–65 (64); tarsus, ♂ 151–162 (157), ♀ 126–136 (131). WEIGHT: (kg, Kenya) ♂ 4·0, 8·0, ♀ 2·6, 3·0.

IMMATURE: undescribed.
DOWNY YOUNG: undescribed.

Field Characters. A fairly large bustard, similar to but somewhat smaller (especially ♀) than Denham's Bustard *N. denhami*; upperparts heavily blotched buff, no rufous on hindneck, closed wing without black and white panel, grey neck separated from white underparts by rufous and black bands; black face and chin of ♂ produce distinctive 'triangular' patch.

Voice. Unknown.

General Habits. Lives in dry lowland habitats, from nearly naked rock desert, open desert with annual grass, to semi-desert savanna and tussocky grassland. Plucks berries, sometimes leaping to reach them. Wary.

Apparently nomadic (the late L. H. Brown, pers. comm.).

Food. Of 3 specimens (Kenya), 1 contained grasshoppers and a mouse, 1 a lizard, insects and vegetable matter, and 1 berries and stones. Small yellow berries also recorded.

Breeding Habits.
NEST: scrape on bare ground.
EGGS: 2. Oval; usually warm buff (sometimes pale clay) with chestnut markings. SIZE: (n = 6) 69·4–75·0 × 52·0–53·0 (72·8 × 52·6). WEIGHT: (estimated) 110.
LAYING DATES: Somalia and Ethiopia, Apr–June; E Africa: Region D, Jan, June, breeding later in rains when grass is tallest.

Reference
Archer, G. and Godman, E. M. (1937).

156 OTIDIDAE

Genus *Chlamydotis* Lesson

Medium-sized bustard with relatively short legs and long tail, crest, and elongated feathers at sides of neck and breast. 1 sp., N Africa, Canary Is., SW and central Asia.

Plate 10
(Opp. p. 161)

Chlamydotis undulata (Jacquin). Houbara; Houbara Bustard. Outarde houbara.

Psophia undulata Jacquin, 1784. Beytr. Gesch. Vög., p. 24; Tripoli.

Range and Status. N Africa, Canary Is., and from Middle East to Outer Mongolia; eastern populations winter NW India, Pakistan, Iran, Iraq, Arabia, and probably occasionally NE Africa.

Resident, sometimes nomadic, and generally widespread, although patchily distributed, throughout most flat to undulating semi-arid and arid zones in Mauritania, S and E Morocco, N-central Algeria, S Tunisia and N Libya, N Sudan and Egypt. Frequent but declining, in places probably drastically (including N Africa) through hunting and habitat changes (Mayaud 1982).

Chlamydotis undulata

Description. *C. u. undulata* (Jacquin): NW and N Africa east to Nile. ADULT ♂: forehead, sides of crown buff with small brown fine speckling; crown a tuft of erectile filamentous white feathers. Nape, throat, foreneck, hindneck and upper breast greyish white with pepper-and-salt brownish find speckling. Sides of upperneck with erectile long black filamentous plumes, sides of lower neck with erectile long white filamentous plumes. Face buffy with some brown streaking, greyish on ear-coverts. Short bristles on lores. Chin whitish. Lower hindneck and mantle buffy light orange-brown with fine brown vermiculations. Rest of upperparts light orange-brown with brown vermiculations; tail similar with 4 pale blue-grey bars and white tips to all but central feathers. Lower breast with long white filamentous feathers, sides of breast as foreneck but tinged buff. Underparts otherwise white. Outer primaries white with brownish black tips; inner primaries and secondaries brownish black faintly tipped white, with whitish bases. Scapulars and tertials buffy light orange-brown with coarse light brown vermiculations. Greater primary coverts blackish brown with light orange-brown bases; median and lesser primary coverts white; alula brownish black. Greater coverts whitish with brown vermiculations; median and lesser coverts light orange-brown with fine brown vermiculations. Underwing white with grey-brown trailing edge; axillaries white. Bill olive-grey; eyes pale yellow; legs and feet pale grey. ADULT ♀: as ♂ but smaller. Crest shorter, tipped and mottled with light orange-brown; black and white neck-plumes reduced in size; 3 bars in tail, tail-tips buffy not white. SIZE: (6 ♂♂, 2 ♀♀) wing, ♂ 354–383 (372), ♀ 333–343 (338); tail, ♂ 185–215 (202), ♀ 138–190 (164); bill, ♂ 35·0–38·5 (37·5), ♀ 29–33 (31); tarsus, ♂ 83–96 (91), ♀ 78–81 (80). WEIGHT: (spring, kg) ♂ c. 3·2, ♂ c. 2·5.

IMMATURE: similar to ♀, but upperparts more heavily marked with dark arrow-shaped blotches; white in wing suffused buff, and crown-tuft poorly developed.

DOWNY YOUNG: light orange-brown with dappled whitish markings, fringed dark brown.

C. u. macqueenii (J. E. Gray): Egypt east of Nile (1 record west of Nile); Sudan (Port Sudan). White crown-tuft with black centre, foreneck clear blue-grey; slightly larger; wing, ♂ (n = 25) 384–426 (399). Resident (breeds Sinai) or rare winter visitor.

Field Characters. A medium-sized bustard, distinguished by uniform pale sandy upperparts, black stripe down side of neck, longish tail with 3–4 broad dark (bluish) bars. White crown-tuft diagnostic but not always obvious even in adults. In flight, has distinctive slow wing-beats, with 'flicked' upstroke; broad black trailing-edge to flight-feathers and black wing-tips, white patch on distal third of wing crossed by black bar, small black carpal patch. Little Bustard *Tetrax tetrax* much smaller, stockier, with short neck and tail, different wing pattern.

Voice. Not tape-recorded. Almost totally silent. Alarmed birds may croak, hiss or whine. ♀ may utter soft 'quop' or 'quip' when returning to eggs.

General Habits. Inhabits arid, open plains and steppe, characterized by shrubby xerophytic and halophytic vegetation, e.g. *Artemisia*, *Suaeda* and *Haloxylon*; in Morocco in areas below 200 mm isohyet; sometimes frequents small cultivations on the edge of plains. Outside breeding season usually in groups of 4–10, occasionally in larger, loose flocks of up to 60, feeding and resting together. Very cursorial, trotting over several km to forage; sometimes remains in small favoured locality (e.g. crop field) to feed. Very retiring and difficult to flush more than once. When flushed, will fly between 20 m and several hundred m or out of sight altogether; may return later to site where first flushed. When disturbed, may also run with head and neck lowered below line of back and tucked into body, keeping small ridges and bushes between itself and intruder; finally squats tight against rock or bush with neck and head on back, or in bush. Flies powerfully with slow 'flicked' wing-beats, legs tucked in, gliding to land. In response to attack from falcon, can eject sticky, liquid faeces at it. When attacked on ground, may respond by spreading wings to show white patches, fanning tail, and drawing head back and bill wide open.

Reportedly both nomadic (in response to rains), e.g. Western Sahara, and site-faithful (always relocated in same area). In Egypt, birds may move inland in winter from coastal breeding areas.

Food. Fruits, seeds, shoots, leaves, flowers, small invertebrates especially ants and beetles, and small reptiles. In spring, more plant than animal food.

Breeding Habits. Probably polygamous or promiscuous (in recent study of Canary Is. race *fuertaventurae*, no evidence of pair-bond was found: D. R. Collins, in prep.); ♀♀ apparently nesting solitarily. ♂♂ use traditional display-areas, possibly maintaining individual territories. ♂ has Trotting-display (**A**): at first stationary, he depresses tail, spreads white crown-plumes, raises black neck-plumes to each side, arches them backwards and ruffles white neck-plumes downward, then curves them forward and up; next, snaps head back onto mantle, pulling white breast-plumes over back, and begins high-stepping trot, moving in zigzags, circles and sometimes long straight courses. Repeated circling may produce trodden patch 1–2 m across. Area over which ♂♂ display variable (6, 16 and 24 ha recorded) distance between displaying birds always over 500 m (D. R. Collins, in prep). In only copulation witnessed, ♂ gave rapidly repeated Trotting-displays as he tried to encircle ♀, then stood still and, with crown,

A

neck and breast-feathers fully erect, repeatedly threw head onto back and thrust it out forward level with body; then turned his forward-thrust head alternately to left and right (at 1 s intervals) and, standing behind or over ♀, pecked at her head and neck for 9 s before suddenly lowering himself onto her, using outstretched wings for balance (D. R. Collins, in prep.).

NEST: shallow unlined scrape, 13–23 cm, in semi-desert; commonly near low bush, sometimes in open; formed by ♀.

EGGS: 2–3, occasionally 4, laid at 1-day intervals. Elliptical; light olive-brown or grey, with darker blotches. SIZE: (n = 60) 58–68 × 43–48 (62 × 45). WEIGHT: (estimated) 67–68.

LAYING DATES: Western Sahara Feb–Mar; Morocco, Mar–May; Algeria, Mar–June; Tunisia, Mar–May. N Africa (country unspecified) also Dec, large chick Jan.

INCUBATION: probably begins before last egg, by ♀ only. Period: *c.* 23 days.

DEVELOPMENT AND CARE OF YOUNG: young squat in response to disturbance, relying on camouflage to avoid detection. Cared for and fed by ♀, although ♂ may attend (*macqueenii*). Fledge at *c.* 35 days; remain with ♀ at least through first autumn.

References
Collins, D. R. (in prep.).
Cramp, S. and Simmons, K. E. L. (1980).

Genus *Ardeotis* Le Mahout

Large bustards with relatively long bills, short backward-projecting crests, uniform brownish grey upperparts, white underparts. Entirely distinct in structure and plumage from *Otis* under which this genus and *Neotis* are sometimes subsumed.

4 spp. forming a superspecies, 1 in India, 1 in Australasia (these 2 closer to each other than to the others), 2 in Africa (1 endemic, 1 also Arabia).

158 OTIDIDAE

Ardeotis arabs superspecies

1 *A. arabs*
2 *A. kori*

Plate 10 **Ardeotis arabs (Linnaeus). Arabian Bustard. Outarde arabe.**
(Opp. p. 161)

Otis arabs Linnaeus, 1758. Syst. Nat. (10th ed.), 1, p. 154; Yemen.

Forms a superspecies with *A. kori*.

Range and Status. Africa and SW Arabia.

Resident and intra-African migrant; from Atlantic to Red Sea, almost entirely between 10°N and 20°N; from Mauritania, Senegambia, Mali, Burkina Faso, Ivory Coast, W Ghana, Niger, Nigeria, and N Cameroon to Chad, Sudan, Ethiopia (including Dahlak archipelago) and Djibouti. Common to abundant, but known to have been much persecuted in past decade and may now be rare in places (e.g. Senegambia). Vagrant Kenya (NW Turkana, 1932) and extreme NW Somalia. In N Africa, formerly resident Morocco (Forest of Marmora, Sous), but only 4 records since 1970; disappearance this century unexplained. Vagrant Algeria (Algiers, 1855).

Ardeotis arabs

Description. *A. a. stieberi* (Neumann): Senegambia to NE Sudan. ADULT ♂: forehead and crown whitish buff with very fine vericulations, sides of crown and straggling crest black. Neck and upper breast plumage filamentous, greyish white with close grey-brown barring. Chin, throat and facial area below ear-coverts similar but creamy white, barring reduced. Lores creamy white with blackish speckling, supercilium creamy white, ear-coverts creamy grey. Upperparts dark orange-brown closely vermiculated with blackish, with pale fringes to some feathers. Tail with greyish base, broad creamy white mottling. Greater primary coverts dark orange-brown with smoky speckling, tipped white. Alula grey-brown with white tip. Greater coverts dark orange-brown at base becoming greyish brown, then white towards tips. Scapulars whitish, tertials likewise but smokier on inner webs. Lesser and median coverts dark orange-brown closely vermiculated with blackish and with conspicuous white tips. Underwing whitish with pale mottling, greyish brown along trailing edges. Axillaries whitish. Bill yellowish, culmen grey-black; eyes pale brown; legs and feet pale yellow or whitish. ADULT ♀: smaller, greyer on upperparts and wings, lacking speckling on lores. SIZE: wing, ♂ (n = 5) 588–690 (634), ♀ (n = 3) 516–554 (530); tail, ♂ (n = 5) 290–330 (310), ♀ (n = 4) 268–277 (273); bill, ♂ (n = 6) 74–90 (82), ♀ (n = 6) 63·5–74·0 (67·9); tarsus, ♂ (n = 5) 182–192 (187), ♀ (n = 4) 166–173 (169). WEIGHT: (kg) ♂ 10·0, 5·7, ♀ 4·5.

IMMATURE: like adult but outer primary pointed, not rounded; younger immature duller, wing-coverts less contrastingly patterned.

DOWNY YOUNG: undescribed.

A. a. arabs (Linnaeus): Ethiopia, Djibouti. Slightly smaller; wing, ♂ (n = 6) 560–629 (602). Upperparts greyish buff, orange-brown almost entirely absent; spotting on median and lesser coverts much reduced.

A. a. butleri (Bannerman): S and SE Sudan, Kenya. Upperparts slightly darker than in *arabs*; greyer on head and neck, reduced area of creamy white on tail.

A. a. lynesi (Bannerman): Morocco (probably extinct). Upperparts darker than *butleri*, spotting on median and lesser coverts and white mottling on secondaries much reduced.

Field Characters. A large bustard, very similar to Kori Bustard *A. kori*, but somewhat smaller and paler; lacks black on wing-coverts and sides of breast. Distinguished from Denham's Bustard *Neotis denhami* by crest, absence of black on closed wing and of rufous on hindneck. In flight shows irregular white barring on wings which in older birds can form an extensive patch.

Voice. Not tape-recorded. General silent. A rasping or honking croak, 'pah pah', when calling during display. Muffled barks in alarm.

General Habits. Inhabits arid environments, including semi-desert, scrub, grassy plains, but also *Acacia* parkland and formerly well gladed *Quercus* woodland (Forest of Marmora). In Ethiopia (Eritrea), occurs in treeless sandy country dominated by *Panicum turgidum*; in Sahelian steppe, (Niger), occurs in short (10 cm) grass with scattered low *Commiphora* and *Acacia* trees (P. J. Jones, pers. comm.). In Chad, closely associated with wadis (Newby 1979). In S Morocco, occurred in wide grassy plains with sandy soil, particularly with dwarf palm *Hyphaene*. Frequently solitary; commonly in pairs or family parties. Drinks frequently and regularly, but also occurs far from water.

Migrates in flocks, generally *c.* 150–450 m above ground, sometimes with Denham's Bustard. In Nigeria, Chad and Sudan and probably across rest of W African Sahel zone, migrates north in June to breed during wet season, returning south around Oct. However, some birds are found N Niger and N Senegambia even in dry season (P. J. Jones, pers. comm.; G. J. Morel, pers. comm.), and in Mali reported to penetrate to 20°N, Feb–Mar (Lamarche 1980).

Food. Mainly insects, especially swarming locusts and grasshoppers, beetles, bugs and caterpillars; also reptiles, nestling birds, rodents, and shoots, leaves, grass heads, seeds and fruits of *Corida sinensis*, *Grewia villosa*, *Salvadora persica* and wild melon *Cucumis*, and gum of *Acacia* (Malzy 1962, Newby 1979).

Breeding Habits. Mating system unknown; biology and behaviour probably similar to Kori Bustard.

NEST: shallow scrape on bare ground, occasionally amid scrub; sometimes lined with vegetation. Solitary nesting ♀♀ encountered every 2–3 km in Sahel zone, Niger (P. J. Jones, pers. comm.).

EGGS: 1–2. Oval; reddish buff to olive-brown, with darker brown streaks. SIZE: (n = 27, *A. a. butleri*) 67·3–79·8 × 51·5–57·0 (73·7 × 53·9). WEIGHT: (estimated) 118.

LAYING DATES: Morocco, Apr; Mauritania, Aug–Sept; Senegambia, July–Sept; Mali, July–Oct; Niger, July–Aug; Chad, 'wet season' (i.e. June–Sept); Sudan, Apr, probably Aug–Oct; Ethiopia, Aug–Oct.

INCUBATION: probably by ♀.

Reference
Cramp, S. and Simmons, K. E. L. (1980).

Ardeotis kori (Burchell). Kori Bustard. Outarde kori.

Otis kori Burchell, 1822. Trav. S. Afr. 1, p. 393, note; Vaal-Orange River confluence.

Forms a superspecies with *A. arabs*.

Range and Status. Endemic resident and intra-African migrant. 2 separate populations: NE Africa, from NW Somalia, central Ethiopia and SE Sudan south through Kenya and NE Uganda to N Tanzania; and southern Africa, in Zimbabwe, Botswana (N to Chobe R.), Namibia, South Africa, S Mozambique and S Angola. Uncommon to locally common, but generally declining, markedly in parts of southern Africa, where

Plate 10
(Opp. p. 161)

Plate 9

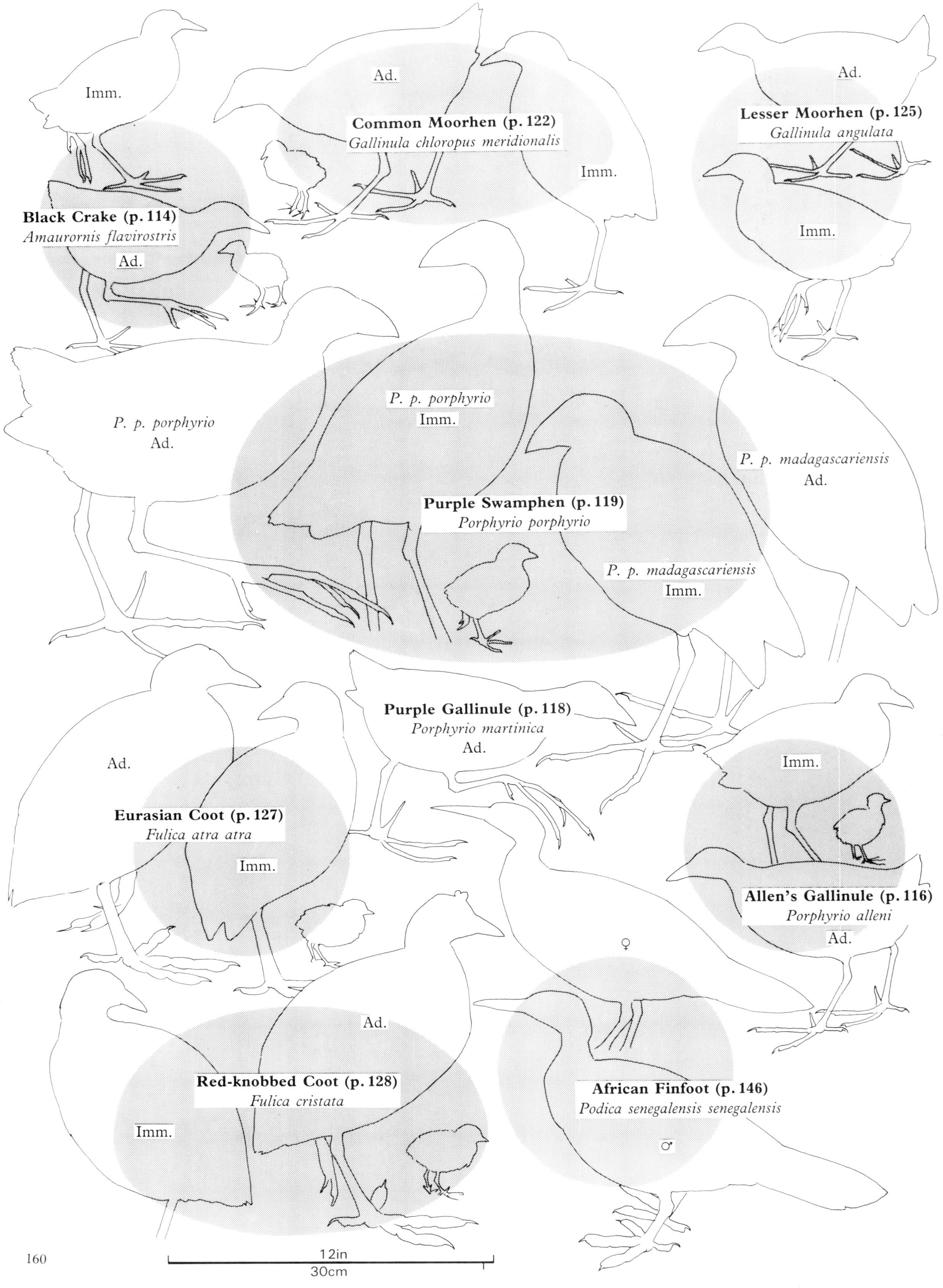

Black Crake (p. 114)
Amaurornis flavirostris
Ad.
Imm.

Common Moorhen (p. 122)
Gallinula chloropus meridionalis
Ad.
Imm.

Lesser Moorhen (p. 125)
Gallinula angulata
Ad.
Imm.

Purple Swamphen (p. 119)
Porphyrio porphyrio
P. p. porphyrio Ad.
P. p. porphyrio Imm.
P. p. madagascariensis Ad.
P. p. madagascariensis Imm.

Purple Gallinule (p. 118)
Porphyrio martinica
Ad.

Eurasian Coot (p. 127)
Fulica atra atra
Ad.
Imm.

Allen's Gallinule (p. 116)
Porphyrio alleni
Ad.
Imm.

Red-knobbed Coot (p. 128)
Fulica cristata
Ad.
Imm.

African Finfoot (p. 146)
Podica senegalensis senegalensis
♀
♂

12in / 30cm

160

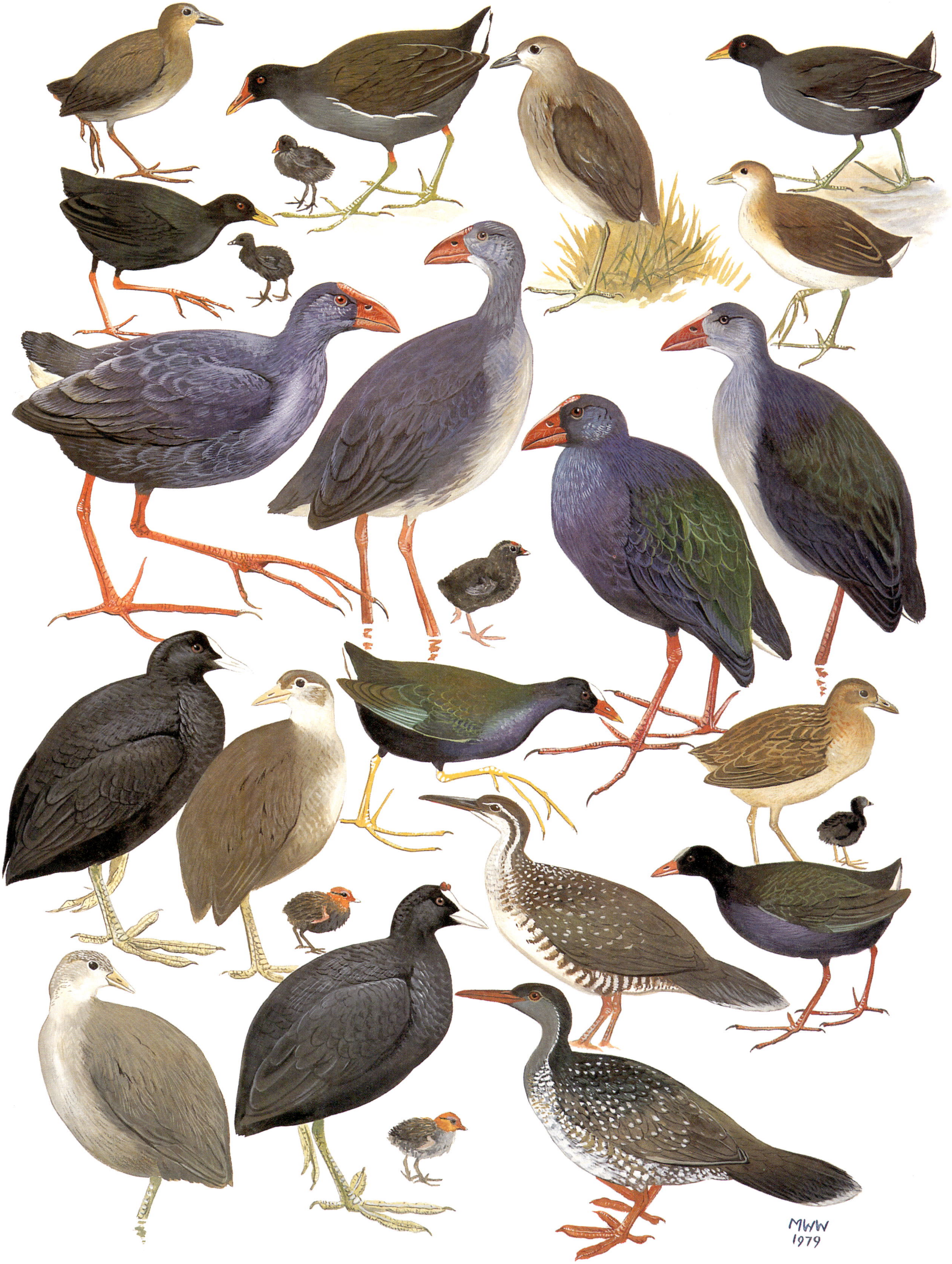
MWW
1979

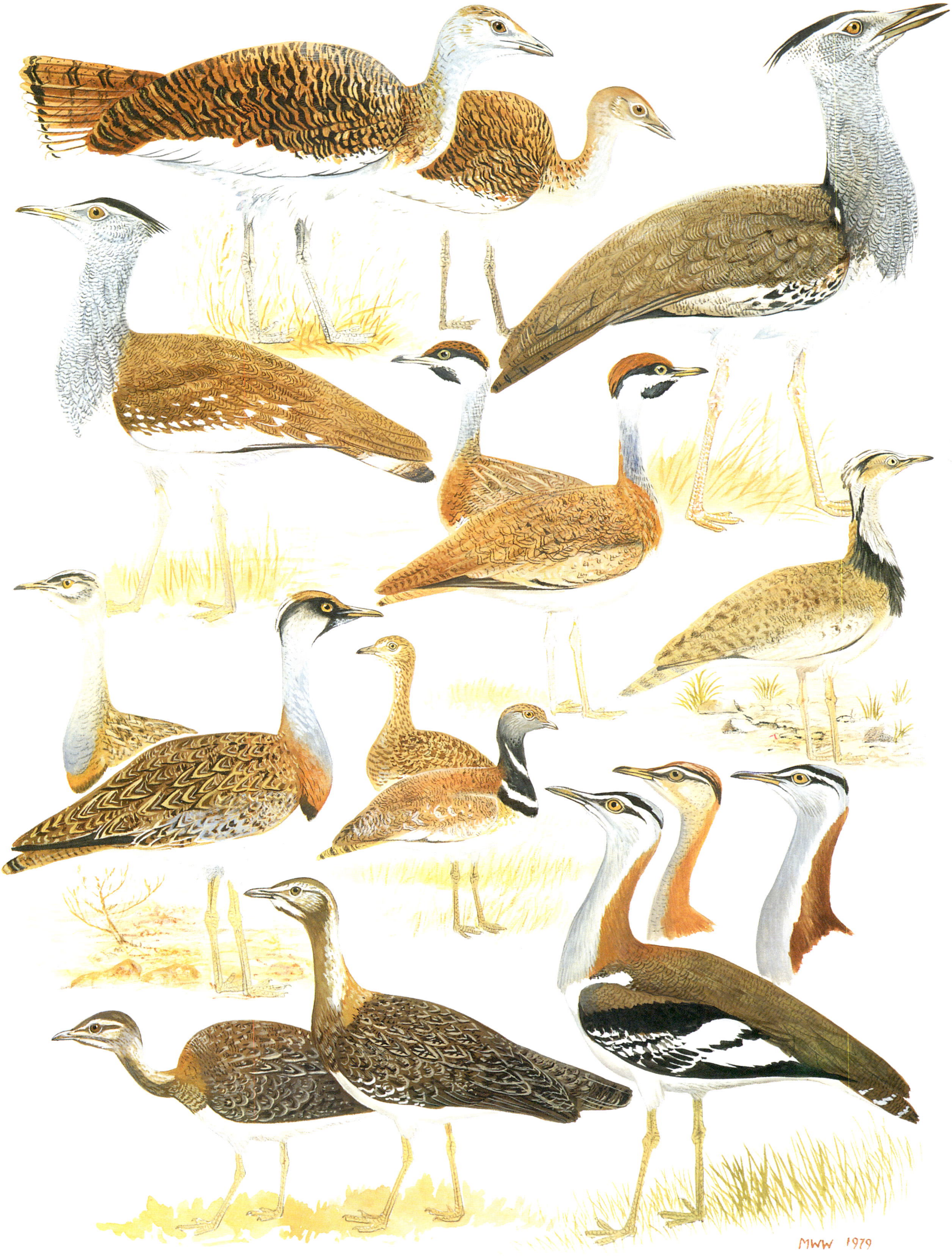
MWW 1979

Plate 10

Great Bustard (p. 164)
Otis tarda tarda
Ad. ♂ non-breeding
Ad. ♀ non-breeding

Kori Bustard (p. 159)
Ardeotis kori struthiunculus
Ad. ♂

Arabian Bustard (p. 158)
Ardeotis arabs stieberi
Ad. ♂

Nubian Bustard (p. 154)
Neotis nuba nuba
Ad. ♀
Ad. ♂

Houbara (p. 156)
Chlamydotis undulata undulata
Ad. ♂

Heuglin's Bustard (p. 155)
Neotis heuglinii
Ad. ♀
Ad. ♂

Little Bustard (p. 149)
Tetrax tetrax tetrax
Ad. ♀
Ad. ♂ breeding

Denham's Bustard (p. 151)
Neotis denhami denhami
Ad. ♀
N. d. jacksoni Ad. ♂
Ad. ♂

Ludwig's Bustard (p. 153)
Neotis ludwigii
Ad. ♀
Ad. ♂

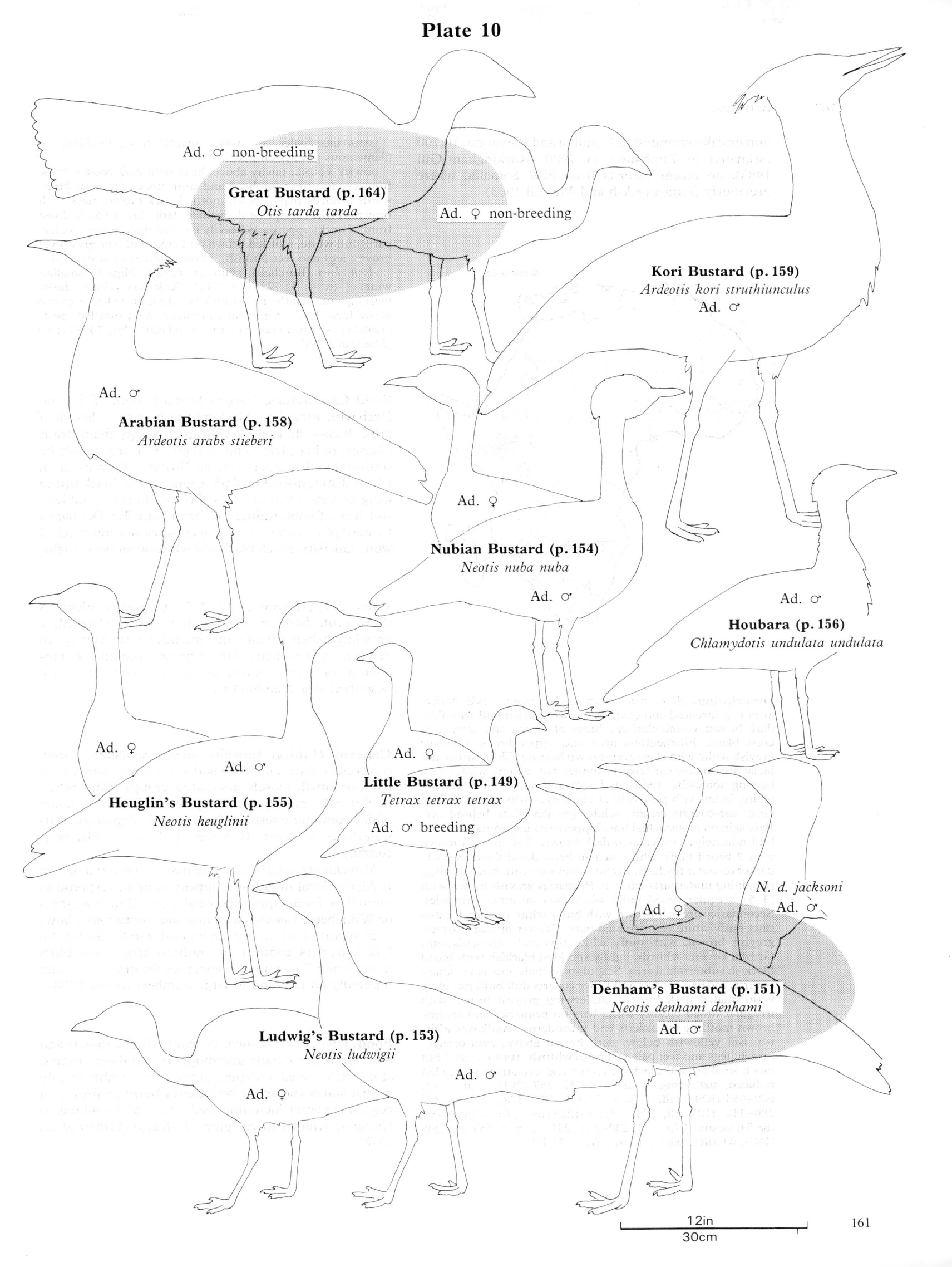

12in / 30cm

numerically strongest in Namibia and Botswana; 10,700 estimated in Zimbabwe in 1980 (Rockingham-Gill 1983); no recent records from NW Somalia, where previously frequent (Ash and Miskell 1983).

Ardeotis kori

Description. *A. k. struthiunculus* (Neumann): NE Africa. ADULT ♂: forehead and centre of crown greyish buff with fine, dark brown vermiculations. Sides of crown and straggling crest black. Filamentous neck and upper breast plumage greyish white with close grey-brown barring. Chin, throat and facial area below ear-coverts similar but creamy white, with barring somewhat reduced. Lores and supercilium creamy white, latter with post-orbital black eye-stripe merging into crest; ear-coverts creamy white, greyish white behind eye. Bare skin on moustachial line. Upperparts including back dull buff minutely vermiculated dark brown. Tail greyish brown with 2 broad buffy white median bars, distal third as back, darker on outer feathers. Breast sometimes with black at sides; remaining underparts whitish. Primaries greyish brown with (3–5) irregular broad buffy white bars on inner primaries. Secondaries greyish brown with buffy white tips, and indistinct buffy white vermiculated bars. Greater primary coverts greyish brown, with buffy white tips and vermiculations. Greater coverts whitish, lightly speckled blackish with broad blackish subterminal area. Scapulars, tertials, median primary coverts, median coverts and lesser coverts dull buff, minutely vermiculated dark brown. Underwing greyish brown with irregular broad creamy white bars on primaries and creamy brown mottling on coverts and secondaries. Axillaries whitish. Bill yellowish below, dark brown above; eyes orange-brown; legs and feet pale yellow to whitish. ADULT ♀: as ♂ but much smaller, and black on crown and eye-stripe somewhat reduced. SIZE: wing, ♂ (n = 3) 752–767 (761), ♀ (n = 13) 600–655 (629); tail, ♂ (n = 3) 370–387 (378), ♀ (n = 13) 280–342 (312); bill, ♂ (n = 3) 95–120 (109), ♀ (n = 15) 81–95 (88·5); tarsus, ♂ (n = 3) 230–247 (241), ♀ (n = 15) 181–205 (190). WEIGHT: (kg) ♂ 10·9, ♀ (n = 2) 5·9.

IMMATURE: paler on crown, mantle more freckled, no filamentous plumes on neck.
DOWNY YOUNG: tawny above, head with dark brown stripe from eye to upper forehead and from upper forehead backwards towards nape; crown mottled dark brown; neck with brown vertical stripe and distinct dark throat-patch down front of neck; upperparts heavily mottled dark brown; underparts dull white, mottled brown on flanks; bill pale grey; eyes brown; legs and feet pinkish. Throat noticeably distended.

A. k. kori (Burchell): southern Africa. Slightly smaller; wing, ♂ (n = 3) 721–769 (742). Back less yellow, dusky markings on mantle and scapulars; black on sides of crown above lores less conspicuous, forehead more mottled, post-orbital eye-stripe reduced or absent. WEIGHT: (kg) ♂ 13·5–19·0 (Maclean 1985).

Field Characters. Largest bustard south of Sahara. Drab with straggling, black-and-grey crest (*c.* length of bill), thick-looking neck with grey shaggy filamentous plumes, rather plain wing in flight. Can really only be confused with similar Arabian Bustard *A. arabs*, from which distinguished by darker upperparts, black tips to wing-coverts, black on sides of breast and on shoulders, and lack of subterminal tail-band. Smaller Denham's Bustard *Neotis denhami* has no crest, rufous hindneck, 3 white tail-bars, much black and white on wings in flight.

Voice. Tape-recorded (C, 217). Generally silent. A short gruff bark or snoring note when alarmed; a growling when threatening intruder near young. In display, a resonant, far-carrying 'voomp voomp-voomp' or 'vum-vum-vum-vum ... vumvum'; also described as a deep loud roar.

General Habits. Inhabits open grasslands, karoo, bushveld and dry, lightly wooded savannas. Solitary, in pairs or small, loosely associated groups. Flies reluctantly; walks rapidly with long strides. In hottest hours of day commonly rests in shade of tree. Regularly visits waterholes to drink, E Africa (J. F. Reynolds, pers. comm.).

Movements poorly documented or understood. In E Africa, local migrations appear to be in response to rainfall or food-supply (especially to follow bush-fires or Wildebeest *Connochaetes taurinus* migrations); however, reported to be a dry season visitor to SE Sudan (G. Nikolaus, pers. comm.). In southern Africa some birds move E or SE to winter at lower levels, in former times reputedly on foot and in large numbers (Snow 1978).

Food. Diet poorly known. Probably chiefly insects and their larvae, especially grasshoppers and dung-beetles; also reptiles, small rodents, carrion (presumably mainly invertebrates and small vertebrates burnt in fires) and vegetable matter including seeds, roots and wild melon *Cucumis*. Known to eat gum of *Acacia* (Urban *et al.* 1978).

A   B

Breeding Habits. Mating system unclear; although birds in pairs and courtship feeding (unrecorded in any other bustard except White-bellied Bustard *Eupodotis senegalensis*) might suggest monogamous system, strong sexual dimorphism (and absence of monogamy in Great Bustard *Otis tarda*, Little Bustard *Tetrax tetrax* and Houbara *Chlamydotis undulata*) indicates more complex system likely. Birds tend to gather in certain areas to display (the late L. H. Brown, pers. comm.) and nest. 2 fighting ♂♂ (**A**) held each other by bill, pushing backwards, forwards and in circles, without using wings or feet; after 30 min of such a fight, 1 ♂ flattened feathers, lowered head and drooped tail, and ran briefly before flying away (Allen and Clifton 1972, Schaller 1973). Displaying ♂ struts back and forth over particular area (commonly, a low hill-top) with tail cocked forward; occasionally stands very upright and gradually inflates neck until it forms a white puffy ball, bulges cheeks, erects crest, holds bill up at angle, opens it (apparently blood-red at gape), fans and cocks tail at various angles (at greatest intensity touching nape), exposing white undertail-coverts, and droops wings so that primaries touch ground (**B**). ♂ may then strut about or stand still, often keeping near small landmark, e.g. bush, anthill, and, when dilation of neck reaches maximum, calls, bill snapping open and shut, neck vibrating. In courtship, ♂ walks slowly around ♀, or stands within 10 m of her, bowing with body tilted forward, neck inflated, head never reaching below level of shoulders (R. Thomson, pers. comm.). Displaying ♂ seen to offer 80 cm long snake to ♀ (Schmidl 1982); captive ♂ repeatedly noted to provision ♀ once both began feeding.

NEST: shallow scrape on bare ground, sometimes with slight bedding of grass, generally near tuft of tall grass, shrub, or low outcrop of rock.

EGG: 1–2. Oval; pale olive streaked greyish and dark brown. SIZE: (n = 5) 76–84 × 53–60 (80·6 × 56·0). WEIGHT: (estimated) 137.

LAYING DATES: Somalia probably Apr–June; Ethiopia, Mar–June (broken egg found in SW, Nov); E Africa: Region A, Jan; Region B, Feb–Apr; Region D, Jan–Apr, June, Nov; Zimbabwe, Sept–Dec, also Apr; South Africa, Sept–Feb; Namibia, Nov–Jan.

INCUBATION: presumably by ♀ only. Period: 4·5 weeks.

DEVELOPMENT AND CARE OF YOUNG: cared for by ♀ only, though ♂ sometimes in attendance. Walking chick keeps under ♀, immediately behind her legs: appears to feed on insects from grass at head-height rather than from ground; receives food from ♀ even when fully feathered.

BREEDING SUCCESS/SURVIVAL: apparently reduced when rains short (eggs and small young Feb–Apr 1966, very wet season; in 2 other years little success in hatching eggs or did not breed at all when rains shorter than usual, Serengeti, Tanzania: Schaller 1973); incubating ♀ killed by jackals (Thomas 1960) and adult killed by Martial Eagle *Polemaetus bellicosus*.

References
Brooke, R. K. (1984).
Rockingham-Gill, D. V. (1983).

Genus *Otis* Linnaeus

Large bustard with stout but short bill, ♂ far larger than ♀ and with moustachial bristles, unique display. 1 sp., N Africa, Europe, Asia.

Plate 10
(Opp. p. 161)

Otis tarda **Linnaeus. Great Bustard. Grande outarde.**

Otis tarda Linnaeus, 1758. Syst. Nat. (10th ed.), 1, p. 154; Poland.

Range and Status. Continental Europe eastward through S Russia and Asia Minor to Mongolia and Manchuria; also Morocco; many populations sedentary, although some winter south to S Europe, Iran, China and possibly Morocco.

Resident N Morocco (near Tangier and in the Habt, the Rhab, and at Moyen Sebou); also reportedly rare Palearctic visitor N Morocco but no proof; vagrant Algeria (Hussein Dey and unspecified localities, undated: Heim de Balsac and Mayaud 1962) and doubtful sight records, Tunisia (Bir Soltane undated, Feriana spring 1983, Rekeb and north of Gabès Apr 1967). Uncommon, with population reduced to 100 or less. Without special protection likely to become extinct in Africa in a few decades, due to continuous disturbance and hunting.

Also winters in breeding range

Description. *O. t. tarda* Linnaeus: only subspecies in Africa. ADULT ♂ (breeding): head and nape bluish grey, feathers darker-tipped and slightly elongated down centre of crown. Chin and upper throat whitish, with greyish white elongated moustachial plumes. Upper hindneck, sides of neck and upper foreneck whitish yellow becoming vinous-chestnut on lower hindneck and rest of neck and breast (with some black barring at sides of lower breast); bare spot behind eye dark grey; bare streak along side of neck violet-black, becoming dark blue-grey when neck inflated. Mantle, scapulars, back and rump boldly barred black and gold-buff. Tail gold-buff, white at base, broad black subterminal bar, white on tips (except central feathers, which have black median bar) and much white replacing gold-buff on outer feathers. Underparts white. Primaries black or dark brown, paler towards bases. Secondaries white with black tips, black more extensive on innermost feathers. Tertials white, innermost with black and gold-buff bars. Greater primary coverts white; alula, median and lesser coverts white suffused with pale grey. Greater, median and lesser coverts gold-buff with irregular black barring. Underwing white with grey-brown trailing edge. Bill grey with dark brown tip; eyes dark brown; feet olive-brown to grey. ADULT ♂ (non-breeding): chestnut on breast replaced by grey and neck appears much sleeker (but still appearing thick) with bare streak absent; moustachial plumes absent. ADULT ♀: like non-breeding ♂, but much smaller and thinner necked (apparently tapering towards head); less white on wing panels, finer barring on tertials and more narrower bars on tail. SIZE: wing, ♂ (n = 10) 598–633 (617), ♀ (n = 14) 475–497 (486); tail, ♂ (n = 9) 222–259 (243), ♀ (n = 13) 208–219 (214); bill, ♂ (n = 14) 32–40 (36·8), ♀ (n = 19) 27–36 (30·8); tarsus, ♂ (n = 15) 145–168 (158), ♀ (n = 19) 118–132 (125). WEIGHT: (E European birds, kg) ♂ AD spring (n = 13) 8·5–18·0 (12·0), ♂ AD winter (n = 11) 5·8–16·0 (8·9); ♀ AD spring (n = 4) 3·5–4·0 (3·8), ♀ AD winter (n = 11) 3·3–5·3 (4·4).

IMMATURE: ♀ as adult ♀, ♂ similar but distinguished by larger size, with broader white panels on closed wing, broad subterminal bar on tail and thicker neck.

DOWNY YOUNG: buff with cryptic mottling of brown blotches bordered black; ♂♂ have broader heads and narrower, darker markings.

Field Characters. A huge bustard, with thickset body, blue-grey head, reddish neck and black and gold barred upperparts. Walks with head held relatively horizontal and often with tail fanned. Broad wings show much white in slow powerful flight.

Voice. Tape-recorded (62, 73). Generally silent. When excited or alarmed, a short, low, grunting bark 'oogh' or 'uf'. In aggression, a grumbling rattle. When raptor overhead, a protracted whine. Food-call of ♀ to chick 'oho-ooho'. Chick has plaintive high descending 'cheeeoo' in distress, a short shriek in alarm, a shrill 'heng' in defensive attack, a rising 'prrip' when relaxed, a short level 'trrip' in greeting, and a brief 'chewyoo' apparently expressing alertness.

General Habits. Inhabits extensive flat or undulating open short grassland, steppe, pastures, and crop-fields, usually well away from trees. Very wary, affected by disturbance, also by deterioration of habitat. Outside breeding season, usually frequents traditional wintering grounds in flocks; flocks single-sexed or with sexes segregated within them. Immature ♂♂ form excitable flocks in spring. Flies powerfully with slow deep wingbeats and no glides, normally below 200 m, usually 30–100 m high. Grabs prey with swift jab of beak, often shaking and sometimes beating it on ground before swallowing.

Generally highly site-faithful with movements occurring between traditional wintering and breeding grounds.

Food. In Europe, young shoots, leaves, flowers, seeds, roots; grasshoppers, beetles and other invertebrates; occasionally amphibians, lizards, eggs and nestlings of ground-nesting birds and small mammals.

Breeding Habits. Some ♂♂ apparently polygamous, others forming no bond with ♀♀, all showing lek-like behaviour. ♀ nests solitarily. ♀♀ reach maturity at 2 years while ♂♂ do so at 4–6 years. ♂♂ utilize traditional display areas, ♀♀ moving onto them for short period each spring to copulate. ♂♂ apparently not strictly territorial (though sometimes the strongest and most successful may be), but move about over general display area, keeping distance from other ♂♂ (presumably through a social rank-order) and displaying at various sites, sometimes in response to ♀'s presence. ♂ has terrestrial display, given mostly in early morning and late afternoon, in which it cocks tail flat on back, inflates gular sac and retracts neck, so that moustachial bristles project upwards around eye and grey stripe of bare skin exposed on neck; simultaneously, wing straightened out, pointed down and back from body (keeping primaries folded so that tips are behind head), lifting secondaries and tertials to create white 'rosettes' at sides of body (**A**). Display stationary, but with some body-trembling, foot-trampling and shifts of position; sustained for several minutes, at end of which bird may only partially relax posture (secondaries and tertials lowered, head raised) before resuming full display. Long pre-copulatory sequence involves ♂ circling ♀ in partial display (head somewhat raised, tail less cocked) and beating on her back with wing, until she squats: he stands over her, pecking at her head, and copulates; insemination occurs very quickly, birds separating and apparently exhibiting no further bond.

NEST: shallow, unlined scrape, 25–35 cm across, in grass or crops, made by ♀ only. Often well away (5–10 km recorded) from ♂ display areas, but may be as close as 60 m.

EGGS: 2–3, rarely 1–4, laid at c. 1–2 day intervals. Elliptical; olive-brown blotched brown. SIZE: (n = 120) 69–90 × 52–61 (80 × 57). WEIGHT: (n = 9) 111–172 (146).

LAYING DATES: Morocco, Apr.

INCUBATION: probably begins with last egg; by ♀ only. Period: c. 25 days; hatching asynchronous.

DEVELOPMENT AND CARE OF YOUNG: young nidifugous and follow ♀ around keeping in contact with peeping calls; bill-fed for first few days; young fledge at 30–35 days; full size at 80–120 days. Age of independence unknown, but probably from 1st autumn, though possibly not until following spring. Immature ♂♂ at least appear to form groups.

BREEDING SUCCESS/SURVIVAL: in 1 population, Spain, young formed only 6% of autumn total (Ena *et al.*, in press). Human disturbance and agricultural intensification are major sources of nesting failure.

References
Cramp, S. and Simmons, K. E. L. (1980).
Gewalt, W. (1959).

Genus *Eupodotis* Lesson

Small to medium-sized bustards, all boldly marked except for the 3 driest-country spp. (*vigorsii*, *rueppellii* and *humilis*), and with penetrating, repetitive calls. Crown of head strongly crested; feathers of lower throat and foreneck conspicuously elongated; tarsus rather long; wing more than 3 times length of tarsus.

5 spp. (*vigorsii*, *rueppellii*, *humilis*, *senegalensis* and *caerulescens*) represent a closely related group (black throat, dark crown, pale sides of head); the others form a 2nd group of 4 black-bellied spp., comprising 2 larger, longer-billed spp. (*melanogaster* and *hartlaubii*) and 2 smaller ones, the pinkish-crested *ruficrista* (possibly representing 3 spp.) and the pinkish *afra*. All 9 are Afrotropical; the Asiatic *Houbaropsis* is sometimes subsumed under *Eupodotis*.

Vigorsii and *rueppellii* form a superspecies, as do *melanogaster* and *hartlaubii*.

166   OTIDIDAE

Eupodotis vigorsii **superspecies**

1 *E. rueppellii*
2 *E. vigorsii*

Eupodotis melanogaster **superspecies**

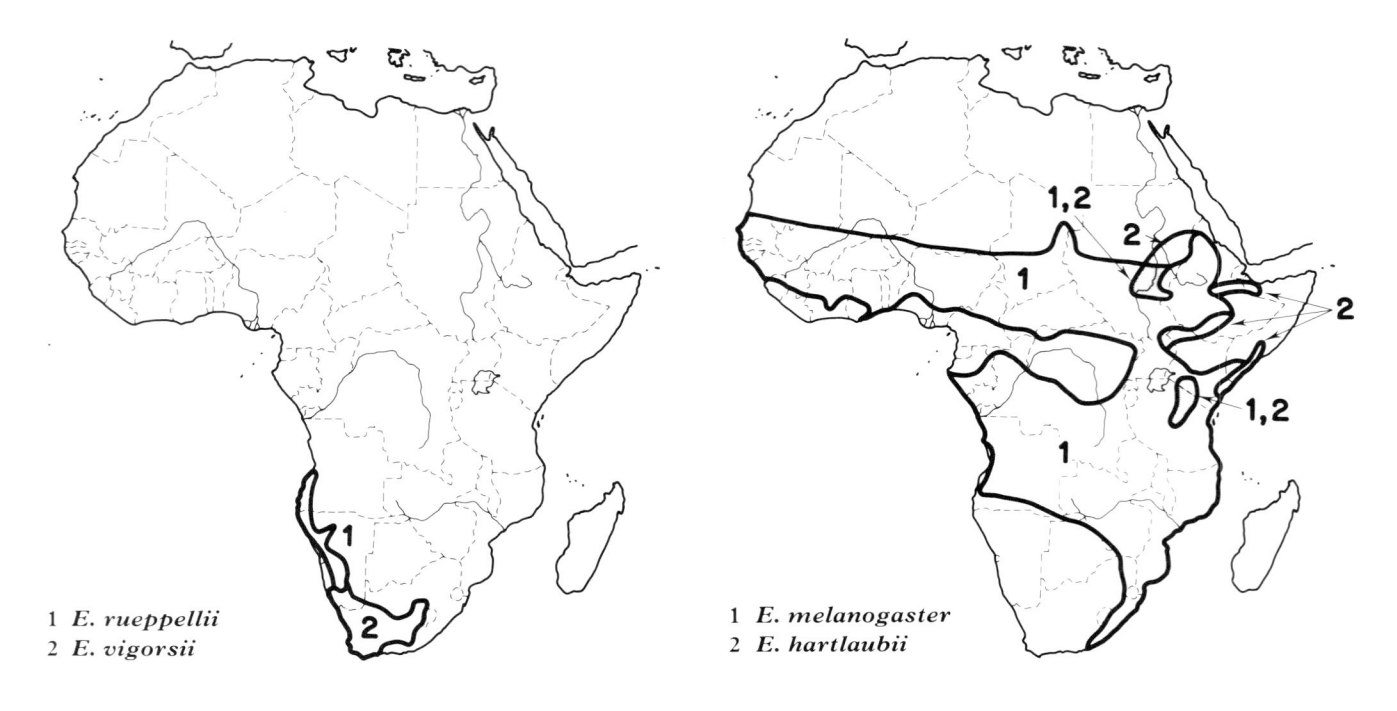

1 *E. melanogaster*
2 *E. hartlaubii*

Plate 11
(Opp. p. 176)

Eupodotis ruficrista (Smith). Crested Bustard; Crested Korhaan. Outarde houpette.

Otis ruficrista Smith, 1836. Rep. Exped. C. Afr., p. 56; Latahoo, Kuruman.

Range and Status. Endemic resident in 3 widely separate areas: (1) in Sahel zone from N Senegambia and S Mauritania through Mali, S Niger, N Nigeria and W Chad to W and SE Sudan; (2) from central, S and E Ethiopia, Djibouti, Somalia, Kenya and NE Uganda into E-central Tanzania from lowlands up to 1400 m; and (3) from central Mozambique, Swaziland and N South Africa across Zimbabwe and Botswana to SW Zambia, N-central Namibia and SW Angola. Common eastern and southern Africa, common to uncommon W Africa, where distribution somewhat patchy; in Mali, common some areas, absent others (Lamarche 1980).

Eupodotis ruficrista

Description. *E. r. ruficrista* (Smith): central and southern Africa. ADULT ♂: forehead and crown dull bluish grey. Buff supercilium with brown flecks extends from before eye to behind crown. Short nuchal crest pinkish brown. Hindneck and sides of neck greyish brown. Ear-coverts grey, buff behind eye; dark grey streak below eye. Rest of face, chin and throat dull white with broad black stripe down throat. Foreneck grey with buff tinge. Mantle pallid brown with blackish mottling and broad buff V-marks. Back, rump and uppertail-coverts palish brown with lighter and darker mottling, tail whitish with dark brown vermiculations and up to 3 indistinct dark bars. Breast grey, sides of breast as mantle with broad white tips forming white patch. Remaining underparts brownish black. Primaries and greater primary coverts dark brown (inner primaries with irregular buff spotting and paler bases);

secondaries similar, with spotting reduced. Scapulars and tertials as mantle but V-markings bigger. Median coverts mottled brown and buff, with whitish tips forming pale bar. Carpal area dark brown; remaining coverts as mantle, but V-markings smaller. Underwing dark grey-brown, some indistinct buff spotting on primaries. Axillaries brownish black. Upper mandible dark greyish, lower mandible paler; eye pale brown; legs and feet dull yellowish green. ADULT ♀: forehead and crown dark brown, flecked buff, with vestigial nuchal crest. Face buff, lightly mottled brown. Chin and throat buffish white. Neck buff, finely vermiculated with brown; rest of upperparts as ♂. Breast buff irregularly barred brown, merging into broad buffish white band on upper belly; rest of undersides brownish black. Wings as ♂, but tips of median coverts buffish, and underwing-coverts paler. SIZE: (12 ♂♂, 12 ♀♀) wing, ♂ 252–279 (264), ♀ 227–279 (255); tail, ♂ 125–144 (136), ♀ 111–139 (130); bill, ♂ 32–37 (33.8), ♀ 28·5–35·5 (32·7); tarsus, ♂ 77–86 (81), ♀ 58–83 (76). WEIGHT: ♂ (n = 3, Namibia) 550–770 (680).

IMMATURE: undescribed.
DOWNY YOUNG: undescribed.

E. r. gindiana Oustalet: eastern Africa. ♂ like ♂ *ruficrista* but crown less bluish, crest more buffy, no subocular streak, black throat stripe extends down neck to belly, more whitish on sides of breast, upperparts darker and less contrasting, tail unbarred, buff spots in wings increased in size and number forming extensive panel (visible on underwing), buff on median coverts suffused with grey. ♀ as ♀ *ruficrista* but pale indistinct buff line down throat and foreneck, breast with whitish ground colour, tail unbarred, wings and back like ♂ *gindiana*.

E. r. savilei (Lynes): W Africa to Sudan. ♂ like ♂ *ruficrista* but smaller: wing, ♂ (n = 9) 216–256 (243); crown suffused pale olive, grey on face and neck much reduced, black throat-stripe broader towards foreneck, black on belly extending onto lower breast and round base of neck, upperparts suffused with light orange-brown (buff V-markings almost absent), buff on median coverts also suffused light orange-brown. ♀ like ♀ *ruficrista* but buff areas all suffused light orange-brown, barring on breast reduced to slight dark flecking, pale area on upper belly more extensive.

Savilei and *gindiana* sometimes treated as specifically distinct (Clancey 1977, Chappuis *et al.* 1979).

Field Characters. Smaller, stockier and shorter-necked than Black-bellied and Hartlaub's Bustards *E. melanogaster* and *hartlaubii*, ♂ with distinctive crest (reduced in ♀). ♂ has buffy face, much less black on chin and foreneck, no white line separating this from greyish neck, race *gindiana* with white band separating grey neck from black belly. ♀ has brown streaked and barred neck with no black line, black belly (belly white in ♀♀ of other 2 spp.). In flight both sexes have dark primaries, dingy underwings. ♀ similar to ♀ Black Korhaan *E. afra*; for differences see that species.

Voice. Tape-recorded (*savilei* 217, 268; *gindiana* 10, 17, 30, 33, 74, B, C, F). Usually silent, except for much repeated (day and night) protracted advertisement call, which varies considerably among the eastern, southern and W African populations. In nominate *ruficrista* ♂ begins advertisement call with a slowly accelerating series (up to 75) of loud tongue-clicks followed without pause by series of long ventriloquial whistles ('kyip') rising in volume and interspersed with tongue-clicks to almost screaming 'keeweep' or 'kee-keeweep', then sometimes fading; repeatedly given but sometimes reaches climax with Rocket-flight (see Breeding Habits). ♂ *ruficrista* also gives a croaking 'wak wak wak' increasing in volume and speed to a deeper 'wuka wuka wuka' when ♀ appears, and a faint peeping in courtship display. In *savilei*, ♂ advertisement call consists of a whistled note followed by a series of slightly lower, short, clear whistles (G. J. Morel, pers. comm.). In ♂ *gindiana* this call begins with frog-like notes but ends with notes close to those of *ruficrista* (Chappuis *et al.* 1979). *Savilei* also has 2 other calls, only known from Niger and Nigeria, 1 a series of short, clear accelerating whistles, the other a series of frog-like notes in the same rhythm (G. J. Morel, pers. comm.). In *ruficrista*, ♀ has a low clucking 'qrock' in alarm, and the chick gives a weak long 'pweeuu'.

General Habits. Lives in arid or semi-arid savanna including thin bush, light woodland and scrub, often in long grass or other cover at the edge of clearings and plains; never found in completely open terrain. In Senegambia occurs by dried pools with thicket cover; in Chad in flat scrub with *Aristida* grass and *Acacia raddiana*. Usually solitary or in pairs; gregarious in Ouadi Rime, Chad (Newby 1979). Secretive, often avoiding detection by remaining motionless in or near thick cover. Reluctant flier; flushes silently, flying fast and low for short distance, lands and runs for a further distance.

Sedentary and site-faithful year-round.

Food. Termites, ants, beetles, grasshoppers, *Acacia* and *Brachystegia* seeds, fruits and gum. Stomach contents of 1 bird, Zambia: 4 centipedes, 3 scarabaeid beetles, 3 tenebrionid beetles, beetle fragments, 1 beetle larva, 1 grasshopper, 5 woody seeds (Benson and Irwin 1964). Of 9 stomachs, Kenya and Namibia, 3 contained insects, 3 beetles, 2 fruits, 2 grains, 2 berries and 1 seeds.

Breeding Habits. Mating system unclear: a dispersed lek system may operate, ♂♂ having regular calling sites (*c.* 100 m²), used in successive seasons (W. R. Tarboton, pers. comm.). In all races except *savilei*, ♂ performs Rocket-flights throughout year (even in moult), but chiefly when breeding; presumably territorial, though also given in response to human intrusion. In this display, bird calls increasingly loudly (see Voice), then runs briefly forward, flies vertically to *c.* 30 m, throws itself on back with feet up, breast-feathers fluffed, rocks forward and then drops vertically down, opening wings and breaking fall at last moment with short winnowing wing-beats, gliding away to land. ♂♂ confront each other in upright posture, walking back and forth together, occasionally coming together with heads down, feathers fluffed, then rearing to kick. When ♀ appears, ♂ calls (see Voice) and may fly over her to adopt upright stance, throat puffed, neck-feathers raised, crest erected; may also droop wings, spread tail, and adopt stiff hobbling gait (**A**). With crest erect also circles bush with nearside wing drooping, offside wing raised, with loud tongue-clicks synchronized with limping gait; and rapidly and rhythmically shakes head while standing hunched by a bush.

NEST: on bare ground, generally in shade or cover of large plant or low bush.

EGGS: 1–2. Roundish oval; glossy; olive-buff with darker brown and grey markings. SIZE: (n = 12) 46·2–57·1 × 38·0–45·2 (50·0 × 42·2). WEIGHT: (estimated) 49.

LAYING DATES: Senegambia, Sept; Mali, Sept–Oct, mainly Oct; Chad, June–Aug (wet season); Ethiopia, Mar–June; Somalia, Apr–June; E Africa: Kenya, Mar–June, Aug; Region D, Jan, Nov; Zambia, Feb (gonads active Nov); Zimbabwe, Sept–Feb; South Africa, Oct–Feb; S Mozambique, Nov.

INCUBATION: probably only by ♀.

DEVELOPMENT AND CARE OF YOUNG: by ♀ only, ♂ not in attendance.

BREEDING SUCCESS/SURVIVAL: birds commonly fall prey to larger raptors, e.g. Tawny Eagle *Aquila rapax* and Pale Chanting Goshawk *Melierax canorus* (W. R. Tarboton, pers. comm.).

References
Chappuis, C. *et al.* (1979).
Kemp, A. C. and Tarboton, W. R. (1976).

Plate 11
(Opp. p. 176)

Eupodotis afra (Linnaeus). **Black Korhaan; Black Bustard. Outarde korhaan.**

Otis afra Linnaeus, 1758. Syst. Nat. (10th ed.), 1, p. 155; Cape of Good Hope.

Range and Status. Endemic resident, Namibia, Botswana, South Africa and lowland Lesotho. Common to locally abundant.

Description. *E. a. afraoides* (Smith): NW to NE Cape Province, Orange Free State, Lesotho lowlands, W Transvaal, SE Botswana. ADULT ♂: crown dark brown with indistinct gold vermiculations, surrounded by thin white line. Rest of head (including vestigial nuchal crest) black, except ear-coverts white. Neck black; white collar on lower hindneck. Mantle and back closely barred dark brown and tawny buff; rump and uppertail-coverts closely barred dark brown and off-white. Tail greyish buff with dark brown vermiculations, all except central feathers with 2 thick black bars (1 subterminal) and white and buff on tips. Underparts black except for patch of white on sides of upper breast adjoining hind collar, and some white on thighs. Primaries, secondaries and tertials brown-black, but inner webs of middle primaries white, forming wing-patch, with white spots or patches on outer webs of inner secondaries forming separate wing-bar. Greater primary coverts brown-black, greater coverts brown-black with white tips, white increasing inwards until all white near body. Median coverts white, lesser coverts, scapulars and tertials barred dark brown and tawny buff; white patch on carpal joint. Underwing white except greater underwing-coverts, secondaries, tertials and tips of primaries grey-brown; axillaries black. Bill yellow, darker on culmen; eye orange-brown; legs and feet orange-yellow. ADULT ♀: much less black than ♂. Crown streaked tawny-buff and dark brown with blackish vestigial crest. Face and neck buff, lightly streaked dark brown; chin and throat to under ear-coverts whitish with some dark brown flecks. Upperparts as ♂ but pattern less regular; tail as ♂. Breast finely barred dark brown fading to whitish on upper belly; flanks and lower belly from thigh brown-black, undertail-coverts brown-black barred buff; rest of plumage as in ♂. SIZE: wing, ♂ (n = 47) 262–308 (281), ♀ (n = 23) 251–298 (270); tail, ♂ (n = 46) 118–162 (131), ♀ (n = 23) 110–137 (125); bill, ♂ (n = 47) 27·5–38·0 (31·3), ♀ (n = 23) 27·3–36·0 (29·5); tarsus, ♂ (n = 46) 78–100 (91), ♀ (n = 23) 79·0–95·1 (87·7). WEIGHT: ♂ (n = 26) 536–851 (716), ♀ (n = 76) 500–878 (669).

IMMATURE: like ♀, but with pale tips to feathers in wings, back and crown.

DOWNY YOUNG: pale tan closely mottled and lined with brown.

E. a. afra (Linnaeus): SW Cape Province to Little Namaqualand, S Karoo to Grahamstown. Darker on upperparts and crown, no white in flight-feathers (above or below).

E. a. etoschae Grote: Ovamboland (Namibia) and Makgadikgadi (Botswana). Upperparts with much paler buff than *afraoides*; ♀ has dark markings much reduced.

E. a. kalaharica (Roberts): central and N Botswana (Kalahari north to L. Ngami and east to Matshakana). ♂ has upperparts intermediate between *etoschae* and *afraoides*, more narrowly barred than latter; white on ear-coverts extends round eye; ♀ more finely marked.

Field Characters. A smallish, stocky bustard; ♂ unique with all black neck and underparts, mainly black head with contrasting white face-patch; strongly barred upperparts separated from neck by white ring, much white in wings in flight. ♀ similar to ♀ Crested Bustard *E. ruficrista* but upperparts blacker and barred, not mottled and streaked, face and chin streaked black, breast strongly barred dark brown and white.

Voice. Tape-recorded (39, 74, F, 262). ♂ gives loud grating 'kr-aaak-a kr-aaak-a ...', persistent and far-carrying, on ground and in flight, day and night, but chiefly when breeding. Also gives a 'kok kok ... kah kah teckok'.

General Habits. Inhabits arid, partially open country, including dry savanna, clearings in thorn scrub and areas with low shrubby patches; also dunes with succulent and other vegetation cover. Largely solitary, rarely in parties. ♂♂ highly conspicuous, noisy and active, at least seasonally; ♀♀ shy, unobtrusive and difficult to flush.
Sedentary.

Food. Invertebrates and vegetable matter.

Breeding Habits. Mating system not known; no evidence of pair-bond; 1 nest found 3 km from nearest ♂. ♂♂ are scattered 300–500 m apart, each displaying intermittently and aerially pursuing each other; ♀♀ rarely seen. ♂ has striking display; he stands on prominent site calling (see Voice), and frequently flies up and circles at *c.* 15 m with exaggerated wing-beats, calling loudly; then descends in fluttering glide with legs dangling, lands and runs off quickly. ♂'s display particularly intense if ♀ flushed; up to 5 ♂♂ display as she flies off. On ground ♂♂ commonly chase each other, with head and neck of pursuer stretched forward, tail laterally fanned (in inverted V).

NEST: scrape on bare ground.

EGGS: 1, occasionally 2. Roundish oval; olive-brown, blotched darker brown and dull purplish. SIZE: (n = 13) 46–63 × 39·8–46·0 (52·6 × 42·8). WEIGHT: (estimated) 52.

LAYING DATES: South Africa, Aug–Mar; Namibia, Oct–Jan.

INCUBATION: by ♀ only.

DEVELOPMENT AND CARE OF YOUNG: ♀ probably raises young independently of ♂.

Reference
Kemp, A. C. and Tarboton, W. R. (1976).

Eupodotis vigorsii (Smith). Karoo Korhaan; Vigors' Bustard. Outarde de Vigors.

Otis vigorsii Smith, 1831. Proc. Comm. Zool. Soc., p. 11

Forms a superspecies with *E. rueppellii*.

Range and Status. Endemic resident (with some movements) in South Africa and S Namibia, from Cape Province, W Orange Free State, north to central Namaqualand; also Lesotho. Generally common.

Description. *E. v. vigorsii* (Smith): SW Cape Province south to Little Karoo, and northeast to Orange Free State and Transvaal. ADULT ♂: face, crown, nape, hindneck, sides of neck, and lower foreneck dull brownish grey with minute black vermiculations; vestigial nuchal crest black. Chin, throat and upper foreneck black with indistinct off-white border. Mantle, back, rump and uppertail-coverts dull brownish grey with minute black vermiculations and reddish tinge; tail similar but with *c.* 3 indistinct narrow dark bars. Breast and flanks dull brownish grey with minute black vermiculations as neck but slightly greyer, merging into off-white belly; undertail-coverts dull brownish grey with minute black vermiculations. Primaries, secondaries, greater primary coverts and alula brown-black with broad dull buff patch in centres; white patch on central primaries. Scapulars and tertials dull brownish grey with minute black vermiculations as mantle but

Plate 11
(Opp. p. 176)

with a few dark irregular blotches. Greater coverts dull brownish grey with minute black vermiculations as mantle but slightly lighter and with dark tips; rest of wing as mantle. Underwing-coverts creamy, separated by black bar along greater coverts from dark brown flight-feathers. Axillaries dark brownish grey with minute black vermiculations as flanks. Bill slaty, yellow at base of lower mandible; eye pale yellow; legs and feet dull yellow. ADULT ♀: as ♂, but black throat patch narrower, smaller and tinged brown; upperparts generally more heavily marked with dark blotches. SIZE: (7 ♂♂, 5 ♀♀) wing, ♂ 318–375 (351), ♀ 316–356 (329); tail, ♂ 141–202 (164), ♀ 139–163 (151); bill, ♂ 34–41 (37·7), ♀ 31·5–37·0 (34·8); tarsus, ♂ 77–96 (88·1), ♀ 78–92 (84·8).

IMMATURE: undescribed.
DOWNY YOUNG: undescribed.

E. v. namaqua (Roberts): NW Cape Province, S Namibia. Upperparts strongly suffused pink to mauve rather than brownish grey; underparts paler, breast more grey than brown.

Field Characters. Medium-sized, rather drab and featureless, greyish brown becoming off-white on belly, pinkish suffusion on upperparts; large black patch (broader and thinly bordered white in ♂) from chin to foreneck, black patch on nape. In flight, upperwing shows white and orange-buff patches on dark primaries, white underwing-coverts contrast with dark flight-feathers. Plumage of both sexes of local race (*barrowii*) of White-bellied Bustard *E. senegalensis* more contrasting and colourful, with reddish buff on neck, breast and wing shoulder, whiter belly, more extensive white on underwing in flight. For differences from similar Rüppell's Korhaan *E. rueppellii*, see that species.

Voice. Tape-recorded (74, F). Noisy, especially in morning when pair utters a croak-like duetting, 'squark' (♂), 'kok' (♀), 'squark' (♂), 'kok' (♀); also written 'kirr-reck arack arack'; this presumably the ventriloquial deep harsh bark, 'waa-u-u' or 'waa-wa-u' of Macdonald (1957). Also vocal at night; voice carries a great distance.

General Habits. Inhabits dry, open, typical karoo country, usually completely treeless but with well spaced shrubs and a stony or gravelly substrate. Occurs in pairs or in small parties. Visits river pools to drink (Macdonald 1957). For data on Kaokoland (Namibia) birds, not distinguished from Rüppell's Korhaan, see under latter.

Movements occur, but pattern not yet clear.

Food. Small invertebrates and their larvae, reptiles; also seeds and other vegetable matter.

Breeding Habits. Mating system unknown. Strongly site faithful and territorial; pair or group defend quite small area (few hundred m radius) for month or more; however, large home range recorded (sightings of an albino up to 8 km apart), within which smaller areas preferred (Quinton 1948). Role of individuals within group unknown; groups may consist of parents and siblings; 1 individual seen settling onto egg in company of 3 other birds (Kemp and Tarboton 1976).

NEST: slight scrape (to remove stones) on bare ground.
EGGS: 1–2. Roundish oval; buffish olive, dappled red-brown and grey. SIZE: (n = 23) 54–69 × 40·0–55·7 (61·3 × 44·0). WEIGHT: (estimated mean) 64.
LAYING DATES: South Africa, Aug–Mar.

References
Macdonald, J. D. (1957).
Quinton, W. F. (1948).
Viljoen, P. J. (1983).

Plate 11
(Opp. p. 176)

Eupodotis rueppellii **(Wahlberg). Rüppell's Korhaan: Rüppell's Bustard. Outarde de Rüppell.**

Otis rüppellii Wahlberg, 1856. Oefv. K. Vet.–Akad. Förh. 13, p. 174; Onanis.

Forms a superspecies with *E. vigorsii*.

Range and Status. Endemic resident, W Namibia and coastal S Angola north to Benguela and plains of Iona National Park; frequent to common.

Description. *E. r. rueppellii* (Wahlberg): Namibia from Windhoek north to SW Angola. ADULT ♂: forehead and crown pinky buff (with minute dark vermiculations) with grey tinge; black vestigial nuchal crest. Face greyish white with faint black flecks along supercilium and moustachial region. Chin and malar region off-white extending under ear-coverts and onto sides of upper neck, joining up below nuchal crest and extending mid-way down hindneck; white on hindneck bordered black. Broad black throat-patch narrowing to black line down foreneck, broadening onto upper breast. Rest of neck buffish grey. Upperparts and tail pinky buff lightly suffused salmon-pink. Pale grey either side of black of upper breast, rest of underparts off-white. Primaries, secondaries, alula and greater primary coverts creamy buff with brown-black tips. Lesser primary coverts white, tipped black. Scapulars pinky buff as mantle, tertials paler; rest of wing as mantle but paler. Underside of primaries brownish black with creamy white on basal half of outer feathers becoming less extensive towards inner wing, forming a large whitish patch; rest of underwing dark brown, extensively mottled pale brown. Bill dark grey, pale yellowish grey at base of lower mandible; eye pale brown; legs and feet pale yellow. ADULT ♀: like ♂ but cheeks appear more mottled, tail often with *c.* 2 faint bars. SIZE: wing, ♂ (n = 18) 312–341 (329), ♀ (n = 13) 293–325 (313); tail, ♂ (n = 20) 132–159 (145), ♀ (n = 7) 135–161 (148); bill, ♂ (n = 9) 35·0–43·2 (37·8), ♀ (n = 7) 33·7–37·5 (35·6); tarsus, ♂ (n = 10) 78·0–89·9 (85·3), ♀ (n = 7) 74·0–85·1 (80·4).

IMMATURE: as adult but head more mottled, more dark markings on back, some bars on tail.
DOWNY YOUNG: undescribed.

E. r. fitzsimonsi (Roberts): Namibia from Maltahöhe north to Windhoek. Like *rueppellii* but darker (browner and more mauve) on upperparts and wings, breast greyer.

Eupodotis rueppellii

Field Characters. The only bustard in its range, but in south adjoins range of larger but very similar Karoo Korhaan *E. vigorsii*; distinguished by white chin and lower face, black of throat extending back to nape patch and down foreneck to centre of breast. In flight, creamy buff wings have black trailing edge and narrow black bars on coverts.

Voice. Tape-recorded (38, 233). Gives croak-like duetting calls, and ♂ warns incubating ♀ with rattling call (Niethammer 1940). Recorded calls virtually identical to *E. v. vigorsii* (Winterbottom 1966); see under that species.

General Habits. Inhabits very arid country, barren plains with thin grass and low shrubs, especially fringing Namib desert; prefers flat, dark, basaltic gravel plains with sparse shrubs and low rainfall. Occurs in pairs or small groups, usually only 3. More reluctant to fly than other South African bustards (except Karoo Korhaan) although readily flies in response to low-flying aircraft (Viljoen 1983.)

Food. Insects including termites, small reptiles, and vegetable matter including seeds.

Breeding Habits. Monogamous, pairs remaining together all year round and bond sustained through incubation; no striking display postures noted (Niethammer 1940).

NEST: on bare ground among small stones.

EGGS: 1–2. Roundish oval; pale pinkish buff, spotted and streaked brown and purplish grey. SIZE: (n = 5) 55·1–61·3 × 38·6–42·2 (57·6 × 40·9). WEIGHT: (estimated mean) 53.

LAYING DATES: probably throughout year (Clancey 1967).

Reference
Viljoen, P. J. (1983).

Eupodotis humilis (Blyth). Little Brown Bustard. Outarde somalienne.

Sypheotides humilis Blyth, 1856. J. As. Soc. Bengal, 24, p. 305; British Somaliland.

Plate 11
(Opp. p. 176)

Range and Status. Endemic resident, N and W-central Somalia and E Ethiopia, apparently having colonized the latter in 1970s (Ash 1977). Common south to 7°N, scarcer as far as 4°N; recent war in Ogaden and prolonged drought in region may have heavily reduced numbers and disrupted breeding.

Description. ADULT ♂: forehead, crown and vestigial nuchal crest light brown with traces of grey, sometimes with black flecking, black tuft on nape beneath nuchal crest. Hindneck and sides of neck grey; foreneck light brown washed grey; lores, supercilium and ear-coverts grey, lightly washed buff; indistinct moustachial stripe creamy white, extending to nape; chin creamy white; throat black with white spots. Mantle, back, rump and uppertail-coverts light orange-brown clouded with pale grey, with minute black vermiculations; tail similar but with 2–3 thin incomplete brown-black bars. Breast light brown washed grey, becoming paler grey (and sometimes more heavily suffused with grey) before merging into creamy white belly; rest of underparts creamy white. Primaries and secondaries brown-black (pale at base of outer webs of primaries), tertials dark brown tipped brown-black; greater primary coverts, greater coverts and alula white with brown-black tips; remaining wing-coverts as mantle but paler;

Eupodotis humilis

scapulars as mantle but with irregular brown-black vermiculations and some brown-black centres to feathers. Underwing silvery white becoming brown-grey at tips of flight feathers; axillaries black. Bill dark brown, pale to dark reddish at base; eye yellow-brown; legs and feet yellow. ADULT ♀: forehead and crown dark brown, mottled straw-buff, with vestigial nuchal crest. Hindneck pale grey-brown with minute brown vermiculations; foreneck creamy buff with brown streaks; face and nape pale buff lightly mottled brown; chin and throat creamy white, extending behind ear-coverts in indistinct line to nuchal crest. Mantle pale brown with regular creamy buff blotching; back, rump and uppertail-coverts light brown with regular dark brown vermiculations. Tail as in ♂ but with up to 4 bars. Breast as mantle, creamy buff blotches growing larger and merging in creamy white belly. Rest of underparts creamy white (sometimes with some dark brown mottling). Wing like ♂ except wing-coverts, which are pale brown with regular creamy buff blotching, and scapulars, which are more finely and densely marked with regular creamy buff arrow-like markings. Bare parts like ♂. SIZE: wing, ♂ (n = 11) 242–260 (253), ♀ (n = 9) 228–253 (246); tail, ♂ (n = 13) 106–141 (115), ♀ (n = 11) 92–134 (109); bill, ♂ (n = 13) 28·0–31·5 (29·6), ♀ (n = 11) 27·5–33·0 (29·5); tarsus, ♂ (n = 13) 56–71 (65), ♀ (n = 11) 59–65 (63).

IMMATURE: undescribed.
DOWNY YOUNG: undescribed.

Field Characters. A small, white-bellied bustard, likely to be confused only with White-bellied Bustard *E. senegalensis*. ♂ rather uniform in appearance except for profusion of black spots on white chin and upper throat, black spot on back of plain head. ♀ has upperparts, breast and crown heavily spotted pale buff; ♀ of sympatric race *canicollis* of White-bellied Bustard much darker, less sandy, with grey hindneck, fine pale vermiculations instead of spots. In flight shows long, narrow pale wing-bar (lacking in White-bellied Bustard) and 2 black spots at carpal joint.

Voice. Not tape-recorded. Call likened to a rattle (Archer and Godman 1937); also rendered 'ka-ki-rak-ka-ki-rak'. Usually given in evening.

General Habits. Inhabits light open thornbush terrain, sometimes also adjacent tussocky plains. Occurs singly or in pairs; fairly tame.

Food. Insects, small molluscs and seeds.

Breeding Habits.
NEST: eggs laid on sandy soil, generally without scrape.
EGGS: 2 (once 3). Clay-buff, streaked umber. SIZE: (n = 12) 43·5–50·6 × 33·7–39·5 (47·0 × 35·8). WEIGHT: (n = 12, estimated) 33.
LAYING DATES: Somalia, Apr–Aug, chiefly May–June.

References
Archer, G. and Godman, E. M. (1937).
Ash, J. S. (1977).

Plate 11
(Opp. p. 176)

Eupodotis caerulescens **(Vieillot). Blue Korhaan; Blue Bustard. Outarde plombée.**

Otis caerulescens Vieillot, 1820. Table Encyc. Orn. 1, p. 334; eastern Karoo.

Range and Status. Endemic resident, South Africa, from E Cape to Orange Free State, interior Natal, and S Transvaal highveld, generally above 1500 m; also Lesotho. Locally common, e.g. 1–1·3 birds/km² in SE Transvaal (in breeding and non-breeding seasons); declined recently in west of range.

Description. ADULT ♂: forehead blackish shading to blue-black over crown, blue-grey on hind crown and vestigial nuchal crest. Neck blue-grey. Broad creamy white supercilium extends to gape; ear-coverts blackish in front of eye merging into blue-grey behind. Chin and moustachial stripe creamy white, latter extending under ear-coverts. Throat and upperneck black. Mantle, scapulars, back and rump buff, densely and finely vermiculated dark brown; uppertail-coverts similar, but light orange-brown replaces buff. Tail light orange-brown with broad brown-black terminal band. Underparts blue-grey, undertail-coverts tinged light orange-brown; some white on thighs. Primaries brown-black with tinges of light orange-brown and blue-grey near base. Secondaries similar but lack light orange-brown. Tertials light orange-brown on outer webs, brown on inner. Greater coverts blue-grey, median and lesser coverts light orange-brown; alula whitish. Underwing-coverts greyish white, flight-feathers grey-brown, whitish at base. Axillaries blue-grey. Bill yellowish; eyes light brown; legs yellowish white. ADULT ♀: similar to ♂, but ear-coverts heavily suffused light orange-brown, creamy white on face tinged buff, and slight buff scaling across upper breast. SIZE: (10 ♂♂, 4 ♀♀) wing, ♂ 315–356 (336), ♀ 325–337 (331); tail, ♂ 135–172 (158), ♀ 145–161 (151); bill, ♂ 24·0–30·5 (28·2), ♀ 26·0–29·5 (28·3); tarsus, ♂ 90–106 (99), ♀ 87–97 (94).

Field Characters. Medium-sized; both sexes easily distinguished by lower neck and entire underparts blue-grey. In flight, conspicuous orange-brown panel on upperwing separated from black primaries by blue-grey band, underwing grey with black tip and white patch in centre, tail orange-brown with blackish terminal band.

Voice. Tape-recorded (74, F). ♂ advertisement call is a clear ringing 'cockow-cockow ...' or 'kuk-pa-wow ...'; or a rasping call like a stylus slipping across record, in tone and volume like Cape Rook *Corvus capensis*. A very noisy species which calls daily throughout the year, most vociferously in early summer. Most calling heard at dawn and dusk; carries 2–3 km (W. R. Tarboton, pers. comm.).

General Habits. Inhabits short sparse highveld grassland, open or with scattered low *Acacia* and termite mounds; prefers closely grazed areas. Visits cultivated fields to forage in winter. Mainly in groups of 2–5, but up to 20 on newly burnt grassland (Maclean *et al*. 1983). Groups roost together at night, sitting in a tight huddle on open ground (W. R. Tarboton, pers. comm.).

Food. Insects and their larvae, scorpions, small lizards and vegetable matter.

Breeding Habits. Mating system and breeding dispersion not known; occurs in parties year-round, with several members giving advertisement call (which is then answered by other parties); suggests group territorialism. In courtship display ♂ approaches ♀ with neck stretched forward, and crown- and throat-feathers raised in ruff. Holding the feathers in this position, ♂ then bobs head rapidly up and down and runs after ♀ (W. R. Tarboton, pers. comm.).

NEST: scrape on ground.

EGGS: 1–3, mostly 2. Oval; pale greenish with streaks. SIZE: (n = 24) 49·5–60·8 × 40·3–46·0 (57·8 × 42·8). WEIGHT: (n = 35, estimated) 58.

LAYING DATES: South Africa, Aug–Feb.

INCUBATION: by ♀ only, though mate or group may visit nest; 2 ♀♀ and 1 ♂ once seen attending a nest (Kemp and Tarboton 1976).

DEVELOPMENT AND CARE OF YOUNG: chicks join adult group on hatching, and may remain with it for up to 2 seasons or more if population densities very high (Maclean *et al*. 1983). Details on fate of young unknown although ♂♂ may stay in family group while ♀♀ disperse (Vernon 1983).

References
Kemp, A. C. and Tarboton, W. R. (1976).
Maclean, G. L. *et al*. (1983).
Vernon, C. (1983).

Eupodotis senegalensis (Vieillot). White-bellied Bustard; White-bellied Korhaan. Outarde du Sénégal.

Plate 11
(Opp. p. 176)

Otis senegalensis Vieillot, 1820. Table Encyc. Meth. Orn. 1, p. 333; Senegal.

Range and Status. Endemic with some local movements. In north, from S Mauritania, Senegambia, Mali and Ivory Coast to Ethiopia, Somalia, and most of E Africa. Distribution patchy in central Africa; chiefly occurs in Gabon, Congo, W Zaïre, central and S Angola and W Zambia (uncommon). In southern Africa occurs chiefly east of 25°E in S Botswana and South Africa to E Cape; also Swaziland and Lesotho. Abundance variable; in Somalia related to altitude, birds commoner at or above 1500 m. Common to abundant W Africa except Gambia where formerly commonest bustard but no recent records (Gore 1981); frequent to common, E. Africa and South Africa.

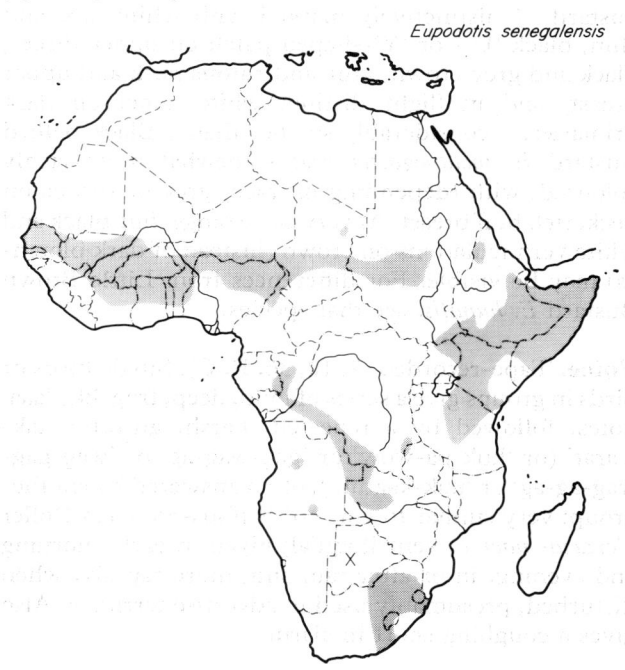

Description. *E. s. senegalensis* (Vieillot): S Mauritania and Senegambia to NW Ethiopia. ADULT ♂: forehead and crown black, centre of hind crown bluish grey, vestigial nuchal crest black. Face, supercilium, chin and upper throat whitish. Black patch on throat extends onto sides of upper neck; whitish patch on lower throat. Neck and upper breast bluish grey, lower hindneck and upper mantle unmarked tawny-buff. Rest of upperparts including scapulars and tertials tawny-buff, minutely vermiculated with brown; tail similar but tinged grey with 2 thin dark bars; scapulars and tertials with coarser vermiculations. Lower breast unmarked tawny buff; rest of underparts whitish, upper belly sometimes with some grey. Primaries and secondaries dark brown with whitish inner webs. Greater primary coverts, greater coverts and alula tawny-buff with dark brown tips; median and lesser coverts tawny-buff. Underwing whitish with darker tips to flight-feathers. Axillaries whitish. Bill yellow, darker on culmen; eye pale brown-grey; legs and feet pale yellow. ADULT ♀: as ♂ but crown dark brown, hind crown finely vermiculated with buff, black on throat absent, foreneck suffused with tawny-buff. SIZE: wing, ♂ (n = 14) 263–287 (276), ♀ (n = 4) 264–276 (269); tail, ♂ (n = 14) 121–145 (129), ♀ (n = 5) 109–132 (120); bill, ♂ (n = 15) 30·0–35·5 (33·0), ♀ (n = 5) 30–33 (31·6); tarsus, ♂ (n = 15) 87–99 (93), ♀ (n = 5) 85–93 (89). WEIGHT: (kg) c. 1·4 kg.

IMMATURE: undescribed.

DOWNY YOUNG: pale sandy with irregular close dark streakings except on belly.

E. s. barrowii (Gray): South Africa, Botswana. ♂ like ♂ *senegalensis*, but tawny-buff extends up back of neck to below nuchal crest, grey extends down centre of breast to white belly, a sometimes indistinct grey-blue spot below eye; back and rump more heavily marked dark brown. ♀ like ♀ *senegalensis* but face and neck much more tawny-buff, back much darker; crown shows dark vermiculations throughout; a small vermiculated patch below eye. Both sexes have obvious dark terminal tail-band.

E. s. canicollis (Reichenow): Ethiopia, Somalia, Uganda, N and E Kenya (south and east to Nakuru, Meru, lower Tana R., Karawa) and NE Tanzania (Arusha, Kilimanjaro and Mpwapwa). ♂ as ♂ *senegalensis*, but larger (wing, 294–326); thick black line from base of upper mandible to below and behind eye, black on throat extending up onto chin; breast all grey; tail, bases of primaries and carpal area suffused grey; axillaries black; belly and tail darker. ♀ as ♀ *senegalensis* but neck and throat with blackish mottling, wings and axillaries as ♂ *canicollis*; upperparts and tail darker than ♂ *canicollis*.

E. s. erlangeri (Reichenow): S Kenya and W Tanzania (west of *canicollis* in Nairobi and Narok to Dodoma, Iringa highlands, Ruaha Nat. Park and Tabora). Like *canicollis*, but darker and less tawny-buff; larger than *senegalensis*; wing, ♂ (n = 8) 299–322 (310).

E. s. mackenziei White: W Zambia, E Angola, S Zaïre. Like *senegalensis* but ♂ with crown mainly grey; ♀ with upperparts intermediate between *senegalensis* and *barrowii*; both with tawny-buff undertail-coverts.

Field Characters. A medium-sized, white-bellied bustard. ♂ distinctively marked with white face and chin, black 'U'- or 'Y'-shaped patch on upper throat, black and grey crown, grey and rufous neck and upper breast, and in flight distinct white panel on dark primaries. ♀ considerably smaller than ♀ Black-bellied Bustard *E. melanogaster* and somewhat more richly coloured, with (depending on race) grey or rufous on neck, rich buff breast. At very close range, fine black and white vermiculations on crown (instead of dark blotches) may be visible. For differences from Little Brown Bustard *E. humilis*, see that species.

Voice. Tape-recorded (2, 10, 36, B, C). Single birds or birds in groups give a series of loud, deep, frog-like 'aaa' notes, followed by a repeated, harsh, guttural 'takwarat' (or 'kuk-pa-wow' or 'gag-wag-ag' or 'wag-gag-wag-ag-ag' or 'kuk-kaatuk'), often answered by another group; very similar to voice of Rufous-crowned Roller *Coracias naevia*; ventriloquial; given in early morning and evening, in or after rain and, more rapidly, when disturbed; presumably used to advertise territory. Also gives a coughing snort in alarm.

General Habits. Inhabits chiefly open grassland and light savanna, commonly occurring in areas of bush or grass near clearings, plains, cultivations, rivers and streams. Visits waterholes (J. F. Reynolds, pers. comm.). Preferred habitat varies in different parts of range. In Chad, adapted to relatively dry conditions, penetrating well into desert in wet season; in Zambia occurs on dry plains and considered the ecological counterpart of the Black Korhaan *E. afra* in Botswana (Benson and Irwin 1967); in South Africa, often keeps to patches of long grass, preferring taller, denser grassland than the Blue Korhaan *E. caerulescens*. Generally occurs in pairs or pairs accompanied by 1 (usual) or 2 (unusual) immatures, presumably offspring of the year (W. R. Tarboton, pers. comm.). Rarely flies far when flushed, but may run through grass on alighting; difficult to relocate.

In Chad moves north in wet season; in South Africa some winter at lower altitudes; other populations resident.

Food. Mainly small invertebrates and vegetable matter including locusts, caterpillars, beetles, spiders, snails, lizards, grass seeds, bulbs, berries and flowers. 1 specimen from Nahud (Kordofan, Sudan) with stomach full of termites; 1 stomach (Senegambia) with 13 scorpions *Buthus* sp. and ants *Messor* sp. (G. J. Morel, pers. comm.).

Breeding Habits. Probably monogamous, with group territorialism (see Voice). In captivity a pair always kept together and called persistently to each other whenever separated; ♀ took food from ♂'s bill. In courtship display, ♂ approaches ♀ with neck stretched forward and crown- and throat-feathers raised in a ruff (W. R. Tarboton, pers. comm.).

NEST: shallow scrape on bare ground; in W Africa, often near or under shelter such as a tree or bush.

EGGS: 1–3, mainly 2. Roundish oval; pale olive, blotched and clouded dark grey and brown. SIZE: (n = 5, *E. s. barrowii*) 50–53 × 41–42 (51·1 × 41·1). WEIGHT: (estimated) 47.

LAYING DATES: Mauritania, July, Sept; Senegambia, Feb, July–Oct; Mali, July–Oct; Nigeria, June, Sept–Oct; Chad, June; Sudan, probably Aug–Sept (2 pullets just able to fly, late Oct); Ethiopia, Mar–June, Dec; Somalia, June; E Africa: Region C, Jan–Feb, Apr–May, Dec; Region D, May, Oct–Nov; Angola, probably Sept–Oct (half-grown young Dec); Zambia, Dec; South Africa, Nov–Feb.

INCUBATION: probably by ♀ only although she may be visited by ♂.

BREEDING SUCCESS/SURVIVAL: burning of long dry grass in acacia veld, South Africa, may seriously affect breeding success (Clancey 1972–1973), but this doubted as breeding occurs outside bushfire periods (W. R. Tarboton, pers. comm.).

References
Johst, E. (1972).
Kemp, A. C. and Tarboton, W. R. (1976).

Eupodotis melanogaster (Rüppell). **Black-bellied Bustard; Black-bellied Korhaan. Outarde à ventre noir.**

Plate 11
(Opp. p. 176)

Otis melanogaster Rüppell, 1835. Neue Wirbelth., Vög., p. 16; Lake Tana.

Forms a superspecies with *E. hartlaubii*.

Range and Status. Endemic resident and partial intra-African migrant. Widely but patchily distributed mainly south of 15°N from Senegambia to Ethiopia and Somalia (very uncommon) south to Angola, Namibia (Caprivi Strip), N Botswana, Zimbabwe, Mozambique, and E South Africa (N Zululand, E and N Transvaal); also Swaziland. Frequent to common although declining locally due to increase in agriculture (E Africa), overgrazing (Zimbabwe) and hunting (Zambia).

Description. *E. m. melanogaster* (Rüppell): from Angola and Zambesi R. north. ADULT ♂: forehead and crown buff, heavily blotched black, less so towards nape. Hindneck and sides of neck buff, finely vermiculated brown. Facial markings variable: lores and chin generally silvery grey, ear-coverts buff tinged with grey; black postorbital supercilium (thinly bordered white above) extends to meet black vestigial nuchal crest. Throat blackish, flecking silvery grey, narrowing to and merging in thin black line down foreneck, latter bordered with white from below ear-coverts to upper breast. Mantle tawny-buff with broad dark brown centres to feathers. Back, rump and uppertail-coverts dark brown speckled with buff. Tail similar but with *c.* 4 regularly spaced dark bars, outer feathers plain dark brown. Underparts black with some white-tipped feathers on sides of breast, a little white on thighs. Outer primary brown-black, others white with brown-black tips; secondaries brown-black, outer ones with white patches on outer webs. Scapulars and tertials tawny-buff with broad dark brown centres to feathers; greater and lesser coverts mainly white, median coverts as mantle but more buff. Underside of flight-feathers as on upperwing, underwing-coverts and axillaries black. Upper mandible dull brown, darker on culmen, lower yellow; eye pale brown; legs and feet dull greyish to yellowish. ADULT ♀: crown, hindneck and upperparts as ♂. Chin and throat off-white, head, neck, breast and flanks and undertail-coverts buff (finely vermiculated brown), belly off-white, tail with no dark outer feathers. Remiges brown-black, upper- and under-surface of inner primaries with off-white spots mid-web, secondaries with buff tips. Axillaries brown-black. Greater and lesser coverts buff with some small dark central spots; scapulars and median coverts like ♂. Underwing-coverts brown-black with white and buff spotting. Bill, eye, legs and feet like ♂. SIZE: (26 ♂♂, 23 ♀♀) wing, ♂ 325–358 (346), ♀ 302–345 (319); tail, ♂ 158–201 (178), ♀ 146–166 (157); bill, ♂ 37–46 (40·5), ♀ 36·5–42·5 (40·1); tarsus, ♂ 120–137 (129), ♀ 119–140 (130). WEIGHT (kg): ♂ 1·8, 2·7, ♀ 1·4.

IMMATURE: like ♀ but wing-feathers edged buff.

DOWNY YOUNG: crown dark brown spotted buff, neck and upper breast buff and black, upperparts brown and black, underparts dull white with black vermiculations on flanks and undertail-coverts.

E. m. notophila (Oberholser): SE Africa, south of Zambesi R. Larger, ♂ (n = 10), wing 357–371 (365); tarsus, 130–157 (140).

Field Characters. A medium-sized black-bellied bustard with slender build, long legs, long, thin neck and rounded head. ♂ distinguished from much smaller and shorter-necked Crested Bustard *E. ruficrista* by lack of crest, dark crown, black line round nape, whiter face, especially ear-coverts and area behind them, more extensive black on chin and upper throat, wider black neck streak; in flight much white in wings, with dark bar along centre of inner wing. ♀ has white belly (black in ♀ Crested Bustard). Browner than very similar Hartlaub's Bustard *E. hartlaubii*, with brown rump, brown tail with distinct dark bars, and less contrasting plumage; for further distinctions see that species. ♀ larger and longer-necked than stockier ♀ White-bellied Bustard *E. senegalensis*, neck uniform dull buff-brown with fine vermiculations; underwing and flight-feathers mainly black with some white barring (underwing and much of flight-feathers white in ♀ White-bellied Bustard).

Voice. Tape-recorded (33, 47, 74, B, C, F). Generally silent. Alarm an intermittent hoarse 'krak'; aggressive ♂ has low growling call; ♂ advertisement call is a short wheezy rising whistle, 'ke-waa-aa-k' ('wheeoo', 'quick'), head retracting sharply onto back for 3–5 s, then a hiccuping or cork-popping 'quok' ('mpok') or 'vook, mmr pick-k-k' (with the 'mmr' guttural and the 'k-k-k' burbling) before head slowly raised back to upright.

General Habits. Inhabits mainly tall grassland with or without trees and bushes; prefers lightly wooded savanna; also uses cultivations, pastures, fallow and old lands, and damper sites like edges of vleis. Generally solitary, occasionally in groups. Up to 2500 m. Often confiding; usually escapes on foot into thicker vegetation. Flight leisurely but powerful, low down (10 m)

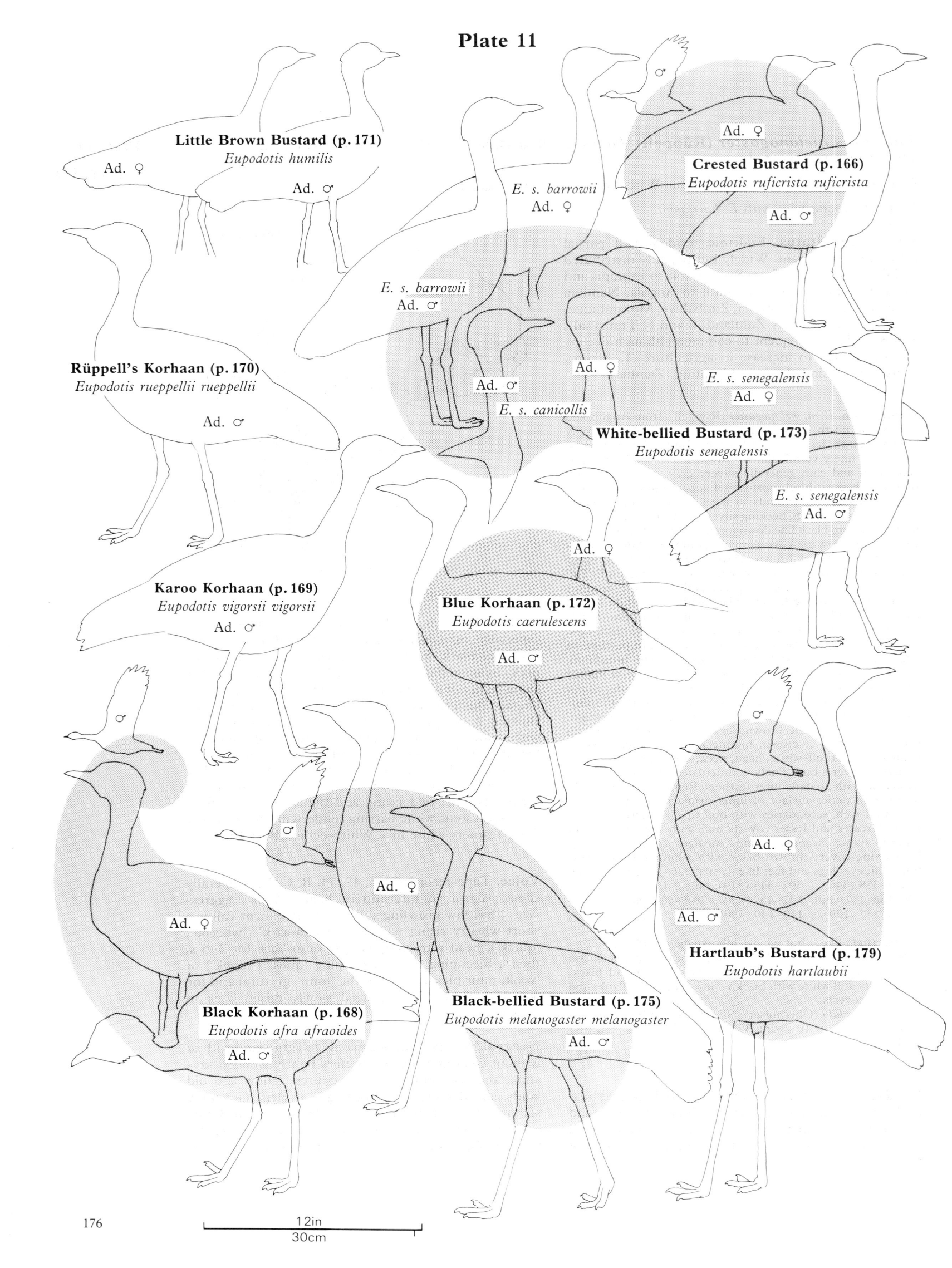

Plate 11

Little Brown Bustard (p. 171)
Eupodotis humilis
Ad. ♀
Ad. ♂

Crested Bustard (p. 166)
Eupodotis ruficrista ruficrista
Ad. ♀
Ad. ♂

E. s. barrowii Ad. ♀
E. s. barrowii Ad. ♂

Rüppell's Korhaan (p. 170)
Eupodotis rueppellii rueppellii
Ad. ♂

E. s. canicollis Ad. ♂
Ad. ♀

White-bellied Bustard (p. 173)
Eupodotis senegalensis
E. s. senegalensis Ad. ♀
E. s. senegalensis Ad. ♂

Karoo Korhaan (p. 169)
Eupodotis vigorsii vigorsii
Ad. ♂

Blue Korhaan (p. 172)
Eupodotis caerulescens
Ad. ♀
Ad. ♂

♂

Hartlaub's Bustard (p. 179)
Eupodotis hartlaubii
Ad. ♀
Ad. ♂

Black Korhaan (p. 168)
Eupodotis afra afraoides
Ad. ♀
Ad. ♂

Black-bellied Bustard (p. 175)
Eupodotis melanogaster melanogaster
Ad. ♀
Ad. ♂

12in / 30cm

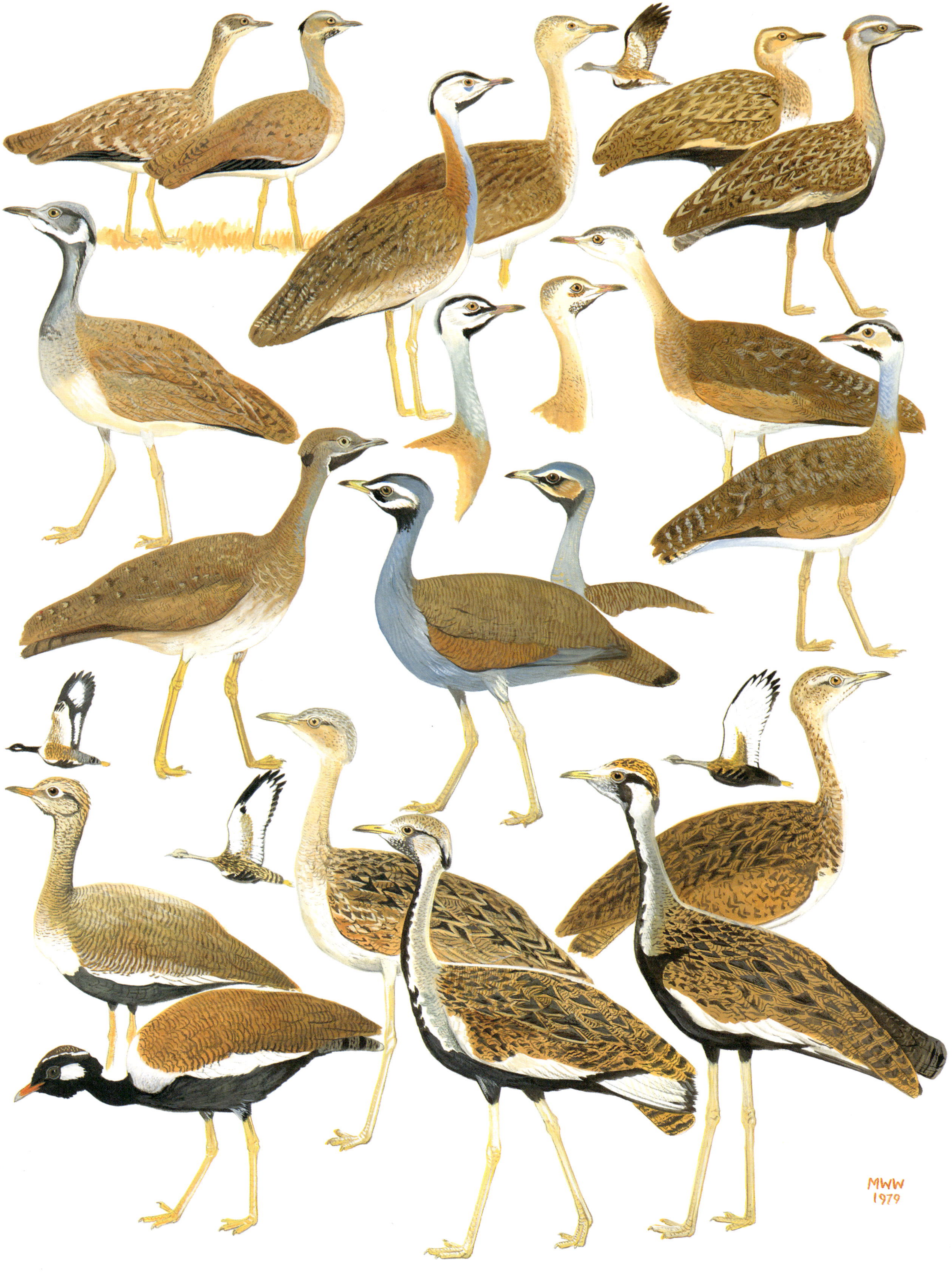
MWW
1979

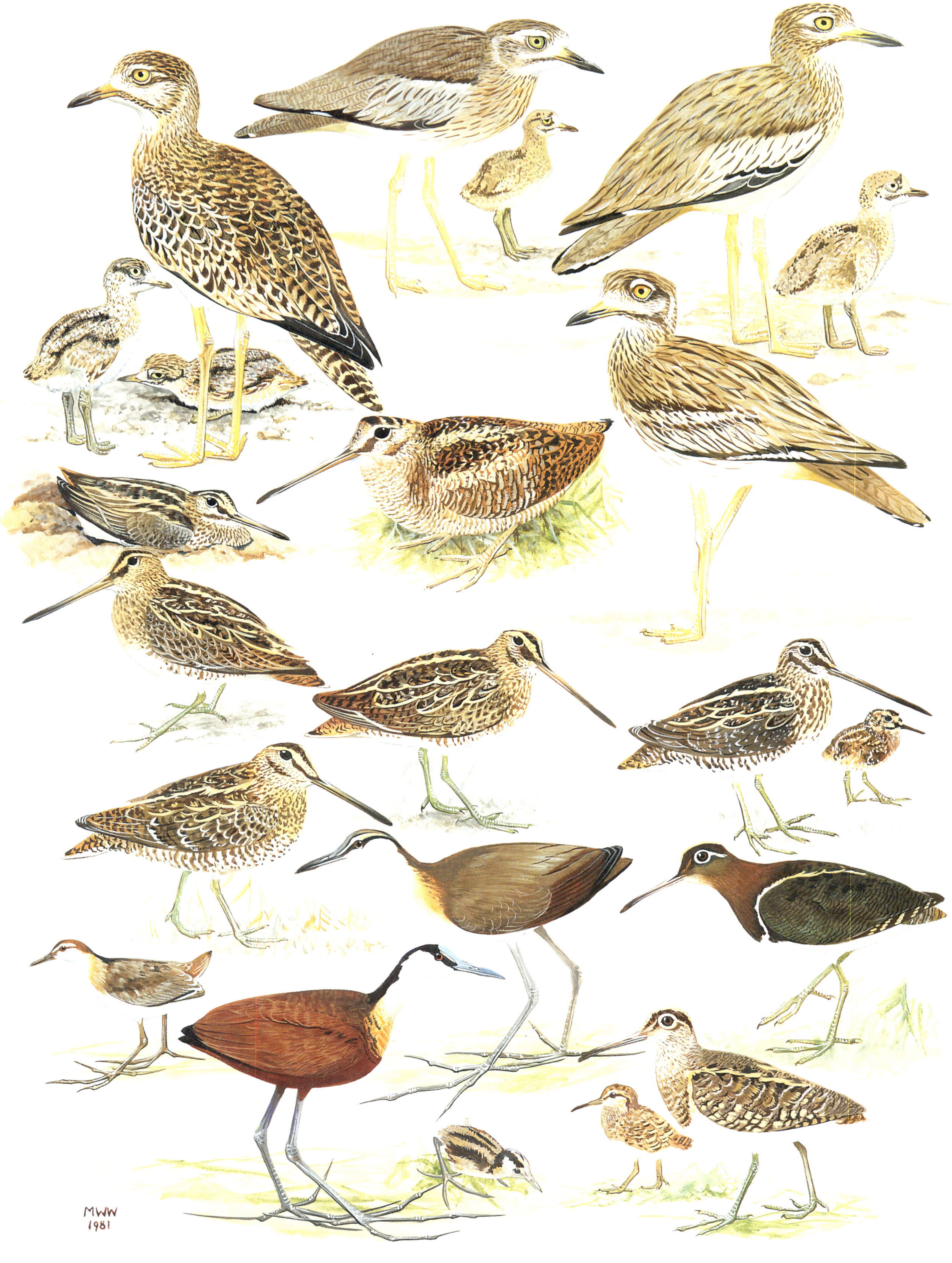
MWW
1981

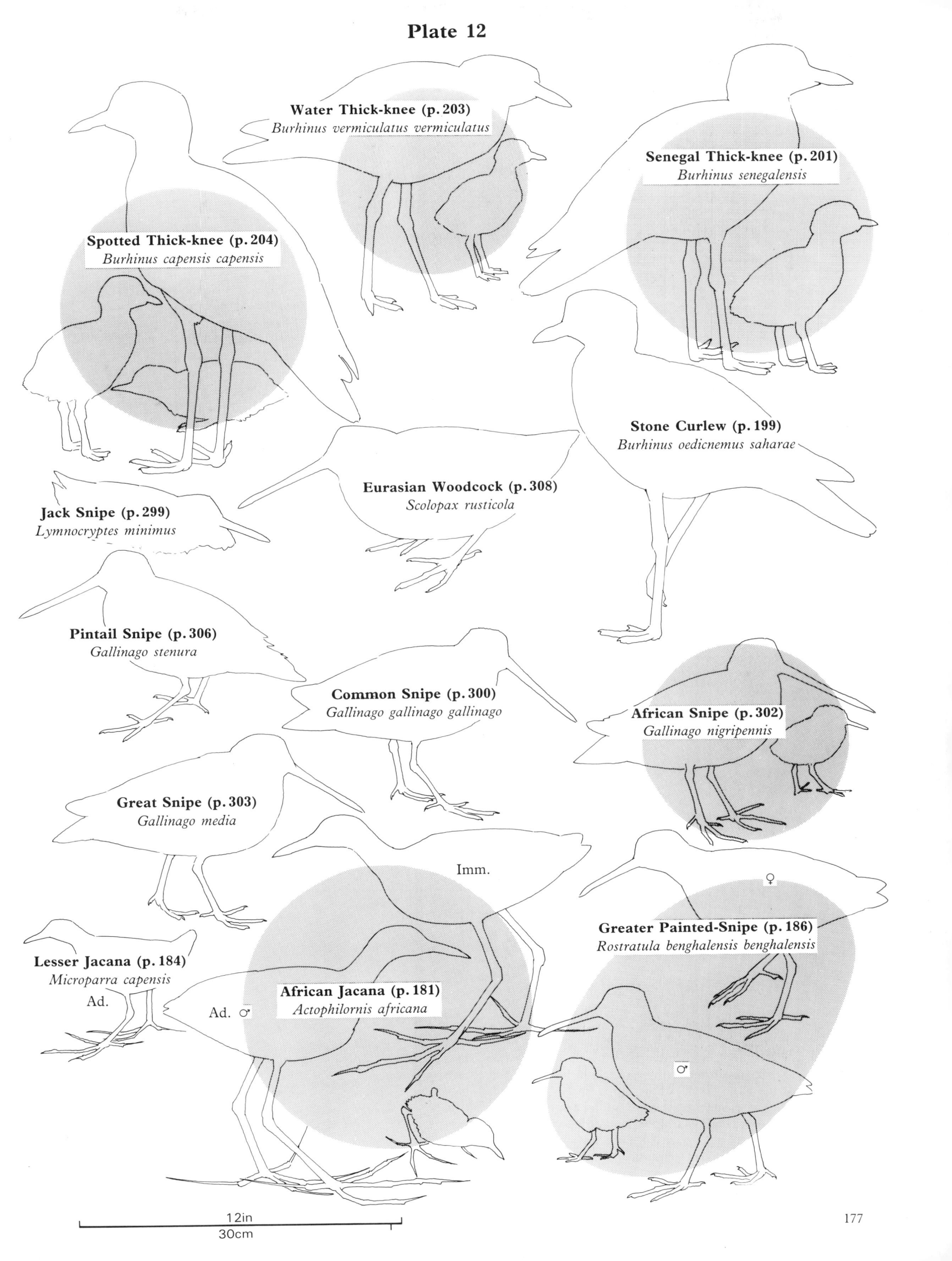

Plate 12

Water Thick-knee (p. 203)
Burhinus vermiculatus vermiculatus

Senegal Thick-knee (p. 201)
Burhinus senegalensis

Spotted Thick-knee (p. 204)
Burhinus capensis capensis

Stone Curlew (p. 199)
Burhinus oedicnemus saharae

Eurasian Woodcock (p. 308)
Scolopax rusticola

Jack Snipe (p. 299)
Lymnocryptes minimus

Pintail Snipe (p. 306)
Gallinago stenura

Common Snipe (p. 300)
Gallinago gallinago gallinago

African Snipe (p. 302)
Gallinago nigripennis

Great Snipe (p. 303)
Gallinago media

Imm.

Lesser Jacana (p. 184)
Microparra capensis
Ad.

African Jacana (p. 181)
Actophilornis africana
Ad. ♂

♀

Greater Painted-Snipe (p. 186)
Rostratula benghalensis benghalensis

♂

12in
30cm

177

178 OTIDIDAE

unless moving between feeding grounds (when up to 165 m). When landing, tends to glide down with wings held in high V.

Largely sedentary, but in Mali and Sudan moves north with rains, June; in Ethiopia numbers increase on high plateau, June; similar movements observed Nigeria. In South Africa, occasionally wanders south and west of breeding range.

Food. Mainly small invertebrates; also vegetable matter including seeds, fruits and palm kernels. Of 8 stomachs, Zaïre, beetles occurred in 7, grasshoppers in 4, bugs and caterpillars in 3, crickets, termites, mantises and centipedes in 2, small seeds in 2 (Chapin 1939). 1 stomach, Tanzania, held 210 beetles, 1 cockroach and 1 large seed.

Breeding Habits. Mating system not known; no clear evidence of monogamy, 1 instance (where 2 chicks attended by 2 ♀♀) of polygamy (Smith 1966); dispersed lek system likely (W. R. Tarboton, pers. comm.). Territorial; nests solitarily. ♂ advertises chiefly by calling (**A**) from a regularly used prominence (see Voice); then takes off and flies about above trees with exaggerated wing-beats (white wings very conspicuous), black chest apparently distended, returning to ground in shallow glide, wings raised in high V. ♂ holds tail well up when aggressive; confronts another ♂ in upright posture; both walk back and forth together, occasionally coming together, head down, feathers fluffed, then rear to kick. 3 birds involved in such disputes, especially when ♀ moving across territorial boundaries. In courtship, ♂ approaches ♀ with small steps, neck upstretched, throat puffed out, neck- and breast-feathers raised to form ruff; when near her, ♂ retracts neck, stretches it horizontally towards her, then raises it to vertical, in repeated circular movement; ♀ crouches, ♂ straddles her, pecking and pulling at head; copulation not always successful (Kemp and Tarboton 1976).

NEST: shallow scrape in bare ground in grass, often near anthill, bush, base of tree, even water.

EGGS: 1–2. Roundish oval; olive-brown, with grey and darker brown blotching. SIZE: (n = 13) 50·0–62·5 × 42·5–53·0 (57·3 × 48·3). WEIGHT: (n = 40, estimated) 77.

LAYING DATES: Senegambia, Aug–Sept; Nigeria, July–Sept (record of still flightless young, Mar, suggests Dec–Jan); Mali, June–Sept; Ethiopia, Apr, Sept; E Africa: Region A, Feb–Apr, June; Region B, Mar–Apr; Region C, Jan, May, Sept (apparently preferring late dry season); Zimbabwe, Oct–Feb; Malawi, Nov–Feb; Zambia, Sept–Mar; Angola, Jan–Mar; South Africa, Oct–Feb.

A

DEVELOPMENT AND CARE OF YOUNG: care of young by ♀ only, ♂ not in attendance.

Reference
Kemp, A. C. and Tarboton, W. R. (1976).

Eupodotis hartlaubii (Heuglin). Hartlaub's Bustard. Outarde de Hartlaub.

Plate 11
(Opp. p. 176)

Otis Hartlaubii Heuglin, 1863. J. Orn., p. 10; Sennar.

Forms a superspecies with *E. melanogaster*.

Range and Status. Endemic resident, with some local movements. E and SE Sudan, Ethiopia, NW and S Somalia to NE Uganda, NW and S Kenya and N Tanzania. Locally common in S Sudan (G. Nikolaus, pers. comm.); generally uncommon elsewhere and rare in Somalia.

Description. ADULT ♂: forehead and crown brown-black spotted with buff, greyish towards nape. Hindneck and sides of neck pepper-and-salt grey. Lores and eye-stripe to behind eye grey; stripe continuous with black (thinly bordered white above) stripe on side of head which extends in thick line to vestigial nuchal crest. Chin and upper throat whitish mottled grey, bordered by thick brown-black line extending from ear-coverts below eye to meet on lower throat where it forms black line down front of neck. Posterior ear-coverts and sides of neck white, extending as thin line bordering dark neck-stripe. Mantle, scapulars and tertials dark brown with thin buff V-markings. Back, rump and uppertail-coverts brown-black with buff peppering. Tail similar but with *c*.4 indistinct black bars, outer feathers all dark. Underparts black with some white-tipped feathers on sides of breast. Outer primary brown-black, others white with brown-black tips; secondaries brown-black with extensive white on outer webs giving a dark bar along centre of inner wing. Greater and lesser coverts mainly white, median coverts as mantle but whitish replacing buff. Underside of flight-feathers as on upperwing, underwing-coverts and axillaries black. Bill yellowish cream, brown on culmen; eye pale hazel; legs and feet dull yellow. ADULT ♀: forehead and crown as ♂, but lacking grey towards nape. Hindneck and sides of neck cream or buff with buff-brown speckles. Face creamy buff with indistinct short dark moustachial streak. Chin and throat creamy white, extending as thin clear line down front of neck, bordered by and merging into creamy buff flecks and short dark streaks. Upperparts as ♂ but slightly lighter. Tail much lighter than in ♂, bars more conspicuous. Breast creamy buff with strong irregular brown barring. Flanks brown-black with large whitish blotches. Belly off-white. Undertail-coverts off-white with irregular dark bars. Outer primary brown-black, others boldly marked with up to 3 white bars; secondaries also with white spotting on outer webs and at tips. Scapulars and tertials as mantle. Greater primary coverts dark brown with white tips, median bar and base; greater, median and lesser coverts as mantle but more buff. Underside of flight-feathers as upperwing, underwing-coverts brown with whitish barring. Axillaries brown-black with white tips. SIZE: wing, ♂ (n = 10) 328–359 (338), ♀ (n = 8) 299–319 (309); tail, ♂ (n = 9) 151–183 (164), ♀ (n = 8) 100–145 (127); bill, ♂ (n = 10) 38·0–45·5 (42·9), ♀ (n = 8) 40·4–45·5 (43·6); tarsus, ♂ (n = 10) 111–131 (124), ♀ (n = 8) 113–128 (120). WEIGHT (kg): ♂ 1·5, 1·6.

IMMATURE: undescribed.

DOWNY YOUNG: ground colour creamy buff with delicate light and dark brown lines on head and neck and mottling on back. Crown dark with light centre, broad throat patch dark with pale brown centre.

Field Characters. Very similar to Black-bellied Bustard *E. melanogaster* but greyer, stockier, plumage more crisply marked. ♂ best distinguished by black lower back, rump and uppertail-coverts, dark tail with only indistinct barring; upperparts black and white with white arrow-shaped marks, more white shows on closed wing; crown with white spots and head pattern different: broad black line behind eye joins narrow black line round nape, broad black line below eye circles down around grey chin. ♀ has similar contrasting plumage on upperparts and crown but lower back to tail grey (brown in ♀ Black-bellied Bustard), tail with black bars; further differs from ♀ Black-bellied by cream line down foreneck, rest of neck streaked or spotted, not vermiculated, breast and upper flanks with blackish spots rather than fine vermiculations.

Voice. Tape-recorded (C, 234, 264). Gives a three-part call that starts with a 'click', is followed 2–3 s later by a 'pop', then followed 1 s later by a deep, drawn out 'boom'. The call, which is repeated several times, is fairly quiet and does not carry far with the first 'click' being heard only at very close quarters (M. Kelsey, pers. comm.).

General Habits. Inhabits drier, more open country than closely related Black-bellied Bustard (e.g. in Ethiopia and Sudan, short-grass *Acacia* savanna rather than tall-grass broad-leaved savanna); also found in thin straggling bush in semi-desert and open grassy plains. Also fond of burnt ground and cultivation. Typically found at low altitudes below 1000 m but up to 1600 m in Somalia. Generally in small groups, sometimes singly. Fairly tame compared to other species.

Mainly sedentary but occurs in Serengeti only Jan–Feb and Sept–Oct.

Food. Orthopterans and other insects; also vegetable matter.

Breeding Habits. Unknown; apparently does not perform display flights (Lynn-Allen 1951).

LAYING DATES: Ethiopia, Apr; E Africa: Region D, Jan, June (breeding in both rains when grass is tallest).

Order CHARADRIIFORMES

A diverse assemblage of small to large mainly wading birds, occurring in marshes, freshwater and marine habitats, and open fields, grasslands and mudflats. Characterized by numerous anatomical features including schizorhinal nostrils, well developed vomers, usually well developed caecae, and absence of crop. Bill long and slender, straight, decurved or recurved. Plumage thick, generally not brightly coloured. Sexes generally alike, although some ♀♀ noticeably brighter in colour than ♂♂ while some ♂♂ spectacularly adorned and unlike duller ♀♀. Some with marked seasonal variation in plumage. Usually 11 primaries with P10 longest and P11 reduced; wing rather long, narrow and pointed. 15–24 secondaries, usually 12 rectrices. All feathers with aftershafts; 15–16 cervical vertebrae; oil gland feathered. Legs generally long, sometimes extremely long; tarsometatarsus unfeathered; anterior toes well developed; hind toe usually small or absent; webs join toes only at base.

Mainly diurnal but some crepuscular or nocturnal. Many migrate long distances. Use bill to probe for or pick up food; eat mainly animal but also vegetable food; some scavenge, eat carrion and pirate for food. Monogamous; some polygamous or polyandrous. Nests on ground, also in burrows, clefts of rocks and on floating leaves; normally with little lining. Usually both parents incubate although in some species only ♂ does. Downy young precocial; mostly nidifugous or semi-nidifugous.

Monophyletic origin with 18 families in 3 suborders, Charadrii (13 families), Lari (4 families) and Alcae (1 family) usually recognized; in Africa 14 families including 9 in Charadrii (Jacanidae, Dromadidae, Rostratulidae, Haematopodidae, Recurvirostridae, Burhinidae, Glareolidae, Charadriidae, Scolopacidae), 4 in Lari (Stercorariidae, Laridae, Sternidae, Rynchopidae), and 1 in Alcae (Alcidae). The 4 families in suborder Lari are sometimes placed in one family Laridae (Wetmore 1960). Classification of order recently reviewed by Cracraft (1981), Sibley and Ahlquist (1972), Storer (1971), Strauch (1978) and Wetmore (1960). Relationship to other orders sometimes controversial but usually thought closest to Gruiformes. Possibly related to Pterocliformes and Columbiformes (Maclean 1967) or Gaviiformes (Storer 1971). May have given rise to Anseriformes (Olson and Feduccia 1980a); Phoenicopteridae may be close to or even belong to the order (being close to Recurvirostridae) (Olson and Feduccia 1980b).

Over 200 spp. world-wide, 149 in Africa.

Suborder CHARADRII

Small to large waders with pervious nares, 15 cervical vertebrae, caecae well developed. Bill as long as or longer than head, usually slender and soft; wing long and pointed in most; tail short. Legs usually long and attached near middle of body; tibia often bare; front toes more or less webbed (palmate or semipalmate); hind toe either small and elevated or absent; oil gland present, always feathered. 3 downy young patterns: (1) pebble-pattern, pale grey above, finely stippled and bordered at sides by dark line down length of back: characteristic of waders nesting in stony or sandy habitats; (2) spotted-pattern, dark above, down densely speckled with pale buff or white tips: characteristic of waders in open tundra; and (3) stripe-pattern, body buff to chestnut above: characteristic of waders in marshes and grasslands. We follow Jehl's (1968) division into superfamilies, 3 in Africa: Jacanoidea (jacanas and painted-snipe), Dromadoidea (crab plovers), and Charadrioidea (oystercatchers, avocets and stilts, thick-knees, coursers and pratincoles, plovers, and scolopacids).

Superfamily JACANOIDEA

Downy young patterned with bold dorsal bands, distinct from striped-pattern; 10 primaries, 15 secondaries, 14–16 rectrices; distinct skeletal and biochemical characters. 2 families, Jacanidae and Rostratulidae.

Family JACANIDAE: jacanas

A distinctive family of extraordinarily long-toed and long-clawed waterbirds, evidently related to waders, particularly painted-snipe (Forbes 1881; Sibley and Ahlquist 1972), but rail-like, and have been treated as gruiform (Lowe 1925). 6 genera, all tropical; 5 are monotypic (1 in Neotropics, 1 in Africa, 3 in SE Asia) while Afrotropical *Actophilornis* has 1 species in Africa and 1 allospecies in Madagascar. There are grounds for treating *Actophilornis* as congeneric with 3 other genera (= *Jacana*, Fry 1983a.) Plumage varied, mainly dark rufous and brown, head and neck black with or without white and gold; most have frontal shield of bright naked skin; all have a small bony knob at carpal joint and very short tail (except pheasant-tailed Asiatic *Hydrophasianus*). All inhabit low green lakeside vegetation and feed mainly while walking over lily leaves. Locomotion deliberate and high-stepping; flight usually low and laboured with legs trailing. Nest a green pad on floating or rooted vegetation; some species (possibly all) polyandrous; eggs beautiful, scrolled (except *Hydrophasianus*) and highly glossed.

Genus *Actophilornis* Oberholser

African and Madagascan jacanas, closely allied, latter with black throat and white hindneck. Leading edge of radius obtusely angled giving forepart of folded wing an angular appearance (Fry 1983b). Naked frontal shield blue-grey, extensive, not frilled. Remex moult synchronous.

Actophilornis africana (Gmelin). African Jacana; Lily-trotter. Jacana à poitrine dorée.

Plate 12
(Opp. p. 177)

Parra africana Gmelin, 1789. Syst. Nat. 1, pt 2, p. 709; Ethiopia.

Range and Status. Endemic, resident. Lakes and marshes throughout sub saharan Africa from sea level up to 2000 m; widespread, abundant or common, but necessarily localized in arid zones bordering Sahara, in Horn and in southwestern Africa; largely absent from equatorial and W African forest zone. Habitat often ephemeral (e.g. flooded land, drying pools), forcing it to disperse far, seasonally or irregularly; *c.* 430 birds on 150 km² Kafue Flats (Zambia); 100–800 on 160 km² Nyl R. flood-plain (Transvaal: W. Tarboton, pers. comm.); varies from none to hundreds on Uaso Nyiro swamp, L. Magadi (Kenya: G. R. Cunningham-van Someren, pers. comm.).

Description. ADULT ♂ (breeding): forehead and forecrown naked, the skin thickened and slightly rugose to form a vivid pale blue frontal shield with straight transverse hind edge well behind level of eyes, up to 22 mm long and 19 mm wide; narrow white supercilium; lores, edge of crown and hind-crown and hindneck glossy black; chin, throat, ear-coverts and sides of neck white, merging through pale yellow to glossy gold on breast; sides of lower neck glossy gold with blackish feather bases showing through. Rest of plumage, above and below, nearly uniform rich chestnut brown (mantle, back and wings glossy); rump, uppertail-coverts and underparts matt chestnut-maroon. Primaries black on both surfaces; highly glossed below; greater under primary coverts chestnut-black; rest of wing chestnut. Bill vivid light blue; eye hazel brown; legs and feet olive brown; nails dark brown. ADULT ♂ (non-breeding): frontal shield dull blue-grey. Sexes alike but ♀ appreciably larger. SIZE: (5 ♂♂, 6 ♀♀): wing, ♂ 144–147 (145·8), ♀ 166–170 (167·5); tail, ♂ 41·5–43 (42), ♀ 46–49 (47); bill to skull, ♂ 31–35 (32), ♀ 37–39 (38); tarsus, ♂ 59–63 (61·2), ♀ 65–72 (68·8); middle toe, ♂ 74–82 (78·7), ♀ 82–91 (89·0); hallux nail, ♂♀ up to 56. WEIGHT: ♂ (n = 14) 115–224 (139), ♀ (n = 7) 167–290 (224) (the heaviest '♂♂' and lightest '♀♀', weighed live, may have been wrongly sexed; ♀♀ may av. over 250 g).

IMMATURE: frontal shield small, grey, *c.* 10 mm long and 10 wide; crown, hindneck and stripe through eye dark brown; supercilium broader than in adult, white in front and buff behind; mantle, back, scapulars, tertials and upperwing-coverts brown, slightly bronzy, with occasional cinnamon feathers (or at least cinnamon undertones), especially on mantle and greater primary coverts; rump brown-cinnamon; tail dark brown; breast white (glossy gold at sides) and remaining underparts white, but flanks rufous (normally concealed beneath folded wing). Hallux nail, 30–45. Acquires adult plumage at 10–12 months by growing cinnamon feathers patchily through brown back and white belly, undertail-coverts being affected last.

DOWNY YOUNG: hatches with sparse down, appearing naked and grey-black with pronounced egg-tooth; when dry, down is thick. Crown rufous with black median line and black lateral lines from eye converging on nape; nape and hindneck black; mantle black in centre with cinnamon line along each side; back and rump cinnamon, with black line along each side meeting above tail and further to each side a broad pale buff line with blackish outer edge (over and behind hip); tail region blackish; flanks and tiny wings tawny buff; forehead, sides of head and entire underparts white; bill pale pink-grey, legs and feet yellowish brown. For 1–2 days after hatching back is tawny cinnamon with the black-bordered pale buff stripe indistinct; foot (tarsus, toes) longer than body and blue-grey; at 24 h tarsus and middle toe 35 mm.

Field Characters. A rich chestnut bird with light blue bill and frontal shield, white foreneck, black hindneck and gold collar, long legs and extraordinarily long toes. Common on low waterside greenery and particularly water-lily fields, walking carefully with deliberate high-stepping gait, often raising wings; noisy; rather shy. Until size of adult, young birds have sandy brown backs, all-white underparts, small grey frontal shield and white eyebrow.

Voice. Tape-recorded (3, 35, 53, 74, C, F). Noisy; calls strident and varied. Mainly a rattling screech on taking off, becoming 'kaaaa kaaka-ka' as it settles; loud scolding and churring notes; piercing shrieks in combat; quieter moans and grating 'kyowrrr, kyowrrr'. ♂, with young threatened by near-by waterfowl, has loud complaining high-frequency trill 'aaaaa'a'a'a'a'a'a'a'a g gh' (rhymes with 'hair'), slightly slowing and falling. Contact call of chick similar but at much greater frequency; sounds very like trill of some bee-eaters, especially Little Green Bee-eater *Merops orientalis*.

General Habits. Inhabits floating and low emergent vegetation at edges of open water, e.g. small and large lakes, dams, sheltered shores and inlets of broad rivers and very large lakes (Chad, Turkana, Victoria); flooded grassland; waterways choked with water-lettuce, water-hyacinth, *Salvinia* or *Elodea*; stagnant marshes with some open water; brackish estuaries; rich fields.

Usually keeps close to overgrown marshy shores where it can take cover, but ♂♂ often venture onto water-lily fields far from shore; sometimes feeds along grassy shores and, if not disturbed, on land, e.g. green pasture 100 m from edge of lake. Perches on nearly submerged hippopotamuses. Pairs or family parties forage more or less dispersed over 50 m of vegetation by moving easily through leaves, walking slowly, pecking at floating leaves and sedge stems; actively flip pieces of greenery over with bill and curl part of large water-lily leaf over to search for small invertebrates on its exposed underside. Leaf often starts to sink beneath bird's weight but it deftly moves on at last moment with a sedate step, sometimes pausing with 1 foot held high, or runs fast to new area a few m away; keeps wings held straight up momentarily before folding them and resuming foraging. Pokes bill into water to catch larvae; picks and pulls at water-lily flowers; occasionally deftly catches flying insect, including bees which are always dipped into water before being swallowed. Territorial encounters are commonplace, with cackling calls and short chases over water-lilies. Flies short distances low over surface, calling and with legs trailing at 45°; flight rather weak; sometimes flies more strongly, 5–10 m high, with legs held straight behind. Occasionally swims; ♂ and young habitually cross stretch of deep open water by swimming. Roosts solitarily or in flocks of up to 50 in sedges and *Papyrus*.

Mainly sedentary, but records at transient pools or in waterless parts of Namibia show that it can move hundreds of km (Winterbottom 1961).

Flight feathers moulted synchronously rendering bird flightless, when it swims and dives to escape detection; swims low in water with neck outstretched on surface and long hindclaw sticking straight up through surface into air (Maclean 1972 and pers. comm.).

Food. Small arthropods (dragonfly nymphs, bees, spiders, crustaceans) and molluscs (probably *Bulinus*, *Biomphalaria*, *Lymnaea*, and gelatinous egg masses); some seeds. Flies (Diptera) commonly taken.

Breeding Habits. Monogamous or polyandrous, sex roles reversed; of 9 breeding groups, 3 = 1♂:1♀; 4 = 2♂♂:1♀; 1 = 3♂♂:1♀; and 1 = 4♂♂:1♀ (W. Tarboton, pers. comm.). Courtship behaviour not elaborate. ♂ solicits ♀ from rudimentary nest platform by pulling at nest material and calling soft 'hkk-hkk' repeatedly; ♀ approaches and ♂ and ♀ walk around each other with heads held low, ♂ then either mounts ♀ or pokes at nest or flies off to pursue rival ♂ (W. Tarboton, pers. comm.). Highly aggressive in breeding season, pair defending territory against intrusion by conspecifics and other species. In polyandrous group each minor ♂ defends small area round nest from other minor ♂♂, but ♀ and major (dominant) ♂ move freely throughout minor ♂♂'s territories. Incubating bird leaves nest and is joined by others to drive off Red-knobbed Coot *Fulica cristata*, Moorhen *Gallinula chloropus*, surface-feeding and diving ducks and Squacco Heron *Ardeola ralloides*, repelling the last after 10 min fight and striking its back in flight.

NEST: flimsy, low, sodden platform of green weed stems drawn together (*Nymphaea*, *Potamogeton*, sedge leaves), floating amongst water-lily leaves or pulled on top of rooted vegetation usually out in open on one or several water-lily leaves, from which nest may become detached if water level rises; or in reedmace bed on waterweeds caught around a few reed stems. Clutch laid in simple depression with a few flattened sedge blades forming perimeter. In Transvaal *Macrophyllum* is most common nest material, then *Polygonum* (n = 150 nests: W. Tarboton, pers. comm.). Sometimes no nest material at all, and eggs simply laid on lily leaf where wetted by wavelets; in Natal nests sometimes placed on small blocks of black peat drifting with water hyacinths backward and forward through 1 km (Miller 1951), without any nest material being added. After a storm, nest and eggs are often moved (once into a higher position on broken-down reeds; once 4 eggs moved to new nest built 4 m away in water-weed bed within 12 h).

EGGS: 2–5, usually 4 (19 out of 28 clutches, Malaŵi; 75 out of 102 clutches, Transvaal: W. Tarboton, pers. comm.). Blunt oval or pyriform; very smooth shell, highly polished, looking varnished or wet; uniform light brown or tan-yellow ground copiously marked with thick dark brown or black vermiform scrolls. SIZE: (n = 100, South Africa) 30·5–37·4 × 21·5–24·8 (33·0 × 23·2).

LAYING DATES: Senegambia all months except Feb, Apr; Ghana (Cape Coast), June–Oct; Nigeria, Apr–Jan, mainly July–Sept; Ethiopia, June–Aug or Sept; E

Africa: Region A, Feb–June, Aug; Region B, Apr–Sept (mainly Aug), Dec; Region C, Mar–May, July, Sept–Nov; Region D, all months (except Mar, Oct), mainly June; Zambia, Nov–June (peak Mar–Apr), Aug; Zimbabwe and Malaŵi, all months (235 records) with peak (53%) Jan–Mar; South Africa, Nov–July; Zaïre, Mar. Breeding season protracted, with peak near end of rainy season.

INCUBATION: by ♂ only, in spells of 15–30 (usually 20) min with breaks of 5 min. Some water-weed added to nest throughout incubation period; eggs once reported to be covered with weeds, and ♀ to visit nest (Phelan 1970). During break ♂ flies well away from nest and feeds; flies back to land some m away then slowly approaches nest foraging (or appearing to) as he moves.

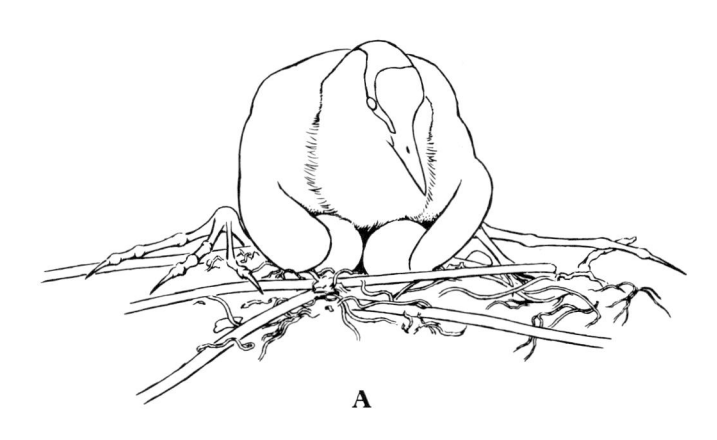

A

Final m of approach is rapid; ♂ straddles clutch in knock-kneed posture with feet wide apart (**A**), fluffs underparts and lowers body carefully onto eggs, quickly settles wrists under sides of breast, and with bill pushes any uncovered eggs under its breast from front and on top of insides of wrists (Fry 1983b); during incubation entire feet are uncovered and lie outside wings. If alarmed, ♂ readily leaves nest, skulking away with head down. If, during break, eggs become hot, ♂ settles on them and immediately rises, crouching or standing up to shade them. When hot, ♂ pants and leaves nest every 10–120 s for 5–10 s excursion close to nest, raising wings vertically each time (Wilson 1974). Despite this, incubating bird is highly cryptic amongst lily leaves showing maroon undersides. ♀ frequently comes near nest, sometimes flicking wings up vertically. Incubation period: 21·5 days (± 1·5 days) to 26 days 9 h (± 9 h). Hatching duration of clutch of 4 variable, 4–24 (?48) h (W. Tarboton, pers. comm.).

DEVELOPMENT AND CARE OF YOUNG: rate of transition from downy to feathered plumage not known (probably 2–3 weeks). Study-skin of ♂ estimated 4 weeks old is 18 cm from bill to tail, bill 30 mm, frontal shield 6 long, tibia *c.* 80, tarsus 57, middle toe 70, hallux nail 30; plumage soft and downy, particularly neck and pale yellow sides of breast, flanks and caudal region; wings cinnamon, feathers tipped whitish, only 38 mm from wrist to end of longest remex. Rectrices 9 mm, soft and downy. Fly when wings *c.* 100–115 long, and ♀♀ (already larger than ♂♂) fly at av. age 35 days (W. Tarboton,

pers. comm.). At 6–7 months still conspicuously smaller than adult; wing 125. Upon hatching eyes are closed and chick is weak for some hours; fairly agile at 24–36 h, haltingly follows parent at 48 h (Cunningham-van Someren and Robinson 1962); or soon becomes active and leaves nest within 2–4 h of hatching (Miller 1951; Tarboton 1976).

As each egg hatches ♂ carries shell away; hatchling lies on side or back kicking feebly, earlier-hatched young being held and brooded between adult's wings and body as he sits or walks around, usually totally concealed except for yellowish toes sometimes dangling down. When adult incubates last egg he takes little notice of latest hatchling squirming a few cm away, and earlier-hatched young (now fluffy, alert and mobile) stay mainly in his wing brood-pouch but from time to time clamber out to peck around edge of nest; they readily hop or clamber back under wing, from in front, until quite concealed (film: G. R. Cunningham-van Someren, pers. comm.). When all hatched, chicks leave nest closely following foraging ♂ who broods them at intervals wherever they happen to be, scooping them

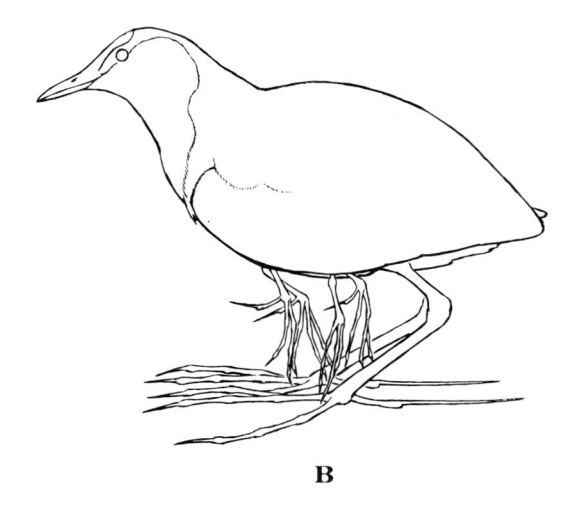

B

under his wings (**B**), walking and foraging some distance, and letting them down onto vegetation. Chicks feed themselves and after a few days are capable lily-walkers in their own right. ♂ leads them from danger and conceals them in vegetation; when he gives alarm call they crouch and 'freeze' or, if caught unaware, dive into water and cling immobile to stem with only bill and face above water. ♂ often gathers 2–4 chicks under wings (1–2 each side) and carries them in crouching posture, their legs dangling, up to 70 m away from danger. He also gives predator-distraction display, jumping up and flying jerkily for few m then collapsing onto vegetation and uttering shrill piping with wings stretched out horizontally, and occasionally fluttering weakly; also walks with weave or stagger, one wing trailing, and collapses on side, pushing himself along by legs with one wing weakly flapping; lasts 5 min (Simpson 1961).

At 12 days chick swims readily, submerged with only head showing; at 3 months a pursued young bird swam for 100 m, only head and threshing legs showing.

184 JACANIDAE

BREEDING SUCCESS/SURVIVAL: in clutches of 4, 1 egg often infertile. Only 16 out of 65 clutches survived to hatching (Transvaal: W. Tarboton, pers. comm.). Chick survival appears to be high.

References
Cunningham-van Someren, G. R. and Robinson, C. (1962).
Hopcraft, J. B. D. (1968).
Steyn, P. (1973).
Postage, A. (1984).
Tarboton, W. (1976).
Wilson, G. (1974).

Genus *Microparra* Cabanis

Monotypic; very small; forehead feathered (lacks naked frontal shield). Radius not angled. Underparts white, plumage remarkably like juvenile African Jacana *Actophilornis africana* of which *Microparra* is probably a neotenic derivative (Fry 1983a). Remex moult ascendant.

Plate 12
(Opp. p. 177)

Microparra capensis (Smith). Lesser Jacana; Lesser Lily-trotter. Jacana nain.

Parra capensis Smith, 1839. Illustr. Zool. S. Afr. Aves, pl. 32; Algoa Bay.

Range and Status. Endemic, resident, localized. From Mali and Burkina Faso to Sudan (Upper Nile Province) and Ethiopia (Rift Valley, rare); in E Africa from Uganda (Murchison Falls National Park) to Rwanda and to Kenyan highlands between 1800 and 3000 m, and from L. Jipe (Kenya/Tanzania border) southwest to Zambia, E Angola, NE Namibia, N Botswana, N and E Zimbabwe, Malawi, Mozambique and (locally in) eastern South Africa to Port Elizabeth. Locally common (e.g. in Mali, Nigeria, Zambia, Okavango delta); probably considerably more widespread in northern tropics than shown in map and almost continuously distributed from Mali to Sudan; reach density of up to 7 birds per 10 ha lake and 19 per 2 km of waterway.

Description. ADULT ♂: plumage soft and silky, rather downy on neck. Forehead golden rufous; crown and nape rich cinnamon or crown shiny violaceous black merging into cinnamon nape (probably an age-related difference); broad white supercilium merging at front with gold forehead; narrow rufous stripe through eye; ear-coverts, chin and throat white; narrow line down hindneck widening on mantle, violaceous black; back and scapulars brown with some violet-black feathers; rump and very short tail rich cinnamon; sides of neck and of breast pale yellow, the feathers thin and with grey bases showing through, edged bright rufous where they conceal bend of folded wing; flanks bright rufous; centre of foreneck, breast and belly white; undertail-coverts white, reaching to or beyond end of tail. Remiges brown-black, outer primaries tipped pale brown, inner ones, secondaries and rather long tertials broadly tipped white; greater coverts pale brown, tipped whitish; other upperwing-coverts brown; axillaries cinnamon, underwing black. Bill pale brownish olive; eye dark brown with narrow yellow orbital ring; legs and feet olive, nails brown. Sexes alike. SIZE (5 ♂♂, 5♀♀): wing, ♂ 82–90 (86·8), ♀ 88–94 (90·6); tail, ♂ 26–31 (28·4), ♀ 27–31 (29·2); bill to feathers, ♂ 14–17 (15·6), ♀ 15–17 (16·2), to skull, ♂ 19–21 (20·0), ♀ 20–22 (20·8); tarsus, ♂ 33–36 (34·4), ♀ 33–36 (34·8); middle toe, ♂ 48–50 (49·3), ♀ 50–55 (52·2); hallux nail, ♂♀ up to 41. WEIGHT: (1 ♀) 41·3.

IMMATURE: like adult but mantle, scapular and back feathers fringed buff; rump and uppertail-coverts black, fringed buff; rectrices dark brown with buff and rufous marks; hallux nail 20–25 mm.

DOWNY YOUNG: like that of African Jacana *Actophilornis africana* (q. v.) (Tarboton and Fry, in press).

Field Characters. Shy; picks its way through water-lily fields like a very small African Jacana *Actophilornis africana*. Rufous hindcrown, flanks and tail, black hindneck, brown back and folded wings, broad white supercilium, narrow stripe through eye, whole underparts white, large straw-coloured area on side of neck and breast. In occasional short flights wings look blackish with white trailing edge and pale brown crescent from wrist to tertials (African Jacana's wings are all blackish brown). Often confused with downy young of African Jacana, which differs in having longer bill and much longer legs, rufous back and wings, smaller white supercilium and stripe through eye inclining to hindcrown (in Lesser Jacana stripe declines to nape.)

Voice. Tape-recorded (233). (a) Quiet, low-pitched 'poop' repeated 5 times per s for 1–5 s; (b) peevish 'see sree shrrr'; (c) soft 'tchr tchr tchr'; (d) muffled 'ti' or 'hli' repeated 2–10 times at rate of 8–10 per s.

General Habits. Inhabits low emergent vegetation and water-lily fields in large expanses of permanent water; swampy edges of rivers, pans, dams and (Natal) coastal lagoons; sometimes small ponds. Avoids shorelines with firm substrate.

Shy, hard to approach closer than 100 m, flying away from open water-lily field or running into cover. Flies readily and strongly, seldom higher than 1 m above surface; on landing momentarily holds wings raised before folding them. Forages by picking its way over low growth, mainly water-lilies, also *Pistia* and *Elodea*, bobbing head, pecking and peering with head down; also wades in shallow water picking insects from emergent grass stems or pursuing them vigorously when they fly (James 1948); preening frequent and varied, including scratching neck with foot (**A**); see also Tarboton and Fry (in press).

Sedentary, seasonal variation in abundance more apparent than real, relating to amount of vegetation cover; but sometimes forced to vacate receding waters whereupon vagrants appear on Transvaal highveld (Tarboton 1976).

Food. Small insects and small pieces of water-plants.

Breeding Habits. Pair breeds solitarily; monogamous and territorial.

NEST: a small, floating accumulation of water-weeds, or clutch simply rests on largely submerged wad of sedges and *Polygonum*.

EGGS: 2–4 (Zimbabwe: one clutch of 2, four of 3, one of 4 eggs); highly glossed; oval-pyriform; tan ground, scrolled and heavily blotched black. SIZE: (n = 18) 23·0–26·9 × 17·5–18·5 (24·7 × 17·9).

LAYING DATES: South Africa (Natal), Nov; Botswana, Mar; Zimbabwe, Feb–Oct, mainly Mar–May with peak Apr; Malaŵi, June, Sept, Oct; Zambia, mainly Feb–Mar, also May, June, Aug; Kenya, May.

INCUBATION and DEVELOPMENT AND CARE OF YOUNG: eggs incubated by ♂ and ♀ alternately and equally; mean duration of 9 spells by ♂ was 31 min and 10 by ♀ 33 min. Bird disturbed from eggs gave distraction display by approaching to 2·5 m of observer and repeatedly sinking in water to belly and flapping wings for *c*. 12 s (Masterson 1969). Young accompany parent until at least half grown and are sometimes carried under its wing.

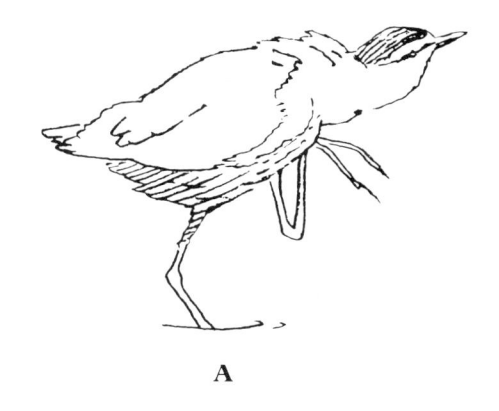

A

References
Fry, C. H. (1983a).
Tarboton, W. and Fry, C. H. (in press).

Family ROSTRATULIDAE: painted-snipe

Medium-sized, plump marsh birds superficially resembling true snipe. Bill long, slender with tip hard, slightly swollen and bent downward. Nostrils in deep, narrow groove extending half way along upper mandible; both mandibles strongly grooved. Eyes large, set well forward for binocular vision. Wings broad, short with extensive yellowish spotting. 10 primaries, *c*. 15 secondaries, 14–16 tail feathers; tail short, round or wedge-shaped. ♀ larger than ♂; sex-role reversal in social behaviour and breeding strategy. Young precocial, nidifugous but initially food-dependent. Striking and complex down pattern of young and 10 (not 11) primaries resemble Jacanidae; some skeletal characters resemble those of rails (Gruiformes, Rallidae).

Two genera, each with one species; *Rostratula* in Old World tropics, *Nycticryphes* in South America.

Genus *Rostratula* Vieillot

Toes not webbed, less decurved bill and tail round (in *Nyticryphes* toes partly webbed between middle and outer toes, bill more decurved, flattened at tip and tail wedge-shaped). Oesophageal crop of ♀ unique in order, with no digestive function; enlarged for use as resonance chamber.

186   ROSTRATULIDAE

Plates 12, 21
(Opp. pp. 177, 352)

Rostratula benghalensis (Linnaeus). Greater Painted-Snipe. Rhynchée peinte.

Rallus benghalensis Linnaeus, 1758. Syst. Nat. (10th ed.), p. 153; Asia.

Range and Status. Africa, Madagascar, and Asia from India to S Manchuria and Japan, Philippines and Australia.

Resident and intra-African migrant, Egypt (Nile Delta) and south of Sahara to Cape except W African forest belt, eastern Somalia, coastal Namibia and western Cape. Mainly uncommon to frequent, occasionally locally common. Resident Egypt; elsewhere with some seasonal movements, possibly at times extensive but poorly documented.

Rostratula benghalensis

Intra-African migrant: see text

Description. *R. b. benghalensis*: only subspecies in Africa. ADULT ♂: similar to ♀ but paler, less brightly coloured; face and upper neck grey-brown and streaked, olive brown neck and upper chest barred; wing-coverts and scapulars noticeably barred. ADULT ♀: crown blackish brown with broad buffy central streak running from forehead to nape. Narrow black circle around eye, surrounded by black-bordered white 'spectacles' that extend to side of nape. Hindneck chestnut with some dark mottling; mantle glossy grey-green, densely and narrowly barred black. Scapulars dark greenish with fine black barring, buff line running posteriorly through scapulars, and white feathers concealed under them. Rest of back, rump and tail grey with white and cinnamon spotting. Lower face, throat, foreneck and upper breast chestnut, shading into black breast band which extends to mantle. Sides of breast black, separated from dark sides of chest by conspicuous white band. Rest of underparts white. Remiges grey-brown, finely vermiculated black and white and spangled with golden buff black-edged spots; upperwing-coverts grey with narrow black bars at base, distally green-buff with golden buff spots enclosed in black lines. Under primary coverts and marginal underwing-coverts grey with black vermiculations; rest of underwing white. Bill greenish yellow or ruddy brown, reddish tip and horn-coloured around nostrils. Eye dark brown; legs and toes light or bright green; toes not webbed. SIZE: wing, ♂ (n = 15) 125–133 (129), ♀ (n = 12) 136–143 (140); tail, ♂ (n = 13) 41–45 (43·2), ♀ (n = 9) 43–51 (45·9); bill, ♂ (n = 15) 46–51 (48·9), ♀ (n = 8) 48–54 (50·6); tarsus, ♂ (n = 16) 42–47 (45·0), ♀ (n = 12) 45–51 (47·9). WEIGHT: ♂♀ (n = 47, Kenya, Sudan) 95–170 (122); ♂♀ (n = 3), South Africa) 113–140 (130); ♂♀ (n = 5, Botswana) 107–131 (117); ♂ (Zambia) 104; ♂ (n = ?, South Africa) 100–118 (115); ♂ (n = 3, Barberspan) 90, 92, 128 (103).

IMMATURE: resembles ♂ but without prominent black band across chest; bill brown. Upperwing-coverts more extensively grey, medium buff at tips; no white feathers under scapulars.

DOWNY YOUNG: ground colour of upperparts pale buff with bold pattern of lines; underparts very pale greyish. Black stripe from base of bill over crown and nape, black stripe through eye, central chestnut stripe on mantle bordered by narrow black stripe.

Field Characters. A strikingly patterned bird with skulking habits. When flushed flies like a rail for a short distance with dangling feet. In flight shows broad rounded wings, rows of large buff spots on remiges, and pale stripe on scapulars. Resembles snipe with long down-curved bill but underparts unbarred. White markings around eye and pale band dividing breast from flanks also diagnostic.

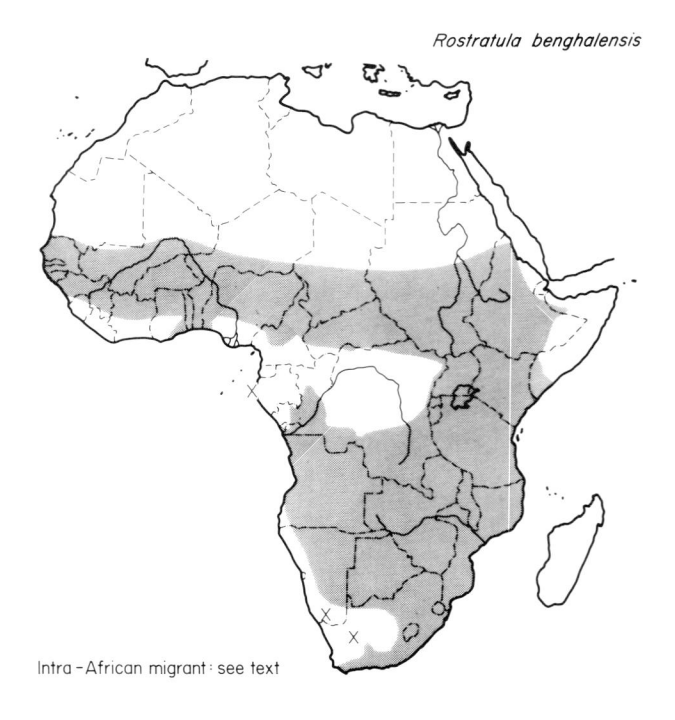

Voice. Tape-recorded (74, 296). ♀'s voice generally deeper and stronger than ♂'s due to longer, convoluted trachea. Both sexes generally silent outside breeding season but give a low 'kek' when flushed. In breeding season (Japan) ♀ calls include: (a) advertising call: 5–6 preliminary low 'vot' notes, followed by up to 50 'kōt' notes in succession; call carries well over 1 km; usually given at night early in breeding season but also by day during laying and incubation; (b) 'booo' call, a soft, mellow sound, when frontal version of spread-wing display given to ♂; (c) 'ko' call, a 'ko-ko- . . . kots-koo', when 2 ♀♀ fighting; and (d) hiss-growl call given in alarm. ♂ calls include; (a) hiss-growl call; (b) squeak call, in response to courting ♀; (c) 'shat' call given in frontal spread-wing display when defending nest; (d) assembly call, a 'jyut-jyut' to summon chicks; and (e) 'jya-jyat' and 'click click' calls used in threat. Young calls include: (a) cheeping calls; and (b) thin, sweet sounding 'pyoh-pyoh'.

General Habits. Inhabits marshes, swamps, thick grass, lake edges and thickly vegetated banks of slow-moving rivers along coastal areas and inland; prefers exposed mud adjacent to cover. Usually singly or in pairs, occasionally in non-breeding season in flocks of 3–15, sometimes 100 (Senegambia, Feb) with ratio of 4–5 ♂♂ to 1 ♀ (although sex ratio of more ♂♂ to ♀♀ may be inaccurate due to similarity of immature ♀ and ♂). Active at twilight but also during moonlit nights and by day. Feeds by probing in soft ground and in shallow water, by scything movement of bill and head. Roosts solitarily in cover day or night; occasionally parties of 3 roost together. When disturbed, 'freezes' for many minutes at a time before flying. When flying, does so for short distance, then alights again. Sometimes bobs hindquarters when foraging or landing; bobs tail like Common Sandpiper *Actitis hypoleucos*.

Undergoes seasonal movements related to dry and wet seasons and availability of suitable habitat, but details poorly recorded in absence of ringing recoveries. Probably undergoes some movements in higher latitudes (but resident Egypt), while resident in lower latitudes of moister vegetational belts. In Nigeria probably moves slowly northward after rains to N Nigeria where remains well into dry season until marshy habitats dry up, then moves south. In Chad, migrates to Sahel zone in wet season to breed in July, then leaves mid-Sept to end Nov. In Senegambia forms flocks up to *c.* 20 Mar–May in few freshwater swamps that remain in dry season. In southern Africa, movements may be extensive but no details. Present Zambia throughout year but mostly Apr–Jan with the peak in any year very much dependent upon water levels.

Food. Invertebrates including grasshoppers, crickets, snails, crustaceans, and earthworms; also seeds.

A

B

Breeding Habits. Sequentially polyandrous mating system with ♀ pairing successfully with at least 2, often 3–4 ♂♂ each year. Solitary nester although nests may be clumped due to polyandrous habit. Bond with each ♂ lasts until eggs laid when ♀ seeks new mate. Occasionally when nesting density low monogamous mating system reported (southern Africa: A. J. Tree, pers. comm.). ♀ defends territory which usually *c.* 200 m in diameter. Main display, performed largely by ♀ but also ♂, is spread-wing display; performed in antagonistic, heterosexual and anti-predator situations, in 2 intensities; (1) lower intensity (**A**) in which stands sideways either with near wing extended, far wing extended and raised vertically, and tail fanned and depressed, or near wing folded, and far wing extended but tail not fanned; and (2) higher intensity (frontal version) (**B**) in which faces opponent or prospective mate with both wings extended and brought forward, tail fanned and raised, body tilted forward, breast lowered and bill pointed down. In courtship ♀ allopreens ♂, then adopts frontal version of spread-wing display and moves around him, given 'booo' call (see Voice); display lasts *c.* 2 min. If ♂ does not respond, ♀ flies over him, feet dangling, lands and courts again. After copulation, pair stand side-by-side and call several times. Sometimes ♀ may open and lift both wings straight upwards.

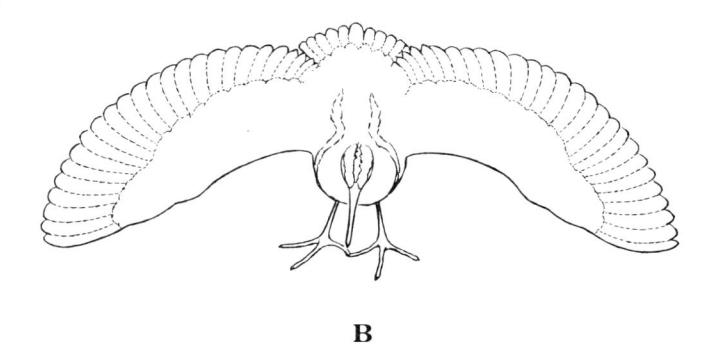

♂ probably builds nest (except in monogamous system when ♀ may help). From a few days before the laying of the 1st egg until the day the 2nd egg laid, ♂ and ♀ stay within 5 m of each other more than 90% of time (Japan). Thereafter time ♀ spends at nest decreases until she does not visit nest except to lay the 4th (last) egg. At this time she seeks new mate.

NEST: on ground in thick marshy vegetation, and well concealed; usually a shallow cup lined with stems and leaves.

EGGS: normally 4 (2–5 recorded); laid at intervals of 24 h. Oval, slightly glossy, light buff-yellow, well marked with black to black brown blotches, spots and thin lines.
SIZE: (n = 140) 33–39 × 22–28 (36 × 26). WEIGHT (n = 140, estimated): 13.

LAYING DATES: Egypt, Apr–May; Senegambia, Jan; Nigeria, Mar–June; Mali, Apr–June; Chad, July, Sept; Ethiopia, Jan, Apr–July, Sept; East Africa: Kenya, Aug–Sept; Uganda, June–July, Region B, June; Region D, Apr–May, July, Sept (in both regions nests chiefly after heaviest rains); Zaïre, Apr, June; Malaŵi, Feb, May–Aug; Zambia, Dec–Apr; Zimbabwe, Jan–Mar, July–Oct, Dec with most Mar; Angola, July–Aug; Namibia, Jan; South Africa: Cape, Aug–Nov, Transvaal, Aug–Apr.

INCUBATION: only by ♂; starts after clutch complete. Period: 15–17 days (Nigeria), 19 days (southern Africa), 16–18 days (Japan). Usually ♂ leaves nest 1.3 times/h for *c.* 10 min (Japan); sometimes he remains on nest for several h, especially during morning; occasionally he stays from nest for as long as 40 min; of 33 nests, on nest 50% of time during day, 89% during night. When leaving nest, ♂ simply flies away. When he returns, he flies to a place *c.* 20 m from nest and then walks to it. When defending nest, may first either creep away cautiously or may perform frontal version of spread-wing display, or fly clumsily as if to attract attention.

DEVELOPMENT AND CARE OF YOUNG: ♂ incubates and rears young to fledging; remains with them for c. 1–2 months after fledging. ♀ not involved in rearing young but may remain in vicinity of brood, perhaps in defence of territory. In monogamous system, ♀ appears to remain with family. ♂ and chicks leave nest within half-day after all chicks hatch; never return to nest. When predator near, young chicks crouch as a group, very close together. Fledging period unknown. ♂ breeds when 1 year old, ♀ when 2 years old.

References
Cramp, S. and Simmons, K. E. L. (1983).
Elgood, J. H. and Donald, R. G. (1962).
Komeda, S. (1983).

Superfamily DROMADOIDEA

Downy young without conspicuous pattern; young remain in nest burrow and are food-dependent; tarsus scutellate. 1 family, Dromadidae.

Family DROMADIDAE: crab plover

A family with 1 genus and 1 monotypic species. Length 35 cm; heavy, black, laterally compressed bill; plumage black and white; sexes alike. Purely sea-coastal, of limited distribution in western Indian Ocean. Affinities uncertain; grey, unpatterned down suggests relationship with gulls, Laridae (Ticehurst 1926), and this and burrow nesting habit suggest affinity with auks, Alcidae (S. Olson, pers. comm.).

Genus *Dromas* Paykull

Plates 13, 20
(Opp. pp. 240, 321)

***Dromas ardeola* Paykull. Crab Plover. Drome ardéole.**

Dromas Ardeola Paykull, 1805. Kongl. Vet.-Akad. Nya Handl. Stockholm 26, pp. 182, 188, pl. 8; Ostindien = India.

Range and Status. Breeds Africa, Persian Gulf, southern coasts Arabian Peninsula; occurs also western Madagascar and most islands northwards to Seychelles, Pakistan and W India. Endemic to western Indian Ocean.

Resident and migrant from other localities of western Indian Ocean. Confined to marine coastlines. Only definitely known to breed in Africa in northern Somalia, in colonies on off-shore islets of Zeyla, Saad Din and Aibat where very abundant Mar–Oct. Occupied breeding colony suspected, Suakim Archipelago, Red Sea, Sudan, June (Moore and Balzarotti 1977, 1983); breeding also possibly Dahlak Archipelago, Ethiopia (Urban and Boswall 1969). Prolonged parental care evident on wintering grounds has led to erroneous reports of local breeding, as in Kenya. Abundant Sept–Apr coasts Tanzania and Kenya; uncommon north to Sudan, south to Natal. Exploitation of breeding colonies by man possible, but no information.

Description. ADULT ♂ (breeding): head, neck, scapulars, back, rump, uppertail-coverts, underparts including underwing and axillaries white. Mantle black, lower feathers elongated and covering back. Inner rectrices pale grey; outer rectrices white on inner webs, pale grey on outer webs; underside of tail wholly white. Tips and outer webs of primaries black, inner webs pale brown; shafts white, dusky at tips. Outer webs of secondaries dark brown, inner pale brown; shafts white, dusky at tips. Tertials pale grey. Greater primary coverts, alula and greater coverts black with concealed inner webs pale grey; median and lesser coverts white. Bill black, paler at base; eye dark brown, anterior and posterior edges of eyelids black; legs slate blue, feet darker. ADULT ♂ (non-

Dromas ardeola

General Habits. Inhabits marine coastal areas, almost entirely tropical, not extending inland more than c. 1 km.

Feeds in inter-tidal zone, usually gregariously, flocks restless and noisy; in Somalia often hundreds together, activity perhaps mainly nocturnal. In non-breeding areas essentially a surface feeder, usually in shallow water, occasionally probing; normal distance between individuals c. 5 m. When prey seen, stabs forward. Traditional sites (e.g. sand-spits, Aldabra) used for roosting, birds congregating from c. 20 km. Sometimes occur in feeding flocks of over 1000 (Providence Group, Aldabra: C. Feare, pers. comm.).

Disperses from breeding grounds in northern Somalia to Aden and southward, regularly to southern Tanzania (Aug–Apr) where some overwinter. Probably migrates in large flocks (over 400, Aldabra, Feb), flying in tight formations, low and fast over water.

Food. Mainly crabs, shells of which are easily broken in powerful bill, downy young being fed on the pulp; also other crustaceans, some molluscs, marine worms and other invertebrates.

breeding): as breeding bird, but crown, nape and hindneck streaked grey. Sexes alike. SIZE: wing, ♂ (n = 9) 202–220 (212), ♀ (n = 12) 206–222 (214); tail, ♂ (n = 9) 65–72 (68), ♀ (n = 12) 62–74 (70); bill from skull, ♂ (n = 9) 63–69 (67), ♀ (n = 12) 61–71 (65); tarsus, ♂ (n = 9) 90–100 (96), ♀ (n = 12) 89–101 (94). WEIGHT: ♂ (n = 2, both immatures) 248, 275; (n = 1) 325.

IMMATURE: 1st plumage—forehead white; crown down to eye and hindneck grey with dark grey shaft streaks. Mantle grey; sides of mantle and scapulars pale grey, tinged brown. Back, rump, uppertail-coverts and underparts white. Grey on rectrices darker than in adult; longest tertials darker, tinged brown. 2nd plumage—mantle darker; lacks brown tinge to sides of mantle and scapulars.

DOWNY YOUNG: down long; upperparts pale grey, darker on lores and around eyes, nape almost white; wings slightly darker grey. Bill horn black; legs light blue-grey, feet darker.

Field Characters. A large, long-legged, rather heavy, white shorebird with build of large plover or thick-knee. Mantle black; in flight, black flight feathers contrast with white of rest of wing. Heavy black bill distinctive. Flies with stiff wings rather like thick-knee.

Voice. Tape-recorded (B, C, 296). Most characteristic call, regardless of age, a constant chatter, with a far-carrying barking 'ha-how' or 'crow-ow-ow'. On breeding grounds a constant chattering 'tchuck-tchuck', continuing well into night, audible c. 1·5 km away; also a shrill 'tchuck-tchuck chuck-chuck-chuck' when returning to breeding burrows. Adults with dependent young give sharp 'keperEP'. Begging young on wintering grounds produce a high, plaintive, twittering whistle, rising in pitch, 'whehehehe . . .'. Downy young have notes recalling those of domestic chickens, and half-grown young soliciting food wheeze constantly, like young *Larus* gulls.

Breeding Habits. Highly colonial, nest-burrows made in coastal sand areas, forming honeycomb effect over an area of 0·8 ha, as in Somalia. Type of mating system uncertain, but monogamous pair-bond assumed; possibly has communal breeding system since groups of up to 10 birds attend nest burrow.

NEST: in chamber at end of tunnel, 120–188 cm long. From entrance, tunnel slopes downwards before rising to chamber; slight hollow excavated for nest, but no lining. Excavation probably by both sexes, using bill and feet. Nest in burrow probably an anti-predator strategy, also protecting egg and chick from heat of sun in open habitat. Same burrows may be used annually, but this requires confirmation.

EGGS: 1, in Persian Gulf rarely 2. Elliptical; smooth or finely pitted; white. SIZE: (n = 12, Somalia) 59·5–66 × 44–47 (63 × 45); very large for size of bird. WEIGHT: (calculated) 45.

LAYING DATES: Somalia May–June (Persian Gulf, last week Apr).

INCUBATION: precise period and role of sexes unknown, although 8 incubating parents taken from burrows in Persian Gulf were all ♀♀.

DEVELOPMENT AND CARE OF YOUNG: young remain in burrows for some time after hatching; food probably brought by both parents for long time after young leave burrow, and even on non-breeding grounds, where parties of 2 adults and 1 begging immature recorded (e.g. Bird Island, Seychelles: Feare 1979).

References
Archer, G. F. and Godman, E. M. (1937).
Cramp, S. and Simmons, K. E. L. (1983).

Superfamily CHARADRIOIDEA

Characters of suborder; variety of downy patterns; 11 primaries with P10 longest, P11 small, 12–22 secondaries, 12 rectrices. 7 families, 6 in Africa.

Family HAEMATOPODIDAE: oystercatchers

Large waders with long, pointed, laterally compressed orange-red bill and black and white or wholly black plumage. Legs reddish, each foot with 3 partly webbed toes; wings long, pointed; tail short. ♀♀ slightly larger than ♂♂, differing especially in length of culmen and weight. Primarily coastal, feeding on intertidal and occasionally terrestrial invertebrates. Have elaborate and highly vocal piping displays, with birds running in pairs or groups,

A

shoulders hunched and bills pointing downwards: in extreme agitation, the wings may be raised above the back during these displays (**A**). Also perform 'butterfly flight', a slow flapping flight usually accompanied by distinctive calls, and 'switchback chases', aerial pursuits by pair with leader alternating whenever birds change direction. Nest a scrape on ground; chicks fed by parents until after fledging. Distributed around much of the world's coasts, but absent from very high latitudes, tropical Africa and southern Asia. 1 genus only, *Haematopus*; 6–11 species worldwide; 1 resident African sp., *H. moquini*, and 1 Palearctic visitor, *H. ostralegus*.

Genus *Haematopus* Linnaeus

Plates 13, 20
(Opp. pp. 240, 321)

Haematopus ostralegus **Linnaeus. Eurasian Oystercatcher. Huîtrier pie.**

Haematopus ostralegus Linnaeus, 1758. Syst. Nat. (10th ed.), p. 152; Europe (restricted type locality Öland Island, Sweden).

Range and Status. Breeds W and central Palearctic; isolated population E USSR; winters W Europe, Africa, Arabia, Persian Gulf, India and China coast.

Palearctic migrant, occasionally oversummering. Occurs around most of African coast except between Cameroon and Namibia (Walvis Bay). Common to abundant Morocco (maximum population *c.* 2500 birds: J. D. R. Vernon, pers. comm.) and Mauritania (midwinter population *c.* 10–11,000 birds), frequent to uncommon Ethiopia and Somalia, uncommon to rare elsewhere. Possible wintering population of several thousand in unsurveyed areas of Guinea-Bissau (W. J. A. Dick, pers. comm.). On eastern seaboard regular on off-shore islands, but at least 7 records inland E Africa and Zaïre.

Description. *H. o. ostralegus* (Linnaeus): N and NW Africa. ADULT ♂ (breeding): upperparts including head and tail glossy black; lower back, rump and uppertail-coverts white. Breast black, rest of underparts white. Bill orange-red; eye red with orange-red orbital ring; legs pink. ADULT ♂ (non-breeding): as breeding but upperparts duller black; neck with irregular white collar. Bill and orbital ring duller orange; bill often slightly dusky towards tip; legs sometimes suffused with grey.

ADULT ♀: as ♂ but bill longer and more pointed. SIZE: wing, ♂ (n = 34) 245–272 (259), ♀ (n = 30) 249–277 (260); tail, ♂ (n = 22)101–118 (107), ♀ (n = 20) 105–119 (110); culmen, ♂ (n = 84) 64–81 (71·4), ♀ (n = 82) 65–87 (79·8); tarsus, ♂ (n = 34) 47–56 (50·6), ♀ (n = 33) 46–54 (49·7). WEIGHT (breeding): ♂ (n = 18) 425–560 (500), ♀ (n = 20) 445–590 (536).

IMMATURE: upperparts brownish black with buff fringes to feathers; broad white neck collar. Bill orange, distal half brown; eye brown, becoming yellowish; legs grey, becoming slate.

H. o. longipes (Buturlin): Ethiopia and Somalia south to South Africa. Upperparts paler: back, scapulars and upperwing-coverts more brownish, less black; bill averaging longer, culmen, ♂ (n = 29) 68·5–81·9 (75·4), ♀ (n = 17) 77·0–90·9 (86·6).

Haematopus ostralegus

Field Characters. A striking large, pied wader. Unique (in Africa) with black upperparts, white underparts, long orange bill and pink or grey legs. White rump and wing-bar conspicuous in flight. Immature browner than adult with bill dull towards tip.

Voice. Tape-recorded (62, 73, 76, C). Usually gives loud, clear, shrill 'klee-eep, klee-eep' or 'kleep-kleep'. When agitated also 'kleep-a, kleep-a' or short, sharp 'pic ... pic'. In flight a lower, quieter 'kip ... kip ... kip ...'. More involved vocalizations, such as given during piping displays, associated with breeding.

General Habits. In Africa generally prefers sandy shores, estuaries and lagoons, but in tropical E Africa appears to favour rocky off-shore islands. Roosts communally at high tide, sometimes with African Black Oystercatcher *H. moquini*.

Scarcity of inland records indicates that it is almost exclusively a coastal migrant. In E Africa most records July–Sept, in South Africa and Namibia, Dec–Mar. Breeding birds from W Palearctic migrate mainly to NW Africa, while those from central Palearctic migrate to NE, E and probably southern Africa. Birds ringed Wales and Netherlands recovered Morocco.

Food. In Mauritania feeds almost exclusively on bivalve mollusc *Arca senilis*. In Europe preys principally on cockles, mussels, limpets, whelks and terrestrial invertebrates.

References
Altenburg, W. *et al.* (1982).
Hockey, P. A. R. and Cooper, J. (1982).

Haematopus moquini Bonaparte. African Black Oystercatcher. Huîtrier de Moquin.

Plates 13, 20
(Opp. pp. 240, 321)

Haematopus moquini Bonaparte, 1856. Comp. Rend. Acad. Sci. Paris 43, p. 1020; Africa (restricted type locality Cape of Good Hope, South Africa).

Range and Status. Endemic to coasts and off-shore islands of southern Africa. Breeds from Namibia (Möwe Bay) to South Africa (Mazeppa Bay, Transkei); in non-breeding season regular as far north as Hoanib Estuary, Namibia, and Bashee River, Transkei. Rare, eastern Transkei and Natal; vagrant Angola (2 records). Records from Senegambia (2 records, 3 birds), Gabon (1), Mozambique (unspecified) and Ethiopia (Dahlak Archipelago) require confirmation. Locally abundant to uncommon. Highest densities—up to 70 birds/km of shore—occur at exposed rocky islands on coast of Namibia and South Africa. Highest densities on mainland occur on sandy beaches, Port Elizabeth area (South Africa). Estimated total population 4780, comprising 1410 on off-shore islands, 3020 on mainland coast and 350 at coastal wetlands. Locally threatened by off-road vehicles and tourist disturbance to mainland breeding areas, heavy inter-tidal exploitation by man for food and bait and introduction of mammalian predators to off-shore islands (Hockey 1984b).

Description. ADULT ♂ (breeding): plumage wholly glossy black. Bill orange-red; fleshy orbital ring orange and swollen; eye red; legs and feet fleshy pink. ADULT ♂ (non-breeding): as

Haematopus moquini

above, but feathers worn and browner after breeding season. Sexes alike, but bill of ♀ longer and more pointed. SIZE: wing, ♂ (n = 64) 265–286 (275), ♀ (n = 64) 265–289 (279); tail, ♂ (n = 13) 104–112 (107), ♀ (n = 8) 101–111 (107); culmen, ♂ (n = 64) 57·7–69·5 (63·2), ♀ (n = 64) 63·6–79·1 (71·6); tarsus, ♂ (n = 64) 50·6–60·8 (56·1), ♀ (n = 64) 52·0–62·0 (57·8). WEIGHT (breeding): ♂ (n = 64) 582–735 (668), ♀ (n = 64) 646–835 (730).

IMMATURE: 1st-year plumage wholly black. Proximal two-thirds of bill orange, distal third brownish; orbital ring burnt orange and much narrower than that of adults; eye reddish brown; legs and feet greyish pink. 2nd-year as adult but with smaller, duller orbital ring and marginally duller distal quarter of bill.

DOWNY YOUNG: grey, with varying amounts of white on belly. Y-shaped black mark on back and dark stripes behind eye and through crown; dark stripe along flanks; legs grey, bill black. Highly cryptic in favoured granite areas.

Field Characters. A large black wader; combination of plain black plumage, pink legs and long orange bill unique in Africa. Immatures black with dull orbital ring, greyish pink legs and distal third of bill brownish.

Voice. Tape-recorded (38, 74, F, 233, 262). Normally loud strident 'kleep', 'kleep-a' or 'klee-eep'. Similar but slower 'kleep-a' given during butterfly displays mainly in pre-breeding and breeding periods. During piping displays, given in territory defence, call much more rapid 'kleepee-kleepee-kleepee-kleepee' varying in volume and speed, often preceded by and interspersed with rapid trilling. In defence of nest and young, call intermediate in speed between butterfly and piping displays, and very loud; also a penetrating 'pic . . . pic . . . pic . . .'. Very soft contact calls given by adults and small chicks. Captive immature birds often give soft chattering calls while resting.

General Habits. Inhabits exposed rocky islands and mixed rocky/sandy shores on mainland, especially sandy beaches where food (*Donax*) abundant.

Forages exclusively intertidally and foraging regime tidally controlled. Opens mussels by jabbing at and severing posterior adductor muscle of gaping individuals. Dislodges limpets normally by a sharp blow to posterior sector of shell and removes flesh with scissoring action (Hockey 1981b). Forages c. 6/24 h (non-breeding, rocky shores): c. 38% of foraging time is at night, with peak activity at low tide; 25–50% of low tide period is spent foraging by day under calm conditions, over 50% during storms.

Roosts communally during non-breeding season with more birds roosting by night than by day. Mean proportion of daylight hours (non-breeding) spent in other activities: resting 37%, preening/bathing 18%, agonistic display 7% (Hockey 1984b).

Before breeding moves to sandy beaches and exposed rocky islands with corresponding decrease in numbers on mainland rocky shores. However, adult population largely sedentary, maximum proven displacement of ringed adult birds from breeding site 15 km.

Food. Mainly gastropods (30 spp. recorded), especially mussels and limpets, notably *Choromytilus meridionalis*, *Perna perna*, *Aulacomya ater*, *Patella granularis* and *P. argenvillei*. Also whelks and winkles, notably *Burnupena* spp., *Oxystele* spp. and *Nucella* spp. Polychaetes eaten, mainly by ♀♀, especially *Pseudonereis variegata* (mussel worm), *Marphysa depressa* and syllids. On sandy shores, mussels, *Donax serra* and *D. sordidus*.

Breeding Habits. Monogamous; solitary nester. Most birds retain feeding territories throughout the year, mate fidelity is high and there are no elaborate courtship displays. Frequency of butterfly displays and switch-back chases (see family description) increases in immediate pre-breeding period.

NEST: a simple scrape in the ground close to shore, usually near rocks; lined to varying extent with shell fragments and stone chips. On hard substrata, edge of nest built up to form cup. Lining added during incubation. Mean internal nest diam. 210 mm, depth 40 mm (n = 95 nests). Modal distance of nests from high water mark at islands 10–30 m, max. 110 m. Nests normally sited adjacent to feeding territories but where nest-sites unavailable may nest up to 350 m away. On islands where breeding densities high, nest-sites as close as 1·5 m. Fresh nest usually made each year, but old nest may be rebuilt where nest-sites limited.

EGGS: 1–2, rarely 3, laid at 2-day intervals; South Africa (SW Cape) mean (n = 46 clutches) 1·74. Slightly pyriform; buffish, washed blue, green or brown and intermediate shades. Spotted and scrolled with black and brown to varying extent. SIZE: (n = 105, South Africa) 55·8–65·2 × 37·9–43·7 (60·7 × 41·0). WEIGHT: (n = 105) 45·0–65·0 (55·8).

LAYING DATES: Oct–Apr, peak Dec–Feb. Breeds slightly later in Namibia than South Africa.

INCUBATION: 1st egg discontinuously incubated, continuously after completion of clutch. Carried out by both sexes. Period: 27–39 (32·1) days.

DEVELOPMENT AND CARE OF YOUNG: newly hatched young weighs c. 40 g (72% of fresh egg weight). Remains in nest area c. 24 h. Bill starts to change colour by 10 days, proximal section becoming brown; by fledging (35–40 days) proximal two-thirds dull orange. Dark areas are first part of body to feather. Remiges and first back feathers appear at c. 14 days, rectrices by 18 days. Tarsus growth complete by 40 days. Culmen fully grown at c. 120 days, wing at 75 days (captive bird). Fledglings weigh c. 450 g. Growth rate: 100 g–11 days, 200 g–17 days, 300 g–23 days, 400 g–32 days. Wing at fledging c. 200 mm, culmen c. 53 mm.

Small chick closely attended by parents and hides under bushes, rocks and in cracks when alarm calls given; freezes in open when no cover available. Fed by both parents above or within intertidal zone depending on exposure and steepness of shore. Swims well and sometimes dives to escape potential predators. 1 adult often guards chick while other adult forages; prey items normally brought complete to chick, then flesh removed, hence formation of characteristic chick 'middens' (piles of emptied shells that collect where the chick is fed).

At islands with high breeding densities all fledged chicks disperse to mainland by mid-July; some disperse in company of parents. Maximum proven dispersal 310 km. Young independent 2–6 months after fledging, all independent by Oct.

BREEDING SUCCESS/SURVIVAL: reproductivity highest on off-shore islands. Estimated 0·3–0·6 young/pair/year on undisturbed islands, much less on mainland. Almost certainly long-lived, as are other oystercatchers. Highest adult mortality occurs during breeding season owing to increased predation. Large mortalities may occur during outbreaks of paralytic shellfish poisoning.

References
Hall, K. R. L. (1959).
Hockey, P. A. R. (1981b, 1983a, b).
Summers, R. W. and Cooper, J. (1977).

Family RECURVIROSTRIDAE: stilts and avocets

Medium-sized pied waders with long, slender bills upturned or straight; legs long (avocets) or extremely long (stilts); hind toe lacking or very small, some webbing between front toes. Gait often brisk with long strides. Cosmopolitan; 3 genera, *Himantopus*, *Recurvirostra* and *Cladorhynchus*, the first 2 each with a single species in Africa.

Possible relationship with flamingoes; Olson and Feduccia (1980b) suggest that flamingoes are derived from this family and not Ciconiiformes (long-legged waders). This proposal based on a number of unique traits flamingoes share with the Australian Banded Stilt *Cladorhynchus leucocephalus* which Olson and Feduccia regard as intermediate between Recurvirostridae and Phoenicopteridae.

Genus *Himantopus* Brisson

Medium-sized shorebirds with extremely long legs (both tibiotarsus and tarsometatarsus long), no hind toe, outer and middle toes with broad web, middle and inner toes with indication of web. Head proportionately small; bill long, straight and slender with long slit-like nostrils, specialized for rapid jaw action and with strong grip. Wings long and pointed, outer primary longest. Tail short, middle and outer pairs of rectrices a little longer than others. ♂ larger than ♀. Flight with fairly quick wing-beats, necks only slightly extended and legs typically extended beyond tail.

We follow Hamilton (1975), Johnsgard (1981) and Snow (1978), and not Mayr and Short (1970), in considering this genus monotypic (1 species with several subspecies).

Himantopus himantopus (Linnaeus). Common Stilt; Black-winged Stilt. Échasse blanche.

Plates 13, 20
(Opp. pp. 240, 321)

Charadrius himantopus Linnaeus, 1758. Syst. Nat. (10th ed.), 1, p. 151; South Europe.

Range and Status. Breeds Africa, N and S America, S Europe, SW Arabia, Madagascar, SW Asia to India, S China, Malay Peninsula, Indonesia, Australia, New Zealand and Hawaii; migratory in northern part of its range, wintering Africa, southern N America and S America.

Resident, Palearctic visitor and possible intra-African migrant. Widely distributed on any area of water from N Africa to Cape. Palearctic visitors winter N Africa and south across Sahara to northern tropics; numbers and southern extent unknown but at least to Senegambia, Chad and N Sudan. N Africa:

resident, uncommon to frequent, Morocco–Egypt (not Libya; very abundant one locality Morocco: several thousand pairs, Iriki 1965–66); Palearctic visitor, common to abundant, especially Morocco (380, Jan, Atlantic coast), Algeria (c. 1000, Apr, Daiet-Tionn 30°05′N, 2°25′W) and Egypt. Some cross Sahara almost anywhere. Western Africa: resident, common only Senegambia and Ghana; Palearctic visitor, common to abundant, coast and inland, Senegambia to Chad; residents and visitors elsewhere frequent to rare. Eastern Africa: locally common to very abundant inland, especially Rift Valley lakes (many hundreds, Kenya: D. Pearson, pers. comm.); locally uncommon on coast. Origin of all N Sudan birds probably Palearctic; most Kenya birds probably Afrotropical (D. Pearson, pers. comm.); origin of birds elsewhere uncertain. Southern Africa: rare to abundant inland and on coast; large numbers South Africa (765, Dec–Jan, southwestern Cape coast).

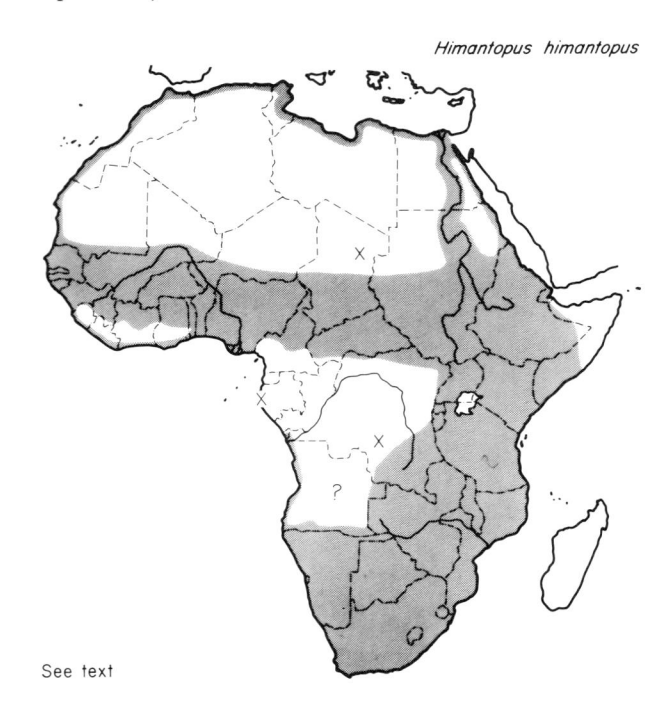

Himantopus himantopus

See text

Description. *H. h. himantopus*: only subspecies in Africa. ADULT ♂ (breeding): crown and nape usually white, sometimes with variable amounts of black. Mantle, scapulars and wings above and below glossy black; tail pale grey-brown, rest of plumage (including axillaries) white. Bill black; eye red; legs and feet vermilion. ADULT ♂ (non-breeding): crown and nape with some grey-brown, legs red; otherwise like breeding ♂. ADULT ♀: like ♂ except mantle and scapulars brown; crown and nape with some dusky tips to white feathers; wing smaller. SIZE: wing ♂ (n = 42) 230–255 (243), ♀ (n = 34) 220–242 (231); tail, ♂♀ (n = 14) 74–91 (81); bill ♂♀ (n = 63) 56–69 (64); tarsus ♂ (n = 43) 107–137 (125), ♀ (n = 32) 100–124 (112). WEIGHT: ♂♀ (n = 20) 112–223 (177); ♂♀ (Kenya, n = 40) 128–238 (D. Pearson, pers. comm.); adults (Zimbabwe, n = 7) 173–194 (A. J. Tree, pers. comm.); ♂ (Botswana, n = 5) 150–198 (170·9), ♀ (Botswana) 154·3, 158.

IMMATURE: juvenile, feathers of upperparts, underparts and underwing-coverts fringed reddish buff; inner primaries and secondaries with broad white tips; legs pale pink. 1st-year winter birds similar to adult ♀ but coverts with some buff fringes.

DOWNY YOUNG: pale buff mottled black above, but neck grey; thin black stripe from bill through lores and behind eye; very pale buff to white below; bill straight; feet slightly webbed with vestigial hind toe; eye pale yellow-grey.

We follow Clancey (1964), Hamilton (1975), Snow (1978), White (1965) and Winterbottom (1962) in considering all stilts in Africa 1 subspecies, *H. h. himantopus*.

Field Characters. Large white shorebird with black back and wings, straight fine black bill and extremely long pinkish red legs which extend far beyond tail in flight. Wings long and triangular in shape in flight. Distinguished from Eurasian Avocet *Recurvirostra avosetta* by straight, not upturned bill, reddish, not blue-grey legs, and lack of white in wings; also by characteristic sharp alarm call.

Voice. Tape-recorded (21, 35, 38, 66, 74). Most common call used in alarm a repeated sharp 'kik-kik-kik-kik' or 'kraak-kraak ... kraak'; sometimes a 'kyip-kyip-kyip' and 'kwit-kwit'. In flight, several may give 'kik-arik' repeated several times. Also gives 'krek' or 'kek' contact call.

General Habits. Inhabits any shallow water, preferring some low vegetation; found in flooded marshes, rivers, freshwater and alkaline lakes, flooded fields, mudflats, coastal lagoons and estuaries but rarely open sea-shore. Usually seen singly or in small parties of 5–10, sometimes in flocks of several hundred to a thousand; often with other waders. Nervous and noisy, alerting other birds to presence of intruder with sharp 'kik-kik' call. Walks deliberately with long strides; occasionally bobs head. Feeds in all habitats from dry land to locally deep water but usually silty margins of water. Frequently wades when feeding, tending to stay in water not deeper than knee; sometimes wades out to belly-deep water. Obtains food by swiftly pecking at surface water, mud or floating vegetation; occasionally immerses head and neck in water to reach food on bottom. Also feeds by probing, moving bill vertically down in a series of short jabs and by sweeping bill from directly in front to a little on one side of the body. Rarely swims. Roosts communally usually at night in small assemblies but sometimes over 1000 and occasionally with other shorebirds; sometimes active at night. Sleeps on one or both legs with head over shoulder or bill tucked behind wings; sometimes rests on tarsus with shank and body upright like stork. Flies with neck only slightly extended, fairly quick wingbeats, and legs held together and extended beyond tail. Flight speed 39–42 km/h (*H. h. mexicanus*: Schnell and Hellack 1978).

Some seasonal movements by Afrotropical breeders likely, e.g. largest numbers Zambia Apr–Oct, with peak July (380: Tree 1969); origin of these birds unknown, but probably from Botswana (A. J. Tree, pers. comm.). Stilt populations south to at least equator (Nigeria to E Africa) increase considerably Oct–Apr possibly due to Palearctic visitors. Most N African residents not sedentary but probably migrate southward after breeding.

Food. A wide variety of insects including flies, water beetles, dragonflies, caterpillars, grasshoppers; also small gastropods, worms, crustaceans, spiders, tadpoles, fish and frog spawn, and occasionally seeds.

Breeding Habits. Seasonally monogamous, often in small colonies, a few pairs to several hundred; occasionally solitary. Courtship in N America and Portugal well documented (Hamilton 1975, Goriup 1982) but detailed accounts lacking for Africa. Early in breeding season 3–4 birds (largely ♀♀) congregate in shallow water, generally face in same direction and make loud calls; 1 or more will then fly 1–5 m over group, flying slowly with legs dangling. Either sex may also assume upright posture with angle of back c. 10° above horizontal and neck and head raised high. They may also assume 'giraffe' posture with back angled c. 50° above horizontal and neck extended up but slightly bent forward; and hunched-run (= crouch-run) posture with neck retracted and back held at slight angle below horizontal. Either sex may perform head-and-legs-down-flight by taking off and hovering 0·5 m or less over another bird with neck extended downwards, tail spread and legs dangling; either mate may also perform butterfly-flight by hovering at height of 5–10 m with neck contracted, tail spread, legs dangling and body held at c. 45° above horizontal. African, but not N American, stilt reported to spring and dance round another, beating wings and jerking bill slightly upwards; eventually staggers around until falls on vegetation with wings half-spread and head raised (Bannerman 1930).

When pairing, ♀ attempts to associate with ♂ in his feeding territory; when he no longer repels her, pair is formed. Before copulating, both approach each other, pecking at ground or water and sometimes also preening breast feathers. When ♀ ready to copulate, stands with head and neck parallel to water. ♂ walks quickly to her side and sometimes around her tail to other side. Each time he reaches her shoulder, he bill-dips and preens his breast or underwing; sometimes he also flicks water on her with his bill. This cycle repeated 2–5 times until he adopts an erect posture prior to mounting. He may then briefly perform giraffe-display, after which he jumps on her back, crouches, opens bill, slightly retracts head, flexes legs so whole length of tarsi rest on ♀'s back, and flaps wings to maintain balance. As ♂ crouches, ♀ swings her head from side to side; as ♂ prepares to copulate, ♀ keeps head inclined to one side, with tail twisted to other side. Copulation brief; as he falls off, ♂'s wing falls across ♀'s back. Each then stands next to the other in upright posture with ♂'s wing over ♀'s back. He next takes wing away, crosses his bill over hers; then both walk c. 1 m either directly forward or in a shallow arc, after which they separate. After a short period of feeding, reunite and perform leaning-ceremony in which both stand side by side in the upright posture, then lean towards, then away from each other by tilting head and neck 3–4 times. They may extend and retract heads rapidly at this time. During courtship, ♂ (and ♀?) may also squat down in a posture resembling incubation. Both sexes build nest, but nest-building behaviour not described.

NEST: a simple scrape on dry ground or, if in damp place, a mound of stems and twigs; occasionally on floating mats of grass or subsurface waterweed (A. J. Tree, pers. comm.). Mound may be surrounded with shallow water and may then be 15–20 cm high, with diameter 30 cm, cavity 15 cm. Lined with a variety of shells, stems, feathers and dry mud. Nest-site close to foraging area; in open area with 360° visibility. In N America 31 nests av. 21·9 m apart, max. 42 m apart (Hamilton 1975); in USSR 51 nests 30 cm to 2·3 m apart (Johnsgard 1981).

EGGS: 3–4, usually 4, once 7 (Every 1974); N Africa mean (n = 23) 3·87; South Africa mean, Natal (n = 13) 3·54; probably laid every 24 h. Pyriform, without gloss, cream-coloured, mottled brown and black. SIZE: (n = 134) 39–49 × 28–33 (44 × 31).

LAYING DATES: Morocco, Apr–July, Sept; Algeria, Apr–May; Tunisia, Mar–Apr; Egypt, Apr–May; Senegambia, May–July; Ghana, Mar, May–July; Ethiopia, Aug–Sept; Somalia, Apr; E Africa: Kenya, May–June; Tanzania, May, Region B, May–June, Region D, Mar–July, mainly late in and after long rains; Zaïre, June; Zambia, May–Oct; Malaŵi, June–July; Mozambique, May, Dec; Zimbabwe, Apr–May, July–Nov with most Aug–Sept; Angola, Aug; Namibia, Jan–Dec; Botswana, Oct; South Africa: SW Cape, June–Nov; Transvaal, May–Dec; Natal, Apr–Sept.

INCUBATION: begins when clutch complete; performed by both sexes. Period 22–27 days. Both parents incubate in fairly short spells. ♀ takes larger share of incubation; appears to do all overnight sitting. At nest-relief, relieving bird moves up behind nest, uttering soft call which cues bird on nest to leave; sometimes off-bird picks up straws and throws them over back as it walks to nest; sitter will do same when departing. Parent returning to nest will soak belly feathers to moisten and cool eggs; bird on nest will raise crown and mantle feathers to help keep cool. When intruder approaches, performs injury-feigning with wings extended from body, tail spread and depressed, neck contracted, and head held close to body; will face intruder and sometimes walk up to it, tipping wings from side to side. Sometimes when performing injury-feigning brings 1 wing to side while extending the other. 1 or both parents will also mob intruder by circling over it at radius of 3–15 m and height of 4–10 m, calling excitedly.

DEVELOPMENT AND CARE OF YOUNG: both parents tend young; normally lead chick away from nest to feeding area within 24 h of hatching. Young will often feed within 20 m of nest until 2 weeks old. Remiges show by 15th day; short distances flown after 5–6 weeks. Fledged when 28–32, sometimes 37 days old; becomes independent c. 2–4 weeks after fledging; breeds when 2 years old.

BREEDING SUCCESS/SURVIVAL: mean brood size 50 pairs (Belgium and Netherlands) 1·4. Oldest ringed bird 12 years, 2 months.

References
Cramp, S. and Simmons, K. E. L. (1983).
Hamilton, R. C. (1975).
Johnsgard, P. A. (1981).

Genus *Recurvirostra* Linnaeus

Medium-sized shorebirds with long legs and long, pointed wings with outer primary longest. Bill, a long, flattened, strongly upcurved, lamellated structure, tapering to a thin point and adapted for gathering large number of small organisms in water and soft mud. Nostrils slit-like. Anterior toes deeply webbed with webs notched in middle; hind toe present but vestigial. Sexes nearly same size. Downy young light tan with darker spots and short stripes. 4 spp., 1 in Africa.

Recurvirostra avosetta Linnaeus. Eurasian Avocet. Avocette élégante.

Recurvirostra Avosetta Linnaeus, 1758. Syst. Nat. (10th ed.), 1, p. 151; Southern Europe (i.e. Italy).

Range and Status. Breeds Africa, Spain, Britain and S Scandinavia through S Europe and S-central Asia to Outer Mongolia and Iran; winters Africa, Middle East, India, Burma and SE China.

Resident, Palearctic visitor, and probably intra-African migrant, widely but locally distributed south to Cape; mainly coastal but also inland, especially alkaline lakes. Northern Africa: rare to uncommon resident Algeria, Tunisia and Morocco; recent status of Rio de Oro and Egyptian resident populations unknown. Palearctic visitor, abundant Atlantic coast Morocco (major wintering population, up to 4000), Algeria (2000–3000) and Tunisia (*c.* 12,000); a few also winter elsewhere. Some oversummer in various localities N Africa; large numbers cross Sahara. Western Africa: Palearctic visitor, frequent Mauritania and Mali; very abundant Senegambian coast (major wintering concentrations, 3000 Cape Verde, 5000 Saloum delta); common to frequent Nigeria, especially last 20 years (180, Borno state, Dec); very abundant Lake Chad (4000, Dec); some oversummer (500 immatures, June, Senegambia). Elsewhere W Africa rare or absent; no resident population known. Eastern Africa: frequent to common N Sudan (along Nile), rare S Sudan and Uganda, common to locally very abundant Rift Valley alkaline lakes Ethiopia to Tanzania (maximum reported 10,000–30,000 L. Manyara, Nov 1940); breeds Kenya, Tanzania and Ethiopia (Awash Valley); wanders eastern Zaïre, Burundi and Rwanda; rare on coast from Red Sea to Tanzania. Southern Africa: breeds South Africa (Cape, Natal, Transvaal), Botswana and Namibia; non-breeding birds very abundant Namibia and south-western Cape coasts (1738 present along *c.* 200 km coastline Durissa Bay-Sandvis, Namibia, Dec–Jan; along southwestern Cape shoreline and estuaries, Nov–Feb, *c.* 2200); locally abundant inland Cape Province (1100, Graaff-Reinet, Van Ryneveld's Pass dam, Aug); large numbers Botswana with some wandering annually to Zambia; elsewhere southern Africa rare to frequent. Origins of major populations in eastern Africa not known but probably include individuals from Palearctic as well as local residents moving from dried up flooded areas; southern African populations probably Afrotropical in origin.

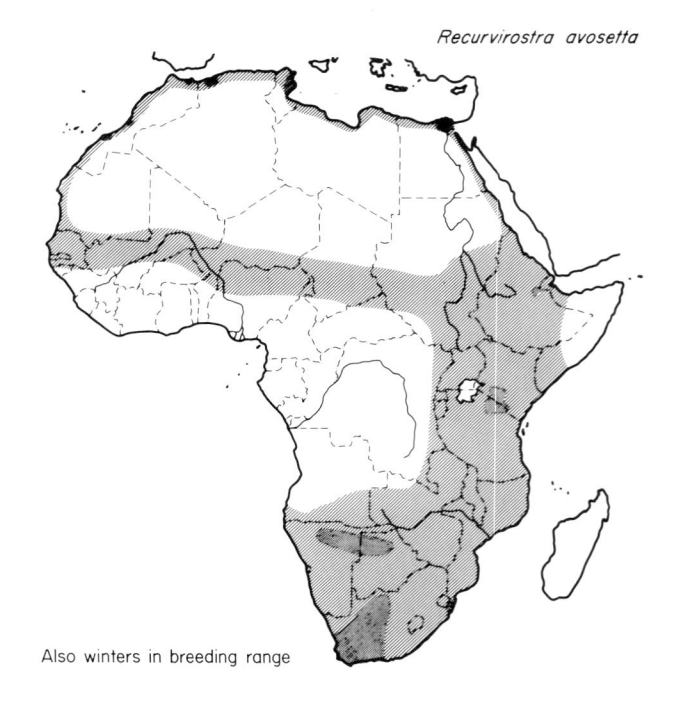

Recurvirostra avosetta

Also winters in breeding range

Description. ADULT ♂ (breeding): forehead, lores, crown to just below eye, nape, hindneck, sides of mantle, upper and inner scapulars black; rest of body plumage white. Central tail-feathers light grey; rest of tail white. Outer 7 primaries black, 3 with large white patch on inner web; secondaries white; primary coverts black with white bases; median and some lesser coverts black; rest of wing white. Bill black; eye brown; legs and feet bluish slate. ADULT ♂ (non-breeding): like breeding ♂ but black on upperparts brownish to pale grey. ADULT ♀: like ♂ except eye hazel; sometimes forehead and crown slightly tinged brown. SIZE: wing ♂♀ (n = 28) 218–238 (227); tail ♂♀ (n = 7) 80–84 (81); bill ♂♀ (n = 17) 75–89 (83); tarsus ♂♀ (n = 27) 75–96 (86). WEIGHT: sex ? (n = 15) 270–390 (319); ♂♀ (Kenya, n = 15) 195–265; ♂ (Botswana) 375·8, 385·9, ♀ (Botswana, n = 4) 202–217 (210·8); ♂♀ (post-juvenile, SW Cape, n = 15) 270–390 (318·7).

IMMATURE: juvenile like adult but black of upperparts brown to grey, many white feathers of upperparts edged brownish buff, 5th primary (counting outwards) with white edge, 4th with large white area; some inner median wing-coverts buff-tipped. 1st-year winter birds, upperparts darker brown but always paler than adult; some median coverts with buff tips.

DOWNY YOUNG: light grey above with sparse black to blackish brown markings; slight black streak from base of bill through eye; white to greyish white below; bill long, slightly upcurved; eye dark brown; feet blue, partly webbed.

Field Characters. Large shorebird with boldly patterned black and white plumage and characteristic recurved bill. Legs shorter and stouter than Common Stilt *Himantopus himantopus*, bluish slate not pink; stance less upright. In flight wings have distinctive black patches separated by white.

Voice. Tape-recorded (3, 38, 62, 73, 76, F). Gives (a) contact call 'cute cute' and 'cuck-cuck'; (b) alarm call, given on ground or in flight, like contact call but as excitement increases, tone becomes harsher, emphasis on 2nd syllable and rate of calling increases; (c) intense alarm or mobbing 'creew'; (d) when bowing (see Breeding Habits) rapid weak 'cwit-cwit-cwit ...' or 'c-c-c-crrrreewer'; (e) when flying at intruder, 'cweet-cweet'; and (f) when calling young, 'crrrr'. Romanian birds (Adret 1982) give 10 calls: (a) chuckle, a single sharp note dry in quality and low-pitched, uttered at nest-relief alternately by both partners until incoming bird replaces partner at nest; (b) triplet call, a series of 3 short repeated notes with 3–6 triplets given in one sequence, uttered as a vocal contact between parent and chicks; (c) parental rallying call, very similar to alarm chirping call (see below), but shorter and slower, uttered when chicks about nest and important in family cohesion; (d) rhythmic call, a series of up to 18 short repetitious notes in a sequence, uttered by an incubating bird when other Avocets fly over nest or when partner approaches nest; (e) mixed call, a 'two-note' call in which 4–5 rhythmic calls are followed by pseudotrill (see below), uttered when defending territory or nest; (f) snarl, a nasal sounding call with an ascending chirp, a sustained chirp, and a whistle, uttered in aggression; (g) coo, a call made of a pulse of rising frequency, a sustained pulse and an ascending pulse which ends abruptly, uttered by foraging non-breeding birds and by ♀ in response to ♂ triplet call; (h) alarm chirp, a call with a short introduction, 2 melodic units and an ascending segment, uttered *c.* 2 times/s by many individuals when mammalian or avian predators at colony; (i) alarm whistle, a whistle with 2 segments, uttered *c.* 2 times/s when avian predators appear in the air; and (j) alarm pseudo-trill, a call with 2 parts, first, the notes of the triplet or rhythmic call, then a short whistle, uttered *c.* 2 times/s, mainly when gulls pass over.

General Habits. Inhabits open flat areas, usually devoid of reeds and shrubs but near water, preferring coastal and inland saline lakes and mudflats; also sand beaches, river deltas, flood-plains; rarely freshwater lakes and rivers. Gregarious, in flocks of 5–30, sometimes several hundred; occasionally singly or in mixed flocks. Walks briskly with body horizontal and neck in gentle curve. When excited or suspicious, bobs head up and down without moving rest of body. Roosts both at night and day, often in close groups; main roosting at night. Most active morning and evening although sometimes active in moonlit nights. Feeds in water 10–15 cm deep in loose flocks scattered over large areas of mud, swinging head from side to side so that upturned end of partly open bill contacts water or mud at each swing; advances after each swing. Prey located by touch; may capture large numbers in short time. Also obtains food by picking up prey with slightly opened bill and by stirring when head and slightly open bill are moved elliptically *c.* 30° above ground in shallow water. Sometimes swims with breast held low in water, tail held high; may then feed by tilting head vertically and raising tail like surface-feeding duck; lifts bill out of water before swallowing food. Sometimes catches fish, walks to shore, repeatedly slams fish to ground for up to 5 min; swallows it, and returns to water to catch another fish (Jacobs 1973). Spends much time resting on open sand, especially in middle of day, often in large flocks. Stands on one or both legs, with bill often tucked into scapulars. In flight legs extend beyond tail. Sometimes flies in V-formation but usually in loose cloud-like flock.

Migrates in large flocks of several hundred; migration patterns in Africa not well known. Southern range of Palearctic birds in Africa not known but at least south to Senegambia in west, N Sudan (Nile and Red Sea coast) in east. Very large flocks in Kenya and Tanzania June–July probably Afrotropical birds possibly from southern populations (e.g. southern origin likely for June flocks at Lake Nakuru: Britton 1980). Other populations make local, seasonal movements caused by drying up of habitat. Source of northern wintering Eurasian Avocets in Africa largely western Palearctic, east to 45°E (Moreau 1972). Birds ringed Denmark, Belgium and Netherlands recovered Morocco, Tunisia and Senegambia; birds ringed France, Germany and Hungary recovered Algeria, Morocco and Tunisia; birds ringed Sweden recovered Morocco and Tunisia; birds ringed Austria and Hungary recovered Tunisia. Earliest arrivals Morocco at end Aug.

Food. Prefers small insects such as larval and adult midges and brine flies associated with salt water; also beetles, small crustaceans, worms, gastropods, sometimes sole (fish) 8–10 cm long (Jacobs 1973) and seeds and small roots.

Breeding Habits. Seasonally monogamous, nesting in colonies of 10–100, sometimes 200 or more pairs; occasionally solitary. Early in breeding season 3–20 birds, typically mainly in pairs, perform grouping ceremony in which they arrange themselves in a circle, point bills to centre, and assume bowing attitude. When grouping, birds may throw straws, peck at water, shake heads, sweep bills as if feeding and, if paired, press their bodies together. When bowing, each bird bends head to ground, stands still for a moment, then tramples feet, moves bill slightly up and down, and makes rapid call (see Voice); bowing also performed when fighting, during nest-relief, when picking up nest material and sometimes before copulation. Either sex may also assume upright posture with head and neck raised; bob head up and down; chase other individuals in crouching attitude with neck drawn back, body sloped forward and wings held away from body; fly at another individual; turn sideways in a ground encounter; spring forward, flapping wings; peck at another bird; sit down; and during course of fight may assume sleeping attitude. Will also attack other shorebirds, gulls, ducks, herons and fish eagles.

When pairing, ♀ attempts to associate with ♂; when ♂ no longer repels her, pair is formed. Before copulating, both birds first dip bills in water and preen. When ♀ is ready to copulate, she stands with legs wide apart and neck forward so her head is flat on water. ♂ then stands beside her, moves behind her and round to her other side; he may move back and forth several times (usually 3–5 times, sometimes up to 25), dipping his bill, shaking and preening while doing so; eventually he jumps sideways on to her. During copulation, which lasts only a few seconds, ♂'s bill is open and his wings are stretched upwards above his back; he does not hold ♀'s bill; ♀ moves head from side to side. After copulation, both cross their bills and run forward 2–10 m. May perform mating ceremony several times a day during courtship and for a few days after clutch complete. After mating ceremony both may feed, rest, drink, bathe or move to dry land to perform nesting display in which both birds may first bow, press bodies together, bow again, then pick up objects and throw them over their heads. ♂ initiates nesting display by sitting on ground and then, leaning on breast with tail pointing into air, scratches ground with feet. ♀ may join him or both may take turns making scrape; sometimes both may simultaneously make separate scrapes; several scrapes made before one is decided upon.

NEST: a simple scrape on dried mud, sand or bare area of soda; sometimes in short grass; diam. (n = 91) 11–13·5, depth 2·5–4 cm, sometimes with edge c. 2–3 cm above ground; lined with various materials including aquatic vegetation and flamingo feathers. May be as close as 20–30 cm apart in dense colonies or as far apart as 62 m.

EGGS: 2–5, usually 3–4; laying interval 1–2 days. Usually ovate, smooth, pale clay-coloured, spotted irregularly with black. SIZE: (n = 330) 46–56 × 33–36 (50 × 34). WEIGHT: (n = 67) 28–40 (32) g.

LAYING DATES: Morocco, Mar–Apr; Algeria, Apr–May; Tunisia, Mar–Apr; Ethiopia (date ?—known to breed but details not published); E Africa: Region C, Jan, Apr, June–July, peaking June early in dry season; Region D, Feb–Aug; Zambia, Aug (1 record); South Africa: Natal, Sept–Nov, Transvaal and Cape, Aug–Oct (also Botswana and Namibia but details not published).

INCUBATION: begins with 2nd and 3rd egg; performed by both parents in approximately equal shares. Period: 23–25 (20–28) days. During incubation, parents change over frequently, often once/h. At nest-relief, sitting bird rises and meets mate; both bow and sometimes throw straws; relieving bird then sits on nest as mate departs. Parent returning to nest will first soak belly to moisten eggs and cool them (Goutner 1984); bird on nest will raise crown and mantle feathers when ambient temperature high. When intruder approaches, may perform injury-feigning with wings extended and tail widely spread and depressed; may sit on ground or walk toward intruder, tipping wings side to side. Sometimes flies around and around intruder, or walks towards it with tail cocked up, wings spread, and bill and head stretched forward. May sometimes protect eggs by standing over them with wings extended and lowered, and tail fanned and elevated.

DEVELOPMENT AND CARE OF YOUNG: young climb out of nest when 1 h old, preen and feed when 1 day old; capable of flight at 30 days although do not usually fly until 6 weeks old. Small young press against ground when intruder present with legs folded and head resting on ground; older young flee. Fully fledged after 32–42 days. Breed when 2 years, sometimes 3 years old.

As soon as young hatch, parents remove eggshells. Both parents tend young until fledged. One or both parents lead young to feeding area. Sometimes (Netherlands) young in groups of 10–20 may be attended by only 1–2 adults.

BREEDING SUCCESS/SURVIVAL: no information on hatching success or mortality for African birds; 1·1 (0·1–3·0) fledged young/pair Suffolk (England) over 29 years. Major predator Lake Manyara (Tanzania) probably Grey-headed Gull *Larus cirrocephalus*. Oldest ringed bird 24 years 6 months.

References
Cramp, S. and Simmons, K. E. L. (1983).
Makkink, G. F. (1936).
Olney, P. J. S. (1970).

Family BURHINIDAE: thick-knees, stone curlews and dikkops

Medium to large plover-like waders with highly cryptic plumages, large heads and eyes, and swollen tibio-tarsal joints. Bill short to fairly long; nostrils long. Neck fairly short; 16 cervical vertebrae. Wings long and pointed; 16–20 secondaries. Tail slightly graduated; 12 feathers. Tarsi long; toes short, only 3 (no hind toe), webbed at base; claw of middle toe broad and dilated on its inner web. Sexes similar or ♀ slightly smaller; no marked seasonal change of plumage. Found in terrestrial and littoral habitats. Can run fast but gait usually more plover-like with short spurts and stops. Often stand with shoulders hunched and will rest squatting on tarsi. Flight deliberate-looking but fast, often low over ground and silent with rapid wing-beats and legs outstretched behind. Nocturnal or crepuscular; in Africa often flushed along roads at night, with nightjars. Chicks precocial but fed by both parents. Superficially resemble bustards but skeletal, biochemical and parasitological characters and down pattern of young evidently suggest they are charadriiforms. 2 genera, *Burhinus* with 7 spp. and *Esacus* with 2 spp.

Genus *Burhinus* Illiger

Plumage vermiculated, spotted or streaked buff and dark above (not plain), whitish with brown streaks below; bill straight, stout and a little shorter than head (not slightly upturned and massive); yellow to green at base, with swollen black tip. 7 spp. in Central and S America, Africa, Europe, Asia and Australia; 4 including 2 endemics in Africa.

Burhinus oedicnemus (Linnaeus). Stone Curlew; Eurasian Thick-knee. Oedicnème criard.

Plates 12, 20
(Opp. pp. 177, 321)

Charadrius Oedicnemus Linnaeus, 1758. Syst. Nat. (10th ed.), 1, p. 151; England.

Range and Status. Breeds W Eurasia, N Africa, Arabia, Iranian region, India, Burma; sedentary except for northern populations which winter in N and E Africa south to Uganda and Kenya.

Resident and Palearctic migrant. Breeds N Africa from Morocco to Egypt; some migrate south to Senegambia, Sudan, Ethiopia, Kenya and N Somalia. Palearctic winter visitor N and E Africa from Morocco to Egypt, mainly to N Uganda, NW Kenya and Ethiopia but recorded south to Elmenteita, Kiambu, and Uaso Nyiro R. in Kenya, Serengeti in N Tanzania, and P. N. de la Garamba et Mauda in Zaïre; also winter Senegambia and Mauritania along Senegal R. and Mali (Sahel region); vagrant Sierra Leone (1, 8°12′N, 10°30′W) and Nigeria (1 sighting Lagos).

Description. *B. o. oedicnemus* (Linnaeus): Palearctic migrant to N Africa, Senegambia, Zaïre, Sudan, Ethiopia, and E Africa. ADULT ♂: crown, nape, hindneck and sides of neck sandy to tawny, streaked with blackish brown; lores, cheeks, superciliary stripe and throat white; malar streak to ear-coverts tawny, streaked blackish brown. Mantle, back and rump sandy to tawny with black shaft streaks; central rectrices sandy to greyish tawny with black shaft streaks; outer rectrices white with brown cross-bars and blackish tip; outermost pair rectrices white with blackish brown tip. Underparts white, streaked dark brown on breast; undertail-coverts light yellowish ochre to tawny. Primaries brownish black, the outermost with white central region, the next with large subterminal white patch and innermost with smaller subterminal white spot; inner primaries tipped and edged with white at end; outer secondaries blackish brown with base of inner web white; inner secondaries sandy to tawny streaked blackish brown; primary coverts brownish black; median wing-coverts greyish brown with blackish brown shaft streaks, the upper row white in centre with dark shaft streaks, blackish brown subterminal bar, and greyish tip; greater coverts pale greyish brown at base, then white with blackish brown tip and white edge, mostly with blackish shaft streaks; coverts form two black and two pale wing-bars at rest (**A**). Bill black with yellow to greenish yellow base; eye yellow; legs and feet yellow. Sexes alike. SIZE: wing, ♂ (n = 10) 234–248 (241), ♀ (n = 10) 237–247 (240); tail, ♂ (n = 13) 115–129 (121), ♀ (n = 13) 113–125 (118); bill, ♂ (n = 14) 37–44 (40·1), ♀ (n = 15) 37–43 (39·3); tarsus, ♂♀ (n = 23) 68–83 (75·7). WEIGHT: (France and central Europe) ♂ (n = 5) 430–502 (475), ♀ (n = 8) 290–535 (449).

IMMATURE: paler than adult; much more tawny and warmer in colour; median coverts and tertials washed with sepia and with darkish subterminal bars; inner medians buffish with distinct bright buff fringes; greater coverts tend to have broad white tips.

DOWNY YOUNG: above greyish buff, finely speckled with black; broken black line across forehead and down centre of crown; two black lines from behind eye down back to tail on either side of midline; black line on each side of rump; black line on wing; below white; bill greenish grey with dusky tip; eye pale yellowish white; legs and feet yellow to grey, soon turning greenish.

B. o. saharae (Reichenow): breeds N Africa where present all year; some migrate to Sudan, Ethiopia, N Somalia, Kenya and Senegambia. Paler, more rufescent, less streaked and smaller than nominate race; wing (n = 20) 228–242 (235).

Burhinus oedicnemus

Also winters in breeding range

B. senegalensis

B. oedicnemus

B. vermiculatus

A

Field Characters. Very similar to Senegal Thick-knee *B. senegalensis* and Water Thick-knee *B. vermiculatus* but folded wing has narrow upper white band bordered above and below by narrow black bands, and below that a broad band of pale grey (**A**). Water Thick-knee lacks lower black band which appears as black streaks only; Senegal Thick-knee has only broad pale greyish band bordered above by narrow black bar.

Voice. Tape-recorded (62, 73, 76). Not described in winter quarters. In breeding areas, calls chiefly at twilight when wailing calls often given in chorus; begins *c.* 30 min after sunset; may involve 4–6 individuals for over 30 min; and may occur throughout hours of darkness but finishes at first light. Calling often occurs in sudden bursts; only calls during day when establishing territory at start of season. Gives the following calls: (a) 'kur-LEE', serves chiefly as aggressive call; (b) 'chrrrwhEE ... chrrrwhEE ... chrrrwhEE', or 'krieehk' alarm calls; (c) 'ker-vic ker-vic' and 'cu-ick' by territorial birds; (d) 'cu-wick ci-wick' nest-scraping call; (e) long series of excited yelping or piping phrases, first rising then falling (up to 80 notes in succession), associated with dancing leaps; (f) 'tuEE tuEE' and 'whee', associated with display; (g) 'quig' between mates to view scrape; (h) 'büde bide' when copulating; (i) hissing call, during distraction display; (j) throaty cluck to call chicks; and (k) 'kuit' call by young. In autumn, gives a musical whistle, a 'tir-whi-whi-whi-whi'.

General Habits. Inhabits open stony ground, short grass plains, semi-desert areas with shrubs, and cultivated fields; usually found away from rivers. Generally gregarious, forming flocks of 10–100 (Morocco), 150 (Tunisia) and 300 (Europe) in non-breeding season; even gregarious in breeding season when several pairs feed communally; sometimes occurs singly. Most active at twilight, but also at night and occasionally by day. In non-breeding season, spends day in flocks; disperses at end of twilight singly or in small groups to feed; reassembles in flock at dawn. During day usually stands or squats under or next to shrub or other shelter, usually in shade. If disturbed crouches flat on ground, or walks with head low and neck stretched forward or runs away; may take flight for a short distance. Flight strong with erratic wing-beats, showing black-and-white wing stripe. When becoming active at twilight, performs wing-waving display by running forward and waving and beating wings. Eventually extends wings, flings them up, leaps into air, makes short flight low over ground, and then lands suddenly in sharp curve. Usually done independently at first although 7–8 birds may become involved, performing group display for *c.* 30 min when all run about, calling, leaping, charging one another and strutting in high-upright display (see Breeding Habits). Whole flock may become involved.

Movements of *B. o. oedicnemus* and *B. o. saharae* poorly known. Substantial numbers of *oedicnemus* cross Mediterranean to winter in N Africa and south of Sahara. Probably crosses Mediterranean and Sahara on wide front although distribution south of Sahara still unclear; known from Senegambia, Mali, Niger, NE Zaïre, Sudan, Ethiopia and E Africa; only in east does it reach equator. Some also cross Red Sea (21°06′N, 38°12′E, Tuck 1964). *Saharae* apparently present all year over much N Africa including coastal plains but not higher plateaux in winter. Partial migrant along Atlantic coast to Senegambia; also reaches Sudan, Ethiopia (Eritrea), N Kenya and N Somalia. *Oedicnemus* arrives Kenya early Oct; Mali Oct–Nov; departs Ethiopia early Mar; Kenya end Mar to end Apr; Mali Apr–May. 1 ringed France, recovered Algeria, Mar; 1 ringed Great Britain, recovered Sierra Leone, Jan.

Food. Mainly insects and their larvae, especially beetles and grasshoppers; also spiders, worms, slugs, crustaceans, small lizards, mice and other small mammals, bird eggs and young, frogs and some seeds.

Breeding Habits. Monogamous, solitary nester; may return to same territory year after year. Often ♂♂ arrive back before ♀♀; pair usually formed on breeding grounds although some arrive with bond already established. Once pair established, ♂ and ♀ usually maintain close contact throughout breeding season vocally and visually. When defending territory, performs high-upright display in which draws up to full height, holds body vertically and tail straight down and fully fanned; folds wings but holds them slightly away from body; and walks stiff-legged. May run at or fly after intruder; sometimes defender and intruder perform this display walking side by side. Usually ♂ does high-upright although ♀ also does so. In courtship ♂ performs quick runs, jumps and skips with wings hanging or half raised, tail spread and moved up and down, neck and head turning and twisting. ♀ may then beg for food and is fed by ♂. Other times ♂ stands erect, runs with neck fully stretched and angled upward, wings hanging and tail raised and spread. In pre-laying period, both perform deep-bowing display, in which ♂ tilts whole body stiffly forward until bill almost touches ground, with back and fanned tail held high in air. At same time ♀ also places bill near ground but does not move rest of body. ♀ then shuffles down on ground at spot indicated by ♂'s bill. Both take turns sitting at this spot, revolving and shuffling, picking up stones and throwing them over shoulder, and sometimes kicking earth back with feet. Sometimes ♂ faces ♀ sitting on scrape; ♀ raises neck at 45°; ♂ then bows in response touching ♀'s bill, then swings into high-upright display. The pair may visit several spots as they perform this display. Both birds also perform neck-arch display, in which both draw themselves up stiffly to full height and, with tail straight down and closed and wings folded, slowly arch their necks with bill pointing down. This display, used in greeting, often followed by copulation. Nests almost always well spaced; some by themselves, others as little as 200 m apart or as far as several km apart (Europe).

NEST: scrape on ground in open, usually on broken, uneven surface; diam. 16–22 cm, depth 5–7 cm, lined with small stones, shells, and pieces of vegetation. Both sexes probably build nest.
EGGS: 2, rarely 3; laid 48 h apart. Oval, smooth, slightly glossed, pale creamy buff to creamy yellow, speckled, spotted, blotched and scrawled with medium to dark brown and some pale purplish grey. SIZE: (n = 27, *saharae*) 50–56 × 35–39 (52 × 37). WEIGHT: 38.
LAYING DATES: Morocco, Algeria, Tunisia, Libya and Egypt, Mar–June.
INCUBATION: begins with last egg, by both parents. Period: 24–27 days. Deep-bowing and neck-arch displays normally not performed once incubation begins.
DEVELOPMENT AND CARE OF YOUNG: downy young precocial, with chicks leaving nest soon after hatching although incapable of much walking for first 24 h. Initially parent brings food to chicks, sometimes also to well grown chicks. Flies at 6 weeks; fledged at 36–42 days; not independent until after fledging. Young squats and freezes to parent's chick-warning call, and at sight of intruder, may lie motionless with head and neck level with ground and eyes open; when 10 days old may escape by running in fast jinking run. When intruders present, parent will depart from nest at earliest sign of disturbance or remain with wings raised and fanned; also may dive attack, and feign partial disablement, fluttering along ground.
BREEDING SUCCESS/SURVIVAL: of 74 eggs over 8 years (England), 77% hatched successfully (annually varied from 54·5 to 100%). Of these eggs, 39% produced young with 0·4–1·5 young/pair (mean 0·8 young/pair).

References
Cramp, S. and Simmons, K. E. L. (1983).
Westwood, N. J. (1983).

Burhinus senegalensis (Swainson). Senegal Thick-knee. Oedicnème du Sénégal.

Plates 12, 20
(Opp. pp. 177, 321)

Oedicnemus senegalensis Swainson, 1837. Bds. W. Afr. 2, p. 228; Senegal.

Range and Status. Endemic resident and intra-African migrant, common, from Senegambia and S Mauritania through S Sahel zone and coastal areas of W Africa (but absent from S Sierra Leone, S Liberia, and SW Ivory Coast) to Sudan, Ethiopia, NE Zaïre, N Uganda and NW Kenya; also Nile Valley north to Delta. Occupies most river systems between equator and Sahara while Water Thick-knee *B. vermiculatus* occupies those south of it. In Zaïre occurs on upper Uelle and adjacent rivers while Water Thick-knee found on lower reaches of same rivers.

Description. ADULT ♂: upperparts from forehead to rump sandy to tawny with dark streaks, very similar to Stone Curlew *B. oedicnemus* except back and scapulars more finely streaked. Tail, central rectrices sandy to greyish tawny with black shaft streaks, outer rectrices white with brown cross-bars and blackish tip; outermost pair white with blackish brown tip. Under-parts white; breast with dark brown streaks. Primaries and secondaries black with white band across outer primaries; inner secondaries tinged grey; greater coverts white with broad black subterminal bar on outer web and narrow white fringes; lesser upperwing-coverts along leading edge black with grey bloom and cinnamon fringes; all other lesser and

Burhinus senegalensis

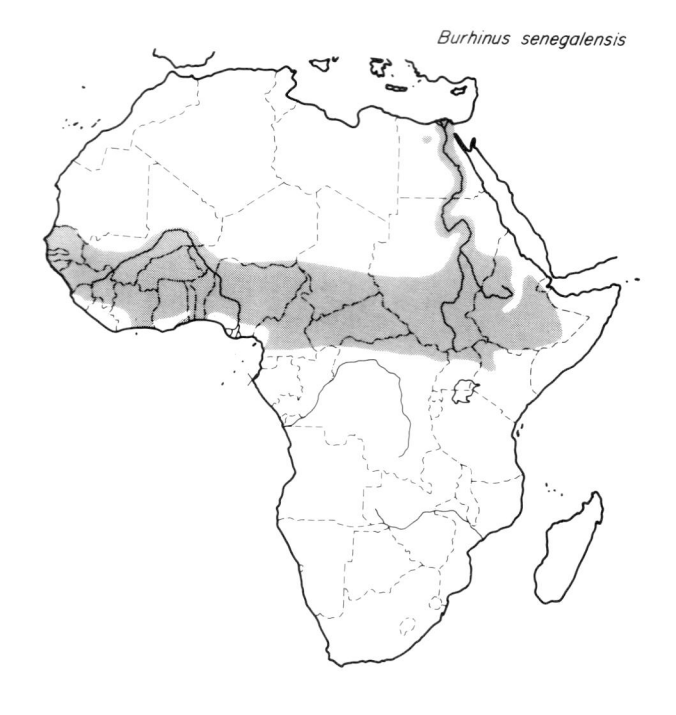

median upperwing-coverts contrastingly pale grey with narrow dark sepia shaft streaks; underwing-coverts and axillaries pale buff to white with sepia shaft streaks near tips. Bill, outer half black, base yellow but less extensive than in Stone Curlew; eye yellow; legs and feet yellow. Sexes alike. SIZE: wing, ♂ (n = 5) 214–234 (221), ♀ (n = 11) 215–231 (221); tail, ♂ (n = 5) 109–119 (114), ♀ (n = 12) 100–117 (110); bill, ♂ (n = 7) 46–48 (47·2), ♀ (n = 11) 44–49 (46·8); tarsus, ♂ (n = 7) 67–76 (71·4), ♀ (n = 10) 67–75 (70·3).

IMMATURE: like adult, but dark shaft streaks on head, neck and chest narrower; median wing-coverts and tertials with distinct buff fringes, tertials more pointed.

DOWNY YOUNG: above greyish buff speckled with black; below white; black lines on forehead, crown, back, and rump (but not on wing); bill greenish grey with dusky tip; eye pale yellowish white; legs and feet yellow to grey.

Sometimes (Vaurie 1965, Prater *et al.* 1977, Cramp and Simmons 1983) divided into 2 races *B. s. inornatus* (E and N Africa) and *B. s. senegalensis* (W Africa), the latter with slightly shorter wing (av. 221 *vs* 214) and grey brown not dark grey to medium brown upperparts. We follow White (1965) in treating it as a monotypic species.

Field Characters. Very similar to Stone Curlew and Water Thick-knee but folded wing with only 1 broad pale greyish band bordered above by narrow black band (see **A**, p. 200), and no white bar on median wing-coverts in flight. Also differs from Stone Curlew in being smaller and inhabiting primarily riverbeds and lake edges, not open areas away from water; less yellow on base of bill. Also differs from Water Thick-knee in not having vermiculation of feathers on posterior upperparts and in having basal part of bill, eye and legs yellow, not green. Voice more nasal and metallic than voice of Stone Curlew and higher pitched than that of Spotted Thick-knee *B. capensis*.

Voice. Tape-recorded (217, 244). Wailing calls similar to Stone Curlew but less strident. Song, given in flight or when settled, a 'pi pi pi-pi-pi-pi-pi-PII-PII-PII-pii-pii'; sometimes gives 'piLI piLI'. Calls often taken up by conspecifics and uttered in chorus.

General Habits. Inhabits sandy country near water, especially sandbanks in rivers with some vegetation cover where roosts during day; also occurs lake shores, cultivated areas, gravel roads and mangroves. In Egypt occurs along shores and banks of Nile and on islands in river while Stone Curlew found in dry situations well away from water. Occurs singly, in pairs or small parties of up to 6, sometimes up to 30 (Ethiopia). Mainly nocturnal but also frequently active during day. Usually forages, sometimes in parties, along water edge although may do so a km or more from water. When disturbed, stands motionless or squats; may run for cover often in zigzags, wings held slightly away from body, neck stretched out or drawn in; occasionally also flies, usually doing so low over ground and landing a short distance away.

In Nigeria makes local movements corresponding with water level changes but not a true migrant. In other parts of range does migrate: in NE Zaïre (upper Uelle) arrives Nov after rains, stays until Apr–May; in Senegambia present Dec–Mar, during and after rains and in Ivory Coast present Nov–Apr.

Food. Mainly insects and crustaceans; also snails, worms and other small invertebrates, tadpoles, frogs and small rodents.

Breeding Habits. Monogamous, solitary to loosely colonial; in Nile Delta nests commonly in small colonies, up to 21 nests on roof of 1 house. Territorial, some pairs using same nesting area from year to year.

NEST: a shallow scrape with little or no lining on bare ground, sandbank, top of flat rock, small rocky islet in stream or flat roof of building. Egg often ringed with tiny pieces of grit, wood, straw, and broken shells.

EGGS: 2, N Africa mean (5 clutches) 2·0. Short oval, smooth, slightly glossy; dull ochre to warm stone brown, lightly marked brown and grey or heavily blotched with sepia at large end. SIZE: (n = 25) 46–51 × 30–34 (49 × 32), (n = 3, Uganda) 49–52 × 32–33. WEIGHT (estimated): 50.

LAYING DATES: Senegambia, Apr–Aug, Nov–Dec; Mali, May–Sept; Nigeria, Mar–Aug, Nov; Egypt, Mar–June; Sudan, Mar–Apr; Ethiopia, Feb–Aug (dry and wet seasons); E Africa: Uganda, Feb–Mar, May–Nov (dry and wet seasons); Kenya, Mar–June, Sept–Nov (wet seasons); Region A, Feb–Mar (at height of dry season when sand banks more exposed); Zaïre Nov–Apr.

DEVELOPMENT AND CARE OF YOUNG: young precocial although remains in nest for 1st few days. If disturbed, freezes with head and neck outstretched on ground. Both parents take care of young; parents considered to carry young (Mackworth-Praed and Grant 1957).

Reference
Cramp, S. and Simmons, K. E. L. (1983).

Burhinus vermiculatus **(Cabanis). Water Thick-knee; Water Dikkop. Oedicnème vermiculé.**

Plates 12, 20
(Opp. pp. 77, 321)

Oedicnemus vermiculatus Cabanis, 1868. J. Orn., 16, p. 413; Jipe near Taita, Kenya.

Range and Status. Endemic resident. Uncommon to abundant along sandy, mangrove-fringed banks of coastal estuaries and rivers from Liberia to Cameroon; rare inland; vagrant Senegambia. Thence, locally common to abundant across central Africa, including Gabon, Congo, Zaïre, Uganda, Kenya and SE Ethiopia, and southward to N Namibia in the west and E South Africa in the east, and down along coastal estuaries and rivers to SW Cape Province. Occupies most of river systems south of equator while Senegal Thick-knee *B. senegalensis* inhabits those north of equator. In NE Zaïre inhabits only lower reaches of Uelle and adjacent rivers; in SE Ethiopia found along Webi Shebelli and Juba rivers.

Description. *B. v. vermiculatus* (Cabanis): S central and E Africa. ADULT ♂: crown to uppertail-coverts pale buffy brown vermiculated with umber and with dark brown or black shaft streaks, broadest on centre of back; outermost 3 pairs of rectrices white, barred and vermiculated with dark brown, tipped blackish and edged with buff; other rectrices buffy grey with subterminal black and white bars, black tip and buffy white margin. Lores and broad streak under eye white; streak immediately below eye to ear-coverts dark brown; streak immediately above and behind eye white. Sides of neck, ear-coverts and malar stripe dark brown streaked with buff; throat buffy white. Breast buffy white boldly streaked with dark brown; abdomen creamy white streaked dark brown on flanks; undertail-coverts rufous-buff. Primaries black, outermost 3 with broad white bar about 54 mm from tip, innermost white basally; secondaries dark grey shading to black at tip, innermost vermiculated with dark brown; tertials and scapulars like mantle; primary coverts and alula blackish brown; secondary coverts grey tipped with black; median coverts buffy grey streaked black; upper lesser coverts dark brown freckled with buff; lower coverts white with black basally, forming distinct wingbar; underwing-coverts and axillaries white tipped with dark brown; bill black, greenish yellow at base; eye pale green; legs and feet light greenish grey. Sexes alike, except bill slightly smaller in ♀. SIZE: wing ♂♀ (n = 16) 191–211 (205); tail, ♂♀ (n = 16) 98–118 (109); bill, ♂ (n = 12) 41–46 (44), ♀ (n = 4) 40–44 (41·3); tarsus, ♂♀ (n = 16) 72–77 (64·8). WEIGHT: ♂ (Botswana) 293, 301, ♀ (Botswana) 308, 315, (Kenya) 320.
   IMMATURE: similar to adult, but grey wing-coverts freckled buffy; more profusely vermiculated on upperparts and tail.
   DOWNY YOUNG: not described.
   B. v. buttikoferi (Reichenow): W Africa from Liberia to Nigeria and Gabon; vagrant Senegambia. Darker, more greyish sepia above; bill longer ♂, 49–54.

Field Characters. More of a water bird than other thick-knees. Very similar to Stone Curlew *B. oedicnemus* and Senegal Thick-knee but on folded wing upper white bar and lower grey bar separated only by streaks, not by black bar (see **A**, p. 200). Also differs from Senegal Thick-knee in basal part of bill, eye and legs green not yellow and feathers of posterior upperparts vermiculated. Easily distinguished from Spotted Thick-knee *B. capensis* by presence of wing-bars (lacking in Spotted).

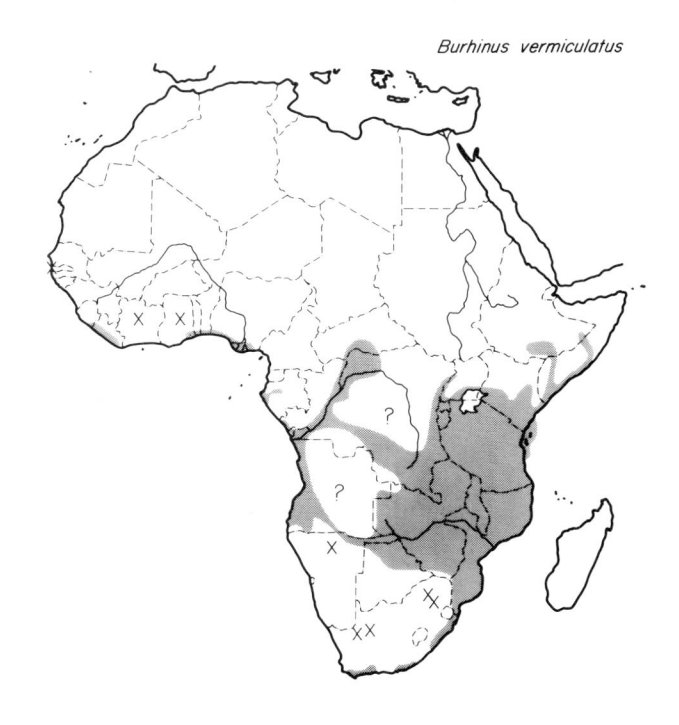

Voice. Tape-recorded (12, 21, 50, 74, F). Loud, rather strident, wild whistling, often in chorus 'ti-ti-ti-ti-ti-tee-tee-teee', drawn out and then dying away towards end.

General Habits. Inhabits riverbanks and lake edges, estuaries, mangrove swamps and sometimes beaches. Found singly or in pairs; also often gregarious, gathering in flocks of 30 or more in non-breeding season. Mainly active at night when flies and calls; more likely to be active and vocal by day than other thick-knees. Sometimes at night forages 1 km or more from water. During day-time roosts by standing or sometimes squatting on ground in light woodland or bush cover adjacent to water habitats; sometimes roosts in open. When disturbed, reluctant to fly; usually runs away with head down; sometimes runs rapidly before taking off; does not fly far. Flies with rapid wing-beats alternated with slower flappings.
   Mainly sedentary; makes only local movements corresponding with fluctuating water levels; moves to higher ground as water rises. In Ivory Coast frequents exposed rocks on rivers in dry season Oct–June; disappears when water rises and rocks are submerged.

Food. Insects, crustaceans and molluscs.

Breeding Habits. Monogamous, solitary nester.
   NEST: a scrape, sometimes lined, on sandbank or shoreline, often next to driftwood or among low shrubs; often close to water.

BURHINIDAE

EGGS: 2 (rarely 1). Rounded oval, buffy white to sandy yellow or cream, irregularly blotched, spotted and speckled with dark brown and black, with underlying violet-grey. Markings tend to be concentrated at thick end. SIZE: (n = 54, southern Africa) 44–54 × 32·7–39 (49·2 × 35·7).

LAYING DATES: E Africa: Kenya and Uganda, Jan, Mar–Dec; Tanzania, Nov; Zanzibar and Pemba, Jan–Oct; Region A, Mar–Apr; Region B, Jan–Apr, June–Nov (with peak at Lake Victoria Aug–Oct in a drier break in main rains of Oct); Region C, July, Sept–Oct, Dec (in dry season); Region D, Mar, June–July, Sept–Nov; and Region E, May, Sept; Zambia, Apr, July–Dec with most Aug–Oct; Malaŵi, Aug–Dec; Zimbabwe, Jan, Aug–Dec; Namibia, Sept; and South Africa: Natal, Mar, Sept; Cape Province, Sept–Jan. In southern Africa breeds in dry season and early rainy season.

INCUBATION: period c. 24 days.

Plates 12, 20
(Opp. pp. 177, 321)

Burhinus capensis (Lichtenstein). Spotted Thick-knee; Spotted Dikkop. Oedicnème du tachard.

Oedicnemus capensis Lichtenstein, 1823. Verz. Doubl., p. 69; Cape of Good Hope.

Range and Status. Breeds Africa and SW Arabian Peninsula.

Resident, frequent to locally common, S Mauritania, Senegambia, S Mali, Upper Volta, N Ghana, Togo, Benin, Nigeria, S Niger, S Chad, N Central African Republic, central and E Sudan, Ethiopia, Somalia, N Uganda, Kenya, Tanzania, E Zaïre, Angola, Zambia, Malaŵi and whole of southern Africa.

Description. *B. c. capensis* (Lichtenstein): S and E Africa to Angola and Kenya (not dry SW parts of South Africa, Namibia and W Botswana). ADULT ♂: upperparts pale pinkish cinnamon with broad blackish shaft streaks; back, rump and uppertail-coverts also barred with blackish brown. Lores, sides of face, around and behind eye white; rest of face, malar stripe and sides of neck pale pinkish cinnamon streaked dark brown; chin and throat white. Breast pale cinnamon streaked blackish brown; abdomen white; undertail-coverts cinnamon; outermost four pairs of rectrices white with bold transverse bars and tips of blackish brown; inner rectrices buffy grey, boldly barred and marbled with blackish brown. Primaries, alula and primary coverts blackish brown; outermost three primaries with broad white subterminal bar; innermost primaries white at base; secondaries dark brown with dusky inner web; tertials like back; wing-coverts like back, but paler and whiter; underwing and axillaries white, streaked dark brown. Bill black, basal third greenish yellow; eye bright yellow; legs and feet yellow with blackish wash on anterior surface. Sexes alike. SIZE: ♂♀ (n = 36) wing, 223–242 (231); tail, 112–138 (123); bill, 34–40·5 (36·8); tarsus, 87–105 (95). WEIGHT: ♂ (Kenya) 400, 425, (Botswana, n = 4) 375–610 (465), (Namibia) 505, ♀ (Kenya) 413, 450, Botswana (n = 9) 375–585 (440·5), (Namibia) 420.

IMMATURE: similar to adult, but more streaked and less spotted dorsally; more finely streaked on breast.

DOWNY YOUNG: forehead whitish grey, bordered behind by black transverse line; rest of upperparts light dirty grey with black tips to down; eyebrow paler grey; indistinct black line down centre of forecrown joins transverse bar on hindcrown; black stripe through eye meets forehead bar; broken black malar stripe meets eye-stripe behind ear-opening; nape lightly mottled with black; two bold black lines on either side of midline of back; bold broken black line on each side of outer lower back from front of thighs to tail; transverse bar across tail; black bar on wing; below dirty white, greyer on breast; bill black; eye very pale yellow or greyish white; legs and feet greyish to yellowish olive. WEIGHT: c. 20 (1st day).

B. c. damarensis (Reichenow): Namibia, W Botswana and SW Angola. Paler and greyer than nominate race with less bold markings.

B. c. maculosus (Temminck): W Africa to Ethiopia, Somalia, N Uganda and N Kenya. Above brighter and more tawny than nominate race.

B. c. dodsoni (Ogilvie-Grant): coastal Ethiopia (Eritrea) to N Somalia (and SW Arabian Peninsula). Paler and more lightly marked than *maculosus*.

Field Characters. Largest African thick-knee and the only one showing no wing-bar at rest or in flight. Upperparts heavily spotted and mottled, not streaked; underparts streaked, especially on breast. In flight wing plain except for white square near tip and white spot on primary coverts.

Voice. Tape-recorded (11, 21, 42, 58, 74, F). Gives a whistled 'ti-ti-ti-teeteetee ti ti ti' growing to crescendo, then dying away towards the end (mainly at night); also harsh growling 'chrrr' anxiety notes, piping alarm notes, and harsh churring notes and 'keh-keeh' when threatening. Young has soft 'chip chip' pleasure call and piercing whistled distress call 'ti ti TEEE'.

General Habits. Occupies a variety of habitats but not waterside; prefers arid areas. Found in savanna and other open woodland, grassland near bush or trees, cultivated lands, stony hillsides, semi-arid scrub, overgrazed and eroded ground, farmyards, large gardens, parks, cemeteries, playing fields, wide beaches and rocky riverbeds. Often found in disturbed weed-grown areas around industrial sites and peri-urban wastelands. Associated with red (laterite) soils in Nigeria. Usually occurs singly or in pairs, and in non-breeding season sometimes in loose flocks of 40–50 at communal daytime roosts. Active mainly at night when often calls, or on dull overcast days. During day roosts, standing or squatting, among stones in the open or under scattered bushes on bare ground. On hot days may lie on ground with legs out behind (**A**). When disturbed will run with head down before taking flight. Sometimes crouches to avoid detection. When flushed, flies with rapid wingbeats, then settles, first spreading its wings and then rapidly folding them.

Subject to local movements (E Africa) but little known.

Food. Insects, crustaceans, molluscs, frogs and some grass seeds.

Breeding Habits. Monogamous, solitary nester.

NEST: shallow scrape lined with small stones, clods, bits of dry plant material, and mammal droppings either in the open on bare ground among grass tufts or low shrubs, or in shade of bush or tree, often next to stone or other object.

EGGS: 2 (rarely 1 or 3), laid at about 48-h interval. Rather elongate oval; pale clay-colour, cream or buff, blotched and spotted with irregular marks of sepia-brown, with smaller underlying marks of ash grey. SIZE: (n = 100, South Africa) 47–58 × 35–41 (52 × 38).

LAYING DATES: Senegambia Apr–July (end dry season, beginning wet season); Mali, May–Oct; Niger, June; Nigeria, Jan, Mar–July (mainly end dry season, early wet season); Chad, Aug; Sudan, May–July (end dry season, early wet season); Ethiopia, Mar–July; Somalia, Mar–June; E Africa: Kenya, Mar–June, Oct–Nov; Tanzania, May, Sept–Dec; Region A, Jan–Feb (dry season); Region C, Jan, May, Sept–Nov (dry season); Region D, Jan–May, Sept–Oct (peaks in rains); Angola, Sept; Malaŵi, Oct; Zambia, July–Oct (end dry season); Zimbabwe, Aug–Dec; South Africa: Orange Free State, Feb; Natal and E Cape, Aug–Dec; W Cape, Sept–Feb.

INCUBATION: starts with 2nd egg or day after clutch complete, by both sexes with ♀ performing most of daytime incubation. Period (n = 2): 24 days. Both sexes defend nesting area by high intensity threatening (**B**), with wings held open, tail raised and fanned, and with calls or harsh notes (see Voice).

A

B

DEVELOPMENT AND CARE OF YOUNG: chicks fed by both parents offering insects in bill, or placing food on ground near chick which then picks it up. Young bird greets parent with head down, wings spread horizontally, tail raised and fanned, and 'chip–chip' greeting calls. Captive young first flew when 8 weeks old.

References
Bigalke, R. (1933).
Broekhuysen, G. J. (1964).
Maclean, G. L. (1966).

Family GLAREOLIDAE: coursers and pratincoles

Small to medium-sized Old World plover-like birds, in 2 subfamilies, the ground-feeding coursers (Cursoriinae) of dry habitats, and the aerial-feeding pratincoles (Glareolinae) associated with water. Included with the coursers is Egyptian Plover *Pluvianus aegyptius* and with the pratincoles the long-legged Australian Pratincole (or Courser) *Stiltia isabella*, intermediate between pratincoles and coursers. Sexes are nearly alike. Bill rather short; nostrils situated at base and not in a groove. Neck short with 15 cervical vertebrae. Wings long and pointed; 14–16 secondaries; 12 tail-feathers. Tarsi with transverse scutellations front and back except in Egyptian Plover. Caeca present; crop absent. Relation to other groups unclear, though seems closest to Charadriidae; many species very plover-like. 17 spp., Old World only, 12 in Africa where 8 endemic.

Subfamily CURSORIINAE: coursers and Egyptian Plover

Ground-feeding terrestrial birds. Several genera of coursers have been recognized; we place all African species in *Cursorius* to emphasize their uniformity. Egyptian Plover is of uncertain affinities, probably closest to coursers; sometimes considered closer to pratincoles (Maclean 1978) or in its own family Pluvianidae (Mackworth-Praed and Grant 1970). 8 spp., 7 in Africa.

Genus *Pluvianus* Vieillot

A strikingly patterned, riverine bird. Structurally similar to *Cursorius* but bill shorter, not decurved; nostrils rounded posteriorly; interorbital septum not extending into base of upper jaw; ectethmoid plate small, occupying medial half of anterior orbit; lachrymal articulating with dorsal swelling of jugal bar (W. Bock, pers. comm.). Tarsi without transverse scutellations front and back. Middle toe claw with flange on inner edge, not pectinate. Down pattern type closest to coursers'. Monotypic, endemic to Africa.

Plates 19, 20

(Opp. pp. 320, 321)

Pluvianus aegyptius (Linnaeus). Egyptian Plover. Pluvian d'Egypte.

Charadrius aegyptius Linnaeus, 1758. Syst. Nat. (10th ed.), p. 150; Egypt.

Range and Status. Endemic, rare to abundant according to local rather than regional conditions; not extensively migratory but makes local movements; lowlands, below 500 m, between 18°N and 10°S except heavily forested regions and areas east of Rift Valley. Formerly abundant along Nile in northern Sudan and Egypt but extinct there since early 20th century. Vagrant Canary Is., Jordan R. valley, Libya.

Description. ADULT ♂: crown, eye-stripe, ear-coverts, sides of neck, hindneck and mantle glossy black; supercilium white, joined around nape; mantle broadly edged with white, posterior mantle feathers long, extending over rump; rump, upppertail-coverts grey; rectrices grey, each with subterminal black band and white tip; underparts buffy white to tawny, palest on chin, upper throat and belly, darkest on flanks and undertail-coverts; black band across breast; remiges white, banded with black basally and/or terminally; outermost primary, alula, lesser and median primary coverts, black; rest of coverts, scapulars and tertials grey; underwing-coverts and axillaries white. Bill black; eye dark brown; legs and feet blue-grey. Sexes alike. SIZE: wing, ♂ (n = 18) 130–143 (135·1); ♀ (n = 16) 127–143 (136·5); tail, ♂ (n = 10) 55–60; ♀ (n = 10) 56–62; bill, ♂♀ (n = 42) 16–19 (17·5), ♂♀ (n = 20) 15–18 (no mean); tarsus, ♂♀ (n = 62) 33–36 (34·4); WEIGHT: ♂♀ (n = 25) 73–92 (81·3).

IMMATURE: resembles adult, but lesser and median coverts rusty brown; body feathers of upperparts with brownish margins.

DOWNY YOUNG: upperparts brownish buff, speckled and blotched with black; post-ocular supercilium white, meeting white patch on nape; eye-stripe black; wings white; underparts white; bill black, lining dark red; eye dark brown; legs and feet blue-grey.

Pluvianus aegyptius

Field Characters. Length *c.* 22 cm; superficially plover-like, but bold pattern of black, grey, white and tawny diagnostic; spread wing pattern unique, largely white with oblique black stripe and black edging.

Voice. Tape-recorded (217, 245). Commonest call a rapid series of loud 'chersk', usually uttered during agitation or aggression, often in flight. Also gives an incisive cluck when agitated near nest, loudest when predator approaches, and this also accompanies visual threat display; a single 'wheep' given as an alerting signal; and sometimes a 'cherk' or 'cluck' during copulation. A 'wheep' followed by a few 'cherks' from bird close to nest usually brings an approach by its mate.

General Habits. Confined to immediate vicinity of rivers, especially those with sand, silt and gravel bars; usually not found around lakes or ponds except in non-breeding season, and not along coasts except where rivers reach sea. In non-breeding season, may form flocks of up to 60; roosting habits not recorded, presumably roosts on riverine rocks or sandbars. Forages mostly above water line on islands in rivers or along banks, occasionally into open ground up to several 100 m away. Forages by: (1) surface-picking, often after stalking or running chase; (2) running after and capturing low-flying insects; may startle insects up by spreading wings; (3) sand-probing, including tossing sand sideways with bill; (4) jumping forward and scratching backward with both feet; (5) stone-turning, like that of turnstones *Arenaria* spp., using bill to flip over stones (up to 70 g) or other objects.

Contention that Egyptian Plover picks food from gaping jaws of crocodile dates back to Herodotus' visit to Egypt in 459 B.C. (his account probably, but not certainly, refers to *Pluvianus*). Brehm in his 19th century work 'Tierleben', romanticized the *Pluvianus*–crocodile association and stated he had seen jaw-picking many times; Meinertzhagen (1959) described jaw-picking by *Pluvianus* (observation undated) and Spur-winged Plover *Vanellus armatus* (in 1907). Both authors apparently relied on memory of times long past; no other naturalists have ever reported seeing jaw-picking.

Makes irregular movements in response to water-level changes; also undergoes some longer movements, e.g. 2 birds ringed Ethiopia (Gambela, Nov and Apr), recovered Sudan (Kosti, Aug, 588 km away and Khartoum, Sept, 840 km respectively). Perhaps a seasonal migrant Nigeria where it leaves southern rivers in wet season, June–Oct, and appears in peak numbers in northern areas of Nigeria. Performs similar behaviour in Chad when it leaves Fort Lamy district May–Sept and makes irregular northward movements to temporary wetlands in Sahel zone.

Food. Primarily insects and other small invertebrates, swallowed whole; also seeds and particles of scavenged fish.

Breeding Habits. Monogamous; solitary nester on islands in rivers. Early stages of pair-formation not described. Pair establishes nesting territory together, becoming extremely aggressive toward other *Pluvianus* and most other species, especially potential predators and competitors for food. Usual territory is all or part of a sand–silt–gravel island formed in river during dry season. Pair attempts to maintain entire island as territory regardless of size, but if strongly challenged may successfully defend only an area with radius of *c.* 15 m around nest. Territorial behaviour and aggressive displays identical in each sex. Immediately threatens and attacks any conspecific bird other than mate if it fails to depart. As intruder approaches, territorial bird first assumes alert posture with plumage compressed, neck extended vertically; next changes to ruffled-out body plumage with neck not extended, facing opponent; this followed by hunched run with head lowered and horizontally directed toward opponent. If intruder does not retreat, territorial bird tilts foreparts down, raises head, suddenly spreads wings fully and tilts them forward, startlingly displaying white and black pattern. (see **A**). If opponent remains, bird attacks with bill-pecking and wing-flapping and pursues retreating intruder in flight. Aggressive intruders respond with same displays and actions; sometimes fighting birds tumble across sand into water. Territorial bird attacks potential predators as above but fully extends and slowly waves wings up and down (**B**); readily threatens and attacks birds as large as storks and eagles.

EGGS: 1–4, laid at 1-day intervals; av. clutch (n = 14, Ethiopia) 2·43 eggs; 4-egg clutches have been recorded only in Egypt (Koenig 1926). Oval; ground colour light yellowish brown, with numerous small spots from reddish brown to grey distributed randomly over entire surface. SIZE: (n = 19) 28·4–33·9 × 23·0–24·4 (30·9 × 23·7). WEIGHT: (n = 19) 8·51–10·48 (9·50).

LAYING DATES: Senegambia Mar–Apr; Sierra Leone, Feb–Mar; Nigeria, Feb–May; Chad, Feb; Sudan,

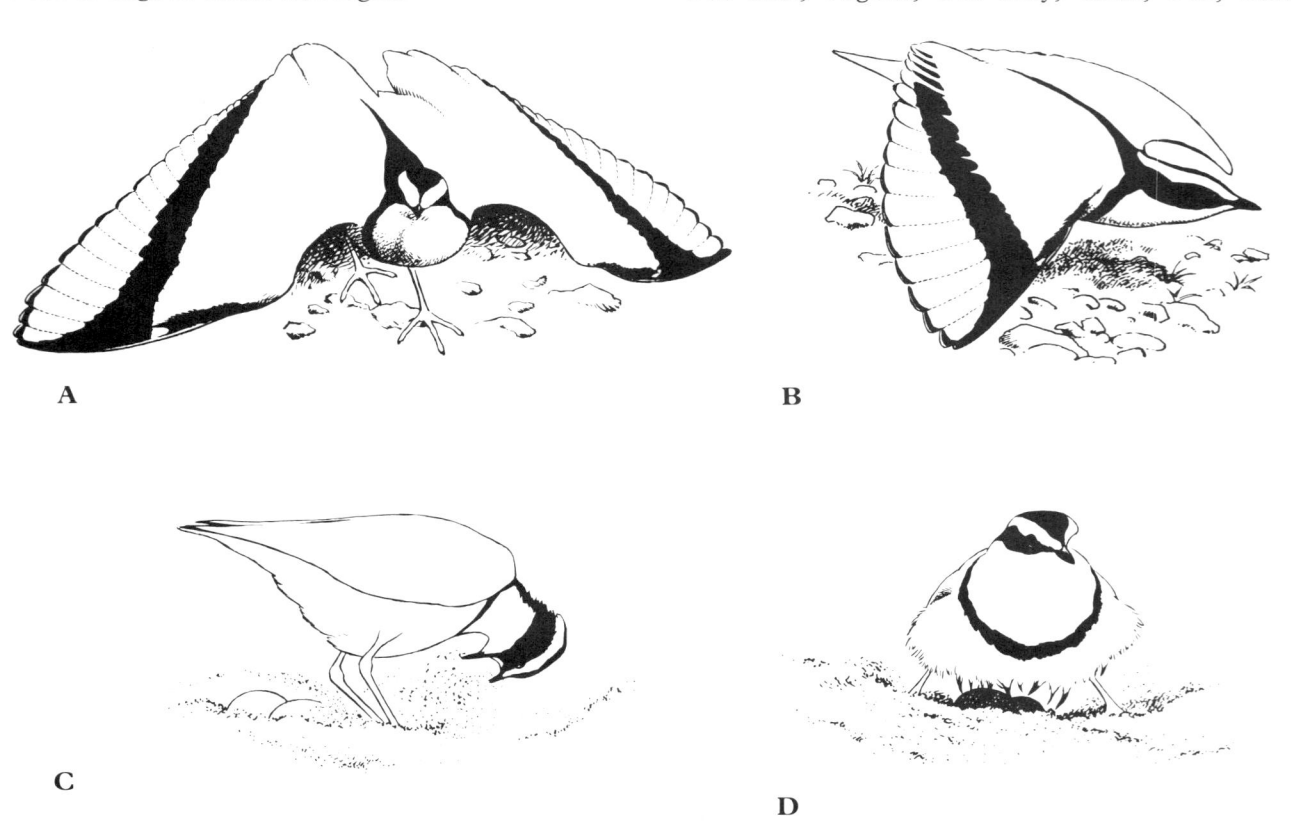

A

B

C

D

Copulations occur from 11 days before to within 2 days after egg-laying. Usually no obvious preliminary display; ♀ sinks into low crouch, wings and tail not spread or quivered, usually without calling; ♂ mounts usually without displaying, wing-spreading, calling or pecking, effects 1 cloacal contact, dismounts after 2–3 s, may then make a few nest-scraping movements; or may run up and half-circle ♀ before mounting; sometimes a few 'cherk' notes or 'clucks' given during copulation by one or both birds.

NEST: a scrape without added material. Pairs begin scrape-making while establishing territories, up to at least 30 days before egg-laying. Many scrapes made, by both sexes; some scrapes repeatedly excavated and 1 eventually used as nest. Bird lowers foreparts, raises head, tilts up hind parts, scrapes backward with feet, settles and rotates while continuing foot-scraping. Whilst scraping, mate may approach it in hunched run, circle or half-circle it, then commence nest-scraping close by. Scrape-making continues to time of egg-laying, and afterward as distraction display. Final scrape, c. 15 cm diam. and 5 cm deep, is fully exposed to sun.

Jan–Apr; Ethiopia, Jan–Apr; Egypt (formerly), Apr (Koenig 1926). Correlated with exposure during dry season of river islands suitable for nesting.

INCUBATION: begins 1st egg, by both sexes. Intervals between nest reliefs 0–4 h in early morning, increasing to 10–20 min during hottest 6 h, declining to 0–3 h in late afternoon. Incubation period: 28–31 (30) days. Whenever bird leaves nest (except when being relieved by mate), tosses sand over eggs with forward/backward motion of bill while standing over nest (**C**), covering them to depth of 2–3 mm (rarely to 10 mm); never uses feet. Adult sits on nest for 1–2 h after sunrise, then leaves nest for 1–2 h after covering eggs with sand. Egg temperature remains within safe range as solar heat is moderate. Bird returns c. 4 h after sunrise when solar heat becomes intense, sits briefly, then goes to river, soaks belly feathers by rapid forward/backward rocking motions ending with quick bill-dip, returns to nest and settles, thoroughly wetting sand and cooling eggs (**D**). Temperature of sunny dry sand now exceeds normal limit of incubated eggs and by midday even shaded air temperature exceeds safe level. Throughout hottest 6 h of day, adults soak feathers in river every few min and return to wet buried eggs which, with shading, keeps them between 34·0 and 41·9°C (mean 37·5°C). From c. 2 h before sunset to c. 1 h after, adults leave nest largely unattended. Returning birds appear to test temperature of sand around buried eggs by probing with opened bill. Adult (sex not determined) sits on nest all night, uncovering eggs to c. two thirds of their depth.

DEVELOPMENT AND CARE OF YOUNG: parents attend nest during hatching. Time from pipping to hatching is 5–18 h. Hatching proceeds with eggs buried in sand; chick gets head above surface before fully emerging from egg, not aided by parents. Weight at hatching (n = 6) 7·0–7·5 (mean 7·23), c. 74% of initial egg weight. Chicks extremely precocial, eyes fully open, capable of rapid running within 1 h after hatching; may go to water up to 30 m away to drink or bathe. Leave nest permanently by end of 1st day. Do not forage during first few days and parents infrequently carry small food items to them in bill. Later accompany foraging adults, which expose food for them.

Recently hatched chicks still in nest with eggs are wetted by adults, and covered with sand if adults leave. Chicks in or out of nest and in danger from predator crouch down and are completely covered with sand by adult, in same manner as eggs. Adults use sand-tossing with bill and scrape-making movements as distraction displays. After danger passes, adult uncovers chick (can also emerge unaided). Parents cover chicks until they are at least 3–4 weeks of age. Young chicks that are covered remain immobile if prodded and scarcely move even if dug out; older chicks remain immobile if prodded and during excavation, then suddenly run off at top speed; buried juvenile may even fly when uncovered. If covered chicks remain in hot sun, adults wet them with soaked belly feathers. Butler (1931) reported an adult regurgitating water on a buried chick, an observation widely quoted but never repeated. Downy chick capable of swimming, even under water, but generally remains on island where hatched until able to fly. Estimated age at 1st flight 30–35 days.

BREEDING SUCCESS/SURVIVAL: Pied Crow *Corvus albus* predator of eggs and perhaps chicks (Ethiopia); probes for them in sand. Black Kite *Milvus migrans* forages constantly over rivers and probably takes some chicks. Eggs and chicks are destroyed if river rises and floods nesting island. In SW Ethiopia almost all eggs hatch unless destroyed by predation or flooding (sample size small).

References
Butler, A. E. (1931).
Howell, T. R. (1979).
Koenig, A. (1926).

Cursorius cursor **superspecies**

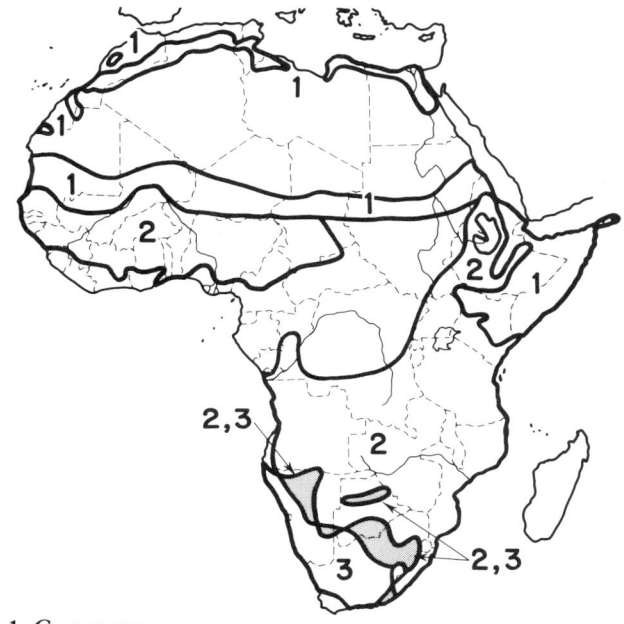

1 *C. cursor*
2 *C. temminckii*
3 *C. rufus*

Genus *Cursorius* Latham

Plover-like birds of arid regions and dry savannas. Legs longer than plovers', bill medium or longish, sometimes used for digging, culmen strongly decurved, without swollen dertrum at tip. Plumage variable, plain or mottled above in adult, always mottled in immature; below usually plain with bands across breast or belly; sometimes lightly streaked on neck and upper breast. Tail square. Legs usually white, but yellow or reddish in 2 spp.; toes short, inner toe absent, middle claw pectinate. Eyes relatively large; diurnal, crepuscular and (1 sp.) nocturnal. Neonatal plumage highly variable, usually complex, with patterns of brown, black and white, rarely simple with black and grey only; in colour and structure like those of sandgrouse (Pteroclidae). 7 spp., 6 in Africa where 5 endemic.

C. cursor and *C. rufus* are very closely allied and comprise an arid-zone superspecies. Between them is *C. temminckii*, whose range is almost exactly parapatric with that of *C. cursor* and nearly so with that of *C. rufus*. Its plumage is very like that of *rufus* and we regard it as a member of the superspecies.

Plates 19, 20
(Opp. pp. 320, 321))

Cursorius cursor (Latham). **Cream-coloured Courser. Courvite isabelle.**

Charadrius cursor Latham, 1787. Gen. Syn. Suppl. 1, p. 293; Kent.

Forms a superspecies with *C. rufus* and *C. temminckii*.

Range and Status. Breeds Africa, Cape Verde and Canary Is., Socotra, Arabian Peninsula, Syria, Iraq, Iran and Afghanistan. Winters in N Africa, Arabia and NW India, also in Sahelian zone from Senegambia to Ethiopia and Somalia, and dry parts of E Africa south to S Kenya.

Resident and intra-African migrant, locally common S Morocco to Egypt, south to Mauritania, Senegambia, Mali, Niger, Chad, N Sudan, N and E Ethiopia, Somalia and NE Kenya. No evidence that it breeds in Central Sahara, but widespread near its borders.

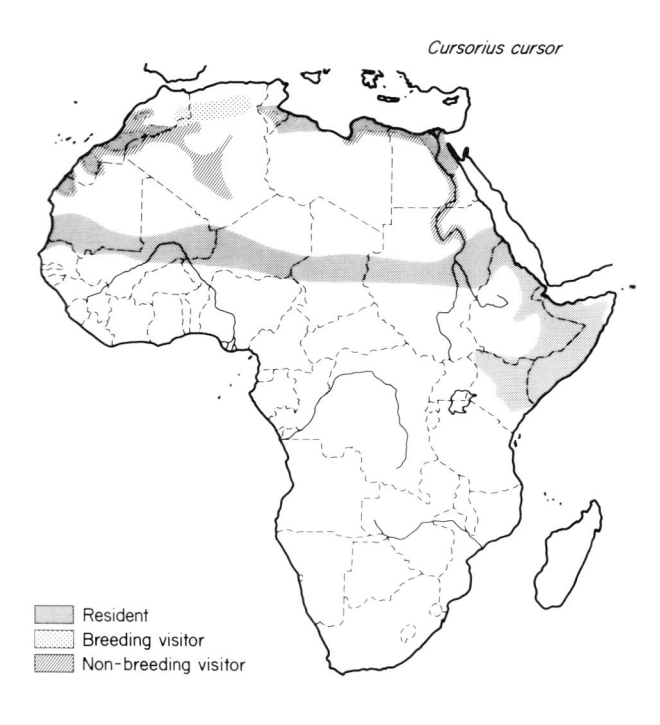

Description. *C. c. cursor* (Latham): N Africa, Socotra. ADULT ♂: forecrown pale sandy, tinged rusty and merging into light grey hindcrown; broad white superciliary stripe, curving down behind eye and meeting on nape where bordered above and below by narrow black stripe; superciliary stripe bordered below by broad black line from eye, narrowing to nape. Hindneck, sides of neck and of face, entire upperparts and tail yellowish sandy buff with slight rufous tinge; innermost rectrices with black subterminal bar or spot and whitish tip; outermost rectrices bordered white on outer web. Chin and throat white merging to pale sandy on foreneck and breast; centre of belly and undertail-coverts white; flanks pale sandy. Primaries black with narrow buff edge to inner web; secondaries blackish with sandy buff outer web and broad white tip; innermost secondaries pale sandy; upperwing-coverts yellowish sandy buff with slight rufous tinge; axillaries and underwing-coverts black. Bill black with white at base of mandible; eye dark brown; legs and feet white. Sexes alike. SIZE: wing, ♂ (n = 13) 155–165 (161), ♀ (n = 12) 148–163 (156); tail, ♂ (n = 14) 56–64 (60·2), ♀ (n = 10) 56–62 (59·4); bill, ♂ (n = 15) 22–26 (23·3), ♀ (n = 11) 21–25 (23·7); tarsus, ♂ (n = 15) 52–60 (55·4), ♀ (n = 11) 51–60 (55·6). WEIGHT: ♂ (Syria, June) 119, (southern Sahara) 102, (Kenya) 93, ♀ (southern Sahara, Jan) 102, 114, (Kenya) 107.

IMMATURE: entire upperparts (including crown) brownish buff with narrow wavy bars of greyish brown; superciliary stripe pale sandy, dark line behind eye narrow and inconspicuous; underparts as in adult, but breast feathers with narrow brown bars; primaries tipped with tawny buff.

DOWNY YOUNG: dense down all over; upperparts finely mottled sandy rufous and white; nape buffy white; dusky brown streak in centre of crown; breast buffy white; throat and remaining underparts white; large-eyed; long-legged.

C. c. somalensis (Shelley): Somalia and E Ethiopia. Slightly darker than nominate race; wing smaller (129–140).

C. c. littoralis (Erlanger): S Somalia to N Kenya. Darker than *somalensis* and more greyish brown.

Field Characters. A pale sandy running bird with black wing-tips and axillaries, long white legs, black and white eye-stripes, but no breast-band. Small-headed and slender. Similar to Temminck's Courser *C. temminckii* but belly pale without black patch in centre.

Voice. Tape-recorded (62, 73). Rather silent; occasionally noisy. Commonest call, a sharp piping 'wit' or 'krit', often repeated 'wit krit', usually in flight, sometimes on ground; a hoarse, croaking 'hark', 'nhark' or 'praak' also repeated, usually in flight; display call 'quit quit whow'; ground call (feeding) 'wut-quoi'; with young 'qua qua', 'weeou', 'cluck'.

General Habits. Inhabits desert, semi-desert, short grassland, arid scrub, cultivated lands, and sandy or stony plains; normally in loose flocks of 5–30. Runs fast, stopping for a few s between runs, either to look about or peck at insects on ground; may also poke into top few cm of soft sand for food. Food items swallowed whole. Often bobs head or tail before running on again. Easily approached and somewhat inquisitive. When disturbed, usually runs a short way before taking wing; may run and crouch, sometimes close to ground, when stretches head forward and closes eyes to slits. When alarmed, stretches upright, cranes neck and appears to stand on tiptoe; then flock may run off together behind a mound or bush, over which intruder is watched with heads stretched up. Flies with curious, jerky flight; groups of 3–4 birds (Sudan, Oct) sometimes tumble about in the air in the cool early morning hours, giving piping calls (see Voice).

C. c. cursor is scarce or absent in N Africa in winter, except in coastal Morocco and Libya; winters mainly south of Sahara, from Mauritania and Mali to Ethiopia where locally common; rare Somalia. Main autumn movement trans-Saharan in 2nd half Sept through Oct. Present Mali Aug–Apr, Niger Oct–Feb, central Chad common Oct–early Mar, Sudan Oct–May (common south to Darfur and Kordofan mid-Oct to mid-Mar), Ethiopia (Eritrea) late Sept–Mar. Migrates north in Sahara in Mar–Apr, some early May; c. 1000 Tunisia in mid–Mar. Breeding populations, Atlantic coastal plains, Morocco, sedentary; present in Algerian Sahara (Beni Abbès) in winter and spring, when breeds, absent in summer (might then have moved north to breed again: Brosset 1957). *C. c. somalensis* is apparently resident; *C. c. littoralis* move southward in Kenya from Sept to May.

Food. Adult and larval insects (grasshoppers, bugs, beetles, ants, flies, caterpillars), small lizards and snails; seeds (recorded in Tunisia and Socotra only).

Breeding Habits. Said to be monogamous, but mating system may in fact be complex (Cramp and Simmons 1983); solitary nester; probable courtship is high, circling flight with quickly beating wings and display calling; this display flight also presumed ♂–♂ aggressive behaviour. Sometimes an adult stands up with chest thrust out, head drawn back, and bill more or less parallel to ground; then with head and legs stationary, moves chest rapidly up and down; this presumed aggressive display performed by several members of a flock or pairs facing each other. When copulating, ♀ first crouches flat; ♂ approaches her with a pattering leg action and upstretched gait, then mounts her with fluttering, outspread wings. After dismounting, ♂ may 'scoop' at ♀ rapidly with his head.

NEST: usually no nest at all; sometimes in scrape which may become slightly hollowed during incubation.

EGGS: 2 (rarely 3). N African mean (100 clutches) 2·0. Short elliptical; cream to yellowish buff, finely spotted and scribbled all over with brown and grey, giving a uniformly darkish effect; sometimes a distinct band of markings at larger end. SIZE: (n = 100, N Africa) 32–39 × 26–28 (35 × 27). WEIGHT (estimated): 14.

LAYING DATES: Morocco, Algeria, Tunisia, Libya and Egypt, Mar–June; Senegambia and Mauritania, Apr; Somalia, May–June; Socotra, Mar–June; E Africa: Kenya, Apr–Sept; Region D, May–Aug (in cool dry season after long rains).

INCUBATION: probably by both sexes, usually $1\frac{1}{4}$–2 h at a time (E Africa); period unknown. Very tame at nest. When intruder present, either steps off nest quietly or performs distraction display with wings held out horizontally and tail fully spread and fanned down. In heat of day incubating bird crouches over its eggs with legs exposed, above or barely touching the eggs. As the ambient temperature increases, the incubating bird performs gular fluttering, spreads its wings slightly and ruffles the feathers on crown and back.

DEVELOPMENT AND CARE OF YOUNG: probably both parents remove hatched shells from nest; either parent will tend young, which remain with parents until after fledging. Young make exploratory trips from nest as soon as dry; eventually parent leads young away from nest 7–8 h later. Initially young are fed bill-to-bill by both parents.

References

Cramp, S. and Reynolds, J. F. (1972).
Cramp, S. and Simmons, K. E. L. (1983).

Cursorius rufus Gould. Burchell's Courser. Courvite de Burchell.

Plate 19
(Opp. p. 320)

Cursorius rufus Gould, 1837. Proc. Zool. Soc. London, p. 81; Potchefstroom.

Forms a superspecies with *C. cursor* and *C. temminckii*.

Range and Status. Endemic resident, nomadic, locally common, Namibia, Botswana, South Africa, SW Angola. In some parts of its range overlaps with Temminck's Courser *C. temminckii* but occupies more arid areas.

Description. ADULT ♂: forehead and forecrown pale sandy, tinged rust and merging into light blue-grey hindcrown and nape; broad white superciliary stripe, curving down behind eye and meeting on nape where bordered above and below by black. Neck, ear-coverts and face light rufous-brown, merging to buffy white on face, chin and throat. Mantle and back light rufous brown; rump and uppertail-coverts buffy grey; outermost rectrices white with black subterminal spot and brown vermiculations basally; remaining rectrices greyish with white tip and black subterminal spot (central pair tipped greyish). Lower throat and breast pinkish buff to light rufous; middle of lower breast dusky brown; flanks, belly and undertail-coverts white. Primaries, alula and primary coverts black; secondaries greyish brown, broadly tipped white and with white inner webs; scapulars and tertials light rufous brown; remaining upperwing-coverts light rufous brown; underwing coverts blackish. Bill black; eye dark brown; legs and toes whitish grey. Sexes alike. SIZE: ♂♀ (n = 13) wing, 132–138 (135); tail, 48–53 (51·2); bill, 21–25·5 (22·9); tarsus, 46·5–51 (48). WEIGHT: adult (Namibia) 75.

IMMATURE: less rufous than adult, mottled and barred with black above; no grey hindcrown and nape; black and white eye-stripes absent; tail mottled with tawny; underparts as in adult, but mottled with black on breast.

DOWNY YOUNG: not described.

Field Characters. Differs from Cream-coloured Courser *C. cursor* (Sahara, E Africa) only in having lower breast brown and demarcated from white belly. Very similar to Temminck's Courser (which it just overlaps) but distinguished by being slightly larger and in having blue-grey, not bright chestnut hindcrown, narrower black eye-stripe (giving less masked effect), uniform pale brown breast terminated below by dark brown bar, uniform whitish belly without black patch, and in flight a pale speculum formed by whitish ends to secondaries. In flight, wing of Temminck's Courser is all dark.

Voice. Tape-recorded (233). A single hoarse 'chuk' on take-off, grunting 'chuk chuk' in flight, and a 3-syllabled 'kok-kok-kwich' or double 'kwirrt-kwirrt' contact call. Voice less harsh than that of Temminck's Courser.

General Habits. Favours very open habitats from short, overgrazed grassland and pastures, and burnt grassland, to semi-desert and true desert; also bare saltpans, sandy or gravelly plains and stony areas dotted with small shrubs. Occurs in pairs or small loose flocks of up to *c*. 10 birds. Forages like other coursers, with quick runs interspersed with pecks at the ground. Nasal salt gland secretes excess sodium from food (insect body fluids) and aids water conservation. When alarmed, bobs tail, stands bolt upright and sways body slowly while holding head still, then runs away. Prefers to escape by running rather than by flying but flight swift and strong. Movements local and highly nomadic.

Food. Insects, especially Harvester Termites *Hodotermes mossambicus*; also seeds.

Breeding Habits. Monogamous; solitary nester.

NEST: on bare ground; sometimes a small patch of soil may be cleared of debris and ringed with small bits of dried plant material, antelope droppings or tiny stones. Shallow scrape may form during incubation.

EGGS: 2 (rarely 1); South Africa mean (12 clutches) 1·9. Rounded oval; creamy white to pale stone or pale fawn, thickly covered with fine speckles and streaks of sepia and black over some grey undermarkings; eggs look almost black at a distance. SIZE: (n = 42, South Africa) 27·8–32·9 × 22·8–26·2 (30·4 × 24).

LAYING DATES: Namibia, Botswana and South Africa, July–Dec (usually in dry season, just before main rains); rarely Feb.

INCUBATION: probably by both sexes; sitting parent performs displacement brooding as distraction display when disturbed from nest. During incubation eggs may be moved several cm as bird rolls them under body on return to nest-site.

Reference
Dixon, J. E. W. (1975).

Plates 19, 20
(Opp. pp. 320, 321)

Cursorius temminckii Swainson. Temminck's Courser. Courvite de Temminck.

Cursorius Temminckii Swainson, 1822. Zool. Ill. 2, text to pl. 106; no type locality (= Senegal).

Forms a superspecies with *C. cursor* and *C. rufus*.

Range and Status. Endemic resident, nomadic, locally common to frequent S Mauritania, Senegambia and S Mali to Ghana and Nigeria, across Sahel to S Sudan and Ethiopia, south to Kenya, Uganda, Tanzania, S and E Zaïre, Angola, Zambia, Zimbabwe, Namibia, Botswana, Mozambique and South Africa. Absent equatorial forest belt and W Cape. Appears to overlap with its allospecies Cream-coloured Courser *C. cursor* and Burchell's Courser *C. rufus*, but ranges may in fact be mutually exclusive.

Description. ADULT ♂: crown bright chestnut, bordered at nape by black band running forward on each side to eye; superciliary stripe white, meeting at nape; neck light greyish brown, paler on foreneck; lores white; sides of face chestnut-buff; chin and throat white, merging with brown of foreneck. Rest of upperparts and two central rectrices light greyish brown; remaining rectrices greyish brown with black subterminal band and white tip; outermost rectrices white with some greyish brown on inner vane. Breast light brown washed with chestnut, merging with deep chestnut on lower breast and belly; central belly patch black, passing down between legs; sides of belly, thighs and undertail-coverts white. Primaries glossy black; outer secondaries black, inner secondaries greyish brown with blackish outer vane and white wedge at tip; innermost secondaries like back; greater coverts black; remaining wing-coverts greyish brown; underwing-coverts black. Bill greyish black (mandible yellowish horn to grey at base); eye blackish brown; legs and feet white. Sexes alike. SIZE: ♂♀ (n = 27) wing, 118–132 (124); tail, 41–50 (45·9); bill, 19–22 (20·2); tarsus, 37–43 (39·9). WEIGHT: adult, (Zambia) 66·5, ♂ (Kenya) 64, 65, (Botswana, n = 5) 68·6; ♀ (Kenya) 70, 71, (Botswana, n = 4) 68–80·5 (74·3).

IMMATURE: crown streaked blackish; superciliary stripe buff; underparts duller than in adult, feathers edged with buff and with dark subterminal band; wing-coverts creamy buff, tipped whitish and with dark brown subterminal band; chest paler than in adult; belly patch smaller.

DOWNY YOUNG: above boldly patterned with patches of golden brown and white, interlaced broadly with black and with a distinct off-white collar; below all white except for buff chest bands; legs and bill grey.

Cursorius temminckii

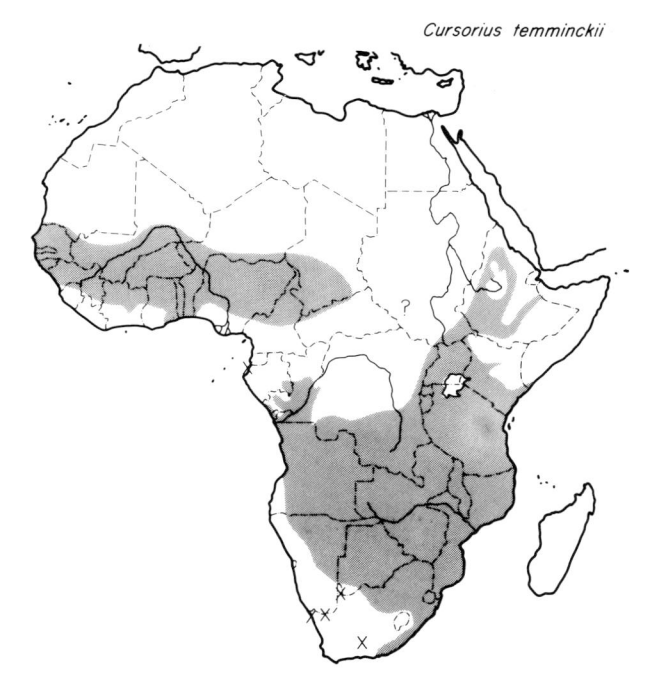

Field Characters. A small courser (length *c.* 20 cm), generally brown above with chestnut breast, dark central belly patch; entire crown bright rusty (not grey as in Burchell's Courser); white eyebrow; and long white legs. Also distinguished by having dark belly patch down abdomen between legs; in Burchell's Courser just dark breast-band.

Voice. Tape-recorded (217). Sharp metallic 'err-err-err' in flight, likened to sound of rusty hinge.

General Habits. Inhabits semi-arid bush savanna, bushveld, short grasslands, ploughed fields and airfields in savanna, burnt savanna grasslands. Occurs in pairs or flocks of around 20, sometimes up to 40. Forages like other coursers by alternate runs and pecks at ground. When feeding, may perform presumed aggressive display by standing upright with chest thrust out, head drawn back and bill held more or less parallel to the ground; then holding head and legs stationary, moves chest rapidly up and down. When alarmed, raises and lowers body by straightening and bending legs while keeping head at one level. Runs and flies fast. Recorded to fly at and disturb pairs of copulating Cream-coloured Coursers (Kenya, J. F. Reynolds, pers. comm.).

Highly nomadic, appearing to breed within days after savanna grasslands burned. Migratory in extreme southern South Africa where present usually Feb–Aug (late summer to winter). In Zaïre (central Katanga) present only Mar–June; in Malawi and Zambia chiefly Apr–Nov in dry season; in northern Tanzania absent Apr–May; and Cameroon, mainly Nov–Mar in dry season. Elsewhere subject only to short movements, either altitudinal or latitudinal.

Food. Insects, molluscs, seeds.

Breeding Habits. Monogamous; solitary nester.
NEST: no nest built; lays eggs on bare ground; sometimes simple scrape formed during incubation.
EGGS: 2. Rounded, pale creamy yellow to white, with dense fine speckles and streaks of blackish brown overlaying pale grey markings. SIZE: (n = 52, South Africa) 25–32·3 × 21·5–24·8 (27·7 × 22·8).
LAYING DATES: Senegambia, Jan–Aug, Nov–Dec; Mauritania, Mar; Nigeria, Feb–June; Mali, May–June; Sudan, Mar–July; Ethiopia, Apr–May, July; E Africa: Region A, Apr; Region B, Aug; Region C, May–Sept, Nov–Dec (with most June–July in dry season); and Region D, Jan, Apr, June, Aug; Malawi, July–Nov with most Aug; Zaïre Aug–Oct; Zambia, June–Oct (most July–Sept); Zimbabwe, May–Dec (most Aug–Oct); South Africa: Transvaal, Natal, Cape, July–Oct. Nests usually in dry season, favouring areas of burnt grass; in southern Africa grass usually burnt July–Sept.
INCUBATION: by both sexes, usually 1¼–2 h at a time (E Africa); period unknown. When incubating, may crouch over eggs with legs exposed, above or barely touching eggs. As ambient temperature increases, performs gular fluttering, spreads wings slightly and ruffles feathers of crown and back; in E Africa 'off-duty' parent seeks shade in middle of day. Anti-predator strategies at nest include flying directly from nest, stepping off nest quietly or feigning injury with wings held out horizontally and tail fully spread and fanned down; sometimes rushes at and pecks at approaching intruder. At nest relief departing bird may peck at ground and pull bits of grit into nest-scrape when mate approaches (J. F. Reynolds, pers. comm.).
DEVELOPMENT AND CARE OF YOUNG: both parents probably remove hatched shells from nest; both attend young. Young remain with parents at least until fledged. Young journey from nest as soon as dry; leave nest with parents 7–8 h after hatching.

Reference
Steyn, P. (1965).

Cursorius africanus Temminck. Double-banded Courser. Courvite à double bande.

Plates 19, 20
(Opp. pp. 320, 321)

Cursorius africanus Temminck, 1807. Cat. Syst. Cab. Orn., p. 175; (Namaqualand).

Range and Status. Endemic resident, occasionally nomadic, common to uncommon, discontinuously distributed in three discrete populations: (a) southern Africa from SW Angola through Namibia and W Botswana to N and W Cape, Karoo, W Orange Free State and W Transvaal; (b) central Tanzania and S Kenya; (c) Ethiopia and N Somalia.

Description. *C. a. africanus*: SW and central Kalahari, N Cape Province and S Namibia. ADULT ♂: crown feathers smoky brown in centre with black subterminal patch and pale sandy buff border; nape, hindneck and ear-coverts sandy brownish, finely streaked with black. Cheeks and superciliary line plain brownish sandy; streak behind eye dusky. Chin, throat and foreneck pale sandy with fine black streaks. Back feathers and upperwing-coverts smoky brown, bordered nar-

Cursorius africanus

rowly with black and fringed broadly with pale sandy; rump and uppertail-coverts white, coverts black at base; central rectrices black with pale sandy border and tip; outermost rectrices white; remaining rectrices black with whitish border and tip. Upper breast pale sandy bordered below by broad black band which extends right around mantle as complete ring; a second similar black breast-band below the first, separated from it by a pale sandy band of somewhat greater width; rest of underparts pale sandy fading to buffy white on belly and undertail-coverts; outer four primaries black with pale brown inner border; inner primaries and outer secondaries bright rusty rufous; inner secondaries and tertials like back feathers; underwing-coverts and axillaries white. Bill black; eye dark brown; legs greyish white. In some individuals the pale sandy feather areas may have a faint rusty tinge. Sexes alike. SIZE: ♂♀ (n = 28) wing, 145–159 (151); tail, 61–73 (65·4); bill, 13–15 (14·1) tarsus, 49–59 (53·7) WEIGHT: adult (South Africa) 97·4; ♂ (Botswana, n = 10) 69–104 (86); ♀ (Namibia) 88, (Botswana, n = 14) 83–101 (90).

IMMATURE: similar to adult, but lacks breast-bands; feathers more shredded in texture and with dark shaft-streaks dorsally, giving more mottled effect.

DOWNY YOUNG: above intricately patterned with various shades of brown (golden to chocolate), black and white, with white occurring as a forehead patch mixed with brown, as a horseshoe shape on nape, as another horseshoe on upper back, and as smaller irregular patches elsewhere. Dark brown to blackish areas concentrated between white and golden-brown areas. Below, variable; may be white, buff or pale golden. Naked areas of skin blackish. Bill black to greyish-brown with black tip; egg-tooth creamy white; eye very dark brown; legs and feet blue grey.

C. a. granti (Sclater): Western South Africa, except N Cape. Mantle dark brown rather than slaty brown; sandy areas more rufous.

C. a. sharpei (Erlanger): Namibia except south. Paler than *C. a. africanus*; wing 138–159.

C. a. bisignatus (Hartlaub): SW Angola. Smaller than *C. a. africanus*; wing 130–138; more rufous above; more finely streaked on throat.

C. a. traylori (Irwin): Botswana. Paler and greyer than *C. a. sharpei*; sandy areas white or off-white.

C. a. gracilis (Fischer and Reichenow): Tanzania and Kenya. Small, wing 126–134, dark and greyish; more rufous on primaries that *C. a. bisignatus*.

C. a. hartingi (Sharpe): Somalia and Ogaden. Dorsal feathers with cinnamon-rufous centre and cream border.

C. a. raffertyi (Mearns): Ethiopia. Similar to *C. a. gracilis*, but mantle darker greyish brown.

Field Characters. Unlike Three-banded Courser *C. cinctus*, inhabits very dry, completely open habitats; further distingusihed by bold double black breast-bands, lack of bold neck stripe and at closer range streaks on brown neck and white legs. In flight black and rufous on remiges, white on rump and outer rectrices obvious. Also distinguished from Three-banded Courser and Bronze-winged Courser *C. chalcopterus* by being smaller and without contrasting head pattern.

Voice. Tape-recorded (296). Calls include: (a) contact call, a plaintive, mellow whistle 'peeu-weee' dropping in pitch in the middle and rising again; (b) alarm call, a sharp 'kikikikik'; (c) anxiety call, a fairly sharp whistled 'pee pee ti-ti-ti-ti-ti'; (d) take-off call, a pleasant, fairly sharp 'wik-wik'; (e) greeting call, a soft, purring 'prrr prrr'; (f) threat call, a grating 'zzzt'; (g) call to summon chick to follow, a soft 'quip quip'; (h) during courtship, ♀ gives high clear 'peee-pi-pi-pi-pi-', ♂ 'trikritkrititkrit ...'; and (i) a 'peep' given by juvenile.

General Habits. Inhabits flat stony and gravelly desert and semi-desert plains, usually with a scattering of low shrubs; also firm sandy soils with sparse tufty grass or thorn scrub, or dry riverbeds in arid country. Partly or largely nocturnal, depending on day-time temperatures (more nocturnal in summer, South Africa). Forages mostly at night, catching prey by sudden jab at the end of a quick run; does not dig with bill. Solitary or in pairs or family groups of 3. Very vocal on moonlit nights. In absence of moon, feeds mainly dawn and dusk. Non-breeding birds spend daytime standing in shade of shrub or bush. Under stress, bobs head back and forth, while independently slowly depressing and relaxing tail. Flies reluctantly. Unlike most coursers, seldom calls in flight.

Some local movements governed by rainfall, but details not recorded. When vegetation becomes too dense after good rain, birds move away to drier, more open areas or overgrazed places.

Food. Insects, especially harvester termites *Hodotermes mossambicus* (Kalahari Desert: Maclean 1967).

Breeding Habits. Monogamous; solitary nester. In courtship ♂ and ♀ walk stiff-legged together, with sideways gait; ♂ then dances in short, hopping steps around ♀ in semicircles causing ♀ to turn and watch his display, during which ♀ calls high clear 'peee-pi-pi-pi-pi' before soliciting ♂ by crouching. With ♀ leading, both may repeat stiff-legged walk after which ♂ again hops around ♀. Copulation not observed, but probably eventually follows after repeated soliciting by ♀ after each walk-and-dance display by the pair. After display, ♀ may side-throw small objects at potential nest-sites, either while standing or while sitting on ground; she may also now and then scrape at soil with bill.

During courtship ♂ also performs upright display (**A**) with body vertical on straight legs, tail depressed, belly feathers slightly fluffed; at this time he gives loud, sharp, metallic, trilling call 'trikritkritkrit ...' which causes body to vibrate noticeably.

NEST: usually lays eggs on bare ground; nest site may become scrape during incubation, and becomes ringed with small stones, bits of vegetation, or antelope droppings collected by sitting bird side-throwing at nest-relief. Nests widely spaced (usually at least 200 m apart, never less than 40 m apart). 60% of all nests in southern Africa found among antelope droppings which confer extra camouflage on egg and sitting adult. Sites always completely exposed for good all-round visibility.

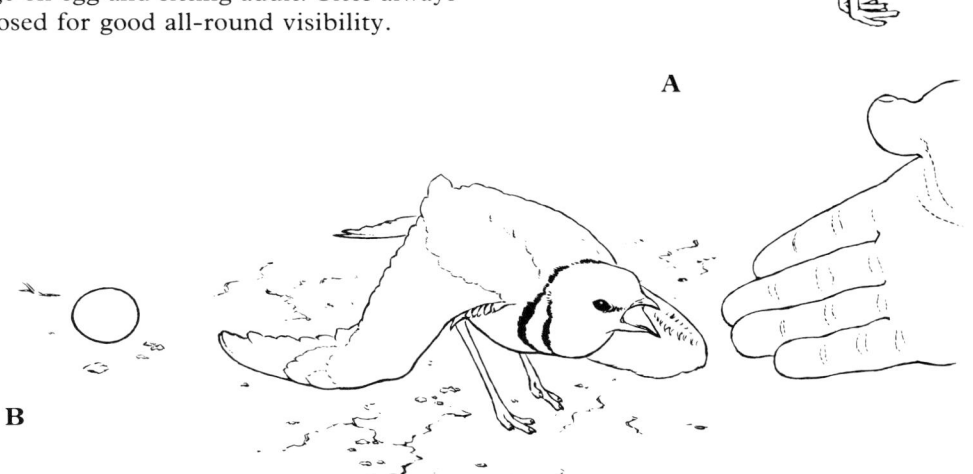

EGGS: always 1. South Africa (86 clutches) 1·0; Somalia (26 clutches) 1·0. Rather rounded and short but slightly pointed at one end; pale yellowish cream to buffy white, finely scrawled and scribbled with sepia and blackish brown, mainly in a zone around the thick end, and with underlying markings of light blue-grey. SIZE: (n = 73, southern Africa) 27·9–35·6 × 23–28·8 (31·4 × 25·5); (n = 26, Somalia) mean 28·0 × 23·5.

LAYING DATES: Ethiopia, Apr, June; Somalia, Feb–July (mainly May–June); E Africa: Tanzania, Nov; Region C, Jan–Feb, Apr–Oct, peaking June–Aug; Region D, Jan–May, July–Aug, Oct–Dec, peaking July–Oct; South Africa, all year with peak Oct–Nov. Most nesting in dry months.

INCUBATION: by both sexes. Period: 26–27 days. Nest-relief occurs every 1·5–2 h throughout the day with extreme intervals of 74 and 218 min. Incubating parent runs away unobtrusively when disturbed by predator; may attempt to fend off or distract non-predator by spreading wings and tail and running at (**B**) or away from intruder, the intensity of display being indicated by the degree of spreading (wings almost fully spread at highest intensity). Besides panting in heat of day, incubating birds lose excess body heat by raising dorsal plumage to allow air to circulate against skin, spreading wings slightly, and exposing legs to air; in E Africa 'off duty' parent seeks shade in middle of day.

DEVELOPMENT AND CARE OF YOUNG: chick takes 3 days to break out of egg. Both parents probably remove hatched shells from nest. Newly hatched chick about 6 cm long, almost naked and very weak, but with eyes open. As down dries, chick looks less naked; leaves nest within 24 h, but moves no more than a few m for the first 3–4 days. Both parents feed chick with small insects brought in bill; chick raises head to take 1 food item at a time. Parents continue to feed chick until it is almost able to fly at about 5–6 weeks. When alarmed chick crouches, but holds head up (typical of coursers, but unlike other Charadrii), sometimes with eyes partly closed. Adult-type vocalizations develop at about flying age. Chest-bands appear at about 3 months of age; full adult plumage at *c*. 4 months. Length of fledging period unknown. Family groups of 3 birds may stay together for several days or weeks until next egg laid, but juvenile not allowed to approach closer than about 6 m to new nest site.

BREEDING SUCCESS/SURVIVAL: egg predation low, e.g. only 4 eggs out of 56 definitely did not hatch (Maclean 1967). Most pairs in South Africa have 2–4 broods every year. Breeding success seems high but no data available.

References
Maclean, G. L. (1967, 1970).

GLAREOLIDAE

Plates 19, 20
(Opp. pp. 320, 321)

Cursorius cinctus (Heuglin). Three-banded Courser; Heuglin's Courser. Courvite à triple collier.

Hemerodromus cinctus Heuglin, 1863. Ibis, p. 31, pl. 1; near Gondokoro, White Nile.

Range and Status. Endemic, resident and intra-African migrant, uncommon to common, NW and SW Somalia, S Ethiopia, SE Sudan, Kenya and NE Uganda to central Tanzania; E and S Zambia, W Zimbabwe, S Angola, N Namibia, N Botswana and N South Africa (Kruger). Northern populations mainly sedentary although with some local movements; southern populations migratory.

Description. *C. c. seebohmi* (Sharpe): southern Africa. ADULT ♂: crown and nape blackish brown with tawny to sandy edges to feathers; hindneck tawny. Superciliary stripe pale tawny becoming white behind eye and meeting at nape; white stripe bordered below with black stripe meeting at nape; black stripe forks at sides of neck to produce black line onto upper breast, forming incomplete collar; lores buffy white; cheeks and ear-coverts bright tawny; chin and throat white, bordered below by chestnut band originating at posterior ear-coverts and spreading in centre of lower throat in V-shape. Mantle, back and upperwing-coverts earth brown with blackish shafts and tawny or buff edges to feathers; uppertail-coverts white (black at base); central rectrices greyish brown fading to blackish brown at end, with white tip; outer vane of second and third rectrices banded black and white; outer vane of outermost rectrices white. Band on upper breast white; rest of upper breast sandy, coarsely streaked with black and bordered below by black band; lower breast and belly, undertail-coverts and underwing-coverts white, with chestnut band separating breast from belly. Outer primaries blackish brown, with white at base of inner vane; inner primaries and secondaries brown with pale sandy border. Bill yellow with black tip; eye-ring yellow, eye brown; legs and feet pale yellow to cream. Sexes alike. SIZE: wing, ♂♀ (n = 4) 162–165, unsexed (n = ?) 164–174; tail, ♂♀ (n = 4) 81–85; bill, unsexed (n = ?) 19·5–20; tarsus, ♂♀ (n = 4) 69–76. WEIGHT: ♂ 125.

IMMATURE: similar to adult but dorsal feather edges paler; lower chestnut band faint or absent; tawny bases to ends of scapulars, innermost secondaries and central rectrices.

DOWNY YOUNG: uppermost parts mottled tan and pale grey with sparse blackish brown patch in centre of head, 2 larger ones on upper back and smaller patches elsewhere; collar white; underparts whitish with faint pale buff chest band; bill dull blackish; eye dark brown; legs and feet grey.

C. c. cinctus (Heuglin): SE Sudan to Somalia and N Kenya. Edges to dorsal feathers paler; more white on rectrices than *C. c. seebohmi*; smallest race; tarsus 57–64.

C. c. emini (Zedlitz): S Kenya, Tanzania, N Zambia. Dorsal feather edges more tawny than in *C. c. seebohmi*; size intermediate; wing 151–166; tarsus 63–73.

Field Characters. A fairly large woodland courser, distinguished from Double-banded Courser *C. africanus* by striped neck pattern, streaked chest and three breast bands.

Voice. Tape-recorded (305). Calls include: (a) soft 'chuick' contact call, (b) a loud whistled 'pieu' alarm call; and (c) 'chuck-a-chuck-a-chuck-a ...' or accelerating 'wick-er-wick-er-wick-er- ...' fading away at end, used as anxiety call.

General Habits. Inhabits dry mopane *Colophospermum mopane* and miombo woodland, thorn savanna and scrub and other dry, open woodlands. Usually seen singly or in pairs when breeding, otherwise in small groups of 5–6 birds. Spends most day-time in shade of bush or small tree; active largely at night. When closely approached, may freeze first before running away or taking off with silent, somewhat owl-like flight; runs less than most coursers. Usually flies less than 20 m; rarely moves far from cover. Often comes onto gravel roads at night.

Mainly sedentary, but apparently migratory in southernmost part of range; present Zimbabwe Apr–Nov. Local movements occur in E and central Africa.

Food. Insects.

Breeding Habits. Probably monogamous; solitary nester.

NEST: usually under shelter of bush or tree; deep scrape filled with loose soil or fine gravel in which eggs are more than half buried during incubation.

EGGS: 2. South Africa mean (2 clutches) 2·0; Somalia mean (5) 2·0; Zambia, 2 adults with 3 chicks (but adults not a pair and chicks could represent 2 different families). Elongated; pale tawny to ochreous with blotches of pale grey and overlying scrawly, scribbly marks of blackish to sepia, evenly distributed. SIZE: (n = 11, Zimbabwe) 36·4–41·2 × 25·5–27 (38·2 × 26·1); (n = 10, Somalia) mean 37 × 25.

LAYING DATES: Somalia, Mar–July, Dec; Ethiopia, Jan–Aug; E Africa: Kenya, Apr–July, Sept–Dec; Tanzania, May–Nov; Region C, May–Oct, Region D, Apr,

June, Sept–Nov; Zambia, May, July–Aug; Zimbabwe, Mar–Apr, June–Nov (most Aug–Oct). Mainly dry season nester.

INCUBATION: by both parents, with 1 bird on nest, usually 1¼–2 h at a time (E Africa), the other in shade of bush or tree a few m away. Period: *c.* 25–27 days. Begins after clutch complete. Sitting parent partly buries eggs in nest scrape by throwing loose soil with bill sideways into it until only about a third of each egg is exposed. Eggs incubated in this position (not uncovered), and may become completely covered with soil as incubation progresses; soil is compacted by body movements of sitting birds. If eggs artificially exposed, parent will again bury them. Bird on nest may not flush until intruder only a few m away.

DEVELOPMENT AND CARE OF YOUNG: both parents probably remove hatched shells from nest, eating small pieces at nest and larger ones some m away; chick leaves nest within 24 h of hatching. Fledged young remain with parents when they incubate 2nd clutch and at one nest until 2nd clutch hatched (E Africa, J. F. Reynolds, pers. comm.).

References
Kemp, A. C. and Maclean, G. L. (1973a, b).

Cursorius chalcopterus Temminck. **Bronze-winged Courser; Violet-tipped Courser. Courvite à ailes violettes.**

Plates 19, 20
(Opp. pp. 320, 321)

Cursorius chalcopterus Temminck, 1824. Pl. col., livr. 50, pl. 298; Senegal.

Range and Status. Endemic, in some areas resident, in others migratory; uncommon to locally common, from Senegambia and Guinea across S Sahel zone to Cameroon, S Chad, Central African Republic, S Sudan, W Ethiopia, Uganda, Kenya, Tanzania, S Zaïre, Angola, Zambia, Zimbabwe, Malawi, Mozambique, N Namibia, N Botswana, Swaziland and E South Africa.

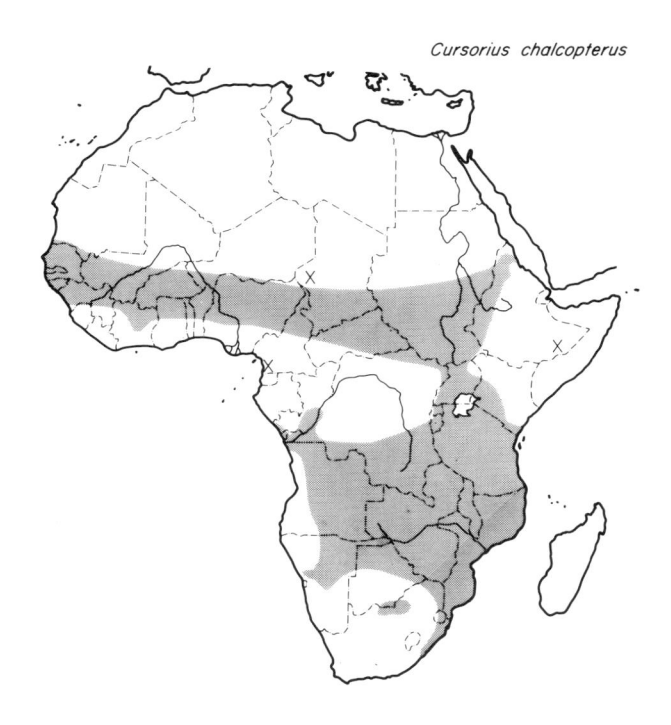

Description. ADULT ♂: forehead buffy white sometimes mottled with sepia and rufous; forecrown sepia; hindcrown, nape and hindneck earth brown; superciliary stripe buffy white, at front joining buffy forehead; lores to ear-coverts blackish brown streaked with buff; short white line curves down behind eye; cheeks and sides of neck to throat white; broad malar stripe brown, sometimes forming brown upper throat by merging in centre. Mantle, back and rump earth brown; uppertail-coverts long and white; tail olive brown, white at base and tip. Breast light brown shading to creamy buff below and bordered below by black band; rest of underparts white; flanks washed buff. Primaries black with purplish gloss, all but outermost tipped with metallic violet, shading to metallic copper and steel blue on black webs; secondaries black to brownish black, white on inner web and glossed coppery green on outer edge; primary coverts black; remaining coverts earth brown, secondary coverts tipped whitish; underwing white, feathers on leading edge blackish tipped with buff. Bill black, purplish red at base; bare eye-ring purplish red; eye dark brown; legs and feet dull red. Sexes alike. SIZE: ♂♀ (n = 8) wing, 176–186 (180); tail, 77–84 (80·4); bill, 20–24 (22·5); tarsus, 71–80 (75·8). WEIGHT: (Botswana) ♂ (n = 6) 117–172 (151), ♀ 148, 157·6, 160.

IMMATURE: dorsal feathers paler and edged with rusty buff; secondary coverts more broadly tipped white; primaries tipped metallic green (not purple); more buffy below; black chest band narrower; malar stripes less extensive.

DOWNY YOUNG: upperparts black with some buffish chestnut patches and a broad white collar; underparts white; bill black; legs dark grey.

Field Characters. The largest courser, plain brown above with brown chest, white belly and single black chest-band; bold black-and-white facial pattern distinctive. In flight white uppertail-coverts, white wing-bar, black remiges with white underwing conspicuous. Distinguished from Crowned Lapwing *Vanellus coronatus*, with which it may be confused, by having plain brown crown, dull red (not bright coral red) legs and distinctive facial pattern.

Voice. Tape-recorded (217). Calls include a thick-knee-like piping 'ji-ku-it', and a harsh, plaintive 'gror-raang' in flight.

General Habits. Inhabits *Acacia* and other savanna, bushveld, thorn scrub, and any woodland except *Brachystegia*. Occurs in pairs or small groups. Strictly nocturnal; rests in shade of trees or bushes by day; often seen on roads ar night. When pursued, flies only a short distance before landing and running away or freezing.

Migratory movements occur but poorly recorded. In E South Africa mainly wet season Sept–May; after nesting moves southward in broad front. In Zimbabwe arrives Aug, leaves Jan–Feb for south. In Zambia mainly in dry season May–Nov with birds arriving apparently from both north and south. In Malaŵi mainly in dry season, particularly May–June. In Zaïre (Katanga) May–Nov. In Tanzania most arrive late Apr to breed during dry season although some as early as Jan–Feb. In Kenya probably mainly a non-breeding visitor with most records Tsavo East N. P. Dec–May. In northern tropics, regularly moves northward Mar–June often with rains, and travels south Oct–Dec in dry season. In Nigeria passage migrant to north in spring, breeds in north in rains, moves south in autumn to *c*. 8°N; in Chad in north to *c*. 15°N in rains.

Food. Insects.

Breeding Habits. Probably monogamous; solitary nester.

NEST: no nest; lays eggs on bare ground.

EGGS: 2–3. Oval, yellowish to buff, heavily blotched with black, sepia and chestnut, with underlying spots and blotches of grey and lilac. SIZE: (n = 31) 33–40 × 25–28 (37 × 27).

LAYING DATES: Nigeria, Jan–Feb; Burkina Faso, May–June; Sudan, Feb–June; Ethiopia, Sept, Nov; E Africa: Kenya, Dec–July; Tanzania, Sept; Uganda, Oct; Region A, Feb; Region B, Feb, Aug; Region C, Jan, Aug–Sept; and Region D, Feb; Malaŵi, Sept–Oct; Zimbabwe, July–Dec (most Sept–Oct); Zambia, May–Oct (most Sept); Mozambique, Aug–Sept; Angola, Aug–Sept; Namibia, July; South Africa: Cape, Sept. Most nesting in dry season; at least in Zimbabwe nests only after grass burnt.

INCUBATION: both sexes incubate. Period: *c*. 25–27 days. In E Africa, 'off duty' parent seeks shade in middle of day; incubating bird on nest $1\frac{1}{4}$–2 h at a time.

DEVELOPMENT AND CARE OF YOUNG: parents with young perform injury-feigning distraction displays.

Reference
Kemp, A. C. (1974).

Subfamily GLAREOLINAE: pratincoles

Highly aerial, and more water-bound than coursers; tern- or swallow-like in air, plover-like on ground. Bill small, but gape wide; feed in air and on ground. Wings long and pointed, tail forked and legs short with hind toe present. Neonatal plumage with dorsal pattern finely stippled making chick look like cluster of pebbles; some with vestige of courser-type pattern. 2 genera, *Glareola* and *Stiltia*; only *Glareola*, with 7 spp., in Africa.

Genus *Glareola* Brisson

Characters as for subfamily. 7 spp., 5 in Africa of which 2 are endemic, 1 is a visitor from Madagascar, 1 from central Palearctic, and 1 from W Palearctic (also breeding in N Africa).

G. nordmanni and *G. pratincola* form a superspecies despite some overlap of breeding ranges. They are very closely allied but differ in depth of tail fork, thickness of tarsus, size of toes and claws, plumage and migratory habits.

Plate 19
(Opp. p. 320)

Glareola pratincola (Linnaeus). **Collared Pratincole; Common Pratincole. Glaréole à collier.**

Hirundo Pratincola Linnaeus, 1766. Syst. Nat. (12th ed.), 1, p. 345; Austria.

Forms a superspecies with *G. nordmanni*.

Range and Status. Africa and S Europe east to central Asia and NW India; also SW Arabia.

Palearctic visitor and intra-African migrant. Breeds N Africa, locally common to rare Morocco (Atlantic coast, Oued Loukos south to Rio de Oro and Mediterranean coast at Sebka Bou Azeg and Moulouya R., total breeding population *c*. 200–300 pairs), Algeria (200–250 pairs), Tunisia (1000–1500 pairs) and Egypt. Breeding birds from N Africa and Europe winter along southern edge of Sahara; not known if they reach equator. Migrants pass through N Africa (Morocco to Egypt) on broad front. Locally abundant; flocks of

3000 noted in migration along Benué R., and 1000 or more along Niger R., Algerian Sahara and S Chad. Locally abundant to rare, S Mauritania, Senegambia and Liberia to lower Niger R. and Chad basin, and from Congo R. to SE Sudan (c. 30,000 moving north, Juba, 3–5 Apr: F. R. Lambert, pers. comm. and *Malimbus* 7, 1985, 136), Ethiopia, W Somalia, south to N Namibia, N Botswana, Zambia, N Zimbabwe, Mozambique and coastal South Africa (Natal). Nomadic, movements considerable, but with no clear pattern. Status in many parts of Africa not well known; may sometimes appear in hundreds at a given locality to breed, then disappear for months or even years.

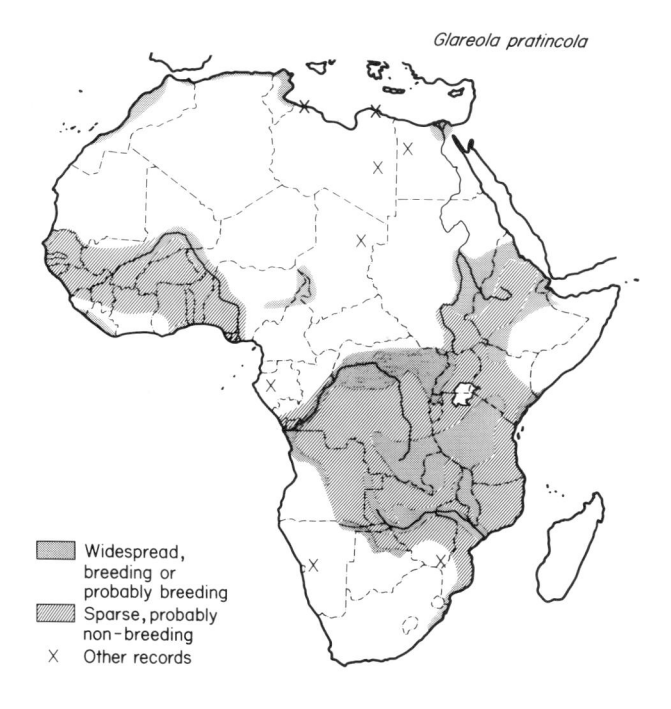

Description. ADULT ♂ (breeding): forecrown to back, sides of face from below lores, around eyes and side of neck earth brown tinged olive. Chin, throat and upper foreneck creamy bordered with a black ring from below eyes to base of foreneck. Rump and uppertail-coverts white; tail forked, outer rectrices white, inner ones black. Breast earth-brown tinged olive; belly and undertail-coverts white. Remiges black, secondaries with white tips; upperwing-coverts earth brown tinged olive; underwing-coverts and axillaries chestnut. Bill black, base red; eye brown; legs and feet black. ADULT ♂ (non-breeding): like breeding ♂ but feathers of forehead, crown and lores fringed rufous-buff. No black on lores, throat spotted and streaked. Sexes alike except ♀ has lores brown. SIZE: wing, ♂ (n = 22) 183–202 (194), ♀ (n = 14) 179–196 (186); tail, ♂ (n = 10) 100–118 (111), ♀ (n = 7) 97–107 (103); bill, ♂♀ (n = 4) 17–18 (17·8); tarsus, ♂♀ (n = 4) 28–31 (29). WEIGHT: ♂ (n = 7) 60–104 (80), ♀ (n = 7) 63–85 (75).

IMMATURE: upperparts and breast mottled with black and buff; blackish streaks on throat; no distinct black ring on throat and neck.

DOWNY YOUNG: upperparts sandy buff, with fine dark stippling giving faint pattern of lines on crown, back and side; underparts pale buff.

Field Characters. Very similar to Black-winged Pratincole *G. nordmanni*, chiefly separable by red-brown underwing-coverts; for further distinctions, see that species. All African pratincoles show conspicuous white rump in flight, but Collared and Black-winged differ from the other species in having deeply forked tail, black ring encircling buff throat. Same size as Madagascar Pratincole *G. ocularis* but considerably larger than Rock and Grey Pratincoles *G. nuchalis* and *G. cinerea*. Madagascar Pratincole also shows red-brown underwing in flight, but has grey-brown throat, rufous belly.

Voice. Tape-recorded (62, 73). Very vocal in groups; calls shrill, harsh, and chattering; given in air and on ground, a raucous, sharp 'karria-krill' and 'kirrie-kirek-karak'. Calls of European breeding birds include: (a) spring-song, a 'kow-ka KILILILI-kow-ka KILI-LILI-kow kow-ah', given on ground or in air; (b) 'tri', 'trrrt' and 'gig gig gig tiririt', given on ground at colony; (c) an accelerating series of gurgling calls given by incubating birds at nest-relief; and (d) a 'kirririk' alarm call.

General Habits. Inhabits flat open areas, dry or wet, muddy or sandy, including grassland, generally not far from inland lakes or rivers or coastal areas. Gregarious, noisy, sometimes in groups of several thousand. Most active dawn and dusk. Roosts by day and night on ground in open. Runs fast; occasionally stops to stretch and stand on tiptoe. Usually forages in flight, hawking for insects; occasionally feeds on terrestrial arthropods. Feeds by day, sometimes on moonlit nights. Sometimes thousands follow swarms of migrating locusts or accompany herds of domestic animals or wild game, feeding in flight on insects disturbed by them. Flight powerful; sometimes spirals up in a great column.

Migrates usually in small flocks, sometimes large ones (Timbuktu, immense flocks, Nov–Apr, highly erratic Apr–May). European and Asian visitors pass through N Africa in spring and early autumn (Morocco, 3rd week Mar–mid May; Tunisia, uncommon passage migrant late Apr–mid May, late Aug–Sept; Libya, uncommon but regular coastal passage migrant Apr–May, Sept–Oct); more numerous in spring than autumn. Some trans-Saharan migrants noted in oases of Morocco, Algeria, Mali and Libya; large numbers use Nile Valley route through Egypt. Only small numbers recorded in desert oases, suggesting many fly at high altitudes. Southern limits of Palearctic visitors in winter unknown due to difficulty in readily separating them from Afrotropical birds. N African birds form large flocks after breeding season (Algeria, near Annaba, formerly Bone, large flock on 22 June increased to 5000 early Aug; all departed south by end Oct). Movements of birds breeding south of Sahara more erratic, but some seasonal movements probably associated with falling water levels (W Somalia, end Aug–early Sept at end of rains; Zambia, large numbers dry season, late Apr–Nov; Ghana, adults depart July in rains, juveniles late Aug; large concentrations build up along Niger R. after rains when water levels fall). 1 bird ringed Spain recovered Senegambia.

Food. Mainly large insects including locusts, click beetles, tiger beetles (Cicindelidae), dragonflies and ephyrid flies (Diptera); also spiders and molluscs.

Breeding Habits. Monogamous, bond probably of seasonal duration. Breeds colonially in groups of 10–100 pairs on deltas, dry mudflats of old flood plains, alkaline and sandflats of lakes, and occasionally coastal reefs (Morocco). Nests sometimes as close to each other as 1–3 m. Breeding birds extremely noisy. Territory defended by running at intruder, then pausing with neck stretched up in High-erect pose, tail fanned and wings held partly open. Defends nest by lowering breast to ground, raising tail and spreading wings wide on ground; and by flight-chasing. When at nest, may jerk tail up and down. Upon alighting, holds wings momentarily over back, showing chestnut underwing. During courtship, pair in air perform spring-song and engage in sudden swoops and climbs. On ground one bird runs to mate, suddenly stops, stands upright before it; both call, run next to each other, stand still again, then stretch necks forward with heads arched down, wing nearest partner drooping slightly and tail often spread. Sometimes one bird circles the other, then approaches and faces it, when both lift their heads, displaying throat-patches, repeatedly raise and lower their tails, and call. Sometimes performs scrape-ceremony, pressing breast down on ground with tail lifted up; the partner stands in High-upright posture close by, then bows, almost touching mate's bill. Courtship feeding may occur before copulation.

NEST: usually none; occasionally eggs laid in tiny hollow or in hoof print; sometimes lined with broken shells.

EGGS: 3 (2–4), laid at 1–2 day intervals. Subelliptical, smooth, pale stone colour to rich buff, marked heavily with blotches, spots and scrawls of black or brown with pale mauve undermarkings. SIZE: (n = 31) av. 32·2 × 23·9; Senegambia (n = 4) 30–31 × 23–25 (30·2 × 24).

LAYING DATES: Morocco, Apr–July; Algeria and Egypt, May; Tunisia, Apr–May; Mali, Apr–May; Senegambia, May–July; Ghana, Apr–July; Chad and Nigeria, Apr–May; Ethiopia, May; East Africa: Region A, May–July; Region B, Apr, Sept, Region C, June–Sept; Region D, Apr–July, with most breeding usually in rains but also in dry season Tanzania; Zaïre, Apr; Angola, Aug; Malâwi, Aug–Oct, mainly Sept; Zambia, June–Nov, mainly Aug–Oct; Botswana, Namibia and Zimbabwe, Aug–Nov; South Africa (Natal), Nov.

INCUBATION: begins when clutch complete; carried out by both parents. Period: 17–19 days.

DEVELOPMENT AND CARE OF YOUNG: young leave nest when 2–3 days old; can fly at 3–4 weeks. Fledging period: 25–30 days. From very early on, young conceal themselves when intruder present. Cared for by both parents, who distract intruder by feigning injury, stretching themselves on ground with wings extended.

Reference
Cramp, S. and Simmons, K. E. L. (1983).

Plate 19
(Opp. p. 320)

Glareola nordmanni Fischer. Black-winged Pratincole. Glaréole à ailes noires.

Glareola nordmanni Fischer, 1842. Bull. Soc. Imp. Nat. Moscow, 15, p. 314, pl. 2; S Russia.

Forms a superspecies with *G. pratincola*.

Range and Status. Breeds SE Europe and SW Asia; winters Africa.

Palearctic visitor, vagrant or locally very abundant, passing largely through Egypt and central Africa (flocks of hundreds S Sudan and W Uganda; E Zaïre, crosses in high continuous flight; W Zambia, flocks of 5000–10,000) to main wintering area from NE Namibia and Botswana south to Orange Free State, S Transvaal, W Natal, W Swaziland and northern Cape. Sometimes also winters in large numbers Ethiopia (c. 500, Itang 8°10′N, 34°14′E, Dec) and Sudan (c. 5000 Baro R. between 8°14′N, 34°01′E and 8°23′N, 33°45′E, Jan); also Nigeria (200, Kainji, Dec–Apr) and Chad (flocks of 100–200 Fort Lamy, Dec). Elsewhere rare to vagrant; recorded Libya, Togo, S Cameroon, Bioko (Fernando Po), Gabon, Somalia, Kenya, Mozambique, Malâwi and Zimbabwe.

Description. ADULT ♂: forecrown to back, side of face from below lores, around eyes and sides of neck earth brown tinged olive. Chin, throat and upper neck creamy bordered with black line from eye to base of foreneck. Rump and uppertail-coverts white; tail forked, inner rectrices black, outer ones white. Breast earth brown tinged olive; belly and undertail-coverts white. Remiges black, secondaries without white tips; upperwing-coverts earth brown tinged olive; underwing-coverts and axillaries pitch black. Sexes alike except ♀ has brown lores. SIZE: wing, ♂ (n = 14) 190–205 (193), ♀ (n = 5) 177–188 (185); tail, ♂ (n = 14) 81–115 (98·9), ♀ (n = 5) 95–115 (102); bill, ♂ (n = 4) 15·5–17 (16·5), ♀ (n = 5) 16–19·5 (17·5); tarsus, ♂ (n = 4) 35–40 (38), ♀ (n = 5) 35–39 (37·2). WEIGHT: (USSR, summer) ♂ (n = 4) 91–105, ♀ (n = 6) 87–99; (Bulgaria, Aug) adult ♀ 105.

IMMATURE: like immature Collared Pratincole *G. pratincola* except axillaries, underwing-coverts and inner lining of wing black; upperparts and chest mottled with black and buff, blackish streaks on throat; no distinct black ring on throat and neck. Birds in immature plumage regularly seen in Africa.

Field Characters. Very similar to Collared Pratincole, frequenting same habitats and sometimes in same flock. Distinguished from it by having black (not reddish brown) underwing; uniform dark upperwing without obvious pale trailing edge on secondaries; upperparts including head darker; tail shorter and less deeply forked; and at rest a more contrasting, pied appearance.

Voice. Tape-recorded (62, 73, C). Very vocal in groups; gives a raucous sharp 'karria-krill'; also 'kirrie-kirek-karak' and 'kirlik kirlik'.

General Habits. Inhabits open barren areas, sparse grass, cleared ground and sandy beaches, often near water. Gregarious, usually in flocks of 10–100, on passage sometimes in thousands. In forest-clad country, birds on autumn migration may descend in ones or twos to rest in clearings and near villages; sometimes large flocks (c. 500) do so. Active at dawn and dusk. Forages usually in flocks, hawking insects; occasionally follows locust swarms. Sometimes feeds on ground. Flocks spiral, whirling up steeply to c. 600 m. Habits similar to Collared Pratincole.

In autumn enters Africa (Egypt and probably Sudan) from Cyprus and from Saudi Arabia (Jiddah, thousands); follows Nile, passing through centre of continent (E Zaïre, W Zambia) before reaching wintering area in southern Africa. Overflies equatorial forests of Zaïre but stops W Zambia. Crosses Cyprus late Aug–early Oct; main movement Zambia Oct–Nov. In spring large numbers Zambia Mar, Zaïre and Sudan Apr.

Food. Insects, mainly locusts and winged termites; also ants, beetles and cockroaches.

Reference
Cramp, S. and Simmons, K. E. L. (1983).

Glareola ocularis Verreaux. Madagascar Pratincole. Glaréole malgache.

Plate 19
(Opp. p. 320)

Glareola ocularis Verreaux, 1833. S. Afr. Quart. J., 2, p. 80; Madagascar.

Range and Status. Breeds Madagascar; spends non-breeding season in Africa and Comoros.

Visitor to eastern Africa, mainly on coast from S Somalia to Mozambique north of Zambezi R. Frequent to locally very abundant on coast and edges of lakes and rivers near coast. Especially abundant coastal Kenya (e.g. Sabaki) where up to 800 regularly rest, sometimes as many as 9000 or more. Inland records erratic; usually rare to uncommon with only 1–6 individuals sighted at a time; sometimes hundreds Lake Victoria. Vagrant Ethiopia (11°44′N, 41°05′E).

Description. ADULT ♂: forehead and crown brown; nape olive-brown, lores darker, ear-coverts blackish brown; rest of upperparts including side of neck olive-brown. A narrow white line below and behind eye, contrasting strongly with dark lores. Chin, throat, foreneck and chest earth-brown. Uppertail-coverts white; outer rectrices white, central ones white with broad dark brown tips. Breast reddish brown to pale chestnut, belly and undertail-coverts white. Remiges dark brown to black with greenish reflections, rest of upperwing dark brown; under primary coverts near bend of wing white; underwing-coverts and axillaries pale chestnut. Bill black, red at base; eye brown; legs black. Sexes alike. SIZE: wing, ♂ (n = 6) 187–210 (196), ♀ (n = 6) 189–210 (198); tail, ♂♀ (n = 12) 66–79 (73·1); bill, ♂♀ (n = 12) 10–13 (11·2); tarsus, ♂♀ (n = 12) 22–25 (23).

IMMATURE: feathers of back edged with rufous; chin lighter than adult, yellowish; upper breast streaked with rufous; line under eye rufous, not white.

GLAREOLIDAE

Field Characters. Same size as Collared *G. pratincola* and Black-winged *G. nordmanni* Pratincoles but tail with only shallow fork and shorter (folded wing-tips extend 50 mm beyond tail). Has dark cap and white facial streak, grey-brown throat without black ring, and rufous belly. Almost completely allopatric with Rock Pratincole *G. nuchalis*, which is very much smaller, lacks rufous belly, and has white collar on hindneck, different habitat.

Voice. Not tape-recorded. Very vocal; a tern-like 'kitt-kirr'.

General Habits. Inhabits exposed mudflats and sand dunes; also edges of lakes and rivers; rarely on exposed rocks. Highly gregarious. Forages in air, usually at twilight, often at considerable height; active during middle of day if weather overcast. Also forages over woodland (e.g. Sokoke Forest, Kenya).

Route to and from Africa not well established but probably enters and leaves via Tanzania. Present in Africa Mar–Sept.

Food. Mainly insects, including hymenopterans, neuropterans and beetles.

References
Britton, P. L. (1977).
Milon, R. *et al.* (1973).

Plate 19
(Opp. p. 320)

Glareola nuchalis **G. R. Gray. Rock Pratincole; White-collared Pratincole. Glaréole auréolée.**

Glareola nuchalis Gray, 1849. Proc. Zool. Soc. London, p. 63; Berber, Sudan.

Range and Status. Endemic resident and intra-African migrant. Locally abundant, in rocky areas of rivers and lakes from Sierra Leone to SW Chad, Zaïre and W Kenya southward to E central Angola, extreme NE Botswana, Zimbabwe and W Mozambique; also SE Tanzania, W Ethiopia and NE Sudan, but largely absent rest of eastern Africa. Sedentary in Gabon and along the northern shore of L. Victoria; migratory in some parts of its range.

Description. *G. n. nuchalis* Gray: L. Chad and central and E Cameroon to Ethiopia, south to Angola, Zimbabwe and Mozambique. ADULT ♂: forehead, crown, nape and hindneck dark sooty brown, lores darker; foreneck and side of neck light sooty brown; white collar from behind eye to hindneck; face and chin dark sooty brown. Mantle, back, and throat to front of belly sooty brown; rump, belly, uppertail- and undertail-coverts and basal part of tail white; outer tail feathers white, broadly tipped brown; rest of tail sooty brown. Primaries and secondaries blackish-brown, basal half of inner webs of secondaries white; rest of wing dark sooty brown. Bill red with tip and culmen black; eye dark brown; legs and feet dull orange to coral red. Sexes alike. SIZE: wing, ♂ (n = 10) 136–158 (143), ♀ (n = 12) 138–158 (150); tail, ♂ (n = 3) 59–63 (61), ♀ (n = 3) 58–61 (60); bill, ♂ (n = 5) all 10, ♀ (n = 2) 10–11 (10·5); tarsus, ♂ (n = 10) 14–19 (16), ♀ (n = 6) 15–18 (16·3). WEIGHT: 43–52.
IMMATURE: slaty grey without collar; feathers of upperparts, breast and tail tipped light rufous. Bill and legs black.
DOWNY YOUNG: pattern quite variable, largely brownish grey above, speckled with dusky; smoke grey below.
G. n. liberiae Schlegel: Sierra Leone to W Cameroon. Collar chestnut, not white (intermediates between *nuchalis* and *liberiae* occur in W Cameroon: Louette 1981).

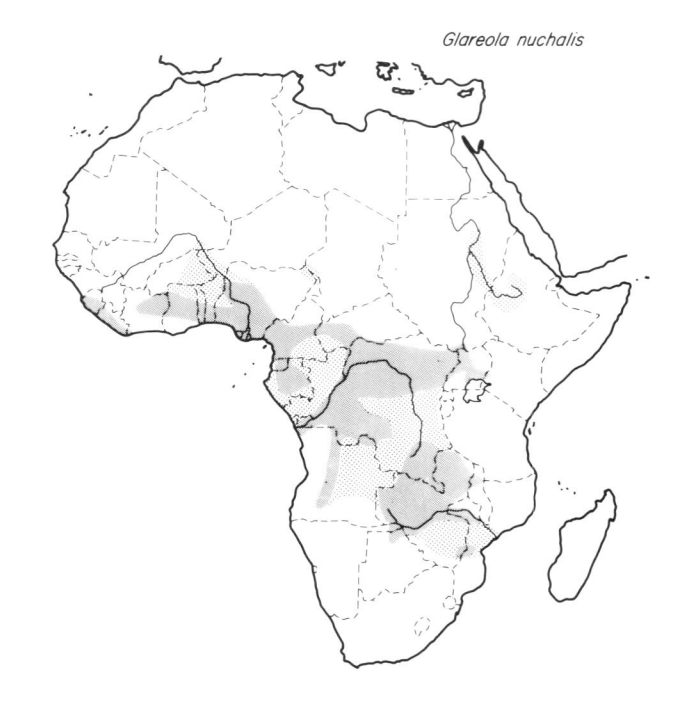

Field Characters. The only pratincole that habitually occurs on rocks in rivers and lakes; grey-brown with rufous or white hind collar. Similar-sized Grey Pratincole *G. cinerea* is much paler grey, lacks collar, and has buffy throat, white belly, different habitat. Much smaller and darker than the other 3 African spp., tail short with very shallow fork.

Voice. Tape-recorded (217). Alarm call, a faint 'kip-kip-kip'; in breeding season both sexes, when sitting or flying, give a long series of short and sharp 'ti-ti-ti- ... te-tic-te'.

General Habits. Inhabits rivers and lakes, especially those with exposed rocks; sometimes mud by lagoons and sand beaches. During low water levels sits on exposed rocks; during high water perches on branches of trees overhanging water. Same rocks and branches perched on daily for years. Gregarious, in groups of 2–10, sometimes up to 100 pairs. Mainly active early morning or late evening when hawks insects; will do so also on cloudy days usually before or after rain. Often accompanies swallows and swifts when feeding. Sometimes hundreds hawk insects at street lamps in evening in centre of towns (e.g. Livingstone, Zambia: Benson *et al.* 1971). Home range of one group of 180 birds (Gabon) *c.* 4 km of large river bed; next group was 50 km away. Outside breeding season, roosts and flies in groups.

Undergoes seasonal movements related to water level changes; departs when area flooded. Regular migrant in Nigeria; on Niger R. seen only between mid Mar and Sept. In Zambia plentiful only July to early Jan with peak Sept when water levels become low. Migrants from NE Namibia have been observed Jan–July in E Zaïre (Clancey 1981). Migrants of the southern race *nuchalis* have been observed in Oct in the range of *liberiae* (Cheke 1982). Extent of migration not known but W African populations may make more extensive N–S migrations.

Food. Insects, especially flies, including tsetse flies (*Glossina*); also beetles.

Breeding Habits. Monogamous, loosely colonial, territories scattered over 4 km or more of river bed; sometimes solitary nester. Strongly defends territory against conspecifics; territory usually *c.* 100 m long. During courtship ♂ and ♀ pursue each other by running or flying; also may hover over territory and exchange calls when flying or on rocks. On landing, crouches submissively and utters a trilling call while mate stretches body upwards so that head is almost vertically above feet. The nuchal collar of the incoming bird is flared at this time.

NEST: none; eggs laid in often precarious situation in shallow depression or crack or on flat top of exposed rock either in centre or periphery of territory.

EGGS: 1–2, laid at 2–day intervals; Gabon mean (71 clutches) 1·56. Oval; stone colour, heavily streaked and blotched brown and grey, sometimes with some black spots and hair-like scrawls; also sometimes creamy white with a ring or cap of grey-brown blotches with lavender undermarkings at blunt end. SIZE: (n = 13) 28–32 × 21–22 (29 × 21·6).

LAYING DATES: Gabon, Dec–Feb, June–Sept (2 breeding seasons correspond to the 2 dry seasons; only 5–30% of the population breeds in each one); Togo, May–July; Nigeria, Apr–June, when water level low, early in rains; Ethiopia, June–Aug; East Africa: Region A, Jan–Mar, Nov, mainly in dry seasons; Region B, Jan–Oct, mainly Aug–Sept late in drier break between rainfall peaks; N Zaïre, Jan onwards; S Zaïre, June, Aug–Nov; Zimbabwe, Aug–Dec (most Sept–Nov); Zambia, Jan, Aug–Dec, mainly Sept.

INCUBATION: carried out by both parents; 1 incubates while other guards territory. Period: 20 days.

DEVELOPMENT AND CARE OF YOUNG: both parents feed young. When parents give alarm call, newly hatched chick 'freezes' on rock; when few days old, swims or runs to cracks in rocks to hide. When begging, chick runs very fast with head lowered toward parent (sometimes colliding with it), then moves in semi-circle around parent, often burying its head in parent's breast; then parent feeds it. Flies when 40–45 days old; remains with parents until *c.* 3 months old.

BREEDING SUCCESS/SURVIVAL: eggs destroyed by flooding, and eaten by Nile Monitor *Varanus niloticus* and Harrier Hawk *Polyboroides typus*. Predation of young rare; of 35, only 2 were killed, by Nile Monitor. Over 9-year period (Gabon) 25 young fledged annually from 180 birds.

References
Brosset, A. (1979).
Cheke, R. A. (1980, 1982).

Glareola cinerea Fraser. Grey Pratincole. Glaréole grise.

Glareola cinerea Fraser, 1843. Proc. Zool. Soc. London, p.26; Niger.

Plate 19
(Opp. p. 320)

Range and Status. Endemic, uncommon to locally abundant, mainly along large rivers from Mali, Niger, upper Volta and Ghana east to S Chad and Central African Republic and south to Gabon, NW Angola and NE Zaïre.

Description. ADULT ♂: forehead to nape, and mantle to rump, pale ashy grey, hindneck and side of neck pale chestnut. White superciliary stripe from above lores to behind eyes; lores and oblique stripe behind ear-coverts and down side of neck black. Rest of face, chin and throat white. Uppertail-coverts white, outer rectrices white with black tips, central ones white with broad black subterminal band and white tips. Breast buff, flanks and belly white. Remiges mainly white with black tips; outer primaries with white confined to inner webs. Underwing-coverts mainly black. Bill black with orange-red base; eye brown to red; legs and feet orange-red. Sexes alike. SIZE: wing, ♂ (n = 23) 138–146 (141), ♀ (n = 5) 139–150 (146); tail ♂♀ (n = 7) 49–59 (54); bill, ♂♀ (n = 7) 7–9 (8); tarsus, ♂♀ (n = 7) 20–24 (21).

IMMATURE: feathers of upperparts edged black and buff; no black on lores or behind ear-coverts; tips of rectrices buff mottled with black.

DOWNY YOUNG: upperparts light greyish buff, finely marked dusky on back, underparts including throat white.

Field Characters. A small, pale grey and white pratincole with buffy throat, found in sandy habitats. Similar-sized Rock Pratincole *G. nuchalis* is much darker and has white or rufous hind collar, white face streak contrasting with dark cap, rocky habitat.

Voice. Tape-recorded (217). Gives a series of 'zi' calls with progressively shorter intervals between each call; also gives an alarm 'tic-tic-tic'; and a modulated series of 'tie-tie-tie'.

General Habits. Inhabits sand bars in slow flowing rivers and lakes; also in forested and open country; occasionally mangroves. Very gregarious, spending most of time sitting or standing on sand bars. Forages mostly in evening by flying over water or along banks of rivers, lakes and estuaries.

After breeding undertakes fairly extensive movements during rains, moving probably along course of rivers to coastal estuaries and along Logone R. to L. Chad. Moves north from Chad basin at end of rains.

Food. Mostly flying insects, including flies, beetles and locusts; also spiders.

Breeding Habits. Probably monogamous; colonial; nests only a few m apart; somewhat territorial.

NEST: a small scrape on sandbank or shingle within 100 m of water, not more than 8 cm wide, without lining or other material. On shingle, eggs often touch or are within few cm of same-sized pebbles, making clutch highly cryptic.

EGGS: 1–2; (Ghana, n = 26), av. 1·85. Broad ovals; variably cream to buff, spotted and mottled with olive-brown and stony grey. SIZE: (n = 61) 25–28 × 20–22 (26·2 × 20·6).

LAYING DATES: Niger, Mar–May; Nigeria, Mar–June; Gabon, Feb; breeds in dry season when water levels are low and sand bars exposed.

Incubation period, role of sexes during incubation, development of young, parent-young interrelationship and breeding success unknown. When intruder approaches, incubating bird may rise, pause a moment, then run rapidly away from nest. Parent performs distraction display by extending both wings full-length, or by shuffling forward on belly.

Reference
Bouet, G. (1955).

Family CHARADRIIDAE: plovers and lapwings

Small to medium waders, rather plump, with large, rounded head, broad forehead, large eyes and short, thick neck. Stand upright, often with head held high; run swiftly, stopping abruptly. Bill short, never longer than head; stout, more or less swollen at tip but without nerve endings at tip, hence locate prey visually rather than by touch when probing. Nostrils slit-like. Wings long; pointed in most, rounded in some; flight strong. Plumage often boldly patterned in brown, olive-grey, black and white. 14–19 secondaries; tail short to medium. Tarsi long with reticulated scales, a few with some transverse scutellae. Most with 3 short toes; hind toe absent in most, vestigial in some. Sexes alike, but ♂ often slightly larger than ♀.

Occupy open terrestrial habitats, and shores of rivers, lakes and seas. Rarely wade when feeding; eat mainly animal matter, some vegetable matter. Often rather gregarious, especially when not breeding. Both ♂ and ♀ care for young. Young precocial, nidifugous, and self-feeding, down short, woolly and densely matted; basically of pebble-pattern type. 2 subfamilies, Charadriinae (ringed plovers) and Vanellinae (lapwings). World-wide except Antarctic. 63 spp.; 29 in Africa.

Subfamily CHARADRIINAE: plovers

Small to medium waders, smaller than lapwings and with different jaw musculature (Burton 1974). Usually 1–2 dark breast-bands on otherwise white underparts; white hind collar; crown and face with distinctive markings. Wing usually pointed; hind toe absent. 39 spp. world-wide, 15 in Africa.

Genus *Charadrius* Linnaeus

Small to medium-sized plovers usually with white and black head markings. Most have one or two breast-bars; some have none; breast-bars usually black but may be white or reddish. Wings pointed, less than 15 cm long. All but 1 sp. lack hind toes. Bob head when alarmed; do not raise wings in aggression. Usually little or no sexual or seasonal variation in plumage. Downy young with spotted dorsal pattern and white nape-patch. 27 spp., 12 in Africa including 7 Palearctic winter visitors of which 2 nest N Africa and 5 Afrotropical spp. of which 1 nests in N Africa.

2 superspecies in Africa, *C. alexandrinus/C. marginatus* and *C. tricollaris/C. forbesi*. The former pair have been considered conspecific (e.g. Peters 1934); however, we follow Vaurie (1964), Clancey (1975), Snow (1978) and Cramp and Simmons (1983) in considering them distinct species because of morphological differences and lack of intergrades; we treat them as allospecies of a superspecies since there is no overlap in their breeding ranges except in Senegambia. *C. tricollaris* and *C. forbesi* have likewise been considered conspecific (e.g. Johnsgard 1981), but we follow Tree (1964) and Snow (1978) in considering them distinct species. In spite of some overlap in their breeding ranges (where *C. forbesi* is apparently rare), their great morphological similarity suggests they have only recently achieved species status, and are best classed as allospecies of a superspecies, with *C. forbesi* replacing *C. tricollaris* in W Africa.

Charadrius tricollaris superspecies

1 *C. forbesi*
2 *C. tricollaris*

Charadrius alexandrinus superspecies

1 *C. alexandrinus*
2 *C. marginatus*

Charadrius dubius Scopoli. Little Ringed Plover. Petit Gravelot.

Charadrius dubius Scopoli, 1786. Del. Flor. et Faun. Insubr., fasc. 2, p. 93; Luzon.

Range and Status. Breeds N Africa, Europe and Asia to Japan and New Guinea; winters Africa, Arabia and S Asia to Malay Peninsula.

Resident and Palearctic visitor, mainly on inland freshwater habitats. Sedentary, common Morocco (nesting on many rivers in north to foot of High Atlas; also in SW on Oued Sous and Massa, and on Dra and Dades Rivers on desert fringe south of High Atlas), frequent to uncommon Algeria and Tunisia, Egypt (Nile Delta) and Libya (suspected breeding Cyrenaica). Winter visitor from Europe and Asia, frequent to locally very abundant, passage migrant N Africa, crossing Sahara in broad front, wintering generally north of equator, south to Rio Muni, Zaïre, Burundi and N Tanzania. Vagrant Zambia; frequent to rare during winter N Africa (e.g. Morocco, along west and north coast in small numbers).

Description. *C. d. curonicus* Gmelin: only subspecies in Africa. ADULT ♂ (breeding): forehead white, forecrown black, separated from brown crown by thin white line which extends back over eyes and ear-coverts. Lores, band below eye and ear-coverts black. Black forecrown band continuous with black around eye. Chin, throat, and foreneck white, white collar around hindneck. Upper breast with black band, continuing as narrow band round neck, separating white collar from brown mantle. Rest of upperparts brown; lower breast and rest of underparts white. Central pair of rectrices ashy brown, 2 outermost pairs white with brown markings, remainder brown tipped white; all but outermost pair with indistinct subterminal black–brown band. Primaries blackish brown, outermost with white shaft; secondaries ashy brown with white tips; coverts brown. Bill black, base of lower mandible yellowish. Eye deep brown, orbital ring bright yellow. Legs and feet dull flesh, sometimes dusky olive, greenish yellow or yellowish. ADULT ♂ (non-breeding): as breeding bird but no black on head, neck or breast; lores and orbital area brown; band on breast brown. ADULT ♀ (breeding): similar but black areas sometimes with some brown; orbital ring less bright; non-breeding ♀ as non-breeding ♂. SIZE: wing, ♂ (n = 36) 112–120 (117), ♀ (n = 13) 112–123 (118); tail, ♂ (n = 6) 53–57 (55·8), ♀ (n = 6) 52–58 (55·0); bill, ♂♀ (n = 59) 11–14 (13); tarsus, ♂♀ (n = 65) 22–26 (24). WEIGHT: ♂♀ (n = 20) 28–39 (32); (Kenya, n = 21) 23–37 (D. Pearson, pers. comm.); (Morocco, n = 20) 26–33 (29·4).

IMMATURE: juvenile, similar to non-breeding adult but breast-band sometimes incomplete; inner median coverts with buff tips. 1st-year winter bird as non-breeding adult but with buff-tipped inner median coverts.

DOWNY YOUNG: forehead buff, black loral stripe; hind crown with black band; white neck collar; upperparts including crown finely mottled orange-cinnamon, light grey and dusky; underparts white to buffy; sides of breast with black half collar; bill black; eye dark brown, orbital ring dusky at hatching, yellow at 10 days; legs and feet dusky; weight 4–6·5.

Field Characters. Very similar to Ringed Plover *C. hiaticula* but smaller and slighter; in breeding plumage distinguished by thin white line separating black frontal band from brown of crown, mostly black bill, flesh to yellowish legs and yellow orbital ring, and in all plumages by lack of white wing-bar in flight. Less maritime than Ringed Plover; prefers freshwater habitats. Call with stress on 1st note; Ringed Plover with stress on 2nd note. Distinguished from Kentish Plover *C. alexandrinus* in non-breeding plumage by yellowish not brownish legs, and lack of wing-bar.

Voice. Tape-recorded (62, 73, 76). Usual call note, 'PEE-oo' with stress on 1st syllable; sometimes also 'cloo'. During aggressive encounters, a fast ringing 'gree-gree-gree ...', a series of low rolling rattles, and a buzzing note especially during aerial chasing. When pair making contact, 'kwee-voo kwee-voo'; when scraping, 'tick tick tick'. When performing butterfly display-flight, a monotonous, rusty 'cree-ah(k), cree-ah(k), cree-ah(k) ...'. Small young give rather high pitched piping; parents call larger young with fast 'pip-pip-pip-pip- ...' or warn them with 'PEE-oo' call.

General Habits. Has distinct preference for freshwater habitats, favouring muddy and sandy shores of rivers and lakes; also residual flood waters, short grassy areas on dry ground near water, airfields, pastures, and around villages. Less commonly inhabits coastal areas including rainwater pools on dry salt-flats bordering mangroves, saltpans, estuaries and creeks. Occurs singly, in pairs, and in groups of 6–12, or more, sometimes with other small waders. Feeds on ground, obtaining food near water's edge on gravel, short grass or mud by running, stopping and dipping body forward to seize prey. Sometimes trembles 1 foot when feeding, probably to bring prey to surface. Non-breeding birds only show aggressive behaviour when individuals come too close to one another, when short chases occur.

1st-year birds wander, many arriving late at European breeding areas. Main route of western Palearctic birds is across W Mediterranean between E Spain and Tyrrhenian Sea to a landfall in NW Africa; subsequent movement across Sahara probably in broad front. Main passage through N Africa Mar–June and July–Aug; present south of Sahara mainly Oct–Apr, e.g. arrives Kenya and Ethiopia end Sept–Oct, departs Mar–Apr; arrives N Sudan from Sept. Birds ringed Chad recovered in Rumania; ringed Kenya, recovered in USSR (Black Sea), Germany and Sudan; ringed Sudan, recovered Germany; ringed Belgium, recovered Senegambia and Tunisia; ringed Britain, recovered Morocco; ringed Germany, recovered Tunisia; ringed Germany, recovered Sudan; ringed Sweden, recovered Libya; and ringed Germany, recovered Algeria and Nigeria.

Food. A wide variety of insects including beetles and dipterans; also spiders, snails, mussels, small crustaceans, worms and seeds.

Breeding Habits. Monogamous, bond lasting for at least 1 brood; solitary nester. ♂♂ arrive first on breeding grounds, ♀♀ following 6–20 days later; rarely paired before reaching breeding area. ♂♂ establish, advertise, and defend territories av. 0·5 ha. Displays include: (1) horizontal-threat display in which bird assumes horizontal position by leaning forward with head depressed into body, white feathers of underparts fluffed out, flank feathers spread over closed wing; then advances toward intruder with black collar especially obvious; (2) chasing, in which bird in horizontal-threat posture raises back feathers, fans tail, and runs at intruder; (3) hunched, in which bird stands or walks about slowly, not necessarily facing opponent, with head held back somewhat, white feathers of underparts and flanks expanded and tail spread and lowered; (4) fighting, in which bird runs at opponent with wings spread; sometimes both birds jump up breast to breast; (5) upright, in which bird expands chest, stretches up head, showing collar, then moves to intruder with legs and feet making rapid 'marking time' movements; (6) butterfly-flight, in which bird flies buoyantly round territory, with wings well extended, arched, beating slowly, and rarely coming above horizontal; sometimes sings and may tilt wings from side to side; (7) stone-tossing, in which standing bird dips head and bill down several times towards ground, sometimes picking up bits of debris and flicking them over shoulder; and (8) flight-threat, in which bird flies toward intruder, uttering aggressive calls with quick wing-beats, and with wings raised in medium V-shape, quivers primaries rapidly, and often expands tail. Defence of territory most persistent before laying of eggs but occurs throughout breeding season. Butterfly-flight display not done after incubation. Once pair formed, both sexes (mainly ♂) defend territory. Sometimes attacks nearby Ringed and Kentish Plovers.

Pair-formation usually takes place in territory. ♂ approaches ♀ in posture similar to horizontal-threat display. ♀ at first may run away; if she remains, ♂ approaches her in upright display, marking time with legs and feet often in one spot. ♀ may then crouch down; ♂ will mount her, both then rapidly wave wings for $c.$ 5s, then, as ♀ shakes herself, ♂ flutters off backward, making cloacal contact as he dismounts. After copulation, both run away and feed separately. Pair-formation also involves scrape-ceremony display in which ♂ forms hollow in ground by (1) leaning forward, (2) rotating on breast while body and tail slanted up (3) scratching backwards with his feet, and (4) sometimes waving spread tail. He then stands up on scrape and, with tail spread, faces away from ♀. ♀ then slips into scrape by passing under ♂'s tail; ♂ then moves away, sometimes stopping to pick up and throw stones over his shoulder toward nest. ♀ may remain in scrape, ♂ may return, or both may go elsewhere and scrape again. ♂ usually makes scrapes, ♀ selects one for laying.

NEST: simple scrape usually on loose sand but sometimes on dried mud, or rock outcropping from sand or mud. One scrape chosen as nest-site; nest sometimes lined with stalks of plants or stones obtained by reaching out from scrape; no materials brought to nest. Nests usually 200 m apart; mimimum $c.$ 56 m.

EGGS: 3–5, usually 4; laid on alternate days; ovate, without gloss; dull bluish green when fresh, rapidly fading to pale brownish buff, marked with small brownish streaks and spots. SIZE: (n = 200) 26–35 × 21–24 (30 × 22). WEIGHT: (n = 1048) av. 7·7.

LAYING DATES: Algeria, Apr–May; Morocco, Mar–Apr; Tunisia, Mar; Egypt, Apr.

INCUBATION: begins between laying of 3rd and 4th egg; both sexes incubate equally; very occasionally by 3rd bird, ♂ or ♀, who also helps take care of young and defend territory. Period: 24–25 days (22–28). During change-over, either parent may toss stones. Either parent may also bring water in belly feathers to eggs (or chicks) for cooling; sometimes covers eggs with earth.

DEVELOPMENT AND CARE OF YOUNG: crack appears on egg 5 days before hatching and unborn young gives faint 'chip' or 'cheap'. Young hatch within a few h of each other, usually during night. Very agile, running and feeding themselves only a few min after hatching. Fledging period: 25–27 (24–29) days. Becomes independent 8–25 days after fledging; may breed at 1 year, usually 2 years.

Parents carry broken eggshells from nest. One parent may brood very small chick on nest at night, especially if 1 egg still not hatched. Parents brood young during 1st week, crouching with young underneath; other parent nearby or off feeding. If predators about, parents call to warn chicks which either freeze or run to them, or parents do distraction displays. In period just before egg-laying, parent runs from nest and crouches as if incubating, then runs and crouches again. During incubation period, runs from nest and crouches but only with legs half bent, then runs with tail cocked and vent showing. With very young chicks, parents call attention to themselves by running with body crouched and tail cocked, then squatting and, with breast pressed to ground, trembling body with tail usually fanned, flicking wings alternately, one raised high, the other held close to body.

BREEDING SUCCESS/SURVIVAL: hatching success 65–85%, fledging success 25·5–64%. Oldest ringed bird 9 years 10 months.

References
Cramp, S. and Simmons, K. E. L. (1983).
Johnsgard, P. A. (1981).
Simmons, K. E. L. (1953b).

228 CHARADRIIDAE

Plates 15, 21
(Opp. pp. 256, 352)

Charadrius hiaticula **Linnaeus. Ringed Plover; Common Ringed Plover. Grand Gravelot.**

Charadrius Hiaticula Linnaeus, 1758. Syst. Nat. (10th ed.), 1, p. 150; Sweden.

Range and Status. Breeds NE Canada east to NE Siberia; winters Africa, S Europe, Persian Gulf and India.

Palearctic visitor throughout, preferring open coastal habitats. Winters abundantly Morocco (21,000 Atlantic coast), Mauritania (136,000 Banc d'Arguin); and Egypt (12,000 Mediterranean coast); uncommon to rare in Sahara, crossing it almost anywhere. Senegambia to Angola, common on passage and in winter along coast, frequent to rare inland. Ethiopia, Kenya and Tanzania, common to abundant in winter on coast and inland, especially on less alkaline Rift Valley lakes; Sudan, abundant along rivers and coast. Zimbabwe, uncommon to frequent with numbers varying annually; Zambia, Malaŵi, Botswana, common to rare passage migrant, mainly on southward passage between mid-Sept and early Dec; Namibia, common to uncommon (0·5% of total wader population along 200 km of coast); South Africa, common to abundant along coast, some inland.

Description. *C. h. hiaticula* Linnaeus: winters Morocco to Mauritania, some east to Libya; also Senegambia, Zaïre. ADULT ♂ (breeding): forehead white, small black band across base of bill; black band on forecrown; rest of crown brown; lores, band under eye and ear-coverts black; white streak extends backwards from above eye. Black forecrown band continuous with black around eye. Chin, throat, foreneck and collar around hindneck white. Black band on upper breast, continued as narrow band round neck, separating white collar from brown mantle. Rest of upperparts brown; lower breast and rest of underparts white. Outer 2 tail-feathers white, rest brown tipped white. Primaries blackish brown, with ends of shafts white; outer secondaries brown tipped white, inner secondaries white; primary coverts black tipped white; rest of wing-coverts and scapulars grey-brown; greater wing-coverts broadly tipped white; rest of coverts with paler whitish fringes. Bill orange-yellow at base, black at tip. Eye dark brown, orbital ring yellow or orange-yellow. Legs and feet bright orange-yellow. Attains breeding plumage before departing north and returns with it in Sept (E Africa). ADULT ♂ (non-breeding): black areas on face, crown, neck and breast become brown; bill dusky yellow at base, black at tip. ADULT ♀: similar but breast-band of breeding birds with many brown feathers; orbital ring dull grey to slightly yellow; non-breeding birds like immature, losing virtually all trace of black. SIZE: wing, ♂ (n = 74) 124–140 (132), ♀ (n = 63) 123–141 (133); tail, ♂ (n = 6) 53–58 (55·0), ♀ (n = 6) 52–57 (54·5); bill ♂ (n = 74) 17–23 (20), ♀ (n = 63) 18–24 (20); tarsus ♂♀ (n = 65) 22–26 (24). WEIGHT: ♂♀ (n = 212) 48–78.

IMMATURE: juvenile, bands on face, crown, neck and breast brown, breast-band often incomplete; upperparts and wing-coverts fringed pale buff; bill black with little yellow at base. 1st-year winter birds as non-breeding adult but with some buff-tipped inner median coverts.

C. h. tundrae (Lowe): winters E Mediterranean and Africa south of Sahara. Slightly darker, browner, less grey; slightly smaller, wing ♂ (n = 152) 122–137 (128). WEIGHT: (Kenya, n = 165) 36–57, typical winter weight 42–50 (D. Pearson, pers. comm.); (Zimbabwe, South Africa, adults/immatures, n = 63) 43–64.

Field Characters. Small, robust plover, very similar to Little Ringed Plover *C. dubius* but larger; no thin white line above black forecrown; bill orange-yellow at base, not mainly black; yellow-orange eye-ring narrow, inconspicuous; legs and feet orange-yellow. In flight, white stripe on wings. Prefers coastal habitats while Little Ringed Plover found mainly inland.

Voice. Tape-recorded (62, 73, 76, C). Ordinary call a liquid, rather whistled 'tee-lee', 'too-it' or 'too-li'; commonly given on winter grounds. In aggression gives 'jhu-weet', 'tjurrr', 'too-dle', guttural croak and short buzz. In display-flight gives a repeated 'tché-rick'.

General Habits. Largely coastal, but also on inland lakes and larger rivers and sometimes in dry open areas during migration. Prefers open shores, most frequently tidal mud- and sandflats; also on coral reefs exposed by tides, rocky and stony shores, lagoons, salt marshes, mud- and sandbanks on rivers and lakes, flooded fields, and short grass. Occurs singly, typically in flocks of up to 50, sometimes 1200–1500. Roosts on bare open shoreline individually or in mixed flocks. Forages by day and night; runs for short distance, pauses in upright or hunched position, probes, then runs and pauses again. Vibrates foot to stir up organisms when feeding.

Displays on wintering grounds include: (1) ground chase, in which bird assumes horizontal compressed attitude, then runs at opponent often with tail fanned; may also fan tail when chasing in flight; (2) threat, leaning forward with bill pointing to ground, then held more or less horizontal, with body slanting forward, and tail fanned and raised; at higher threat intensity tail fanned and elevated, sometimes placed sideways, back feathers raised, black markings standing out conspicuously, breast expanded and sometimes head stretched up and legs and feet making rapid 'marking-time' movement; (3) flight-threat, flying at intruder with quick wing-beats; and (4) butterfly-flight, in wide circles with wings extended, arched, beating slowly, rarely above horizontal.

Main migration route of *tundrae* is E coast and Rift Valley, and across Central Africa to Namibia and South African coasts; also probably from E Mediterranean across Sahara to Gulf of Guinea. Arrives Egypt (Hurghada) early Sept, leaves late May. Adults arrive Kenya Sept, young birds mainly Oct; main departure Apr. Marked passage Ethiopia Apr–early May. Arrives South Africa late Sept–Oct, departs Apr. Passage Tunisia in autumn from end July–Aug onwards. *Hiaticula* migrates along Atlantic coast to Mauritania; also crosses Sahara. Main passage Morocco mid-Aug–Nov and Apr–early June. Birds recovered Morocco, ringed Greenland, Belgium, France, Germany, Iceland, Norway, Sweden, Finland, Estonia, Poland, Algeria, Great Britain; Tunisia, ringed Estonia; Senegambia, ringed Greenland, Norway, Sweden, England, Germany; Mauritania, Ghana, and Benin, ringed England; and South Africa, ringed France and USSR. A few, presumably 1st-year birds, oversummer almost anywhere in Africa; in E Africa mainly on coast.

Food. Small crustaceans, worms, small snails and clams, beetles, larvae, flies and various other insects, millepedes, and some vegetable material.

References
Cramp, S. and Simmons, K. E. L. (1983).
Johnsgard, P. A. (1981).
Simmons, K. E. L. (1953a).

Charadrius pecuarius Temminck. Kittlitz's Plover; Kittlitz's Sand-Plover. Pluvier pâtre.

Charadrius pecuarius Temminck, 1823. Pl. col., livr. 31, pl. 183; Cape of Good Hope.

Plates 15, 21
(Opp. pp. 256, 352)

Range and Status. Africa, Madagascar.
Resident, widespread, partly migratory, inland as well as on coast, from Mauritania, Nile Delta and Sudan south to Cape; locally abundant to rare.

Description. *C. p. pecuarius* Temminck: only subspecies in Africa. ADULT ♂ (breeding): forehead white; forecrown with blackish bar followed by narrow white bar, rest of crown brown with sandy tips to feathers. Black stripe from bill through eye to side of neck, continuing as collar across upper mantle, separated from crown by white superciliary stripe which continues round hindneck to form white collar. Mantle dark grey-brown, rest of upperparts sooty brown, the feathers with sandy rufous margins. Central pair of tail-feathers blackish, others progressively whiter towards side of tail, outer 1–2 pairs completely white. Face, chin and upper throat white; rest of underparts light yellow-ochre, paler belly and lower flanks. Primaries and secondaries blackish; inner primaries with white edges at base, outer greater coverts with small white tips; lesser coverts blackish brown, rest of upper coverts lighter; axillaries white, underwing-coverts white flecked brown. Bill black; eye dark brown with black eyelids; legs and feet black, sometimes greenish grey. ADULT ♂ (non-breeding): like breeding ♂ except eye-stripe browner, frontal bar lacking, underparts much paler buff. ADULT ♀: like ♂ except black band across forecrown narrower. SIZE: wing ♂ (n = 31) 99–112 (106), ♀ (n = 15) 101–111 (105), ♂♀ (n = 90) 96–116 (105); tail ♂♀ (n = 20) 35–44 (40); bill ♂♀ (n = 74) 15–23 (19); tarsus ♂♀ (n = 82) 26–33 (29). WEIGHT: Zimbabwe, Botswana, Namibia, adult (n = 590) 30–40, immature (n = 540) 29–40; South Africa, coastal and Transvaal, adult (n = 72) 32–44, immature (n = 56) 29–42 (A. Tree, pers. comm.); Barberspan, adult (n = 426) 19–49 (34), juvenile (n = 251) 20–44 (33); adult/immature Zambia (n = 96) 25–44 (32); Kenya (n = 42) 23–40 (D. Pearson, pers. comm.).

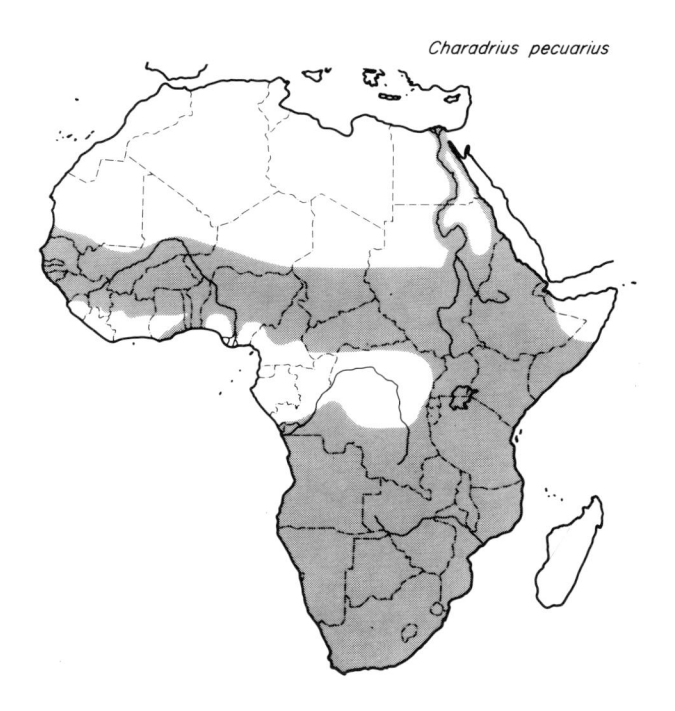

IMMATURE: like adult ♂ above but lacks black face marks, upperparts brown broadly edged pale buff-brown; collar on hindneck buff; underparts whitish, tinged buff-brown on breast.

DOWNY YOUNG: upperparts grey to white, mottled black with dark centre line on back; underparts white, weight (n = 14) 4–6 (5·5).

Field Characters. Distinguished from other small plovers by buffy underparts and conspicuous black line down side of neck, and from all except White-fronted Plover *C. marginatus* by lack of breast-bands or patches; latter has paler upperparts, pure white underparts. In flight, pale greater and median coverts contrasting with dark remiges and leading lesser coverts give characteristic wing pattern; also short narrow wing-bar formed by white tips of outer greater coverts and white edges at base of inner primary webs.

Voice. Tape-recorded (264). Calls include 'pipip', gentle 'towhit', 'tip-peep', 'trit-tritritritrit', 'perrup', and 'kich-kich-kich'. Also gives an alarm 'chirrt' or 'prrrt'. During disputes, ♂ gives a buzzy call. When feigning injury, both give 'cheep-cheep'; when inviting young to brood, 'chip-chip'. Warning note to young, 'trr-trr'.

General Habits. Prefers dry ground with very short grass or dried mud, usually but not always near water; also inhabits edges of dams, lakes, rivers, tidal mudflats and dry saltflats, but avoids sandy or rocky coasts. Like Forbes' Plover *C. forbesi*, found consistently away from water. Usually occurs in pairs but gregarious in non-breeding season, in small flocks of up to 20, sometimes more (e.g. 11–100 Zambia (Kafue) with maximum 270: Blaker 1966). Feeds, often in groups of 2–5, by running rapidly, stopping suddenly to peck at object, then running on again. Sometimes stands erect with 1 foot vibrating on surface, lunging forward to catch prey. Often forages in small mixed flocks in association with *Calidris* spp.; feeds also on moonlit nights until *c.* 23.00 h. Roosts mainly back from water on broad, fairly bare open shoreline, sometimes in mixed flocks; individuals also roost separately.

Undergoes some seasonal movements that are not well documented. In Botswana mainly summer visitor, while in Zimbabwe and Zambia mainly dry season visitor Apr–Dec with entire Zambian population leaving during high floods. The few ringing recoveries indicate birds from Harare move to Bulawayo and presumably on to Botswana, Namibia and Cape Province where they join sedentary coastal populations (Tree 1976, 1978). In Ghana, present on coast only Oct (Macdonald 1979). Movements not studied in detail but seem to be regulated by seasonal rainfall with birds leaving during rains and flooding.

Food. Insects, especially small beetles, insect larvae, spiders, crustaceans and molluscs; most food small but seen to take cricket *c.* 40 mm long (Tree 1974).

Breeding Habits. Monogamous, solitary nester or in loose colonies. Pairing takes place 3–4 weeks before territories occupied. Pair defends territory until soon after young hatch; also copulate and feed in territory during incubation. Territory may extend along 60–70 m of shore and occupy 3600–4200 m². Early in breeding season individuals of a flock suddenly fly or run at other members, chasing them a short distance; sometimes 2 birds may run towards and even fly at each other, nearly colliding in mid-air (Clark 1982a). When chasing intruder, pursuer runs at but stops short of it, then adopts an upright posture with legs almost straight and head up. Courtship behaviour includes Scrape-ceremony display in which 1 bird of pair places breast on sand, rotates on breast with body slanting up at *c.* 30° from horizontal and tail somewhat raised, and kicks out sand with a backward and forward movement of legs, ankles sticking out behind (**A**). ♂ usually takes initiative in making scrapes although both may take turns making them before selecting one. Occasionally either bird picks up pebble and carries it to nest or throws it over back with backward jerk of head. Either bird may also break off fragments of dead vegetation round perimeter of hollow and flick them in its direction. Copulation usually occurs on or near scrape, sometimes before final choice of nest-site has been made. When preparing to copulate, ♀ stands with head lowered slightly or skulks along on bent legs with tail down, away from scrape. ♀ may do this 2–3 times before ♂ responds. When responding, ♂, in stiff upright stance with feathers puffed out, approaches ♀ from behind, often high-stepping. ♀ lowers her body slightly. He may first stroke ♀'s back with his feet or immediately run forward and jump on her back. He may stand on her back up to 150 s, but usually only a few s; he then crouches, copulates, grabs her neck with his bill, and pulls both of them over backwards. ♂ ends up on his back with wings spread and ♀ on top of him with legs in air. Copulation occurs 6–11 days before egg-laying (2 pairs: Clark 1982a). No 'butterfly' song flight recorded.

NEST: simple scrape in sand or dried mud in open terrain where birds can see intruders; usually within *c.* 50–100 m of water, although may be several km away; 10–15 cm in diameter, 2·5 cm deep; lined with small pebbles, pieces of shell, animal dung and bits of vegetation. Most 20 m apart but some as close as 9 m. Old scrape may be used, probably by same pair (Hall 1959).

EGGS: 1–3, usually 2, laid at 1–2 day intervals; Zimbabwe, Zambia and Malaŵi (n = 34) av. 1·82; South Africa, W Cape (n = 466) av. 1·86; Natal (n = 362) av. 1·81; overall (n = 852) av. 1·84. Pointed ovals; creamy buff, marked with fine lines of sepia and heavy lines and spots of very dark brown or black. SIZE: (n = 301) 28–34 × 19–24 (32 × 22).

LAYING DATES: Egypt, Apr; Mali, Feb–Mar; Senegambia, Feb–July, Oct; Ghana, June–Sept; Nigeria, May–Aug, Nov; Ethiopia, Mar–Sept; E Africa: Kenya,

Apr–June, Nov; Tanzania, Feb–Mar, May–June, Nov, Dec; Region A, Jan–Sept; Region B, Apr–Sept; Region C, Feb, Apr–July, Sept; Region D, mainly in rains but also in dry season; Zaïre, Aug–Sept at end of dry season; Zambia, Apr–Nov; Malâwi, Aug–Oct; Zimbabwe, Mar–Nov but mostly July–Nov; Mozambique, June–July; Angola Apr, Aug; Namibia, July–Sept; Botswana, Nov–Jan; South Africa: W Cape, all year but mostly Sept–Jan; Natal, July–Feb but mostly July–Oct.

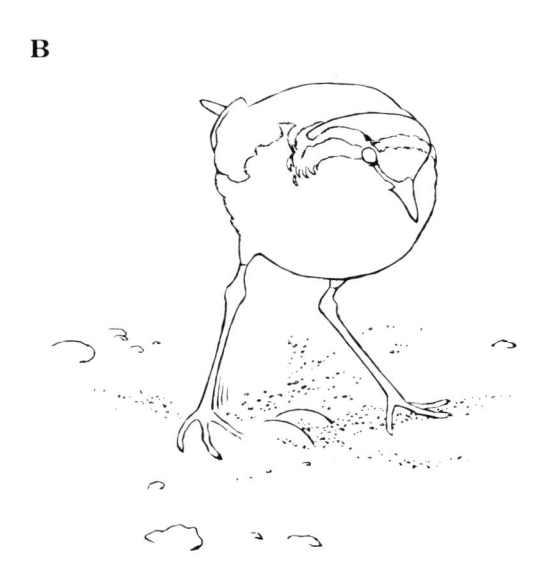

B

INCUBATION: begins when clutch complete; shared by both parents with ♂ at night, ♀ by day (Clark 1982a). Period: 22–28 days. When leaving nest at night or when leaving on its own initiative in early morning or late evening, bird does not cover eggs; when leaving during other times of day to feed, covers eggs with sand by kicking feet inward alternately (**B**). When approached by any potential predator, bird quickly covers eggs with sand. May move 2–3 times round nest when covering it, taking 3–90 s. Eggs two-thirds or completely covered. Also covers newly or partly hatched chicks. When returning to nest, uncovers eggs before incubating. Does not cover nest during normal nest-relief. During nest-relief incoming bird may relieve mate on nest without ceremony or mate may stand up and make side-throwing movements before departing. When hot, incubating bird places legs alongside of body, raises crown and mantle feathers and crouches over eggs. Occasionally relieving bird first soaks belly plumage in water to help cool eggs. Either parent removes eggshells from nest-site; will transfer eggs from 1 scrape to another if 1st scrape flooded; moves eggs up to 30 cm (Kutilek 1974). Sometimes leaves eggs unattended up to 5–7 h (Clark 1982a).

DEVELOPMENT AND CARE OF YOUNG: wander from nest within few h of hatching or brooded in nest for up to 24 h. Do not feed for first 24 h. Either parent, but usually ♀, broods young at frequent intervals. Parent calls young when inviting them to brood or when danger present (see Voice). Chicks fly when 26–32 days old but brooded in captivity for 42 days. Some breed 1st year.

When intruder present, either parent may perform: (1) injury-feigning, by first lying flat on ground, waving wings about in helpless manner, holding 1 or both wings up, and fanning tail; then running away or towards intruder; (2) 'rodent running', in which it runs fast with head held low, tail drooped and spread, and wings spread; and (3) false brooding. When young feed, parents defend immediate area against conspecifics and Three-banded Plovers *C. tricollaris* by running towards them and putting them to flight.

BREEDING SUCCESS/SURVIVAL: hatching success, southern Africa, 52% eggs hatched (Blaker 1966); Botswana, usually less than 2 young/pair raised successfully (Wilson 1981); Zimbabwe, 77% hatching success, minimum fledging success 0·5 young/pair (Tree 1974); Kenya (Nakuru) 19% eggs hatched (62 nests with 112 eggs over 3 years); mortality due mainly to flooding but also to motor vehicles and predators.

References
Clark, A. (1982a).
Conway, W. G. and Bell, J. (1968).
Cramp, S. and Simmons, K. E. L. (1983).
Hall, K. R. L. (1958, 1959, 1965).

***Charadrius tricollaris* Vieillot. Three-banded Plover; Treble-banded Plover. Pluvier à triple collier.**

Plates 15, 21
(Opp. pp. 256, 352)

Charadrius tricollaris Vieillot, 1818. Nouv. Dict. Hist. Nat., 27, p. 147; Africa (restricted type locality, Cape Town, Grant, *Ibis*, 1915, p. 57).

Forms a superspecies with *C. forbesi*.

Range and Status. Africa and Madagascar.

Resident, with some local movements, from Gabon, S Zaïre, E Africa, E Sudan, Ethiopia and N Somalia south to Cape. Also N Cameroon and N Nigeria (only non-breeding visitor); vagrant Ghana and Mali. Generally common throughout southern Africa, frequent to uncommon in northern part of range; uncommon to vagrant W Africa.

Description. *C. t. tricollaris* Vieillot: only subspecies in Africa. ADULT ♂: forehead white; crown, hindneck and nape dark brown; white stripe from forehead over eye and around nape. Side of face and throat greyish white to grey. Mantle and back olive-brown; tail, outer feathers white, central pair brown, rest brown and white, becoming more white laterally with darker subterminal bar and white tips. Black band across lower throat and upper breast separated by white band from a 2nd black band; rest of underparts white. Upperside of wings

Charadrius tricollaris

Also winters in breeding range

largely dark brown; inner primaries tipped white; secondaries, primary coverts and alula black tipped white. Underside of wings white. Bill reddish pink with black tip; eye hazel brown with large red eye-ring; legs and feet purplish to pink. Sexes alike. SIZE: ♂♀ (n = 48), wing, 104–119 (110); tail, 56–67 (62); bill, 14–18 (16); tarsus, 21–24 (23). WEIGHT: (n = 11) 25–38 (31) (Skead 1977), (n = 55 post-juveniles) 25–43 (34) (Summers and Waltner 1978); (n = 8, Kenya) 27–39 (D. Pearson, pers. comm.); (n = 90, Zimbabwe, adults) 29–39, (n = 690, Zimbabwe, immatures) 27–37 (A. Tree, pers. comm.); (n = 147) 25–43 (32·2) (G. Maclean, pers. comm.); (n = 92, Barberspan, adults) 25–38 (31·2).

IMMATURE: incomplete brown upper chest-band, lower band incomplete but with some black feathers; head more uniform in colour; red eye ring less noticeable; upperparts with feathers edged buff.

DOWNY YOUNG: boldly patterned; upperparts sandy speckled black, black median stripe on back; white below.

Field Characters. A distinctive plover, distinguished from all except Forbes' Plover by double breast-band, grey face, red eye-ring and pink-based bill. Separated from Forbes' Plover *C. forbesi* by white forehead connecting with white superciliary stripe, grey throat, narrower and darker breast-bands and pink legs. In flight shows white wing-bar, white outer tail-feathers and subterminal black bar and terminal white bar on all tail-feathers.

Voice. Tape-recorded (38, 74, 217, 262, 305). When alarmed, a loud 'wik-wik', 'twi-twi', 'wick-wick', 'tweep-tweep' and 'tsip-tsip'. In flight a high-pitched 'slee', 'peep', 'seep' and a loud 'tiuu-t-tiuu-t'. When chasing another bird, a rattling trill 'kee-kee-kirra-kirra'. Before copulating, ♂ gives a guttural, subdued 'krrr'; 'krrr' also given when pair makes scrape and when performing distraction display.

General Habits. Occurs usually in pairs or small flocks of 6–10, up to 20, rarely 40. Prefers firm shorelines and gravelly areas near water; also muddy areas associated with lakes and rivers; less attracted to salt water where found on tidal pools and lagoons but rarely on open seashore. Usually seen running along water's edge or making short flights; flight erratic with jerky wing beats; usually flies only short distance before alighting; upon alighting bobs tail. When giving alarm call, inclines body forward, bobbing up and down with each call. When feeding, darts forward in short spurts with head down, stopping now and then to peck at ground. Pecks vary from 10 to 40 pecks/min. Sometimes before pecking, trembles foot by standing on one leg and vibrating the ground with the toes of the other (Freeman 1970, Clark 1982b). Flies about a good deal at night, probably to forage. Roosts away from water on broad, open shoreline with gravel or stone, in mixed flocks or individually.

Some movements, poorly known. In Nigeria mainly present at end of rains and through most of dry season Aug–Feb (Hall 1976); in Ethiopia present Mar–Apr and July–Dec (Ash 1972); in E Zaïre and Rwanda influx of birds Mar (Curry-Lindahl 1981). Movements complex in southern Africa: Zambia, dry season visitor; Zimbabwe, population moves out with rains but partially replaced by visitors from elsewhere; Botswana, present all year but commoner in rains; Namibia, peak numbers May–Sept; South Africa, Cape coast, large number in winter rains, Orange Free State, large numbers winter (Tree 1980a). Birds ringed Goreangah Dam (22°37′S, 17°08′E) recovered 260 km west at Walvis Bay (Tree 1980a).

Food. Terrestrial and aquatic insects, insect larvae, crustaceans, small molluscs and worms.

Breeding Habits. Monogamous; solitary nester. Early in breeding season pair establishes a territory, usually along 80–100 m of shoreline, where both mates feed and build nest; may establish 2 territories some distance apart, 1 for nesting and 1 for feeding (Clark 1982b). Either mate defends territory by running at intruder in a crouched position with head held low and flank feathers fluffed; pursuer may give rattling trill and may also fly at intruder. When threatened by gull or large plover, may turn its back and fan tail but will not usually move away. When disturbed, will also bob body up and down, usually more than once. When several conspecifics arrive in feeding area, either mate runs alongside intruders, giving rattling trill and fluffing flank feathers over edge of wing. Sometimes in response to mate, conspecifics or other species, may perform Jig display by quickly stepping from side to side, flirting wings, exposing white belly and flank feathers and, very occasionally, stretching wings vertically. During courtship ♂ performs Vent display by squatting at angle with breast touching ground and tail thrust upwards and sometimes fanned. As ♂ performs this display, ♀ stands facing his tail. Maintaining Vent display, ♂ nest-scrapes with legs moving backwards alternately; sometimes he may scrape by pushing chest into sand, moving from

B

A

side to side and turning around. ♂ then moves off scrape, maintaining the elevated tail, and picks up and discards fragments around the scrape. Mate will take his place, scrape briefly, lift her body slightly, pick at fragments, then move away. Either mate brings and drops small pebble or twig on nest; sometimes throws it over head with backward jerk of head. Several scrapes may be made before 1 is chosen. ♀ initiates copulation by adopting an upright stance with tarsal joint straightened and jerking her body up and down; she may perform this several times in succession. ♂ runs up behind her; then 'high-steps' behind her and makes 'krrr' sound (**A**). ♀ lowers body (**B**), and ♂ steps up on her back. 1st copulation usually followed by 1 or more further copulations. After copulation, ♀ may bob, ♂ may perform Jig and Vent displays. Copulations may occur over several days (up to 19 days, Leeupan: Clark 1982b).

NEST: a simple scrape in sand or dried mud, always close to water; sometimes on hard rock; frequently lined with bits of plant material, dried mud or pebbles.

EGGS: 1–2, usually 2, probably laid at 2–4 day intervals; Zimbabwe mean (89 clutches) 1·89; South Africa, W Cape (69 clutches) 1·94, Natal (14 clutches) 1·54; overall (172 clutches) 1·88. Regular oval; cream-white, heavily covered all over with fine black lines. SIZE: (n = 67) 27–33 × 21–24 (30 × 22).

LAYING DATES: Ethiopia, May, Sept–Oct, Dec; Somalia, May; E Africa: Region A, Apr; Region B, Apr–June; Region C, Apr, June–Oct; Region D, all year with preference Apr–June, during rains; Region E, July; Burundi, May–July; Zambia, May–Oct; Malaŵi, July–Oct; Zimbabwe, Jan, Mar–Dec with most July–Sept; Mozambique, May–June; Namibia, Jan, July, Sept–Dec; Botswana, July; South Africa: Natal, July–Dec, mainly Sept–Dec; SW Cape, all year but mainly July; Transvaal, mainly July–Sept.

INCUBATION: begins probably when clutch complete; performed by both parents. Period: 26–28 days (Tyler 1978, Blaker 1966). Incubating bird sometimes allows intruder to approach within 3 m before slipping off nest and running quickly away. Does not cover eggs before leaving. When incubating, displays by raising posterior part of body, sometimes cocking tail up and showing white undertail-coverts. Occasionally soaks belly plumage to help cool eggs. When temperature high, incubating bird will place legs alongside body, raise crown and mantle feathers, or crouch over eggs. Occasionally removes broken egg from nest and transfers remaining egg to new nest c. 20–30 cm from original site (Maclean 1965).

DEVELOPMENT AND CARE OF YOUNG: young agile, feeding themselves as soon as they are dry. Small young brooded often (during 2 h mid-morning, young c. 2 days old brooded on average 2·8 min every 6 min: Clark 1982b). Before flying, some may remain in a limited area of c. 15 m² around nest while others wander 50 m or more in an evening when feeding (A. Tree, pers. comm.). Fly at 21–22 days, fully fledged at 30–32 days but sometimes remain with parents until 40–42 days old (Tyler 1978). On hearing alarm call, hide or freeze (froze in 3/40 observed disturbances: Tyler 1978).

Both parents brood chicks against intruders or rain until 24 days old. When intruder present, parent may approach it, sometimes calling; then feigns injury by hiding in a crouch with head jerking up and down, tail raised, spread, and fanned to one side, and wings held out. Sometimes drives away other birds, e.g. Kittlitz's Plover *C. pecuarius*, Wood Sandpiper *Tringa glareola* and Squacco Heron *Ardeola ralloides*.

BREEDING SUCCESS/SURVIVAL: 1 pair in Ethiopia multiple-brooded with 3 clutches within 6 months, the 2nd being a repeat clutch (Tyler 1978). In Zimbabwe frequently double-brooded, once triple-brooded (1st brood fledged, 2nd lost when chick half grown, 3rd fledged: Tree 1974). ♀ frequently lays again well before 1st brood fully fledged. Breeding success, Botswana, usually less then 2 young/pair raised successfully (Wilson 1981). Lives at least 10 years (A. Tree, pers. comm.).

References
Clark, A. (1982b).
Tree, A. (1974).
Tyler, S. (1978).

Plates 15, 21
(Opp. pp. 256, 352)

Charadrius forbesi (Shelley). Forbes' Plover. Pluvier de Forbes.

Aegialtis forbesi Shelley, 1883. Ibis, 1883, p. 560; Shonga, Niger River.

Forms a superspecies with *C. tricollaris*.

Range and Status. Endemic, seasonally migratory, locally common to uncommon, in muddy and grassy habitats from Senegambia to N Angola and east to SW Sudan, W Uganda, W Tanzania and N Zambia. More common in western part of range; uncommon to rare in extreme eastern and southern parts of range.

Description. ADULT ♂ (breeding): forehead light brown, white superciliary stripe from eye around nape. Crown, hindneck, mantle and back dark brown; tail dark brown centrally, white with dark bars laterally, all rectrices with subterminal black bar and, except for central ones, tipped white. Side of face grey; throat grey to brown becoming dark brown on upper breast and forming dark band, separated by white band from another broader dark brown band; rest of underparts white. Wings dark brown, secondaries with narrow white tips; axillaries white. Bill dark brown, pinkish at base; eye brown with large red eye-ring; legs and feet pale yellowish to purple-brown. ADULT ♂ (non-breeding): like breeding ♂, but somewhat lighter brown. Sexes alike. SIZE: wing ♂♀ (n = 8) 121–133, ♂, 125, 130, 132, ♀, 129, 134; tail ♂♀ (n = 8) 63–71, ♂, 71, 74, 77, ♀, 74, 76; bill ♂♀ (n = 8) 16–18, ♂, 17, 19, 19, ♀, 18, 19; tarsus ♂♀ (n = 8) 29–32, ♂, 27, 30, 33, ♀, 32, 33. WEIGHT: (Liberia) ♂ 46·4, ♀ 47, 48, 49.
    IMMATURE: not described.
    DOWNY YOUNG: not described.

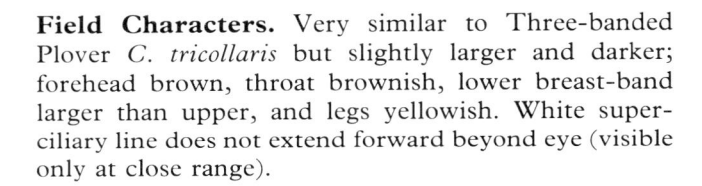

Field Characters. Very similar to Three-banded Plover *C. tricollaris* but slightly larger and darker; forehead brown, throat brownish, lower breast-band larger than upper, and legs yellowish. White superciliary line does not extend forward beyond eye (visible only at close range).

Voice. Not tape-recorded. Calls include single 'peeuw', repeated at infrequent intervals, repeated plaintive 'per-ooh', and, when displaying in flight, 'pleuw-pleuw-pleuw-pleuw . . .'.

General Habits. Usually solitary or in pairs; sometimes small flocks of 15–20. Inhabits open grassy plains, bare ground, recently burnt areas, open places in forest belts, and edges and muddy areas by lakes and rivers; prefers drier areas than Three-banded Plover. In breeding season found in rocky hills and slopes with granite outcrops. When alarmed, bobs head up and down and calls, and flies off in fast, erratic zig-zag movements. When feeding, runs quickly, then pauses for long periods.
    Seasonally migratory in Nigeria, gathering at beginning of dry season (Sept) in flocks in grasslands, open areas in forests, rice fields, and by lakes and rivers. With first rains (Mar) moves to rocky hills and slopes to breed. Present Ivory Coast only Nov–Apr.

Food. Beetles, grasshoppers and other insects including larvae; also small molluscs, crustaceans and worms.

Breeding Habits. Monogamous, solitary nester. During courtship one member of pair may fly into air, drop quickly down on shallow wing-beats calling 'pleuw-pleuw . . .', level off, land and repeat the performance.
    NEST: a scrape on top of granite outcrop or in gravel stream bed; sometimes made of small pebbles *c.* 0·5 cm in diam. Same nest-site may be used year after year.
    EGGS: 2–3; clutch (Nigeria, n = 18) av. 1·5 (Brown 1948). Ovate to somewhat pyriform with slight gloss; creamy coloured, marked with blotches and spots of reddish brown; some grey undermarkings but not hair lines or other linear markings as in Three-banded Plover. SIZE: (n = 3) 31–32 × 23.
    LAYING DATES: Ghana, July–Aug; Nigeria, Mar–Aug.
    INCUBATION: probably begins when clutch complete; performed by both parents. Period: at least 22 days (Brown 1948).
    DEVELOPMENT AND CARE OF YOUNG: both parents attend young; usually 1 parent at a time broods them. Parents carry out injury-feigning. Young freeze when intruder present.
    BREEDING SUCCESS/SURVIVAL: possibly double brooded (Nigeria).

References
Brown, L. H. (1948).
Tree, A. (1964).

Charadrius alexandrinus Linnaeus. Kentish Plover; Snowy Plover.
Pluvier à collier interrompu.

Charadrius alexandrinus Linnaeus, 1758. Syst. Nat. (10th ed.), 1, p. 150; Egypt.

Plates 15, 21
(Opp. pp. 256, 352)

Forms a superspecies with *C. marginatus*.

Range and Status. Breeds Africa, Mediterranean basin, Europe, central Asia and E China, Japan and Indonesia, western North America, US Gulf coast, West Indies, coasts of Peru and Chile; Eurasian population winters Africa, Mediterranean basin and India; other populations winter Central and South America and SE Asia.

Resident and Palearctic visitor, mainly on coasts, south to Cameroon, N Zaïre and Somalia. Resident, rare to common Mediterranean (Morocco to Egypt) and W African coasts south to Senegambia (southernmost west coast breeding localities: 90 km south of Dakar, 110 km southeast of Dakar and Khalissaye N. P., Casamance (*c.* 12°30′): G. Morel and A. Sanya, pers. comm.); common to locally abundant Red Sea and Gulf of Aden coasts south to N Somalia and Socotra. Palearctic visitor, very abundant to common, Tunisia (17,000), Morocco (10,000, Atlantic coast), Algeria (1500, Oranie), Mauritania (18,000, Banc d'Arguin), Egypt (50,000+), Sudan coast and in north (many hundreds along Nile to at least 30 km south of Khartoum), Ethiopia and Somali coasts (flocks of 80 Somalia). Elsewhere N Africa, frequent to rare winter visitor; common passage migrant with some crossing Sahara on broad front. Elsewhere south of Sahara locally frequent to rare, mainly coastal although recorded inland in Mali, Chad, Nigeria, Cameroon, Zaïre, Sudan, Ethiopia and Kenya. Precise southern limits unclear, with most records south of equator not reliable due to confusion with White-fronted Plover *C. marginatus*; known to occur as far south as Magadi, Kenya, 1°52′S, 36°17′E.

Description. *C. a. alexandrinus* Linnaeus: only subspecies in Africa. ADULT ♂ (breeding): forehead white; forepart of crown black (not reaching eye); rest of crown and hindneck light buff to rufous. Lores, eye and upper ear-coverts form black stripe, separated from crown by white stripe. Hindneck with white collar. Black patch on sides of breast sometimes extends narrowly around hindneck directly below white collar. Mantle, back, rump and central tail-feathers pale brown; outer tail-feathers white. Underparts white. Primaries dark brown, most with white shafts; secondaries pale brown, white toward tip; primary coverts dark brown, inner ones tipped white; greater coverts pale brown tipped white; rest of coverts ash brown. Bill black; eye dark brown, eye-ring black; legs and feet dark grey to black, occasionally olive or yellowish grey. ADULT ♂ (non-breeding): like breeding ♂ but patches on head, face and breast brown; coverts grey-brown. ADULT ♀ (breeding): like ♂ but head, face and breast patches brown with a few black feathers; white on forehead usually less extensive; only edge of crown rufous. ADULT ♀ (non-breeding): white collar and white eye-stripe narrower than in non-breeding ♂. SIZE: wing, ♂ (n = 53) 102–115 (111), ♀ (n = 26) 105–116 (112); tail, ♂ (n = 6) 41–48 (45·3), ♀ (n = 6) 39–48 (43·8); bill, ♂♀ (n = 59) 14–17 (15); tarsus, ♂♀ (n = 60) 26–30 (28). WEIGHT: ♂♀ (winter) (n = 22) 27–41 (37).
IMMATURE: juvenile, breast patches pale brown; upper parts and coverts grey-brown fringed sandy buff. 1st-year winter bird like non-breeding adult but with pale buff-fringed inner median coverts.
DOWNY YOUNG: forehead and hindneck white; rest of upperparts, including crown and back, pale grey with fine stippling; discontinuous black line between crown and white nape; underparts white; bill black, eye dark brown, legs and feet bluish grey.

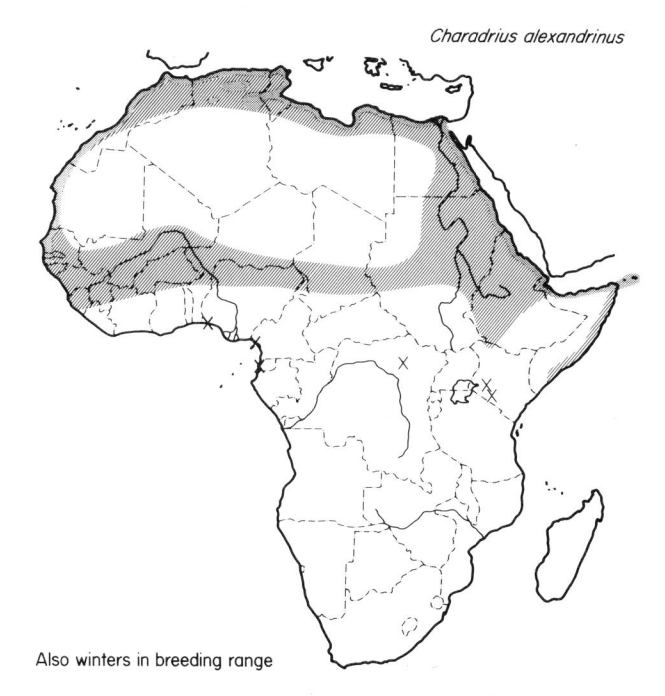

Field Characters. Easily confused with very similar White-fronted Plover; distinguished by black or dark grey legs (not pale, buff or olive grey), conspicuous breast-patches (not poorly defined or absent), slightly larger size with stronger legs and larger feet, upperparts without dorsal rusty tinge, whiter underparts, and shorter tail (usually less than 45 mm). ♀♀ of both species very similar but White-fronted ♀ has some black separating white forehead from grey crown. At all ages distinguished from Ringed Plover *C. hiaticula* and Little Ringed Plover *C. dubius* by paler upperparts, incomplete breast-band, broader white sides of tail and black legs; adults separated by rufous crown, less black on cheeks and all black bill. In flight separated by white wing-bar from Little Ringed Plover. Non-breeding adult and immature very similar to very young juvenile Chestnut-banded Plover *C. pallidus* but larger (wing 111 *vs* 102 mm), legs darker and breast marks different.

Voice. Tape-recorded (62, 73, 76, C). A low-pitched 'poo-et', 'chu-wee' or 'chu-uu-ee'; 'wee-it'; 'prrr'; soft 'hwick'; soft and fluty 'uit' or 'wit-wit-wit'; and song-rattle, a hard fast mechanical trill 'dwee-dwee-dwee ...', which at lower intensity becomes a liquid 'cher-wee'. Also gives 'kittup' alarm call. Song when in flight, a loud sharp 'TJEKke-TJEKke ...'.

General Habits. Prefers sea coast habitats, especially flat open areas including sand and shingle beaches, estuaries, fringes of lagoons, and tidal mudflats; inland found predominantly along saline lakes and mudflats and on steppes with limited vegetation; sometimes also along rivers. Occurs singly, in pairs or in small groups of up to 20 or 30, sometimes in larger flocks of up to 260; often associates with other waders. Runs about with neck drawn into shoulders like other small plovers, moving as quickly as or faster than Little Ringed Plover. When foraging, runs quickly, stops, dips bill into ground or pecks at prey, then runs on again. Sometimes trembles one foot on sand or shakes leg, presumably to make nearby prey reveal itself. Roosts often in large mixed flocks.

Most Palearctic migrants winter south of Sahara Sept–May. Most migrants pass through N Africa Mar–Apr and July–Nov, although a few remain to winter along coast and inland in Saharan oases. Extent and timing of migration (if any) of resident African population not known; it may be sedentary. Birds ringed Austria, Belgium, East Germany, France and Italy recovered Morocco; birds ringed Tunisia recovered Algeria. Main passage Morocco in Sept, Banc d'Arguin into Nov with largest numbers in Oct.

Food. Insects, including sand flies, beetles, sandhoppers, and aquatic species; spiders; small crabs and other crustaceans; worms; small molluscs; bits of sea weed; and grit.

Breeding Habits. Seasonally monogamous although some bonds last up to 6 years; rarely ♂ incubates alternately at 2 nests, each with different ♀. Solitary nester. ♂♂ establish territories; pair-bond established in neutral feeding grounds; once bond formed both sexes, although mainly ♂, defend territory, sometimes expelling larger species such as Eurasian Avocet *Recurvirostra avosetta*. When defending territory, may chase, threaten, fight or make display flights. Displays include: (1) Chasing display in which either ♂ or ♀ in horizontal position, with head depressed into body, feathers raised and tail fanned, runs at intruder, sometimes facing it with open tail tilted sideways; (2) erect, Puffed-breast display in which ♂ stands up at angle of 50°, increases its size by puffing out and extending its white breast, flank and belly feathers over its side, and brings its neck well down into a hunched pose; (3) Butterfly-flight display in which ♂ flies about territory, slowly beating extended and arched wings, and producing its trilling 'dwee ...' song; performed apparently only by ♂; and (4) Stone-tossing display in which standing ♂ or ♀ picks up small pebble or bit of debris and flicks it over shoulder with quick sideways jerk of head.

Pair-formation and pre-copulatory behaviour involves Scrape-ceremony display in which ♂ makes scrape by scratching backwards with feet until slight impression formed; may make several scrapes. If ♀ approaches ♂, he stands at edge of newly formed scrape with bill pointed into it, tail raised and fanned, wing nearest ♀ slightly drooped, and wing farthest from her raised and slightly spread. ♀ then enters scrape by passing under his raised and spread tail. During pair-formation and before copulation, ♂ may walk towards ♀, raising his feet very high at each step; he may also mark time in one spot. If ♀ receptive, ♂ mounts her and may sometimes remain on her back 1–5 min before making cloacal contact. After copulation, ♂ may grasp nape feathers of ♀ and pull her over backwards. Copulation activity may occur for several days; ♂ makes scrapes; 1–4 scrapes made before ♀ settles on one.

NEST: a simple scrape, sometimes lined with bits of grass, leaves, pebbles, or broken shells; av. diam 6–9 cm, depth 1·5–3 cm; nests usually solitary although sometimes close together (Somalia, several within 1 ha; Kansas, some 15–20 m apart, 17 nests on av. 85 m apart; Europe, shortest distance 0·8 m apart, many 2–5 m, most 21 m, some 80 m).

EGGS: usually 3, sometimes 2, rarely 4; laid at *c*. 2-day intervals. Oval; smooth, dull; cream to grey ground colour, washed with green and marked with darker specks, spots and streaks. SIZE: (n = 230) 30–35 × 22–25 (33 × 23). WEIGHT: (n = 221) 8–11 (9).

LAYING DATES: N Africa: Morocco and Libya, Mar–June; Egypt Mar–Apr; recorded also Algeria and Tunisia, presumably Mar–June; Mauritania, Apr–Aug; Senegambia, Mar–July; Sudan (Port Sudan, no date, June?: Cave and Macdonald 1955); Ethiopia, Feb–June; Somalia, Apr–July; Socotra, Mar–May.

INCUBATION: begins when clutch complete; by both sexes in equal amounts with ♀ by day, ♂ by night. Period: 24–27 days (Europe 26·3 days: Rittinghaus 1961; Kansas 25·5 days: Boyd 1972). Change-over during incubation characterized by symbolic nest-building. Relieving bird may have its breast wet to moisten eggs and keep them cool. Eggs often 20–80% buried with sand; sand sometimes scratched on eggs by parents to protect them against sun rather than predators.

DEVELOPMENT AND CARE OF YOUNG: young tended by both sexes; after hatching led away to feeding grounds; fledge at *c*. 27–31 days; become independent shortly after fledging. Sometimes breed when 1 year old, normally at 2 years.

When predators near young, parent performs distraction display, usually after eggs hatch. Runs from nest, crouches, faces predator and fans and depresses tail and quivers wings; may also raise both wings high and fully extended and wave them up and down.

BREEDING SUCCESS/SURVIVAL: of 168 eggs laid, 94% hatched (W Germany); 51 pairs, mean 2·3 young per pair (W Germany); oldest ringed bird 10 years.

References
Cramp, S. and Simmons, K. E. L. (1983).
Johnsgard, P. A. (1981).
Simmons, K. E. L. (1953).

Charadrius marginatus Vieillot. **White-fronted Plover; White-fronted Sand-Plover. Pluvier à front blanc.**

Plates 15, 21
(Opp. pp. 256, 352)

Charadrius marginatus Vieillot, 1818. Nouv. Dict. Hist. Nat. 27, p. 138; Cape Peninsula.

Forms a superspecies with *C. alexandrinus*.

Range and Status. Africa and Madagascar.

Resident, largely sedentary although with some seasonal movements, mainly in sandy coastal areas from Senegambia to Somalia south to Cape; also inland locally along sandy shores of larger rivers and lakes. In northern part of range uncommon to frequent (Ethiopia, only Rift lakes, vagrant coast) to locally common (Somalia, recorded as far north as 5°30′; Nigeria, Ghana, Mali) to abundant (Kenya and Tanzania coasts, flocks of 70; inland, numerous only on L. Turkana and L. Malaŵi). In southern Africa one of the most common resident coastal shorebirds (South Africa, 7900 along 1126 km open shoreline; Namibia, flocks of up to 375, June–Aug, Sandvis, 3059 along 200 km shoreline Durissa Bay–Sandvis, Dec–Jan). Northernmost known breeding sites in Senegambia (*c.* 13°30′N), Mali (*c.* 17°N), Nigeria (*c.* 13°N) and Somalia (*c.* 5°N).

Description. *C. m. marginatus* Vieillot: coasts, SW Cape to southern Angola. ADULT ♂ (breeding): forehead white, forecrown black (not reaching eye); rest of crown and hindneck grey with tawny wash. Dark streak from base of bill through lores and eye to ear-coverts; separated from crown by white streak continuous with forehead. Lower hindneck with white collar. Mantle and back grey to warm brown, feathers fringed pale brown to off-white; rump rusty; central tail-feathers brown; rest of tail white. Pale rusty patch on side of breast; chest and upper belly sometimes washed tawny; rest of underparts white. Primaries and secondaries brown with bases, inner webs and shafts white. Coverts brown, edged sandy brown, greater coverts tipped white. Bill black, eye brown, legs and feet pale grey to olive grey or buff. ADULT ♂ (non-breeding): like breeding ♂ but dark streak through eye and forecrown more brownish; lateral breast patches less pronounced. ADULT ♀: like ♂ but less black on forecrown. SIZE: ♂♀ (n = 24) wing, 103–115 (110); tail, 45–54 (50); bill, 16–18 (17); tarsus, 23–29 (25); (n = 46) wing, 101–112; tail, 46–55; bill, 14–17·5; tarsus, 25–29 (A. Tree, pers. comm.). WEIGHT: (n = 262) av. 48·3; (n = 46) 42–55.

IMMATURE: like non-breeding adult but no black on head; lores dark brown; little or no rusty patch on side of breast.

DOWNY YOUNG: above grey, with black mottling, forming lines down centre of head and back; below white.

C. m. arenaceus Clancey: coasts, SW Cape to southern Mozambique. Similar to nominate *marginatus* but upperparts browner, fringing of mantle feathers, scapulars and tertials rustier, cinnamon-buff, not off-white. SIZE: (n = 82) wing, 102–115; tail, 46–58; bill, 14·5–18; tarsus, 25–30. WEIGHT: (n = 82) 40–55 (A. Tree, pers. comm.).

C. m. mechowi (Cabanis): coastal Angola and Mozambique northward; also inland from Zimbabwe and Botswana northward. Tawnier than *arenaceus*; smaller. SIZE: (n = 47) wing, 92–103; tail, 41–49; bill, 13–16; tarsus, 24–27. WEIGHT: (n = 47) 27–40 (A. Tree, pers. comm.).

Field Characters. Easily confused with Kentish Plover *C. alexandrinus*, but slightly smaller, tail proportionately longer; upperparts with rusty wash; underparts less white; breast-patch rusty, indistinct; legs grey. ♀ similar to ♀ Kentish Plover but has some black separating forehead from forecrown. Distinguished from Kittlitz's Plover *C. pecuarius* by different head and wing patterns, paler upperparts, usually white underparts (rather than buff-tinted), and, in flight, legs not extending beyond tail. May be confused with immature Ringed Plover *C. hiaticula* but lacks dark breast-patches and pale buff fringes to feathers of upperparts.

Voice. Tape-recorded (262, 264). Produces soft 'wit', 'woo-et', 'twirit', and 'tirit-tirit'; also a plaintive 'pi-peep'. When chasing intruders from territory, gives harsh 'chiza-chiza-chiza', followed by 'purrr' or 'squeak'. Incubating birds give short 'clirrup' or 'clup' when alarmed. Other alarm calls include 'chirrrt', 'kittup' and a low, drawn-out 'pirr' or 'churr'. Makes 'croo' call when visiting nest before clutch complete.

General Habits. Occurs usually in pairs or small parties although sometimes forms larger flocks in non-breeding season (375, Namibia: Berry and Berry 1975). Prefers sandy coastal seashores; found also on rocky coasts, coastal and tidal mudflats, saltpans, estuaries, and larger rivers and lakes where mainly on exposed sandy beaches but also mudflats. Forages on beaches in intertidal areas mainly during low tide, during daytime as well as night, and in inland areas during moonlit nights (A. Tree, pers. comm.); also on coastal dunes. Sometimes forages in mixed flocks with Sanderlings

Calidris alba and Curlew Sandpipers *C. ferruginea*. Feeds by running quickly, stopping suddenly to peck at food, then running on again. May vibrate toes of one foot on ground, presumably to disturb insects (Steyn 1966). Also may fly up into air from the beach and hawk flying termites (C. J. Skead, pers. comm.). Roosts mainly away from water on broad, open shorelines, sometimes in mixed flocks; individuals may roost separately.

In southern Africa adults of coastal populations sedentary, often maintaining territories on sections of beaches or on sand dunes; much dispersal of immatures. Inland populations in central and southern Africa migratory, with birds from southern Africa moving to east coast as far south as E Cape; those in central Africa present on sandy rivers May–Dec, moving to coast when rivers flood Dec–May, although in years of poor rainfall and no floods, all remain inland (Tree 1969, 1977a, 1980b). Elsewhere, small numbers of non-breeding birds with flocks of up to 70 occur seasonally on E African coast (Britton 1980); some movements with changing water levels Nigeria (Elgood 1982); present all year Mali, with population increase Dec–Jan (Lamarche 1980).

Food. Insects including sandflies, grasshoppers, termites, mosquito pupae and *Cheirocephalus* larvae; also gastropods (*Assiminea* spp., *Marginella capensis*), bivalves (*Nucula* spp.), crabs (*Cleistostoma edwardsii*), isopods (*Exosphaeroma* spp.), other small crustaceans, and worms. Feeds also on insects blown into water and washed up on beach; consumes 5·69 kg/km/year of insects on beaches of E Cape (South Africa) (McLachlan *et al.* 1980); consumes 3·77 kg insects/year or 10·33 g/day (A. Tree, pers. comm.).

Breeding Habits. Monogamous, pair bond lasting for several months; solitary nester. 2 territories: 1, a section of sand where nest located; the other, an intertidal sandflat where bird feeds; maintained all year although those with territories sometimes join flocks of non-territorial birds. Nesting territories 0·6–2·5 (1·6) ha in size (SW Cape: Summers and Hockey 1980). Defends territories against neighbours and other species such as Sanderling and Kittlitz's Plover by running at intruder with head held low or flying low over sand straight at it. Also adopts aggressive posture in which it assumes horizontal posture by leaning forward with head depressed into body, feathers of flanks fluffed over closed wing. Produces harsh notes (see Voice) when chasing intruders. Pair-formation displays not described but probably similar to those of Kentish Plover.

NEST: a simple scrape in sand, usually on beach near high water mark or on dunes or riverine sandbanks; occasionally quarries. Frequently lined with pebbles, shell fragments, dried seaweed or small twigs. Some as close as 16·2 m apart (Summers and Hockey 1980); on Zambezi, 1981, 1 pair/2 ha of sandbank (A. Tree, pers. comm.). Pair may dig 1–2 scrapes before selecting one.

EGGS: 2–4, usually 2, laid at 2–4 day intervals; mean clutch sizes, Zimbabwe, Zambia and Malaŵi (n = 17) 2·53; South Africa, Natal (n = 296) 1·95, W Cape (n = 318) 1·94, SW Cape (n = 26) 2·1, overall (n = 657) 1·95. Pointed ovals; creamy buff, marked with fine lines of dark brown scattered over surface. SIZE: (n = 150) 30–37 × 21–25 (33 × 23). WEIGHT: (n = 35) 47–59 (53).

LAYING DATES: Senegambia, May–July; Ghana, Feb–Sept; Nigeria, Feb–May; Somalia, May–July; E Africa: Tanzania (coast), Dec; Zanzibar, May–July; Region D, May–July, Dec; Region E, Aug; Zambia, June–Oct; Malaŵi, Aug–Oct; Zimbabwe, May–Nov with most Aug–Sept; Mozambique, Oct; Namibia, all year with peak Jan; South Africa, SW Cape, all year, mainly Aug–Sept; Natal, all year, mainly July–Sept.

INCUBATION: probably begins when clutch complete, performed equally by both parents. Period: 26–29 days, up to 33 (McLachlan and Liversidge 1978). Pair may make brief visits to nest before clutch complete. Parent partly covers eggs with sand very early in morning, leaves them covered during day, uncovers them at night. Broods sand and eggs during day, uncovers eggs at night. Degree of egg-covering increases as incubation proceeds (Liversidge 1965); probably serves as thermoregulatory and concealing device. In 23% of nests, eggs may not be covered during the day (Hall 1960). Parent covers eggs with sand by rapid kicking motion. When very hot, bird preparing to brood first wades into shallow water and fluffs feathers until soaked; then settles over eggs, wetting them (Reynolds 1977).

DEVELOPMENT AND CARE OF YOUNG: young very agile, feeding themselves a few h after hatching. Fledging period: 35–38 days. Parents with chicks, or with eggs containing calling chicks, perform distraction displays if predators nearby; displays include: (1) 'rodent-running' in which bird runs fast with head held low, tail drooped and spread, and wings spread; and (2) injury-feigning in which it runs along or crouches on sand, flaps one or both wings, and holds partly spread tail either up or down (**A**).

A

BREEDING SUCCESS/SURVIVAL: pair may make several breeding attempts within breeding season if clutches lost. In SW Cape pairs reared an average of 0·14 fledged young 1976–76 and 0·08 fledged young 1976–77 (Summers and Hockey 1980). In 56 nests, W and E Cape, Natal and Zimbabwe, 37·5% of eggs hatched and 28% of hatched young survived (Blaker 1966). Poor nesting success SW Cape probably due mainly to mongooses (Grey Mongoose *Herpestes pulverulentus* and Water Mongoose *Atilax paludinosus*) and also Bokmakierie *Telophorus zeylonus*. Main agents of egg destruction elsewhere in southern Africa include floods, high tides, and motor vehicles; some eggs also destroyed by African Black Oystercatcher *Haematopus moquini* (Blaker 1966).

Av. life expectancy 7·5 years (Summers and Hockey 1980), but can live 9 years or more (Tree 1980b).

References
Kieser, J. A. and Liversidge, R. (1981).
Liversidge, R. (1965).
Summers, R. W. and Hockey, P. A. R. (1980).

Charadrius pallidus Strickland. Chestnut-banded Plover; Chestnut-banded Sand-Plover. Pluvier élégant.

Plates 15, 21
(Opp. pp. 256, 352)

Charadrius pallidus Strickland, 1852. Contr. Orn., p. 158; Walvis Bay.

Range and Status. Endemic; some local movements. Rift Valley alkaline lakes in S Kenya and N Tanzania (Magadi, Natron, Manyara, Lagarja, Masek, near Dodoma, and Bahi Swamp 6°14'S, 35°18'E) and southern Africa, mainly coastal, from SW Angola to Mozambique (32 km north of Beira) but also inland north to N Namibia and central Botswana. In E Africa, abundant to very abundant; southern Africa, common to abundant west and southwest coasts (Namibia, 4534 on *c*. 200 km coast Durissa Bay-Sandvis, Dec–Jan; 1909 in Walvis Bay), locally common to rare inland. Vagrant Zambia (Akatiti Dam 13°10'S, 28°24' E) and Zimbabwe (Victoria Falls, Bulawayo, Limpopo River). We follow Britton (1980) and Dowsett (1977) in not accepting records from Lakes Baringo and Nakuru (Kenya) and Katwe Salt Lake (Uganda).

Description. *C. p. pallidus* Strickland: southern Africa. ADULT ♂ (breeding): forehead white; forecrown with narrow black and chestnut bands separated from eye by small amount of white; rest of crown and nape greyish brown, side of neck pale chestnut. Narrow black line from bill to eye; rest of face, chin and throat white. Mantle, back and rump pale greyish brown; centre of tail brown to sandy grey, rest white. Underparts white except for pale chestnut band extending across chest to meet grey upperparts. Primaries dark brown; secondaries grey with white terminal bands; primary coverts dark brown with white tips and white edging; rest of upperwing-coverts greyish brown; axillaries and underwing-coverts white. Bill black; eye dark brown; legs and feet olive to greenish grey. ADULT ♂ (non-breeding): like breeding ♂ but black on head browner, less conspicuous. ADULT ♀: similar to ♂ but lacks black markings on forehead and in front of eye. SIZE: ♂♀ (n = 13) wing, 100–105 (102); tail, 39–44 (41); bill, 13–14 (13); tarsus, 26–29 (28). WEIGHT: ♂♀ (n = 27) 28–39 (35·1) (G. Maclean, pers. comm.); ♂♀ (Kenya, n = 204) 20–37 (D. Pearson, pers. comm.); southern Africa, ♂ (n = 7) 33–44 (37·0), ♀ (n = 5) 39–43 (40·5) (A. Tree, pers. comm.); ♂♀ (SW Cape, n = 10) 29–39 (36·0); Botswana, ♂ (n = 5) 32·9–37·7 (36·7), ♀ 32·9, 32·9, 34.

IMMATURE: no black on forehead, chestnut on sides of neck or black mark through eye; smaller chestnut band on chest. In 1st (juvenile) plumage lacks band but has 2 ashy or smoky grey patches on sides of breast, buff edges to feathers of back and wings. Bill dark brown.

DOWNY YOUNG: upperparts grey mixed with black markings, black lines down centre of crown and back; underparts white; bill black, eyes dark brown, and legs and feet olive grey.

C. p. venustus Fischer and Reichenow: Kenya and Tanzania. Smaller: wing, ♂ (n = 5) 85–91 (88·8), ♀ (n = 5) 89–94 (90·8); mantle darker and greyer. WEIGHT: ♂, 25.

Charadrius pallidus

Also winters in breeding range

Field Characters. A small, very pale plover of alkaline areas with conspicuous chestnut band across chest in adult. Adult ♂ distinguished by black line between bill and eye and black and chestnut bands on forecrown. In flight, remiges contrastingly darker than rest of wing and back, faint white wing-bar, and outer tail-feathers white. Resembles a pale White-fronted Plover *C. marginatus* or Kentish Plover *C. alexandrinus* but smaller and more chubby with proportionately longer and paler legs. Very young immature (juvenile) has grey patches on sides of breast; very difficult to distinguish from immature and non-breeding adult Kentish Plover and Mongolian Plover *C. mongolus*. However, Chestnut-banded smaller (wing *pallidus* av. 102, *alexandrinus* av. 111, *mongolus* av. 128).

Voice. Not tape-recorded. Call notes include 'tsk, tsk', a long piping whistle, a quiet 'chuck' when landing or taking off, a 'chee-wich-ii-you' when defending territory, and a deeper but similar 'chuck' when warning young.

Plate 13

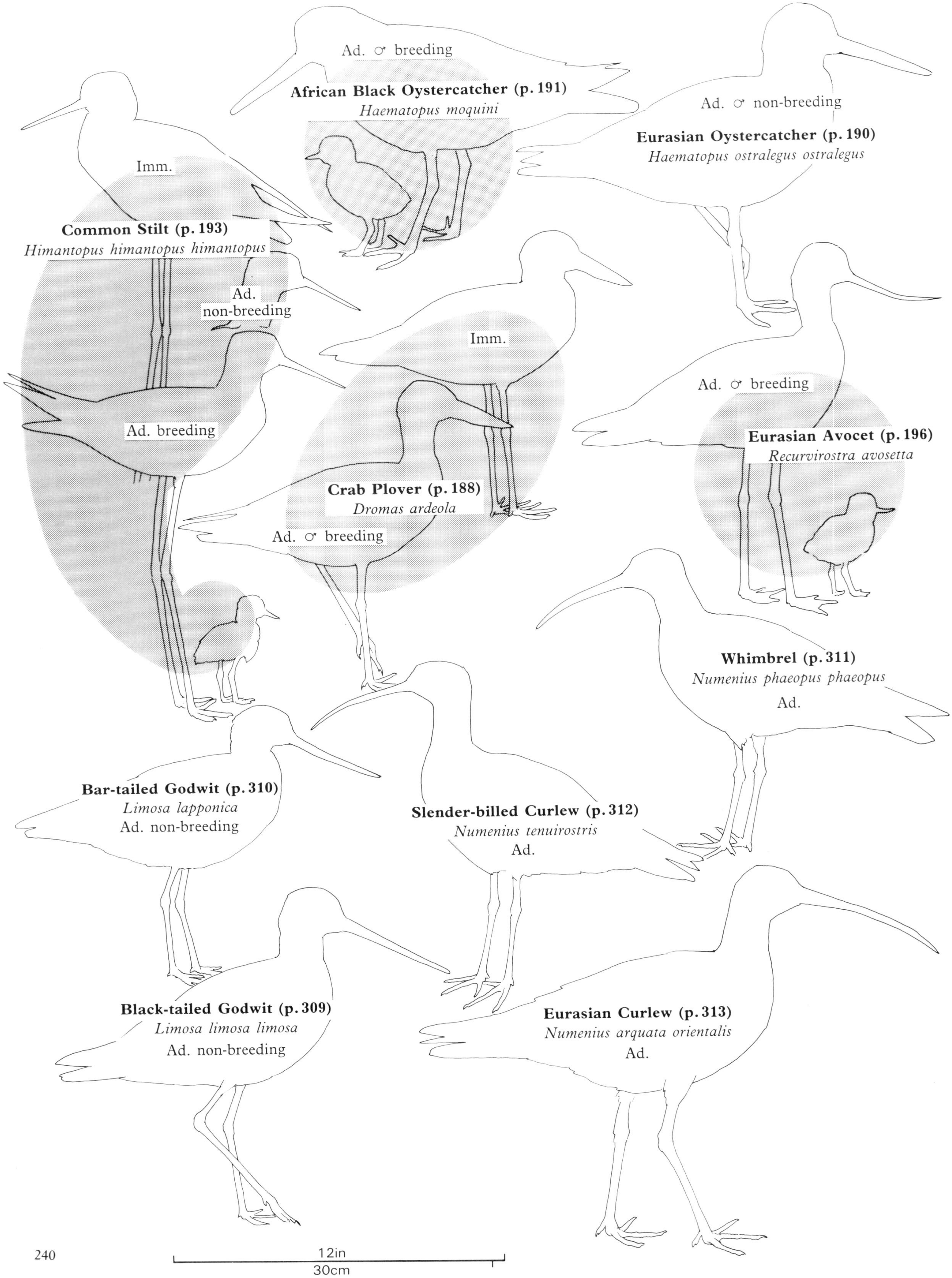

African Black Oystercatcher (p. 191)
Haematopus moquini
Ad. ♂ breeding

Eurasian Oystercatcher (p. 190)
Haematopus ostralegus ostralegus
Ad. ♂ non-breeding

Common Stilt (p. 193)
Himantopus himantopus himantopus
Imm.
Ad. non-breeding
Ad. breeding

Crab Plover (p. 188)
Dromas ardeola
Imm.
Ad. ♂ breeding

Eurasian Avocet (p. 196)
Recurvirostra avosetta
Ad. ♂ breeding

Whimbrel (p. 311)
Numenius phaeopus phaeopus
Ad.

Bar-tailed Godwit (p. 310)
Limosa lapponica
Ad. non-breeding

Slender-billed Curlew (p. 312)
Numenius tenuirostris
Ad.

Black-tailed Godwit (p. 309)
Limosa limosa limosa
Ad. non-breeding

Eurasian Curlew (p. 313)
Numenius arquata orientalis
Ad.

12in / 30cm

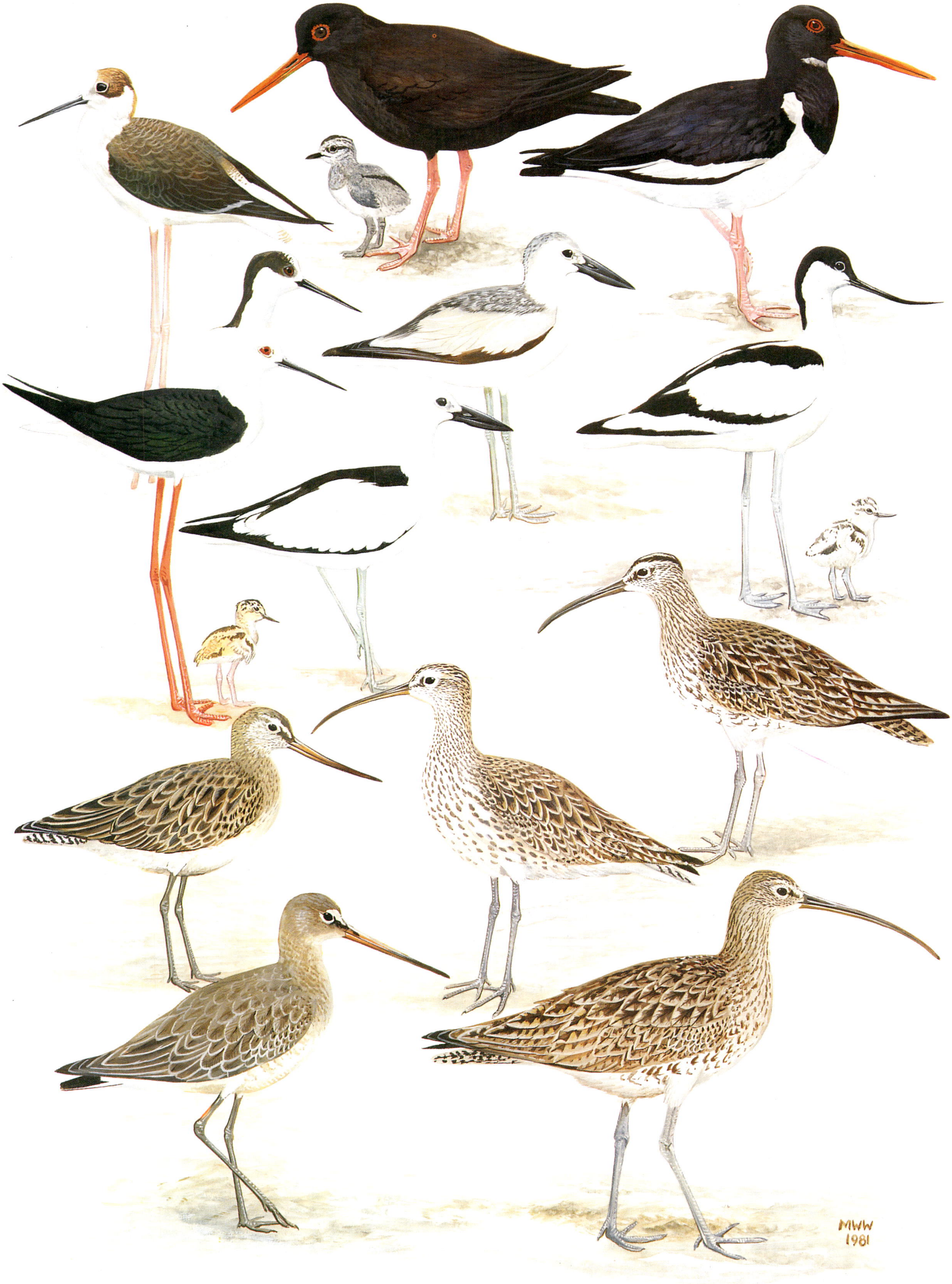
MWW
1981

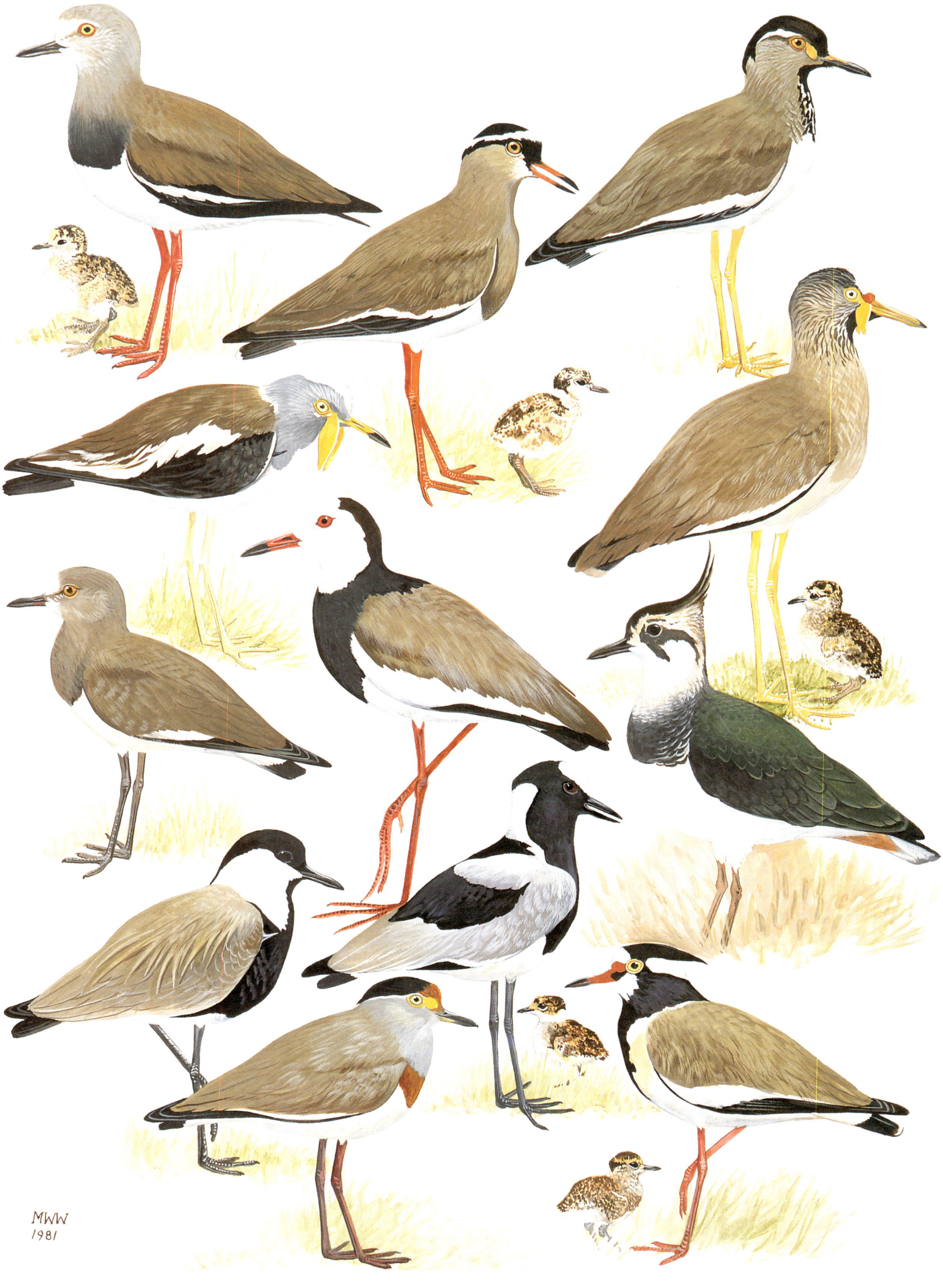
MWW
1981

Plate 14

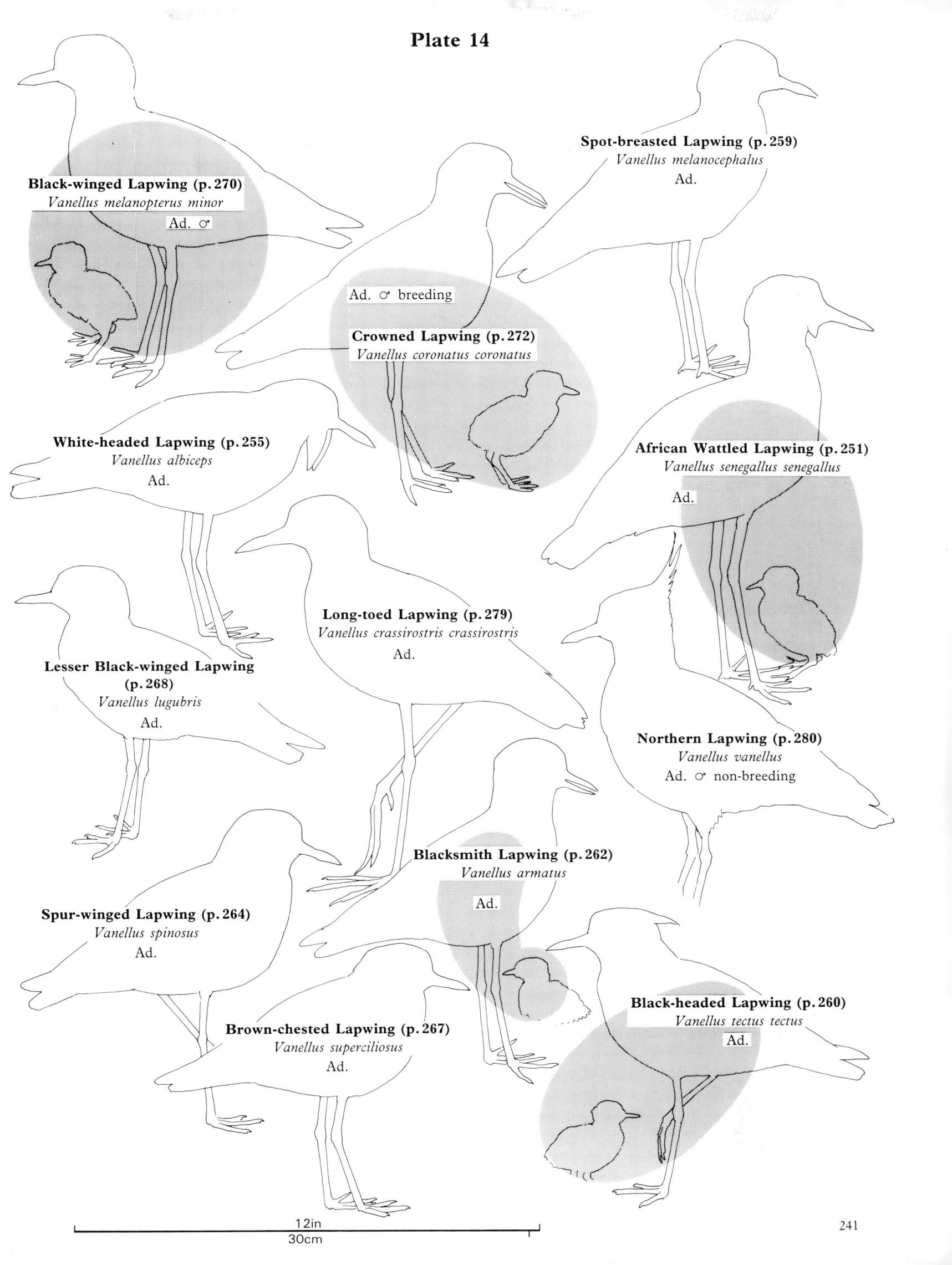

Spot-breasted Lapwing (p. 259)
Vanellus melanocephalus
Ad.

Black-winged Lapwing (p. 270)
Vanellus melanopterus minor
Ad. ♂

Ad. ♂ breeding

Crowned Lapwing (p. 272)
Vanellus coronatus coronatus

White-headed Lapwing (p. 255)
Vanellus albiceps
Ad.

African Wattled Lapwing (p. 251)
Vanellus senegallus senegallus
Ad.

Lesser Black-winged Lapwing (p. 268)
Vanellus lugubris
Ad.

Long-toed Lapwing (p. 279)
Vanellus crassirostris crassirostris
Ad.

Northern Lapwing (p. 280)
Vanellus vanellus
Ad. ♂ non-breeding

Spur-winged Lapwing (p. 264)
Vanellus spinosus
Ad.

Blacksmith Lapwing (p. 262)
Vanellus armatus
Ad.

Brown-chested Lapwing (p. 267)
Vanellus superciliosus
Ad.

Black-headed Lapwing (p. 260)
Vanellus tectus tectus
Ad.

12in / 30cm

General Habits. Occurs singly, in pairs, and in flocks of up to 50 or 60; sometimes in mixed flocks with other *Charadrius* and *Calidris* spp. Prefers saline pans, alkaline lakes, and coastal lagoons and estuaries; rarely in freshwater habitats. Shy; flies fast and low. Calls extensively in early morning; as sun rises, ceases to call but continues to feed usually until 0900 h, sometimes to 1200 h (Kenya). Feeds along water's edge, sometimes springing into air to catch insects in flight (Jeffery and Liversidge 1951).

Both races possibly partly migratory or at least with some local movements associated with drying up of breeding habitat. In E Africa movements probably up and down Rift, being most numerous Lake Manyara July–Sept (Morgan-Davies 1960) but more information needed. In southern Africa wanders locally (e.g. Zambia: Taylor 1982; Zimbabwe: Perry 1975, Pollard 1980), some inland birds dispersing to coast after breeding season to mix with sedentary coastal populations (Tree 1980a). Large numbers Namibia coast Dec–Jan (Whitelaw *et al.* 1978).

Food. Insects and small crustaceans.

Breeding Habits. Monogamous, solitary nester. Pair establishes territory *c.* 20 × 100 m, and defends it until young capable of flight. As many as 70 nests along edges of 1 saltpan (Jeffery and Liversidge 1951). ♂ (and ♀?) defends territory by flying or running at intruder, with head down between the shoulders, and at same time giving 'chee-wich-ii-you' call. Courtship behaviour includes Scrape-ceremony display with either bird pushing breast into sand, rotating breast, and pushing out sand with legs. Upon arrival of ♀, ♂ may stand on edge of scrape with bill open, pointed downward and vibrating. ♂ and possibly ♀ throw small pebbles with tip of bill over shoulder towards nest. ♂ usually selects scrape site although ♀ will do so. Several scrapes made before one chosen. At time of copulation, ♂ approaches ♀ from rear, performs 4–6 'high-steps' directly behind her, hops on her, copulates, then falls backwards pulling her over him.

NEST: a shallow scrape on bare calcareous soil, mudflat or stony slope, within 50 m of water, 5 cm in diam and 1 cm deep; lined with small white quartz chips, bleached grass and fish bones, and small gastropod shells.

EGGS: usually 2, probably laid at 2–4 day intervals; mean clutch size South Africa (W Cape), (n = 13) 1·77 (Blaker 1966). Obtuse one end, acute the other; greyish cream lightly flecked with fine streaks and spots of brownish black. SIZE: (n = 42) 29–33 × 22–24 (32 × 23).

LAYING DATES: E Africa: Kenya and Tanzania, Apr–Oct; Region C, July; Region D, May–Sept, with peak late in long rains and after them up to July; Mozambique, Sept; Namibia, July; South Africa, Mar–May, Sept–Jan, with most Nov–Dec.

INCUBATION: probably begins when clutch complete; incubation period unknown. Performed by both parents, by ♀ apparently during daylight hours, by ♂ usually during evening. When temperature increases, incubating bird places legs alongside body, crouches over eggs, and raises crown and mantle feathers; may leave eggs unattended for long periods, morning or evening. Relieving bird occasionally soaks belly to help cool eggs. Incubating bird very cautious; at first sign of danger, moves several feet away from nest to resettle among stones, without covering eggs before departing. When intruder close, leaves by flying low over ground with slow wing-beats, remaining hidden until some distance away.

DEVELOPMENT AND CARE OF YOUNG: when 10 days old, markings down side of body more pronounced; upperwing buff, bill brown. Behaviour of young bird not described. When chick quite young, parent holds it between wing and body so chick's feet just touch ground. When intruder near chick, either parent may perform broken wing distraction display, fluttering along the ground away from chick. Parents defend young until capable of flying. Length of fledging period unknown.

BREEDING SUCCESS/SURVIVAL: nest mortality and breeding success unknown. Possibly double-brooded (1 nest used in Oct prepared for use again Nov: Jeffery and Liversidge 1951). Known to live at least 11·5 years (Tree 1980b).

References
Jeffery, R. G. and Liversidge, R. (1951).
van Someren, V. G. L. (1956).

Plates 15, 20
(Opp. pp. 256, 321)

Charadrius mongolus Pallas. **Mongolian Plover; Lesser Sand-Plover. Pluvier mongol.**

Charadrius mongolus Pallas, 1776. Reise versch. Prov. Russ. Reichs, 3, p. 700; Kulussutai, probably on the Onon River, Siberia, *fide* Ridgway, 1919, p. 134.

Range and Status. Breeds central Asia and NE Siberia; winters Africa, Madagascar, Comoros, Persian Gulf and S Asia to Australia.

Palearctic visitor to coasts of eastern Africa from Sudan to South Africa. Locally common to abundant, Ethiopia to Tanzania; frequent to rare southern Africa south to eastern Cape; vagrant SW Cape, Namibian coast and Egypt (Suez). Inland, uncommon or vagrant; recorded Burundi, Ethiopia, Kenya, Somalia, Tanzania, Uganda, Zaïre and Zambia.

Charadrius mongolus

IMMATURE: juvenile, warm tinge to hindneck, bright buff fringing to feathers of upperparts, warm wash on breast. 1st-year winter birds like non-breeding adult but inner median coverts with buff fringes.

Field Characters. Very similar to Great Sand-Plover *C. leschenaultii* but slightly smaller with smaller, weaker bill, shorter, dark grey not greyish green legs and, in breeding plumage, broader breast-band and blacker forehead. Differs from Kentish Plover *C. alexandrinus* in non-breeding plumage by larger breast-patches, lack of white collar, generally larger size and proportionately longer legs (see also Great Sand-Plover, below).

Voice. Tape-recorded (73). Fairly quiet in non-breeding season. In flight a quite dry 'trr-trr-trr', 'trik' or 'drrit'; also 'pip-ip' and 'chittick'.

General Habits. Prefers mainly sandflats and mudflats of coastal bays and estuaries; also sand dunes near coast; sometimes along shores of lakes. Occurs singly or in flocks of up to 100 or more. Forages during day, sometimes at night, by making quick runs, then stopping to capture food. Mixes with Great Sand-Plovers when feeding and roosting.

Passage migrant, mainly autumn Ethiopia; thousands winter and many oversummer Kenya coast (Ash 1980b). Most adults arrive Kenya Sept, young birds Oct, somewhat later than Great Sand-Plover, and depart end Apr–early May. In South Africa, assumes full summer plumage before departing with some showing traces as early as Feb. In E Africa all returning adults are in non-breeding plumage; rarely assumes summer plumage before migrating northward. Many 1st-year birds retain non-breeding plumage and oversummer in E Africa.

Description. *C. m. pamirensis* Richmond: only subspecies in Africa. ADULT ♂ (breeding): lores, most of forehead, forecrown and patch through eye to ear-coverts black; some small white spots in front of eye. Rest of crown greyish brown with some buff, hindneck rufous, nape and rest of upperparts greyish brown. Chin, throat and foreneck white. Broad rufous breast-band; rest of underparts white. Middle rectrices deep grey-brown, others gradually paler, outermost with white outer webs. Primaries blackish brown, inner ones edged white basally on outer webs; secondaries blackish brown with white bases and narrow white tips; primary coverts blackish brown; tertials and rest of wing-coverts greyish brown, greaters broadly tipped white, forming pale bar. Underwing-coverts and axillaries whitish. Bill black; eye dark brown; legs greyish, sometimes with green tinge, feet with darker toes. ADULT ♂ (non-breeding): greyish brown above, mantle feathers and scapulars with buff edges. Forehead white; lores and ear-coverts brown; distinctive white superciliary stripe. Lateral brown-grey breast-patches, sometimes meeting in centre; rest of underparts white. ♀ very similar but duller black on head mixed with grey-brown; non-breeding ♀ like non-breeding ♂. SIZE: wing, ♂ (n = 17) 124–132 (128), ♀ (n = 7) 128–134 (131); tail, ♂ (n = 6) 46–54 (49·5), ♀ (n = 26) 44–47 (45·3); bill, ♂ (n = 18) 15–18 (16·9), ♀ (n = 8) 16–18 (16·9); tarsus, ♂ (n = 18) 32–35 (33·9), ♀ (n = 8) 32–34 (33·4). WEIGHT: (n = 1) 54 Oct, (n = 1) 89 Mar (Waltner 1981); (Kenya, n = 91) 40–81, typical wintering weight 45–60 (D. Pearson, pers. comm.).

Food. Insects, crabs, small molluscs, marine worms, amphipods.

References
Cramp, S. and Simmons, K. E. L. (1983).
Johnsgard, P. A. (1981).

Charadrius leschenaultii Lesson. Great Sand-Plover; Greater Sand-Plover. Pluvier de Leschenault.

Plates 15, 20
(Opp. pp. 256, 321)

Charadrius Leschenaultii Lesson, 1826. Dict. Sci. Nat., ed. Levrault, 42, p. 36; Pondicherry, India.

Range and Status. Breeds Armenia east to Mongolia; also Jordan; winters Africa, Persian Gulf, Malagasy Region, and S Asia to N Australia.

Palearctic visitor, mainly to eastern Africa from Egypt to South Africa. Common to abundant coastal Egypt (Red Sea), Sudan, Ethiopia, Somalia and E Africa; frequent to uncommon E coast southern Africa to E. Cape (Cape St Francis); rare on coasts of SW Cape and Namibia (north to about Walvis Bay). Frequent to rare Mediterranean coast west to Tunisia (Djerba, Korba). Uncommon to vagrant inland, recorded Burundi, Egypt, Kenya, Malaŵi, Tanzania, Uganda and Zambia. Reported breeding NE Africa (Egypt–Somalia: Ticehurst 1929, Meinertzhagen 1930,

244 CHARADRIIDAE

Charadrius leschenaultii

Archer and Godman 1937, Harrison 1963), but records not confirmed and based on uncertain identification of eggs and chicks from Somalia and lack of definite proof of breeding in Egypt (Vaurie 1964).

Description. *C. l. columbinus* Wagler: Red Sea, Gulf of Aden, vagrant Lake Chad. ADULT ♂ (breeding): forehead white (sometimes with some median black feathers), separated from ashy grey to rufous crown by thin black bar. Hindneck and sides of neck rufous; rest of upperparts ashy brown. Thin black line from bill through eye, broadening on ear-coverts; rest of face, chin, throat and foreneck white. Pale chestnut band across breast; rest of underparts white. Tail brown centrally, outer rectrices white. Primaries blackish brown, inners edged white basally on outer webs; secondaries blackish brown with white bases and narrow white tips; primary coverts blackish brown; tertials and rest of wing-coverts greyish brown, greaters broadly tipped white, forming pale bar. Underwings pale. Bill black, eye dark brown, legs and feet olive or yellowish grey, sometimes light brown, darker on joints. ADULT ♂ (non-breeding): above ashy brown; forehead and superciliary stripe white; lores and ear-coverts brown; chin, throat, and foreneck white. Sometimes a slight pale collar around hindneck. Greyish brown lateral breast-patches sometimes joining a thin grey line across centre of breast; rest of underparts white. ♀ very similar but in breeding plumage often duller with brownish black replacing black. Non-breeding ♀ like non-breeding ♂. SIZE: wing, ♂ (n = 15) 137–149 (144), ♀ (n = 27) 135–150 (143); tail, ♂ (n = 6) 47–52 (49·0), ♀ (n = 6) 42–53 (48·2); bill, ♂ (n = 27) 21–24 (22·6), ♀ (n = 43) 20–24 (22·6), bill depth ♂ (n = 21) 4·6–5·3 (5·0), ♀ (n = 43) 4·7–5·4 (5·1); tarsus, ♂ (n = 22) 36–38 (37·0), ♀ (n = 39) 34–38 (36·6). WEIGHT: (unsexed) 122, 130 Mar, 81 July (Waltner 1981); 99 May (A. Tree, pers. comm.); (Kenya, n = 92) 57–107, typical wintering weight 65–80 (D. Pearson, pers. comm.).

IMMATURE: juvenile, warm tinge to hindneck, bright buff fringing to feathers of upperparts, warm wash on breast. 1st-year winter birds like non-breeding adult but inner median coverts with buff fringes.

C. l. crassirostris Severtzov: eastern and southern Africa. Bill heavier than *columbinus*; ♂ (n = 20) 23–26, ♀ (n = 32) 22–27; depth ♂ (n = 20) 5·4–6·3 (5·7), ♀ (n = 32) 5·4–6·4 (5·9).

Field Characters. Easily confused with Mongolian Plover *C. mongolus*, especially in non-breeding plumage, but with longer, heavier and bulbous bill; legs longer and more greenish or greyish green; head and body larger, particularly crown and eyes; upperparts often paler, sandier grey brown; forehead, superciliary stripe and tail usually with more white; and habit of feeding in more scattered, larger flocks with other species, and call a trill, not 'chitik'. In breeding plumage chestnut breast-band less extensive than that of Mongolian Plover. Differs from Caspian Plover *C. asiaticus* in having slightly longer, stouter bill, black not white lores, prominent wing-bar, tail darker than rump with white edges, narrower breast-band, less slender build and more upright stance, smaller eyes and different habitat. In winter, has clearly defined chest-patches like Mongolian Plover while Caspian Plover has dusky wash right across breast.

Voice. Not tape-recorded. In flight, sometimes gives short 'drrit' or 'treep' and a quiet dry trill 'chirrirrip'; also clear 1- or 2-note whistle.

General Habits. Prefers littoral habitats, favouring tidal and estuarine sand- and mudflats; also dunes near coast. Typically gregarious, in groups frequently of 2–50, sometimes 1000 when roosting; occasionally singly. Keeps to itself when foraging but often roosts with Curlew Sandpipers *Calidris ferruginea*, Sanderling *C. alba* and longer-legged species such as Grey Plover *Pluvialis squatarola*. Commonly in mixed flocks with Mongolian Plovers, when individual identification difficult. Feeds in typical plover stop–run–peck manner; may also probe. When threatening, stands upright, exposing white abdomen; wings held out horizontally. Aggressive behaviour includes chasing or flying at opponent which either runs off in crouched posture or ducks while attacker sails over it.

Migrates mainly along coast; present Egypt Red Sea shoreline Aug–Apr; arrives Sudan to Tanzania late Aug–Sept with main concentration arriving later, sometimes as late as Nov. Juveniles seen late Sept–Nov, especially on Red Sea. In E Africa, all returning adults in breeding plumage; large proportion assume nuptial plumage Mar–Apr before departing in 2nd half Apr–early May for Palearctic breeding grounds; in South Africa only a few assume nuptial plumage before heading northward. One bird ringed Nairobi, captured Sudan (19°08'N, 31°17'E); bird ringed Sudan (Suakin) in Sept, recovered on breeding grounds Syria (D. Pearson, pers. comm.). Hundreds, i.e. significant proportion of wintering birds, perhaps all or most non-adults, summer in winter quarters on E African coast and occasionally inland (P. Britton, pers. comm.)

Food. Crustaceans, molluscs, worms and insects.

References
Cramp, S. and Simmons, K. E. L. (1983).
Johnsgard, P. A. (1981).

Charadrius asiaticus Pallas. Caspian Plover. Pluvier asiatique.

Charadrius asiaticus Pallas, 1773. Reise versch. Prov. Russ. Reichs, 2, p. 715; southern Tatar steppes.

Plates 16, 21

(Opp. pp. 257, 352)

Range and Status. Breeds central Asia; winters Africa and Sinai.

Palearctic visitor to eastern and southern Africa from E Egypt and E Sudan (Red Sea) south to SW Angola and Cape, mainly in short grasslands. Winters in 2 main areas: (1) upland plains of SW Kenya and N Tanzania, and (2) Botswana, N and E Namibia, and South Africa (N Cape and Orange Free State). Wintering birds, Kenya and Tanzania, locally very abundant (many thousands in flocks of up to 150, Serengeti and adjacent areas of SW Kenya and N Tanzania; many hundreds Amboseli, Tsavo East and central Tanzania; elsewhere only odd birds or parties). Wintering birds South Africa, common. On passage rare to uncommon Egypt and Sudan; common to very abundant Ethiopia and Somalia, Kenya and Tanzania coasts (also L. Turkana); common N and W Uganda; abundant Zambia (mainly in W, 2000+ Lochinvar) and common Malaŵi. Frequent to rare elsewhere in range; vagrant Nigeria, Cameroon, Mali.

Description. ADULT ♂ (breeding): crown, nape, mantle, and back sandy brown; uppertail-coverts light brown edged ash white. Forehead, lores, superciliary stripe, cheeks, chin, throat and foreneck white; streak behind eye on ear-coverts black. Lower foreneck and chest bright chestnut, separated by narrow black line from white underparts. Tail brown; central feathers pointed and larger, outer ones lighter. Primaries black-brown with white shafts, outer webs with some white near base. Rest of upperwings brown, outer greater coverts tipped white; underwing-coverts mixed brown and white. Bill black, eye dark brown, legs and feet grey-green. ADULT ♂ (non-breeding): similar to breeding ♂ but head and face buffy, band on chest greyish brown; underparts white to buffy. ADULT ♀ (breeding) very similar to breeding ♂ but chestnut breast-band less distinct, black breast-band less distinct or absent; non-breeding ♀ like non-breeding ♂. SIZE: wing, ♂ (n = 71) 137–159 (148), ♀ (n = 38) 136–154 (146); tail, ♂ (n = 15) 48–56 (51), ♀ (n = 8) 51–61 (52·5); bill, ♂ (n = 35) 19–28 (23), ♀ (n = 20) 18–23 (21); tarsus, ♂ (n = 35) 33–44 (39), ♀ (n = 20) 37–45 (41). WEIGHT: ♂ (n = 5) 67–89 (80), ♀ (n = 3) 66–75 (70); all ages (n = 51) 59–78 (A. Tree, pers. comm.).

IMMATURE: juvenile, like non-breeding adult except that feathers of upper parts more distinctly edged sandy buff; underparts washed sandy buff. 1st-year winter bird like non-breeding adult but with buff-fringed inner median coverts.

Field Characters. Medium-small plover resembling Mongolian Plover *C. mongolus* but unlikely to occur with it since habitat very different. In breeding plumage has less complex head pattern than Mongolian Plover and black band separating chest-band from white underparts. In winter, patch behind eye bigger, darker, superciliary stripe broader. In all plumages has less compact shape, thinner bill, usually complete grey-brown breast-band, and lack of conspicuous wing-bar and white at side of rump and tail. Stance more upright, like golden plovers (*Pluvialis*).

Voice. Tape-recorded (F). Gives a flight note 'kwhit', a low double whistling 'ku-wit', 'ptrrwhit', a soft piping and a shrill whistle.

General Habits. Prefers short grasslands and other open areas, often far from water; particularly partial to recently burnt grasslands; not normally common on coast except when migrating (although overwinters on coastal dunes, Somalia). Gregarious, in small flocks of up to several hundred. Squats close to ground when difficult to see even on bare ground. Moves like a courser, standing upright, then runs; often runs rather than flies when disturbed. Wanders a great deal, rarely remaining long in one locality (e.g. Serengeti, appears in same area for 1–2 weeks, moves on but then returns again several weeks later). Active at night. Flies with very rapid wing-beats. Feeds in typical plover run–stop–peck manner.

Migratory routes not well understood. Autumn migrants probably move across Red Sea to pass through Ethiopia, Kenya, Tanzania, Malaŵi and Zambia to Botswana, Namibia and South Africa. Northbound migrants make fewer stops, probably first drifting to Angola, then moving to NE Africa (A. Tree, pers. comm.); recorded W and N Uganda, N and coastal Kenya, Somalia and Ethiopia. In Somalia Aug–Apr, once June. In Ethiopia Aug–Apr with peak autumn passage Aug–Sept. Most reach E Africa Oct, with many

♂♂ arriving later; leaves in Mar (L. Turkana Oct and Apr, Uganda Mar). Reaches Tanzania (Serengeti) 1st half Aug with most arriving Oct; moults into breeding plumage Jan–Feb. In Zambia Aug–Jan, with most Sept–Dec. In southern Africa mainly Sept–Mar (Orange Free State Aug–Feb). May assume breeding plumage before leaving for breeding areas; in southern Africa assumed Dec–Feb. Not known to summer in Africa.

Food. Almost entirely insects, including grasshoppers and beetles; also some grass seeds.

References
Bannerman, D. A. (1961).
Cramp, S. and Simmons, K. E. L. (1983).
Johnsgard, P. A. (1981).

Plates 16, 21
(Opp. pp. 257, 352)

Charadrius morinellus Linnaeus. Eurasian Dotterel. Pluvier guignard.

Charadrius Morinellus Linnaeus, 1758. Syst. Nat. (10th ed.), 1, p. 150; Sweden.

Range and Status. Breeds Scotland to NE Siberia, also NW Mongolia, central Europe, and Italy; winters Mediterranean basin to Persian Gulf.

Palearctic visitor, locally common to abundant from Morocco to Egypt (e.g. several hundred in Constantine, Algeria; 200, Tripoli airport, Libya); frequent to uncommon along Red Sea coast south to N Sudan; vagrant Mauritania.

Description. ADULT ♂ (breeding): crown and upper nape uniform black, sometimes blackish brown, occasionally with some white feathers; forehead with white and brown feathers. Broad white stripe starting over or before eye, running to lower nape, meeting in V. Lores with brown and white feathers, cheeks white with some brown feathers, ear-coverts and side of neck ash brown; chin and throat white. Hindneck, mantle, scapulars, back, rump and uppertail-coverts ash brown. Tail ash brown with broad subterminal black band; 3 outer pairs of rectrices broadly tipped white. Ash brown upper breast separated from orange-cinnamon lower breast by narrow sepia line and a broader white line. Flanks orange cinnamon; upper belly chestnut, lower belly black; vent and undertail-coverts white. Primaries sepia with white tips on inner feathers, secondaries ash brown with white tips; wing-coverts ash brown; underwing-coverts and axillaries pale grey. Bill black, eye brown, legs and feet dull yellow. ADULT ♂ (non-breeding): general pattern similar to breeding bird but crown brown, rest of upperparts greyish brown, sometimes faint white line across greyish brown breast, and belly white. ADULT ♀ (breeding): often indistinguishable from ♂; in general crown purer black, less streaked; superciliary stripe wider. ADULT ♀ (non-breeding): like ♂. SIZE: wing, ♂ (n = 82) 137–163 (150), ♀ (n = 59) 142–162 (153); tail, ♂ (n = 25) 63–72 (67), ♀ (n = 23) 64–74 (69); bill, ♂ (n = 23) 14–17 (16), ♀ (n = 22) 15–18 (16); tarsus, ♂ (n = 26) 34–38 (36), ♀ (n = 23) 35–40 (38). WEIGHT (breeding birds): ♂ (n = 10) 88–116 (100), ♀ (n = 5) 99–130 (117).

IMMATURE: like non-breeding plumage but feathers of upperparts with broad buff fringes, breast brownish buff with indistinct white line, inner wing-coverts with distinctive brown areas.

Field Characters. A medium-sized plover with pointed wings, fairly short legs and bill; in non-breeding plumage not unlike a small golden plover *Pluvialis* spp. or Caspian Plover *C. asiaticus*, but has white superciliary stripe forming a V on back of nape, and white breast-stripe. Birds in non-breeding plumage most common; occasionally birds in breeding plumage seen in Africa. Breeding birds distinctive with white breast-stripe, lower breast chestnut, belly black. In flight, axillaries grey, no wing-stripe.

Voice. Tape-recorded (73, 76). In flight a soft 'ter-tee', 'ter-too' or 'peep-peep-peep'. Individuals in feeding flocks often give a 'kwip kwip', 'kwip-ter ... tee-ta' and subdued 'wā-wā'; also a sharp 'ting'.

General Habits. Favours steppes, plains and other stony or sandy areas with scanty vegetation; also visits sea-shore habitats. Occurs singly or in small flocks of less than 20, occasionally 30–50. Roles of sexes reversed with ♀ dominant. Has typical plover gait; runs quickly, pauses, then pecks at food; sometimes probes for food; occasionally vibrates foot on bare soil presumably to expose prey. Tends to feed at night or very early in morning. Flight swift, strong with quick wing-beats.

Birds ringed Norway, captured Morocco and Algeria; ringed Scotland, captured Morocco and Algeria; ringed Sweden and Austria, captured Libya; ringed Finland, captured Algeria and Tunisia.

Food. Principally insects and spiders; also molluscs, worms and plant material.

References
Cramp, S. and Simmons, K. E. L. (1983).
Nethersole-Thompson, D. (1973).

Genus *Pluvialis* Brisson

Chunky, medium-sized plovers with extensive pale spotting on upperparts. Wings pointed and long, nearly 4 times as long as tarsus. Tail short, square-tipped and barred. Bill slender, more than half length of tarsus. Tarsus with hexagonal scales all round. Front toes webbed at base; hind toe small or absent. Raises wings, ruffles back feathers, and lowers head in aggressive situations; does not bob head. Sexes alike. 3 spp., Holarctic, all wintering in Africa.

Pluvialis dominica (P. L. S. Müller). Lesser Golden Plover. Pluvier fauve.

Plates 16, 20
(Opp. pp. 257, 321)

Charadrius Dominicus P. L. S. Müller, 1776. Natursyst., suppl., p. 116; Hispaniola.

Range and Status. Breeds N Siberia to NE Canada; winters Africa, Asia, Australia, southern Pacific and South America.

Palearctic visitor mainly eastern Africa. Locally common on coasts from Ethiopia to S Somalia (max. 38 Sept Assab and 138 Mar near Mogadishu). Uncommon inland Sudan (Nile Valley), Ethiopia and Somalia, inland and coastal Kenya and coastal Tanzania. Rare Zambia and South Africa; vagrant Algeria, Burundi, Ghana, Namibia, Nigeria, Senegambia, Sierra Leone, Tunisia and Zaïre.

Description. *P. d. fulva* (Gmelin): eastern Africa. ADULT ♂ (breeding): upperparts from crown to back blackish brown with extensive golden and white spotting; rump and uppertail-coverts barred yellowish buff and dark brown. Forehead, area above eye, sides of neck and breast, undertail-coverts, and thighs white. Lores, sides of face, chin, throat, foreneck, breast and abdomen black. Tail dusky, barred pale greyish or yellowish. Primaries and primary coverts dusky to blackish; some primary shafts partly white, some inner primary webs with some white; secondaries dusky tipped with some white; rest of upperwing dusky to black, broadly barred and notched yellow. Underwing and axillaries grey to buffish grey. Bill black; eye dark brown; legs and feet greyish black; hind toe missing. ADULT ♂ (non-breeding): upperparts including crown and nape golden buff with dark brown centres to feathers; forehead cream to buff; chin and throat white to buff white; ear-coverts dusky; whitish superciliary stripe; neck, breast and flanks pale brown edged yellow to white; rest of underparts white. Wing and tail like breeding plumage. Sexes alike. SIZE: wing, ♂ (n = 34) 153–174 (163), ♀ (n = 24) 152–173 (163); tail, ♂ (n = 17) 55–62 (58), ♀ (n = 15) 52–64 (57); bill, ♂ (n = 17) 27–31 (30), ♀ (n = 15) 26–31 (29); tarsus, ♂ (n = 17) 37–43 (40), ♀ (n = 15) 37–44 (40). WEIGHT: Wake Island, Apr, ♂ (n = 16) 122–193 (152), ♀ (n = 11) 123–172 (154), Aug, ♂ (n = 8) 115–146 (133), ♀ (n = 3) 126–138 (133), Dec, ♂ (n = 9) 121–141 (132), ♀ (n = 3) 123–133 (128).

IMMATURE: juvenile, like non-breeding adult but upperparts brown-black with some golden and white; brown-tipped white feathers on flanks and abdomen, which are retained in some 1st-year winter birds until Mar.

P. d. dominica (P. L. S. Müller): sight records (e.g. Field 1974) indicate vagrant to rare, coasts W and NW Africa. Larger, less golden yellowish plumage, nape paler than crown and unstreaked; upperparts edged buff-yellow to grey; throat, neck and breast streaked or mottled grey-brown; belly almost grey. SIZE: wing, ♂ (n = 20) 176–191 (183), ♀ (n = 16) 176–193 (185); tail, ♂ (n = 16) 63–71 (67), ♀ (n = 16) 62–76 (66).

Field Characters. Very similar to Greater Golden Plover *P. apricaria*, though their ranges do not overlap in Africa; slightly smaller and longer-legged, but best distinguished in flight by grey rather than white axillaries and underwing-coverts. For distinction from Grey Plover *P. squatarola*, see that species (p. 249).

Voice. Tape-recorded (62, 73). Gives a melodious 'tu-ee' or 'tee-tew'; also a repeated 'chu-leek', 'too-lick', 'klee-yee'. Sometimes also a 'pfeeb' and 'deedleck'.

General Habits. Found mainly in muddy open freshwater habitats close to coastal areas; also tidal flats, lagoons, dunes, and ploughed fields; occasionally inland water habitats. Sometimes found in shallow tidal lagoons surrounded by mangroves where it rests and drinks but flies to nearby freshwater habitats to feed (Ash 1980b). Occurs singly or in flocks of 20–100+, often closely packed. Occasionally flocks with other waders. When feeding, runs, pauses, turns head close to ground and pecks at food. Does not forage as fast as Grey Plover (Burton 1974). Individuals defend feeding areas in non-breeding range. Feeds day or night. Flight powerful and agile, with deliberate buoyant wing-beats;

248 CHARADRIIDAE

travels up to 113 km/h.

Returns Africa Aug–early Sept, leaves late Apr–early May, once July (Somalia).

Food. Beetles, flies, bees, caterpillars, grasshoppers, locusts, larvae, spiders, earthworms, slugs, small molluscs, crustaceans and miscellaneous plant material including seeds; rarely very small fish and lizards.

References
Ash, J. S. (1980b).
Dowsett, R. J. (1980).
Kieser, J. A. (1981).

Plates 16, 20

(Opp. pp. 257, 321)

Pluvialis apricaria (Linnaeus). Greater Golden Plover. Pluvier doré.

Charadrius apricarius Linnaeus, 1758. Syst. Nat. (10th ed.), 1, p. 150; southern Sweden.

Range and Status. Breeds N Europe east to western Siberia; winters N and NW Africa, S Europe and Mediterranean basin to Caspian Sea.

Palearctic visitor to coasts of N and NW Africa from Senegambia to Egypt; frequent to locally abundant Morocco to Libya, uncommon to rare Egypt, Mauritania and Senegambia; vagrant São Tomé. Largest numbers recorded Morocco, Jan, 14,000–17,000. Records from eastern Africa (e.g. Somalia) probably all refer to Lesser Golden Plover *P. dominica*.

Description. ADULT ♂ (breeding): upperparts from crown to tail including scapulars and larger wing-coverts blackish brown with many bright ochre yellow to golden spots and bars; feathers of upperparts notched and tipped golden to give spangled effect. Forehead white, separated from bill by narrow black to greyish black band; area above eye, sides of neck and breast, thighs and vent white; undertail-coverts white with some dusky brown and golden barring. Lores, side of face, chin, throat, foreneck, breast, flanks and abdomen black. Tail black-brown barred and tipped golden. Primaries dusky, inner primaries with faint white tips; secondaries dusky with white bases; rest of upperwing black-brown notched and edged golden. Underwing and axillaries all white. Bill black; eye dark brown; legs and feet greyish black to greenish grey; hind toe missing. ADULT ♂ (non-breeding): upperparts, wing and tail similar to breeding plumage; sides of head and neck, breast and flanks pale brown, edged white; rest of underparts white. Sexes alike. SIZE: wing, ♂ (n = 54) 177–197 (187), ♀ (n = 25) 170–195 (185); tail, ♂ (n = 6) 68–72 (70·3), ♀ (n = 6) 69–77 (71·2); bill, ♂♀ (n = 82) 21–25 (22·5); tarsus, ♂♀ (n = 87) 38–43 (40·3). WEIGHT: sex? (n = 5) 200–210 (203).

IMMATURE: like non-breeding adult but upperparts pale with duller yellow spots; brown-tipped white feathers on flanks and abdomen.

Field Characters. Medium-sized, plump, round-headed plover with yellow-brown upperparts and, in breeding plumage, face, throat, and underparts black bordered by white. Very similar to Lesser Golden Plover but slightly larger, proportionately shorter legs, smaller bill, upperparts more spangled with black and gold; best field character is colour of axillaries and underwing: white in this species, grey in Lesser Golden Plover. Also similar to Grey Plover *P. squatarola*, especially in non-breeding plumage; for distinction see under that species (opposite).

Voice. Tape-recorded (62, 73, 76). A melancholy liquid 'tluit' or 'lui' repeated 2 or 3 times.

General Habits. Frequents moist grasslands, high steppes, ploughed fields, and various freshwater habitats; also tidal areas and salt marshes. Usually gregarious outside breeding season, in flocks varying from small groups to 2000; also occurs singly. When feeding, runs, pauses, turns head close to ground, and pecks at food. Locates food by seeing and also hearing. Does not forage as fast as Grey Plover (Burton 1974); sometimes performs foot-trembling. Feeds day or night. Flies quickly with swift wing-beats.

Returns to Africa Oct (occasionally Sept), leaves late Mar–early May; Morocco, main autumn passage early

Nov (some Sept onwards), main spring passage mid Feb (some as early as early Feb and as late as 1st week May); Algeria Oct (occasionally Sept) to Feb, peak numbers Jan; Tunisia mainly late Oct–Mar; Libya mainly Oct–Apr. Birds ringed Norway and Denmark, captured Morocco and Algeria; ringed Britain, recovered Morocco; ringed Iceland and Netherlands, captured Morocco.

Food. Earthworms, caterpillars, grasshoppers, locusts, beetles, larvae, spiders, small molluscs, crustaceans and miscellaneous plant material, including seeds.

References
Cramp, S. and Simmons, K. E. L. (1983).
Johnsgard, P. A. (1981).

Pluvialis squatarola (Linnaeus). Grey Plover; Black-bellied Plover. Pluvier argenté.

Plates 16, 20
(Opp. pp. 257, 321)

Tringa Squatarola Linnaeus, 1758. Syst. Nat. (10th ed.), p. 149; Sweden.

Range and Status. Breeds N Russia east to NE Canada; winters Africa, Madagascar, S Asia, Australia, central Pacific islands, southern North America and South America.

Palearctic visitor throughout Africa in aquatic habitats, mainly coastal shores and tidal mudflats where locally very abundant to uncommon, Aug–Apr. Large wintering concentrations Tunisia (21,000), Morocco (Atlantic coast *c.* 20,000), Mauritania (Banc d'Arguin 24,000), Egypt (400), Kenya (*c.* 1000), South Africa (SW Cape *c.* 4600), and Namibia (200 km coastline Durissa-Sandvis, *c.* 4600). Uncommon to rare inland; in Botswana, Burundi, Chad, Congo, Ethiopia, Ghana, Kenya, Malaŵi, Nigeria, South Africa, Sudan, Tanzania, Uganda, Zaïre, Zambia, and Zimbabwe.

Pluvialis squatarola
Migrants cross Africa

Description. ADULT ♂ (breeding): upperparts from crown to tail including scapulars pale grey with brownish black spots; forehead, area above eye, sides of neck and breast white; lores, face, chin, throat, foreneck and underparts to belly black. Rump white, sometimes with darker barring; tail barred black-brown and white. Flanks, vent, and undertail-coverts white. Primaries blackish brown with white shafts, inner ones tipped and edged white on outer webs; primary coverts blackish brown tipped white; secondaries greyish brown edged white; rest of upperwing-coverts greyish brown, greaters edged and tipped white, medians and lessers edged and notched white. Underwing-coverts white; axillaries black. Bill black; eye dark brown; legs and feet black to greyish black; small hind toe present. Breeding plumage usually partly, occasionally fully, assumed before spring departure; most returning adults still have partial breeding plumage in Sept (E Africa). ADULT ♂ (non-breeding): upperparts including tertials and wing-coverts greyish brown, edged greyish white; forehead, lores, area above eye, cheeks, lower neck, breast and flanks with brownish markings; throat and chin white with fewer brownish markings; rest of underparts white. Rest of wings and tail as in breeding plumage. Sexes alike. SIZE: wing, ♂ (n = 154) 188–206 (197), ♀ (n = 83) 187–205 (198); tail, ♂ (n = 12) 69–82 (76), ♀ (n = 6) 68–77 (73·3); bill ♂♀ (n = 106) 25–34 (29), ♂ (n = 29) 25–30 (29), ♀ (n = 21) 26–31 (29); tarsus, ♂ (n = 29) 45–52 (48), ♀ (n = 22) 46–51 (48). WEIGHT: ♂ (n = 5) 170–224 (188), ♀ (n = 2) 194, 209, sex ? (n = 7) 138–188 (164), (Kenya n = 114) 160–287, typical winter weight 180–230 (D. Pearson, pers. comm.), ♂♀ (n = 31) 154–321 (A. Tree, pers. comm.).

IMMATURE: juvenile as non-breeding adult but upperpart feathers, tertials, and wing-coverts with bold golden buff on whitish lateral spotting; underparts white, but with brownish barring on breast and flanks. 1st-year winter birds retain distinct spotting; pale areas of juvenile inner median coverts lost to leave dark central wedges. Underparts white with darkest fringes lost although central wedges often retained.

Field Characters. A medium-sized, plump, large-headed plover with relatively short bill. Resembles Greater and Lesser Golden Plovers *P. apricaria* and *P. dominica* but distinguished from them in flight at all ages by black axillaries, white wing-bar, white rump and pale tail. At rest appears larger and stouter. Bill stouter than that of Lesser Golden Plover. Breeding birds have grey, not golden upperparts; non-breeding birds harder to distinguish but generally paler and greyer, less brown; also more uniform, less mottled. Immature more heavily spotted, like golden plovers; easiest told apart in flight.

Voice. Tape-recorded (62, 73, 76). A loud 3-syllable, musical trill with middle syllable lower, higher pitched than in golden plovers' call, 'tee-oo-ee' or 'plee-uu-wee'; also 'tlui-tlui'. Flocks often rather silent; occasionally individuals on wing give soft mellow 'quu-hu' or low, almost growling note.

General Habits. Mainly in coastal saltwater habitats, especially lagoon mudflats but also rocky and sandy shores, mudflats along seashore, salt marshes, dunes near shore and estuaries; occasionally freshwater pools near sea. Often occurs singly or in small parties of 2–3, sometimes in large flocks of up to 1000. Usually remains apart but will flock with other species. When feeding, runs, pauses in an upright attitude, turns head close to ground, and pecks at food. Often combines pecking with a sideways flicking movement. Forages faster than Greater Golden Plover, averaging 23·2 pecks/min (Burton 1974). Often feeds at night as well as by day. Will swim when necessary. Flight swift, up to 80 km/h (Summers and Waltner 1979), with much agile turning and twisting; flies usually directly along coastline with larger flocks spreading out in wide V-shapes (Ash 1981).

Enters Africa early Aug, with main influx Sept, following W African coast (occasional birds inland waters Morocco); in eastern Africa some fly south over Red Sea and Rift Valley lakes to reach southern Africa. Sometimes stops in southward passage; crosses Sahara. Adults with much breeding plumage mostly return Kenya late Aug–early Sept; young birds arrive Oct. Follows same routes when northbound; leaves late Apr–early May, many in partial or full breeding plumage (Kenya). Return trip to Arctic probably with few stops. Leaves South Africa from Feb–Mar, Morocco from early Apr. Birds ringed Russia, Norway, England and Finland, recovered Morocco; ringed in Britain, recovered from Ghana; ringed Rumania, recovered South Africa; ringed Norway, recovered Senegambia; ringed Namibia and South Africa, recovered Russia. Some summer African coasts; most 1st-year birds in Kenya appear to oversummer.

Food. Crustaceans including sand and mysid shrimps, barnacles, crabs, prawns; small clams and snails; polychaetes and other segmented worms; insects; occasionally fish; and seeds.

References
Cramp, S. and Simmons, K. E. L. (1983).
Dowsett, R. J. (1980).
Johnsgard, P. A. (1981).

Subfamily VANELLINAE: lapwings

Medium-sized plovers, larger than typical plovers Charadriinae and with different jaw musculature (except *Hoploxypterus*). Bony protuberance at carpal joint (proximal end of carpometacarpus) blunt and vestigial in some species, long pointed spur in others. Many have colourful facial wattles and bare skin round eye, some have crests. Plumage very variable but primaries black, many species with broad white patch or stripe on secondaries and greater coverts, all except *Vanellus leucurus* with broad black subterminal band on tail. Wing usually rounded, flight more buoyant than in Charadriinae, wing-beats slower.

24 spp., 23 in *Vanellus* (14 in Africa), 1 in *Hoploxypterus*. Bock (1958) placed all lapwings in *Vanellus*, but Burton (1974) showed that *cayanus* lacks specialized jaw musculature of other species; it is also smaller, with unique plumage pattern, and we retain it in *Hoploxypterus*. 2 spp., *gregarius* and *leucurus*, are often put in separate genus *Chettusia*, to which Glutz *et al.* (1975) added *coronatus*, *melanopterus* and *crassirostris*; however, *crassirostris* is highly specialized for aquatic life and often retained in monotypic genus *Hemiparra*, e.g. by Johnsgard (1981), chiefly because of its long toes, and has no obvious close relative. 6 genera recognized by Cramp and Simmons (1983) but generic diagnoses not given. Until further evidence is available we prefer to place all species except *Hoploxypterus* in *Vanellus*.

Genus *Vanellus* Brisson

Large, wing longer than 150 mm. Jaw musculature distinct from Charadriinae. Scapulars not black with white inner edges (special feature of *Hoploxypterus*).

Relationships within *Vanellus* are obscure. Grouping of African species generally follows that of Snow (1978), except that we consider *V. lugubris* and *V. melanopterus* to form a superspecies, in spite of a limited area of seasonal overlap, because of great similarity in plumage. In grouping the 3 Eurasian spp. not treated by Snow (*V. gregarius*, *V. leucurus*, *V. vanellus*) it is assumed that lapwings arose in tropical Africa and later invaded Eurasia, just as *V. spinosus* is currently invading SE Europe via the Middle East; its allospecies *V. duvaucelii* in India and SE Asia is presumably the product of an earlier invasion. *V. gregarius* with its striped head may be closest to *V. coronatus* whereas

V. leucurus, with which it is often linked in *Chettusia*, has lost both head-stripes and tail-band, so a progression *coronatus—gregarius—leucurus* seems reasonable. *V. coronatus* is also linked to V. *melanopterus*/V. *lugubris*, all 3 having formerly been joined in *Stephanibyx*. The rather plain *superciliosus* may possibly be part of this assemblage although its affinities are very uncertain. *V. vanellus* is not close to any African species and is placed on its own at the end, next to the equally distinctive *crassirostris*.

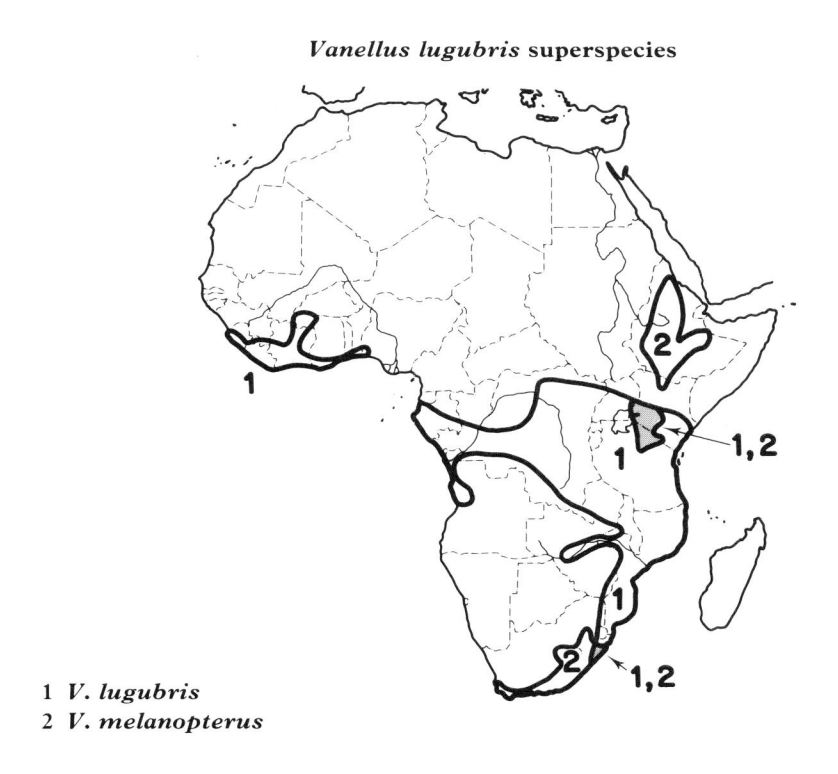

***Vanellus lugubris* superspecies**

1 *V. lugubris*
2 *V. melanopterus*

Vanellus senegallus (Linnaeus). African Wattled Lapwing; Wattled Plover. Vanneau du Sénégal.

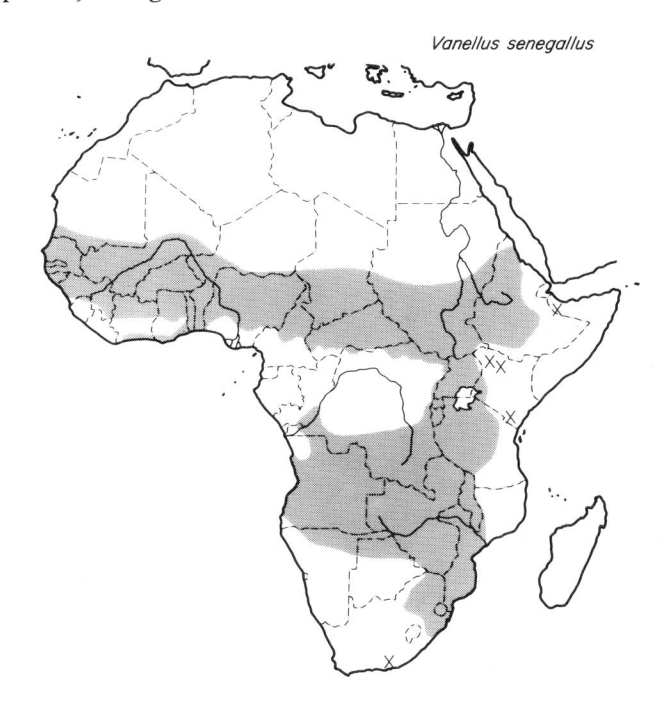

Plates 14, 20
(Opp. pp. 241, 321)

Parra senegalla Linnaeus, 1766. Syst. Nat. (12th ed.), 1, p. 259; Senegal.

Range and Status. Endemic resident and intra-African migrant. Widespread in savanna country north of equator from SW Mauritania and Senegambia to central Ethiopia, N and E Zaïre and Uganda. In Kenya only in W highlands from Uasin Gishu and Nandi to Mara. Occurs south and east of lowland forest belt in Zaïre, common in E highlands, Rwanda and Burundi, extending east through W Tanzania to Mt Meru, Serengeti, Tabora, Dodoma and Kilombero R. Throughout Angola (except Luanda and Cuanza Norte), Zambia (local), Malaŵi and Zimbabwe (scarce in some drier W areas and along Limpopo), extending to N Namibia (especially Caprivi), N Botswana (to Okavango, Makgadikgadi), S Mozambique (north at least to Zambezi and Malaŵi border at L. Chilwa, possibly further), Transvaal (except extreme W) and most of Natal, becoming rare in south. Vagrants in NW Somalia, N Kenya (Lossodok Hills, North Horr), SE Kenya (Tsavo West Nat. Park) and E Cape Province (Port Elizabeth). Seldom very numerous in spite of wide range; generally frequent to common, but concentrations of up to 50 in Zambia and 150 in Rwanda.

Description. *V. s. lateralis* (A. Smith) (including '*solitaneus*'): S and E Zaïre (L. Edward), central Uganda (Masindi, Mengo, Mt Elgon) and Kenya to Namibia and South Africa. ADULT ♂: feathers at base of upper mandible brown; forehead and forecrown white, sides of crown and hindcrown blackish grading to grey-brown on nape and hindneck, latter with dark shaft streaks. Face finely streaked black and grey, sides of neck and foreneck, throat, and upper breast more broadly streaked black and white; chin white, centre of throat black. Upperparts, including scapulars, tertials and lesser upperwing-coverts, greyish olive-brown, paler on wing-coverts. Uppertail-coverts and tail white, latter with broad black subterminal band, broadest in centre, narrower on outer feathers; central feathers tipped brown, rest tipped white. Breast and upper belly brownish grey, separated from white flanks, lower belly and undertail-coverts by blackish band. Primaries black, base of inner webs white; secondaries white tipped black; median coverts white with pale grey bases, greater coverts white, primary coverts black. Axillaries and underwing-coverts white. Bill yellow with black tip; large wattles in front of eyes, upper portion red, pendent portion yellow; narrow eye-ring yellow; eyes pale grey; legs and feet yellow. Sexes alike. ♀ said to have less black on centre of throat but this character very variable. SIZE: (8 ♂♂, 8 ♀♀) wing, ♂ 223–244 (236), ♀ 226–249 (237); tail, ♂ 91–100 (96·0), ♀ 88–102 (93·6); bill, ♂ 32–36 (34·0), ♀ 32–35 (33·9); tarsus, ♂ 87–93 (89·5), ♀ 82–91 (86·6); wing-spur, ♂ (n = 5) 10–12·5 (11·5), ♀ (n = 5) 6–9 (7·6). WEIGHT: ♂ (n = 4) 209–240 (224), ♀ (n = 4) 197–248 (224); 1 unsexed, 270; 1 juv ♂ 246.

IMMATURE: like adult but small amount of white on forehead, forecrown dark brown with white bases to feathers, hindcrown brown, not black; throat off-white with faint dusky streaks; wattles and wing-spurs very small.

DOWNY YOUNG: speckled buff and black. Forehead white, black patch on crown and back; narrow white collar round hindneck; underparts yellowish white. Bill slate grey, legs very pale yellow.

V. s. senegallus (Linnaeus): Senegal to Somalia, south to NE Zaïre (L. Albert) and Uganda north of Masindi, Mengo and Mt Elgon. Chin black, no black band on belly, tip of bill mainly yellow (only tip of culmen black). There is a west–east cline in size: largest birds (Ethiopia and Somalia) have been described as separate race *major*.

Field Characters. A large lapwing appearing more or less uniform grey-brown at rest; in flight shows white patch on inner wing contrasting with black primaries and grey-brown lesser coverts. White-headed Lapwing *V. albiceps* shares white crown and yellow face wattles but has pure white underparts, white collar on hindneck, much more white on wings in flight, and different habitat.

Voice. Tape-recorded (15, 17, 42, 58, 74, C, F). In agitation or excitement, including during fights, a series of very rapid, high 'peep-peep's which gradually change into staccato 'yip-yip'. Alarm, 'ke-WEEP, ke-WEEP' as bird flushes; more intense form, 'kee-up, keeup, tejreep' Loud 'peep' given by bird landing in territory, usually answered by mate. When approaching nest, both birds give soft 'peep' at intervals. High-pitched piping made by both sexes before copulation, twittering by ♂ during copulation. ♂ clucks softly when nest-building. Adults attending chicks keep up constant soft 'peep-peep'. Calls a great deal at night.

General Habits. Demonstrates great ecological plasticity in choice of habitats; in some areas occupies same habitat all year, elsewhere opportunistic, changing habitats seasonally. Generally associated with water, in marshes and in damp grass or on bare muddy or sandy ground beside lakes, ponds, rivers and streams, but also in areas of dry, short grass, including airports, and among sparse vegetation and dry savanna far from water. Feeds and sometimes nests on ploughed land and dry, weedy, fallow fields; after rains moves in to flooded rice fields, inundated grassland and temporary pools, wading in water several cm deep; after fires occupies burnt grassland. In Akagera Nat. Park, Rwanda, associated with termite mounds, which it uses as look-out perches. Sometimes forced by dominant sympatric congeners, e.g. Crowned Lapwing *V. coronatus* and Spur-winged Lapwing *V. spinosus*, into marginal habitat, such as long grass by lake-shore, to which it is able to adapt, thereby reducing competition (Vande weghe and Monfort-Braham 1975). Generally in lowlands but up to 2200 m in E Africa and Ethiopia.

Rather more solitary than other lapwings, found singly, in pairs or small groups, but larger flocks form in non-breeding season in newly available habitat such as burnt grassland or temporary waters, and may form mixed flocks with congeners such as Crowned Lapwing and Long-toed Lapwing *V. crassirostris*. Often fairly tame and tolerant of man when not nesting, occurring and even nesting not far from human habitation, but during breeding season one of the noisiest birds.

Daily activity includes a considerable amount of time spent on preening. In 100 min of observation, bird fed for 60 min, preened for 20 min, rested for 20 min (Little 1967). Preening accompanied by comfort movements, e.g. fluffing out feathers, spreading and shaking tail, scratching head (leg over wing), standing on 1 leg while stretching the other backwards, raising wings vertically. Bathes by wading into water up to belly, dipping head into water, squatting and beating wings vigorously, also scooping up water with head and neck and letting it run over back. Fluffs out feathers and shakes body to dry, then preens carefully. When resting often squats or stands with head on back, bill tucked into feathers. When foraging does not use quick run-and-stop of other lapwings; instead walks slowly and deliberately, pauses with leg raised, takes 1 step forward and grabs prey from ground, often with small jump, both feet together. Also probes with bill into base of grass tuft. Only one instance noted of 'foot-trembling' while foraging (Little 1967): after lifting and then dropping large cricket-like insect bird stood with 1 leg raised and trembling, then stood on this leg and raised the other which likewise trembled. Watched insect for a few s, then picked it up and swallowed it. When drinking dips head into water with bobbing, pecking movement. In bad weather, crouches behind grass tuft, molehill or other natural object with legs half bent, head and neck drawn in; sometimes squats on tarsi. On approach of storm 1 pair called to each other and came together; when close, 1 bird 'marked time' with high-stepping gait otherwise seen only just before copulation. Alert; constantly scans sky for danger; when raptor spotted sinks and squats motionless on ground, incubating bird remains huddled on nest. Flight strong; on alighting raises wings vertically before folding them, displaying striking black and white pattern.

Resident in some areas, migratory elsewhere but often with no definable pattern; movements frequently nomadic and opportunistic. In W Africa moves north in wet season, e.g. in Mali to Sahel, in Chad to 13°30'N, but in Nigeria while some birds move north others also move south, e.g. to Lagos, so no clear pattern of seasonal movements (Elgood 1982). In Ivory Coast present all year in wet areas but only Dec–Apr in savannas. 5 birds seen extreme N Liberia (Nimba) Oct apparently on migration (Colston and Curry-Lindahl, in press). In Cameroon migrates according to the water level, and in Zaïre moves locally during rains, e.g. in Uelle and Ubangi where birds absent June–Oct, moving to Sudan. Present and moulting in non-breeding season, Aug–Nov, in W Sudan (Darfur). In Akagera Nat. Park (Rwanda) apparently resident; and partial resident E Africa, but also moves to temporary waters and flooded fields. Seasonal visitor to various localities in Tanzania, e.g. in dry season to Ufipa Plateau, and to flooded rice fields near Tabora. In Ethiopia occurs around Addis Ababa July–Oct. Moves when habitat flooded by rains in Zambia but no regular migration. Much seasonal movement in Zimbabwe, with possibility of 2 different populations, 1 arriving Mar–Apr, while local birds nomadic during rains (Irwin 1981). Resident in northern half of Kruger Park, South Africa, vagrant in summer to southern half; migrant to highveld areas of E Transvaal, arriving first week Sept and leaving end Mar. Of 191 birds ringed Zimbabwe, 8 retrapped 1 or more seasons later.

Food. Chiefly insects: grasshoppers, locusts, beetles (including dung beetles and weevils), crickets, termites and various aquatic insects; also worms and some grass seed. Food sometimes regurgitated in pellets (Brooke 1967): (n = 23) length 37–60 (46), diam.15–21 (18), weight (n = 21) 1–5 (2·9). Pellets contained coarse grass and many fragments of large and small grasshoppers, dung beetles and a few termites.

Breeding Habits. Monogamous; pairs form before arrival at South African breeding grounds, E Transvaal. Possible courtship behaviour exhibited by paired birds includes loud calling, side-throwing movements and the clucking call associated with scrape formation; also swooping courtship flight in air. In Rwanda 3–4 pairs found nesting close together in small colony (Vande weghe and Monfort-Braham 1975), but highly territorial in South Africa where defends feeding territory. Nest usually placed just outside this territory but young brought to territory after hatching; size of 1 territory in vlei, 330 m × 100–200 m (Little 1967). Fighting takes place when conspecific enters territory during breeding season; birds (a) strike each other with wings, either from sideways or frontal position, (b) face each other with raised wings, 1 rushes at and jumps over the other without striking, and (c) dive-bomb opponent from air; these actions accompanied by much loud 'yip-yip' calling and use of threat postures, especially high-intensity neck stretch in which bird stretches neck high, raises striped neck-feathers and crown-feathers, puffs out breast, bobs slightly. Fighters sometimes break off and redirect aggression at predator (buzzard, mongoose). Fighting almost all by ♂, but ♀ sometimes helps ♂ drive off intruder. Fighting often followed by copulation. Towards end of breeding season reaction to invader much less intense: bird hopped about on ground, displacement-preened, adopted threat-crouch, then suddenly flew up with high-pitched calling, flew down at intruder with rolling motion, circled and swooped again, swinging from side to side; intruder flew off (Little 1967). Before nesting season feeding territory only partly defended; some birds tolerated, but Blacksmith Lapwing *V. armatus* and several small passerines driven off. Dogs and mongoose mobbed but little attention paid to man. Threat postures towards intruders include (a) low intensity: slight crouch, head hunched into shoulders, bill horizontal and pointed at intruder, bird runs at intruder (effective with small birds); (b) also low intensity: bird bends neck to ground and places head parallel to ground (used in situation where bird withdrawing from fight or human approaching nest); (c) more aggressive: ground-pecking threat where bird bends forward, puffs out head and neck feathers and pecks stiffly at ground; (d) high-intensity neck stretch—see above. Displacement activities during encounters include yawning, preening, side-throwing of nest material, and giving the clucking call connected with nest-building.

Copulation takes place within feeding territory. Birds approach each other calling with necks stretched forward, rumps raised, legs bent; ♀ walks ahead of ♂ and crouches, raising tail and lowering wings to expose white rump; ♂ steps onto her back, spreads wings slightly, both birds raise and spread tails. Sometimes much wing-flapping, and twittering call from ♂. As ♂ dismounts he raises wings vertically (sometimes sideways), fans tail, stretches head up, ruffles feathers, preens, and moves off to feed. ♀ usually does not stretch wings but walks forward, ruffles feathers and preens. Copulation sometimes preceded by displacement nest-building behaviour; ♂ may also approach ♀ with crouch threat or goose-stepping walk. Copulation can occur up to time 1st eggs laid but not during incubation; frequently occurs after intruder driven off. In 1 instance ♀ solicited from intruder while mate was away; intruder attempted copulation, then flew off (Little 1967).

Scrape ceremony performed mainly by ♂, though ♀ may take part later. Bird first side-throws nest material such as grass stems and bits of dry dung while giving clucking call, then crouches on ground, raises rump and tail, spreads wings slightly, and pushes breast into ground with rotary movement of tail while kicking legs backward in 'running' motion; then stands up and as it moves away pecks at ground and side-throws material. Several scrapes made; ♂ leads ♀ round scrapes, ♀ chooses 1 for nest-site.

NEST: shallow depression in ground, diam. *c.* 15 cm, depth *c.* 2·5 cm, lined with grass stems, small sticks, pebbles and pieces of dry dung; lining on bottom of nest sometimes thin, sometimes thick with eggs one-third embedded; much material around rim of nest; adds to lining during incubation. Situated in open where bird has unobstructed view, typically on bare ground or in short grass (including burnt grassland), also among weeds in fallow fields, and not infrequently near road or human habitation; usually within 100 m of water. Sites also include gravel ridge in flooded quarry, patch of *Helichrysum* on bare stony ground, and among boulders on ridge above stream (Little 1967).

EGGS: 2–4 (usually 2–3 in equatorial regions, 4 in south) laid at 1–2 day intervals. Deep cream to pale or warm buff, putty or pale olive-brown, blotched and marked with black and dark brown, sometimes heavily, especially at large end; some eggs almost entirely covered with black. SIZE: (n = 11, Uganda) 44·0–46·8 × 30·8–33·7 (45·3 × 32·1); (n = 42, southern Africa) 44·9–53·8 × 32·9–36·0 (49·3 × 35·2); (n = 23, Transvaal) 44·0–51·6 × 31·0–34·5 (48·2 × 33·3). WEIGHT: 21·2–25·4. Eggs in nest usually rest with sharper ends pointing inwards.

LAYING DATES: Senegambia, Jan, May–Aug, Oct–Nov; Ghana, May; Nigeria, Mar–June; Zaïre: Shaba, June–Dec, Virunga Nat. Park, Apr–May (rains), Upemba Nat. Park, begins with rains (Sept–Oct), young flying by Dec–Jan; Rwanda, Apr, Aug–Nov, both dry and wet seasons; E Africa: Region A, Jan–May, July–Aug, peak Mar–Apr; Region B, July–Sept, mid-rains; Region C, Aug–Dec, peak Aug–Oct in dry season; Ethiopia, Mar–June; Angola (♀ with developed ovary Aug); Zambia, July–Nov, peak Sept; Malaŵi, Sept–Oct; Zimbabwe, July–Jan, peak Sept–Oct; South Africa: Natal, July–Dec, Transvaal, Sept–Oct.

INCUBATION: begins with 3rd egg; by both sexes, shared equally except during first few days when more by ♂. Period: 30–32 days. In heat of day incubating bird sometimes squats over eggs with legs bent, feathers fluffed out, rather than brooding them; also recorded soaking belly and wetting eggs and newly hatched young (Wright 1963). Incubating bird sits still for hours, occasionally turning eggs or leaving nest briefly to preen; sits facing wind; keeps watching sky for raptors, presses itself onto ground if one spotted. When ♀ is incubating, ♂ stands guard nearby. Changeovers take place every 40–45 min, usually silent, but birds sometimes give soft 'peep'; variable behaviour noted by Little (1967): (a) ♀ flew off nest to join ♂; 10 min later ♂ walked slowly back to nest by devious route, (b) both arriving and leaving birds side-throw material, (c) changeover with element of threat—♂ approached nest, gave 'upright threat', ♀ stood up and pecked at ground, made side-throwing movements, (d) ♀ approached nest, incubating ♂ refused to move; only after ♀ made several approaches, including pecking at ground, did ♂ move.

During incubation other birds generally tolerated in territory, including Crowned Lapwing, but dogs and mongoose swooped at with excited screeching. Domestic animals ignored unless very close to nest; cow and calf threatened with spread wings, head and neck stretched, sheep 0·3 m from nest threatened by both birds with spread wings and fanned tail, much calling; sheep did not move until birds pecked at its head and almost struck its eye. On approach of person slips quickly and silently off nest, stands *c*. 30 m away, mate sometimes gives alarm call and swoops; closer to time of hatching reaction increasingly aggressive, similar to that when chicks hatch.

DEVELOPMENT AND CARE OF YOUNG: adults eat eggshells. At 1 week has general shape of adult, buff-brown with white neck and underparts; at 2 weeks white crown-patch starting to appear, bill longer, yellow with black tip; at 3 weeks no white on nape, neck and back same uniform grey-brown as adult; white crown distinct, tiny yellow wattles, below as adult (brown breast, white belly), tail-feathers growing. At 4 weeks black and white wing and tail pattern starting to develop, clear by 5th week. Fledging period: *c*. 40 days.

Chicks sometimes remain near nest after 24 h but usually leave nest on 1st day and go to feeding territory. Continuously tended by parents (somewhat more by ♀ since ♂ more active on guard duty) but never fed by them. Tending bird does some brooding, especially in bad weather when chicks up to 2 weeks old brooded, but mostly stands nearby, often on molehill or other eminence for better view, and keeps up constant low 'peep-peep'. Bird not on duty feeds nearby but outside feeding territory. Changeover initiated by feeding bird which flies towards mate and chicks with loud call to which other bird replies. Intervals between changeovers up to 95 min (Little 1967). Chick very independent at 1 week, seen attempting to preen even though it could hardly balance on 1 leg, managed to scratch bill with foot; at 2 weeks pulled up and swallowed earthworm, also made comfort movements, stretching leg back, raising stumpy wings, in spite of difficulty balancing. Chicks silent. Parents tend chicks until they can fly.

Behaviour of adults towards intruders more aggressive than during incubation. Harriers, ibises, guineafowl and lapwings attacked; waders dispersed with low-intensity threatening run. Distraction display to dogs in addition to swooping—1 wing extended and dragged. Flies up and calls loudly on approach of human, but near human habitation can become habituated to cars and pedestrians and eventually ignores them.

BREEDING SUCCESS/SURVIVAL: 4 out of 6 nests destroyed, Transvaal; causes unknown. Newly hatched chicks died when exposed to sun for 10–15 min while parent away from nest (Wright 1963). 1 nest destroyed by hailstorm. Ringed bird retrapped in Zimbabwe 8 years later.

Reference
Little, J. de V. (1967).

Vanellus albiceps **Gould. White-headed Lapwing; White-crowned Plover. Vanneau à tête blanche.**

Vanellus albiceps Gould, 1834. Proc. Zool. Soc. Lond., p. 45; Niger River or Fernando Po.

Plates 14, 20
(Opp. pp. 241, 321)

Range and Status. Endemic resident and intra-African migrant. Occurs throughout most of the river systems of W Africa, breeding west at least to Niokolo-Koba Nat. Park (SE Senegambia) but scarce and irregular visitor to lower Senegal R.(St Louis, Richard-Toll) and lower Gambia R. (to Farrafenni). From Nigeria extends east and south to SW Sudan (Bahr-el-Ghazal), Zaïre (Lualaba R. to Kongolo; Lulua, Kabinda and Kwango Districts) and NW Angola (Cuanza Sul). Common on rivers of SE Tanzania (Ruaha, Rufiji, Luwegu, Kilombero), and on Luangwa and Zambezi Rivers, in Zimbabwe also on some Zambezi tributaries. In Mozambique occurs from Zambezi to Limpopo R., extending inland along Limpopo to Beit Bridge, and to Sabi and other rivers of SE Zimbabwe. In Kruger Park (NE Transvaal) known from Olifants, Levubu and Letaba Rivers. Isolated occurrences L. Kivu (Kisenyi, Rwindi Plain), Serengeti (L. Lagarja) and Namwala (Kafue Flats), Zambia, possibly migrants. Vagrant to Bioko (Fernando Po) and SW Transvaal (Potchefstroom). Frequent to common.

Description. ADULT ♂: forehead and broad stripe over crown to nape white, narrowly bordered dark grey; sides of crown, facial area, and sides of neck and throat grey, continuing as broad collar around hindneck which becomes blackish where it meets narrow white band around upper mantle. A small white spot below and just in front of eye, continuing back as narrow rim under eye. Chin, centre of throat and entire underparts white. Mantle, inner scapulars and tertials earth-brown; back white crossed by narrow black band; rump, uppertail-coverts and basal half of tail white, rest of tail black. 3 outer primaries black with pale shaft and whitish base to inner web; rest of primaries and secondaries white. Outer scapulars white, forming line separating back from black lesser and median upperwing-coverts; greater coverts, primary coverts, axillaries and underwing-coverts white. Bill yellow, distal third black; long wattles (*c.* 40 mm) hanging down from in front of eye yellow with greenish base; eye, legs and feet pale yellow. Sexes alike. SIZE: (8 ♂♂, 5 ♀♀) wing, ♂ 210–228 (218), ♀ 212–224 (218); tail, ♂ 95–102 (99·4), ♀ 95–101 (98·4); bill, ♂ 31–36 (33·4), ♀ 32–34 (33·0); tarsus, ♂ 70–78 (75·1), ♀ 70–77 (74·2); wing-spur, ♂ 18–24 (20·1), ♀ 20–22 (20·6). WEIGHT: 1 unsexed, Zambia, 201.

IMMATURE: like adult but centre of crown brown, brown back faintly barred darker brown, black wing-coverts tipped brown or white, webs of scapulars mottled, spurs and wattles small.

DOWNY YOUNG: upperparts speckled buff and black, specks larger on head. Narrow black band round nape, white collar on hindneck; underparts white. Bill horn, tip black; wattles 0·25 cm long, pale buff; legs grey.

Field Characters. A medium-large, pale lapwing with distinctive combination of white crown and collar, grey face and neck, long yellow wattles, and all-white underparts. In flight, black wing-coverts and outer primaries contrast with white of rest of wing. Riverine habitat also distinctive. Browner-looking African Wattled Lapwing *V. senegallus* shares white crown, yellow wattles and yellow legs, but has streaked neck, greyish underparts, no white collar, and much less white in wing in flight.

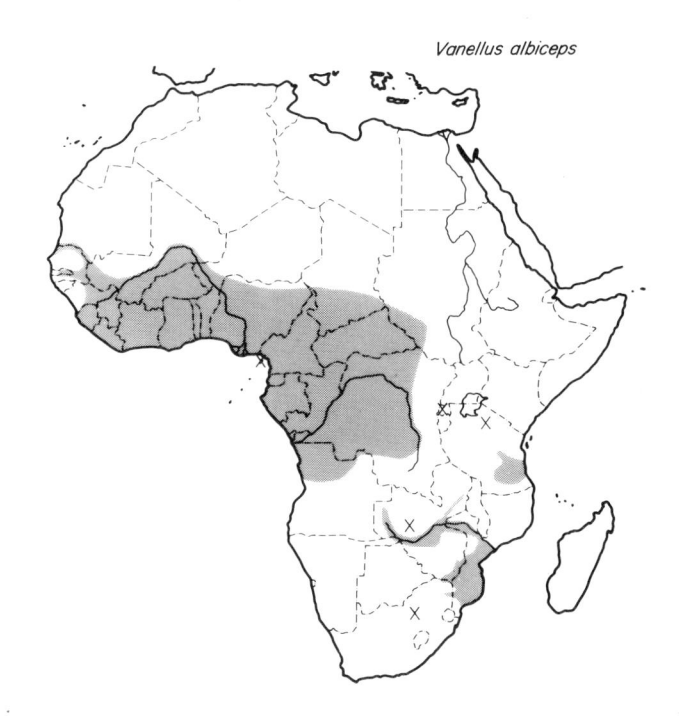

Voice. Tape-recorded (21, 35, 74, F). A single, high-pitched, ringing 'pew' or 'peeuw', often repeated, with a quality reminiscent of an oystercatcher *Haematopus*. Incubating bird gave subdued 'wheet'.

General Habits. Inhabits sandy and gravelly rivers, both in open country and in forest, nesting on riverbeds at time of low water. Particularly partial to large rivers with islands in the middle, but during floods occurs on smaller streams, pans and lagoons. Sometimes occurs on lake-shores, e.g. L. Kariba, and forages for worms in damp grassy places.

Occurs in pairs or parties of 6–12, occasionally up to 30. Flight buoyant and graceful; calls a lot in flight. Alert, perching on logs and rocks while watching for danger. Like the Egyptian Plover *Pluvianus aegyptius*, rumoured to pick teeth of crocodiles, but no proof, though certainly lives near crocodiles.

Migratory to some extent through most of its range since it must vacate its riverine habitat when rivers flood. Distance travelled depends on degree of flooding and availability of suitable habitat nearby. A regular migrant in most of W Africa, e.g. absent Ivory Coast, Aug–Sept, returning late Oct. Leaves S Nigeria June–Oct and moves north beyond breeding range to Sokoto and Yankari Game Reserve. Present in Park W, Niger, only in dry season, and rare dry season visitor to lower Gambia R. South of equator birds more sedentary, e.g. Zambia, where leave rivers only at peak of

Plate 15

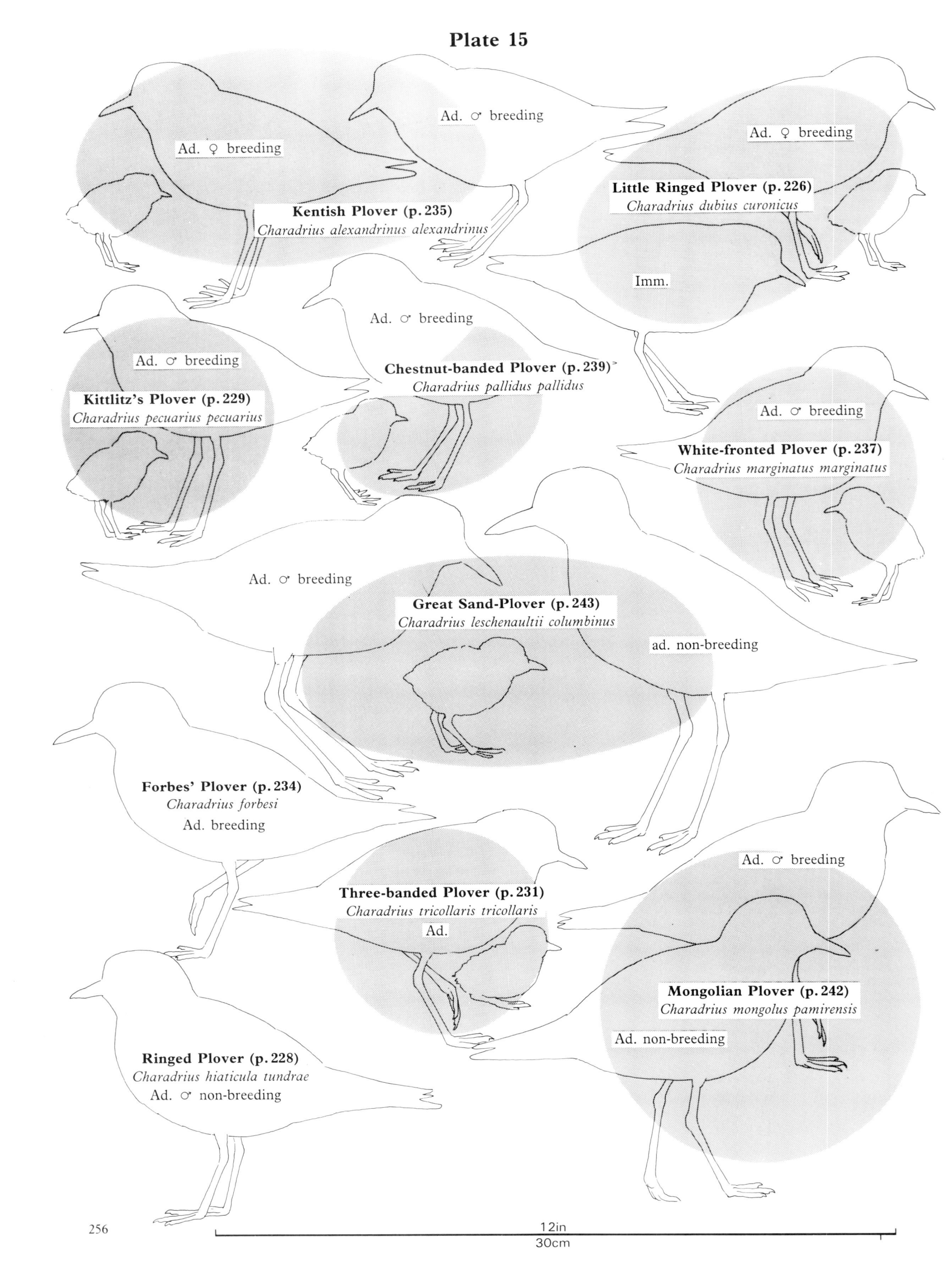

Kentish Plover (p. 235)
Charadrius alexandrinus alexandrinus
Ad. ♀ breeding
Ad. ♂ breeding

Little Ringed Plover (p. 226)
Charadrius dubius curonicus
Ad. ♀ breeding
Imm.

Kittlitz's Plover (p. 229)
Charadrius pecuarius pecuarius
Ad. ♂ breeding

Chestnut-banded Plover (p. 239)
Charadrius pallidus pallidus
Ad. ♂ breeding

White-fronted Plover (p. 237)
Charadrius marginatus marginatus
Ad. ♂ breeding

Great Sand-Plover (p. 243)
Charadrius leschenaultii columbinus
Ad. ♂ breeding
ad. non-breeding

Forbes' Plover (p. 234)
Charadrius forbesi
Ad. breeding

Three-banded Plover (p. 231)
Charadrius tricollaris tricollaris
Ad.

Mongolian Plover (p. 242)
Charadrius mongolus pamirensis
Ad. ♂ breeding
Ad. non-breeding

Ringed Plover (p. 228)
Charadrius hiaticula tundrae
Ad. ♂ non-breeding

12in
30cm

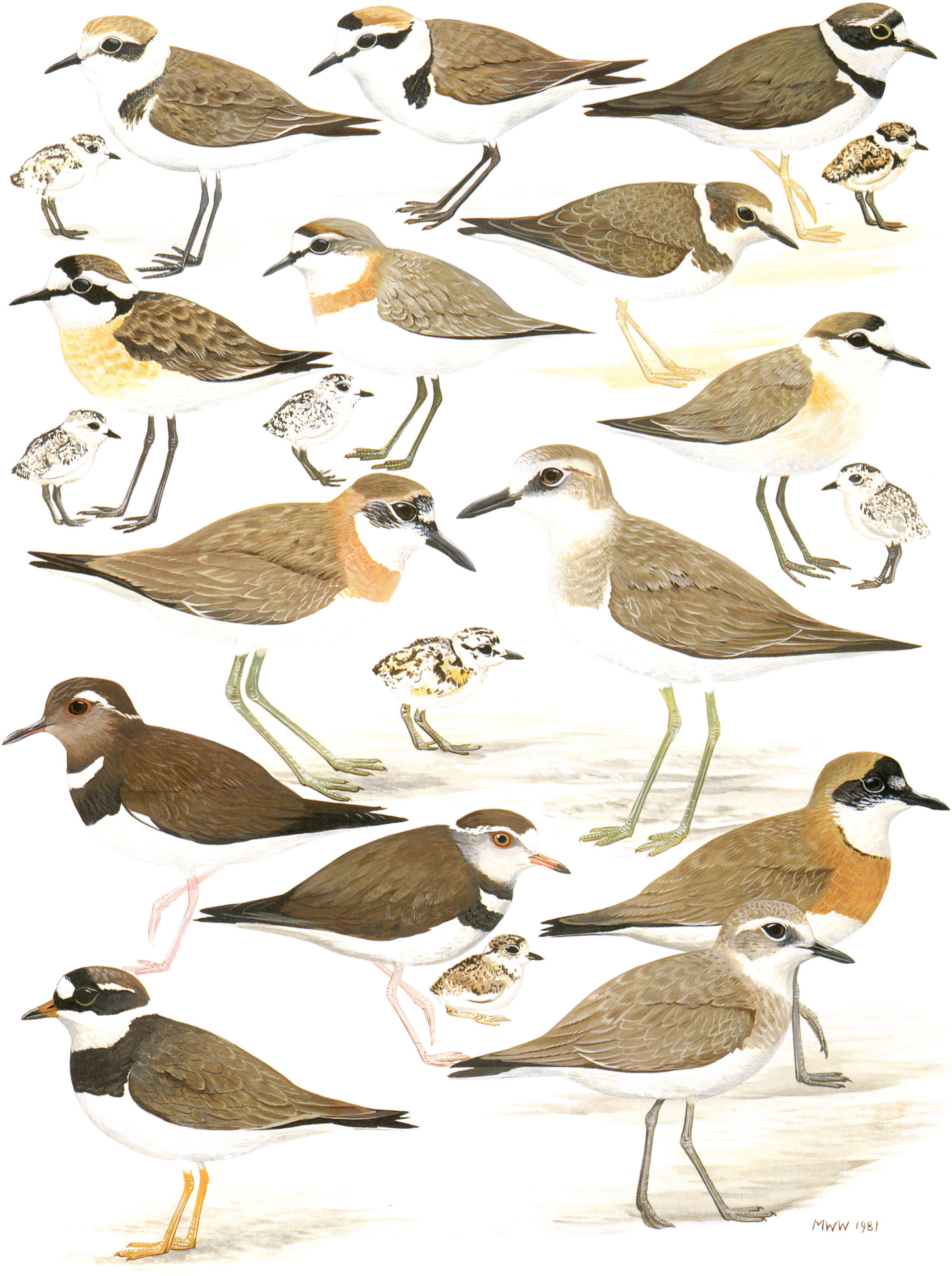
MWW 1981

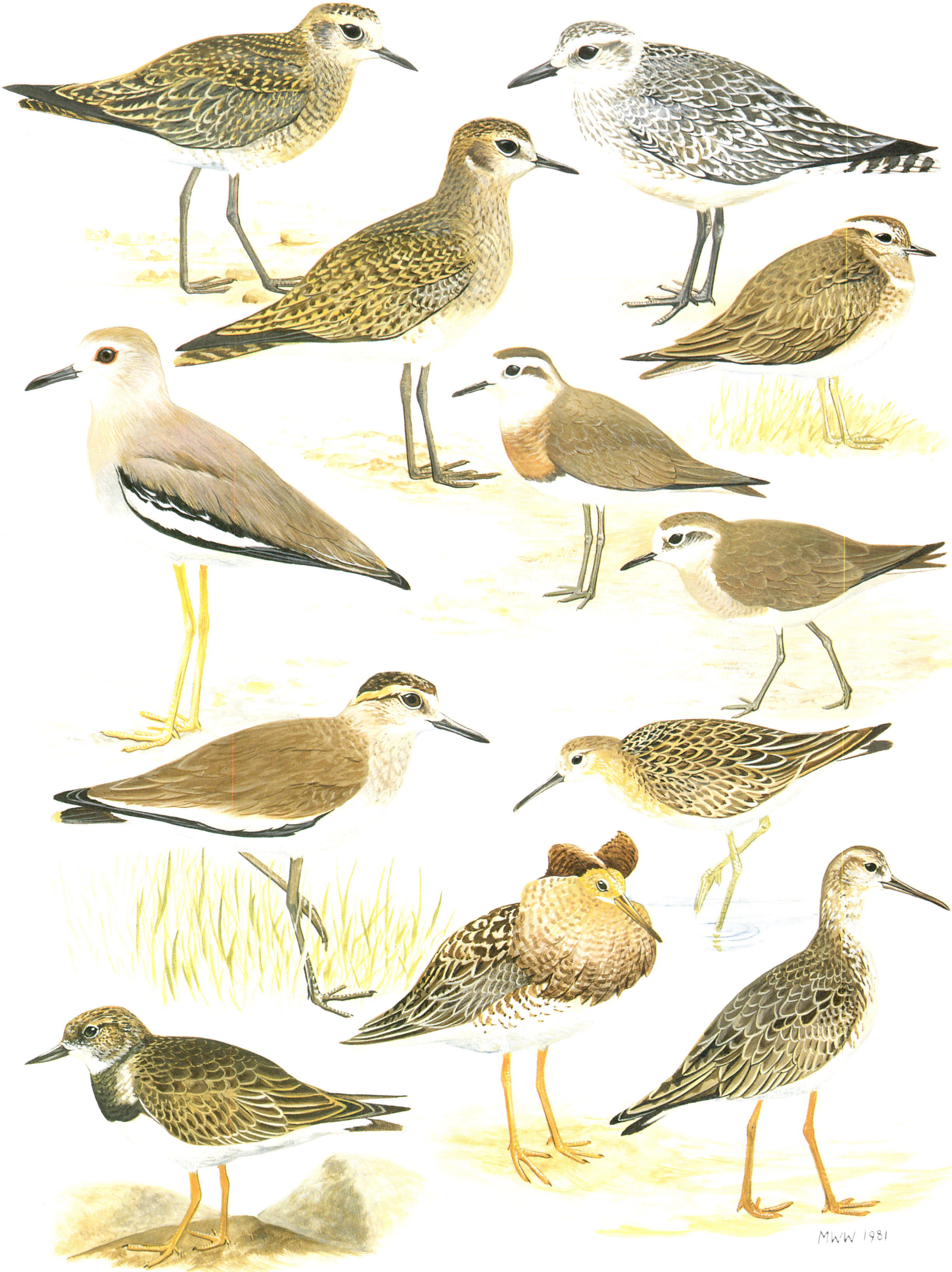
MWW 1981

Plate 16

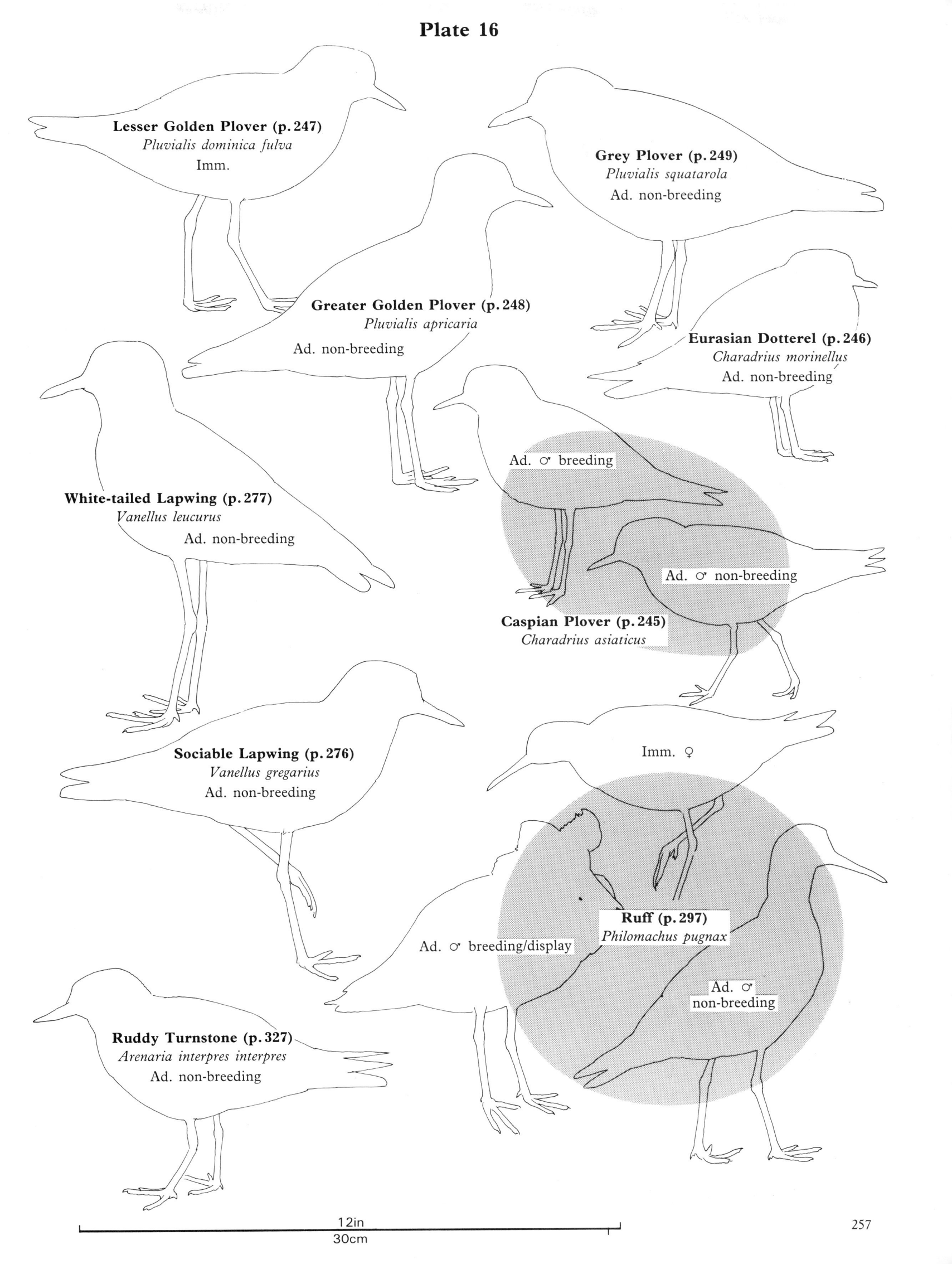

Lesser Golden Plover (p. 247)
Pluvialis dominica fulva
Imm.

Grey Plover (p. 249)
Pluvialis squatarola
Ad. non-breeding

Greater Golden Plover (p. 248)
Pluvialis apricaria
Ad. non-breeding

Eurasian Dotterel (p. 246)
Charadrius morinellus
Ad. non-breeding

Ad. ♂ breeding

White-tailed Lapwing (p. 277)
Vanellus leucurus
Ad. non-breeding

Ad. ♂ non-breeding

Caspian Plover (p. 245)
Charadrius asiaticus

Sociable Lapwing (p. 276)
Vanellus gregarius
Ad. non-breeding

Imm. ♀

Ad. ♂ breeding/display

Ruff (p. 297)
Philomachus pugnax

Ad. ♂ non-breeding

Ruddy Turnstone (p. 327)
Arenaria interpres interpres
Ad. non-breeding

12in
30cm

floods, moving to ponds on drier ground, and Zimbabwe, where move only short distance at times of heaviest rain. However, 10 birds in N Tanzania (Serengeti) Feb were presumably migrants.

Food. Beetles (including weevils), mantids, mutillid wasps and other insects; worms, molluscs, crabs and other small crustaceans; small fish.

Breeding Habits. Monogamous; territorial. Solitary nester in most of range but loose colonies in Zaïre. In Nigeria nests at least 400 m apart; breeding density South Africa, at limit of range, 1·46 pairs/km of river. Territory on Ruaha R., Tanzania, consisted of sandy island 91 m × 45 m (Reynolds 1968).

Conspecifics excluded from territory; some birds of other species tolerated, e.g. Grey Pratincole *Glareola cinerea* and African Skimmer *Rynchops flavirostris*, but most driven away. Especially aggressive towards Egyptian Plover and Three-banded Plover *Charadrius tricollaris*. Usually calls loudly on appearance of human intruder and flies about overhead; in Zaïre whole colony joins in mobbing. Also has Distraction Display, with outstretched wings, and false-broods. However, during incubation sitting bird sometimes leaves nest silently, walks a few m, stands still or feeds, mate remaining quietly nearby. Passing African Fish Eagles *Haliaeetus vocifer* and Bateleurs *Terathopius ecaudatus* regularly mobbed, but Vervet Monkeys *Cercopithecus aethiops* and Yellow Baboons *Papio cyanocephalus* only 27 m away from nest ignored. After chicks hatch all birds driven away, even Ring-necked Dove *Streptopelia capicola*.

Nuptial flight recorded (Lippens and Wille 1976) but not described. Copulation noted during incubation period, also just as chicks hatching after pair drove off Three-banded Plover.

NEST: shallow, circular scrape in sand or shingle, diam. 12–18 cm, sometimes unlined but usually lined with small sticks and pebbles; chiefly in open but once in grass near tree trunk (Reynolds 1968), 1–18 m from water's edge; sometimes in bed of dry watercourse. Eggs sometimes partly buried.

EGGS: 2–4 (once 5); mean (32 clutches) 3·1. Laying interval not recorded but chicks hatch at intervals of c. 24 h. Ovato-pyriform; unglossed, shell thin; creamy, pale or brownish buff, stone or clay-coloured, blotched and spotted brown and black, markings sometimes large and coalescent, sometimes a few black or yellow-brown hair lines; undermarkings ashy or mauve. SIZE: (n = 6, Cameroon) 41·8–43·7 × 30·2–31·7 (42·8 × 31·0); (n = 12, Nigeria) 40·6–46·2 × 29·3–31·2 (43·6 × 30·4); (n = 14, southern Africa) 41·2–47·6 × 28·0–31·9 (43·7 × 30·3). WEIGHT: (estimated) 20·4.

LAYING DATES: Senegambia, Feb; Ivory Coast, Feb–May; Nigeria, Feb–May; Cameroon, Dec–Jan (♀ with egg in oviduct May); Chad, (♀ in breeding condition Feb); Zaïre, Dec–Feb, Aug; Tanzania, July–Oct, peaking Sept at end of dry season, (young June); Zambia, July–Oct, peak Sept; Zimbabwe, July–Nov, peak Sept; Mozambique, 'breeds from Sept' (Clancey 1971); South Africa, Oct.

INCUBATION: by both sexes. Period: over 25 days. Amount of time eggs incubated during day varies greatly; in some cases rarely left uncovered, especially in heat of day, but at 1 Zambia nest eggs uncovered for large periods of time in sunlight and ambient temperature of 105°F (Bainbridge 1965); in heat of day birds belly-soaked and wetted eggs but only stayed on nest few min, then resumed feeding and preening; wetted eggs every 15 min. On cloudy days eggs not wetted. Eggs sometimes covered with small fragments of aquatic plants when birds away from nest. Incubating bird pants but remains sitting on eggs, does not crouch over them. At another nest, Zimbabwe, bird incubated for 45 min in heat of day, panting vigorously, then walked to edge of lake, squatted in shallow water and soaked belly. In this posture (**A**) wings and tail held high, unlike in bathing. Bird got up, moved, and squatted several times, then stood up and preened. On this occasion did not return to nest, so objective of belly-soaking presumably to cool itself; later eggs were wetted (Begg and Maclean 1976).

DEVELOPMENT AND CARE OF YOUNG: eggshells carried some way from nest after chicks hatch. Young self-feeding within 24 h, occasionally brooded, tended by

A

both parents. In nest of C/4 1st 2 chicks hatched tended by 1 parent while other brooded 1 newly hatched chick and 1 well chipped egg (Reynolds 1968); no brood division, however, since parents regularly exchanged roles, sitting bird leaving nest to tend older chicks while mate returned to nest to brood.

BREEDING SUCCESS/SURVIVAL: One nest destroyed by elephant. Nests on sandbanks frequented by crocodiles, hippopotami and ungulates seemed likely to be trampled on, but 1 found which was surrounded by recent buffalo tracks except for an untrodded area of *c.* 50 cm around nest, suggesting owner successfully deterred buffalo (Tarboton and Nel 1980).

References
Bainbridge, W. R. (1965).
Reynolds, J. F. (1968).
Serle, W. (1939).

Vanellus melanocephalus (Rüppell). Spot-breasted Lapwing; Spot-breasted Plover. Vanneau d'Abyssinie.

Lobivanellus melanocephalus Rüppell, 1845. Syst. Uebers. Vog., N.–O. Afr., p. 115, pl. 44; Simien Mts, Ethiopia.

Plates 14, 20
(Opp. pp. 241, 321)

Range and Status. Endemic resident in highlands of Ethiopia, from Simien Mts and highlands north of Gondar south in western highlands at least to Shoa District, and in highlands SE of Rift Valley in Arussi and Galla Districts. Locally frequent to common, but because of its extremely restricted range it should be watched closely for any signs of decline.

Description. ADULT ♂: forehead, crown and nape black, feathers of nape elongated, forming short crest; narrow white supercilium from just before to well behind eye. Face, sides of neck and hind neck grey-brown, becoming warmer earth brown on sides of breast and upperparts, including scapulars, tertials and lesser upperwing-coverts. Uppertail-coverts and tail white, latter with black subterminal band, broadest in centre, narrowing towards edge of tail, on outer feathers only *c.* 7 mm. Chin and throat black, centre of breast white with short, dark brown streaks and spots, rest of underparts white. Primaries and secondaries black; inner secondaries with some white on inner webs and bases of outer webs, increasing inwardly. Median upperwing-coverts earth brown tipped white; greater coverts white; alula, lesser and median primary coverts earth brown, greater primary coverts black; axillaries and underwing-coverts white. Bill black, basal third of upper mandible and base of lower mandible yellow; small wattle in front of eye rich yellow; narrow eye-ring dark yellow; eyes yellowish grey; legs and feet pale yellow. Sexes alike. SIZE: wing, ♂ (n = 1) 233, ♀ (n = 3) 232–237 (234), unsexed (n = 3) 233–235 (234); tail, ♂ (n = 1) 104, ♀ (n = 4) 82–95 (89·0), unsexed (n = 4) 92–97 (95); bill, ♂ (n = 1) 25, ♀ (n = 4) 25–26 (25·3), unsexed (n = 3) 21–26 (23·7); tarsus, ♂ (n = 1) 63, ♀ (n = 4) 60–62 (61·0), unsexed (n = 4) 55–65·5 (59·6). No wing-spur but small knob *c.* 3 mm. WEIGHT: unsexed, 199, 228.

IMMATURE: undescribed.
DOWNY YOUNG: undescribed.

Field Characters. A medium-large lapwing with black crown and throat, small yellow wattles in front of eye, and distinctive spotted breast. Isolated from most Ethiopian lapwings by montane habitat; most likely to overlap with Black-winged Lapwing *V. melanopterus* which has plain grey head and throat without wattles, unspotted grey and black breast, and, along highland streams, with Spur-winged Lapwing *V. spinosus* which has white face, mainly black underparts. African Wattled Lapwing *V. senegallus* has black throat surrounded by black and white streaking, but has plain grey breast, no black on crown, long yellow wattles hanging down below eye, longer, yellow bill.

Voice. Tape-recorded (C). Similar to that of other lapwings; has a double note, 'ku-eep' or 'pewit', and a longer call, 'kree-kree-kre-krep-kreep-kreep'.

General Habits. Inhabits highland grassland, moorlands with giant lobelia, giant heath, *Alchemilla* and tussock grass, marshes, streams and damp meadows; partial to cattle pastures and often occurs around domestic stock. Altitudinal range 1800–4100 m.

Normally occurs in pairs or small parties but outside breeding season also found in large flocks. Noisy; not really shy but restless and quarrelsome, often on the move; seldom found with other species. Travels a considerable distance during the day to places frequented by domestic animals. Forages in typical *Vanellus* fashion, with a short run, pause, and examination of ground. Flight even more erratic than other lapwings, and tends to fly closer to the ground.

Some local seasonal movements likely. In locality north of Gondar, a few present all year but large numbers appear during rainy reason (Olson 1976).

Food. Undescribed.

Breeding Habits.
NEST: only example recorded was shallow scrape in patch of grass and moss on 1 m² rock island in shallow moorland pool, in an area of small lakes and giant lobelias.

EGGS: 4 (from same nest). Brownish blue to smoke grey, heavily marked with black.
LAYING DATES: Apr (Bale Mts); also breeds Aug in Shoa district (Urban 1978).

References
Von Heuglin, T. (1869–1873).
Urban, E. K. (1978).

Plates 14, 20
(Opp. pp. 241, 321)

Vanellus tectus (Boddaert). **Black-headed Lapwing; Blackhead Plover. Vanneau coiffé.** Plates 14, 20

Charadrius tectus Boddaert, 1783. Table Pl. enlum., p. 51; Senegal, *ex* Daubenton, pl. 834.

Range and Status. Endemic resident, common across most or all of Sahel zone from Atlantic to Red Sea. Occurs S Mauritania north to 18°8′ N, Senegambia (throughout) and W Guinea-Bissau, through Mali (common except in south), extreme N Ghana (Tumu), Niger (north to Agadès), Nigeria (sparse resident, south to Kaduna and Bauchi Plateau), N Cameroon (and vagrant on Adamawa Plateau) and Chad (common, extending north to Ennedi) to Central African Republic (Bamingui-Bangoran Nat. Park, rare), central and southern Sudan, Ethiopia (except highlands and Danakil desert) and W Somalia. In E Africa found in N Uganda south to Murchison Falls Nat. Park and Karamoja, NW Kenya south to Kerio Valley and Baringo, and, after gap, in E Kenya west and south to Isiolo, Kajiado, Tsavo West Nat. Park and Malindi. Single record Tanzania (Ardai Plains, Arusha). 1 extralimital record, from Wadi Araba, Israel-Jordan border.

Description. *V. t. tectus* (Boddaert): Senegambia to Ethiopia, Uganda, NW Kenya south to Baringo. ADULT ♂: forehead white, lores, crown and short crest black; white stripe from behind eye round nape, continuing forward as narrow rim under eye broadening below lores into white chin and throat. Cheeks and broad band round hindneck, down sides of neck and round throat black, continuing as broad stripe down centre of breast, lowest feathers long and loose. Upperparts, including scapulars, tertials and most of upperwing-coverts light earth brown, paler on upper mantle where it meets black hindneck. Uppertail-coverts and tail white, latter with broad subterminal black band, broadest in centre, narrowing towards outer feathers, and narrow white tips. Sides of breast pale earth brown, becoming whitish where they meet the black neckband; rest of underparts white. Outer half of primaries and secondaries black, basal half white; primary coverts and outer half of greater coverts white; axillaries and underwing-coverts white. Bill dull red, distal third black; small red wattle above lores; eyes yellow; legs and feet carmine. Sexes alike. SIZE: (8 ♂♂, 8 ♀♀) wing, ♂ 188–197 (191), ♀ 184–198 (190); tail, ♂ 82–91 (85·9), ♀ 80–91 (85·5);

bill ♂ 23–26 (24·3), ♀ 22–25 (23·3); tarsus, ♂ 56–63 (58·9), ♀ 52–60 (56·6). No spur on wing, small knob *c.* 2 mm. WEIGHT: ♀ in breeding condition, Kenya, 100.

IMMATURE: as adult but crown feathers tipped buff, feathers of upperparts with narrow buff tip and subterminal dark band.

DOWNY YOUNG: top of head, mantle, rump and wings spotted black and yellow; short black stripe down centre of crown; black band round nape; hindneck, cheeks and underparts white. Bill dark grey; eye dark; legs and feet grey.

V. t. latifrons (Reichenow): Somalia, Kenya east of the Rift. Smaller (wing, 170–175); white band on forehead broader. WEIGHT: ♂ (n = 6) 113–120 (114); ♀ (n = 6) 99–120 (113).

Field Characters. A rather small, dry-country lapwing with a distinctive head pattern combining red wattles beside white forehead, black crown and crest separated by narrow white line from black face and neck. Black streak down middle of breast unique in lapwings. Tame and approachable, unlike most species.

Voice. Tape-recorded (217, 264). Like that of Spur-winged Lapwing *V. spinosus* but with less disagreeable tone. On approach of intruder gives a harsh 'kwairr'; on closer approach this becomes a shriller 'kiarr', and birds finally take flight with piercing 'kir-kir-kir'.

General Habits. Inhabits dry ground, bare or with covering of short grass, in semi-arid regions; found in open desert with annual grass and on grassy plains, also in bushed grassland and in open, bare areas in thorn scrub. Occurs near rivers, ponds and waterholes, not on mud but on hard ground nearby; also occurs far from water. Partial to vicinity of human habitation, found on airfields, race courses, polo grounds, football fields and tilled ground, nesting in gardens and near buildings.

Occurs in groups of 6–10, sometimes up to 40. Usually tame and curious, allowing observer to approach within a few m even when nesting, but sometimes mobs intruders even in non-breeding season. Spends day in shade of tree or bush if available; more active in early morning and evening, very active at night, when it feeds and often calls. Wags tail on alighting after short flight.

Chiefly sedentary throughout its range but in Mauritania moves north in rains, and a few local movements noted Mali and Nigeria during the rains. In Ethiopia found mainly below 1200 m, but groups wander to higher ground in Tacazze Valley and Tana Basin, 1500–2000 m, Nov–May (Olson 1976).

Food. Insects, including ant-lion larvae, and snails.

Breeding Habits. Monogamous; breeds singly or in loose colonies; 14 pairs/ha Chad, 7 nests within 300 m Kenya. Pair in garden very tame while nesting (Greenham and Greenham 1975), but colonial birds aggressive, mobbing and dive-bombing intruder (North 1936); at beginning of laying period, birds still without eggs attacked intruder, those with eggs did not.

NEST: shallow depression or scrape in sand or earth, encircled by ring of small stones or gravel, with or without lining; diam. 15–17 cm; depth 2·5 cm; footprints of cattle or buffalo also used. In colony at Garissa, Kenya, 2 types of nests: 'hidden', lined with tiny pebbles, pieces of earth, twigs, dung and grass seeds, with eggs half buried in lining; and 'open', unlined or lined with a few small stones, eggs not buried (North 1936). Of 20 nests, 11 were 'open', 7 'hidden', 2 'hidden' at first, later 'open' (North 1936). Situated in bare area or short grass, including clearing in bush as small as 10 m²; often close to road, footpath or building.

EGGS: 1–6 laid at less than 24 h intervals; Senegal mean (117 clutches) 2·2. Clutches of 5 and 6 probably from 2 ♀♀ laying in same nest (Etchécopar and Morel 1960). Subpyriform; light stone or clay colour to warm buff, sometimes greenish brown or light yellowish buff, irregularly marked with blotches and spots of blackish, dark brown and chestnut, with light grey undermarkings. SIZE: (n = 23, Kenya) 32–37 × 24–25 (34 × 25). WEIGHT: (estimated) 12·3.

LAYING DATES: Senegambia, Jan–Aug, Dec; Mali, Feb–May, Nov–Dec; Niger, Mar; Nigeria, Mar–May, July, Nov; Chad, Mar–Apr, during time of greatest heat; Sudan, Mar–Apr; Ethiopia, Feb–Aug; E Africa: Region A, Jan, Mar, June, Sept, Dec, mainly in dry months; Region D, Jan, Mar–Oct, peaking at height of main rains.

INCUBATION: begins with 1st egg; by both sexes. Period: 26 days. During heat of day incubating bird squats on tarsus beside nest and shades eggs rather than sitting on them, fluffing out body feathers and extending wings. Eggs may be left for fairly lengthy periods in early morning, evening, and at night but never unattended during heat of day. Sitting bird often gapes to lose heat; often calls to mate, who stands in shade nearby. No ritual at changeover; incubating bird moves off nest and pecks at ground, mate comes and sits on eggs. Human intruders sometimes tolerated, but incubating bird may slip off nest at first sign of intruder, walk short way and start displacement feeding. When herd of domestic animals approaches nest, birds stand firm and shriek loudly, pecking at legs of animals coming too close.

DEVELOPMENT AND CARE OF YOUNG: hatching interval less than 24 h; adults make soft clucking noises to newly hatched chick; eggshells carried some way from nest. Chicks leave nest within 24 h of hatching and are active up to 25 m from nest; run like adults, with short run, stop, pause, run on again. Nest area deserted within 24 h of hatching of last chick. Some adults with young allow close human approach, others call loudly, walking around with head held low, crest flat on back, displacement feeding; false copulation may also take place (North 1936).

BREEDING SUCCESS/SURVIVAL: nests trodden on by domestic animals; a few nesting adults caught with snares laid around eggs. Seen mobbing Red-necked Buzzard *Buteo auguralis* (Mundy and Cook 1972).

References
Greenham, R. and Greenham, L. M. (1975)
North, M. E. W. (1936).

Plates 14, 20
(Opp. pp. 241, 321)

Vanellus armatus (Burchell). Blacksmith Lapwing; Blacksmith Plover. Vanneau armé.

Charadrius armatus Burchell, 1822. Travels, 1, p. 501, note; Klaarwater (= Griquatown), Cape Province.

Range and Status. Endemic resident and partial local migrant; occurs from N-central and SE Kenya (Maralal, Tsavo East Nat. Park) south across central Tanzania to SE Zaïre (extending at least to Upemba Nat. Park), Zambia, Malaŵi, central Mozambique and Angola (south of line from Cuanza R. to Lucuano and L. Cameia), and thence throughout southern Africa to Cape except for extreme desert region of W Namibia. Vagrant Burundi (Katumba). Has greatly extended range in southern Africa in recent years. In Zimbabwe range expanded following introduction of artificial water supplies; in South Africa formerly rare south of Orange R. and Natal, now breeds south to Cape Town. Since 1970 has invaded coastal belt of SE Cape (C. J. Skead pers. comm.). Common to locally abundant; non-breeding flocks of 500 Zambia (Kafue Flats, Liuwa Plain); large resident population Barberspan (South Africa), counts up to *c.* 400.

Description. ADULT ♂: forehead and forecrown white; hindcrown, nape, sides of head and neck, chin, throat and breast glossy black. A broad white patch on hindneck; back and outer half of scapulars glossy black; uppertail-coverts, rump and tail white, tail with broad terminal wedge of black, broadest on central feathers (*c.* 35 mm), narrowest on outer ones (*c.* 12 mm), which are narrowly tipped white. Underparts below breast white. Primaries black with narrow black bases; secondaries black with broader white bases, the white increasing in extent on inner feathers until innermost mostly white. Inner scapulars, tertials and wing-coverts light grey, secondary coverts tipped white; alula and primary coverts black, lesser primary coverts mottled grey. Axillaries and underwing-coverts white. Bill, legs and feet black; eyes ruby-red. Sexes alike. Leucistic individuals sometimes occur. SIZE: (7 ♂♂, 7 ♀♀) wing, ♂ 196–215 (207), ♀ 200–219 (207); tail, ♂ 83–92 (87·7), ♀ 79–91 (83·6); bill, ♂ 26–29 (27·7), ♀ 25–30 (27·6); tarsus, ♂ 66–81 (72·1), ♀ 71–77 (73·0); wing-spur, ♂ (n = 7) 8–13 (10·3), ♀ (n = 6) 6–11 (8·5). WEIGHT: ♂ (n = 13, Botswana) 120–194 (162), ♀ (n = 11, Botswana) 150–213 (167); unsexed (n = 47, Kenya) 138–197 (164), (n = 264, South Africa) 114–211 (156).

IMMATURE: as adult but forehead and crown buff or brownish with a few black feathers, chin off-white, buff tips to black feathers, dark tips to some grey feathers of upperwing.

DOWNY YOUNG: crown and broad stripe down centre of back dark brown; rest of upperparts mottled dark brown and buff; white collar round neck; underparts white, breast mottled black and buff.

Field Characters. A medium-sized lapwing with distinctive combination of white crown, black face and neck, white patch on hindneck, black back and grey upperwing-coverts. Long-toed Lapwing *V. crassirostris* shares white crown and black nape but has white face and throat, black hindneck, grey-brown back and large white patch on wings. Spur-winged Lapwing *V. spinosus* has reverse head pattern, with black crown, white face and neck; also grey-brown back and upperwing-coverts.

Voice. Tape-recorded (15, 35, 38, 74, B, C). Best known call is 'tink-tink' or 'click-click' of alarm, likened to noise of blacksmith's hammer striking anvil; repeated at varying intensities and frequencies according to degree of anxiety. Given at low intensity by bird making aggressive bowing movements, high intensity while flying up and over intruder. When young recently hatched, a screech added to high-intensity tinking; during incubation, other calls directed at intruder are repeated: 'chi-uk, chiuk', 'whit-it', and 'ki-ko-i, koi, koi'. Territorial call (uncommon), 'cheu-whic-i-u'; excitement call during pair-formation or after chasing other birds, 'kerweek'; a twitter given during copulation; low 'chuck-chuck' by parent as chick pips egg and low 'tink-tink' to newly hatched young; at higher intensity, young freeze in response to call.

General Habits. Inhabits dry ground near water, occurring beside rivers, lakes, dams, ponds, lagoons, waterholes and sewage farms. Also open dry or marshy grassland near swamps and on mudflats and soda flats, and seasonally occupies temporary vleis and shallow flood-plains. Forages on ploughed land, in fields of vegetables and lucerne, and where cattle grazing and dung abundant.

Usually occurs singly or in pairs, but in non-breeding season sometimes in large flocks of up to 100 (500 recorded). An alert, noisy and nervous bird, flushing with loud warning cries at first sign of intruder; flight dashing and panicky. Stands on small hummock of earth or grass above ground level to obtain better view of surroundings. Forages like other lapwings (makes short run, stops, leans body forward, points at prey with bill, steps forward quickly and stabs at ground), but also has other methods. Walks forward very slowly, then makes rush of 3–4 steps ending with violent stab at ground; sometimes, especially on wet surfaces, forward leg trembles rapidly as foot touches ground; occasionally feeds from surface of shallow water while wading. Hunts for insect larvae in cattle dung, turning over cowpat with sharp lateral movement of bill; after seizing grub drops it and picks it up head first before swallowing it (van Someren 1956). Bathes by wading into water to top of legs, dipping head into water (sometimes shaking it from side to side) and waggling tail. Also splashes water under wing with leg, flutters wings to splash water over whole body. Then leaves water and preens at edge. In evening flies from one feeding ground to another. At Barberspan, roosts in large flocks on islands.

Mainly sedentary but subject to local movements in some parts of range. In Zaïre makes considerable movements after breeding season; in Malaŵi local movements likely, due to flooding of habitat; numbers in Zimbabwe fluctuate seasonally, influx of birds in dry season to suitable habitats; at Barberspan, influx in winter, possibly of birds that bred locally.

Food. Small molluscs, crustaceans, worms and insects.

Breeding Habits. Monogamous; territorial. Nests widely spaced at South African sites, usually 400 m apart, sometimes only 200 m, minimum territory size c. 3·6 ha (Hall 1959). Pairs form while in non-breeding flocks. Courtship behaviour often involves 3–5 birds, with much excited action and calling. Behavioural components include: (1) 1 bird running closely followed by another, (2) excited flying around, (3) running on ground in upright posture, (4) long-distance flying in which pair take off calling and fly out of sight, later returning to flock, (5) single bird pivots on one spot in upright stance, sometimes bowing, or turning head from side to side, (6) several birds peck violently at ground or run rapidly about, sometimes picking up object in bill and shaking it; later, pair may copulate. Copulation also occurs while birds still in flocks, just before breeding season, and continues after birds paired off, up to time that eggs are laid. ♀ walks in front of ♂ who is in upright position with neck stretched, then crouches displaying white rump and back feathers; ♂ mounts for c. 5 s without flapping or extending wings, during which twitter call given (by which sex?).

Intraspecific aggression rare in flocks but some evident just before breeding season, including bowing, standing in stiff-necked upright position, running in hunched threat position (body horizontal, bill pointed towards opponent, head and neck extended), and chasing in air. In South Africa no aggression shown to conspecifics while nesting; group interactions similar to those of courtship, including copulation, noted when intruding pair landed near nest of incubating pair (Hall 1964). However, some intraspecific territoriality in Kenya, birds running at each other giving territorial call 'cheu-whic-i-u' (van Someren 1956).

Very little interspecific territoriality with Spur-winged Lapwing *V. spinosus* in areas of breeding overlap, but very aggressive to birds of other species on its territory, including gulls, other waders and even passerines; larger birds attacked from air or on ground in hunched position, smaller ones sometimes dispersed simply by fluffing out feathers.

NEST: usually a shallow depression in ground, with variable lining of small pebbles, earth pellets, mud flakes, dung, bark and small pieces of wood (including matchsticks), dry grass, plant stems, dead water-weeds and other debris, (occasionally unlined). Footprint of Wildebeest *Connochaetes taurinus* (and doubtless other animal tracks) also used as site, and 1 nest was unlined scrape in middle of cowpat. Some nests constructed, e.g. pile of sticks, platform of pebbles, mound of grass; nests in damp grassy sites often with perimeter woven of grass stems. Situated on dry ground or in short grass close to water, sometimes at edge of water or on islet, occasionally up to 60 m away. Site must afford incubating bird all round view, i.e. not be obscured by vegetation, bank or other objects. For 2nd brood uses new site, 50–150 m away in same territory.

Nest-scrape made by both sexes, by pressing breast against soil and making twisting movements, with tail raised and wings slightly spread. Several scrapes often made, 1 eventually chosen for nest. Scrape-formation and nest-building accompanied by ceremonial crouching postures and side-throwing of nest material.

EGGS: 1–4; mean (130 clutches) 3·1. Exceptional clutches of 5 and 6 probably from 2 ♀♀ laying in same nest. Pyriform; cream, buff, sandy buff, light grey or olive, fairly evenly blotched and spotted black, dark brown or dark grey-green over slate grey spots. SIZE: (n = ?, E Africa) 39–41 × 27–30; (n = 19, Zimbabwe) 37·4–41·9 × 27·4–30·7 (39·0 × 29·2); (n = 100, southern Africa) 37·0–42·5 × 27·0–30·7 (39·6 × 29·0). WEIGHT: (estimated) 16·5. Double brooded; interval between hatching of 1st clutch and completion of 2nd clutch 50–55 days.

LAYING DATES: E Africa: Kenya, Mar–Sept, Dec–Jan; Region C, Jan–Feb, Apr–Sept, mainly early in dry season; Region D, all months except Feb, peaking May–July in cool dry season just after long rains; Malaŵi, May; Zambia, Mar–Oct; Zaïre, May–Oct (dry season); Angola, Aug, probably Apr; Namibia, Feb, May–July, Sept–Nov; Botswana, Mar–Dec; Zimbabwe, all months, mainly Apr–Sept; South Africa: S-central Transvaal, July–Sept, exceptionally Apr, Dec; Barberspan, all months except Feb, June, peaking Sept; Natal, June–Nov; Cape Province, Feb, Apr–Nov, peaking July–Aug.

INCUBATION: begins with last egg; by both sexes. Period: 23–31 days. Nest-reliefs occur every 20–80 min, occasionally 90 min or more. Changeover initiated either by incubating bird flying or walking away (nearby partner walks rapidly to eggs) or by relieving bird flying or walking close to sitter; sitting bird sometimes side-throws nest material while walking away after being relieved, and incubating bird sometimes side-throws material on or near nest. In heat of day stands over eggs or newly hatched chicks rather than brooding them. Incubation behaviour seen to change in response to changing air and soil temperature (Brown 1972); bird incubated in cold early morning, left eggs for long periods when it became warm, and in heat of day partly buried them under animal droppings and dry grass. Incubating pair indulged in ceremonial scraping and nest-building behaviour (Hall 1964).

Sitting bird reacting to person jumps up and walks rapidly away, head low to ground, or flies off without calling. Usually returns quickly when person out of sight, e.g. in car, but even car sometimes treated with suspicion, birds circling and examining it. If observer still visible but at considerable distance bird returns very slowly, stretching neck high to look around, false-preening, scratching under wing with foot, displacement-feeding. Distraction or threat behaviour towards humans rare during incubation, though dogs mobbed.

DEVELOPMENT AND CARE OF YOUNG: weight of chicks at hatching (n = 6) 11·2 ± 0·6; fully fledged at 41 days. Eggshells removed soon after hatching. Young leave nest within few h of hatching; 1st young hide close to nest until last egg hatched. Young stay very close to parents for first few days, not left alone for more than few min, parents maintain contact with low 'tink' call. Parents share brooding; brooding reliefs occur at intervals of 3–97 min; young freeze on approach of predator. After few days young move considerable distance from nest but remain with parents within territory; young of 1st brood remain with parents during incubation of 2nd brood, only driven away when chicks of 2nd brood hatch, at age of 2·5 months, having been fledged for 30 days. Parents swoop at human intruder with tinking call, land close by and make aggressive bowing movements, displacement feed, false-brood, and tear at leaves of plants; dive-bomb dogs with loud shrieks, try to distract them with false-brooding. Attacks decrease in intensity as young become older and more mobile. In response to attack on hatching egg by safari ants *Dorylus*, parent carried egg away from nest and dropped it in water. Ants then left nest, which contained 3 cracked eggs (Reynolds 1984).

BREEDING SUCCESS/SURVIVAL: in 10 nests in Nakuru Nat. Park, Kenya, 40% of eggs hatched; known causes of egg loss mainly flooding during rainstorm, secondarily destruction by automobiles and abandonment. In South Africa nests waterlogged, eggs taken by Pied Crow *Corvus albus*, gulls, coots and jackal, eggs and young trampled by cattle. Chicks eaten by safari ants. Adults apparently distasteful since aposematic behaviour (maximizing conspicuousness to predator) successful in repelling 6–8 attacks by Lanner Falcon *Falco biarmicus* (Thomas 1983). Birds responded to attack by running into water, turning to face falcon, spreading wings downwards and forwards to show bold plumage pattern, all the time calling loudly; when falcon flew low (1 m) overhead they merely ducked.

References
Brown, L. H. (1972).
Hall, K. R. L. (1959, 1964).
Reynolds, J. F. (1984).
Thomas, D. H. (1983).

Plates 14, 20
(Opp. pp. 241, 321)

Vanellus spinosus (Linnaeus). Spur-winged Lapwing; Spur-winged Plover. Vanneau éperonné.

Charadrius spinosus Linnaeus, 1758. Syst. Nat. (10th ed.), 1, p. 151; Egypt.

Range and Status. Africa, Israel, Jordan, Iraq, Syria, Turkey, Greece.

Resident, local intra-African migrant, and possible Palearctic migrant. Breeds Egypt from Nile delta, Wadi Natrun and Fayoum along Nile to Sudan where widespread except in desert. Found throughout Ethiopia, Kenya and Uganda but no breeding records Tanzania, where records believed to refer to wanderers (Britton 1980). Very common on Rift Valley lakes of W Uganda and E Zaïre, becoming uncommon to rare in Rwanda and Burundi, south to delta of Ruzizi R. and north shores of L. Tanganyika. Elsewhere in Zaire confined to dry country of Uelle District. Occurs throughout W Africa except in forest, extending into the Sahel zone; in Mauritania found north to c. 18°N. Common to abundant nearly everywhere. 4 records on Red Sea coast from Hurghada to Port Sudan, possibly migrants.

Description. ADULT ♂: forehead, lores, crown down to level of eye and nape black with slight gloss, nape feathers somewhat elongated. White patch from base of bill and below eye down sides of neck and around hind neck, also continuing as narrow crescent from lower neck onto breast. Mantle, back, scapulars, tertials, lesser and inner median upperwing-coverts grey-brown; uppertail-coverts and rump white, tail black. Chin, throat, breast, upper belly and lower flanks glossy black; upper flanks, lower belly and undertail-coverts white. Remiges black; white band across upperwing from wrist across outer median to greater coverts; primary coverts black, lessers mottled white; alula black, much of inner webs white. Axillaries and underwing-coverts white. Bill, legs and feet black; eye crimson to reddish purple. ADULT ♀: like ♂ but eye somewhat duller and browner. SIZE: wing, ♂ (n = 18) 201–220 (209), ♀ (n = 11) 193–206 (202); tail, ♂ (n = 12) 89–98 (92·8), ♀ (n = 9) 86–98 (91·7); bill, ♂ (n = 13) 27–32 (28·9), ♀ (n = 13) 26–30 (28·3); tarsus, ♂ (n = 13) 66–78 (70·7), ♀ (n = 13) 66–72 (69·1); wing-spur ♂ (n = 11) 8–12 (10·3), ♀ (n = 7) 5–10 (7·0). WEIGHT: (Kenya) ♂ (n = 4) 142–175 (161), ♀ 142, 177; unsexed (n = 6) 127–159 (148).

IMMATURE: like adult but feathers of crown, upperparts, and upperwing-coverts narrowly edged buff, black of underparts duller. Eyes dark brown to brownish crimson.

DOWNY YOUNG: crown and upperparts cinnamon-buff mottled grey, with black streaks on forehead, crown, upper back, sides of lower back, wings and base of tail; white collar on hind neck bordered above by black band; underparts buffish white. Bill blue-grey or dark grey, tip dark horn-brown to black; eye brown; feet greenish grey to olivaceous lead grey.

Vanellus spinosus

Field Characters. A medium-sized lapwing distinguished by black crown and throat contrasting with white face and neck, and lack of face wattles. Similar-sized Blacksmith Lapwing *V. armatus* has black back, black and grey wings and reverse head pattern, white crown contrasting with black sides of head and neck.

Voice. Tape-recorded (10, 62, 73, C, 217, 264). Contact and territorial call, a loud, screeching 4-note (sometimes 3–5 note) phrase, 'ti-ti-TURR-it', with accent usually on 3rd note, also rendered 'did-he-do-it', 'charadlio', or 'di-dridri-dri-drit'. Alarm, a high-pitched 'tick' or 'kick', very like that of Blacksmith Lapwing but less metallic. Birds threatening predator near nest gave continuous twittering and chipping; twittering call during copulation, and ♂ also squawks; cooing sounds made during distraction display.

General Habits. Occurs in wide variety of waterside habitats: bare ground by lakes and rivers, mudflats, sandflats, shortgrass meadows, dry burnt grassland, rice fields and other cultivation, both dry and flooded, soda flats by alkaline lakes, sandy and gravel beaches, dunes, occasionally on open shore (Kenya). Tends to prefer dry ground but also found in ponds and seasonal marshes; seldom far from water.

Occurs in pairs or parties of up to 15, occasionally up to 50 or even 200 (Mali). Flocks of 5–15 seen to fly in formation and land at same time (Dragesco 1960). Flies strongly but never for great distance. Generally shy and vigilant but sometimes allows close approach. When first disturbed on ground runs a short way, stops and bobs body up and down, wags and depresses or partly fans tail; when observer too close flies up and wheels overhead calling loudly and alerting neighbourhood to presence of intruder. At rest, often stands with head hunched into shoulders. Forages mainly at edge of water, taking a few quick steps and stabbing at prey. Also flushes prey from ground with foot: stands on 1 leg, moves other foot rapidly 5–6 times back and forth over ground, *c.* 2–3 cm above it; then runs forward 6–7 steps and repeats procedure (Clifton 1972). Grasshoppers, other insects and a small lizard were flushed and caught in this way. The lizard, caught by its body, dropped its tail; bird ate body then tail. Probably forages at night, when flies from one feeding ground to another, and frequently flies around calling.

African breeders mainly sedentary, e.g. in Nigeria, where not affected by changing water levels. In Mauritania and Chad some birds move north into drier country during the rains. Present all year in Akagera Nat. Park, Rwanda, but commonest Aug–Nov. Appears irregularly in certain parts of Uelle District, Zaïre; some local migration in E Africa. In Egypt, wanders somewhat in winter. Palearctic birds leave northern breeding grounds in winter, some may reach Africa. Possible autumn migrants noted N end of Suez Canal but no proof of influx into Nile delta. However, birds recorded from Hurghada and other points on Red Sea coast possibly come from Middle East via Gulf of Aqaba and coast of Sinai, where small passage noted in autumn.

Food. Chiefly insects and their larvae, especially beetles (Tenebrionidae, Carabidae, Silphidae, Dytiscidae, Dryopidae, Gyrinidae), also grasshoppers, flies, midges (Chironomidae), termites, ants, spiders, centipedes, millepedes; also some crustaceans, molluscs, small lizards and a few grass and other seeds (Gramineae, Convolvulaceae, Papilionaceae).

Breeding Habits. Monogamous; flocks break up into pairs at beginning of breeding season. Pair-bond maintained throughout year and for several years, perhaps for life. Territorial; territory occupied either for whole year or for breeding season only. Nests singly or in loose colonies or associations. 30 nests on 1 sandspit, Egypt; in Uganda, 1 pair every 200 m each side of track for 1·6 km (breeding density on strip of Greek coast 1 pair/2·6 ha).

Most other birds driven from territory, especially waders, e.g. Ringed, Kittlitz's and Kentish Plovers *Charadrius hiaticula*, *pecuarius* and *alexandrinus*, Egyptian Plover *Pluvianus aegyptius*, Marsh Sandpiper *Tringa stagnatilis* and Little Stint *Calidris minuta*. Like Blacksmith Lapwing, has particular antipathy towards Common Stilt *Himantopus himantopus*. Wagtails *Motacilla* either driven off or ignored. Very aggressive towards African Wattled Lapwing *V. senegallus* nesting nearby on lake-shore, Rwanda, with continuous pursuits on ground and in air, Spur-winged Lapwing dominant in all observed encounters (Vande weghe and

Monfort-Braham 1975). However, in area of breeding sympatry with closely related Blacksmith Lapwing at Amboseli, Kenya, low level of aggression shown (Reynolds 1980). Blacksmith pair twice landed in territory of incubating Spur-winged; on first occasion Spur-winged aproached them with forward threat crouch; the Blacksmiths ignored it and wandered away. On second occasion Blacksmiths landed near nest, sitting Spur-winged went into threat crouch at nest, then flew towards the Blacksmiths and landed just in front of them. Neither species displayed, and the Blacksmiths flew off.

Though territories distinct, boundary disputes not common, and conspecifics often tolerated in territory. In dispute with neighbours birds bob heads, puff out head-feathers, adopt high-upright posture (standing tall with neck stiff), run at each other and may fight briefly, striking with wings. Loser adopts appeasement position with legs bent, head lowered to ground, wings raised and held half open above body. Victorious pair often copulates. Much loud calling accompanies such encounters.

In courtship display ♂ adopts horizontal posture, head drawn into shoulders and level with back, and circles ♀ with slow, stiff, deliberate steps for many min. This leads to mating ceremony in which pair stand side by side, bob heads, turn to face each other; ♂ circles ♀ in high-upright posture, ♀ stretches forward, ♂ runs to her, jumps on her back and copulates. After copulation ♂ often resumes circling. Other copulatory behaviour similar to that of Blacksmith Lapwing (Hall 1965). Copulation also takes place during scrape ceremony, which likewise similar to that of Blacksmith Lapwing, performed by either bird: slow, crouching walk followed by pressing of breast to ground, tail elevated, scraping movements with feet, pecking at ground and side-throwing of material. ♂ takes lead in forming several scrapes, ♀ follows him and chooses one for nest.

For further examples of displays and postures used during courtship and aggressive activity see Cramp and Simmons (1983).

NEST: shallow scrape, unlined or lined with dry grass, pieces of reed or other plant material, dung, bits of earth and other debris, sometimes with rim of earth and small shells; natural hollow in rock lined with earth or pebbles; on mudflat, sometimes a fairly substantial structure of dead grass and reed stems. Interior diam. 9 cm, exterior 10–11 cm. Situated in open on sandy riverbed, mudflat or other bare, dry ground, usually near water, also on small sandy or rocky islet in lake or river. 14 nests, Uganda, located 6–75 m from water's edge. Sometimes placed in dry grassland near reeds, and 1 recorded built on pieces of elephant dung (Verschuren 1978); another placed on shore among small boats in fishing village. ♂ makes several scrapes, ♀ selects one.

EGGS: 1–5, usually 3–4; laying interval varies, perhaps with latitude; said to be 2 days, perhaps sometimes 1 (Glutz et al. 1975), but longer in Uganda where clutch of 4 contained 1 fresh egg, 2 half incubated, 1 almost ready to hatch (McInnis 1933). Pyriform or pointed oval; smooth, not glossy; deep cream, creamy yellow, pale umber, buff, putty, stone, clay or olive, thickly spotted (sometimes blotched or lined) with black and sepia over grey, pale purple or lilac. SIZE: (n = 36, Egypt and Nile Valley) 37·5–43·0 × 27·0–29·0 (n = 14, Uganda) 33·9–41·7 × 22·2–28·3 (36·0 × 26·6); (n = 13, Nigeria) 36·4–39·1 × 27·0–28·6 (37·7–27·6). WEIGHT: (estimated) 16·4. Double brooded; also, replacements laid after egg loss.

LAYING DATES: Egypt, Apr–Aug; Sudan, Apr; Senegambia, all months except Sept–Oct; Guinea, Mar; Mali, Feb–July; Niger, June; Nigeria, Mar–May, during rains (chick Aug); Chad, Jan–Feb; Ethiopia, Jan, Mar–Sept; E Africa: Region A, Jan–Oct, peak Apr; Region B, Feb, Apr, Aug; Region D, Jan–May, Aug–Sept, Nov–Dec, peak May; Region E, Apr (peaks indicate laying in main rains); Zaïre, Apr–June; Rwanda, Sept–Nov.

INCUBATION: partial brooding begins with 1st egg, continuous brooding with last egg; by both sexes, though ♀ does somewhat more. Period: 22–24 days. Nest-relief performed without ceremony, sometimes with side-throwing by one or both birds of nest material which is later incorporated into nest by sitting bird. Pre-laying behaviour: ♀ brooded on empty nest for 2 h before egg laid, fluffed out feathers in hot sun, ♂ standing nearby under bush; ♂ relieved her for few min, ♀ stood beside him; ♀ indicated desire to return; ♂ got up, removed piece of earth from nest lining, ♀ sat back on nest; 1 h later observer found 1 egg in nest, 75% buried in lining, presumably in hole made by ♂ (North 1936). Birds sometimes wetted underparts just before incubating. Incubating bird sits with head tucked into back, sometimes with head and neck stretched forward, chin on ground.

In reaction to human intruder, incubating bird slips off nest and either circles with loud cries, landing at different spots and taking off again, or simply stands 50–75 m away without calling. No distraction display used. Nest approached in vehicle to within 10–20 m before bird stood up and moved away, sometimes in horizontal crouch posture, with either stiff-legged walk or wobbly 'rodent-run' (Hall 1965).

DEVELOPMENT AND CARE OF YOUNG: young tended by both parents; leave nest immediately after hatching, sometimes led to water by parents. Soon able to feed themselves; brooded and shielded from sun by parents. When predator appears young chicks freeze, older ones run for cover; in 1 case young ran into vegetation and crouched on ground, heads stretched out in front, and did not move even when picked up (Guichard 1947). Young fledged at 7–8 weeks, independent soon thereafter, but sometimes remain with parents until next breeding season when driven away as if they were intruders. In case of 2 broods, ♀ incubates 2nd clutch while ♂ tends chicks of 1st brood.

Reaction of parents to predators much more violent than during incubation. Human intruder dive-bombed, birds flying overhead in agitated manner and calling loudly; also posturing on ground, including grovelling on bellies in sand. In distraction display birds walk rapidly away leaning forward, head between shoulders, rear end elevated, giving cooing call; then sit on tarsi or on ground as if incubating. Raptors, crows and gulls attacked in the air; in reaction to Marabou *Leptoptilos crumeniferus* either stood facing it with open wings and spread tail, or walked away from it in distraction display. See also Cramp and Simons (1983).

BREEDING SUCCESS/SURVIVAL: nests and eggs frequently trampled by cattle and game animals. In Israel, av. of 2 young raised per brood (S. Su-Aretz *in* Cramp and Simmons 1983).

References
Cramp, S. and Simmons, K. E. L. (1983).
Dragesco, J. (1960).
Hall, K. R. L. (1965).

Vanellus superciliosus (Reichenow). Brown-chested Lapwing; Brown-chested Wattled Plover. Vanneau caronculé.

Plates 14, 20
(Opp. pp. 241, 321)

Lobivanellus superciliosus Reichenow, 1886. J. Orn. 34, p. 116, pl. 3; Marungu, west of Lake Tanganyika.

Range and Status. Endemic resident and intra-African migrant. Only proven breeding records are from Nigeria, but probably breeds in narrow band just north of forest from Togo to Uelle District, NE Zaïre. Known from E Ghana, Nigeria north to Zaria and Maiduguri, Cameroon north to Maroua, and Zaïre, where widespread in east. Non-breeding birds occur south of forest in S and SE Zaïre, south to Shaba Province (Kasaji) and NW Zambia (Mwinilunga); 1 record from Congo (Brazzaville); regular also in W Uganda (Rift Valley region), Rwanda, Burundi, and W Tanzania. Has occurred W Kenya (Kisumu area), N Tanzania (Serengeti, 1 record), possibly regular that far east, but single bird from extreme SE Tanzania (Mikindani) probably vagrant. Generally uncommon to rare, though concentrates in some numbers locally in a few areas of non-breeding range.

Description. ADULT ♂: forehead and lores pale rufous, extending back to just above eye; crown and nape black, feathers of latter slightly elongated to form short, bushy, pointed crest (usually laid flat). Cheeks, ear-coverts, sides of neck, hindneck, and upper breast grey; throat pale grey, chin white. Very thin pale line extending back from eye separates black crown from grey face. Upperparts, including scapulars, tertials and lesser coverts, olive-brown, darker on lower back and rump. Uppertail-coverts and tail white, latter with broad black terminal band, broadest on centre, becoming somewhat narrower towards outer feathers; outermost pair pure white. Broad band of chestnut across lower breast, rest of underparts white. Primaries black; secondaries white, outer ones tipped black, the black decreasing inwardly, innermost with some brown, grading into brown tertials. Median coverts pale brown with whitish fringes; greater coverts white; lesser and median primary coverts dark brown, alula and greater primary coverts black. Axillaries and underwing-coverts white. Bill black; eyes yellow or yellowish brown; small yellow wattle in front of eye, extending back to just above eye; legs and feet dark red. Sexes alike. SIZE: (2 ♂♂, 5 ♀♀) wing, ♂ 179, 186, ♀ 178–190 (185); tail, ♂ 72, 72, ♀ 68–72 (70·3); bill, ♂ 21, 21, ♀ 20–22 (20·8); tarsus, ♂ 57, 60, ♀ 52–55 (53·6); unsexed (n = 2): wing, 184, 186; tail, 72, 73; bill, 21; tarsus, 53, 56. WEIGHT: (estimated) 150.

IMMATURE: rufous on forehead more extensive, reaching forecrown and extending back as broad supercilium to sides of nape; crown brown, some feathers with black bases showing through, producing mottled effect; brown feathers of upperparts and wings with pale rufous margins; grey areas of face, neck and breast browner, feathers of lower breast dark brown with whitish fringes and wash of chestnut, producing mottled effect. SIZE: 1 ♀, wing 176, tail 66, bill 21, tarsus 53.

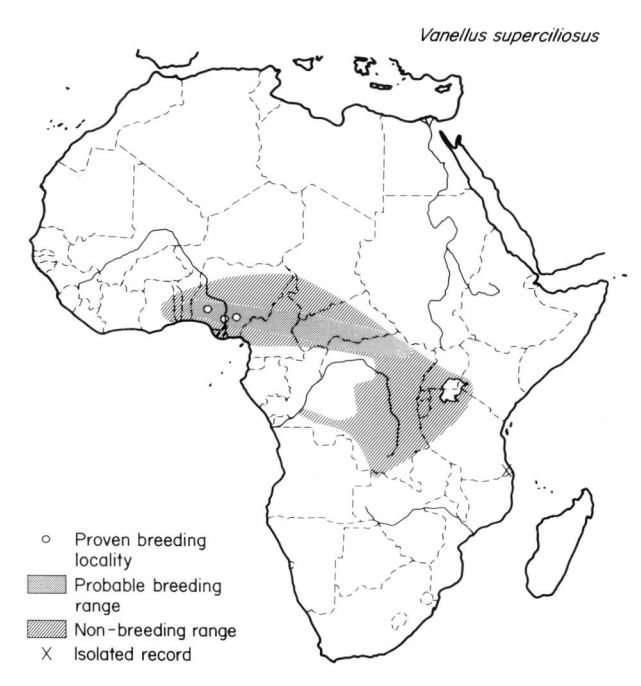

DOWNY YOUNG: upperparts variegated buff and dark brown, dark patch on occiput, white patch on nape. Chin and throat greyish white, breast and belly white, pectoral band brown. Bill blackish; eye brown, well formed eye wattle creamy yellow; feet fleshy brown.

Field Characters. A small lapwing with black crown, rufous forehead, yellow face wattles, grey neck and broad chestnut breast-band unique in lapwings. Similar-sized Lesser Black-winged Lapwing *V. lugubris* has grey neck and breast but blackish breast-band, grey crown, white forehead and no wattles; Spur-winged Lapwing *V. spinosus* has black crown but white face and neck, black underparts.

Voice. Tape-recorded (217). 3 high-pitched, harsh calls in quick succession, given in flight; said to have squeaky quality, like rusty hinge.

General Habits. Inhabits grasslands, open, bare areas by rivers and lakes, and cleared ground, including football fields and lawns. Partial to recently burnt grassland but less dependent on it than other lapwings, occurring also in thicker grass with an admixture of stunted bushes and scrub. In Nigeria inhabits 'orchard bush', and in Rwanda (Akagera Nat. Park) lives in both grassy savanna and wooded savanna with *Acacia gerrardii* and *Dychrostachys cinerea*, in both plains and hills. On migration in Zaïre occurs in cleared areas within forest, and once found in mangrove forest (Rio del Rey, Cameroon).

Occurs in flocks of up to 30 on migration and on wintering grounds; exceptionally 100 (Rwanda). Regularly associates with Lesser Black-winged Lapwing, and mixed flocks do not separate in flight. Probes loose soil for insects. In Rwanda commonest when other lapwings nesting but little interspecific aggression noted. Tends to be wary and restless; flight rapid, intermittent; flies around at night, calling.

A transequatorial migrant; present in known or presumed breeding range in dry season, Nigeria late Nov–early June, Togo Jan, N Zaïre Jan–Apr; passes through E and central Zaïre in Nov–Dec (all adults, presumably heading for breeding areas) and July–Aug (adults and immatures). Occurs in S Zaïre July–Oct, Rwanda Sept–end Nov, E Africa July–Dec, mainly Aug–Oct; single Zambia record Oct. Premigratory accumulation of fat noted in month before birds left Nigerian breeding grounds (Serle 1956).

Food. Ants and other insects.

Breeding Habits. Monogamous; birds already paired on arrival at Nigerian breeding grounds Dec. Territorial; nests placed on freshly burnt ground; where this is restricted, at beginning of burning season, 3 pairs found breeding within 200 m radius, but territory presumably larger when more habitat becomes available (Serle 1956).

Courtship behaviour (Clarke 1936): ♀ ran about chased by 2 ♂♂; all kept stopping to peck nervously at grass. If ♂♂ got too close, ♀ flew ahead a few m. After 45 min 1 ♂ flew off. Roles were then reversed; remaining ♂ ran ahead and appeared to feed while ♀ followed, uttering a low call and crouching before him; ♂ paid no attention, walked away. Finally ♂ ran around her and they mated twice in 5 min; both times ♀ rose and ran off twittering. 3 evenings later frequent mating was again noted, initiative always being taken by ♀.

NEST: shallow circular depression on ground in rather stony red lateritic soil, with thick lining of small red stones and anthill fragments which half bury the eggs, and rim built up above ground level. Diam. 115–220, depth *c*. 25. Situated in grassy savanna or orchard–bush savanna, sometimes near building.

EGGS: 2–4; mean (8 clutches) 3·5. Pyriform; smooth, without lustre; ground colour ranges from cream-buff to warm buff and pinkish brown (exceptionally olive-clay), the reddish colour matching that of surrounding soil; marked with blotches and spots of blackish, chocolate-brown and red-brown, with ashy and pale mauve undermarkings; last egg laid somewhat paler than others. SIZE: (n = 27, Nigeria) 28·0–40·5 × 25·5–33·8 (36·3 × 26·7). WEIGHT: (estimated) 13·4.

LAYING DATES: Nigeria, Jan–Feb; Zaïre (♀ with somewhat enlarged ovaries Jan), probably breeds Dec–June during dry season.

INCUBATION: begins with 1st egg; by both sexes. Changeover quick and without ceremony, lasting 2–3 s. Period: at least 24 days. Incubating bird restless, often rising from eggs, turning around and settling in new position. Frequently reaches out to pick pebbles from ground to add to nest; sometimes leaves nest for few moments to fetch pebble out of reach. When person appears in distance quietly runs off eggs and either stands motionless nearby or displacement-feeds, pecking at ground; sometimes leaves vicinity of nest entirely. Very occasionally (possibly as eggs come close to hatching?) flies about approaching human, calling shrilly, or runs around with wings drooped (Serle 1956).

DEVELOPMENT AND CARE OF YOUNG: downy young said to behave like Northern Lapwing *V. vanellus* (Serle 1956); squats motionless on ground but can also run very fast. Dark patch on occiput makes it visible in the field. Both parents attack human picking up chick, also give broken-wing distraction display. Observer placed young under perforated metal food cover near house, which was close to nest-site; within 5 min young had established vocal contact with parents, after another 5 min 1 adult was sitting beside container (Clarke 1936). Breeding grounds deserted in month before departure; family parties unite to form flocks of up to 50.

References

Clarke, J. D. (1936).
Serle, W. (1956).

Plates 14, 20
(Opp. pp. 241, 321)

Vanellus lugubris (Lesson). Lesser Black-winged Lapwing; Senegal Plover. Vanneau demi-deuil.

Charadrius lugubris Lesson, 1826. Dict. Sci. Nat., ed. Levrault, 42, p. 36; locality unknown = Senegal, designated by Grant, Ibis, 1915, p. 56.

Forms a superspecies with *V. melanopterus*

Rang and Status. Endemic resident and intra-African migrant, with wide range but very patchy distribution. 1 record Senegal, small numbers S Mali, otherwise in W Africa mainly in coastal savannas from Sierra Leone to extreme SW Nigeria, with population north of forest in Guinea-Bissau and Ivory Coast.

Reported from Yankari Game Reserve and Malamfatori in E and NE Nigeria. Apparently absent Chad, Cameroon and Central African Republic, but occurs along N edge of forest in NE Zaïre and in forest–savanna mosaic from Gabon across S–central Zaïre, where locally very common. Widespread E Africa, also on Latham, Pemba, Zanzibar and Mafia Is. Widespread but uncommon Malaŵi, very local and uncommon Zambia, chiefly around L. Mweru, L. Bangweulu and in Luangwa Valley. Widespread and not uncommon Mozambique extending to low-lying areas of E Zimbabwe, where otherwise known only from Wankie Nat. Park and Middle Zambezi Valley (Mupata Gorge). Also occurs from lowveld of E Transvaal to E Swaziland, Zululand and (rarely) N Natal. Disjunct population in W Angola in Luanda, Cuanza Norte and W Malanje (Cangandala Nat. Park). Small numbers in Zambia possibly due to competition with Crowned Lapwing *V. coronatus* (Benson *et al.* 1971).

Description. ADULT ♂: forehead and chin white; crown dark brownish grey, becoming lighter on lores, facial area and nape and medium grey on throat, sides of neck and hindneck. Upperparts, including scapulars, tertials and upperwing-coverts greyish olive-brown with faint greenish gloss, becoming darker on rump; tail white with blackish subterminal wedge, broadest on inner pair of feathers, narrowing to small dark spot on outer feathers; outermost pair completely white. Breast grey with lower border of blackish feathers narrowly tipped white, rest of underparts white. Primaries black, innermost 3 tipped white; secondaries and tips of greater coverts white. Axillaries white; underwing-coverts mottled brown and white. Bill black tinged red at base; eye-ring dark reddish purple; eyes orange-yellow to pale yellow with inner ring of brown; legs and feet purplish black. Sexes alike, ♀ somewhat smaller. SIZE: (5 ♂♂, 7 ♀♀) wing, ♂ 175–182 (178), ♀ 162–179 (171); tail, ♂ 60–67 (64·8), ♀ 58–66 (61·9); bill, ♂ 21–23 (22·2), ♀ 20–22 (20·7); tarsus, ♂ 59–65 (62·4), ♀ 57–60 (59·0). No spur on wing; small knob 2–3 mm. WEIGHT: ♂ (n = 2) 107, 112, ♀ (n = 5) 100–120 (113).

IMMATURE: like adult but with buff tips to feathers of upperparts, including uppertail-coverts and tail, primaries, scapulars, tertials and upperwing-coverts.

DOWNY YOUNG: above, mottled yellowish and blackish brown; white neck-band; below, greyish white, slightly spotted on breast.

Field Characters. A small lapwing with plain grey head, neck and breast, unmarked except for white forehead. Can only be confused with similar Black-winged Lapwing *V. melanopterus*, which is larger, with more white on forehead (extending to forecrown), paler grey crown, browner upperparts, broader black breast-band, different wing pattern. In flight, Lesser Black-winged shows pure white secondaries without black terminal bar, less white on coverts; at rest, shows pure white patch on centre of folded wing; Black-winged has much longer white wing-stripe with upper edge partly barred with black.

Voice. Tape-recorded (10,74, C, F). Normal call a clear, piping 'tlu-wit' or 'thi-wit'; a plaintive 3-syllable whistle 'ti-ti-hooi' often given, especially by migrants on moonlit nights; also has wailing alarm call.

General Habits. Inhabits dry, lightly wooded savanna, open grassland with bushes and scrub, patches of burnt grass in *Acacia* woodland, dry, almost bare, short grassland, and open, dry ground of various types; especially partial to newly burnt grassland and croplands with growth of new grass. Also occurs on ground cleared for cultivation, farmland, pastures and airfields, beside large lakes and rivers, occasionally on seashore (Zanzibar).

Gregarious, forming flocks of up to 20, more usually 5–10. May form loose association with Blacksmith Lapwing *V. armatus*. Forages by making quick, short run and then pecking at ground. Active at night, when feeds and migrates. Not particularly shy; tends to walk away on approach of observer or stand still. On close approach bobs up and down and makes many short, quick runs, and if pressed finally flushes with wailing alarm call, flies short distance in leisurely manner.

Makes regular movements in some areas. In Zaïre migrates in large flocks by night, moving south to north Feb–Apr and north to south Apr–Oct (Lippens and Wille 1976), perhaps indicating movement between northern and southern edges of main forest block. Movement also noted between lakeside habitat and higher plateaux in Upemba Nat. Park (Verheyen 1953). In E Africa seldom, if ever, resident (Britton 1980). Main centres of abundance are W Uganda, L. Victoria basin and on coast, but wanders up to 3000 m in E and W Kenya. Makes regular seasonal appearance in certain areas, e.g. coastal Tanzania and Tsavo East Nat. Park June–Aug, coastal Kenya Apr–July, W Kenya (Siaya) Nov–June (mainly Jan–Mar), L. Victoria basin (Uganda) Sept–May, N Tanzania (Arusha Nat. Park) Jan–Apr. Breeding resident Zanzibar, not migrant (Pakenham 1979). In Malaŵi present every month except Apr and June; no evidence of regular migration but known to wander. Zambia records all Apr–Oct except 1 in Jan, thought to indicate that it is only a visitor (Tree 1969), but habitat largely inaccessible during rains, and it may in fact be sedentary (Benson *et al.* 1971). All Zimbabwe records are July–Nov, so may be dry season visitor. In South Africa, regular non-breeding summer visitor to Kruger Nat. Park, rare in winter; adults arrive accompanied by immatures (Kemp 1974). Movements may be dictated in part by brush fires, causing flush of new grass suitable for nesting.

Food. Small invertebrates, insects and their larvae, especially beetles (including dung beetle larvae), and small grass seeds.

Breeding Habits. Monogamous; territorial.

NEST: a circular scrape or simple depression on ground, lined with a few stalks of dry grass; usually sited on burnt ground among new grass, also on bare spots in grassy areas or on ploughed land; sometimes no nest at all, eggs laid on bare ground.

EGGS: 2–4, usually 3. Pyriform; buff or dull olive-brown heavily spotted with black, dark brown and yellow-brown over grey. SIZE: Uganda (n = ?) 37 × 27; (n = 5, southern Africa) 33·0–35·6 × 25·3–27·0 (34·6 × 26·3). WEIGHT: (estimated) 13·7.

INCUBATION: period, 18–20 days.

LAYING DATES: Sierra Leone, Apr; Nigeria, Sept; Zaïre, May–Nov; Burundi, Aug–Nov; E Africa: Uganda, Jan, Sept; Kenya, Apr; Tanzania mainland Feb, Oct; Zanzibar, Sept; Region A, Apr; Region B, Jan, July–Sept; Region C, Feb, Sept–Nov; Region D, May, Aug–Oct; Region E, Mar (dry season preferred in Regions B, C, and D); Angola (Cabinda) Aug; Zambia, Aug–Oct; Malaŵi, Aug; Mozambique, Sept; Zimbabwe, Aug, Nov; South Africa: Natal and Zululand, June–Oct.

DEVELOPMENT AND CARE OF YOUNG: parents vociferous in defence of young or eggs, running about in agitated manner. Nestlings remain with parents for *c.* 2 months.

References
Irwin, M. P. S. (1976).
Vincent, J. (1934).

Plates 14, 20
(Opp. pp. 241, 321)

Vanellus melanopterus (Cretzschmar). Black-winged Lapwing; Black-winged Plover. Vanneau à ailes noires.

Charadrius melanopterus Cretzschmar, 1829. *In* Rüppell's Atlas, Vög., p. 46, pl. 31; Djedda, Arabia.

Forms a superspecies with *V. lugubris*.

Range and Status. Africa and S Arabia.

Resident and intra-African migrant, with disjunct populations in NE, E and South Africa. In Ethiopia occurs mainly at 1800–3000 m, extending west to lower altitudes in extreme E Sudan. In E Africa confined to highlands of Kenya and N Tanzania, from Trans-Nzoia and Laikipia districts to Mbulu and Mt Kilimanjaro, and intervening Serengeti, Athi, and Ardai plains. Wanderers recorded in Tanzanian lowlands from Taveta, Dar es Salaam and Rukwa (but see Vesey-Fitzgerald and Beesely 1960); sight record from S Tanzania highlands (Njombe) may represent unknown breeding population. Breeds South Africa in highlands of E Transvaal (north to Zoutpansberg), interior Natal, Orange Free State (Bloemfontein) and E Cape Province west to Mossel Bay, once in Zululand (Lake St Lucia); also in W Swaziland; winters mainly on coastal plains from S Mozambique to E Cape; wanderers to W Cape. Common to locally abundant. Post-breeding flocks of over 1000, perhaps up to 10,000, assemble before migration in W Kenya.

Description. *V. m. minor* (Zedlitz): Kenya, Tanzania, South Africa. ADULT ♂: forehead, part of forecrown, loral area and narrow supercilium white; crown and nape slate-grey, hindneck and sides of neck light grey, becoming somewhat paler and browner on face and grading to white on chin and throat. Upperparts, including scapulars, tertials, lesser and all but tips of median coverts earth brown, darker on rump; uppertail-coverts and tail white, latter with broad blackish subterminal band, broadest in centre, narrower at sides. Foreneck and upper breast grey, broad band across lower breast blue-black; rest of underparts white. Primaries blue-black; basal half of secondaries white, outer half black, the black decreasing inwardly until 2 innermost feathers completely white. Median coverts tipped white with narrow black subterminal band; greater coverts white; alula and greater primary coverts black, median primary coverts earth brown. Axillaries and underwing-coverts white. Bill black, tinged red at base; eye-ring purplish red; eyes pale yellow; legs and feet dark purplish red. ADULT ♀: as ♂ but head and neck duller and browner, forehead off-white, white not extending to forecrown; breast grey-brown, band across lower breast dull black without blue gloss, primaries dull black. Somewhat smaller. SIZE: (7 ♂♂, 7 ♀♀) wing, ♂ 202–217 (210), ♀ 97–216 (205); tail, ♂ 70–75 (73·0), ♀ 66–79 (70·9); bill, ♂ 24–27 (26·1), ♀ 25–27 (26·4); tarsus, ♂ 54–64 (60·1), ♀ 55–61 (58·3). No wing-spur; small knob c. 2 mm. WEIGHT: ♂ 163, 170; ♀ 170.

IMMATURE: as adult ♀ but head, neck and breast light brown with hint of grey on crown; forehead light brown; dark band across lower breast narrow and indistinct; feathers of upperparts narrowly edged buffy.

DOWNY YOUNG: top of head and upperparts buff with black spots; white collar on hindneck; face and underparts white, face and flanks washed buff.

V. m. melanopterus (Cretzschmar): Sudan, Ethiopia, Somalia. Larger; wing (♂, ♀) 214–230.

Field Characters. A medium-sized lapwing with white forehead, unstriped grey head and neck, black breast-band and no wattles. Lack of complex head and neck pattern distinguishes it from all resident African lapwings except Lesser Black-winged *V. lugubris*; latter is smaller and has less white on forehead, narrower black breast-band, olive tone to upperparts, blackish legs, no red eye-ring, and different voice; the 2 spp. are largely allopatric both geographically and ecologically, but may sometimes overlap in central Kenya and NE South Africa.

Voice. Tape-recorded (3, 10, 74, C, F). A variety of loud cries and screams, including harsh 'tlu-wit' and 'che-che-che-chereck'. On flushing bird typically gives a series of rapid staccato notes with a tinny, piercing quality, 'taytaytaytaytaytaytay', which may be followed by a double 'too-ghrweet' (C. J. Skead, pers. comm.). Common call, especially when breeding or in flight at night a plaintive, protesting 'titihoya'. Alarm call while bird squatting on ground, a squeaky, grating 'ch,rrr' (C. J. Skead, pers. comm.).

General Habits. Breeds in grassland, typically where grass very short, on highland plateaux, mountain slopes, and at lower elevations on open plains and dry savanna, sometimes in long grass; partial to areas frequented by game animals, also to recently burned ground with carpet of new green grass. In winter also occupies fallow fields, meadows, coastal flats and airfields.

Gregarious, occurring in flocks of up to 50 in non-breeding season. Noisy and restless; when approached, at first runs away, then bursts into flight with loud cries of alarm. Forages like Crowned Lapwing *V. coronatus*; quick run forward, stops with body leaning forward, bill pointed at target, sometimes 1 leg held off ground; strikes at prey with 5–6 rapid jabs (C. J. Skead, pers. comm.).

No movements noted Ethiopia, but altitudinal migrant E and South Africa. In highlands of W Kenya, 2750–3050 m, first birds appear late Jan, numbers increase slowly to Mar–Apr; breeding takes place Mar–July, birds depart Aug–Sept; later seen 1000 m lower down on plains c. 65 km to the south, where they may spend rest of year (Sessions 1967). In South Africa most move from highland breeding areas to coastal plains in non-breeding season, where present especially Apr–Aug; some may remain on breeding grounds, and some non-breeding birds remain on coast during breeding season. Migrates at night as well as by day.

Food. Molluscs, worms, beetles and their larvae, flies and other insects; sometimes small fish.

Breeding Habits. Monogamous; territorial. Sometimes breeds in small, loose colonies. Very aggressive toward human intruders, constantly flying around, swooping down to ground and dive-bombing them, while screaming and giving 'titihoya' call. Drags wings in distraction display; also runs up to intruder with wings raised and fully extended. In post-copulatory display (**A**), birds run side by side with wing away from partner raised, outer half of wing remaining folded (Maclean 1972).

A

NEST: simple scrape in ground c. 7·5 cm diam. lined with straw, grass roots, plant stems, sometimes sheep droppings and bits of cattle dung; situated among grass, especially new grass on burnt ground; also on bare earth on newly ploughed land, where birds quick to lay on soil just turned by tractor (Sessions 1967). Dung evidently a desirable component; nest recorded in hollow scraped out in centre of dry cowpat (Cyrus 1982). Eggs sometimes partly buried, with only round ends showing.

EGGS: 2–4, usually 3. Slightly glossy; pale olive-brown to buff with spots and blotches of black and brown over grey, densest at broad end. SIZE: (n = 44, southern Africa) 36·5–46·5 × 26·7–31·3 (41·5 × 29·4); (n = 3, Kenya) 35–39 × 26–28. WEIGHT: (estimated) 18·2.

LAYING DATES: Ethiopia, Apr–July; E Africa: Kenya, Jan, Mar–July, Sept, Nov–Dec; Tanzania, Sept–Dec; Region A, Apr–July, in main rains, usually early in rains before ground really wet; Region C, Feb, Sept–Nov, mostly in dry season; Region D, Feb–July, Sept, Nov–Dec, peaking Feb–Mar at end of hot dry season when new short grass present after early showers ('grass rains'); South Africa, July–Oct.

References
Cyrus, D. P. (1982).
Sessions, P. H. B. (1967).
Vincent, A. W. (1945).

Plates 14, 20
(Opp. pp. 241, 321)

Vanellus coronatus (Boddaert). Crowned Lapwing; Crowned Plover. Vanneau couronné.

Charadrius Coronatus Boddaert, 1783. Table Pl. enlum., p. 49; Cape of Good Hope, *ex* Daubenton, pl. 800.

Range and Status. Endemic resident and intra-African migrant. Occurs from Ethiopia (common to abundant below 2400 m, absent Eritrea), Somalia (common and widespread), extreme SE Sudan and NE Uganda (Kidepo Valley Nat. Park) to Kenya and Tanzania (reasonably common resident and wanderer, chiefly at middle elevations but up to 3000 m; absent lush coastal areas, extreme W Kenya, and much of central and SE Tanzania). An apparently disjunct population in SW Uganda, extreme E Zaïre, Rwanda and Burundi, where locally common. Found in SE Zaïre (Tanganyika District to Kundelungu Mts) and adjacent parts of Zambia but absent much of NE and NW Zambia, only becoming common in SW. In Malaŵi present apparently only in central provinces and extreme N. Range more or less continuous from central Angola, SW Zambia, Zimbabwe (throughout) and SW Mozambique to Cape; absent from parts of Karoo and highlands of Lesotho.

A common and familiar bird, extending its range due to activities of man in clearing bush and providing open habitat, and especially in overgrazing by domestic stock; now often an indicator of mismanaged land (Irwin 1981).

Description. *V. c. coronatus* (Boddaert): Ethiopia, SE Sudan and E Africa to Angola (except SW), Zimbabwe, and South Africa (central Transvaal, Cape Province south of Orange R.). ADULT ♂ (breeding): forehead, lores, feathers around base of bill and broad supercilium black, extending behind eye to join on nape. Centre of crown black encircled by broad white ring. Chin and throat white, extending up as narrow wedge bordering black feathers at base of bill to just below front of eye; rest of face, neck, and upper breast light grey-brown, paler on lower throat, darker on hindneck. Upperparts, including scapulars, tertials, lesser and median upperwing-coverts greyish earth brown; uppertail-coverts and tail white, latter with broad subterminal black band becoming somewhat narrower on outer feathers. Breast grey-brown, darkening into narrow black lower border; rest of underparts white. Primaries black with white bases; secondaries mainly white with black tips, extent of black decreasing inwardly (outer feathers have distal half black, innermost nearly all white); primary coverts, tips of median coverts and greater coverts white, which together with base of secondaries form broad white band across entire wing. Outer webs of tertials margined with white; alula dark brown, paler on inner web. Axillaries and underwing-coverts white. Bill pinkish red, outer half black; eyes yellow; legs and feet pinkish red, upper surfaces of toes darker. ADULT ♂ (non-breeding): like breeding bird but eyes brownish orange. Sexes alike, ♀ somewhat smaller. SIZE: (8 ♂♂, 8 ♀♀) wing, ♂ 192–200 (196), ♀ 185–201 (195); tail, ♂ 84–93 (87·9), ♀ 79–88 (83·3); bill, ♂ 29–34 (31·1), ♀ 29–32 (30·4); tarsus, ♂ 65–74 (70·0), ♀ 62–70 (65·5). No wing-spur; small carpal knob *c.* 3 mm. WEIGHT: ♂ (n = 10, Botswana) 68 (? imm.)–192 (171); (n = 3, E Africa) 156, 170, 175; ♀ (n = 11, Botswana) 126–188 (156), (n = 1, Namibia) 155; (n = 4, E Africa) 155–200 (170); unsexed (n = 2, South Africa) 148, 158; 1 imm ♂ 150.

IMMATURE: similar to adult but lores and feathers around base of bill off-white, entire crown dark brown with pale edges to feathers; brown feathers of upperparts and wings with pale buffy fringes and dark brown subterminal bar; no black border to lower breast. Eye brown; legs pale orange.

DOWNY YOUNG: upperparts deep buff broadly spotted with black; black on crown forms ring on nape, hindneck whitish, unspotted; underparts white. Bill dark slate with white egg-tooth; eye black or brown; tibiotarsus grey, tarsometatarsus pinkish grey, small black claws on alulae.

V. c. demissus Friedmann: Somalia. Brown feathers of upperparts, wings and breast paler, broadly tipped whitish or buffy, giving overall sandy tone to plumage.

V. c. xerophilus Clancey: SW Angola, Namibia, Botswana, dry areas of W Transvaal and extreme W Zimbabwe, and Cape Province north of Orange R. Doubtfully distinct. White on throat more extensive, extending to lower face; brown areas of plumage somewhat paler and greyer. Dark centre of crown often reduced, replaced by white.

Field Characters. A medium-sized brownish lapwing with red and black bill, red legs, prominent pale eye and black and white striped head pattern unique in genus. The 2 'black-winged' lapwings (Black-winged *V. melanopterus*, and Lesser Black-winged *V. lugubris*) are similar above and below but have unpatterned grey heads, white foreheads. In NE Africa might be confused with Sociable Lapwing *V. gregarius*, q.v.

Voice. Tape-recorded (10, 27, 58, 64, 72, 74). Screech given in flight, a loud, grating, rolling 'kerrrritt' or 'kerrrreet', rising in pitch, somewhat reminiscent of a francolin; often followed by some short, sharp single notes, 'kerritt-rit-rit-rit'; or single notes may be given alone, 'rrit-rrit-rrit'; uttered by both sexes, during 'normal' flight (including on moonlit nights) and during 'back-and-forth' display flight; intensified when danger approaches. On ground, lowest-intensity alarm call is low, growling 'krrt'; medium intensity call, a quiet 'skrik' or 'skwirrk', becoming louder ad more anxious as intruder comes closer; if flushed, gives the screech. In display flight gives a double note 'chikka, chikka, chikka ...', and the same notes but in a quieter tone given when 2 groups join together. Double note, 'snatcha', given in series of 7–8 by pair or party, usually on ground, sometimes in flight; at any time of year, but not common. Incubating birds give soft 'kruit' at change-over; during copulation, calls include a quiet crowing noise, 'tik-a-tik', 'krrr, krrr' and a deep throaty note. An apprehensive 'ghreet, ghreet' made by birds walking cautiously around cobra. Chicks give a quiet squeak while still inside egg, quiet 'oo-oo' just after hatching, and older but non-flying young call 'to-yeet' when disturbed.

Also reported to give a double 'kievi' or 'ker-krit', sometimes a triple 'kiverkie' or 'kievietje' (the Afrikaans name), but context not clear. Calls a great deal at night as well as by day.

General Habits. Inhabits dry, open country of all kinds, both treeless and lightly bushed: grasslands, pastures, shortgrass *Acacia* savanna, dry open thornbush, lightly wooded savanna, bare areas in bushveld, *Colophospermum* shrublands; readily occupies habitat opened up by man: ploughed land, cultivation, fallow fields, airports, golf courses, roadsides, school playing fields, open areas at edges of towns. Particularly partial to recently burnt grasslands, where it nests when the new grass appears. Tolerates semi-arid and even desert conditions, including dunes, in Somalia and Namibia, and common in Kaisut Desert, Kenya (Lewis 1981). Avoids damp and clay soils and lush grasslands in high rainfall areas; not associated with water, though sometimes found on dry ground beside lakes, and vacates seasonal grasslands when grass grows too long.

Gregarious, occurring in pairs or small flocks, typically of 10–40, occasionally larger (100 or even 150), especially just after breeding season. Flocks are very loose associations, which fragment and recombine; they range widely and have no territory, though there may be a 'headquarters' area where they rest during the day (Skead 1955). At Barberspan (South Africa) loafs and roosts during day at waterside or on islands. Besides standing around and resting, midday activities include preening, squatting on ground, running slowly about with neck tucked into shoulder in hunched-up attitude, head-bobbing, drooping tail-feathers and wagging tail briskly, stretching with wings raised and neck forward, and shuffling to ground, as if settling on eggs. Pecks idly at ground for food, and from time to time cocks head and looks briefly into sky for predators. Bathes by bending down until breast touches water, plunges head in and out of water and flaps wings to splash water onto body. On leaving water stretches wings over back, gives a little hop, and then preens. When apprehensive draws itself up, thrusts out breast, gives single 'kwerrt' while bobbing head. Prefers to run rather than fly on approach of danger, slinking away either silently or with a 'krrr' of apprehension. Calls become louder as intruder gets closer; occasionally bird will turn and threaten intruder with a crouching walk. Most active and noisy at dusk, and also flies about and calls on moonlit nights. When foraging, runs short distance, stops suddenly with body poised, neck stretched forward and downward, 1 leg half raised; then makes rapid forward dart towards food, stabbing at it 3–4 times with bill; movements clean and precise.

In off-season display, small parties fly back and forth over distance of up to 300 m at altitude of 10–60 m for 4–5 min, with slow, deliberate wing-beats, giving screeching flight call and a rapid, stuttering 'chikka, chikka, chikka ...'. Several groups may be in the air at once. Then they descend to earth in series of crazy, high-speed side-steps, swerves and swoops, likened to similar flight of Northern Lapwing *Vanellus vanellus* (Archer and Godman 1937). Display flight takes place usually in late afternoon or evening, often on moonlit nights, sometimes in morning (Skead 1955).

Daily activity in Kenya recorded by van Someren (1956): during day flocks break up into small bands, birds resting or idly picking up a few insects. More active in late afternoon, moving to ground frequented by cattle and game where dung plentiful, insects abundant. Before sunset birds congregate and perform communal display: first they run around each other, making little hops into the air; then, as though at given signal, they raise wings straight up and fan tail. Wings held up for quite a time, producing display of black and white, then brought to rest, when birds merge with background. This display apparently a preliminary to evening feed and flight. Flight destination not known but some birds fly up from lower plains to feed in higher pastures where cattle feed; still there in early morning, but gradually drift back in small groups to lower ground during day. Flocks seen flying high during day, and similar phenomenon noted in Serengeti, where flocks often encountered 150 m in air (Turner 1964), possibly all connected with regular daily movements.

Performs regular movements connected to changes in habitat. Follows dry season fires which produce preferred breeding habitat of burnt grassland. In Akagera Nat. Park, Rwanda, isolated individuals appear at tiny fires Jan–May but bulk of breeding population arrives during main fire season June–July (Vande weghe and Monfort-Braham 1975). In Burundi breeds on plains of Ruzizi R. June–Nov, then disappears, returning Apr–May (Gaugris 1979). In E Africa, both wanderer and resident. Mainly a migrant in Zambia, present during dry season Apr–Nov; most have left breeding areas by end Nov, some linger into Dec, very scarce Jan–Mar when very little suitable habitat even in dry SW (Taylor 1979). Absent Malaŵi at height of rains. In Zimbabwe present in wettest areas (Mashonaland) only when grass dry or burnt. In South Africa, resident Kruger Nat. Park but numbers decrease during rains (Kemp 1974); in S–central Transvaal leaves breeding areas when grass long and wanders in flocks (Tarboton 1974); in Orange Free State sedentary in drier western areas but in east only plentiful in summer, disappearing from Apr to July/Aug.

Food. Insects and their larvae, especially termites (Isoptera: Hodotermitidae), grasshoppers, crickets, beetles (Coleoptera: Curculionidae, Tenebrionidae, Cerambycidae, Scarabaeidae) and ants; possibly earthworms.

Breeding Habits. Generally monogamous but 1 case of polygamy recorded (Hanley 1983). Non-territorial, social, semi-colonial nester; in Zimbabwe 5 nests spaced randomly on 11 ha and 6 nests in adjacent triangular paddock 41 × 37 × 9 m, with no animosity between pairs and flock of 20 non-breeding birds also in paddock; on disused airport, 12 nests on 10 ha, closest 25, 26 and 50 m apart (Moore and Vernon 1973). 3 nests within 25 m of one another producing communal 'early warning' system: when 1 bird called on approach of danger, all 3 birds left nests (Ade 1979). In South Africa, 3 nests 50 m apart, 5 pairs on 120 ha (Skead 1955). Non-breeding groups of 3–10 remain in between territories of nesters, Rwanda (Vande weghe and Monfort-Braham 1975). Other conspecifics tolerated in close proximity even when eggs hatching or chicks present; Black-winged Lapwing frequently tolerated in South Africa, where seen bathing with party of Crowned Lapwings, and 1 flew in to join Crowned Lapwing with chick in protest at intruder (Skead 1955). Partly territorial in Rwanda: on arrival, when habitat restricted and population dense, much intraspecific aggression, including aerial pursuits and displays on ground. As fires open up more habitat, birds disperse and aggression diminishes. In interspecific conflict, dominant over African Wattled Lapwing *V. senegallus* and Lesser Black-winged Lapwing; on arrival sometimes forms mixed flocks with latter without apparent competition but rivalry soon develops, only diminishing as fire opens up more ground.

Copulation often performed casually, in presence of other birds. Examples: (a) ♂ approaches ♀ with deep, throaty call; ♀ crouches, ♂ mounts, squats for few moments, then dismounts; ♀ stands up and raises 1 wing over back; (b) ♂ probes cowpat, ♀ joins him, then solicits in crouching position; ♂ mounts, both wag tails; ♂ stays 3 s, dismounts and raises both wings; ♀ does not raise wings; (c) ♀ crouches, ♂ nonchalantly mounts and copulates; dismounts after 2 s, both raise wings; (d) ♀ squats, stands up, rotates on spot like bird about to settle on eggs, squats again; ♂ approaches with legs and body stiff, mounts from behind; afterwards ♂ stretches wings, ♀ just walks off (Skead 1955). Sometimes vocal during copulation, 1 bird making quiet crowing noise, other 'tik-a-tik'. 'False copulation' observed after birds have mobbed dog: ♂ mounts but does not squat or copulate. Also seen in group of 15 birds standing around idly in non-breeding season: 1 bird gives 'zikker' call, others answer and birds strut about calling; 1 bird approaches another which crouches and solicits copulation; 1st bird climbs on its back and squats but does not copulate. Squatting bird then stands up, holds head erect and slowly raises wings, holding them up for 1–2 s (C. J. Skead, pers. comm.).

Highly demonstrative to intruder after chicks hatched, flying around, calling loudly and dive-bombing invader, even striking with wing. In threat display on ground (**A**) raises fully opened wings over back and rushes at invader with threatening calls.

Distraction displays include spreading wings and tail, both in standing and crouching positions; wings may be half closed, or dropped and trailed. Squats on ground with neck stretched forward and shuffles along with wings arched and tail fanned; bobs head, displacement feeds. With ungulates, waits on nest until last possible moment then flashes 1 or both wings or flies up shrieking, thus frightening animal away from nest. Calf frightened by call of 'kehk, kehk, kehk' from crouched position. On approach of Blue Cranes *Anthropoides paradisea* bird crouched then turned away from them and walked off, as if to lead them away from nest; same method used on person (C. J. Skead, pers. comm.). Sometimes staggers back and forth feigning illness, and pair may confuse intruder by separating and running in different directions. 'False brooding' strategy noted on approach of raptor: sitting bird rises from nest and stands 1 m away from it, while sentinel 20 m away sits down and pretends to be brooding, drawing attention of raptor to fake nest-site (Ade 1979).

NEST: a simple scrape excavated by bird or mere depression in ground, sometimes on bed of small pebbles, lined with material in immediate vicinity— dead grass, roots, seeds, dry leaves, small sticks, earth pellets, pebbles, shell fragments, animal droppings, lid of trapdoor spider's nest (C. J. Skead, pers. comm.); little lining to start with, more added during incubation. As nearby material used up, birds bring it from farther away. More distant material not taken straight to nest; bird first picks it up and flicks it back in direction of nest with brisk sideways movement of head, later retrieves it and takes it to nest. Likewise, sitting bird may pick up scrap of material and flip it over its back. By the time chicks hatch, nest-scrape is full of material; up to 1340 fragments counted in 1 nest (C. J. Skead, pers. comm.). Diam. 11·5–13·5 cm, depth 2·5–3·8 cm. Nest on low-lying ground built up with pellets of sludge in anticipation of rain (Hanley 1983). Site in open or near stone, dung or low shrub, affording sitting bird uninterrupted distant view; often 10–25 m from shade tree used by partner on sentinel duty; sometimes re-used annually.

EGGS: 2–3, occasionally 4 (records of 5 eggs are product of 2 ♀♀), laid at 1-day intervals; mean (143 clutches) 2·6. 1 joint nest contained 3 eggs of Crowned Lapwing and 4 of African Wattled Lapwing (Vincent 1945). Pyriform; ground colour pale to dark buff, stone, brown, olive or yellow-brown, with spots, speckles, blotches and scrolls of black, sepia and brown over grey or pale purple, heaviest at thick end. A clutch in Zimbabwe exactly matched colour of dry earth background, black blotches and scrolls matched burnt grass, producing very effective camouflage (Ade 1976). SIZE: (n = ?, Somalia 34·5–39 × 26·5–27·5 (37 × 27); (n = 150, southern Africa) 36·2–44 × 26–31·4 (40·1 × 29·0). WEIGHT: (estimated) 17·0. Double brooded.

LAYING DATES: Ethiopia, Jan–Sept; Somalia, Mar–June, exceptionally Jan, Aug; E Africa: SW Kenya, Mar–May, Sept–Nov; NW Tanzania (Serengeti), all months, peak in dry season May–Sept; Region B, Feb, July, Aug; Region D, all months, peak Apr–June, during and just after main rains; Zaïre, Apr–May, at end of wet season but grass kept short by herbivores (Lippens and Wille 1976); Rwanda, Aug–Dec, peaks Sept–Oct; Burundi, June–Sept; Zambia, June–Oct, peaking Aug, exceptionally Apr; Malawi, Aug; Mozambique, June–Nov; Zimbabwe, all months except Jan–Feb, mainly July–Nov, peak Aug–Oct; Namibia, Aug–Dec, Mar; Botswana, Aug–Jan, Apr; South Africa: Transvaal, June–Nov, exceptionally Apr; Natal/Zululand, June–Nov; W Cape, all months but nearly all Aug–Feb; N Cape (Gemsbok Park), Oct–Feb, May.

INCUBATION: begins probably with 1st egg; by both sexes. Period: 28–32 days. 4 changeovers noted between 08·14 h and 17·30 h at irregular intervals (Ade 1979). Changeover initiated by either bird uttering a soft 'kruit'; 1 moves off nest as other moves towards it, they pass not less than 2 m from eggs. No display at changeover; both hold heads low, move slowly and cautiously; bird approaching eggs checks sky for predators. Sitting bird regularly adds material immediately around nest to lining, especially on arrival or departure. From time to time sitting bird stands up and turns around, or leaves nest and preens nearby for several minutes; in heat of day stands up to shade eggs, sometimes pants. Mate often stands guard nearby but sometimes not present. At beginning of incubation eggs may be left unattended for considerable periods, sometimes with no bird in sight; sitting bird may leave nest and squat nearby, allowing eggs to incubate in sun. Eggs most often unattended in cool of evening or morning, when both parents feed, preen and exercise, and brooded most continuously during hottest part of day. As hatching time approaches, less time spent away from nest.

On first sign of danger incubating bird slips quickly and quietly off nest in a crouch, assumes normal stance when *c.* 10 m from nest and displacement-feeds or just stands around. At beginning of incubation not particularly demonstrative if intruder comes close to nest: moves back towards nest, stands near intruder or squats with chin on ground; also bobs head, wags tail and moves about uttering a single low 'skreet'; this attracts mate, who returns and both circle intruder. As hatching time approaches, reacts much more strongly, using aggressive or distraction displays described above.

DEVELOPMENT AND CARE OF YOUNG: chick 25% grown: body and wing-coverts well feathered, down remaining on back and rump; remiges and rectrices emerging from quills. Black band on occiput, blackish band on breast; bill pink at base, tip black; legs pale pink. Chick as large as adult but unable to fly (*c.* 1 month old): down remaining on rump, otherwise well feathered; no red on bill, legs greenish yellow. When able to fly, white speckles on wing-coverts, legs pale red. ♀ known to breed at 16 months, ♂ at 18 months (Tree 1981).

Squeaking and movement detectable inside eggs up to 48 h before hatching. 3 eggs hatch within 1–2 days. Chicks leave nest within 24 h of hatching; first to hatch do not wait until all eggs hatched but follow 1 parent away from nest while other parent broods remaining eggs. Eggshells removed from nest soon after hatching, either eaten or carried well away from nest. Both parents feed and tend young; lead them away from nest-site after hatching, sometimes to considerable distance, but if not disturbed whole family may remain near nest until young are adult size. Parents show food to young, attracting them with characteristic posture and call to food source, typically dung pat containing insect larvae (Walters 1982). Young also crowd round bill of adult as if waiting for food. On warm days chicks rest in shade of parent's body, moving when parent moves. Brood remains together for feeding but divides on approach of danger, parents moving in different directions, each with part of brood. Parent may then sit brooding young while watching sky; or chicks squat and remain motionless, neck stretched out and chin resting on ground. Chicks sometimes forget proper drill and run to parents instead of 'freezing'. When young full grown, brood division ceases. Young remain with parents until fully grown; family may mix with other families even before young can fly, regularly does so when young flying. At end of breeding season large mixed flocks of adults and full-grown young occur; last groups to leave breeding areas in Rwanda composed solely of young of the year.

BREEDING SUCCESS/SURVIVAL: 57 out of 65 eggs hatched, Zimbabwe (Ade 1979); of 790 eggs, South Africa, 54·3% hatched, 30·9% reared to maturity (Hanley 1983). Recorded cases of adult mortality include being run over by cars (frequents roads at night, blinded by headlights), legs broken by golf balls, legs badly misshapen and covered with nodules when attacked by Scaly Leg Mite *Knemidokoptes mutans*. Many eggs destroyed and young die in brush fires; young recorded being attacked by drongos, 2 later disappeared, possibly eaten by drongos (Hanley 1983). Side-striped Jackal *Canis adustus* believed to have eaten chick; cow nearly trampled on egg and chick after adult failed to drive it off (Ade 1979). Other animals attacked and chased on breeding grounds (predation not proven) include Slender Mongoose *Herpestes sanguineus*, dogs, Hamerkop *Scopus umbretta*, Secretary Bird *Sagittarius serpentarius*, Lanner Falcon *Falco biarmicus* and Pied Crow *Corvus albus*; cautious treatment of cobra (see above) also suggests it is a predator.

References
Ade, B. (1979).
Hanley, A. W. D. (1983).
Moore, R. and Vernon, C. J. (1973).
Skead, C. J. (1955).

Plates 16, 20
(Opp. pp. 257, 321)

Vanellus gregarius (Pallas). Sociable Lapwing; Sociable Plover. Vanneau sociable.

Charadrius gregarius Pallas, 1771. Reise versch. Prov. Russ. Reichs, 1, p. 456; Volga, Jaiku and Samara.

Range and Status. Breeds SW and central Russia; winters NE Africa, Middle East (chiefly Iraq) and N India. Vagrant in Europe west to Spain, Britain and Ireland.

Palearctic migrant, wintering mainly in N Ethiopia (Eritrea) and Sudan, south to Darfur (El Fasher, Zalingei, Bahr 'el Arab), Kordofan, Sennar, Rahad and Atbara R. near Ethiopian border. Formerly passed through and wintered in Egypt in small numbers, but very few recent records; 1 seen in Fayoum 15 Nov 1977 (Short 1977). 1 record NE Somalia (Sheikh) but possibly more frequent since it occurs in SW Arabia. Vagrant Morocco (34 km south of Tangiers, Dec: Giraud-Audine and Pineau 1972). Scarce and local in Sudan; numbers vary from year to year in Eritrea, absent some years, 1 year flock of *c*. 150 recorded (Smith 1957).

Description. ADULT ♂ (non-breeding): crown grey-brown, some feathers with narrow buff fringes; forehead, lores and broad white line encircling crown off-white. Incomplete narrow black eye-ring in front of and behind eye, not quite meeting in middle, extending behind eye as thin line across ear-coverts and around nape below white stripe. Cheeks off-white grading to pale brown on ear-coverts and sides of throat and neck, with variable amount of fine dark streaking; chin and centre of throat white, becoming pale brown on lower throat. Upperparts, including scapulars, tertials, lesser, median and base of greater coverts greyish brown variably tipped whitish to rufous buff; uppertail-coverts and tail white, latter with broad black subterminal band decreasing in width outwardly; 2 outer pairs white (t 5 sometimes with a little black). Breast whitish or pale grey with variable amount of dark brown V-shaped markings; rest of underparts white. Primaries black with concealed white bases, secondaries and tips of greater coverts white, alula and primary coverts black; axillaries and underwing-coverts white. Bill and legs black; eyes dark brown. ADULT ♂ (breeding): crown and nape black with bluish sheen, forehead and stripe encircling crown pure white; broad streak from lores through eye and onto ear-coverts black, not continuing round nape; cheeks and ear-coverts pale buff, chin and throat white. Sides of neck, hindneck and upperparts, including scapulars and tertials, pale grey-brown. Upper breast pale grey-brown tinged with lilac, becoming darker on lower breast; belly black with blue sheen, bordered below by chestnut band. Rest of underparts, wings and tail as non-breeding. ADULT ♀ (non-breeding): like ♂. ADULT ♀ (breeding): like breeding ♂ but crown with some grey feathers, belly less black, often dark grey, chestnut band less extensive and mixed with white. SIZE: wing, ♂ (n = 24) 200–220 (206), ♀ (n = 22) 197–210 (204); tail, ♂ (n = 22) 76–91 (84·2), ♀ (n = 19) 76–87 (81·9); bill, ♂ (n = 23) 26·2–32·4 (29·4), ♀ (n = 22) 25·8–31·3 (28·7); tarsus, ♂ (n = 25) 56·5–64·5 (59·8), ♀ (n = 21) 55·3–63·1 (59·2). No wing-spur; small knob c. 2 mm. WEIGHT: ♂ (n = 2), winter 206, 227, spring 245, 260; ♀ spring (n = 3) 200–252 (217), age and season unknown (n = 4) 180–195 (187).

IMMATURE: similar to non-breeding adult. Crown either as adult or darker grey-brown with some black spotting; forehead and superciliary stripe buff; feathers of upperparts and wings more extensively and broadly tipped whitish or buff; breast and sides of neck more extensively streaked and mottled dark brown.

Field Characters. A medium-sized lapwing with distinctive breeding plumage of black crown, black and white striped face, pale grey breast and upperparts and chestnut-bordered black belly which easily separates it from other species. Drab winter adults and especially the more heavily marked immatures somewhat reminiscent of Grey and Golden Plovers *Pluvialis* spp. when standing with wings folded, but legs longer, eye-stripe more pronounced, upperparts and wings less heavily marked. In flight, black and white wing and tail pattern quickly dispels any similarity. Sympatric with similar-sized White-tailed Lapwing *V. leucurus*, which differs in having all-white tail, unmarked pale head and pale legs.

Voice. Tape-recorded (62, 73). Rasping, grating notes 'jek', 'krek', 'etch-etch-etch' given mainly by breeding birds but also by vagrants in Europe. In winter mainly silent, but sometimes a short, shrill, piping whistle.

General Habits. Occurs mainly in dry country but also in damp situations: dry plains, shortgrass savanna, highland grassland, *Panicum* steppe, burnt grassland, sandy wastes; ploughed land and grainfields; also river edges, irrigated fields and coastal saltmarsh.

Gregarious on migration and in winter quarters; typically in groups of 5–20, once 150. Single individuals in Egypt and Europe often join flocks of Northern Lapwing *V. vanellus*; bird in France followed them everywhere, seemed anxious when they flew off, assumed upright stance and quickly flew to rejoin them (Grisser 1983); it was also somewhat aggressive towards them, chasing those that came too close. British bird more independent of its flock, associating loosely with them and alternately aggressive toward and tolerant of them (Riley and Rooke 1962). Forages in plover manner with quick runs and stops, but runs shorter than other species; more active and faster moving than Northern Lapwing. Seen to lean forward and remain for several s in watchful pose, with bill c. 5 cm from ground (Grisser 1983); also, when feeding on ploughed land, trampled ground with 1 foot at a time while making rapid, bobbing body movements (Riley and Rooke 1962). Generally rather shy. Flight like Northern Lapwing but more direct, less wavering, with more frequent wing-beats. Tends to fly low over ground for short distances but at considerable height for journeys of 1 km or more.

Present Eritrea Nov–Mar, N-central Sudan (Khartoum) Oct–Mar; arrives W Sudan (Darfur) end Oct. Single Somalia bird Nov.

Food. Mainly insects and their larvae, especially grasshoppers, locusts, beetles and spiders; a small amount of plant material also taken.

Reference
Cramp, S. and Simmons, K. E. L. (1983).

Vanellus leucurus (Lichtenstein). White-tailed Lapwing; White-tailed Plover. Vanneau à queue blanche.

Plates 16, 20
(Opp. pp. 257, 321)

Charadrius leucurus Lichtenstein, 1823. *In* Eversmann's Reise von Orenburg nach Buchara, p. 137; between the Kuwan and Ian Daria, Turkestan.

Range and Status. Breeds SW USSR and Middle East, west to Turkey. Winters NE Africa, Middle East, N India. Vagrant Europe west to Sweden and Britain.

Palearctic migrant. Formerly uncommon winter visitor N Egypt (not west of Alexandria), recorded Nile delta Nov–Dec, Fayoum Feb–Mar (Meinertzhagen 1930). Recent records from Nile in Middle and Upper Egypt, Tahta–Kom Ombo, Sept–Oct and Feb–Mar indicate passage and possible occasional wintering. Frequent N Sudan in winter on Red Sea coast and rivers south to below Khartoum on Blue and White Niles, and 1 record Kordofan (Um Ruaba), but not yet recorded Ethiopia. 2 records from L. Chad (Malamfatori, Nigeria and Djiboulboul, Chad: Vieillard 1972). Vagrant to Tunisia (Gabes) and Libya (Benghazi: Baker 1982).

Description. ADULT ♂ (non-breeding): forehead, lores, chin and throat whitish; face and sides of neck pale brown; crown, hindneck and upperparts, including scapulars, tertials, lesser and inner median coverts greyish brown, with lilac wash on back, rump and wing-coverts. Tips of lower rump feathers, uppertail-coverts and tail white. Upper breast light brown, lower breast grey with white fringes to feathers, belly and flanks pinkish buff, paling to off-white on undertail-coverts. Primaries black with concealed white bases; secondaries white, outer ones tipped black; tips and concealed inner webs of tertials white; greater and outer median coverts grey-brown

Vanellus leucurus

with white tips and black subterminal band, amount of white on tips increasing outwardly; lesser and median primary coverts white, alula and greater primary coverts black. Axillaries and underwing-coverts white. Bill black; eye-ring bright red to pink; eyes red-brown, hazel, orange-brown or sepia; legs and feet bright lemon-yellow, often tinged green at joints. ADULT ♂ (breeding): similar but forehead, sides of head, chin and throat washed with buff, crown and hindneck browner, upperparts more washed with lilac, no pale fringes to grey feathers of breast. Sexes alike. SIZE: wing, ♂ (n = 22) 172–186 (179), ♀ (n = 19) 170–184 (177); tail, ♂ (n = 20) 67–81 (74·0), ♀ (n = 19) 67–77 (72·1); bill, ♂ (n = 24) 27·2–31·4 (29·4), ♀ (n = 22) 26·6–31·0 (29·1); tarsus, ♂ (n = 21) 68·0–76·3 (73·5), ♀ (n = 25) 65·9–74·4 (71·0). No wing-spur; small knob *c.* 2 mm. WEIGHT: ♂ May 139; ♀ Feb 99, 198, Apr 107, May 141, Oct 114, 123. ♂♀ (n = 7, Asia) 99–198 (132).

IMMATURE: differs from adult as follows: crown blackish brown, feathers narrowly fringed buff; upper mantle pale grey-brown, feathers fringed pale buff; rest of mantle, scapulars and lesser upperwing-coverts dark brown with pale brown fringes; back, rump and uppertail-coverts white, feathers fringed pinkish buff. Central tail-feathers with dark subterminal band and buff or pale brown tips, rest of tail feathers faintly tipped grey-buff. Upper breast pale grey mottled buff-brown, rest of underparts pinkish buff, wearing to pale grey or off-white. Tertials pale grey-brown with lilac gloss, dark brown subterminal band and buff or off-white tips; lesser and median upperwing-coverts with dark brown subterminal mark.

Field Characters. A medium-sized, plain lapwing in pastel shades of grey, lilac, buff and pale brown, with long yellow legs; whitish face and sides of head give white-headed appearance. Distinguished from similar-sized Sociable Lapwing *V. gregarius* by lack of head-stripes, unmarked underparts, yellow (not black) legs and, in flight, all white tail. Similarly plain and long-legged Caspian Plover *Charadrius asiaticus* is smaller, browner, with dark legs and typical *Charadrius* flight pattern (pale wing-bar separating brown coverts from darker remiges, white-edged brown tail).

Voice. Tape-recorded (70, 73). Common call rendered 'pi-wick', 'kwie-wuk' or 'kie-witt', like Northern Lapwing *V. vanellus* but less plaintive, more subdued. When flushed, a harsh 'kwett-kwett'. Excited birds in flight give rapid 'wik-wik-wik ...' Also a soft whistle.

General Habits. Mainly aquatic, inhabiting rivers, drainage ditches, coastal lagoons, ponds, marshes. Noted feeding in stream of sewage effluent, Sudan (Pettet 1982). Also occurs on river banks and other dry ground near water, but avoids the dry, open country occupied by Sociable Lapwing.

Occurs singly, in pairs or small flocks outside the breeding season. Not particularly shy. Flight like Northern Lapwing but more powerful. Sometimes bobs head like *Tringa* spp. On dry ground feeds like other lapwings, taking 3–4 steps, pausing, leaning forward and pecking at ground. Wades in both deep and shallow water, taking insects from water surface, in deep water also immersing head. Uses 'foot-pattering' method not unlike that recorded for some other plovers; bird wading in muddy water pauses, stretches body forward somewhat, puts 1 foot forward and pats mud under water 4–6 times, then stoops to inspect surface, often picking items from it (Pettet 1982). Regularly forms feeding association with Common Stilt *Himantopus himantopus*, joined also by Marsh Sandpiper *Tringa stagnatilis*; follows stilts, actively striding about in water and picking items from surface presumably stirred up by stilts; from number of items taken this method apparently more productive than feeding alone (Pettet 1982).

Present Sudan Oct–Mar. Autumn migration on Nile noted Egypt Oct 4–10; total of 11 between Tahta and Kom Ombo, including flock of 7 near Qena (Hume 1983). Early spring migrants noted mid Feb–mid Mar on Nile: 2–5 between Qena and Kom Ombo (Quinn 1982, 1983; Welch 1983). Not yet recorded on Red Sea coast in Egypt, but since it winters on Red Sea coast in Sudan and small numbers pass through Gulf of Aqaba on migration it may possibly follow route suggested for Spur-winged Lapwing *V. spinosus*, crossing Red Sea to land near Hurghada and continuing on down coast.

Food. Insects and their larvae, including beetles and grasshoppers; worms, snails.

References
Cramp, S. and Simmons, K. E. L. (1983).
Pettet, A. (1982).

Vanellus crassirostris **(Hartlaub). Long-toed Lapwing; White-winged Lapwing. Vanneau à ailes blanches.**

Chettusia crassirostris Hartlaub, 1885. J. Orn. 3, p. 427; Nubia.

Range and Status. Endemic resident and partial intra-African migrant. Distribution patchy, dictated by specialized habitat. Apparently isolated populations at L. Chad, NW Angola (floodplain of R. Cuanza) and E Kenya (lower Tana R. and adjacent coast). Main range from S Sudan (Bahr-el-Ghazal, Sudd) and extreme W Ethiopia (Baro R. area) through E Zaïre, Uganda (except extreme NE), W Kenya (north to Baringo, east to L. Naivasha, Amboseli, L. Jipe), Rwanda, Burundi, N and W Tanzania (to Rukwa, Kilombero R. and Tarangire Nat. Park), Zambia (local) and E Angola to Malaŵi (fairly common below 600 m), S Mozambique (north to Zambezi, Quelimane and L. Chilwa), Zimbabwe (Zambezi, L. Kariba, lower Sabi R.), E Namibia (Caprivi), N Botswana (Okavango, L. Ngami) and Zululand, with a single record for extreme NE Transvaal (Pafuri). Frequent to common, with local concentrations (97 at Kafue Flats, Zambia).

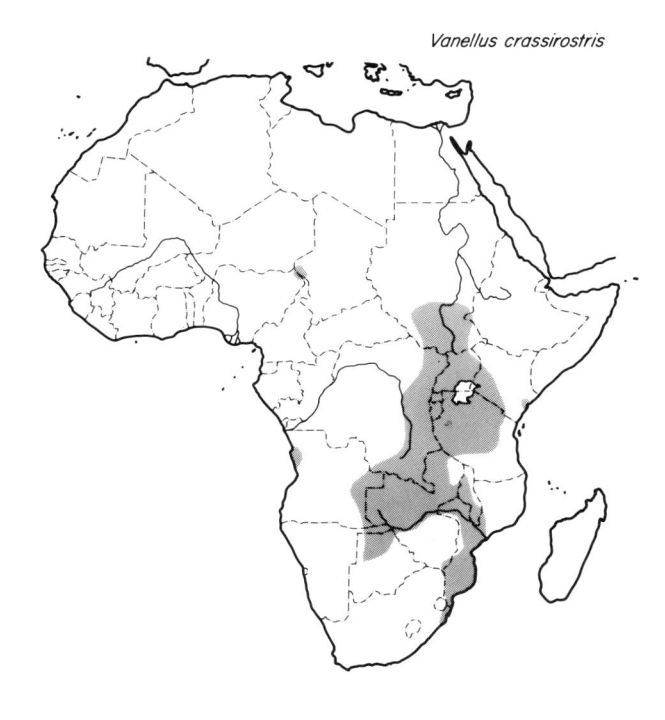

Description. *V. c. crassirostris* (Hartlaub): Nigeria and Sudan to E Zaïre, Uganda, Kenya and N Tanzania, intergrading with *leucoptera* in Tanzania, SE Zaïre and N Malaŵi. ADULT ♂: forehead, forecrown, facial area from lores to behind eye and ear-coverts, chin, throat and sides of neck white; hindcrown, nape, hindneck and broad collar across lower neck joining breast and upper belly black with blue gloss. Upperparts dull earth brown, scapulars with darker tips, rump darker brown, lower feathers blackish tipped white. Uppertail-coverts and basal third of tail white, rest of tail glossy blue-black. Lower belly and undertail-coverts white. Remiges black, tertials dark olive-brown with bronze wash, upperwing-coverts, including primary coverts, white, alula black. Axillaries and underwing-coverts white. Bill purplish pink, outer third black; narrow ring of bare skin round eye purplish pink to crimson; eyes crimson; legs and feet dull red, frontal scutes blackish. Sexes alike. SIZE: (3 ♂♂, 5 ♀♀) wing, ♂ 209–212 (211), ♀ 198–211 (207); tail, ♂ 99–102 (101), ♀ 90–99 (94·8); bill, ♂ 30–33 (31·6), ♀ 32–35 (34·0); tarsus, ♂ 76–83 (80·3), ♀ 78–87 (81·4); wing-spur, ♂ 5–6 (5·3), ♀ 3–5 (4·0). WEIGHT: 1 ♂ (Uganda) 170, 1 ♂ (Zaïre) 225; 1 ♀ (Uganda) 170, 1 ♀ (Zaïre) 187; unsexed (Kenya) 162, 191; unsexed (Zambia) 186, 204; imm. 140.

IMMATURE: not described.

DOWNY YOUNG: not described.

V. c. leucoptera Reichenow: S Zaïre, Zambia and Malaŵi to South Africa. Secondaries and all but 3 outer primaries white.

Intermediate birds from Tanzania, SE Zaïre and N Malaŵi were formerly separated as *hybrida*.

Field Characters. A medium-large lapwing with distinctive combination of white crown, face and neck, black back of head and neck, black breast, brown upperparts, large white wing patch conspicuous in flight, and reddish legs. Confined to aquatic habitats from which other lapwings largely absent.

Voice. Tape-recorded (264, 296). Alarm, a loud clicking 'kick-kick-kick'; also a loud, plaintive, whistled 'wheet' on being flushed.

General Habits. Typically inhabits floating vegetation on lakes, ponds, canals and other stagnant water; wide variety of plants utilised includes water-lilies *Nymphaea*, also *Salvinia molesta*, *Typha latifolia*, *Ludwigia*, *Cyperus*, *Pistia*, *Leersia*, *Oryza*, *Panicum repens*, *Vossia*, *Wolffia* and *Azolla*; in Sudan, clumps of 'sudd' and rotting vegetation. Also occurs in extensive swamps, water meadows, on grass-covered floodplains and in short grass near water. In dry season feeds on exposed mud at edges of ponds and small waterways; in wet season occupies inundated grassland, also flooded rice fields. Noted paddling in shallow water just outside reeds fringing shore, L. Chilwa (Vincent 1934).

In pairs or small parties; larger concentrations in dry season in flooded grassland or at contracting waters: flocks of 10–20, Rwanda, and 30–40 SE Zaïre (Ruwet 1965). Forages mainly on surface of floating vegetation, where its long toes enable it to run about with the agility of a jacana, but also on mud; associates with Spur-winged Goose *Plectropterus gambensis*, catching insects when goose turns over dead vegetation.

Largely sedentary, moving only when habitat dries up, but some movements between habitats noted Rwanda (Akagera Nat. Park): in July–Aug leaves permanent swamps and moves into inundated grassland beginning to dry out. Said to be only migrant in Upemba Nat. Park, Zaïre (Verheyen 1953).

Food. Aquatic insects, maggot-like insect larvae, dragonfly nymphs, beetles, ants and small snails.

Breeding Habits. Monogamous; territorial. Territories small at Amboseli, Kenya, 0·15–0·72 ha (mean 0·29); no distinction between nesting and foraging areas (Walters 1982). In territorial quarrels with neighbours, mainly uses swooping attack from air without striking opponent, but sometimes birds strike each other with wings; when attacked, territory holder rarely retreats more than 5 m. Very aggressive towards birds of other species in territory after chicks hatch; av. of 1 attack every 8·3 min, Kenya (Walters 1979). African Jacana *Actophilornis africana* accounted for 76% of its attacks, Blacksmith Lapwing *V. armatus* and Squacco Heron *Ardeola ralloides* for further 16%. Harrier *Circus* sp. and White-browed Coucal *Centropus superciliosus* harshly attacked; others chased were Sacred Ibis *Threskiornis aethiopica*, sandpipers, Black Crake *Amaurornis flavirostris*, Wattled Starling *Creatophora cinerea* and Long-tailed Fiscal *Lanius cabanisi*. Ducks and grebes entered territory with impunity. Most intruders are swooped at from air but sometimes strikes jacanas with wings. In Namibia nesting birds attacked eagles, gulls and lapwings and ignored other waterfowl (Saunders 1970). In Rwanda tolerated African Wattled Lapwing *V. senegallus* (Vande weghe and Monfort-Braham 1975). Approach of humans to nest elicited broken-wing display—bird left nest and crawled and fluttered across surface of weeds.

NEST: varies with site. On mudbank exposed at low water, shallow scrape in wet mud lined with aquatic plants; in short grass near water, shallow depression lined with grass; on floating vegetation, cup of grass or other vegetation; in shallow swamp, platform of damp mosses, weeds and debris built up so that eggs 5–10 cm above water level.

EGGS: 2–4. Pyriform; earth brown, buff, clay, putty, dull grey-green, olive-green or dark olive-khaki, heavily blotched, spotted and streaked with dark brown and black over slate. SIZE: (n = 4, Tanzania) 42–44 × 29–30; (n = 8, Namibia) 42·8–44·7 × 29·0–31·7 (43·6 × 30·4); (n = 5, South Africa) 40·0–47·8 × 27·3–33·5 (44·9 × 31·0). WEIGHT: (estimated) 19·1–20·0.

LAYING DATES: Ethiopia, Apr; Rwanda, Sept–Dec; E Africa: Region B, Mar, May–Sept (both in dry season and in drier break after heaviest rain, peaking June); Region D, Jan, July; Region E, July; Angola, July; Malaŵi, June, Aug–Oct; Zambia, July–Sept; Mozambique, Sept; Namibia, July; Zululand, Oct.

INCUBATION: period *c.* 30 days. Incubating bird usually slips off nest at first sign of human intruder and runs some distance away, but can be approached in boat to 5 m; bird ran off, later returning to nest by fits and starts, with plover-like crouching run interspersed with frequent stops and displacement activity, rearranged eggs before sitting down; mate remained on nearby vegetation, giving intermittent 'kick-kick' call (Saunders 1970).

DEVELOPMENT AND CARE OF YOUNG: young fledge at *c.* 2 months. Tending parent inactive, not staying close to young and rarely calling to them, often stationary; young call frequently, enabling parent to monitor their location (Walters 1982). Distance of foraging unfledged chick from tending adult varies with size of brood, chicks of smaller broods remaining closer (within 10 m of adult 85% of time in broods of 1, 35% of time in broods of 2, 15% of time in broods of 4) (Walters 1979). On approach of person, parent brooding 4 newly hatched chicks on nest ran off with them into dense sedge bed; mate hovered overhead with short wingbeats, both birds called (Saunders 1970).

References
Saunders, C. R. (1970).
Walters, J. R. (1979).

Plates 14, 20
(Opp. pp. 241, 321)

Vanellus vanellus (Linnaeus). Northern Lapwing. Vanneau huppé.

Tringa Vanellus Linnaeus, 1758. Syst. Nat., (10th ed.), 1, p. 148; Europe, Africa. Restricted type locality Sweden.

Range and Status. Breeds Palearctic from N Africa and W Europe to N Iran, central USSR, Manchuria and N China; winters south to N Africa, Middle East, N India, S China, S Japan.

Resident and Palearctic migrant. Many summer NW Morocco, where small numbers breed (Moulay Bousselham (70–80 pairs), Sidi Jmil, Merdja Zerga, Ras-el-Daoura). Widespread in the winter from Morocco to Egypt, mainly on coastal plains, numbers fluctuating widely according to weather conditions in Europe; in Morocco, up to 100,000 winter in Le Rharb and 8500 along Atlantic coast south regularly to the Sous and Oued Massa, irregularly to Nouadhibou and Baie de l'Etoile, Mauritania (Port Etienne, Nouakchott), and delta of R. Senegal; a few in mountains, to 1800 m. Common on coastal plains of Algeria, less so on Hauts-Plateaux; a few winter in desert oases south to Beni Abbès, In Salah and Aïn Amenas. In N Tunisia abundant in some winters, scarce in others; less numerous farther south, irregular south of Sfax and Gabes; noted inland at Gafsa. Scarce in coastal Libya (regular Tripolitania, irregular Cyrenaica); 2 desert records from Fezzan (Brak, Ashkidah). In Egypt common in Nile delta, Wadi Natrun, Fayoum and along Nile to

Asyut and Abu Tieg; smaller numbers farther south. Said to winter in Siwa Oasis (Meinertzhagen 1930) and 1 found dead Qattara Depression (El Moghra), Apr 1977 (Goodman and Ames 1983). In Sudan, rare in extreme north, once at Kosti. (1 rejected South African record: *Ostrich* **29**:27.)

Vanellus vanellus

Description. ADULT ♂ (breeding): forehead brownish black; crown and long thin crest glossy blue-black, stripe from eye round nape white, sometimes pale buff on nape; lores, area around eye and stripe under eye to ear-coverts black, sometimes continuing as indistinct dark line separating nape from ash brown hindneck; white patch from cheeks to sides of upper neck. Mantle and back glossy green, scapulars similar but tipped with bronze and purple; centre of lower back and rump dark ash-brown tinged greenish; uppertail-coverts red-brown. Base and sides of tail white, outer half black, often with greenish gloss, narrowly tipped buff. Chin, throat and breast glossy blue-black; rest of underparts white except for cinnamon undertail-coverts. Primaries black, outer 3 with pale brown tips and subterminal whitish spots, 4th black or with pale tip to outer web only, outer half of shafts whitish. Secondaries black with white bases; upperwing-coverts green with variable amount of purple gloss, outer ones dark brown. Axillaries and underwing-coverts white except for black under primary coverts. Bill black; eyes dark brown; legs and feet brownish flesh. Hind toe present. ADULT ♂ (non-breeding): like breeding but face and neck pattern different. Buff stripe from base of bill over and behind eye and round nape; blackish area around and under eye, extending as streak across ear-coverts and short moustachial stripe; cheeks and sides of upper neck buff, chin and throat off-white, sometimes with a few black feathers. Feathers of hindneck, mantle, scapulars, and inner upperwing-coverts variably tipped buff; feathers of breast variably tipped white, producing mottled appearance. ADULT ♀ (breeding): like breeding ♂ but duller; crown and breast tinged brown, latter often with pale tips to some feathers, not solid black; upperparts more olive, less green; black areas on face broken up with variable amounts of white, especially around eye. Chin and throat usually with a few white feathers. Crest shorter. ADULT ♀ (non-breeding): as non-breeding ♂ but black and green areas duller, browner, less glossy; black patches on face less extensive and browner. Outer 4 primaries more pointed, pale patch near tips less extensive and differently shaped. SIZE: wing, ♂ (n = 52) 220–240 (229), ♀ (n = 21) 214–231 (224); tail, ♂ (n = 49) 95–112 (106), ♀ (n = 21) 92–111 (101); bill, ♂ (n = 67) 21·7–26·4 (24·1), ♀ (n = 45) 21·9–25·7 (23·9); tarsus, ♂ (n = 58) 45–50 (47·4), ♀ (n = 39) 44·5–50 (47); crest, ♂ (n = 26) 61–114 (91), ♀ (n = 10) 57–75 (65). No spur on wing; small knob *c*. 2 mm. WEIGHT: (Europe) Feb–Mar ♂ (n = 25) 128–287 (192), ♀ (n = 5) 139–261 (189); Apr–June ♂ (n = 32) 140–242 (211), ♀ (n = 40) 180–317 (226); Sept–Dec ♂ (n = 16) 135–330 (254), ♀ (n = 11) 151–316 (233).

IMMATURE: like non-breeding ♀ but crown duller and browner, crest very short, black areas on face small, mainly around eye; feathers of upperparts and upperwing-coverts edged buff, breast-band narrower, browner, many feathers edged white.

DOWNY YOUNG: top and sides of head and upperparts buff irregularly mottled and streaked black; nape off-white separated from crown by narrow black ring around hindcrown. Chin greyish, throat off-white, breast dark brown, rest of underparts white. Eyes dark brown, bill dark violet-grey, legs and feet flesh-grey, grey or blue-grey.

Field Characters. A very distinctive medium-sized lapwing with broad, rounded wings, long crest, black and white face pattern and cinnamon undertail-coverts. In flight appears uniform blackish above except for white rump and wing-tips; from below, black underwing contrasts with white underwing-coverts, black breast contrasts with white belly. Spur-winged Lapwing *V. spinosus* has brown upperparts, black underparts, no crest.

Voice. Tape-recorded (62, 73, 76). 3-part territorial call ('song') given in song flight starts with 'peeerr-willup-o-weep', the 'willup' component having a light, bubbly texture; continues with a double 'weep-weep', and ends with 'peeeyuweet'. Contact/alarm calls 'pee', 'peewit', 'peeawee' and variations depending on context and degree of alarm. Alarm call in flight, 'cheew-ep'.

General Habits. Inhabits rivers, lakes, fresh and salt marshes, drainage ditches, estuaries, mudflats; very partial to damp grassland, and often on agricultural land—pastures, stubble fields, plough, irrigated fields. In Moroccan mountains occurs in clearings in cedar forest and other woodland and on moorlands (K. D. Smith *per* J. D. R. Vernon, pers. comm.).

Gregarious outside breeding season. Adults and juveniles usually join together in post-breeding flocks but sometimes form separate flocks. Flocks may have mid-day roost at waterside for loafing and other activities, including drinking, bathing, preening and sleeping. Roosts communally at night at traditional site, often feeding intensively on way to roost. Forages on ground, mainly during day but also at night, especially in moonlight; when moon full, feeds at night and sleeps by day; makes short walk or run, stops, bends forward, pauses, probes in ground with bill. Locates prey by both sight and sound. Foot-pattering or trembling (patting ground rapidly with foot) frequent. Generally rather shy but less so while breeding. Flight airy, with slow, erratic wing-beats.

In Morocco most migrants arrive mid-Oct and leave late Feb (earliest 8 Sept, latest 10 May). Cold weather in Europe may bring additional migrants in mid-winter (60 seen arriving from Spain, Jan: Smith 1965). Algeria: present Sept–Apr, mainly Oct–Feb. Tunisia: present mid-Oct to late Mar, a few until mid-May. Libya: present Oct–early May. 16 at Nouakchott (Mauritania) Dec 1973–Jan 1974 (11 died) (J. P. Gee 1984).

Birds recovered Morocco ringed in Finland, Sweden, Norway, Denmark, Estonia, Lithuania, Hungary, Czechoslovakia, Germany, Austria, Switzerland, Italy, Netherlands, Belgium, France, England; recovered Algeria ringed in Sweden, Norway, Hungary, Netherlands, Belgium, Spain; recovered Tunisia ringed in Hungary. African recoveries, confirmed by those in Europe, indicate strong southwesterly movement of European birds to Iberian peninsula and Morocco; some also pass through Italy and cross central Mediterranean to Africa (Imboden 1974). Most recoveries are of young birds; others aged 1–9 years.

Food. Mainly small invertebrates living on or in ground: insects and their larvae, including dragonflies, mayflies, crickets, grasshoppers, cicadas, butterflies, moths, ants, flies, beetles; also spiders, earthworms, snails and other molluscs, a few crustaceans, frogs, small fish, seeds and other plant material.

Breeding Habits. Monogamous, but promiscuous early in breeding season (and bigamous where breeding density high). Pair-bond lasts only 1 season. Nests singly or in loose group of 3–10 pairs. Tends to be colonial, with poorly defined or partly overlapping territories where food plentiful, nests more widely spaced and boundaries fixed where food scarce. Nests in colony placed 10–150 m apart. Breeding density in good habitat, Europe, c. 1 pair/ha, but very variable, range 0·1–11·0 pairs/ha.

♂ establishes and defends territory. Neighbouring ♂♂ run at each other in forward-hunched posture, legs bent, body bent low and forward, neck retracted, crest lowered; also use high-upright posture, facing opponent with crest raised, breast puffed out, wings slightly raised. Fighting rare but birds sometimes strike each other with wings or bill. Aerial pursuits common, and one may dive-bomb another on ground. Song-flight over territory includes rolling, twisting and diving. For details of these and of courtship displays see Cramp and Simmons (1983). ♂ performs scrape ceremony with breast pressed into ground, wings held out from body, tips pointing upward, tail bobbing up and down, legs kicking, body pivotting; ♀ often stands nearby. Then ♂ (and ♀) may side-throw nest material. Many scrapes made, ♀ chooses 1 just before laying.

NEST: shallow scrape, diam. 9–15 (12) cm, depth of cup 3–7 (5) cm; lined with grass, leaves and other vegetation; on open ground, usually slightly above surrounding level on mound or tussock; may be re-used in successive years.

EGGS: 1–5, usually 4, laid at intervals of 1–5 days (mean 2·8); mean (558 clutches) 3·85. Pyriform; smooth, without gloss; brown, olive, umber, clay or stone spotted, streaked and blotched with black. SIZE: (n = 500) 39–53 × 30–36 (47 × 33). WEIGHT: 22–29 (25–26). Single brooded but replacements laid after egg loss.

LAYING DATES: Morocco, Mar–June.

INCUBATION: begins with last egg; by ♀ at night and ♀ or ♂ by day. Period: 24·7–34·0 days (mean 28·1). Changeover fast, without ceremony; relieving bird lands some distance from nest and walks or runs to it, mate may fly away from nest or walk away side-throwing material.

DEVELOPMENT AND CARE OF YOUNG: chicks weighed c. 18 g for first 4 days (Scotland: Redfern 1983); at 1 week weight 30 g, at 2 weeks 55 g, at 3 weeks 90 g, at 4 weeks 130 g. Remiges start appearing at c. 2 weeks. Fledging period: 29–42 (35–40) days. First breeds at 2 (sometimes 1) years.

Chicks remain in nest until all eggs are hatched, keep close to nest on 1st day out and 50–125 m from nest after 1st week. Self-feeding within a few h of hatching; tended by both parents, mainly ♀ (♂ usually on guard duty); brooded for long periods in first few days; brooding diminishes after 1st week. Young freezes when parent gives alarm call; anti-predator behaviour of parents includes mobbing and injury-feigning. Young independent soon after fledging, though may remain with parents for another week.

BREEDING SUCCESS/SURVIVAL: hatching success, Europe, 54–90%; fledging success, Switzerland (n = 1620), 16·4%; losses mainly caused by predators (crows, gulls, foxes, dogs), farming operations and bad weather. Mortality of grown young, Sept–Mar, 30·4–57·5% (39·7%); thereafter, annual mortality 32·3% (Glutz et al. 1975). 1 bird lived 21 years (Dejonghe and Czajkowski 1983).

References
Cramp, S. and Simmons, K. E. L. (1983).
Imboden, C. (1974).
Redfern, C. P. F. (1983).

Family SCOLOPACIDAE: sandpipers and snipe

Small to medium large wading birds; neck and legs typically rather long, and bill long and fine. Lower part of tibia bare, hind toe reduced and occasionally absent. Wings long, usually pointed. Tail short, square or rounded. 11 primaries, outermost much reduced; usually 12 rectrices.

Mostly long distance migrants, spending short breeding season at high northern temperate or arctic latitudes, wintering south to equator and beyond. Terrestrial birds of open habitat, associating with sea-shores and with inland marshes and wet places. Flight usually strong and rapid, with quick wing-beats, wings angled and legs projecting back close beneath body. Mostly gregarious, associating in mixed species gatherings. Often fly in close ranked, highly co-ordinated flocks. Feed in shallows or along shore's edge, less commonly in deeper water whilst swimming. Food secured by deep probing in mud or sand, by snatching or skimming from water, by picking from surface of mud or vegetation, or by foraging among debris or beneath stones. Post-nuptial moult complete. Pre-nuptial moult partial, involving body and head plumage, and usually tertials, rectrices and some inner wing-coverts. Breeding and non-breeding plumages often differ conspicuously; juvenile plumage usually distinct from either. Some species assume full breeding plumage at end of 1st year; others not until end of 2nd year. Important identification features include calls, and pattern of wings, rump and tail in flight.

In Africa, 15 genera, 42 spp. including 1 breeding resident and 1 Palearctic visitor with very small N African breeding population. 41 regular or vagrant non-breeding visitors of which 32 from Palearctic (at least 2 also from Greenland) and 9 as stragglers from Nearctic (1 probably also from E Siberia).

Note on immature plumages of Palearctic Scolopacidae: young birds in Africa in Sept have full juvenile plumage, but this is replaced during Oct–Nov on body and head (sometimes also inner wing-coverts and tertials) by 1st winter plumage like that of non-breeding adult. 1st summer plumage, acquired by partial moult Jan–June, may partly or fully resemble adult breeding plumage. Although juvenile flight feathers are typically retained throughout 1st year of life, outer primaries of smaller species are commonly replaced in tropical and southern Africa at 6–10 months.

Subfamily CALIDRIDINAE

Stints and 'peep' sandpipers of the genus *Calidris* and allies. Small to medium-sized and rather short-legged waders, mostly gregarious species of the water's edge, highly active and feeding largely by picking. Typically rather patterned above, especially in richly coloured breeding plumages.

Genus *Calidris* Merrem

Large genus of small to medium-sized waders, most with short, slender bills and shortish legs. Wings long and pointed; tail square, middle rectrices often projecting slightly beyond the rest. All except Sanderling *C. alba* with small hind toes. ♀♀ slightly larger and usually with longer bills than ♂♂. Most with dull greyish or brownish non-breeding plumage, but striking breeding plumage, rich chestnut or tawny above, mottled or streaked blackish; breeding plumage of many species seen in Africa for 2–4 months, except in south. Juveniles dark above with prominent creamy or rich buff edging and fringing. In flight, show narrow whitish wingbar and dark back, and most have dark centre to rump. Rectrices rather plain, central feathers dark, outers pale.

18 migrant spp. of which 13 reach Africa, 10 from the Palearctic region, 2 as stragglers from the Nearctic and 1 probably from both regions. 2 reaching Africa, *C. minuta* and *C. ruficollis*, form a superspecies.

Calidris tenuirostris **(Horsfield). Great Knot. Grand bécasseau maubèche.**   **Not illustrated in colour**

Totanus tenuirostris Horsfield, 1821. Trans. Linn. Soc. London, 13, p. 192; Java.

Range and Status. Breeds NE Siberia; winters S and SE Asia to Australia.

Vagrant with a single record Morocco (near Agadir, 27 Aug 1980).

Description. ADULT ♂ (breeding): forehead to nape, sides of head and neck and hindneck boldly streaked black and whitish; no distinct supercilium; mantle feathers black, fringed whitish or cinnamon; new scapulars black, fringed cinnamon and with broad chestnut subterminal bar, but some (remaining non-breeding plumage) dull grey; back and rump dark grey fringed white; uppertail-coverts white with large dark spots; rectrices dark grey. Underparts white, foreneck with bold black spots, breast heavily scaled black and tinged cinnamon laterally, flanks and sides of belly and undertail-coverts blotched black. Primaries and primary coverts blackish; secondaries dark grey; greater coverts grey, and these and inner primary coverts with white tips, forming inconspicuous wing-bar. Some tertials new, black edged chestnut, but most tertials old, and these and rest of upperwing greyish; axillaries and underwing-coverts white, marked grey. Bill blackish, tinged olive at base; eye brown; legs and feet dark olive-grey. ADULT ♂ (non-breeding): differs from breeding bird in having upperparts grey, narrowly streaked black, all tertials grey, and markings on white uppertail-coverts fewer and greyer; streaking on head and neck fine; spotting below fine and confined to foreneck, breast and flanks. Sexes alike. SIZE: wing, ♂ (n = 14) 179–193 (186), ♀ (n = 21) 181–198 (191); tail, ♂♀ (n = 38) 61–70 (65·1); bill, ♂♀ (n = 31) 39–47 (43·3); tarsus, ♂♀ (n = 47) 32–38 (34·7). WEIGHT: (20 New Guinea, Sept–May) 125–230.

IMMATURE: juvenile differs from non-breeding adult in having top of head boldly streaked blackish, mantle and scapulars blackish, fringed pale buff, and tertials and inner wing-coverts fringed pale buff.

Field Characters. Structure and tail pattern recall Red Knot *Calidris canutus* but larger and less dumpy, with longer thinner bill, slightly decurved at tip, and longer wing. In all plumages, supercilium and wingbar less pronounced than in Red Knot, but white uppertail band contrasts strongly with dark tail. Similar to Red Knot in non-breeding plumage, but less barred and more distinctly spotted below. Distinctive in breeding plumage with black streaking and blotching above and below, and heavy marking forming distinct breastband; chestnut confined to scapulars and tertials, appearing as clear patches as feather tips wear. Juvenile more heavily streaked on head and blacker above than in Red Knot, with round spots below (**A**).

Voice. Not tape-recorded. Rather silent. A low 'knut-knut' or 'chukker-chukker-chukker' in flight.

General Habits. Inhabits coastal mudflats and sandbanks in non-breeding season, often gregarious in Asian winter quarters. Flight and feeding movements similar to those of Red Knot.

Food. Invertebrates including crustaceans, annelids, molluscs and insects.

Reference
Lister, S. M. (1981)

Plates 18, 21

(Opp. pp. 305, 352)

Calidris canutus (Linnaeus). Red Knot. Bécasseau maubèche.

Tringa canutus Linnaeus, 1758. Syst. Nat. (10th ed.), p. 149; Sweden.

Range and Status. Breeds North America, Greenland, Spitzbergen and high arctic USSR; winters North and South America, Africa, W Europe and SE Asia to Australia.

Palearctic visitor to coastal mudflats and beaches. Locally very abundant Morocco to Guinea-Bissau, especially Mauritania (Banc d'Arguin, 300,000 + wintering). Also winters southern Africa (5000–10,000, Walvis Bay area, Namibia, to Natal, of which up to 4000 Langebaan Lagoon and 3000+ Cape Cross–Sandvis, Namibia, Dec–Jan). Rare to uncommon elsewhere W Africa. Frequent to common on passage Mediterranean coasts Morocco–Tunisia and Egypt; small numbers winter Tunisia. Vagrant or rare E and NE African coasts. Recorded inland as vagrant or rare Algeria, Libya, Sudan, Mali, Botswana and South Africa.

Description. *C. c. canutus* (Linnaeus): only subspecies in Africa. ADULT ♂ (breeding): forehead to nape and mantle blackish brown, feathers edged chestnut; scapulars blackish, edged and spotted chestnut, tipped whitish; supercilium and cheeks chestnut; lores, ear-coverts, sides of neck and hindneck chestnut, narrowly streaked blackish; back greyish brown, feathers edged white; lower rump and uppertail-coverts white, barred and marked dark brown; rectrices greyish brown. Underparts chestnut, but vent intermixed with white and undertail-coverts white, narrowly streaked brown. Primaries blackish brown, inners edged white on outer webs; primary coverts and alula blackish brown; secondaries and outer greater coverts greyish brown; inner primary and outer greater coverts tipped white; in some individuals, tertials and inner greater coverts as scapulars, and a few median and lesser coverts edged chestnut; otherwise tertials and rest of upperwing-coverts greyish; axillaries and underwing-coverts white, marked brown. Bill black; eye brown; legs and feet

Calidris canutus

olive-green. ADULT ♂ (non-breeding): differs from breeding bird in having upperparts, tertials and inner wing-coverts grey, feathers with darker shaft streaks and lighter edges; indistinct whitish supercilium; loral streak grey; ear-coverts, cheeks and sides of neck white, narrowly streaked greyish brown; underparts white, sides of breast greyish, foreneck, breast and flanks barred greyish brown. ADULT ♀: as ♂, but in breeding plumage upperparts less chestnut, with more greyish edging; underparts paler with more white feathers and some dusky barring; lower belly and vent white. Pre-nuptial moult Feb–early May usually well advanced before departure from Africa. Post-nuptial body moult mainly after arrival in Africa; wing-moult July–Nov (W Africa), Oct–Feb (South Africa). SIZE: wing, ♂ (n = 29) 160–176 (168), ♀ (n = 17) 167–177 (171); tail, ♂♀ (n = 39) 57–67 (62·2); bill, Siberian ♂♂ (n = 48) 33–37 (34·7), ♀♀ (n = 41) 34–40 (36·6), Greenlandic birds smaller ♂ (n = 28) 29–36 (32·6), ♀ (n = 18) 31–37 (34·2), mean for c. 1000 South African ♂♀ 35·4, and for 820 Mauritanian ♂♀ 35·3; tarsus, ♂♀ (n = 34) 29–33 (31·7). WEIGHT: (n = 1700+, South Africa) 93–212, typical wintering weight 110–150.

IMMATURE: juvenile plumage like non-breeding adult, but crown and nape streaked dark brown and buff; mantle and scapulars greyish brown, feathers fringed creamy with dark subterminal borders; sides of face and neck, hindneck and underparts suffused buff; tertials and inner greater and median coverts as mantle. 1st winter body plumage (as non-breeding adult) acquired Oct. Breeding plumage not usually acquired until 2nd year.

Field Characters. Large stocky *Calidris* with short dark bill and greenish legs, feathers with pale edges producing scaly appearance. In flight shows bold wing-bar, whitish rump and pale tail. In breeding plumage, with chestnut underparts and dark richly marked upperparts, can be confused only with Curlew Sandpiper *C. ferruginea*, which has white rump contrasting with darker tail; distinguished by larger size, paler shorter legs and short straight bill.

Voice. Tape-recorded (62, 73, 76, C). A low-pitched monosyllabic 'k-nut', often by many birds in concert, and a low flight call, 'quick-ick'.

General Habits. Occurs on entirely coastal habitats, mainly estuaries and tidal mudflats. Gregarious; often in flocks of thousands. Birds feed close together, heads down, walking unhurriedly and probing rapidly and continuously. Flight fast and powerful; often flies in long flocks to and from tidal roosts.

Flocks return NW Africa Aug–Sept and South Africa mainly Nov; spring departure from south mid Apr, from north late Apr–early May. 1st year birds commonly oversummer. Migration is by very long flights between few strategic staging areas. African wintering birds originate Siberia, reaching west coast via NW and W Europe, and southern Africa probably all part of same movement, perhaps across open Atlantic. Main return movement from south again apparently via W Africa and W Europe rather than by more direct NNE route across Africa and through Middle East. Autumn passage birds ringed Norway, E Germany, Poland and British Isles recovered Mauritania, and autumn birds ringed British Isles, E Germany and Belgium recovered South Africa. British ringed autumn birds also recovered Senegambia, Liberia, Gabon, Congo, and Mozambique, and a Norwegian ringed bird found Ivory Coast. Birds ringed Morocco and Mauritania Sept–Nov recovered France and W Germany in May, and South African ringed birds found Sweden, W Germany and Denmark in autumn and twice France in May. A juvenile ringed Mauritania (Sept) recovered South Africa following June.

Food. In coastal wintering areas, mainly molluscs, polychaetes and crustaceans.

References
Dick, W. J. A. *et al.* (1976).
Summers R. W. and Waltner, M. (1979).

SCOLOPACIDAE

Plates 18, 21
(Opp. pp. 305, 352)

Calidris alba (Pallas). Sanderling. Bécasseau sanderling.

Tringa alba Pallas, 1764. In Kroeg's Cat. Adumbr., p. 7; coast of North Sea.

Range and Status. Breeds high arctic from NE Greenland and Spitzbergen to N Taimyr and New Siberian Is.; also North America; winters Africa, Madagascar, W Europe, S and E Asia to Pacific islands, Australasia and coasts of North and South America.

Palearctic visitor, widespread on coasts throughout. Winters commonly to abundantly on Atlantic and Indian Ocean seaboard (e.g. 34,000 Banc d'Arguin, Mauritania; 25,000 on 250 km Namibian coast; 15,000 1100 km SW Cape shoreline; 2000–5000 on 200 km Kenya coast: D. J. Pearson, pers. obs.); less commonly and more locally Mediterranean and Red Sea coasts. Rare to uncommon inland, with records Morocco, Libya, Ghana and Nigeria, and from S Sudan, Ethiopia and E Africa through Zambia and Malawi to Namibia and South Africa.

Calidris alba

Main wintering range
Rare to locally frequent on passage

Description. ADULT ♂ (breeding): forehead to nape, mantle and scapulars blackish, broadly edged pale chestnut and tipped whitish; hindneck and sides of neck chestnut, mottled darker; lores and sides of face pale chestnut, spotted and streaked brown; back, centre of rump and uppertail-coverts black, tipped and fringed chestnut and grey; sides of rump and uppertail-coverts white. Rectrices pale greyish brown, centre pair edged chestnut. Chin whitish; foreneck to upper breast pale chestnut, streaked brown; rest of underparts white. Primaries blackish brown, outer webs of inner feathers white basally; primary coverts, alula and secondaries blackish brown; tertials and inner wing-coverts blackish brown edged chestnut; outer median and greater coverts grey; primary and greater coverts with broad white tips; axillaries and underwing-coverts white. Bill black; eye dark brown; legs and feet black. ADULT ♂ (non-breeding): differs from breeding bird in having top of head, hindneck, mantle and scapulars ashy grey with dark shaft streaks; forehead and sides of face white, lores and ear-coverts streaked dusky; underparts white, sides of breast mottled grey; tertials and inner wing-coverts grey with dark mesial streaks; lesser coverts blackish grey. ADULT ♀: as ♂, but in breeding plumage upperparts edged greyer and inner wing-coverts greyer. SIZE: wing, ♂ (n = 30) 120–131 (125), ♀ (n = 34) 119–131 (127); tail, ♂ (n = 53) 48–56 (50·9), ♀ (n = 30) 48–55 (51·6); bill, ♂ (n = 18) 21–26 (24·0), ♀ (n = 19) 22–27 (25·2); tarsus, ♂ (n = 17) 22–27 (24·2), ♀ (n = 19) 24–27 (24·7). WEIGHT: (n = 1000+, South Africa) 42–103; typical wintering weight 45–60.

Post-nuptial body moult July–Aug, mostly before arrival Africa; wing moult Africa late Sept–Feb. Pre-nuptial moult begins Feb–Mar, but chestnut feathers gained after leaving Africa.

IMMATURE: juvenile plumage same as non-breeding adult, but top of head and upperparts blackish brown, feathers tipped and edged pale buff and white; hindneck ashy; forehead and supercilium whitish; lores and ear-coverts dusky; rest of head and underparts white, breast suffused creamy buff; tertials, inner greater and inner median coverts broadly fringed and spotted pale buff or whitish; inner medians whitish with dark shaft streaks and thin dark fringes; lesser coverts dark brown. Replaced by 1st winter plumage (as non-breeding adult) on body Sept–early Nov. 1st summer usually as non-breeding adult; few attain breeding plumage before 2nd year.

Field Characters. Compact, whitish medium-sized *Calidris* with short stout bill, short dark legs, prominent dark wing shoulder and conspicuous white bar along middle of upperwing. In breeding plumage (rarely well developed in Africa) mottled chestnut upperparts and chestnut head and breast well demarcated from white belly. In juvenile, whitish forehead, face and underparts contrast with blackish cap and upperparts.

Voice. Tape-recorded (62, 73, 76, C). A sharp distinctive 'chick', especially in flight and when alarmed.

General Habits. Almost entirely confined to coasts, especially sandy beaches, but also rocky shores and coral; scarce on creeks and mudflats. Inland records mainly from sandy or stony shores of larger freshwater and soda lakes, or riverbanks. On coast usually in parties of 5–20, sometimes over 200 together; usually associates with Great Sand-Plover *Charadrius leschenaultii*, Mongolian Plover *C. mongolus* and Curlew Sandpipers *Calidris ferruginea*, or with Ruddy Turnstones *Arenaria interpres*, but often commonest wader on open shores. Extremely active; feeds mainly by probing, interspersed with short fast runs, body horizontal and head held low; also picks items from surface of recently exposed beach, characteristically following receding wave to take food then retreating rapidly before next breaker. May roost in hundreds on beach at high tide. Flight quick and agile, typically in compact flocks low over breakers.

Arrives Morocco from late July onwards, but scarce elsewhere N Africa until late Sept–Oct. 1st adults reach tropics early Aug, but main arrival E and W Africa late Aug–Sept and Cape Sept–Oct. Young birds arrive E

Africa Sept–Oct and Cape late Oct–Dec. Main departure South Africa and E Africa is late Apr–early May. Spring passage N African coast late Mar–May. Adults commonly 60–90% above typical winter weight before departure South Africa. 1st year birds oversummer commonly African coasts south to Cape, often congregating into small flocks in more sheltered sites. Inland occurrences are mainly late Aug–Nov, suggesting trans-African passage to southern and southwestern coasts. African birds probably all of Palearctic rather than Greenland origin. Birds ringed South Africa recovered June on Yenisei delta (72°N, 84°E) and several on autumn passage Black and Caspian Seas; others recovered May Tunisia, Britain, Netherlands, and (once) S Russia indicating return to Siberia via W Sahara, Mediterranean and NW Europe. Birds ringed Britain, recovered Algeria, Tunisia, Western Sahara, Mauritania, Senegambia, Ghana and South Africa.

Food. Small crustaceans, molluscs and marine worms; also insects, especially adult and larval dipteran flies, and sometimes plant material.

References
Dowsett, R. J. (1980).
Summers, R. W. and Waltner, M. (1979).

Calidris ruficollis (Pallas). Rufous-necked Stint; Red-necked Stint. Bécasseau à col roux.

Plate 18
(Opp. p. 305)

Trynga ruficollis Pallas, 1776. Reise Versch. Prov. Russ. Reichs, Vol. iii, p. 700; Kulussutai, E Siberia.

Forms a superspecies with *C. minuta*.

Range and Status. Breeds NE Siberia and Alaska; winters mainly S China to the Andaman Islands and south to Australasia; also Africa.

Palearctic visitor to Indian Ocean and South African coasts, rare Natal, vagrant Somalia, Kenya, Mozambique and Cape Province. Perhaps more widespread on E African seaboard than current records show.

Description. ADULT ♂ (breeding): forehead white; crown, nape, mantle and scapulars blackish brown, broadly edged chestnut and tipped whitish; hindneck pale chestnut; back, centre of rump and uppertail-coverts blackish edged chestnut; sides of rump white. Central rectrices blackish brown, narrowly edged chestnut; rest of tail brownish white. Throat, sides of neck, foreneck and upper breast bright chestnut (the last with small brown spots); rest of underparts white. Primaries blackish, inner feathers with whitish outer edging; primary coverts, alula and secondaries blackish brown; tertials and inner wing-coverts greyish brown; inner primary and outer greater coverts tipped white; underwing-coverts light brown, fringed white, axillaries white. Bill black; eye blackish brown; legs and feet black. ADULT ♂ (non-breeding): differs from breeding bird in having top of head, hindneck, sides of neck, mantle and scapulars greyish brown, feathers with dark shaft streak; back, centre of rump and uppertail-coverts brown edged greyish; foreneck and upper breast white with fine greyish brown streaking; tertials and inner wing-coverts as mantle; central rectrices dark brown. Sexes alike. SIZE: wing, ♂ (n = 21) 98–108 (103), ♀ (n = 25) 96–111 (105); tail ♂ (n = 16) 40–46 (42.6), ♀ (n = 12) 40–45 (42.3); bill, ♂♀ (n = 48) 16–20 (17.4); tarsus, ♂♀ (n = 47) 18–21 (19.1). WEIGHT: (n = 88, wintering Australia) 17–28.

IMMATURE: juvenile plumage as non-breeding adult, but upperparts dark brown with narrow pale buff fringes; centre of breast washed buffish, unstreaked; tertials, upperwing-coverts and central rectrices fringed pale buff.

Field Characters. Breeding bird distinguished from Little Stint *C. minuta* by chestnut throat and upper breast; tertials and wing-coverts also usually greyer. In non-breeding plumage, extremely difficult to distinguish from Little Stint; usually subtly greyer above, bill fractionally more robust; wing and tail longer, legs slightly shorter, to give more tapered body shape; dark centres to upperpart feathers typically narrower than Little Stint's. Juvenile generally duller and more uniform looking than in Little Stint, and lacks clear V-pattern on mantle.

Voice. Not tape-recorded. A thin, squeaking 'chit' or 'week' and a short, rolled 'tirriw'; also 'tirriw-chit-chit'. General tone weaker and less penetrating than in Little Stint.

General Habits. Feeding and flocking behaviour much as Little Stint, but more a bird of seacoasts in non-breeding season, preferring mudflats and sandspits to marshy puddles.

Food. Insect larvae and small crustaceans.

Reference
Sinclair, J. L. and Nicolls, H. (1976).

288 SCOLOPACIDAE

Plates 18, 21
(Opp. pp. 305, 352)

Calidris minuta (Leisler). Little Stint. Bécasseau minute.

Tringa minuta Leisler, 1812. Nachtr. zu Bechst. Naturg. Deuschl., p. 74; near Hanau, Germany.

Forms a superspecies with *C. ruficollis*.

Range and Status. Breeds in tundra, NE Norway to Siberia at 150°E; winters Africa, Madagascar, Mediterranean basin and S Asia.

Palearctic visitor, abundant to very abundant on mudflats and shallow water habitats inland, and on coastal creeks and estuaries. Winters with probably a few hundred thousand on freshwater and soda lakes in Ethiopia and E Africa (e.g. 10,000–20,000 S Kenya, 7000 on 5 km W shore L. Turkana, also large numbers locally in N tropical belt and in South Africa (6000 SW Cape). Mainly on passage N African coast.

Description. ADULT ♂ (breeding): forehead white; crown, nape, mantle and scapulars blackish brown, broadly edged chestnut and tipped whitish; hindneck and sides of neck pale chestnut; back, centre of rump and uppertail-coverts blackish edged chestnut; sides of rump white. Central rectrices blackish brown, edged chestnut; rest of tail brownish white. Foreneck and upper breast pale chestnut or whitish spotted with brown; rest of underparts white. Primaries blackish, inner feathers with whitish outer edging; primary coverts, alula and secondaries blackish brown; tertials and inner wing-coverts blackish brown edged chestnut; outer median and greater coverts greyish brown; inner primary and outer greater coverts tipped white; underwing-coverts light brown, fringed white, axillaries white. Bill black; eye blackish brown; legs and feet black. ADULT ♂ (non-breeding): differs from breeding bird in having top of head, hindneck, sides of neck, mantle and scapulars greyish brown with darker feather centres; back, centre of rump and uppertail-coverts brown edged greyish; foreneck and upper breast white with fine greyish brown streaking; tertials and inner wing-coverts as mantle; central rectrices dark brown. Sexes alike. SIZE: wing, ♂ (n = 45) 92–99 (95·9), ♀ (n = 49) 96–103 (98·7); tail, ♂ (n = 16) 38–42 (39·6), ♀ (n = 14) 38–42 (40·3); bill, ♂ (n = 81) 16–19 (17·3), ♀ (n = 99) 17–20 (18·5); tarsus, ♂ (n = 40) 19–23 (20·9), ♀ (n = 50) 20–22 (21·5). WEIGHT: (n = 10,000, Africa) 15–43, typical wintering weight 18–25.

Pre-nuptial moult completed Africa, Feb–early May. Post-nuptial body-moult Aug–Oct, mainly in Africa; wing-moult Africa, typically Sept–Feb, but earlier in north.

IMMATURE: juvenile plumage as non-breeding adult, but feathers of underparts dark brown fringed rich buff and whitish, edges of outer mantle feathers forming prominent whitish lines converging posteriorly; hindneck greyish; centre of breast washed buffish, unspotted; tertials, inner greater coverts, median and lesser coverts and central rectrices fringed rich tawny buff or cinnamon. Replaced on body by 1st winter plumage (as non-breeding adult) Oct–early Nov. 1st winter wing-moult usually complete Dec–Apr. Full breeding plumage attained in 1st summer.

Field Characters. A small, compact-bodied *Calidris*, with short fine bill and dark legs. In flight shows narrow wing-bar extending onto primary coverts, dark rump centre and greyish outer tail. In breeding plumage (Aug–Sept and Apr–May, less frequent further south) mottled chestnut upperparts and white throat and underparts are distinctive. In juvenile plumage whitish forehead, face and underparts contrast with dark upperparts, and a V-shaped pattern formed by pale mantle edges is prominent. Distinguished from Temminck's Stint *Calidris temminckii* by dark legs, greyish outer tail, more mottled upperparts and whiter breast; also by call and more gregarious habits. From Rufous-necked Stint *C. ruficollis* in breeding plumage by pale foreneck and upper breast. For distinctions from Long-toed Stint *C. subminuta* and small Nearctic calidrids see under those species.

Voice. Tape-recorded (62, 73, 76, 233, 264). A high-pitched 'chit' or 'chi-chi-chit'; flocks produce a twittering chorus when disturbed.

General Habits. Frequents muddy lake-shores, edges of small dams and pools, floodwater margins and sandbanks along rivers. On coast, mainly confined to shallow lagoons, saltpans, creeks and estuaries. Young birds more partial to temporary floods and rain puddles than adults. Usually in small groups, but many hundreds may occur together. Frequently consorts with Curlew Sandpipers *C. ferruginea*. Rather tame, usually feeds busily with head down, picking continuously from side to side at mud or water surface. Rests in compact flocks, usually on ridge or sandbank. Flight fast and agile, often in tight flocks of tens or hundreds, birds flashing white underparts simultaneously as they turn and bank together.

Adults arrive N Africa late July–Aug. Most reach tropical wintering areas Aug–Sept, and Cape Oct. 1st year birds appear at equator by early Sept, but most arrive Oct, and reach Cape Nov–Dec. Adult passage Kenya Aug–early Sept; 1st year passage mainly Oct.

Southward passage Zambia late Aug–Nov; Malawi Aug–Oct. Birds leave Cape early Apr, and southern, central and W Africa late Apr–early May. Many remain Ethiopia and E African lakes to mid and even late May. Commonly 60–100% above usual winter weight before departure from South Africa and Kenya. Little northward passage noted south of Sahara, but substantial movement N Africa late Apr–early June. Very few birds oversummer in Africa.

Birds ringed in Senegambia and Tunisia recovered from central Mediterranean to Black Sea and from Finland to USSR (Kara Sea). Birds ringed southern Africa recovered on passage Kazakhstan and S Caspian and once in June NE European Russia. Several birds ringed Zimbabwe and 1 ringed Zambia recovered Kenya Rift Valley lakes in Aug–Oct or Mar–May. Birds ringed Kenya recovered Kazakhstan (Aug), Zimbabwe and E Zaïre. Birds ringed Britain recovered Morocco and Tunisia respectively.

Food. Chiefly insects, especially larvae and adults of flies; also small crustaceans and molluscs.

References
Middlemiss, E. (1961).
Pearson, D. J. (1984).
Pearson, D. J. *et al.* (1970).

Calidris temminckii (Leisler). Temminck's Stint. Bécasseau de Temminck.

Plates 18, 21
(Opp. pp. 305, 352)

Tringa temminckii Leisler, 1812. Nachtr. zu Bechst Naturg. Deuschl., p. 78; near Hanau, Germany.

Range and Status. Breeds N Scandinavia and across N USSR; winters Africa, and S and SE Asia.

Palearctic visitor, locally common and sometimes abundant on freshwater marshes and lake margins south to Mali, N Nigeria, N Zaïre, Burundi and Kenya; rare N Tanzania and vagrant Zambia and Namibia. Mainly a passage migrant in N Africa and Ethiopia (Eritrea) but small numbers winter W Morocco, Tunisia and Egypt.

Description. ADULT ♂ (breeding): top of head, hindneck and some mantle and scapular feathers blackish brown, tipped and edged warm buff; rest of mantle and scapulars greyish brown; indistinct supercilium; lores and sides of face and neck pale buff, streaked dusky; back and centre of rump and uppertail-coverts mixed grey-brown and blackish with warm buff edging; sides of rump and uppertail-coverts white. Outer 3 pairs of rectrices white, rest greyish brown, central pair darker. Foreneck and upper breast greyish with darker streaking and some warm buff edging; rest of underparts white. Primaries blackish brown, inners with whitish outer edging; primary coverts and alula blackish brown, inner primary coverts broadly tipped white; secondaries brown, fringed white; some tertials and inner wing-coverts olive-brown, edged buff; rest of tertials and upperwing-coverts greyish brown, greaters broadly tipped white; axillaries white; underwing-coverts white mottled dusky. Bill dark brown; eye dark brown; legs olive-green or yellowish. ADULT ♂ (non-breeding): similar to breeding plumage but uniform greyish brown above including tertials and inner wing-coverts; white below, upper breast suffused with greyish brown and lacks streaking. Sexes alike. SIZE: wing, ♂ (n = 31) 94–103 (97·6), ♀ (n = 14) 92–102 (98·1); tail, ♂♀ (n = 20) 42–48 (45·6); bill, ♂♀ (n = 30) 16–18 (16·9); tarsus, ♂♀ (n = 34) 17–18 (17·5). WEIGHT: (n = 39, Kenya) 17–26, typical wintering weight 19–22.

IMMATURE: juvenile as non-breeding adult but upperparts, tertials and upperwing-coverts browner and fringed with buff; coverts with thin subterminal dusky lines; breast washed brown.

Field Characters. A small *Calidris*, greyish brown or brown above, with short pale legs. Shows dark rump centre and narrow wing-bar in flight, but has distinctive white outer tail feathers. Separated from Little Stint *C. minuta* also by plainer upperparts, more pronounced breast-band and call; legs typically more flexed, and flight more jerky and erratic, often with flickering wing action recalling Common Sandpiper *Actitis hypoleucos*.

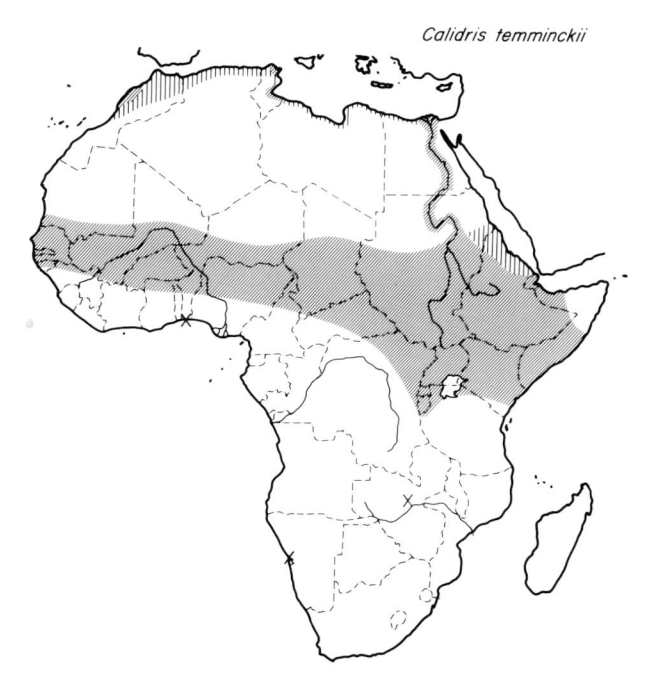

Voice. Tape-recorded (62, 73, 76). A quiet twitter, 'titititititi', usually given when flushed.

General Habits. Typically an inland species, frequenting ditches, edges of marshes and floodpools, and muddy lake margins, especially where cover afforded by low plants. Usually singly or in 2s and 3s, but flocks up to 20–30 on migration. Walks briskly when feeding, head down, picking continuously from mud surface and vegetation. Tends to tower when flushed.

Southward passage Morocco and Tunisia from late July to early Oct; Mali Aug–Oct; Ethiopia (Eritrea)

Sept–early Oct. A few reach tropics by Aug, but most from Oct. Departs from winter quarters mainly late Mar–Apr, but some linger tropics into May. Spring passage Morocco end Mar–early June, NE Africa and Ethiopia (Eritrea) Apr-mid May. Birds ringed S France (Aug) and Scandinavia recovered Benin (Mar) and Tunisia (Feb) respectively.

Food. Chiefly insects; also annelids, crustaceans, small molluscs and some vegetable matter.

Plate 18
(Opp. p. 305)

Calidris subminuta (Middendorf). **Long-toed Stint. Bécasseau à longs doigts.**

Tringa subminuta Middendorf, 1851. Siber. Reise II, p. 222; Udskii Ostrog.

Range and Status. Breeds E Siberia, probably also boreal zone of central and W Siberia; winters S and SE Asia; vagrant Africa.

Palearctic vagrant Ethiopia (L. Abiata, Jan 1964), Kenya (Naivasha, Apr 1969 and Feb 1974, Nakuru, May 1970 and Nairobi, Feb 1984), Mozambique (Maputo, Dec 1976–Jan 1977), and South Africa (Steynrus Dam, Jan 1967).

Description. ADULT ♂ (breeding): forehead to nape, mantle and scapulars blackish brown, feathers edged tawny, tipped greyish white; prominent whitish supercilium; lores dusky; sides of face and neck and hindneck buffish white streaked dark brown, cheeks tinged warm brown; back and centre of rump and uppertail-coverts blackish edged warm buff; sides of rump and uppertail-coverts white. Rectrices brownish white, central 2 pairs blackish brown edged buff. Underparts white, foreneck, breast and flanks streaked and mottled brown. Primaries, primary coverts, alula and secondaries blackish brown; tertials, inner wing-coverts and new lesser coverts brown, edged warm buff; rest of wing-coverts greyish brown, greaters tipped white; axillaries white; underwing-coverts white mottled brown. Bill black; eye brown; legs and feet greenish or yellowish. ADULT ♂ (non-breeding): as breeding bird, but upperparts, tertials and innermost wing-coverts dark brown with greyish brown fringes; sides of face and neck, and breast, more lightly streaked. Sexes alike. SIZE: wing, ♂ (n = 22) 90–97 (93·5), ♀ (n = 30) 91–100 (95·4); tail, ♂♀ (n = 10) 36–38 (37); bill, ♂♀ (n = 24) 17–19 (18·0); tarsus, ♂♀ (n = 26) 20–23 (21·4). WEIGHT: (n = 3, Kenya) 20, 28, 33.

IMMATURE: juvenile as non-breeding adult, but top of head blackish brown edged rich buff, and mantle same with fringes of white forming fairly distinct V laterally; tertials and upperwing-coverts fringed pale buff; underparts whitish, breast suffused greyish brown and faintly streaked.

Field Characters. Very small *Calidris* distinguished from Little Stint *C. minuta* by pale legs; also by longer necked appearance, more prominent supercilium, dark top of head contrasting with paler hindneck and in flight by shorter wing-bar. Distinguished from all small calidrids by longer toes. In non-breeding plumage, darker and more mottled above than Little Stint, and more prominently streaked on head, neck and upper breast, with rather sharp demarcation from rest of white underparts. Brightly coloured and stongly patterned in breeding plumage, but face and upperparts darker than in Little Stint, and breast marking heavier. Distinguished from Temminck's Stinct *C. temminckii* in all plumages by well patterned appearance, greyish outer tail feathers and call (see below).

Voice. Tape-recorded (F). Gives a purring 'chrrrp'.

General Habits. Mainly a freshwater species found on muddy puddles and shallow marshy lake edges, usually in small parties. Feeding and movements much as Little Stint, but adopts pronounced alarm posture with long neck stretched.

Food. Insects; also small molluscs and crustaceans.

Not illustrated in colour

Calidris fuscicollis (Vieillot). **White-rumped Sandpiper. Bécasseau de Bonaparte.**

Tringa fuscicollis Vieillot, 1819. Nouv. Dict. d'Hist. Nat. nouv. ed. xxxiv, p. 146; Paraguay.

Range and Status. Breeds N Canada, winters South America; vagrant Africa.

Nearctic vagrant to Namibia (Hoamib R. mouth, Dec 1981) and South Africa (Cape Town, Dec 1979; Port Alfred, 2 Oct 1983).

Description. ADULT ♂ (non-breeding): upperparts greyish brown, feathers with darker centres; sides of face and neck white, streaked dusky; back and rump dark brown, edged greyish, lateral feathers white; uppertail-coverts white. Rectrices dull brown, centre 2 pairs darker. Foreneck and upper breast greyish brown, finely streaked dusky; flanks washed greyish brown; rest of underparts white. Primaries and alula dark brown; primary coverts same, inners narrowly tipped white; secondaries dark brown, tipped white; tertials and rest of upperwing-coverts greyish brown, outer greaters tipped white; axillaries white, underwing-coverts mainly white. Bill blackish, dull greenish at base; eye dark brown; legs and feet blackish. Sexes alike. SIZE: wing, ♂ (n = 29) 120–129 (123), ♀ (n = 29) 118–127 (124); tail, ♂♀ (n = 24) 50–54 (51·2); bill, ♂ (n = 15) 22–24 (22·8), ♀ (n = 19) 23–26 (24·5); tarsus, ♂♀ (n = 35) 23–25 (24·2). WEIGHT: (n = 13, North America, June–Aug) 32–51; (n = 20, Venezuela, Nov–Apr) 27–42.

Field Characters. Medium-sized *Calidris* with dark legs and straight dark bill, showing white across uppertail-coverts in flight. Curlew Sandpiper *C. ferruginea* is larger, with longer curved bill and proportionately much longer neck and legs. Distinguished from Dunlin *C. alpina* at rest by shorter finer straight bill and noticeably longer wings, well overlapping tail. Greyer and more uniform above in non-breeding plumage (**A**) than Baird's Sandpiper *C. bairdii*.

Voice. Tape-recorded (62,73). Rather silent; in flight, a high-pitched 'treet'.

General Habits. Occurs mainly on coastal mudflats, creeks and saltwater lagoons, but also on inland pools and marshes. Often singly or in small parties, but at times in large flocks. Flight and feeding behaviour much as in Dunlin *C. alpina*.

IMMATURE: juvenile as non-breeding adult, but top of head, mantle, scapulars, tertials, inner wing-coverts and central rectrices dark brown, edged rich tawny; median and lesser wing-coverts dark brown, fringed buff or whitish; breast suffused with buff and narrowly streaked brown.

Food. Annelids, crustaceans, insects, molluscs and seeds.

Calidris bairdii (Coues). Baird's Sandpiper. Bécasseau de Baird.

Actodromas bairdii Coues, 1861. Proc. Sci. Philad., p. 194; Great Slave Lake, Canada.

Plates 18, 21
(Opp. pp. 305, 352)

Range and Status. Breeds arctic North America and easternmost Siberia; winters South America; vagrant Africa.

Nearctic vagrant with single records Senegambia (Bund Road, Nov 1976; Gandiol, 16 Dec 1965) and Namibia (Walvis Bay, 24 Oct 1863).

Description. ADULT ♂ (non-breeding): top of head light buff, streaked dark brown; mantle and scapulars greyish brown with darker centres, edged warm buff; supercilium whitish; lores dusky; sides of face and neck buff, narrowly streaked dark brown; back, rump and uppertail-coverts dull brown edged buff; lowermost lateral uppertail-coverts dull white. Rectrices greyish brown, centre pair darker. Foreneck and upper breast buff, narrowly streaked dark brown; rest of underparts white. Primaries and alula blackish brown; primary coverts same, inners narrowly tipped white; secondaries dark brown tipped white; tertials dark brown edged sandy buff; rest of upperwing-coverts greyish brown, greaters tipped white, medians and lessers edged pale buff. Bill black; eye dark brown; legs and feet black or dark olive. Sexes alike. SIZE: wing, ♂ (n = 21) 119–131 (126), ♀ (n = 17) 123–135 (128); tail, ♂♀ (n = 10) 51–54 (52·3); bill, ♂♀ (n = 23) 21–24 (22·8); tarsus, ♂♀ (n = 25) 21–24 (22·5). WEIGHT: (n = 3, Peru) 32, 35, 38; (n = 38, Alaska, May) 32–48.

IMMATURE: juvenile as non-breeding adult, but crown to nape, mantle and scapulars dark brown with broad sandy buff fringes; back and rump blackish, edged sandy; breast sandy buff with streaking inconspicuous; tertials, inner greater, median and lesser coverts dark brown, fringed sandy buff.

Field Characters. Medium-sized *Calidris*, adult rather mottled brown above and buffish across breast, with short dark legs and straight or slightly drooping fine dark bill. Shows dark rump and uppertail-coverts in flight, and short narrow wing-bar. At rest distinguished from Dunlin *C. alpina* by shorter, straighter and finer bill, buff-toned plumage and long wings projecting beyond tail. Juvenile appears very scaly on mantle.

Voice. Tape-recorded (63, 73). Flight note, a reedy 'chirr'.

General Habits. Frequents marshy pools inland; also brackish and saltwater lagoons. Usually singly or in small parties. Feeds actively, picking from surface like a stint.

Food. Insects, crustaceans and vegetable matter.

292 SCOLOPACIDAE

Plates 18, 21

(Opp. pp. 305, 352)

Calidris melanotos (Vieillot). Pectoral Sandpiper. Bécasseau tachète.

Tringa melanotos Vieillot, 1819. Nouv. Dict. d'Hist. Nat. nouv. ed. xxxiv, p. 462; Paraguay.

Range and Status. Breeds arctic coast E Siberia, Alaska and NW Canada, wintering E Australia, New Zealand and South America; vagrant Africa.

Vagrant from Nearctic and perhaps also from E Palearctic; recorded Morocco (4 times), Libya (1), Kenya (2), Zambia (2), Zimbabwe (3), Botswana (3), Namibia (1) and South Africa (10).

Description. ADULT ♂ (breeding): forehead to nape, mantle and scapulars blackish brown, feathers fringed bright cinnamon; long narrow buffish supercilium; ear-coverts, cheeks, sides of neck and hindneck pale cinnamon, streaked blackish brown; back, rump and uppertail-coverts blackish with narrow cinnamon fringes, sides of rump and uppertail-coverts white. Central rectrices blackish, fringed cinnamon, rest pale brown. Chin white; foreneck and upper breast creamy, boldly streaked or mottled blackish brown, sharply demarcated from white lower breast, flanks, belly and undertail-coverts. Primaries, alula and primary-coverts dark brown, last with small white tips on inners; secondaries dark brown, tipped and edged white; greater coverts dark brown, tipped white; tertials and rest of upperwing-coverts blackish brown, fringed bright cinnamon; axillaries white, underwing-coverts mainly white. Bill blackish, base olive green; eye dark brown; legs dull yellow or brownish. ADULT ♂ (non-breeding): like breeding adult, but upperparts, tertials and wing-coverts less blackish, feather fringes pale greyish buff; streaking on foreneck and breast narrower. Sexes alike. SIZE: wing, ♂ (n = 30) 138–149 (144), ♀ (n = 20) 126–135 (130); tail, ♂ (n = 14) 57–65 (61·3), ♀ (n = 17) 51–60 (53·3); bill, ♂ (n = 29) 27–32 (29·3), ♀ (n = 22) 24–29 (27·4); tarsus, ♂ (n = 31) 27–31 (28·8), ♀ (n = 20) 25–28 (26·6). WEIGHT: (N South America, late Aug–Sept) ♂ (n = 2) 79–83 (81·1), ♀ (n = 7) 41–61 (52·2).

IMMATURE: juvenile as breeding adult, but streaking on foreneck and breast narrower, hindneck greyish, and outer feathers of mantle and scapulars with buffish white outer edges.

Field Characters. Rather large, stocky *Calidris* with yellow legs and base to bill greenish. Heavily streaked neck and upper breast end sharply against white lower breast and belly. Neck somewhat longer than smaller stints. In flight appears uniform above, with dark rump centre and inconspicuous pale wing-bar. Juvenile has prominent pale lines bordering mantle.

Voice. Tape-recorded (62, 73). A reedy 'krrik-krrik' when flushed.

General Habits. Favours marshy habitat and wet grassland, but also lake edges, dams and saltpans. May associate with parties of smaller calidrids. Feeds on mud or in shallow water, walking slowly and picking and probing continuously. Flight fast with strong deliberate wing-beat, and at times quite agile.

African records all months Sept–May. The number of occurrences in eastern and southern parts of the continent suggests many are of Siberian rather than Nearctic origin.

Food. Insects, small crustaceans, annelids and vegetable matter.

Plates 18, 21

(Opp. pp. 305, 352)

Calidris ferruginea (Pontoppidan). Curlew Sandpiper. Bécasseau cocorli.

Tringa ferruginea Pontoppidan, 1763. Danske Atlas, 1, p. 624; Iceland and Christiansø.

Range and Status. Breeds tundra zone USSR from Yenisei to New Siberian Is.; winters Africa, Madagascar and S Asia to Australia.

Palearctic visitor, abundant to very abundant on coast throughout, and locally common to very abundant on muddy lake-shores inland. Some hundreds of thousands winter along Atlantic coast from Mauritania southwards, and S Red Sea and Indian Ocean coasts (wintering counts of 150,000+ Banc d'Arguin, Mauritania; tens of thousands Archipel des Bijogos, Guinea–Bissau; 21,000 Namibia on 200 km coast; 56,000 SW Cape; 5000–10,000 estimated Kenya coast). Inland, occurs mainly on passage, especially Mali, Nile Valley and E African lakes; winters locally and in relatively small numbers. Abundant both passages N African coasts, where a few winter SW Morocco and S Tunisia.

Description. ADULT ♂ (breeding): crown, nape, mantle and scapulars dark brown, broadly edged chestnut, tipped whitish; hindneck chestnut streaked darker; sides of face and neck deep chestnut red; back and rump dark brown edged grey;

uppertail-coverts white barred black. Rectrices greyish brown, centre pair dark brown edged chestnut. Chin whitish; throat to belly deep chestnut red, feathers edged white, sometimes barred dark grey subterminally and intermixed with white on lower breast, belly and flanks; undertail-coverts white barred dusky. Primaries, primary coverts and alula blackish; secondaries blackish brown; inner primaries narrowly edged whitish and inner primary and greater coverts tipped white; a few (new) tertials and inner wing-coverts as mantle; rest of upperwing-coverts greyish brown; axillaries white, underwing-coverts white, leading feathers mottled dusky. Bill blackish, slightly decurved; eye dark brown; legs and feet black. ADULT ♂ (non-breeding): differs from breeding bird in having upperparts greyish brown, feathers medially dusky and fringed whitish; lores dusky; ear-coverts, cheeks and sides of neck streaked dusky; rest of head including supercilium white; uppertail-coverts white; underparts white, foreneck and upper breast with fine dusky streaking; tertials, inner wing-coverts and central rectrices greyish brown. ADULT ♀: as ♂, but in breeding plumage has slightly paler edging above, and is paler below with more white feathers, and chestnut more confined to distinct bars. SIZE: wing, ♂ (n = 10) 125–130 (131), ♀ (n = 20) 125–136 (131); tail, ♂ (n = 12) 42–49 (45·4), ♀ (n = 19) 43–50 (45·8); bill, ♂ (n = 35) 33–39 (36·0), ♀ (n = 16) 35–42 (39·4); tarsus, ♂ (n = 31) 27–32 (29·3), ♀ (n = 15) 29–31 (29·7). WEIGHT (n = 2500, South Africa) 36–117, typical wintering weight 48–64; (n = 2000, Kenya) 32–86, typical wintering weight 42–56.

Breeding plumage in both sexes partially and sometimes fully acquired in Africa, Feb–May. Post-nuptial body-moult July–Sept, partly or mainly before arrival in Africa; wing-moult Africa Sept–Feb.

IMMATURE: juvenile plumage as non-breeding adult, but upperparts dark brown fringed pale buff; hindneck greyish; underparts suffused pale cinnamon buff, breast finely streaked brown; tertials and upperwing-coverts brown, fringed pale buff or whitish. Body feathers replaced by 1st winter plumage (as non-breeding adult) late Sept–Oct. Juvenile outer primaries usually moulted Feb–June. 1st summer plumage as non-breeding adult; breeding plumage acquired before 2nd summer.

Calidris ferruginea

Field Characters. Medium-sized *Calidris* with long slightly decurved bill and rather long dark legs. White rump and narrow white wing-bar prominent in flight. Distinctive chestnut breeding plumage on head and underparts usually only partial in Africa, Apr–May and July–Sept. Distinguished from non-breeding Dunlin *C. alpina* by more decurved bill, longer legs, white rump, more pronounced supercilium and less heavy streaking below.

Voice. Tape recorded (62, 73, 76, 264). A soft 'chirrup' in flight, mainly when disturbed.

General Habits. Particularly favours coastal estuaries, creeks and mud or sandflats; also frequents exposed coral, rocky shores and tidewrack on sandy beaches. Inland, mainly on muddy edges of freshwater and soda lakes; also at dams, irrigation and small flood patches. Gregarious; on coast usually in parties or larger flocks up to several hundreds, consorting with Sanderling *C. alba*, Red Knot *C. canutus*, Great Sand-Plover *Charadrius leschenaultii* and Mongolian Plover *C. mongolus*; typically the most abundant migrant wader on muddy shores and coastal wetlands. Inland, usually in small parties, outnumbered by Little Stint *C. minuta* and Ruff *Philomachus pugnax*, but in hundreds locally on passage. Usually feeds in shallow water or on exposed tidal flats, walking purposefully, head down, pausing frequently to make single or repeated probes or occasionally to pick from surface. At Langebaan, South Africa, birds foraged 55–65% of daytime during Sept–Feb, but up to 80% of daytime during June–Aug. Adults foraged more rapidly and more successfully than 1st-year birds (Puttick 1979). Wing-beat rather deliberate and powerful, but flight can be fast and flocks perform agile aerial manoevres. Coastal birds roost at high tide on sandspits and beaches, or among mangroves.

Adult passage N Africa mid July–mid Oct, young birds late Aug–Nov. Adults reach equator end July–early Aug, with large flocks there late Aug; they arrive South Africa mainly Sept. Young birds reach E Africa from Sept (occasionally end Aug) and Cape from mid Oct. Main passage period E Africa mid Aug–Nov (adults mid Aug–Sept), Zambia late Aug–Nov, Malawi Aug–Oct. Most adults leave Cape and E African coasts late Apr, South African birds accumulating large and E African birds moderate fat reserves; departure W Africa late Apr–early May. Marked northward passage E African Rift Valley late Apr–late May, and N African coast mid Apr–early June. Most 1st year birds apparently oversummer in Africa.

Birds ringed on passage in W Europe recovered NW Africa south to Mauritania and once Ivory Coast. Birds wintering in W Africa also arrive via S Europe, and recoveries suggest this is favoured return route. Birds ringed Tunisia, mostly recovered Black Sea area; ringed South Africa, recovered mainly Black and Caspian Seas areas but also Zaïre (May) and arctic coast USSR at 131°E (June); ringed Kenya, recovered Caspian area (2). Much migration to and from Guinea coast and South Africa is transcontinental, but there is also a continuous southward movement of adults along E African coast Aug–early Sept.

Food. Marine worms, molluscs, crustaceans, insect larvae and occasional vegetable matter. At Langebaan, South Africa, main items were nereid worm *Ceratonereis* and hydrobiid gastropod *Assimina*; also amphipods, small crabs and dipteran larvae. Adults took larger prey than did young birds, and range of items selected by ♂♂ and ♀♀ also differed significantly (Puttick 1978).

References
Elliott, C. C. E. *et al.* (1976).
Wilson, J. R. *et al.* (1980).

Calidris maritima (Brünnich). Purple Sandpiper. Bécasseau violet.

Plates 18, 21
(Opp. pp. 305, 352)

Tringa maritima Brünnich, 1764. Orn. Borealis, p. 54; Christiansø Is., off Bornholm.

Range and Status. Breeds N and NE Canada, Greenland, Iceland and N Scandinavia to arctic USSR at 100°E; winters Atlantic North America, W and NW Europe; vagrant N Africa.

Vagrant N Morocco (Tangier).

Description. ADULT ♂ (non-breeding): forehead to nape, mantle and scapulars sooty brown, feathers edged greyish; sides of head sooty brown; back, rump and central uppertail-coverts blackish brown, edged whitish, last with slight purple gloss; sides of rump and uppertail-coverts white. Rectrices greyish brown, 2 central pairs darker. Underparts whitish, but foreneck and upper breast sooty brown, lower breast heavily mottled and flanks and undertail-coverts streaked sooty brown. Primaries blackish brown, inners tipped whitish and margined white on outer webs; primary coverts and alula sooty brown, former tipped white; secondaries blackish brown tipped white, inners mainly white; tertials and rest of upperwing-coverts sooty brown, fringed greyish, greaters broadly tipped white. Bill blackish brown, base yellow; eye brown; legs and feet greenish yellow. Sexes alike. SIZE: (Scottish wintering birds) wing, ♂ (n = 10) 122–130 (127), ♀ (n = 10) 124–134 (129); tail, ♂ (n = 10) 53–56 (54·7), ♀ (n = 10) 55–58 (56·3); bill, ♂ (n = 9) 23–30 (26·4), ♀ (n = 12) 28–34 (30·2); tarsus, ♂♀ (n = 10) 20–23 (21·8). WEIGHT: (Scotland, Oct–Mar) 51–81.

IMMATURE: juvenile as non-breeding adult but upperparts, tertials and upperwing-coverts dark brown with creamy fringes, mantle slightly glossed; breast more streaked.

Field Characters. A robust, medium-sized *Calidris*, dark sooty brown above and heavily marked on breast and flanks, with short yellowish legs, and rather long, slightly decurved bill with yellowish base. In flight shows dark rump centre, narrow wing-bar and pale trailing edge to secondaries.

Voice. Tape-recorded (62, 73, 76). A low 'weet-wit'; rather silent.

General Habits. Frequents rocky shores, typically in small parties. Rather tame. Walks briskly on rocky coasts picking and foraging among seaweed and debris, occasionally wading in shallow water. Flight fast and direct.

Food. Insects, crustaceans, molluscs and small fish.

Calidris alpina (Linnaeus). Dunlin. Bécasseau variable.

Plates 18, 21
(Opp. pp. 305, 352)

Tringa alpina Linnaeus, 1758. Syst. Nat. (10th ed.), p. 149; Lapland.

Range and Status. Breeds E Greenland and from W and N Europe across N USSR; also W and N Alaska and N Canada; winters Africa, Europe, SW and E Asia and North America.

Palearctic visitor, wintering commonly to very abundantly Atlantic coast south to Guinea, and on Mediterranean and Red Sea coasts. Main Atlantic concentration Banc d'Arguin, Mauritania, estimated at 800,000; wintering counts 100,000–200,000 Tunisia (M. A. Czajkowski, pers. comm.). Vagrant to rare Nigeria and Somalia. Inland, uncommon to frequent Morocco and Algeria, mainly on passage; small flocks reported Mali and common along Nile south to Jonglei Province, Sudan (M. Rae, pers. comm.); otherwise vagrant to rare south to Burundi, Uganda, and Kenya, and South Africa (recorded once Kruger N. P. and twice from Cape).

Description. *C. a. alpina* (Linnaeus): NE Africa. ADULT ♂ (breeding): forehead to nape, mantle and scapulars blackish brown edged tawny; hindneck and sides of neck greyish brown streaked darker; lores and sides of face buff streaked dusky; back and centre of rump and uppertail-coverts blackish brown, edged tawny, sides of rump and uppertail-coverts white. Rectrices greyish brown, centre feathers darker. Foreneck and breast suffused buff and heavily streaked dark brown; belly black; rest of underparts white. Primaries blackish brown, outer webs of inners edged white; alula and primary coverts blackish brown, latter with white tips on inners. Secondaries blackish brown, tipped white, innermost mainly white; tertials and upperwing-coverts greyish brown, greaters with broad white tips; axillaries white, underwing-coverts mainly white. Bill black, slightly decurved; eye brown; legs and feet black. ADULT ♂ (non-breeding): as breeding bird, but upperparts greyish brown, feathers medially darker; ear-coverts, cheeks and sides of neck white, streaked greyish brown; streaking on breast less prominent and belly white.

Calidris alpina

ADULT ♀: as ♂, but in breeding plumage hindneck washed warm brown. SIZE: wing, ♂ (n = 55) 109–120 (115), ♀ (n = 30) 112–123 (117); tail, ♂ (n = 55) 42–58 (49·2), ♀ (n = 30) 44–57 (49·3); bill, ♂ (n = 48) 27–35 (31·1), ♀ (n = 30) 29–36 (32·9); tarsus, ♂ (n = 53) 23–26 (24·4), ♀ (n = 28) 23–27 (25·6). WEIGHT: (n = 88, Sudan, Nov–Dec) 31–55, mostly 35–48.

Pre-nuptial body-moult in Africa Feb–Apr; post-nuptial body- and wing-moult completed Africa Aug–Nov.

IMMATURE: juvenile plumage as non-breeding adult but upperparts blackish brown edged tawny buff; sides of face, foreneck and upper breast washed buff; streaking and spotting on breast heavier and extending to flanks; tertials and upperwing-coverts greyish brown, fringed buff. Body feathers replaced by 1st winter plumage (as non-breeding adult) Aug–Nov. Breeding plumage acquired in 1st year.

C. a. schinzii (Brehm): NW Africa. Breeding plumage with upperpart edging more yellowish buff than nominate race, belly patch smaller with more white feathers intermixed. SIZE: wing, ♂ (n = 226) 105–123 (112), ♀ (n = 171) 109–124 (115); tail, ♂ (n = 225) 40–58 (46·8), ♀ (n = 173) 41–55 (47·5); bill, ♂ (n = 218) 23–36 (28·7), ♀ (n = 161) 27–36 (31·7); tarsus, ♂ (n = 220) 22–27 (24·1); ♀ (n = 169) 23–28 (25·1). WEIGHT: (Mauritania, Oct–Nov, mainly *schinzii*) 26–59.

C. a. arctica (Schiøler): forms small percentage of NW African wintering population. Breeding plumage with upperpart edging more greyish brown or buffish than nominate, breast streaking fainter, belly patch smaller. Generally smaller with shorter bill. SIZE: wing, ♂ (n = 52) 107–117 (112), ♀ (n = 32) 112–122 (116); bill, ♂ (n = 51) 23–29 (26·1), ♀ (n = 31) 27–32 (29·5); tarsus, ♂ (n = 52) 22–24 (22·8), ♀ (n = 32) 22–25 (23·4).

Field Characters. Medium-sized *Calidris* with medium to longish black bill, slightly decurved at tip, and black legs. In flight shows dark rump centre and prominent narrow wing-bar. In breeding plumage (Aug–Sept and Apr) upperparts richly coloured and belly with diagnostic black patch. Distinguished from non-breeding Curlew Sandpiper *C. ferruginea* by dark rump centre, pronounced streaking on breast (flanks also in juvenile), less pronounced supercilium and shorter legs. Broad-billed Sandpiper *Limicola falcinellus* is smaller and shorter legged, with characteristic face pattern (see that species) and greyer non-breeding plumage.

Voice. Tape-recorded (62, 73, 76, C). Flight call a distinctive 'treerr'.

General Habits. Occurs mainly on estuaries and muddy sea coasts, also coastal lagoons, marshes and saltpans; inland on dams and lake edges. Gregarious; usually in parties or larger gatherings up to thousands. Associates freely with other small calidrids. Feeds chiefly on exposed tidal mud, walking steadily with head down and forward, picking and probing methodically. Small parties are rather tame, but large tidal roosts less easy to approach. Flight fast and agile, large flocks banking and wheeling rapidly.

1st birds return to Africa late July. Adult arrival and passage Morocco late July–Aug, juvenile passage Sept–Oct; main influx Tunisia Oct, Egypt Sept, Sudan and Ethiopia (Red Sea) late Sept–Oct, Mali Oct. Depart from wintering areas mainly Mar–early Apr, with passage Morocco and Mediterranean coast end Mar–late May. A few oversummer anywhere in main African range. Main wintering grounds of *schinzii* appear to be NW Africa. Moroccan and Mauritanian birds mainly this race, together with small contingent of nominate *alpina*, whilst contribution of Greenland *arctica* remains unclear. Many Moroccan and Mauritanian recoveries of birds ringed British Isles (passage), France and S Scandinavia, and 1 ringed Iceland recovered Morocco. Birds ringed Britain also recovered Western Sahara and Senegambia. Tunisian ringing indicates many birds came from Baltic area, but some (presumably nominate) recovered on passage Black Sea. Racial identity of birds wintering Senegambia and Mali unknown. Nominate birds wintering NE Africa presumably derived from NE Russia and Siberia.

Food. Inland, mainly insects, especially dipteran flies and beetles; in coastal areas, intertidal invertebrates, mainly polychaete worms, molluscs and crustaceans, including small crabs, shrimps and sandhoppers; commonly also vegetable matter.

Reference
Pienkowski, M. W. and Dick, W. J. H. (1975).

SCOLOPACIDAE

Genus *Limicola* Koch

Like small, short-legged *Calidris*, but bill long, flattened between nostril and tip, tip decurved. Plumages as typical *Calidris*. 1 sp., visiting Africa from Palearctic.

Plates 18, 21
(Opp. pp. 305, 352)

Limicola falcinellus (Pontopiddan). Broad-billed Sandpiper. Bécasseau falcinelle.

Scolopax falcinellus Pontopiddan, 1763. Danske Atlas, 1, p. 623; Denmark.

Range and Status. Breeds N Scandinavia to E Siberia; winters Africa, Iraq to India and SE Asia to Australia.

Palearctic visitor, mainly coastal, wintering probably Egypt, regularly at one site Kenya (up to 65 birds: D. J. Pearson pers. obs.) and in small numbers Ethiopia (Eritrea), South Africa (Natal) and Namibia. On passage regular in small numbers Tunisia, rare Libya, vagrant Morocco. Also recorded rarely coastal Somalia and Tanzania, and inland from Mali, Nigeria, Chad, Sudan, Ethiopia, Burundi, Uganda, Kenya, Tanzania, Malaŵi, Zambia, Zimbabwe and South Africa (Cape). Other E or NE African wintering sites probably yet to be discovered.

Description. ADULT ♂ (breeding): forehead to nape, mantle and scapulars blackish brown, feathers edged bright tawny, tipped whitish; white supercilium bifurcated behind eye; lores and ear-coverts dusky, streaked tawny; hindneck and sides of face and neck white, streaked and spotted dark brown; back, rump and uppertail-coverts blackish brown tipped tawny, sides of uppertail-coverts white. Rectrices greyish brown, central pair blackish, edged tawny. Chin whitish; foreneck, breast and flanks tinged tawny, heavily streaked and spotted dark brown; rest of underparts white. Primaries, secondaries and alula blackish brown; primary coverts same, inners tipped white; tertials and inner greater coverts like mantle; rest of upperwing-coverts dull brown, outer greaters broadly tipped white; axillaries white, underwing-coverts mainly white. Bill blackish, distinctively broad from base to tip and decurved near tip; eye dark brown; legs and feet dark brown or dark greenish. ADULT ♂ (non-breeding): differs from breeding bird in having upperparts, tertials and inner wing-coverts greyish with lighter edges; small lesser coverts blackish brown. Sexes alike. SIZE: wing, ♂ (n = 13) 102–109 (105), ♀ (n = 17) 103–114 (110); tail, ♂ (n = 12) 36–40 (37·9), ♀ (n = 15) 37–42 (38·9); bill, ♂ (n = 11) 28–33 (30·4), ♀ (n = 15) 29–35 (33·2); tarsus, ♂ (n = 13) 20–22 (20·9), ♀ (n = 17) 20–22 (21·5). WEIGHT: (n = 5, Kenya) 28, 30, 31, 32, 34.

IMMATURE: juvenile plumage similar above to breeding adult, but sides of face and neck, foreneck and breast white with buff suffusion, streaking light and confined to foreneck and upper breast; median coverts broadly edged pale buff.

Field Characters. A small *Calidris*-like wader, rather larger than Little Stint *C. minuta*, with longish, heavy-looking bill kinked downward at tip and short, rather dark legs. In breeding and juvenile plumages dark and richly coloured above with prominent pale V on mantle. Distinguished from Dunlin *C. alpina* by distinctive forked supercilium, bill shape and often by greenish leg colour; also in flight by fainter wing-bar. In non-breeding plumage upperparts are paler grey than in most calidrids, but streaking is quite prominent on neck and upper breast. Dark wing shoulder generally hidden when feeding.

Voice. Tape-recorded (62, 73, 76). A twittering 'tri-tri-trit-tri-trit' when flushed, but generally rather silent.

General Habits. Confined to tidal mudflats and muddy lakeshores, often curiously localized. Found singly or in small flocks, often loosely dispersed with other calidrids. When feeding, walks briskly, picking from side to side and occasionally probing, and frequently making short runs. Flight much as in Little Stint.

Present at wintering sites mainly Sept–early Apr, early adults from Aug. Passage records NE Africa to mid May. Inland records mainly late Aug–Nov, most presumably passage, whilst single birds N Nigeria (Aug) and central Chad (Sept) suggest Saharan crossing from central Mediterranean.

Food. Annelids, small molluscs, insects and vegetable matter.

Reference
Dowsett, R. J. (1980).

Genus *Tryngites* Cabanis

Differs from *Calidris* in having black speckling on inner webs of primaries, secondaries and many wing-coverts, and completely buff underparts in all plumages; also in lacking noticeable wing-bar. Frequents drier grassland habitats than *Calidris*. ♂ larger than ♀. 1 sp., straggling to Africa from Nearctic.

Tryngites subruficollis (Vieillot). Buff-breasted Sandpiper. Bécasseau rousset.

Plates 18, 21
(Opp. pp. 305, 352)

Tringa subruficollis Vieillot, 1819. Nouv. Dict. d'Hist. Nat., nouv. ed., p. 465; Paraguay.

Range and Status. Breeds arctic Alaska and W Canada; winters South America, vagrant Africa.

Nearctic vagrant, with records Egypt (El Quseir, Feb 1928), Tunisia (Sidi Mansour, Dec 1963), Sierra Leone (Freetown, Nov 1973) and Kenya (L. Turkana, Dec 1973).

Description. ADULT ♂: forehead to nape, mantle and scapulars blackish, broadly edged warm buff; sides of face warm buff; sides of neck and hindneck the same, spotted dusky; back, rump and uppertail-coverts as mantle. Central rectrices blackish brown; rest pale buff, tipped whitish, freckled and sub-terminally bordered blackish. Chin to breast and flanks warm buff, tipped whitish, sides of breast spotted blackish; rest of underparts light buff. Primaries brown, becoming blackish near tip; tips white and inner webs finely barred and freckled dark brown; alula and primary coverts similar, but pale inner webs plain; secondaries and outer greater coverts as primaries, inner webs boldly marked; tertials, inner greater, median and lesser coverts dark brown edged sandy buff; axillaries and small underwing coverts buff; larger underwing-coverts patterned as secondaries. Bill black; eye dark brown; legs and feet yellowish. SIZE: wing, ♂ (n = 20) 133–140 (137), ♀ (n = 19) 124–132 (128); tail, ♂ (n = 10) 57–62 (59·8), ♀ (n = 10) 52–56 (54·3); bill, ♂ (n = 16) 19–21 (20·0), ♀ (n = 19) 18–20 (18·6); tarsus, ♂ (n = 21) 31–33 (31·9), ♀ (n = 19) 27–31 (29·0). WEIGHT: (n = 5, Surinam) 43–67.

IMMATURE: juvenile as adult, but with narrower white fringes to upperparts and wing-coverts; whitish tips to tertials; median coverts with blackish subterminal bars.

Field Characters. A distinctively-shaped sandpiper, with small, round, pigeon-like head, longish neck, short dark bill and yellow legs, mottled buffish brown above and with whole underparts and sides of face buff. Resembles juvenile Ruff *Philomachus pugnax*, but smaller and distinguished by leg colour, uniform rump and lack of conspicuous wing-bar. Wing-feathers appear pale above, bordered by dark trailing edge, while underwing is noticeably white.

Voice. Tape-recorded (70, 73). A low 'chwup' when flushed. Rather silent.

General Habits. In Nearctic frequents coastal and inland marshes as well as drier grasslands. Usually in parties or small flocks on migration, where found on airports, ploughed fields, and dry mud and grass at edge of marshes, avoiding water. Often very tame. Carriage and movement rather plover-like, long neck stretched when alarmed. Often runs rather than flies when approached, but flight strong and agile.

Food. Mainly insects.

Genus *Philomachus* Merrem

Similar to *Calidris*, but considerably larger, with bill more tapering, neck rather long. ♂ much larger than ♀, and adopts brightly coloured and extremely variable breeding plumage with 'ruff' and ear-tufts, and lores and sides of face bare. 1 sp., visiting Africa from W, central and E Palearctic.

Philomachus pugnax (Linnaeus). Ruff. Chevalier combattant.

Plates 16, 21
(Opp. pp. 257, 352)

Tringa pugnax Linnaeus, 1758. Syst. Nat. (10th ed.), p. 148; Sweden.

Range and Status. Breeds Scandinavia to easternmost Siberia; a few also temperate central and W Europe; winters Africa, SW Europe and S Asia.

Palearctic visitor, widespread on shallow inland lakes and floodland. Locally very abundant, although rare to uncommon in humid forested regions. Very large numbers winter N tropics: counts of over 1,000,000 Senegal R. delta and NE Nigeria; hundreds of thousands Niger inundation zone Mali, L. Chad and Sudan. Also tens of thousands in Ethiopia, Zaïre/Uganda border lakes,

Kenya and N Tanzania, and further concentrations south to Cape Province. Mainly on passage N Africa, but also winters with up to 4000+ in Tunisia.

Philomachus pugnax

Description. ADULT ♂ (2nd or 'spring' non-breeding plumage): top and sides of head, sides of neck and hindneck typically warm brown, mottled darker, with little indication of supercilium; in some, neck or whole head and neck white; mantle and scapulars blackish brown, barred and variegated rich tawny; back and centre of rump and uppertail-coverts blackish brown; sides of rump and uppertail-coverts white. Inner rectrices barred black and tawny; outer 3 pairs greyish brown. Underparts white, foreneck, breast and flanks strongly mottled blackish brown. Primaries, secondaries, primary coverts and alula blackish brown; tertials, inner greater and inner median coverts and some lesser coverts richly marked as mantle; rest of upperwing-coverts greyish brown; axillaries white; underwing-coverts brown, broadly fringed white. Bill dark brown, often with orange-red, pinkish or yellowish base; eye brown; legs and feet orange or pinkish. ADULT ♂ (1st or 'winter' non-breeding plumage): as 2nd non-breeding plumage, but upperpart feathers greyish brown with darker centres; face and sides of neck whiter; underparts white with only faint mottling on foreneck and breast; tertials, inner wing-coverts and inner rectrices greyish brown. ADULT ♂ (breeding): differs from 2nd non-breeding plumage in having upperparts finely streaked and vermiculated in various combinations blue-black, white, rufous and buff; face covered with orange caruncles; head with 'eared' crest, blue-black or white; feathers of throat and sides of neck long and stiffened into erectile 'ruff', white, tawny or buff, plain or vermiculated with black; breast and flanks heavily scaled black. ADULT ♀: bill lacks coloured base. Breeding and non-breeding plumages as adult ♂ 2nd and 1st non-breeding plumages respectively. SIZE: ♂ much larger than ♀. Wing, ♂ (n = 56) 176–196 (189), ♀ (n = 25) 151–162 (156); tail, ♂ (n = 67) 62–70 (65·9), ♀ (n = 23) 51–60 (54·3); bill, ♂ (n = 53) 31–39 (35·3), ♀ (n = 25) 28–33 (30·6); tarsus, ♂ (n = 59) 46–54 (50·5), ♀ (n = 24) 38–44 (41·0). WEIGHT: ♂ (n = 400+, Africa) 121–234, typical wintering weight 140–180; ♀ (n = 5000+, Africa) 66–150, typical wintering weight 80–105.

♂ 2nd non-breeding plumage and ♀ breeding plumage acquired by partial moult Africa, Dec–Mar and Jan–Apr respectively. ♂ breeding plumage developed from 2nd non-breeding plumage by moult of head, neck, breast, flanks and upperparts Apr–May after departure from Africa. ♂ 1st non-breeding plumage and ♀ non-breeding plumage acquired by complete moult July–Nov and Aug–Jan respectively; wing-moult mainly or entirely in Africa.

IMMATURE: juvenile plumage as adult ♀ (non-breeding), but upperparts and upperwing-coverts dark brown with extensive pale buff edging; underparts white, breast washed buff and streaked brown. Replaced on body by 1st winter plumage (as non-breeding adult ♀) late Sept–Oct. Some 1st year birds moult juvenile outer primaries Feb–Mar. Plumage as breeding adult ♀ acquired 1st year, Jan–May. Legs greyish green or yellowish green in juvenile, becoming brown during 1st year and fully orange at 15–24 months.

Field Characters. A plump-bodied, short-billed generally brown wader, suggesting a large, long-legged *Calidris*. Orange legs of adult are distinctive, and white-necked ♂ is unmistakable. Distinguished in flight by slow wing-beat and by short narrow wing-bar and oval white patches on either side of dark rump centre. Other common waders of similar size have a wholly white or whitish rump. ♂ with ruff rarely seen in Africa, but traces of ♂ breeding plumage common in newly arrived Aug birds.

Voice. Tape-recorded (62, 73, 76). A rarely heard flight note 'too-i'. Generally silent.

General Habits. Frequents marshy and muddy lake borders, floods, irrigation and rice stubble; and away from water on wheat fields and grassland. Generally uncommon in coastal intertidal habitats, but in South Africa roosts on estuaries and saltings. Usually in flocks; hundreds or many thousands. ♀♀ appear to outnumber ♂♂ south of the Sahara by about 10 to 1. Flocks usually of mixed sex, but ♂♂ sometimes segregated in deeper water; parties of ♂♂ may fly together. Rather tame and sluggish. Carriage rather erect when standing, especially when alarmed, with long neck stretched upwards. Flight strong, with slow, regular wing-beats often interrupted by gliding. Walks with deliberate gait, and feeds with body horizontal and head down, picking at mud or water surface or taking insects or seeds from grass; commonly also feeds from surface of deeper water whilst swimming. Concentrates at night at specific feeding sites during dark as well as moonlit conditions; also gathers in thousands to roost on mudbanks.

Adults arrive Tunisia from late July, and many moult there before moving south. Some adults reach tropics from late July and South Africa from mid Aug, but most arrive there late Aug–Sept, ♂♂ slightly before ♀♀. Adult passage Kenya Aug–Sept. 1st year birds occur Senegambia from mid Aug and Kenya from late Aug; continue to pass south through continent for many weeks with main influx Kenya Oct–Nov and Cape Province Nov–Dec. Most birds leave South Africa late Feb–early Apr. ♂♂ leave Kenya late Mar–early Apr, ♀♀ mostly late Apr–mid May. Main departure Senegambia late Apr. Spring passage from N African coast mainly

Mar–early May, but birds begin to appear Tunisia late Feb. Most 1st-year birds migrate, but some ♀♀ oversummer with flocks of 100+ June–July, throughout subsaharan Africa.

Senegal and Niger inundation zones appear to be main wintering areas for European birds which cross Sahara; however, single Sudanese recoveries of birds ringed Sweden and Finland, and one ringed SW Sudan recovered Algeria. Birds ringed Britain recovered Morocco and Mali. Siberian birds reach W Africa; one Senegal ringed bird found on the Ob, and Tunisian and Nigerian birds as far east as the Lena. E and S African birds apparently all Siberian, with northern recoveries from Kenyan and South African ringing all 70–160°E.

Food. Insects, worms, crustaceans, grain and grass seeds.

References
Morel, G. J. and Roux, F. (1966).
Pearson, D. J. (1981).

Subfamily GALLINAGININAE

Snipes and dowitchers, long-billed waders with skull structure adapted for deep probing with head partly submerged.

Genus *Lymnocryptes* Boie

Similar to *Gallinago*, but smaller, bill higher at base; outer secondaries pointed; 12 rectrices, inner ones long and pointed. 1 sp., visiting Africa from W and central Palearctic.

Lymnocryptes minimus (Brünnich). Jack Snipe. Bécassine sourde.

Scolopax minima Brünnich, 1764. Orn. Bor., p. 49; Christiansø Is., off Bornholm.

Plate 12
(Opp. p. 177)

Range and Status. Breeds N Scandinavia and N USSR to Taimyr; also E Siberia; winters Africa, W and S Europe, and central and S Asia.

Palearctic visitor to marshes and swamps, locally frequent to common south to Senegambia, N Nigeria, Chad and S Sudan; rare to uncommon Ethiopia, and south to N Cameroon, Uganda and central Kenya, though subject to occasional influxes; vagrant Somalia, Tanzania and Zambia.

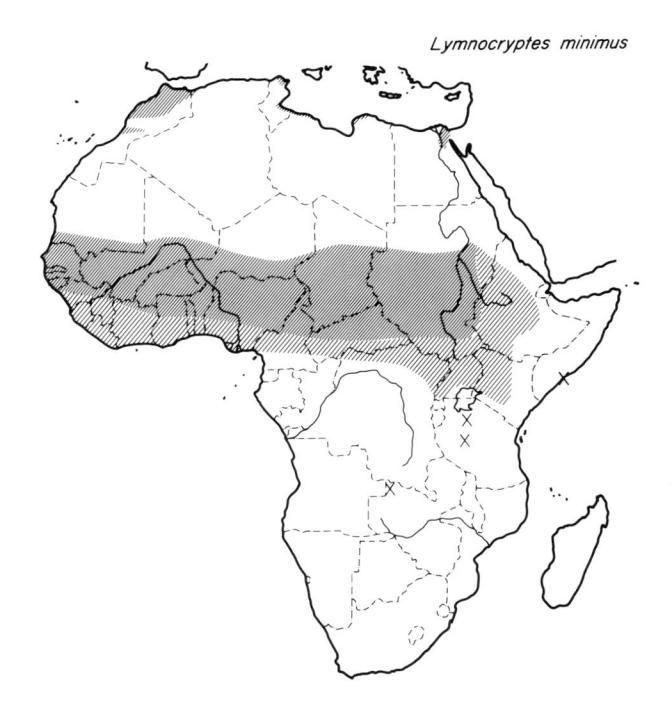

Description. ADULT ♂: black median band on forehead to crown, broadening on nape; broad buff supercilium, dark feather tips immediately above eye forming additional narrow stripe parallel to crown; blackish brown lines from lores to eye and base of lower mandible through ear-coverts; otherwise sides of face buffish white; sides of neck buffish streaked brown; hindneck and upper mantle dark brown, marked buff and tipped whitish; lower mantle and scapulars black glossed purple and green, spotted and marked warm buff; broad buff outer margins at sides of mantle and scapulars forming parallel longitudinal body stripes; lower mantle feathers particularly long; back and rump black glossed purple, tipped white. 12 rectrices, dark brown edged and marked buffish, centre 2 pairs darker, longer and pointed. Belly, vent and undertail-coverts white; rest of underparts suffused warm buff, foreneck and breast heavily streaked and flanks marked dark brown. Primaries dark brown, inners tipped white; primary coverts and alula same tipped white; secondaries dark brown, broadly tipped white; tertials dusky brown tipped white, outer webs marked black; greater coverts dark brown broadly tipped white, inners barred blackish and buff; median and lesser coverts blackish edged and tipped buff; axillaries and underwing-coverts white marked dusky. Bill dark brown; eye dark brown; legs and feet pale greenish. SIZE: wing, ♂ (n = 29) 110–121 (116), ♀ (n = 22) 107–119 (111); tail, ♂ (n = 40) 45–53 (46·8), ♀ (n = 29) 46–52 (48·7); bill, ♂ (n = 26) 38–42 (40·3), ♀ (n = 19) 39–43 (40·7); tarsus, ♂ (n = 26) 23–25 (23·7), ♀ (n = 19) 23–25 (23·8). WEIGHT: (n = 18, Sudan and Kenya) 46–58.

IMMATURE: juvenile and other 1st year plumages not distinguishable from adult.

Field Characters. Much smaller and shorter-billed than Common Snipe *G. gallinago*, with buffish hindneck, gloss on back, scapulars and rump, and all-dark wedge-shaped tail (of 12 feathers). Has white trailing edge to inner wing, white belly and prominent buff body stripes like Common Snipe, and general colour similar, but head pattern differs, with 'double' supercilium and no pale crown-stripe. Distinguished also by behaviour (see General Habits).

Voice. Tape-recorded (62, 73). Low weak flight call rarely heard.

General Habits. Favours waterlogged, tussocky sedge and flooded grassland; also saltings. Usually singly, but sometimes in groups up to *c.* 5. Behaviour on ground much as Common Snipe but even more reluctant to fly; camouflage astonishing (Beere 1981); when feeding, 'bounces' rhythmically on flexed legs. Takes food from surface and by probing. Rises silently, more slowly than Common Snipe, without twisting, and usually flies no more than 50–100 m before pitching; flight rather jerky, bill angled downward.

Recorded in Africa end Sept–early May (in tropics mainly Nov–Feb). Birds from W and central Europe recovered Morocco (5), Algeria (1) and Tunisia (1); Maghreb probably an important wintering area.

Food. Mainly insects and their larvae, small land and water molluscs and annelids; also seeds and other vegetable matter.

Genus *Gallinago* Brisson

Mostly medium-sized waders. Bill very long, upper mandible soft and flexible behind flattened tip. Eyes large and high in skull. Legs rather short, toes long and slender, hind toe well developed. Wings usually pointed in migratory species, rounded in more sedentary species. Tail usually rounded; 14–26 rectrices, in some species lateral ones narrow and stiff. Upperparts intricately patterned buff, blackish and tawny, with pale longitudinal lines on upper body and stripes on head. No distinct breeding plumage. More crepuscular or nocturnal than most waders, resting by day amongst thick marsh vegetation.

13 spp. world-wide; breed all continents, but most species Palearctic and Neotropical. In Africa, 1 resident and 3 visitors from W and central Palearctic. 2, *G. gallinago* and *G. nigripennis*, are best regarded as members of a superspecies.

Plate 12
(Opp. p. 177)

Gallinago gallinago (Linnaeus). **Common Snipe. Bécassine des marais.**

Scolopax gallinago Linnaeus, 1758. Syst. Nat. (10th ed.), p. 147; Sweden.

Forms a superspecies with *G. nigripennis*.

Range and Status. Breeds Azores, Iceland, and widely through Europe, USSR and North America; winters W and S Europe, Africa, S and E Asia and Central and South America.

Palearctic visitor, wintering commonly to very abundantly in marsh and floodlands south to N Zaïre and W Tanzania, and rarely to N Zambia and N Malaŵi.

Description. *G. g. gallinago* (Linnaeus): only subspecies in Africa. ADULT ♂: 2 blackish brown lines across top of head separated by narrow buff central stripe; broad buffish supercilium; dark brown streak from lores to eyes; sides of face buff, heavily marked lower ear-coverts forming darker line below eye; cheeks, sides of neck and hindneck buff, streaked blackish brown; mantle and scapulars blackish brown, feathers tipped and finely marked tawny and buff, those at sides of the scapular tracts with broad golden buff outer borders forming 2 parallel stripes; back and rump olive brown, tipped whitish; uppertail-coverts barred warm buff and blackish brown. Rectrices usually 14, occasionally 12 or 16; outer pair whitish, tinged russet, barred dark brown on both webs; rest blackish, barred laterally pale brown, tips buff with blackish subterminal band. Belly and vent white; rest of underparts pale buff, foreneck and breast streaked and barred, and sides of flanks barred dark brown. Primaries dark brown; primary coverts and alula dark brown tipped white; secondaries dark brown, broadly tipped white; tertials blackish brown, finely marked and edged rich tawny and buff; rest of upperwing-coverts blackish brown, outer greaters and lessers tipped buff or white, inner greaters and medians irregularly marked, edged and tipped buffish and tawny; buff tips to median and lesser coverts divided by dark terminal shaft streaks; axillaries and underwing-coverts white lightly barred olive brown. Bill dark brown, pale at base; eye brown; legs and feet pale greenish. Sexes alike. SIZE: wing, ♂ (n = 30) 129–142 (136), ♀ (n = 23) 129–141 (135); tail, ♂ (n = 10) 51–55 (53·3), ♀ (n = 10) 50–54 (51·7); bill, ♂ (n = 27) 60–70 (65·7), ♀ (n = 33) 55–74 (65·1); tarsus, ♂ (n = 30) 30–35 (31·6), ♀ (n = 23) 30–34 (32·4). WEIGHT: (n = 97, Kenya and Sudan) 72–145; typical wintering weight 80–110.

IMMATURE: juvenile similar to adult but median and lesser coverts with broad buff tip, without median shaft streak, but with dark terminal fringe; narrower white tips to secondaries. Most juvenile upperwing-coverts replaced by adult type feathers during 1st autumn and winter.

Field Characters. A long-billed snipe, dark above but with much rich tawny patterning, and prominent pale stripes; belly more conspicuously white than Great Snipe *G. media*, with less barring on flanks. In flight shows rufous-tipped tail with little white at sides, and broad white trailing edge to inner wing; remiges more pointed than African Snipe *G. nigripennis* (see **A, B**). Lacks barred upperwing of Great Snipe and has different call and zig-zag flight when flushed. Less blackish above than African Snipe, from which best distinguished by pointed wings and by not fluttering wings in flight. Distinguished from Jack Snipe *Lymnocryptes minimus* by larger size, longer bill, buff sides to tail and different head pattern; also by voice and twisting flight.

Voice. Tape-recorded (62, 73, 76). Flight note, characteristically given when flushed, a harsh explosive 'scaap'.

General Habits. Frequents permanent and temporary swamps, marshy edges of lakes, dams and waterholes, and flooded sedge and grassland. Sometimes in ones and twos, but usually larger parties. Tens, and sometimes hundreds, may be scattered in a small area. Mainly crepuscular, spending the day resting in tussocky grass or marsh vegetation. Usually feeds in shallow water, probing deeply and vigorously in soft or semi-liquid mud, whilst almost stationary or moving slowly forward; sometimes up to belly in water, head often immersed to eye level. Food often swallowed without withdrawal of bill. Walks deliberately, with long bill angled downwards. Body more upright and neck stretched when alert, but crouches on approach of danger. Usually flushed singly, twisting rapidly close to ground, also towering in air, then flying several hundred metres before pitching. Small groups fly together to and from feeding areas.

1st birds reach Morocco and some penetrate to S Sudan by end Aug, but main influx N Africa late Sept–early Oct. Passage Ethiopia Oct and Mali Nov, and main arrival south of Sahara Oct–early Nov. Northward movement from end Feb; most leave Africa Mar, but a few remain in tropics to late Apr and N Africa to May. Exceptionally N Africa June–July. Many recoveries NW Africa (mainly Morocco) of birds ringed Scandinavia, and W and central Europe; lack of ringing recoveries indicates origin in USSR for birds crossing Sahara.

Food. Annelids, insects and their larvae, small molluscs and crustaceans; also spiders, small frogs and plant material.

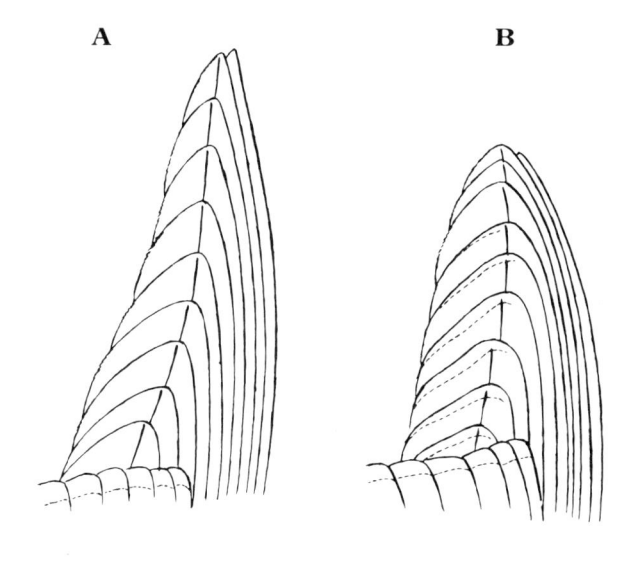

A   **B**

G. gallinago   **G. nigripennis**

302 SCOLOPACIDAE

Plates 12, 21
(Opp. pp. 177, 352)

Gallinago nigripennis Bonaparte. African Snipe. Bécassine africaine.

Gallinago nigripennis Bonaparte, 1839. Icon. Faun. Ital. Ucc. fasc. 25; Cape of Good Hope.

Forms a superspecies with *G. gallinago*.

Range and Status. Endemic resident; locally common to abundant Ethiopia; W and central Kenya to N Tanzania; S Sudan and Uganda to Rwanda, Burundi, E Zaïre, W Tanzania and Malaŵi; Angola and N Namibia to Botswana, Zambia, Zimbabwe, and Mozambique; S and E South Africa.

Description. *G. n. aequatorialis* Rüppell: Ethiopia to E Zaïre, W Tanzania, Malaŵi, E Zimbabwe and N Mozambique. ADULT ♂: 2 black lines across top of head separated by narrow buff central stripe; broad buffish supercilium; black streak lores to eye and another below eye; sides of face buff; cheeks sides of neck and hindneck buff, streaked blackish; mantle and scapulars black, feathers tipped and finely marked tawny and buff, those at sides of scapular tracts with broad golden buff outer edges forming 2 parallel stripes. Back and rump blackish brown tipped whitish; uppertail-coverts barred blackish and warm buff. 16 rectrices, inner 4 pairs blackish, barred laterally pale brown and tipped buff with blackish subterminal band; next barred blackish and pale brown; outer 3 pairs whitish with a few brown bars or centre marks (see **A**). Belly and vent white; rest of underparts pale buff, foreneck and breast boldly streaked and barred and sides of flanks barred blackish. Primaries blackish, inner feathers tipped white; primary coverts and alula same tipped white; secondaries blackish broadly tipped white; tertials and inner greater and median coverts blackish finely and regularly barred warm buff across both webs; outer greater coverts blackish, tipped white and barred buff; lesser coverts tipped white; axillaries and underwing-coverts white lightly barred blackish. Bill flesh brown, darker at tip; eye brown; legs and feet yellowish brown or greenish. Sexes alike. SIZE: (E Africa) wing, ♂ (n = 6) 122–134 (130), ♀ (n = 8) 129–138 (132); tail, ♂♀ (n = 22) 53–60 (56·6); bill, ♂ (n = 7) 64–74 (70·7), ♀ (n = 8) 74–80 (77·5); tarsus, ♂ (n = 7) 34–37 (35·6), ♀ (n = 8) 36–38 (36·8);

A

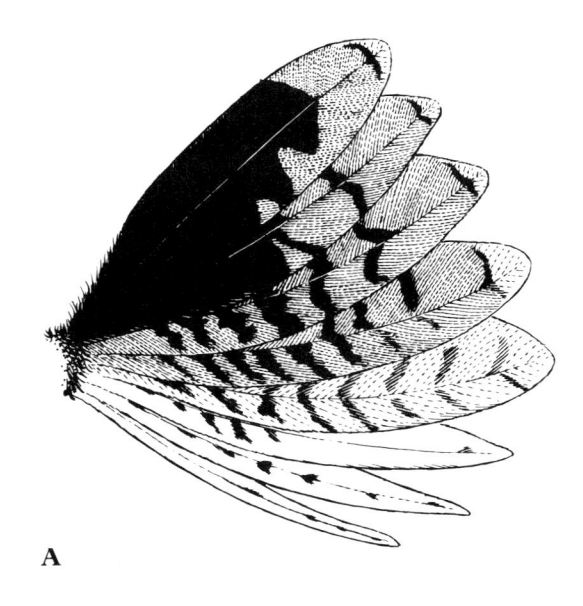

(southern Africa) wing, ♂♀ (n = 10) 129–138 (133); bill, ♂♀ (n = 22) 73–89 (80·4); tarsus, ♂♀ (n = 10) 35–39 (37·4). WEIGHT: (n = 35, Kenya) 95–130; (n = 8, southern Africa) 104–147.

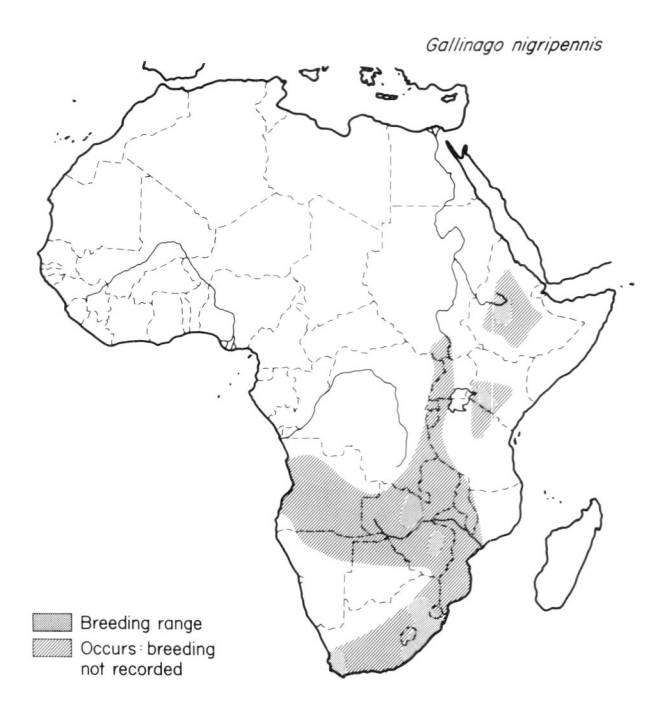

Gallinago nigripennis

Breeding range
Occurs: breeding not recorded

IMMATURE: much as adult, but juvenile less strongly marked on body, wings more mottled and barred.
DOWNY YOUNG: ground colour rich tawny. Black spot above base of bill. Black streak through lores; malar streak, streak on chin and band across foreneck. White streak below eye and white band across forehead. Irregular black and white bars across crown and nape. Black dorsal stripes. Sides of body with white tips giving superimposed stippled effect.
 G. n. nigripennis (Bonaparte): South Africa and S Mozambique. Head stripes and ground colour of upperparts and wings more brownish black than in *aequatorialis*, longitudinal body stripes more tawny, face more heavily marked, tips of small wing-coverts more buffish. SIZE: similiar to *aequatorialis*; wing, ♂♀ (n = 25) 126–141 (132); bill, ♂♀ (n = 25) 70–86 (75·5).
 G. n. angolensis (Bocage): Angola and Namibia to Botswana, W Zimbabwe, Zambia and S Katanga. Coloration as nominate, but bill longer. SIZE: wing, ♂♀ (n = 10) 130–142 (136); bill, ♂♀ (n = 40) 88–103 (93·2).
 The close affinity of the African Snipe *G. nigripennis* and the Common Snipe *G. gallinago* is indicated by similarity of call, and of general and breeding behaviour, and the 2 have sometimes been treated as conspecific (e.g. Voous 1960, Dowsett 1978). However, the extent of difference in plumage pattern, especially in the tail, and the major difference in wing structure seem amply to justify separate specific treatment as given here.

Field Characters. Slightly larger than Common Snipe but longer billed. At rest, rather blacker above, heavily streaked neck and breast contrasting more with white belly. In flight shows broad white trailing edge to secondaries and prominent body-stripes, like Common Snipe, but differs in having white sides to outer tail, broader rounder wings (**B**, p. 301), and white tips to alula and primary and greater coverts, more conspicuous against generally blackish upperwing. Flight rather slower than in Common Snipe with characteristic fluttery wing action. Distinguished from Great Snipe *G. media* by wing shape, longer bill, white belly, and lack of bold spotting on upperwing-coverts; also by harsh call and more explosive take-off when flushed.

Voice. Tape-recorded (10, 74, B, C, 296). A harsh 'scaap' on rising, much as Common Snipe. Mechanical 'drumming': see below.

General Habits. Inhabits highland bogs, swampy lake edges and ditches and seasonally flooded grassland; in E and NE Africa mainly above 1500 m, occasionally to 4000 m; in S also at low altitudes. Usually in small numbers and somewhat scattered. Behaviour on ground similar to Common Snipe. Mainly crepuscular or nocturnal feeder, favouring muddy areas with scattered low vegetation. Typically flushed singly; flight zig-zagging, usually pitching again after 100–200 m.

Subject to local movements, often associated with drying of temporary floods. In E Africa nests typically at 1800–2700 m, descending somewhat in non-breeding season.

Food. Mainly annelids and insect larvae; also small molluscs and crustaceans.

Breeding Habits. Monogamous; solitary, or several pairs associated in same small marsh. Aerial drumming display over nesting territory by night and day to end of incubation. Bird flies in wide circle *c.* 10 m above ground, repeatedly stooping almost to ground with tail fanned, outer feathers vibrating to produce sound 'whu-whu-whu-whu-whu-whu', similar to but lower pitched than that of displaying Common Snipe.

NEST: pad of grass leaves, hidden in tussock, surrounded by moist or flooded ground.

EGGS: 2–3. Pale greenish or buffish, strongly spotted and blotched blackish brown, pale brown and slate grey, marking heavier at large end. SIZE: (n = 53, South Africa) 37–45 × 27–32.

LAYING DATES: E Africa: during or just after rains, Kenya, mainly Apr–Oct; Region A, Apr–July, Sept–Oct; Region B, Nov; Region D, Mar–May; Ethiopia, July, Sept; Zambia and Malaŵi, mainly Apr–June; South Africa, Apr, June–Oct.

Gallinago media (Latham). Great Snipe. Bécassine double.

Scolopax media Latham, 1787. Gen. Syn. Bds. Suppl. 1, p. 292; England.

Plate 12
(Opp. p. 177)

Range and Status. Breeds Scandinavia through boreal zone USSR to Yenisei; winters entirely Africa.

Palearctic visitor in smaller numbers than formerly, uncommon most areas, but regular and common to abundant locally on marsh and grassland. Widespread on autumn passage in N tropics from Senegambia to Ethiopia, with thousands N Nigeria and Niger inundation zone. Some remain W Africa, but most apparently move on to S Zaïre, W and S Tanzania, Angola, Zambia and Malaŵi with occasional records south to N Namibia, Natal, and E Cape Province. Several recent Jan–Feb records Kenya. Uncommon to rare N Africa, mainly on spring passage, though recent Moroccan records mostly Nov–Feb.

Description. ADULT ♂: 2 blackish brown lines across top of head separated by narrow buff central stripe; broad buffish supercilium; dark streak from lores to eye and another below eye; sides of face pale buff, spotted and scaled blackish brown; mantle and scapulars blackish brown, finely barred buff, feathers at sides of scapulars with buff outer edges forming 2 parallel stripes; back and rump olive brown, tipped whitish; uppertail-coverts barred buff and blackish brown. Rectrices 14–18 (usually 16); centre 2 pairs blackish, barred laterally pale brown, tips buff with blackish subterminal band; remainder with extensively unmarked white at tip, outer 3 pairs also white basally with a few narrow brown bars. Centre of belly white; rest of underparts pale buff, foreneck, breast, flanks, sides of belly, tibiae and undertail-coverts barred dark brown.

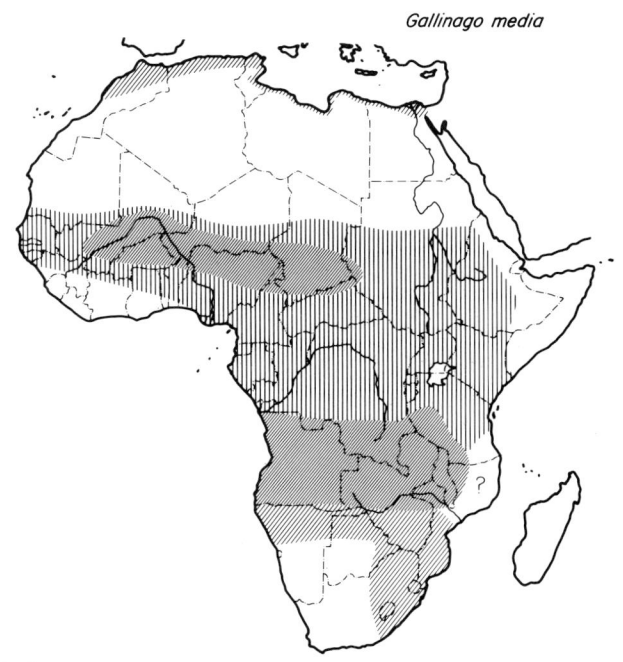

Primaries blackish brown; alula, primary coverts and greater coverts blackish brown, broadly tipped white; secondaries dark brown, narrowly tipped white; tertials dark brown, much barred, variegated and tipped buff; rest of upperwing-coverts dark brown, edged and flecked buff and tipped white, broadly on outer medians; axillaries and underwing-coverts white, barred blackish. Bill dark brown, pale at base; eye dark brown;

Plate 17

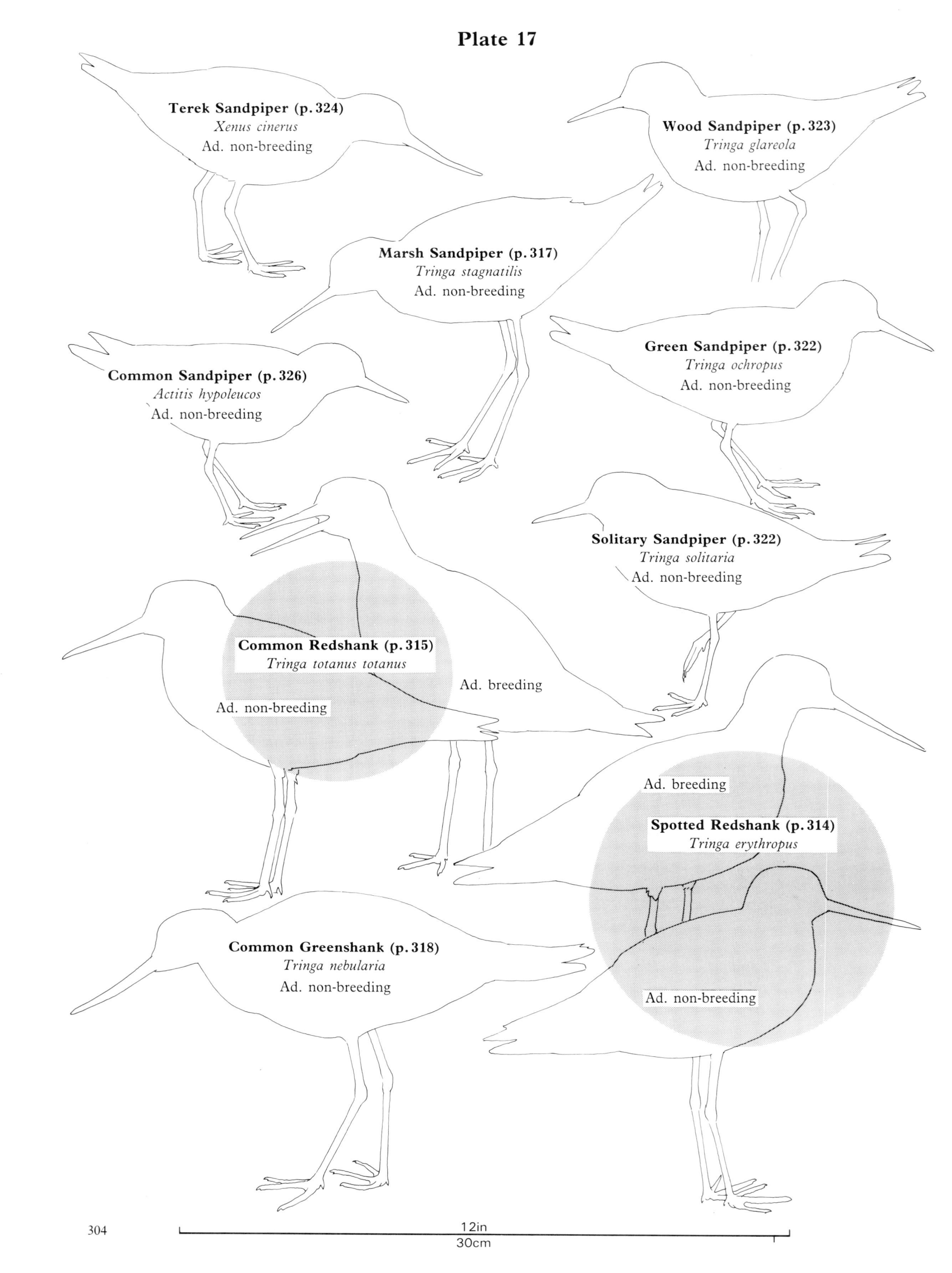

Terek Sandpiper (p. 324)
Xenus cinerus
Ad. non-breeding

Wood Sandpiper (p. 323)
Tringa glareola
Ad. non-breeding

Marsh Sandpiper (p. 317)
Tringa stagnatilis
Ad. non-breeding

Common Sandpiper (p. 326)
Actitis hypoleucos
Ad. non-breeding

Green Sandpiper (p. 322)
Tringa ochropus
Ad. non-breeding

Solitary Sandpiper (p. 322)
Tringa solitaria
Ad. non-breeding

Common Redshank (p. 315)
Tringa totanus totanus
Ad. non-breeding
Ad. breeding

Ad. breeding

Spotted Redshank (p. 314)
Tringa erythropus
Ad. non-breeding

Common Greenshank (p. 318)
Tringa nebularia
Ad. non-breeding

12in / 30cm

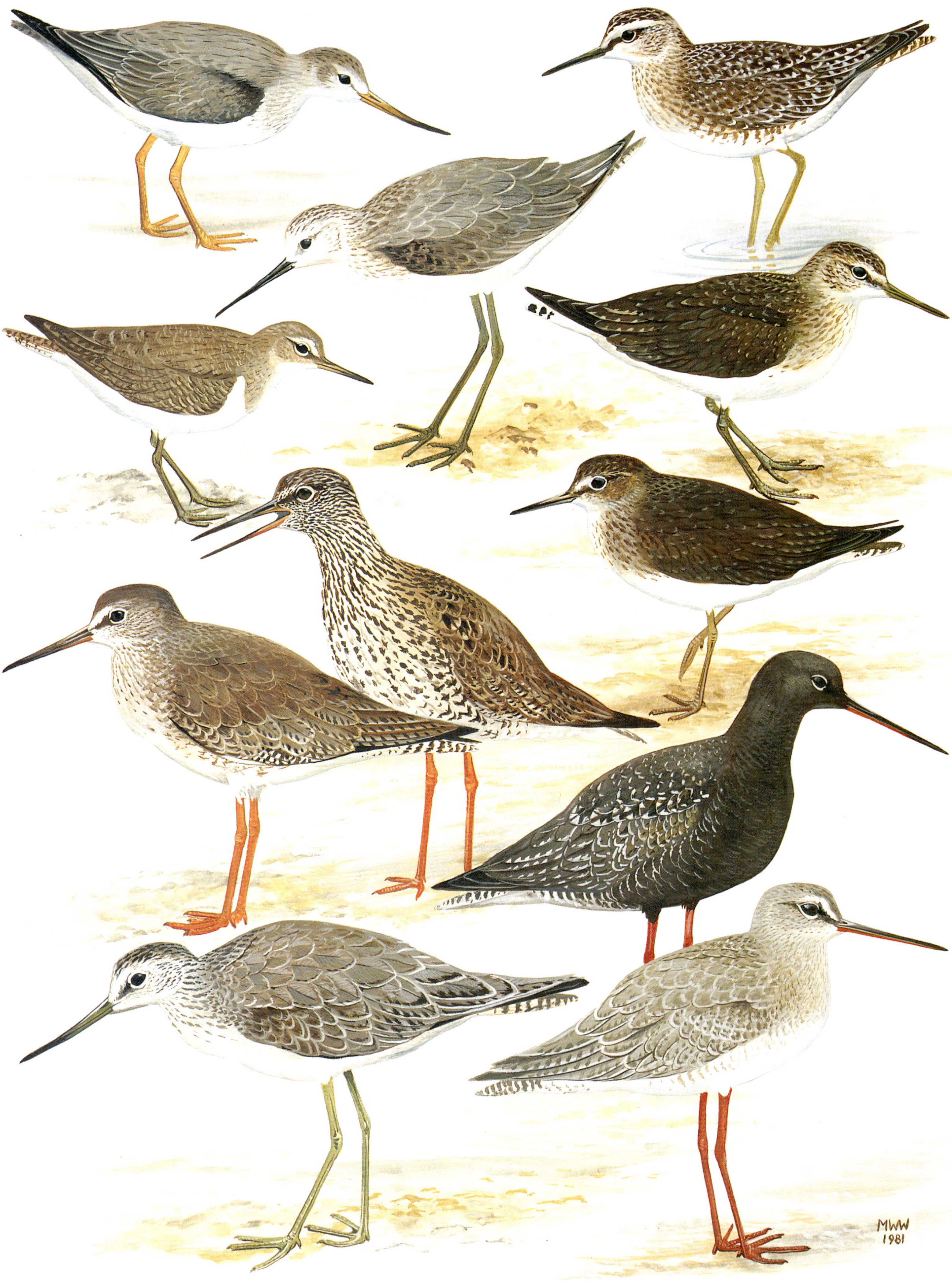
MWW
1981

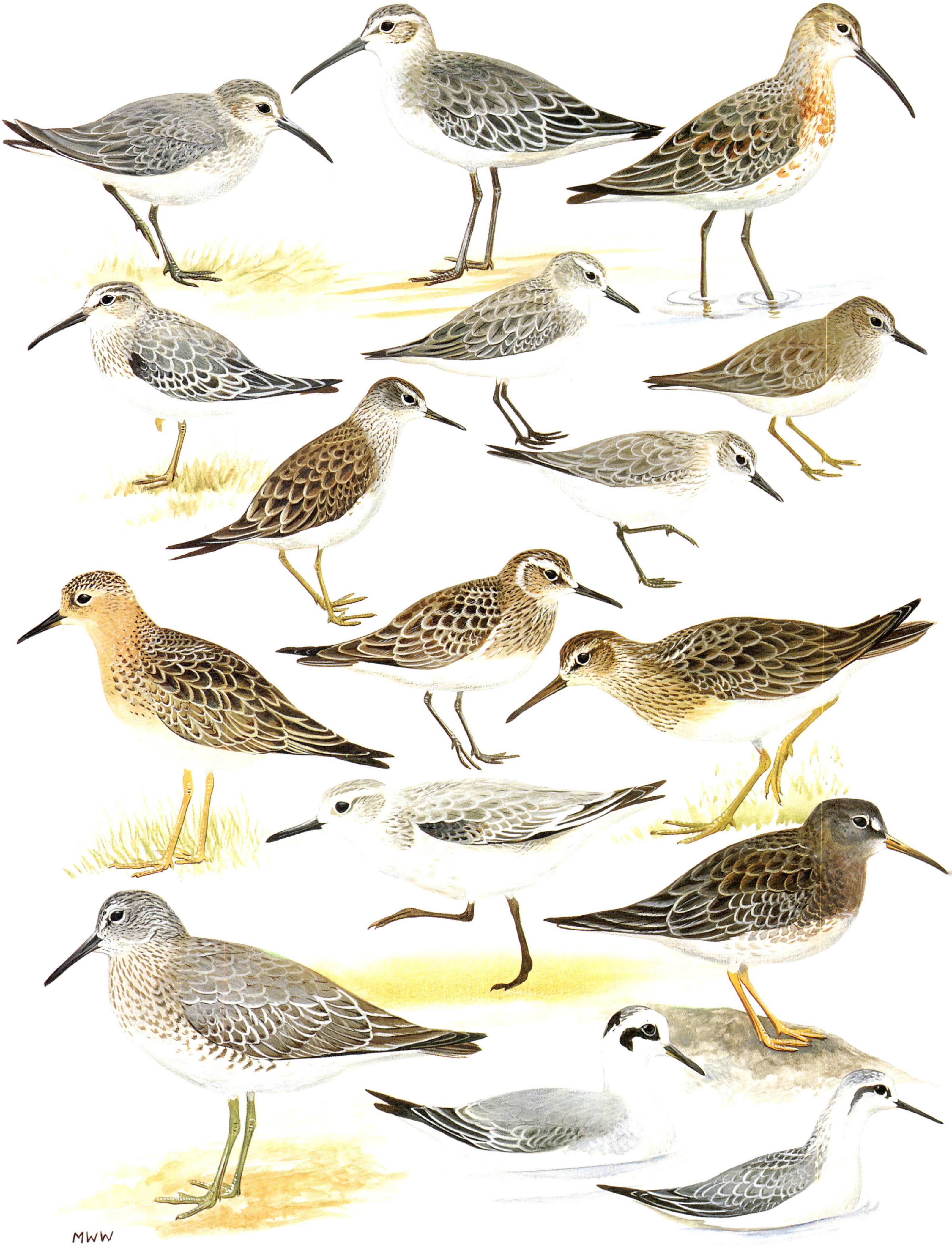
MWW
1981

Plate 18

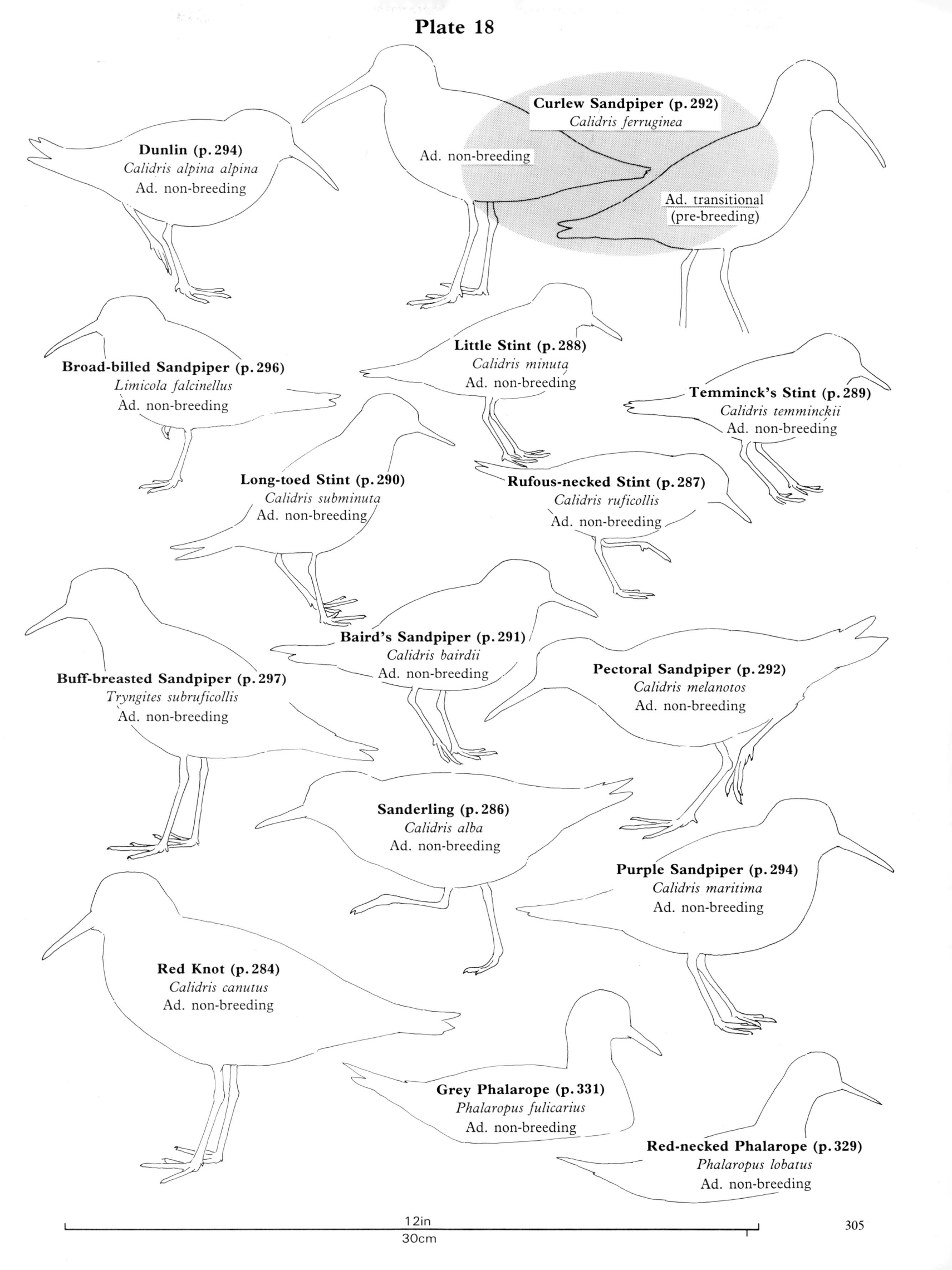

Dunlin (p. 294)
Calidris alpina alpina
Ad. non-breeding

Curlew Sandpiper (p. 292)
Calidris ferruginea
Ad. non-breeding
Ad. transitional (pre-breeding)

Broad-billed Sandpiper (p. 296)
Limicola falcinellus
Ad. non-breeding

Little Stint (p. 288)
Calidris minuta
Ad. non-breeding

Temminck's Stint (p. 289)
Calidris temminckii
Ad. non-breeding

Long-toed Stint (p. 290)
Calidris subminuta
Ad. non-breeding

Rufous-necked Stint (p. 287)
Calidris ruficollis
Ad. non-breeding

Buff-breasted Sandpiper (p. 297)
Tryngites subruficollis
Ad. non-breeding

Baird's Sandpiper (p. 291)
Calidris bairdii
Ad. non-breeding

Pectoral Sandpiper (p. 292)
Calidris melanotos
Ad. non-breeding

Sanderling (p. 286)
Calidris alba
Ad. non-breeding

Purple Sandpiper (p. 294)
Calidris maritima
Ad. non-breeding

Red Knot (p. 284)
Calidris canutus
Ad. non-breeding

Grey Phalarope (p. 331)
Phalaropus fulicarius
Ad. non-breeding

Red-necked Phalarope (p. 329)
Phalaropus lobatus
Ad. non-breeding

12in
30cm

305

legs and feet pale greenish. Sexes alike. SIZE: wing, ♂ (n = 49) 142–154 (147), ♀ (n = 23) 145–155 (149); tail, ♂ (n = 11) 52–57 (54·4), ♀ (n = 10) 50–57 (52·9); bill, ♂ (n = 43) 58–67 (62·1), ♀ (n = 24) 59–72 (66·7); tarsus, ♂ (n = 52) 35–39 (36·9), ♀ (n = 27) 36–40 (38·3). WEIGHT: (n = 7, Sept–Nov, Sudan) 145–180, (n = 5, May, W Tanzania) 191–255.

IMMATURE: juvenile plumage as adult but browner above, longitudinal stripes narrower and paler, white tips to upperwing-coverts usually less clear and white outer rectrices barred to near tips. By midwinter most upperwing-coverts and all rectrices usually replaced by adult type.

Field Characters. A bulky looking snipe with distinctive bold spotting on wing-coverts, forming pale bars in flight along tips of greaters and outer medians. Differs from Common Snipe *G. gallinago* also by boldly barred flanks, white sides to tail, less conspicuous back stripes, shorter bill and more spotted, less streaked appearance of neck and breast. African Snipe *G. nigripennis* also has much white in tail, but has a very long bill, no barring on flanks, more uniform upperwings and rounded wing-tip. Great Snipe rises more silently and flies off more slowly and directly than either Common Snipe or African Snipe.

Voice. Tape-recorded (62, 73). Mainly silent, but may utter monosyllabic croak when flushed.

General Habits. Occurs mainly on lake edges and seasonally flooded areas, especially in short sedge (B. Stronach, pers. comm.); sometimes also in short grass away from water, particularly on passage. Usually singly or in small loose groups up to *c*. 15. Feeding mainly crepuscular. Habits on ground much as common Snipe. When flushed rises rather slowly, usually silently, without twisting; typically flies low and steadily a few hundred m before pitching, with bill nearly horizontal.

1st birds arrive south to Nigeria mid Aug, but most reach N tropics late Aug–Oct, leaving again from Nov. Passage in Ethiopia mid Sept–early Oct, E Africa Oct–early Dec. Main arrival southern Africa end Oct–Dec, departing again Apr; common W Tanzania to mid May. A few passage birds E Africa and N African coast (where very rare autumn) Apr–May, but remarkably little trace of return movement through Africa. Birds ringed Ryazan, USSR (54°45′N, 40°50′E) recovered Nigeria (Sept), Congo Republic (Dec) and Zaïre (Jan); 1 ringed S Sudan (Sept) recovered Kharkov, USSR. A Finnish ringed bird recovered N Namibia in July was presumably oversummering.

Food. Mainly annelids; also small molluscs, insects and vegetable matter.

Plate 12
(Opp. p. 177)

Gallinago stenura (Bonaparte). **Pintail Snipe. Bécassine à queue peinte.**

Scolopax stenura Bonaparte, 1830. Annali di Storia Naturale Bologna, 4, p. 335; Sunda Archipelago.

Range and Status. Breeds NE Russia and through boreal zone of Siberia; winters S and SE Asia.

Palearctic vagrant recorded twice Socotra (Jan 1899), once S Somalia (Juba R., Mar *c*. 1920) and five times Kenya (Naivasha, Jan 1969 and Jan 1982, Mombasa, Sept 1981–Jan 1982, Thika, Sept–Nov 1982, and Shombole Swamp, Jan 1983).

Description. ADULT ♂: 2 dark lines along top of head separated by narrow buff central stripe; broad buffish supercilium; dark brown streak from lores to eye and another below eye; sides of face buff; cheeks, sides of neck and hindneck buff, streaked dark brown; mantle and scapulars dark brown, finely marked and spotted buff, scapulars with broader warm buff barring, and pale outer edges of outer feathers forming parallel longitudinal body-stripes; back and rump brown tipped whitish; uppertail-coverts barred brown and buff. 24–28 (usually 26) rectrices, of which outer 6–8 pairs reduced to pin shapes 1–2 mm wide; central and other large feathers blackish brown, barred paler on outer web and broadly tipped buff, pin feathers tipped white (see **A**). Centre of belly white; rest of underparts buff, foreneck and breast streaked, and flanks, sides of belly and tibiae barred dark brown. Primaries dark brown; primary coverts and alula same tipped white; secondaries dark brown with only very narrow white tips; tertials and outer greater coverts barred dark brown and tawny buff; inner greater, median and lesser coverts brown, tipped and barred buff, tips of median and lesser coverts divided by dark terminal shaft streak; axillaries and underwing-coverts heavily barred dark brown and white. Bill dark brown, paler at base; eye brown; legs and feet pale greenish. Sexes alike. SIZE: wing, ♂ (n = 26) 128–142 (134), ♀ (n = 14) 130–140 (136); tail ♂♀ (n = 10) 41–46 (44·0); bill, ♂ (n = 25) 56–64 (60·5), ♀ (n = 14) 59–69 (64·9); tarsus, ♂ (n = 25) 30–34 (31·1), ♀ (n = 16) 32–36 (33·3).

A

IMMATURE: in juvenile, the broad buff tips to median and lesser coverts lack dark shaft streaks. Other 1st year plumages not distinguishable from adult.

Field Characters. Very similar to Common Snipe *G. gallinago* in size and build. Differs in having shorter bill and in flight, upperparts paler, less richly coloured, with body-stripes less prominent; upperwing paler and lacking noticeable white inner trailing edge; flanks and underwing more barred; and feet trailing more noticeably beyond very short tail. Also rises more directly and slowly than Common Snipe; flight less powerful.

Voice. Tape-recorded (C). Call when flushed a reedy 'krreck' or 'squik', weaker and lower pitched than in Common Snipe.

General Habits. Frequents ditches and flooded sedge and grassland, but also drier grassy areas.

Food. Mainly insects, molluscs and annelids; occasionally crustaceans; also seeds and other vegetable matter.

Reference
Taylor, P. B. (1984).

Genus *Limnodromus* Wied

Medium-sized waders with long bill, soft and flexible behind broadened tip. Wings long and pointed; tail square. Legs rather short, hind toe well developed, partial web between middle and outer toe, shorter one between middle and inner. Greyish or brownish above in non-breeding plumage; striking breeding plumage, dark and patterned above, chestnut below. 3 migrant spp., breeding North America and Asia; 1 vagrant to Africa from central Palearctic.

Limnodromus semipalmatus (Blyth). Asiatic Dowitcher. Limnodrome semipalmé.

Macrorhamphus semipalmatus Blyth, 1848. Jour. Asiat. Bengal, 17, pt. 1, p. 252; Calcutta.

Not illustrated in colour

Range and Status. Breeds W and central Siberia; winters SE Asia; vagrant Africa.

Palearctic vagrant, recorded once in Kenya (Nakuru, 20–21 Nov 1966).

Description. ADULT ♂ (non-breeding): forehead to hindneck, mantle and scapulars dull brown with paler edgings; prominent whitish supercilium; lores dusky; sides of face and neck white, streaked dusky; back dark brown, fringed whitish; rump and uppertail-coverts white, barred and marked dark brown. Rectrices dark brown, barred and variegated whitish, except sometimes on central pair. Underparts white, chin to foreneck finely streaked, and breast, flanks and undertail-coverts lightly barred greyish brown. Primaries blackish brown, inners broadly fringed white on outer webs; alula and primary coverts blackish brown, the latter tipped white on inners; secondaries greyish brown, fringed white, inners barred white on inner webs; outer greater coverts as inner secondaries; tertials greyish brown; rest of upperwing-coverts the same, fringed white, darkest on lessers; axillaries and underwing-coverts white, marked brown. Bill blackish; eye brown; legs and feet blackish. Sexes alike. SIZE: wing, ♂♀ (n = 18) 174–188 (179); tail, ♂♀ (n = 10) 61–65 (63·1); bill, ♂ (n = 14) 75–82 (80·1), ♀ (n = 9) 78–82 (82·9); tarsus, ♂ (n = 14) 46–53 (50·7), ♀ (n = 9) 48–54 (50·8).

IMMATURE: juvenile differs from non-breeding adult in being darker above with broad buff fringing on mantle, scapulars, tertials and wing-coverts, and tinged tawny buff on foreneck, breast and flanks.

Field Characters. Very long-billed wader, size of large *Tringa*, with structure and coloration recalling Bar-tailed Godwit *Limosa lapponica*. Differs, however, in its rather smaller size; straight all-black bill, distinctly swollen at tip; shorter-looking neck and legs; and whitish underwing with dusky patch at bend of wing and dusky tips to flight feathers. Rump and uppertail-coverts are barred and there is no clear white above. Paler area of secondaries often extends to primaries as whitish bar (**A**). Puppet-like feeding action (see below) is quite different from that of a *Tringa* or a godwit.

A

Voice. Tape-recorded (F). Generally silent, but has a quiet, monosyllabic 'chewsk' and quiet, plaintive, mewing 'miau'.

General Habits. Frequents marshes and mudflats in non-breeding season, usually singly or in small parties. Stands with neck tucked in, bill angled downward; gait unhurried; feeds in wet mud or shallow water, probing repeatedly with quick deep strokes, neck rigid and body pivoting about legs. Flight strong with steady wing-beats, bill slightly depressed.

Food. Larvae and small worms.

[*Limnodromus spp.* Long-billed Dowitcher *L. scolopaceus* reported from Morocco (2, Oued Massa, Apr 1981: J. D. R. Vernon, pers. comm.) and Short-billed Dowitcher *L. griseus* from Ghana (1, Cape Coast, Oct 1976) but insufficient details for them to be accepted as certain species identification. 1 unidentified *Limnodromas* sp. also reported Senegambia (Faraja, Dec 1978: Smalley 1979).]

Subfamily SCOLOPACINAE

Medium-large waders with long stout bill, upper mandible soft and flexible behind tip. Eye very large and placed high and very far back in skull. Wings relatively short and rounded; tail short and rounded. Legs short and stout, tibiae entirely feathered, toes long, hind toe small; plumage soft and thick, intricately patterned black. Resorting to cover by day. 1 genus only, *Scolopax*; 5 spp. worldwide; 1 visits N Africa from W Palearctic.

Genus *Scolopax* Linnaeus

Scolopax rusticola Linnaeus. Eurasian Woodcock. Bécasse des bois.

Scolopax rusticola Linnaeus, 1758. Syst. Nat. (10th ed.), p. 146; Sweden.

Range and Status. Breeds from temperate Europe through Russia and Siberia to Manchuria; also Azores, Madeira and Canary Is., and Himalayas; winters Europe, N Africa and central, S and SE Asia.

Palearctic visitor, wintering NW Africa south to High Atlas, near Tunisian and Libyan coasts, and in Nile delta. Locally common to abundant.

Description. ADULT ♂: forehead to forecrown greyish buff, narrowly barred blackish; 3 broad blackish transverse bands across hindcrown, nape and hindneck, separated by narrow bands of tawny buff; sides of face pale buff, with broad blackish streak across lores to eye, and narrower streak below ear-coverts; mantle and scapulars intricately barred and patterned warm buff, rich tawny and black, lower scapulars broadly tipped greyish buff, and greyish buff outer webs forming broad pale V on outer edge of mantle; back, rump and uppertail-coverts rich tawny, narrowly barred blackish. Rectrices black, notched tawny, tipped greyish buff on upperside, bright white on underside. Below, chin whitish; rest of underparts pale buff, narrowly barred dusky, suffused browner on foreneck and upper breast, and with patches of rich tawny feathering and broader barring at sides of lower throat and sides of breast. Primaries, primary coverts, alula and secondaries dark brown, tipped paler, deeply notched tawny; tertials tawny, barred dark brown, finely banded blackish and tipped buff; inner greater and inner median coverts as tertials, broadly tipped buff; outer greater, outer median and lesser coverts barred rich tawny and brown; axillaries and underwing-coverts buff barred brown. Bill flesh brown, darker at tip; eye dark brown; legs and feet greyish flesh. Sexes alike. SIZE: wing, ♂ (n = 13) 195–206 (199), ♀ (n = 8) 185–202 (196); tail, ♂ (n = 53) 80–90 (85·4), ♀ (n = 50) 75–87 (81·0); bill, ♂ (n = 115) 64–80 (71·5), ♀ (n = 90) 63–81 (74·7); tarsus, ♂ (n = 13) 34–37 (35·1), ♀ (n = 9) 34–37 (35·6). WEIGHT: (n = 109, Morocco, Dec–Jan) mean 296, (n = 1200+, Ireland, Feb) 205–420 (307).

IMMATURE: juvenile often indistinguishable from adult, but tawny pattern on rectrices more prominent, and pale tips to undersides of rectrices usually greyish white.

Field Characters. Medium-large russet-coloured wader with long straight bill, but heavy build, broad rounded wings and slow wing-beat give appearance quite different from any snipe in flight. On ground distinguished from snipe by black transverse bars on nape, by marbled patterning of rich browns, grey and black on upperparts, and by fine barring on whole underparts.

Voice. Tape-recorded (62, 73). A thin 'tzizik', carrying some distance, uttered on feeding flights at dusk.

General Habits. Frequents open woodlands, especially cork oak, and rough ground with scattered scrub growth; also along coast in wadis with thick cover and in luxuriant gardens. Solitary, almost always flushed singly. Rises with considerable wing noise, often twisting amongst trees. Clear of cover, flight is rather slow and wavering, often quite high, bill angled downwards and tail spread. Crepuscular, usually resting amongst cover by day and flying to feeding areas at dusk. Feeds by walking a few steps then probing deeply all round in moist soil, generally swallowing prey without withdrawing bill.

Reaches Africa end Oct–Nov, departs late Feb–Mar, occasionally early Apr. Birds ringed Scandinavia recovered Algeria and Tunisia.

Food. Earthworms, insects and their larvae, arachnids, freshwater molluscs, seeds and grass roots.

Subfamily TRINGINAE

Medium to large waders, with long legs and necks and slender build. Adapted to wading and feeding in deeper water than most other species.

Genus *Limosa* Brisson

Medium large waders. Bill very long, straight or slightly recurved. Wings long and pointed; tail almost square. Legs long with long bare lower tibia; hind toe small, small web between bases of middle and outer toes. ♀♀ with longer bills and larger than ♂♂. Greyish brown non-breeding plumages replaced by striking breeding plumages, darker above, rich chestnut below. 2 Nearctic spp. and 2 Palearctic (1 also breeding Alaska); 2 visit Africa from W and central Palearctic.

Limosa limosa (Linnaeus). Black-tailed Godwit. Barge à queue noire.

Scolopax limosa Linnaeus, 1758. Syst. Nat. (10th ed.), p. 147; Sweden.

Plates 13, 20
(Opp. pp. 240, 321)

Range and Status. Breeds W and central Europe to central and E Asia; winters Africa, W and S Europe and S and SE Asia to Australia.

Palearctic visitor, common to very abundant N Africa and N tropics on freshwater margins and floodland. Winters mainly Senegambia (tens of thousands), Niger inundation zone (up to 40,000+), L. Chad, N Sudan (20,000+), and W Morocco (10,000–15,000). Thousands probably also spend the winter around Gulf of Guinea, and small flocks regular to N Zaïre and Keyna; vagrant to rare further south as far as Cape.

Description. *L. l. limosa* (Linnaeus): only subspecies in Africa. ADULT ♂ (breeding): forehead to crown blackish brown, edged chestnut; supercilium pale chestnut; lores dusky; sides of face and neck and hindneck pale chestnut, slightly spotted and streaked; mantle and scapulars with some feathers blackish brown, notched, barred and fringed chestnut, others dull brown; back and rump dark brown; uppertail-coverts white. Rectrices blackish with white bases. Foreneck, breast, and sides of belly and flanks chestnut, feathers tipped white and barred dusky; rest of underparts white with dark barring. Primaries blackish brown, inners white basally; primary coverts blackish brown, inners broadly tipped white;

secondaries blackish brown, basal halves white; some tertials and inner wing-coverts as scapulars; rest of upperwing-coverts brown, greaters broadly tipped white; axillaries and underwing-coverts white. Basal half of bill pinkish brown, tip brown; eye dark brown; legs and feet blackish. ADULT ♂ (non-breeding): differs from breeding bird in having upperparts rather uniform greyish brown; narrow white supercilium; sides of face and neck, foreneck, breast and flanks light greyish brown; rest of underparts white; all tertials and inner wing-coverts as mantle. ADULT ♀: as ♂ but breeding plumage usually paler, less chestnut on head and neck; upperparts with more dull brown feathers, underparts white. Breeding plumage partially acquired Africa Feb–Mar. Post-nuptial body-moult before arrival in tropical Africa but wing-moult commonly completed in winter quarters. SIZE: wing, ♂ (n = 67) 188–228 (207), ♀ (n = 41) 205–231 (219); tail, ♂ (n = 19) 71–81 (76·1), ♀ (n = 19) 76–87 (79·1); bill, ♂ (n = 63) 79–123 (92), ♀ (n = 41) 95–122 (107); tarsus, ♂ (n = 70) 64–78 (74), ♀ (n = 40) 73–88 (81). WEIGHT: (n = 17, Morocco, Aug–Sept) 198–285.

IMMATURE: juvenile as non-breeding adult but mantle, scapulars and upperwing-coverts darker brown with tawny buff fringes; neck and breast tawny buff. Some birds assume breeding plumage in 1st summer; others moult again into non-breeding plumage.

Field Characters. Medium-large, with long slender neck, long straight bill, pale on basal half, and long dark legs. Taller than Bar-tailed Godwit *L. lapponica* with longer legs; in flight separated by broad white wing-bar and dark rump contrasting with white uppertail-coverts and black tail-tip. Belly white in breeding plumage.

Voice. Tape-recorded (62, 73, 76). A monosyllabic 'kik' or 'teuk'; also a louder flight call 'wicka-wicka-wicka', especially from flocks.

General Habits. Inhabits marshy pools, rivers and lake edges, and receding floodwaters; also coastal lagoons, estuaries and salt pans especially on migration. Gregarious; usually in parties or small flocks, flocking in main wintering areas. Feeds in more leisurely way than Bar-tailed Godwit, usually in deeper water, with prolonged and vigorous probing, head completely immersed. Gait measured and movements generally graceful. Flight swift, often erratic, wing-beat relatively rapid, frequently in coordinated flocks.

Many arrive Morocco July–Aug (earliest end June) with passage to Oct; passage Tunisia late July–Sept and across Chad mainly mid-late Sept. 1st adults reach Senegambia early Aug with juveniles by mid Aug, but main arrival across N tropics Sept–Oct, some moving away from drying habitat in Sahel about Dec. Spring departure in west mainly late Feb–Mar; in east Mar–early Apr. Birds ringed W and central Europe recovered mainly Morocco–Sierra Leone; those wintering Mali eastwards presumably from unringed USSR population. Scarcity in Mauritania suggests birds reach Senegambia by crossing the Sahara.

Food. Insects, crustaceans, small fish and amphibians, molluscs and annelids. Also plant material, particularly seeds of rice and other grasses.

Reference
Morel, G. J. and Roux, F. (1966).

Plates 13, 20

(Opp. pp. 240, 321)

Limosa lapponica (Linnaeus). Bar-tailed Godwit. Barge rousse.

Scolopax lapponica Linnaeus, 1758. Syst. Nat. (10th ed.), p. 147; Lapland.

Range and Status. Breeds from N Norway through USSR to W Alaska; winters Africa, W and S Europe, SW, S and SE Asia to Australia.

Palearctic visitor to coastal mudflats, locally very abundant SW Morocco to Guinea Bissau, especially Banc d'Arguin, Mauritania (500,000 wintering). Locally and in small numbers on Gulf of Guinea coasts, but up to 2000+ reach Namibia and common W Cape Province. Small numbers winter Morocco to Tunisia, Red Sea and N Somalia, but uncommon and local Indian Ocean coasts where mainly an autumn passage migrant. Inland, recorded rarely Morocco, Nigeria, Chad, Sudan, Ethiopia, Kenya, Burundi, Zambia, Malaŵi, Botswana and South Africa.

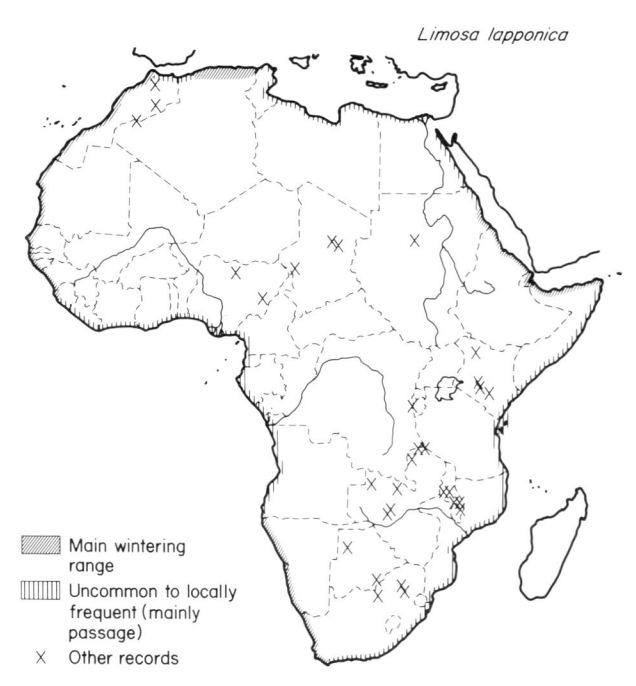

Description. *L. l. lapponica* (Linnaeus): only subspecies in Africa. ADULT ♂ (breeding): forehead to nape blackish brown, edged chestnut; supercilium and cheeks chestnut; lores dusky; ear-coverts, sides of neck and hindneck pale chestnut, streaked dusky; mantle and scapulars blackish brown, edged and notched chestnut; back, rump and uppertail-coverts white marked brown. Rectrices and posterior uppertail-coverts barred brown and white. Whole underparts chestnut, slightly streaked on sides of breast and foreneck. Primaries, primary coverts and alula blackish brown; secondaries dark brown edged white; tertials, inner greater and inner median coverts as scapulars; rest of upperwing-coverts greyish brown fringed whitish; axillaries and underwing-coverts white, streaked and barred brown. Bill blackish brown, pinkish on basal half, slightly recurved; eye brown; legs and feet dark grey. ADULT ♂ (non-breeding): differs from breeding bird in having upperparts, tertials and inner wing-coverts pale greyish brown, streaked darker brown; supercilium white; sides of head and neck white, streaked greyish brown; underparts white, foreneck and breast washed pale brown, faintly streaked dusky, flanks and undertail-coverts barred brown. ADULT ♀: as ♂ but in breeding plumage upperpart feathers edged paler, more buffish; chestnut below paler, usually confined to breast and flanks and intermixed with buffish, sometimes virtually absent; streaking on neck and breast heavier; eye-stripe pale and sides of face, neck and foreneck suffused light buff; belly and undertail-coverts mainly white. Pre-nuptial moult commonly well advanced before departure from Africa. Post-nuptial wing moult and completion of body-moult usually in winter quarters. SIZE: wing, ♂ (n = 64) 200–221 (211), ♀ (n = 31) 214–231 (222); tail, ♂ (n = 22) 67–77 (72·1), ♀ (n = 23) 72–81 (75·8); bill, ♂ (n = 62) 69–87 (79), ♀ (n = 30) 86–108 (99); tarsus, ♂ (n = 65) 46–53 (50), ♀ (n = 28) 52–59 (55·4). WEIGHT: (n = 15, Kenya and South Africa) 200–490.

IMMATURE: juvenile plumage as non-breeding adult, but upperparts browner, feathers with buff fringes; wing-coverts edged and tertials notched buff; breast and flanks strongly washed buff. Some birds acquire partial breeding plumage in 1st year.

Field Characters. Distinguished from Black-tailed Godwit *L. limosa* by slightly upturned bill, shorter legs (not projecting beyond tail in flight), barred tail, whitish rump patch extending up back, and lack of wing-bar. In breeding plumage ♂♂ also differ from Black-tailed Godwit by unbarred chestnut below extending to undertail-coverts.

Voice. Tape-recorded (62, 73, 76, C). Flight note a low 'kirruc kirruc', often heard from flocks, although single birds usually silent.

General Habits. Favours estuaries and offshore mudbanks, but also occurs on sandy shores and coral. Inland occurrences mainly on muddy lake edges. Often singly or in small groups, but gregarious on main wintering grounds, with flocks of up to many thousands. Usual gait a brisk walk. Feeds in shallows or on exposed tidal mud, probing deeply and vigorously all round every few steps, occasionally burying whole head. On Banc d'Arguin, $36·0 \pm 0·9$ probes per min. in foraging ♂♂ ($3·3 \pm 0·2$ successful), and $35·4 \pm 1·4$ probes per min. ($2·0 \pm 0·2$ successful) in ♀♀ (NOME 1980). May wade more deeply and feed by moving bill from side to side through water. Flight fast with steady powerful wingbeats. Flocks typically in loose lines and skeins. Returns NW Africa from late July, but mainly late Aug–Oct; to Red Sea and Somalia mainly from Sept. Returns early to South Africa, and coastal movement Kenya and Tanzania mainly late Aug–early Sept. Wintering areas vacated mid Apr–early May, but a few oversummer. Inland records mainly Aug–Nov, suggesting overland migration for some SW African wintering birds. 1 bird ringed Cape, recovered Iranian Caspian; 1 ringed Great Britain, recovered W Sahara.

Food. Mainly annelids and molluscs; crustaceans and fish fry also recorded. Main prey items on Banc d' Arguin were small polychaetes (especially *Nereis*), large polychaetes (mainly *Marphysa*) and bivalves (*Loripes*).

Genus *Numenius* Brisson

Medium to large waders with long decurved bills. Wings long and pointed; tail rather rounded. Legs rather long, hind toe well developed. ♀♀ larger than ♂♂, typically with much longer bills. Plumage streaked, mottled and barred dark brown and buff or whitish; no distinct breeding dress. 8 spp., Holarctic, 3 visiting Africa from Palearctic.

Numenius phaeopus (Linnaeus). Whimbrel. Courlis corlieu.

Scolopax phaeopus Linnaeus, 1758. Syst. Nat. (10th ed.), p. 146; Sweden.

Plates 13, 20
(Opp. pp. 240, 321)

Range and Status. Breeds Iceland to W Siberia; also NE Siberia and NW and N Nearctic; winters Africa, Madagascar, SW Europe, S and SE Asia to Australasia and South America.

Palearctic visitor, frequent to abundant. Winters all Atlantic, Red Sea and Indian Ocean coasts, but mainly Mauritania and Ethiopia (Eritrea) southwards. Regular passage migrant N African coasts. Inland, rare to uncommon Niger delta, Mali, and E and S central Africa, mainly on southward passage, but frequent all months L. Victoria, Uganda; also recorded Sudan, Ghana, Nigeria, Namibia and Transvaal.

Description. *N. p. phaeopus* (Linnaeus) (including the form '*alboaxillaris*'): throughout Africa ('*alboaxillaris*' recorded Mozambique and Natal). ADULT ♂: top of head and nape dark brown with pale medial stripe and whitish supercilium; lores and ear-coverts dusky; cheeks, sides of neck and hindneck whitish, streaked with dark brown; mantle and scapulars dark brown, feathers edged paler, scapulars notched pale brown; back and rump white with light streaking; uppertail-coverts white, finely barred dark brown. Central rectrices greyish, rest white, all barred dark brown. Chin and throat whitish; foreneck and breast washed pale buff, streaked dark brown; flanks barred brown; rest of underparts white. Primaries dark brown, inners tipped and spotted buffish on outer webs; primary coverts dark brown, inners tipped white; secondaries

312 SCOLOPACIDAE

Numenius phaeopus

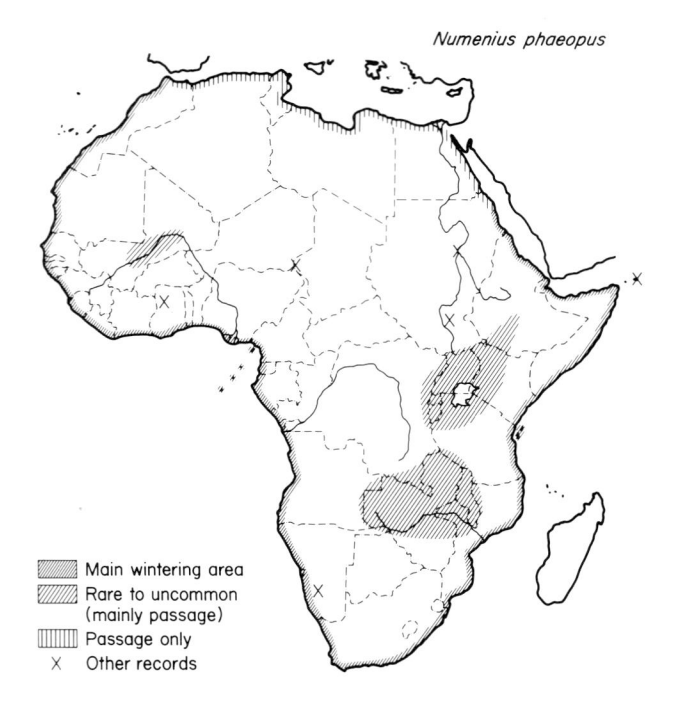

- Main wintering area
- Rare to uncommon (mainly passage)
- Passage only
- X Other records

and greater coverts dark brown with pale tips; tertials and inner wing-coverts as scapulars; rest of wing-coverts light brown, edged and notched whitish; axillaries and underwing-coverts white, flecked and barred dark brown (axillaries, upperwing-coverts and rump of '*alboaxillaris*' pure white). Bill blackish, strongly decurved; eye brown; legs and feet dark grey. Sexes alike. SIZE: wing, ♂ (n = 16) 239–255 (246), ♀ (n = 12) 232–265 (253); tail, ♂ (n = 19) 92–103 (97·2), ♀ (n = 18) 92–102 (98·2); bill, ♂ (n = 16) 76–92 (82), ♀ (n = 12) 76–99 (84); tarsus, ♂ (n = 16) 53–64 (58), ♀ (n = 11) 56–65 (61). WEIGHT: (n = 20, South Africa and Kenya) 315–475.

IMMATURE: juvenile plumage as adult, but scapulars, tertials and coverts with large buff lateral spots.

N. p. variegatus (Scopoli): SE African coast (irregular). Back and rump more heavily streaked and barred than in nominate; axillaries and underwing-coverts more heavily marked, and flanks, sides of body and undertail-coverts more strongly barred.

N. p. hudsonicus (Latham): Sierra Leone (once). Differs from nominate in being generally paler buff above, rump and back buff streaked brown, underwing and axillaries buff marked brown.

Field Characters. Differs from Eurasian Curlew *N. arquata* in being smaller and rather darker, and in having dark cap with pale central crown-stripe and pale supercilium; also by shorter bill, hooked nearer tip, and distinctive tittering call (see Voice). Gait and wing-beats quicker than Eurasian Curlew.

Voice. Tape-recorded (62, 73, 76, C, 264, 296). Utters a whinnying 'tititititititit', usually when flushed and commonly when migrating.

General Habits. More exclusively coastal than Eurasian Curlew, frequenting rocky and coral shores and sandflats as well as estuaries, mangroves and lagoons. Habits much as those of Eurasian Curlew, but forms more compact tidal gatherings, associating loosely with other species such as Bar-tailed Godwit *Limosa lapponica* and Grey Plover *Pluvialis squatarola*. Rarely probes deeply when feeding.

Most return to Africa late July–early Oct, quickly reaching Cape. Southward passage in Morocco mainly Sept, Tunisia July–Aug; passage offshore Kenya late Aug–early Sept. Most inland records mid Aug–Oct, suggesting some transcontinental passage to southern and W African coasts. Spring departure from late Mar, but mainly Apr; heavy offshore northward movement along coast of Somalia Apr–early May. Many over-summer Africa. Birds ringed Europe and NW European Russia recovered Senegambia, Ghana and Congo Republic; populations wintering southern and E Africa presumably of E Russian and Siberian origin.

Food. Crustaceans, shellfish and annelids; inland, commonly also insects and plant material.

Plates 13, 20
(Opp. pp. 240, 321)

Numenius tenuirostris Vieillot. Slender-billed Curlew. Courlis à bec grêle.

Numenius tenuirostris Vieillot, 1758. Nouv. Dict. d'Hist. Nat., nouv. ed., 8. p. 802; Egypt.

Range and Status. Breeds N Kazakhstan and W Siberia; winters N Africa and 1 recent record Iraq, Jan.

Palearctic visitor, formerly wintering regularly N African coastal sites. Now very reduced, rare to uncommon and present status uncertain. No recent records Egypt or Libya, but several Dec–Feb reports Tunisia (max. 32) and 1 record of 3 Algeria. Morocco apparently now main wintering area with scattered records of parties up to 120 Atlantic coast, mainly Nov–Mar; 1 report of 500–800 in SW (Blondel and Blondel 1964); irregular migrant in NE late July–Aug and Apr–May.

Description. ADULT ♂: top of head dark brown, broadly streaked pale buff; hindneck, sides of face and neck pale buff, streaked brown; mantle and scapulars dark brown, broadly edged buff, some scapulars with lateral barring; back and rump white; uppertail-coverts white barred dark brown. Rectrices white barred dark brown. Chin and throat white; foreneck and breast whitish or creamy, streaked dark brown; rest of underparts white with large heart-shaped dark brown spots on lower breast, flanks and sides of belly. Primaries dark brown, outers blackish; inner webs distally mottled and barred white, outer webs of inner feathers also broadly tipped and deeply notched white; primary coverts dark brown, inners broadly notched and tipped white; alula, secondaries and greater coverts brown, tipped and deeply notched white on both webs; tertials and inner wing-coverts as scapulars; lesser coverts brown, edged whitish; axillaries and underwing-coverts white. Bill brown, strongly decurved; eye dark brown; legs slate grey. Sexes alike. SIZE: wing, ♂ (n = 12) 242–259 (251), ♀ (n = 5) 258–274 (262); tail, ♂ (n = 15) 87–99 (93·6), ♀ (n = 5) 96–108 (102); tarsus, ♂ (n = 21) 59–66 (63·4), ♀ (n = 12) 64–69 (66·1); bill, ♂ (n = 21) 68–78 (72·9), ♀ (n = 12) 82–96 (89·9). WEIGHT: (♂, France, Sept) 360; (juv. ♀, Italy) 255.

IMMATURE: juvenile similar to adult but sides of belly and flanks streaked and barred, lacking heart-shaped markings.

Field Characters. Size of Whimbrel *N. phaeopus*, from which distinguished by lack of dark cap and head-stripes, more gradually curving, tapering bill with finer tip, and generally paler plumage. In flight, pale secondaries contrast markedly with dark wing-tips. Spotting on breast and flanks diagnostic when present. Otherwise from Eurasian Curlew *N. arquata* by smaller size and finer bill.

Voice. Not tape-recorded. Call similar to that of Eurasian Curlew but notes higher in pitch and shorter.

General Habits. Appears to prefer saline or alkaline situations in non-breeding season, but also occurs in damp meadows. Movements faster than those of Eurasian Curlew and said to feed more often whilst wading.

Food. Insects, small molluscs, crustaceans and worms.

Numenius arquata (Linnaeus). Eurasian Curlew. Courlis cendré.

Plates 13, 20
(Opp. pp. 240, 321)

Scolopax arquata Linnaeus, 1758. Syst. Nat. (10th ed.), p. 145; Sweden.

Range and Status. Breeds W and central Europe, and through Russia to central Asia and N China; winters Africa, Madagascar, W and S Europe, S and SE Asia.

Palearctic visitor, wintering commonly to abundantly in estuarine habitats on all coasts except Somalia to Tanzania where local and mostly uncommon to frequent; 14,000 counted Mauritania (Banc d'Arguin) and 10,000–12,000 Tunisia. Inland, locally uncommon to frequent on lakes and rivers especially on southward passage.

Description. *N. a. orientalis* (Brehm): E and southern Africa; also west to Nigeria and Sierra Leone. ADULT ♂: top of head streaked dark brown and buff; hindneck, sides of face and neck pale buff, streaked brown; pale patch above and below eye; mantle and scapulars dark brown, broadly edged buff, some scapulars with lateral barring; back and rump white with inconspicuous blackish streaks; uppertail-coverts white barred blackish brown. Rectrices tinged brown, especially centres, and narrowly barred blackish. Chin and throat white; foreneck and breast buff streaked blackish brown; rest of underparts white, flanks and undertail-coverts finely streaked blackish. Primaries blackish brown, inner webs distally mottled and barred white, inner feathers also tipped white and notched white on outer web; primary coverts blackish, inners broadly tipped and notched white; alula, secondaries and greater coverts blackish, tipped and deeply notched white; tertials and inner wing-coverts as scapulars; lesser coverts dark brown edged white; axillaries and underwing-coverts white. Bill dark brown; strongly decurved; eye brown; legs and feet greenish grey. Sexes alike. SIZE: wing, ♂ (n = 11) 284–305 (293), ♀ (n = 15) 297–322 (310); tail, ♂ (n = 10) 106–115 (111), ♀ (n = 10) 109–116 (112); bill, ♂ (n = 14) 123–146 (135), ♀ (n = 16) 138–184 (170); tarsus, ♂ (n = 14) 78–85 (82), ♀ (n = 17) 79–91 (86). WEIGHT: (n = 1, Kenya) 590.

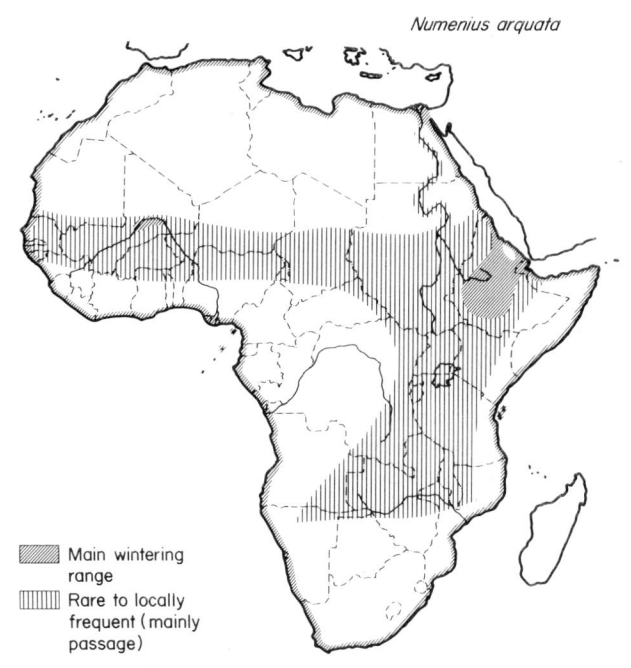

Main wintering range

Rare to locally frequent (mainly passage)

IMMATURE: juvenile like adult but upperparts with more extensive buff edges.

N. a. arquata (Linnaeus): N and W Africa; also South Africa. Top of head and upperparts darker than in *orientalis*, with edging narrower; dark markings on back and rump more prominent; more heavily streaked below including belly, and flanks usually barred; axillaries and underwing marked and barred brown. Bill (mean 23 ♂♂, 116; 15 ♀♀, 153) and tarsus (mean 23 ♂♂, 75; 16 ♀♀, 83) shorter than in *orientalis*. WEIGHT: (n = 1, Morocco) 522; (respective ranges for 266 ♂♂ and 243 ♀♀, Netherlands, Aug–Apr) 500–1010 and 700–1360.

Field Characters. Very large, with long legs and very long sturdy decurved bill. Uniform brown above in flight except for whitish rump patch extending in V up back. Distinguished from Whimbrel *N. phaeopus* by larger size, lack of supercilium and crown-stripe, and longer bill. Calls distinctive (see Voice).

Voice. Recorded (62, 73, 76, C, 264). Commonest notes in flight and when feeding, 'quoi-quoi' and more long-drawn 'coor-wi'; also 'kyoi-yoi-yoi-yoik' when moving whilst feeding, and a hoarse 'kahiyah-kahiyah' when flushed.

General Habits. Inhabits mainly estuarine mudflats and saltings, and coastal mangroves, but on migration also on lagoons, beaches and sand dunes, and inland on river banks, lake-shores and grasslands. In ones and twos or loose flocks. Scattered widely when feeding, typically wading and securing prey by pecking, jabbing or deep probing, sometimes using whole bill length. Gait slow and deliberate, body horizontal and neck somewhat retracted. Flight quite fast with steady wing-beats, flocks usually in straggling lines.

Birds return Morocco from late July, present in force most wintering areas Aug–Mar or early Apr, but many oversummer. Inland, a few birds on larger lakes all months, but most records Aug–Nov. Regular autumn occurrence in Sahel indicative of broad front trans-Saharan migration. Swiss and Hungarian ringed birds recovered Morocco and Algeria, but Banc d'Arguin perhaps usual southern limit for European nominate birds.

Food. In littoral zone, shellfish, small crabs, shrimps, small fish and annelids. Inland takes adult and larval insects, occasionally small vertebrates and vegetable matter.

Genus *Tringa* Linnaeus

Medium small to medium large waders, mostly rather slender with long necks and long slender bills, straight or slightly recurved. Legs usually long, pale or brightly coloured, hind toe present, small web between bases of middle and outer toes, trace also between middle and inner toes. Wings long and pointed, tail square; upperwing uniform. Rump and often back white; tail barred. Greyish or dull brown above in non-breeding plumage, whitish below; breeding plumage usually similar, although feathers more strongly marked. Most have loud ringing calls.

10 migrant spp. breeding Palearctic or Nearctic, 8 of which have occurred Africa.

Plates 17, 21
(Opp. pp. 304, 352)

Tringa erythropus **(Pallas). Spotted Redshank. Chevalier arlequin.**

Scolopax erythropus Pallas, 1764. In Kroeg's Cat. Adumbr., p. 6; Holland.

Range and Status. Breeds N Scandinavia and N boreal zone of USSR to E Siberia; winters Africa, W and S Europe, and S and E Asia to Japan.

Palearctic visitor to inland marshes and floodland and coastal saltpans south to N Zaïre, Burundi and N Tanzania; vagrant Angola (Cabinda), Zimbabwe and Mozambique. Uncommon to frequent throughout much of winter range, but abundant to very abundant locally. Regular on Ghana coast with up to 2000+ Mar; 100+, Aweil rice scheme, SW Sudan (G. Nikolaus, pers. comm.); up to 160 W Kenya (D. Pearson, pers. obs.). Mainly a passage migrant N Africa but hundreds winter Morocco (J. D. R. Vernon, pers. comm.). Apparently ranges further south than formerly, for present status as locally regular and frequent to common Kenya and Uganda contrasts with total lack of records before 1953.

Description. ADULT ♂ (breeding): head, neck and upper breast sooty black, feathers narrowly edged white; mantle and scapulars sooty black, tipped and spotted white or pale buff; back and rump white; uppertail-coverts black, barred white or pale buff. Rectrices greyish brown with incomplete black bars, centre pair spotted buff laterally. Flanks black, narrowly barred white; belly and vent black, tipped white; undertail-coverts barred black and white. Primaries blackish, inners narrowly edged white and spotted blackish on outer web; primary coverts and alula blackish tipped white; secondaries greyish brown, edged and irregularly barred white; tertials, inner greater coverts and inner median coverts black, tipped and spotted white; some lesser coverts as head and neck; rest of upperwing-coverts pale greyish brown, edged and barred white; axillaries and underwing-coverts white. Bill dark brown, dusky red at base; eye brown; legs and feet dark red. ADULT ♂ (non-breeding): as breeding bird but crown, nape, hindneck, sides of neck and mantle grey, tipped white; scapulars grey, spotted dusky and white; conspicuous broad white supercilium and dark loral streak; cheeks and ear-coverts white, spotted and streaked pale ash-brown; under-parts white, foreneck and upper breast suffused ashy and faintly streaked dusky; sides of breast and flanks barred ashy and white; tertials, inner greater and inner median coverts as mantle; all lesser coverts pale greyish brown, edged white; legs orange-red. Sexes alike. Breeding plumage largely acquired Africa, Feb–Apr; post-nuptial body- and wing-moult complete before arrival in tropics. SIZE: wing, ♂ (n = 19) 158–173 (167), ♀ (n = 16) 162–178 (170); tail, ♂ (n = 10) 62–69 (65·1), ♀ (n = 10) 62–68 (66·2); bill, ♂ (n = 17) 54–62 (57·4), ♀ (n = 17) 55–62 (58·9); tarsus, ♂ (n = 20) 52–61 (58·6), ♀ (n = 18) 54–60 (56·9). WEIGHT: (n = 2, Kenya) Jan 133, late Mar 157.

IMMATURE: juvenile plumage as non-breeding adult but upperparts darker and browner, spotted white; coverts with large lateral buff spots; underparts heavily washed and barred greyish brown. Replaced on body Oct–early Nov by 1st winter plumage which is like non-breeding adult; spotting on coverts worn by Dec. Many 1st summer birds attain breeding plumage.

Field Characters. Medium large sandpiper with long, slender neck, long straight bill with red base, and long red legs. Pale grey above in non-breeding plumage and white below, with prominent eye-streak and supercilium. Black breeding plumage diagnostic. White rump and back show in flight, and white mottling on secondaries and inner primaries gives broad paler trailing edge to wing. Non-breeding birds distinguished from Greenshank *T. nebularia* by red legs and base to bill, dark eye-streak and paler upperparts. From Redshank *T. totanus* by greyer plumage, more slender build and lack of clear-cut white trailing edge to inner wing. Feeds gregariously, often spinning on water like phalarope. Single birds commonly fly high, when identified by distinctive voice.

Voice. Tape-recorded (62, 73, 76, C). A distinctive 'chuwet' or sometimes an abbreviated 'tewt', usually in flight.

General Habits. Favours inland freshwater habitats, typically marshy lake edges, floodland and irrigation; often occurs on quite small dams and pools; locally on saltpans and soda lakes. Often seen singly, although commonly also in parties of up to 20 and exceptionally over 100.

Feeds by picking and probing, and also by chasing with head partly submerged. In deeper water often in compact parties, feeding with whole head and neck immersed; swims readily, up-ending like a surface-feeding duck. Shy and easily flushed; flight fast and erratic.

Moulting birds present Tunisia June–July. Passage continues N Africa generally to Nov with strong coastal movements W Morocco late Aug–early Nov. Present in tropics mainly Oct–Apr. Return movement N Africa late Mar–early May. Few records of oversummering. Passage birds ringed Britain, Germany and Denmark recovered respectively Morocco, Algeria and Tunisia. A Swiss-ringed passage bird found Senegambia is sole recovery south of Sahara.

Food. Aquatic insects, small crustaceans and molluscs, small fish and amphibians.

Tringa totanus (Linnaeus). Common Redshank; Redshank. Chevalier gambette.

Plates 17, 21
(Opp. pp. 304, 352)

Scolopax totanus Linnaeus, 1758. Syst. Nat. (10th ed.), p. 145; Sweden.

Range and Status. Breeds N Africa and from Europe, including Iceland, through Russia and central Asia to Tibet and Mongolia; winters Africa, W and S Europe, and S and SE Asia.

Mainly a Palearctic visitor; also breeds Tunisia (200–250 pairs: M. A. Czajkowski, pers. comm.). Winters commonly to abundantly Mediterranean and Red Sea coasts, Somalia and Atlantic coasts south to Sierra Leone. Large wintering populations Mauritania (200,000), Morocco (5000–7000: J. D. R. Vernon, pers. comm.), Tunisia (5000+ Jan: M. Smart, pers. comm.) and Egypt (3000–4000). Locally frequent to common Ivory Coast–Cameroon and Somalia–Kenya. Uncommon to frequent inland in N tropics, where regular along Nile to Jonglei Province (Sudan) and on passage in Mali and N Nigeria. Rare to uncommon in central, southern and rest of E Africa, mainly on coast.

Description. *T. t. totanus* (Linnaeus): N and NW Africa. ADULT ♂ (breeding): forehead to nape and hindneck blackish brown, edged buff; lores dark brown; whitish supercilium inconspicuous; sides of face and neck whitish, spotted and streaked dark brown; mantle and scapulars brown, streaked and barred dark brown, fringed buff; back and rump white; uppertail-coverts barred brown and white, centre pair suffused greyish brown; underparts white, sides of breast suffused brown, chin to breast heavily streaked and spotted, flanks and undertail-coverts barred dark brown. Primaries, primary coverts and alula dark brown; secondaries white; tertials, inner greater and inner median coverts as scapulars; rest of upperwing-coverts dark brown, edged and tipped whitish; axillaries and underwing-coverts spotted and barred brownish. Bill horn, base orange; eye brown; legs and feet orange-red. ADULT ♂ (non-breeding): as breeding bird, but upperparts drab brown, feathers darker medially; underparts whiter, streaking and barring less heavy; tertials and inner wing-coverts plainer, as upperparts. Sexes alike. Most moult

Tringa totanus

into breeding plumage in African winter quarters, but already in non-breeding plumage on return to Africa. SIZE: wing, ♂ (n = 56) 149–161 (156), ♀ (n = 46) 151–167 (159); tail, ♂ (n = 56) 56–67 (62·4), ♀ (n = 46) 58–69 (62·8); bill, ♂ (n = 28) 34–43 (40·4), ♀ (n = 23) 39–44 (41·4); tarsus, ♂ (n = 28) 41–51 (46·4), ♀ (n = 23) 44–52 (46·2). WEIGHT: (n = 162, S France, Aug–Mar) 78–176; (n = 600+, Morocco/Mauritania, Aug–Nov) 62–141.

IMMATURE: juvenile as non-breeding adult but upperparts warmer brown, edged and spotted buff, scapulars barred dark laterally; sides of neck, foreneck and breast light buff, heavily streaked; tertials and inner wing-coverts as scapulars, median and lesser coverts fringed buff. Legs pale orange-yellow. Body plumage replaced Sept by 1st winter plumage like non-breeding adult.

DOWNY YOUNG: black median line from base of upper mandible across centre of crown, and black line from above eye encircling rear of crown; rest of crown warm buff; black line from lores through eye to sides of nape, above this a broad buff band. Broad black band down centre of upperparts, divided by narrow median buff line in back region; broad black band on either side of back, and black lines across wing and along thighs; blackish patch also above exposed tibia; rest of upperparts warm buff. Breast suffused buff; rest of underparts white. Legs greyish.

T. t. ussuriensis Buturlin: NE, E and southern Africa. Breeding bird paler than *T. t. totanus* and feather edging more chestnut; streaking below less extensive, throat whiter. Non-breeding bird paler, more greyish above than *totanus*, underparts usually whiter. Slightly larger than nominate: (16♂♂, 16♀♀) wing, ♂ 152–170 (161), ♀ 156–167 (161); bill, ♂ 38–47 (42·7), ♀ 40–48 (41·8); tarsus, ♂ 45–55 (50·7), ♀ 47–55 (48·7). WEIGHT: (n = 8, Sudan/Kenya, Sept–May) 100–183.

T. t. robusta (Schiøler): Morocco. Breeding plumage darker, more heavily marked than *T. t. totanus*. Wing larger: ♂ (n = 13) 160–175 (167), ♀ (n = 16) 163–175 (170).

Field Characters. Medium-sized *Tringa* with red-based bill and orange-red legs. In flight, white back and rump like other species but broad white trailing edge to inner wing diagnostic. Smaller than Spotted Redshank *T. erythropus*, with shorter bill and legs, legs more orange; in non-breeding plumage browner, breast more streaked, eye-stripe less pronounced. Ruff *Philomachus pugnax*, also generally brown and often with bright orange legs, differs in having dark secondaries, back and rump centre. Also distinguishable by excitable behaviour. Performs deep suspicious bobbing, nervous, shy; yelping call nearly always given when flushed.

Voice. Tape-recorded (38, 62, 73, 76). A whistling 'tuuuu', and a triple 'tu-hu-hu' with 1st note higher commonly given in flight. Alarm call a persistent 'tewk-tewk-tewk ...'. On breeding grounds a repeated 'tut-tut-tut-tut-tut ...' accompanying song flight or from vantage point, and a yodelling 'taweeo-taweeo ...'.

General Habits. Frequents estuaries, mudflats, coastal lagoons and saltmarsh; inland on open lake-shores and pools. Often singly or in small groups, but forms loose flocks on coast, sometimes many hundreds together.

Walks briskly, feeding by picking and intermittently probing in shallows or exposed tidal mud. Noisy and restive. Flight strong but rather erratic, wing-beats quick but slightly jerky.

Adults moult Tunisia June–July; arrive Morocco July–Aug, Red Sea late Aug–Sept, and Gulf of Guinea coast mainly from late Sept. Inland passage W Africa Oct. Crosses Sahara on broad front. Overwintering birds depart Tunisia from Feb, NW African and Gulf of Guinea coasts Mar–early Apr and Red Sea mainly Apr. Commonly oversummers Africa. Ringing indicates birds wintering NW Africa derived mainly from W Europe, but 1 recovery Iceland to Morocco. 17 birds recovered further south in Africa (Mauritania, Senegambia, Liberia, Sierra Leone, Ivory Coast and Ghana) include 10 ringed W Europe, also 4 from Sweden, 2 from Norway and 1 from Finland. Lack of ringing recoveries suggests birds wintering from Nigeria eastwards and southwards are from little-ringed European Russian and Siberian populations.

Food. In estuarine habitats principally polychaetes, crustaceans and small molluscs. Inland mainly insects, and diets may include vegetable matter; also arachnids and occasionally small fish and amphibians.

Breeding Habits. Monogamous; territorial; 1st breeding usually in 2nd year, occasionally in 1st year. Display by individual pairs or small flocks. In song flight, by both sexes, birds alternately rise a few m, wings depressed and fluttering rapidly, head and neck held up, then descend on slanting glide, legs hanging down and tail fanned. This flutter and descending action often performed over one spot, but also during extended flights. Before coition ♂ usually chases ♀ with tail fanned. If she responds and stops, he approaches with high stepping gait and wings raised and beating rapidly, usually from behind, then mounts and, if successful, copulates with wings fluttering continuously. Afterwards ♂ may perform bowing action round ♀, tail again fanned.

NEST: usually solitary or in small loose colonies on low coastal islands and saltmarshes in Tunisia; in Europe in low-lying rough pasture or saltings near coast; also on edges of high moorland. Usually concealed in long grass tuft with entrance at side; lined with dry grasses; cup 8–12 cm across (mean 23 nests Europe 10·3 cm), 2–7 cm deep (mean 5·2 cm).

EGGS: in Europe usually 4, occasionally 3 and rarely 5, laid at 2-day intervals; W Germany mean (379 clutches) 3·95. Oval; pale greyish or buffish white, blotched or spotted dark or reddish brown, mainly at large end. SIZE: (n = 100, Britain) 41·5–48·4 × 28·5–33·1 (45·2 × 31·6).

LAYING DATES: Tunisia, Apr.

INCUBATION: begins on completion of clutch; by both sexes. Period (Europe): 22·5–24·5 days.

DEVELOPMENT AND CARE OF YOUNG: young leave nest soon after hatching, tended by both parents; fly after 23–35 days.

BREEDING SUCCESS/SURVIVAL: no data Africa; from 1047 eggs at W German site in 1955–58, 73% hatched (2·99 per nest); c. 50% of chicks hatched were estimated to have fledged. Mortality of fledged young during 1st year estimated Europe at 55%; av. annual mortality of 2–3 year-old birds 27–30%. European ringed birds recovered after up to 16 years (Glutz von Blotzheim 1977).

References
Goss-Custard, J. D. (1969).
Grosskopf, G. (1958–59).
Hale, W. G. (1971).

Tringa stagnatilis (Bechstein). Marsh Sandpiper. Chevalier stagnatile.

Plates 17, 21
(Opp. pp. 304, 352)

Totanus stagnatilis Bechstein, 1803. Orn. Taschenb. 2, p. 292; Germany.

Range and Status. Breeds SE Europe, central and S USSR east to Mongolia; winters Africa, Madagascar, S and SE Asia to Australasia. Palearctic visitor throughout Africa, mainly to lakes, rivers and flooded ground; also locally to coastal mudflats and lagoons. Widespread and locally abundant winter visitor from N tropics to Cape. One of the commonest waders on Rift Valley freshwater and soda lakes Ethiopia, Kenya and N Tanzania (e.g. 940 southern part Kenya Rift Valley: Pearson and Stevenson 1980; 1100 along 5 km western shore L. Turkana: D. J. Pearson, pers. obs.). Uncommon forested central and W Africa but frequent in e.g. Nigerian savannas. Regular on passage N Africa though uncommon Morocco and Algeria; a few winter S Tunisia and Egypt. Common Red Sea coast.

Description. ADULT ♂ (breeding): forehead to nape, hindneck, sides of neck and face buffish white, speckled and streaked black; mantle and scapulars pale greyish buff with broad black streaks and barring; back, rump white; uppertail-coverts white with black centre markings. Rectrices white, lightly variegated and barred blackish. Underparts white, breast boldly spotted with black, flanks slightly barred. Primaries, primary coverts and alula blackish; secondaries grey, fringed white; tertials, inner greater coverts and inner median coverts greyish buff, streaked and barred black; outer greater and outer median coverts greyish; lesser coverts dark brown; axillaries and underwing-coverts white. Bill black; eye brown; legs and feet olive-green to yellowish brown. ADULT ♂ (non-breeding): as breeding bird, but crown, nape, hindneck, mantle and scapulars pale grey with narrow black medial streaks and whitish fringes; forehead and face white, ear-coverts and cheeks lightly streaked brown; tertials and inner wing-coverts as mantle. Sexes alike. Breeding plumage acquired Africa Jan–Mar; post-nuptial body-moult and most or all of wing-moult usually completed before arrival in Africa. SIZE: wing, ♂ (n = 32) 131–145 (139), ♀ (n = 28) 135–148 (142); tail, ♂ (n = 37) 52–63 (56·9), ♀ (n = 27) 53–61 (57·8); bill, ♂ (n = 37) 38–45 (40·0), ♀ (n = 35) 36–45 (40·9); tarsus, ♂ (n = 41) 47–57 (51·6), ♀ (n = 37) 47–57 (51·5). WEIGHT: (n = 1000+, Kenya) 44–99; typical wintering weight 55–75.

IMMATURE: juvenile plumage as non-breeding adult, but crown, nape, mantle, scapulars, tertials and lesser, median and inner greater wing-coverts dark brown, spotted and fringed buff. Replaced on body in Sept by 1st winter plumage like that of non-breeding adult. Juvenile outer primaries sometimes moulted Feb–Mar. 1st summer plumage same as breeding adult but moult less complete and many winter feathers retained.

Field Characters. Smallish sandpiper resembling miniature Greenshank *T. nebularia*, but more slender and graceful, with bill fine and straight, and greenish or yellowish legs proportionately longer. Secondaries paler than in Greenshank, and dark carpal area more contrasting. Breeding bird paler and more spotted-looking than Greenshank; non-breeding bird whiter about head. Similar body size to Green Sandpiper *T. ochropus* and Wood Sandpiper *T. glareola*, but easily distinguished by much longer bill and legs, white rump patch extending in V-shape up back, and whitish underwing.

Voice. Tape-recorded (2, 62, 73, 76, B, C). A single 'cheeoo', usually when flushed, less ringing and fluty than call of Greenshank; also a sharp 'chik'. Several birds disturbed together can recall sound of a flock of Wood Sandpipers.

General Habits. Frequents muddy and marshy lake-edges and irrigation, sometimes also small pools and riverbanks; confined on coast to creeks, lagoons and estuaries. Usually singly or in small parties, often associating with Ruff *Philomachus pugnax* or Wood Sandpiper. Monospecific flocks sometimes exceed 300.

Movements quick and agile; feeds by wading in shallow water, frequently changing direction and picking repeatedly from side to side. When flushed often flies short distance only, with rather weak-looking action.

Adults moult Tunisia late June–July, and passage continues N Africa to Oct. Crosses Sahara on broad front. Most adults reach tropical wintering areas Sept. Young birds reach equator from end Aug, but main influx Sept–Oct. Spring departure from wintering areas mainly late Mar–mid Apr. Passage Tunisia Mar–early Apr, but continuing to early May in NE Africa. Many 1st-year birds oversummer in Africa. Wintering areas W Africa far to west of breeding range and westernmost passage localities Europe and N Africa; a SW–NE crossing of Sahara is thus indicated. Birds ringed SE Zaïre and Kenya recovered USSR in Kazakhstan and Tyumen region respectively; another bird ringed Kenya (Apr) recovered South Africa (Jan).

Food. Mainly insect larvae; also small molluscs and crustaceans.

Reference
Pearson, D. J. *et al.* (1970).

Plates 17, 21
(Opp. pp. 304, 352)

Tringa nebularia (Gunnerus). **Common Greenshank; Greenshank. Chevalier aboyeur.**

Scolopax nebularia Gunnerus, 1767. Leem. Beskr. Finm. Lapp., p. 251; Norway.

Range and Status. Breeds NW Europe east through boreal zone of USSR to Kamchatka; winters Africa, Madagascar, Mediterranean basin, and S and E Asia to Australasia.

Palearctic visitor, widespread throughout on coasts and inland. Common to abundant winter visitor N tropics to Cape; also locally NW Africa east to Tunisia, Lower Nile Valley and Red Sea coasts. Mainly on passage N Africa when common to abundant all coasts.

Description. ADULT ♂ (breeding): forehead, crown, nape and hindneck greyish white, heavily streaked dark brown; mantle feathers blackish brown, edged white; scapulars greyish brown, streaked and barred dark brown; sides of face and neck white, streaked dark brown; back and rump white; uppertail-coverts white barred and variegated brown. Rectrices white with incomplete narrow brown bars. Underparts white, foreneck, upper breast and flanks spotted and streaked dark brown. Primaries, primary coverts and alula blackish; secondaries greyish brown, fringed whitish; tertials, inner greater coverts and inner median coverts streaked and barred blackish; rest of upperwing greyish brown, darkest on lesser coverts; axillaries and underwing-coverts white, narrowly barred dark brown. Bill slightly upturned, blackish, greyish green at base; eye brown; legs and feet pale olive-green. ADULT ♂ (non-breeding): as breeding adult, but crown, nape and hindneck greyish brown, edged white; mantle and scapulars greyish brown with narrow shaft streaks and white fringes; forehead and sides of face white, lores, cheeks and ear-coverts streaked dusky; streaking below confined to sides of neck and breast; tertials and inner wing-coverts as mantle. Sexes alike. Pre-nuptial moult in Africa Jan–Mar; post-nuptial body-moult before return to tropics, but wing-moult usually completed in Africa Aug–Dec. SIZE: wing, ♂ (n = 15) 176–194 (186), ♀ (n = 17) 184–204 (192); tail, ♂ (n = 20) 70–81 (77·1), ♀ (n = 8) 71–82 (76·9); bill, ♂ (n = 16) 51–58 (55·2), ♀ (n = 24) 51–63 (55·8); tarsus, ♂♀ (n = 85) 57–69 (62·6). WEIGHT: (n = 400+, E and southern Africa) 121–305; typical wintering weight 145–195.

IMMATURE: juvenile plumage as non-breeding adult, but upperparts dark brown, broadly edged buff; wing-coverts with broad buff edging and tertials and inner greater coverts with brown lateral barring; marking below confined to sparse speckling and barring on sides of breast. 1st winter plumage as non-breeding adult, replaces juvenile plumage on body Sept–Oct. Some birds moult juvenile outer primaries Feb–Mar. 1st summer plumage as breeding adult but many winter feathers retained.

Field Characters. A large sandpiper with long, rather stout, slightly upturned bill and long greenish legs; greyish brown above and white below in non-breeding plumage. In flight, white back forms prominent wedge between dark scapular tracts. In breeding plumage browner above and less white about head and neck. Distinguished from Marsh Sandpiper *T. stagnatilis* by larger size and stouter bill; from similarly sized Spotted Redshank *T. erythropus* by greenish legs, upturned bill without red at base, rather darker upperparts and lack of prominent eye-streak.

Voice. Tape-recorded (38, 58, 74, B, C, F). Highly distinctive; a fluty, far-carrying 'tew-tew-tew', mainly in flight.

General Habits. Frequents open muddy or rocky lake-shores and larger rivers, estuaries, sandflats and exposed coral. Normally solitary or in small parties well spread out whilst feeding. May form larger flocks on migration, and gatherings up to 100+ are not unusual at high tide or night roosts.

Shy and easily flushed. Flight powerful and agile. Feeds by wading steadily in shallow water with head held up, bending frequently to probe or pick from surface; also by wading with head forward, scything from side to side with bill submerged.

Some adults arrive Tunisia late June–July and moult; passage N Africa generally July–Oct. Crosses Sahara on broad front. Adults reach winter quarters as far distant as South Africa from late July, but most arrive Aug–Sept. Young birds appear in tropics from end Aug and South Africa from early Sept, but main influx probably Oct. Full wintering numbers not established in South Africa until Dec. Departure from south mainly late Feb–early Apr, and from equatorial latitudes late Mar–late Apr. Spring passage N Africa Mar–early May. Many birds remain in Africa throughout 1st summer. Birds ringed W and central Europe recovered in Mali, Ivory Coast, Ghana and N Zaïre; 1 ringed Italy recovered Nigeria and 1 ringed Morocco recovered Finland. Birds wintering E and southern Africa presumably from unringed USSR populations.

Food. Insects and their larvae, crustaceans, molluscs, annelids, fish fry and small amphibians.

Reference
Tree, A. J. (1979).

Tringa melanoleuca (Gmelin). Greater Yellowlegs. Grand chevalier à pattes jaunes.

Scolopax melanoleuca Gmelin, 1789. Syst. Nat., 1, ii, p. 659; Labrador.

Not illustrated in colour

Range and Status. Breeds Alaska and N Canada; winters North and South America.

Vagrant, recorded once South Africa (Cape Province near Cape Town, 2 Dec 1971).

Description. ADULT ♂ (non-breeding): forehead to nape, hindneck, mantle and scapulars greyish brown edged paler; scapulars also notched blackish brown; some streaks on upperhead and hindneck and white spots on mantle and scapulars; prominent whitish supercilium; loral streak greyish brown; sides of face and neck white, streaked greyish brown; back and rump dark brown, feathers tipped white; uppertail-coverts white, lowermost barred greyish brown. Central 2 pairs of rectrices greyish, rest white; all barred and marked dark greyish brown. Underparts white, sides of breast greyish brown, foreneck and breast lightly streaked and flanks lightly barred greyish brown. Primaries, primary coverts and alula blackish brown, inner primaries broadly tipped and notched white; secondaries dark brown, fringed, broadly tipped and notched white; tertials, greater coverts and median coverts like scapulars, but greater coverts and median coverts more broadly fringed and notched white; lesser coverts dark brown, fringed white; axillaries and underwing-coverts barred dark brown and white. Bill black, basal third grey or greenish; eye brown; legs bright orange-yellow. Sexes alike. SIZE: wing, ♂♀ (n = 9) 186–200 (193); tail, ♂♀ (n = 30) 71–83 (77); bill, ♂♀ (n = 9) 50–60 (55·9); tarsus, ♂♀ (n = 9) 59–66 (62·3). WEIGHT: (n = 87, Venezuela) 110–220.

IMMATURE: juvenile as non-breeding adult, but much darker and browner; upperparts, tertials and upperwing-coverts extensivly spotted buff; neck and breast more heavily streaked, flanks prominently barred.

Field Characters. Larger and more robust than Lesser Yellowlegs *T. flavipes*; bill usually longer, 1·2–1·7 times head length (from base of bill to back of skull), versus 1·0–1·3 times head length in Lesser; also stouter, and usually with slight upturn (less pronounced than in Greenshank *T. nebularia*), though sometimes straight Pale base of bill more extensive than in Lesser. I winter plumage the 2 yellowlegs are almost identical, but in summer Greater has much more heavily marked underparts, belly barred and streaked (white in Lesser). Caution: lone winter bird with intermediate-length straight bill can be difficult to identify. Resembles Greenshank in structure and size but browner, more spotted-looking, with yellow legs and white rump not extending up back.

Voice. Tape-recorded (C). Loud ringing call of 2 or more syllables, 'wheu-wheu-wheu', reminiscent of Greenshank.

General Habits. In Nearctic wintering and passage birds frequent on inland lakes and marshes and coastal lagoons and mudflats. Less gregarious and less tame than Lesser Yellowlegs. Flight fast and well sustained. Feeds by picking at water surface and skimming with bill; also by seizing swimming prey.

Food. Insects, crustaceans, molluscs, arachnids, small fish.

Reference
Wilds, C. (1982).

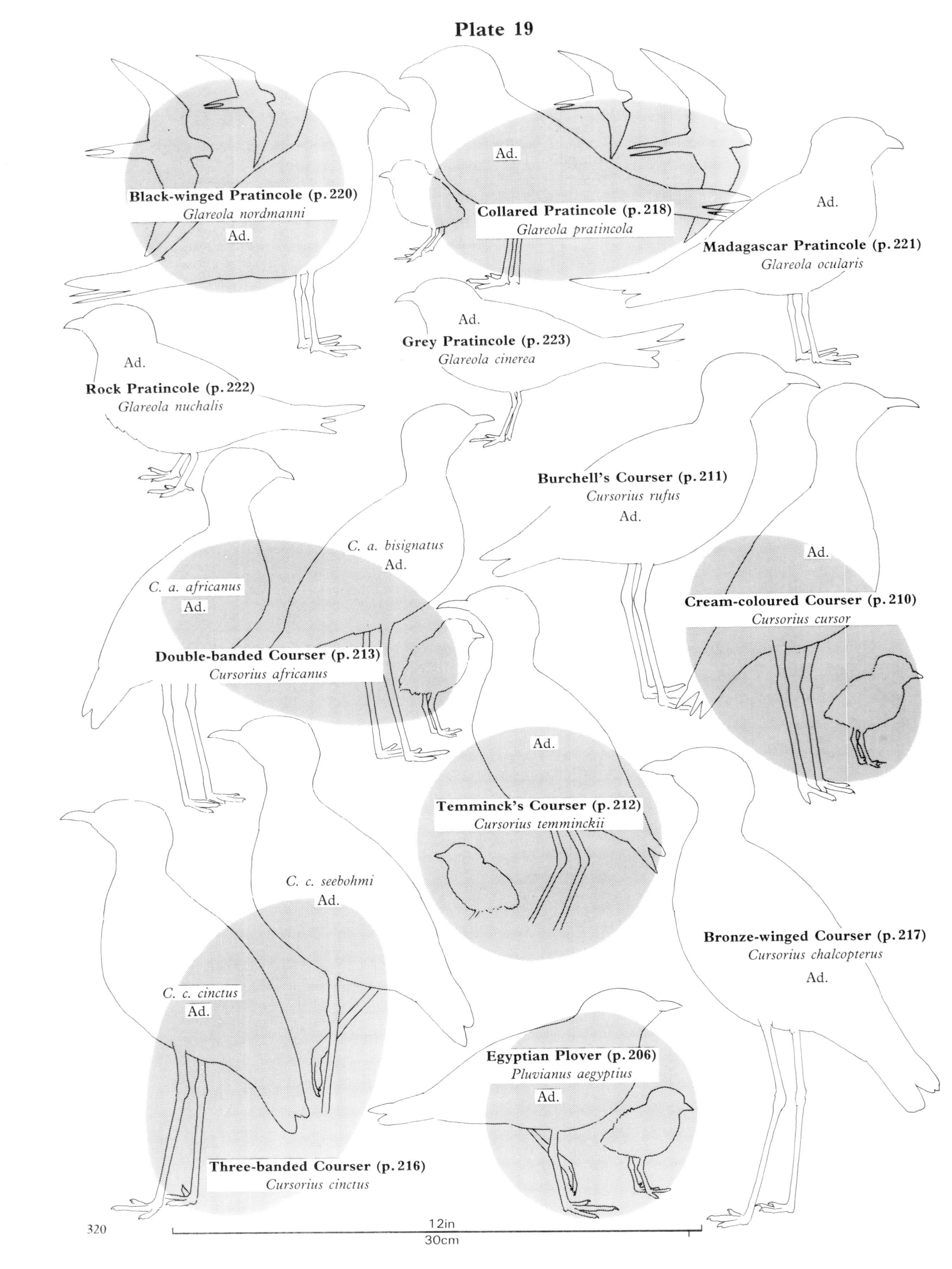

Plate 19

Black-winged Pratincole (p. 220)
Glareola nordmanni
Ad.

Collared Pratincole (p. 218)
Glareola pratincola
Ad.

Madagascar Pratincole (p. 221)
Glareola ocularis
Ad.

Rock Pratincole (p. 222)
Glareola nuchalis
Ad.

Grey Pratincole (p. 223)
Glareola cinerea
Ad.

Burchell's Courser (p. 211)
Cursorius rufus
Ad.

C. a. bisignatus
Ad.

C. a. africanus
Ad.

Double-banded Courser (p. 213)
Cursorius africanus

Cream-coloured Courser (p. 210)
Cursorius cursor
Ad.

Temminck's Courser (p. 212)
Cursorius temminckii
Ad.

C. c. seebohmi
Ad.

C. c. cinctus
Ad.

Three-banded Courser (p. 216)
Cursorius cinctus

Bronze-winged Courser (p. 217)
Cursorius chalcopterus
Ad.

Egyptian Plover (p. 206)
Pluvianus aegyptius
Ad.

12in / 30cm

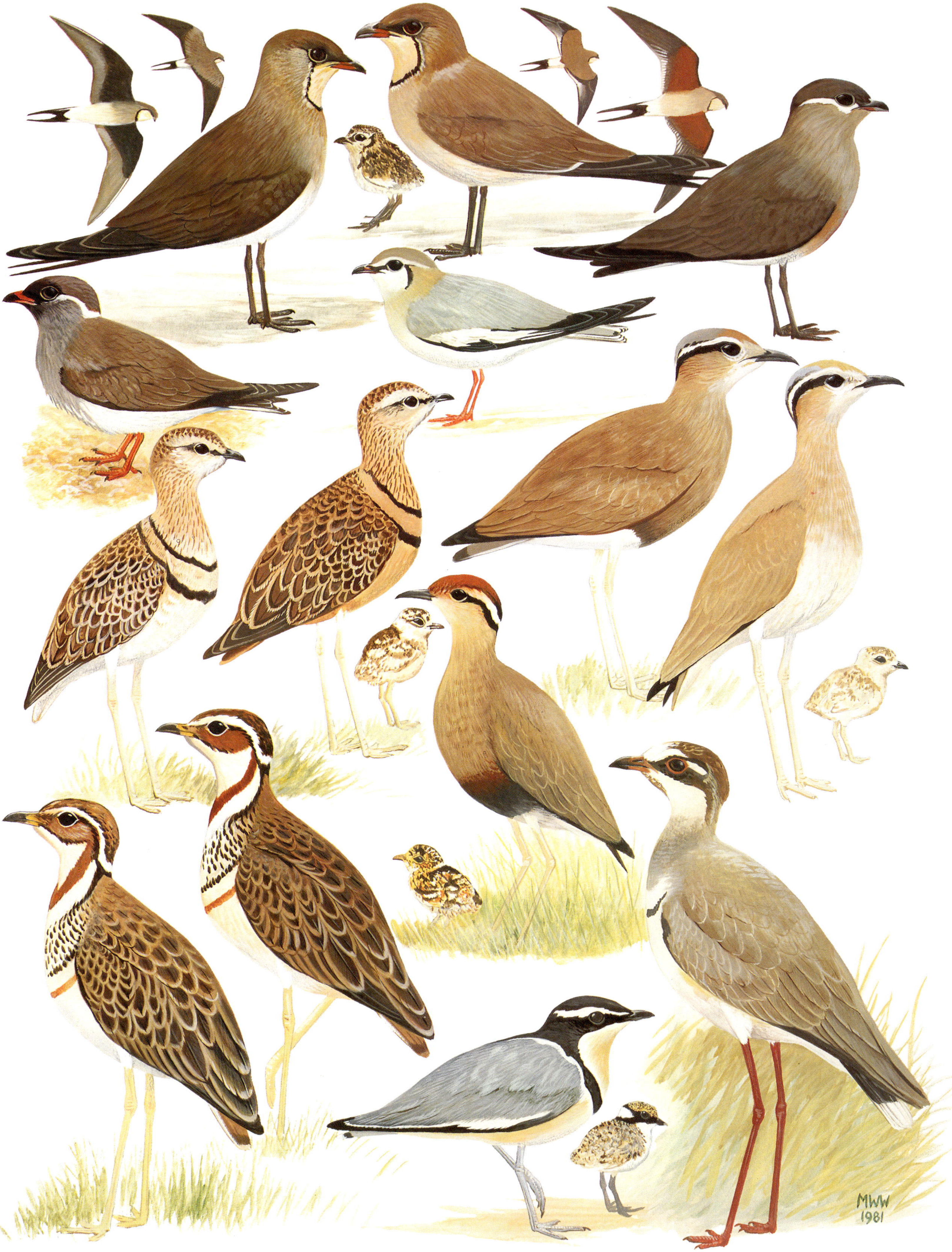
MWW
1981

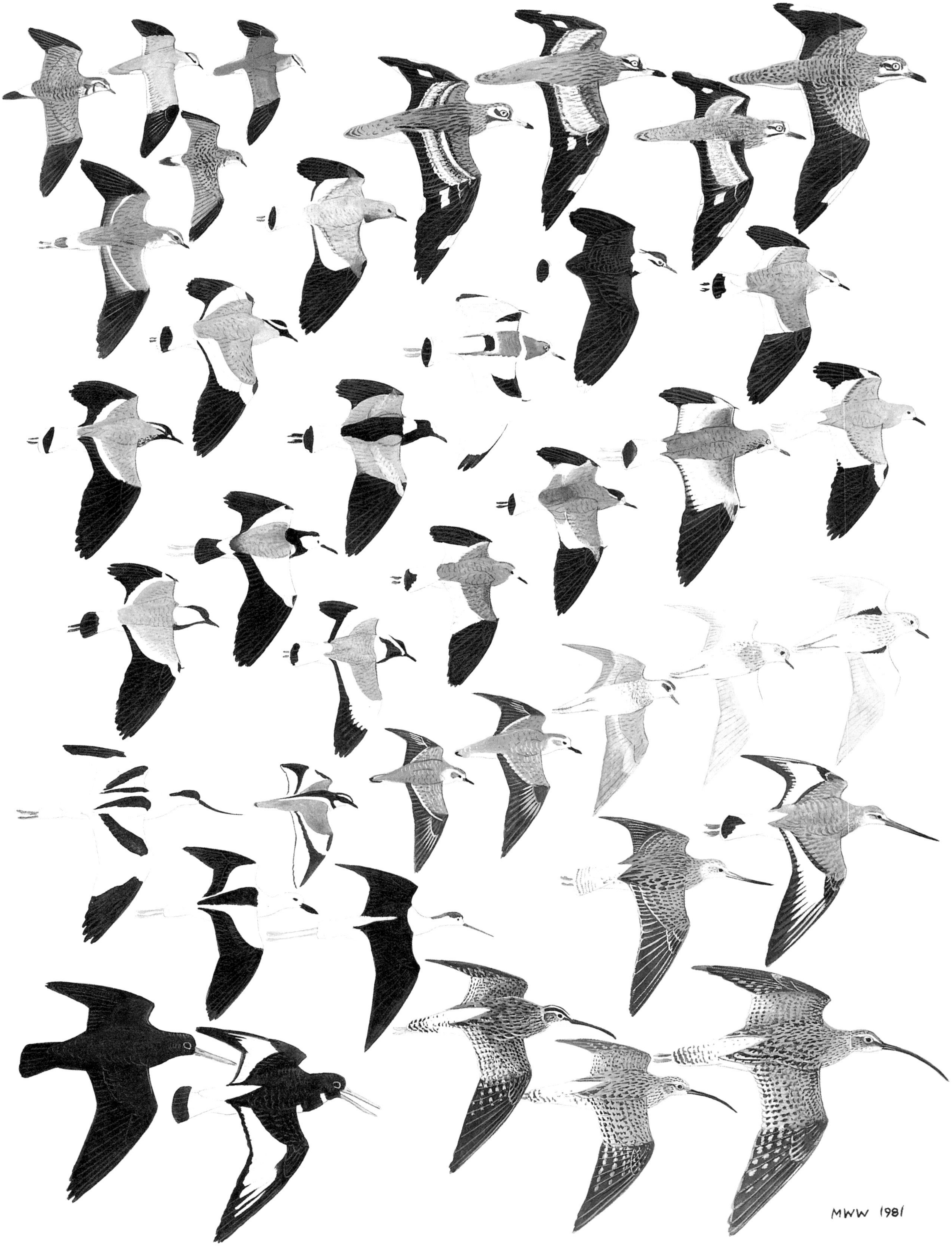
MWW 1981

Plate 20

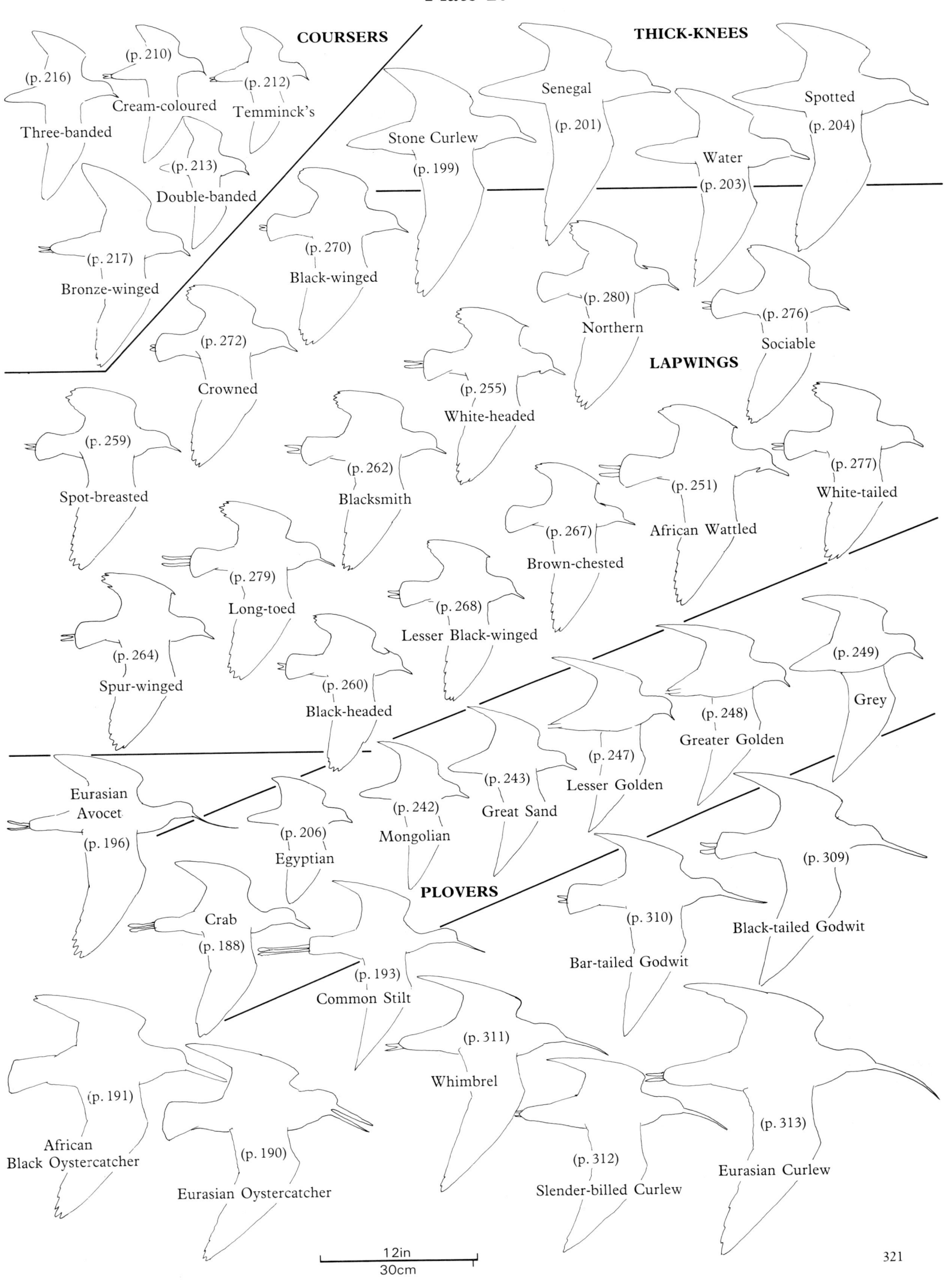

COURSERS

- Three-banded (p. 216)
- Cream-coloured (p. 210)
- Temminck's (p. 212)
- Double-banded (p. 213)
- Bronze-winged (p. 217)

THICK-KNEES

- Stone Curlew (p. 199)
- Senegal (p. 201)
- Water (p. 203)
- Spotted (p. 204)

LAPWINGS

- Black-winged (p. 270)
- Crowned (p. 272)
- White-headed (p. 255)
- Northern (p. 280)
- Sociable (p. 276)
- Spot-breasted (p. 259)
- Blacksmith (p. 262)
- Brown-chested (p. 267)
- African Wattled (p. 251)
- White-tailed (p. 277)
- Long-toed (p. 279)
- Lesser Black-winged (p. 268)
- Spur-winged (p. 264)
- Black-headed (p. 260)

PLOVERS

- Eurasian Avocet (p. 196)
- Egyptian (p. 206)
- Crab (p. 188)
- Common Stilt (p. 193)
- Mongolian (p. 242)
- Great Sand (p. 243)
- Lesser Golden (p. 247)
- Greater Golden (p. 248)
- Grey (p. 249)
- Bar-tailed Godwit (p. 310)
- Black-tailed Godwit (p. 309)
- African Black Oystercatcher (p. 191)
- Eurasian Oystercatcher (p. 190)
- Whimbrel (p. 311)
- Slender-billed Curlew (p. 312)
- Eurasian Curlew (p. 313)

12in / 30cm

321

Plate 21
(Opp. p. 352)

Tringa flavipes (Gmelin). Lesser Yellowlegs. Petit chevalier à pattes jaunes.

Scolopax flavipes Gmelin, 1789. Syst. Nat. 1, ii, p. 659; New York.

Range and Status. Breeds Alaska, N and NW Canada; winters central and South America.

Vagrant, recorded Nigeria (Lagos, 15 Feb–18 Mar 1969), Zambia (21 Jan–18 Feb 1979) and Zimbabwe (31 Dec 1979) and South Africa (Cape, Velddrif, 2 Aug 1983).

Description. ADULT ♂ (non-breeding): forehead to nape, hindneck, mantle and scapulars greyish brown edged paler; scapulars also notched blackish brown; prominent whitish supercilium; loral streak greyish brown; sides of face and neck white, streaked greyish brown; back and rump dark brown, feathers tipped white; uppertail-coverts white, lowermost barred greyish brown. Central 2 pairs of rectrices greyish, rest white; all barred and marked dark greyish brown. Underparts white, sides of breast greyish, foreneck and breast lightly streaked and flanks lightly barred greyish brown. Primaries, primary coverts and alula blackish brown; secondaries dark brown, fringed white; tertials, greater coverts and median coverts as scapulars; lesser coverts dark brown, fringed white; axillaries and underwing-coverts barred dark brown and white. Bill black; eye dark brown; legs and feet bright yellow. Sexes alike. SIZE: wing, ♂ (n = 10) 149–168 (161), ♀ (n = 17) 155–169 (162); tail, ♂♀ (n = 10) 54–59 (57·2); bill ♂♀ (n = 37) 33–40 (36·6); tarsus, ♂♀ (n = 36) 47–58 (52·3). WEIGHT: (n = 52, Surinam) 60–106.

IMMATURE: juvenile as non-breeding adult, but darker and browner; upperparts, tertials and upperwing-coverts extensively spotted pale buff; neck and upper breast streaked brown and heavily washed grey.

Field Characters. Slightly larger than Marsh Sandpiper *T. stagnatilis* and different in shape—less slender-looking, legs shorter, less spindly, bill fine but shorter. Legs bright yellow, never greenish. At rest lacks dark carpal patch of Marsh Sandpiper, and in flight white of rump does not extend in wedge up back. Among Palearctic waders most resembles Wood Sandpiper *T. glareola* (flight pattern almost identical) but larger, greyer, bill usually longer than head (same length as head or shorter in Wood), legs longer, brighter yellow; wing-tips projecting well beyond tail-tip; call very different. For distinction from Greater Yellowlegs *T. melanoleuca* see account for that species.

Voice. Tape-recorded (62, 73). A sharp, yelping 'tu-tu', shorter, less ringing and fluty than call of Greater Yellowlegs.

General Habits. On passage and in winter frequents lakes, marshes and wet grassland inland; also coastal lagoons and mudflats. Often forms flocks of up to 100 or more. Movements quick and graceful; feeding behaviour and flight resemble those of Marsh Sandpiper.

Food. Insects, crustaceans, molluscs and fish fry.

Plates 17, 21
(Opp. pp. 304, 352)

[*Tringa solitaria* Wilson. Solitary Sandpiper. Chevalier solitaire.]

[Recorded once Angola (Cabinda: Mackworth-Praed and Grant 1970) but not fully substantiated; no details on locality and date of collection available.]

Plates 17, 21
(Opp. pp. 304, 352)

Tringa ochropus Linnaeus. Green Sandpiper. Chevalier culblanc.

Tringa ochropus Linnaeus, 1758. Syst. Nat. (10th ed.), p. 149; Sweden.

Range and Status. Breeds from Baltic and E Europe through boreal region of USSR to Sea of Okhotsk; winters Africa, SW and S Europe and S and SE Asia to Philippines.

Palearctic visitor, wintering on freshwater streams, pools and river edges throughout; locally frequent to abundant from Mediterranean basin to Zaïre and Zambia; rare to uncommon in Angola and south of Zambezi. Mainly on passage N Africa. Many passage records at Saharan oases.

Description. ADULT ♂ (breeding): forehead to nape, hindneck, mantle and scapulars dark olive-brown with fine spotting; whitish supercilium; lores, ear-coverts, cheeks and sides of neck heavily streaked olive-brown; back and rump olive-brown edged white; uppertail-coverts white. Rectrices white, distal half of inner 4 pairs and tip of 5th broadly barred blackish. Underparts white, foreneck and upper breast heavily streaked and flanks barred dark brown. Primaries, primary coverts and alula blackish brown; secondaries brown; rest of upperwing dark olive-brown, tertials, inner greater coverts and inner median coverts with small white lateral spots; axillaries and underwing-coverts dark brown, narrowly barred white. Bill blackish brown, tinged green at base; eye dark brown; legs and feet dark olive-green. ADULT ♂ (non-breeding): as breeding bird but more uniform above, spotting inconspicuous; sides of face and neck and breast less heavily streaked. Sexes alike. Pre-nuptial moult Africa Jan–Apr; post-nuptial moult completed Africa Aug–Dec. SIZE: wing, ♂ (n = 23) 137–147 (143), ♀ (n = 27) 140–155 (147); tail, ♂ (n = 22) 54–59 (56·9), ♀ (n = 15) 55–61 (58·1); bill, ♂ (n = 25) 31–38 (34·5), ♀ (n = 29) 31–37 (34·5); tarsus, ♂ (n = 26) 32–37 (33·9), ♀ (n = 31) 31–36 (33·9). WEIGHT: (n = 41, Kenya) 66–120; typical wintering weight 70–90.

IMMATURE: juvenile plumage as non-breeding adult but upperparts, including wing-coverts, with more distinct buff spots.

Field Characters. Small sandpiper with short straight bill and dark green legs. Rather uniform dark olive-brown upperparts and dark underwing contrast in flight with square white rump patch and white belly. Darker

Migration only

and a little larger than Wood Sandpiper *T. glareola*, with darker legs and larger rump patch; whiter below with dark marking on breast more confined to sides; supercilium narrower and indistinct behind eye.

Voice. Tape-recorded (62, 73, 76, C). Flight call a ringing 'clewit-wit-wit', or, when flushed, 'clewit-lewit'.

General Habits. Occurs in sheltered habitats in a variety of country including wooded lake-edges, small dams and ponds, riverbanks, forest pools and streams, temporary inundations and small ditches and rain puddles; often near villages and cultivation; occasionally also coastal creeks. Usually solitary, but forms small parties on migration. Ones and twos remain attached to same site for many weeks.

Flight erratic with jerky wing-beat. When flushed rises high, calling repeatedly, after zig-zagging close to ground at take-off. When feeding walks quickly, picking from shallow water, mud or vegetation. Bobs and moves tail like Common Sandpiper *Actitis hypoleucos*.

On passage N African coast July–early Nov. Crosses Sahara on broad front. Present N and equatorial tropics mainly late Aug–early Apr, and in southern Africa Oct–Mar. Return passage N Africa late Feb–early May. A few remain in tropics June–July. Birds ringed Belgium and W Germany and 4 ringed Fennoscandia recovered Morocco; 1 ringed Byelorussia found Tunisia.

Food. Mainly insects and their larvae; also small molluscs and crustaceans.

Tringa glareola Linnaeus. Wood Sandpiper. Chevalier sylvain.

Tringa glareola Linnaeus, 1758. Syst. Nat. (10th ed.), p. 149; Sweden.

Plates 17, 21
(Opp. pp. 304, 352)

Range and Status. Breeds N Europe through boreal zone of USSR to E Siberia and Kamchatka; winters Africa, Madagascar, S and E Asia to Japan and Australasia.

Palearctic visitor, abundant to very abundant throughout in swamp and marshland. Inland, the most widespread wintering Palearctic wader from N tropics to Cape, with pronounced southward passage at many localities. Probably many tens of thousands lake-edges and swamplands of central Africa (e.g. 30,000 west shore, L. Edward, Zaïre: Curry-Lindahl 1981). Abundant autumn and spring passage migrant N Africa, and more locally on Red Sea coast. A few overwinter in N Africa.

Description. ADULT ♂ (breeding): forehead to nape and hindneck dark brown edged whitish; prominent whitish supercilium; lores, ear-coverts, cheeks and sides of neck streaked brown; mantle and scapulars dark brown, barred and spotted pale brown and white; uppertail-coverts white, rearmost feathers marked dark brown. Rectrices white with dark brown barring, confined to outer webs in outer 2 pairs. Underparts white, foreneck and breast washed brown and heavily streaked and spotted dark brown, flanks barred brown.

Migration only

Primaries, primary coverts and alula dark brown; secondaries dark brown, edged white; tertials, inner greater and inner median coverts dark brown with bold paler barring and white lateral spotting; rest of upperwing-coverts dark brown; axillaries and underwing-coverts white, barred brown. Bill blackish brown, greenish at base; eye dark brown; legs and feet olive-green or yellowish green. ADULT ♂ (non-breeding): as breeding bird but plainer brown above, with narrower pale edges and less conspicuous spotting; streaking below faint and confined to breast and sides of neck. Sexes alike. Pre-nuptial moult Africa Jan–Apr; post-nuptial body- and wing-moult completed Africa Aug–Jan. SIZE: wing, ♂ (n = 23) 120–131 (126), ♀ (n = 12) 123–131 (127); tail, ♂ (n = 26) 46–53 (48·3), ♀ (n = 12) 47–52 (49·1); bill, ♂♀ (n = 41) 27–31 (28·7); tarsus, ♂♀ (n = 41) 32–40 (37·0). WEIGHT: (n = 250+, Kenya) 43–87; typical wintering weight 50–65.

IMMATURE: juvenile plumage as non-breeding adult, but with clear buff spotting on upperparts and sides of tertials and upperwing-coverts; breast heavily washed brown, with pale mottling. Many birds moult juvenile outer primaries Jan–Mar. 1st-year birds acquire full breeding body plumage.

Field Characters. Small sandpiper with short straight bill, greenish legs, spotted brownish plumage, prominent supercilium and square white rump patch. More slender than Green Sandpiper *T. ochropus*, with longer, paler legs, browner, more heavily marked breast and upperparts, paler underwing and smaller rump patch.

Voice. Tape-recorded (38, 58, 74, B, C, F). Flight call 'chiff-iff', 'chiff-iff-iff', usually given when flushed, and occasionally from ground. Also a persistent 'chip' when alarmed. Display call 'tuleea-tuleea ...' sometimes heard.

General Habits. Inhabits marshy areas including permanent swamp, flooded grassland, irrigation, sedgy dams, lake-edges and muddy creeks. Usually in small scattered parties, but 20–50 often flock together and concentrations may exceed 1000 on migration. Commonly consorts with Ruffs *Philomachus pugnax*, Marsh Sandpipers *T. stagnatilis* and Common Snipe *Gallinago gallinago*.

Flight fast and erratic, but less jerky than in Green Sandpiper. Bobs tail on landing. Feeds in shallow water or from floating vegetation, walking rapidly with head up and forward, picking frequently from water surface. Perches commonly on logs, tree roots and other vantage points above water. Chasing and display calling recorded in Zaïre and South Africa Mar–Apr.

Autumn passage N African coast July–Oct. Crosses Sahara on broad front. Most adults return to tropics late July–early Sept. Young birds appear from late Aug, but most arrive and pass through in Sept–Nov. Peak overall numbers in E and central Africa Aug–Oct. Adults commonly reach South Africa from late Aug, but full wintering numbers present only from Dec. Most birds leave winter quarters Apr, although some remain in equatorial and N tropics to mid May. Evidence of movement in Zambia early Apr. Spring passage in N Africa Mar–early June. Some birds oversummer in N tropics. Of birds ringed in Scandinavia and W and central Europe, 8 recoveries in NW Africa and over 35 in tropical W Africa (Senegambia, Guinea-Bissau, Mali, Burkina Faso, Niger, Liberia, Ivory Coast, Togo, Benin, Nigeria, Cameroon, Gabon). 1 Swedish-ringed bird recovered in Egypt and 2 Finnish-ringed birds recorded in Zambia. 1 Senegambia-ringed bird recovered Italy, 3 Kenya-ringed birds recovered N European Russia at 44–49°E, and Kenyan and S Sudan birds found in W Siberia at 64°E and 70°E respectively. 1 birds ringed Kenya (Sept) found South Africa (Nov) 9 years later.

Food. Aquatic insects, insect larvae, small molluscs and crustaceans, and occasionally small fish.

Genus *Xenus* Kaup

Allied to *Tringa*, but characterized by long recurved bill, very wide at base. Legs short, toes joined by short web at base, hind toe well developed. 1 sp., visiting Africa from Palearctic.

Plates 17, 21

(Opp. pp. 304, 352)

Xenus cinereus (Güldenstadt). Terek Sandpiper. Bargette de Terek.

Scolopax cinerea Güldenstadt, 1775. Nov. Comm. Petrop. 19, p. 473; Terek River, SE Russia.

Range and Status. Breeds S Finland to central Siberia; winters Africa, Madagascar, and S and SE Asia to Australia.

Palearctic visitor, wintering commonly to abundantly S Red Sea and Indian Ocean coasts south to Natal (e.g. regularly 1000+ on 200 km Kenya coast: D. J. Pearson, pers. obs.; 3200 Inhaca Is., Mozambique, Nov: Waltner and Sinclair 1981). Locally frequent to common E and SW Cape coast; uncommon to rare Namibia. On passage, common to abundant on Sudan Red Sea coast; rare but probably regular in autumn in Tunisia, and rare in Libya and Egypt. Rare to vagrant W Africa with records in Angola (including Cabinda), Togo, Nigeria and Senegambia, and inland records in Mali, N Nigeria and Chad. Rare to uncommon passage migrant eastern and southern inland lakes and rivers, with records in S

Sudan, Ethiopia, Kenya and Uganda south to Zambia, Botswana and South Africa. Small groups winter regularly on Ruzizi Flats, Burundi.

Xenus cinereus

Main wintering areas
Rare to uncommon (mainly passage)

Description. ADULT ♂ (breeding): crown, nape, hindneck and mantle greyish brown, feathers with dark medial streaks; forehead and face whitish, lores, ear-coverts and cheeks flecked dusky; scapulars greyish brown with broad blackish centres; rump and uppertail-coverts greyish brown, narrowly barred whitish. Rectrices grey fringed white. Underparts white, sides of breast greyish and foreneck finely streaked brown. Primaries blackish brown; primary coverts, alula and lesser coverts dark brown; secondaries brown, apical half of all except outermost feathers white; median and greater coverts greyish brown, the latter tipped white; tertials greyish brown with dark shaft streaks; axillaries and underwing-coverts white. Bill blackish, grading to orange-brown basally; eye brown; legs and feet orange-yellow. ADULT ♂ (non-breeding): as breeding bird but dark markings above confined to narrow inconspicuous shaft streaks. Sexes alike. Breeding plumage acquired Africa Feb–Apr; wing-moult Africa Sept–Feb. SIZE: wing, ♂ (n = 33) 129–142 (134), ♀ (n = 21) 131–140 (136); tail, ♂ (n = 54) 49–58 (52·6), ♀ (n = 40) 50–59 (53·3); bill, ♂ (n = 34) 43–52 (46·2), ♀ (n = 19) 42–52 (48·0); tarsus, ♂♀ (n = 65) 26–32 (28·3). WEIGHT: (n = 350+, E and South Africa) 48–120; typical wintering weight 60–80.

IMMATURE: juvenile plumage as non-breeding adult, but scapulars with blackish V-markings, and tertials and upperwing-coverts with buff fringes and brown subterminal bar. 1st-year birds usually moult outer primaries in south, Jan–May.

Field Characters. Easily distinguished from other small-medium waders by long, slightly upturned bill and rather short orange-yellow legs. Dark carpal area striking at rest. Broad white bar on rear edge of innerwing and grey-brown rump and tail distinctive in flight.

Voice. Tape-recorded (62, 73, 76, C, 296). A carrying 'du-du' or 'du-du-du', usually in flight.

General Habits. Frequents coastal sandbars, exposed coral and muddy lagoons and estuaries. Inland records mainly from muddy lake or river edges. Usually singly or in parties up to 20 consorting freely with species such as Curlew Sandpiper *Calidris ferruginea* and Great Sand-Plover *Charadrius leschenaultii*. Up to 300 together in tidal roosts. Feeds along water's edge, running actively and probing intermittently and usually deeply with long bill held at flat angle. Often moves rapidly with body held low and bill horizontal. When stopping, sticks or jabs bill down, coming to abrupt halt (almost 'tripping over' its own bill). Occasionally bobs like Common Sandpiper *Actitis hypoleucos*.

Adults arrive on wintering grounds Aug–Sept (Oct in south). Marked passage Red Sea and E African coasts late Aug–early Sept. Spring departure late Apr–early May. Many 1st-year birds oversummer Africa. Inland records mainly Aug–Dec, and a few Apr–May, suggesting transcontinental passage to S and SW African coasts. W African records also spring and autumn, and these, together with Tunisian passage birds, perhaps indicative of migration route of most westerly breeding birds. 1 bird ringed South Africa recovered June in N European Russia; another in Apr in Ethiopia.

Food. Small crustaceans, marine worms and molluscs.

Reference
Waltner, M. A. and Sinclair, J. C. (1981).

Genus *Actitis* Illiger

Differs from *Tringa* in having longer, more rounded tail and prominent wing-bar; also in unusual flight, flickering wing-beats alternating with gliding. 1 superspecies with 2 allospecies, 1 breeding in Palearctic and 1 in Nearctic; the former visits Africa.

326 SCOLOPACIDAE

Plates 17, 21
(Opp. pp. 304, 352)

Actitis hypoleucos (Linnaeus). Common Sandpiper. Chevalier guignetta.

Tringa hypoleucos Linnaeus, 1758. Syst. Nat. (10th ed.) p. 149; Sweden.

Range and Status. Breeds Europe and throughout S boreal and steppe zones of USSR; winters Africa, Madagascar, SW Europe, S and SE Asia to Australasia.

Palearctic visitor to coasts and inland waters. Winters commonly to abundantly south of Sahara throughout. Common to very abundant on passage N Africa, and winters locally in small numbers Morocco–W Libya and Nile Delta. Numerous passage records Sahara oases.

Description. ADULT ♂ (breeding): top of head olive-brown, feathers streaked dark brown, edged buff; ill-defined whitish supercilium; lores olive-brown; ear-coverts and sides of neck olive-brown, streaked dark brown; mantle and scapulars olive-brown, heavily streaked and marked dark brown, scapulars tipped buff; back, rump and uppertail-coverts olive-brown, finely barred dark brown, lateral uppertail-coverts with blackish and white barring. Rectrices olive-brown, with incomplete dark barring, all except centre pair with broad white tips, outer 2 pairs edged and broadly notched white. Underparts white, sides of breast washed greyish olive, foreneck and breast streaked dark olive-brown. Primaries and alula dark brown; primary coverts same tipped white; secondaries dark brown, tipped white, with broad white band across middle of both webs; tertials and inner greater coverts as mantle; median and lesser coverts olive-brown, finely barred and freckled darker; axillaries white, underwing-coverts mainly white. Bill dark brown; eye dark brown; legs and feet light greenish. ADULT ♂ (non-breeding): similar to breeding bird, but upperparts plainer, streaking inconspicuous; streaking on neck and breast less extensive. Sexes alike. Pre-nuptial moult Feb–Apr in Africa; post-nuptial wing- and body-moult in Africa, mainly Oct–Jan. SIZE: wing, ♂ (n = 15) 107–115 (112), ♀ (n = 12) 109–116 (113); tail, ♂ (n = 24) 48–57 (53·3), ♀ (n = 15) 50–57 (53·4); bill, ♂♀ (n = 33) 22–27 (24·8); tarsus, ♂♀ (n = 36) 22–25 (23·6). WEIGHT: (n = 80+, Kenya) 33–69; typical wintering weight 40–50.

Field Characters. Confusion unlikely with any other wader in Africa. A small *Tringa*-like species with short straight bill and short greenish legs, olive-brown above and whitish below with greyish breast patches. In flight dark rump centre, white edges to broad graduated tail and broad bar on middle of upperwing are clear characters. Call and flickering flight action also distinctive (see Voice and General Habits).

Voice. Tape-recorded (62, 73, 76, F, 233). Flight note, often given when flushed, a high-pitched 'twee-wee-wee'. Alarm call, usually from ground, a more long-drawn 'tweee'. Display call, occasionally heard in winter quarters, in flight or on ground, a repeated 'twee-wit-it-it-it'.

General Habits. Frequents wide range of habitats inland, sandy, rocky or wooded lake-shores, riverbanks, small dams, pools and ditches and temporary puddles in savanna, cultivation and even primary forest; rocky or sandy sea-shores, tidal creeks and estuaries; few at soda lakes except on passage. Usually solitary, individuals often maintaining winter territories with calling, threat postures and flight attacks at boundaries. In aggressive behaviour, E Africa, 2 birds suddenly run together between intervals of feeding, face each other with heads lowered and stretched forward, and then begin jumping into the air with half-opened wings. In autumn, E Africa, apparent courtship activities observed in pairs included wing-raising, tail-fanning and lowering, squatting and chasing with head down and back humped (Pakenham 1939). May occur in small parties, especially on migration, and in gatherings of 20–30 (sometimes 100+) at night roosts.

Feeds mainly by picking along water's edge. Runs quickly, pausing frequently with tail moving up and down in characteristic action, head bobbing. Usual flight highly distinctive, low and direct, flickering wing-beats alternating with gliding on bowed wings.

Crosses Sahara on broad front. 1st adults penetrate south to equator by mid-July, but major influx from Sahel to Cape is during Aug (adults) and Sept (young birds). Marked passage N Africa July–Sept, Mali Aug–Sept, Chad and Kenya especially late Aug–early Sept, Zambia Aug–late Oct. Leaves wintering areas late Mar–Apr; northward passage Ethiopia mid Apr and N African coast to early May. Oversummers in some northern areas (Ethiopia, Mali) but very rarely further south. No observations in recent decades to suggest any African breeding despite reports of eggs and young in Uganda and Kenya respectively early in century (van Someren and van Someren 1911; Jackson 1938). European-ringed birds recovered Morocco; 14 from Scandinavia and W and central Europe found further south, all in W Africa (Guinea-Bissau, Guinea, Mali, Sierra Leone, Ivory Coast, Ghana, Nigeria). Birds ringed Zambia (2), Zimbabwe, Zaïre and Kenya recovered European Russia at 30°–57°E. 1 ringed South Africa (Dec) recovered S Sudan following Apr.

Food. Chiefly free-flying adult invertebrates, particularly insects; also spiders, crustaceans, molluscs, annelids, snails, frogs and toads, small fish and occasionally plant material including seeds.

References

Pearson, D. J. (1977).
Simmons, K. E. L. (1951).

Subfamily ARENARIINAE

Thickset, medium-sized waders, with short, pointed bill and short legs. Wings long and pointed; tail rather rounded; small hind toe present. Black and white patterned plumage, *A. interpres* with chestnut in breeding dress. 1 genus only, *Arenaria*; 2 migrant spp., 1 breeding N Holarctic and 1 N Nearctic; the former visits Africa.

Genus *Arenaria* Brisson

Arenaria interpres (Linnaeus). Ruddy Turnstone; Turnstone. Tournepierre à collier.

Tringa interpres Linnaeus, 1758. Syst. Nat. (10th ed.), p. 148; Gotland Is., Sweden.

Plates 16, 21
(Opp. pp. 257, 352)

Range and Status. Breeds Arctic North America, Greenland and USSR, and Scandinavia south to Baltic; winters Africa, Madagascar, W Europe, S and SE Asia to Australasia and Pacific islands, and North and South America

Palearctic visitor, mostly frequent to abundant and wintering on coasts throughout; many reach southern Africa (e.g. 7000+ counted on 250 km Namibian coast). Rare to frequent inland, chiefly on autumn passage, with records Libya, Mali, Nigeria and Chad, and many from S Sudan, Ethiopia and E Africa south through Zambia, Malawi and Zimbabwe to South Africa.

Description. *A. i. interpres* (Linnaeus): only subspecies in Africa. ADULT ♂ (breeding): forehead white, crown to nape white streaked blackish; hindneck white extending as collar around sides of neck to sides of upper breast; lores, supercilium and ear-coverts white; narrow black line from base of upper mandible to eye and blackish patch below eye; the last continuous with narrow black line to base of lower mandible, broad black band towards nape, and black foreneck and centre and lower sides of breast; chin and rest of underparts white. Mantle and scapulars blackish, variably marked and intermixed chestnut, mainly on upper mantle and V-shaped area surrounding lower mantle; back white; lower rump and most of uppertail-coverts black; lowermost uppertail-coverts white. Rectrices blackish, bases and tips white, central pair broadly tipped chestnut; outers mainly white. Primaries blackish brown, inners with white tips and basal half of inner web white. Primary coverts blackish brown, inners tipped white; secondaries blackish brown with white tips and bases; tertials and inner greater and median coverts brown, notched and margined chestnut; outer greater coverts blackish brown, broadly tipped white; innermost lesser coverts and outermost scapulars forming white patch on side; rest of median and lesser coverts dark brown; axillaries and underwing-coverts white. Bill black; eye dark brown; legs and feet orange. ADULT ♂ (non-breeding): as breeding bird, but forehead to hindneck dusky, streaked brownish; mantle and scapulars blackish brown, edged pale brown and tawny; sides of face and neck dusky brown, paler above and behind eye and on sides of neck; chin white, foreneck and breast blackish brown, paler at bend of wing; tertials and inner wing-coverts (except innermost lessers) as mantle; legs dull orange. ADULT ♀: as ♂, but breeding plumage duller with darker crown; hindneck, collar and facial markings tinged buff; chestnut in mantle, scapulars and wing-coverts less bright; black on face and breast tinged brown. SIZE: wing, ♂ (n = 23) 147–163 (155), ♀ (n = 13) 150–164 (158); tail, ♂ (n = 22) 57–66 (61·5), ♀ (n = 18) 59–66 (62·3); bill, ♂ (n = 31) 20–25 (22·0), ♀ (n = 24) 20–25 (22·0); tarsus, ♂ (n = 30) 25–27 (25·7), ♀ (n = 24) 24–27 (25·6). WEIGHT: (n = 200+, South Africa) 81–185; typical wintering weight 90–120.

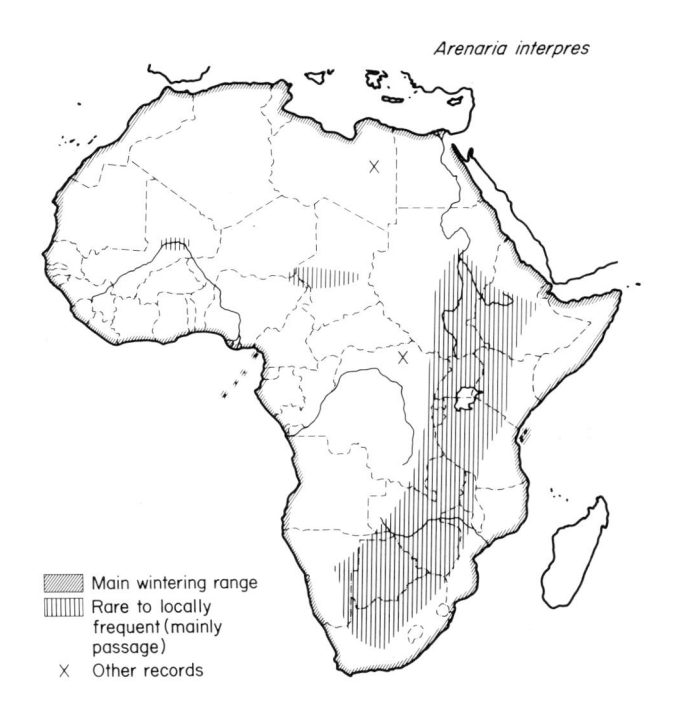

Arenaria interpres

Main wintering range
Rare to locally frequent (mainly passage)
X Other records

IMMATURE: juvenile as non-breeding adult, but upperparts, tertials and upperwing-coverts fringed warm buff; pectoral band narrower and browner; legs yellowish brown. Breeding plumage partly assumed by 1st summer ♂♂, which resemble adult ♀♀. Little breeding plumage in 1st summer ♀♀.

Field Characters. Plump, medium-sized wader with broad blackish breast band and dark brown upperparts (non-breeding), short orange legs and short, dark, pointed bill. Distinctively patterned in flight with white lower back, uppertail-coverts, tail-tip, wing-bar and sides of scapulars contrasting with rest of dark upperparts. In breeding plumage has striking black and white head pattern and much chestnut on upperparts.

Voice. Tape-recorded (62, 73, 76, C). Flight call a hard 'kitititit'.

General Habits. Inhabits rocky shores where often the most abundant wader; also commonly associates with Sanderling *Calidris alba* on sandy beaches and coral; scarce on mudflats and estuaries. Occurs inland on open lake-shores and river edges. Usually in small scattered parties, but in larger flocks on migration and sometimes 100+ in tidal roosts. Forages actively on rocks and along shore, searching under stones and debris and among seaweed. Flight can be fast and strong although usually low and over short distances on feeding area.

Main arrival tropics Aug–Sept, Cape late Sept–Oct. Many inland records in Aug–Nov, indicating transcontinental flights to southern and SW African coasts. Departs wintering areas mainly Apr, southern birds fattening extensively before migration. Passage in N Africa to mid May. 1st summer birds commonly remain in Africa. Ringing evidence indicates that birds wintering W Africa south to Gulf of Guinea are derived mainly Scandinavia and NW Russia, but some birds of Greenlandic origin also reach Morocco and Mauritania. E and southern African wintering birds presumably derived NE Russia and Siberia.

Food. Chiefly invertebrates, especially insects, crustaceans and molluscs; also scavenges.

References
Dowsett, R. J. (1980).
Summers, R. W. and Waltner, I. M. (1979).

Subfamily PHALAROPODINAE: phalaropes

Small- to medium-sized waders with slim graceful bodies. Neck slender; head small with narrow forehead. Bill usually shorter or only slightly longer than head and either needle-like or broad, deep and with depressed tip. Tongue with well developed musculature. Legs relatively short, tarsi laterally compressed. Hind toe relatively long, raised. Front toes with small basal webs and scalloped lateral lobes; arrangement different in each species. Body feathers thicker than in calidridine waders. Wings pointed, narrow; flight low, fast, direct. ♀ larger and more brightly coloured than ♂ (in breeding plumage); associated with reversal of usual sexual roles; polyandrous mating in at least 2 species with ♂ incubating eggs and brooding chicks. Expert swimmers, readily distinguished from other waders by ability to land on water and by spinning and dipping action when foraging on water. Spinning action performed to stir up edible particles from bottom in shallow water, to make prey more visible, and to bring food items within reach of bill. Spin in shallow and deep water.

1 genus; 3 spp. with *lobatus* and *fulicarius* breeding in Holarctic and wintering in tropical waters including Africa; and *tricolor* breeding in Nearctic, wintering in South America and vagrant in Africa and Europe.

Genus *Phalaropus* Brisson

Not illustrated in colour

***Phalaropus tricolor* (Vieillot). Wilson's Phalarope. Phalarope de Wilson.**

Steganopus tricolor Vieillot, 1819. Nouv. Dict. d'Hist. Nat., 32, p. 136; Paraguay.

Range and Status. Breeds North America from British Columbia and California to Manitoba and Kansas; also Minnesota and S Ontario; winters from Texas to South America, mainly pampas of Argentina. Vagrant Africa and Europe although increasing number of records from W Europe since the first in Sept 1954.

Vagrant to Morocco (1, Merdja Zerga, 4 Oct 1963; 1, Merzouga, end Apr 1980; and 1 Oued Massa, 22 Sept 1981); Namibia (1, Swakopmund, 2 Apr 1983), and South Africa (Natal: 1, Umvoti Estuary, 28 Dec 1982–1 Jan 1983; Cape: 1, Muizenberg, 9 Jan 1977; 1, Velddrif, 21 Oct 1979; 1, Paternoster, 20–22 Nov 1983).

Phalaropus tricolor

IMMATURE: as winter adult but all upperparts brown in tone with buff edges to mantle, scapulars, tertials, back, rump and inner median coverts.

Field Characters. Distinguished from Red-necked and Grey Phalaropes *P. lobatus* and *P. fulicarius* by being larger (20% larger than Red-necked, 15% than Grey) and by having long, needle-shaped bill, square white rump, pale grey tail and no wing-bar. In winter plumage lacks dark eye-patch and has unstriped grey back like Grey Phalarope but contrastingly darker, plain wings. Distinctive breeding plumage has black stripe through eye and down neck and russet base of neck. When foraging on land may be mistaken for sandpiper *Tringa* spp. Similar to Wood Sandpiper *Tringa glareola* but larger and paler, without pale speckled upperparts, barred tail, or clouded and streaked breast, and Marsh Sandpiper *T. stagnatilis* but bill and legs shorter and white rump not extending up back.

Voice. Tape-recorded (62, 73). Usually silent on passage and in winter; occasionally gives soft nasal grunt or subdued quack, 'aaugh' or 'ork'.

General Habits. Inhabits freshwater marshes, shallow lakes, mudflats and coastal tidal pools; more prone to remain inland than other phalaropes, usually avoiding open coastal areas and rarely found at sea. Less gregarious than other phalaropes although occasionally in large flocks in winter. Typically forages in small groups, by spinning in tight circles; also forages by wading in shallow water or by walking on shoreline, constantly sweeping bill sideways, often with neck extended; runs and picks up food, often with legs bent in half-crouching posture.

Description. ADULT ♀ (breeding): forehead and crown pale bluish grey, black eye-stripe beginning with lores and continuing down side of neck. Area between eye and crown and on cheek below eye-stripe white; chin and throat white. Foreneck cinnamon, sides of neck chestnut with 2 chestnut stripes longitudinally across scapulars. Central hindneck pale blue-grey; rest of hindneck white. Mantle pale blue-grey, rest of back slate-grey, uppertail-coverts and rump white; rectrices mouse grey tipped white. Breast creamy buff, rest of underparts white. Upperwing dark grey, the coverts and tertials with pale margins. Underwing white. Bill black, eye dark brown to black, legs and feet black. ADULT ♀ (non-breeding): upperparts uniform pale-grey, underparts white, legs and feet yellow, yellow-green or yellowish flesh. ADULT ♂ (breeding): much duller than ♀ with no chestnut on scapulars. ADULT ♂ (non-breeding): like non-breeding ♀. SIZE: wing, ♀ (n = 34) 128–143 (136), ♂ (n = 28) 120–129 (125); tail, ♀ (n = 10) 50–58 (53·6), ♂ (n = 10) 48–52 (50·1); bill, ♀ (n = 10) 32–36 (33·6), ♂ (n = 10) 29–32 (30·7); tarsus, ♀ (n = 10) 32–36 (34·0), ♂ (n = 10) 31–34 (32·4). WEIGHT: breeding season, Alberta, ♀ (n = 53) 55–85 (68·1), ♂ (n = 100) 30–64 (50·2); Netherlands, May, ♀ 78·3.

Food. Mostly flies; also other insects, crustaceans, spiders and seeds of aquatic plants.

Reference
Cramp, S. and Simmons, K. E. L. (1983).

Phalaropus lobatus (Linnaeus). Red-necked Phalarope; Northern Phalarope. Phalarope à bec étroit.

Plates 18, 21
(Opp. pp. 305, 352)

Tringa Lobata Linnaeus, 1758. Syst. Nat. (10th ed.), p. 148; Hudson Bay.

Range and Status. Breeds N Holarctic, winters at sea mainly off coasts of W South America, S Arabia and W Pacific, generally nearer equator than other phalaropes; some also winter E Africa.

Palearctic visitor, locally common to abundant offshore waters Kenya (on passage mainly Nov and Mar, many small parties between Shimoni and Malindi beyond reef probably this species although only 3 definite records along coast: 1 inland Diani, 50 off Diani, 1 near Mombasa) and Somalia including Gulf of Aden (100,000 one transect off S Arabia, 21 Jan 1954; thousands north coast Somalia, spring). Frequent to uncommon inland Kenya (L. Turkana, Ferguson's Gulf, 30 annually especially Oct and Apr) and Ethiopia

(Rift Valley lakes especially Basaaka and Koka Reservoir, c. 10 annually, 2 Dec–25 Apr). Elsewhere vagrant to rare; recorded Morocco (Tangier, Dayet, Ifrah, Qualidia, Chafferine Is., Villa Cisneros), Tunisia (5 coastal, 1 inland records), Libya (Tripoli), Egypt (Suez), Ethiopia (Massawa), Kenya (Nakuru), Uganda (Rwenzori N. P.), Tanzania (Momella lakes, Tabora, L. Masek), Burundi (Ruzizi mouth), Zaïre (L. Edward), Zimbabwe (Bulawayo, Sabi R.), Namibia (Walvis Bay, Swakopmund) and South Africa (Cape, Strandfontein, Cape St Francis, Graaf Reinet, East London; Natal, Mont-aux-Sources; Transvaal, Vanderbijl Park).

Phalaropus lobatus

See text

Description. ADULT ♀ (breeding): forehead, crown, lores and cheeks dark bluish slate; chin and throat white. Bright white marks above and below eye. Central hindneck dark slate-grey. Streak from behind eye down along side of neck deep rufous-cinnamon, extends across foreneck and upper breast. Mantle and scapulars dark bluish slate with feathers fringed cinnamon. Back and rump dark grey with narrow off-white feather fringes; uppertail-coverts dull black with cinnamon-buff fringes and narrow white tips; lateral feathers of rump and uppertail coverts largely white. Tail feathers dark grey with narrow white edges. Lower breast, sides of breast and flanks slate-grey, feathers fringed white; rest of underparts white. Primaries and secondaries dull black, shafts white, feather edges narrowly margined white. Upperwing-coverts dark grey, greater coverts broadly tipped white, forming broad wing-bar. Underwing-coverts and axillaries white. Bill black; eye dark brown; legs slate-grey to greyish black. ADULT ♀ (non-breeding): head and neck white except for dull black streak from just in front and below eye to lower ear-coverts; rest of upperparts mainly grey. Grey patch at side of breast, rest of underparts white. Wings like breeding bird. ADULT ♂ (breeding): smaller and duller, forehead, crown, lores and cheeks dark slate-grey, mantle and scapulars black, cinnamon mainly confined to side of neck. ADULT ♂ (non-breeding): like non-breeding ♀. SIZE: wing, ♀ (n = 12) 109–117 (114), ♂ (n = 11) 104–114 (108); tail, ♀ (n = 21) 48–53 (50·4), ♂ (n = 13) 45–53 (47·9); bill, ♀ (n = 25) 20–23 (21·4), ♂ (n = 22) 19–23 (21·1); tarsus, ♀ (n = 24) 19–22 (20·3), ♂ (n = 13) 19–22 (20·4). WEIGHT: Kenya (inland) 28·2, 28·8, 30·7, 34·7; Iran (passage) 20–37 (26·5).

IMMATURE: as non-breeding adult in winter but upperparts darker, more brown-black; crown, centre of hindneck and eye-panel also brown-black; rest of head and underparts white. Wings and tail like adult except larger coverts with paler margins. Eye and bill like adult, foot yellow-buff, legs slate-grey.

Field Characters. Smaller than Grey Phalarope *P. fulicarius* (20% shorter wings, 10% shorter tail), with all-black, needle-thin bill. In non-breeding plumage upperparts darker, wing-stripe more contrasting, back striped, not uniform. In breeding plumage, has only a patch of russet on side of neck; Grey Phalarope has underparts entirely russet. Distinguished from Wilson's Phalarope *P. tricolor* by being smaller (wing-span 30% smaller) and in non-breeding plumage by having dark eye-patch, dark rump and wing-stripe. In breeding plumage, has russet patch on foreneck and upper breast rather than black stripe on side of neck. In flight may be confused with small *Calidris* spp. but head smaller, chest relatively broader and tail longer.

Voice. Tape-recorded (62, 73, 76, C). Usually silent in non-breeding season although may produce a 'twit' or 'whit', lower pitched than Grey Phalarope; sometimes also gives a low cluck, 'prek', 'kirk', 'chek' and 'cherrp'.

General Habits. Winters mainly at sea, often far from shore, in groups of several tens, sometimes in flocks of several thousand. Occasionally visits inland water habitats, including fresh and alkaline lakes, pools and sewage treatment ponds. Feeds by spinning, up to 46 spins per min, faster than Grey Phalarope but slower than Wilson's Phalarope. Obtains food by: (1) picking it off water surface or from emergent vegetation; (2) seizing it from just below surface using rapid forward lunge of head; (3) up-ending; and (4) seizing flying insects, sometimes with short flutter-leap. Occasionally wades or walks when it pecks or probes for food. Rarely perches; flies mostly close to ground or water.

Migratory patterns in Africa little known; wintering birds in Arabian Sea probably spread west to Gulf of Aden by late Oct; also to Somalia and Kenya coasts and inland along northern Rift Valley. No regular wintering area yet known in Atlantic.

Food. Chiefly invertebrates, especially dipteran flies and their larvae; also other insects, crustaceans, annelids and molluscs. Rarely small fish and some plant material.

Reference
Cramp, S. and Simmons, K. E. L. (1983).

Phalaropus fulicarius **(Linnaeus). Grey Phalarope; Red Phalarope. Phalarope à bec large.**   Plates 18, 21

Tringa fulicaria Linnaeus, 1758. Syst. Nat. (10th ed.), 1, p. 148; Hudson Bay.   (Opp. pp. 305, 352)

Range and Status. Breeds N Holarctic, winters mainly off west coasts of Africa and South America.

Palearctic visitor, locally abundant, mainly in upwellings of Guinea and Canary currents off bulge of W Africa between Tropic of Cancer and 7–8° N latitudes, especially abundant off Mauritania (3000, 18°00'N, 17°30'W, Feb; 2500, 24°45'N, 16°15'W, Nov; 2500, 29°00'N, 14°30'W, Nov). Also locally abundant in northern winter in Benguela current off Namibia and South Africa ('large' flocks, 26°05'S, 14°30'E, 31°47'S, 15°43'E, Feb) but details of exact range imperfectly known. Vagrant to rare elsewhere inshore and inland; recorded Morocco (several records including 40 Larache, Feb; 61 Mehdia, Jan; and 70 Villa Cisneros, May), Tunisia (1, Thyra, 24 Apr), Libya (1, 12 km north of Benghazi, 26 Dec), Mauritania (1, Banc d'Arguin), Senegambia (5, 70 km southeast Dakar, 5 Aug; 200, 14°00'N, 18°30'W, Feb; large flock, Banjul, 27 Mar), Sierra Leone (numbers?, Freetown), Ghana (1, Kumasi, 20 Nov; 1, 50 km S Kumasi, 3 Apr), Nigeria (12, Lagos, 30 Mar), Cameroon (1, Bitye, Mar), Namibia (several records inshore, 1 Okahandja), Botswana (1, Tshono Pan, 3 Mar), South Africa (several records inshore and inland), Zimbabwe (1, Birchenough Bridge, Sept; 1, Bulawayo, Nov; 1, Hartley, 1 Nov) and Kenya (1, L. Elmenteita, 17 Feb; 2 L. Nakuru, 15 Feb, 28 Mar; 1 L. Turkana, 12–13 Apr). Occurs in Gulf of Aden and Red Sea only as a vagrant.

Description. ADULT ♀ (breeding): forehead, crown, lores and chin black, centre of nape grey. Large oval white patch on side of head, surrounding eye and extending to sides of nape. Side of neck and underparts deep, uniform rufous-chestnut. Mantle and scapulars black; back dull black; rump and uppertail-coverts cinnamon-rufous with central feathers streaked black. Tail, central feathers dull black, outer ones grey. Primaries and secondaries dark grey, nearly black on outer primaries; shafts white. Secondaries tipped white. Greater upperwing-coverts dark grey with distal half white, forming wing-bars; underwing-coverts and axillaries white. Bill yellow with orange tinge to base and black on tip; eye dark brown; legs and feet pale blue, feet with yellow lobes. ADULT ♀ (non-breeding): forehead, crown and lores white, crown with some grey. Small area around eye and to nape black. Mantle and back pale blue-grey with feathers narrowly bordered white. Underparts white. Wing and tail as in breeding adult. Bill black with brown or grey tinge; eye dark brown; legs and feet grey, feet with yellow-grey lobes. ADULT ♂ (breeding): slightly duller than ♀; otherwise similar except belly whitish. ADULT ♂ (non-breeding): like non-breeding ♀. SIZE: wing, ♀ (n = 55) 130–143 (137), ♂ (n = 68) 124–135 (129); tail, ♀ (n = 17) 59–74 (66·9), ♂ (n = 13) 57–65 (61·7); bill, ♀ (n = 17) 21–24 (22·8), ♂ (n = 12) 19–22 (21·6); tarsus, ♀ (n = 5) 21–24 (22·1), ♂ (n = 5) 22–24 (22·5). WEIGHT: Canada, Alaska and Spitsbergen, migrants May–June, ♀ (n = 4) 51–66 (57·1), ♂ (n = 4) 37–50 (42·4).

IMMATURE: like non-breeding adult except crown, mantle and back darker brown to black; inner coverts margined buff rather than white.

Field Characters. A typical phalarope but with a bulkier body and rather shorter, stouter bill. Breeding adult unmistakable with all-russet underparts and white eye-patch, but separation from other phalaropes in winter difficult except at close range. Distinguished from Red-necked Phalarope *P. lobatus* in winter plumage by paler, more uniform, unpatterned grey back, stronger, more robust flight and slightly larger size. Distinguished from Wilson's Phalarope *P. tricolor* in winter plumage by dark eye-patch, shorter bill, dark rump, pale wing-stripe and smaller size.

Voice. Tape-recorded (62, 73, 76, C). Commonest call is short, sharp, high-pitched, whistling 'pit' or 'wit' given in alarm, also a chirruping 'zhit', both being shriller than in Red-necked Phalarope. Birds feeding in a flock also give excited twittering 'PHEErreep-PHEErreep-PHEErreep ...'.

General Habits. The most oceanic of all phalaropes, inhabits mainly plankton-rich areas of open ocean in flocks of mixed age and sex, typically c. 20 but up to several hundred, sometimes several thousands; very occasionally occurs singly. Found inshore or inland only under stress of weather. Flies swiftly, skimming water and following trough of waves; makes frequent short flights. Forages while swimming (although on breeding grounds also does so by wading or walking and pecking at food). Feeds by: (1) spinning (39 spins/min) and pecking outwards at food particles; (2) picking prey off surface; (3) quickly lunging forward; (4) up-ending;

and (5) seizing flying insects, often in short flutter-leaps. Sometimes rests and feeds on floating mats of seaweed.

Details of migratory routes to main concentrations off coast of W Africa not known since no ringing results. However, many pelagic records spanning N Atlantic suggest major southeast autumn movement by Nearctic population to winter off W Africa. Also some numbers seen flying southeast between St Helena Island and African mainland (17°S 2°E to 8°S 5°W) in Aug suggest movement toward Benguela current. Leaves South African seas Mar, W African seas Apr.

Food. Mainly insects and larvae including dipteran flies, caddis flies, beetles, bugs; also molluscs, crustaceans including amphipods, annelids, spiders, mites, small fish and plant material.

Reference
Cramp, S. and Simmons, K. E. L. (1983).

Suborder LARI

Mainly aquatic birds, some partly terrestrial, some highly pelagic in non-breeding season. Bill relatively short, stout and variable in shape; nostrils perforate (with no median septum). Wings very long, quite narrow and pointed. Legs rather short and stout and attached near centre of body; tibiae partly bare; tarsus with overlapping scales in front, small irregular plates elsewhere; hind toe usually small and elevated, sometimes rudimentary or absent; anterior toes fully webbed. Plumage predominantly black, white and grey. Feeding methods include scavenging, plunge-diving, rodent-hunting, pirating, and skimming. 4 families, Stercorariidae (skuas), Laridae (gulls) Sternidae (terns) and Rynchopidae (skimmers), all occurring in Africa.

Family STERCORARIIDAE: skuas

Oceanic gull-like birds with mainly brown plumage. Bill dark, stout, hooked at tip with complex horny sheath covering most of base and nostrils. Neck strong and thick. Tarsus and feet much as in gulls, but rougher, claws strong, hooked and sharp; 19–23 secondaries; 12 rectrices, central pair slightly or much elongated. Wings long and pointed; flight strong and agile. Feed by scavenging, killing some small animals, but mainly by harrying gulls and terns until they drop or disgorge food (kleptoparasitism). Sexes similar and ♀ slightly larger.

6 spp. in 2 genera, all breeding at high latitudes, and wintering at sea in lower latitudes; 5 visit African waters.

Genus *Stercorarius* Brisson

Small, slim skuas with central tail feathers elongated and inconspicuous white wing-flash. Each species with 2–3 colour phases and seasonal changes; non-breeding plumage resembles immature with central rectrices not elongated. Plumage of young birds extensively barred below. Strong, agile fliers, associating at sea with foraging flocks of terns and gulls. 3 spp.

Plate 22
(Opp. p. 353)

***Stercorarius pomarinus* (Temminck). Pomarine Skua; Pomarine Jaeger. Labbe pomarine.**

Lestris pomarinus Temminck, 1815. Man. d'Ornith., p. 514; Arctic regions Europe (*ex.* Brisson).

Range and Status. Breeding distribution circumpolar; migrates at sea and along coasts of Africa, Europe, E Asia, Australasia, North America and western South America; winters southern oceans and coasts.

Palearctic visitor; main wintering area zone of upwelling off W African coasts between 8° and 20°N, common to locally abundant (500 around fishing fleet 20°N, 17°W; flocks of 50 off Senegambia with 30% juveniles, 80% of adults in light phase plumage). Elsewhere, off-shore waters and coasts Morocco to W Nigeria, N Namibia to Cape (east to Mossel Bay), and Red Sea to Gulf of Aden. Morocco, usually uncommon to frequent, Atlantic coast only, mostly 1–2 but once 100+ off Ifni, Oct; Rio de Oro, large concentrations; Nigeria, rare, Lagos, 7 records, 1 inland record 10°N, 4°30'E; Gulf of Guinea, rare; Angola, vagrant, 1 record, mouth of Cunene R.; Namibia, uncommon on coast, mostly pairs or singles but off-shore Benguela Current several hundred around fishing boats 18°07'S, 11°39'E and on 1 day *c.* 100, from Lüderitz to Mercury Is.; uncommon in W South Africa, vagrant to E South Africa (Durban). Egypt (Mediterranean coast) uncommon on passage. Red Sea, uncommon but regular. Djibuti, coast, rare (1 record Obock). Somalia coast,

Stercorarius pomarinus

Main wintering range

Passage and other wintering areas

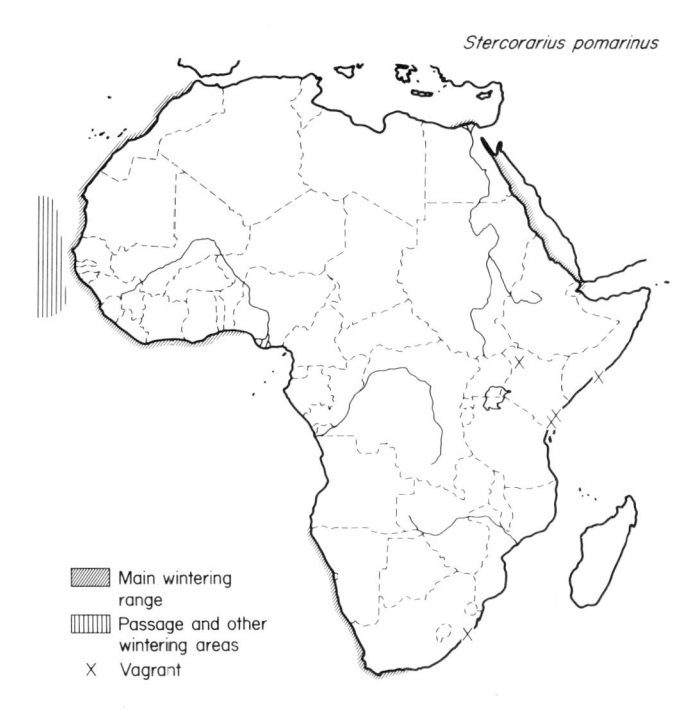

X Vagrant

rare (1°57'N, 45°11'E, 10°25'N, 51°20'E). Kenya coast, rare to uncommon (2 records Mombasa, several observations Sabaki), vagrant inland (L. Turkana).

Description. ADULT ♂ (non-breeding): two distinct colour phases. Light phase: forehead, crown, nape, lores and under eye dark brown, extending as dusky hood to chin and throat; some feathers of head edged pink-buff. Ear-coverts dark brown; hindneck and side of neck barred grey-brown, white and pink-buff, sometimes feathers with straw yellow tips. Throat white, sometimes with dark bars or spots; feathers with some straw yellow tips. Upperparts black-brown, with some feathers of mantle and scapulars broadly edged pink-buff. Uppertail-coverts with black and white bars. Tail brown-black; central tail feathers slightly elongated. Underparts largely white; feathers of breast and flanks with dark edges and subterminal bars, forming dark band across upper breast. Undertail-coverts with black and white bars. Upperwings blackish brown; bases of outer 5–8 primaries white; underwing blackish brown but with more extensive white crescent on outer primaries. Bill dull brown to buff or yellow at tip, black at base; eye black; legs black, blue-grey scales on upper tarsus. Dark phase: upper and underparts smoky black; only bases of primaries and shafts of outer primaries white. ADULT ♂ (breeding): light phase: head with sooty black cap, rest of head and entire neck yellowish white. Upperparts brownish black, tail black; basal two-thirds of rectrices including entire outer rectrices paler, central feathers twisted, spatulate-shaped and project 17–20 cm. Underparts mostly yellowish white; breast-band and undertail-coverts dark brown. Wings as non-breeding ♂, but more extensive white crescent on outer primaries. Dark phase: as non-breeding ♂ but cheeks and hind collar with variable amounts of brownish yellow. Sexes alike. SIZE: wing, ♂ (n = 17) 354–374 (363), ♀ (n = 11) 363–382 (373); wing-span, ♂♀ 1250–1380; tail, (outer feather) ♂ (n = 50) 116–134 (124), ♀ (n = 36) 121–136 (127), (from tip of central pair to tip of outer tail-feather) ♂ (non-breeding adult, n = 3) 38–57 (49), ♂ (breeding adult, n = 9) 75–105 (95), ♀ (non-breeding adult, n = 5) 32–53 (40), ♀ (breeding adult, n = 9) 65–111 (92); bill to feathers, ♂ (n = 31) 38–42 (40), ♀ (n = 23) 39–44 (40·9); tarsus, ♂ (n = 52) 50–56 (53·7), ♀ (n = 36) 53–58 (55). WEIGHT: (breeding adult,

N Holarctic) ♂ (n = 73) 542–797 (648), ♀ (n = 52) 576–917 (740).

IMMATURE: 1st winter, light phase: head dark grey-brown to dark slate-grey, sometimes dull brownish white with indistinct darker cap; mantle and back brownish, sometimes grey or slaty in tone, feathers with narrow paler edges forming indistinct transverse bars; rump and uppertail-coverts broadly barred; tail black with short blunt central projections, not twisted; underparts dusky or brownish white with grey-brown breast-band; some barring on flanks, undertail-coverts and axillaries; wings as adult; bill dark horn-brown or dark leaden grey, tip black. 1st winter, dark phase: uniform sooty black; underwing-coverts, tail-coverts and axillaries barred. 2nd winter, light phase: like non-breeding adult but underparts and underwing-coverts heavily barred; tarsus with some black. 2nd winter, dark phase: like 1st winter except crown slightly darker than upperparts. 3rd winter, light phase: crown and lores dull black, hindneck, sides of head, chin and throat white, usually streaked dull black; sides of head and hindneck slightly tinged yellow; upperparts black-brown, uppertail-coverts black, barred white; underparts heavily barred dull black and white, belly whitish. 3rd winter, dark phase: as 2nd winter. 4th winter: as non-breeding adult but tail shorter, less twisted; some wing-coverts partly barred; tarsus sometimes with some blue-grey.

A

Field Characters. Blunt spatulate-tipped central tail feathers diagnostic if present (Arctic Skua *S. parasiticus* has short pointed ones, Long-tailed Skua *S. longicaudus* has long pointed ones and *Catharacta* skuas have none). Largest *Stercorarius* skua, size between Great Skua *C. skua* and Arctic Skua. About two-thirds size of Great Skua, but lacks nape hackles and has smaller white wing-patches. Plumage variable, from all-pale to all-dark below, but white wing-flashes more obvious than in Arctic Skua. Appears broader and heavier than Arctic and Long-tailed Skuas with wider head, proportionately longer, more distinctly hook-tipped bill, deep chest, broad breast-bar, thick barrel-shaped body and longer broad full tail (see **A**). Wings with broader bases; flight steadier and slower, with shallow wing-beats; wings appearing less angled than Arctic Skua. Immatures difficult to distinguish from other immature skuas; separable from Arctic Skua by lack of dark streaks on head and nape, heavier bill, dark barring on underwing, chest and vent, more conspicuous barring on upper- and undertail-coverts and more white on upper surface of primaries.

Voice. Tape-recorded (62, 73). Generally silent on migration and wintering grounds. Low 'gack', hawk-like 'kek' or low 'hek'; harsh gull-like 'yowk'; and sharp 'which-yew'.

General Habits. In non-breeding season, mainly pelagic in cool upwelling systems. In winter usually solitary or in small groups but sometimes up to 500 near trawlers. Gregarious on migration, in 'spring' flocks of scores of birds; in 'autumn' in flocks rarely more than 8. Soars, sometimes high (like migrating *Accipiters*). Catches fish by dipping-to-surface; occasionally sub- merges briefly in search of food; also scavenges, and pursues other birds with swift and agile flight; kleptoparasitizes Sabine's Gull *Larus sabini*, Sooty Terns *Sterna fuscata* and other small gulls and terns (J. C. Sinclair, pers. comm.).

Origin of wintering birds not known; passes along Morocco's Atlantic coast to main wintering area off Senegambia, a few birds continuing to Gulf of Guinea east to Nigeria and south across equator into Benguela Current. Birds in Red Sea and Gulf of Aden probably from upwellings in Arabian Sea. Present Gulf of Guinea, Oct–Mar; Nigeria, Aug–Mar; SW coasts South Africa, Dec–Apr; and Kenya, Dec–Mar.

Food. In non-breeding season mainly fish; also marine invertebrates, carrion and birds.

References
Cramp, S. and Simmons, K. E. L. (1983).
Harrison, P. (1983).

Plate 22
(Opp. p. 353)

Stercorarius parasiticus (Linnaeus). Arctic Skua; Parasitic Jaeger. Labbe parasite.

Larus parasiticus Linnaeus, 1758. Syst. Nat. (10th ed.), 1, p. 136; north of Tropic of Cancer to Europe, North America and Asia.

Range and Status. Circumpolar; winters coasts of Africa, S Asia, Australasia, and North and South America.

Palearctic visitor, main wintering area upwelling zone in Benguela Current off Namibia and W South Africa; common to locally abundant ('many' adults and imma- tures 22°S, 9°45′E; 400 including 'many' immatures 17°30′S, 5°E; 50 on 1-day trip from Lüderitz to Mercury Is.). Also occurs coast and off-shore waters from W Libya, to Morocco; Atlantic Morocco to W Nigeria, Angola to Cape; also east to Natal and S Mozambique, and from Egypt (Suez) to Somalia (Cape Guardafui). Libya, rare; Tunisia and Algeria, uncommon; Morocco, uncommon on Mediterranean coast, common to locally abundant in autumn passage on Atlantic coast (1000 in 2·5 h, Ifni), uncommon in winter and rare in summer (2, July); Mauritania, uncommon; Senegambia, un- common autumn passage, rare winter visitor; Guinea, Ghana and Togo, uncommon; Nigeria (Lagos), rare; Gulf of Guinea (3, 2°N, 3°25′E), rare; Angola, frequent to common; Namibia, frequent to common (also see above); South Africa, common especially Cape Pro- vince winter, other seasons uncommon, E Coast, un- common; S Mozambique, rare. Egypt (Suez southward) frequent; Red Sea and Gulf of Aden, uncommon. Vagrant, S Somalia coast (1°54′N, 45°05′E; 1°57′N, 45°11′E; 2°12′N, 45°37′E), inland Ethiopia (L. Awasa, L. Abiata), Kenya (L. Turkana, probably this species, on coast at Malindi and at Shimani 4°39′S, 39°23′E), Zaïre (7°24′S, 25°38′E), South Africa (Shiyanemane, Kruger National Park).

Description. ADULT ♂ (non-breeding): 2 colour phases with intermediates. Light phase: forehead, crown and just below eye dull to sooty brown, forehead becoming pale buff with wear; sides of head and neck pink-buff or brown-buff variably streaked or dotted with dark brown. Mantle, scapulars, and back to uppertail-coverts black, shafts and bases of four outer primaries white; other upperwing-coverts dark brown, some of feathers edged buff or white. Tail wedge-shaped; dark slate-grey to dark brown, darker toward tip and with white base; 2 central rectrices long and tapered, projecting. Throat and chin uniform pink-buff. Breast, flanks, vent and undertail-coverts pink-buff; barred black patches at side of breast often form breast-band; belly white with some black barring; undertail-coverts barred black and off-white. Remiges and greater coverts black; shaft and bases of four outer primaries white; other upperwing coverts dark slate-grey to dark brown. Eye dark brown; bill black, tinged brown, olive, grey-green or slate at base; legs and feet black. Dark phase: completely brown-black except for white shafts of outer primaries; black cap; faint slaty tinge from mantle to rump; underparts sometimes sooty grey to sooty brown. Upperparts sometimes narrowly edged or dotted dark rufous-brown; underparts including vent and tail-coverts sometimes sooty; vent and tail coverts sometimes edged dark rufous-brown. Intermediate phases: whitish buff cheeks, hind collar, chin and throat; upper- and underparts brown (but much variation). ADULT ♂ (breeding): light and intermediate phases: like non-breeding ♂, but upperparts darker brown without feathers edged buff or white, cap on head more distinct, sides of head and hindneck white with feathers tipped golden yellow; flanks with less barring. Dark phase: like non-breeding ♂ but with golden brown hackles on hindneck and sides of head and upperparts, vent and tail-coverts without feathers edged or dotted dark rufous-brown. Sexes alike. SIZE: wing, ♂ (n = 123) 290–345 (322), ♀ (n = 119) 290–340 (326); wing-span, ♂♀ 1100–1250; tail (outer feather), ♂ (n = 38) 109–122 (115), ♀ (n = 48) 108–125 (116), (from tip of central pair to tip of outer tail-feather) ♂ (n = 27) 65–105 (83), ♀ (n = 26) 65–90 (78); bill to feathers, ♂ (n = 34) 29–34 (31), ♀ (n = 46) 30–34 (31); tarsus, ♂ (n = 34) 41–47 (44), ♀ (n = 46) 42–47 (44). WEIGHT: ♂ (n = 108) 306–585 (429), ♀ (n = 125) 315–636 (462).

IMMATURE: 1st winter, light phase: head, neck, chin and throat warm buff with some black streaks; rest of upperparts black, feathers tipped warm buff. Tail like adult, but 2 central feathers project only c. 10–15 mm. Underparts black-brown mixed with dark brown and rufous-buff barring. Wings like adult except underwing-coverts and axillaries barred brown, grey or buff. Bill blue-grey, tip black; eye like adult; legs black, some blue-grey, feet blue-grey. 1st winter, dark phase: like adult but head cap less defined; chin, throat and sides of face tipped tawny and buff, upperparts with indistinct rufous and buff barring; underparts uniform blackish to brown with rufous and buff barring; underwing-coverts and axillaries barred dark brown, rufous and white; base of bill pale blue-grey to pinkish grey. 1st winter, intermediate phase: like dark phase but underparts paler, more apparent barring. 2nd winter, all phases: like 1st winter, but light and intermediate phases with whiter underparts with more barring, and indistinct breast-band. 3rd winter, all phases: like non-breeding adult but with underwing-coverts, axillaries and upper- and undertail-coverts barred; some with blue-grey spots on tarsus. 4th winter, all phases: like non-breeding adult but some with a few dark bars on underparts and a few with marks on tail-coverts; central rectrices longer each year.

Field Characters. Medium-sized, smaller than Pomarine Skua *S. pomarinus* but usually larger than Long-tailed Skua *S. longicaudus*. Much more lightly built than Great Skua *Catharacta skua*. On water rides high with tips of tail and wings raised well up. Underparts variable at all ages, from all pale to all dark. More buoyant flight than Pomarine Skua, wings appearing more angled and less broad at base; also has thinner, shorter, less contrasting bill, dark cap with some white on forehead, partial rather than complete breast-band, less pronounced white wing-flash, especially on upper surface, and sharply pointed rather than spoon-tipped central tail-feathers (see **A**). Distinguished from Long-tailed Skua by generally larger head and body, longer bill, fuller chest with a partial breast-band, shorter broader wings, more white on primaries (upper surface), and tail shorter and narrower at base, without long whipping tail streamers; also flight more purposeful and direct; less buoyant and tern-like; wings not appearing as far forward. At rest, wing and tail-feathers more or less equal; in Long-tailed Skua wings extend at least 4 cm past outer tail-feathers. Immatures distinguished from immature Long-tailed Skua by heavier body, richer plumage with tawny buff barring and noticeable white across bases of primaries on both surfaces of wings.

Voice. Tape-recorded (62, 73, C). Usually silent when migrating and in wintering areas; makes a repeated nasal squealing 'eee-air'.

General Habits. Mainly pelagic; sometimes frequents estuaries and sewage outfalls (Cape Town: Furness 1983). Generally solitary; also in pairs or small groups, in flocks of a few hundred on migration when accompanying migrating terns and gulls, especially Arctic Tern *Sterna paradisaea*. Flight purposeful and rather falcon-like, with jerky wing-beats and low glides. Away from breeding grounds, obtains food mainly by chasing gulls and terns. Searches on its own but usually within view of other skuas; sometimes in pairs or in groups of up to 5. When attacking, flies low over water at high speed and approaches victim from behind and below. Kleptoparasitizes Common Tern *S. hirundo*, Arctic Tern, Sandwich Tern *S. sandvicensis*, Sabine's Gull *Larus sabini* and Hartlaub's Gull *L. hartlaubii*. Also scavenges galley refuse from ships; sometimes catches fish by dipping-to-surface. At sewage outfall (Cape Town) inactive 90% of time; remainder of time kleptoparasitizes gulls and terns.

'Autumn' passage largely late Sept–Nov on-shore and off-shore; 'spring' passage Mar–Apr mainly out to sea. Morocco, autumn mid Aug–Oct on Atlantic coast; Tunisia, Sept–May; Senegambia, mid Aug–Mar; Nigeria, Aug–Mar; Namibia, end Sept–Feb; South Africa, Sept–Jan; and Kenya, Sept–Oct, Mar. In Benguela Current immatures arrive late Sept, most immatures Oct–Nov. 1 bird (age ?) ringed Finland, recovered Egypt; 1 3rd-year bird ringed Sweden, recovered Guinea (autumn); Scottish-ringed birds: juveniles recovered Algeria (Aug), western Sahara (Dec), Ghana (Dec), Zaïre (inland, Dec); 2nd-year birds recovered Togo (6°08′N, June), Angola (Oct), Ghana (Jan); 3rd-year birds recovered western Morocco (autumn); older birds recovered Morocco (Nov), Mauritania (Jan), Ghana (no date) and Angola (Nov). Origin of Benguela Current and South African birds unknown.

Food. Mainly fish; actual composition hardly known.

References
Cramp, S. and Simmons, K. E. L. (1983).
Furness, B. L. (1983).

Plate 22
(Opp. p. 353)

Stercorarius longicaudus Vieillot. **Long-tailed Skua; Long-tailed Jaeger. Labbe à longue queue.**

Stercorarius longicaudus Vieillot, 1819. Nouv. Dict. d'Hist. Nat., 32, p. 157; northern Europe.

Range and Status. Circumpolar; migrates off coasts of W and SW Africa, Europe and Pacific and Atlantic coasts of North and South America; wintering grounds poorly known but mainly southern oceans.

Palearctic visitor; main wintering area, upwelling zone off Namibia and W South Africa in Benguela Current, common to locally abundant (220, 26°36′S, 14°23′E 17 Dec; 220, 25°12′S, 13°39′E 25 Jan). Elsewhere uncommon to vagrant, Morocco (1, Tangier Oct; 2 old records 1846, 1858); Senegambia (1, 14°20′N, 17°44′W 20 Feb; 1, off Daker 18 Mar; 3, Dakar 20 May); Togo (1, 7°48′N, 1°18′E 24 Aug); Nigeria (1, Lagos 2 Dec); Gulf of Guinea (2, 0°17′N, 3°24′E 5 Apr); Angola (6, Baia dos Tigres; 1, Porto Alexandre); Namibia (6, Lüderitz bay 16 Dec); South Africa (1, Cape coast; 1, Agulhas Bank July; 1, Kalahari Gemsbok National Park, N Cape c. 26°59′S, 20°18′E 28 May, c. 550 km inland). Vagrant Kenya (2, 25km north of Malindi 19 Oct; 1, Ferguson's Gulf, L. Turkana, 3°31′N, 35°55′E).

Description. ADULT ♂ (non-breeding): 2 colour phases with intermediates (dark and intermediate phases very rare). Light and intermediate phases: forehead, crown to just below eye, and lores dark brown, feathers fringed off-white to grey; rest of head and neck white, dotted and tipped dusky grey to brown. Rest of upperparts pale slate-grey; mantle with pink-buff tips; uppertail-coverts barred pale buff and black-brown. Tail pale slate-grey, feathers darker toward tips; central pair sharply pointed and 40–100 mm longer than others. Throat white; rest of underparts variably barred brown and white; dark grey band across breast; some bars on flanks, lower vent and undertail-coverts. Remiges black, rest of upperwing-coverts slate-grey; underwing-coverts and axillaries uniform grey. Bill black, slightly tinged dark olive-grey at base; eye dark brown; legs and feet black, legs sometimes spotted grey-blue or mainly grey-blue. Dark phase: uniform dark slate-grey all over. ADULT ♂ (breeding): light and intermediate phases: like non-breeding ♂ but forehead, crown and lores black, rest of head and neck white tinged yellow, throat white sometimes tinged yellow. Mantle without pink-buff tips, uppertail-coverts pale slate-grey. Tail streamers up to c. 180 mm longer than others. Breast white; breast-band indistinct or lacking; rest of underparts slate-grey; no dark barring on flanks, vent or undertail-coverts. Dark phase: like non-breeding ♂. Sexes alike. SIZE: wing, ♂ (n = 36) 292–318 (306), ♀ (n = 38) 294–323 (309); wing-span, ♂♀ 1050–1170; tail (outer feather), ♂ (n = 73) 104–121 (112), ♀ (n = 55) 104–121 (111), (from tip of central pair to tip of outer feather), ♂ (n = 31) 151–246 (178), ♀ (n = 29) 135–216 (174); bill to feathers, ♂ (n = 75) 26–31 (29), ♀ (n = 53) 26–31 (28); tarsus, ♂ (n = 75) 39–46 (43), ♀ (n = 54) 39–45 (43). WEIGHT: ♂ (n = 26) 236–343 (280), ♀ (n = 18) 258–358 (313).

IMMATURE: 1st winter, light and intermediate phases: forehead and crown mottled grey-brown and white; crown darker than rest of head; hindneck, side of head, chin and throat white, streaked grey-brown. Head of light phase sometimes strikingly yellowish white. Rest of upperparts black-brown, each feather fringed white, forming thin transverse bars;

uppertail-coverts barred black-brown and white. Tail black-brown, tips of feathers edged white; central tail-feathers 18–32 mm beyond others, tips blunt or rounded, not pointed. Underparts white, flanks, vent and undertail-coverts with some black bars; intermediate forms with more barring. Remiges dull black, shafts of only outer 2 primaries white; inner webs of primaries white; underwing-coverts and axillaries dark grey and white. Base of bill pink and white, tip black; legs pale blue-white to pink. 1st winter, dark phase: mostly blackish brown; chin throat, and sides of neck and hindneck paler yellowish brown, upperparts faintly barred buff; undertail-coverts barred brown and white. Subsequent immature plumages: (light and intermediate phases) like adult but uppertail-coverts narrowly barred white and brown, underparts with partial breast-band, white on tips of primaries reduced or lacking, and underwing-coverts with some barring; (dark phase) as adult.

Field Characters. Smallest skua. Flight silhouette like Arctic Skua *S. parasiticus* and also petrels particularly if seen head on over sea surface or in wave trough (see **A**). Diagnostic features of breeding adult are very long tail streamers, no white on long narrow pointed wings, and no breast-band (**B**); also neat appearance, slender build, small-headed look, small bill and small cap, greyish mantle, slim chest, wedge-shaped long tail (even without long streamers), and flight light, floating, graceful and tern-like. Much less variable than other skuas. When resting on sea, wings extend at least 4 cm beyond outer tail-feathers. Immatures difficult to distinguish from other immature skuas; useful characters include body shape and flight action (see above), greyish and less yellow-brown colour, pale barring on rump and uppertail-coverts, whitish transverse bars on upperparts and conspicuous black and white barring on undertail-coverts and axillaries. See also Pomarine Skua *S. pomarinus* and Arctic Skua.

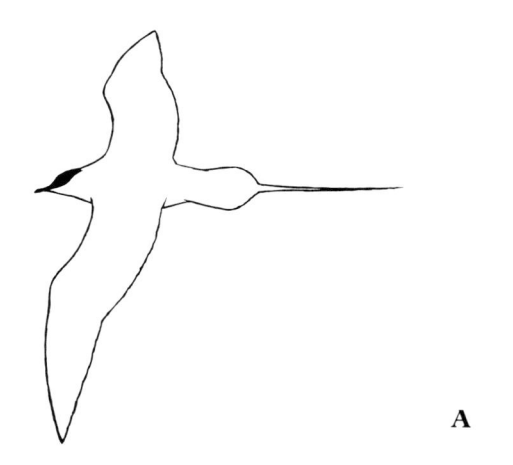

Voice. Tape-recorded (62, 73). Silent when migrating and in winter areas; various calls at breeding colony (see Cramp and Simmons 1983).

General Habits. Pelagic; rarely on coast. Solitary, or occasionally up to 5 except in spring migration when occurs in groups of up to 40, occasionally hundreds; and in wintering ground in Benguela Current when 100 or more, mainly immatures, may congregate. Flies higher than other skuas, up to *c.* 250 m; less ready to approach ships and trawlers. If flying into high waves, either alternates careening on fixed wings with flapping and gliding, or hugs wave troughs. Hovers freely; chase-flights very agile. Regularly settles on sea. Probably less piratical than other skuas although off Namibia occasionally kleptoparasitizes Common Tern *Sterna hirundo*, Arctic Tern *S. paradisaea* and Sabine's Gull *Larus sabini*. Off western Cape (South Africa) forages in flocks of Sabine's Gulls; sometimes obtains food from water surface in mid-flight; also scavenges offal from trawlers and feeds on carrion.

Apparently winters only south of equator. Arrives in Benguela Current off Namibia late Sept with numbers building up to Nov; departs Apr, sometimes May; sometimes oversummers (1 adult, Agulhas Bank, South Africa, July). Morocco, Oct; Senegambia, Feb–Mar; Nigeria, Dec; Gulf of Guinea, Apr; and Kenya, late Aug, Oct. Origin of wintering birds in Africa unknown.

Food. Marine fish, offal, carrion and refuse; composition hardly known outside breeding season.

References
Cramp, S. and Simmons, K. E. L. (1983).
Lambert, R. (1980).

STERCORARIIDAE

Genus *Catharacta* Brünnich

Large, heavy-bodied, brown skuas with conspicuous white area at base of primaries. Wings broad but pointed at tip; tail wedge-shaped, two central tail-feathers not or barely elongated; 1–2 colour phases. Plumage of young birds not barred below.

Taxonomy complex. We recognize 1 superspecies of 3 spp.: *chilensis* and *maccormicki* (monotypic), and *C. skua* with 4 races (*skua* of Northern Hemisphere, *antarctica* of Falkland Is. and S Argentina, *hamiltoni* of Gough and Tristan da Cunha, and *lonnbergi* of Antarctic Peninsula, subantarctic islands and Australasian region. The last 3 races are sometimes split from *skua* to form *C. antarctica* (Brooke 1978). 2 spp. occur in Africa.

Plate 22
(Opp. p. 353)

Catharacta skua Brünnich. Great Skua; Brown Skua. Grande Labbe.

Catharacta Skua Brünnich, 1764. Ornithol. Bor., p. 33; Färoes = Iceland.

Forms a superspecies with *C. maccormicki*.

Range and Status. Breeds NW Europe, southern tip of South America, Atlantic subantarctic islands, Antarctic Peninsula; northern population winters Atlantic, south to NW and W Africa and N Brazil; southern population winters southern oceans north to Africa, Australasia and southern South America.

Palearctic visitor Tunisia (uncommon) to Senegambia (common in upwelling zone off Senegambia); vagrant Libya (Tripoli) and Nigeria (Lagos). Antarctic visitor to coasts from S Angola to S Mozambique (frequent to common); vagrant Pagalu, Kenya (Kiunga, 1°45′S, 41°29′E) and Somalia (1, Mallable, 2°12′N, 45°37′E 1 June 1979). Seen (race?) Equatorial Guinea.

Description. *C. s. skua* Brünnich: coasts NW and W Africa south to Senegambia. ADULT ♂ (non-breeding): forehead, crown to eye, lores and neck dark brown, with some pale streaks at centre of feathers. Rest of upperparts black-brown, each feather with narrow cinnamon-rufous central streak at tip. Tail blackish brown, off-white near base, wedge-shaped. Chin and throat uniform dark grey-brown; rest of underparts cinnamon-brown, feathers with pale grey sides and bases, giving mottled appearance; breast, flanks, undertail-coverts often darker grey. Remiges and greater upperwing-coverts blackish brown; primaries with much white at base; outer webs of outer primaries whitish, decreasing towards inner primaries; bases of secondaries with some white. Underwing-coverts grey; smaller and inner median ones dark grey-brown with cinnamon streaks. Axillaries black-brown, each feather cinnamon-rufous at tip. Bill blackish grey, sometimes with darker tip; eye brown; legs and feet blackish grey. ADULT ♂ (breeding): head with indistinct cap, feathers of sides of head and hindneck elongated with pale yellow to brown-yellow shaft streaks forming hackled mane; head feathers rufescent. Feathers of upperparts with broad cinnamon-rufous central streak at tip. Sexes alike except that ♀ has longer, paler yellow hackles, and broader, paler streaks on upperparts. SIZE: (21♂♂, 21♀♀) wing, ♂ 393–431 (412), ♀ 416–425 (422), span, ♂♀ 1320–1400; tail, ♂ 140–162 (150), ♀ 144–156 (149); bill to feathers, ♂ 48–50 (49), ♀ 44–52 (50); tarsus, ♂ 67–71 (69), ♀ 63–73 (68). WEIGHT: ♂ (n = 21) 1250–1500 (1339), ♀ (n = 21) 1420–1650 (1490).

IMMATURE: much like adult but cap less distinct, head without golden shaft streaks, upperparts more uniform brown with paler rufous edges to feathers. Underparts more uniform, sometimes rufous-cinnamon; upperwings with less white at bases of primaries; legs and feet with some blue-grey dots.

C. s. lonnbergi Mathews (= *C. s. madagascariensis*): coasts of S Angola south and east to S Mozambique; also Kenya and Somalia. Wing relatively shorter and more rounded, bill stronger, plumage more uniform dark brown. SIZE: wing, ♂ (n = 7) 401–406 (407), ♀ (n = 8) 397–416 (407).

Field Characters. A large, stout-billed, broad-winged and relatively short-tailed skua, like a large gull but with heavier, more barrel-shaped body (see **A**). In distance, appears uniformly dark but with striking large white patches on both surfaces of primaries; closer up, plumage has mottled brown appearance. Tail short and slightly rounded. Also has hunched appearance with head sunk into shoulders. Size variable; some as small as Pomarine Skua *Stercorarius pomarinus*, but differs in having longer bill, bulkier head, thicker neck, broader wings with almost rounded tips and shorter tail. Distin-

General Habits. Highly pelagic; rarely on coasts. Singly or in groups of up to 4; much less gregarious than other skuas even when migrating. Rests on sea. Flight direct and purposeful, hugging wave contours, but in migration may soar well above sea. Follows ships. Feeds by scavenging, dipping-to-surface, surface-seizing, and food-piracy. Usually initiates pursuit-chases with sneak low level attack, often grabbing victim's wing or tail and pulling bird into sea. Off South Africa and Namibia, kleptoparisitizes mainly Little Shearwater *Puffinus assimilis* and Cory's Shearwater *Calonectris diomedea*, also harries Great Shearwater *P. gravis* and Atlantic Petrel *Pterodroma incerta* and attacks and kills Broad-billed Prion *Pachyptila vittata*, Kerguelen Petrel *Pterodroma brevirostris*, Soft-plumaged Petrel *P. mollis* and Hartlaub's Gull *Larus hartlaubii* (Sinclair 1980); elsewhere harries shearwaters, gannets, gulls and terns.

Northern population present NW and W Africa Sept–Mar/Apr with some 1st-year birds oversummering. 32 ringed Scotland, recovered NW and W Africa (Morocco, Algeria, Tunisia, Mauritania, Senegambia: 11 Jan–Feb, 8 Mar–Apr, 4 May–June, 2 July–Aug, 5 Sept–Oct, 2 Nov–Dec). Southern population present off African waters all year but mainly Apr–Sept. 1 ringed Marion Is. recovered Namibia.

Food. Fish, crustaceans and cephalopods; also small birds, carrion and trawler offal.

Reference
Cramp, S. and Simmons, K. E. L. (1983).

guished from South Polar Skua *Catharacta maccormicki* by uniform plumage (not 2-toned with upperparts darker than underparts); immatures with rufous (not dark grey) underparts. South Polar Skua frequently has large pale patch on nape formed by golden hackles. For further differences, see that species.

Voice. Tape-recorded (62, 73). Generally silent outside breeding season; harsh nasal 'skeerrr', deep barking 'uk-uk-uk' and guttural 'tuk-tuk-tuk' when attacking.

Catharacta maccormicki (Saunders). South Polar Skua; McCormick's Skua. Skua de McCormick.

Plate 22
(Opp. p. 353)

Stercorarius maccormicki Saunders, 1893. Bull. Brit. Orn. Club 3, p. 12; Possession Island.

Forms a superspecies with *C. skua*.

Range and Status. Breeds Antarctic; 'winters' north to N Atlantic, N Indian and S Pacific oceans.

Rare, South Africa: 2 obtained off Cape St Francis, Cape Provine, 16 May 1963; also a few sight records off Western Cape. Status in Benguela and Agulhas currents uncertain but probably very rare (J. Sinclair, pers comm.). Vagrant, Somalia (1, Hal Hambo, 1°54′N, 45°05′E).

Description. ADULT ♂: 2 colour phases with intermediates. Light phase: forehead and cheeks pale brown, crown dark brown, nape and hindneck greyish light brown, sometimes nearly greyish white forming partial collar; nape and sides of neck with variable amount of golden hackles. Rest of upperparts uniform brownish black with a few feathers with pale tips. Tail blackish, wedge-shaped and short. Underparts pale brownish grey, upper breast streaked with straw yellow. Remiges brownish black; shafts and bases of primaries white; rest of upperwing and underwing blackish brown. Bill blackish grey; eye dark brown; legs and feet blackish grey. Intermediate phase: as light phase except head, hindneck and underparts uniform straw-brown to buff-brown. Dark phase: uniform black-brown except for golden hackles on head and

neck; nape somewhat paler; pale grizzled area at base of bill; and underparts slightly lighter than upperparts. Sexes alike.
SIZE: wing, ♂ (n = 14) 380–406 (390), ♀ (n = 12) 372–401 (388); tail, ♂ (n = 10) 145–152 (149), ♀ (n = 10) 137–154 (146); bill to feathers, ♂ (n = 12) 43–50 (46), ♀ (n = 12) 43–50 (46); tarsus, ♂ (n = 12) 59–67 (63), ♀ (n = 12) 60–68 (65).
WEIGHT: (Antarctic adult, sex?, n = 4) av. 1224.

IMMATURE: light and intermediate phases: similar to adult but head and neck grey or dusky greyish brown; no golden hackles; chin and throat paler; hindneck uniform paler greyish buff, continuing on foreneck as unmarked pale colour; upperparts greyer, less brown, sometimes with light grey or buffish edges to scapulars, mantle, back and upperwing-coverts; upper- and underwing with less white than adult. Bill pale blue, tip black; legs and feet blue. Dark phase: like adult except for blue base to bill and blue legs.

Field Characters. Resembles Great Skua *C. skua* but smaller, with short slender bill, and plumage varying from greyish white to blackish. Light and intermediate phases show striking contrast between uniform blackish upperparts and paler head and underparts (see **A**); Great Skua's head, nape and body are heavily marked with tawny streaks and spots; upper- and underparts are same colour. In flight pale body contrasts with dark underwings. Light phase birds have pale collar on hindneck. Dark phase birds very difficult to distinguish from Great Skua, but slight contrast between upper- and underparts and only few golden hackles on neck. Immatures have paler tips on feathers of upperparts, paler hind collar, and greyer (not rufous) underparts. Distinguished from adult *Stercorarius* spp. by being larger and having large white wing-flashes and wedge-shaped tail without elongated central feathers.

Voice. Not tape-recorded. Seldom vocal outside breeding season; harsh nasal 'skeeer'; deep barking 'uk-uk-uk' and guttural 'tuk-tuk-tuk' when attacking; calls higher-pitched than those of Great Skua.

General Habits. Pelagic; usually solitary but feeds communally at trawlers, when quarrels noisily. Flight sustained and powerful; impressively agile when hunting. Dips-to-surface for fish and crustaceans; kleptoparasitizes gulls, cormorants and terns; predatory on other birds; also scavenges and eats offal.

Food. Fish, crustaceans and probably cephalopods; also birds and trawler offal; little known outside breeding season.

Reference
Brooke, R. K. (1978).

Family LARIDAE: gulls

Medium-sized to large seabirds, less slender and mostly larger than terns; bill shorter than head, slender in small species, stout with pronounced gonys and end of upper mandible decurved in larger species; wings long and narrow, distinctly angled in flight, exceeding tail when closed; tail of Sabine's *Larus sabini* markedly forked, others square, slightly rounded or slightly cleft; feet webbed.

Most mature in 3 or 4 years; early plumages at least partly brownish and cryptic. Adults of most species white below, wings and mantle grey or blackish; head white or with seasonal hood of black, brown or grey. Variations in bill and leg colour with age and season less marked than in terns.

Habitually rest and feed on water. Flight strong and buoyant with regular, usually leisurely, wing-beats. Several are long-distance migrants. Walk with body almost horizontal. Adaptable inshore and terrestrial feeders; some small species also successful aerial (insect) feeders. Various feeding strategies; fish usually scavenged rather than caught. Often associates with man and increasing locally in response to increased or modified food supply (refuse, fish offal). Most species are typically coastal. In Africa, inland breeding and migrant populations mostly on large waters.

Usually noisy and gregarious, nesting in often dense colonies; all monogamous and defend a nest-site territory; displays elaborate. Based on ritualized displays, African breeding species are either 'large white-headed gulls' (Herring *L. argentatus*, Kelp *L. dominicanus*, Audouin's *L. audouinii*, and the aberrant Hemprich's *L. hemprichii* and

White-eyed *L. leucophthalmus*), or 'masked gulls' (Grey-headed *L. cirrocephalus*, Slender-billed *L. genei* and Hartlaub's *L. hartlaubii*). Hood may not be a reliable character since it was lost comparatively recently by Hartlaub's Gull, secondarily developed (along with dark plumage) by Sooty and White-eyed (Fogden 1964). Displays detailed below are representative of both groups; there is comparatively little variation within each. Long Call and related patterns are the most useful displays for taxonomic purposes.

Behaviour patterns include:

(1) Large white-headed gulls. (i) Long Call: head jerked into Head-down position (see **A**), then thrown up with a jerk into Throwback position (**B**), head and bill pointing upwards, often beyond vertical; series of loud, screaming calls begin, head and neck relax gradually, carpals raised; aggressive, in defence of territory. (ii) Mew Call (**C**): head lowered, neck arched, often accompanied by peculiar slow gait; associated with nest-relief and nest-building. (iii) Upright Postures: in aggressive form (**D**), neck is swollen, carpals raised, bill depressed; in anxiety form (**E**) (often with Alarm Call), neck is thin, carpals not usually raised, bill horizontal or upwards. (iv) Hunched (Begging) Posture: neck withdrawn, bill pointing forwards and slightly upwards; mostly ♀ and large young perform this. (v) Head Tossing (**F**): head and bill flicked upwards; bill (slightly open) moves through about 90° until almost vertical; soft (Begging) call given with each toss; usually in Hunched Posture by ♀, in Semi-Upright by ♂, primarily sexual.

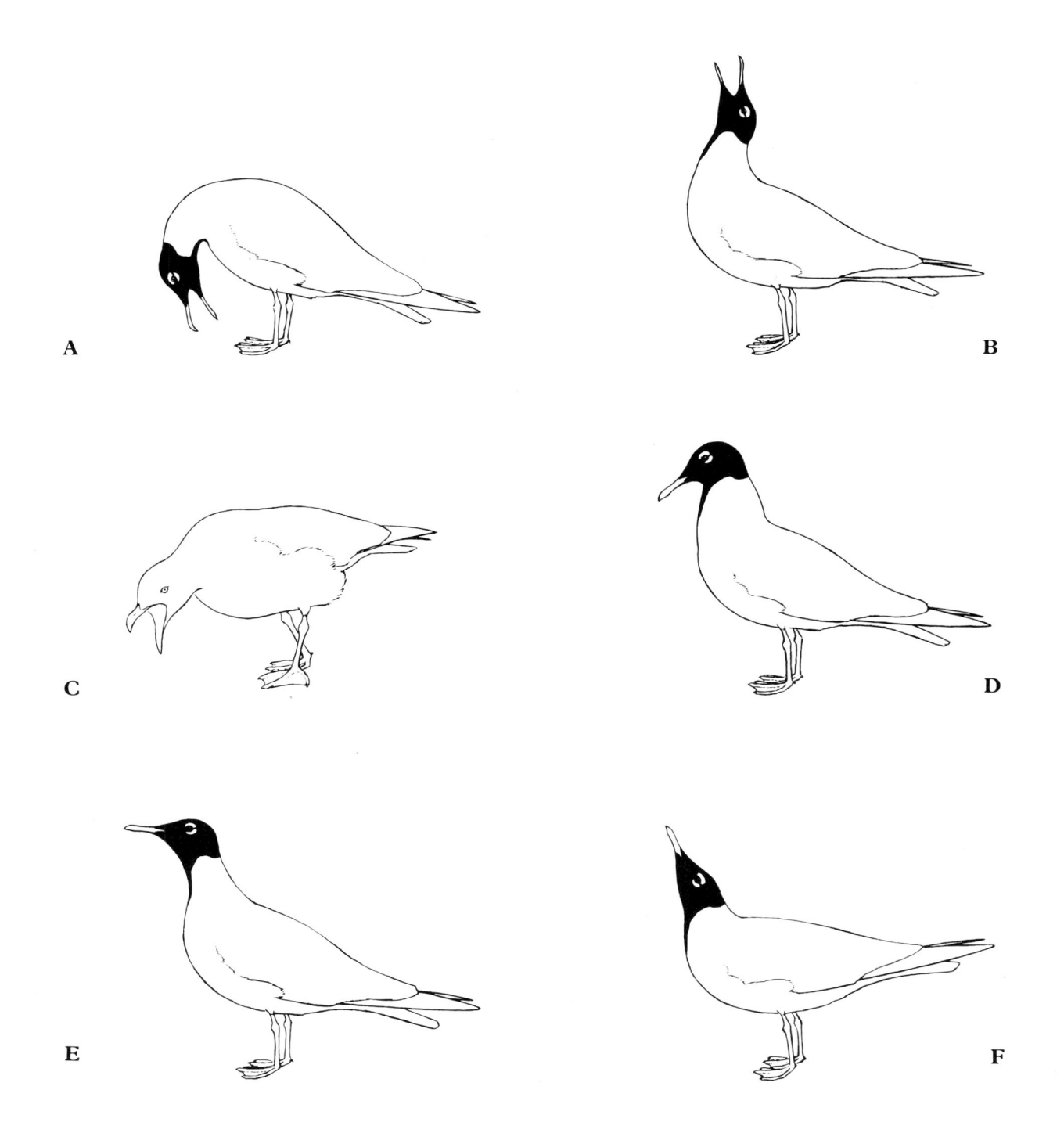

342 LARIDAE

(vi) Choking (**G**): body tilted forward to *c.* 45° below horizontal, carpals usually raised, wings sometimes spread; part of meeting ceremony during pair formation and establishment of nest-site, uncommon once incubation begins. (vii) Pecking-Into-Ground: defence of territory, often involves Grass Pulling at territory boundaries to settle disputes.

(2) Masked gulls. Several postures similar to those of white-headed gulls, Choking especially conservative. (i) Long Call: often less extreme than Group (1), e.g. head of Hartlaub's moves from about 60° below horizontal to 60° above (**H–L**), final position termed Forward Posture rather than Throwback. (ii) Facing Away (**M**): a speciality of this group, serves to exaggerate contrast of dark hood and pale nape, usually superimposed on Upright, head turned away from bird at which display is directed; movement of head jerky (Head Flagging), bill often depressed; prominent during pair formation, mutual displays at nest, and immediately after copulation; much less developed in Group (1).

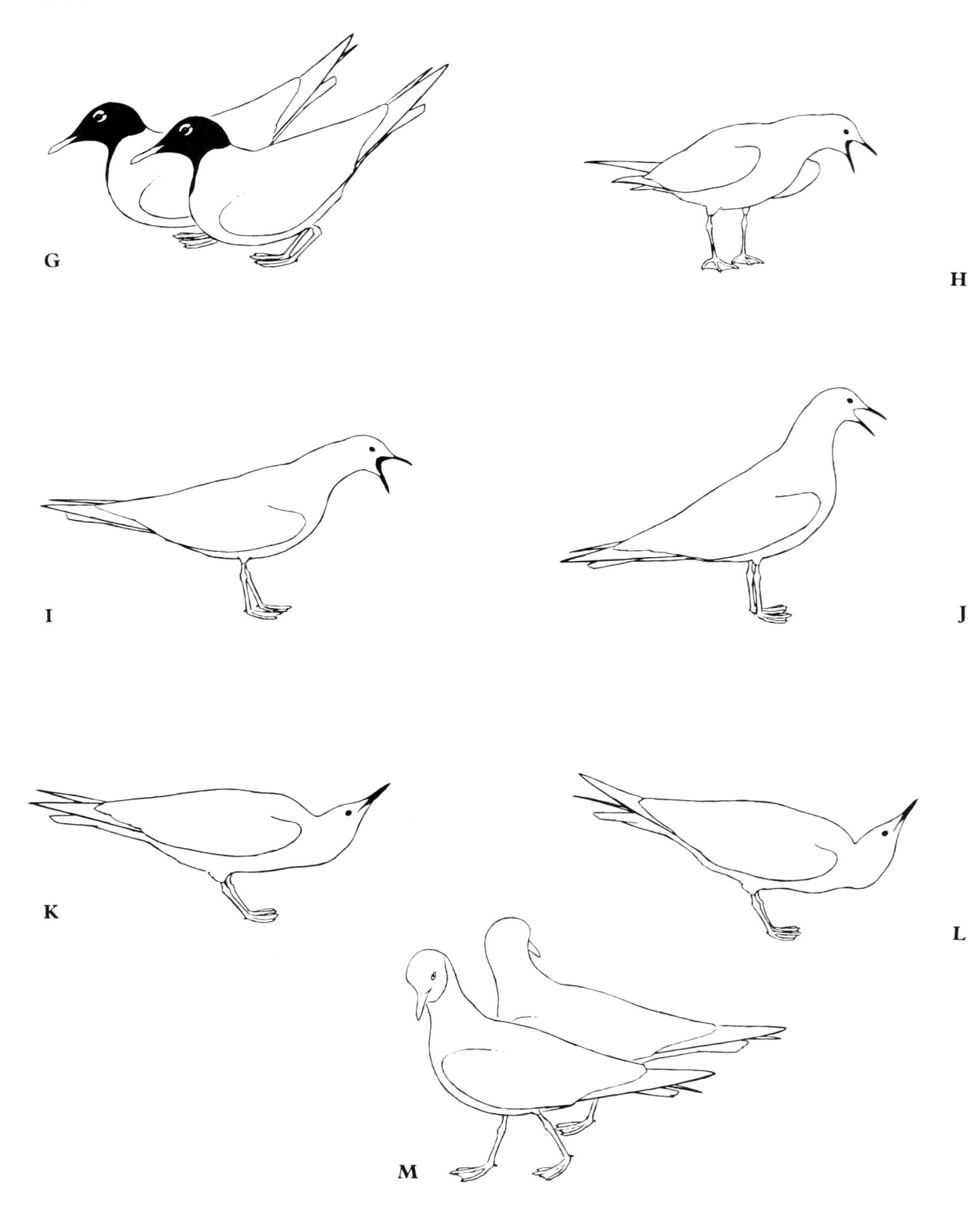

In Africa, 2 genera, 21 spp., of which 6 are resident, 7 are migratory from Palearctic, 2 have both resident and Palearctic wintering populations, 1 is migratory from Nearctic and 5 are vagrant.

Note: certain terms used in the plumage descriptions have been taken from Grant (1982); these are: ear-spot (well defined area of dark feathers on ear-coverts, appearing as dark spot behind eye; feature mainly of hooded species in non-breeding plumages), eye-crescent (semi-circular dark area immediately in front of eye) and mirror (rounded white area near tip of otherwise black outermost 1 or 2 primaries). Juvenile and immature plumages are described with the juvenile plumage being the 1st full plumage following the nesting or downy young stage, and the immature plumage being one or more recognizable plumages following the juvenile plumage.

Genus *Larus* Linnaeus

Small to large gulls. Bill stout or slender, upper mandible slightly longer than lower, somewhat bent down at tip; no cere, nostrils linear ovals about half-way along bill; lower third of tibia bare of feathers, tarsus longer than middle toe with claw, transverse scutes in front; hind toe short but well developed with claw.

Cosmopolitan; mostly Holarctic genus of 40 spp.; 20 in Africa with 5 Afrotropical, 7 visitors, 3 visitors with Afrotropical populations and 5 vagrants.

The genus separates into well defined groups, usually described with reference to ancestral morphological characters (e.g. masked), though behavioural studies and distribution patterns are at least as important in clarifying membership of superspecies. The two 'masked' species (*cirrocephalus* and *hartlaubii*) are virtually allopatric members of a superspecies, along with the migrant *ridibundus*. The black-backed *dominicanus* and the closely related *fuscus* form a well marked superspecies, which includes also *argentatus*. Amongst still larger species, the migrant *marinus* and *hyperboreus* are members of a superspecies. Smaller than other large white-headed gulls, *audouinii* and *canus* are perhaps best regarded as members of a diverse superspecies in which some authorities also include *delawarensis*.

Larus cirrocephalus superspecies

Larus argentatus superspecies

1 *L. cirrocephalus*
2 *L. hartlaubii*
+ *L. ridibundus* (**non-breeding visitor**)

1 *L. argentatus*
2 *L. dominicanus*
+ *L. fuscus* (**non-breeding visitor**)

344 LARIDAE

Plates 24, 27
(Opp. pp. 369, 416)

Larus hemprichii Bruch. Hemprich's Gull; Sooty Gull. Goéland d'Hemprich.

Larus hemprichii Bruch, 1853. J. Orn., p. 106; Red Sea.

Range and Status. Coasts of NE Africa, Arabia and Persian Gulf to W Pakistan.

Resident and partial intra-African migrant, abundant to uncommon. Breeds mostly north of Gulf of Aden in Red Sea north to Wadi El Gamal Island (Egypt); also breeds Gulf of Aden at Zeyla (Somalia); only 50–100 breeding pairs south of Gulf of Aden, in Somalia (Brava) and Kenya (Kiunga). Non-breeders eastern seaboard south to Tanzania-Mozambique border. Scarce winter visitor Egypt north to Gulf of Suez, once 3 immatures inland Kenya (L. Turkana, Apr).

Description. ADULT ♂ (breeding): forehead and lores to hindneck, ear-coverts, chin, throat and foreneck very dark brown, shading to blackish on hindneck and bib; lower hindneck, sides of neck and crescent above eye white; mantle, back and scapulars brown washed dark grey; rump, uppertail- and undertail-coverts, tail and belly white; breast-band and flanks grey-brown; underwing dull brown, coverts and axillaries blackish brown; remiges blackish, all except outer primaries tipped white. Bill yellow or greenish yellow, blackish band before red tip to bill, extreme tip often yellowish; eye dark brown; legs yellow or greenish yellow. ADULT ♂ (non-breeding): much as breeding but head paler brown, collar on hindneck less defined or lacking, bare parts duller. Sexes alike except ♀ somewhat smaller. SIZE: (10 ♂♂, 14 ♀♀) wing, ♂ 330–352 (343), ♀ 318–340 (330); tail, ♂ 120–132 (127), ♀ 114–126 (120); bill, ♂ 47–52 (50·1), ♀ 44–52 (46·9); tarsus, ♂ 53–59 (55·0), ♀ 48–56 (52·1). WEIGHT: (Kenya) ♂ 510, ♀ 400; (Egypt) ♂ (n = 9) 491–640 (558), ♀ (n = 6) 412–522 (461).

IMMATURE: head pale brown, shading to brown on nape and whitish on chin and face; whitish crescent above eye, dark eye-crescent; mantle, back, scapulars, breast-band, flanks and underwing grey-brown; rump, uppertail- and undertail-coverts and belly whitish. Tail mostly black, narrow white terminal fringe, white basally on inner webs of outermost rectrices. Secondaries and outer primaries blackish (distinctly paler when worn), inner primaries and tertials paler; inner wing with narrow white trailing edge, scaly coverts. Bill and legs greyish or blue-grey, legs a shade darker, bill tipped black; eye dark brown. By 2nd winter resembles non-breeding adult; head and bill little different from younger immature; body plumage patchy and browner; white trailing edge to wing less pronounced; tail patterned black, grey and white.

JUVENILE: much as immature; body paler (darker than head); rump and uppertail-coverts pale grey-brown or whitish.

DOWNY YOUNG: upperparts and wings brownish white, underparts and head white; bill dull pink tipped black.

Field Characters. A medium-sized dark gull, larger, heavier and browner than White-eyed Gull *L. leucophthalmus*, wing-tips more rounded in flight. Adult hood like that of White-eyed, extending as bib onto upper breast, but dark brown, not glossy black. White crescent distinct above eye, faint below eye. Young birds of all ages with only small area of white confined to crescent above eye; White-eyed young with bold crescentic white above and below eye. Mantle and wing-covert feathers of young birds of all ages have more obvious scaly pattern than White-eyed; less contrast between whitish lower belly and dark head, chest, underwing and upperparts than White-eyed. Bill deeper, straighter and heavier than White-eyed, basally yellow, not red (adult) or greyish with dark tip, not blackish (immature). Flies with broad easy strokes like other large gulls; flight not graceful or tern-like as White-eyed.

Voice. Tape-recorded (C). Gives (a) Long Call: loud, screaming 'kioow', usually repeated 12–16 times, becoming progressively shorter and lower pitched. (b) Alarm Call: rhythmic, repeated, short, hard 'ke' (usually 3–6); more deliberate, less staccato 'koup-koup-koup'. (c) Attack Call: harsh 'kek-kek-kek-kyaaa', ending in loud, frenzied scream (at lowest point of swoop). Calls mostly given at colonies, but Long Call common, especially in flocks circling at roost and in food squabbles.

General Habits. Feeds singly or in small parties in all littoral habitats and to 10 km or more beyond reef. Mostly a scavenger; picks up floating items in flight or alights alongside. Forages with waders on exposed flats, commonly foot-paddling in mud. Fishes large shoals with White-eyed Gull and mixed flocks of terns, by plunge-diving from surface and air. Kleptoparasitism uncommon but very successful: of 2 or 3 gulls chasing Swift Tern *Sterna bergii* at Brava, all but 1 attempt successful; 32/38 attempts at Sacred Ibis *Threskiornis aethiopica* at Kiunga successful. Takes eggs and young of terns. Several roosts of 500 or more Kenya and Tanzania, especially near harbours and fish markets; now much commoner than early this century.

Partial migrant and wanderer, mostly southwards. South of 2°S, June–Sept numbers (few adults) only 10–20% of Oct–May numbers.

Food. Fish (mostly offal), invertebrates, eggs and young of terns, eggs of White-eyed Gull, turtle hatchlings. In Egypt, fish include *Amblyglyphidodon flavilatus* (Pomacentridae) 60–70 mm long and cyprinodontids 110 mm long (S. M. Goodman and R. W. Storer, pers. comm.).

Breeding Habits. Typically solitary nester, 1–3 pairs per island, nests as far apart as possible on exposed promontories at edge of island, once 14 incubating pairs on 1 ha island; small loose colonies on larger islands in Gulf of Aden and Red Sea, 1 or 2 pairs on most islands in Dahlak and Suakin Archipelagoes. Performs displays typical of a large white-headed gull; dark plumage and hood are secondary characters. Long Call, Alarm Call, Begging Call, Hunched and Head Tossing postures, Choking in meeting ceremony, frequency of Upright display, quality of voice and poor development of aerial behaviour patterns all very similar to Herring Gull *L. argentatus*; Ground Pecking, Forward and Mew Call absent. Facing Away highly developed and similar to masked gulls; usually superimposed on Upright posture, deliberate and precisely orientated; bill depressed sufficiently for hood to be virtually hidden, while collar exaggerated in area and conspicuous; prominent at meeting ceremony and immediately after copulation. Pair-formation takes place away from nesting area, especially on beaches. Nest-site territoriality minimal. Long Call by overflying bird often prompts Long Call from incubating or off-duty ♂. Neither sex responds aggressively to other adults on ground in vicinity, even when 2 neighbours alighted 30 cm away and preened for 2 min.

NEST: scrape or depression in coral, usually fairly deep, lined with mangrove leaves, grass, feathers, small stones; under mangrove bushes on bed of leaves or in scrape in loose sand under *Suaeda* bushes where eggs protected from White-eyed Gull and/or exceptionally high temperatures (NW Somalia).

EGGS: 2–3, rarely 1; Kenya (31 clutches) av. 2·42. Subelliptical; cream to cream-buff with distinct spots, blotches and lines, separate and evenly distributed, in various shades of yellow-brown, deepening to chocolate brown; also pale purple undermarkings. SIZE: (n = 50) 53·0–63·0 × 37·0–44·5 (57·8 × 41·5). WEIGHT: 53 g (estimated).

LAYING DATES: Sudan, Ethiopia, Somalia and Kenya, June–Sept (especially July).

INCUBATION: carried out by both parents. Period: at least 25 days. Individual incubation period lasts 3–4 h (sometimes 5 h: same ♂ sat 0900–1735 h, 0845–1700 h, almost continuously but for 1 h break midday). Bird on nest may pant, apparently distressed by heat. Both sexes leave nest for brief intervals, circling before alighting on sea to sip water; most nest-reliefs take place during these intervals, relieving bird (often sitting quietly nearby) moving onto vacant nest. Some nest-reliefs involve display, including Long Call, Choking, carrying off leaves and twigs; waiting bird may push other off nest. If incubating bird otherwise disturbed often picks up vegetation (displacement nest-building).

DEVELOPMENT AND CARE OF YOUNG: chick stimulates parent to regurgitate by stretching upwards, spreading wings, squeaking and pecking at bill; parent gives notes of Long Call.

BREEDING SUCCESS/SURVIVAL: direct human predation insignificant; dark plumage has cryptic value; nests inconspicuous, well spaced, often hidden, usually far outnumbered by conspicuous nests of terns and White-eyed Gulls.

References
Cramp, S. and Simmons, K. E. L. (1983).
Fogden, M. P. L. (1964).

Larus leucophthalmus Temminck. White-eyed Gull. Goéland à iris blanc.

Plates 24, 27
(Opp. pp. 369, 416)

Larus leucophthalmus Temminck, 1825. Pl. Col. livr., pt. 62, plate 366; Red Sea.

Range and Status. Endemic to Red Sea and Gulf of Aden; abundant to common. Breeds Somalia (thousands Aibat and Saad Din Islands off Zeyla), Ethiopia (1400 adults, spread over 10 islands, Dahlak Archipelago; 9 clutches collected Fatmah Island), Sudan (200, Mukawar Island), Egypt (up to 50 pairs, Hurghada, Tiran). Vagrant Kenya (L. Turkana, 4–14 Apr 1975) and N Egypt (Port Said, 3, 28 June 1977); reported occurrence Mozambique (Beira) and South Africa (Port Elizabeth and Natal) erroneous.

Description. ADULT ♂ (breeding): forehead and lores to hindneck, ear-coverts, chin, throat and foreneck glossy black; lower hindneck, sides of neck and broad crescents above and below eye white; mantle, back and scapulars dark grey, breast-band and flanks paler grey; rump, uppertail- and undertail-coverts, tail and belly white; underwing dull grey-brown, coverts and axillaries blackish brown; remiges blackish, all except outer primaries tipped white. Bill red, tipped black; legs yellowish; eye dark brown, orbital ring red. ADULT ♂ (non-breeding): similar to breeding ♂ but black areas of head flecked white, collar on hindneck less defined, and bare parts duller. Sexes alike except ♀ somewhat smaller. SIZE: (5 ♂♂, 6 ♀♀) wing, ♂ 318–332 (325), ♀ 305–312 (309); tail, ♂ 115–125 (120), ♀ 107–115 (111); bill, ♂ 44–52 (48·8), ♀ 43–46 (44·5); tarsus, ♂ 45–50 (47·8), ♀ 44–47 (45·2). WEIGHT: (Egypt) ♂ (n = 10) 325–415 (357), ♀ (n = 9) 275–355 (303).

IMMATURE: head brown, shading to whitish on face and throat, faintly streaked on crown and around ear-coverts; blackish mask through eye to nape, whitish crescents above and below eye; mantle, scapulars, breast-band, flanks and underwing grey-brown; rump and uppertail-coverts greyish; belly and undertail-coverts whitish. Tail black, usually lacking white terminal fringe; small white area basally on inner web of outermost rectrices. Primaries and secondaries blackish, narrow white trailing edge to inner wing; tertials and wing-coverts grey-brown with pale fringes. Bill black, brownish at base of lower mandible; eye dark brown; legs greenish grey. By 2nd winter resembles non-breeding adult; head browner, tail usually patterned black and grey, broken subterminal band, rump usually greyish.

JUVENILE: much as immature; mask reduced and dusky, breast-band and flanks brownish.

DOWNY YOUNG: upperparts and wings brownish white, underparts and head white, bill black.

Larus leucophthalmus

Field Characters. A medium-sized gull, resembling Hemprich's *L. hemprichii* but smaller, greyer, hood blacker, breast glossy black. White crescents above and below eye, giving 'white-eyed' look; Hemprich's has only inconspicuous crescent below eye. Bill slimmer, somewhat decurved, almost tern-like; red (not yellow) in adults, blackish (not grey) in immatures. Head pattern similar to vagrant Franklin's Gull *L. pipixcan*, but larger, darker, wing-tips lacking black and white pattern of Franklin's, bill longer. (See Hemprich's Gull account for details of juvenile plumage.)

Voice. Tape-recorded (73, 233). Gives (a) Long Call, a screaming 'kioow', usually repeated 12–16 times, like Hemprich's but somewhat less harsh and deep. (b) Alarm Call, like Hemprich's, a 'ke' repeated usually 3–6 times and 'koup-koup-koup', but less harsh and deep.

General Habits. Occurs inshore and offshore; scarce in major ports, scavenges far less than Hemprich's; takes eggs and young of Lesser Crested Tern *S. bengalensis*. Usually in small flocks, often associates with Hemprich's, frequents dumps and village refuse heaps Egypt; follows ships in Suez Canal. Unlike most dark or hooded gulls and terns, fishes by plunge-diving, mostly whilst swimming; requires waters with superabundance of fish; follows fish shoals and schools of whales off Ethiopia.

Occurs throughout range all year; pronounced southward and eastward shift in winter (Nov–Mar) when scarce north of Hurghada (Egypt). Occupies breeding islands in Somalia in late Apr.

Food. Fish, marine invertebrates, birds (eggs and chicks) and plants. In Egypt, fish include *Scarpus* sp. 110 mm long and the plant *Nitraria retusa* (S. M. Goodman and R. W. Storer, pers. comm.).

Breeding Habits. Breeds in colonies on offshore islands, usually in small groups, sometimes spread over several islands. Entire colony sometimes attacks intruders threatening young (Dahlaks). Only 1 instance of courtship observed, late in season 1 km or more from nests (Clapham 1964): presumed ♀ in Hunched position, ♂ Long Called in flight and landed 5 m from ♀; ♀ flew to ♂, resumed hunched position, ♂ picked up piece of seaweed, walked 3 m away from ♀; ♀ followed, still in Hunched position, started to Head Toss and to utter Begging Call when near to ♂, postures virtually identical to Hemprich's.

NEST: on sand, a conspicuous ring of twigs, seaweed and debris, no bedding or foundation for eggs; on rocky islands, small pads of vegetable matter by rocks; scrapes among impenetrable *Euphorbia* clumps.

EGGS: 2–3 (usually 2). Subelliptical; cream-white to pale olive-buff ground, with small and evenly distributed brown spots, blotches and stripes; ash-grey or purple undermarkings. SIZE: (n = 12, Somalia) 52–56·5 × 36·5–40 (54 × 39).

LAYING DATES: Somalia, Ethiopia, Sudan, Egypt, early June–late July, extends to Sept in north (Egypt).

DEVELOPMENT AND CARE OF YOUNG: immatures leave Dahlaks very soon after fledging, probably going to sea.

BREEDING SUCCESS/SURVIVAL: predation of eggs by humans and Hemprich's Gull, Somalia.

References
Clapham, C. S. (1964).
Cramp, S. and Simmons, K. E. L. (1983).

Larus ichthyaetus Pallas. Great Black-headed Gull. Goéland ichthyaète.

Plates 22, 27

Larus ichthyaetus Pallas, 1773. Reise versch. Prov. Russ. Reichs, pt. 2, p. 713; Caspian Sea.

(Opp. pp. 353, 416)

Range and Status. Breeds from southern Russia (Crimea) east to NW Mongolia. Winters south to E Mediterranean, Red Sea, Persian Gulf and coasts of India and Burma.

Palearctic winter visitor, common to uncommon, mostly inland. Annual flocks on Ethiopian Rift Valley lakes (winter counts 32–111, 1973–76); Egypt (up to 50, but numbers reduced in recent decades). Vagrant Tunisia, Sudan, Somalia, Kenya and Uganda.

Description. ADULT ♂ (breeding): forehead and lores to upper nape, ear-coverts, chin and throat black; broad white crescents or oval patches posteriorly above and below eye; lower nape, hindneck, rump, uppertail-coverts, tail and underparts white; mantle, back, most of inner wing and inner primaries pale pearl-grey, thin whitish fringes to scapulars. All remiges broadly tipped white; outer web of outermost primary wholly black, remainder of leading edge of wing narrowly fringed white, subterminal black on primaries decreasing inwards to small spot on 6th; remainder of outer wing white; underwing white except for black tip and some dusky marks on coverts. Bill yellow or orange-yellow, black subterminal band, reddish or orange-red tip; eye dark brown, orbital ring red; legs yellow or greeish yellow. ADULT ♂ (non-breeding): like breeding ♂ except head mostly whitish, crescents above and below eye strikingly white against dark patches before and behind eye; dark streaking often extends diffusely over crown; nape and hindneck streaked or spotted blackish, concentrated in patch on lower hindneck. Bill yellowish, subterminal band black, tip pale; orbital ring dull; legs dusky, greenish or yellowish. Sexes alike except ♀ somewhat smaller. SIZE: (9 ♂♂, 8 ♀♀) wing, ♂ 470–500 (483), ♀ 422–468 (451); tail, ♂ 180–203 (190), ♀ 171–185 (177); bill, ♂ 58–65 (61·7), ♀ 50–60 (55·9); tarsus, ♂ 74–83 (78·5), ♀ 65–76 (71·2). WEIGHT: (USSR, Mongolia, Pakistan) ♂ (n = 8) 1130–2000 (1580), ♀ (n = 8) 960–1500 (1220).

IMMATURE: body much as non-breeding adult; markings on nape and hindneck browner, often extending onto sides of breast as prominent shoulder patches; uniform grey of mantle and back somewhat darker, usually with some brown feathers; broad, blackish subterminal band on tail, outer web of outermost pair of rectrices wholly white. Outer wing and secondary-bar blackish brown; greyish or grey-brown mid-wing panel, brownish carpal-bar; underwing has dusky and greyish markings, typically forming lines along wing. Legs, orbital ring and base of bill greyish; bill sometimes yellowish, always with clear-cut black tip or broad subterminal band. Partial or full hood usually acquired by 2nd summer (sometimes 1st); blackish tail-band still prominent; wing more like adult, secondary-bar and carpal-bar faint and broken, more black and less white on outer wing; bill and legs yellowish. By 3rd summer resembles breeding adult; more black on wing-tip, faint or mottled tail-band usually present.

Field Characters. Larger than Herring Gull *L. argentatus*; about size of Great Black-backed Gull *L. marinus*; wings appear longer and narrower in flight than either. Black hood, frequent from Jan onwards, unique amongst large gulls; after breeding season birds of all ages have dusky or blackish around eye and on ear-coverts, and hindcrown to lower hindneck, concentrated at base of neck and behind eye. Legs long; bill mostly pale, black at or near tip; rather long and heavy, accentuated by long sloping forehead which peaks well behind eye. At rest, primaries extend well beyond tail, giving body attenuated appearance. Mantle and mid-wing panel of immatures uniform grey; rump white, tail mostly white with broad black subterminal band.

Voice. Tape-recorded (70, 73). Non-breeders mostly silent. Loud, raucous 'kraa-a' rather like Common Raven *Corvus corax*.

General Habits. Frequents beaches on large lakes and coasts; usually solitary or in parties of 2–3; never in flocks. In Kenya mostly associates with Herring Gull; invariably successful in squabbles for fish remains at Malindi. Often follows fishing boats returning to port; picks up garbage and takes food from other gulls, terns, grebes and coots. Remarkably timid in comparison with other gulls.

Present Dec–Apr, mostly adults and sub-adults. Some in 1st summer plumage Uganda Apr–Sept 1966.

Food. Fish (mostly scavenged), also mammals, insects; less often crustaceans, reptiles and birds and their eggs.

References
Ash, J. S. and Ashford, O. M. (1977).
Cramp, S. and Simmons, K. E. L. (1983).

348 LARIDAE

Plates 23, 27
(Opp. pp. 368, 416)

Larus melanocephalus Temminck. Mediterranean Gull. Mouette mélanocéphale.

Larus melanocephalus Temminck, 1820. Manuel d'Ornith., pt. 2, p. 777; Adriatic Sea.

Range and Status. Breeds France, Italy, Hungary, Greece, Black Sea, Sea of Azov and Turkey. Winters mainly central Mediterranean; also Black Sea, Sea of Azov, Adriatic, E and W Mediterranean, and Atlantic coast Morocco and Iberia, straggling north to British Isles.

Palearctic winter visitor, abundant to rare. Whole Mediterranean coast; Atlantic coast of Morocco; Gulf of Suez. Tunisia, 8000 Jan–Feb 1978, mostly in southeast; usually 50 or fewer elsewhere. Vagrant Mauritania, Senegambia, Sudan and Kenya.

Description. ADULT ♂ (breeding): forehead and lores to hindneck, ear-coverts, chin and throat black; white crescents above and below eye; mantle, back and wings pale pearl-grey; rump, uppertail-coverts, tail, underparts, secondaries and underwing white; primaries pale pearl-grey, shading to white at tips, outermost with thin black line on outer web. Bill scarlet; eye dark brown, orbital ring red; legs scarlet. ADULT ♂ (non-breeding): like breeding ♂ but head white with blackish markings, usually a patch of fine streaks behind eye extending diffusely onto rear crown; eye-crescent blackish. Bill and legs orange, red or blackish; bill with dark tip or indistinct subterminal mark, often yellowish at extreme tip. Sexes alike except ♀ somewhat smaller. SIZE: (9 ♂♂, 12 ♀♀) wing, ♂ 291–311 (303), ♀ 282–296 (289); tail, ♂ 118–127 (123), ♀ 113–120 (117); bill, ♂ 33–38 (35·5), ♀ 31–36 (33·4); tarsus, ♂ 48–53 (51·1), ♀ 47–51 (48·3).

IMMATURE: head and body same as non-breeding adult. Outer webs of outer 5 or 6 primaries and their coverts mainly black; others mainly grey, subterminal black marks decreasing in size inwards; inner webs of outer primaries with extensive white, nearly to tips, outermost 1 or 2 sometimes wholly black; secondaries black fringed white; carpal-bar brown or ginger-brown; tertials dark centred, whitish fringes; greater coverts mainly pale grey, forming contrasting mid-wing panel; tail white, black subterminal band. Bill blackish, base usually paler (buff or yellowish to orange or red); legs vary, blackish or grey to orange or red. 1st summer, head markings more extensive, sometimes like breeding adult; by 2nd summer resembles breeding adult, some with white flecking on otherwise black head.

JUVENILE: like early immature; head with buff wash and partial hood; brown markings most prominent on ear-coverts and hindcrown, separated from hindneck by having white collar; dark eye-crescent; thin white crescents above and below eye; mantle and hindneck rich brown or ginger-brown, mantle with scaly pattern; underparts and rump whitish, sides of breast washed buff. Bill blackish, paler at base; legs blackish.

Field Characters. Adults at all seasons have diagnostic 'white-winged' appearance, mantle and wings very pale pearl-grey, only outer web of outer primary black. Hood black, not brown, extending further down on neck than Black-headed Gull *L. ridibundus*. Further differs from Black-headed in shorter, stouter bill, longer legs, strutting gait, slower take-off from water; in flight appears heavy-bodied and bull-necked, wings less angled, wing-beats stiffer, tail less spread. 1st winter birds resemble Mew Gull *L. canus* but have paler mantle and underwing, more contrasting upperwing pattern, also heavy, blackish bill, blackish legs, blackish eye-patch, trace of hood. 2nd winter birds have whitish wings of adult with diagnostic wing-tip pattern—outer primaries have black subterminal marks, white tips.

Voice. Tape-recorded (62, 73, 217). Raucous 'kae-aeh', 'kieu', 'kau-kau' or 'kiau'; rather tern-like, harsher than Black-headed, different tone.

General Habits. Frequents coasts, favouring estuaries, saline lagoons and other sheltered habitats; seldom far inland or far from water. Thousands winter in shallow inshore waters SE Tunisia; on Atlantic feeds mostly on flat seaweed-covered shelves. Of 10 feeding in Algeria, 4 ate fish scraps, 2 accompanied Black-headed Gulls patrolling near beach, 3 harassed Manx Shearwaters *Puffinus puffinus*, 1 followed trawler; methods similar to Black-headed, foraging on foot, paddle-feeding, plunge-diving and feeding in air.

Many ringed Black Sea recovered Libya to NW Morocco, mostly Oct–Apr. Up to 50 annually L. Tunis (N Tunisia) July–Aug, probably 2nd-year birds.

Food. Small fish, molluscs, insects, earthworms, berries.

Larus atricilla Linnaeus. Laughing Gull. Goéland atricille.

Larus atricilla Linnaeus, 1758. Syst. Nat. (10th ed.), p. 136; Bahamas.

Not illustrated in colour

Range and Status. Breeds North and South America, winters eastern seaboard from South Carolina and Gulf coast to Brazil, western seaboard from southern Mexico to Peru; vagrant Europe and N Africa.

Vagrant Morocco (1 adult, south of Agadir, 25 Apr 1978), Senegambia (2, Banjul, Apr 1981). 1 (possibly this species) paired with Grey-headed Gull *Larus cirrocephalus*, nested Saloum Delta, Senegambia, 1983 (Erard *et al.* 1984).

Description. ADULT ♂ (breeding): forehead and lores to nape, ear-coverts, chin and throat slaty black; prominent white crescents above and below eye; hindneck, rump, uppertail-coverts, tail, underparts and underwing-coverts white. Mantle, back and wings dark grey, broad white trailing edge to secondaries and inner primaries; outer 4 or 5 primaries black, black extending along leading edge of wing to near carpus; 5th to 6th or 8th narrowly tipped white. Bill and orbital ring dull red; eye brown; legs dull red. ADULT ♂ (non-breeding): head white except for indistinct partial grey hood, darkest in patch on ear-coverts and extending over hindcrown; eye-crescent blackish; hindneck and sides of breast suffused grey; white tips to primaries more extensive; bill black or blackish brown, with small red line on ridge near tip; legs blackish or grey. Sexes alike except ♀ somewhat smaller. SIZE: (16 ♂♂, 10 ♀♀) wing, ♂ 308–330 (321), ♀ 295–326 (312); tail, ♂ 113–133 (127), ♀ 115–123 (119); bill, ♂ 37–44 (40·2), ♀ 35–41 (38·9); tarsus, ♂ 50–54 (51·6), ♀ 46–55 (49·7). WEIGHT: (Florida, unsexed, n = 39) 270–400 (325).

IMMATURE: forehead, lores, chin, throat and crescents above and below eye whitish; rest of head grey, darkest from eye-crescent and ear-coverts to nape; hindneck, mantle, back, breast-band and flanks uniform or mottled dark grey; belly, rump, uppertail-, undertail- and underwing-coverts dull white, last with extensive dusky markings; wing-coverts brown or grey-brown. Rectrices with dull grey outer webs, whitish inner webs, broad black subterminal band, thin white terminal fringe. Outer primaries and their coverts wholly blackish, thin white fringes at tips from P8–P6 inwards, dull grey on both webs increasing from P7–P5 inwards; secondaries black, edged and tipped white; variable number of secondaries with blackish on inner webs, forming partial subterminal bar; flanks suffused grey; tail mainly white, or with grey at base, often with narrow, broken subterminal band of black or grey; bill and legs blackish until 2nd summer.

Field Characters. Medium-sized, slightly smaller than Mew Gull *L. canus*. Adults have distinctive combination of dark grey upperwings and blackish wing-tips without white 'mirrors'; 1st winter birds with dark grey mantle and mainly sooty black wings also unlike any Old World gull. Can only be confused with Franklin's Gull *L. pipixcan*, but larger, with longer, slightly drooping bill, flatter head, and longer, narrower wings which extend further beyond tail when perched. 1st winter birds differ from Franklin's in less distinct hood, grey-brown breast and broader tail-band; winter adults also have less distinct hood than Franklin's. Breeding adults of both species have slaty black hood with prominent white crescents above and below eye, pattern very similar to Mediterranean Gull *L. melanocephalus*, but last easily distinguished by white wings.

Voice. Tape-recorded (73). Commonest note a whining 'ke-ruh'; also a high, clear, repeated 'hah', resembling human laughter.

General Habits. Frequents coastal habitats, typically inshore; occasionally found inland on rivers and lakes. Follows ships, especially fishing vessels, for offal and refuse; also circles over bays taking flying insects or picking them from water surface. Searches seashores, flats and riverbanks for food; sometimes follows ploughs and after heavy rains visits fields to get earthworms. Takes food from other colonial birds, even pelicans.

Adult south of Agadir (25 Apr) was part of heavy northward passage of gulls, mostly Lesser Black-backed Gulls *L. fuscus*.

Food. Mostly small fish; also invertebrates, carrion, refuse.

Plates 23, 27
(Opp. pp. 368, 416)

Larus pipixcan Wagler. Franklin's Gull. Mouette de Franklin.

Larus pipixcan Wagler, 1831. Isis von Oken, p. 515; Mexico.

Range and Status. Breeds North America, winters Central and South America; vagrant Europe, Africa.

Vagrant Mozambique (Beira, twice), Senegambia (Banjul) and South Africa (Port Elizabeth, Ungeni estuary, Durban, Langebaan). 1, paired with Grey-headed Gull *Larus cirrocephalus*, nested Saloum Delta, Senegambia, 1983 (Erard *et al.* 1984).

Description. ADULT ♂ (breeding): forehead and lores to nape, ear-coverts, chin and throat slaty black; thick white crescents or oval patches posteriorly above and below eye; hindneck, rump, uppertail-coverts, underparts and underwing-coverts white, underparts with pale pink flush; mantle, back and wings dark grey, broad white trailing edge to wings; tail grey centrally, white terminal fringe and sides. Ends of outer 5 or 6 primaries white with variable amount of black subterminally; basal portion of all remiges dark grey. Bill red, usually with dark subterminal band; eye brown, orbital ring red; legs red. ADULT ♂ (non-breeding): forehead, chin and throat whitish; eye-crescent, ear-coverts, hindcrown and nape blackish brown, forming half-hood; lacks pink flush. Bill blackish or brownish, tipped red; legs blackish or dull red. Sexes alike except ♀ somewhat smaller. SIZE: (14 ♂♂, 12 ♀♀) wing, ♂ 263–286 (277), ♀ 262–283 (273); tail, ♂ 98–111 (104), ♀ 97–107 (101); bill, ♂ 29–34 (30·7), ♀ 27–33 (29·2); tarsus ♂ 41–45 (42·5), ♀ 39–44 (41·4). WEIGHT: (Minnesota) ♂ (n = 29) 220–335 (281), ♀ (n = 11) 250–325 (279).

IMMATURE: hindneck blackish, mantle dark grey, latter sometimes finely streaked brown; tail white with black subterminal band, broadest in centre, usually not extending to outermost pair of rectrices. Outer primaries and their coverts mainly black, with grey on outer webs increasing inwards from P3–P4, and black decreasing to subterminal band on P6–P7; primaries usually with small white tips; secondaries grey-brown with blackish centres and white tips; prominent white trailing edge to inner wing, tertials and coverts brownish. Bill and legs blackish, sometimes shade lighter at base of bill. By 1st summer somewhat resembles breeding adult; hood fully or partially developed, most with white flecking on forehead and throat; black wing-tip and leading edge of outer wing, all remiges tipped white, black on primaries decreasing in extent inwards to subterminal mark on P7–P6; dark centres to some secondaries form indistinct bar; subterminal spots on rectrices form indistinct partial band centrally.

Field Characters. Smaller than Black-headed Gull *L. ridibundus* and Laughing Gull *L. atricilla*. Most likely to be confused with Laughing, but has short bill, rounded head, short neck, less pointed wings and buoyant flight. Adult easily distinguished by wing-tip pattern: white tips and black subterminal marks, separated from grey upperwing by white band. Immatures more difficult, but hood darker and more pronounced, breast whitish, upperwing greyer, outer tail feathers often white to tip (in Laughing they have dark subterminal band). Has been confused with White-eyed Gull *L. leucophthalmus*, but latter has dark upperwing with blackish tips, long decurved bill and yellow legs.

Voice. Tape-recorded (70, 73). Feeding note an incessant, shrill 'kuk', sometimes varied with plaintive 'weeh-a'; also a repeated 'po-lee' (resembling Eurasian Curlew *Numenius arquata*).

General Habits. Frequents mostly coastal habitats (although nests exclusively in freshwater marshes). Forages widely, feeding inshore or inland on irrigated fields; often hovers or soars when searching for food. Sometimes flocks; leaving resting places to mount in circles to upper air space.

Present Dec–Apr; probably arrives via S Atlantic (specimen Tristan da Cunha) although Banjul (Senegambia) bird more likely via W Europe.

Food. Mostly invertebrates, particularly insects; also fish.

Larus minutus Pallas. Little Gull. Mouette pygmée.

Larus minutus Pallas, 1776. Reise versch. Prov. Russ. Reichs, pt. 3, p. 702; Siberia (Berezof).

Plates 23, 27
(Opp. pp. 368, 416)

Range and Status. Breeds Europe and Asia from central Sweden to L. Baikal; also North America; winters British Isles, Mediterranean, Black and Caspian Seas; vagrant Africa south of Sahara.

Palearctic winter visitor, locally common N African coast. Annual Mediterranean, Egypt to Morocco, varying greatly from year to year. Algeria, hundreds feeding at sea off Oran, only 40 (45% adult) in shore-based counts whole coastline, several inland at L. Boughzoul; 900 (3 flocks) Tunisia Jan 1975, few other years. Erratic Atlantic coast Morocco; very small numbers elsewhere Atlantic. Vagrant Senegambia, Sierra Leone, Nigeria (3–5, Lagos, 1969–70), Angola and Kenya (60, of which half adults, L. Turkana, Jan 1979).

Description. ADULT ♂ (breeding): forehead and lores to lower nape, ear-coverts, chin and throat black; underparts and hindneck white flushed pink; rump, uppertail-coverts and tail white; mantle, back and wings pale grey, prominent white trailing edge and tip to wings above and below; much of underwing blackish, forepart greyer, axillaries pale grey. Bill dark reddish brown; eye and orbital ring blackish brown; legs scarlet. ADULT ♂ (non-breeding): head white with blackish eye-crescent and ear-spot, grey crown and nape; body lacks pink flush; bill blackish, legs dull red. Sexes alike. SIZE: (16 ♂♂, 9 ♀♀) wing, ♂ 210–230 (221), ♀ 212–227 (221); tail, ♂ 85–95 (91·9), ♀ 89–97 (91·8); bill, ♂ 21–25 (23·2), ♀ 22–24 (23·3); tarsus, ♂ 25–28 (26·5), ♀ 25–29 (26·5). WEIGHT: (Europe) ♂ (n = 8) 82–117 (99), ♀ (n = 4) 93–108 (98).

IMMATURE: crown blackish; nape and hindneck uniform with pale grey mantle, back and scapulars, grey extending to sides of breast. Underwing whitish except for blackish tip and leading edge. Tail white, black subterminal band; white terminal fringe narrow or absent; outermost 1 or 2 pairs rectrices sometimes wholly white. Outermost 6–8 primaries black on outer web and tip, decreasing in extent inwards; inner webs all primaries mainly white; secondaries and outer webs inner primaries grey, broadly tipped white, secondaries with blackish centres forming subterminal bar. Distinctive W pattern of blackish brown on wings in flight formed from outer coverts of outer wing and extensive carpal-bar (to inner greater coverts and tertials, often almost continuous across back). Bill, eye and orbital ring blackish; legs pale flesh or reddish. By 1st summer, partial or full hood of grey, brown or black sometimes acquired, grey on hindneck lost; tail-band and carpal-bar often faded to pale brown, band often broken in centre, tail sometimes wholly white. Hood usually fully developed or white-flecked by 2nd summer; axillaries and some underwing-coverts still whitish.

JUVENILE: much as immature; distinctive pattern of blackish brown on crown, mantle, back and scapulars, whitish fringes giving scaly pattern (especially on scapulars); white collar across nape.

Field Characters. Smallest gull, wing-span 20–30% less than Black-headed *L. ridibundus*. Contrasting blackish underwing and pale grey upperwing distinctive in adults and subadults; black hood in summer. In immature combination of small size and striking dark inverted W pattern across wings distinctive (but see Black-legged Kittiwake *Rissa tridactyla*). Dainty, tern-like in flight.

Voice. Tape-recorded (62, 73). Non-breeders mostly silent. Commonest note 'kek-kek-kek', lower pitched than Black-headed; also gives 'kuk-kuk-kuk' and 'truk' when alarmed.

General Habits. Frequents mainly coastal habitats; prefers sandy or muddy beaches, especially where freshwater streams or sewage outlets reach the sea; also found offshore but rarely inland. Often gregarious, in flocks of 10–20; sometimes hundreds or thousands congregate briefly at favourable sites or during bad weather. Often accompanies terns in small foraging flocks; feeds much like a tern, moving slowly forward with bill pointed downwards and periodically dipping-to-surface of land or water to snatch up prey. Also lands on water to pick up prey; sometimes submerges head.

Present Mediterranean Dec–Mar; marked Apr–May passage in Algeria and Tunisia; immatures sometimes oversummer.

Food. Chiefly small fish, marine invertebrates and insects.

Reference
Cramp, S. and Simmons, K. E. L. (1983).

Plate 21

PLOVERS

- Kittlitz's (p. 229)
- White-fronted (p. 237)
- Little Ringed (p. 226)
- Caspian (p. 245)
- Buff-breasted Sandpiper (p. 297)
- Ruff (p. 297) ♂ ♀ (p. 297)
- Kentish (p. 235)
- Three-banded (p. 231)
- Eurasian Dotterel (p. 246)
- Ringed (p. 228)
- Chestnut-banded (p. 239)
- Forbes' (p. 234)
- Lesser Yellowlegs (p. 322)
- Solitary Sandpiper (p. 322)
- Green Sandpiper (p. 322)
- Marsh Sandpiper (p. 317)
- Spotted Redshank (p. 314)
- Common Greenshank (p. 318)
- Common Redshank (p. 315)
- Wood Sandpiper (p. 323)
- Common Sandpiper (p. 326)
- Curlew Sandpiper (p. 292)
- Sanderling (p. 286)
- Red-necked Phalarope (p. 329)
- Grey Phalarope (p. 331)
- br. (p. 327) non-br. (p. 327) Ruddy Turnstone
- Broad-billed Sandpiper (p. 296)
- Dunlin (p. 294)
- Pectoral Sandpiper (p. 292)
- Terek Sandpiper (p. 324)
- Baird's Sandpiper (p. 291)
- Little Stint (p. 288)
- Temminck's Stint (p. 289)
- African Snipe (p. 302)
- ♂ (p. 186)
- ♀ (p. 186)
- Greater Painted-Snipe
- Purple Sandpiper (p. 294)
- Eurasian Woodcock (p. 308)
- Red Knot (p. 284)

12in / 30cm

MWW 1981

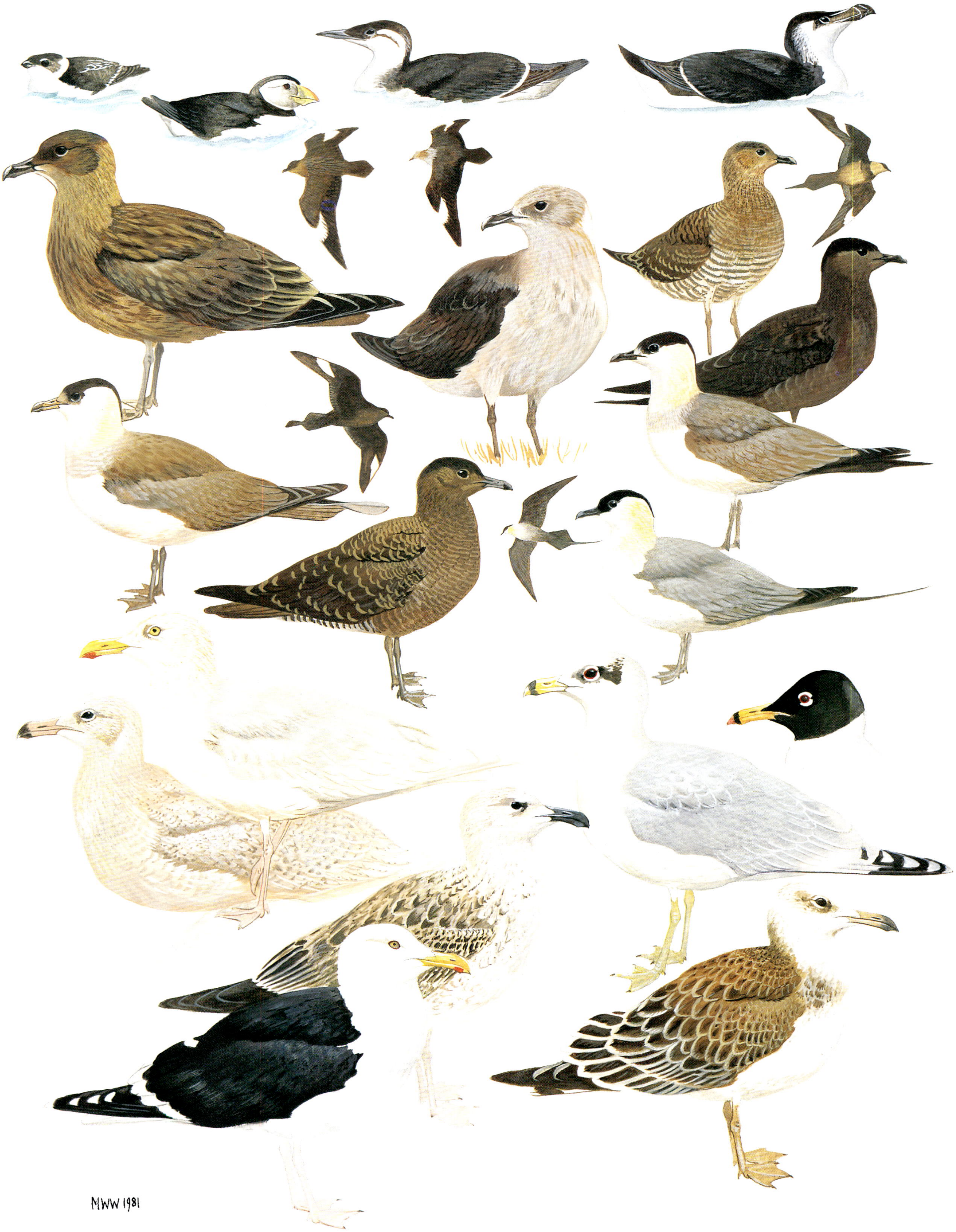
MWW 1981

Plate 22

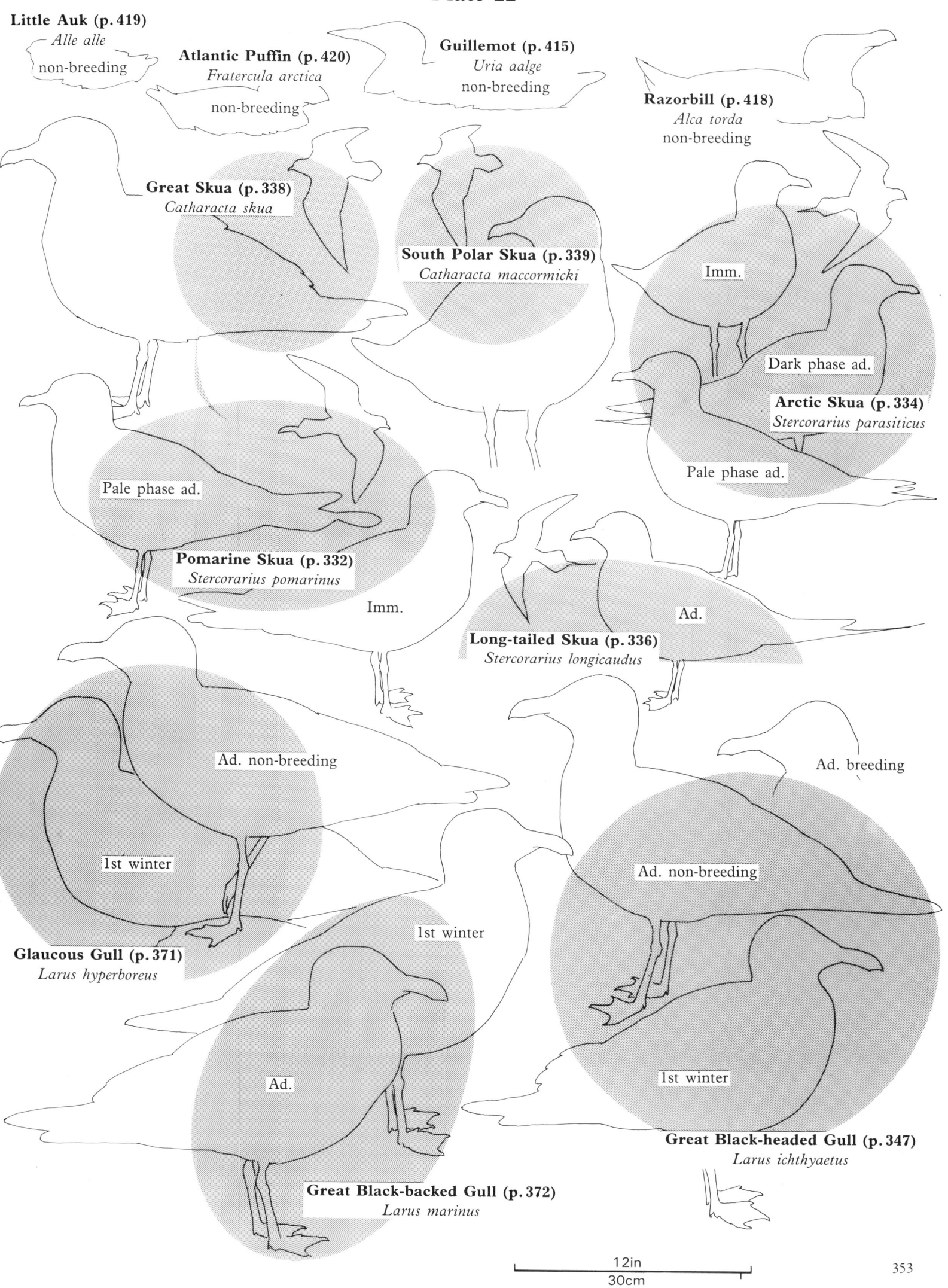

Little Auk (p. 419)
Alle alle
non-breeding

Atlantic Puffin (p. 420)
Fratercula arctica
non-breeding

Guillemot (p. 415)
Uria aalge
non-breeding

Razorbill (p. 418)
Alca torda
non-breeding

Great Skua (p. 338)
Catharacta skua

South Polar Skua (p. 339)
Catharacta maccormicki

Imm.

Dark phase ad.

Arctic Skua (p. 334)
Stercorarius parasiticus

Pale phase ad.

Pale phase ad.

Pomarine Skua (p. 332)
Stercorarius pomarinus

Imm.

Ad.

Long-tailed Skua (p. 336)
Stercorarius longicaudus

Ad. non-breeding

Ad. breeding

1st winter

Ad. non-breeding

1st winter

Glaucous Gull (p. 371)
Larus hyperboreus

Ad.

1st winter

Great Black-backed Gull (p. 372)
Larus marinus

Great Black-headed Gull (p. 347)
Larus ichthyaetus

12in / 30cm

353

354 LARIDAE

Plates 23, 27
(Opp. pp. 368, 416)

Larus sabini Sabine. Sabine's Gull. Mouette de Sabine.

Larus Sabini Sabine, 1819. Trans. Linn. Soc. London, 12, p. 522; Greenland.

Range and Status. Breeds Alaska, N Canada, Greenland to Spitsbergen, Taimyr Peninsula and New Siberian Islands. Winters Pacific and Atlantic south to southern Africa.

Nearctic winter visitor, abundant to uncommon, mainly cool Benguela current off South Africa and Namibia, 18–37°S. Passage Morocco to Liberia, S Angola; scarce Agulhas Current (SE Africa) north to 29°S and Gulf of Guinea east to Nigeria; vagrant Congo, Cameroon, Mozambique, Somalia.

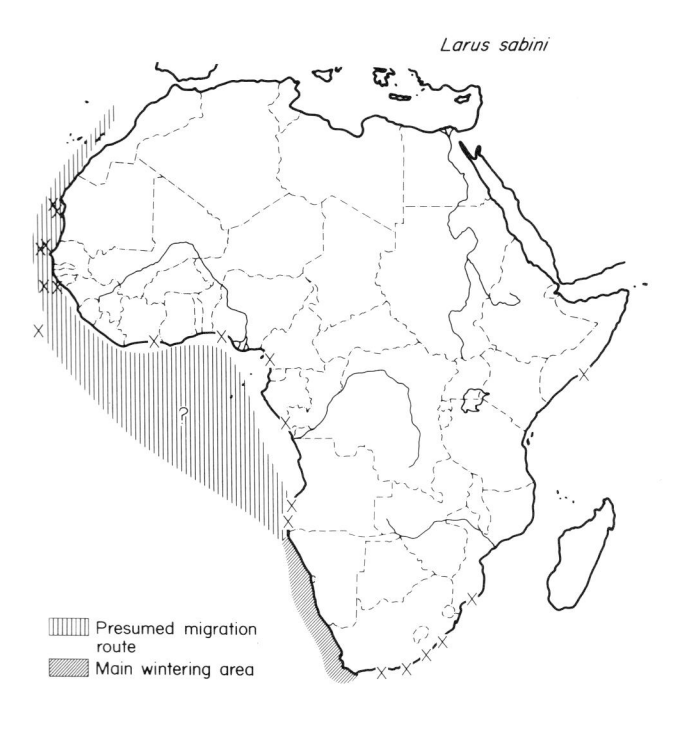

Description. ADULT ♂ (breeding): forehead and lores to nape, ear-coverts, chin and throat dark grey; hindneck, rump, uppertail-coverts, tail and underparts white; underwing mostly white, wing-tip black. Mantle, back, inner wing-coverts and innermost secondaries grey; wing-tip and leading edge of outer wing mostly black, rest of wing white; extensive white on inner webs and tips of otherwise black outer 5 primaries, visible as lines when wing fully spread. Bill black, tipped bright yellow; eye blackish brown, orbital ring red; legs blackish or dark grey. ADULT ♂ (non-breeding): head white with blackish eye-crescent and variable pattern of blackish grey, usually confined to nape and hindneck, sometimes extending to ear-coverts, crown and sides of neck; outer primaries often brownish and lacking white tips. Sexes alike except ♀ somewhat smaller. SIZE: (15 ♂♂, 10 ♀♀) wing, ♂ 256–284 (267), ♀ 245–267 (260); tail ♂ 113–126 (120), ♀ 108–131 (118); bill, ♂ 22–28 (25·5), ♀ 23–28 (25·2); tarsus ♂ 31–38 (33·7), ♀ 31–35 (33·7). WEIGHT: (South Africa) ♂ 164, 174, ♀ 168, 170.

IMMATURE: some inner wing-coverts grey-brown with blackish subterminal crescents and whitish or brownish fringes, giving scaly appearance; black terminal band on tail broadest in centre, accentuating marked fork; black of wings and tail worn and faded; bill black, legs dull flesh or blackish.

By 1st summer resembles adult; hood intermediate between adult breeding and non-breeding plumage; occasionally some subterminal dark marks on tertials and tail.

JUVENILE: wings and tail much as young immature, but fresh and not faded; sometimes has dusky bar across underwing-coverts; forehead, narrow eye-ring, lores, chin and throat white; eye-crescent blackish, remainder of head grey-brown with whitish fringes to feathers; scaly, grey-brown innerwing-coverts extend across mantle and back, most prominent on scapulars. Bill black; eye brown, orbital ring blackish; legs pinkish flesh or greyish flesh.

Field Characters. Small, with shallowly forked tail, short dark bill tipped yellow in adult. Breeding adult has diagnostic combination of dark grey hood and black legs. In flight distinguished at all ages by contrasting tricoloured wing pattern, with large white triangle almost separating black wing-tip from inner wing and mantle (grey in adult, brown in immature). Mainly pelagic.

Voice. Tape-recorded (73). Gives 3 calls: (a) very high, plaintive, twittering 'vihihihi'; (b) very high 'hrier-hrier', hirikri' or 'grih'; (c) very short, clicking or snapping 'tsett'. Different calls not obviously associated with different behaviour or activity.

General Habits. Usually found within 40 km of shore in shelf-edge zone, although occurs up to 150 km offshore. Occurs singly or in flocks up to 50, sometimes in flocks up to 2000, especially around offshore fishing fleets. Graceful and agile, often hovering; takes most food from surface in flight; escapes attacking skuas by keeping close to surface although 25% attacks by Arctic Skua *Stercorarius parasiticus* successful. Associates with Grey Phalaropes *Phalaropus fulicarius*. When migrating, flies low over water, settling frequently. Roosts in tight flocks on water at night, bill often tucked in feathers of back; not attracted to trawler searchlights.

Follows Canary/Senegal upwellings with bulk of migrants passing rapidly through tropics to winter in cool Benguela Current off Namibia and South Africa. In W Africa, southward passage late Aug–Nov; northward passage mid Mar–June with most 1st half May; rare Dec–mid Apr, some immatures in June. Present southern Africa Sept–May, most numerous Jan–Feb.

Food. Fish offal, small fish and cyprid larvae.

References
Cramp, S. and Simmons, K. E. L. (1983).
Lambert, K. (1975).
Zoutendyk, P. (1968).

Larus hartlaubii Bruch. Hartlaub's Gull. Mouette de Hartlaub.

Larus Hartlaubii Bruch, 1853. J. Orn., 1, p. 102; Cape of Good Hope and the Indian coasts.

Forms a superspecies with *L. cirrocephalus* and *L. ridibundus*.

Plates 24, 27
(Opp. pp. 369, 416)

Range and Status. Endemic resident on coasts of southern Africa, very abundant to uncommon. Breeds Namibia and W Cape, from Swakopmund to Dyer Island. Limited dispersal, rare east of Mossel Bay (Cape); vagrants east to Natal, Atlantic north to *c*. 22°S. Numbers increasing at traditional (Robben Island, 400 pairs 1953, 4309 pairs 1979) and man-made sites.

Description. ADULT ♂ (breeding): head, underparts, rump, uppertail-coverts and tail white, some birds with lavender-grey ring from upper nape down sides of neck and across throat; mantle, back and much of wing pale grey. Outermost 3 primaries (P10–P8) mostly black, mirrors (outermost 2) and tips white; primaries P9–P5 with basal white (mostly outer webs), increasing in extent inwards, white tips prominent when fresh, otherwise mostly black; inner primaries grey. Underwing appears dusky, white translucent spots towards darker wing-tip. Bill, legs and orbital ring red to maroon, eye brown. ADULT ♂ (non-breeding): head wholly white; bill and legs blackish. Sexes alike except ♀ somewhat smaller. SIZE: (5 ♂♂, 8 ♀♀) wing, ♂ 276–281 (280), ♀ 260–275 (266); tail, ♂ 103–110 (107), ♀ 101–110 (105); bill, ♂ 31–37 (33·8), ♀ 30–35 (32·5); tarsus, ♂ 44–46 (45·0), ♀ 41–45 (43·2). WEIGHT: ♂♀ (n = 18, South Africa) 235–340 (292).

IMMATURE: inconspicuous dusky eye-crescent and light brown mottling on mantle and back; brown and blackish spots on inner webs form indistinct, broken subterminal band on tail; outer rectrices often wholly white. White in wing confined to outer webs of middle primaries, ill-defined apical spots on inner primaries, variable subterminal patches on outermost 2. Grey-brown tertials and subterminal spots on secondaries form prominent band on inner wing, band of light brown from carpal joint across upper wing formed by tips of coverts. Bill, eye and orbital ring dark brown or blackish, eye brown. At age 1 year resembles non-breeding adult, bare parts darker.

JUVENILE: much as immature; head and upperparts mostly brown and buff; recently fledged birds in colony vary (Sinclair 1977), perhaps different ages.

DOWNY YOUNG: buff mottled black; darker shade of buff and more profusely marked above, fewest markings on throat and belly; bill flesh, legs yellowish flesh.

We follow Johnstone (1982) and give specific status (*L. hartlaubii*) rather than subspecific status (*L. novaehollandiae hartlaubii*) to African populations of this gull.

Field Characters. Only similar gull in its restricted range is Grey-headed *L. cirrocephalus*, from which it must be distinguished with great care. Somewhat smaller; bill slightly finer; white-headed adults no problem, but in breeding season some birds have pale grey dark-bordered hood, closely resembling Grey-headed. Wing-tip pattern very similar, but eye brown (whitish in Grey-headed). Immatures very similar but Hartlaub's lacks dark ear spot, has darker bill.

Voice. Tape-recorded (C, 262, 305). Harsh, loud 'kwarrk' or 'kwarrr'. Alarm call, staccato 'kekekek' or 'kakakakaka'.

General Habits. Inhabits coastal areas, within 3 km of land. Feeding methods similar to Black-headed, foraging on foot, paddle-feeding, plunge-diving, and feeding in air. Hundreds feed together on shallow inshore waters, beaches, inter-tidal zone, especially when concentrations of kelp fly larvae and amphipods present. Adapting increasingly to man-modified terrestrial habitats, particularly at Cape Town. Scavenges rubbish tips, abattoirs, sewage works, harbours; pulls earthworms from ground water-logged by sprinklers; aerial-feeds at night on insects attracted to street lights, mostly crickets, some moths. Kleptoparasitic chases by Arctic Skua *Stercorarius parasiticus* 25% successful.

Mostly sedentary; birds marked Rondevlei recovered up to 120 km north.

Food. Mostly marine invertebrates and fish including Lantern Fish *Lampanyctodes* sp. and Pelagic Goby *Sufflogobius bibarbatus* (Walter 1984); also insects and their larvae, worms, and offal.

Breeding Habits. Displays detailed under family description (p. 341). Colonial, mostly 10–1000 pairs, often on islands with Swift Tern *Sterna bergii* and other colonial species; also nests on man-made islands (sewage, salt works) and roofs of buildings. Nearest-neighbour distance usually 1–2 m, density about 0·5 (0·13–1·0) nests/m², small territory defended, mostly by ♂ (area defended around chicks if moved). Interbreeds with Grey-headed, sometimes producing young (Sinclair 1977).

NEST: slight hollow, lined with plant material, seaweed, shells, feathers; some scant, others bulky and solid. Usually on ground, but 20–50 cm up on densely matted sclerophyllous scrub, also on buildings. Ground sometimes bare or with only sparse cover of beach halophytes; preferred sites usually associated with substantial vegetation.

EGGS: 1–5, usually 1–3, av. (1962 clutches) 1·80. Seasonal decrease in clutch-size SW Cape attributed to diverse age composition, some breeding birds less than 6 years old. Subelliptical, cream to olive or pale green, fairly evenly marked with spots, blotches and streaks of brown and black, undermarkings grey. SIZE: (n = 29) 48·9–56·9 × 36·1–40·3 (52·3 × 37·9). WEIGHT (estimated): 40.

LAYING DATES: Namibia and South Africa, Feb–Sept, most Apr–June.

INCUBATION: carried out by both sexes. Period: c. 25 days.

DEVELOPMENT AND CARE OF YOUNG: within 12 h of hatching 24 young weighed 22–28; by day 4, 41–44; by day 5, 52–66; by day 6, 94. Feather growth begins with scapulars, initially mottled brown and cryptic. Fledging period c. 40 days. Post-juvenile plumage of hybrids unknown.

BREEDING SUCCESS/SURVIVAL: 23 out of 36 chicks (28 broods) were killed within few days of hatching, 17 on single day, mostly by Sacred Ibises *Threskiornis aethiopica* and Kelp Gulls *L. dominicanus*. Mole Snake *Pseudaspis cana*, 60 cm long, jabbed at by adult when 30 cm from nest, ate recently hatched chick in nest.

References
Sinclair, J. C. (1977).
Tinbergen, N. and Broekhuysen, G. J. (1954).

Plates 24, 27
(Opp. pp. 369, 416)

Larus cirrocephalus Vieillot. Grey-headed Gull; Grey-hooded Gull. Mouette à tête grise.

Larus cirrocephalus Vieillot, 1818. Nouv. Dict. d'Hist. Nat., 21, p. 502; Brazil.

Forms a superspecies with *L. hartlaubii* and *L. ridubundus*

Range and Status. Breeds Africa and South America; some non-breeding birds Madagascar.

Resident, very abundant to uncommon. Coastal W Africa from Mauritania to Togo, inland along major rivers from Mali to L. Chad. More widespread from Nile valley in Sudan, and Ethiopia to Cape, breeding mostly inland. 2500 or more pairs on single islands in Kenya, Uganda and Senegambia. (Mixed pair of Grey-headed Gull × Franklin's Gull *L. pipixcan* reported Senegambia, Saloum delta, 1983; also Grey-headed × *Larus* sp. (possibly Laughing Gull *L. atricilla*); Erard *et al*. 1984.)

Description. *L. c. poiocephalus* Swainson: only subspecies in Africa. ADULT ♂ (breeding): forehead and lores to nape, ear-coverts, chin and throat dove-grey, shading to whitish on forehead and chin; mantle, back and inner wing pale grey; lower nape, hindneck, rump, uppertail-coverts, tail and underparts white, sometimes with faint pink flush. Black wing-tip with prominent white mirrors (both webs) outermost 2 primaries (P10–P9) and indistinct white tips (P8–P7 to P4–P3); white bases to outer primaries (mostly outer webs) form large white wedge; outer primaries mostly blackish, grey increasing in extent inwards, secondaries pale grey. Underwing dusky, few translucent spots towards tip when spread. Bill red; eye pale yellow or whitish, orbital ring red; legs red. ADULT ♂ (non-breeding): much as breeding but hood paler, less defined, sometimes with darker ear-spot; lacks pink flush, bill with subterminal dark tip. Sexes alike except ♀ somewhat smaller. SIZE: (10 ♂♂, 8 ♀♀) wing, ♂ 302–325 (317), ♀ 285–321 (299); tail, ♂ 117–132 (125), ♀ 114–123 (118); bill, ♂ 39–41 (40·4), ♀ 35–37 (36·1); tarsus, ♂ 50–54 (52·1), ♀ 45–52 (47·5). WEIGHT: ♂♀ (n = 49, South Africa) 211–377 (278); (Botswana) ♂ 347, 352, ♀ 260, 291, 321; (E Africa) ♂ 325, 384, ♀ 280, 310, 320.

IMMATURE: similar to non-breeding adult; mantle sometimes with few brown feathers, tail with blackish subterminal band; head mostly white, indistinct dusky eye-crescent and ear-spot; hindneck greyish, paler than mantle. Outermost 2 or 3 primaries wholly black, white on inner webs at base of P8–P7 to P6–P5 forms patch in middle of outer wing; innermost 4 or 5 primaries mostly grey, blackish tips join with blackish secondaries to form dark trailing edge; most primaries with tiny white spots at tips, increasing inwards; pale grey mid-panel to wing separates brown carpal-bar from trailing edge. Bill and legs flesh, extensive dark tip to bill, eye brown. Little change at age 1 year; grey of head usually more extensive, some with adult hood and white hindneck.

JUVENILE: wings, uppertail-coverts, tail and underparts much as immature, brown and blackish areas darker and more defined; ear-spot and eye-crescent darker, partial hood of grey-brown on ear-coverts, crown, upper nape; hindneck and thin crescents above and below eye whitish; feathers of mantle, back and scapulars brown fringed paler, scaly effect most pronounced on lower back; grey-brown sides of breast, pale grey rump.

DOWNY YOUNG: buff, becoming greyer, with rather ill-defined brown markings; legs flesh, bill dusky.

Field Characters. Winter adult differs from winter Black-headed Gull *L. ridibundus* by pale eye, indistinct pale grey hood, darker upperwing with more extensive blackish tips, small white mirrors, darker underwing without pale leading edge on outer half; carriage more upright, legs longer, wings broader and less angled, bill longer and heavier. Immature resembles Black-headed, with dark ear-spot and dark-tipped pinkish or yellowish bill, but has solid dark wing-tips without white panel on outer primaries, narrower tail-band. Profile, with sloping forehead and longish bill, similar to Slender-billed Gull *L. genei*, but larger, with different wing pattern and lacks breast bulge; bill and legs without orange tone. For differences from Hartlaub's Gull *L. hartlaubii*, see that species.

Voice. Tape-recorded (35, 36, 38, 74, C, F). Noisy, similar to Black-headed, but with slight twanging quality. Some notes of Long Call short and simple, others longer, slightly quavering. Repeated, staccato 'karr' during perch disputes. ♀ gives 'gwow' during courtship feeding.

General Habits. Inhabits coasts and wetlands, especially estuaries, harbours, large freshwater and alkaline lakes. Gregarious; mostly feeds on ground and in air, often dips-to-surface. Fishes inshore sardine shoals in South Africa; feeds on cichlid fish, insect larvae and molluscs in drying pools in Zaïre. Regularly harasses surfacing Long-tailed Cormorants *Phalacrocorax africanus* Zaïre; once robbed a Lesser Crested Tern *Sterna bengalensis* Mozambique. Scavenges in harbours, shoreline, rubbish tips and places with fish scraps. Voracious predator at water-bird colonies; one in full flight swallowed a week-old Long-tailed Cormorant chick.

Recoveries of birds ringed in Transvaal to southeast (coastal Natal, S Mozambique) and southwest, mean distance 330 km; also Zimbabwe, W Zambia, Botswana, coastal Namibia and S Angola (2000 km). Numbers varied 14–874 during year St Lucia (Natal), being lowest during summer. Tropical breeders wander; has reached S Spain; 8000 Lake Nakuru (Kenya) Feb, unknown origin.

Food. Insects, molluscs, fish, and birds' eggs and young.

Breeding Habits. Colonial nester on islands or floating vegetation. Displays hardly recorded for wild birds, most observations on breeding birds in captivity, London (A. S. Richford *in* Cramp and Simmons 1983). Cries of alarm and feigned attacks considered prelude to breeding; in 'panic' flights of pre-breeding flocks, whole flock takes flight, calling, returns to original site after short circular flight. Unritualized supplanting attacks and fights in defence of nest-site territory fairly frequent. When rushing at opponent in Aggressive Upright, wings commonly opened almost fully and body tilted backwards. Long Call given from Oblique posture in which body slightly inclined and carpal joints lowered away from body; neck arched, bill pointed slightly downwards. Immediately before attack, bill may be slightly raised, wings extended and raised. Forward posture often follows Oblique posture: body tilted forward up to 15° below horizontal, neck bent backwards and upwards, bill level or pointing slightly upwards; feathers of back may be ruffled. Choking usually performed near nest; body inclined at 45° or more to ground, neck vertical, bill pointed downwards, wings sometimes raised and half extended. Pair may Choke together at nest, side by side, particularly in response to intruding bird which may also Choke before moving away. Fighting birds jump at each other, trying to get on top and pecking downwards; wings not used for striking, though flapping movements sometimes make contact; one may grab other by wing or base of bill and both birds grapple for a time. In courtship feeding, ♀ approaches ♂ in Hunched posture, giving rather strangled, clipped 'gwow' call; back feathers may be ruffled and Head Tossing may accompany call; ♂ regurgitates onto ground, sometimes with brief head toss. Courtship feeding with ♀ touching crop of ♂ forms immediate preliminary to copulation.

NEST: varies from shallow scrape to well-constructed cup. In 1 Mauritanian colony, nests varied from substantial structure of grass, twigs and seaweed to unlined scrape in sand. Often elaborate and camouflaged in tall, thick vegetation; in short grass in Natal (St. Lucia). Floating nests are always substantial, made from lily leaves, sedges; often built on closely growing group of water lilies.

EGGS: 1–4, usually 2–3; E Africa mean (312 clutches) 2·4. Subelliptical, cream to olive-brown, blue-green or rich brown; darker, evenly distributed spots, lines and blotches, mostly dark brown or black; purple undermarkings. SIZE: (n = 100, Uganda) 49·2–60·6 × 32·3–41·4 (53·9 × 37·3); (n = 100, Senegambia) mean 49·7 × 35·3. WEIGHT(estimated): 41.

LAYING DATES: Mauritania, May–July; Senegambia, May–Aug; Guinea; Apr–Sept; Ethiopia, June–Aug; Kenya, Mar–Sept; Uganda, May–Aug; Tanzania, Aug–Sept; Zambia, May–July; South Africa, Feb–Nov. Onset varies from year to year (Guinea; St Lucia), seldom well synchronized.

INCUBATION: carried out by both sexes, off-duty bird often stands 1 or 2 m away. Pieces of withered lily leaves brought repeatedly from shore, deposited at edge of floating nest, carefully arranged and banked up by incubating bird (often standing) until only head and wing-tips visible above edge (Kenya).

Plates 23, 27
(Opp. pp. 368, 416)

Larus ridibundus Linnaeus. Black-headed Gull; Common Black-headed Gull. Mouette rieuse.

Larus ridibundus Linnaeus, 1766. Syst. Nat. (12th ed.), p. 225; England.

Forms a superspecies with *L. cirrocephalus* and *L. hartlaubii*.

Range and Status. Breeds Eurasia from Iceland to northern Mongolia; winters south to Azores, northern Africa, Persian Gulf, India, Malaysia, China, Japan and Philippines.

Palearctic winter visitor, locally abundant on coast, uncommon inland except Mali and E Africa. Widespread and numerous Mediterranean, Red Sea, Atlantic coast south to Mauritania: Jan–Feb counts Tunisia 45,000, Algeria 13,800, Egypt 72,700, Mauritania 275 (Banc d'Arguin); rather uncommon to south. Increasing but still rather erratic Senegambia to Nigeria (as many as 45, Kazaure, Dec 1977) and Mali, NE Africa south to Tanzania and Burundi; up to 3000 Sudan (Jebal Aulia dam), flocks of hundreds (few adults) annual E Africa since 1971, mostly coast and Rift Valley lakes, especially where fish offal plentiful. Vagrant Chad, Gabon, Mozambique and Zimbabwe. (Normally does not nest in Africa, although 1 pair with a nest reported from Bird Island, Saloum N. P., Senegambia, May 1982: Depuy 1983.)

Description. ADULT ♂ (breeding): forehead and lores to nape, ear-coverts, chin and throat chocolate-brown, thin white eye-ring; hindneck, rump, uppertail-coverts and tail white; mantle, back, rump and most of wing pale grey; underparts white, sometimes with pale pink flush; underwing-coverts white or very pale grey. Outer 6–8 primaries tipped black; white on outer primaries and their coverts appears as pronounced wedge from wing-tip to carpus, above and below; other primaries appear mainly blackish from below. Bill and eyelids dark red; eye brown; legs dark red. ADULT ♂ (non-breeding): like breeding ♂ but head mostly white, dusky eye-crescent, prominent blackish ear-spot (often more extensive); pale grey tips to middle primaries, dark tip to bill, lacks pink flush. Sexes alike except ♀ somewhat smaller. SIZE: (7 ♂♂, 5 ♀♀) wing, ♂ 284–315 (300), ♀ 280–297 (288); tail, ♂ 113–124 (118), ♀ 104–117 (109); bill, ♂ 31–37 (33·6), ♀ 30–33 (31·5); tarsus, ♂ 43–47 (44·7), ♀ 42–47 (44·2). WEIGHT: (Eurasia) ♂ (n = 45) 235–400 (301), ♀ (n = 21) 190–314 (254).

IMMATURE: body much as non-breeding adult, tail with black subterminal band, usually some brown marks on mantle. Secondaries mainly blackish, forming dark subterminal trailing edge; pale grey mid-wing panel, contrasts with brown carpal-bar; black on outer primaries, mostly outer webs; white wedge less pronounced than adult. From below, remiges appear mainly blackish, narrow white translucent leading edge to outer wing. Bill dark flesh or yellowish flesh with extensive dark tip; eye dark brown; legs dull yellowish flesh. Little different in 1st summer; most have white-flecked full hood, wings and tail worn and faded, bill and legs more orange. Striking ginger-brown juvenile plumage partly or completely lost by the time Africa reached.

Field Characters. Hood of breeding adult chocolate-brown, not black (though may appear black at a distance), and reaching only to nape, not to hindneck as in Mediterranean Gull *L. melanocephalus* and Little Gull *L. minutus*. Immature similar to immature Slender-billed Gull *L. genei* but has rounded head, shorter neck and bill, and dark-tipped pinkish or yellowish bill without orange tone. For distinctions from Grey-headed Gull *L. cirrocephalus* see under that species.

Voice. Tape-recorded (62, 73, C). Various harsh notes including 'krriaeh' (in flight), 'kwarr', 'kwup'.

only 0·07 young reared per nest Saloum (Senegambia) 1977.

Reference
Cramp, S. and Simmons, K. E. L. (1983).

DEVELOPMENT AND CARE OF YOUNG: young run freely within day or so of hatching. At 10 days, will run rather than hide, wings outspread; plunge into water, swim well.

BREEDING SUCCESS/SURVIVAL: temporarily abandoned eggs frequently eaten, and exposed young birds killed by neighbours. Human predation (eggs) serious locally;

General Habits. Frequents inshore tidal waters, especially inlets and estuaries with extensive sandy or muddy beaches; also found inland usually along margins of ponds, lakes and slow-moving rivers, but also grasslands, ploughed land, refuse dumps and sewage farms. In Algeria, of 6780 birds counted, 22% found at ports, 21% rubbish dumps, 20% coastal plains and wetlands, 15% cultivated sites, 11% sewage outflows, 7% beaches, and 4% urban areas, with some penetrating Sahara; of these, immatures rather than adults tended to travel inland with proportions of immatures at rubbish dumps increasing as day progressed. Gregarious throughout year, with pairs, siblings and perhaps groups from 1 colony sometimes remaining together on migration and in winter quarters. Defends feeding territories in winter quarters. Roosts communally at night, on raised vantage point or in open area. When foraging: (1) follows plough, by walking, then making short catch-up flights; seizes prey while walking or makes fluttering leaps; (2) flies with steady wing-beats, dipping down to pick up food; (3) hovers over hedges and treetops, then feeds on fruit; (4) circles and rises rapidly to catch insects; (5) dips-to-surface; (6) surface-plunges (submerges, often completely); (7) surface-feeds; (8) paddle-feeds; (9) runs or swims with much of head submerged; (10) practices intra- and inter-specific food piracy; and (11) scavenges, often following ships.

Arrives (mostly adults) Tunisia from mid-July, Atlantic and Red Sea from Sept, tropics mostly Nov–Apr. Not known if reaches tropics via Sahara. Recoveries Mauritania and Senegambia from Great Britain, Norway, France and W Germany; Morocco from Europe (north and east to Finland and Latvia); Algeria mostly from France and Switzerland; Tunisia from W and central Europe (especially Czechoslovakia); Libya from Denmark; and Egypt from Yugoslavia and USSR. Birds in E Africa probably from eastern USSR. Oversummers most areas, some in adult-like plumages.

Food. Largely animal material including fish, crustaceans, molluscs, insects, earthworms; also refuse and vegetable matter.

Reference
Cramp, S. and Simmons, K. E. L. (1983).

Larus genei Brème. **Slender-billed Gull. Goéland railleur.**

Larus genei Brème, 1840. Rev. Zool. for 1839, p. 321; Sardinia.

Plates 23, 27
(Opp. pp. 368, 416)

Range and Status. Breeds W and N Africa, Spain, Black and Caspian Seas, Sea of Azov, Persian Gulf to Pakistan; ranges throughout Mediterranean and N Africa.

Resident and Palearctic winter visitor, very abundant to uncommon, mostly coastal. Breeds Senegambia and Mauritania (3000 pairs), Morocco (few), Tunisia (200 pairs), Egypt (200–400 pairs). Breeding population increasing parts W Africa but overall situation northern Africa precarious. Black Sea breeders winter Egypt to Tunisia; Tunisia flocks 2000 Aug, 15,000 Jan–Feb; Egypt 6000 on lakes Port Said Jan–Feb, 2000 July–Aug, 1240 Lake Qarun (200 km from coast) Jan; Libya flocks Oct–Apr. Vagrant winter Algeria, Morocco; also vagrant Nigeria, Ethiopia, Sudan, Kenya.

Description. ADULT ♂ (breeding): head, rump, uppertail-coverts and tail white; mantle, back and most of wing pale grey. Underparts pink. Outer 6–8 primaries with black tips; outer 5 and their coverts otherwise white, forming conspicuous wedge on leading edge of outer wing. From below, inner primaries appear blackish, contrasting with white wedge, rest of underwing whitish or pale grey. Bill dark red or orange-red, looking blackish at distance; eye white or pale yellow; legs paler than bill, tone more orange. ADULT ♂ (non-breeding): much as breeding ♂ but underparts white, pale grey ear-spot usually present. Sexes alike except ♀ somewhat smaller. SIZE: (16 ♂♂, 8 ♀♀) wing, ♂ 290–320 (304), ♀ 278–310 (293); tail, ♂ 112–125 (117), ♀ 110–122 (114); bill, ♂ 35–46 (42·2), ♀ 37–44 (40·1); tarsus, ♂ 47–54 (52·1), ♀ 46–54 (48·6). WEIGHT: (unsexed, n = 8, USSR, Arabia) 243–350 (299).

Also winters in coastal breeding range

IMMATURE: head, upperparts and underparts mostly as non-breeding adult; indistinct dark eye-crescent, pale grey ear-spot. Blackish secondaries and brown carpal-bar separated by pale grey mid-wing panel; blackish areas at tips of inner primaries indistinct; outer wing has adult pattern, but outer web of outer primaries often black, breaking up white wedge; from below, wedge contrasts with mostly blackish remiges. Bill pale orange-flesh, dark tip small or lacking; eye appears dark (pale at close range); legs pale orange-flesh. Little change 1st summer; worn and faded, ear-spot usually lacking, carpal-bar hardly darker than grey of upperparts, eye usually pale by late summer.

JUVENILE: wing and tail similar to young immature; pale buff and grey head markings form indistinct partial hood; mantle, back and sides of breast grey-brown.

DOWNY YOUNG: upperparts, chin and throat beige to grey-brown, typically streaked blackish, sometimes virtually plain, often barred blackish on back of head. Rest of underparts whitish, underwing streaked blackish; bill and legs brownish flesh.

Field Characters. Size, immature plumage, and wing pattern at all ages similar to Black-headed Gull *L. ridibundus*, but bill and legs usually with orange tone, bill seldom with dark tip, and adult eye pale. Ear-spot never distinct; breeders have head wholly white and underparts flushed pink (as late as Oct). Long neck, markedly elongated flat forehead, small head and long, thin bill give characteristic profile. When neck retracted (in flight and at rest) thick base gives pronounced bulge above breast. Swimming stance distinctive, neck tipped forward, bill below horizontal. Similar to Grey-headed Gull *L. cirrocephalus*, but Grey-headed lacks pronounced bulge above breast and bill and legs without orange tone; wing pattern very different.

Voice. Tape-recorded (62, 73). Repeated 'ka' or 'kra' of Long Call deeper and more mellow than Black-headed, nasal inflection recalls larger gulls.

General Habits. Frequents mostly coastal habitats, preferring lagoons and salt-extraction pans with high salt content. Gregarious when fishing, paddles rapidly with neck fully stretched, angled 45° to surface; fish caught with rapid lunge or plunge. Forages on foot in very shallow water, 'ploughing' sediment with bill; sometimes individuals or groups of up to 14, and often with Little Egrets *Egretta garzetta* wade with heads held high and necks outstretched, shepherding small fish into shallow water before plunging after them (Libya: N. E. Baker, pers. comm.). Also takes fish scraps near settlements and colonies of piscivorous birds. Rests on water.

Some dispersal of W African breeding birds but destination unknown. Winter quarters of increasing, oversummering population in Algeria also unknown. Flocks arrive Tunisia Aug, maximum numbers Jan–Feb; some also oversummer Egypt.

Food. In W Africa, young were fed mostly fish, some crustaceans. Winter, Tunisia, 1571 individuals ate fish, 135 *Artemia salina*, 14 *Gammarus*, 16 larval *Ephydra* and 6 debris floating in mouth of spate river.

Breeding Habits. Colonial, nesting on islands, coastal and inland seasonal salt lakes. In subcolonies, Senegambia, 105 nests in 8 groups, 5–20 nests in each group, group centres 10–50 m apart. Territoriality and ceremony minimal; gives ground to neighbouring Gull-billed Tern *Gelochelidon nilotica*. In courtship both sexes perform Upright posture with long neck especially emphasized and carpals held well away from body. Also gives Long Call like other masked gulls. Apparently does not perform Forward posture and Choking display. During pair formation several birds in Upright posture advance in tight group, sometimes giving Long Call and Facing Away; may follow this with aerial pursuits. May perform courtship feeding before copulating, with ♀ Begging with Head Tossing. All adults in colony rise when disturbed, soon return.

NEST: deep scrape in sand, pad of vegetable matter within and overlapping depression, often ringed with white excreta; nests 20–50 cm apart. Nests very conspicuous, apparently attractive to Grey-headed Gull and Senegal Thick-knee *Burhinus senegalensis* (both laid in occupied nests Senegambia).

EGG: 1–3, usually 2; Senegambia mean (105 clutches) 2·15. Subelliptical; cream, buff, white (washed blue) or pale grey, spots and blotches of pinkish brown, reddish, dark brown and blackish, undermarkings lavender. SIZE: (n = 100, Senegambia) mean 53·8 × 38·3. WEIGHT (estimated): 45.

LAYING DATES: Tunisia, Apr–June; Egypt, May–June; Mauritania, Apr–June; Senegambia, May–July.

INCUBATION: carried out by both sexes beginning with last egg. Period: 25–28 days.

DEVELOPMENT AND CARE OF YOUNG: recently hatched young carefully shaded or brooded; fed infrequently. After a few days, it leaves nest (and often nesting area) to seek regular hiding place, fed by parent frequently ($\frac{1}{2}$ h intervals). Solicits food by striking adult's lower mandible; food is regurgitated in thick saliva, often falls to ground, eaten from ground. Form waterside crèches of 20–200 young, accompanied by up to 10 adults; assemblages difficult to observe, tend to disperse if approached. Fledging period: 35–40 days.

BREEDING SUCCESS/SURVIVAL: predation by fishermen locally serious. In Egypt all eggs taken in 1979; greatly prized Saloum (Senegambia), only 0·09 young reared per nest 1977. Increasing Banc d'Arguin (Mauritania), 770–870 pairs 1964, 1733 pairs 1974. Overall situation precarious N Africa.

References

Cramp, S. and Simmons, K. E. L. (1983).
Dragesco, J. (1961).
Gowthorpe, P. (1979).
de Naurois, R. (1969).
Isenmann, P. (1976).

Larus audouinii Payraudeau. Audouin's Gull. Goéland de Audouin.

Plates 23, 27
(Opp. pp. 368, 416)

Larus audouinii Payraudeau, 1826. Ann. Sci. Nat. 8, p. 462; Sardinia and Corsica.

Forms a superspecies with *L. canus* and *L. delawarensis*.

Range and Status. Endemic to Mediterranean islands and coasts; winters mainly Mediterranean, some along Atlantic coast of Morocco.

Resident and partial migrant, locally abundant, but mainly uncommon; 2700 pairs (70–75% of known population) breeds Algeria and extreme E Morocco; in 1978, 500 pairs (8 sites) Algeria, with largest colony 96 pairs; increasing at Chafarinas Is. (Morocco) where 500 pairs in 1966, 2220 pairs in 1981; 12 colonies on 2 islands in 1982 varied 10–606 pairs (mean 142); possibly breeds Tunisia. Partial migrant, wintering largely NW Morocco to Libya: winter counts 108 Atlantic Morocco, 824 Algeria (mostly west), 141 Libya; 82–85% adults Algeria and Libya, 50% or fewer Atlantic; 1st winter birds especially rare Algeria. Rare N Tunisia, up to 36 extreme SE Tunisia. Up to 20 oversummering Atlantic include adults which might breed Mogodor Island (Morocco). Probably overlooked Atlantic south of 30°N, 4 records Senegambia Jan–May. Also vagrant Egypt (200 km inland), Mauritania.

Description. ADULT ♂ (breeding): head, underparts, uppertail-coverts and tail white; mantle, back, rump and much of wing pale pearl-grey; some pearl-grey on crown, nape, flanks, underwing and base of tail. Underparts white. Black confined to outer primaries, decreasing in extent inwards to P5, forming clear-cut wing-tip above and below; all remiges tipped white, small white mirror on inner web of outermost; secondaries and inner primaries markedly translucent. Bill red (often looking blackish) with single or double black subterminal band and yellowish tip; eye brown, orbital ring red; legs greyish or olive, soles yellow. Non-breeders hardly differ. Sexes alike, ♂ with larger bill. SIZE: (5 ♂♂, 3 ♀♀) wing, ♂ 370–400 (389), ♀ 382–402 (389); tail, ♂ 138–151 (147), ♀ 140–158 (148); bill, ♂ 43–53 (48.9), ♀ 43–47 (45.3); tarsus, ♂ 52–60 (58.2), ♀ 57–60 (58.5). WEIGHT: 500–600.

IMMATURE AND JUVENILE: hindneck, upper mantle, chin to belly, most of head and most of rump uniform grey-brown; dusky eye-crescent, thin white crescents above and below eye; face, crown and nape whitish or cream, producing white-capped effect; rest of mantle and back dark brown, paler fringes forming scaly pattern; lower belly to undertail-coverts and distinctive U-shaped area on rump and uppertail-coverts white; tail mainly black, white terminal fringe, base greyish; outer wing blackish, white bar on mainly blackish rear half of inner wing; elsewhere wing browner and scaly; ill-defined dark and light bars along underwing, narrow white trailing edge except for mainly dark tip. Bill black, sometimes pale at base and tip; eye dark; legs dark grey. By 1st summer head and underparts mainly white, wings and tail worn and faded; mantle paler, less scaly. By 2nd summer tail has neat black subterminal band; secondary bar and whole outer wing blackish, inner primaries mostly grey; remiges except outermost 1 or 2 with white tips; rest of inner wing mostly pale grey, variable brown markings.

DOWNY YOUNG: head and upperparts buff-grey, irregularly mottled blackish brown; underparts grey-white or buff-white; legs pale grey, bill pale flesh, base greyer.

Field Characters. Like Herring Gull *L. argentatus*, but somewhat smaller, slimmer build, more graceful flight, more upright stance, markedly sloping and elongated forehead. Grey of upperparts slightly paler; black wing-tip lacks prominent mirrors; bill red and black, eye brown legs greyish. Immature wing pattern similar to Lesser Black-backed Gull *L. fuscus*, but head and upper mantle unstreaked grey-brown, whitish forehead and crown, no dark ear-patch, distinctive U-shaped area of white at base of tail; the last retained when older (at least through 1st winter), mantle partly clear grey, bill mostly pale, blackish at or near tip.

Voice. Tape-recorded (62, 73). Varied repertoire, given mostly by breeders. Repeated, staccato, goose-like 'gug' or 'leuk' (usually 4) fading in intensity; long, hoarse greeting call 'ki-aou'; loud 'geeaak' in feeding flight, resembles child's cry; subdued 'criek-criek' or 'dog-dog' at nest.

General Habits. Mostly feeds 3 km or more offshore, in calm or moderately calm water, often at night. Skims within 30 cm of surface, rather like Cory's Shearwater *Calonectris diomedea*, maintains momentum by 'rowing' action of primaries, with level glides of up to 100 m; picks food from just below surface, settles briefly (remiges not folded) or lunges, hardly pausing. Associates with feeding Gannets *Sula bassana*; fishes in surf (using rocks as vantage points), scavenges for fish remains; joins Herring Gull and shearwaters at surface shoals and trawlers; flies over fields, often chasing flying insects; sometimes preys heavily on migrant small birds when fish difficult to catch. In winter, 92% at estuaries or sandy beaches Algeria, commuting to open sea; most Atlantic birds at outflows of non-tidal rivulets, also flood water near Tangier; mostly small groups (once 380 Algeria), very pugnacious when bathing, snapping at larger gulls if too close.

Food. Of 120 Apr–June pellets from Morocco, 87% contained fish, 5% insects, 13% birds; 214 identified fish were of 17 spp., 73% were sardines *Sardina pilcharda* of which most were 10–17 cm long (max. 22 cm); insects were winged ants (40%), locusts, mole-crickets, and beetles (20% each); 68% of pellets 30 Apr–9 May contained 1–6 birds (mostly 1–3) including 12 Pied Flycatchers *Ficedula hypoleuca*, 9 Woodchat Shrikes *Lanius senator*, and 30 other birds up to about 60 g. Also molluscs, crustaceans, squid, and geckos.

Breeding Habits. Nests in colonies on offshore islands. Nests usually $0.2/m^2$ but some 30–40 cm apart in crowded colonies (Chafarinas Is., 1976, 1 colony, 1000 pairs, 63 nests in 10 m × 10 m quadrant). Sometimes nests placed amongst Herring Gull nests (Algeria, 1978, 2 compact groups total 50 pairs amongst 150–200 pairs Herring Gulls), and immatures in 2nd summer plumage defend territories with rudimentary nests at colony edge (Algeria, 1978, 3 pairs of 56 pairs did so). Behaviour typical of 'large' gull (e.g. staccato Alarm Call, Upright and Hunched postures); some displays scarcely ritualized (e.g. Choking, Head Flagging); probably closest to Mew Gull *L. canus* (Witt 1977b).

NEST: foundation of grass, mostly sticks from shrubbery, also grass with seaweed and flight feathers often added. Internal diam. (n = 15) 14–19·5 (16·5 cm); external 19–28 (23·5); preferred site is scattered medium-sized bushes (*Withania*, *Lycium*, *Salsola*), building on exposed soil in between. Sometimes in shrubbery (Algeria); where colonies mixed, earlier-breeding Herring Gull occupies island summits.

EGGS: 1–4, laid on alternate days, mean 2·6 (n = 198). Subelliptical; pale blue to olive or dark beige, rather sparingly marked with blotches and spots of brownish grey, brown or black. SIZE: (n = 125, Morocco) 55·8–70·1 × 39·1–48·5 (62·8 × 44·2). WEIGHT: means Morocco, fresh (n = 16) 66·8, 25–50% incubated (n = 22) 63·4, >50% incubated (n = 21) 59·4.

LAYING DATES: Morocco and Algeria, late Apr–mid May, well synchronized. Ideal feeding conditions (small fish close to surface, long periods of low winds) usual from late May, necessary when feeding young.

INCUBATION: carried out by both parents; begins with 2nd, sometimes 3rd egg. Period: 21–25 days.

DEVELOPMENT AND CARE OF YOUNG: at hatching, mean weight 49, tarsus 22·1 mm, bill 16·6 mm (2 birds). Newly hatched chicks remain in nest 1st day; later hide nearby, called out by parents for food.

BREEDING SUCCESS/SURVIVAL: eggs still taken for sale from Chafarinas Island(s) despite status as national game refuge and military restrictions. 1% of eggs taken by Herring Gulls nesting nearby. Adequate vegetation especially important (0·9 fledged per nest in ideal sites, 0·3 where only smaller and much scattered bushes). Some found dead in offshore fishing lines. 3 unhatched eggs 1978 contained 1·94 ppm DDE, 0·76 ppm mercury, levels unlikely to influence hatching success. Most mortality of young after 1 month and *c.* two-thirds fledged.

References

Cramp, S. and Simmons, K. E. L. (1983).
Jacob, J. P. and Courbet, B. (1980).
de Juana, E. *et al.* (1979 and in press).
Witt, H. H. *et al.* (1981).

Not illustrated in colour

Larus delawarensis Ord. Ring-billed Gull. Goéland à bec cerclé.

Larus delawarensis Ord, 1815. *In* Guthrie, Geogr., ed. 2 (Am.), 2, p. 319; Delaware River, below Philadelphia, Pennsylvania.

Range and Status. Breeds North America from S Alaska, east to James Bay and south to about Oregon, Colorado, Wisconsin, S Quebec and NE Newfoundland; winters North America south to Mexico, Cuba and Bahamas.

Vagrant, Atlantic coast Morocco (Essaouira, Aug 1982).

Description. ADULT ♂ (breeding): head, rump, uppertail-coverts, tail, underparts and underwing-coverts white; mantle, back and wings pale grey; clear-cut black wing-tip confined to 5 or 6 outer primaries; white tips to primaries, large white tips towards tip of outermost 2, sometimes reduced or absent; broad white trailing edge and thin white leading edge to wing. Bill bright yellow, broad subterminal black band; eye lemon-yellow, orbital ring red; legs yellow, slight green tinge. ADULT ♂ (non-breeding): much as breeding ♂ except head with fine dark streaks and spots, most dense on hindneck, forming noticeable pattern of diffuse spots, base of neck pale grey-white; spotting extends to sides and sometimes centre of lower throat and upper breast; eye-crescent dusky. Yellow of bill duller, more green-yellow; orbital ring brown-black. Sexes alike except ♀ somewhat smaller. SIZE: (12 ♂♂, 4 ♀♀) wing, ♂ 369–401 (382), ♀ 360–377 (368); tail, ♂ 141–158 (148), ♀ 137–145 (140); bill, ♂ 38–44 (41·0), ♀ 36–39 (37·2); tarsus, ♂ 56–60 (58·3), ♀ 53–55 (53·8). WEIGHT: (USA) ♀ 496.

IMMATURE: body much as non-breeding adult, except that head, neck and breast are generally more heavily marked; streaks and spots form grey-brown area at base of hindneck; flanks and sometimes undertail-coverts barred. Primary coverts and most of outer 5 primaries black-brown; other primaries mostly grey-white, dark brown subterminal areas; secondaries dark brown, narrow white edges, pale grey bases; greater coverts grey, edged pale buff; median and lesser coverts dark brown, pale grey bases; underwing-coverts white, variable amount of narrow brown edging near tip. Tail white, tinged grey, finely speckled brown; broad, blackish subterminal bar, reduced and often broken into few narrow bars on outer feathers. Bill with bright flesh-pink base, otherwise dark; eye dark brown; legs dull flesh-pink. By 2nd winter, body like non-breeding adult but flanks sometimes spotted; tail mostly white, subterminal black spots sometimes forming almost continuous band; innermost 3 primaries generally grey with white tip, others grey and black, brownish virtually absent, outermost with full or partial mirror; secondaries like adult.

JUVENILE: much as young immature except ear-coverts fairly uniform grey-brown; mantle and scapulars brown with broad buff margins.

Field Characters. Medium-sized; larger than Mew Gull *L. canus*, smaller than Herring Gull *L. argentatus*, bill and head shape more like latter. Overall resembles Mew Gull but upperparts paler, bill banded or dark tipped, wing mirrors smaller or absent, noticeable spotting on nape and breast; more pronounced wing pattern (immatures).

Voice. Tape-recorded (C). In non-breeding season sometimes gives soft, mellow 'kowk'.

General Habits. Inhabits mostly coastly areas including harbours; also refuse dumps and sewage outlets. Feeds on tidal flats; follows plough and ships. A buoyant flier; catches insects on the wing.

Food. Invertebrates, fish, refuse and carrion.

Reference
Hoogendoorn, W. (1982).

Larus canus Linnaeus. Mew Gull; Common Gull. Goéland cendré.

Plates 23, 27
(Opp. pp. 368, 416)

Larus canus Linnaeus, 1758. Syst. Nat. (10th ed.), p. 136; Sweden.

Forms a superspecies with *L. audouinii*.

Range and Status. Breeds NW North America, Iceland, British Isles to L. Baikal; winters south to California, Mediterranean, N Africa and Persian Gulf, east to Japan, S China and Indochina.

Palearctic winter visitor, uncommon; very small numbers Mediterranean east to Egypt and Atlantic coast of NW Morocco. Vagrant Mauritania and Senegambia.

Description. *L. c. canus* Linnaeus: only subspecies in Africa. ADULT ♂ (breeding): head, rump, undertail-coverts, tail, underparts and underwing-coverts white; mantle, back and wings blue-grey; clear-cut black wing-tip confined to 5 or 6 outer primaries; white tips to all primaries except outermost, and large white mirrors towards tip of outermost 2; broad white trailing edge and thin white leading edge to wing. Bill yellowish green; eye brown, orbital ring red; legs yellowish or greenish. ADULT ♂ (non-breeding): head with fine dark streaks and spots, most dense on hindneck; eye-crescent dusky, thin white crescents above and below eye; white tips to primaries more prominent. Bill with greyish base and faint dark subterminal band. Sexes alike except ♀ somewhat smaller. SIZE: (7 ♂♂, 9 ♀♀) wing, ♂ 335–385 (357), ♀ 320–365 (341); tail, ♂ 124–148 (139), ♀ 125–147 (133); bill, ♂ 31–38 (34·4), ♀ 30–37 (32·6); tarsus, ♂ 48–58 (52·3), ♀ 48–54 (50·2). WEIGHT: (Norway) ♂ (n = 75) 325–475 (413), ♀ (n = 73) 300–480 (360).

IMMATURE: ear-coverts and crown densely streaked grey-brown; grey-brown mottling on hindneck and sides of breast; mantle grey, some dark marks on rump and undertail-coverts. Most of primary coverts and outer 3–5 primaries wholly blackish brown; grey on outer webs of other primaries, forming pale division between primaries and blackish, tipped white, secondaries; carpal-bar light brown; greater coverts grey-brown, forming paler panel between carpal-bar and secondaries; dark fringes to coverts and axillaries form lines on mainly whitish underwing. Tail white with broad, blackish brown subterminal band. Bill pale flesh or yellowish flesh, tipped black; legs flesh-pink. By 2nd winter, much as adult except black wing-tip extends to P5–P3 and along leading edge of wing onto outer primary coverts; primaries lack prominent white tips, mirrors smaller; rarely some brown on coverts and small amount of black on tail.

JUVENILE: much as young immature; breast-band, flanks and hindneck grey-brown; mantle buff, feathers fringed paler; bill blackish, base usually flesh.

Field Characters. Medium-sized; a smaller, more delicately built version of Herring Gull *L. argentatus*, but smaller, yellow-green bill lacks red spot, legs yellow-green, larger white mirrors in black wing-tips. Immature resembles immature Mediterranean Gull *L. melanocephalus*, but mantle darker, upperwing pattern less contrasting, head streaking finer, without blackish eye patch or trace of hood; bill slender, with pink base and dark tip; legs pinkish.

Voice. Tape-recorded (62, 73), Non-breeders rather silent, commonest call a whistling 'keee-ya'.

General Habits. Mostly coastal, especially in estuaries and harbours; twice on rubbish tip Tunisia, few at L. Qarun (Egypt) 200 km from Mediterranean coast. Often associates with Black-headed Gull *L. ridibundus*; almost exclusively a ground-forager or swimmer (paddle-feeding); sometimes scavenges and steals food. Occasionally live fish constitute the main food source.

Present Nov–Feb. Birds ringed Sweden and Finland recovered NW Morocco.

Food. Invertebrates, carrion, refuse and fish.

Reference
Cramp, S. and Simmons, K. E. L. (1983).

Plates 24, 27
(Opp. pp. 369, 416)

Larus dominicanus Lichtenstein. **Kelp Gull; Dominican Gull. Goéland dominicain.**

Larus dominicanus Lichtenstein, 1823. Verz. Doubl., p. 82; Brazil.

Forms a superspecies with *L. argentatus* and *L. fuscus*.

Range and Status. Southern Hemisphere; breeds on coasts South America, southern Africa, SW Madagascar, Australia, New Zealand and oceanic islands.

Resident, coasts of southern Africa, uncommon to very abundant. Breeds Namibia and South Africa (Cape), non-breeders range north to Angola (Luanda), South Africa (Natal) and Mozambique (Delagoa Bay) where uncommon. 11,200 breeding pairs at 55 sites southern Africa, 57% of birds near Cape Town; largest concentration (2982 pairs) on flat terrain at Dassen Island. Rare nester Northern Hemisphere (Senegambia, Saloum R. delta: Oiseaux Is., 28–30 May 1983, 1 nest with 1 egg (Erard *et al.* 1984) and Terema Is., 26–30 June 1980, 1 nest with 2 eggs (Dupuy 1984)).

Description. *L. d. vetula* (Bruch): resident southern Africa. ADULT ♂ (breeding): head, underparts, rump, uppertail-coverts and tail white; mantle, back and much of wing slaty to sooty black, appearing black; prominent white trailing edge to wing, narrower white leading edge to inner wing. Black on outer primaries decreasing inwards to small subterminal spot on P5–P4; prominent white mirror on outermost, sometimes another (much smaller) on 2nd outermost primary; secondaries greyer than primaries, white tips broader. Underwing white with subterminal grey trailing edge merging into blackish outer primaries. Bill yellow, spot on gonys scarlet or orange-red; eye dark brown, orbital ring orange-red; legs olive-grey or olive-yellow. ADULT ♂ (non-breeding): crown narrowly streaked darker, legs sometimes bluish grey. Sexes alike except ♀ somewhat smaller. SIZE: (13 ♂♂, 25 ♀♀) wing, ♂ 419–452 (429), ♀ 395–438 (414); tail, ♂ 156–173 (166), ♀ 145–166 (156); bill, ♂ 54–61·5 (57·4), ♀ 50–54·5 (52·5); tarsus, ♂ 66–75 (68·6), ♀ 60·5–70 (65·1). WEIGHT: ♂ 1060, 1086, 1096, ♀ 870, 966, 970.

IMMATURE: head and underparts mostly streaked dark grey-brown, forehead, chin and throat whiter, dusky ear-patch and eye-crescent; flanks and sides of breast washed and mottled blackish grey. Mantle, back and scapulars grey-brown, feathers edged pale buff; uppertail- and undertail-coverts and rump whiter, coarsely barred blackish brown; tail otherwise mostly blackish, basal half of outermost pair rectrices and tail fringed whitish. Wing mostly blackish brown above and below, tips to greater coverts and secondaries form double white bar. Bill and eye blackish, legs pale flesh-brown or dusky brown. By 3rd winter much as non-breeding adult; some dark markings on rectrices, streaks on head more extensive and pronounced, primary mirrors smaller.

JUVENILE: much as young immature; body darker and browner, head and underparts more dusky.

DOWNY YOUNG: grey, mottled dark brown; bill and legs dark.

L. d. dominicanus Lichtenstein: scarce non-breeding visitor W Cape, mostly in austral winter. Eye pale yellow; somewhat smaller than *vetula*, crown flatter; some at least with white diamond-shaped or triangular patches on wing near carpus. SIZE: (14 ♂♂) wing, 372–423 (404), bill, 46–54 (50·5).

Field Characters. Marginally larger than Lesser Black-backed Gull *L. fuscus* but more heavily built, wings extending only short distance beyond tail when perched, giving more compact, less elongated profile. Head more massive, bill much stouter (length of culmen less than 3 times depth at base); eye dark in most birds, legs olive-grey to olive-yellow, not bright yellow. Juveniles extremely similar; best told by structural characters.

Voice. Tape-recorded (38, 74, C, F). Similar to Lesser Black-backed. Thorough study New Zealand (Fordham 1963) distinguishes 10 main calls. Lack of good data Africa.

General Habits. Generalist, coastal forager, mostly between high tide line and edge of continental shelf. Finding food apparently not difficult, non-breeders spending only 4 h/day foraging, accessible dead fish frequently ignored. Individuals tend to specialize in particular techniques. Treads in shallows (sandy substrate) and scavenges the inter-tidal zone (rocky shores) for bivalve molluscs, probably basic food, particularly *Choromytilus meridionalis*, *Macoma ordinaria* and *Donax* spp. Molluscs dropped from flight on hard surfaces to break shells; 19/21 adults succeeded at 1st attempt to break molluscs dropped on concrete breakwater from mean height of 3·9 m. Feeds up to 3 km inland, seeking terrestrial snails; locally at rubbish tips and abattoirs to 14 km inland. Regular (has bred) on freshwater lakes Cape Flats; once on dam 27 km inland. Soars in updraught to catch insects in bill (soaring with legs extended in slip-stream, bill open in ambient temperature of 35°C, apparently to dissipate heat). Kleptoparasitizes African Black Oystercatcher *Haematopus moquini*, chasing those with opened *Choromytilus* but not those with small *Aulocomya*; also Great Cormorant *Phalacrocorax carbo*, Cape Cormorant *P. capensis*, Long-tailed Cormorant *P. africanus*, Crowned Cormorant *P. coronatus* and Osprey *Pandion haliaetus* for fish, latter only successful at feeding perch where gulls more often took discarded pieces. Initiated or joined in 16% of chases of small gulls and Common Terns *Sterna hirundo* by Arctic Skua *Stercorarius parasiticus* off South Africa. Feeds on swarming crustacea, catches crayfish *Jasus lalandii*. Predator of eggs and young, including its own species; killed healthy adult Hartlaub's Gull *L. hartlaubii* by repeated blows to head, attacked wounded migrant waders *Calidris* spp. Perhaps competes with juvenile Cape Vultures *Gyps coprotheres* for offal at sea, though gulls rarely eat large carrion, except eyes and protruding tongue of recently dead sheep. Follows trawlers to 150 km offshore.

Recoveries from near Cape Town suggest limited dispersal, 115 km north, 100 km south.

Food. Mostly marine organisms including molluscs, crustaceans and fish; also birds (eggs, young and adults), carrion and offal. Breeding birds in Namibia caught fish *Thyrsites atun*, *Coracinus capensis*, *Xiphurus capensis*; cephalopods; bivalves; crayfish of carapace length up to 70 mm (mean 56); bones, skin and fur of Cape Fur Seal *Arctocephalus pusillus*.

Breeding Habits. Colonial nester; usual nearest neighbour distance 0·2–0·6 m, overall density 4 pairs/m^2. Similar nearest neighbour distances in different colonies and habitats indicates strong social attraction. Most larger colonies are on larger islands; size of colony also influenced by food availability, level of human disturbance, competition for space from other colonial seabirds (63% of islands shared, mostly Cape and Bank Cormorants *Phalacrocorax capensis* and *neglectus*, Jackass Penguin *Spheniscus demersus* and Cape Gannet *Sula capensis*), and by types of terrain and vegetation.

Displays very similar to Herring Gull *L. argentatus* (Fordham 1963). Mate recognition is accompanied by Facing Away when one or both turn heads away, followed by preening and relaxation. Choking is associated with nest-building and aggression; Long Call and Oblique posturing are associated with defence of food. Food begging by ♀, and feeding of her by ♂, involves ♂ regurgitating onto ground. Copulation is preceded by Head Tossing, often followed by preening. Grass Pulling is a common territorial feature, deterring potential intruders, and often precedes ritual fighting. Hostile actions include dives in flight, Choking, and Upright, in which bird tenses body, extends neck, and bends head somewhat downwards while wings are folded but held high against the body. For further details of breeding biology see Williams *et al.* (1984).

NEST: scrape, built up with grass, twigs, large feathers, small stones and shells, jetsam, kelp, seal hair. Sites include ledges on almost vertical cliffs, scattered large boulders, troughs in mainland dunes, low islands in lakes, wooden platforms, exceptionally buildings. Prefers flat, stable areas with some cover (rocks or vegetation) avoiding areas devoid of or with too much vegetation. Intermediate cover provides adequate protection for chicks yet allows visibility and escape for parents; local preference for sites near walls and prominent boulders (vantage points).

EGGS: usually 2–3, sometimes 1, means 1·47–2·1 (6 colonies). Subelliptical, pale green, turquoise or brown-ochre, markings dark brown or black. SIZE: (n = 100) 62·5 – 76·9 × 44·5–51·8 (68·6 × 47·6). WEIGHT: c. 90.

LAYING DATES: Namibia and South Africa, late Sept–early Jan, most 1st clutches Oct–Nov; earlier South Africa than Namibia; also Senegambia, May–June.

INCUBATION: period 30–31 days.

DEVELOPMENT AND CARE OF YOUNG: young leave nest when few days old; fledging period 35–40 days.

References
Brooke, R. K. and Cooper, J. (1979a, b)
Crawford, R. J. M. *et al.* (1982).

Plates 24, 27
(Opp. pp. 369, 416)

Larus fuscus Linnaeus. Lesser Black-backed Gull. Goéland brun.

Larus fuscus Linnaeus, 1758. Syst. Nat. (10th ed.), p. 136; Sweden.

Forms a superspecies with *L. dominicanus* and *L. argentatus*.

Range and Status. Breeds Iceland, British Isles, N Europe east to NW USSR; also Spain and Portugal; winters Palearctic, Africa, Arabia and Persian Gulf.

Palearctic migrant, uncommon to very abundant, coasts and inland. Winters widely north of equator, in Morocco (on passage, 5500 in 5·5 h, 29 Oct 1962), Algeria (1800 whole coast), Tunisia (numerous, parties up to 180), Egypt (70 Jan–Feb 1979), Mauritania (24,000 wintering Banc d'Arguin), Senegambia (3550 of which 90% adult), Nigeria (400 Lagos); smaller numbers Upper Niger Inundation Zone and inner Gulf of Guinea. In eastern Africa passage migrant along Nile and Red Sea; regular on coast south to Mozambique, largest numbers Rift Valley lakes and L. Victoria (typically scores or hundreds, 1000 L. Turkana Mar–Apr). Also Malaŵi, Zambia, Zaïre, Angola, NE Namibia, Botswana, Zimbabwe and South Africa (Natal and Transvaal). Increasing Nigeria, Somalia, probably elsewhere. Numbers oversummer throughout, some in adult-like plumage. (Normally does not nest in Africa although 2 nests with eggs reported from Saloum N. P. (13°39′N, 16°39′W), Senegambia, June 1980: G. J. Morel, pers. comm.; Morel and Morel 1982.)

Description. *L. f. fuscus* Linnaeus: E seaboard and E Africa west to Algeria, Niger and Gulf of Guinea. ADULT ♂ (non-breeding): head, underparts, rump, uppertail-coverts and tail white; mantle back and much of wing blackish grey, often appearing black and hardly contrasting with black wing-tip; all remiges with white tips; prominent white trailing edge, indistinct white leading edge. Black on outer primaries decreasing inwards, usually to small subterminal spot on P4; white mirror on outermost primary (P10), sometimes another (much smaller) on 2nd outermost primary (P9). Underwing white, with broad, dark grey subterminal trailing edge, merging with blackish outer primaries. Bill yellow, reddish spot near gonys, whitish tip; eye pale yellow, orbital ring red; legs yellow or cream-yellow. ADULT ♂ (breeding): upperparts and wings with brownish cast, white tips to primaries reduced or lacking. Sexes alike except ♀ somewhat smaller. SIZE: (13 ♂♂, 8 ♀♀) wing, ♂ 415–438 (426), ♀ 394–410 (402); tail, ♂ 152–169 (160), ♀ 142–159 (150); bill, ♂ 49–55 (52·4), ♀ 45–48 (46·2); tarsus, ♂ 58–66 (63·6), ♀ 57–60 (59·1). WEIGHT: (Kenya) ♂ 822, 850; ♀ 624; (Norway) ♂ (n = 62) 750–1045 (872), ♀ (n = 64) 545–840 (662).

IMMATURE AND JUVENILE: head and underparts mostly streaked dark grey-brown, face and nape paler, prominent dark patch on ear-coverts, blackish eye-crescent; blackish mottling on flanks and sides of breast, belly paler; mantle and back grey-brown, feathers edged paler, giving scaly effect; rump, uppertail-coverts and base of tail whitish with darker streaking, contrasting with back; tail with broad clear-cut blackish subterminal band and series of narrow blackish bars; median and lesser coverts much as upperparts, wing otherwise mostly blackish brown; black outer greater coverts and secondaries form double black bar on inner wing; underwing uniform blackish brown, 'window' effect in middle wing slight or lacking. Bill blackish; eye dark brown; legs dull flesh. By 1st summer, head and underparts whiter, mantle and upper wing-coverts rather faded. First signs of adult plumage in 2nd winter; upper mantle showing uniform ash-grey or black, inner wing-coverts losing barring; tail whiter, band narrower; head mainly dusky, but face, rump and body mostly white. Standing 2nd summer bird as adult, but some brownish on upperparts and wings; tail mainly white, subterminal bar narrow and broken; bill yellowish, subterminal dark area; eye usually pale; legs yellowish.

L. f. graellsii Brehm: Atlantic coast south to Guinea and Nigeria, also inland W Africa and Mediterranean east to Tunisia. Upperparts and wings paler slate-grey than nominate race. SIZE: similar but wing shorter, ♂ (n = 11) 394–428 (412).

Field Characters. Slightly smaller than Herring Gull *L. argentatus*, considerably smaller than Great Black-backed Gull *L. marinus*, with slimmer build and more elongated appearance at rest produced by proportionately longer wings. Adults told from Great Black-backed by yellow legs, less white in wing-tips, and (*L. f. graellsii*) contrast between slate-grey back and black wing-tips. In flight overhead, told from Herring by dusky trailing edge to underwing. Juvenile much darker than Great Black-backed, with slim bill; somewhat darker than Herring with contrasting white rump, solid dark primaries (lacking pale 'window' of Herring and Great Black-backed), greater coverts as well as secondaries dark, forming broader dark trailing edge to wing; bill blackish without pale base. For distinctions from Kelp Gull *L. dominicanus*, see that species.

Voice. Tape-recorded (62, 73). Gives 'keew', 'kleew', 'kyi-ki-ki ...' and 'kiaou', all similar to Herring Gull but somewhat deeper and louder.

General Habits. Frequents both coastal and inland habitats, including lagoons, estuaries, harbours, seashore, and large lakes and rivers (especially Nile and Niger). Gregarious, although tends not to mix with other species of gulls (Algeria, only 4 with 1800 Herring Gulls at rubbish dump while 380 with 40 Herring Gulls 2 km away at Chelif estuary). Feeds in flocks of hundreds at rubbish dumps or over shoals of fish at sea; also feeds singly or in small flocks. Feeds by dipping-to-surface, surface-plunging, and shallow-plunging; also takes insects in air, picks up food from ground, scavenges food and steals it both intra- and interspecifically. Roosts communally on salt or fresh water. Normally migrates singly or in small groups of less than 10.

Migrates along Atlantic coast from Morocco to Senegambia with heavy passage Morocco Sept–Oct. Some movements along Mediterranean coast east to Tunisia with pronounced westerly movements in Nigeria Mar–Apr. Not known if crosses Sahara. Undergoes extensive movements along Nile and Red Sea to East African Rift Valley with main passage Sept–Oct and Mar–Apr. Most NW Africa recoveries 1st winter birds (83%). Recoveries of *graellsii* from Morocco, Tunisia, Algeria, Western Sahara = Morocco, Mauritania, Senegambia, Sierra Leone, Nigeria and Mali; of nominate race from Zaïre, Angola, Chad, Mali, Nigeria, Benin and Cameroon.

Food. Small rodents, birds and their young, fish, beetles, flies and larvae, ants, moths, crustaceans, molluscs, segmented worms, starfish, seaweed, grain, berries and rubbish.

Reference
Cramp, S. and Simmons, K. E. L. (1983).

Larus argentatus **Pontoppidan. Herring Gull. Goéland argenté.** Plates 24, 27

Larus argentatus Pontoppidan, 1763. Danske Atlas, pt. 1, p. 622; Denmark. (Opp. pp. 369, 416)

Forms a superspecies with *L. fuscus* and *L. dominicanus*.

Range and Status. Breeds North America, Iceland and British Isles east to NE Siberia and W Manchuria; also Iberia, N Africa. Winters Panama, N Africa, India, China, Philippines and Indochina.

Resident and Palearctic winter visitor, mainly coastal; uncommon to very abundant. Breeds and mostly winters Tunisia to Western Sahara: 7000 breeding pairs mostly Mediterranean including 3500 pairs Morocco (Chafarinas), 12,000 winter Algeria; slight increase Tunisia (Kneiss Is. 200 pairs 1971, 400–500 pairs 1976–77). Also winters E Mediterranean, NE Africa, and Djibouti to Tanzania (Dar-es-Salaam), Nov–Mar. Marked increase Kenya (Malindi): 5 in 1973–74, 30 in 1977–78, 62 in 1979–80, 280 in 1980–81; 60% or more adult, some immatures oversummer. Up to 10 regular Kenya (L. Turkana), only singly elsewhere far inland. Some from Tunisia wander annually Libya, Aug–Mar. Vagrant Nigeria, Burundi, Uganda, South Africa (Natal).

Description. *L. a. michahellis* Naumann: NW Africa. ADULT ♂: head, underparts, rump, uppertail-coverts, tail and most of underwing white; mantle, back and much of wing pale grey; prominent white trailing edge, indistinct leading edge to inner wing. Black on outer primaries, decreasing inwards to small subterminal spot on 6th, forms clear-cut black wing-tip above and below; white mirrors on outermost 2 usually prominent. Bill yellow, spot near gonys orange or red, tip whitish; eye pale yellow, orbital ring red; legs yellow. Non-breeders hardly differ. Sexes alike except ♀ somewhat smaller. SIZE: (6 ♂♂, 5 ♀♀) wing, ♂ 408–463 (439), ♀ 387–437 (420); tail, ♂ 158–177 (170), ♀ 155–171 (162); bill, ♂ 52–60 (55·7), ♀ 47–54 (49·4); tarsus, ♂ 63–70 (65·0), ♀ 57–64 (61·6). WEIGHT: (S France) ♂ (n = 80) 1040–1500 (1275), ♀ (n = 80) 800–1400 (1033).

IMMATURE AND JUVENILE: head and underparts streaked grey-brown, face and nape distinctly paler, dark ear-patch, blackish eye-crescent. Mantle, back, inner wing-coverts and much of tail rust-brown; rump, uppertail-coverts and base of tail whiter; subterminal tail-band clear-cut, blackish. Secondaries and outer wing mostly blackish brown; paler inner primaries form pale 'window', prominent above and below; thin white trailing edge to inner and middle wing; underwing mostly pale grey-brown. Bill blackish, eye dark brown, legs

Plate 23

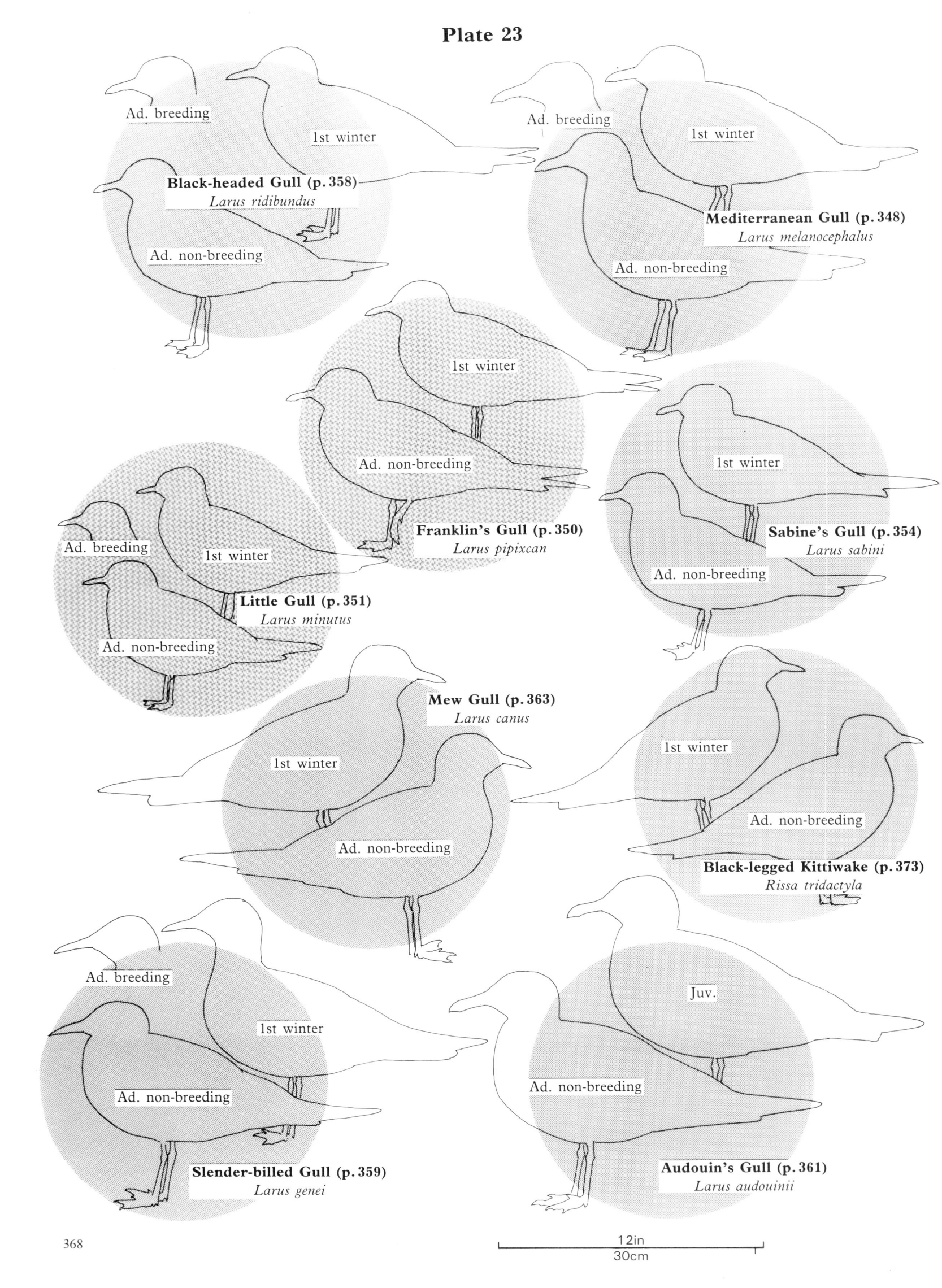

Ad. breeding
1st winter
Black-headed Gull (p. 358)
Larus ridibundus
Ad. non-breeding

Ad. breeding
1st winter
Mediterranean Gull (p. 348)
Larus melanocephalus
Ad. non-breeding

1st winter
Ad. non-breeding
Franklin's Gull (p. 350)
Larus pipixcan

1st winter
Sabine's Gull (p. 354)
Larus sabini
Ad. non-breeding

Ad. breeding
1st winter
Little Gull (p. 351)
Larus minutus
Ad. non-breeding

Mew Gull (p. 363)
Larus canus
1st winter
Ad. non-breeding

1st winter
Ad. non-breeding
Black-legged Kittiwake (p. 373)
Rissa tridactyla

Ad. breeding
1st winter
Ad. non-breeding
Slender-billed Gull (p. 359)
Larus genei

Juv.
Ad. non-breeding
Audouin's Gull (p. 361)
Larus audouinii

12in
30cm

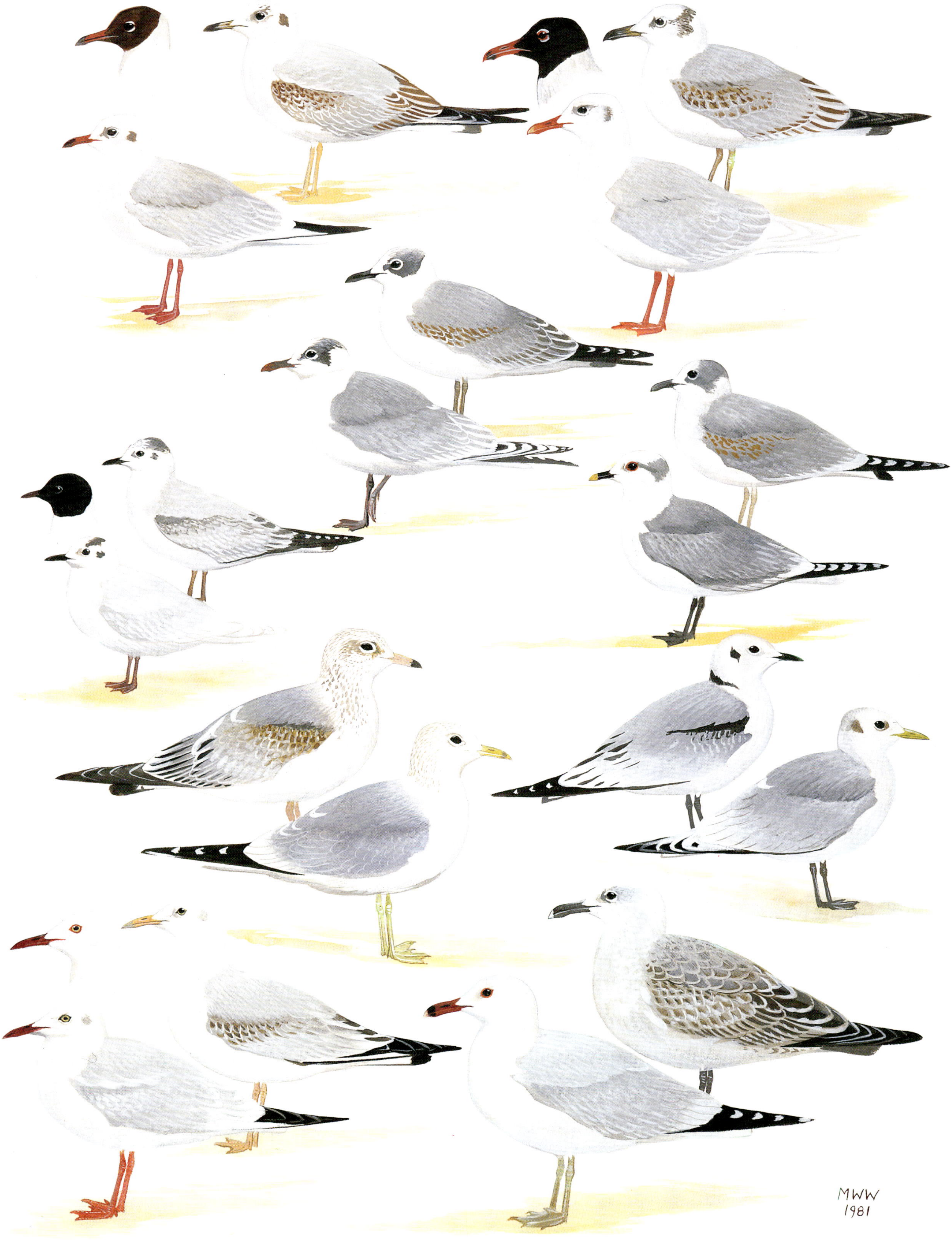
MWW
1981

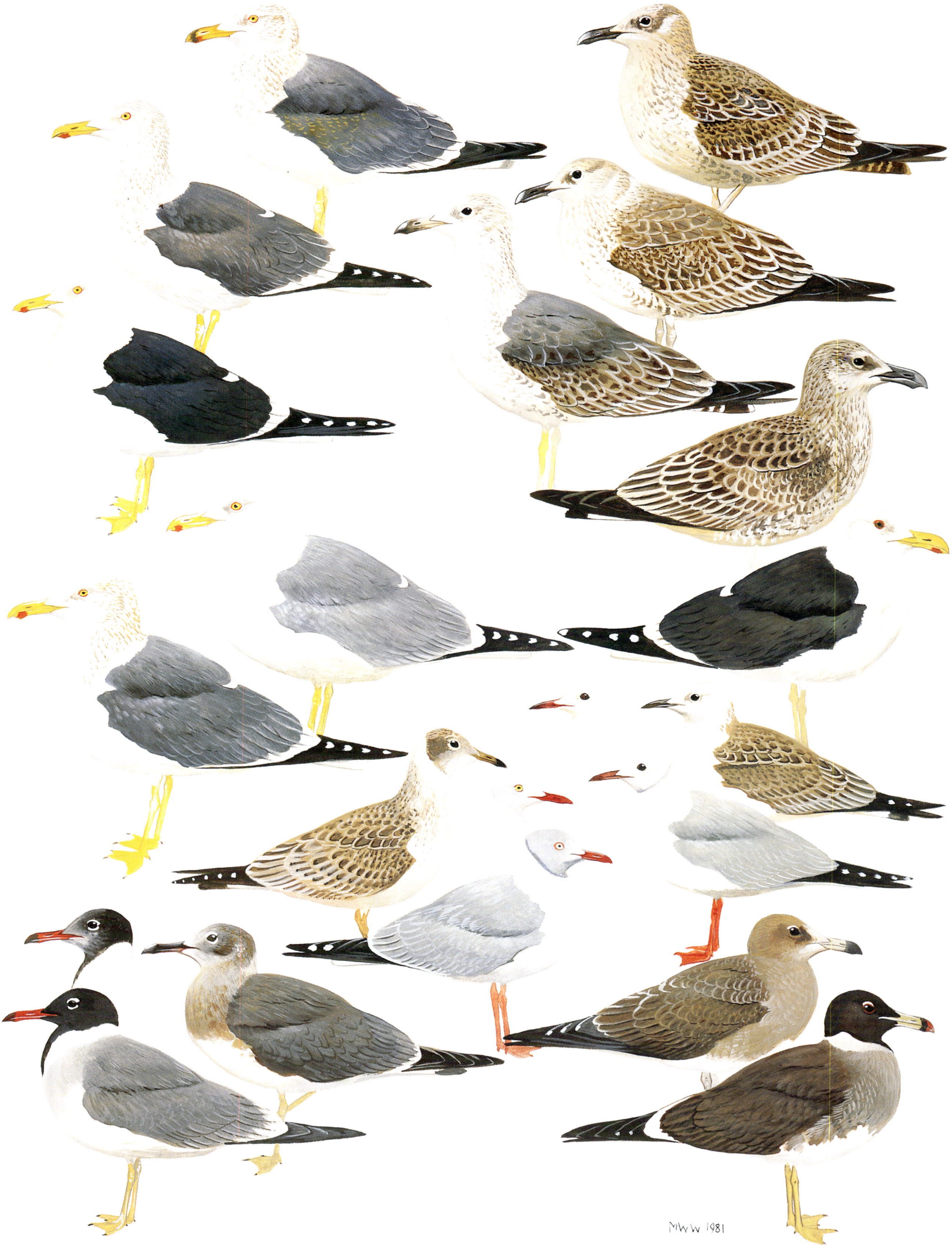
MWW 1981

Plate 24

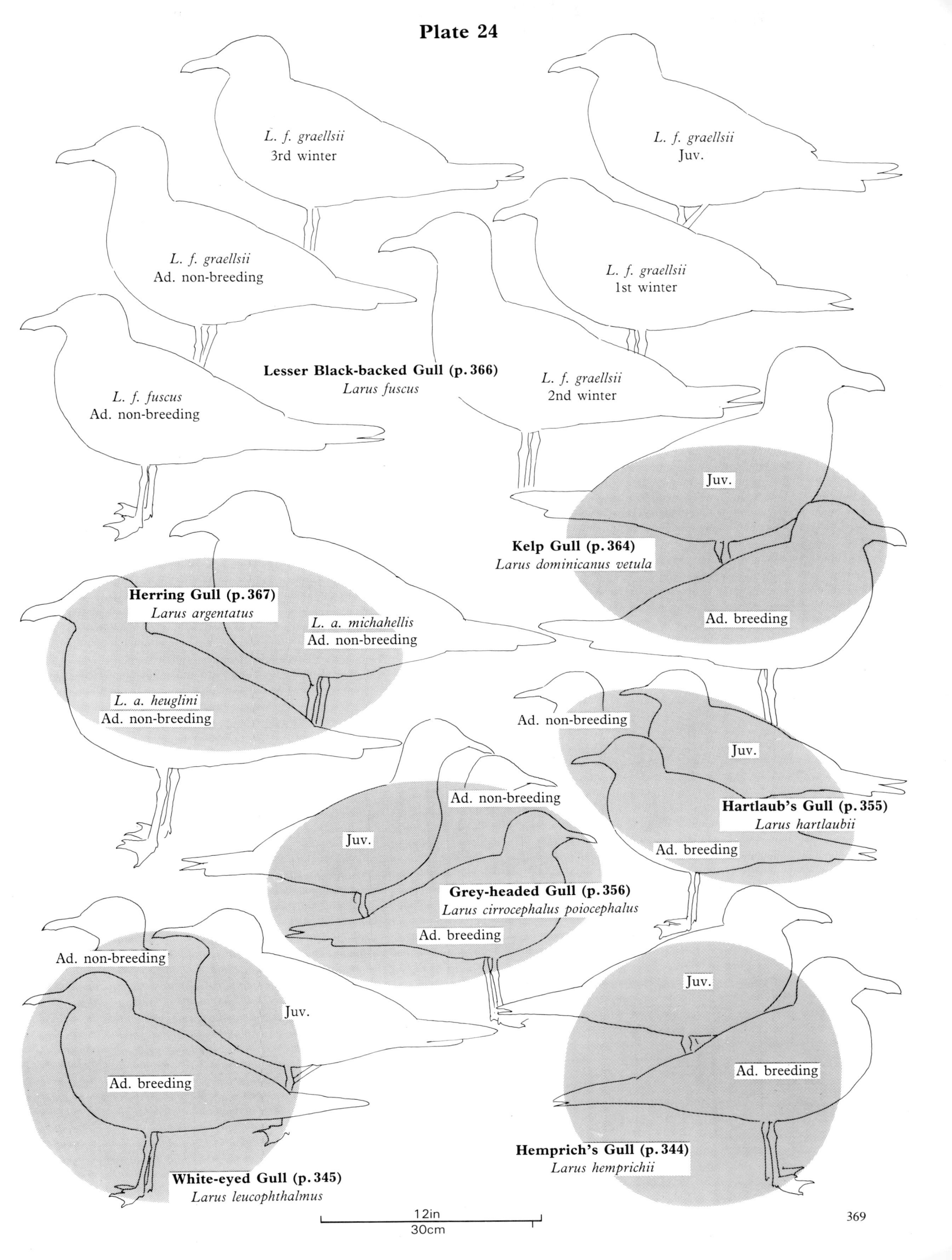

L. f. graellsii
3rd winter

L. f. graellsii
Juv.

L. f. graellsii
Ad. non-breeding

L. f. graellsii
1st winter

Lesser Black-backed Gull (p. 366)
Larus fuscus

L. f. fuscus
Ad. non-breeding

L. f. graellsii
2nd winter

Juv.

Kelp Gull (p. 364)
Larus dominicanus vetula

Ad. breeding

Herring Gull (p. 367)
Larus argentatus

L. a. michahellis
Ad. non-breeding

Ad. non-breeding

L. a. heuglini
Ad. non-breeding

Juv.

Hartlaub's Gull (p. 355)
Larus hartlaubii

Ad. breeding

Ad. non-breeding

Juv.

Grey-headed Gull (p. 356)
Larus cirrocephalus poiocephalus

Ad. breeding

Ad. non-breeding

Juv.

Juv.

Ad. breeding

White-eyed Gull (p. 345)
Larus leucophthalmus

Ad. breeding

Hemprich's Gull (p. 344)
Larus hemprichii

12in / 30cm

369

dull flesh. By 1st summer, head and underparts strikingly white; wings and tail worn and faded, some pale at base of bill. By 2nd summer, mantle and back mostly uniform pale grey; wings and tail faded, pale areas whitish, contrasting with mantle; bare parts much as adult, bill with subterminal dark area.

DOWNY YOUNG: down long and soft, fine silky tips; upperparts and throat buff-grey, spotted and streaked blackish brown on head and throat, larger blotches on upperparts less regular; underparts buff-white; bill blackish tipped pink, legs dusky pink.

L. a. cachinnans Pallas: winter visitor Egypt. Adult slightly paler than *michahellis*; less black and more white on wing-tip; legs yellow. SIZE: wing, ♂ (n = 10) 435–470 (450).

L. a. heuglini Bree: winter visitor NE Africa, erratic inland and to south. Upperparts slate-grey, non-breeders with extensive dusky streaking on nape and hindneck. Immatures paler and greyer than *michahellis*, tail-bar less defined, bill pale at base, legs yellow. SIZE: wing, ♂ (n = 28) 435–465 (450).

L. a. argentatus Pontoppidan: winter vagrant Morocco; Dutch ringed bird recorded Agadir, sight record Massa. Similar to *michahellis* but mantle marginally paler, legs flesh-pink. SIZE: wing, ♂ (n = 23) 391–461 (425).

Many adults Kenya paler than *heuglini*, legs pale flesh to yellow, hindneck less streaked, apparently *taimyrensis* Buturlin. Uncommon winter visitor Senegambia probably *atlantis* Dwight, winter specimens Cape Verde Islands; adult smaller, darker, more heavily streaked than *michahellis*.

Field Characters. Larger and more heavily built than Mew Gull *L. canus*, wings broader with less white in black tips, red spot on heavy orange-yellow bill, legs yellow or pink. Larger and heavier than Audouin's Gull *L. audouinii*, which lacks white mirrors in wing-tips and has red and black bill, brown eye and grey legs. Juvenile paler than Lesser Black-backed Gull *L. fuscus*, with narrower dark trailing edge to wing, pale 'window' on inner primaries, pale brown rump, pinkish base to bill; smaller and less heavy than juvenile Great Black-backed, with much brown on head, plainer, browner back, more brown on tail.

Voice. Tape-recorded (62, 73, C). Throughout year gives socially contagious, clear, loud, incessant 'keew' or 'kleew'; shrill, triumphant clamour in 3 phases, low 'aou-ao', subdued, high'kyi-ki-ki....', and a series of loud, resounding screams 'kiaou'. Breeding birds also give plaintive 'maaeaeae-moaeh-miaou' (Mew Call); repeated, rhythmic 'huoh' (Choking); hoarse, rhythmic 'hahaha-haehaehaehae' (Alarm Call); and soft 'klee-ew' (Begging).

General Habits. Mainly marine, but rarely found out of sight of land; prefers rocky coasts but also inhabits beaches, marshes and other flat areas; sometimes found inland at freshwater and terrestrial sites including grasslands, rubbish dumps and farmland. In tropics usually found inshore or at offshore fisheries; in N Africa prefers rubbish dumps and other terrestrial habitats near coast. Gregarious, usually in flocks of up to several hundred, but occurs singly or in pairs. Feeding methods include dipping-to-surface, surface-plunging, surface-seizing, shallow surface-diving, foot-paddling (presumably to disturb prey) and aerial pursuit of insects. Follows ships and trawlers, picking up offal, and ploughs, taking earthworms and insects. Also scavenges at rubbish dumps and sewage outlets. Steals from seabirds including own species (e.g. as many as 30 individuals chase Audouin's Gulls at sunrise, with one third of the chases being successful and each chase lasting 5–85 s (n = 22)). Eats eggs and young of all sizes and kills adult birds as large as Eurasian Coot *Fulica atra*. Flies strongly; swims buoyantly.

Adults in Mediterranean basin largely sedentary, present all year at colonies, whereas some immatures disperse, e.g. 1st year birds from Tunisia recovered Mediterranean France, Italy, Morocco and Algeria. Origin of birds reaching Senegambia and Gulf of Guinea as far as Nigeria not known but possibly from breeding colonies in NW Africa and SW Europe. Origin of E African visitors unknown.

Food. Fish, fish waste, earthworms, small molluscs, insects, starfish, crustaceans, frogs, birds (eggs, young and adults), small mammals, plant material, rubbish and offal. Of 65 pellets from Morocco (May) 50% contained fish, 50% insects, 30% molluscs, 13% mice and none contained plant material, birds or garbage (Witt *et al*. 1981).

Breeding Habits. Colonial; in 38 sites Algeria, 1–350 pairs (mean 67); 25 sites on islands, 11 on mainland cliffs, 2 on both; at crowded sites nests regularly spaced 1 m apart; largest colonies on islands. Displays detailed under family (large white-headed gulls).

NEST: N Africa, varies from simple depression with few twigs to substantial saucer of grasses lined with leaves, feathers, shell fragments; internal diam. 18–24 cm (mean 21·5), external 28–49 (34·0) (n = 29).

EGGS: 1–3, Algeria mean (182 clutches) 2·7, Morocco mean (279 clutches) 2·0. Subelliptical, colour and pattern very variable, usually various shades of buff or olive, markings dark brown. SIZE: (n = 68, Morocco) 63·9–77·8 × 46·2–52·3 (70·8 × 49·8). WEIGHT: (n = 131, Aegean) mean 72·7.

LAYING DATES: Morocco, Algeria, Tunisia, late Mar–early June (mostly mid Apr–mid May).

INCUBATION: carried out by both parents with ♀ probably taking larger share at least to start; begins with 1st egg. Period: 27–30 days.

DEVELOPMENT AND CARE OF YOUNG: young precocial, leaves nest within 2–3 days after hatching but stays within territory. Fledging period: 35–45 days. Cared for and fed by both parents about equally. May be fed by parents on territory for up to 30 days after fledging.

BREEDING SUCCESS/SURVIVAL: of 610 eggs laid (Wales), 64% hatched. Survival in 1st year, Britain 70–83%, North America 38–73%; survival 2nd year, Britain 93%; adult, Britain 90–94%, North America 80–96%. Overall breeding success related to nest spacing and density, weight of egg and chick at hatching and time within season. Main egg and chick losses due to predation by gulls (including own species) and weather. Oldest ringed bird 31 years, 11 months.

References
Jacob, J. P. and Courbet, B. (1980).
de Juana, E *et al.* (1979 and in press).
de Naurois, R. (1969).
Witt, H. -H. *et al.* (1981).

Larus hyperboreus Gunnerus. Glaucous Gull. Goéland bourgmestre.

Plates 22, 27
(Opp. pp. 353, 416)

Larus hyperboreus Gunnerus, 1767. *In* Leem's Beskr. Finn. Lapper., pt. 1, p. 226 (footnote); Norway.

Forms a superspecies with *L. marinus*.

Range and Status. Holarctic, wintering mostly north of 40°N. Vagrant Mediterranean region, NW Africa, Azores, Madeira.

Vagrant Morocco; specimen of immature Tangier, and sightings of immatures Agadir (Apr 1977) and Larache (Apr 1980).

Description. *L. h. hyperboreus* Gunnerus: only subspecies in Africa. ADULT ♂ (non-breeding): head, rump, uppertail-coverts, tail, underparts and underwing white, head and upper breast with variable dense brownish or orange-brown streaking; eye-crescent dusky, whitish crescents above and below eye. Mantle, back and wings pale grey; narrow white leading edge and broad white trailing edge to wings in flight. Primaries and secondaries pale grey broadly tipped white; shafts straw. Bill yellowish with orange-red spot near gonys, whitish at extreme tip; eye pale yellow, orbital ring yellowish; legs pink. ADULT ♂ (breeding): head and underparts wholly white, bare parts brighter. Sexes alike except ♀ somewhat smaller.

SIZE: (11 ♂♂, 10 ♀♀) wing, ♂ 435–477 (459), ♀ 430–450 (439); tail, ♂ 180–210 (197), ♀ 182–200 (188); bill, ♂ 57–67 (62·6), ♀ 56–61 (58·3); tarsus, ♂ 69–77 (72·6), ♀ 64–73 (68·1). WEIGHT: (USSR) ♂ (n = 14) 1232–2180 (1680), ♀ (n = 10) 1221–1733 (1580).

IMMATURE: head, underparts and underwing pale brownish grey or buff, shading to whitish on chin around base of bill, darkest on belly, undertail-coverts strongly barred; eye-crescent dusky, whitish crescents above and below eye; streaking on head fine and inconspicuous, mottled or faintly barred on sides of breast and flanks; mantle, back, scapulars and wings pale buff barred brownish, rump and uppertail-coverts strongly barred; tail pale grey or buff with whitish and dark mottles and bars; remiges grey-brown or buff tipped and fringed white. Bill pale flesh with sharply demarcated black tip; eye dark brown; legs pale flesh. Over 4 year period plumage becomes generally whitish or paler, more uniform, less barred. Head and underparts of 2nd winter sometimes rather coarsely streaked; by 2nd summer whole plumage generally faded or whitish, mantle with variable amount of clear grey, eye pale.

Field Characters. Large and stout, size between Herring Gull *L. argentatus* and Great Black-backed Gull *L. marinus*. Pale plumage and absence of black on wings and tail at all ages diagnostic. Iceland Gull *L. glaucoides*, recorded as vagrant Madeira but not in Africa, is very similar but smaller and slighter; bill shorter, thinner, mainly dark in juvenile. Flight heavy, powerful, wing-beats comparatively slow.

Voice. Tape-recorded (62, 73). Rather silent. Commonest note similar to 'kyow-kyow-kyow' of Herring, more shrill.

General Habits. Frequents inshore and offshore habitats; follows boats and frequents fisheries and harbours; inclined to be solitary, but sometimes associates with Herring Gull (e.g. Madeira). Predator, kleptoparasite and scavenger, though less fierce and rapacious than Great Black-backed.

Food. Carrion, birds, fish, invertebrates, refuse.

Reference
Cramp, S. and Simmons, K. E. L. (1983).

372 LARIDAE

Plates 22, 27
(Opp. pp. 353, 416)

Larus marinus Linnaeus. Great Black-backed Gull. Goéland marin.

Larus marinus Linnaeus, 1758. Syst. Nat. (10th ed.), p. 136; Sweden.

Forms a superspecies with *L. hyperboreus*.

Range and Status. Breeds eastern coastal North America, Greenland, Iceland, British Isles east to Scandinavia; winters south to Great Lakes and coastal Georgia; and coasts of N and W Europe; vagrant to rare N Africa.

Palearctic migrant; formerly vagrant Tunisia to NW Morocco; a few now regular each year Morocco, especially Agadir.

Description. ADULT ♂ (non-breeding): head mostly white, few dark streaks (mainly around eye and on nape), eye-crescent dark; mantle, back and most of wing blackish grey, looking black at distance; rump, uppertail-coverts, tail and underparts white. Prominent white trailing edge to wing, narrow leading edge to inner wing; black on outer primaries decreasing inwards, usually to subterminal mark on P5. White tips largest on outer primaries, merging with large mirror on outermost to form extensive white end; large mirror on P2. Underwing mostly white; dark grey subterminal trailing edge merges with blackish undersides of outer primaries. Bill pale yellow, red spot near gonys; eye pale yellow, orbital ring red; legs flesh. ADULT ♂ (breeding): head wholly white. Sexes alike except ♀ somewhat smaller. SIZE: (8 ♂♂, 7 ♀♀) wing, ♂ 463–498 (481), ♀ 454–491 (466); tail, ♂ 183–211 (198), ♀ 181–209 (189); bill, ♂ 57–72 (64·8), ♀ 57–66 (60·7); tarsus, ♂ 72–85 (79·8), ♀ 74–81 (75·7). WEIGHT: (Britain, Norway) ♂ (n = 142) 1290–2150 (1740); ♀ (n = 172) 1140–1868 (1470).

IMMATURE AND JUVENILE: head and upper breast mostly white, prominent blackish eye-crescent, some dark streaking especially on nape and around eye; rest of underparts coarsely streaked grey-brown; mantle, back and inner wing-coverts blackish, feathers marked whitish, forming barred pattern on wing, more chequered elsewhere; rump, uppertail-coverts and much of tail whitish streaked darker, contrastingly paler than back; blackish subterminal band on tail typically diffuse and broken; remiges and outer wing mostly blackish brown, paler inner primaries form rather indistinct pale 'window' from above and below; underwing grey-brown and blackish brown. Bill black, whitish at tip; eye dark brown; legs dull flesh. By 2nd summer, standing bird resembles adult; some yellow in bill, eye pale. In flight, brownish and grey-brown inner wing contrasts with mainly blackish mantle and back; outer wing mainly blackish brown, usually with trace of mirror. Subterminal blackish band on tail still distinct. By 3rd winter resembles adult; mirrors on outer primaries smaller, tail with faint, broken band.

Field Characters. Much larger than Lesser Black-backed Gull *L. fuscus*, legs pink. Larger than Herring Gull *L. argentatus*, though some individuals of some races of Herring may approach it in size; head proportionately larger, bill heavier. Larger, dark-winged races of Herring (*heuglini, taimyrensis*) have yellow legs. White wing-tip of adult diagnostic (lacking in Lesser Black-backed and Herring). Juvenile distinguished from juvenile Herring and Lesser Black-backed by white head, diffuse dark tip to tail, chequered back pattern produced by blackish feathers with white edges, heavy black bill.

Voice. Tape-recorded (62, 73). Gives 'keew', 'kleew'. kyi-ki-ki ...' and 'kiaou', all very similar to Herring Gull but lower and more powerful.

General Habits. Inhabits inshore and offshore areas, sometimes 150 km from land; follows boats and frequents fisheries and harbours; found also at sewage outlets and garbage heaps. Largely solitary or in loose groups although roosts at night and loafs during day in flocks, sometimes in association with other gulls. Dominant over other gulls. Flies strongly; swims readily even on rough seas far from land. Catches live prey, scavenges and robs other birds of food; kills prey by stabbing with bill and vigorously shaking it. Sometimes catches birds in air; also catches flying ants and dipterans. Opens molluscs and thick-shelled eggs by dropping them from air.

Food. Fish, frogs and toads, birds (eggs, young and adults), mammals including mice, rabbits and sheep (weak young and adults), insects including dipterans, beetles and ants, starfish and relatives, small crustaceans, segmented worms, small molluscs, some plant material, carrion and refuse.

Reference
Cramp, S. and Simmons, K. E. L. (1983)

Genus *Rissa* Stephens

Medium-sized, marine gulls, similar to *Larus* but tarsus shorter than middle toe with claw; hind toe vestigial, usually without claw. Bill slender, gonys not pronounced. Nests in dense and numerous colonies on ledges of steep cliffs or sea caves. Differences in habits from most gulls correlated with cliff nesting habit; e.g. food regurgitated from ♂ bill to ♀ bill directly rather than dropped onto ground; also broad nape mark of young bird used as a signal to the parents if young about to fall from cliff nest-site. Holarctic, 2 spp., 1 visiting Africa.

Rissa tridactyla (Linnaeus). Black-legged Kittiwake. Mouette tridactyle.

Plates 23, 27
(Opp. pp. 368, 416)

Larus tridactyla Linnaeus, 1758. Syst. Nat. (10th ed.), p. 136; Great Britain.

Range and Status. Breeds eastern North America, Greenland, Iceland, British Isles, Spain, Portugal and northern Europe east to NE Siberia and coasts Alaska. Winters northern oceans Asia south to Japan and coasts to *c.* 30°N, sometimes further.

Palearctic winter visitor, rare, but probably annual Tunisia to Mauritania: single adults Algeria and Western Sahara, otherwise immatures Dec–Feb, mostly singles but up to 15 Western Sahara, 5 inshore Mauritania (Banc d'Arguin). Vagrant Senegambia, Guinea, Egypt, South Africa (twice west of Cape Town).

Description. ADULT ♂ (non-breeding): underparts, rump, uppertail-coverts, tail and most of head white; rear crown to hindneck grey, merging with ill-defined, often crescent-shaped dark grey or blackish ear-spot which often extends over rear of crown; ill-defined dusky eye-crescent; underwing whitish, wing-tip black; mantle, back and wings dark grey, shading to white on trailing edge of wing and immediately before black wing-tip. Outermost longest primary (P 10) has outer web wholly black; clear-cut black triangle across both webs of P10–P7, subterminal black on P6 and sometimes P5; white tips from P7 inwards; remiges otherwise pale grey or whitish. Bill greenish yellow or yellow, sometimes whitish at tip; eye blackish brown, orbital ring orange-red or red; legs blackish or dark grey. ADULT ♂ (breeding): head (including hindneck) wholly white, plumage more immaculate. Sexes alike except ♀ somewhat smaller. SIZE: (12 ♂♂, 10 ♀♀) wing, ♂ 295–322 (305), ♀ 285–314 (280); tail, ♂ 124–136 (129), ♀ 113–140 (123); bill, ♂ 33–39 (35.7), ♀ 31–36 (33.6); tarsus, ♂ 32–36 (33.6), ♀ 31–34 (32.6). WEIGHT: (USSR) ♂ (n = 355) 305–512 (420), ♀ (n = 225) 305–525 (390).

IMMATURE: head, upperparts, underparts, underwing and tail much as non-breeding adult; dark grey or blackish collar on hindneck; black terminal band on tail, broadest in centre, thus accentuating slightly forked tail. Black on primaries extends inwards (as subterminal spots) to P5/P4; mainly blackish outer coverts of outer wing combine with broad blackish carpal-bar to give distinctive inverted W pattern on upperwing in flight, broken across back. Bill, legs, eye and orbital ring blackish, bill sometimes paler at base. Late in 1st summer somewhat resembles non-breeding adult; W pattern and tail-bar retained but less obvious, carpal-bar faded to pale brown; ear-spot and collar on hindneck faded but often conspicuous.

Field Characters. Smaller than Mew Gull *L. canus*, with short black legs, solid black wing-tip without white mirrors, and band of white separating grey wing from black wing-tip; in winter dark ear-spot and sometimes nape but no brown streaking on head. Immature has dark inverted W pattern on upperparts similar to that of much smaller Little Gull, but dark band does not join across lower back, and Kittiwake has black collar on hindneck, pale crown, Little has whitish hindneck, dusky cap; also red legs. Immature might be mistaken for Sabine's Gull *L. sabini*, having similar white 'triangle' and black wing-tip, but in Sabine's mantle and forward portion of inner wing concolourous, without contrasting dark band.

Voice. Tape-recorded (62, 73). Gives 'Kitti-wa-a-k' call, but non-breeders mostly silent. Alarm note raucous 'gok-hagagaga'.

General Habits. Mostly frequents offshore habitats; often follows ships. Leisurely, buoyant flight in calm conditions; distinctive combination of deep, stiff wingbeats and accomplished shearing on strong wings. Most food taken from surface in flight; also plunges from air or surface, often completely immersing.

3 birds ringed Britain recovered Morocco (2) and Algeria (1); wreck of large numbers northern coast of Morocco Dec 1956–Jan 1957.

Food. Mostly fish (scavenged and caught) and pelagic invertebrates.

Reference
Cramp, S. and Simmons, K. E. L. (1983).

Family STERNIDAE: terns

Medium-sized seabirds, more slender and mostly smaller than gulls; wings narrow and pointed; bill long and pointed; tail long, usually forked, often with outermost rectrices elongated. Accomplished and elegant fliers; most at home on the wing; many undergo long-distance migrations. Feet webbed, yet seldom or never swim; walk well. Mostly marine, inshore feeders, roosting on land; several breed in freshwater habitats (rivers, marshes), but winter at sea.

Noddies *Anous* and immature Sooty Tern *Sterna fuscata* are blackish above and below; all 3 *Chlidonias* are partly black or blackish in breeding plumage. All others are white or grey below, grey or blackish above at all ages and seasons. Have various feeding strategies, in or over water; most dip-to-surface or plunge-dive for fish, squid, crustaceans, amphibians; some (especially *Chlidonias* and *Gelochelidon*) also catch insects in air, sometimes far from water.

Calls raucous; especially vocal at breeding colonies. Displays are elaborate, superficially similar in many species, often well documented, few data Africa. In species accounts, most displays are detailed under representative species, e.g. Royal Tern *S. maxima* for crested species, Roseate *S. dougallii* for typical medium-sized *Sterna*. Panics or social flights are common tern displays in which some or all members of a colony take part, flying swiftly and silently towards the sea in a dense flock. After a short time, the birds begin to call again and return individually to their previous activities in the colony. This behaviour is probably adapted to protect adults against aerial predators. Different intensities of this same group of responses have different names, e.g. dreads or upflights.

In the tern accounts most measurements of Palearctic breeders are from Cramp and Simmons, *Birds of the Western Palearctic*, Vol. IV, 1985. Also juvenile and immature plumages are both described, the juvenile plumage being the first full plumage following the nestling or downy young stage, and the immature plumage being one or more recognizable plumages followng the juvenile plumage. The term '*portlandica*', now used for subadult plumage phase of several *Sterna* terns, was originally used for the first summer plumage of the Arctic Tern *S. paradisaea*.

7 genera, 43 spp. world-wide. In Africa 4 genera, 23 spp. including 1 endemic. (White Tern *Gygis alba*, recorded 'Cape Seas', doubtfully from African waters and not included in this book.) Many are mainly or wholly migratory from Palearctic, breeding ranges fragmented in tropics. Important breeding populations in Africa, including Damara Tern *Sterna balaenarum* in South Africa and Namibia, Roseate Tern *S. dougallii* in Kenya (largest known in world), Royal and Caspian Terns *S. maxima* and *S. caspia* in Senegambia and Mauritania, Sooty Tern *S. fuscata* in Gulf of Guinea and Somalia, Kenya and Tanzania coasts, White-cheeked Tern *S. repressa* in the Red Sea and NE Africa south to Kenya, Lesser Crested Tern *S. bengalensis* and Swift Tern *S. bergii*; in the Red Sea and NE Africa, and Brown Noddy *Anous stolidus* in Gulf of Guinea.

Genus *Gelochelidon* Brehm

Monotypic; a comparatively large tern (length 40 cm, wing-span 85 cm); bill stout, deep, shorter than head; gonys sharply sloped up from prominent angle on lower mandible; tip of lower mandible sharply pointed, that of upper more obtuse. Tail short, forked, outermost rectrices same length summer and winter, much less than twice length of innermost. Tarsus fairly long, slightly exceeding length of middle toe with claw. Walks better than most terns.

Gelochelidon nilotica (Gmelin). **Gull-billed Tern. Sterne hansel.**

Sterna nilotica Gmelin, 1789. Syst. Nat. 1, pt 2, p.606; Egypt.

Plates 25, 28
(Opp. pp. 400, 417)

Range and Status. Africa, Eurasia, Australia, North and South America.

Resident and Palearctic winter visitor, abundant to uncommon. Resident W and NW Africa: Mauritania (Banc d'Arguin and mainland near Nouakchott, 1600–1900 pairs), Senegambia (100–200 pairs, Senegal delta), Morocco, Algeria, Tunisia (irregular, small numbers); breeds mostly islands off coast, also inland. Palearctic visitors mostly Sept–Apr; passage migrant Mediterranean, Atlantic and eastern seaboards; also inland crossing Sahara. Winters widely tropical Africa, uncommon N Africa, mainly inland; in numbers south to Upper Niger Inundation Zone, N Nigeria, Chad, Sudan and E Africa; very small numbers Gulf of Guinea, SW Tanzania and Zambia, vagrant to south.

Description. *G. n. nilotica* (Gmelin): only subspecies in Africa. ADULT ♂ (breeding): forehead, crown, nape, hindneck, upper part of lores, sides of head to level of lower part of eye glossy black. Outer web of outer pair of rectrices usually almost white, rest of tail and wing-coverts as upperparts. Primaries with tips and outer webs pearl-grey, broad blackish line along shafts on inner webs, base of inner webs pale grey extending towards tips in wedge; secondaries grey, edged and tipped white; underwing white. Bill black; eye dark brown; legs and feet black, soles usually paler. ADULT ♂ (non-breeding): grey of plumage a shade paler than summer. Forehead white; crown and nape whitish to pale ash-grey, with or without fine black shaft-streaks; hindneck whitish; lores white, slightly speckled black; area around eyes and streak behind eyes grey-black, feathers edged white. Sexes alike except ♀ somewhat smaller. SIZE: wing, ♂ (n = 35) 309–341 (326), ♀ (n = 24) 307–333 (319); tail, ♂ (n = 22) 123–143 (132), ♀ (n = 14) 118–136 (127); bill, ♂ (n = 36) 38–42 (39·8), ♀ (n = 28) 35–40 (37·7); tarsus, ♂ (n = 34) 33–38 (34·8), ♀ (n = 27) 31–35 (33.4). WEIGHT: ♂♀ (n = 13, E Africa) 142–268 (192), ♀ (Zambia) 207. Little sexual dimorphism in 7 ♂♂ and 3 ♀♀ breeding Mauritania; wing (318) and tail (127) rather short (de Naurois and Roux 1974).

IMMATURE: like non-breeding adult except retains juvenile rectrices and dark-tipped primary coverts; primaries blacker and worn; legs dark red-brown. In 1st summer, varying amount of black on crown and nape, never wholly black, and most flight feathers worn (moulted by Aug/Sept).

JUVENILE: resembles immature, but upperparts buff streaked dark brown and blackish; remiges and rectrices grey, latter with brown wedge-shaped marks at tips and slight buff tinge. At fledging, bill pale orange with black tip, darkening to blackish by Sept; legs and feet dull flesh, soles brownish.

DOWNY YOUNG: down longish, soft, with fine hair-like tips. Upperparts buff or greyish buff, mainly streaked and spotted brown-black, distinct pattern of longitudinal streaks on crown and mantle; underparts white, tipped buff below and in front of eye, throat grey. Bill pink to orange, often tipped dusky; legs orange-yellow to brownish.

Field Characters. A comparatively large tern similar in size and proportions to paler Sandwich *Sterna sandvicensis* and darker Lesser Crested *S. bengalensis*. With heavy black bill, slow steady beats of markedly long wings, and tail less forked than most terns, it is somewhat gull-like; immature and non-breeding birds are especially so with head mainly white. Stands higher than terns of similar size. At distance, especially on wing, appears whiter than most terns. Voice distinctive, very different from *Sterna* terns.

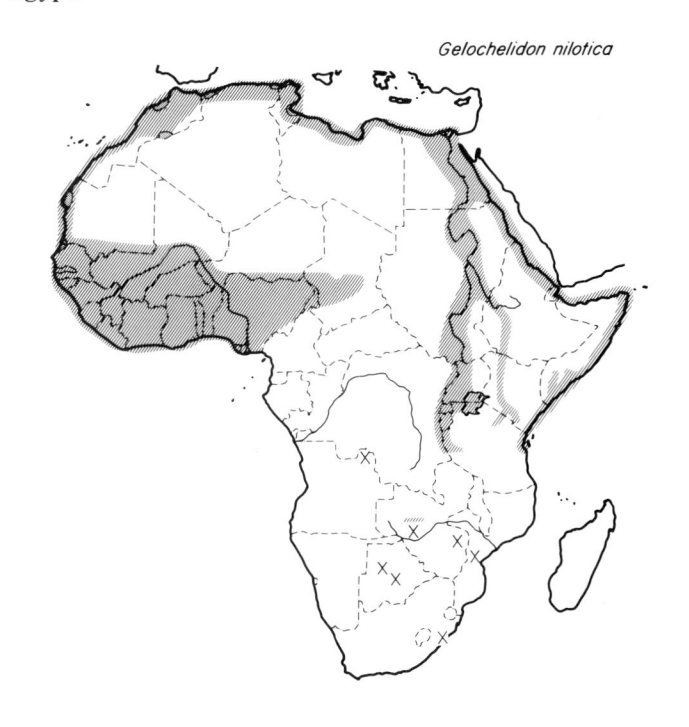

Voice. Tape-recorded (62, 73, C). Both sexes highly vocal at breeding colony; gives various sharp and metallic calls, and somewhat liquid 'ger-erk' or 'ger-vik'. Young give piping, high 'pee-eep' or faster 'pe-pe-eep'. Not very vocal away from colony. Commonest metallic calls given year-round are a repeated 'kaak', and a softer, rasping 'char-ock', 'tirr-uck' or 'ka-huk'.

General Habits. Widespread and numerous inland on larger lakes and rivers, and extensive inundation zones; also coasts (estuaries, creeks, lagoons, saltpans and extensive sandflats) especially in Somalia and N Kenya. Typically near water, flying slowly at height 3–6 m into wind, head down, dropping repeatedly to near ground for insects. Often feeds over lakes, far from shore, but seldom plunges into water. Sometimes catches insects at higher levels (20 m or more); attracted to grass fires where catches insects and perhaps small vertebrates.

Mainly winter visitor, Sept–Apr. Immatures oversummer in small numbers in main wintering areas;

some may migrate northwards 1st summer; occasional birds in apparently adult plumage E Africa, June–July, probably 2nd summer. Banc d'Arguin vacated July–Aug; hundreds there Sept–Oct perhaps are mainly European breeders on passage. Some Atlantic coast passage W Africa, perhaps reaching interior via Senegal R. Heavy passage Tunisia Aug–Sept; regular parties Chad (Ouadi Rime-Ouadi Achim Faunal Reserve) from 9 Aug (mainly Sept), and birds SE Morocco Apr, suggest numbers cross Sahara. Inland and coastal passage in east.

Food. In Africa, flying insects (especially Orthoptera, Coleoptera, Odonata); in Europe also small mammals, young birds, lizards, frogs, tadpoles and worms.

Breeding Habits. Monogamous; territorial, nesting in colonies. Display and posturing much as other terns (see Common Tern *S. hirundo*); e.g. in high flight, pairs rise in circling sweeps to 200 m or more, glide down close together, finally plunging at great speed. Nesting territories of about 3 m² established only 1 or 2 days before 1st egg laid. At Banc d'Arguin it is loosely colonial, with nests frequently spaced 10 m or more apart in open dune sites between scattered clumps of halophytic herbs. In Senegal delta, spacing comparatively dense with nests often only 1 m apart; nests there with Slender-billed Gull *Larus genei*, Grey-headed Gull *L. cirrocephalus* and Caspian Tern *S. caspia* in fairly thick herbaceous vegetation on small islands, mainly near shore; also on edge of large colony of Slender-billed Gulls. Ground predators deterred by group attack; after noisy chase flock hovers over predator, birds dive individually, then swoop up to the flock; resembles typical feeding flight except calling throughout. Banc d'Arguin birds particularly wary; Senegambia birds unusually aggressive, frequently striking intruders.

NEST: shallow scrape or natural hollow in sand or bare ground, scantily lined with grass or other vegetation; some nests on dense carpet grass. Nest scraped out by both adults.

EGGS: 2–4, usually 2; subelliptical, ochreous to cream-white or brownish, spotted various shades of darker brown and grey. SIZE: (n = 24, NW Africa) mean 50·4 × 35·5. WEIGHT (estimated): 32.

LAYING DATES: Morocco, Algeria and Tunisia, late Apr–June; Mauritania, May–June, Sept; Senegambia, May–July.

INCUBATION: begins with 1st egg, sometimes 2nd; carried out by both parents with frequent nest-reliefs. Period: 22–23 days. Ritualized sideways-throwing (shells or other small items thrown towards nest from up to 1 m away) and sideways-building (material placed on nest rim) at nest before and during incubation, similar to *Sterna*. Much calling accompanies nest-relief, relieving bird walking to nest from 2–4 m away.

DEVELOPMENT AND CARE OF YOUNG: fledging period 28–30 days; young apparently no longer dependent by time winter quarters reached.

References
de Naurois, R. (1969).
Dupuy, A. R. (1975).

Genus *Sterna* Linnaeus

Small to large terns; several medium-sized species with outermost rectrices of deeply forked tail elongated as streamers (often seasonal); tail of *S. caspia* (sometimes placed in monotypic genus *Hydroprogne*) short, only slightly forked. Largest species is *caspia* (length 55 cm, wing-span 135 cm), smallest are *albifrons*, *saundersi* and *balaenarum* (length 21 cm, wing-span 50 cm). Bill of *caspia* very stout and deep, structure similar to *Gelochelidon* but pointed, slender, slightly decurved, and angle of lower mandible far less prominent. Tarsus short, sometimes equal to or longer than middle toe with claw usually somewhat shorter. Most have bill length about equal to length of head or longer. All are white and various shades of grey or black; red and/or yellow on bill and legs, and black on crown varies markedly with age and season.

Cosmopolitan; of 31 spp., 16 in Africa, including 1 endemic. Several Palearctic species have fragmented breeding populations in Africa, usually of same subspecies.

There are very few superspecies in *Sterna* or in the Sternidae as a whole; *saundersi* and *albifrons* are usually regarded as allospecies in a superspecies (Vaurie 1965). Further research will probably reveal that *balaenarum* belongs in this superspecies, and that *vittata and paradisaea* form a superspecies.

Sterna albifrons superspecies

1 *S. albifrons*
2 *S. saundersi*

Sterna caspia Pallas. Caspian Tern. Sterne caspienne.

Sterna caspia Pallas, 1770. Nov. Comm. Acad. Sci. Petrop. 14, pt. 1, p. 582; Caspian Sea.

Plates 26, 28
(Opp. pp. 401, 417)

Range and Status. Africa, Eurasia, Australasia, North America and Madagascar.

Resident and Palearctic winter visitor. Breeds Mauritania (1200–1800 pairs Banc d'Arguin, 100 Aftout), Senegambia (1500 between Gambia and Casamance rivers, up to 300 Senegal delta; smaller numbers Saloum delta, Bijolo Is.), Guinea-Bissau (400–600 Bijagos Islands), South Africa (100 L. St. Lucia); smaller numbers also Tunisia, Egypt (Gulf of Suez and Red Sea), and southern Africa (Zambezi delta, Transvaal, Cape, Namibia). Breeding Kenya (L. Turkana), Nigeria (Niger delta), Ethiopia (Dahlak Islands) unsubstantiated. Palearctic visitors common south to Gulf of Guinea and Kenya, mainly inland; uncommon coastal Gabon, Cameroon, Tanzania, and inland Zaïre. Passage migrants from Palearctic mainly cross Sahara or follow Nile and eastern seaboard. Large numbers winter Mediterranean, including 800 Tunisia Jan–Feb 1978.

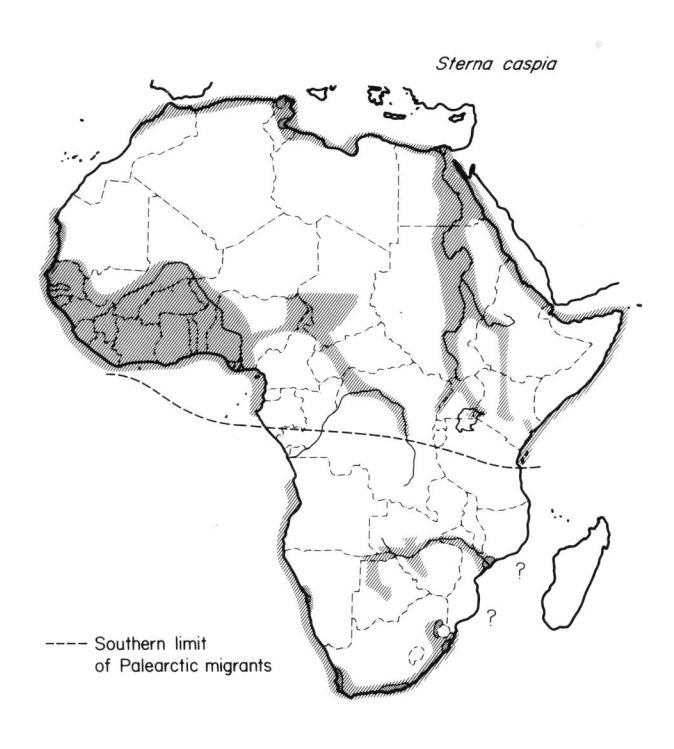

Description. ADULT ♂ (breeding): forehead, crown, nape, upper lores, upper ear-coverts, line under eye glossy black; hindneck white; upperparts (including secondaries and most wing-coverts) ash-grey, rump, uppertail-coverts and tail paler, sometimes almost white. Primaries with tips, coverts, outer webs, and margins of inner webs silver-grey; rest of inner webs dark grey, bases paler, shafts straw-white. Bill vermilion, tipped dusky; eye dark brown; legs black. Southern African breeders differ from Palearctic birds in not assuming complete breeding plumage over head, ♂ bill perhaps heavier (Clancey 1971); ♂ and ♀ of the only pair in Senegal delta 1972 had cap markedly streaked white (Latour 1973). ADULT ♂ (non-breeding): mainly as breeding except forehead, crown, lores and nape white, closely streaked black; ear-coverts and below eye mostly blackish, with varying amount of white. Sexes alike except ♀ smaller. SIZE: wing, ♂ (n = 14) 404–441 (422), ♀ (n = 14) 387–429 (412); tail, ♂ (n = 13) 130–155 (141), ♀ (n = 13) 125–147 (135); bill, ♂ (n = 14) 69–79 (72·5), ♀ (n = 14) 62–73 (67·4); tarsus, ♂ (n = 15) 45–50 (46·9), ♀ (n = 15) 42–47 (44·7). WEIGHT: (Palearctic) ♂ (n = 13) 600–780 (688), ♀ (n = 4) 500–640 (581); (Namibia), unsexed 690.

IMMATURE: mainly as non-breeding adult except head more streaked and mantle and scapulars duller grey. Rectrices with blackish distal portions, tips buff; outer webs with some grey; proximal part inner webs whitish, 2 central pairs with brown-black subterminal bands. Primaries brown-black, outer webs washed silver-grey, inner feathers edged and tipped white; secondaries blackish grey, tips and most of inner webs white; primary coverts blackish, tipped whitish. In 1st summer resembles breeding adult although some dull grey in tail and secondaries.

JUVENILE: remiges, rectrices and underparts as immature. Forehead, crown, nape and upper lores brown-black, streaked grey-buff; ear-coverts, below eye and lower lores blackish, fringed grey; hindneck white, mottled dark grey. Mantle and scapulars grey, tipped buff, subterminal brown spots; back and rump paler grey; uppertail-coverts almost white, with dark brown subterminal marks. Wing-coverts grey, edged and tipped buff; lessers with dark centres. Bill paler, more orange; dusky tip more defined.

DOWNY YOUNG: some fine brown-black markings above, head unmarked, browner than Gull-billed Tern *Gelochelidon nilotica*; underparts white, often tinged buff, throat and chin dusky. Bill orange-red, usually tipped black; legs brownish.

Field Characters. A very large tern, larger than many gulls, with crest on nape, heavy red bill, heavy build, powerful flight, and only slightly forked tail. More gull-like than other terns except Gull-billed. Primaries dusky towards tip above, appearing mainly blackish below. Dark underwing distinguishes it from noticeably smaller Royal Tern *S. maxima*, which has more forked tail, more obvious crest, and less heavy, orange-red or orange-yellow bill. Raucous call diagnostic and frequent year-round.

Voice. Tape-recorded (62, 73, C, 262). In flight, gives deep, loud, raucous 'kaah-aah' or 'kraak-kraak' that resembles call of Grey Heron *Ardea cinerea*. Also gives softer, repeated, gull-like 'kuk-kuk-kuk'. Young give shrill whistling 'klee-eep'.

General Habits. Found on coasts, larger lakes and rivers, and extensive inundation zones. Seldom gregarious except when nesting and roosting; usually occurs singly, often flying (sometimes soaring) at height of 30 m or more. Mainly feeds inshore in calm, shallow water; on coasts frequents estuaries, creeks, and lagoons. At L. St. Lucia (South Africa), fishes parallel to shore, 5–20 m above water, often making repeated transects; when prey sighted, hovers, wings flexed inwards, dives head first, submerging completely, dives often aborted at last moment; prey swallowed on wing, head first; fishes throughout day, regularly 30 km from colony, most actively early morning (Whitfield and Blaber 1978). Aerial-skimming recorded (Géroudet 1965).

African breeders probably disperse widely. At L. St. Lucia, numbers decline in austral summer, varying 6–341 over year. Small flocks regular Zambia, Botswana, Angola probably from south. Banc d'Arguin mainly vacated Oct. Most European breeders probably cross Sahara to reach tropical W Africa by Oct, since very soon after dispersal of flocks on passage that congregate in SE Tunisia, numbers appear in Mali and Chad. No ringing recoveries from NW Africa, and little evidence of coastal passage from there. Many recoveries of Baltic and Black Sea breeders in winter quarters: bulk Upper Niger Inundation Zone (Mali), SE Tunisia, and Ghana coast, south to Zaïre (5°S, 25°E). Passage Red Sea and Somalia Aug–Oct and Mar. Regular flocks coastal Kenya Aug–Mar (mainly Dec–Feb), also L. Turkana. Eastern birds mainly of uncertain origin, although Sudan and Ethiopia recoveries from Baltic and USSR (45°N, 81°E). Some remain in Africa for 1st and 2nd summers but not in main wintering area.

Food. Small and medium-sized fish; also flying insects, birds' eggs and young, and dead fish from nets. At L. St. Lucia, 17 fish spp. 5–242 g (most 10–30), pelagic and demersal; 3 most frequent (*Johnius belengerii*, *Sarotherodon mossambicus*, *Clarius gariepinus*) perhaps taken mainly as carrion.

Breeding Habits. Monogamous; territorial. Most African breeders loosely colonial in small colonies on coastal islands (offshore, estuaries, pans); occasional solitary pairs nest inland and coast. At Banc d'Arguin, most colonies 20–40 pairs (largest 400), nests 2–15 m apart; 15 pairs Cape, nearest neighbour distance 0·65–2·90 m (mean 1·23). Some W African breeders especially aggressive, rising off nest to harass intruders 300 m away. Display and posturing much as other terns; closer to that of Royal *S. maxima* than to Common *S. hirundo*. Bergman (1953) distinguished display flight and fish flight. Former corresponds to high flight of smaller species though less elaborate; participants (2 or 3) climb gradually, apparently not in circles; switch from ascent to descent not abrupt; no long final downward rush, but downward glides and horizontal pursuit continue until flight ends low over water and partners separate, the same bird in lead throughout. Thus resembles pursuit flight of gulls, with alternating swoops and soars, and without prolonged circling ascent. 'Fish-flight' is less frequent and less conspicuous than among smaller species.

NEST: deep scrape, scant lining, sometimes decorated with coloured shells, red locust legs, scales and bones of fish; 15 nests Cape 220–290 mm (246), diam. 25–50 (40) deep, lined with dead vegetation.

EGGS: 1–3, usually 2, laid at 2–3 day intervals; Guinea-Bissau mean (37 clutches) 1·54; Cape (14) 1·86; N Africa (6) 2·33. Subelliptical; greenish to stone-buff, spotted sparingly blackish brown, ashy or dark purple. 2

types (South Africa): green with heavy dark markings (48%), pale turquoise with few purplish markings (35%). SIZE: (n = 23, Cape) 58·7–69·0 × 40·4–46·0 (65·7 × 43·9); (n = 25, Senegambia) mean 64·7 × 45·3. WEIGHT: (n = 19, Cape) 52–72 (63·3).

LAYING DATES: Tunisia and Egypt, Apr–May; Mauritania, Banc d'Arguin mainly Apr–June, also Nov–Dec, Aftout, Dec–Jan; Senegambia, Apr–July; Guinea-Bissau, Dec–Jan; South Africa, Natal, L. St. Lucia, June–July, Transvaal, Mar–Sept, Cape, Nov–Jan; Mozambique (Zambezi delta), Nov–Jan; Namibia, Dec–Mar. At L. St. Lucia breeding coincides with decreasing lake levels, and increase in water transparency due to decreased wind speeds which produced optimal fishing conditions and secure (island) site.

INCUBATION: begins with 1st egg; carried out by both parents with frequent nest-reliefs. Period: 22–24 days.

DEVELOPMENT AND CARE OF YOUNG: young remain in or near nest 1–7 days; thereafter regularly use refuge sites on shore. At Banc d'Arguin parents make a nest-scrape on ground to shelter young against sun. Fledging period 28–35 days. Young fed for as long as 6–8 months after fledging.

BREEDING SUCCESS/SURVIVAL: predation by jackals, dogs, humans and probably Kelp Gulls *Larus dominicanus*. Chicks of solitary Transvaal pair hatched July 1969 did not survive long; chicks fledged from replacement clutch hatched Oct 1969. 15 pairs Cape reared 16 young.

References
Cramp, S. (1985).
Dragesco, J. (1961).
Dupuy, A. R. (1975).
Hockey, P. A. R. and Hockey, C. T. (1980).
de Naurois, R. (1969).
Whitfield, A. K. and Blaber, S. J. M. (1978).

Sterna maxima Boddaert. Royal Tern. Sterne royal.

Sterna maxima Boddaert, 1783. Table Planch. Enlum. p. 58; Cayenne, *ex* Daubenton, pl. 988.

Plates 26, 28
(Opp. pp. 401, 417)

Range and Status. Africa, North and South America.

Resident and probably intra-African migrant western seaboard, Morocco (Tangier) to Namibia (Walvis Bay); very abundant to frequent Mauritania to N Angola, uncommon elsewhere. 5 known breeding stations: Mauritania (3000–6000 pairs annually Banc d'Arguin), Senegambia (up to 12,000 pairs, Saloum, Casamance and Senegal deltas, Bijolo Islands).

Description. *S. m. albididorsalis* Hartert: only subspecies in Africa. ADULT ♂ (breeding): forehead, upper lores, crown and nape black; elongated nape feathers form distinct crest; hindneck, sides of head, lower lores, underparts and most of underwing white; elsewhere very pale grey, white at bases of inner primaries, tips and fringes of these and all secondaries white, tail paler, appearing almost white. Worn outer primaries contrastingly blackish. Bill orange-red; eye dark brown; legs and feet black, soles paler. ADULT ♂ (non-breeding): much as breeding except tail shorter, head mainly white; small blackish spots and streaks on crown and nape; crest short and black, feathers edged white. Sexes alike, except ♀ somewhat smaller. SIZE: wing, ♂ (n = 7) 354–366 (360), ♀ (n = 10) 346–358 (352); tail, ♂ (n = 6) 158–173 (166), ♀ (n = 7) 151–174 (165); bill, ♂ (n = 9) 64–69 (66·3), ♀ (n = 11) 61–65 (63·0); tarsus, ♂ (n = 13) 32–35 (33·5), ♀ (n = 12) 31–34 (32·2). WEIGHT: ♂♀ (n = 4, Ghana) 308–345 (324).

IMMATURE: much as non-breeding adult, except faint carpal-bar, formed by dark grey lesser coverts, which is usually lost 1st summer, sometimes visible until early 2nd (when resembles breeding adult, although feathers of forehead and crown have white fringes). Legs yellowish and black, becoming black.

JUVENILE: resembles immature; black feathers of head greyer, fringed white; mantle white, feathers with grey centres, brownish subterminal markings, buff tips; remiges, rectrices and rest of upperparts grey, rectrices and much of upperparts marked as mantle; inner webs of outer primaries grey-brown, white basal wedges, inner primaries tipped and fringed white; most wing-coverts very pale grey, lesser coverts contrastingly darker. Bill and feet various shades of yellow.

DOWNY YOUNG: 100 Senegambia, similar age, half with bill blackish, feet black, half with bill red, feet orange. In America, young are extremely variable; throat, head and upperparts various shades of buff, dark markings sometimes dense; underparts white or pale buff. Some resemble Caspian Tern *S. caspia*, but middle toe with claw shorter than tarsus, and bill more slender.

Field Characters. Large, crested sea-tern very similar to Swift Tern *S. bergii* except paler in all plumages and bill and head somewhat heavier. Breeding adult with orange-red rather than pale yellow or greenish yellow bill, no white band on forehead; non-breeding adult much whiter than Swift. Juvenile Swift with more brown and blackish on upperparts and head, tail darker. Confusion with Caspian Tern possible at a distance, when best character is much whiter underwing of Royal. Caspian larger (though this not obvious at distance) with less graceful, gull-like shape and different call; bill of Royal less deep, less blunt, never red or tipped dusky. Larger and paler above than Lesser Crested Tern *S. bengalensis* which has similar orange bill; bill longer, wing-beats slower.

Voice. Tape-recorded (63, 73, C). Varied repertoire, resembling Sandwich Tern *S. sandvicensis*, though generally deeper. Commonest year-round flight calls are high-pitched, shrill 'tee-err' and loud, strident 'krryuk'.

General Habits. Occurs in numbers in inshore upwelling zone off Senegambia, but typically found on more sheltered waters of lagoons, creeks, harbours and the open shore. Gregarious at roost; usually feeds singly or in small groups. When feeding, flies usually 5–10 m above surface, plunge-diving; aerial-skimming and dipping-to-surface for fish offal from boats. Sometimes individuals in flock steal food from conspecifics; usually avoids food-pirates by flying faster than them.

Most move south after breeding; Mauritania (Banc d'Arguin), few individuals present Sept-Mar although occasionally in some winters (Jan–Feb) 2000–3300 present. Birds ringed Mauritania recovered Senegambia, Sierra Leone, Nigeria, Liberia, Ivory Coast and Ghana. Bulk probably winter Gulf of Guinea; westerly passage of adults in Nigeria and Ghana, Mar. Small numbers Atlantic coast Morocco, mainly June–Oct, also twice Jan at 34°N.

Food. Pelagic fish; 15 from 5 stomachs W Africa, 3·3–18·5 cm (mean 6·7), Clupeidae, Mugilidae, Carangidae, Pomadasyidae, Ephippidae.

Breeding Habits. Monogamous, territorial, forming dense colonies on offshore sand islands. Performs fish-offering ceremonies, spiralling aerial flights involving 3 or more birds and pre-copulatory strutting displays, all of which persist throughout egg-laying. In fish-offering, one bird flies round, calling loudly, usually with fish cross-wise in bill; another joins it and flies in front; rear bird bends head down, other stretches neck; fish not transferred on wing, sometimes on ground, accompanied by strutting; not eaten by recipient. In aerial flights, gliding more spectacular than medium-sized terns (e.g. Common *S. hirundo*): makes fast descent from 100 m or more on gentle incline, with wings raised and each bird on independent zig-zag path, often crossing flight-line of other(s). When strutting, ♂ and ♀ face each other, wings spread, carpus almost touching ground, necks stretched, crest raised; ♂ then walks in circle round ♀. Alarm-calls and aggressive posture given by Banc d'Arguin adults when Grey-headed Gulls *Larus cirrocephalus* fly over colony, yet Royal often breeds adjacent to Caspian colonies. At least 5–6 pairs/m² at massive colony on small island of Saloum. Many small colonies Banc d'Arguin, yet closely packed, 5–8 pairs/m²; often disjunct subcolonies, e.g. eggs laid 3 weeks later in 2 new small colonies 100 m from main colony.

NEST: shallow scrape or natural hollow in sand, occasionally lined.

EGGS: almost invariably 1, rarely 2. Subelliptical; stone background, brown spots and blotches mainly broad end (Bijolo); cream, numerous brown-red spots (Saloum). SIZE: (n = 2, Bijolo) 67 × 46, 69 × 46. WEIGHT: (n = 25, nominate race) 58–70 (64); (n = 6, Saloum) 53–56 (54·5).

LAYING DATES: Mauritania, Apr–July; Senegambia, May–July.

INCUBATION: by both parents. Period: 25–30 days (nominate race 28–35 (30–31)).

DEVELOPMENT AND CARE OF YOUNG: small chicks weigh 45–55 g (Saloum, Dupuy 1975). Do not crèche until 15 days, remaining in or near nest 7 days (when may group). 1 parent forages, other remaining with small young, often shading it under wing in heat of day. Straying chicks occasionally dragged along ground by neighbour; if parents intervene, aerial chase involving 3–6 birds may result. At colony in North America, 20-day chick attacked and killed young half its size (Buckley and Buckley 1969). Young of different ages and from different subcolonies form different crèches, probably to minimize trampling. Parents feed only own young in crèches; recognition likely by plumage of young and individual calls. Fledging period about 30 days; however, parental care for 5–8 months at least. First breed probably when 3–4 years old.

BREEDING SUCCESS/SURVIVAL: at crowded Saloum colony of 10,000 pairs, many young wandered into shallows where taken by patrolling sharks (Dupuy 1975). Predation likely by jackals and Caspian Terns (Dragesco 1961, de Naurois 1969). Accidental ingestion of crude oil particles killed immatures (Lagos: Wallace 1973).

References
Cramp, S. (1985).
Dragesco, J. (1961).
Dupuy, A. R. (1975).
de Naurois, R. (1969).

Sterna bergii Lichtenstein. Swift Tern; Greater Crested Tern. Sterne huppée.

Sterna bergii Lichtenstein, 1823. Verz. Doubl., p. 80; Cape.

Plates 26, 28
(Opp. pp. 401, 417)

Range and Status. Breeds Africa, Persian Gulf and Malagasy Region to tropical Pacific and Australia.

Resident and partial intra-African migrant, Namibia, South Africa and throughout eastern seaboard. Breeds Sudan (Suakin, 400 pairs or less) to NW Somalia (Zeila, 200 pairs or less); Tanzania (Latham Island, 200–500 pairs); Namibia and South Africa (Luderitz to Algoa Bay, colonies mostly 100–200, total 4600 pairs).

Description. *S. b. bergii* Lichtenstein: Namibia and Cape; non-breeders southern Africa to S Mozambique. ADULT ♂ (breeding): crown, nape and much of forehead black, feathers of hindcrown and nape elongated into crest. Centre of forehead, band between upper mandible and sides of head, hindneck, underparts, axillaries and underwing-coverts white. Mantle, back, rump, uppertail-coverts and most wing-coverts pale blue-grey; inner wing with white leading edge. Elongated outermost pair rectrices white; others with outer webs pale grey, inner webs paler. Primaries silver-grey, margins and tips of inner webs white. Secondaries with outer webs and coverts grey, tipped white, inner webs mainly white. Bill bright yellow, paler and tinged greenish later in season; eye dark brown; legs and feet black, soles yellowish. ADULT ♂ (non-breeding): similar to breeding adult except forehead and crown broadly edged white; ear-coverts, cheeks and sides of neck freckled slate-grey; crest feathers shorter and rounder, sometimes with fine whitish streakings; outermost pair of rectrices grey tipped white; bill pale greenish yellow. Sexes alike. SIZE: ♂♀ (n = 10) wing, 347–375 (359); tail, 164–184 (175); bill, 57–66 (60·7); tarsus, 28–31 (29·6).
IMMATURE: as non-breeding adult except dusky carpal-bar formed by dark grey centres to lesser coverts, some grey-brown juvenile wing-coverts retained, and rectrices grey, broadly tipped whitish.
JUVENILE: forehead and crown brownish black, edged white; crest short, dark brown; hindneck and ear-coverts white, blotched dark brown; white mottled brownish black round eye, merging to uniform brown over ear-coverts; upperparts and wing-coverts grey, heavily mottled and barred black and white, some brown and buff edges; rectrices dark grey, edged white, primaries mainly blackish. Bright yellow legs recorded South Africa; otherwise bill and legs as immature.
DOWNY YOUNG: down long and coarse. Colour of upperparts and head variable, cream, buff or grey; dark brown and buff bases to down; mantle and wing-feathers of older birds with large dark brown centres, rich buff borders, especially marked on wings; ear-coverts mainly brown; underparts cream-white. Bill pale yellow, legs greyish yellow or greenish yellow.

S. b. thalassina Stresemann: breeds Tanzania; non-breeders to Kenya and S Somalia. Upperparts and tail paler grey, white band on forehead broader than nominate, face always white; distinctly smaller, bill shorter, stubbier. SIZE: ♂♀ (n = 8) wing, 322–350 (332); ♂♀ (n = 10) bill, 54·5–59 (56·0). WEIGHT: ♂ 325, 345, 350, ♀ 350.

S. b. velox Cretzschmar: breeds eastern seaboard, Sudan to NW Somalia; non-breeders throughout Red Sea, and south to Kenya. Upperparts and tail slate-grey, darker than nominate, forehead band broader, face always white; larger, bill especially long and slender. SIZE: ♂♀ (n = 12) wing, 354–381 (366); bill, 59–68 (63·9). WEIGHT: ♂ 375, 380, ♀ 340, unsexed 360.

S. b. enigma Clancey: breeds Zambezi delta; non-breeders probably south to Natal. Dimensions and bill colour as nominate, paler above, less white on forehead, non-breeding head as Royal Tern *S. maxima*.

Field Characters. Large sea-tern, similar to Royal Tern but bill somewhat decurved and pale greenish yellow, never orange-yellow; upperparts and tail darker, *velox* especially so. Distinctive mottled and barred upperparts of juvenile retained for some months. Wear exaggerates '*portlandica*' plumage of immature which has inner wing with 3 dark bands (leading edge, greater coverts, across secondaries). Lesser Crested Tern *S. bengalensis* much smaller, slimmer, more graceful, and orange or orange-yellow bill, not pale yellow or greenish yellow. Flight fast, powerful with long, sweeping wing action; wings sharply angled at carpus.

Voice. Tape-recorded (F). Often silent, despite varied repertoire. At nest, gives cawing 'korr' on ground, softer 'wep-wep' in flight. Fishing flocks give continuous grating 'kik-kik-kik' and harsh 'krow'. Other calls include rasping 'kerrack' and high-pitched screaming 'kree-kree'.

General Habits. Occurs in offshore and coastal water habitats. Fishes singly or in small parties, mainly inshore in sandy, rocky, coral coast and estuary habitats; often flocks with other feeding terns. Feeds mostly by plunge-diving; grips fish behind head, and usually swallows them on wing, head first. Also dips-to-surface; younger birds mostly feed by contact-dipping, usually incompletely submerging themselves. Regularly bathes at water's edge.

Partial migrant, though small numbers remain all months throughout range. In Red Sea and Gulf of Aden numbers reduced in Sept–May; Egypt present mostly Apr–Aug; Kenya mainly Jan-June; and mainland Tanzania mostly Dec–Feb. Non-breeding birds from

Namibian and South African colonies move along Namibian and South African coast to S Mozambique; those from Zambezi delta colony probably move south at least to Beira and Natal; some of those breeding in Tanzania move north to Kenya and Somalia; and those breeding in Red Sea regular throughout Red Sea and south to Kenya in non-breeding season.

Food. Pelagic fish, often 10–15 cm including Hake *Merluccius* sp. and Pelagic Goby *Sufflogobius barbatus* (Walter 1984); in Somalia mainly Clupeidae.

Breeding Habits. Monogamous, territorial; forms dense colonies on small coastal islands; sometimes on offshore islands or in coastal pans or sewage works. Nest-site established very soon after arrival at colony; pairs often formed long beforehand. In Mozambique (Inhaca Island) many pre-breeding displays and copulations observed, Oct-Nov, though nearest colony 600 km to north. In Somalia (Brava), pairs seen chasing each other at dusk, making hoarse calls, impressive swoops and dives, May-June, though nearest colony 2000 km to north. Often nest in subcolonies of 10 or fewer pairs; spacing very regular Sudan, 7–8 pairs/m² (R. J. Moore, pers. comm.). Associated with much larger colonies of Lesser Crested and Sooty Terns *S. fuscata*, Masked Booby *Sula dactylatra* and Hartlaub's Gull *Larus hartlaubii*. Aerial attack poorly developed, and vulnerable despite large size. Displays, postures and crèche like Royal's; aerial flights as high as 250 m.

NEST: shallow scrape, rarely lined, on open flat site, average 30 cm apart.

EGGS: 1–2, usually 1; 1% or fewer nests have 2, probably laid by 2 birds. In some subcolonies eggs laid 1 week later than in other subcolonies (Sudan). Subelliptical, variable, olive-yellow to buff-cream, blotched, spotted and scrawled black, brown and grey. SIZE: (n = 250, *velox*) 53·8–68·6 × 40·0–45·1 (62·1 × 41·7); (n = 88, *bergii*) 55·4–66·8 × 39·3–45·3 (62·1 × 43·0). WEIGHT: (n = 5, Seychelles) 50–57.

LAYING DATES: Ethiopia and Sudan, June–July; Somalia, Aug; Tanzania, Oct–Nov; Mozambique, Jan; South Africa and Namibia, Feb–June.

INCUBATION: period not recorded.

DEVELOPMENT AND CARE OF YOUNG: individual chicks recognized by parents after 2 days, when scatter or form crèche. Fledging period 35–40 days. Young fed whole fish. Usually remain at colony 1 month or less after fledging; dependent 4 months at least. Juveniles contact-dip to pick up pieces of weed which are sometimes repeatedly dropped and retrieved; accompany parents on fishing excursions over shallow water, returning to shore to be fed.

BREEDING SUCCESS/SURVIVAL: continual harassing by Hemprich's Gulls *Larus hemprichii* causes significant losses of eggs (Sudan). Sacred Ibises *Threskiornis aethiopica* take eggs and downy young. Kelp Gulls *Larus dominicanus* also prey on eggs but are beaten off by the terns (South Africa: J. Cooper, pers. comm.).

References
Clancey, P. A. (1975).
Cramp, S. (1985).

Plates 26, 28
(Opp. pp. 401, 417)

***Sterna bengalensis* Lesson. Lesser Crested Tern. Sterne voyageuse.**

Sterna bengalensis Lesson, 1831. Traité d'Ornith., p. 621; India.

Range and Status. Breeds southern Mediterranean (Libya), Red Sea and Gulf of Aden; also Persian Gulf and India to Australia. Winters E and W coast of Africa and Madagascar.

Resident and intra-African migrant, very abundant to uncommon. Breeds many sites eastern seaboard, sometimes thousands, Red Sea coast Egypt to NW Somalia (Zeila); in non-breeding season south to Natal, vagrant E Cape. Also breeds E Libya (2000 pairs), spends non-breeding season Senegambia.

Description. ADULT ♂ (breeding): forehead, upper lores, crown and nape black; underparts, ear-coverts, lower lores, axillaries, underwing-coverts, hindneck and upper mantle white, latter washed grey; lower mantle, scapulars and wing-coverts pale blue-grey; rump, uppertail-coverts and tail paler, some rectrices almost white. Primaries with outer webs silver-grey, inner webs white, narrow dark grey shaft-line; secondaries as back, fringed and tipped white. Bill orange-yellow; eye dark brown; legs and feet black, soles yellow. ADULT ♂ (non-breeding): mainly as breeding except forehead and lores white, black spot in front of eye; crown white, mottled greyish black; upper ear-coverts, hind crown and nape black; bill yellow. Sexes alike except ♀ somewhat smaller. SIZE: wing, ♂ (n = 13) 287–315 (298), ♀ (n = 5) 277–308 (292); tail, ♂ (n = 6) 149–173 (157), ♀ (n = 3) 153–172 (159); bill, ♂ (n = 13) 49–56 (53·6), ♀ (n = 6) 47–57 (50·9); tarsus, ♂ (n = 13) 24–27 (25·4), ♀ (n = 6) 24–27 (25·7). WEIGHT: (Kenya) ♂♂ 185, 190, ♀♀ 205, 235; (Sudan) ♂ (n = 1) 160; (Egypt) ♀ 204.

IMMATURE: as non-breeding adult except remiges and rectrices darker grey; indistinct carpal-bar across forewing, formed by dark grey centres to lesser coverts; bill pale yellow.

JUVENILE: much as immature: white of forehead and crown tinged buff, mottling less contrasting; feathers of mantle and wing-coverts with brown subterminal marks; feet orange-yellow to yellow, black at joint, later patterned black.

DOWNY YOUNG: head and upperparts pale grey to pale buff, varying amount black markings above, mantle distinctly streaked; bill olive-yellow or black; legs orange-yellow, black developing at joint.

Monotypic (Vaurie 1965) although Mediterranean birds larger, bill heavier.

Field Characters. About size of Sandwich Tern *S. sandvicensis*; smaller and more slender than other African crested terns. Separable from Swift Tern *S. bergii* by orange or orange-yellow, straight (not decurved) bill, and in breeding season, narrower white band between bill and crown. Bill colour similar to Royal Tern *S. maxima*, from which it must be distinguished with care. Royal is larger, heavier, bill more massive, wingbeats deeper, tail somewhat shorter.

Voice. Not tape recorded. Varied repertoire including high-pitched 'krr-eep' and twittering 'kee-kee-kee'. Commonest year-round calls are rasping 'kirrik', less ringing than Sandwich, and softer 'kriik' whilst fishing.

General Habits. Occurs mainly inshore on sandy and coral coasts and estuaries, in both deep and shallow water. Gregarious, regularly found in feeding flocks with other terns 3–6 km off coast of Kenya. Flight fast, 4–10 m above water. Most fish caught from vertical plunge-dive; bird usually submerges completely. Swallows food in flight; also dips-to-surface and readily settles on water. At roost associates especially with Swift and Sandwich Terns.

Libyan breeders migratory, passage along Mediterranean May–July, late Aug–Nov on way to Senegambia. Small numbers present most months Morocco; Atlantic coast passage mainly Oct. Present throughout year along whole eastern seaboard. Partial migrant, numbers reduced Sept–Apr Red Sea and Gulf of Aden, probably wholly migratory Gulf of Suez. Kenya to South Africa (Natal) mainly present Dec–Mar; immatures oversummer in small numbers. Rare inland E Africa, Aug–Sept; also Burundi (L. Tanganyika) Feb 1982 and Sudan (Khartoum) June 1982.

Food. Mostly small pelagic fish, also crustaceans. In Egypt fish include *Diplodus noct* (Sparidae), 65 and 95 mm long.

Breeding Habits. Monogamous, territorial; nests in dense colonies on small offshore islands, sitting birds almost touching one another (Ethiopia, Dahlak Islands). Displays, postures and crèche behaviour like Royal's. Actively drives away gulls in vicinity of colony. Leaves nest when human intruders still distant, wheeling noisily but not striking them.

NEST: shallow scrape or depression in sand; unlined. In Sudan, prefers open sandy sites at highest point of islet, either in centre of islet surrounded by vegetation, or on highest part of beach.

EGGS: almost invariably 1. Subelliptical, white to buff, variably marked black or dark brown spots, blotches and smudges, underlain ash-grey. SIZE: (n = 40, Indian Ocean) 48·0–55·0 × 33·4–37·0 (52·0 × 35·9). WEIGHT (estimated): 35 g.

LAYING DATES: Libya, July–Aug; Egypt, Sudan and Ethiopia, June–Aug, mainly July; Somalia, Aug.

INCUBATION: by both parents. Period: 21–26 days.

DEVELOPMENT AND CARE OF YOUNG: cared for and fed by both parents, remaining in or near nest for few days, then form crèche on shore or scatter. In crèche watched over by several adults, fed by own parents. Several thousand on Dahlak Islands with ages varying from 1 week to within 1 week of flying, in number of disjunct droves, each attended by own group of adults. Once crèche is established, used as roost by some young of all ages. If crèche approached, attendant adults wheel, giving alarm-calls; some dive at intruder, others circle above fleeing chicks. Fledging period: 30–35 days. Adult continues to feed young for up to 5 months after fledging; young still dependent by time winter quarters reached.

Sterna sandvicensis Latham. Sandwich Tern. Sterne caugek.

Plates 26, 28
(Opp. pp. 401, 417)

Sterna sandvicensis Latham, 1787. Gen. Synopsis Birds, suppl. 1, p. 296; England.

Range and Status. Breeds N Africa, European coast, Black and Caspian Seas, and North America; winters African coasts, Persian Gulf, NW India and northern South America.

Resident and Palearctic winter visitor. Breeds Tunisia, uncommon (juveniles, age 25–30 days, SE Tunisia 11 July 1959, Castan 1961; bred Tunisia and Algeria in 19th century). Palearctic visitor, very abundant to uncommon, wintering throughout western and northern seaboard, mainly in tropics where generally common. In Mediterranean commoner in winter than summer (920 Egypt Dec 1979–Jan 1980, 1000 Tunisia

Jan–Feb 1978, 1800 Algeria Dec 1977–Jan 1978); common Mauritania (35,000 Dec 1978). Also winters and on passage E and NE Africa, Sudan to Mozambique, uncommon.

Sterna sandvicensis

Also winters in breeding range

Description. *S. s. sandvicensis* Latham: only subspecies in Africa. ADULT ♂ (breeding): forehead, lores, crown to lower part of eye and nape black; nape feathers elongated and pointed; hindneck and upper mantle white; lower mantle, back, scapulars and most wing-coverts pale ash-grey; rump, uppertail-coverts, tail and tips of scapulars white. Underparts white, sometimes tinged pink. Primaries with outer webs and coverts silver-grey, shafts white; inner webs white, with grey-black shaft-line; secondaries with inner webs white, outer webs pale ash-grey. Bill black tipped yellow; eye dark brown; legs and feet black, soles yellow. ADULT ♂ (non-breeding): mainly as breeding except forehead and lores white; small black spot in front of eye, few black streaks under eye; fore-part of crown white, finely streaked black; nape and rest of crown blackish, streaked white, nape feathers less elongated; pink on underparts slight or absent. Tail with trace of grey, usually on innermost pair rectrices. Bill all black or with horn tip. Sexes alike except ♀ somewhat smaller. SIZE: wing ♂ (n = 21) 302–317 (309), ♀ (n = 16) 294–320 (304); tail ♂ (n = 7) 148–172 (159), ♀ (n = 4) 142–158 (147); bill, ♂ (n = 21) 53–58 (55·5), ♀ (n = 16) 49–56 (53·1); tarsus, ♂ (n = 21) 26–28 (26·9), ♀ (n = 16) 24–27 (25·9). WEIGHT: ♂♀ (n = 13, South Africa) 193·5–228 (215).

IMMATURE: mainly as non-breeding adult except tail with tips buff mottled brown-black; rest (sometimes whole) central pair of rectrices white, others with outer webs and distal portions of inner webs grey, brown-black towards tips. Outer webs of remiges darker than adult, extending to inner webs of inner feathers, innermost barred black-brown.

JUVENILE: remiges and rectrices as immature. Forehead, upper lores, crown, nape and line under eye black, feathers edged and tipped buff; mantle, back, scapulars and greater coverts buff, widely barred brown-black; median coverts ash-grey washed buff, with crescent-shaped blackish penultimate bars; lesser coverts dark grey; uppertail-coverts buff-white with black marks.

DOWNY YOUNG: head, throat, wings and upper mantle buff to grey, with matted appearance; varying amount of blackish on crown and sides of head, black line along wing; rest of upperparts buff mottled blackish; underparts except throat white; bill and legs varying shades of grey.

Field Characters. A large sea-tern, pale ash-grey above, white (sometimes tinged pink) below with size and proportions of Lesser Crested Tern *S. bengalensis*. Flight and distinct crest very like Lesser Crested, but bill mostly black (not orange-yellow). At distance could be confused with Gull-billed Tern *Gelochelidon nilotica* but Gull-billed has shorter, broader, less pointed wings, heavier body, and often shorter tail, which together with gull-like flight suggest small gull. Several medium-sized terns (e.g. Common *S. hirundo*) have mainly black bills in some plumages; Sandwich has heavier build, less buoyant flight, less deeply forked tail, longer bill, and distinctive call; also only Roseate *S. dougallii* appears as white.

Voice. Tape-recorded (62, 73). Common calls include loud and strident 'kirrick' and rising 'kirr-quit'; also gives alarm note, 'gwit' or 'gwut'.

General Habits. Occurs inshore along sandy beaches, mangrove-fringed mudflats, rocky shores, and estuaries; often offshore when migrating. Gregarious, especially noisy at roost. Plunge-dives from 5–10 m above water surface, remaining submerged or partly submerged longer than smaller sea-terns, with which it often feeds. Dipping-to-surface and shallow dives (1·5–3 m) frequent in W Africa; also takes fish escaping from nets; for immatures, dipping-to-surface is 1st step in learning plunge-diving.

In 1st year, winters mainly Senegambia to Angola, also southern Africa to Natal; adults and older immatures usually winter south of equator. On passage most follow Atlantic coast, but early Apr birds SE Morocco suggest some cross Sahara. Most spend 1st summer in Africa; minority do so in 2nd summer and occasionally so in 3rd summer. Many juveniles reach tropical W Africa during Sept, including British bird recovered Ghana 2 Sept; Oct recoveries south to Angola. Preponderance of Dec recoveries of British juveniles between equator and 10°N suggests most winter northern tropics, but recoveries include 63 in Angola in 1st year of life (Langham 1971). Wintering population Tunisia (July–Mar) originates in Black Sea (numerous recoveries). Origin of birds wintering E and NE Africa unknown but probably breed Caspian Sea. Annual flocks along SE coast of southern Africa north to Natal arrive from Atlantic. Birds ringed Great Britain recovered Morocco, Western Sahara, Mauritania, Senegambia, Guinea-Bissau, Guinea, Sierra Leone, Liberia, Ivory Coast, Ghana, Togo, Benin, Nigeria, São Tomé, Cameroon, Gabon, Zaïre, Congo, Angola, Namibia, South Africa and Mozambique; 1 ringed Denmark recovered South Africa (L. St. Lucia).

Food. Small pelagic fish; mean length (134 sightings) Sierra Leone 5 cm, 4 fish from stomachs 3–10 cm (mean 7·1). Sardines important, mainly *Sardinella aurita* in Ghana, *S. eba* recorded in Sierra Leone.

Breeding Habits. Monogamous; breeds in dense colonies, though numbers small in Mediterranean where species is marginal. Displays, postures and crèche behaviour like Royal's *S. maxima*.

NEST: shallow hollow scratched in sand; scant lining.
EGGS: 1–3, usually 2. Slightly pyriform; cream to pale brown, sometimes unmarked, usually marked densely brown or grey. SIZE: (n = 20, Mediterranean) mean 51·5 × 36. WEIGHT: 35 (estimated).

LAYING DATES: Tunisia, probably May.
INCUBATION: begins with 2nd egg; carried out by both parents with frequent nest-reliefs. Period: 23 days.
DEVELOPMENT AND CARE OF YOUNG: young form crèche at 7–15 days; fledging period *c.* 35 days. Inefficient feeding makes fledged birds strongly dependent on parents; become self-sufficient during 1st winter.

References
Campredon, P. (1978).
Cramp, S. (1985).
Dunn, E. K. (1972a).
Elliott, C. C. H. (1971).
Langham, N. P. E. (1971).

Sterna sumatrana Raffles. Black-naped Tern. Sterne diamante.

Sterna sumatrana Raffles, 1822. Trans. Linn. Soc. London 13, pt. 2, p. 329; Sumatra.

Plates 26, 28
(Opp. pp. 401, 417)

Range and Status. Breeds tropical Indo-Pacific including Indian Ocean from Amirantes and Aldabra to Chagos Islands.

Vagrant South Africa and Mozambique; 4 immatures at Umvoti river mouth, Jan–Mar 1976, 1 netted 30 Mar; other Natal records Umvoti and Umdloti river mouths Oct and Dec 1978 and Feb 1981; adult with immature Mozambique (Inhaca Island) 11 Nov 1976.

Description. ADULT ♂ (non-breeding): forehead and crown white; broad black streak through eye extends to lores, narrowing to a point, and behind eye encircles crown, broadening to cover whole nape; feathers of nape slightly elongated. Hindneck white, not clearly differentiated from very pale grey mantle, back, uppertail-coverts and wing-coverts. Tail white, 4 innermost rectrices very pale grey. Underparts white, variably suffused pink; underwing-coverts and axillaries white. Primaries white with very pale grey centres, shafts white, outer web of outermost contrastingly black or blackish; secondaries somewhat greyer. Bill black (variable yellowish tip indistinct or absent); eye dark brown; legs and feet black, soles flesh. Little seasonal variation. Sexes alike except ♀ somewhat smaller. SIZE: wing, ♂ (n = 10) 216–229 (221), ♀ (n = 9) 209–226 (217); tail, ♂ (n = 10) 143–182 (156), ♀ (n = 6) 142–166 (151); bill, ♂ (n = 9) 32–37 (35·1), ♀ (n = 8) 32–36 (34·1); tarsus, ♂ (n = 10) 19–21 (19·6), ♀ (n = 9) 18–20 (19·1).

IMMATURE: forehead and crown buff streaked dusky; nape, ear-coverts and lores brownish black; hindneck, lower back and tail whitish, rest of upperparts very pale grey, lesser coverts dark grey or blackish; underparts white, sometimes suffused pale pink; primaries dark grey, shafts white; bill and feet blackish.

JUVENILE: forehead, crown and nape buff or brownish, mottled dark grey-brown and blackish, blacker on nape; upperparts mainly pale grey, feathers of mantle, scapulars and secondaries with blackish subterminal crescents, lesser coverts dark grey or blackish; tail pale grey, tipped blackish; underparts white; primaries dark grey, with broad white stripe on inner webs of outer 4; bill yellowish, legs browner.

On measurements, bird netted by Sinclair (1977) assigned to subspecies *mathewsi*, breeding W Indian Ocean; considered monotypic by Vaurie (1965).

Field Characters. Adult active, all white sea-tern with long tail streamers; black of nape, eye-stripe, outer web of leading primary, bill and legs contrasting and prominent. In other plumages, resembles somewhat larger, more heavily built Roseate Tern *S. dougallii*, but fast shallow wing-beat is striking. At distance resembles White Tern *Gygis alba*.

Voice. Not tape-recorded. Gives high-pitched 'keeyik' or 'kee', often repeated 3 or 4 times; also lower, sharp, alarm call 'chit-chit-chit-rer'.

STERNIDAE

General Habits. Prefers shallow, calm water inshore, especially inside reef; also feeds offshore. Feeds at or near surface, using variety of techniques, but usually dipping-to-surface; occasionally swims like noddies *Anous* to snatch prey from surface; submerges from both swimming position and plunge-dive.

Food. Small fish and invertebrates.

Reference
Sinclair, J. C. (1977).

Plates 26, 28
(Opp. pp. 401, 417)

Sterna dougallii Montagu. Roseate Tern. Sterne de Dougall.

Sterna dougallii Montagu, 1813. Orn. Dict: Suppl. (not paged); Scotland.

Range and Status. Breeds Africa, Eurasia, Australia, North and South America, and Malagasy Region.

Resident, possibly intra-African migrant and Palearctic winter visitor, very abundant to uncommon. Palearctic birds winter W Africa from Mauritania to Nigeria, especially south of 12°N. Small numbers regular Mediterranean east to Libya, once Suez; formerly bred SE Tunisia. Breeds E Africa, 8000 pairs, mainly Kenya (Kiunga Islands), with more erratic breeding (in hundreds or thousands) to south in Kenya and Tanzania; also S Somalia. In South Africa (Cape Province), breeding confined to islands in Algoa Bay (70–80 pairs); formerly Dyer Island and near Cape Town. In non–breeding season leaves breeding colonies but extent of movement unknown; only rare visitor SE Africa (Mozambique, Natal) and probably mainly vagrant elsewhere away from countries where it breeds. Reported occurrence in numbers and breeding NE Africa apparently erroneous.

Description. *S. d. dougallii* Montagu: range of species in Africa except 1 individual Natal. ADULT ♂ (breeding): forehead, crown and nape black, hindneck white with slight grey tinge; rest of upperparts very pale grey, rump and uppertail-coverts paler. Tail, outer pair rectrices greatly elongated and white, central pair ash-grey with white tip, rest same but with white inner webs. Underparts white with rosy wash; in E African breeders rosy wash indistinct or absent after laying, whereas acquired by Cape breeders only by the time chicks hatch. Primaries slightly darker than upperparts; white on inner edge of inner webs extending to tips; secondaries with white tips and inner webs, inner feathers with dark grey central patch. Bill of African breeders more red than in temperate populations with base red, rest black (sometimes only distal quarter), red becoming more extensive as season progresses and fading to pink towards end; eye dark brown; legs red. ADULT ♂ (non-breeding): tail without streamers; forehead and forecrown white, sometimes with few blackish tips and spots; nape and back of crown black-brown; underparts without rosy wash; bill black, legs orange-red. Sexes alike except ♀ somewhat smaller. SIZE: wing, ♂ (n = 19) 225–242 (233), ♀ (n = 12) 228–235 (231); tail, ♂ (n = 13) 165–205 (180), ♀ (n = 11) 158–201 (178); bill, ♂ (n = 20) 37–40 (38·8), ♀ (n = 12) 35–39 (36·8); tarsus, ♂ (n = 16) 19–21 (20·2), ♀ (n = 10) 19–21 (19·6). WEIGHT: ♂♀ (n = 27, South Africa) 102·5–136·5 (112·2).

IMMATURE: resembles non-breeding adult but forehead, lores and hindneck white freckled brown; crown, nape and ear-coverts dark sooty brown or blackish, forecrown streaked brown, black and white; some black around eye, mainly in front. Mantle pale blue-grey; back, rump and uppertail-coverts ash-grey; most wing-coverts pale blue-grey tipped whitish, lessers dark grey fringed whitish. Outer pair of rectrices white; next 2 pairs with outer webs dark grey, fine blackish line near tip, inner webs white; next 2 pairs of rectrices with ash-grey outer webs and dark brown band at tip, inner webs white; central pair of rectrices ash-grey with dark brown bar at tip. Bill and legs black, latter becoming orange-red.

JUVENILE: like immature but mantle feathers have brown-black crescent-shaped bars at tips; back, rump and uppertail-coverts faintly mottled brown.

DOWNY YOUNG: down appears coarse and matted, buff or grey, heavily marked brown-black. Chin, throat and sides of head often blackish, varying grey-brown to nearly black; breast and belly white or buff-white; bill blue- to flesh-grey, tip dark red-brown to blackish; legs varying shades grey, blackish when older.

S. d. bangsi Mathews: Natal (recorded once). SIZE: wing ♂ (n = 15) 213–229 (221). Bill sometimes wholly red.

Subspecies limits poorly defined, e.g. breeding bird collected Tanzania most closely matches *bangsi* from Sri Lanka (Thomas and Elliott 1973).

Field Characters. Graceful medium-sized sea-tern. Breeding birds distinguished from all except vagrant Black-naped Tern *S. sumatrana* by rosy wash on breast; distinguished from Black-naped by being larger. In all plumages adult appears much whiter than Common *S. hirundo*, Arctic *S. paradisaea*, Antarctic *S. vittata* and White-cheeked Terns *S. repressa*. At rest, long and deeply forked tail projecting well beyond wing-tips distinguishes it from all but Antarctic. Also smaller than all of these, with flight more buoyant and with usually shallower and stiffer wing-beats. At all ages, no dark line along trailing edge of outer wing (*cf.* Common, Arctic). Mantle of recently fledged individuals boldly marked, noticeably brown and buff. Like other species, immature has dark carpal-bar, but this and other grey areas somewhat paler; more extensive areas of blackish on head, forehead never pure white and sometimes only slightly paler than crown. 'Aak' and 'chew-ik' calls diagnostic. Black-billed birds have been confused with larger Sandwich Tern *S. sandvicensis*. Immature readily confused with Black-naped Tern.

Voice. Tape-recorded (62, 73, C). Especially vocal at breeding colonies. Most frequent year-round call a harsh 'chew-ik' or 'chiv-ik', varying in pitch and intensity, sometimes described as liquid or musical. Alarm-calls loud, most distinctive a guttural, rasping 'aak' or 'kraak'. When defending nest, a chattering 'kekekekek'.

General Habits. Inhabits open shore on sandy, rocky or coral coasts, sometimes resting and feeding in sheltered estuaries and creeks. Scores or hundreds feed together, usually with other terns, in deep and generally clear water, 0·5–3 km offshore; breeders at Algoa Bay feed mainly within 300 m of colony in water *c.* 20 m deep, frequently with Jackass Penguin *Spheniscus demersus*. Feeds on congregations of small pelagic fish, usually caught by diving deeply, from *c.* 2 m or more above water surface; also dips-to-surface for prey driven there by larger pelagic fish. Fast flyer, commutes regularly up to 10 km to favoured feeding areas. Kleptoparasitism sometimes regularly organized; victimizes Common, Arctic and Sandwich Terns. Rests mainly on offshore banks and reefs.

Movements of African breeders little known; most leave Kenya Nov–May; recent Algoa Bay (South Africa) records May–Feb. Palearctic birds visit mainly Oct–Apr. Leaves European waters rapidly, most reach tropics Sept–Oct; by Nov all ringing recoveries from countries south of 10°N. Immatures remain in winter quarters until 2nd and often 3rd summer. Recoveries suggest movement northwards within tropics, one-third of those in 2nd summer being north of 10°N, most northerly record off Western Sahara (22°N), most Ghana (*c.* 5°N) (Langham 1971). Birds ringed Great Britain recovered Morocco, Western Sahara, Mauritania, Senegambia, Sierra Leone, Liberia, Ivory Coast, Ghana, Togo and Nigeria.

Food. Small pelagic fish and probably crustaceans. South Africa (Algoa Bay), 11 species of fish identified, mostly Ratfish *Gonorhynchus gonorhynchus* (mean length 83 mm), but Cheilodactylidae and Clupeidae (39–88 mm) comprised 40% of total (Randall and Randall 1978).

Breeding Habits. Monogamous, breeding in dense colonies on small offshore islands; territory 0·2–4·0/m^2 (mean 0·4 South Africa, 0·7 Kenya); few colonies used annually, though probably annual Kiunga (Kenya) where numbers can be as little as 1% of those of previous year (Britton and Brown 1974). Arrive at colony 15–45 days before 1st eggs laid. Pair-formation involves both aerial and ground displays. In 'low flight', ♂ carries fish noisily around colony and often is joined by excited group of up to 6 other birds. In 'high flight', several birds ascend rapidly to 100 m or higher, and 1st pair at highest point displays during descent. 'Dreads' and 'panics' involve only part of colony. Ground display, with drooping wings, elevated tail and upstretched neck, often includes more than 2 birds; postures comparatively exaggerated, whereas aerial display less elaborate than those performed by Common and Arctic, both of which are more pugnacious towards intruders at colony. Courtship feeding and presentation of fish to mate on nest infrequent. Establishes and defends territory around nest after egg laid; quarrels continuously with neighbours (mainly conspecifics) throughout season; mean nearest neighbour distance 0·66 m (Randall and Randall 1981).

NEST: shallow scrape (up to 1 cm deep) or natural hollow, with scant lining of grass or succulents, or bed of small stones; typically in short grass or halophytic herbs, but also bare coral, rock-strewn slopes, dense growths, or tangles of *Commiphora* and *Mesembryanthemum*; 1st egg may be laid almost without any nest lining, material being collected around it later.

EGGS: 1–3, laid at 2–3 day intervals; Kenya mean (1188 clutches) 1·43; South Africa (37) 1·32. Pyriform, light brown background, dark brown and black spots, underlain by purple blotches. SIZE: (n = 23, Kenya) 38·0–46·2 × 28·3–31·1 (42·8 × 29·8); (n = 28, South Africa) 39·8–44·5 × 28·6–31·5 (42·1 × 30·1). WEIGHT: 20 (estimated). In South Africa, 1st egg laid has significantly greater volume than 2nd.

LAYING DATES: E Africa, mainly July–Aug (especially mid to late Aug), but also Tanzania Sept–Oct and Somalia June; South Africa, June–Oct. In Kenya lays during rough weather season when islands least accessible; laying often well synchronized, and may coincide with neap tides (Whale Island, 3°24′S, 36°E) (Britton and Brown 1974).

INCUBATION: begins with 1st egg; carried out by both parents with frequent nest-reliefs. Period: *c.* 25 days.

DEVELOPMENT AND CARE OF YOUNG: at hatching, mean weight 16·2 g, mean bill length 9·3 mm (12 chicks). Newly hatched chicks remain in nest, constantly attended by both parents; when small, camouflage rather than hiding used to escape predators; leave nest at 7–12 days, hide under rocks or vegetation, and move as much as 10 m in a day (Randall and Randall 1981). Fledging period: 23–30 days. Most chicks dependent for several weeks, probably months, after fledging.

BREEDING SUCCESS/SURVIVAL: human predation major cause of mortality: trapping W Africa, on line and hook or in baited snares, believed to be principal factor responsible for rapid decline Europe; taking of eggs for food, all breeding areas, can cause catastrophic losses, e.g. complete failure 1000 pairs Kenya. Eggs and chicks taken by Hemprich's *Larus hemprichii* and Kelp Gulls *L. dominicanus*, eggs by Brown-necked Raven *Corvus ruficollis*; many chicks die from exposure, drenching and intra-specific aggression. No evidence of significant losses from pollution; inexplicable desertion on massive scale Whale Island perhaps tick infestation as in Seychelles (Bourne *et al.* 1977). Rearing success, 0·09–0·40 chicks per pair Algoa Bay; too low to sustain already reduced South African population (Randall and Randall 1981).

References
Britton, P. L. and Brown, L. H. (1974).
Cramp, S. (1985).
Randall, R. M. and Randall, B. M. (1978, 1981).

Plates 26, 28
(Opp. pp. 401, 417)

Sterna hirundo Linnaeus. Common Tern. Sterne pierregarin.

Sterna hirundo Linnaeus, 1758. Syst. Nat. (10th ed.) p. 137; Sweden.

Range and Status. Breeds Africa, Eurasia, North America; winters Africa, South America, Madagascar, India, New Guinea, Solomon Islands and Australia.

Resident, intra-African migrant and Palearctic winter visitor, mainly coastal; very abundant to uncommon. Annual breeding Mauritania (Banc d'Arguin, 200–900 pairs) and SE Tunisia; erratic Western Sahara (Puerto Cansado), Senegambia (Saloum delta), Guinea-Bissau (Bijagos Islands), Nigeria (Dodo R. mouth) and Libya. Palearctic visitors winter all coasts; inland in Kenya (Rift Valley lakes) and Burundi (L. Tanganyika); passage Sudan (Nile); vagrant Uganda, Zambia and Malaŵi.

Description. *S. h. hirundo* Linnaeus: breeds W and NW Africa, common winter visitor to coasts throughout. ADULT ♂ (breeding): forehead, crown to lower part of eye and nape black; hindneck, mantle, back, scapulars and wing-coverts blue-grey; rump, uppertail-coverts, lower part of lores, sides of head, chin, throat, neck, underwing-coverts and undertail-coverts white; axillaries washed grey, rest of underparts grey tinged mauve. Inner webs of rectrices white, also outer webs of central pair, outer webs of outer feathers grey. Primaries with outer webs and tips grey; shaft and much of inner web white, although most inner webs of outer primaries more grey-black than white. Secondaries grey, tips and part of inner webs white; primary coverts silver-grey, lesser coverts along edge of wing white. Bill scarlet, tipped black; eye dark brown, legs vermilion. ADULT ♂ (non-breeding): remiges and rectrices much as breeding, grey areas of tail darker, no rectrices wholly white. Forehead and hindneck white, latter tipped grey; forecrown white, feathers with grey-brown centres; nape and rest of crown black-brown, lores white, spot in front of eye and speckles under eye black. Upperparts paler, less blue; rump and uppertail-coverts grey, lesser coverts dusky grey; underparts white with trace of grey. Bill black, sometimes crimson at base; legs red or red-brown. Sexes alike except ♀ somewhat smaller. SIZE: wing, ♂ (n = 73) 257–287 (272), ♀ (n = 39) 259–290 (270); tail, ♂ (n = 13) 148–168 (157), ♀ (n = 11) 148–170 (158); bill, ♂ (n = 66) 35–40 (37·1), ♀ (n = 36) 32–37 (35·2); tarsus, ♂ (n = 45) 19–22 (20·2) ♀ (n = 36) 19–21 (19·8). WEIGHT: ♂♀ (n = 323, South Africa) 93–155 (124); (Egypt) ♂ (n = 1) 103, ♀ (n = 1) 81.

IMMATURE: much as non-breeding adult, but mantle duller grey, rump whiter, greater coverts ash-grey, lesser coverts dark brown or blackish, primary coverts dull grey tipped white; tail feathers with outer webs darker and browner, tips white, and subterminal brown bar. Bill blackish, base flesh; legs orange.

JUVENILE: underparts, tail and wing much as immature, wing-coverts (except lessers) tipped brown or buff. Forehead white, washed buff; forecrown similar, streaked black-brown; feathers of hindneck white tipped buff, with subterminal brown bar; back and rump ash-grey, sides of rump and lateral uppertail-coverts white tipped buff.

DOWNY YOUNG: down longish, soft, with hair-like tips. Upperparts buff to brown or grey, marked black-brown, forehead often unmarked dusky; face buff, throat brown or dusky; rest of underparts white; bill pink to orange or scarlet, tipped black; legs pink to orange.

S. h. tibetana Saunders: regular Natal in numbers, once Malaŵi, presumably eastern seaboard to north. Breeding plumage darker and duller than nominate; in other plumages white upper part of crown and hindneck less extensive, crown streaking finer, upper parts darker and more blue. Ratio bill to wing small, bill ♂♀ (n = 26) 29–36 (33·7) (Clancey 1976).

Field Characters. A medium-sized sea-tern. In breeding plumage, very like Arctic Tern *S. paradisaea*, but body paler, facial streak less contrasting, bill proportionately longer and never all red, legs longer, tail streamers shorter, rump less contrasting, flight more buoyant. In other plumages easily confused with Arctic, Antarctic *S. vittata*, White-cheeked *S. repressa* and Roseate Tern, *S. dougallii* and individuals are often left indeterminate. Distinguished from them by dark on primaries forming dark wedge on upperwing and to a lesser extent on underwing. Also rump pale grey, tail mainly grey-white while Arctic and Antarctic with contrasting white rump. Less robust than White-cheeked with wings and legs longer; also less robust than Antarctic with bill rather shallower, and appearing lighter and less vivid red. Dark carpal-bar particularly prominent; primaries not so white or translucent as Arctic. Scaling on mantle of juvenile more conspicuous than juvenile Arctic, less than juvenile Roseate or Antarctic.

Voice. Tape-recorded (62, 73, C). Varied repertoire, mainly harsh and strident. Most distinct year-round calls are 'keee-yaah' and 'keee-rrr', accent on 1st syllable; also 'kik-kik-kik' in fishing parties and migrant flocks. When defending nest, rattling 'kek-kek-kek', and harsh, screaming 'karr'.

General Habits. Occurs in inshore and offshore habitats. Gregarious. Usually plunge-dives for food from about 3 m. In W Africa, eats fish offal thrown by fishermen, and dips-to-surface for fish escaping from nets or driven to surface; scavenges on foot. Feeds inshore and offshore with flocks regularly seen at upwellings 600 km off Guinea where remains dawn to dusk; presumably roosts on sea or is airborne all night. Also roosts on sea out of sight of land along South African coast (150, roosting for 2 h, Natal: Sinclair 1982). Some lose food to skuas; off South Africa, 17% of kleptoparasitic chases by Arctic Skua *Stercorarius parasiticus* successful.

Main passage Palearctic birds late Aug–Oct, late Mar–Apr. Spends little time in winter quarters (Kasparek 1982); lingers on southbound migration, but NW African coast traversed quickly during spring (estimated 80–120 km/day on average). Birds from S and W Europe winter mainly north of equator, Mauritania to Nigeria, some Angola, Namibia and South Africa; others winter Angola, Namibia, South Africa and Mozambique but predominantly on west coast. Ringing recovery New York to Ivory Coast. In east, common on passage Egypt, abundant Somalia, only small passage Red Sea. Large flocks Kenya of uncertain origin, probably include many *tibetana*; recoveries Tanzania from Austria and West Germany, and 2 Somalia birds ringed Black Sea. Inland passage probably regular, especially Rift Valley: airstrike Oct over inland Sudan, where also Norwegian bird recovered Nile; almost annual Burundi (L. Tanganyika) since 1975, and several times Rift Valley lakes Kenya. Immatures oversummer throughout; most probably spend 2nd summer in tropics. Birds ringed Great Britain, recovered Morocco, Algeria, Western Sahara, Mauritania, Senegambia, Guinea, Sierra Leone, Liberia, Ivory Coast, Ghana, Togo, Benin, Nigeria, Gabon, Angola, Namibia, and South Africa.

Food. Mainly fish. In Sierra Leone, Anchovy *Engraulis guineensis* and Mugilidae frequent; in Ghana sardines important, especially *Sardinella aurita* and *S. eba*; also locally crustaceans probably important.

Breeding Habits. Monogamous, mostly paired on arrival at nest-sites. In Africa, loosely colonial in small colonies on offshore islands, rarely on mainland; Banc d'Arguin population spread over 7 islands, nests usually 1 m or more apart. Aerial displays usually involve 2 birds, not always an established pair (Cullen 1960). The high flight begins with a circling ascent, wings angled back, flight jerky. After 1 or 2 min, as high as 200 m, the pursuer (flying slightly above its partner) initiates the descent by swooping low over the leader. During this and subsequent passes, the pursuer overtakes. The descent involves a fast glide, steep at first. The low flight (or fish flight) usually involves an unmated territorial ♂, often carrying a fish; flight mostly level, also a glide with wings locked in upwards 'V' position; ♂ tries to get ♀ to engage in a pass. Two main ground postures: bowing, with back slanting, tail cocked at about 45°, neck arched, bill pointing obliquely down; and stretching, with neck stretched up, bill nearly vertical, tail elevated (less than in bow). Both ground postures involve held-out, drooping wings, and head thrown up (then held vertical).

NEST: shallow scrape in sand, bare ground or very short grass; sometimes lined with grasses and herbs.

EGGS: 1–3, laid at 1–2 day intervals; Tunisia mean (9 clutches) 2·7, Senegambia (Saloum) 1·55 (11 clutches) perhaps low due to disturbance. Subelliptical, buff to grey-green or brown, blotched and spotted dark brown and ash-grey; variable colour said to act as natural camouflage at Saloum. SIZE: (n = 24, Tunisia) 36·5–43·5 × 29–32; (n = 15, Saloum) mean 40·5 × 29·5. WEIGHT (estimated): 20.

LAYING DATES: Senegambia (Saloum, Casamance), Guinea-Bissau (Bijagos) and Mauritania (Banc d'Arguin), Apr–June, mainly May–June with Saloum a little behind Bijagos, perhaps 1–2 weeks ahead of Banc d'Arguin; Tunisia and Libya Mar; Nigeria, July.

INCUBATION: begins with 1st egg; carried out by both parents with frequent nest-reliefs. Period: 25 days.

DEVELOPMENT AND CARE OF YOUNG: fledging period *c.* 25–30 days. Young rely on camouflage rather than cover. Dependent on parents for only short time after fledging.

BREEDING SUCCESS/SURVIVAL: low success rate of 12% at Saloum probably due to human predation; little evidence of predation at other colonies. Opportunistic feeding behaviour makes them susceptible to capture. Mass mortality in viral epizootics South Africa, notably in 1961 (Rowan 1962).

References
Clancey, P. A. (1976).
Cramp, S. (1985).
de Naurois, R. (1969).

STERNIDAE

Plates 26, 28
(Opp. pp. 401, 417)

Sterna paradisaea Pontoppidan. Arctic Tern. Sterne arctique.

Sterna paradisaea Pontoppidan, 1763. Danske Atlas 1, p. 622; Denmark.

Range and Status. Breeds Arctic and subarctic regions Eurasia and North America; winters in Antarctic Ocean south to 74°S.

Holarctic winter visitor, mostly Palearctic and mostly passage, abundant to uncommon, along western seaboard and all coasts South Africa; also Mozambique, Somalia and Algeria. Vagrant inland Cameroon, Sudan, Somalia and South Africa.

Description. ADULT ♂ (breeding): forehead, crown to lower part of eye and nape black; hindneck, mantle, back, wing-coverts and scapulars blue-grey, latter broadly tipped white; rump, uppertail-coverts, lower part of lores, sides of head, chin, throat, neck, underwing-coverts, undertail-coverts and most of tail white; rest of underparts pale slate-grey. Outer webs of 2 outermost pairs rectrices and all remiges grey, primaries with shaft and most of inner web white; some grey-black on most inner webs; secondaries with tip and most of inner web white. Bill red, tip of upper mandible rarely black; eye dark brown; legs red. ADULT ♂ (non-breeding): remiges, rectrices and most upperparts as breeding, but forehead, forecrown and hindneck white, mottled grey; nape and rest of crown black-brown, lores white, spot in front of eye and speckles under eye black; lesser coverts brownish or dusky; underparts whitish. Bill black, reddish at base; legs and feet dark brown-red or blackish, soles paler. Sexes alike except ♀ somewhat smaller. SIZE: wing, ♂ (n = 20) 270–290 (279), ♀ (n = 16) 261–288 (274); tail, ♂ (n = 10) 190–214 (198), ♀ (n = 9) 170–204 (180); bill, ♂ (n = 19) 31–35 (33·0), ♀ (n = 20) 29–33 (30·8); tarsus, ♂ (n = 20) 15–17 (15·9), ♀ (n = 19) 15–17 (15·6). WEIGHT: ♂♀ (n = 39, South Africa) 76–120 (94); (Sudan) ♂♀ (n = 1) 80.

IMMATURE: much as non-breeding adult, except varying amounts of black on forehead and forecrown, lesser coverts slate-grey, underparts and inner webs of outer rectrices often partly grey, and outer webs of inner rectrices ash-grey.

JUVENILE: much as immature, lacking most of the brown and buff of Common Tern *S. hirundo*: forehead, forecrown and broad band on hindneck white, rest of crown contrastingly black. Bill black by Aug–Sept; legs darken from orange-red to dark brown-red or blackish by Oct–Nov.

Field Characters. Medium-sized tern, readily confused with Common, Roseate *S. dougallii*, Antarctic *S. vittata* and White-cheeked *S. repressa* Terns. In breeding plumage underparts darker than Common, white rump contrasting with back. More robust Antarctic and White-cheeked are still darker below, white cheek more contrasting; white rump of Antarctic contrasts with back like Arctic; White-cheeked is uniformly dark above. At rest, tail of Antarctic (and much paler Roseate) projects well beyond wing-tips; Arctic's about equal to wings. Antarctic has markedly heavy, wholly red bill. Legs of Arctic distinctly shorter than Common, similar to White-cheeked. Further differences detailed under these species, especially Common and Antarctic.

Voice. Tape-recorded (62, 73, C). Varied repertoire; voice resembles Common Tern but pitched somewhat higher. Most distinct year-round call is 'keee-yaah', with accent usually on drawn-out 2nd syllable.

General Habits. Occurs both inshore and offshore, but more pelagic than Common. At sea rests on floating objects. Crustaceans caught by dipping-to-surface; most fish caught by plunge-diving from 1 to 6 m (mean 3 m), after searching upwind and hovering; immersion usually complete.

Ringed immatures have reached Liberia and Cameroon from Europe by Aug; some winter equatorial region (e.g. Nigeria), many more South Africa (predominantly E Cape and Natal), and some penetrate Antarctic region with bulk of adults; South African recoveries include birds from Canada, Greenland and NW Russia (Elliott 1971, Langham 1971, E. K. Dunn, pers. comm.). Adults on passage reach southern Africa by Sept, most Oct–Nov; return passage Mar–May, mainly breeding plumage. Atlantic coast passage further offshore than Common, includes many Nearctic birds which have crossed Atlantic to Europe. Many immatures oversummer; ringing recoveries of 1st summer birds within 10° of the equator indicate northward movement, probably followed by a return south for 2nd winter (Langham 1971); breeds 3rd–4th summer. Recent records Somalia mainly passage, also oversummering. Inland records, mostly adults May. Birds ringed Great Britain recovered Morocco, Senegambia, Guinea, Sierra Leone, Liberia, Ivory Coast, Ghana, Benin, Nigeria, Gabon, Cameroon, Congo, Angola, Namibia and South Africa.

Food. Fish and crustaceans, proportions uncertain.

Sterna vittata Gmelin. Antarctic Tern. Sterne couronnée.

Sterna vittata Gmelin, 1789. Syst. Nat.1, pt. 2, p.609; Kerguelen.

Plates 26, 28
(Opp. pp. 401, 417)

Range and Status. Breeds on islands of southern oceans, mainly subantarctic, including Ascension, St Helena, Tristan da Cunha, Gough, South Georgia and Bounty Islands; ranges in non-breeding season to southern Africa, South America and New Zealand.

Non-breeding visitor South Africa, Cape to Natal, mostly W Cape, abundant to uncommon. Census of 12 day-roosts Cape, 2148 birds, including 1200 at Dassen Is.; probably accounts for most birds spending austral winter South Africa (Cooper 1976).

Description. *S. v. tristanensis* Murphy: regular Cape, once Natal. ADULT ♂ (breeding): forehead, crown to lower part of eye and nape black; hindneck, mantle, back and wing-coverts blue-grey; rump, uppertail-coverts, lower part of lores, sides of head, chin, throat, neck, underwing-coverts and undertail-coverts white; rest of underparts pale slate-grey. Tail very pale grey or whitish, tip and outer web of outermost pair of rectrices slate-grey. Primaries with shaft and about half of inner web white, other half dark grey; outer webs grey, those of outer primaries darker, outermost blackish; secondaries grey, tips and most of inner webs white. Bill red; eye dark brown; legs orange-red. ADULT ♂ (non-breeding): remiges, rectrices and most upperparts as breeding but forehead and crown pale grey, feathers broadly tipped grey-black, especially on crown; nape and rear crown blackish; lores white, heavily streaked and spotted blackish brown towards eye. Underparts often as breeding, but with varying amounts of white, usually on throat and upper breast. Bill with varying amount of blackish near base; legs red-brown. Sexes alike except ♀ somewhat smaller. SIZE: wing, ♂ (n = 11) 245–274 (258), ♀ (n = 16) 236–272 (255); tail, ♂ (n = 7) 166–196 (180), ♀ (n = 10) 162–189 (173); bill, ♂ (n = 11) 36–39·5 (38·1), ♀ (n = 12) 33–36·6 (34·9); tarsus ♂ (n = 11) 19–22 (20·2), ♀ (n = 14) 18–21·4 (19·8). WEIGHT: (n = 85, South Africa) 105–160 (128); (Tristan da Cunha) ♂ (n = 10) 120–159 (140), ♀ (n = 10) 125–160 (139).

IMMATURE: much as non-breeding adult but lesser coverts dusky, forming rather indistinct carpal-bar; tail shorter, more strongly marked grey; some juvenile barring retained on mantle, back, secondary coverts and rectrices; underparts white; bill and legs blackish.

JUVENILE: abrasion of dark tips to white feathers of underparts and rump mostly complete before arrival in Africa. Forehead and forecrown white, with varying amount of black or brown streaks; crown and nape mostly blackish, washed and tipped buff; upperparts and wings grey, strongly barred dark brown, blackish and tawny, rump whitish. Rectrices mainly grey, outer webs darker; 3 innermost pairs broadly barred dark brown, others barred near tips.

S. v. georgiae Reichenow: regular Cape, perhaps more frequent than *tristanensis* (R. K. Brooke, pers. comm.). Small, tarsus especially short; ♂♀ (n = 14) wing, 246–266 (257); tail, 117–141 (130); tarsus, 15·9–17·8 (16·8).

S. v. vittata Gmelin: vagrant, Cape and Natal. Darker grey than *tristanensis*, tail streamers shorter and broader; ♂♀ (n = 7) wing, 260–286 (272); tail, 135–157 (149).

Field Characters. Medium-sized sea-tern. In breeding plumage resembles Arctic Tern *S. paradisaea* but bill heavier, wholly red, tone more vivid; somewhat larger, body more robust; darker, underparts hardly paler than upperparts, white cheek more contrasting; tail streamers shorter, almost white; tarsus (at rest) usually longer, and tail usually projects well beyond wings. Less seasonal variation than other grey *Sterna* spp.. Juvenile with distinctly barred upperparts, partly retained when immature. Equally dark White-cheeked Tern *S. repressa* has uniform upperparts; darker, more slender bill; and shorter tail. Common Tern *S. hirundo* is paler, less contrasting rump and facial streak. Still paler Roseate Tern *S. dougallii* has uniform upperparts, slight build and slender bill; despite graceful flight, confusion with Roseate unlikely, since juvenile Roseate with patterned upperparts and adult with long tail streamers.

Voice. Tape-recorded (C, 301). Shrill, high-pitched 'trr-trr-kriah', similar to Arctic.

General Habits. Occurs mostly on low-lying rocky headlands, sometimes sandy beaches; preference for headlands probably related to offshore feeding, mainly in cold water (Benguela Current); also feeds inshore. Roosts in groups of 10–1200. Flight graceful, undulating. Feeds in small parties, hovering 2–15 m above water, dropping to catch small prey items at or near surface. Inshore in South Africa, follows and feeds over Jackass Penguins *Spheniscus demersus*. Scavenges intertidal zone for stranded marine life; follows ships for scraps.

Arrival and departure dates at main roosts South Africa 15–31 May, 1–20 Oct (Cooper 1976). A few immatures occur there in other seasons.

Food. Small fish and crustaceans. In South Africa, probably mainly fish; at higher latitudes mainly crustaceans.

Reference
Cooper, J. (1976).

Plates 26, 28
(Opp. pp. 401, 417)

Sterna repressa Hartert. White-cheeked Tern. Sterne à joues blanches.

Sterna repressa Hartert, 1916. Nov. Zool. 23, p. 298; Persian Gulf.

Range and Status. Africa and Arabia to India.

Resident, partial migrant and visitor from Persian Gulf to eastern seaboard south to Kenya. Many breeding sites Egypt (Tiran Island, Abu Mingar) to N Kenya (Lamu Archipelago); uncommon Kenya south of breeding sites and Red Sea north of Suakin Archipelago (Sudan); vagrant Tanzania and Natal. Somalia residents uncommon and restricted; intermixed with non-breeding visitors from Persian Gulf.

Description. ADULT ♂ (breeding): forehead, upper lores, crown to lower level of eye and nape black; upperparts and tail dark ash-grey, outer web of outermost rectrix silvery (blackish when worn). Underparts ash-grey (tinged lilac when fresh); white patch from lower lores, gape and chin to sides of nape; undertail-coverts and areas bordering white cheeks paler than rest of underparts. Primaries silver-grey, inner webs with paler wedges at base. Secondaries ash-grey, fringed white on outer web; axillaries and underwing-coverts white tinged grey; white line on leading edge of wing. Bill scarlet, tipped black; darkens to blackish from about July, and many incubating birds have red restricted to gape and base of lower mandible. Eye dark brown; legs red or orange-red. ADULT ♂ (non-breeding): forehead and lores white; crown grey, mottled black at rear; nape and ear-coverts black. Underparts, including undertail-coverts and sides of head, white tinged grey. Remainder of body, wings and tail as breeding. Bill black, red tinge often visible at base. Sexes alike. SIZE: wing, ♂ (n = 11) 232–248 (240), ♀ (n = 5) 236–249 (243); tail, ♂ (n = 9) 131–154 (142), ♀ (n = 4) 132–140 (136); bill, ♂ (n = 11) 34–37 (35.7), ♀ (n = 5) 34–37 (35.1); tarsus, ♂ (n = 11) 18–20 (18.6), ♀ (n = 4) 18–19 (18.3). WEIGHT: (Egypt ♂ (n = 9) 78–105 (90), ♀ (n = 3) 87–92 (90).

IMMATURE: much as non-breeding adult but conspicuous carpal-bar formed by dark grey or brownish lesser coverts along white leading edge of wing; juvenile outer primaries retained until about 1 year, worn and blackish. At age 2 years resembles breeding adult, tail somewhat shorter, trace of carpal-bar, some white on forehead and underparts.

JUVENILE: much as immature with upperparts and tail paler ash-grey; underparts white, much of plumage elsewhere mottled buff or brown. Legs and base of bill pale red at fledging, soon darkening.

DOWNY YOUNG: usually black or blackish on chin, throat and lores, otherwise extremely variable. Above whitish or buff to sandy brown or bluish grey; blackish markings, sometimes absent, vary greatly in size and pattern. Underparts mainly white or buff. Bill red with dusky tip; legs flesh.

Field Characters. Medium-sized sea-tern. Size and bill as Common Tern *S. hirundo*, but shorter-winged, shorter-legged, and appearing rather stout on the wing. In breeding plumage, superficially resembles Whiskered Tern *Chlidonias hybridus*, although tail longer and more deeply forked, wings longer, and habitat exclusively marine. Darkest and most uniformly coloured of medium-sized congeners, rump uniform with mantle, not white as in Arctic Tern *S. paradisaea* and Antarctic Tern *S. vittata*. In other plumages, dark, uniform upperparts and tail, contrasting with white underparts; underwing predominantly grey. Many birds have patchy grey and white underparts.

Voice. Tape-recorded (271). Distinctive year-round flight call, a loud, harsh 'kee-errr', accent on variable 2nd syllable; also rendered 'kerrit' and 'kee-leek'. Breeding calls mainly resemble Common Tern; include piercing 'kee-kee-kee' in defence of nest.

General Habits. Occupies inshore and offshore habitats. Feeds mostly within 3 km of land. In Kenya, usually feeds with other terns beyond reef, regularly 5–10 km from land; Red Sea (Ethiopia) winter flocks feed offshore, apparently following schools of whales. Seldom submerges, most food obtained by contact-dipping. Settles on water; often stands in shallow water. Large numbers, many immature, associate with breeding colonies of Saunder's Tern *S. saundersi* on coastal plain Somalia, resting and roosting with them; all rise up and circle together if colony disturbed.

Present Red Sea mainly Apr–Sept; Somalia and Kenya all months; largest flocks Kenya Mar–June. Persian Gulf visitors on Somalia coast mainly Sept–Apr.

Food. Small fish and invertebrates driven to surface by predatory fish. In Egypt fish include *Atherinomorus lacunosus* (Atherinidae) (n = 8, 35–70 (51) mm long) and *Spratelloides* sp. (Clupeidae) (6, *c.* 45 mm long) (S. M. Goodman and R. W. Storer, pers. comm.).

Breeding Habits. Monogamous, nesting colonies on small islands. Numbers in African colonies comparatively small, usually 10–200 pairs, often split into subcolonies which lay up to 7 days apart. At Kiunga Islands, Kenya, mainly confined to one edge of large

Roseate Tern *S. dougallii* colony. On Red Sea islands (Sudan) forms smaller colonies than other terns, often at edge of Lesser Crested Tern *S. bengalensis* colonies. At nest, less pugnacious than Common and Arctic Terns, wheeling overhead and calling; reluctant to swoop or strike. Displays hardly documented; said to resemble Common Tern.

NEST: shallow scrape or natural depression, with scant lining. Siting and construction extremely varied in Sudan: may be natural depression in sand, or made from various combinations of seaweed, twigs and coral rubble; or sometimes a built-up structure 5 cm or more high (R. J. Moore pers. comm.).

EGGS: 1–3, usually 2; 4 in nest Kiunga Islands laid by 2 birds; Kenya mean (304 clutches) 1·72, varying 1·47 to 1·94 at 4 subcolonies. Subelliptical, extremely varied, pale buff to brown, blackish and dark brown markings, underlain by lilac or grey blotches. SIZE: (n = 23, Kenya) 35·8–44·7 × 27·3–30·4 (41·5 × 29·0). WEIGHT: (n = 20, Kenya) 14–18 (16·0), variation within clutches as much as 2·5.

LAYING DATES: Egypt and Ethiopia, June–Aug, mainly July; Sudan, May–June; Somalia, July–Sept; Kenya, July–Oct, mainly Aug.

DEVELOPMENT AND CARE OF YOUNG: fledged birds gather in very tight groups at water's edge, begging from adults; presumably dependent for some weeks after fledging.

BREEDING SUCCESS/SURVIVAL: at Kiunga Islands (Kenya) chicks taken by Yellow-billed Egrets *Egretta intermedia*, Sacred Ibis *Threskiornis aethiopica* and Yellow-billed Stork *Mycteria ibis*; small flock of latter wiped out small colony.

References
Britton, P. L. and Brown, L. H. (1974).
Cramp, S. (1985).

Sterna anaethetus Scopoli. Bridled Tern; Brown-winged Tern. Sterne bridée.

Plates 25, 28
(Opp. pp. 400, 417)

Sterna anaethetus Scopoli, 1786. Del., Faun. et Flor. Insubr. 2, p. 92; Philippines.

Range and Status. Breeds widely tropical and subtropical Indo-Pacific and Atlantic, including Africa.

Resident and partial intra-African migrant, abundant to uncommon. Breeds Egypt, Sudan, Ethiopia, NW Somalia, N Kenya in east, Western Sahara (Virginie), Mauritania (Banc d'Arguin, 1500 pairs), Pagalu (Gulf of Guinea, 200 pairs) in west; unsuccessful attempts Senegambia and S Somalia. Disperses widely offshore after breeding; seldom recorded other coasts; uncommon Tanzania, Mozambique, Natal, Togo; vagrant Cape (32°48′S, 17°55′E). Banc d'Arguin population apparently increasing, 1480 pairs in 1974 almost double 1960 estimate (Trotignon 1976).

Description. ADULT ♂ (breeding): forehead white, extending in stripe over and beyond eye; crown, nape and stripe from ear-coverts to base of bill black; hindneck pale grey, merging into dark grey-brown of upperparts and tail. Wings mainly as upperparts, some coverts darker, leading edge of inner wing strikingly white. Outermost pair of rectrices greatly elongated, outer web and basal half of inner web white; others with white bases to inner webs. Underparts white, breast and belly washed grey; underwing-coverts white, underwing grey with paler tip. Primaries black-brown, bases pale brown; outer 3 with inner webs whitish, outermost whiter. Bill black; eye dark brown; legs black. ADULT ♂ (non-breeding): little seasonal variation; black areas of head black-brown, crown lightly streaked whitish; upperparts with whitish tips; tail less deeply forked. Sexes alike except ♀ somewhat smaller. SIZE: wing, ♂ (n = 13) 255–274 (264), ♀ (n = 15) 242–263 (256); tail, ♂ (n = 9) 168–205 (179), ♀ (n = 6) 164–201 (177); bill, ♂ (n = 8) 40–46 (42·6), ♀ (n = 9) 38–41 (39·5); tarsus, ♂ (n = 13) 20–22 (21·1), ♀ (n = 15) 19–22 (20·5). WEIGHT: ♂♀ (n = 69, Seychelles) mean 95·6.

IMMATURE: as non-breeding adult; wings and tail browner, pale tips, crown more streaked, 'bridle' less defined.

JUVENILE: forehead and stripe over eye white; crown, nape and lores streaked grey-white and brown; hindneck black-brown, few whitish streaks; upperparts and wing-coverts dark brown, feathers broadly tipped white or cream, especially mantle and back. Rectrices dark brown, tipped whitish, shorter and blunter than older birds.

DOWNY YOUNG: down of head and upperparts long and fine, buff or grey, speckled dark brown and blotched darker grey; underparts white or buff; eye, bill and legs black, legs tinged blue.

Field Characters. A medium-sized sea-tern, only confused with adult Sooty Tern *S. fuscata*, but paler, less uniform above; white eyebrow narrower and longer, extending *c*. 1 cm behind eye; broad pale neck collar, contrasting with crown; smaller and shorter-winged. Juvenile has pattern similar to adult; upperparts paler, feathers broadly tipped white or cream; head not black and hardly contrasts with mantle; immature Sooty mainly dark brown.

Voice. Not tape recorded. Vocal at colony; commonest note a barking staccato 'wep-wep' or 'wup-wup'. When fishing, gives variety of grating cries 'kr-arr', 'ka-karr', 'karr-kow', 'k-rrr'; also low subdued 'kwit'. At roost commonly gives mournful 'kree-err', and during courtship gives 'kek-kek', 'greer-greer' and 'mer-er-er' (see Breeding Habits).

General Habits. Offshore rather than pelagic, usually within 50 km of land. Often inshore in Red Sea, even harbours; occasionally blown inland Ethiopia in squally weather. Reluctant to rest on water, commonly perches on driftwood, buoys, masts and rocks. Regularly roosts in low bushes, rocks, sandbanks, posts, and tall trees; lack of inshore records in Kenya indicates marine roosting. Often gregarious at roosts. Feeds individually, in small parties or large flocks; joins in feeding associations with other terns. Flight markedly buoyant and graceful, with exaggerated wing-beats. Often flies within 1 m of water, hovers frequently, dips-to-surface for prey; also plunge-dives with complete immersion. Comfort movements typical of genus: 1 wing and leg stretched over fanned tail, frequent preening; head twisted to one side to rub on shoulder; scratching of head or beak with foot often done whilst flying.

Atlantic breeders probably move south to upwelling areas off Senegambia, or remain in Gulf of Guinea. Occurs seasonally in Red Sea and Somalia, Apr–Nov; appears in large flocks offshore Ethiopia 2nd week Apr, moving inshore 3rd week, colonies form in May. Small numbers regular off Kenya all year. Occasional Tanzania, including bird ringed as nestling Seychelles Nov 1973 recovered off Pemba May 1974. Several birds Natal Dec 1973–Mar 1975, 1st recorded southern Africa, perhaps associated with drastic food shortages in Seychelles or elsewhere in Malagasy Region. Massive wrecks of Sooty Terns from Natal to Maputo (Mozambique) associated with Jan 1976 cyclone in Mozambique channel included scattering of Bridled Terns.

Food. Small fish and invertebrates (including crustaceans) taken mainly at surface. Seychelles, by weight 95% fish, mostly Mullidae; also squid, water-bug *Halobates*. Most fish taken in Australia 2–8 cm.

Breeding Habits. Monogamous, breeding in loose colonies, mostly 30–400 pairs, on small offshore islands; nests 1–5 m apart, sometimes only 30 cm. W Africa, 1 pair/5–6 m²; Kenya, sometimes forms small, compact colonies among far more numerous Roseate Tern *S. dougallii*. In courtship display, bodies of ♂ and ♀ canted towards each other, wings folded and drooping outwards; goose-step side by side, or circle one another, head averted slightly from partner; use bills to pick up and manipulate stones and vegetation. During 'parade', short note 'kek' or 'kuk'; changes to growling 'greer-greer' during mutual nodding and bill fencing. 'Dreads' and 'panics' involve only part of colony, triggered by 'mer-er-er' from single bird, making violent evasive movements in fast flight, as though avoiding hawk. Birds nearby follow. Interspecific territoriality involves raised crest and braying call, bill opened very wide.

NEST: shallow scrape, depression or crevice, little or no lining. Site very varied, usually hidden and shaded; in broken terrain (e.g. fallen rocks, ridges in coral), concealed in scrubby vegetation, or under old nests of Long-tailed Cormorant *Phalacrocorax africanus*.

EGGS: 1, rarely 2 (probably 2 birds: C. J. Feare, pers. comm.). Subelliptical, buff or cream, with dark brown or red-brown speckles, spots and blotches. SIZE: (n = 108, Indian Ocean) 40·4–46·0 × 28·5–33·1 (44·0 × 31·2); (n = 6, Banc d'Arguin) mean 45 × 32·5. WEIGHT: (n = 6, Seychelles) mean 20.

LAYING DATES: Egypt, Sudan and Ethiopia, June-July; Somalia, June-Aug; Kenya, July–Aug; Pagalu, Oct–Nov; Mauritania, May–July, mainly late May–mid June; Western Sahara, May-early June. Laying often well synchronized. No evidence of subannual breeding (*cf*. nesting every 7½–8 months Seychelles: Diamond 1976).

INCUBATION: carried out by both parents. Period: 28–30 days. Nest-site territoriality after laying involves off-duty bird standing on nearby vantage point (rock, coral outcrop, shrub). Intruding Hemprich's Gulls *Larus hemprichii* often successfully driven away, Sudan; wandering chicks of gulls and other terns may be attacked violently and lifted bodily, Banc d'Arguin. Not very pugnacious towards humans; sometimes defaecates on them, never strikes them. Often performs broken-wing display after returning from initial flight.

DEVELOPMENT AND CARE OF YOUNG: young cared for and fed by both parents; leaves nest after 3 or more days; escapes predators by hiding, usually in nearby vegetation. Wandering chicks of conspecifics seldom attacked. Fledging period: 55–63 days. Leaves colony area *c*. 35 days after fledging, possibly independent.

BREEDING SUCCESS/SURVIVAL: jackals take toll of nestlings and incubating adults, Banc d'Arguin; probably responsible for destruction of Senegal delta colony July 1974 (Dupuy 1975). Difficult for humans to locate eggs and young at most sites; losses from disturbance rather than direct predation.

References
Cramp, S. (1985).
Nicholls, G. H. (1977).

Sterna fuscata **Linnaeus. Sooty Tern. Sterne fuligineuse.**

Sterna fuscata Linnaeus, 1766. Syst. Nat. (12th ed.), p. 228; Santo Domingo.

Plates 25, 28
(Opp. pp. 400, 417)

Range and Status. Breeds and ranges tropical and subtropical Indo-Pacific and Atlantic, including Africa.

Resident, intra-African migrant and visitor from New World tropics, abundant to uncommon. Colonies of 5000–200,000 pairs Principe (Tinhosas Islands, Gulf of Guinea), NW Somalia (Mait Island), Kenya (Tenewe Islands), and Tanzania (Latham Island); also breeds in small numbers S Somalia and Senegambia. Otherwise mainly pelagic, dispersing widely after breeding; on west coast, within Gulf of Guinea; on east coast, Gulf of Aden to Natal. Birds breeding New World tropics and semi tropics (Florida) disperse across tropical Atlantic to Gulf of Guinea; large numbers may be involved. Most records SE Africa associated with cyclones, south to E Cape, sometimes far inland (Malaŵi, Zimbabwe, Transvaal); vagrant Tunisia.

Description. *S. f. fuscata* Linnaeus: only subspecies in Africa. ADULT ♂: forehead and superciliary stripe to middle of orbit white; crown, nape and broad line through eye black. Upperparts, wings and tail mainly black-brown, inner webs of remiges and rectrices paler and greyer; outermost pair of rectrices greatly elongated and mostly white. Underparts white, washed ash-grey on flanks, belly and undertail-coverts. Bill black; eye dark brown; legs black. No seasonal variation other than wear (especially rectrices). Sexes alike except ♀ somewhat smaller. SIZE: wing, ♂ (n = 16) 280–304 (294), ♀ (n = 13) 276–297 (287); tail, ♂ (n = 10) 161–194 (174), ♀ (n = 8) 154–184 (174); bill, ♂ (n = 16) 40–46 (42·3), ♀ (n = 9) 38–43 (40·2); tarsus, ♂ (n = 18) 22–25 (23·6), ♀ (n = 12) 21–25 (23·1). WEIGHT: ♂♀ (n = 450, Seychelles) date of laying, 148–240 (201); later in season ♂ (n = 21) 165–224 (172), ♀ (n = 16) 157–225 (189).

IMMATURE: brown-black above, dark brown below; feathers of back, mantle, scapulars, rump, uppertail-coverts, wing-coverts, tertials and rectrices with variable cream-white tips, broader from mantle to uppertail-coverts, those of scapulars still broader and whiter; underparts dark brown washed grey, undertail-coverts pale grey. From age 1–2 years, resembles adult; upperparts mainly black-brown, mantle washed grey, some whitish fringes. 30% of 3 year olds and 5% of 5 year olds visiting colonies retain some dark speckling below.

JUVENILE: much as young immature except somewhat paler, more olive.

DOWNY YOUNG: down with rather long rami, hair-like tips short, filaments joined together at tips; appears rather coarse. Breast and belly white; otherwise brown-black speckled grey-white and grey-black, with varying amount of buff.

Field Characters. A quite large sea-tern resembling Bridled Tern *S. anaethetus*, but larger, longer-winged and darker; white eyebrow broader, less tapered, not extending behind eye; bounding flight, frequent soaring. Younger birds all dark, resembling noddies *Anous* at distance; tail forked (not wedge-shaped), vent pale grey, and upperparts with white flecks and spots.

Voice. Tape-recorded (62, 73). Flight call nasal 'ker-wacky-wack' carrying far over water; often given at night. Alarm calls include croaking 'krark' and more prolonged neighing 'kreeaa'.

General Habits. Mainly pelagic, believed to sleep on wing; rests on flotsam; reluctant swimmer. Usually gregarious. Flock of 5000 in Gulf of Guinea mistaken for cloud at distance, displaying remarkable cohesion even whilst fishing; wheeling and dipping in tight formation. Flocks often include small numbers of other terns, especially Bridled Tern and Brown Noddy *Anous stolidus*. Plunge-dives infrequent and shallow, usually dips-to-surface; most fish and squid caught at or above surface whilst escaping from shoals of larger predatory fish. Preponderance of squid in diet suggests feeding activity concentrated at dawn and dusk; bathypelagic fish, which ascend to surface at night, in diet suggests it also feeds at night.

Makes long-distance movements to seasonally productive areas, especially convergence zones. Gulf of Aden and Indian Ocean breeders believed to move south with shifting belt of SE trades, Oct-May. Present Somalia, north and east coasts, Mar–Nov. Many birds from largest African colony (Tinhosas) may spend whole year in Gulf of Guinea; flocks include up to 20% immatures. New World birds from Florida move to southern Caribbean, then disperse across tropical Atlantic to Gulf of Guinea Oct–Nov, travelling 11,000 km by supposed route via Brazil; immatures and adults appear to winter in separate areas. 1st year birds wintering in Gulf of Guinea may remain there up to their 6th year; numerous recoveries Sierra Leone to Cameroon, mostly Oct–Jan, oldest bird 25 months (Robertson 1969).

Food. Small fish and invertebrates; squid can form 50% of diet by weight.

Breeding Habits. Monogamous, nesting mostly in massive dense colonies, on small offshore islands. On the ground, both sexes stretch wings vertically, making sharp V over back; head thrust forwards horizontally. 'Parade' is characteristic display in which one circles the other; both droop wings, arch neck, turn head slightly away, and take small running steps; used especially at nest changeover, but seen throughout breeding period. Parading birds may break off to attack nearby conspecifics. Bowing and 'high flight' also common (see Common Tern *S. hirundo* account for description of these).
  NEST: shallow scrape or depression, little or no lining; in sand, short grass, shallow soil, coral, on rock. Sites usually flat and low-lying, segregated from other species; single nest Senegal delta in Gull-billed Tern *Gelochelidon nilotica* colony, in hollow between sand dunes, concealed by large branch (Dupuy 1979).
  EGGS: invariably 1. Subelliptical, pale or deep buff, blotched or spotted chestnut-brown or blackish brown, with lilac undermarkings. SIZE: (n = 100, Atlantic Ocean) 44·5–56·0 × 31·8–38·1 (50·5 × 35·0); (n = 42, Indian Ocean) 47·5–58·0 × 34·7–38·1 (51·8 × 36·0). WEIGHT: fresh eggs Seychelles, means 31·8 (n = 74), 36·2 (n = 117), generally lighter as season progresses.
  LAYING DATES: Tanzania (Latham Island), Oct–Nov; Principe, Senegambia, Somalia and Kenya, late June–early Aug, mainly July. Laying often synchronized. Seychelles, replacement eggs laid 12–13 days after loss of 1st; some individuals lay 3 eggs in a season.
  INCUBATION: carried out by both parents, with frequent nest-reliefs; often accompanied by parade; shifts lengthen as season progresses. Period: 28 days. At large dense colonies, defends nest-site territory as far as it can reach whilst sitting. Incubating birds are not very pugnacious in defence of nest except at hatching when will wheel noisily in dense cloud over colony, 100 m or more high. Many reluctant to leave vicinity of nest; readily caught by hand whilst incubating or hovering around intruder's head.
  DEVELOPMENT AND CARE OF YOUNG: young grow rapidly first 30–35 days. Fed mainly fish and squid, size of items not increasing as chicks grow; those fed solely on squid lose weight; often given drinks, not only by own parents; chicks hatched at 29–34 g increase in weight more rapidly than 20–23 g group, heavy chicks also heavier when last caught (186 g, n = 29; 172 g, n = 28); wing of flying juvenile 4% shorter than adult, bill 20% shorter. Fledging period: 8–10 weeks. Period of dependence uncertain, but remain near colony 2–3 weeks after fledging. Return to colony at night to be fed by parents, after spending day alone at sea. Also fed over colony with adult flying above juvenile, latter giving begging call; adult regurgitates food and passes it directly to juvenile's bill. Near colony, juveniles engage in erratic darting movements close to water surface, dipping-to-surface to pick up weed. These movements during learning period strikingly similar to adults' darting in erratic flight over shoals of predatory fish (Feare 1976a).
  BREEDING SUCCESS/SURVIVAL: hatching success 75% at colony centre, only 10% at edges due to increased predation. Overall chick mortality 27%, early hatched chicks surviving better. Most losses due to pecking by adults; rain increases mortality, especially of pecked chicks. Mass desertion of eggs and young probably due to heavy infestation with *Ornithodoros capensis* ticks infected with Soldado virus; high incidence of congenital defects and plumage deficiencies in such colonies in Seychelles (Feare 1976a).

References
Clancey, P. A. (1977).
Cramp, S. (1985).
Feare, C. J. (1976a).

Plates 26, 28
(Opp. pp. 401, 417)

Sterna balaenarum (Strickland). Damara Tern. Sterne des baleiniers.

Sternula balaenarum Strickland, 1852. Jardine's Contr. Orn., p. 160., Damaraland.

Range and Status. Resident, endemic to western seaboard and Cape, abundant to uncommon. Namibia breeding population 1000–2000 pairs, mainly north of Swakopmund; far smaller numbers S Cape (east to Algoa Bay), N Cape (4 colonies Orange R. south to Kleinzee: J. Cooper, pers. comm.), formerly near Cape Town. Probably breeds S Angola (150 at Cunene R. in breeding season, Nov–Jan). Partial migrant, flocks of up to 250 Nigeria (Wallace 1973), 30 or 40 Ghana since 1967 (Sutton 1970), up to 75 Togo 1976–77 (Browne 1980), possibly increasing W Africa. Published records tropical Indian Ocean erroneous. Development, especially tourism, mining and off-road vehicles, seriously disturbs nesting habitat of this rare and vulnerable species.

Description. ADULT ♂ (breeding): forehead, crown, lores, upper ear-coverts and hindneck black; mantle, back, scapulars, wing-coverts, rump, uppertail-coverts and tail pearl-grey; outer web of outermost pair rectrices white tipped grey; underparts white, washed grey; axillaries and underwing white. Primaries with white shafts; 3 outermost primaries with outer webs and shaft-line of inner web dark grey, rest of inner web white; others, and secondaries, grey, white margins to inner webs. Bill black, some yellowish at gape; eye dark brown; legs yellowish.
  ADULT ♂ (non-breeding): much as breeding except upperparts and tail paler grey, underparts white; forehead white, crown pale grey; nape, upper ear-coverts and spot in front of eye brownish black, many crown feathers tipped same; bill wholly blackish, legs ochre or dusky, toes and back of tarsus yellowish. Sexes alike except ♀ somewhat smaller. SIZE: wing, ♂♀ (n = 6) 164–175 (171); tail, ♂♀ (n = 6) 68–74 (72); bill,

Sterna balaenarum

♂♀ (n = 7) 27 – 32 (29·3); tarsus, ♂♀ (n = 7) 14 – 16 (14·7). WEIGHT: (n = 11, South Africa) 51 – 66 (58·7).

IMMATURE: much as non-breeding adult except crown whiter, less flecked; lesser coverts dusky, showing as indistinct carpal-bar; primaries darker grey; base of lower mandible horn or dusky.

JUVENILE: forehead and crown rich buff or tawny, crown flecked brown; contrasting dark brown or blackish stripe through eye, from lores to ear-coverts; upperparts and tail fawn, distinct U-shaped subterminal dark brown markings, mainly on mantle and back; scapulars and wing-coverts partly coloured as mantle, also partly grey, carpal-bar indistinct; underparts white, breast and flanks washed fawn, indistinct greyish crescent-shaped markings; primaries blue-grey, edged silver-white, shafts dark, coverts grey. Upper mandible and tip of lower mandible ash-grey, rest of lower mandible dull orange-ochre; eye dark brown; legs dull orange.

DOWNY YOUNG: down of head and upperparts pale fawn, covered with small black speckles, some black patches; underparts white; bill blackish, white egg tooth prominent; legs yellow; fresh feathers of mantle and scapulars rich tawny, soon fading to pale fawn.

Field Characters. Small sea-tern with rapid and erratic flight, easily confused with Little Tern *S. albifrons* but bill black (not yellow with black tip), longer, thinner and slightly decurved with pronounced gonys; black on head more extensive, extending from forehead to hindneck and down to below eye; head larger; body stockier and heavier; and flight more powerful, not so jerky; also adult appears white in flight in all seasons. Forehead and crown of immature and non-breeding adult whiter, less marked than Little, eye-stripe generally more pronounced; rump and tail uniform with mantle; carpal-bar indistinct or absent. Upperparts of juvenile strongly barred brown and buff, while upperparts of juvenile Little less so; also juvenile Little's eye-stripe less pronounced or absent on lores, extending across nape. (For other differences, see accounts of Little and Saunders' *S. saundersi* Terns.)

Voice. Tape-recorded (252). Frequently gives harsh and rapid calls especially when fishing, wheeling over disturbed nest, or when flushed from roost. Main calls include high-pitched 'tsit-tit', repeated several times, and sharp metallic 'tit-tit'.

General Habits. Inhabits estuaries, creeks, bays, harbours, lagoons, pans and surf zone on open shore. Seldom joins in feeding assemblages with other terns, although roosts communally on sandbanks and beaches. Usually feeds alone, individuals spaced 10–50 m apart; plunge-dives after hovering; small parties occur, but not for communal fishing (J. Cooper, pers. comm.). In Cape Town harbour, at refuse from ships (where small fish feeding), 5 of 7 dives successful; associates closely with Cape Cormorant *Phalacrocorax capensis*: 2 terns follow underwater movements from hovering position, catch fish disturbed by cormorant, plunges often coinciding with cormorant's surfacing.

Occurs in numbers all months Namibia, with most adults present in austral winter Mar–Oct. Absent Cape austral winter (J. Cooper, pers. comm.). Nigeria (Lagos) recorded all months Apr–Jan; flocks July–Oct, 72–81% adult; uncommon Nov–Jan, mainly immatures. Togo (Lomé) flocks July–Oct; Angola (Cabinda, Luanda) flocks May–Oct, mainly immatures.

Food. Small fish, also squid and probably crustaceans. Fish fed to chicks Namibia, 1·5–12·5 cm long, weighing 2–30 g; included *Mugil richardsonii*, *Engraulis japonica*, also squid and larval blenny.

Breeding Habits. Monogamous; semi-colonial, forming loose aggregations of 4–60 pairs, usually 10 or fewer, sometimes nests singly. In colony of 9 pairs, Namibia, nearest distance to neighbour 32–96 m (mean 57); 9 pairs Algoa Bay (Cape) mean 185 m. No territorial interactions. Foot-shuffling by ♀ immediately prior to copulation; pebble-flicking used by ♂ and ♀ as displacement activity when disturbed near nest. Namibia breeders are seldom aggressive. Disturbed bird leaves nest early, flies overhead noisily; does not mob or harass intruders. Usually returns within 2 min of intruder's departure, lands few metres away, pauses, then walks to nest. Marked fidelity to colony site in subsequent seasons, even though coast in vicinity appears visually similar. Recent breeders Cape mobbed humans, Kelp Gulls *Larus dominicanus* and White-fronted Plovers *Charadrius marginatus* whenever they approached nest-site.

NEST: normally some sort of scrape, decorated with pebbles, pieces of shell or pieces of lichen around rim. ♂ and ♀ actively scrape, shuffling with both feet, slowly turning in chosen spot, flicking pebbles and other small objects from scrape with bill. Cape breeders lay near sea, mean distance 159 m (n = 10) from spring high tide level. In Namibia, eggs are laid up to 3 km from sea, on firm stony rather than sandy substrate in sites with good visibility, on gravel plains, hardened surface of dry saltpans, bare rocky areas, and slack areas between dunes. Nests located mostly on featureless gravel plains sometimes 1 km from sea or at edge of Namib Desert. Although *Arthraerura* and *Zygophyllum* shrubs and *Parmelia* and *Teloschites* lichens are found widely in nesting habitats, nest-sites not associated with them or with large stones.

EGGS: invariably 1. Buff, with even scattering of dark brown speckles, underlain with lighter brown and purple blotches. SIZE: (n = 48) 31·3–35·8 × 23·0–25·2 (33·3 × 23·9). WEIGHT: fresh 7·8, 8·2; incubated 7·8, 9·5.

LAYING DATES: Namibia and South Africa, Nov–Jan, peak mid Dec, once late Feb, probably repeat laying.

INCUBATION: carried out by both parents with frequent nest-reliefs. Period: 18–22 days. Incubate 94% of daylight h; periods eggs unattended 1·5–66 min (mean 5·2); parent always on nest sunset and sunrise. Hatching can take 6 h or more. Small fragment of shell membrane often adheres to chick's back, sometimes remaining there for a few days.

DEVELOPMENT AND CARE OF YOUNG: weight of chick at hatching 6·5 g, bill 7·7 mm; 7 days, 19·5 g, bill 10·7, wing 29; 15 days, 36·0 g, bill 15·4, wing 61; flies when 6 g lighter than adult, bill 19. Chick weighing 40 g ate 30 g fish. Young brooded in nest for first 2 days, then leaves nest, often moving several metres away, actively seeking concealment among stones or bushes. Incubating adults peck others' chicks if they wander too close, either while standing beside chick or swooping from above, pecking at each pass. Parents do not intervene, standing 10 m away, calling excitedly. Fledging period: 20 days. After fledging, young still accompany adults, apparently remaining dependent for 2–3 months.

BREEDING SUCCESS/SURVIVAL: hatching success 33–72%; young fledged from 16–35% of eggs laid. In Namibia, main predator is Black-backed Jackal *Canis mesomelas*; shoreline regularly patrolled by this and other predators. Siting of colonies inland perhaps reduces losses.

References
Clinning, C. F. (1978a).
Frost, P. G. H. and Shaughnessy, G. (1976).
Randall, R. M. and McLachlan, A. 1982.

Plates 26, 28

(Opp. pp. 401, 417)

Sterna albifrons Pallas. Little Tern. Sterne naine.

Sterna albifrons Pallas, 1764. Cat. Adumbr., p. 6; Holland.

Forms a superspecies with *S. saundersi*.

Range and Status. Breeds and winters Africa, Eurasia and Australasia; also winters Madagascar.

Resident, partial intra-African migrant and Palearctic winter visitor, very abundant to uncommon. Breeds mainly on N and W African coasts north of equator; largest coastal populations (hundreds of pairs) Tunisia, Mauritania, Senegambia, Ghana, Nigeria, Cameroon; also inland W African major rivers and Kenya (L. Turkana, thousands). Partial intra-African migrant, especially W African inland breeders; vagrant Lake Naivasha (Kenya) and Lake Tanganyika (Burundi). Passage and wintering of Palearctic breeders exclusively coastal; winters Senegambia to Gulf of Guinea in west, Somalia to South Africa (Natal) in east; commonest migrant tern some sites Mozambique and Natal.

Description. *S. a. albifrons* Pallas: breeds Egypt to Morocco; winters Red Sea to Natal, Senegambia to Gulf of Guinea. ADULT ♂ (breeding): forehead and above eye white; crown, nape and lores black; hindneck, uppertail-coverts and tail whitish, rump pale grey; rest of upperparts (including wing-coverts) blue-grey; underparts white. Remiges mainly blue-grey, inner webs edged white; outermost 2 or 3 primaries with tips, outer webs and shaft-line on inner webs blackish grey, shafts brown. Bill yellow, tipped black; eye dark brown; legs yellow or orange-yellow, claws black. ADULT ♂ (non-breeding): much as breeding except crown ash-grey tinged brownish; nape to behind eye brown-black, flecked white; blackish patch in front of eye; rest of head and underparts white; rump and uppertail-coverts only slightly paler than mantle; lesser coverts blackish grey; bill black, legs yellow-brown or dusky. Sexes alike except ♀ somewhat smaller. SIZE: wing, ♂ (n = 19) 176–186 (181), ♀ (n = 20) 167–180 (176); tail, ♂ (n = 10) 77–98 (85), ♀ (n = 10) 69–84 (78); bill, ♂ (n = 15) 28–33 (30·4), ♀ (n = 19) 27–31 (28·6); tarsus, ♂ (n = 12) 16–18 (16·8), ♀ (n = 18) 16–18 (16·6). WEIGHT: (n = 8, South Africa) 48·3–56·7 (53·2); (Egypt) ♂ 41·5, 44·5, ♂ 46·4.

IMMATURE: as non-breeding adult except rectrices mainly ash-grey, dark brown subterminal markings, outermost white; remiges, lesser coverts and leading edge of wing darker grey.

JUVENILE: much as immature except forehead, lores and crown sandy buff, last streaked blackish; nape and sides of crown brown-black, narrowly fringed buffish; hindneck buffish white; mantle and scapulars sandy buff, U-shaped brown subterminal bars; back, rump and uppertail-coverts ash-grey tipped buff or white; leading edge of wing blackish grey; bill horn-brown to blackish, base dull yellow.

DOWNY YOUNG: down short with fine hair-like tips; pale sandy brown, with variable amount of darker streaking, often as 2 or 3 parallel streaks; white below; bill flesh-grey, tipped dark brown; legs flesh.

S. a. guineae Bannerman: coastal Mauritania and from Ghana to Gabon; also larger rivers W Africa, Lake Chad, Lake Turkana (Kenya). Rump, uppertail-coverts and tail strikingly white; little or no dusky or black on bill of breeding adult. SIZE: ♂♀ (n = 18) wing, 155–169 (163).

S. a. sinensis Gmelin: Ghana, presumably vagrant. Shaft outermost primary white; tail long, more deeply forked. SIZE: ♂ (n = 10) wing, 182–192 (186); tail, 95–140 (109).

Field Characters. A small tern, similar to Saunders' Tern *S. saundersi* and Damara Tern *S. balaenarum*. Saunders' hardly distinguishable in non-breeding or immature plumage. Breeding Saunders' has white of forehead more restricted than Little, not extended posteriorly at sides to behind eye (**A**); upperparts paler more chalky grey; tail whiter; 3 outermost primaries mostly black, with black and white portions clearly demarcated; legs yellowish olive or olive. Damara lacks white forehead and mostly yellow bill of breeding Little and Saunders'; head of Damara more robust, bill longer, gonys more pronounced (see Damara for other differences). All 3 have proportionately longer bill and shorter tail than larger congeners, wings narrower, flight more fluttering. All *Chlidonias* are larger with shorter bill, smaller head, more leisurely flight.

Voice. Tape-recorded (62, 73, C). Gives various calls, more liquid and less harsh than most terns; include 'kit-kit', 'kruit', 'queet', the latter often given as prolonged, excited trill.

General Habits. Favours inshore shallow water of creeks, harbours, lagoons and saltpans, but regularly joins in feeding flocks with other terns up to 15 km offshore; seen feeding with Common Tern *S. hirundo* 600 km off Guinea. Occurs inland on larger rivers and lakes with exposed beaches. Hovering flight 3–6 m above water surface, often alters height by sudden drop; feeds mostly by plunge-diving, immersion often complete; dips-to-surface to take prey with backward flick of bill; takes insects from ground and vegetation, rather like *Chlidonias*. Nigeria, recorded in tree 15 km from nearest water (Ogun R., mainly dry, July); flew back to tree after being flushed.

Heavy passage Egypt, Tunisia, Morocco Aug–Oct, peak Atlantic coast Sept. Common Senegambia from early Sept; very small numbers Palearctic breeders recorded Nigeria where passage westerly Apr–May. Heavy passage Mauritania from mid Apr; Egypt heaviest May. Recoveries of W European breeders Morocco, Senegambia, Ivory Coast, Ghana; fledgling *sinensis* ringed Java, recovered Ghana. Most visitors along east coast (Indian Ocean) from Caspian–Aral Seas via Persian Gulf; no evidence of passage in Red Sea south of Egypt. Movements of tropical breeders little known. Migratory inland W Africa, where breeding sites flood; Gulf of Guinea breeders mainly resident, numbers swollen seasonally, including 1000 Nigeria July. Some residents from L. Turkana (Kenya) probably also migratory although destinations unknown. Non-breeders throughout year most tropical coasts; confusion with Saunders' Tern likely NE Africa. Some breed 2nd summer, but usually do so when older.

Food. Small fish (*c.* 5 cm long) and invertebrates (mostly crustaceans, insects).

Sterna saundersi

Sterna albifrons

Breeding Habits. Monogamous; semi-colonial, loose aggregations of 6–150 pairs, sometimes nests singly. In Senegal delta, 100 pairs in area 500 m square, 1972, many only 4–6 m apart; 100 pairs at lagoon, 1979, 100 m or more apart. Displays much as Common Tern. Courtship flights involve curious, deliberate, hesitating

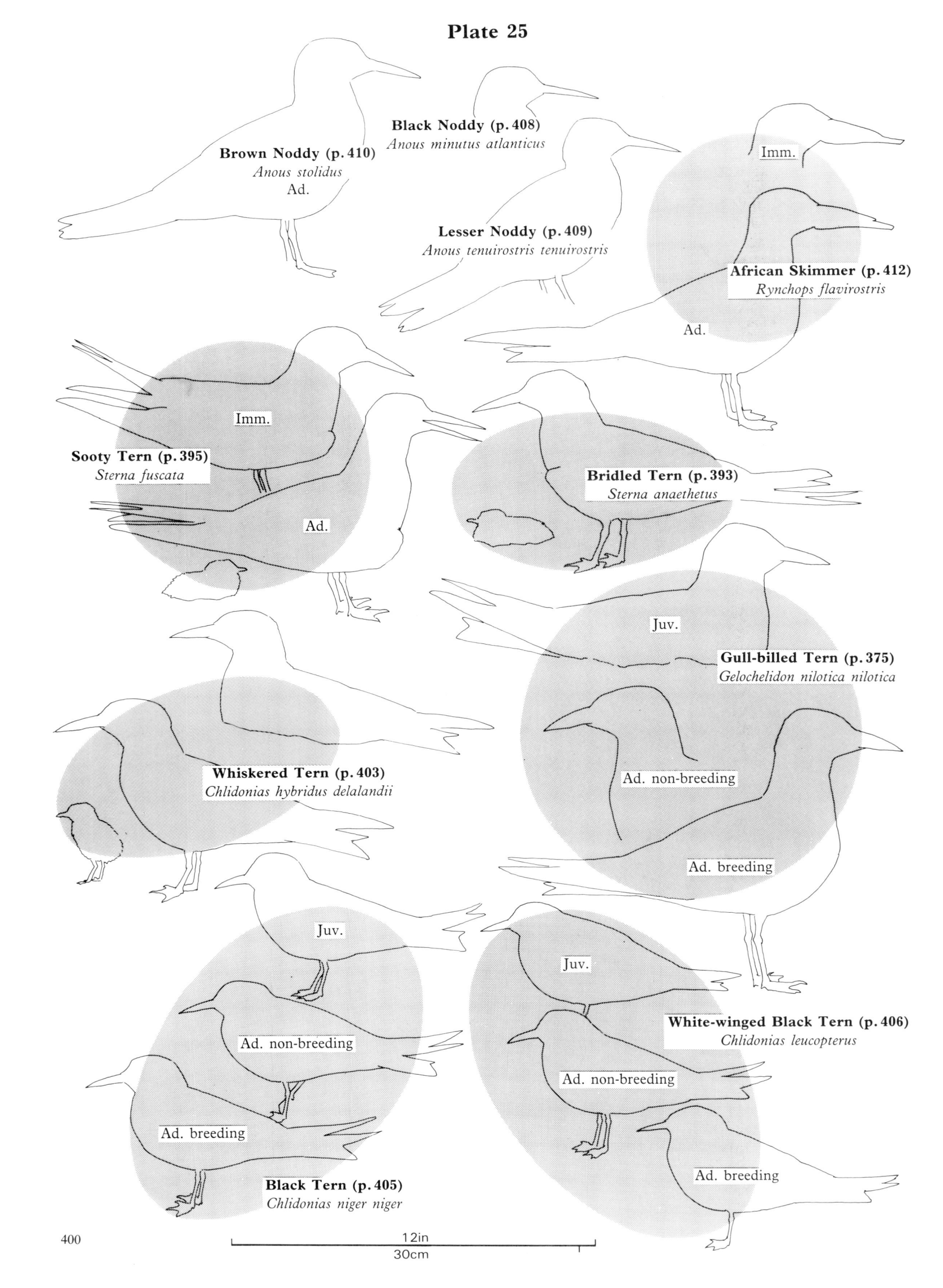

Plate 25

Brown Noddy (p. 410)
Anous stolidus
Ad.

Black Noddy (p. 408)
Anous minutus atlanticus

Lesser Noddy (p. 409)
Anous tenuirostris tenuirostris

African Skimmer (p. 412)
Rynchops flavirostris
Imm.
Ad.

Sooty Tern (p. 395)
Sterna fuscata
Imm.
Ad.

Bridled Tern (p. 393)
Sterna anaethetus

Gull-billed Tern (p. 375)
Gelochelidon nilotica nilotica
Juv.
Ad. non-breeding
Ad. breeding

Whiskered Tern (p. 403)
Chlidonias hybridus delalandii

Black Tern (p. 405)
Chlidonias niger niger
Juv.
Ad. non-breeding
Ad. breeding

White-winged Black Tern (p. 406)
Chlidonias leucopterus
Juv.
Ad. non-breeding
Ad. breeding

12in / 30cm

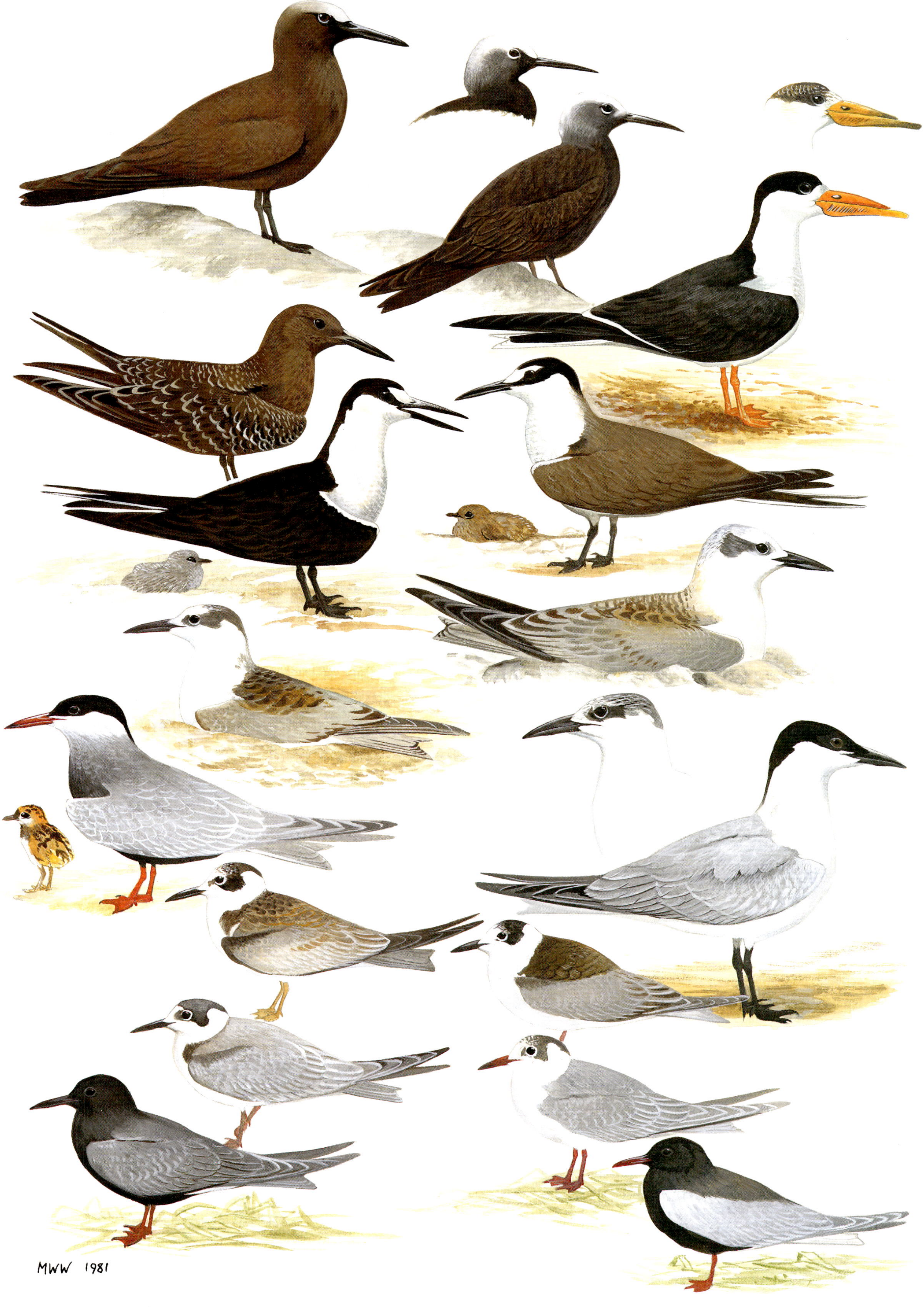
MWW 1981

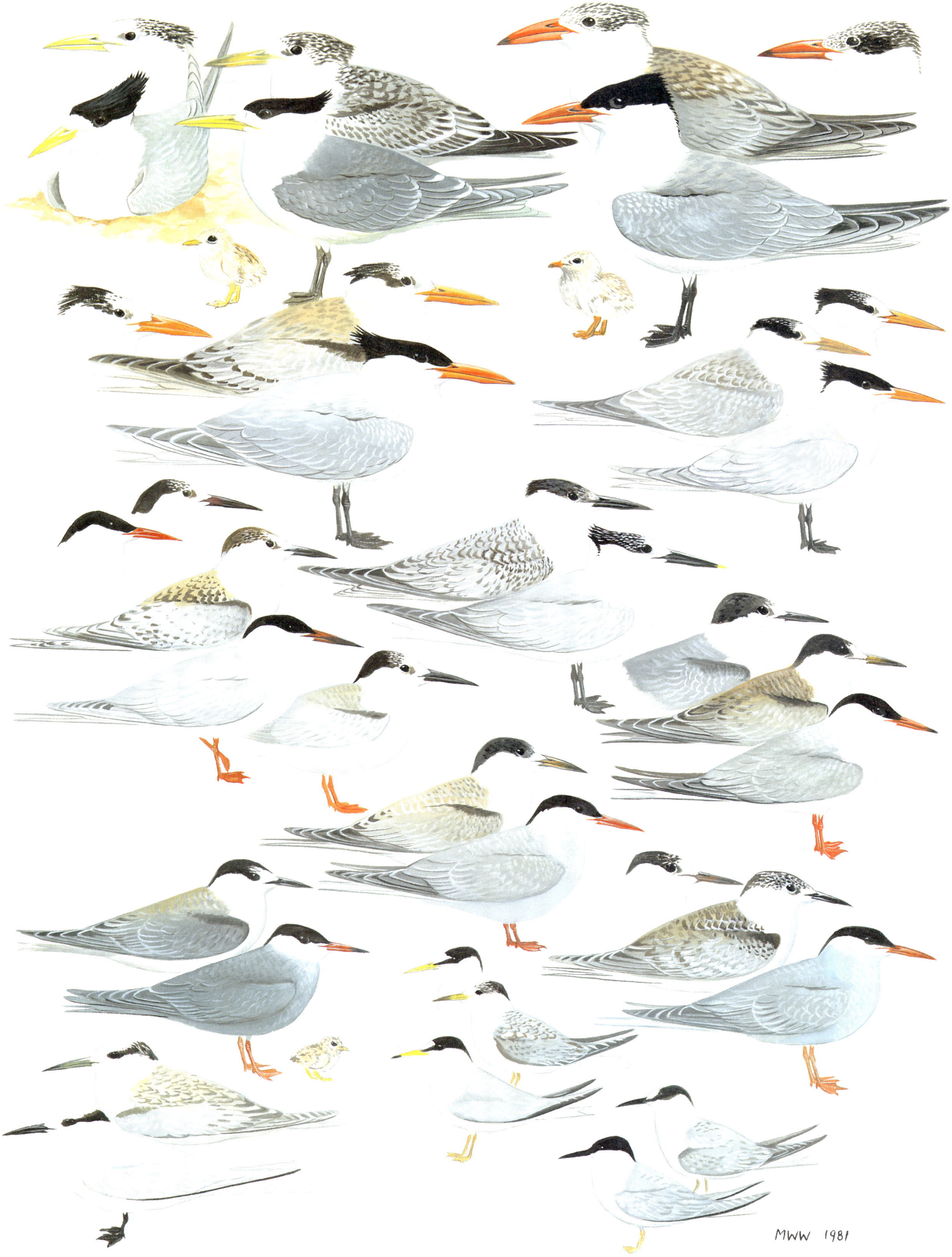
MWW 1981

Plate 26

Swift Tern (p. 381)
Sterna bergii
- *S. b. bergii* Ad. breeding
- *S. b. velox* Ad. breeding
- Ad. non-breeding
- Juv.

Caspian Tern (p. 377)
Sterna caspia
- Ad. breeding
- Ad. non-breeding
- Juv.

Royal Tern (p. 379)
Sterna maxima albididorsalis
- Ad. breeding
- Ad. non-breeding
- Juv.

Lesser Crested Tern (p. 382)
Sterna bengalensis
- Ad. breeding
- Ad. non-breeding
- Juv.

Roseate Tern (p. 386)
Sterna dougallii
- Ad. breeding *S. d. korustes*
- Ad. non-breeding
- Juv.

Sandwich Tern (p. 383)
Sterna sandvicensis sandvicensis
- Ad. non-breeding
- Juv.

Common Tern (p. 388)
Sterna hirundo hirundo
- Ad. breeding
- Ad. non-breeding
- Juv.

White-cheeked Tern (p. 392)
Sterna repressa
- Ad. breeding
- Juv.

Arctic Tern (p. 390)
Sterna paradisaea
- Ad. breeding
- Juv.

Little Tern (p. 398)
Sterna albifrons
- Ad. breeding

Antarctic Tern (p. 391)
Sterna vittata
- Ad. breeding
- Ad. non-breeding
- Juv.

Black-naped Tern (p. 385)
Sterna sumatrana
- Ad. non-breeding
- Juv.

Saunders' Tern (p. 402)
Sterna saundersi
- Ad. breeding
- Juv.

Damara Tern (p. 396)
Sterna balaenarum
- Ad. breeding
- Juv.

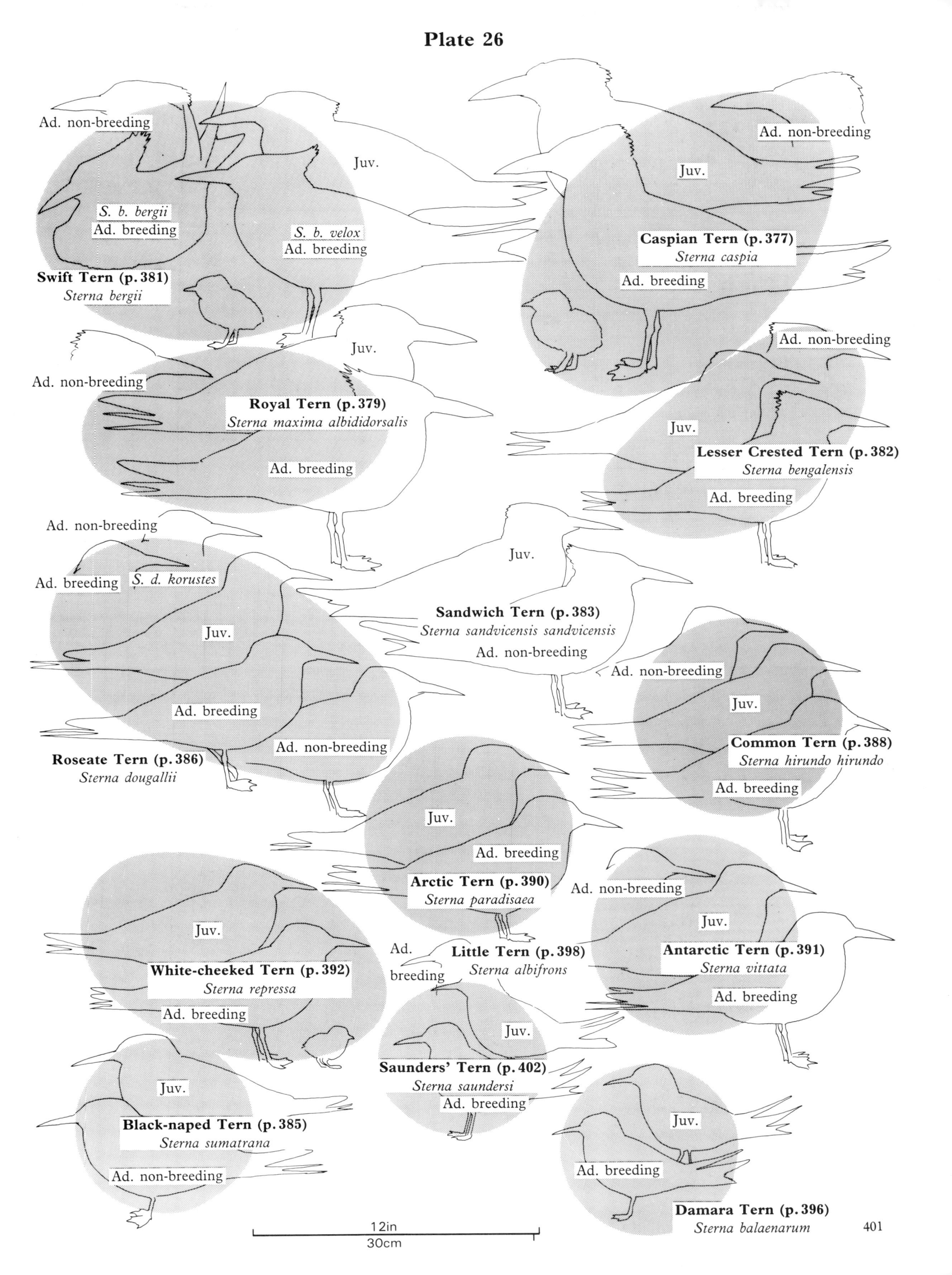

12in / 30cm

wing-strokes, with wings curved back; high flight involves shorter, less frequent glides, wings often set in V; fish presentations performed rapidly.

NEST: shallow, circular scrape without lining. Senegal delta, 5 cm deep, 15 cm diameter; Cameroon, 10 cm diameter, in sand or shingle.

EGGS: 1–3, usually 2, laid 1–2 day intervals. Subelliptical; Cameroon, buff to olivaceous or pale greenish grey with speckles, spots and small blotches of various shades of brown, and some irregular scrawls, usually at heavier larger end, underlain ashy or violet; Mauritania, dark and remarkably homogeneous, markings, black or lavender. SIZE: (n = 22, W Africa) 28·2–31·9 × 21·3–23·6 (29·9 × 22·6). WEIGHT: 8–11.

LAYING DATES: Egypt, Libya, Tunisia, Algeria, Morocco, May; Mauritania, Senegambia, Ghana, Nigeria, Cameroon, Apr–Aug, mostly May–June. Dates Kenya (L. Turkana) population unknown.

INCUBATION: carried out by both parents with frequent nest-reliefs. Period: 20–22 days.

DEVELOPMENT AND CARE OF YOUNG: leaves nest when few days old; fledging period 19–20 days. Mainly relies on camouflage to escape predators. Family groups on passage; dependent for unknown period after fledging.

BREEDING SUCCESS/SURVIVAL: many eggs destroyed as rivers rise W Africa. Young eaten by canines Banc d'Arguin (Mauritania). African populations stable, perhaps increasing in Tunisia.

References
Clancey, P. A. (1982).
Cramp, S. (1985).

Plates 26, 28
(Opp. pp. 401, 417)

Sterna saundersi Hume. **Saunders' Tern. Sterne de Saunders.**

Sterna Saundersi Hume, 1877. Stray Feathers 5: 324–326; Karachi.

Forms a superspecies with *S. albifrons*.

Range and Status. Indian Ocean, Persian Gulf and Red Sea east to India, south to Malagasy Region.

Resident SE Somalia, Sudan, Socotra, common to uncommon; partial migrant south to Tanzania (Dar es Salaam, Kilwa) and north to Egypt–Sudan border.

Description. *S. s. saundersi* Hume: only subspecies in Africa. ADULT ♂ (breeding): forehead white; crown, nape and lores black; hindneck, uppertail-coverts and tail whitish, rump pale grey; rest of upperparts (including wing-coverts) chalky grey; underparts white. Remiges mainly blue-grey, inner webs edged white; outermost 3 primaries mostly black (including shaft), black and white zones of inner webs markedly contrasted; no frosting patina to other primaries. Bill yellow, tipped black; eye dark brown; legs brownish yellow, yellowish olive or olive. ADULT ♂ (non-breeding): much as breeding except crown ash-grey tinged brownish; nape to behind eye brown-black, flecked white; blackish patch in front of eye; rest of head and underparts white; rump and uppertail-coverts only slightly paler than mantle; lesser coverts blackish grey; bill black; legs brownish or dusky. Sexes alike except ♀ somewhat smaller. SIZE: (10 ♂♂, 10 ♀♀) wing, ♂ 164–174 (168), ♀ 157–170 (165); tail, ♂ 66–99 (76), ♀ 55–73 (66); bill, ♂ 27–31 (29·6), ♀ 26–30 (28·3); tarsus, ♂ 16–19 (17·4), ♀ 14–18 (15·9). WEIGHT: ♀ 40, 42, 45.

IMMATURE: as non-breeding adult except rectrices mainly ash-grey, dark brown subterminal markings, outermost white; remiges, lesser coverts and leading edge of wing darker grey, dark area on lesser coverts somewhat broader than in adult.

JUVENILE: much as immature except forehead, lores and crown sandy buff, last streaked blackish; nape and sides of crown brown-black, narrowly fringed buffish; hindneck buffish white; mantle and scapulars sandy buff; back, rump and uppertail-coverts ash-grey tipped sandy buff or white; leading edge of wing blackish grey; bill horn-brown to blackish.

DOWNY YOUNG: down short with fine hair-like tips; pale sandy brown with darker streaking; white below; bill flesh-grey, tipped dark brown; legs flesh.

Sterna saundersi

Also winters in breeding range

Field Characters. Small tern, chalky grey above, white below. Similar to Little Tern *S. albifrons* and Damara Tern *S. balaenarum*. Saunders' and Little hardly distinguishable in non-breeding or immature plumage. Breeding Saunders' has white of forehead more restricted than Little, not extended posteriorly at sides to behind eye (**A**, p. 399); upperparts paler grey, tail whiter, 3 outermost primaries mostly black, legs yellowish olive or olive rather than yellow or orange-yellow. Damara lacks white forehead and mostly yellow bill of breeding Little and Saunders' (see Damara for other differences). All 3 have proportionately longer bill and shorter tail than larger congeners, wings narrower, flight more fluttering. All *Chlidonias* are larger with shorter bill, smaller head, more leisurely flight.

Voice. Not tape-recorded. Gives various calls, more liquid, less harsh than most terns. When chasing in courtship, gives a staccato 'kitik'. Calls often given as prolonged, excited trill.

General Habits. Favours inshore shallow water of creeks, harbours and lagoons, but regularly joins in feeding flocks with other terns up to 15 km offshore. Hovering flight 3–6 m above water surface, often alters height by sudden drop; feeds mostly by plunge-diving, immersion often complete; dips-to-surface to take prey with backward flick of bill.

Extent and regularity of movements uncertain; many sightings are of Little or Saunders', usually not distinguished. Non-breeders throughout year Somalia and Kenya.

Food. Small fish (*c.* 5 cm long) and invertebrates (mostly crustaceans).

Breeding Habits. Monogamous; semi-colonial, loose groups of 5–30 pairs; sometimes nests singly. Displays much as Common Tern *S. hirundo*. Courtship flights involve curious, deliberate, hesitating wing-strokes. Sometimes presumed ♂♂ offer small fish to mates, with wing-lifting and crests raised; fish often not accepted; birds returning from sea with fish chase each other in pairs, giving staccato 'kitik'. Vociferous when defending territory; may mob intruders; does not settle at nest whilst intruder in vicinity.

NEST: shallow, circular scrape, without lining. In Somalia (Brava) built only on hard, stony, red sand, or loose drifts of white sand 1 m high in extremely desolate site on slope of sandhills, 30 m above sea level, 2 km from sea; only 1 of 6 nests in scant vegetation, despite perennial sandstorm at ground level. Similar inland sites with numerous small colonies along 150 km stretch of coastline near Mogadishu (Somalia), and 2 km inland on Socotra.

EGGS: 1–3, usually 2, laid 1–2 day intervals. Subelliptical; stone background, dark brown markings, underlain grey; consistently paler, less spotted than Little Tern. SIZE: (n = 11, Somalia) 30·4–35·9 × 22·3–24·1 (32·6 × 23·5).

LAYING DATES: Somalia, Sudan, Socotra, June–Aug.

INCUBATION: carried out by both parents with frequent nest-reliefs. Period: *c.* 20 days.

DEVELOPMENT AND CARE OF YOUNG: leaves nest when a few days old; mainly relies on camouflage to escape predators. Fledging period: *c.* 20 days.

Reference
North, M. E. W. (1945).

Genus *Chlidonias* Rafinesque

3 small and delicate, dusky-plumaged freshwater-breeding 'marsh' birds. Bill slender, only slightly curved, shorter than head; legs small and feeble, webs of front toes much incised, appear only half webbed; tail less than half length of wing, only slightly forked, lateral rectrices not elongated into streamers. Seasonal variation especially striking. All 3 spp. visit Africa; 1 breeds there.

Chlidonias hybridus (Pallas). Whiskered Tern. Guifette moustac.

Plates 25, 28

Sterna hybrida Pallas, 1811. Zoogr. Rosso–Asiat. 2, p. 338; Volga.

(Opp. pp. 400, 417)

Range and Status. Africa, Europe, S Asia to Australia; also Madagascar.

Resident and Palearctic winter visitor, abundant to uncommon. Breeds N Africa: traditional sites mostly drained (Mees 1977); S Morocco (Iriki, 100 pairs 1966), Tunisia (L. Kelbia, 60 pairs 1977), Algeria (oversummering several sites 1974–78, some bred). Breeds E and southern Africa: Kenya (L. Naivasha; also flocks in breeding dress L. Jipe and dams north of equator, perhaps extending range), Tanzania, Malaŵi, Zimbabwe, Botswana, South Africa (Cape, Transvaal); population small, largest colony 60 pairs (Malaŵi); wanderers Zambia, Angola, Zaïre. Palearctic birds widespread north of equator; largest concentrations Egypt (24,000 counted 2 lakes Nile delta, Jan–Feb 1979); common to abundant Mali, coastal Senegambia. Some Atlantic passage, including 60 moving south Angola, Oct; heavy passage Egypt (Nile); Aug–Sept flocks (up to 200) Chad and numbers SE Morocco Apr suggest substantial passage across Sahara to winter tropical W Africa. Uncommon to rare Sudan (Khartoum and south), Uganda, Rwanda, Burundi, Ethiopia, Somalia and Kenya (N and coast). Vagrant Djibouti (2, Apr, L' Escale).

404 STERNIDAE

Chlidonias hybridus

Also winters in breeding range

Field Characters. In non-breeding and immature plumages, confusion with Black *C. niger* and White-winged Black *C. leucopterus* Terns likely. Overall impression more like medium-sized *Sterna*; wing-beats more regular and deeper; larger, bulkier, longer-winged, tarsus 50% longer than Black; longer, heavier bill, flatter head; collar on hindneck indistinct or absent; mantle and back of juvenile more variegated. Lacks dark shoulder smudge of Black, white rump of White-winged Black. In all plumages confusion possible with *Sterna* spp., especially White-cheeked Tern *S. repressa*, but best distinguished by subtle differences of structure, flight action and feeding behaviour, preference for wetlands; shorter, less forked tail; smaller size; less attenuated structure. Feeding behaviour much as Black Tern; beats upwind low over water or vegetation, turns to fly high and quickly downwind; then turns again to repeat process.

Voice. Tape-recorded (62, 73). Repertoire varied; calls given mostly while breeding, harsher than congeners. Alarm call, often repeated, loud, rasping 'kerch'. Harsh breeding calls include 'kirirìrick', 'kek', 'kee-kee'; and soft, low 'kura-kura-kiu'.

General Habits. Inhabits wetlands, especially marshes. Often gregarious; flies low into wind, 2–4 m above water surface or aquatic vegetation, periodically dipping-to-surface or surface-plunging from hovering position; also pursues insects in flight, especially those disturbed by domestic stock or labourers in fields. Perches on posts and telegraph wires. In Upper Niger Inundation Zone, only occurs where areas of deep water, apparently dependent on fish. Seen to pluck Arum Frogs *Hyperolius horstocki* from lilies, Cape; then flew over dam, frog in bill, swooping to dip bill in water (Craig 1974).

Palearctic winter visitors follow Nile, Atlantic coast, or cross Sahara to Chad and Mali. Passage mainly Aug–Oct and Apr. Birds ringed France and Spain recovered W Morocco, Senegambia and Ghana. Breeding and movements of *delalandii* erratic, mainly influenced by water levels. Age of 1st breeding and extent of oversummering of Palearctic birds not known.

Food. Mostly insects and small fish; sometimes frogs. In Zaïre, feeds on fry of cichlid fish Aug–Nov, insects over marsh vegetation June–July. Insects and worms brought to very small young; when older mostly small fish and amphibians.

Breeding Habits. Monogamous; colonies typically number 10–30 pairs; breeds on seasonal pans with abundant floating vegetation, areas of open water 0·6–1·5 m deep; colonies often split into asynchronous-laying subcolonies; nests 3–18 m apart (mean 10–12 m). Displays much as Black Tern, similar to *Sterna* (Swift 1960). Courtship includes 'high flight' which involves up to 300 birds (almost the whole colony, Camargue); ascent to 100 m or more involves tight, noisy group with rather jerky flight; split up before descent, gliding down silently and rapidly with

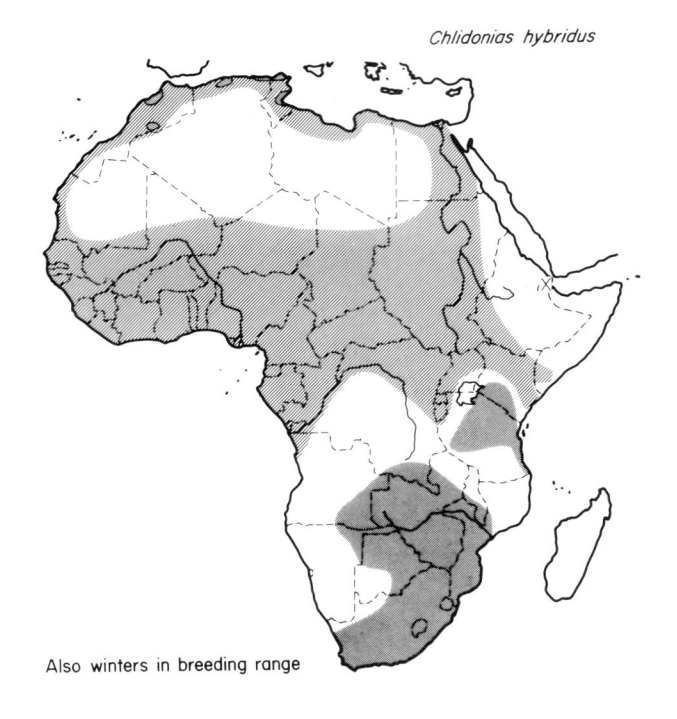

Description. *C. h. hybridus* (Pallas): N Africa south to equator. ADULT ♂ (breeding): forehead, crown and nape black; rest of upperparts, wing-coverts and innermost secondaries ash-grey; broad white stripe from gape and lower part of lores through ear-coverts to nape. Chin pale slate-grey, becoming darker on breast, blackish on flanks and belly; undertail-coverts and underwing-coverts white; axillaries and outer secondaries pale slate-grey. Rectrices as upperparts, tipped white, outermost pair with outer webs whitish. Primaries with outer webs and tips pearl-grey; inner webs mostly darker, bases white, extending as wedge towards tip; shafts white. Bill and legs dark crimson; eye dark brown or reddish brown. ADULT ♂ (non-breeding): upperparts, tail and wing as breeding. Forehead and hindneck white, latter tinged grey; crown streaked black and white; nape and spot in front of eye black; rest of head and underparts white. Bill black, tinged red-brown at base; legs red-black or red-brown. Sexes alike except ♀ somewhat smaller, especially bill. SIZE: wing, ♂ (n = 8) 231–250 (242), ♀ (n = 12) 228–238 (232); tail, ♂ (n = 17) 85–101 (88), ♀ (n = 7) 81–95 (87); bill, ♂ (n = 8) 30–33 (31·6), ♀ (n = 11) 26–30 (28·5); tarsus, ♂ (n = 9) 23–24 (23·3), ♀ (n = 11) 21–24 (22·6). WEIGHT: (Botswana) ♂ (n = 4) 80–108 (99), ♀ (n = 3) 97–115 (109).

IMMATURE: much as non-breeding adult except some remiges and rectrices with brown tips, few brown feathers on crown. 1st summer resembles breeding adult.

JUVENILE: much as immature except crown black, lightly streaked grey, buff and white; hindneck white tipped dark slate; mantle feathers tipped black-brown, narrowly edged buff.

DOWNY YOUNG: above buff-brown or chestnut-brown, streaked and mottled black; underparts white, chin and throat dusky brown; forehead and lores blackish; bill blackish, base vinaceous; legs grey-pink.

C. h. delalandii (Mathews): Kenya, Tanzania, and central Africa south to Cape. Breeding adult darker than nominate; grey areas of plumage dark slate-grey, belly black; white streak below eye narrower; indistinguishable in other plumages. SIZE: wing, ♂ (n = 6) 225–243 (237), ♀ (n = 8) 222–240 (232). WEIGHT: South Africa, ♂ (n = 4) 80–108 (92·5), ♀ (n = 4) 97–115 (107).

wings stiff; some glide for further 7–8 min after majority have returned to colony. Copulations on newly completed nest platforms.

NEST: bulky floating structure, free or anchored to underwater weeds, 80% of structure underwater; tapers from base 50–75 cm wide to top 20–25 cm wide, rising 8–15 cm above water level. *Cyperus* up to 30 cm long, built up loosely in criss-cross pattern (Tanzania); ill-formed cup, 12 cm in diameter, mainly grass and *Cyperus* leaves. Sinking nests built up so as to keep eggs just above water. Nests built on floating rafts of vegetation (*Polygonum*, *Aponogeton*) much less bulky, shallow platform, rudimentary cup.

EGGS: 1–3, usually 2, 1 probably incomplete; mean (68 clutches) 2·56; laid on consecutive days. Subelliptical, buff to olive-brown, blotched and spotted dark brown, with mauve undermarkings; fresh eggs may be distinctly green, fading after 1st week. SIZE: (n = 12, South Africa) 36·5–41·4 × 26·8–29·0 (38·7 × 27·6). WEIGHT: 16 (estimated).

LAYING DATES: Tunisia, Algeria, Morocco, May–June; Kenya, May–July; Tanzania, Dec–Feb, May–June; Malaŵi, Aug–Sept; Botswana, Zimbabwe, South Africa, Oct–Apr, mainly Jan–Feb.

INCUBATION: begins with last egg; carried out by both parents with frequent nest-reliefs. Period: 18–22 days. No fixed ritual at nest-relief; sometimes involves stoop and/or erect neck, often effected with no ceremony at all. Postures of incubating birds include stoop (chest horizontal, close to nest, rear half of body angled upwards at 60°) and erect neck (neck stretched forwards and upwards, bill as much as 50° above horizontal).

Territorial disputes more frequent after hatching. Aerial chase and threat from bird at nest usually involves erect neck posture; on ground sometimes involves ruffling of head and body feathers and lifting of wings, ruffled white cheeks contrast markedly with black crown. Dives at human intruders, frequently defaecates, will strike and draw blood. Chases away Grey-headed Gull *Larus cirrocephalus* and Red-knobbed Coot *Fulica cristata*. Period of incubation lasts on average 25 min in Europe; in Africa lasts mostly 10–15 min with birds showing signs of distress, with bill wide open after 10 min. Shorter incubation period and deliberate egg-wetting activity during incubation shift by bird paddling at edge of nest; probably result from high temperature and exposure to sun. Mate feeds brooding bird and often stands along side for much of shift. Mate hovers over nest prior to changeover; sitting bird gives short harsh call, neck outstretched, head flat on nest.

DEVELOPMENT AND CARE OF YOUNG: young able to swim competently, and climb back onto nest platform, few hours after hatching. Leaves nest within few days, often returns to be fed. Shrill wheezing call to solicit food. Fledging period: 20–23 days. Dependent for unknown period after fledging, probably several weeks.

BREEDING SUCCESS/SURVIVAL: many losses of eggs due to flooding; 5 fledged from 30 eggs, Cape.

References
Cramp, S. (1985).
Fuggles-Couchman, N. R. (1962).
Steyn, P. (1960).
Tarboton, W. R. *et al.* (1975).

Chlidonias niger (Linnaeus). Black Tern. Guifette noire.

Sterna nigra Linnaeus, 1758. Syst. Nat. (10th ed.) p. 137; Sweden.

Plates 25, 28
(Opp. pp. 400, 417)

Range and Status. Breeds Europe, W Asia, North America; winters Africa and South America.

Palearctic winter visitor, largely coastal; Egypt to Morocco, entire western seaboard, and around Cape to Natal. Main winter concentrations Senegambia to Namibia with bulk Gulf of Guinea to Namibia (Nigeria, Lagos, 4000 Jan; Namibia, Walvis Bay, 900 Feb; offshore in Gulf, thousands Dec). Also winters Egypt (uncommon, mostly Nile delta), rare Tunisia and Morocco. Heavy coastal passage Tunisia to Ivory Coast; major concentrations Mauritania (Banc d'Arguin, 100,000 Sept, thousands Oct–Nov, few wintering); southward passage Angola (mid Oct, 2000, largest flock 350); annual Chad and oases N Africa, suggests some cross Sahara. Vagrant Somalia, Burundi, Kenya, Botswana, Transvaal; access to E and SE Africa probably via Atlantic.

Description. *C. n. niger* (Linnaeus): only subspecies in Africa. ADULT ♂ (breeding): forehead, crown, hindneck, sides of head and chin to belly black; rest of upperparts slate-grey, hindneck and upper mantle merging; undertail-coverts white, underwing-coverts ash-grey, axillaries slate-grey. Bill black, cutting edges red towards base; eye dark brown; legs dark red-brown or black tinged reddish. ADULT ♂ (non-breeding): wings and tail as breeding except remiges darker slate when worn. Forehead, forecrown and broad collar on hindneck white; rest of crown and upper ear-coverts sooty-black; small patch of dark slate on each side of upper breast; small black patch in front of eye; lores, sides of head and rest of underparts white; axillaries and underwing-coverts pale ash-grey. Moulting birds may be strikingly pied on head and underparts. ADULT ♀ (breeding): underparts and sides of head dark slate-grey, chin and throat usually paler grey; black of head not extending to upper mantle; otherwise sexes alike except ♀ somewhat smaller. SIZE: wing, ♂ (n = 49) 210–226 (218), ♀ (n = 36) 204–224 (213); tail, ♂ (n = 11) 83–94 (89), ♀ (n = 10) 82–91 (88); bill, ♂ (n = 47) 26–30 (27·8), ♀ (n = 36) 25–28 (26·5); tarsus, ♂ (n = 39) 15–18 (16·4), ♀ (n = 38) 15–17 (16·3). WEIGHT: ♂♀ (South Africa) 59, 73; ♂ (n = 9, Europe) 63–88 (73), ♀ (n = 4, Europe) 71–73 (72).

IMMATURE: much as non-breeding adult except blackish cap less compact, downward extension of dark cheek patch especially obvious; patches on sides of upper breast often joined to upper mantle; underwing-coverts and axillaries white, contrasting with greyer undersurface of remiges. Older birds have patchy '*portlandica*' wing-pattern; strikingly pale panel between dusky lesser coverts and secondaries; outerwing darker when worn; rump, uppertail-coverts and tail rather pale. Birds resembling breeding adults with paler or partly pied underparts probably 2nd summer.

JUVENILE: much as younger immature; forehead and hindneck washed pale brown; mantle and back dusky, mottled brown and grey; legs and base of bill paler, soon darkening to adult colours.

Field Characters. Breeding adult has grey tail, narrow leading edge of white on upperwing, visible on bend of wing at rest; easily distinguished from White-winged Black Tern *C. leucopterus* which has white tail and much white on upperwing, red bill and legs, and black body. The two are easily confused in other plumages. Size and proportions are similar, but wings of Black narrower, tail longer and more forked, bill longer and finer, and build slimmer. Dark smudge at shoulder at all ages, usually prominent, distinguishes it from congeners; dusky band across leading edge of wing and upper mantle; rump and tail uniform grey, also diagnostic. Juvenile's mantle somewhat darker than wings, but lacks contrasting black 'saddle' of White-winged Black Tern. Whiskered Tern *C. hybridus* larger, more robust, bill heavier; overall impression on wing more like medium-sized *Sterna*. Unlike congeners, Black Tern is mainly marine in Africa; habits and habitat much as Common Tern *S. hirundo*.

Voice. Tape-recorded (62, 73). Mainly silent in Africa. Flight calls, mostly spring passage, quiet 'kik-kik' and sharper 'teek-teek'.

General Habits. Mostly marine; inshore, favours estuaries, lagoons and other sheltered habitats. Large flocks regular offshore; in Gulf of Guinea, flocks of thousands 400–600 km from land, winter and on passage; flocks change direction abruptly. Feeds in groups or singly, flying into wind, *c.* 2 m (0·5–4·5) above water surface, bill pointing vertically downwards, hovers on fluttering wings. Mostly dips-to-surface; also surface-plunges, and hawks for aerial insects. Joins with other terns over shoals of predatory fish and schools of dolphins; on passage follows shoals of fish. Inshore Ghana, flock of 150 resting on sea, also rests regularly on flotsam. Takes waste thrown from fishing boats and fish factories. Also takes fish from seine nets; frequents sewage outlets and accumulations of flotsam.

Passage late July–Nov, late Mar–mid May, mostly offshore, rarely far inland; mainly along western seaboard. Birds ringed Mauritania Oct–Nov recovered Ghana (2, following spring) and Poland (July); ringed Morocco Sept, recovered Togo (following spring); ringed USSR, recovered Tunisia; ringed Namibia, recovered USSR (52°N, 62°E). Some oversummer in tropics: Nigeria, 1000–2000, mainly immatures, June–July; Senegambia and Togo, flocks arrive July, mainly winter and immature plumage; also oversummer on occasion Algeria.

Food. Mainly marine fish; also crustaceans and insects. In Sierra Leone and Ghana takes fish (*Sardinella eba*, *Brachydeuterus auritus*, *Engraulis guineensis*) and insects (Isoptera, Formicidae, Gerridae).

Reference
Cramp, S. (1985).

Chlidonias leucopterus (Temminck). White-winged Black Tern; White-winged Tern. Guifette leucoptère. Plates 25, 28

Sterna leucoptera Temminck, 1815. Man. Orn. (1st ed.), p. 483; South Europe.

Range and Status. Breeds SE Europe and central Asia; winters Africa, Persian Gulf and India to Australia.

Palearctic winter visitor, very abundant to uncommon; wetlands throughout N Africa to Cape. Abundant in winter Senegambia, Nigeria and Sudan to South Africa, particularly so E Africa (tens of thousands, mostly Rift Valley soda lakes) and Mali (Upper Niger Inundation Zone, very abundant). Heavy passage Nile Valley, Red Sea and Chad. Rare most coasts, sometimes with large flocks of Black Terns *C. niger* (Nigeria, Angola, Namibia).

Chlidonias leucopterus

Description. ADULT ♂ (breeding): head glossy black or blue-black; mantle, back and scapulars blackish slate; chin to belly, axillaries and most underwing-coverts black. Rump, uppertail-coverts, tail, undertail-coverts and lesser coverts white. Primaries with tips and outer webs pearl-grey, inner webs with white bases, extending towards tip as wedge; shafts whitish; secondaries and greater coverts slate-grey, median coverts paler. Bill dark red; eye dark brown; legs orange-red. ADULT ♂ (non-breeding): forehead and forecrown white, rest of crown streaked grey-black; black spot in front of eye; blackish ear-coverts. Upperparts and tail pale slate-grey, upper mantle darker, rump and hindneck white, outermost pair of rectrices with whitish outer webs; rest of head and entire underparts, including axillaries and underwing-coverts, white; remiges and wing-coverts pearl-grey to slate-grey, lesser coverts darkest. Legs darker red or blackish; bill black. Moulting birds may be strikingly pied. ADULT ♀ (breeding): head less glossy; innermost pair rectrices grey, others mottled ash-grey; otherwise sexes alike except ♀ somewhat smaller. SIZE: wing, ♂ (n = 24) 208–221 (215), ♀ (n = 31) 203–221 (212); tail, ♂ (n = 12) 72–85 (79), ♀ (n = 11) 69–84 (78); bill, ♂ (n = 28) 25–28 (25·9), ♀ (n = 34) 23–26 (24·7); tarsus, ♂ (n = 21) 19–22 (20·0), ♀ (n = 28) 18–21 (19·6). WEIGHT: ♂ (n = 23, Afrotropical) 49–78 (65·4), ♀ (n = 29, Afrotropical) 50–79 (61·2).
IMMATURE: much as non-breeding adult except tips of some remiges and rectrices brown; wing-coverts brown, lessers especially dark. Black underwing-coverts of breeding adults or other signs of partial breeding plumage occur in some 1st summer birds.
JUVENILE: much as immature except black-brown 'saddle' on mantle and back, contrasts with white hindneck and silver-grey wings; hindcrown blackish.

Field Characters. Breeding adult unmistakable; easily distinguished from Black Tern *C. niger* by white tail, white upperwing, blackish bill and blackish legs. In other plumages lacks dark shoulder smudge of Black Tern; white rump contrasts with darker mantle and tail; broad white collar on hindneck. Black-brown 'saddle' of juvenile striking and diagnostic. Whiskered Tern *C. hybridus* larger, more robust; gives impression of *Sterna*, upperside far more uniform. Faster and more agile than Black Tern; wing-beats shallower, flight more erratic; more prone to settle on water.

Voice. Tape-recorded (62, 73). Alarm call, loud churring 'kerr' or 'kreek-kreek', more frequent and piercing than Black Tern.

General Habits. Frequents wetlands; mostly rare on coasts; flock at sea 1·5 km off Natal exceptional. Feeding methods more varied than Black Tern: dips-to-surface from hovering position; skims insects from water surface and aquatic vegetation; wades in search of food; hawks for aerial insects, including grasshoppers at grass fires or disturbed by livestock; and follows plough. Forages far from water on passage and in winter. Feeding varies seasonally, Zaïre: Nov, catches insects (abundant after 1st rains) over grass plains; Dec–Jan, eats fry of cichlid fish trapped in residual pools; Feb–Mar, forages over newly flooded plains, often taking insects from aquatic plants. In Kenya, small flocks follow predatory tiger fish *Hydrocyon* spp., catching minute fry jumping out of water to escape; hovers within 10–15 cm of water surface while waiting. Also in Kenya flocks of thousands feed on concentrations of Army Worms *Laphygma exempta*; in South Africa plague of American bollworms *Pectinophora gossypiella* in growing maize eradicated by flock. Flocks in breeding dress Apr–May wheel high in tight formation.

Mostly arrives via Nile, Rift Valley and Red Sea, mid Aug–Oct. Uncommon on passage coastal Libya to Morocco, Apr–May, Aug–Sept. Apparently crosses Sahara to winter tropical W Africa: Chad, commonest tern, from late July, up to 500 from mid Aug; Senegambia, annual northwards passage Apr, 500 km inland; SE Morocco uncommon late Apr. Adults reach southern Africa early Sept. Immatures spend 1st summer in winter quarters; oversummering birds resembling fully moulted breeding adults are probably 2nd summer birds.

Food. Invertebrates (mainly insects) and fish; occasionally amphibians. At L. Kariba (Zimbabwe), along shore feeds almost exclusively on insects (dragonflies, grasshoppers, bugs, beetles, termites) while on open water feeds mainly on introduced sardines *Limnothrissa miodon*, but also *Haplochromis* and *Synodontis*. Feeds on small fish. L. Tanganyika (*Limnothrissa*, *Stolothrissa*) and L. Victoria (abundant whitebait). In Kenya Army Worm caterpillars; in South Africa, bollworms.

References
Begg, G. W. (1973).
Cramp, S. (1985).

Genus *Anous* Stephens

Dark, fairly uniform plumage, little variation with age or season. Tail rounded and heavy, 3rd or 4th pair of rectrices (from outside) longest. Tarsus much shorter than middle toe with claw. Bill long and strong. 3 spp., all occur Africa, 2 as breeders. *A. minutus* and *A. tenuirostris* are best regarded as allospecies which form a superspecies (Serventy *et al.* 1971).

Plates 25, 28
(Opp. pp. 400, 417)

Anous minutus Boie. Black Noddy; White-capped Noddy. Noddi noir.

Anous minutus Boie, 1844. Isis von Oken, col. 188; New Holland = Raine Island, Australia.

Forms a superspecies with *A. tenuirostris*.

Range and Status. Breeds widely tropical and subtropical Atlantic and Pacific.

Abundant resident Gulf of Guinea, breeding Principe (Tinhosas Islands, 5000 pairs) and Pagalu (70,000 pairs); populations mainly sedentary. Elsewhere W Africa, Senegambia to Ghana, uncommon to rare; vagrant South Africa (E Cape, Feb 1975).

Description. *A. m. atlanticus* (Mathews): only subspecies in Africa. ADULT ♂: forehead and crown white washed grey, merging to slate-grey round margins of cap; lores and narrow line round eye black, lower eyelid and centre of upper eyelid white; remiges and rectrices black; elsewhere sooty-black, slight umber tinge on back and wing-coverts, more apparent with wear. Bill black; eye dark brown; legs black tinged reddish. Sexes alike except ♀ somewhat smaller. SIZE: wing, ♂ (n = 9) 225–240 (231), ♀ (n = 7) 214–230 (224); tail, ♂ (n = 9) 110–120 (116), ♀ (n = 7) 111–124 (116); bill, ♂ (n = 9) 42–48 (46·0), ♀ (n = 8) 41–45 (42·5); tarsus, ♂ (n = 9) 22–23 (22·5), ♀ (n = 8) 21–22 (21·6). WEIGHT: Ascension, mean 120.

IMMATURE AND JUVENILE: much as adult except plumage somewhat browner. Juvenile with less white on crown, upperwing-coverts and secondaries tipped with buff. Immature similar to juvenile but without pale feather tips to secondaries and coverts.

DOWNY YOUNG: down short and soft, uniform dark brown; top of head contrastingly pale ash-grey or grey-white.

Field Characters. Difficult to separate from Lesser Noddy *Anous tenuirostris* and Brown Noddy *A. stolidus*. At distance, all three noddies might be confused with small dark phase *Stercorarius* skua or immature Sooty Tern *Sterna fuscata*. Flight slower; Brown Noddy especially lethargic, akin to idly foraging gull. All lack wing panel of skuas. Sooty Tern has dashing flight and forked tail typical of *Sterna*, paler underwing and vent, white flecks and spots above, while noddies have heavy rounded tails, often ragged with central cleft. Black Noddy and somewhat smaller Lesser Noddy very similar, but Black with heavy and slightly forked tail rather than rounded tail with cleft, and with more sharply defined blackish lores, giving clear-cut separation from pale cap. Combination of smaller size and more contrasting, whiter cap distinguishes Black and Lesser Noddies from Brown Noddy; immature Brown can have cap almost uniform with body, cap of others contrasts markedly at all ages. Wing-beats of Black and Lesser more rapid, less planing, flight more direct; bill finer, proportionately longer, gonys more pronounced, head proportionately smaller; dark areas of plumage more uniform, less brown, lacking 2 shades of upperwing (remiges darker, coverts paler) typical of Brown.

Voice. Tape-recorded (C). Mainly silent away from colonies and communal roosts. Alarm-note a harsh staccato rattle, also a sharp 'kerr'; flight call, also used when perched, more sustained crackling 'krikrikrik', beginning slowly, speeding up, ending slowly. In courtship, ♂ gives 'kedrutt-urrutt' (see Breeding Habits). Young and ♀♀ soliciting food give sibilant, piercing 'see-ew'.

General Habits. Occurs in offshore, sometimes inshore habitats. Usually feeds in flocks, 1–3 m above water surface; sweeps low and snatches prey from surface, splashes, does not plunge deeply. Also feeds on surface like gull, occasionally hovers; rests on flotsam, buoys, masts. Often associates with Brown Noddy at feeding areas and roosts.

Anous minutus

Also winters in breeding range

Food. Small fish and squid driven to surface by predatory fish, proportions uncertain. At Christmas Island, Pacific Ocean, smaller prey than larger Brown Noddy and Sooty Tern; takes few squid, but squid 25% of diet by volume.

Breeding Habits. Monogamous; dense colonies offshore islands. Pagalu colony consists of c. 10% Brown Noddies, all closely packed; 1 tree had 143 nests, 1–9 m above ground, most 4·5–5·5 m. In courtship display at nest-site, ♂ bobs head slightly, calling 'kedrutt-urrutt'; caresses head and neck of ♀ with bill, then both touch each other's bill; ♀ makes rattling guttural response, bobbing head, bill open wide; ♂ feeds her regurgitated food. Nest-building begins soon after mating. When bringing nesting material, ♂ stands on ♀'s back and passes stick or weed to her; she works it into the nest. Structure shaped by pushing with breast, especially if seaweed is wet. Intention movement preceding departure from nest involves 'snaking' of neck (rather as Guillemot *Uria aalge*), and flight takes place after bird has reached forward to full extent of neck. In aggressive territorial display, ♂ raises crest slightly, gives rattling call, beak slightly open, tongue arched; this is followed by forward thrust (neck held stiffly) and bowing. In all displays the orange tongue, white crown and black and white markings around eye feature prominently. 'Dreads' are common. Much preening activity to eliminate hippoboscid flies, by which heavily parasitized, Pagalu.

NEST: Pagalu, built of coarse grasses; flimsy; sits on ledge of small cliff or rock-strewn ground under boulder overhang; nearby tussocks and low bushes avoided. In much of range, nest more substantial, with leaves and other plant material (especially seaweed) cemented with excreta; placed in trees; diameter c. 15 cm, depth of cup c. 6 cm; leaves and seaweed droop as much as 15 cm below nest rim.

EGGS: invariably 1. Subelliptical, white to pink-buff, with spots and irregular blotches of sepia and blackish, mainly near broad end, some eggs almost unmarked. SIZE: (n = 12, Ascension) 45·6–50·0 × 30·8–33·6 (47·6 × 32·4). WEIGHT: (n = 11, Pacific) mean 23·7.

LAYING DATES: Pagalu Aug–Oct.

INCUBATION: carried out by both parents with frequent nest-reliefs; relieving bird nibbles bill of sitting bird, sitting bird then nibbles bill of relieving bird. Period: 35 days. Incubating birds reluctant to leave nest when disturbed; readily caught by hand; some have to be eased off nest to inspect eggs.

DEVELOPMENT AND CARE OF YOUNG: fledging period: 42–52 days. For unknown period after fledging, juveniles return to colony to be fed by adults in late afternoon, spending day alone at sea.

References
Ashmole, N. P. (1962).
Clancey, P. A. and Wooldridge, T. (1975).

Anous tenuirostris (Temminck). Lesser Noddy. Noddi marianne.

Plate 25
(Opp. p. 400)

Sterna tenuirostris Temminck, 1823. Pl. col., livr. 34, pl. 202; Seychelles.

Forms a superspecies with *A. minutus*.

Range and Status. Breeds tropical and subtropical Indian Ocean.

Seasonal visitor E Africa, common to uncommon. Recorded in numbers in recent years, perhaps regular, apparently breeders from Malagasy Region. Seasonally numerous S Somalia (Apr–Oct, especially Mogadishu) and Kenya (erratic, most records Aug–Dec 1976); 5 sight records South Africa (Natal, 2 sites, May 1979); once N Somalia.

Description. *A. t. tenuirostris* (Temminck): only subspecies in Africa. ADULT ♂: forehead and crown pale grey shading to darker grey at nape and margins; lores and narrow line round eye black, lower eyelid and centre of upper eyelid white; remiges and rectrices blackish; elsewhere dark brown, can appear charcoal grey in strong sunlight; slight umber tinge on back and wing-coverts, more apparent with wear. Bill and legs black; eye dark brown. Sexes alike except ♀ somewhat smaller. SIZE: wing, ♂ (n = 4) 216–220 (218), ♀ (n = 3) 202–209 (205); tail, ♂ (n = 3) 112–113 (112), ♀ (n = 3) 106–111 (108); bill, ♂ (n = 3) 36–43 (39·7), ♀ (n = 3) 33–42 (37·7); tarsus, ♂ (n = 3) 22, ♀ (n = 3) 22. WEIGHT: ♂ 97·4, ♀ 110, 115, 120.

IMMATURE AND JUVENILE: both much as adult but with plumage somewhat browner, and pale cap more contrasting due to absence of darker grey at nape and margins.

Field Characters. Difficult to separate from Black Noddy *A. minutus* and Brown Noddy *A. stolidus*, unless good view or 2 together. At distance, all might be confused with small dark phase *Stercorarius* skua or immature Sooty Tern *Sterna fuscata*. Smaller with more contrasting white cap and greyer plumage than Brown Noddy. Difficult to distinguish from Black Noddy in field (see Black Noddy for details).

Voice. Tape-recorded (217). Various rattling and purring alarm calls at colonies; otherwise mainly silent.

General Habits. Occurs offshore, sometimes inshore. Usually feeds in flocks, 1–3 m above water surface, sweeps low and snatches prey from surface, splashes, does not plunge deeply. Also feeds on surface like gull, occasionally hovers. Rests on flotsam, buoys, masts. At high tide, flock of 110–220, Kenya, roosted daily on fish trap and nearby beach. Often associates with Brown Noddy at feeding areas and roosts.

Food. Small fish and squid driven to surface by predatory fish; proportions uncertain.

References
Ash, J. S. (1980).
Britton, P. L. (1977).

Plates 25, 28
(Opp. pp. 400, 417)

Anous stolidus (Linnaeus). Brown Noddy; Common Noddy. Noddi brun.

Sterna stolida Linnaeus, 1758. Syst. Nat. (10th ed.), p. 137; West Indies.

Range and Status. Breeds and ranges widely tropical and subtropical Atlantic and Indo-Pacific.

Resident, very abundant, breeds Gulf of Guinea at Principe (Tinhosas Is., tens of thousands) and Pagalu (thousands); NW Somalia (Mait Is. 20,000 or more Nov 1847, Nov 1942, May 1979); Tanzania (Latham Island, thousands); and Red Sea in Sudan (Barra Musa Saghir, Suakin Archipelago, 300 pairs). Also breeds in hundreds Kenya, São Tomé and probably S Somalia. Elsewhere W Africa, uncommon and erratic Senegambia to Cameroon; seldom recorded away from Gulf of Guinea islands; not associated with concentrations of Sooty Terns *Sterna fuscata* at sea. Non-breeders widespread in east, 12°N to 7°S; common Somalia north and east coasts, mainly Apr–Sept, uncommon Kenya throughout year (exceptional concentrations, up to 1000 together, Aug–Dec 1976). Vagrant South Africa (sick bird Durban Jan 1969, assigned to nominate race; sight record W Cape 1977, and unidentified noddy, probably Brown, inland at Rondevlei, Cape).

Description. *A. s. stolidus* (Linnaeus): islands in Gulf of Guinea, vagrant mainland W Africa, Senegambia to Cameroon; also South Africa, Natal (Durban). ADULT ♂: forehead and forecrown grey-white, merging to lavender-grey on rear crown and nape, bordered by narrow white line from base of bill to above eye; lores and narrow line round eye black, lower eyelid and centre of upper eyelid white. Rectrices black-brown, remiges black. Elsewhere, including underwing-coverts and axillaries, dark sooty-brown; much of body plumage tinged grey when fresh, most marked on head and chest. Bill black; eye dark brown; legs black, brown tinge to tarsus and toes, webs sometimes ochre. No seasonal variation, but rather drab when worn, body more uniform dark brown, crown greyer, less contrasting. Sexes alike except ♀ somewhat smaller. SIZE: wing, ♂ (n = 6) 273–294 (282), ♀ (n = 5) 266–281 (274); tail, ♂ (n = 6) 144–157 (152), ♀ (n = 5) 134–152 (142); bill, ♂ (n = 5) 42–48 (44·2), ♀ (n = 4) 39–43 (40·7); tarsus, ♂ (n = 6) 25–27 (25·5), ♀ (n = 5) 25–26 (25·6). WEIGHT: ♂♀ (n = 10, Ascension) 160–205 (186).

IMMATURE AND JUVENILE: much as adult except body somewhat darker, grey tinge less obvious; forehead greyer, rear crown browner; cap separated from black lores by broken grey-white line; in worn plumage especially drab, cap almost uniform with body.

DOWNY YOUNG: down short and soft; uniform brown, shade varies individually, dark brown to paler grey-brown; always paler than Black Noddy *A. minutus*. Forehead, crown and nape either uniform with body or paler ash-grey, forehead whiter; cap contrasts less than Black Noddy. Bill black; legs and feet dark brown, webs paler.

A. s. plumbeigularis Sharpe: eastern seaboard, Sudan to Tanzania. Forehead and crown deeper blue-grey, less ashy; black area on face more extensive; grey tinge to fresh plumage more plumbeous. SIZE: ♂♀ (n = 11) wing, 266–294 (278); tail, 134–157 (147).

Field Characters. A dark sooty brown tern with grey-white cap and wedge-shaped tail; difficult to separate from Black Noddy and Lesser Noddy *A. tenuirostris* unless good view of two together; then, larger size, less contrasting grey-white (not white) cap, and slightly browner colour distinguishes Brown from others. At distance, all 3 might be confused with small dark phase *Stercorarius* skua or immature Sooty Tern *Sterna fuscata* (see Black Noddy for details).

Voice. Tape-recorded (62, 73, B, C). Mainly silent away from colonies and communal roosts. Alarm note a low, harsh, croaking 'kar-r-rk' or 'karr-karr', rather like Common Raven *Corvus corax*; given in flight and at rest, more frequently at night. Also a shrill 'pee', 'pay-ee' or 'pay-lili-li'. At colony 'kuk-kuk-kuk' and 'kree-aw' used in territorial disputes with neighbours; also cackling and purring notes during courtship (see Breeding Habits).

General Habits. Pelagic, occurring mainly offshore, sometimes inshore. Less pelagic than often supposed, though more so than Black Noddy. Feeds in flocks (sometimes singly) by day and moonlight. Generally flies 1·5–3 m above water surface, sometimes hovers. Sweeps low to drink, bathe and snatch prey from surface; splashes, does not plunge deeply. Feeding flocks of up to 1000 non-breeders regular near Mogadishu (Somalia). In Kenya, joins with Roseate Tern *Sterna dougallii*, Saunders' Tern *S. saundersi* and other terns in feeding flocks over shoals of predatory fish, 3–15 km offshore; also feeds whilst swimming in small groups, sometimes with Black Noddy and Hemprich's Gull *Larus hemprichii*, riding buoyantly like small gull. At Mait Island (Somalia), large concentrations feed just offshore, but feeds up to 50 km from colonies. Readily settles on rigging of fishing boats, as well as on flotsam and buoys; reluctant to rest on water for prolonged periods. Infrequent use of roosts on beaches, reefs and fishtraps Kenya suggests some roost at sea.

Food. Small fish and squid driven to surface by larger predatory fish. At Christmas Is., Pacific Ocean, eats larger prey than Black Noddy, and greater proportion of squid; kinds and sizes of prey much as Sooty Tern.

Breeding Habits. Monogamous, breeds in dense colonies on offshore islands. Often in mixed colonies with Black Noddy (Tinhosas Islands and Pagalu) and Sooty Tern (Latham Island, Tanzania; Tenewe Islands, Kenya). In 'high flight' wing-beats slowed down and curiously exaggerated. 'Dreads' frequent, birds rising silently, often triggered by alarm-call from other species; birds return noisily, performance sometimes repeated at intervals of less than 1 min, usually involving only minority of individuals. Ground displays given before and during incubation. Pair also displays by stretching neck forward, bill open, white forehead and eyelids and orange mouth prominent; fence with bills, nodding, bowing and jerking head backwards and upwards repeatedly; display more extensive early in season. Birds also circle each other, wings drooped, primaries on or close to ground; seaweed and sticks often presented in pre-laying ceremonies. Presumed ♀ begs from mate, picks up and eats disgorged food; accompanied by cackling and purring notes; ♀ later preens head of mate. In aggressive display, head thrust forward, bill open, accompanied by croaking call; this often followed by actual conflict with bills locked. Open mouth display, with tongue arched, used to threaten birds approaching too closely. Swoops at human intruders.

NEST: a substantial structure of sticks and seaweed usually placed on a tree or shrub; c. 25 cm in diameter; depth variable, typically c. 15 cm. In West Indies built in depression on rock, 20–30 cm in diameter; in Sudan, amongst 3 m high shrubbery. Elsewhere Africa, built on coral islets, rocky promontories and slopes, small cliff ledges and rock-strewn ground, with or without low creeping vegetation (e.g. *Euphorbia*). In Tanzania (Latham Island) made of small stones and bones; in Kenya (coral islets), lined with seaweed or waterworn shell fragments, with most nests exposed to sea. Both sexes build nest; ♀ fed when building.

EGGS: invariably 1. Subelliptical, white to greyish or pinkish, sparsely spotted and blotched reddish brown, dark brown or purple, with lavender undermarkings. SIZE: (n = 100, Indian Ocean) 47·0–55·2 × 32·8–37·6 (51·8 × 35·8). WEIGHT: (n = 32, Ascension) 26·0–41·5 (33·5).

LAYING DATES: Sudan, June; Somalia, breeding season uncertain, perhaps all year; Kenya, June–Sept, mainly Aug; Tanzania, Nov; Pagalu, Aug–Oct.

INCUBATION: carried out by both parents with frequent nest-reliefs. Period: 32–36 days.

DEVELOPMENT AND CARE OF YOUNG: fledging period: 40–56 (46·5) days (n = 66, Hawaii). For 100 days or more after fledging, juveniles return to colony at night after day alone at sea, many emitting high, whispering calls associated with soliciting of food from parents (Hawaii). Fledged birds have wing, tail and bill 2–10% shorter than adult; bill not fully grown until c. 1 year.

References
Ash, J. S. (1980).
Cramp, S. (1985).
Woodward, P. W. (1972).

Family RYNCHOPIDAE: skimmers

Skimmers are large tern-like birds with several unique attributes: lower mandible (bone and particularly its rhamphotheca or horny covering) much longer than upper, striated (**A**), most of it less than 1 mm wide, and flexible towards tip (**B**—view from below); flight-feeding by 'skimming', with lower mandible slicing through water surface; bone and muscle adaptations in head and neck associated with snap-closing bill mechanism (Zusi 1962); and pupil which becomes cat-like vertical slit in bright light (Wetmore 1919, Zusi and Bridge 1981). Striations on mandible are shallow grooves probably serving to counteract tendency for it to be pushed downward in water during skimming (Schildmacher 1931). Skimmers are related to gulls and terns (Moynihan 1959, Schnell 1970) but certain behaviours suggest that they are no closer to latter than to former (Sears *et al.* 1976).

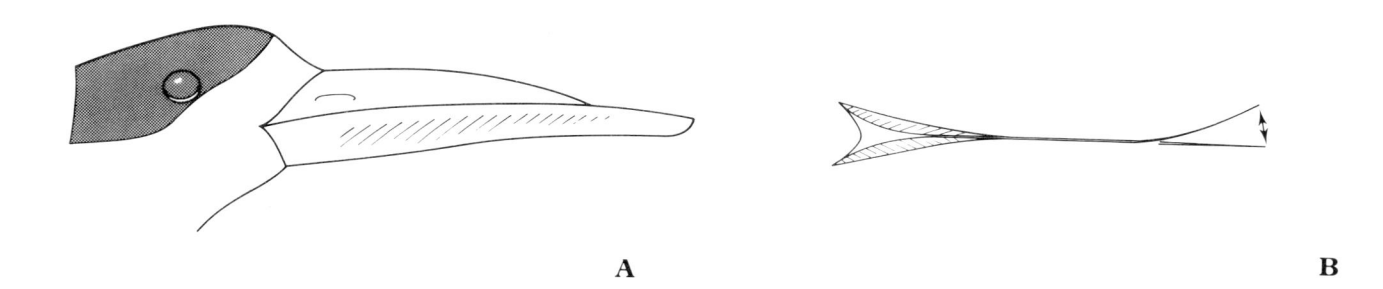

A **B**

1 genus; conventionlly 3 spp. are recognized, 1 in American tropics and subtropics, 1 in Africa and 1 in India, but differences between them trivial and they might even be thought conspecific. We regard them as a superspecies, with the African bird systematically intermediate (black hindneck agrees with American and bill colour with Indian allospecies). At present only American species known to skim sand, have vertical pupil, carry fish cross-wise in bill, and feed chick by regurgitation onto ground, and chick to partly bury itself (Terres 1980); but African and Indian species may prove to have similar behaviour when better known.

Skimmers nest on hot dark or blindingly white tropical river sandbars (and sea beaches to lat. *c.* 40°, Americas), and migrate along rivers according to height and turbulence of water; high water prevents nesting and turbulence prevents feeding.

Genus *Rynchops* Vieillot

Plates 25, 28
(Opp. pp. 400, 417)

Rynchops flavirostris **Vieillot. African Skimmer. Bec-en-ciseaux d'Afrique.**

Rhynchops flavirostris Vieillot, 1816. Nouv. Dict. d' Hist. Nat., 3, p. 338; Senegal.

Range and Status. Endemic resident and intra-African migrant; as breeding bird, practically confined to broad rivers with large sandbars: mid Senegal R. and mid Niger R. (Mali, L. Debo downward to Katcha, Nigeria), lower Benue, Logone and Chari rivers (Nigeria, Cameroon, Chad), Congo R. from mouth (Zambi) to Lukolela and probably its main tributaries (Zaïre), Nile and tributaries from Butiaba (Uganda: L. Albert) to about Dongola (N Sudan), W and SW Ethiopia, N Kenya (L. Turkana: Central Island, Allia Bay and formerly Ferguson's Gulf), S Tanzania (Ruaha, Kilombero, Rufiji and Ruvuma rivers), all of Zambezi R. and major tributaries (Zambia, Zimbabwe, Mozambique) except Kariba and Cabora Bassa sections; Kafue and Luangwa rivers (Zambia) and Shire/Mwanza confluence (Malaŵi); Okavango, Cunene and lower Cuanza rivers (Angola, NE Namibia, NW Botswana breeding south to L. Ngami); Lundi R. (SE Zimbabwe), Save and lower Limpopo R. (Mozambique); nested L. St Lucia (Natal) in 1943 and 1944 but no bird seen Natal in 1970s.

Widespread in non-breeding season. Regular in Senegal R. delta, Gambia R., saltpans coastal Ghana (20–70 birds), coastal Nigeria, lower Logone R. (Gamsay, 200), Nile north to S Egypt (Kôm Ombo, 40, Oct 1979); Uganda, Murchison Falls National Park, 300; Kenya, Ferguson's Gulf 1000, L. Nakuru, L. Naivasha, L. Shakababo (up to 70); Zaïre, Congo R. up to Lisala; Oubangui-Uele rivers from Irebu to Niangara, Aruwimi R. up to Banalia, upper Lualaba R. (Kadia, Kiabo) and hundreds at mouths of Rutshuru and Rwindi rivers; Tanzania, L. Rukwa once 1500, Jan; Angola, Cunene and lower Cuanza rivers. Vagrant to Red Sea coast; Somalia; South Africa, Durban, Vaal R., Potchefstroom.

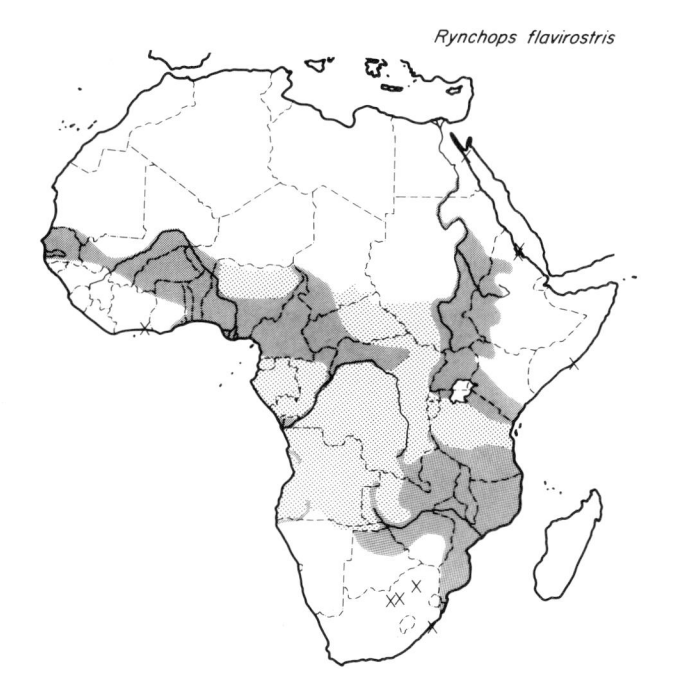

Rynchops flavirostris

Frequent to common on breeding rivers but local; 35 adults on 35 km Luangwa R. in breeding season, mainly in 1 colony of 5–10 nests, but only 3 birds on 210 km stretch downstream which appeared equally suitable (Attwell 1959). On 85 km Shire R. (Malaŵi) *c.* 200 birds, many breeding on one 20 km stretch (Hanmer 1982), *c.* 5 pairs on each of 4–5 small islands in Congo R. mouth. Greatest known concentration, 50 pairs Central Island, L. Turkana, where much disturbed by people, and protection, if not too late, long overdue. At Lochinvar (Zambia) up to 600 occur every dry season, Apr–June to Oct–Jan, presumably breeding locally (Taylor 1979).

Description. ADULT ♂ (breeding): crown including eye region, mantle to rump and central rectrices blackish brown; uppertail-coverts dark brown in centre, white at sides; scapulars, all upperwing-coverts and remiges blackish brown, inner primaries tipped pale brown, secondaries whitish on inner edge and broadly tipped white; rest of plumage white, sharply demarcated on head and neck from dark brown; but outer rectrices off-white with brown shafts. Tail forked, outermost rectrix pointed and 20–38 mm longer than rounded innermost pair. Underwing: axillaries white, coverts brownish white, secondaries grey with white tips, primaries blackish. Eye dark brown; bill orange-red, extremely flattened laterally, upper mandible yellowish at tip, lower mandible *c.* 24 mm longer, the 'overshot' part being somewhat translucent, yellowish with whitish tip; 5 × 60 mm area on side of lower mandible with *c.* 20 parallel diagonal striations; legs and feet bright orange-red, nails black. ADULT ♂ (non-breeding): hindneck dusky white, forming collar; black on head and neck not sharply demarcated from white; back brown rather than blackish; bill less bright red. Sexes alike. SIZE: (10 ♂♂, 10 ♀♀) wing, ♂ 311–374 (336), ♀ 321–361 (334); wing-span *c.* 106 cm; tail, ♂ 105–127 (115), ♀ 102–120 (108); upper mandible to feathers, ♂ 52–71 (62·6), ♀ 49–64 (55·3); lower mandible, tip to jaw symphysis, ♂ 68–96 (83·0), ♀ 58–104 (73·5); bill depth at nares, ♂ 20·9–23·9 (22·25), ♀ 19·3–21·4 (19·6); tarsus, ♂ 26–31 (28·2), ♀ 25–31 (26·4). WEIGHT: 10♂, 111–204 (164).

IMMATURE: forehead whitish; crown, nape and hindneck brown mottled whitish; patch through eye (lores, ear-coverts) dark brown, scarcely mottled; feathers of remainder of upperparts brown broadly fringed pale buff; tail dark brown, outer rectrices fringed buff; primaries, secondaries and tertials broadly edged pale buff forming long crescents and V-marks against blackish primaries in folded wing. White parts as in adult but chin, under lores, and throat with soft brown freckles. Bill yellowish with dusky distal half, lower mandible longer than upper but shorter than in adult; legs and feet dusky yellow.

DOWNY YOUNG: down thick, buff-white above with irregular peppering of small black dots mainly on crown and rump, white below; eye black; bill pale yellow; fleshy feet dull purple.

Field Characters. A long-winged, black-and-white river bird, in shape rather like large tern, with long red-orange bill and short red legs. Usually seen resting in small, huddled flocks on sandbar, or 'skimming'—flying low over still water with head down and mandible cutting surface. A solitary recently fledged bird, brown above and white below, could be mistaken for a dark-backed tern but distinguished by its long yellowish lower mandible and yellowish legs. Folded wings project 12 cm beyond tail; tail deeply forked.

Voice. Tape-recorded (234, 264, 296). Sharp, loud 'kip-kip-kip' and longer, tern-like harsh 'kreeep'. Calls infrequently, mainly on wing; but noisy when breeding. Soft shrill cooing (courtship).

General Habits. Breeding habitat is broad, meandering rivers with huge dry sandbars more or less bare of vegetation; sometimes sandy lakeshores. Has nested sandy sea-coast (Zululand). Also, in non-breeding season, estuaries, coastal lagoons, industrial saltpans and (vagrant birds) sewage ponds, dams and swamps. Requires expanse of calm water for feeding. Mainly in hot lowlands, below 1800 m. Roosts close to water's edge on hot, blinding white sand.

Spends most of time resting on sandbar, in small flock huddled together all facing same way, not calling, occasionally preening itself (long mandible evidently no impediment). Wings hang loosely, the 'hand' not tucked into breast feathers. Flight buoyant with slow (sometimes swift) deep beats, body rising slightly with each downstroke. Movements of flying flock synchronized, all birds rising, falling and wheeling in unison. Disturbed roosting bird settles soon in same place or far away, or it may take a few 'skimming' flights over water. Comfort behaviour tern-like; stretches with legs bent, head held very low with chin almost against ground, wrists held high, wing-tips crossing above tail and touching ground. Does not swim.

Normally feeds for short period towards dusk, on moonlit nights, and in early morning; birds in foraging flocks well separated. Flies just above water, with forepart of body tilted down, opens bill *c.* 45° wide and lower mandible slices surface, leaving wake. Mandible cuts water at 45° angle (so that striations on side of bill

A

are parallel to surface) and is immersed for two-thirds of its length; tip of upper mandible passes few mm above surface. Bird 'skims' for 50–100 m, then turns and often skims same path again in other direction. Long wings held high, the beat regular and shallow, tips not touching water. From time to time upper mandible snaps shut onto lower, neck momentarily bends down and bill moves back in water, as bird strikes prey; feeding continues with barely discernible break in skimming flight (**A**).

Migrations are evidently dictated by need for calm weather and still, not very shallow water; in Nigeria vacate nesting sandbars in June, July and Aug, soon after breeding and when rivers rising, and disperse up to 600 km in all directions, mainly downstream (southward on Niger R. toward coast, northward on Benue/Logone rivers to L. Chad). L. Chad vacated with onset of strong winds Nov–Dec, when birds return to nesting areas. Hundreds, present L. Edward (Zaïre) and Murchison Falls National Park (Uganda) late Dec–mid Apr, are non-breeders probably from Zambia and S Tanzania (Britton and Brown 1974). Main arrival Zambia from Mar to July; departure Nov, vagrants Dec–Jan (Taylor 1979). Present Zimbabwe Apr–Dec.

Food. Fish up to 8 cm long (Chapin 1939).

Breeding Habits. Monogamous, colonial. At colony on Central Island, L. Turkana, a few birds present all year but most appear 1–2 weeks before laying. They arrive at night and are active and noisy until early morning when each day they leave (Modha and Coe 1969). When permanent residence established, much aerial chasing and calling. After chasing and 'skimming' a ♂ landed on beach, kept up cooing call, opening and closing bill and jerking head from side to side; ♀ landed nearby, ♂ walked to her jerking his head and copulated for 7 s.

NEST: simple unlined deep scrape in sand; diam. (n = 6) 220–280 (260) mm; depth 40–45 (42) mm; sides steep, sometimes nearly perpendicular. Formed by bird pivoting around its breast, tail held high, scrabbling sand backward with feet. Substrate varies from fine white to coarse blackish sands. Scratch marks in sand around nest with radiating grooves said to be where incubating bird rests bill (Chapin 1939). Density: in 3 colonies 10, 25 and 37 nests averaged one per 35, 150 and 85 m^2; 2–14 m apart. Sited within 28 m of waterline, but falling water may increase distance considerably.

EGGS: 1–5; in Kenya 1 (15 clutches), 2 (13) and 3 (7), av. clutch size 1·8; 2–3 eggs usual Zaïre and 3–4 Nigeria and Natal, 5 once (Natal, Beven 1944). Laid generally at night, at intervals from 9 h to 5 days; clutch of 3 laid usually in 4–5 days. May be double brooded (Zambia, Pitman 1932). Ground colour buff, heavily blotched light or dark brown or purplish brown overlying thick spots of slate-grey or black-purple. SIZE: (n = 22, Kenya) 35·7–43·1 × 26·4–30·0 (39·4 × 28·2). WEIGHT: (n = 6, Kenya) 15·1–17·5.

LAYING DATES: Senegambia, Apr; Mali, May–June; Nigeria, Apr–June; Ethiopia, May–June; Congo, June; Zaïre Mar; E Africa: Region B Apr, Region C May–July, Region D Mar–May; Angola, Aug; Zambia, May–Oct (11 clutches May–June, 67 July, 44 Aug, 27 Sept, 19 Oct): Zimbabwe, July–Oct (mainly Sept, 24 out of 61 clutches); Botswana, Aug–Oct; Malaŵi, Sept–Oct; Natal, Oct. On lakes with stable water level breeds in rains (Brown and Britton 1980) but on rivers nests when water falling, i.e. in dry season.

INCUBATION: by both sexes, usually together at nest during night and nearly all of day. Non-incubating bird restless until it relieves mate. Nest-reliefs more frequent as air temperature increases; at 35·5°C, 8 nest-reliefs in 30 min. After 3–4 min sitting bird shows heat stress by moving in nest, opening bill and tossing head from side to side; if not relieved it leaves nest but soon returns. At nests on black sand, nest-reliefs every c. 60 s, mate gently pushing incubating bird off nest and replacing it; displaced bird then flies low over water, dips feet 4–5 times probably splashing its breast, and returns to sit on or stand over and wet eggs (Turner and Gerhart 1971, Dowsett 1975). On cool cloudy days all incubating birds in colony may 'panic' like terns, flying away together, soon returning. Period: c. 21 days.

DEVELOPMENT AND CARE OF YOUNG: chick weighs 9·7g at hatching; with egg-tooth; lower mandible slightly hooked at tip and longer than upper, at 4 days 2 mm longer. Rapid elongation of lower mandible does not begin until chick fully fledged. Chick leaves nest within day of hatching, then wanders all over colony area. 1 or both parents attend it closely for several days, feeding and constantly shading it. Chick led to water's edge as day heats up, and into grass in evening. Chick a few days old actively follows parents about beach, washes in surf, drinks, and digs shallow resting place in moist sand where it lies, with parents nearby. Squats down and 'freezes' when danger approaches. Parents mob and chase away small reptiles and most other bird species. No distraction display; but curious floppy flight over water with legs trailing then feet scrabbling surface was either distraction display or breast-wetting. Parent flies to water to wet breast feathers and ♂ and ♀ brood chick alternately; disturbed brooding adult said to pick chick up in bill and fly off with it for 60–200 m (Roberts 1976). Young fledges in 5–6 weeks or less (Hanmer 1982). Flying young continue to be fed by parents.

BREEDING SUCCESS/SURVIVAL: of 35 nests (62 eggs) on Central Island, L. Turkana, 20 (28 eggs) destroyed in 19 days by storms and predators equally. Main predator Sacred Ibis *Threskiornis aethiopica*; others, ravens *Corvus rhipidurus*, gulls *Larus cirrocephalus*, *L. fuscus*, herons *Ardea goliath*, and cobras *Naja nigricollis*. 21 small chicks reduced to *c.* 7 by such predators in 5 days.

Untended chick soon dies of heat and drowns in rain or when washed from shoreline resting place (Modha and Coe 1969). Elsewhere Nile Monitors *Varanus niloticus* and probably crocodiles prey on eggs and young, and small colonies obliterated by hippopotamuses.

References
Cramp, S. (1985).
Modha, M. L. and Coe, M. J. (1969).

Family ALCIDAE: auks

Medium to small, short-necked seabirds with compact, streamlined bodies. Thick, stiff, black and white plumage; moult twice a year; flightless during autumn moult. Wings short and narrow; 11 primaries with outermost reduced, 15–19 secondaries. Short, rounded tail of 12–16 rectrices. Bill strong. 3 front toes, long (especially middle), united by membranes; hind toe absent or rudimentary; strong claws; tarsometatarsus unfeathered. Sexes alike.

Float high on water; dive swiftly; swim quickly under water with wings. Take-off laboured, wings flapped for considerable time with feet touching water. Flight direct, fast. Legs stretched backward in flight. Sociable all year. Feed near surface and in deep water on fish and invertebrates; on land only during nesting season; stand and walk upright.

Occur chiefly in northern regions of northern hemisphere; 12 genera, 20 spp.; 4 genera and spp. reach N Africa, mainly W Mediterranean and Atlantic coast of Morocco, in Palearctic winter.

Genus *Uria* Brisson

Bill long, tapered toward tip, not laterally compressed; nostrils slit-like; tail with 12–14 rectrices, each comparatively broad and not pointed at tip; tarsus length *c.* 25% of body length. 2 spp., 1 occasionally wintering N African coastal waters.

Uria aalge (Pontoppidan). Guillemot; Common Murre. Guillemot de Troïl.

Colymbus aalge Pontoppidan, 1763. Danske Atlas, 1, p. 621, pl. 26; Iceland.

Plate 22
(Opp. p. 353)

Range and Status. Breeds N Atlantic and Arctic Lapland and Baltic to Nova Scotia; also N Pacific coasts and islands N Japan and Bering Sea to Washington and California. Most winter at sea near breeding range, some disperse south to Iberian Peninsula, western Mediterranean, California, and New England.

Palearctic visitor, uncommon to rare, Morocco (20, Straits of Gibraltar, 1 May 1977).

Description. *U. a. albionis* Witherby: only subspecies in Africa. ADULT ♂ (breeding): head, neck and throat chocolate brown to black; rest like non-breeding plumage. ADULT ♂ (non-breeding): upperparts from crown to tail and including wings dark grey brown; face white with narrow dark brown stripe from eye across ear-coverts. Chin, throat and foreneck white. Some feathers of throat and sides of head tipped blackish, occasionally making dark bands across throat. Flanks streaked brown; underparts white. Secondaries with

Plate 27

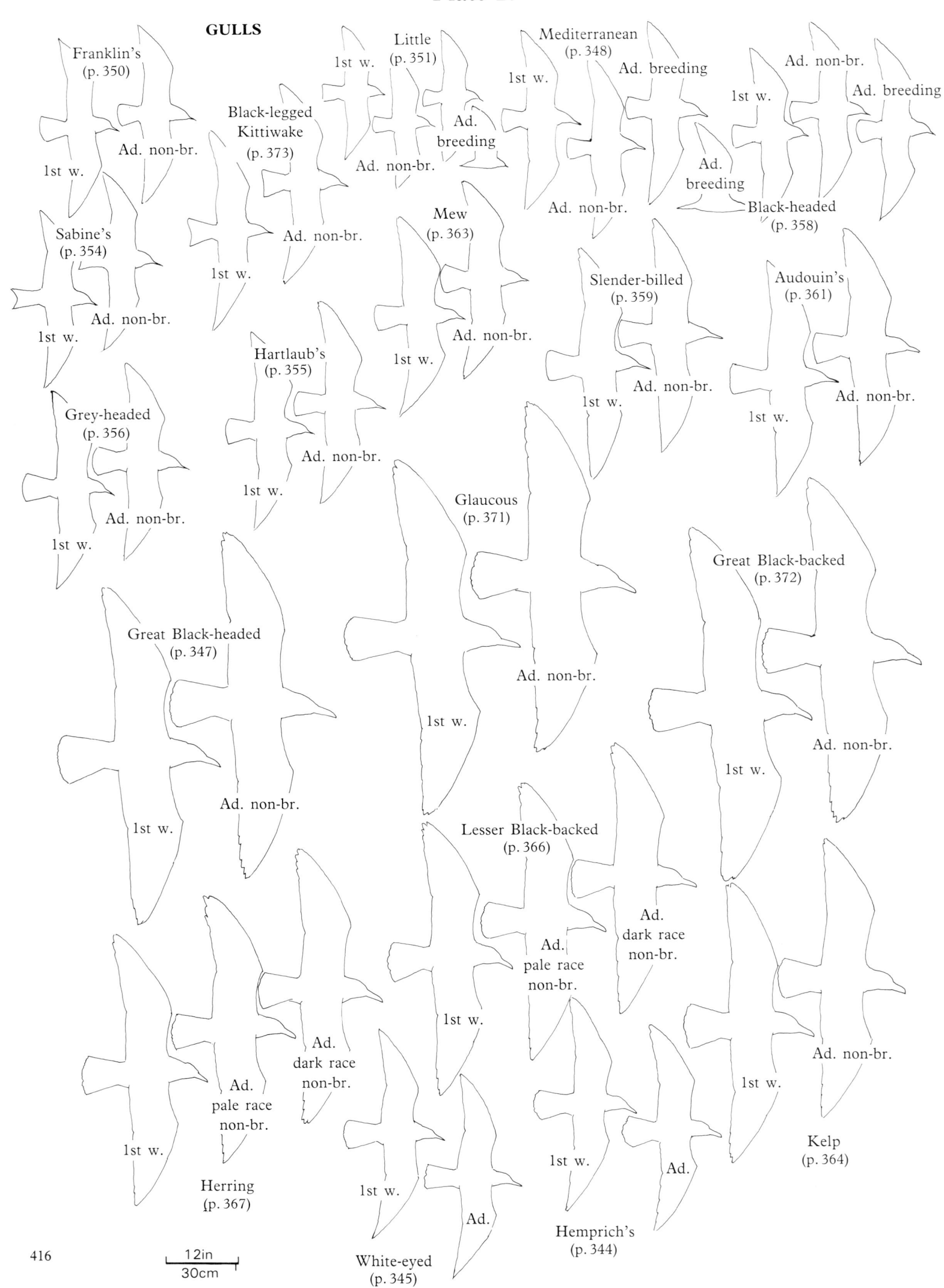

GULLS

Franklin's (p. 350) — 1st w., Ad. non-br.

Sabine's (p. 354) — 1st w., Ad. non-br.

Black-legged Kittiwake (p. 373) — 1st w., Ad. non-br.

Little (p. 351) — 1st w., Ad. breeding, Ad. non-br.

Mediterranean (p. 348) — 1st w., Ad. breeding, Ad. non-br.

Black-headed (p. 358) — 1st w., Ad. breeding, Ad. non-br.

Mew (p. 363) — 1st w., Ad. non-br.

Hartlaub's (p. 355) — 1st w., Ad. non-br.

Grey-headed (p. 356) — 1st w., Ad. non-br.

Slender-billed (p. 359) — 1st w., Ad. non-br.

Audouin's (p. 361) — 1st w., Ad. non-br.

Great Black-headed (p. 347) — 1st w., Ad. non-br.

Glaucous (p. 371) — 1st w., Ad. non-br.

Great Black-backed (p. 372) — 1st w., Ad. non-br.

Herring (p. 367) — 1st w., Ad. pale race non-br., Ad. dark race non-br.

White-eyed (p. 345) — 1st w., Ad.

Lesser Black-backed (p. 366) — 1st w., Ad. pale race non-br., Ad. dark race non-br.

Hemprich's (p. 344) — 1st w., Ad.

Kelp (p. 364) — 1st w., Ad. non-br.

416

12in / 30cm

MWW
1981

MWW
1981

Plate 28

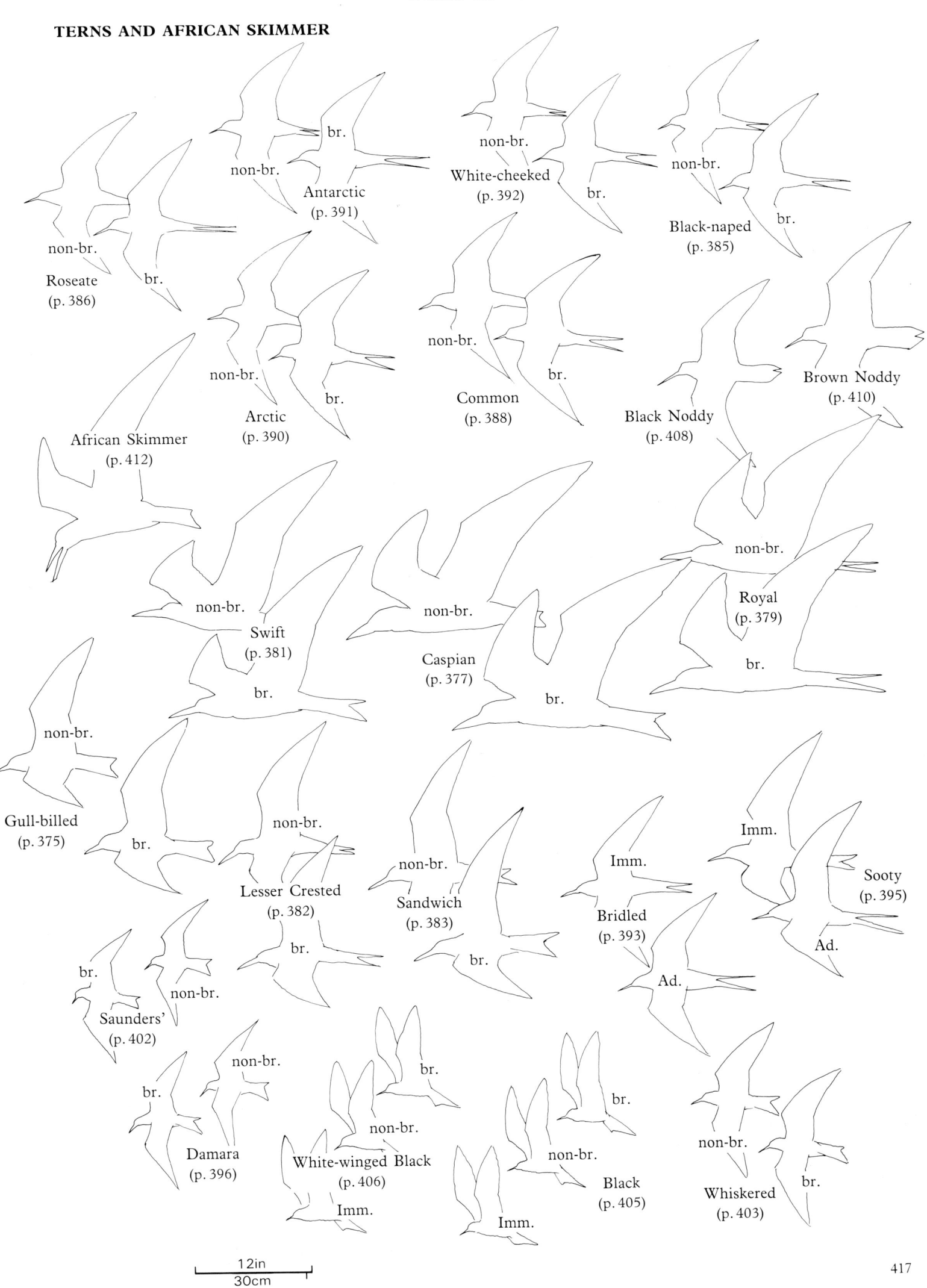

TERNS AND AFRICAN SKIMMER

non-br.
br.
Roseate
(p. 386)

non-br.
br.
Antarctic
(p. 391)

non-br.
br.
White-cheeked
(p. 392)

non-br.
br.
Black-naped
(p. 385)

non-br.
br.
Arctic
(p. 390)

non-br.
br.
Common
(p. 388)

Black Noddy
(p. 408)

Brown Noddy
(p. 410)

African Skimmer
(p. 412)

non-br.
br.
Swift
(p. 381)

non-br.
br.
Caspian
(p. 377)

non-br.
br.
Royal
(p. 379)

non-br.
br.
Gull-billed
(p. 375)

non-br.
br.
Lesser Crested
(p. 382)

non-br.
br.
Sandwich
(p. 383)

Imm.
Ad.
Bridled
(p. 393)

Imm.
Ad.
Sooty
(p. 395)

br.
non-br.
Saunders'
(p. 402)

br.
non-br.
Damara
(p. 396)

br.
non-br.
Imm.
White-winged Black
(p. 406)

br.
non-br.
Imm.
Black
(p. 405)

br.
non-br.
Whiskered
(p. 403)

12in / 30cm

417

ALCIDAE

Uria aalge

white tips. Bill black; eye brown-black; legs and feet ochre to brown-black. In 'bridled' form a narrow white ring around and line back from eye. Sexes alike. SIZE: wing, ♂ (n = 30) 189–205 (197), ♀ (n = 4) 191–212 (198); tail, ♂ (n = 12) 41–48 (45), ♀, 44, 47, 51; bill, ♂ (n = 12) 41–49 (45), ♀ 43, 46, 46; tarsus, ♂ (n = 12) 33–37 (35), ♀ 33, 37, 38. WEIGHT: ♂ (n = 50) 825–1185 (1028); ♀ (n = 29) 820–1112 (978).

IMMATURE: like non-breeding adult but often more white showing across back of neck; less streaked on flanks.

Field Characters. A medium-sized, dark brown and white auk; dark stripe across white face distinguishes it from other alcids in winter. Resembles Razorbill *Alca torda* but has longer, narrower, pointed bill, thinner head and neck, and, when seen closely, dark brown, not black upperparts. In flight, no white wing-bar.

Voice. Tape-recorded (62, 73). Usually silent in winter; very noisy in breeding season.

General Habits. Winters at sea, rarely within sight of land. Often seen in small flocks, flying in lines low above water. Congregates in large numbers on feeding grounds. Occasionally forages singly. Takes off and flies in large flocks. Flight straight and relatively rapid. In severe storms spends a great deal of time under water rather than flying. Dives well, ranging as deeply as other auks; remains submerged up to 68 s. Often leaps from water when surfacing from a dive. Moults at sea and is then flightless. Flocks in late spring largely separated into either adults or immatures.

Food. Pelagic fish; also bottom-dwelling fish, crustaceans, polychaetes, molluscs and algae.

Reference
Tuck, L. M. (1960).

Plate 22
(Opp. p. 353)

Genus *Alca* Linnaeus

Bill strong and laterally compressed, decurved at tip; nostrils slit-like; tail with 12–14 rectrices, each pointed at tip; tarsus short (*c.* 28% of body length). 1 sp., occasionally wintering N Africa in coastal waters.

Alca torda Linnaeus. Razorbill. Petit Pingouin.

Alca Torda Linnaeus, 1758. Syst. Nat., (10th ed.) 1, p. 130; southern Sweden.

Range and Status. Breeds coasts NE North America east to N Europe; winters NE North America, N and W Europe.

Palearctic visitor, uncommon to rare Morocco to Tunisia; vagrant Egypt.

Description. *A. t. islandica* C. L. Brehm: only subspecies in Africa. ADULT ♂ (breeding): upperparts black; conspicuous white line from base of bill to eye; sides of face, chin, throat, fore and side of neck chocolate brown, underparts white. Bill blacker, less greyish than in non-breeding ♂, enlarged near base with scale plates, upper and lower mandible grooved. ADULT ♂ (non-breeding): lores, ear-coverts and upperparts, including head, nape, hindneck, and tail black, tinged brown; nape and hindneck with some white feathers; sometimes with thin white line from base of bill to eye. Chin, throat, foreneck and underparts white. Underwing-coverts grey-brown; secondaries tipped white; rest of wings black. Bill greyish black crossed by vertical white curved line; eye dark brown; legs and feet black. Sexes alike. SIZE: wing, ♂ (n = 22) 181–202 (190), ♀ (n = 9) 179–206 (194); tail, ♂ (n = 14) 71–84 (78), ♀ (n = 9) 70–87 (78·2); bill, ♂ (n = 14) length 30–35 (33), depth 21–23 (22), ♀ (n = 10) length 27–35 (31·5), depth 14–24 (18·8); tarsus ♂ (n = 14) 30–34 (32), ♀ (n = 11) 28–34 (30·4); WEIGHT: (*A. t. torda*) ♂ (n = 11) 662–800 (727), ♀ (n = 8) 631–810 (717).

IMMATURE: similar to non-breeding adult but upperparts with more brown, white tips to secondaries narrower and bill smaller, not crossed with white line.

Blacker, less brown than Guillemot *Uria aalge* and with heavier head and thicker neck. White wing-bar noticeable in flight. When swimming, bill and tail tilted upwards; tail pointed.

Voice. Tape-recorded (62, 73). Usually silent at sea; weak whistle and protracted growl at breeding grounds.

General Habits. Winters far out to sea, occasionally seen in inshore waters. Often in small flocks flying in line close to water; sometimes occurs singly. Dives frequently, remaining under water up to 22 s. In African waters end Sept to early May. 4 birds ringed Great Britain, recovered Algeria (1, Dellys, Dec; 1, near Algiers, Mar), Tunisia (1, 36°47'N, 10°11'E, Jan) and Morocco (1, 31°30'N, 9°47'W, Dec).

Food. Mainly fish; also molluscs, crustaceans, and polychaetes.

Field Characters. A medium-sized black and white bird, distinguished from other auks by short, deep, laterally compressed bill crossed by white line in adults.

References
Bédard, J. (1969).
Cramp, S. (1985).

Genus *Alle* Link

Bill stout and short; nostrils rounded; 12 rectrices; tarsus little shorter than middle toe with claw, being 31·5% of body length. Throat pouch present. One of the smaller auks. 1 sp., occasionally wintering N African coastal waters.

Alle alle (Linnaeus). Little Auk; Dovekie. Mergule nain.

Alca Alle Linnaeus, 1758. Syst. Nat. (10th ed.), 1, p. 131; Scotland.

Range and Status. Breeds eastern Ellesmere I. east to N Russia; winters south to Atlantic coast of North America, southern Greenland, Iceland and Scandinavia, sometimes reaching Florida and western Mediterranean.

Palearctic visitor, uncommon to rare Morocco.

Description. *A. a. alle* (Linnaeus): only subspecies in Africa. ADULT ♂ (breeding): like non-breeding adult but whole of head, neck and upper breast chocolate brown. ADULT ♂ (non-breeding): upperparts from forehead to tail glossy blue-black to brown black; lores and anterior ear-coverts brown-black; rest of sides of head and neck to side of nape white. Sides of upper breast light brown, sometimes forming incomplete necklace; rest of underparts white, except flanks streaked black. Wings brown-black, secondaries tipped white, scapulars edged white, forming narrow white line. Bill black; eye black to hazel brown; legs and feet brown to slaty grey, webs and joints black. Sexes alike. SIZE: wing, ♂ (n = 32) 118–133 (125), ♀ (n = 33) 120–132 (125); tail, ♂ (n = 12) 32–38 (35), ♀ (n = 10) 29–38 (32·5); bill, ♂ (n = 12) 14–16 (15), ♀ (n = 10) 12–15 (14·2); tarsus, ♂ (n = 12) 19–21 (20), ♀ (n = 10) 18–22 (19·9). WEIGHT: (n = 21, AD, sex?) 338–452 (386).

IMMATURE: similar to adult but upperparts browner, less glossy; feathers of scapulars edged darker brown.

Plate 22
(Opp. p. 353)

420 ALCIDAE

Field Characters. A very small auk not more than 20 cm in total length. The combination of small size, chubby appearance and short stout bill distinguishes it from other auks.

Voice. Tape-recorded (62, 73). Silent at sea; constant, shrill, laughing chatter at breeding colony.

General Habits. Spends non-breeding season on high seas, usually far from land, in small flocks, sometimes in large assemblages. Usually flies in flocks, close to water; flight rapid. Dives well, remaining submerged up to 63 s, av. 33 s. Rises easily and directly from water. Can hold food in throat pouch. Moves fairly well on land, walking on toes.

Food. Mainly planktonic crustaceans, less frequently molluscs and polychaetes; very occasionally fish.

References
Cramp, S. (1985).
Evans, P. G. H. (1981).

Genus *Fratercula* Brisson

Bill large, stout, exceptionally deep, laterally compressed, strongly decurved and with strong folds and horny plates in breeding season. Nostrils slit-like, tail with 14–16 rectrices. Claws strongly curved. 2 spp., 1 occasionally wintering N Africa in coastal waters.

Plate 22
(Opp. p. 353)

Fratercula arctica (Linnaeus). Atlantic Puffin. Macareux moine.

Alca arctica Linnaeus, 1758. Syst. Nat. (10th ed.), 1, p. 130; Norway.

Range and Status. Breeds E Canada and Maine to Britain, Scandinavia and N Russia; winters south to middle Atlantic coast of USA, Iberian Peninsula and western Mediterranean.

Palearctic visitor, uncommon to rare, Morocco (mainly on passage, rare winter); vagrant Algeria (2 recoveries of ringed birds 1978; 4 individuals sighted since 1972: E. Johnson, pers. comm.); no recent records Tunisia; occasionally large numbers migrate along Mediterranean coast of Morocco; rarely inland (Lac de Tunis).

Description. *F. a. grabae* (C. L. Brehm): only subspecies in Africa. ADULT ♂ (breeding): like non-breeding ♂ except side of head, including lores and around eyes, dusky to whitish grey; forehead at base of bill sometimes ashy, chin darker grey. Rim at base of bill dull yellow; feathered proximal part with blue-grey plates separated by yellow ridge from bright red distal part; tip of bill, junction of jaw, and inside mouth yellow. Orbital ring vermilion; blue-grey line extending behind eye; small blue horny plates above and below eye. ADULT ♂ (non-breeding): forehead, crown and nape grey-black; rest of upperparts including tail black to brown-black; chin, throat and sides of head including lores, around eye, ear-coverts and side of nape grey to dusky black; black extends forward at base of neck to centre of throat to form complete black collar. Flanks dark brown; rest of underparts white. Primaries and secondaries black with inner webs brown, upperwing-coverts black, underwing-coverts brown. Bill laterally compressed and expanded vertically; culmen arched, curving down to decurved tip; proximal half flattened and grey-brown, distal half yellow with grooves and ridges curving back to base. Eye grey-brown, hazel to white with red orbital ring; legs and feet yellow. Sexes alike. SIZE: wing ♂ (n = 10) 152–164 (159), ♀ (n = 43) 152–164 (157); tail, ♂ (n = 10) 42–56 (49·7), ♀ (n = 10) 43–50 (46·0); bill, length ♂ (n = 10) 44–49 (46), ♀ (n = 43) 41–46 (43), greatest depth ♂ (n = 10) 33–38 (36), ♀ (n = 43) 31–37 (34); tarsus, ♂ (n = 10) 23–28 (25·4), ♀ (n = 10) 21–27 (24·3). WEIGHT: (*F. a. arctica*) ♂ (n = 18) 466–586 (511), ♀ (n = 19) 445–565 (498).

IMMATURE: like non-breeding adult except sides of head duller ash grey, lores blackish, collar brown; shallow and small bill dark brown with 0–2 transverse grooves; culmen sloping but not arched.

Field Characters. A small chunky auk, with large head, large triangular bill, pale cheeks and dark collar on foreneck. Bill red, blue and yellow in summer, less colourful and slightly smaller in winter; immature bill darker and still smaller. Appears large-headed in flight; wings pointed and narrow with no white wing-bar.

Voice. Tape-recorded (62, 73). Usually silent although makes growling 'ow' or 'arr' notes at breeding sites and croaking sound at sea or during mating.

General Habits. Spends non-breeding season at sea, seldom near shore, in small or large flocks. Moults at sea when loses bright breeding adornments. Flies low, just above water in short lines. Takes off from water with difficulty, rising after much flapping of wings and rapid skimming over water. Moves fairly well on land, walking on toes. Feeds in large flocks. Courtship activities including bowing, billing, head flicking and pair-formation, occur at sea.

Mainly seen on western passage in Straits of Gibraltar in Spring (1 Apr–6 May); 2 Scottish ringed birds recovered Altantic shore, Morocco; birds ringed Great Britain and Sweden, recovered Algeria (Algiers and Cape Djenet).

Food. Mainly fish; also invertebrates including crustaceans and molluscs; very occasionally algae.

References
Cramp, S. (1985).
Harris, M. P. (1980).
Lockley, R. M. (1962).

Order PTEROCLIFORMES

Pigeon-like, desert-dwelling ground birds, probably as closely related to Charadriiformes as to Columbiformes (Voous 1973, see p. 22). A single family of 16 spp., confined to Afrotropics, southern Palearctic and western Oriental Regions. Most live in arid habitats and fly regularly to distant water.

Highly nomadic, some migratory. Food mainly seeds, especially of legumes.

Family PTEROCLIDAE: sandgrouse

2 genera. *Pterocles* inhabits hot deserts, with 6 spp. endemic to Africa and 6 shared between Saharan N Africa and Asia; also 1 endemic to India and 1 to Madagascar (other genus, *Syrrhaptes*, inhabits cold Asiatic deserts).

Compact birds, pigeon-sized (150–400 g), cryptically coloured in dull shades of olive, khaki, ochre, rufous, chestnut and black; some with white markings. Sexually dimorphic, ♂♂ and sometimes ♀♀ having 1 or more breast-bands in most species. Most ♀♀ barred dorsally and sometimes ventrally with black; ♂♂ of *P. bicinctus*, *P. quadricinctus*, *P. indicus* and *P. lichtensteinii* (subgenus *Nyctiperdix*) likewise barred above and below and have black and white forehead patterns.

Genus *Pterocles* Temminck

Hind toe rudimentary and legs short and feathered in front to toes (in *Syrrhaptes* hind toe absent and legs and toes completely feathered). Belly feathers modified to carry water, the barbules being tightly spiralled and holding water by capillarity between terminal filaments, which uncoil during immersion (not so in *S. tibetanus*).

Rectrices 14–18, tail rounded or with elongate central rectrices. Bill short, culmen decurved; nostrils covered with bristly feathers. Plumage with dense underdown, even on apteria. Aftershaft present. 11 primaries; wings long and pointed; flight strong and direct. Crop well developed. Clutch of 3 cryptically coloured eggs laid in scrape on ground; eggs elongate, equally rounded at each end. Nest-scrape may be lined with small stones or bits of dry vegetation, or it may be unlined. Chicks precocial with characteristically patterned natal down, except in *P. lichtensteinii* which has uniform brown down.

Unlike most wader chicks, sandgrouse chicks crouch with head up. Young not fed by parents, but are provided with water from soaked belly plumage of ♂. Food mostly dry seeds. Drinking flights characteristically occur in morning after sunrise (most species) or in evening just after dusk (subgenus *Nyctiperdix*); morning-drinking species may also drink in late afternoon. Individuals may not drink daily, although flocks occur daily at watering places. Drink by sucking up water and raising head to swallow. Gregarious, especially at watering places.

Immatures of some species pass quickly through 2 distinct plumages ('1st plumage' and '2nd plumage' in the species texts), but age determination has not yet proved possible (Kalchreuter 1979).

Affinities within *Pterocles* are not easy to discern. *P. quadricinctus* forms a superspecies with the Indian *P. indicus* and is closely allied with *P. bicinctus* and the very desert-adapted *P. lichtensteinii* and probably with *P. decoratus*. *P. coronatus*, *P. senegallus*, *P. exustus* and probably *P. namaqua* comprise a species group. The remaining 3 African spp. are independent: the Palearctic *P. alchata* and *P. orientalis*, and the alluvial-plains-dwelling *P. gutturalis*.

Plate 29
(Opp. p. 448)

Pterocles alchata (Linnaeus). Pin-tailed Sandgrouse. Ganga cata.

Tetrao alchata Linnaeus, 1766. Syst. Nat. (12th ed.), 1, p. 276; SW Europe (Bogdanov, 1881, Bull. Acad. Imp. Sci. St Petersburg 27, col. 165–167).

Range and Status. N Africa, SW Europe, Middle East to W Pakistan, Afghanistan and Aral Sea region, some wintering NW India.

Resident and nomadic in Morocco, Algeria, Tunisia and Libya. Probably breeds throughout S and SE Morocco; absent from extreme NE plains; occurs in semi-arid steppe north of High Atlas, with tens of thousands on high plateaux in winter and large-scale movements into Algeria (Figuig) in early Apr (J.D.R. Vernon, pers. comm.). Invasions in extreme NW Morocco, e.g. in 1965 (Pineau and Giraud-Audine 1974) probably of European origin. In Algeria common between Biskra, Béni-Ounif and Ouargla; thousands on high plateaux in summer and winter; probably moves somewhat south in winter (Ledant et al. 1981). In Tunisia scarce resident between Chott region and Sahara, formerly in Gafsa region; few breeding records; numbers fluctuate; autumn and winter visitor probably from central Sahara often in large flocks Oct (max. 1000, Apr, Oct) (Thomsen and Jacobsen 1979). In Libya scarce but widespread in NW; southern and eastern limit obscure but probably about El Hammam (29°N, 15°45′E); probably extends south to Erg of Oubari in Fezzan; large flocks sometimes water in Wadi Zemzem (Bundy 1976). Vagrant Egypt (2 records, before 1917).

Description. *P. a. caudacutus* Gmelin: only subspecies in Africa. Range of species except Iberia. ADULT ♂ (breeding): crown, nape, back, uppertail-coverts and rectrices greenish yellow; upper back mottled with yellow; rump, uppertail-coverts and rectrices barred black; central rectrices elongate; sides of head deep orange-buff with black line through eye; superciliary stripe yellowish buff; chin and throat black; upper breast greenish ochre; lower breast deep rufous-buff bordered above and below by narrow black bands; rest of underparts white; primaries grey; secondaries yellowish green; secondary coverts yellowish green edged with black; primary coverts yellowish green; minor coverts on bend of wing vinaceous to chocolate, edged with white forming a white band in flight; underwing-coverts white; bill dull brown or slate; eyes brown; feet dusky yellow. ADULT ♂ (non-breeding): in early post-nuptial moult similar to breeding ♂ but back, scapulars and tertials barred with black and clay colour ('zebra' feathers); in late post-nuptial moult these feathers are plain greenish grey; in pre-nuptial moult they are greenish with large yellow terminal spot. ADULT ♀: upperparts yellowish buff closely barred black and grey; lesser coverts edged with black; supercilium, sides of face and throat band dull ochre; chin and throat white; upper breast ochre-yellow bordered below with narrow black band; 2 black bands on breast, upper one passing up sides of neck and behind ear-coverts to meet a black line behind eye; rest of underparts white; bill, eyes, legs and feet as in breeding ♂. SIZE: wing ♂ 213–224, ♀ 194–231; tail, ♂♀ 140–190; bill, ♂♀ 13–15; tarsus, ♂♀ 25–29. WEIGHT: (48 ♂♂, 23 ♀♀, Spain) ♂ 250–408 (315), ♀ 207–374 (302) (M. A. Casado, pers. comm.).

IMMATURE (1st plumage, 5 weeks old): feathers of upperparts and breast concentrically marked with black and buff, with golden yellow border; rectrices barred black and buff; belly white. (2nd plumage, 3 months old): upperparts and breast rich buff finely barred with black ('zebra' feathers); throat, rest of underparts and eyebrow white.

DOWNY YOUNG: crown rust with black edges to forehead feathers and whitish line down mid-crown; white horseshoe extends from base of bill above and below bright rust-brown eye-patch; dark malar stripe; whitish lines on back bordering 4 rust-brown patches to form double figure-of-8 pattern; all brown patches separated from white lines by narrow black edging; irregular brown patches on posterior flanks; underparts creamy white; bill pale yellowish; eye brown; legs and feet pale yellowish.

Field Characters. The only sandgrouse with pure white belly; combined with white underwing, white wing-bar, and long central tail feathers, it gives unique appearance in flight. At rest, maroon breast-band and green upperparts of ♂ and black-banded yellow-buff breast of ♀ readily distinguish it from other N African sandgrouse.

Voice. Tape-recorded (62, 73). Far-reaching 'catar catar', 'crau crau'; alarm call 'twoi, twoi, twoi'; threat call 'ok-ok-ok-ok-ok-kurrrrr'.

General Habits. Inhabits arid and semi-arid steppes, but avoids severe desert. Prefers Mediterranean steppe with winter and spring rainfall of c. 200 mm a year. Sparsely vegetated stony ground, dry plains and plateaux, mudflats, coastal wadis.

Apart from flights to water, flies little by day; birds rest when it is very hot and fly only short distances before landing. In late afternoon small flocks (up to 20 birds) perform flight manoeuvres of short duration: these are probably non-breeding birds, since ♂♂ predominate. In evenings adults and chicks dustbathe in sun-warmed earth, often turning onto backs with feet pointing straight up. Large numbers roost together on ground.

Out of breeding season forages in small groups (2–3) which gather into flocks when flying to drink. Flock size depends on conditions: after good rainfall flocks may number hundreds or thousands. Drinks daily, earlier on hot than cool mornings. Drinking flocks seldom exceed 15–25 birds (but hundreds in Iraq). Drinking time occurs from 1–2 h after sunrise and is of c. 2 h duration; flocks mix with Black-bellied Sandgrouse *P. orientalis* (Morocco). If suddenly frightened, tail is raised and fanned.

Not regularly migratory; rather, nomadic throughout range (but see Range and Status).

Food. Exclusively small seeds. In captivity takes no animal matter at all but grazes short turf and eats shepherd's purse *Capsella bursa-pastoris*, dandelion *Taraxacum* spp., hops *Humulus lupulus*, sorrel *Rumex* spp., clover *Trifolium* spp. and chickweed *Stellaria* spp. Moves onto farmland in early summer to eat wheat and oat grains, lentils and other cereals (Spain, 41 plant spp. identified: Casado, Levassor and Parra 1983).

Breeding Habits. Monogamous; solitary nester. Sometimes nests in small colonies, but nests not very close together. Both sexes reproductively mature in 2nd spring after hatching and make scrapes as part of pair-bond reinforcement.

Display of ♂ is stiff-legged walk with lowered, fanned tail, a jump of 1–2 m into the air, followed by copulation. No corresponding display by ♀.

NEST: bare scrape of 113–135 (125) mm diam., usually unlined, but sometimes contains a few small stones; either close to large stone or bush, or in open without notable landmarks; often in natural depression or hoofprint up to 70 mm deep.

EGGS: 2–4, usually 3; N Africa mean (11 clutches) 2·7, Iraq (16 clutches) 2·9. Laid mid-morning or late afternoon at 24–48 hour intervals. Eggs equally rounded at both ends, long elliptical; smooth and somewhat glossy: shades of buff, variably spotted, blotched or speckled with browns and underlying marks of grey. SIZE: (n = 30) 42–50 × 27·5–31·7 (47 × 30).

LAYING DATES: throughout African range mainly late Apr to June towards end of rains; birds may be double-brooded in some seasons.

INCUBATION: by ♀ in day and by ♂ at night. Period: 19–25 days (usually 21–21·5 days) and 17 days in captivity in warm weather. Steady incubation starts with 3rd egg but both sexes sit irregularly on incomplete clutch. ♀ sits from c. 09·00 h to c. 17·00 h, ♂ for rest of time (3 nests, Morocco); in captivity, ♀ sits from 06·00 h to 20·00 h, ♂ overnight. Both sexes drive conspecifics from nest with lowered head and outstretched neck, accompanied by threat call. Incubation is continuous and sitting bird faces into wind; leaves nest when animal disturbance is 80–200 m away. At nest-relief ♀ leaves nest as soon as ♂ flies in and lands, but in evening ♂ waits for ♀ to get to nest before he leaves.

DEVELOPMENT AND CARE OF YOUNG: day-old chick weighs 17–18 g. By 4 weeks young are fully feathered and able to fly; central rectrices elongate at about 4 months. Chicks leave nest as soon as they are dry, or at most within 1st day after hatching. Led by both parents and feed by themselves from the start, but must have seeds shown to them by pecking action of parents. Chicks stay very close to parents for first few days, especially in hot sunshine. In captivity chicks are watered by belly-soaked ♂ each morning and evening, but in wild apparently only in morning. ♂ has a characteristic upright watering posture while chicks drink from central groove of his belly plumage. In captivity chicks can drink from an open pan of water, so may do so in the wild if standing water available.

BREEDING SUCCESS/SURVIVAL: very low in Iraq, where eggs destroyed by ungulates and predators (Marchant 1961).

References
Casado, M. A., Levassor, C. and Parra, F. (1983).
Cramp, S. (ed.) (1985).
Marchant, S. (1961, 1962).
Thomas, D. H. and Robin, A. P. (1977).
von Frisch, O. (1969, 1970).

Plate 29
(Opp. p. 448)

Pterocles exustus **Temminck and Laugier. Chestnut-bellied Sandgrouse. Ganga à ventre brun.**

Pterocles exustus Temminck and Laugier, 1825. Planch. Color. d'Ois. 5, Plates 354, 360; Senegal.

Range and Status. Africa, Arabia to India. Accidental central Europe.

Resident, nomadic or migratory; Egypt, and sahelian savannas from Mauritania to Red Sea and Gulf of Aden coasts (Sudan, Ethiopia, Somalia), also Kenya and NE Tanzania. Widespread, frequent to locally abundant. In Mauritania from Senegal valley to 18°N; SE Senegambia, and very common in N from coast to upper Senegal R. in groups of 20–250 (de Smet and van Gompel 1980); huge flocks in N-central Mali—30,000 visit lakes at Dimamou (15°26′N, 01°48′W) daily from Dec to Apr, 50,000 at L. Kabara (15°48′N, 04°33′W) in late Mar, 40,000 at 0·5 ha pool at Tahabanet, and similar concentrations Dec–June elsewhere in Sahel zone (e.g. Faguibine, Anderamboukane: Lamarche 1980); in central Chad the commonest and most widespread sandgrouse (see General Habits); locally abundant in Sudan; in lowland Ethiopia frequent to abundant; in Kenya locally common below 1500 m, and widespread in S Kenya and adjacent Tanzania (see Britton 1980); in Somalia abundant, but absent in NE and south of 3°N (Ash and Miskell 1983).

Description. *P. e. exustus* Temminck and Laugier: Mauritania to Sudan. ADULT ♂: forehead to uppertail-coverts brownish sandy to olive-grey; ear-coverts, chin and throat bright ochre-yellow washed rusty on throat; breast pale sandy to pinkish grey with narrow (3–4 mm wide) black pectoral band; belly burnt umber-chestnut sometimes merging to black in centre and merging into brownish sandy-pink anteriorly; undertail-coverts cream; feathering on tarsus cream to chestnut; central pair of rectrices sandy to olive-grey with long chocolate brown terminal filaments; remaining rectrices olive-grey to brown, some barred yellowish, all with sepia subterminal bar and broad buff tip. Primaries sepia, 5 innermost with white tips; secondaries blackish brown to sandy; scapulars and tertials dusky brown with buff tips and black terminal fringes; primary coverts chocolate brown; median and larger lesser coverts yellowish buff with narrow chocolate terminal bands and the larger ones with pearly white subterminal spot; remaining lesser coverts sandy; underwing dusky brown. Bill slate blue with darker tip; naked orbital ring pale greenish; eyes brown; feet light blue-grey to blackish grey. ADULT ♀: crown and nape yellow-buff streaked finely with sepia; sides of neck, foreneck, chin, throat and breast yellow-buff heavily streaked on breast with sepia, and with 2–3 narrow sepia pectoral bands; mantle, back and uppertail-coverts yellowish sandy with short, wavy, transverse sepia bars; rectrices sandy barred with sepia, central pair with elongate terminal filaments; belly dark brown with yellowish barring on flanks and tarsus, often uniformly black in centre; undertail-coverts and tarsal feathers cream to chestnut; bare parts as in ♂. SIZE: (♂♀) wing, 168–189; tail, 80–85, or to end of pin 120–145; bill, 13–15; tarsus, 22–27. WEIGHT: ♂ (n = 72) 174–295 (202), ♀ (n = 64) 170–225 (191).

IMMATURE (1st plumage): no sexual difference. Dorsally yellowish with rusty wash and very fine, often obscure, dark vermiculations; throat and breast like back, more finely banded; no breast-bands; belly as in adult but banded on flanks and tarsis with rust; undertail-coverts cream with irregular dark spots, often with downy remains on tips of feathers; rectrices shorter than in adult, with rusty terminal band, central pair not elongated; remiges shorter and narrower than in adult, rusty towards tips, with dark marbling; inner secondaries, coverts, scapulars and mantle sandy with rusty centres to feathers, and fine dark banding. (2nd plumage): above olive-grey with obscure dark banding towards tips of feathers only, and feathers with pale edges, especially on back and mantle; primary coverts boldly edged brown; breast greyish pink with obscure black vermiculations; no breast-band; belly as in adult; scapulars more boldly marked (especially in ♀) than in 1st immature plumage, but not rusty.

DOWNY YOUNG: dark streak above bill separated from lores by creamy white line; patch around eye golden buff edged with black; 2 transverse patches on crown golden buff bordered black and separated by creamy white lines; 2 lateral patches on nape golden buff bordered black and separated by creamy white line; line down centre of back white; 4 transverse white lines on back form 4 pairs of patches of golden buff edged with black; irregular band around flanks golden buff edged with black, meeting across rump and separated from dorsal patches by white lines; wings mostly creamy buff with patches of golden buff and black on proximal regions; underside creamy white with brownish patch on upper breast; bill and bare eye-ring lead blue; eyes hazel; feet dusky pink; claws white.

P. e. olivascens Hartert: SE Ethiopia, Kenya, Tanzania. More olive above and ♀ more closely barred than *exustus*; sometimes poorly differentiated. WEIGHT: ♂ 200, ♀ (n = 3) 170–200 (182).

P. e. floweri Nicoll: Egypt (Faiyum to Luxor). (Now probably extinct—S. M. Goodman, pers. comm.).

P. e. ellioti Bogdanow: N Ethiopia, Somalia, SE Sudan. Paler and brighter, and ♂ more rufous below than *exustus*. WEIGHT: ♂ (n = 18) 170–220 (183), ♀ (n = 15) 140–213 (171).

Field Characters. Combination of dark belly and underwing and long tail diagnostic. Spotted Sandgrouse *P. senegallus* also has long tail but less black on belly, pale underwing. ♂ has yellowish wings and narrow black breast-band like Black-bellied Sandgrouse *P. orientalis*, but face yellowish, without black on throat, tail long. ♀ also not unlike ♀ Black-bellied, but dark belly barred pale, breast streaked, not spotted, face yellow.

Voice. Tape-recorded (2, 10, 63, 73, C). Deep, fairly musical 'gutter-gutter' or 'gouta-gouta'. At waterhole, a musical 'creen' as a continuous low murmur.

General Habits. Inhabits bare semi-desert and arid scrub; also cultivated fields and grassland. In flocks of 5–30 birds, usually 15–20. When disturbed, crouches or creeps silently away until hidden. On take-off wings make loud clattering. Feeds during cooler morning and late afternoon hours; mostly inactive at midday. Drinks 16 km or more away, in large flocks (see Range and Status) 2–3 hours after sunrise, on hot days as early as 08.00 h, on cool or cloudy days not until 10.00 h. In very hot weather a few may drink also shortly before sunset. In Serengeti (Tanzania) associates with Yellow-throated Sandgrouse *P. gutturalis*. They fly high, calling continuously, circle waterhole and land 20–30 m away before running down to drink. Sometimes they land further away and then fly in to drink. For drinking behaviour in relation to Yellow-throated Sandgrouse, see that species. ♂♂ in flocks often threaten each other and spar or fight.

Movements in Chad (Newby 1979) probably typify entire range: in dry season common in Kanem and Mortcha (belt across Chad between 14° and 15° N); in wet season, especially after early heavy rains from July, many move north into S Wadi Rime-Wadi Achim Reserve (14°–16° N); as wet season progresses they move further north until Sept when a 2nd incursion arrives in south, all moving north again by the hottest months.

Food. Hard seeds, mainly of legumes, grain, and sometimes young shoots and insects. In Chad feeds on long trails of black ants (Newby 1979). In E Africa feeds mainly on seeds of *Trianthema*, *Indigofera* and *Heliotropium* (and in India mainly of *Tephrosia*, *Indigofera*, *Crotalaria* and *Phaseolus*: up to 10,000 seeds per crop).

Breeding Habits. Monogamous, solitary nester.

NEST: shallow, unlined scrape in open desert, sometimes a natural hollow cleared out by the birds.

EGGS: 3, rarely 2. Elongate and equally rounded at each end. Creamy buff, greyish to light olive, heavily spotted and blotched with brown and pale purplish grey to give a drab effect. SIZE: 34–38 × 24·5–25·5 (36·1 × 25·2). WEIGHT: 7·6–9·0.

LAYING DATES: Senegambia, Feb–Nov, mainly Mar–July (late dry season and 1st rains); Mali, Mar–June; Sudan, Apr–May, and young just able to fly Aug; Ethiopia, Apr–June (*exustus*) and Mar, May and

Sept (*ellioti*); East Africa, Region C, May, July–Aug; Region D, Feb–Mar, May–July (mainly May); Tanzania (Serengeti), young chicks Aug–Sept (Schmidl 1982).

INCUBATION: period reportedly 20 days.

DEVELOPMENT AND CARE OF YOUNG: birds breed at end of 1st year. Nothing further known.

References
Christensen, G. C. and Bohl, W. H. (1964).
Cramp, S. (ed.) (1985).
Kalchreuter, H. (1979, 1980).

Plate 29 (Opp. p. 448) *Pterocles namaqua* (Gmelin). Namaqua Sandgrouse. Ganga namaqua.

Tetrao namaqua Gmelin, 1789. Syst. Nat. 1(2), p. 754; Namaqualand.

Range and Status. Endemic resident; locally nomadic, southern populations migratory, especially in Karoo. From SW Angola to Namibia (except extreme NE), Botswana (except extreme N), Cape Province (except extreme S and E), Orange Free State and W Transvaal. Vagrant to Natal and Zimbabwe. Common to locally abundant. Now absent from former range in Orange Free State and Lesotho.

Description. ADULT ♂: crown buffy olive grading to olive-yellow on hindneck and sides of neck; face light buffy brown washed with olive-yellow; chin, throat and malar region bright rusty yellow; lower hindneck and mantle light brownish olive; upper back, scapulars and tertials brownish olive, with apical 3rd of each feather dull buffy, and an apical spot of metallic bluish silver surrounded by dark brown; back, rump and uppertail-coverts light brownish olive, each feather paler in the centre, the brownish olive fading to paler toward tail; breast light ochre yellow washed increasingly with greyish and bordered below with an upper band of white and a lower band of deep red-brown; belly grading from greyish chestnut to blackish to paler posteriorly with paler feather centres; tarsal feathers and undertail-coverts pale buff; tail olive-brown broadly tipped with white and mottled on the outer vane with pale buff, except for central 2–3 pairs; central rectrices greatly elongated, pale brownish olive with elongated tip black and base pinky buff. Primaries dark brown, paler on inner web, with shafts of first 2 white and innermost tipped white on inner web; secondaries dark brown narrowly tipped white; primary coverts dark brown; secondary coverts light brownish olive with broad white tips; remaining wing-coverts light brownish olive, each feather with a pale buff or cream terminal spot edged on inner vane with dull chestnut; many upper coverts spotted with metallic silver as on mantle, scapulars and tertials; axillaries light brown; underwing pale buffy brown. Bill light greyish horn, darker at tip; bare skin around eye yellow; eye dark brown; legs and toes pinky grey or buff.
ADULT ♀: generally mottled above, barred on belly and streaked on breast, with yellow face and throat. Crown and nape olive-yellow streaked black; hindneck and sides of neck pale buff streaked blackish brown; face and throat pale buff finely speckled with black; mantle, scapulars, back, rump and uppertail-coverts olive-yellow with narrow shaft streaks and broad wavy bars of blackish brown, scapulars with subterminal cream spot and terminal chestnut tip; breast pinkish buff with elongate spots and bars of dark brown giving a streaked effect; belly dull whitish barred with dark brown, centre of belly washed sepia; tarsal feathers and undertail-coverts pale buff; tail more heavily barred than in ♂; remiges as in ♂; tertials barred olive-yellow and greyish black; wing-coverts barred olive-yellow and black, tipped with cream; bare parts as in ♂. SIZE: (10 ♂♂, 10 ♀♀) wing, ♂ 167–179 (173), ♀ 163–173 (167); tail, ♂ 89–124 (112·5), ♀ 88–106 (95.6); bill, ♂ 11–13 (11·9), ♀ 11–13·5 (12·2); tarsus, ♂ 21–24 (22·6), ♀ 20–23 (21·8). WEIGHT: ♂(n = 13) 166 – 191 (180), ♀ (n = 15) 143–192 (172).

IMMATURE: similar to ♀, but upperparts more rufous tinged, barred with sepia and each feather fringed and tipped with whitish; throat and breast pale pinkish buff, barred with light brown and tipped with off-white; belly pale reddish brown, unmarked in ♂, mottled and barred with dark brown in ♀; tail as in adult ♀, but paler and lacking elongate central rectrices; primaries brown, paler at tip, mottled and barred brown and dusky; wing-coverts as upperparts but broadly tipped and fringed with buffy white.

DOWNY YOUNG: crown and nape yellowish brown with white bars outlined with black; face sandy with white vertical line in front of eye edged with black and running to throat; lines above and below eye whitish edged with black; back yellowish brown divided by creamy white lines into double figure-of-8 pattern, all lines edged with black (**A**); underparts pale buff, darker on breast; wings patterned like back. Bill black; eye brown; legs and feet dusky or greyish.

Field Characters. The only southern African sandgrouse with long pointed central tail feathers. At rest ♂ distinguished by plain olive-yellow head and neck, white and chestnut breast bands. ♀ lacks chestnut belly of ♀ Yellow-throated Sandgrouse *P. gutturalis*; similar to ♀ Double-banded Sandgrouse *P. bicinctus* but has yellow throat, streaked breast. Highly characteristic 3-

note flight call, well expressed by Afrikaans name 'kelkiewyn'.

Voice. Tape-recorded (35, 38, 39, 74, F). Flight call has slightly nasal initial rising inflection, then falls on slightly drawn-out final note: 'ki-ki-keeu'; it is a contact call, and may be given by incubating bird of either sex to its mate flying over. It also attracts birds on ground to join flocks flying to waterholes; in captivity it has been heard as a duet between ♂♂ (D. H. Thomas, pers. comm.). Take-off call, also sometimes given when landing, is rapid burst of staccato notes 'kip-kip-kip-kip'. Flock of birds on ground keeps up a continuous muttering 'kip kip kip' or rich, low-pitched 'ca-caaa-c'. When flushed from nest or young, parents have strident alarm/anxiety call 'ki-keeee', which develops into high-pitched churring very like voice of Suricate *Suricata suricatta* (possibly of adaptive significance to the birds); may be accompanied by injury-feigning distraction displays. Parents call to chicks with soft 'quip quip', similar to notes uttered by flocks at waterhole. Chick has 2-syllabled distress call 'ch-wrrr', with slightly rolling quality.

General Habits. Inhabits stony and gravelly desert and open semi-desert with sparse scattering of low shrubs or grass tufts; sometimes arid sandy savanna with denser vegetation, like the Kalahari. Occurs typically where av. annual rainfall is < 300 mm. Non-breeding birds occur in flocks of 3–4, seldom more than 20–30 but sometimes several hundred. Feeds mostly in early morning and late afternoon, lying up during day in any shade or sitting quietly out in open. When feeding, takes small steps and pecks frequently at ground, often calling quietly.

Flocks fly 60+ km each day to waterholes, calling 'ki-ki-keeu', travelling *c.* 50–100 m high, and joined by other birds *en route* so that 300–400 birds arrive together. Drinks mainly in morning; but individual bird may not need to drink daily (except perhaps in very hot weather) and may do so only every 3–5 days. Seldom alights at edge of water; early arrivals gather on high ground 200 m away, and after half-hour fly down to drink in huge numbers, often thousands. Later arrivals alight closer to water, wait for several min, then run to water's edge to drink. A few birds drink in late afternoon or just before dusk. ♂♂ take av. 9·6 draughts per drink, ♀♀ 9·4; at each draught bird sucks once, raises head and swallows with bill open. Intake is *c.* 10–12 ml water per drink (av. 2·3 ml per draught). Prefers water with molarity <50 mM (about half the plasma molarity); does not drink water saltier than *c.* 250 mM. Drinking takes 10–15 s, then bird takes off directly or first walks up to 60 cm away from pool.

In early morning birds disperse to feed before flying to water. After drinking, flocks split up into smaller groups or pairs as they return to feeding grounds. In late evening birds that have fed together during day fly in groups to stony areas where each makes shallow roosting scrape in which it spends the night (but birds in captivity roost in close huddle). In favoured areas same roosting scrape probably used on successive nights. In captivity birds droop wings and huddle together at temperature >35°C (Thomas and Maclean 1981).

Birds on ground react to those flying low overhead or to any other sudden movement above, by crouching and raising fanned tail toward stimulus (**B**); at high intensity head is lowered (**C**). Bird threatened by aggressor adopts same posture towards it. Reaction to predator is to crouch without fanning tail. Flies fast and can outfly a falcon in level flight; cruises usually at *c.* 60 km/h.

B

C

Nomadic throughout range; southern populations seem to winter further north, moving from Karoo to N Cape, Botswana and Namibia.

Food. Small hard seeds. Staple in Kalahari is *Lophiocarpus burchelli* seeds (1 chick's crop contained 1400) and in Namib *Tephrosia dregeana*, *Cleome diandra* and *C. luederitziana* seeds. Known high values of gross energy and crude protein content of these seeds indicate selection of a nutritious diet. Birds' crops from Namib Desert contained very small proportion of insects and mollusc shell fragments (perhaps eaten incidentally) and gravel taken as gastroliths.

Breeding Habits. Monogamous; solitary nester. In courtship ♂ chases ♀ on ground with head low and tail raised. Sometimes 2 or more ♂♂ chase ♀ and may bowl her over. Territoriality weakly developed. Adjacent nests may be as little as 25 m apart, but birds do not form colonies.

NEST: scrape 10–12 cm wide and *c*. 10 mm deep in soil, usually next to small shrub, grass tuft or stone, sometimes flanked or surrounded by several such objects; occasionally completely exposed on gravelly soil. Scrape lined with small stones, clods of earth or dry plant material accumulated during incubation by side-throwing, usually at nest-relief.

EGGS: 2–3, laid at 24-h intervals; southern Africa mean (62 clutches) 2·9. Elliptical, pale greenish stone, greyish green, olive brown or pinkish stone with blotches or spots of pale to dark brown or reddish brown and underlying spots or smears of pale grey; pinkish eggs have reddish brown markings, greenish eggs have olive-brown markings. SIZE: (n = 190) 31–42·7 × 22·5–27·8 (36·3 × 25·2). 1st and last eggs of clutch same size; weight 11–12·5.

LAYING DATES: probably depend on rainfall: throughout range eggs found in every month of year, mainly in July–Nov (73% of 175 clutches, McLachlan 1985).

INCUBATION: sitting starts with 1st egg laid and proper incubation with 2nd or 3rd egg. ♂ usually attends nest (at least by day) until clutch is complete, then ♀ incubates from 08·00 h to 18·00 h in summer months (*c*. 09·00 h–15·00 h in winter); her spell seldom exceeds 10 h. ♂ incubates rest of the time, in 14 h spell in summer (18 h in winter). Period: 21–23 days, normally 21. Ambient day-time temperature in nesting habitat often exceeds 45°C. Incubating ♀ faces sun to reduce exposure, or into any wind. At soil temperature of 45°C parent keeps eggs at 30–35°C, thermoregulating by gular fluttering and raising mantle feathers, and (in Namib where cool onshore breezes occur) by raising body to shade clutch while resting on lowered wings. At air temperatures >40°C soil temperature exceeds 50°C, and bird has difficulty keeping eggs below 45°C; eggs must tolerate such heat. Adult disturbed at nest usually departs long before danger is close; sometimes sits tight and leaves only at last moment, injury-feigning and anxiety-calling.

DEVELOPMENT AND CARE OF YOUNG: eggs hatch synchronously or over 2–3 days. 1st chick remains in or near the nest until the last can run. Parents drop eggshells several m from nest. Hatchlings feed themselves within 24 h, eating only seeds usually shown to them by parents pecking at selected items. When chicks are very small, ♂ broods them while ♀ flies to drink in the morning; on her return ♂ flies to drink while ♀ broods them. Before soaking, (**D**) ♂ may rub belly on sand, probably to loosen plumage and remove waterproofing. When ♂ returns chicks run out to take water from his soaked belly plumage while he stands in upright watering-posture; when they have drunk he dries plumage by rubbing it on sandy soil. After being watered family moves off together, usually 1 chick to each parent. In hot weather chick keeps in shade by walking under parent's tail, even while feeding. As young grow, ♂ and ♀ fly to water together, leaving young crouched in shade of plant. Apart from water-transport, roles of sexes are about equal in care of young. ♀ transports water at times, probably only when ♂ has been lost. Chicks fly at about 4 weeks but remain with parents for at least 2 months, and are still brought water by ♂ well after they can fly.

BREEDING SUCCESS/SURVIVAL: hatching success 68% (n = 69 eggs). 1 chick usually lost early and parents rarely accompanied by more than 2 young. Not over, and probably well under, 23% of eggs laid produce flying young. In Kalahari Bat-eared Foxes *Otocyon megalotis* and Black-backed Jackals *Canis mesomelas* prey on eggs and Greater Kestrels *Falco rupicoloides* and Pied Crows *Corvus albus* on chicks. Main avian predators of adults at waterholes are Lanner Falcons *F. biarmicus*.

References
Clancey, P. A. (1979).
Dixon, J. E. W. (1976, 1978).
Maclean, G. L. (1976).
Thomas, D. H. and Maclean, G. L. (1981).
Thomas, D. H. *et al.* (1981).

Pterocles orientalis (Linnaeus). Black-bellied Sandgrouse. Ganga unibande.

Tetrao orientalis (Linnaeus), 1758. Syst. Nat. (10th ed.), 1, p. 161; 'In Oriente' = Anatolia, *ex* Hasselquist.

Plate 29
(Opp. p. 448)

Range and Status. NW Africa, Canaries, S Iberia, Cyprus and Middle East to S central Asia and W Pakistan, some wintering in Egypt and from Mesopotamia to NW India.

Resident, nomadic, and Palearctic visitor. Breeds Morocco to Libya. In Morocco breeds in northeast (Trifa, Angad, Hautes Plateaux) and Erfoud, where sedentary, and probably in mid west where shown on map; occurs in semi-arid steppe north of High Atlas in NW (J. D. R. Vernon, pers. comm.). In Algeria uncommon to frequent, absent from NE (Ledant *et al.* 1981). In Tunisia scarce but evenly distributed central plains between Kelbia region and chotts; in north mountain plains to Medjerda valley; in flocks up to 20, rarely 50 (Thomsen and Jacobsen 1979). In Libya frequent in N Tripolitania south to 31°N and east to 15°E; winters in Wadi Kaam June–Sept; a few winter in Cyrenaica from Tobruk to Egypt (Bundy 1976). Regular non-breeding migrant to Egypt, presumably from SW Asia.

Description. *P. o. orientalis* (Linnaeus): only subspecies in Africa. Fuerteventura, Morocco and Iberia to Caucasus. ADULT ♂: crown, nape, hindneck and upper back grey tinged russet; sides of head grey; chin, upper throat and sides of neck chestnut; back and rump grey with rich buff at base and tip of each feather; uppertail-coverts blackish edged with yellow; lower throat black, forming triangular collar; breast pale grey crossed below by narrow black band; belly black, with broad pale pinkish band between it and breast-band; undertail-coverts, thighs and feathering on tarsus white; remiges greyish brown with white bases; scapulars and inner wing-coverts grey with large basal and terminal areas of yellow ochre; remaining wing-coverts yellow ochre; alula grey; underwing-coverts white; central rectrices barred grey and buff, tipped grey-green and narrowly edged buff; remaining rectrices similar, darker towards outermost and tipped white; bill leaden with dark tip; eyelids lemon yellow; eye brown; legs and feet greyish to brownish with darker claws. ADULT ♀: pale fawn with black belly. Upperparts pinkish grey, streaked black on head, nape and mantle, tinged rufous on back, and barred black on back and rump; chin and throat ash grey with narrow black crescent on throat; breast and upper belly yellowish to pinkish grey; sides of head and throat streaked black; breast spotted with black, and with black band as in ♂; rest of underparts and remiges as in ♂; scapulars and inner wing-coverts like back; median wing-coverts yellow ochre; secondary coverts grey; central rectrices like back; succeeding rectrices tipped white and darker towards outside; bare areas as in ♂. SIZE: wing, ♂ (n = 38) 224–256 (237), ♀ 220–235 (226); tail ♂♀ 101–128; bill, ♂♀ 10–14; tarsus, ♂♀ 24–28. WEIGHT: ♂ (n = 4) 480–550 (514), ♀ (n = 4) 410–465 (434).

IMMATURE: similar to adult, but greyer.

DOWNY YOUNG: above ochre-tawny with black tips to most of down; bold cross-shaped pattern of buffy white lines dividing tawny back into bold double figure-of-8 pattern, and meeting with buffy white lines laterally; throat and neck whitish; rest of underparts pinkish buff, faintly barred on flanks with greyish; bill light grey; down on legs buff; bare areas on legs and feet buffy pink.

Field Characters. Black belly, black and white underwing, short tail and large size distinctive in flight. Chestnut-bellied Sandgrouse *P. exustus* also has dark belly but tail long, underwing all dark. At rest ♂ told from Spotted Sandgrouse *P. senegallus* by black throat, black band across breast and contrasting yellow wings, ♀ by dark band below breast spots, more extensive black on belly.

Voice. Tape-recorded (62, 73). Deep, mellow 'cudrrrii' and 'djur-djur-djur'; also clucking calls, a soft double chuckle, 'catarr' repeated several times, a deep 'tchourou' and a musical 'churr-churr-rur-rur'.

General Habits. Inhabits semi-arid steppe north of Atlas Mts in west and along Mediterranean coastal strip to Egypt (with rainfall of 250–350 mm, mainly in Mar–Apr; summer, June–Sept, is dry). Summer breeding range lies roughly between 20°C and 30°C July isotherms, where temperature often 40°C, sometimes 55°C.

During most of day single bird or pair stays in feeding areas; at 07·00–09·30 h flocks of 15–25 fly to drink at waterholes (Morocco, May), sometimes with Pin-tailed Sandgrouse *P. alchata*. After drinking, return to feeding areas where spend rest of day. Reluctant to fly in heat of day even when disturbed, and normally seek shade to stand in.

Sedentary or nomadic from Morocco to Libya, but further east a winter visitor about Oct–Mar (origin unknown).

Flocks fly daily at least 30 km to water (although individual birds may not drink every day), usually 1–2 h after sunrise. Drinking period seldom lasts more than 2 h. In India several thousand birds drink during that time. Preferred drinking places are clear of vegetation. Flock wheels around waterhole once or twice before pitching in steeply to land a short distance away. Waits there a moment or two, even 30 min, before running or flying down to water to drink; or may land at water, drink and take off.

Food. (Data mainly from India.) Mostly small grass and weed seeds (30,000 seeds found in 1 bird), and fallen grains in cereal crops. Favourite foods are tiny seeds of legumes *Indigofera linifolia* and *I. cordifolia* (in 92% of birds' crops, forming 21% by volume). Other preferred seeds are of *Panicum, Heliotropium, Cyamopsis, Gynandropsis* and the legumes *Phaseolus, Tephrosia, Melilotus* and *Astragalus*. Seeds of *Artemisia, Salsola* and *Alhagi* also recorded. Predominance of legumes in diet may reflect their high protein content. Evidently also eats some insects (mainly termites, beetles and their larvae).

Breeding Habits. Monogamous; solitary nester. Courtship unknown.

NEST: exposed shallow scrape on ground, usually unlined.

EGGS: 2–3, usually 3; N Africa mean (17 clutches) 2·6. Equally rounded ellipses; ground colour pale to deep buff or greyish, sometimes with greenish tinge, smudged, blotched and spotted with brown and grey.
SIZE: (n = 44, N Africa) 40·5–52·5 × 29·5–34·2 (47·0 × 32·0); (n = 78, India) 47·5 × 32·3.

LAYING DATES: N Africa, mid Apr–July, after rains and into dry season. Double-brooded (in Israel), with young as late as Aug.

INCUBATION: carried out by ♀ during day and by ♂ at night, starting with 1st egg. Period 21–22 days.

DEVELOPMENT AND CARE OF YOUNG: ♂ provides young with water from soaked belly feathers. Nothing further known.

References
Ali, S. and Ripley, S. D. (1969).
Christensen, G. C. *et al.* (1964).
Cramp, S. (ed.) (1985).

Plate 29 *Pterocles senegallus* (Linnaeus). Spotted Sandgrouse. Ganga tacheté.
(Opp. p. 448)

Tetrao senegallus Linnaeus, 1771. Mantissa: 526; Senegal, *errore*, Algeria accepted as type locality by Hartert, 1924, Nov. Zool. 31: 7.

Range and Status. Africa, Israel and Arabia to Afghanistan and Pakistan; winter visitor to W Pakistan and NE India. Accidental S Europe.

Resident, nomadic, partially migratory, Morocco to Mauritania, Mali and Libya and Egypt to Somalia. In Morocco patchily distributed in subdesert steppe, uncommon, little evidence of breeding; winters N to Sous and Dades valleys and NE to Berguent (J. D. R. Vernon, pers. comm.); in Algeria frequent to common, in flocks up to 200, north to Chegga, Ksours Mts, Taghit and Beni Abbès (Ledant *et al.* 1981); in Tunisia irregular and rare breeding resident, greatest flock 300 (Thomsen and Jacobsen 1979); in Libya widespread as mapped, little evidence of breeding, greatest flock 250, appears locally in coastal zone to drink (Bundy 1976); in Mauritania occurs mainly to 14°23'W, between 18° and 19°44'N (P. W. P. Browne, pers. comm.); but also in Banc d'Arguin National Park (Lunais 1984); in Mali occurs only in Saharan zone in east, in groups of 10–15, breeds (Lamarche 1980); in Chad known only from Borkou and Ennedi; in Egypt and Sudan widespread north of Khartoum; in Ethiopia frequent to abundant in NW and NE, including Dahlak islands; also N Somalia, where evidently increasing (Ash and Miskell 1983).

Also winters in breeding range

Description. ADULT ♂: crown pale rufous-buff; broad line from bill over eye and around nape and hindneck pale ashy grey; rest of upperparts sandy or pinkish buff; rectrices blackish, central pair elongated; chin, throat and sides of foreneck deep orange-yellow; upper breast pale buffy grey merging with buff underparts; centre of belly and undertail-coverts black. Primaries blackish with buff tips and buff outer vane; secondaries buff with brown subterminal spot; scapulars and tertials buff with large brown subterminal patch; greater coverts buff with brown base; median coverts brown with buff tip; lesser coverts sandy buff. Bill bluish grey; eye brown, orbital skin yellow; legs and feet pale grey, claws blackish.

ADULT ♀: above and below buff; crown streaked and spotted with black; rest of upperparts and breast spotted with black; sides of foreneck washed with yellow, belly black; primaries as in ♂, rest of upper surface of wing spotted with black; bill, eye, feet and claws as in ♂. SIZE: wing, ♂ 190–208, ♀ 176–197; tail, ♂ 80–90, or to end of pin 127–167; bill ♂♀ 12–15; tarsus, ♂ 23–26. WEIGHT: ♂♀ 250–340; ♀ max. 255.

IMMATURE: upperparts sandy with darker crescentic bars and somewhat vermiculated streaks, central rectrices barred, shorter than in adult; underparts sandy with small horseshoe marks on breast; centre of belly black.

DOWNY YOUNG: hatchling light greyish brown, obscurely mottled.

Field Characters. ♂ rather pale and uniform, not unlike Crowned Sandgrouse *P. coronatus* but with long tail, black on belly and no black on face or throat. ♀ heavily spotted above and on breast; ♀ Black-bellied Sandgrouse *P. orientalis* also has spotted breast but is barred above and more extensively dark on belly; also has short tail.

Voice. Tape-recorded (73, 217, 246). Staccato calls, mainly in flight, distinctive and liquid 'cuito cuito', 'wittu wittu', 'waku waku', or 'cuddle-cuddle'. Alarm call a soft 'hu hu'.

General Habits. Inhabits open, flat, stony desert, especially with isolated patches of vegetation. Prefers hammadas (stony plains) with light grey or brown background, dotted with sandy patches and small vegetated wadis; unlike Crowned Sandgrouse (with which it sometimes flocks) avoids more sparsely vegetated areas with larger rocks and stones where wadis are scarce.

At daybreak flocks move out to feed for 1–2 h; a few birds fly around area 06·30–08·00 h giving flight call until one after another leaves ground and whole flock flies to waterhole, usually only few km away. Flight speed said to be *c.* 100 km/h (George 1976) but that estimate probably too high. On arrival at water, flocks of up to 60 circle several times then land a few m from shore, still calling loudly; several hundred congregate. They wait for min or two, taking off at the slightest disturbance. If at ease, run to water and drink by immersing bill to level of eyes, raising head between draughts to swallow. After several draughts, in <15 s, run back to landing place, then fly back to feeding areas. Very few drink in evening. In Egypt gathers in large flocks with Crowned Sandgrouse to feed on raw grain spilled from lorries trading between Red Sea ports and Nile valley markets (S. M. Goodman, pers. comm.).

Non-breeding birds roost in flocks of up to 50, usually those that fed together during day, on open hammadas near their feeding areas, but not in wadis (where likely to be preyed on by foxes and jackals). Each makes a scrape 1–6 cm deep in which to sleep, *c.* 1 m from adjacent scrapes.

Food. Small hard seeds. In Morocco feeds largely on seeds of *Euphorbia guyoniana* (in non-breeding season, George 1969), and *Asphodela tenuifolia* (Thomas and Robin 1977). In India, grass and weed seeds as well as 'quantities of insects' (Ali and Ripley 1969).

Breeding Habits. Monogamous; solitary nester. Non-breeding flocks disperse into pairs few weeks before egg-laying begins, but no marked displays evident; ♂♂ merely become less aggressive toward ♀♀. Flights in which several ♂♂ may chase a ♀ may constitute courtship.

NEST: small unlined scrape on ground; formed by ♀ excavating with feet and ventral plumage; also in hoofprint or other natural depression in soil. Sited among conspicuous stones about same size as bird, usually on flat ground.

EGGS: 3; N Africa mean (20 clutches) 3·0. Interval between eggs 24–48 h. Elliptical, equally rounded at each end; buff or *café-au-lait*, sparingly spotted and scrawled with light brown, reddish brown and underlying purplish grey. SIZE: mean 41·7 × 28·0 (Harrison 1975); 40·9 × 28·4 (Ali and Ripley 1969).

LAYING DATES: Morocco and Tunisia, Mar–July; Sudan, Mar, May. In Morocco may be double brooded in years with good rainfall (D. H. Thomas, pers. comm.).

INCUBATION: starts with 1st egg. Period: 29–31 days (n = 3). Only by ♀ until clutch is complete. Between 06·30 h and 08·00 h ♂ calls ♀ and both fly to drink, sometimes even when clutch complete. Usually ♀ takes over from ♂ on complete clutch after she has drunk; ♂ waits on nest until ♀ alongside. In evening ♂ relieves ♀, which may leave while ♂ is still many m away.

DEVELOPMENT AND CARE OF YOUNG: downy chick at 4–6 days sandy-buff with typical pattern in brown, but less distinctly marked than congeners. Leaves nest on day last chick hatches. ♀ makes roosting scrape that evening in which she broods chicks; ♂ makes his own roosting scrape nearby. Both parents lead chicks to suitable feeding places in morning and indicate food by pecking.

When chicks *c.* 3–4 days old ♂ takes over brooding from ♀ *c.* 1 h after sunrise. ♀ flies to waterhole; on her return ♂ flies to drink. On return he locates ♀ by clear 'wit-wit-wit' contact calls *c.* 3 s apart in flight; ♀ on ground responds with same call. ♂ stands upright in watering posture, using tail as prop, and chicks drink from his soaked belly plumage by placing bills in ventral groove. Oldest chick said not to drink until 3 days old, but that seems unlikely. After watering chicks, ♂ rubs belly feathers in sand (presumably to dry them). After 4 days, parents leave chicks crouched at suitable hiding place under shrub and fly together to drink. When parents return (after up to 2 h absence) they utter contact calls and chicks answer only their own parents' calls. Both parents perform injury-feigning distraction displays in presence of danger near eggs or young.

Chicks <4–6 days old in greyish brown down always try to hide among brown stones; later, in sandy yellow down, hide in small pockets of yellowish drift sand, forming a scrape with body movements until half body hidden in sand.

Whole family feeds together in morning and late afternoon, remaining inactive during heat, *c.* 11·00–17·00 h. They cover about 200 m/day in search of food. At 5–6 weeks young can fly with parents to waterhole.

BREEDING SUCCESS/SURVIVAL: seldom are more than 2 young reared from brood; Lanner *Falco biarmicus* and Barbary Falcon *F. peregrinus pelegrinoides* main avian predators.

References
Cramp, S. (ed.) (1985).
George, U. (1969, 1970, 1976).
Marchant, S. (1961).
Thomas, D. H. and Robin, A. P. (1977).

Plate 29
(Opp. p. 448)

Pterocles coronatus Lichtenstein. Crowned Sandgrouse; Coronetted Sandgrouse. Ganga couronné.

Pterocles coronatus Lichtenstein, 1823. Verz. Doubl., p. 65; Nubia.

Range and Status. Africa and Middle East to W Pakistan.

Resident, locally nomadic and partly migratory, from Morocco through N and central Sahara to Nile, south to Tibesti (Chad); Red Sea coastal records as shown on map (Goodman and Watson 1983). Inhabits the severest desert of all sandgrouse species, hence range poorly known. In Algeria frequent and widespread, south to Tassili and Hoggar; the commonest sandgrouse in Ksours Mts (Ledant *et al.* 1981). In Tunisia occurs mainly between Chebika and Metlaoui and between Kebili and El Hamma, but even there erratic and a rare breeder (Thomsen and Jacobsen 1979). In Libya, the least numerous sandgrouse, scarce north of 32°N in Tripolitania, in Cyrenaica widely distributed between Jebel Akhdar and the sand sea of Kalanshu, and in Fezzan widespread, uncommon or locally common, absent from Oubari and Mourzouk sand seas (Bundy 1976). Recorded in Aïr (Niger), Hoggar (Algeria, locally abundant, Meinertzhagen 1934) and Tibesti (Chad) massifs in central Sahara; in Tibesti very common on Wadi Tao plains and at Guelta Sao (Guichard 1955). In N central Chad (Wadi Rime-Wadi Achim reserve) scarcer than Lichtenstein's Sandgrouse *P. lichtensteinii* (Newby 1979). Occurs in Nile delta and E Egypt (including Wadi Kansathrope and Wadi Umm Taqhir) and fairly common around Dongola (Sudan). Recorded Mopti (Mali); may occur in Kordofan (Sudan).

Description. *P. c. coronatus*: only subspecies in Africa. ADULT ♂: forehead white, bordered by black stripe from base of bill to front of crown; sides of face white; crown pale rufous surrounded by pale blue-grey stripe from above and behind eye to nape; chin and upper throat black, joining black forehead stripes at gape; rest of throat, cheeks and hindneck ochre-gold; mantle, back, uppertail-coverts rufous-sandy; rectrices rufous-sandy with white tip and black subterminal band; central rectrices slightly elongated; underparts sandy, washed grey on breast, rufous or pinkish on belly; feathering of tarsus white. Primaries brown with whitish outer border and white shafts; secondaries rufous-sandy with brownish inner webs; scapulars, tertials and wing-coverts rufous-sandy washed grey at base and with pale sandy tips; greater underwing-coverts greyish brown; rest of underwing white washed with sandy. Bill bluish grey with blackish tip; narrow bare orbital ring bluish grey; eye brown; legs and feet bluish grey to greenish grey. ADULT ♀: chin, throat, ear-coverts (and sometimes nape) ochre-yellow, often finely spotted with black; rest of plumage sandy-buff, closely barred all over with black, most coarsely on mantle. SIZE: wing, ♂ (n = 5) 193–205, ♀ (n = 6) 183–195; tail, ♂ (n = 5) 79–86, ♀ (n = 6) 75–83; bill, 13–14; tarsus, ♂ (n = 5) 22–24, ♀ (n = 6) 23–24. WEIGHT: (n = 1) 300.

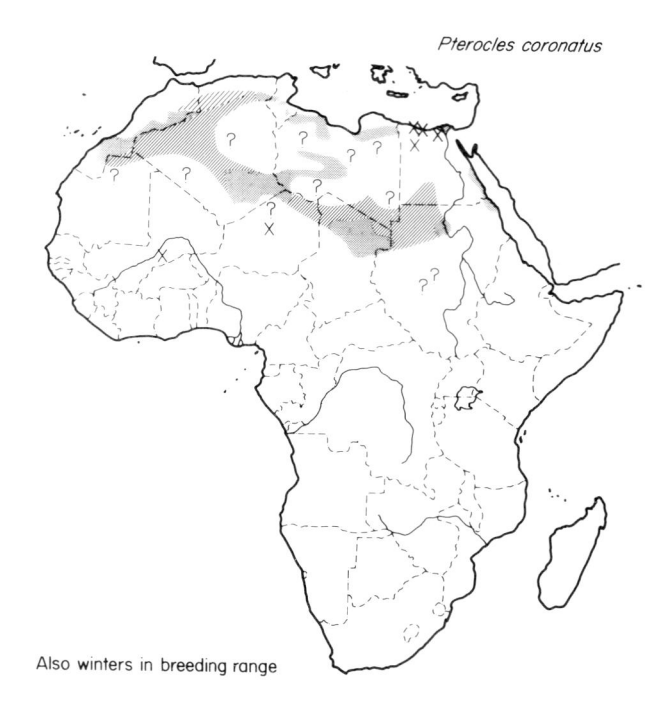

Also winters in breeding range

IMMATURE: similar to adult ♀ but throat whitish; generally more rufous-buff; barring coarser, little or none on belly. Secondary-coverts and central rectrices of ♂ buff with black vermiculations.

DOWNY YOUNG: pattern above like that of chick of Namaqua Sandgrouse *P. namaqua*; rufous-buff above, whitish below; dark patch around eye; for 1st few days head is conspicuously whitish.

Field Characters. A smallish, short-tailed sandgrouse. ♂ similar in overall colour to ♂ Spotted Sandgrouse *P. senegallus* but lacks long tail, black on belly; ♀ very like ♀ Lichtenstein's Sandgrouse but has unspotted yellow throat. Voice distinctive.

Voice. Tape-recorded (217, 282). Unlike other sandgrouse, a soft 'kla kla kla' or 'cha-chagarra', distinctly staccato. Alarm call a soft disyllabic 'hu hu' repeated 2–3 times. Flocks flying around waterholes call continuously.

General Habits. Inhabits the most arid regions of Sahara, especially dark reddish brown stony areas. Apparently not as gregarious as other sandgrouse; flocks number 6–30 birds. Flies to drink in the mornings only;

said to drink also at dusk (Pakistan). Where range overlaps that of Spotted Sandgrouse, 2 spp. time their morning drinking flights together, between 06·00 h and 09·00 h, starting about an hour after sunrise and arriving synchronously at waterhole. At other times flies little or not at all, especially in hot weather. Presence of a Lanner *Falco biarmicus* at usual waterhole caused sandgrouse to fly several dozen km to next one. In Egypt large flocks gather on the Safaga–Qena and Quseir–Qift roads to feed on grain spilled copiously from fleets of lorries (S. M. Goodman, pers. comm.).

In E Morocco extends north in winter (Oct–Mar) to Berguent; in Chad moves from 15°40'N (southern limit) to Tibesti and Erdi regions in hottest months (Apr–June) (Newby 1979).

Food. In N Africa almost exclusively seeds of *Asphodela tenuifolia* where this plant extremely abundant 6 months after heavy rain (Thomas and Robin 1977).

Breeding Habits. Monogamous; solitary nester. Courtship and territorial behaviour unknown.

NEST: scrape or unlined natural hollow in sand, sometimes with ring of small stones around rim.

EGGS: 2–3, usually 3. Long, elliptical with equally rounded ends; smooth and moderately glossy; cream to pale drab yellowish, evenly spotted and blotched with light brown and pale purplish grey. SIZE: (n = 9, N Africa) av. 39·4 × 27·4.

LAYING DATES: Morocco, Algeria, May–July; Chad, Mar.

INCUBATION: ♀ incubates by day, showing no signs of heat stress even at ambient air temperatures of 40–50°C. ♂ incubates from about 17·00 h and probably through night. 1 ♂ reported taking over incubation about 1 h before sunset with well soaked belly plumage (Thomas and Robin 1977) but could have been incidental result of getting wet while drinking. Period unknown.

DEVELOPMENT AND CARE OF YOUNG: young are provided with water from soaked belly feathers of ♂ but (unlike congeners) chicks are watered both morning and afternoon (George 1976); during watering ♂ always stands with back to sun so that chicks and wet belly are shaded.

References
Cramp, S. (ed.) (1985).
George, U. (1970, 1976).
Thomas, D. H. and Robin, A. P. (1977).

Pterocles decoratus Cabanis. Black-faced Sandgrouse. Ganga à face noire.

Pterocles decoratus Cabanis, 1868. J. Orn. 1868, p. 413; Lake Jipe, Kenya.

Plate 29
(Opp. p. 448)

Range and Status. Endemic resident. S, SE and NE Ethiopia where frequent, S Somalia (common and widespread), NE Uganda (Kidepo Valley Nat. Park, Karasuk hills, S Mt Elgon), throughout Kenya below c. 1600 m and locally common, and most of E Tanzania (see Britton 1980).

Description. *P. d. decoratus* Cabanis: SE Kenya and NE Tanzania. ADULT ♂: forehead black with small buff patch over nostrils; supercilium and line over forecrown white; crown, nape and hindneck khaki streaked with black; eye-stripe black; ear-coverts yellowish buff; chin buff; centre of throat black bordered by pale buff merging to buff on face; mantle khaki narrowly barred sepia; back, rump and uppertail-coverts khaki coarsely barred brownish black and brown, except for broad buffy-khaki feather-tips giving a chequered effect; tail khaki boldly barred brownish black; tips of rectrices buffy with broader blackish subterminal band; upper breast khaki, lower breast white, separated by brown-black band; belly very dark brown with narrow whitish tips to feathers; undertail-coverts khaki-buff with broad brownish black chevrons. Primaries dark brown, innermost with narrow whitish tips; secondaries dark brown; scapulars, tertials light khaki, each feather with 2 broad blackish bars; primary coverts dark brown-black; remaining coverts light khaki coarsely and sparingly barred blackish; underwing whitish washed with khaki, merging to very dark brown on axillaries and flanks; feathering on legs pale buff. Bill deep yellow; bare skin around eye dull yellow; legs and feet dull orange-yellow. ADULT ♀: similar to ♂, but more closely barred dorsally, and barred on upper breast; no black face mask, eye-stripe or breast-band; chin yellowish; face and sides of neck finely spotted with blackish. SIZE: wing, ♂♀ 154–175; tail, ♂♀ 60–75; bill, ♂♀ 13–15; tarsus, ♂♀ 25. WEIGHT: ♂ (n = 35) 160–216 (187), ♀ (n = 34) 149–210 (178).

IMMATURE (1st plumage): similar to adult ♀ but feathers downy; dorsally, scapulars, inner secondaries and secondary coverts rusty with narrower, darker barring; primaries with rusty tips, bordered whitish; tail more irregularly barred, with streaks at base. (2nd plumage): ♂ and ♀ like respective adults, but feathers mostly edged whitish; ♂ has indications of facial pattern, but no breast-band.

DOWNY YOUNG: sandy to gold and grey above, with bold black mottling.

P. d. ellenbecki (Erlanger); NE Kenya to NE Uganda, S Somalia and S Ethiopia. Paler, yellower, less brown and less heavily barred. WEIGHT: ♂ (n = 14) 130–190 (156), ♀ (n = 17) 140–170 (154).

P. d. loveridgei Friedmann: SW Kenya and interior Tanzania. Paler than nominate race, more greyish sandy and less rusty than other races.

Field Characters. Broad white band across upper belly distinguishes both sexes from all sympatric sandgrouse. ♂ Lichtenstein's Sandgrouse *P. lichtensteinii* has black and white face pattern but no black on throat and no pronounced supercilium; both sexes of Lichtenstein's have finely barred, not chequered upperparts; ♀ lacks yellow-buff on face.

Voice. Tape-recorded (36, B, C, 244, 264, 296). A repeated 3-note whistle of 2 long and 1 short syllables; also calls 'chucker chucker chucker' and a series of short notes 'quit-quit-quit-quit-quit'; also a clucking note while feeding.

General Habits. Inhabits dry savanna, thorn bush, semi-desert scrub and coastal dunes. Often frequents bare areas, road verges, and small open spaces in thick bush. Usually singly or in pairs, also groups of 4–12, gathering into larger flocks to drink. Flies to drink in 1st half of morning, travelling high up, calling, and on arrival circles round a few times before swooping down.

Food. Dry seeds, mainly of legumes: *Trianthema salsoloides*, *Indigofera* spp. and *Heliotropium undulatifolium*; roots.

Breeding Habits. Monogamous; solitary nester.
NEST: unlined scrape on bare sandy or stony ground.
EGGS: 2 (Serengeti, Tanzania), usually 3 elsewhere. Long oval, almost equally rounded at both ends; glossy; greyish buff, spotted or blotched with reddish brown and mauve-grey undermarkings. SIZE: 33–36 × 25–26.
LAYING DATES: E Africa: Region C, June–July; Region D, Jan–Apr, June–Sept (mainly June–Aug).

References
Kalchreuter, H. (1979, 1980).

Plate 29
(Opp. p. 448)

Pterocles lichtensteinii Temminck. Lichtenstein's Sandgrouse. Ganga de Lichtenstein.

Pterocles lichtensteinii Temminck, 1825. Pl. col. Livr. 60, pl. 355; Nubia.

Range and Status. Africa, S Jordan, Arabia, Socotra, Baluchistan (Iran/Pakistan) to W Sind.

Resident, somewhat nomadic, S Morocco (rare; known only in stone deserts south of Jebel Bani), Mauritania, Senegambia (once, N'Goui), central Mali (rare; Kami, Bankas), E Mali (Adrar des Ifoghas, common, flocks up to 20) and adjacently in Aïr Massif (Niger), also Hoggar Mts and Tassili (Algeria) where locally common in N foothills (Meinertzhagen 1934), N Chad (Tibesti, drinking flocks of hundreds: Simon 1965), central Chad (frequent Wadi Rime-Wadi Achim Reserve: Newby 1979), N and extreme SE Sudan (frequent to locally common), SE Egypt, Ethiopia (uncommon to frequent), N Somalia, NE Uganda (Karamoja), and N Kenya south to L Bogoria (Hannington) and lower Tana R. (uncommon). In Morocco and probably elsewhere, occurrence strictly associated with that of main food plant, *Acacia sayal*.

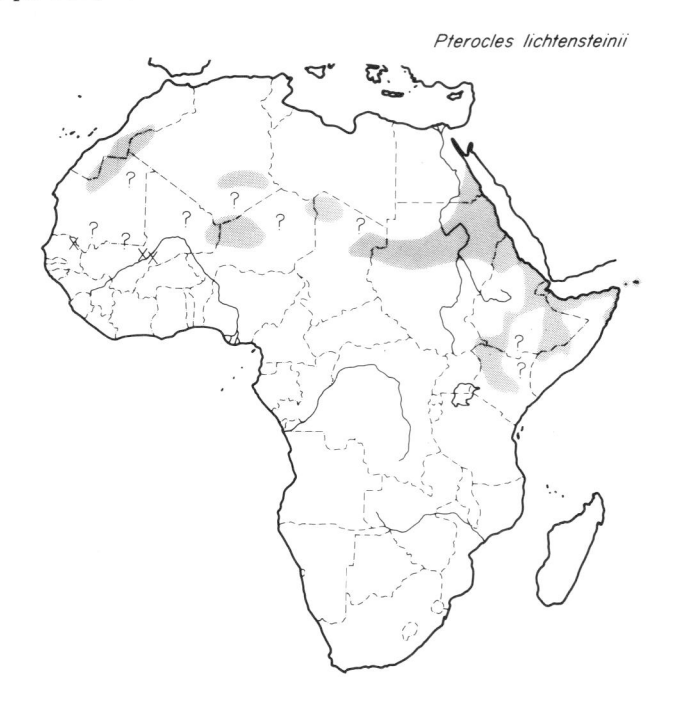

Description. *P. l. targius* Geyr von Schweppenburg: Sahara and Sahel from Chad west to Morocco and Mauritania. ADULT ♂: forehead and forecrown white, divided transversely by broad black bar; hindcrown, nape and sides of neck rufous-buff streaked with black; small black patch behind and above each eye, with a white patch behind it; face and ear-coverts buff with fine black shaft streaks; chin and throat rufous-buff with border of small black spots, mantle, back, rump and uppertail-coverts rufous-buff heavily barred with black; tail rufous-buff, more coarsely barred with black than rump and with broad black subterminal band and deep rufous-buff or yellow-ochre tip; base of foreneck and upper breast buff with black barring, bordered below by narrow black band; lower breast dull yellow-ochre, divided in centre by narrow transverse band of deep blackish chestnut and bordered below by another narrow black band; belly whitish barred with black; undertail-coverts yellow-ochre with sparse heavy black bars. Primaries blackish brown with narrow white margins; outer web of outer secondaries whitish with 1–5 oblique black) bars and inner web blackish brown; remaining secondaries evenly barred blackish brown and rufous; scapulars deep rufous-buff with coarse black barring and yellowish buff tips; primary coverts blackish brown; remaining coverts rufous-buff barred with black and fringed with yellowish buff; underwing light brown; axillaries greyish, mottled with brown at tips. Bill yellow to orange; bare skin around eye yellow; eye brown; legs and feet chrome yellow with whitish feathering on tarsus. ADULT ♀: crown pale rufous-buff narrowly streaked black; nape, sides of neck, chin and throat buff finely spotted black; rest of upperparts as well as breast pale rufous-buff narrowly barred black; belly whitish narrowly barred black; remiges and rectrices and bare parts as in ♂. SIZE: wing, ♂♀ 180–195 (188); tail, ♂ 74–90, ♀ 69; bill, ♂♀ 14–16; tarsus, ♂ 25–30 (26), ♀ 25.

IMMATURE: similar to ♀, but more closely barred above and below.

DOWNY YOUNG: warm donkey brown, slightly paler ventrally; but lores, supercilium, ear-coverts and under eye light chocolate brown, bordered below by pale malar stripe and above by pale lateral crown-stripe running from bill to nape; pale spot below and behind eye; bill, naked orbital ring, legs and feet light slate grey (Thomas and Robin 1983).

P. l. lichtensteinii Temminck: SE Egypt, Sudan except SE, N Ethiopia, N Somalia. More yellowish buff, less rufous.

P. l. sukensis Neumann: S Ethiopia, SE Sudan, NE Uganda, N Kenya. Darker than *lichtensteinii*, with heavier black barring. WEIGHT: ♂ (n = 11) 175–250 (217), ♀ (n = 7) 190–230 (207).

Field Characters. Fine barring above and below in both sexes, buff breast of ♂ bisected by narrow black band is distinctive. ♂ has black and white forehead and face pattern not unlike ♂ Black-faced Sandgrouse *P. decoratus*, but lacks black throat and supercilium; throat and breast barred, not buff. ♀ similar to ♀ Crowned Sandgrouse *P. coronatus*, but face and throat spotted, not plain yellow, barring finer, broader pale bands on wing evident at rest.

Voice. Tape-recorded (5, B, C, 258). A liquid whistle 'quitoo-quitoo, quitoo-quitoo'. Flocks at water have constant musical chatter termed 'creening'. On drinking flights a soft 'wheet ... wheet ... wheet' flight call.

General Habits. Prefers rocky or scrubby country, hillsides, bushy wadis with *Acacia* in arid areas. Does not occur in open desert. Normally in pairs or small groups, seldom more than 4 birds together. Spends much of day in shade of rock or plant; active mainly at night, especially in drinking habits. Pants by gular fluttering. Flocks fly just above bush and tree height to drink after dusk and again before daybreak, when can be detected by flight calls and rustling sound of wings. Lands *c.* 30 m from water and runs to edge, taking off directly after drinking. Main food seeds have low salt content and sandgrouse also drinks water of low salt content; its relatively small kidneys perhaps reflect adaptation to low salt intake, but good water conservation because of relatively few glomeruli (kidney units) per unit bodyweight.

Food. In NW Africa mainly seeds of *Acacia sayal*, *A. radiata* and *Asphodela* spp. 10 crops yielded 3674 acacia seeds (Arabia); also feeds on *Cassia* seeds.

Breeding Habits. Monogamous; solitary nester.

NEST: unlined scrape on ground usually among scattered trees or rocks; sometimes under shelter of shrub.

EGGS: 2–3. Elliptical; olive, brownish olive, pinkish or very pale buff, blotched, spotted or sparsely marked with reddish brown and underlying purplish grey. SIZE: 42·0 × 26·3.

LAYING DATES: Somalia, Feb, July; Ethiopia, June; East Africa, Region A, May–June; Region D, Sept.

DEVELOPMENT AND CARE OF YOUNG: young leave nest soon after hatching. In heat of sun, chicks walk in parents' shadows. When alarmed, young crouch in shade of rock or bush.

References
Cramp, S. (ed) (1985).
Thomas, D. H. and Robin, A. P. (1977, 1983).

Pterocles quadricinctus Temminck. Four-banded Sandgrouse. Ganga quadribande.

Plate 29

Pterocles quadricinctus Temminck, 1815. Pig. et Gall. 3, p. 252; Senegal.

(Opp. p. 448)

Range and Status. Endemic, resident and migrant. Widespread from Senegambia to Ethiopia and NW Kenya; in central latitudes of this range resident, in southern parts a breeding dry-season visitor and in northern parts a non-breeding wet-season visitor. Locally common Senegambia; common Mali (abundant Bougoni Mts), northernmost record Diré (16°16′N); abundant N Ivory Coast (dry season only); frequent to common Nigeria, south to Ilorin (08°30′N); common in Chad north to 16°N in groups of 4–6; frequent in Ethiopia; uncommon N Uganda and NW Kenya south to 2°N.

Description. ADULT ♂: forehead and forecrown white divided by broad transverse black bar; short black bar over bare eye-patch; rest of crown rufous-buff to khaki, heavily streaked with black drop-shaped marks; nape, hindneck, sides of crown, sides of neck and face sandy buff with olive wash; chin creamy buff; upper throat and cheeks washed sandy, merging to deep khaki-buff on throat and upper breast; mantle, back and scapulars yellowish buff, with basal region of feathers heavily barred black and rufous giving a chequered effect; uppertail-coverts and tail banded ochre-yellow and black; breast khaki-buff bordered below by narrow bands of chestnut, white and black in that sequence; belly finely barred black and white, blackish toward centre; feathering of tarsus white;

undertail coverts barred ochre-yellow and black. Primaries, outer secondaries and primary coverts dark brown; inner secondaries brown with yellowish buff and white edging;

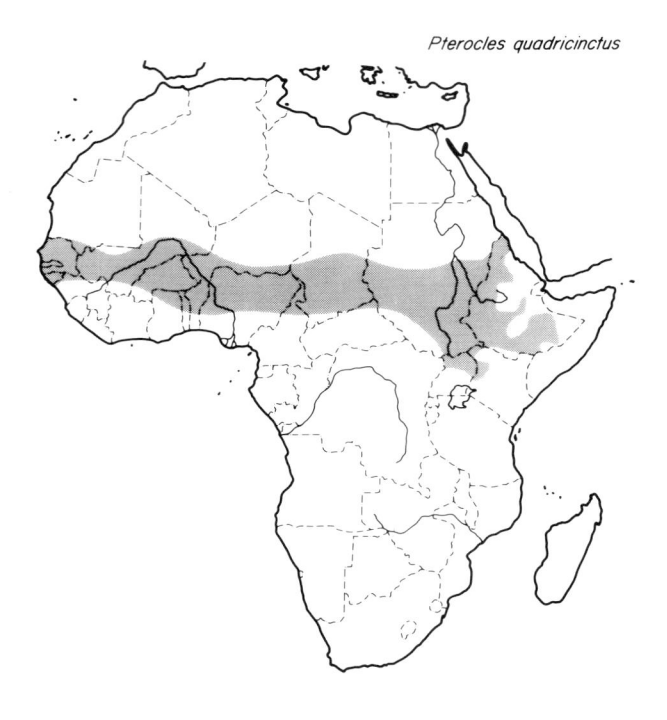

Pterocles quadricinctus

innermost secondaries barred rust and black with ochre-yellow edging; remaining wing-coverts plain ochre-yellow; underwing-coverts and axillaries grey. Bill dull yellow; bare skin around eye yellow; eye dark brown; legs and feet yellow. ADULT ♀: general appearance similar to ♂ but without black and white forehead pattern and breast bands. Entire crown rufous-buff streaked black; nape rufous-buff spotted black; hindneck rufous-buff barred black; foreneck and face buff fading to whitish on chin and throat; rest of upperparts as in ♂, but paler and more closely barred; breast khaki-buff or rufous-buff without bars; belly and tarsal feathers as in ♂. Primaries, secondaries and primary coverts as in ♂; remaining wing-coverts deep buff barred black; underwing barred greyish brown and white (larger coverts brownish grey); bill yellowish with dark tip, or mostly black; eye and surrounding bare skin, legs and feet as in ♂. SIZE: wing, ♂ 175–182, ♀ 176–184; tail ♂ 75–80, ♀ 72–78; bill ♂♀ 13–15; tarsus, ♂ 25–26, ♀ 23–25.

IMMATURE: more rufous than adult; black barring finer; feathers of mantle, back and scapulars tinged with buff; primaries broadly tipped with rufous or whitish buff.

DOWNY YOUNG: dorsal pattern similar to that of most sandgrouse chicks, but with fewer white lines dividing the dark areas posteriorly, so that the figure-of-8 is single on upper back and not double over whole back: eyebrow broadly white, brown eye-patch confined to area below eye (from figures in Gore 1981 and Fjeldså 1976).

Field Characters. A smallish, short-tailed sandgrouse with boldly chequered upperparts, plain buff breast and finely barred belly. ♂ has black and white forehead pattern similar to that of Black-faced Sandgrouse *P. decoratus* and Lichtenstein's Sandgrouse *P. lichtensteinii*, but lacks black throat of former, barred throat and breast of latter; white breast-band much narrower than that of Black-faced, and bordered above by chestnut band. ♀ has unique combination of plain buff breast and barred belly. Wing-tips black in flight.

Voice. Tape-recorded (217, 267). Sounds like 'wur-wulli' (Hausa name, Nigeria) or 'pirrou-ee'; a soft or shrill, piping whistled call in flight. Flocks at water twitter loudly. Gabbling call when flushed.

General Habits. Inhabits dry, open or partly wooded savannas, cultivation, open and bushy grasslands, dissected pasture, cotton soils, and scrub by coastal dunes. Mainly nocturnal; largely inactive by day, resting in groups of 2–4. When flushed by day zig-zags away and seldom flies far before settling again. From *c.* 20 min before sunset to *c.* 50 min after, flies low to water to drink in groups of up to 12 birds, gathering at some waterholes in large numbers. After drinking, groups scatter to feed until late at night. May also drink before daybreak (Lewis *et al.* 1984).

After breeding in Senegambia north to 15°30′N, migrates north to Senegal R. valley July–Oct. In Nigeria resident in central latitudes; in south a dry-season visitor, e.g. in Borgu and at Zaria Oct–May; in NE a wet-season visitor to Sahel zone, Apr–Oct, locally abundant from mid–May; passage noted Mar–Apr and Oct. Similarly in Mali and Chad a rains visitor to Sahel zone, in Chad occurring north to 15°50′N, June–Oct.

Food. Almost certainly seeds.

Breeding Habits. Monogamous; solitary nester.

NEST: a scrape, bare, or lined with a little dry grass, on bare stony soil or shingle, or among broken scrub and bushes, sometimes under a shrub.

EGGS: 2–3; E Africa mean (3 clutches) 2·7. Elliptical, rounded equally at each end, clay-pink or salmon-buff, spotted and flecked with orange-brown or pale brown and underlying mauve; highly cryptic among dry leaves of *Bauhinia* trees, where often laid. SIZE: (W Africa) 36·4–38·0 × 26·2–27·9; (n = 2, E Africa) 38·7–40·5 × 27·6 (39·6 × 27·6).

LAYING DATES: Senegambia, Nov–June, mainly Mar; Nigeria, Nov–Jan; Cameroon (Benue), chicks Jan; Sudan, Feb–Mar; Ethiopia, Feb–Mar, possibly also July, Aug and Dec; East Africa, Region A, Jan.

Nothing further known.

Pterocles bicinctus Temminck. Double-banded Sandgrouse. Ganga bibande. Plate 29

Pterocles bicinctus Temminck, 1815. Hist. Nat. Pig. et Gall 3, p. 247; Great Fish River, Great Namaqualand. (Opp. p. 448)

Range and Status. Endemic resident. SW Angola (coastal desert north to Benguela), S Angola, Namibia, N Cape Province, Transvaal, Botswana, Zimbabwe, S Zambia, W Mozambique and extreme S Malaŵi. Common.

Description. *P. b. bicinctus*: NW Cape, Namibia, extreme S Angola, Botswana. ADULT ♂: forehead and forecrown white divided by transverse black band; crown and nape deep rufous-buff streaked with black; hindneck and sides of neck deep olive-buff; lores black, continuous with bar across forecrown; short black line above eye; rest of face, chin and upper throat creamy buff; lower throat rufous-buff, with olive-buff edges to feathers; mantle, scapulars and tertials blackish brown with white tips to feathers, barred with pale rufous; back and rump barred greyish brown and rufous; uppertail-coverts deep buff barred greyish brown; upper breast pinkish buff merging with rufous-buff of lower throat and bordered below by double band of black and white; lower breast and belly barred whitish and blackish brown; undertail-coverts deep buff barred with dark brown; feathers of tarsus white with darker bands; tail deep buff banded with brownish black, all rectrices except central pair tipped buff. Primaries blackish brown tipped with white; secondaries blackish brown with basal half of outer vane paler; primary coverts brownish black; median and secondary coverts greyish brown, barred and apically spotted with white; remaining coverts rufous-buff fringed olive-buff; underwing greyish brown, tipped paler; axillaries grey. Bill yellowish to reddish with dark tip; bare skin around eye yellow; eye dark brown; legs and feet dull yellowish to brownish. ADULT ♀: crown and nape light rufous-buff streaked black; hindneck and sides of neck pale buff streaked black; face creamy buff faintly mottled brown; chin and upper throat creamy buff; mantle, scapulars and tertials light rufous-buff heavily barred blackish brown and tipped whitish; back, rump and uppertail-coverts light rufous-buff heavily barred blackish brown; tail barred buff and black, tipped with pale buff and with wavy black subterminal bar; lower throat and upper breast pinkish buff mottled and barred with wavy bands of brownish black; rest of underparts off-white or greyish white finely barred black; undertail-coverts deep buff barred blackish brown. Primaries, secondaries and primary coverts brownish black; remaining coverts pale pinkish buff barred dark brown and apically spotted white. Soft parts as in ♂. SIZE: (10 ♂♂, 10 ♀♀) wing, ♂ 172–187 (181), ♀ 171–185 (177); tail, ♂ 73·5–86·5 (80·6), ♀ 76–82·5 (79·1); tarsus, ♀ 22–26·5 (23·5), ♀ 22–26 (23·5); culmen, ♂ 14–16 (14·9), ♀ 12·5–15 (13·8). WEIGHT: ♂ (n = 9) 215–250 (234), ♀ (n = 19) 210–280 (239).

IMMATURE: similar to ♀ but upperparts more pinkish fawn and less barred, with speckles and coarse vermiculations of black and edges of buff; chin to mid-breast pale pinkish buff washed sandy, each feather with subterminal bar of dark brown; rest of underparts off-white finely barred black; primaries blackish brown tipped pinkish buff; secondaries blackish brown; primary coverts blackish brown; tertials pinkish fawn edged whitish, and with bold blackish brown barring on outer vane; median and secondary coverts brown at base, buff over apical half; lesser coverts pale pinkish fawn fringed pale buff; tail as in adult ♀.

DOWNY YOUNG: crown elaborately patterned in dark brown patches outlined with black and separated by broad creamy white lines; face rich brown with white forehead stripe running down to lores and up as white eyebrow to top of crown; 2 narrow white lines extending horizontally behind eye and outlined in black, the lower joining its opposite number on throat; rest of upperparts patterned in deep rusty brown patches outline with black and separated with creamy white lines to form bold double figure-of-8 pattern **(A)**; centre of chin white outlined with black; side of chin and upper throat rich brown; rest of underparts off-white; wings patterned like upperparts; bill blackish horn with paler tip; eye dark brown.

P. b. ansorgei Benson: SW Angola. Smaller and paler than *bicinctus*.

P. b. multicolor Hartert: Transvaal, Mozambique, Malaŵi and Zambia. Richer and more rufous than *bicinctus*, smaller in east and larger in west of range.

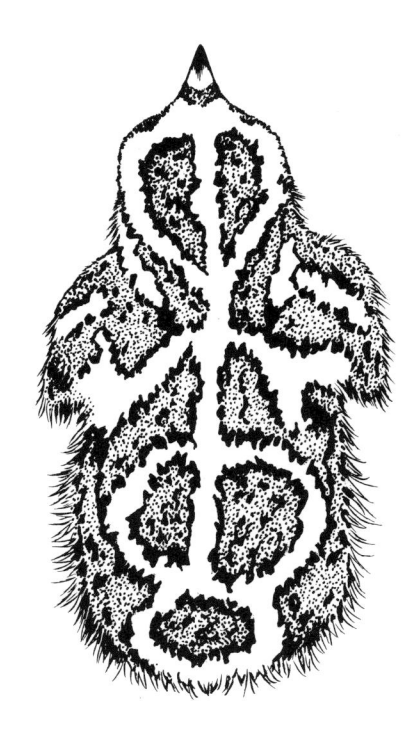

A

Field Characters. Tail short and blunt. ♂ is only sandgrouse in its range with black and white forehead, black and white breast-band and barred belly. ♀ similar to ♀ Namaqua Sandgrouse *P. namaqua* but underparts entirely barred, including throat; ♀ Namaqua has plain yellow throat, streaked breast, long pointed tail.

Voice. Tape-recorded (21, 35, 74). In flight a harsh 'chuck chuck'. On ground a set phrase of 7–10 syllables with accent on 2nd and final ones, 'pitee *you're* iti purpur' or 'oh *no*, he's gone and done it *again*!', given in slightly rasping whistle uttered repeatedly by birds congregated at waterhole at dusk. Also has churring, somewhat growling anxiety note sounding very like Suricate *Suricata suricatta*, and which may represent mimicry of adaptive significance to sandgrouse (D. H. Thomas, pers. comm.).

General Habits. Inhabits *Acacia* and other wooded savannas, bushveld mopane woodland from fairly moist in east to very arid in Namibia and Angola; prefers rocky areas in drier country, such as hill slopes or gravel plains with tussocky grass and scrub. Crepuscular, mostly inactive by day, when usually found in pairs or small groups (rarely of up to 50 birds) in shade of plants or rocks. Drinks at night; flies low over ground or treetops, almost silently, arriving at water well after sundown and staying there until at least 1 h after dark, even on the darkest nights. Flocks of 10–20 land a few m from water, wait, then run to water, drink, take off and fly directly to feeding grounds. Feeds at night if moonlit, otherwise usually during early morning and late afternoon. Experiments suggest it may not need to drink daily; in captivity birds huddle together at temperature > 35°C (Thomas and Maclean 1981).

Food. Dry seeds of grasses and herbs, including those of exotic weed *Bidens bidentata*.

Breeding Habits. Monogamous; solitary nester.
NEST: scrape in soil, sand or gravel, usually lined with few bits of dry plant material; placed under bush or between grass tufts or in open.
EGGS: 2–3; Namibia mean (17 clutches) 2·6. Elliptical; pale salmon pink, cinnamon-pink or pinkish buff, spotted with purple-brown or reddish brown and slate-grey. SIZE: (n = 40) 31·1–39·9 × 24·5–27·8 (37·3 × 26·7).
LAYING DATES: mainly in dry winter months, throughout range, Apr–Oct, particularly May–July; in Namibia also Nov–Dec; Angola, Oct; Zimbabwe, Apr–Oct.
INCUBATION: a captive ♂ occupied nest scrape for several days before eggs laid; then infertile eggs were incubated for 33 days, by ♀ from 19.00–20.00 h to *c*. 13.00 h next day and by ♂ in afternoon only (unlike other sandgrouse) (D. H. Thomas, pers. comm.). In Namibia ♀ seems to incubate until *c*. 09.30 h and ♂ for rest of daylight hours (R. A. C. Jensen, pers. comm.).

Reference
Thomas, D. H. *et al.* (1981).

Plate 29
(Opp. p. 448)

Pterocles gutturalis **Smith. Yellow-throated Sandgrouse. Ganga à gorge jaune.**

Pterocles gutturalis Smith, 1836. Rep. Exped. Expl. Centr. Afr., p. 56; Kurrichane (Zeerust), western Transvaal.

Range and Status. Endemic, sedentary and partially migratory. 2 populations: one (*gutturalis*) breeds S Zambia and N Botswana (mainly Caprivi Strip) and 'winters' in W Zimbabwe, SE Botswana, N and W Transvaal and N Cape; other (*saturatior*) is resident from extreme NE Zambia through interior of Tanzania (east to Kilingali, Fuggles-Couchman 1984) and Kenya to S, central and N Ethiopia. Frequent or locally common; abundant in some E African grasslands.

Description. *P. g. gutturalis* Smith: Zambia to Transvaal. ADULT ♂: crown and nape greyish yellow merging into dark olive-buff on hindneck and sides of neck; face creamy buff; superciliary stripe creamy buff, narrowly edged with black above; lores black; chin and throat creamy buff, bordered below by black band extending to sides of neck; mantle and scapulars dull greyish washed yellowish, tinged blackish slate subterminally; back, rump and uppertail-coverts greyish olive; tail wedge-shaped, black broadly tipped with buffy white and with outer vanes barred with buffy white; innermost pair of rectrices greyish olive; breast pale dull grey, feathers fringed olive–buff; rest of underparts chestnut brown, vermiculated with black on flanks; central belly feathers pale greyish near tip with black outline to grey area; undertail-coverts edged with rust. Primaries brownish black with pale shafts and with tip and inner vane edged whitish; secondaries blackish brown; tertials like scapulars; primary coverts brownish black; secondary coverts, median coverts and lower lesser coverts light grey, broadly tipped clay colour; remaining coverts dull grey washed yellowish; underwing brownish black. Bill pale bluish grey; bare skin around eye grey; eye dark brown; legs and toes pinkish brown (tarsal feathers buffy brown). ADULT ♀: crown and nape blackish brown, feathers edged with pale buff; hindneck brownish black, feathers edged ochre-yellow; face and superciliary stripe creamy buff; lores black; chin, throat and sides of neck creamy buff; mantle, back, rump and uppertail-coverts mottled and barred blackish brown and olive-buff; upper breast light ochre-yellow with shaft streaks and barring of dark brown; lower breast and remaining underparts chestnut heavily barred black; undertail-coverts greenish brown fringed black; remiges brownish black. Primaries with pale shaft and off-white edging; primary coverts brownish black; remaining wing-coverts as back, somewhat greyer on greater coverts; underwing brownish black; tail as in ♂, but central rectrices barred black across both webs; bare parts as in ♂. SIZE (10 ♂♂, 10 ♀♀): wing, ♂ 208–228 (217), ♀ 205–220 (213); tail, ♂ 83·5–94·0 (88·1), ♀ 75–88 (81·1); bill, ♂ 13·5–16 (14·6), ♀ 13·5–16 (14·9); tarsus, ♂ 28–32 (29·9), ♀ 28–31 (29·2). WEIGHT: ♂ (n = 3) 340–345 (342), ♀ (n = 6) 285–400 (336).

IMMATURE: similar to adult ♀, but upperparts with smaller spots and narrower bars of olive-buff; primaries broadly tipped and edged on outer vane with olive-buff.

DOWNY YOUNG: crown and back boldly and elaborately patterned in rufous-brown patches edged with black and separated by white lines to form double figure-of-8 pattern on back and white cross on crown; superciliary stripe white edged black; malar stripe and ear-coverts rufous-brown edged with black and interspersed with white; below greyish white, washed smoky brown on breast; wings patterned like upperparts.

P. g. saturatior Hartert: Zambia to Ethiopia. Wing-coverts with brighter, more cinnamon margins.

Field Characters. A large sandgrouse with wedge-shaped tail lacking long central rectrices. ♂ has yellow throat, black collar and lores, and chestnut belly, plain grey back and black underwing. ♀ has black lores and chestnut belly, unlike any other ♀ sandgrouse in southern part of its range; in north might be confused with ♀ Chestnut-bellied Sandgrouse *P. exustus*, but latter has unmarked yellow face, pale band between breast and belly, and pin tail. Throat is buff (brighter yellow in ♀ Namaqua Sandgrouse *P. namaqua*). Orbital skin grey (yellow in other sympatric species).

Voice. Tape-recorded (2, 25, 34, 36, 244, 277). Take-off call is harsh 'glock-glock-glock'. Flight call is 'twet-weet, twet-weet' or 'wha-ha, wha-ha', with 2nd and 4th syllables slightly higher and more accentuated.

General Habits. Inhabits short-grass plains near swamps and rivers; also recently burnt ground and cultivated fields; 800–2000 m altitude. Usually in pairs or small groups which feed conspicuously in the open and freeze when disturbed. Takes off with noisy wing-beats. Flocks of 10–50 congregate into hundreds before flying to drink from 07.00 h to 10.00 h, sometimes later; flocks with Namaqua and Burchell's Sandgrouse *P. burchelli* in southern and with Chestnut-bellied Sandgrouse *P. exustus* in E Africa. In Serengeti (Tanzania) flock of 200 Yellow-throated and Chestnut-bellied Sandgrouse congregates near water, then single-species groups fly in to drink, a flock of 1 sp. constantly displacing drinking flock of the other (Mungure 1974). Drinking flocks split up into small parties or pairs when flying back to feeding grounds immediately after drinking.

P. g. gutturalis moults when breeding (Apr–Sept). Present in S Zambia mid Apr to mid Jan but frequent and regular only in Apr–Oct (Liuwa plains, Kafue Flats, Lusaka etc.); numbers decrease (Liuwa) from July and breeding grounds are vacated in Oct, with southward passage through S Zambia in flocks of hundreds and up to 1000 in Oct–Dec (Taylor 1979). *P. g. saturatior* is resident throughout its range but some movements occur, with influx of breeding birds at Rukwa (Tanzania) in dry season (Britton 1980).

Food. Seeds of *Sesbania*, *Crotolaria*, *Cassia*, *Achyranthes aspera*, *Leersia hexandra*, *Rottboelia exaltata* and *Bidens* (Zambia). Also fallen grains of oats, wheat, barley and sorghum.

Breeding Habits. Monogamous, solitary nester. Bobbing movements of paired birds facing each other seem to constitute courtship behaviour.

NEST: scrape or natural hollow (e.g. hoofprint), unlined or scantily lined, in open among grass tussocks and stubble tufts.

EGGS: 2–3, usually 3. Elliptical; smooth and glossy; light brown to pale buff, spotted and blotched with light umber, reddish brown and pale mauve. SIZE: (Zambia) 42·0–43·0 × 27·5–33·3.

LAYING DATES: mainly dry season. Zambia, May–Sept with peak in July; Zimbabwe, Sept; Tanzania, (Serengeti), July–Aug, young chicks June–Aug; East Africa, Region C, Apr–Sept (mainly July); Region D, June–Aug (mainly Aug); Ethiopia July–Sept.

INCUBATION: carried out by both sexes.

DEVELOPMENT AND CARE OF YOUNG: young can fly when only about half size of parent (Clancey 1967). Water transported to young in belly feathers of ♂. Nearly full grown young (Serengeti, mid Oct) would be 2–4 months old.

References
Brooke, R. K. (1968).
Mungure, S. A. (1974).
Taylor, P. B. (1979).

Plate 29
(Opp. p. 448)

Pterocles burchelli Sclater. Burchell's Sandgrouse; Spotted Sandgrouse. Ganga de Burchell.

Pterocles burchelli Sclater, 1922. Bull. Brit. Orn. Club 42, p. 74; near Griquatown, N Cape.

Range and Status. Endemic, resident and nomadic. SE Angola, N and E Namibia, Botswana (except extreme E), W Transvaal, W Orange Free State and N Cape. Common.

Description. ADULT ♂: crown and nape dark olive-brown, feathers edged with buffy ochre to give streaked effect; face and superciliary stripe light blue-grey; ear-coverts and sides of neck buffy ochre streaked dark brown; lores, chin and upper throat light blue-grey; upperparts olive-brown at base of feathers, orange-yellow on apical half; on mantle and scapulars, each feather with white spot on each vane where olive-brown and orange-yellow meet; on rump and uppertail-coverts 2–3 white spots on both vanes of each feather; tail wedge-shaped, dark grey, broadly tipped white, with black subterminal bar and barred on both vanes pinkish buff; central rectrices with ochre tip and bolder barring; breast bright cinnamon washed ochre, each feather subterminally spotted white and faintly banded greyish; belly greyish rust with lower breast feathers having broad whitish centres; undertail-coverts buff; tarsus feathered buff. Primaries blackish brown tipped white mainly on inner web, and shafts of first 3 or 4 white; secondaries blackish brown; tertials like scapulars but white spots elongate and diffuse; primary coverts blackish brown; upper lesser wing-coverts light olive-brown washed yellowish; remaining coverts greyish olive-brown, broadly tipped orange-yellow grading to deep rufous, each feather boldly spotted white on each vane as on mantle; underwing deep rufous grading to greyish at base of primaries. Bill blackish; bare skin around eye yellow; eye dark brown; legs and feet yellowish pink. ADULT ♀: upperparts similar to ♂; face, superciliary stripe, chin and throat deep yellowish buff with some rusty spots on throat; breast pinkish cinnamon washed ochre, each feather laterally and subterminally spotted white, becoming white barring on lower breast; belly buffy white lightly barred light brown, speckled and vermiculated dusky; undertail-coverts deep buff; wings and tail as in ♂, but tips of coverts less rufous and tips of rectrices more buffy; underwing light buffly brown; bare parts as in ♂. SIZE (10 ♂♂, 10 ♀♀): wing, ♂ 163–175 (169), ♀ 163–173 (167); tail, ♂ 66–75 (70·8), ♀ 63–72 (66·7); bill, ♂ 11–12·5 (11·8), ♀ 9·5–12 (10·9); tarsus, ♂ 23–29 (27·3), ♀ 25–29 (26·7). WEIGHT: ♂ (n = 4) 180–200 (192), ♀ (n = 4) 160–185 (171).

IMMATURE: similar to ♀; breast buffy, feathers with paler tips and marked with subterminal and median chevrons of light brown; white spots on wings duller and more diffuse, feathers barred buff; some innermost secondaries barred on outer vane buff; tips of primaries white, extending onto outer vane.

DOWNY YOUNG: forehead pale golden buff anteriorly and white posteriorly, narrowly bordered black, light golden brown at sides and behind, narrowly bordered black; crown white in centre, bordered black, golden brown at sides, flecked and bordered black; hindneck white in centre, golden brown at sides; sides of neck mainly light golden brown flecked white; face and lores white, bordered behind by golden brown stripe, then a white stripe, both narrowly bordered black; ring around eye golden brown, extending as stripe to sides of neck; chin white; throat white in centre, light golden buff at sides; back boldly marked with white double figure-of-8 pattern, bordered black and filled in with golden brown; tail golden brown flecked black; chest dull greyish buff with long white filaments; belly and feathering of legs pale dingy buff; upper arm white with bold black blotches; forearm brown bordered with black and white; hand light golden buff. Bill dusky; legs and feet yellow.

Field Characters. Bright salmon-rufous underparts and bold white spots above and below are distinctive; face and throat grey in ♂, yellow in ♀. Looks dumpy and rather short-tailed, especially in flight; on ground is longer-legged than other sandgrouse.

Voice. Tape-recorded (74, 233, 262, 282). A 2-syllabled mellow flight call 'kwok-wok' or 'chok-lit', the 2nd note slightly higher and more accented than the 1st. On ground, a rapid cheeping 'ch pt, ch pt pt', rolling *en masse*. On take-off the call is a rapid 'wok-wok-wok-wok-wok-wok' until fully airborne, when flight call takes over.

General Habits. Largely confined to red Kalahari sands with good cover of grass 30–50 cm tall, often mixed with shrub or scrub; a bird of dry savanna rather than true desert. Seldom seen away from waterholes; behaviour secretive. When disturbed, either crouches or sneaks away between grass tufts in hunched posture. Largely inactive in hot midday hours. Usually found singly or in pairs, or in loose groups of 3–4. Drinks in morning (never in afternoon or evening) from 2–4 h after sunrise, when flocks number a few tens, up to 50, sometimes with Namaqua Sandgrouse *P. namaqua* or Yellow-throated Sandgrouse *P. gutturalis*. Flies to water up to 80 km daily. Flocks fly high (up to 300 m) giving occasional 'kwok-wok' calls, land at water's edge or even on surface, drink quickly and depart at once.

Those settling on water float low, with wings held partly open to aid buoyancy, then take off without difficulty like a teal. Occasionally flock lands a few m from water and after few min runs down to drink. Takes av. 7 draughts, dipping bill at each draught to suck in water, then raising head to swallow with bill open. Drinks 10–15 ml of water in just a few s. Some birds dustbathe before drinking, often rolling over onto their backs, with legs in the air (**A**). Being long-legged it runs fast and easily, adopting a nearly vertical stance. Excessively shy at all times and appears to have an anxious disposition; mist-netted birds often die within seconds of release.

Food. Seeds; 16 crops in Namibia (Etosha National Park) contained seeds and 1 chloropid fly (perhaps eaten incidentally). In Kalahari mainly seeds of chenopod *Lophiocarpus burchelli*.

Breeding Habits. Monogamous; solitary nester.
NEST: shallow scrape in sandy soil, sparsely lined with dry plant material or sometimes unlined, usually placed next to a grass tuft or shrub.
EGGS: 3, rarely 2. Elliptical; pale cream, olive-cream or buff with elongate smears of dark olive-brown and underlying blotches of greyish mauve; some eggs lightly marked and look pale. SIZE: (n = 15) 34·5–38·8 × 23·4–27·3 (36·7 × 25·7).
LAYING DATES: throughout range, Apr–Oct, after rains and in dry winter season.

DEVELOPMENT AND CARE OF YOUNG: both adults accompany chicks after they leave nest. ♂ gathers water for chicks in belly feathers. Rocks body while soaking feathers in water (**B**).

References
Dixon, J. E. W. (1978).
Macdonald, J. D. (1957).
Maclean, G. L. (1968).

Order COLUMBIFORMES

A distinctive order of about 290 species in a single family, practically cosmopolitan. Sandgrouse have often been included, but opinion is growing that they are quite unrelated. Columbiformes may be allied distantly with Psittaciformes (Sibley and Ahlquist 1972).

Family COLUMBIDAE: pigeons and doves

Medium-sized arboreal and terrestrial birds which eat grain, fruit and some other plant materials. Body plump and compact, with very small head and rather short neck and legs. Bill short, with bare cere. Plumage soft in texture and colours and very dense, often shot with metallic reflections, particularly on neck and, in some terrestrial species, wings. Oil-gland naked or absent. Mostly sexually monomorphic and all seasonally monomorphic. 11 primaries (1 much reduced), 12 secondaries, 12–14 rectrices; usually diastataxic; primaries moult descendently, taking several months. Solitary or highly gregarious. Flight strong; most African species are not markedly migratory but all have pronounced daily roosting movements. Food ingested whole; seeds never husked; stored in capacious crop, then ground up in powerful gizzard.

Monogamous. Nest a simple, insubstantial platform of twigs, in tree, shrub, rock crevice or ledge, building, or rarely on ground. Eggs 1–2 (rarely 3), white; protracted breeding seasons with many successive clutches. Sexes take equal share in nest-building, incubation and raising young. Young fed at first by regurgitation of parental crop secretion, 'pigeons' milk', a whitish fluid comparable with mammal milk in composition. Young or 'squabs' hatch blind and nearly naked, then grow long, sparse, yellowish down. Nidicolous; juvenile leaves nest at three-quarters of adult size and is only capable of flight several days later. 42 genera in 4 subfamilies; 2 subfamilies represented in Africa with 6 genera and 37 species (26 endemic).

Subfamily TRERONINAE: fruit pigeons

A large, heterogeneous and perhaps polyphyletic group of strictly arboreal, essentially Paleotropical pigeons which eat small fruits. Medium-sized, mostly with beautiful plumages of soft greens, yellows and mauves. Bill and legs short, usually brightly coloured; eyes and ceres often brilliant. About 120 species in 10 genera, much the largest being the Oriental *Ptilinopus*, to which *Treron*, the only genus represented in Africa, is probably closely allied.

Genus *Treron* Vieillot

21 species of green fruit pigeons. 14 rectrices; 8th primary notched on inner vane; upper part of tarsus feathered. Gizzard hard, strong and muscular; gut long; may digest seeds and flesh of fruit. Somewhat gregarious, feeding on figs and buds, rarely coming to ground. Lack cooing notes typical of other pigeons; songs are complex series of whistles, barks, growls, throaty chuckles and clicks of several seconds' duration. Otherwise quiet; move sedately even when feeding, and, like parrots, can be very hard to pick out in green trees. When startled sit quiet, or burst from tree with swishing wings; flight strong, some species making creaking sound with wings.

Green pigeons range throughout subsaharan Africa, Gulf of Guinea Is., E African coastal islands, Comoro Is., Madagascar and SW Arabia; they are highly polytypic, varying in plumage tone, tail colour, cere width, and soft part colours. Several species have been recognized, and all have been treated as a single superspecies. We follow Goodwin (1983) in recognizing 1 superspecies (*T. calva*, *T. sanctithomae*, *T. pembaensis*, *T. australis* (Madagascar) and *T. phoenicoptera* (Orient)) and an 'independent' species (*T. waalia*). *T. waalia* is little better differentiated than an allospecies, and although it overlaps geographically with *T. calva* it does not do so locally. *T. waalia* inhabits dry wooded savannas and *T. calva* forest; they 'overlap' only where gallery forests penetrate savannas along rivers.

***Treron calva* superspecies**

1 *T. calva*
2 *T. sanctithomae*
3 *T. pembaensis*

Treron calva **(Temminck). African Green Pigeon. Pigeon vert à front nu.**

Columba calva Temminck, 1809. In Knip, Les Pigeons, p. 35; Loango: French Congo.

Forms a superspecies with *T. sanctithomae* and *T. pembaensis*.

Plate 32
(Opp. p. 465)

Range and Status. Endemic resident, widespread in forests and mesic savanna woodland from Senegambia, S Mali and Liberia to S Nigeria, Cameroon, S Central African Republic, Uganda and S Kenya, ranging south to N Namibia, N and E Botswana, Zimbabwe, Mozambique, and South Africa (Transvaal, Natal, E Cape with records from SW Cape). From Uganda ranges in narrow corridor to central Ethiopia. Bioko, Principe and Zanzibar Is., recorded also Mafia Is. From sea level up to 2100 m. Frequent to locally abundant, but declining in some areas because of habitat destruction and hunting (e.g. Gambia, Principe Is.).

Description. *T. c. nudirostris* (Swainson): Senegambia, Guinea. ADULT ♂: entire head, breast and belly yellowish green (belly slightly yellower); lower hindneck and upper mantle olive-green tinged grey; mantle, back, rump and uppertail-coverts olive-green. Tail above pale grey, concealed proximal two-thirds of all rectrices except middle pair dark grey; below, stronger contrast between blackish proximal two-thirds and silver-grey tips. Flanks grey tinged olive; vent feathers olive green fringed creamy white; thighs bright yellow; short undertail-coverts olive green broadly edged cream, long ones chestnut with pale tip. Primaries grey-brown, distal part of outer web finely edged yellow, 8th with deep indentation on inner web; secondaries grey-brown with outer web conspicuously edged yellow near end; tertials olive-green with narrow yellow edge near tip; outer greater wing-coverts brown broadly edged pale yellow, making conspicuous wingbar; inner greater coverts olive-green edged yellow; rest of wing-coverts like back, with mauve patch near wrist; underwing grey. Base of bill red, tip pale grey; cere 7·5–9·0 mm long in midline; eye pale blue, sometimes with concentric rings of mauve, red and brown; legs and feet yellow. Sexes alike. SIZE: (13 ♂♂, 8 ♀♀) wing, ♂ 154–164, ♀ 150–161; tail, ♂ 85–90, ♀ 80–87; bill, ♂ 22–28, ♀ 21–26; tarsus, ♂♀ 20–23. WEIGHT: ♀, 210.

IMMATURE: similar to adult but duller; mauve patch absent.

NESTLING: skin pink, covered with sparse straw-coloured down; bill flesh-coloured. Wings and breast turn greenish after a few days.

T. c. sharpei (Reichenow): Sierra Leone to central Nigeria. Olives slightly greyer, less yellow, than in *nudirostris*. Cere 12–16 mm long, extending well up forehead, and up to 15 mm wide.

T. c. calva (Temminck): SE Nigeria and Bioko to NW Angola and Zaïre (Kasai, Manyema). Darker and greener than *sharpei*, grey hindneck collar less pronounced, cere wider near forehead; Bioko population slightly larger (wing, 155–175).

T. c. virescens Amadon: Principe Is. Head and underparts deeper and duller green than *calva*; bill more slender.

T. c. uellensis (Reichenow): N Zaïre and S Central African Republic (Ubangi-Chari) and Sudan (Bahr-el-Ghazal) to W Uganda (L. Albert); also SW Ethiopia. Like *sharpei* but larger (wing, 157–171); cere 15–17 mm long and more expanded than in *sharpei* (except in Ethiopia, where population is probably taxonomically distinct). WEIGHT: ♂ (n = 17) 160–250 (218), ♀ (n = 6) 130–225 (179).

T. c. gibberifrons (Madarasz): SE Sudan (Imatong Mts, intervening between W and NE parts of range of *uellensis*) to L. Victoria basin (Kenya west of Rift Valley, Rwanda, Burundi, NW Tanzania south to Kigoma and east to Mwanza, and Zaïre: Shinyanga, Kivu). Like *sharpei* but yellower; grey hindneck collar more pronounced. Legs and feet red.

T. c. brevicera Hartert and Goodson: Kenya east of Rift Valley to near coast, NE Tanzania (Crater Highlands, Kilimanjaro, Iringa). Distal one-third of tail paler than in *nudirostris*, *sharpei* or *calva*, and head and underparts brighter and yellower, particularly hindneck; hindneck collar clear grey, contrasting sharply with bright yellow-olive hindneck and with dull olive back. Legs and feet red. Cere only 6–8 mm long in midline. Larger than *sharpei* (wing, 160–179).

T. c. wakefieldii (Sharpe): coastal Kenya and NE Tanzania (Lamu to Usambara). Like *gibberifrons* but tail green (not grey); smaller (wing, 147–164). WEIGHT: ♂ (n = 5) 175–220 (200), ♀ (n = 5) 180–225 (196).

T. c. schalowi (Reichenow): Zaïre (S L. Tanganyika to Katanga), SW Tanzania, Malawi south to Nkata Bay, Zambia west of Luangwa R., Zambezi valley east to about Tete, Mozambique; E Angola (west to 21°E), NE Namibia, N Botswana, NW Zimbabwe. Like *gibberifrons* but tail green (not grey); larger (wing, 171–185).

T. c. ansorgei Hartert and Goodson: W Angola (south of Cuanza R.). Like *gibberifrons* but greener.

T. c. vylderi Gyldenstolpe: NW Namibia, east to Grootfontein. Like *gibberifrons* but greyer. WEIGHT: ♂ 195, 200, ♀ (n = 4) 175–220 (196).

T. c. delalandii (Bonaparte): SE Tanzania (Njombe, Songea) to South Africa, west to range of *schalowi*. Head, neck, breast and most of belly grey-green (not green or yellow-green); hindneck collar grey, boldly contrasted with nape and mantle; remaining upperparts and tail bright yellowish green; tail with 2 cm wide whitish terminal band, sometimes also affecting central rectrices. Large: wing, ♂♀ 172–192. WEIGHT: ♂ (n = 8) 230–242, ♀ 209, 247.

T. c. granti (van Someren): lowland E Tanzania (Kilosa and Dodoma to Pangani and Rovuma). Like *delalandii* but belly yellower; smaller (wing, 161–171).

Field Characters. A parrot-like gregarious tree pigeon with very distinctive voice. Distinguished from *Poicephalus* parrots by small bill, grey-tipped tail, red or yellow feet, and voice. The only green pigeon south of equator. In north, all races sympatric with Bruce's Green Pigeon *T. waalia* have green head, neck and underparts (with just a little yellow on extreme lower belly); for further distinctions see under Bruce's Green Pigeon.

Voice. Tape-recorded (5, 21, 36, 44, 47, 74). Up to at least 20 high-pitched, fluty, whistling trills, followed by lower-pitched harsh creaking, barking or growling notes, or throaty chuckles, ending in 3 quiet popping clicks. Song lasts 5–15 s and may be repeated several times per min. Sometimes up to 5 whistling trills interpolated before final clicks. Several birds in perched flock often sing in chorus.

General Habits. Occurs in dense woodland and forest savanna mosaic zones, coastal dunes, gallery, primary and secondary forests, *Ficus*, *Brachystegia* and *Combretum* woodlands; clearings, open cultivated country in vicinity of fruit trees, gardens. Shy, but less so when nesting aggregations occur near human habitations in rural areas. In parties of 3–50; seldom solitary; flock forages 7–18 m high in leafy canopy. Stocky and short-legged but nimble, clambering among thin branches like parrot and sometimes dangling head-down to pluck fruit. Blends into foliage and difficult to spot, particularly in heat of day when inactive and silent. When disturbed flies fast out of far side of tree, swooping down then rising to perch in canopy of distant tree, without gliding. Rarely takes food from ground, with waddling walk. Does not drink.

Moves locally in all parts of range, probably related to fruit availability, but not migratory.

Food. Figs; also fruits of *Musanga acropioides* (Germain *et al.* 1973), *Syzigium cordatum*, *Diospyros mespiliformis*, *Podocarpus* sp., *Melia azederach*, *Duranta repens*, *Myrica conifera*, *Mimusops* sp., *Cassina schlechteri* and *Pappea capensis* (Rowan 1983). Cultivated fruits include mulberry *Morus*, peach *Prunus* and (captive birds) raisins, bananas and pawpaw. Takes some cereal and wild grass seeds (*Loudetia simplex*) from ground. Buds of mangroves.

Breeding Habits. Monogamous. Nest solitarily, but around villages in concentrations of up to 11 nests in 5·5 ha (nests 18–270 m apart, Kruger National Park, South Africa: Tarboton and Vernon 1971). Courtship not studied; some postures are: (a) quick shuddering to-and-fro movement of head, head held horizontally and mandibles rapidly open and close (threat, and appeasement); (b) 2 birds turn away and erect tails to show bright undertail-coverts (Goodwin 1983); (c) fanned tail moves slowly up and down (aggressive, not reproductive?); (d) 7 birds half-raised wings, holding posture for 5–10 s; (e) ♂ said to bow to ♀ on ground, like domestic pigeon (Rowan 1983).

NEST: flat platform, diam. 13–15 cm, depth 3–4 cm; 1 made of 26 thin petioles 7–20 cm long and 28 forked twigs 7–15 cm long, interlocking to form strong latticework. Sited 2–21 m high (median 5 m, n = 97 nests) in tree: in fork, on horizontal or gentle sloping branch, on old pigeon nest, or in dense growth of parasitic *Viscum* or *Loranthus*. 85% (n = 48 nests) are among foliage and 15% in bare trees. Nest material gathered in living tree *c.* 60 m distant (not from ground) by 1 bird; other bird (♀?) remains on site, receives material from mate and works it into platform. Much wastage; ♂ (?) brought *c.* 60 twigs in 6 h, helping mate to incorporate larger ones, but only 12–15 were successfully used. During that period nest was attended continuously and neither bird fed (Rowan 1983).

EGGS: 1–2, laid at 24–48 h interval; in central and W Africa 1 egg (n = 10 clutches, Gabon: A. Brosset, pers. comm.), in southern Africa 1 (18% of 133 clutches) or 2 eggs (82%); white, smooth, matt or slightly glossy, oval. SIZE: (n = 30, southern Africa) 28·0–33·5 × 22·0–26·5 (30·9 × 24·0).

LAYING DATES: Senegambia, Jan–May; Nigeria, Jan–Aug (peak Jan–Apr); Cameroon, Jan–Sept, Nov; Principe, Mar; Zaïre, July–Dec; Gabon, every month; E Africa: Region A, Feb–June, Sept, Dec; Region B, Jan, Mar–June, Sept, Nov–Dec; Region C, June, Sept, Dec; Region D, Jan, Feb, June, July, Aug, Oct–Dec; Region E, Mar, Apr, Sept, Nov; Malaŵi, Sept–Oct; Zambia, July, Sept–Nov; Zimbabwe, Jan, Feb, May, July–Dec; Mozambique, Sept–Dec; South Africa, Aug–Dec.

INCUBATION: carried out by both sexes; parent sits very closely day and night and can be touched by person. Period: 13–14 days. Material can be added to nest during incubation (and nestling) period.

DEVELOPMENT AND CARE OF YOUNG: 2 chicks hatch at interval of 3–24 h. Shells remain in nest or fall below. Primary quills sprout on day 3; eyes open on day 5. Iris then black, and bill dark brown with light brown base. Well feathered by day 8/9 except that breast still downy. Nestling period: 12–13 days. Fledged young probably roosts near nest for first few nights. Adult disturbed from nest with chick dropped to ground with fanned wings and tail then flew up to alight nearby; at another nest adult fled with very slow emphatic wing-beats—distraction flight?

BREEDING SUCCESS/SURVIVAL: egg rolls out and nest sometimes deserted when incubating bird disturbed; nests destroyed by high winds. Adult very susceptible to shock death (e.g. gunshot noise: van Someren 1956).

References
Rowan, M. K. (1983).
Tarboton, W. R. and Vernon, C. J. (1971).

Treron sanctithomae (Gmelin). São Tomé Green Pigeon. Pigeon vert de São Tomé.

Plate 32
(Opp. p. 465)

Columba s. thomae Gmelin, 1789. Syst. Nat. i, pt. 2, p. 778; São Tomé.

Forms a superspecies with *T. calva* and *T. pembaensis*.

Range and Status. Endemic resident, São Tomé Is., where common. Also (formerly) neighbouring Rollas Is., where exterminated by forest destruction (R. de Naurois, pers. comm.).

Description. ADULT ♂: entire head, neck, mantle, breast and upper belly dark greenish grey; lacks grey 'shawl' on mantle; back, scapulars, rump, uppertail-coverts and wing-coverts dark olive green; tail black-brown with slaty grey 3-cm tip. Following features all like African Green Pigeon *T. calva*: bright yellow lower belly, vent and thighs; wing and purple bend of wing; and buff-edged rufous undertail-coverts. Sexes alike. SIZE: ♂♀ (n = 3) wing, 164–179; tail, 84–88; tarsus, 24; bill massive, hooked and parrot-like, 9–10·5 mm deep, culmen curved, cere 2·5–3·5 mm long, hooked tip projects 2·5–5·0 mm.

IMMATURE and NESTLING: undescribed.

Field Characters. Only 'green' pigeon on São Tomé Is. A large, plump pigeon, predominantly grey, with grey tail, rufous-chestnut undertail-coverts, and stout bill with red cere.

Voice. Not tape-recorded. Not described.

General Habits. Inhabits plantations and forest, up to 1250 m. Feeds in canopy down to 12 m above ground but rests higher, *c.* 20 m. Shy. Vocal. Wings make swishing noise in flight.

Food. Unknown.

Breeding Habits: 1 nest 2·5 m up in cocoa tree, Jan, held 1 fresh egg.

446 COLUMBIDAE

Plate 32
(Opp. p. 465)

Treron pembaensis Pakenham. Pemba Green Pigeon. Pigeon vert de Pemba.

Treron pembaensis Pakenham, 1940. Bull. Br. Orn. Club, 60, p. 94.

Forms a superspecies with *T. calva* and *T. sanctithomae*.

Range and Status. Endemic resident, restricted to well timbered parts of Pemba Is., mainly in north. Frequent to common.

Description. ADULT ♂: differs from São Tomé Green Pigeon *T. sanctithomae* as follows: head slightly greyer; rufous undertail-coverts lack buff edges; bill not robust and parrot-like but slender, like that of African Green Pigeon *T. calva*. Sexes alike. Slightly smaller than African Green Pigeon; wing, ♂♀ 162–169.
IMMATURE and NESTLING: undescribed.

Field Characters. Only 'green' pigeon on Pemba Is.; a large, plump pigeon, predominantly grey, with grey tail, dark chestnut undertail-coverts and red cere.

Voice. Not tape-recorded. Usual call is 'kiu, tiu, kiuriu, kiwrikek, wrikek' followed by soft 'krrr-rrr-rrr', very like voice of African Green Pigeon (Pakenham 1943). Also subdued 'guweck' or 'kuwok', and a rapidly repeated castanet-like snapping of bill. Sings throughout year and all day, mainly near dawn and dusk.

General Habits. Inhabits forested and well timbered parts of Pemba, perching high in tall trees (kapoks, mangos, *Areca*, *Pterocarpus*). Feeds in canopy, where very hard to detect. Shy; but during breeding season much less so and then inhabits gardens, nesting close to dwellings and on branches overhanging much-used paths.

Food. 1 crop contained 106 young fruits of betel palm *Areca catechu*.

Breeding Habits.
NEST: flimsy, untidy, asymmetric platform of twigs (of *Tamarindus*, *Pterocarpus*, *Millingtonia*, *Cassia siamea*) in *Pterocarpus*, mango, *Millingtonia* or *Cassia javanica*, often near end of bough overhanging path. 2–3 nests are half-built and aborted before final site chosen. 1 bird carries material, other stays by nest and works material in (once 1 bird both carried and incorporated material).
EGGS: 2.
LAYING DATES: Oct–Feb; ♂ in breeding condition June.
BREEDING SUCCESS/SURVIVAL: nest and eggs often destroyed by wind.

Reference
Pakenham, R. H. W. (1979).

Plate 32
(Opp. p. 465)

Treron waalia (Meyer). Bruce's Green Pigeon. Pigeon vert waalia.

Columba waalia F. A. A. Meyer, 1793. System. Sum. Über. Zool. Entdeck. p. 128; near Lake Tana, NW Abyssinia.

Range and Status. Africa and SW Arabia.
Resident in dry, wooded savannas from Senegambia to 48°E in coastal N Somalia. Ranges north to 15°N, and in Ethiopia (Red Sea hills) to 18°N; and south to near border of rain forest zone in W Africa and to 2°N in central Uganda; from sea level up to 2000 m. Also Socotra Is. Frequent to locally abundant, but was evidently far more abundant early in century. Shot for sport.

Description. ADULT ♂: entire head, neck and breast light grey with faint greenish tinge; mantle, back, scapulars, rump and uppertail-coverts olive green; belly bright yellow, sharply demarcated from breast. Central rectrices grey above, others slate grey with 25 mm grey terminal band. Flanks grey, washed olive; vent feathers whitish with greenish grey streaks; undertail-coverts chestnut broadly edged cream or tawny, the longest almost reaching tip of tail. Primaries brown, 2 outermost boldly and others narrowly edged pale yellow, the 8th deeply notched; secondaries brown, boldly edged pale yellow (visible in folded wing); tertials olive green; lesser and median wing-coverts mauve (but innermost ones olive like mantle); outermost wing-coverts broadly edged pale yellow; underwing grey. Bill whitish, cere blue; iris blue usually with red outer ring; legs and feet orange. ADULT ♀: like ♂ but mauve wing-patch smaller. SIZE: (10 ♂♂, 10 ♀♀) wing, ♂ 171–182, ♀ 172–176; tail, ♂♀ 95–110; bill, ♂ 15–18, ♀ 14–16; tarsus, ♂ 25–28, ♀ 25–26. WEIGHT: (n = 1, Ghana) 268.
IMMATURE: duller, mauve wing-patch incomplete.
NESTLING: not known.

Field Characters. A medium-sized, fast-flying green pigeon replacing African Green Pigeon *T. calva* in drier, more open country of subsaharan savanna belt. Head, neck and breast much greyer than African Green Pigeon, belly bright yellow, extreme lower belly white and legs orange, a useful character when bird seen from below in typical canopy view. African Green Pigeon has green belly, becoming yellowish on extreme lower belly, and yellow legs.

Voice. Tape-recorded (44). Song lasts 9–10 s. Starts with rapid cracking notes (like knocking a wooden box, or a creaking door opened slowly) (2 s); then high-pitched fluty and modulated whistles (2 s); then a short growl (1 s); and ends with 6–8 sharp unequally spaced grunts or yaps (4 s).

General Habits. Inhabits dry wooded grasslands, *Acacia* and *Commiphora* savannas, lowland riverine and *Podocarpus* forests; closely associated with *Ficus* sp.; occurs in food trees in villages and towns (e.g. Ndjamena, Chad). Usually in flocks of less than 10 birds, sometimes of 50; formerly in hordes, mixing with African Green Pigeons. Spends most of time foraging or sitting quietly in canopy of food tree, where very difficult to spot. Rather shy; when person approaches either flies out of tree at 1st alarm, with creaking wings, or remains very still, only the head moving a little to peer between leaves. But, particularly in villages, tolerates some human activity immediately beneath tree. When not anxious, rather more conspicuous, making foliage move by tugging at fruits, calling while foraging, moving within canopy and flying freely in small flocks between adjacent trees. Never feeds on ground, and in more mesic parts of range never comes to ground at all; but in arid country drinks, presumably on ground, at sunset before going to roost (Salvan 1968).

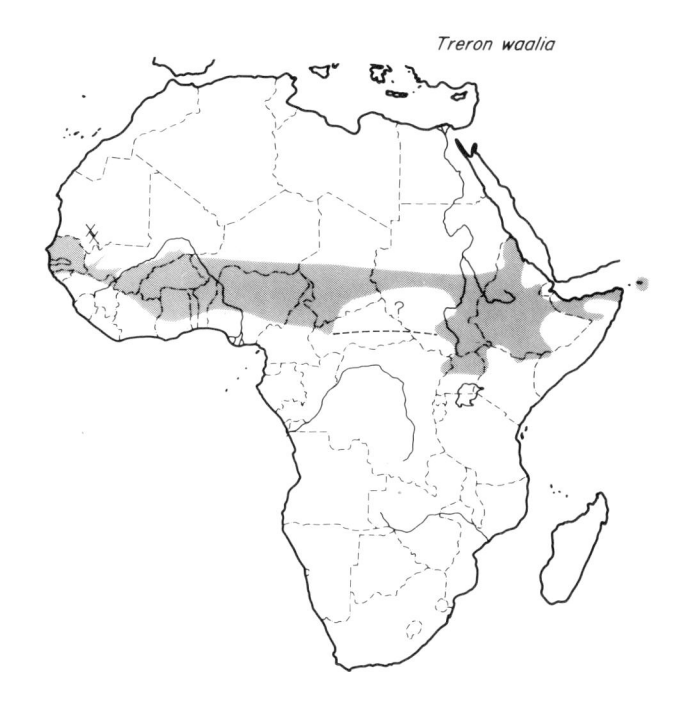

Makes some local movements in relation to rains (Mali) or to ripening of fruits (Gambia, Nigeria, Ethiopia). Figs are most important food, and sometimes converges in large flocks onto ripe fig trees.

Food. Fruits of *Ficus*, *Podocarpus*, *Zizyphus*; said to favour guttapercha tree *Ficus platyphylla* and (in Sudan) gemeiza *F. sycamorus*.

Breeding Habits. Evidently monogamous.
NEST: flimsy platform in thick foliage, often at end of wide spreading branch; usually at least 3 m above ground. Sometimes in bare tree, or low down (*Acacia*).
EGGS: 1–2. Glossy white. SIZE: (n = 3) 30·0–30·9 × 21·6–22·5 (30·5–22·0).
LAYING DATES: Gambia, Dec–Apr; Mali, Dec–June; Ghana, Apr–Sept; Nigeria, Mar–Apr; Sudan, Jan–Apr; Ethiopia, Nov–May; Somalia, Mar–May; Socotra, Jan–Feb; E Africa: Region A, Apr–May.
INCUBATION: incubating bird shuffles to rotate itself on nest, to keep its cocked tail towards observer on ground below (undertail-coverts cryptic, resembling piece of bark: Walsh 1981).

Subfamily COLUMBINAE: pigeons and doves

Cosmopolitan, about 165 species; larger ones are 'pigeons' and smaller ones 'doves', but the terms are not used consistently. Mostly seed-eaters, feeding on ground; some so adapted to living on ground that they somewhat resemble quails; others more strictly arboreal. Heterogeneous; broadly, 3 groups may be recognized, all represented in Africa: (a) quail-doves and bronzewings, cosmopolitan, probably primitive, terrestrial; plumage often green and rufous, but the 6 species in Africa, in *Oena* and the endemic genus *Turtur*, are brown; (b) rock doves and wood pigeons, mainly *Columba*, a cosmopolitan genus of 52 species with 15 in Africa (12 endemic); and (c) turtle doves *Streptopelia*, an advanced, essentially Paleotropical genus of 15 species, 11 in Africa (6 endemic). Many species in groups (b) and (c) have feathers at sides of neck in diagonal rows, raised during excitement to give distinct furrows or pleats (at other times not apparent).

Vocal repertoire limited and unvarying, mostly cooing notes in advertising and display; crop inflated during cooing; often a distinctive alighting note; noisy wing-clatter when taking off. Courtship displays include allopreening, bowing, display flight, and billing before copulation; display flight has several slow beats with widely spread wings in upward flight, loud claps produced by wing-tips meeting below body, followed by long glide.

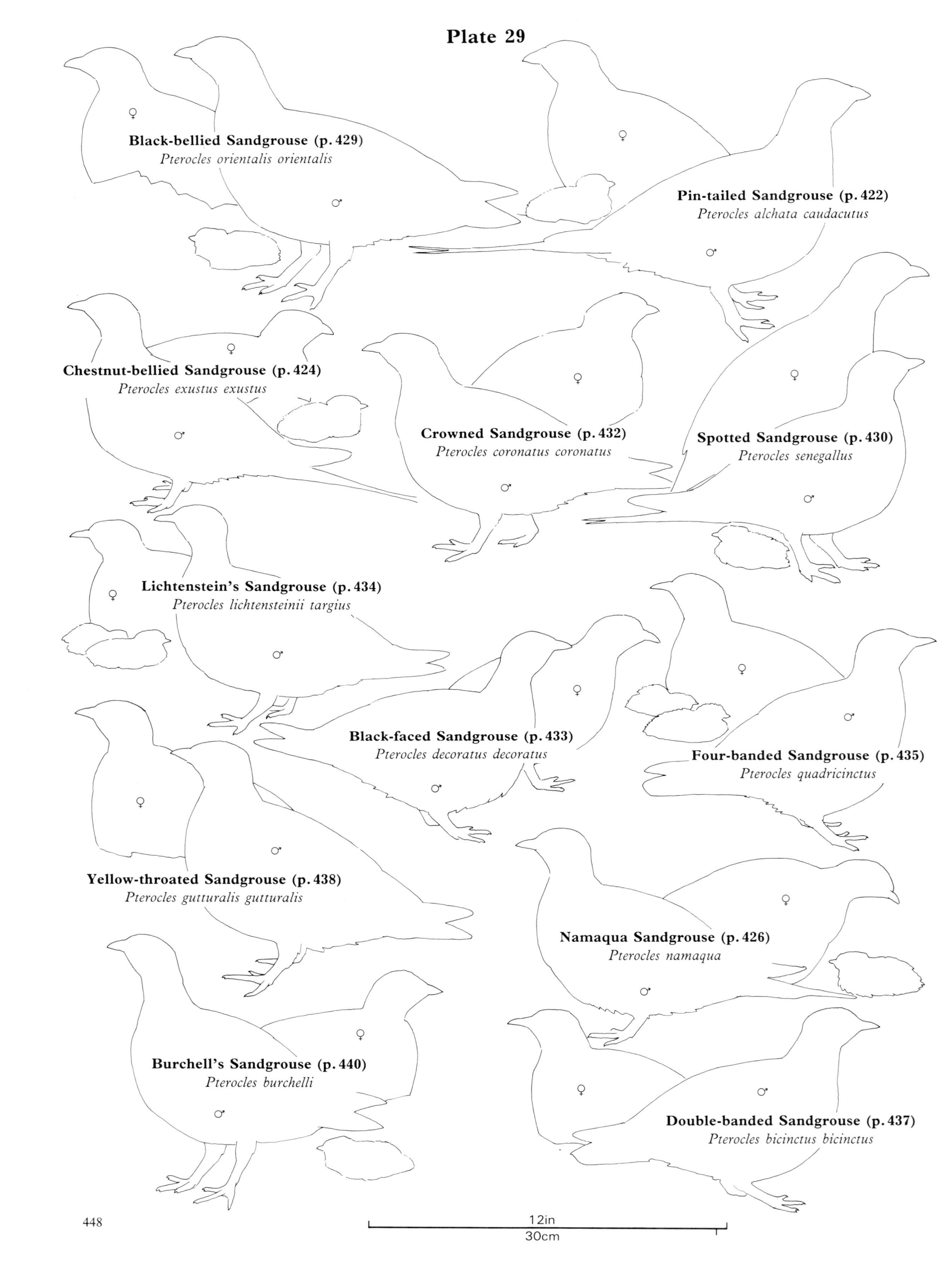

Plate 29

Black-bellied Sandgrouse (p. 429)
Pterocles orientalis orientalis

Pin-tailed Sandgrouse (p. 422)
Pterocles alchata caudacutus

Chestnut-bellied Sandgrouse (p. 424)
Pterocles exustus exustus

Crowned Sandgrouse (p. 432)
Pterocles coronatus coronatus

Spotted Sandgrouse (p. 430)
Pterocles senegallus

Lichtenstein's Sandgrouse (p. 434)
Pterocles lichtensteinii targius

Black-faced Sandgrouse (p. 433)
Pterocles decoratus decoratus

Four-banded Sandgrouse (p. 435)
Pterocles quadricinctus

Yellow-throated Sandgrouse (p. 438)
Pterocles gutturalis gutturalis

Namaqua Sandgrouse (p. 426)
Pterocles namaqua

Burchell's Sandgrouse (p. 440)
Pterocles burchelli

Double-banded Sandgrouse (p. 437)
Pterocles bicinctus bicinctus

12in
30cm

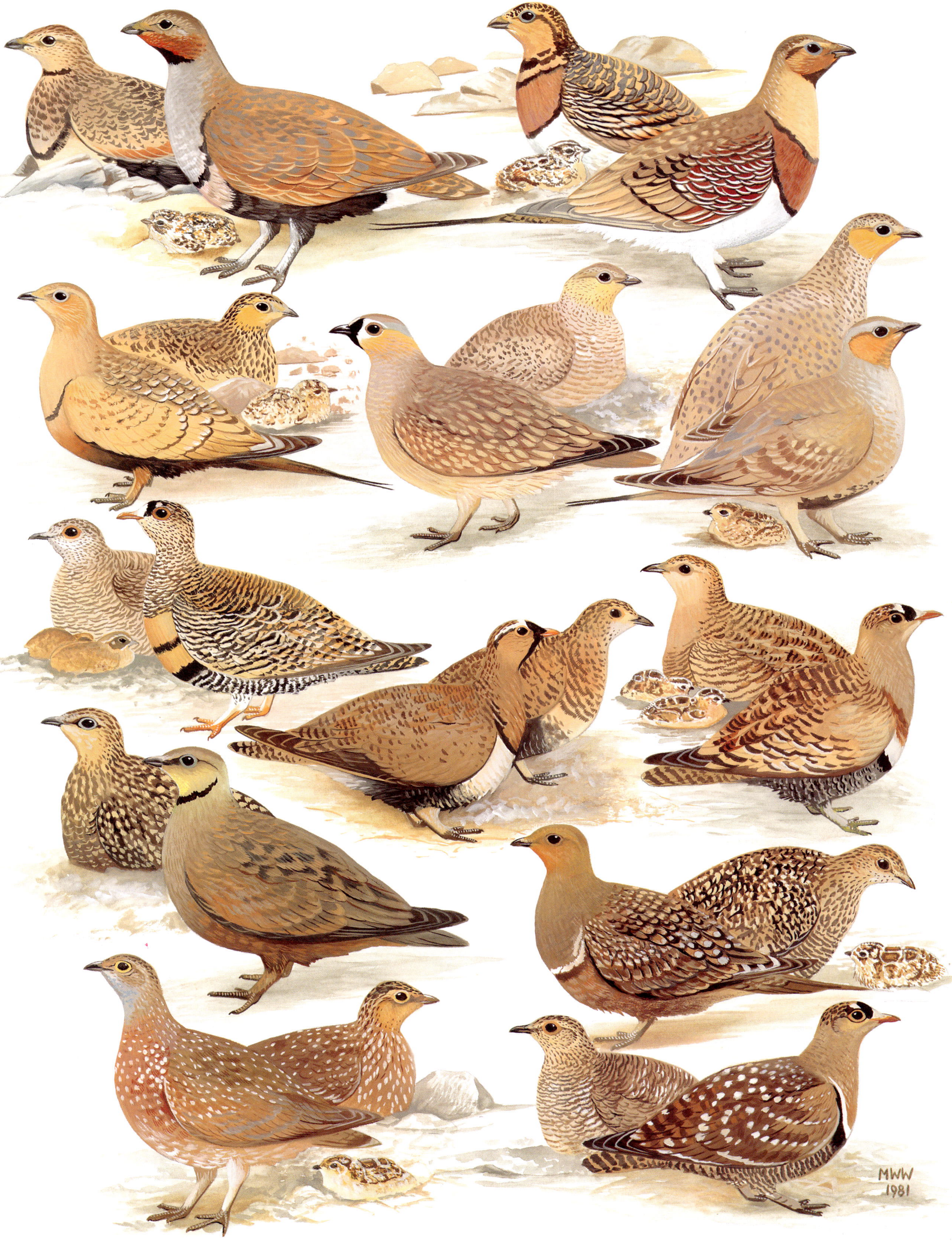
MWW
1981

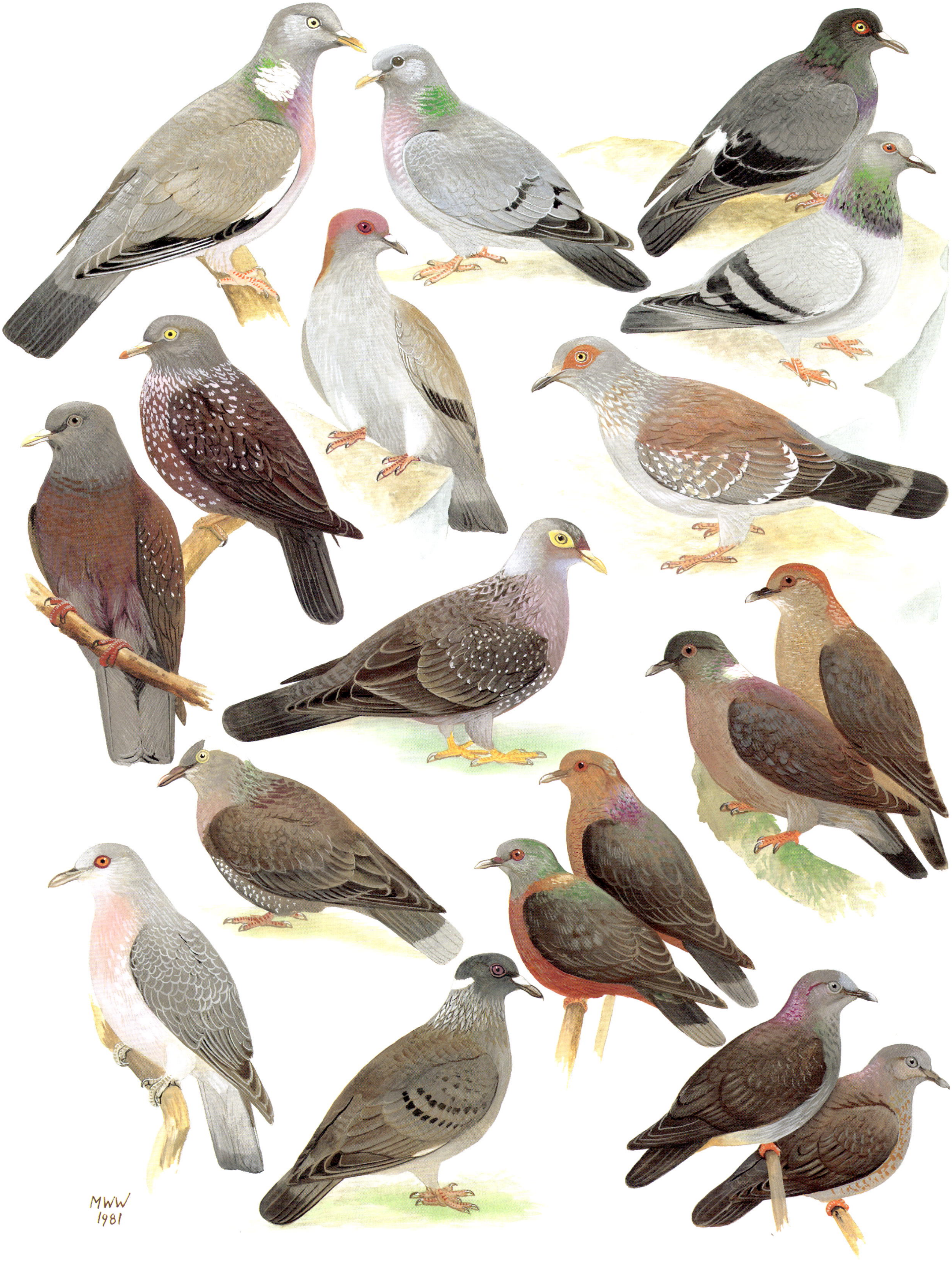
MWW
1981

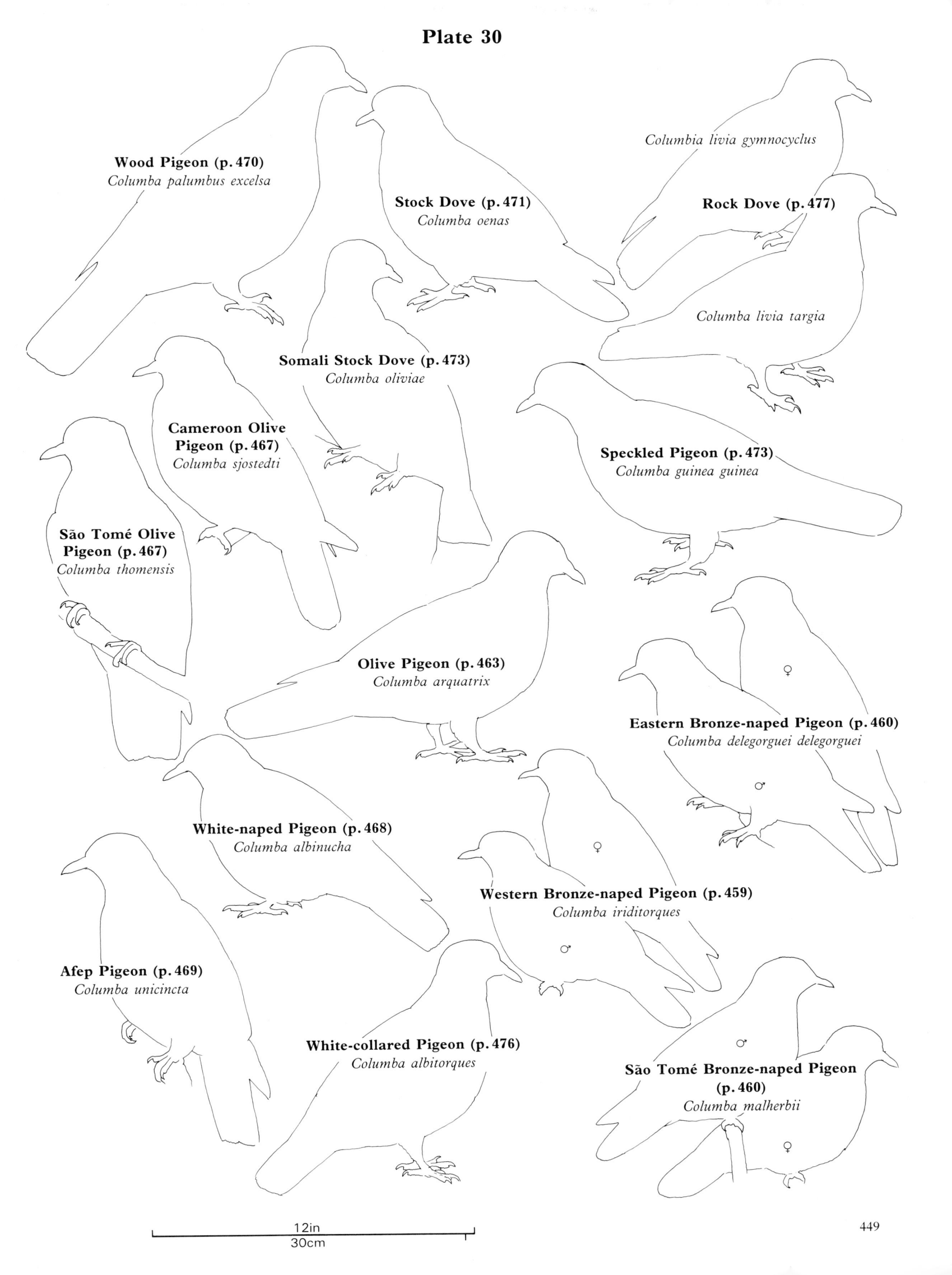

Plate 30

Wood Pigeon (p. 470)
Columba palumbus excelsa

Stock Dove (p. 471)
Columba oenas

Columbia livia gymnocyclus

Rock Dove (p. 477)

Columba livia targia

Somali Stock Dove (p. 473)
Columba oliviae

Cameroon Olive Pigeon (p. 467)
Columba sjostedti

Speckled Pigeon (p. 473)
Columba guinea guinea

São Tomé Olive Pigeon (p. 467)
Columba thomensis

Olive Pigeon (p. 463)
Columba arquatrix

Eastern Bronze-naped Pigeon (p. 460)
Columba delegorguei delegorguei

White-naped Pigeon (p. 468)
Columba albinucha

Western Bronze-naped Pigeon (p. 459)
Columba iriditorques

Afep Pigeon (p. 469)
Columba unicincta

White-collared Pigeon (p. 476)
Columba albitorques

São Tomé Bronze-naped Pigeon (p. 460)
Columba malherbii

12in / 30cm

450 COLUMBIDAE

Genus *Turtur* Boddaert

Endemic. Small, relatively weak-flying terrestrial doves, feeding on ground, and flying, perching and nesting close to ground. Wings and tail rather short. Plumage brown and pale grey with rufous primaries, 1 species with white face and underparts, another largely cinnamon, all with *c*. 5 conspicuous metallic green, blue or violet spots on inner greater and median wing-coverts and with broad black-white-black bands across rump. Monomorphic. Mainly sedentary. Song a protracted series of coos gradually tailing away.

The 5 species embrace a dry-woodland superspecies (*T. abyssinicus*, *T. chalcospilos*) closely related and broadly overlapping with a moist-woodland species (*T. afer*), and 2 independent forest species.

***Turtur abyssinicus* superspecies**

1 *T. abyssinicus*
2 *T. chalcospilos*

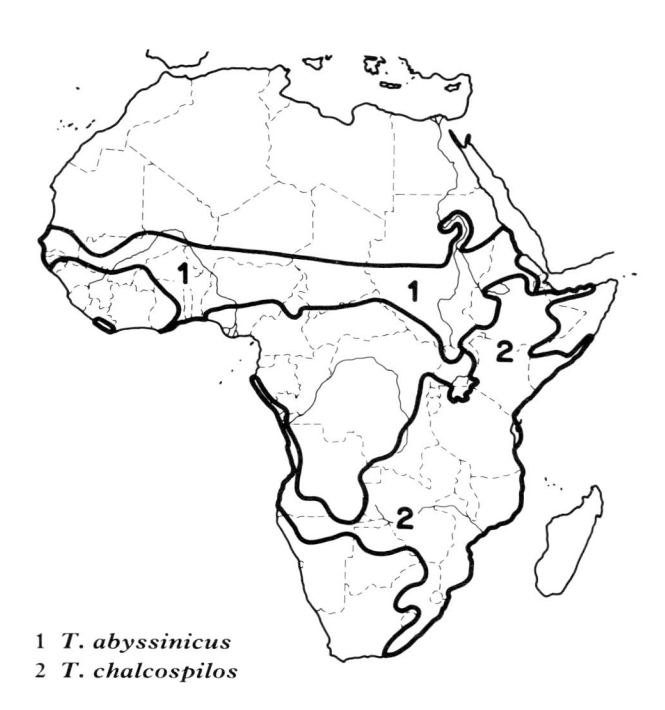

Plate 32
(Opp. p. 465)

***Turtur brehmeri* (Hartlaub). Blue-headed Wood Dove. Tourtelette demoiselle.**

Chalcopelia brehmeri Hartlaub, 1865. J. Orn., p. 97; Gabon.

Range and Status. Endemic resident. Rain forest of Sierra Leone, Liberia, Ghana; Nigeria, Gabon, N Congo, and equatorial Zaïre east to Rift Valley; and SW Zaïre and Angola (Cabinda). Uncommon to frequent.

Description. *T. b. brehmeri* (Hartlaub): S Cameroon (south of 3°30′N, also near Douala) and Congo basin. ADULT ♂: crown blue-grey, nape slightly darker; hindneck chestnut tinged light blue; forehead, face and chin light blue; foreneck chestnut with vinous tinge; all other upperparts chestnut with vinous sheen but rump somewhat paler (in some specimens with indistinct brown bars); uppertail-coverts and tail dark rufous, 2 outermost rectrices with basal half grey-blue, with 15 mm blackish subterminal bar and rufous tip. Underparts rufous, belly and undertail-coverts slightly darker. Remiges rufous-brown, primaries with tawny area on inner web and secondaries with paler inner web; tertials and inner greater coverts with 4–6 gold-copper patches. Underwing pale rufous; tail below dark rufous, outer 2 feathers with grey-brown base and rufous tip. Bill dark red, tip greenish; eye dark brown; legs and feet dark red. Sexes alike. SIZE: wing, ♂ (n = 7) 129–139 (133), ♀ (n = 6) 124–136 (131); tail, ♂ (n = 7)

94–112 (100), ♀ (n = 4) 94–103 (99); bill to skull ♂ (n = 7) av. 20·8, ♀ (n = 3) 19–20 (19·8); tarsus, ♂♀ 22–27. WEIGHT: ♂ (n = 7) av. 133 ± 2, ♀ (n = 7) 92–132 (112).

IMMATURE: forehead, face and chin pale cinnamon brown; upperparts with indistinct black barring.

NESTLING: clad in long, soft cream-coloured down.

T. b. infelix (Peters): Sierra Leone to coastal Cameroon (ascending 130 km up rivers). Upperparts and foreneck with less of a vinous sheen; wing-spots metallic green.

Field Characters. Rufous-chestnut plumage with pale blue head is unique among doves. Very large-eyed (Fry 1984).

Voice. Tape-recorded (44, C, 226). Up to 40 hollow coos or 'poos' in 10–15 s, starting slowly and quietly, accelerating with slight increase in volume, then descending down scale. Hardly separable from song of Tambourine Dove *T. tympanistria*; last notes of *T. brehmeri* song sharper.

General Habits. Inhabits lowland primary forest and old secondary growth; never in cultivation or much-disturbed parts of forests. Solitary; in pairs; or (much less often) in small parties. Forages among leaves on forest floor or by track; often perches in understorey, within 3 m of floor. Flight fast, low and precise.

Food. Seeds; also slugs, insects and their larvae.

Breeding Habits. Monogamous.

NEST: a platform of twigs and roots of 1–2 mm diam., with shallow cup, built on top of dead dry leaves; 1 15 × 20 cm diam.; placed on 1 cm diam. horizontal branch 2·5–5·5 m above ground in leafy undergrowth of primary forest.

EGGS: 1–2. Dark buff with olive gloss. SIZE: (n = 3) 27·4–30·0 × 20·0–21·4 (28·3–20·8).

LAYING DATES: Cameroon, Aug; Gabon, Dec–May; Zaïre, July–Oct.

DEVELOPMENT AND CARE OF YOUNG: at 1 nest 2 same-sized young fed by plunging bills into left and right sides of parent's mouth at same time. At c. 1 week, chicks have yellowish skin, the pterylae looking blackish with rows of small slaty quills, entirely but sparsely covered with blond down, quite thick on nape (Brosset 1976).

Turtur tympanistria (Temminck). Tambourine Dove. Tourtelette tambourette.

Plate 32
(Opp. p. 465)

Columba tympanistria Temminck *in* Knip, 1810. Les Pigeons, p. 80; restricted type locality, Gamtoos River, South Africa.

Range and Status. Africa and Comoro Is.

Common resident in forest from S Senegambia to SE Kenya, south to W and N Angola, E Zimbabwe, Natal and S Cape Province. Bioko, Zanzibar, Pemba and Mafia Is. Also west-central Ethiopia, S Sudan, N Uganda and S Somalia (lower Juba R.). Up to 900 m (Cameroon), 1150 m (Natal), 1800 m (Bioko and Zimbabwe), 2000 m (E Zaïre), 2200 m (Malaŵi), 2500 m (E Africa) and 3200 m (Ethiopia). Also in gardens and castor-oil plantations in South Africa where increasing; rubber, cocoa and kola plantations in W Africa. Elsewhere, probably affected adversely by forest destruction. Locally common. 20 pairs in 40 ha forest (Malaŵi).

Description. ADULT ♂: crown very dark brown, nape dark brown; sides of neck greyish, foreneck white; forehead and face pure white with narrow black line from gape to eye and black spot on ear-coverts; rest of upperparts, including wing-coverts, uniform dark brown; rump a little paler, crossed by 2 ill-defined blackish bars; uppertail-coverts with maroon tinge; tail with central pair of rectrices dark chestnut-brown, 2 outer ones grey with subterminal black bar and greyish tip, and remaining ones dark brown with blackish terminal third. Underparts white, brownish flanks and brown undertail-coverts. Primaries mostly chestnut, 10th with distal 2 cm of inner web strongly attenuated; secondaries with outer web brown and decreasing amounts of chestnut on inner web; tertials brown; c. 6 metallic green patches on innermost wing-coverts and tertials. Underwing chestnut, with dark brown rectrices, 2 outermost ones with greyish tip. Bill deep purple with black tip; eye dark brown; legs and feet purplish red. ADULT ♀: forehead greyish, breast and sides of foreneck grey; upperparts somewhat paler brown and wing-spots almost black, barely glossy. SIZE (W Africa): wing, ♂ (n = 18) 110–117 (110·5), ♀ (n = 6) av. 109·5 ± 2·1; tail, ♂♀ 72–85, ♂ (n = 9) av. 69·7 ± 1·7, ♀ (n = 6) av. 66·8 ± 2·5; bill to skull, ♂ (n

Turtur tympanistria

= 9) av. 17·8 ± 1·0, ♀ (n = 6) av. 17·2 ± 0·8; tarsus, ♂♀ 20–21. WEIGHT: (E Africa) ♂ (n = 38) 51–79 (69·6), ♀ (n = 23) 52–77 (65·6); (W Africa) ♂ (n = 9) av. 64·6 ± 7·3, ♀ (n = 6) av. 60·2 ± 2·8; (southern Africa) ♂ (n = 20) 64·8–84·9 (73·8), ♀ (n = 11) 59·2–77·3 (67·4).

IMMATURE: brown above, most feathers barred rufous and brown; underparts freckled with grey, middle of belly white.

NESTLING: dark brownish flesh colour with pale creamy white wiry down on head, scapulars, and dorsum (van Someren 1956).

Field Characters. Usually seen fleetingly, flying low down near observer in dense undergrowth or when flushed: a small, dark brown dove with white face and underparts, and chestnut wings. Solitary; flight quiet and fast without weaving.

Voice. Tape-recorded (4, 14, 30, 44, 74, 75). Song, distinctive of genus, is series of 20–40 coos delivered in up to 16 s; starts with *c.* 5 muffled notes uttered rapidly (almost a short trill), then increases in volume with remaining coos about equally spaced and slightly descending scale, and ends with long series of brief coos gradually tailing off in same pitch. Song rather sharper than in other *Turtur* species (except Blue-headed Dove *T. brehmeri*), and more regular, lacking hesitation.

General Habits. Inhabits montane, lowland and riparian fringing forests; gallery and secondary forests, coastal dune thickets, well-timbered gardens and plantations (castor-oil, cocoa, rubber and kola). Deep in main body of forest, also near large clearings and border of primary and secondary forests (A. Brosset, pers. comm.), on ground and in low and mid-height in vegetation. Elusive. Solitary or in pairs, seldom up to 6 together. Feeds on forest floor usually under leafy cover, sometimes on tracks and verges; mixes with Lemon Doves *Columba larvata*. Drinks regularly (at least in drier months) at midday and in early afternoon; sunbathes on forest floor. Rises with clatter then flies fast, silently, and with precision through undergrowth.

Sedentary; but in Natal (Mtunzini) markedly commoner July–Aug than in rest of year, suggesting an influx.

Food. Partial to seeds of castor-oil *Ricinus*; seeds of *Sorghum*, *Eleusine*, *Albizzia*, *Celtis*, *Croton*, *Neoboutonia* and *Polyscias*; mulberries; fruits of *Trema orientalis*, *Syzygium* and *Solanum*. On Nyika Plateau (Zambia/Malaŵi) eats seeds of *Neoboutonia* and *Polyscias* (Dowsett-Lemaire 1983). Frequently eats invertebrates, including termites and small molluscs.

Breeding Habits. Monogamous, territorial. Minimum nest territory size, 2·3 ha (Malaŵi: F. Dowsett-Lemaire, pers. comm.). No courtship known (van Someren 1956).

NEST: frail platform, diam. 9 cm, of thin sticks lined with rootlets in fork near trunk of small tree; or in bush, or near end of horizontal branch in tangle of creepers. In southern Africa 13 nests in riverside thickets, 2 in forest and 1 in fallen mangrove tree; 1–10 m (median 2·5 m) above ground. Once used an old nest of Laughing Dove *Streptopelia senegalensis*. ♂ carries twigs to ♀ at nest; ♀ arranges them. A pair took a week to complete nest, building until 11.00 h and again in late afternoon.

EGGS: 1–2; southern Africa, 3 clutches of 1 and 18 of 2. Creamy white. SIZE: (n = 14) 23·0–27·0 × 17·7–19·5 (24·8 × 18·3).

LAYING DATES: Sierra Leone, Jan–Apr; central Africa, all months; Gabon, Sept–Feb; S Zaïre, Apr–May; E Zaïre, May, Oct, Nov; E Africa: Regions A, B, Jan–Mar, May–Nov; Region C, July; Region D, Apr–July, Nov; Region E, May, Aug–Oct; Malaŵi, Feb, June, Oct–Dec; Nyika plateau (Zambia/Malaŵi), Aug–Sept; Zimbabwe, Jan, Mar–May, Sept–Nov; Natal Sept–Mar, May.

INCUBATION: carried out by ♀ and ♂, mainly ♀. Incubating ♀ fed by ♂ (by regurgitation).

DEVELOPMENT AND CARE OF YOUNG: hatchlings brooded closely by ♀; once ♂ arrived, took eggshells from ♀ and carried them off. Both parents brood, constantly at first, later intermittently. Young fed 4 times daily, about 07·30–08·30, 11·00–12·00, 16·30 and 18·00 h; either both young feed at once by plunging bills into parent's mouth (time: 3 min), or chicks fed in turn (time: 10 min). Arriving parent alerts chicks with short call. Any berry dropped in nest is promptly eaten by parent. When young, chicks are briefly brooded after feeding. During long brooding spell, chicks preened by ♀ to remove feather sheaths. Chicks well feathered at 2 weeks, when bill becomes narrower and harder. Faeces voided over side of nest. 3 nestling periods recorded as 13–14, 20 and 22 days. Young can fly strongly at 14 days (but may not normally do so).

BREEDING SUCCESS/SURVIVAL: 20% of birds survive at least 3 years; 1 lived at least 7 years.

References
Rowan, M. K. (1983).
van Someren, V. G. L. (1956).

Turtur afer (Linnaeus). Blue-spotted Wood Dove; Red-billed Wood Dove. Tourtelette améthystine.

Plate 32
(Opp. p. 465)

Columba afer Linnaeus, 1766. Syst. Nat. (12th ed.), p. 284; Senegal.

Range and Status. Endemic resident and partial migrant; common in evergreen woods and forests, up to 2000 m, throughout Africa between 10°N and about 15°S except for equatorial latitudes east of 30°E (where sparse or absent). Ranges north to 11°N in Nigeria, 13°N in Chad, 15°N in Senegambia and 17°N in Ethiopia, and south to S Angola, E Zimbabwe and S Mozambique. Common to abundant in Ethiopia, widespread and common in Uganda and around L. Victoria, but elsewhere in E Africa range fragmented (S Somalia, S Kenya, Tanzania) and bird only locally common (Britton 1980). Zanzibar and Pemba Is. In most of E Africa segregated from Emerald-spotted Wood Dove *T. chalcospilos* by altitude or habitat.

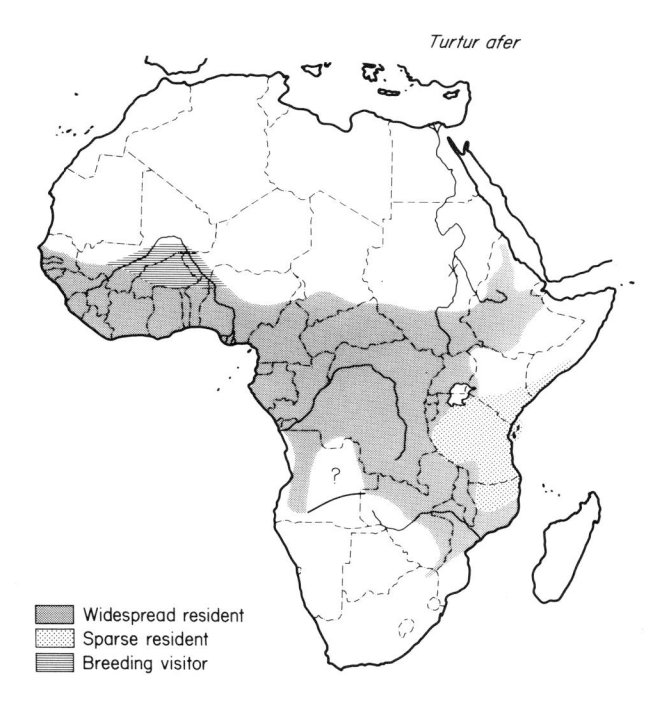

Widespread resident
Sparse resident
Breeding visitor

Description. ADULT ♂: differs from Emerald-spotted Wood Dove (*q. v.*) as follows: wing-spots amethyst blue and smaller; outer web of outermost rectrix pale grey at base (not white); bill wine red at base, orange-yellow at tip. Commonly cited difference in shade of upperparts and underparts (paler in Emerald-spotted) does not hold good throughout range and is unreliable. Sexes alike. SIZE: (n = 5, southern Africa) wing, ♂ 107–112 (109), ♀ 109–112 (110); tail, ♂♀ 79–87 (83); bill, ♂♀ 15·5–17 (16·3); tarsus, ♂♀ 17·5–19 (18·3); (n = 11, W Africa) wing, ♂ 104–116 (107), ♀ 103–112 (105); tail, ♂♀ 64–87; bill, ♂♀ av. 15; tarsus, ♂♀ 15–20. WEIGHT: (W Africa) ♂ 53, 60, ♀ 54, 56, ♂♀ (n = 8) 57·5–74 (66·5); (E Africa) ♂ (n = 3) 61–66 (64), ♀ (n = 3) 53–67 (60). Some W African variation in size and shade (Rand 1949).

IMMATURE: like adult but browner; feathers of upperparts tipped and barred buff; wing-spots smaller and duller.
NESTLING: not known.

Field Characters. The common and ubiquitous ground-foraging dove of secondary forests and mesic woods. 5–6 shiny blue spots on wing; black and white banded rump; rich rufous remiges apparent in flight. Distinguished from overlapping Black-billed Wood Dove *T. abyssinicus* (northern tropics) and Emerald-spotted (eastern Africa) by blue wing-spots (not green) and red and yellow bill (not black or dark red). See also Emerald-spotted Wood Dove.

Voice. Tape-recorded (2, 10, 44, 74, C, F). Song like that of Emerald-spotted Wood Dove (*q. v.*) but of fewer coos (8–12) and shorter duration (8–14); also pitch perhaps lower (500–600 Hz), 1st coo shorter, and song less irregular (Chappuis 1974).

General Habits. Inhabits forest edges, moist woodlands, gallery forests, secondary bush, clearings in montane forests, dense evergreen thickets in *Acacia* and *Combretum* savannas; gardens, farms, mangroves and *Eucalyptus* plantations. Forages on ground, solitarily or in pairs, sometimes several together with other species of doves. Other habits same as Emerald-spotted Wood Dove (*q. v.*).

Resident; but probably moves south in Ghana as dry season progresses (Taylor and Macdonald 1978), and in Mali a wet season visitor as far north as 17 °N (Lamarche 1980).

Food. Seeds of grasses (A. Brosset, pers. comm.); also castor-oil *Ricinus* seeds and emerging termites.

Breeding Habits. Single description of courtship flight (Bannerman 1931): ascends rapidly and steeply from perch to 7 m, claps wings under body, then glides down on outstretched, rigid wings with tail fanned.

NEST: slight platform of twigs, rootlets, tendrils and wool-like material, diam. 8–20 cm, placed 1–3 m above ground on leafy tree stump, in bush, mango or other large tree. 2 instances of building nest on old thrush nest.

EGGS: 2. Pure white or creamy-white. SIZE: (n = 9) 21·5–25·0 × 17·5–19·9 (23·8 × 18·4).

LAYING DATES: Gambia, Oct–Mar, Aug; Nigeria, Oct–Mar; Gabon, Oct–Mar, June; Ethiopia, Jan, May, June; East Africa: Region A, Jan; Region B, every month; Region D, Aug; Rwanda, every month; Zimbabwe, Apr, Aug–Sept, Dec; Tanzania (Pemba), Aug; Zambia, Mar, Aug–Oct; Malaŵi, Mar–Apr, June, Sept.

Plate 32
(Opp. p. 465)

454   COLUMBIDAE

Turtur abyssinicus **(Sharpe). Black-billed Wood Dove. Tourtelette d'Abyssinie.**

Chalcopelia abyssinica Sharpe, 1902. Bull. Br. Orn. Club, 12, p. 83; Kokai, Eritrea.

Forms a superspecies with *T. chalcospilos*.

Turtur abyssinicus

Range and Status. Endemic resident, perhaps partial migrant. Dry sudanian and sahelian savannas from Senegambia to Ethiopia (north to 16°N in Mali, 17°N in Ethiopia and 18°N in Nile valley), south to 'Dahomey gap' coast (Ghana, Togo, Benin, W Nigeria, perhaps only seasonally) and to N Cameroon, N Central African Republic and N Uganda. From sea level to 1800 m. Common to abundant, reaching density of 5/ha (riparian woods, N Senegambia).

Description. ADULT ♂: differs from Emerald-spotted Wood Dove *T. chalcospilos* (*q. v.*) as follows: crown and nape blue-grey shading to whitish on forehead; hindneck brown tinged grey; sides of neck and foreneck pale grey-pink; face greyish; chin and throat whitish; mantle, back, scapulars and all inner wing-coverts pale earth brown; rump and uppertail-coverts slightly paler than back; 2 conspicuous blackish bars *c.* 5 mm apart and 5–7 mm wide divide rump from back; a much narrower black bar near base of uppertail-coverts, which are broadly tipped black. Central pair of rectrices grey-brown, others grey with 2 cm-broad black tip; black below, outermost shorter, above and below with white outer web and black tip edged white. Breast and belly pale greyish pink; vent whitish; undertail-coverts black; flanks washed rufous. 3 outermost primaries broadly edged brown on outer web, others rufous with broad brown tip and small extension onto outer web; 7th–9th primaries emarginated; innermost secondaries grey-brown tinged rufous, others a little greyer than mantle; 3 innermost wing-coverts and 3 tertials with dark blue metallic patch on outer web 10–15 mm long, sometimes partly blackish; only a small portion of patch visible on folded wing; outermost wing-coverts paler and greyer than back; underwing rufous. Bill black with dull carmine base; eye dark brown, large; legs and feet purple. Sexes alike. SIZE: wing, ♂ (n = 15) 107–113, ♀ (n = 10) 108–112; tail, ♂♀ 80–90; bill, ♂♀ 13–14; tarsus, ♂♀ 18–20. WEIGHT: ♂♀ (n = 86, Senegambia) 54–78 (66); ♂♀ (n = 22, Ghana and Nigeria) 51–75 (62·2).
IMMATURE: most feathers barred buff or rufous on dusky ground.
NESTLING: undescribed.

Field Characters. See Emerald-spotted Wood Dove. Paler brown and rather greyer than Emerald-spotted (which overlaps in Ethiopia and N Uganda) and bill black (not dull red and black); in Ethiopia Black-billed inhabits woodland and forest, Emerald-spotted thornbush. Distinguished from Blue-spotted Wood Dove *T. afer* by having black (not red and yellow) bill, green (not blue) wing-spots, and arid- rather than mesic-woodland habitat.

Voice. Tape-recorded (44, 267). Song like that of Emerald-spotted Wood Dove (*q. v.*) but rather longer, with 24–27 coos lasting 15–18 s.

General Habits. Inhabits arid *Acacia* and *Combretum* savanna woodland, bushes around desert wells, shrubby banks of drying watercourses, well-wooded edges of marshes, cultivation, clumps of shade trees, edges of dense woods, and also impenetrable shady green thickets. In Ethiopia, heavier woodland and forest.
Habits same as Emerald-spotted Wood Dove (*q. v.*). Absence of breeding records south of 10°N in Ghana, Togo, Benin and Nigeria suggests that birds recorded there are non-breeding migrants. In north of range concentrates in thicker vegetation nearer water as dry season advances.

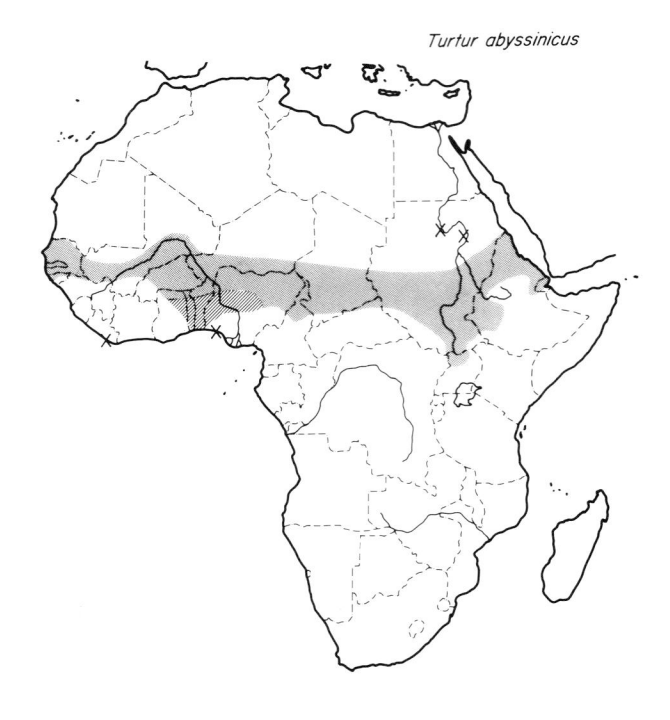

Food. Grass and herb seeds weighing 0·2–0·5 mg. In Senegambia eats at least 63 seed species, mainly of grasses *Panicum laetum* (56% of total food weight), *Brachiaria* spp. (30%) and *Dactyloctenium aegyptium* (8%). Occasionally eats much larger grains of *Sorghum*, *Pennisetum* and *Oryza*. Full crop contains 2·0 g seeds (n = 9) usually of 4–7 spp.

Breeding Habits. Monogamous.
NEST: platform of thin sticks, more or less hidden in foliage of *Balanites*, *Acacia nilotica*, *Zizyphus mauritiaca*, *Salvadora*, *Boscia* and *Maytenus*, at 1·0–2·5 (av. 1·55) m above ground (n = 12, N Senegambia).
EGGS: 2. Brownish cream. SIZE: (n = 7) 20·5–23·5 × 15·8–17·4 (21·9 × 16·6).
LAYING DATES: N Senegambia, Aug–Nov (peak Oct) and Feb–Apr; Nigeria, Oct–Mar; Sudan, Sept–Oct, Jan.

Turtur chalcospilos **(Wagler). Emerald-spotted Wood Dove. Tourterelle émeraudine.**

Columba chalcospilos Wagler, 1827. Syst. Av. Columba, sp. 83; East Cape Province.

Forms a superspecies with *T. abyssinicus*.

Plate 32
(Opp. p. 465)

Range and Status. Endemic resident. Widespread in shady savanna woodland from SE Sudan, Ethiopia and W Somalia, south through E Africa to N Namibia, N Botswana, Transvaal and Natal. Zanzibar and Mafia Is. Also in coastal belt from Gabon to SW Angola. Up to 2000 m. Common to abundant; often in suburbs and gardens.

Description. ADULT ♂: crown bluish grey inclining to brownish blue on nape; hindneck earth brown; sides of neck and foreneck dull light mauvy pink; face pinkish grey; chin and throat pinkish white; mantle earth brown; back and rump tawny brown to light brown; back, rump and uppertail-coverts crossed with 2–4 conspicuous black bars, the upper ones 7–8 mm wide and 10 mm apart and lower ones much narrower and 15 mm apart, lowest forming tip of uppertail-coverts, which are light brown. Tail with basal half of central pair of feathers pale brown, terminal part darker; other rectrices grey with broad black tip, outermost with narrow whitish tip and white basal half to outer web; below, black except for white basal half of outer web to outermost rectrix. Underparts pale dull mauvy pink. Major part of remiges rufous (hardly visible on closed wing, conspicuous in flight); 4 outermost primaries with inner web rufous, outer web and terminal band dark brown; inner primaries rufous except for broad brown tip; secondaries dark brown with some rufous at base; 3 outermost tertials and 3 greater coverts earth brown with large oblong metallic green patch on outer web (patches sometimes golden or bluish green); other coverts uniform brown. Underwing rufous, tips of primaries brownish. Bill black, reddish at base (southern Africa) or dull red with black tip (E Africa); eye brown; legs and feet purplish red. Sexes alike. SIZE: wing, ♂ (n = 7) 111–116 (113), ♀ (n = 6) 108–112 (110); tail, ♂♀ 75–89 (83.5); bill, ♂♀ 17–19.5 (18.2); tarsus, ♂♀ 17–19 (18.0). WEIGHT: (E Africa) ♂ (n = 17) 50–70 (56.2), ♀ (n = 9) 52–63 (55.8); (n = 22, South Africa) ♂♀ 52.8–70.7 (64.1).

IMMATURE: like adult but browner, barred buff or tawny particularly on upperparts; wing-spots smaller and duller.

NESTLING: pale brownish flesh, with long, yellowish hair–like down on head, scapulars and dorsal tracts.

In southern Africa adults have plumage cline, from paler birds in arid Namibia to darker ones south of Limpopo R.

Field Characters. A small, brown, ground dove of savanna woodland, with paler head and breast and darker tail, 5–6 shiny emerald spots on wing and black and white banded rump. Remiges rich rufous, hardly apparent when perched but very obvious in flight. Stumpy, rather short-tailed; solitary. Walks in dappled shade with nodding head; shy, flies low and direct but not far. Common; unlikely to be confused with forest congeners (*q. v.*), but very similar to Black-billed Wood Dove *T. abyssinicus* (overlapping with it in Ethiopia and N Uganda) and to Blue-spotted Wood Dove *T. afer* (which broadly overlaps other 2 species almost everywhere).

Bill dusky carmine red with black tip. Distinguished from Black-billed by bill colour and plumage (Black-billed is paler brown), and from Blue-spotted Wood Dove by wing-spot colour, bill colour (red with yellow tip in Blue-spotted) and habitat (Blue-spotted in denser, evergreen cover).

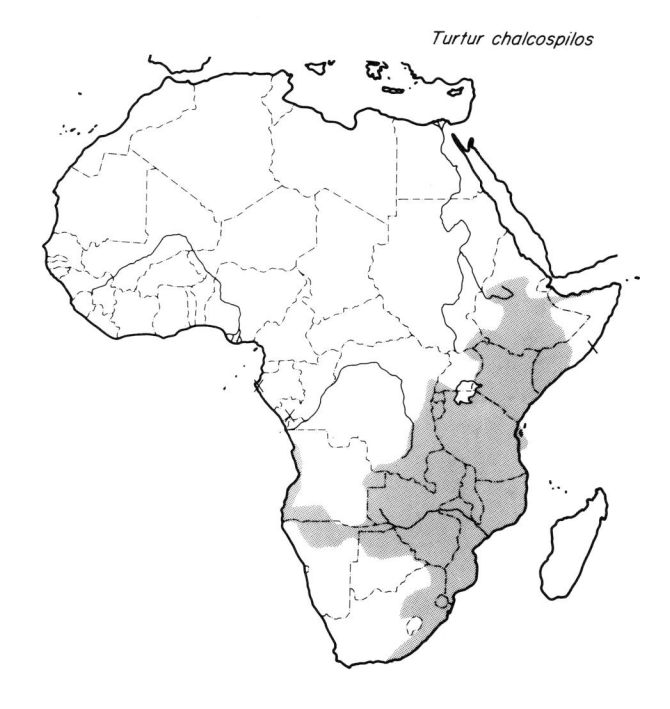

Songs of all 5 *Turtur* spp. are practically indistinguishable, although quite different from other doves (see Voice).

Voice. Tape-recorded (7, 12, 13, 49, 51, 69). Song is unvarying phrase of 13–15 soft, hollow coos and lasts 12–14 s. Song starts with 2 quiet, muffled notes; then consists of coos uttered hesitatingly and irregularly, singly or in pairs, at first in same pitch (*c.* 600 Hz) then slowly descending scale, suddenly accelerating to fast, regular rhythm; volume diminishes and last notes are hardly audible. Irregular part of song lasts *c.* 10 s and final, accelerated, regular part 3.5–4.0 s. Song may be repeated 2–3 times per min for minutes on end. Whole effect is hesitantly mournful; sometimes transliterated as 'My mother is dead ... dead ... another ... brother is dead, oh boohoohohohoho'.

General Habits. Inhabits deciduous woodland thickets, *Acacia* savanna, lowland riverine and coastal forests and open cultivated areas. Most numerous below 1600 m. Solitary or in pairs, never in flocks; forages on ground with restless zig-zag walk, in early morning or on cloudy days well out in open, but at other times keeping in or near dappled shade of trees and thickets. Never feeds in trees. Readily perches in trees, usually within 3 m of ground and in shade. Leaves perch silently but rises from ground with clatter of wings;

flight fast, direct and silent; flies low down and often for only short distance. Raises and lowers tail quickly, once or twice, upon alighting. Drinks mid-morning or midday.

Sedentary; but varying numbers at some southern African localities suggest local movements in relation to weather.

Food. Small seeds or herbs and grasses; also a few small invertebrates (molluscs, termites).

Breeding Habits. Monogamous, pair-bond persisting for several years. Territorial, pair regularly nesting in same spot (van Someren 1956). During courtship ♀ sits hunched on horizontal branch with bill tucked into breast feathers; ♂ preens her nape, gives series of jerky nods (like feeding movements but not touching ground) at rate of 2/s, then mounts and copulates; then both birds briefly preen themselves (Rowan 1983).

NEST: small platform of twigs, rootlets and grass stems, diam. c. 10 cm. Sited on stump, in bush or tree, at 0·5–6·0 m, usually 2 m, above ground (n = 75), often among creepers; once on open sand (N. E. Baker, pers. comm.). Same nest-site often re-used in following year.

EGGS: 1–2 (8 clutches of 1 egg; 76 of 2). Creamy white.

SIZE: (n = 18) 19·5–24·5 × 14·5–19·5 (23·2 × 18·1).
WEIGHT: (n = 2) 4.

LAYING DATES: Ethiopia, Apr–June, Nov; East Africa: Region C, Jan, Mar–May; Region D, Jan, Apr–June, Aug–Sept; Region E, May–June, Aug–Sept, Dec; Malaŵi, Feb–Apr; Zambia, Feb–Apr, Aug–Sept; Zimbabwe (n = 123 clutches), Jan–May 45%, June–Aug 12%, Sept–Oct 31%, Nov–Dec 12%; Mozambique, Oct–Dec; Namibia, Mar, May, Aug; South Africa: E Cape, Mar–Apr, Sept; Natal, Oct–Feb.

INCUBATION: carried out by both sexes, mainly ♀. Period: 17 days.

DEVELOPMENT AND CARE OF YOUNG: fledging period: (n = 2) 16 ± 1 days and 18 days. Parent disturbed with young in nest gave distraction display, fluttering along ground with dragging wings. Young fed 5 times daily.

BREEDING SUCCESS/SURVIVAL: chicks eaten by Boomslang *Dispholidus typus*, shrikes and mongooses, and mortality rate very high (van Someren 1956). 1 lightweight bird had high parasitaemia (Peirce 1984).

References
Rowan, M. K. (1983).
van Someren, V. G. L. (1956).

Genus *Oena* Swainson

Monotypic. A very small dove of dry woodlands and subdesert steppe, closely resembling *Turtur* in possessing rufous primaries, metallic violet wing-spots and banded rump; but has long, attenuated tail and is strongly dimorphic, ♂ having orange bill and black face and breast. Migratory.

Plate 32
(Opp. p. 465)

Oena capensis (Linnaeus). Namaqua Dove; Long-tailed Dove. Tourtelette à masque de fer.

Columba capensis Linnaeus, 1766. Syst. Nat. (12th ed.), p. 286; Cape Peninsula, South Africa.

Range and Status. Africa, SW Arabia and Madagascar; vagrant Jordan.

Resident and intra-tropical migrant. Common and widespread throughout subsaharan Africa except for rain forest zone, up to 3000 m but commoner below 1800 m; Zanzibar, Socotra and Dahlak Is. Probable breeding resident Jebel Elba, SE Egypt (Goodman and Atta in press). Vagrant Algeria (Tarhit) 1942, Agadir 1981, and Egypt (Kom Ombo) 1971. Breeds in drier savannas and partially migrates to winter in mesic, low-latitude open country. Common to abundant; flocks of hundreds occur in Kalahari sandveld and in sahelian zone.

Description. ADULT ♂: forehead, lores, chin, throat and breast black; rest of head and neck grey, pale next to black mask and browner on hindneck; mantle to rump light earth brown, upper rump crossed with 2 blackish and 1 whitish band; uppertail-coverts pale grey-brown, with blackish tips. Tail long and steeply graduated, central rectrix pointed, twice length of outer one, dark silver grey with blackish end; outer rectrix mainly white, intervening ones grey with distal half or third black with whitish tip. Belly white, undertail-coverts black. Primaries, greater primary coverts and outer secondaries rufous with sharply defined blackish border; inner secondaries mainly grey-brown; remaining coverts pale grey, with 3–5 large amethyst spots on outer webs near tertials. Underwing mainly rufous, undertail mainly black. Bill yellow-orange with purple proximal half; eye dark brown; legs and feet vinous red. ADULT ♀: like ♂ but with total absence of black mask; instead, face whitish, chin, throat and breast pale grey-brown; bill brown; outer wing-coverts duller grey. SIZE: (southern Africa) wing, ♂ (n = 42) 94–117 (108), ♀ (n = 28) 97–111 (105); tail, ♂♀ 121–160; bill, ♂♀ 13–15 (14·8); tarsus, ♂♀ 14–16 (15·5). WEIGHT: (southern Africa) ♂ (n = 39) 29–47 (40·2), ♀ (n = 27) 39–45 (39·4); (W Africa) ♂ (n = 70) 28–45 (38), ♀ (n = 30) 31–42 (35); (E Africa) ♂ (n = 11) 29–41 (36·3), ♀ (n = 7) 30–43 (35·6).

IMMATURE: upperparts heavily spotted, each feather golden buff and crossed by several narrow brown V-shaped bars.

NESTLING: skin blackish purple; short, creamy white hair-like down.

Field Characters. Much the smallest African dove. Plumage distinctive: ♂ grey with black mask, black and white banded rump, long pointed tail, white belly contrasting sharply with black undertail-coverts, mainly rufous remiges (not visible in perched bird), shiny blue spots near tertials, and yellow and purple bill. ♀ has the same features except that throat and breast are grey (lacks black mask) and bill is brown. Habits like those of *Turtur* doves but more gregarious and forages on unshaded ground; readily distinguished from them by small size and long graduated tail.

Voice. Tape-recorded (44, 74, B, C, F). Advertising song a quiet, frequently repeated, very short double 'hoo' followed by longer, plaintive 'woooo', slightly rising, lasting less than 1 s, medium-pitched. ♂ walking behind ♀ utters 'kuk-warr', the 'kuk' sharp and 'warr' long, deep and throaty. Also, series of 6–12 quiet 'du' notes at 1 s intervals in descending pitch (Rowan 1983).

General Habits. Inhabits *Acacia* savannas and thornveld, bushed grasslands, scrub, erosion gullies, cultivation, open places in riparian vegetation, dry pans and open savanna in gardens and around villages. Forages mainly singly or in pairs; hundreds may assemble in open places with other species of doves. Usually tame. Rises with clatter of wings, then flight is direct, fast (64 km/h), low and silent; often lands on ground or in low vegetation after only 5–50 m. On alighting quickly raises tail then lowers it more slowly. Forages exclusively on ground, often where earth is very hot, walking easily and sometimes making rapid short run. Perches in open, on lower branches, thorny hedges and dead wood usually within 2 m of ground. Drinks mainly between 11.00 h and 14.00 h at edge of lake or at tiny puddle.

Makes both local and migratory movements. In dry season tends to concentrate near water; in Sahelian zone large flocks fly 25 km to arrive in late morning at wells, where they stay during hottest hours; parties of 20–30 often remain on moist ground and sunbathe collectively. In dry season moves daily between Zanzibar and mainland. In drier parts of range disperses widely into dry bush as soon as ponds refill in early wet season.

Markedly migratory in northern tropics: resident in sahelian zone but further south a breeding dry-season visitor to sudanian and guinean savannas; in Nigeria present in sudanian zone Sept–June, in northern guinean savanna zone Sept–May, and in southern guinean zone Oct–Apr, with southernmost records (Ilorin, Mkar) occurring in arid 'harmattan' weather, Dec–Feb (Elgood *et al.* 1973). Bird ringed in Namibia Nov recovered Bulawayo, Zimbabwe, 1000 km to east in July.

Food. Mainly minute seeds. In Senegambia seeds of 68 spp., including 8 sedges, 10 grasses, 3 cereals and *Pennisetum* meal. Seeds of monocotyledonous plants predominate and annual percentage varies with amount of rain from 50% of diet by weight (*Dactyloctenium*, rhizomes of *Fimbristylis*) to 90% (*Panicum laetum*, 80%; *Dactyloctenium*, 7%; *Pennisetum*, *Sorghum* and *Oryza*, 3%). Other important seeds are dicotyledons *Gisekia*, *Cleome*, *Heliotropium* and *Sesuvium*; rarely eats broken *Tribulus* seeds. Most wild seeds weigh less than 1 mg, many 0·1–0·5 mg. Max. dry crop content 2·09 (*Gisekia* seeds) and 3·0 g (*Panicum* seeds). Rarely eats small invertebrates (insects, snails).

Breeding Habits. Monogamous, weakly territorial. Singing ♂ lowers head and sometimes raises closed tail. In display flight ♂ rises steeply then glides down with tail spread wide; 2–3 ♂♂ may chase ♀ in flight. On ground ♂ walks very close behind ♀, stops to utter 'kuh-warr', then catches her up with short bounding flight.

NEST: frail circular platform of rootlets, tendrils and small twigs with grass lining, diam. 5·5–8·5 cm (n = 2). Sited close to ground in bush; of 244 nests in Senegambia, 60% less than 50 cm above ground in piles of dead wood and low branches of small trees (*Commiphora*, *Acacia senegal*) and 40% from 50 cm to 3 m high mainly in *Balanites* and *Acacia raddiana* (median height, 60 cm); of 167 nests in southern Africa 4% on ground, 60% in low bushes (*Acacia*, *Lycium*, tamarisks), 15% in piles of brushwood and dead branches, 10% on low leafy cut stumps, and remainder on broken-down reeds, grass tufts and mealie stalks, in hedges, on a wood railing and in an old tin (median height, 1 m). Nest often fully exposed to sun. Built by both sexes. Rarely, same nest re-used up to 6 months later.

EGGS: 1–4, laid at 1–2 day intervals; southern Africa (187 clutches): 10 clutches of 1 egg, 173 of 2, 3 of 3, 1 of 4. Elongated ellipse; pale yellow, cream or buff. SIZE: (n = 49, southern Africa) 19·4–24·5 × 14·9–17·1 (21·1 × 15·9). WEIGHT: (n = 2) 3·2.

LAYING DATES: Senegambia, every month; Nigeria, Oct–Apr (dry season over whole range); Chad, July–Feb; Ethiopia, Feb–Sept, ?Nov; Zambia, Feb–May, July–Nov; Malawi, Feb, Apr–July; E Africa: Region A, Mar, Sept; Region B, Mar–May; Region C, Aug–Sept; Region D, May–July, Sept; Zimbabwe, every month, (n = 151 clutches) 64% Aug–Oct (mainly Sept), 13% Nov–Mar, 16% Apr–May, 7% June–July; South Africa, every month, mainly Oct (and Feb after rains in N Cape).

INCUBATION: begins with 1st egg; carried out by both sexes, ♂ in 1 bout of c. 5 h ending about 11.00 h (South Africa, Kalahari Gemsbok Park; Senegambia) or about 09.30–15.30 h (South Africa, Barberspan) and ♀ in unbroken spell for rest of day and night. Incubating ♂ turns eggs, and flies at encroaching Namaqua Dove to drive it away. Slightly alarmed ♂ crouches low over eggs with head down, concealing black throat, with rump feathers raised. ♂ and ♀ with eggs and nestlings have distraction display, creeping along ground with fluttering wings and spread tail. Eggs hatch at 1-day intervals.

DEVELOPMENT AND CARE OF YOUNG: hatchling brooded for 4–5 days by both sexes. Size doubles by day 5; wing-quills appear on day 6 and eyes open in days 4–7. Young fed by both ♂ and ♀. Nestling period: 16 days (n = 3).

BREEDING SUCCESS/SURVIVAL: of 69 eggs (35 clutches) 62% hatched and of 43 chicks (23 broods) 48% fledged (30% of eggs laid).

References
Cramp, S. (ed.) (1985).
Rowan, M. K. (1983).

Genus *Columba* Linnaeus

Wood pigeons and rock doves. Medium or large, mainly arboreal pigeons, heavy-bodied, less gregarious than *Streptopelia*; most monomorphic, with dark grey and vinous plumage, partly iridescent, sometimes (not in Africa) strongly patterned, often with bright red, yellow or bluish bare parts (legs, bill, skin around eye); some have white flashes on neck or wing; most have grey tail with blackish terminal band. Several African species purplish. Tail square or slightly rounded. Tarsus shorter than middle toe with claw; upper tarsus concealed by plumage. Eat grain but mainly small hard fruits and small leaves.

African species fall into 5 groups: bronze-naped pigeons of the *C. iriditorques*/*C. delegorguei*/*C. malherbii* superspecies and their terrestrial relative *C. larvata*; olive pigeons *C. albinucha* and the closely allied *C. arquatrix*/*C. sjostedti*/*C. thomensis* superspecies; wood pigeons *C. unicincta* and *C. palumbus*; stock doves *C. oenas* and *C. oliviae*, which are intermediate between wood and rock pigeons; and rock pigeons *C. livia*, *C. guinea* and *C. albitorques*. *C. albinucha* is very like the *C. arquatrix* superspecies and may belong to it; although broadly sympatric in E Zaïre, the 2 birds may in fact be separated there altitudinally. *C. larvata* was formerly placed in *Aplopelia* Bonaparte, but we follow Fry *et al.* (1985) in treating it in *Columba*.

Columba iriditorques superspecies

Columba arquatrix superspecies

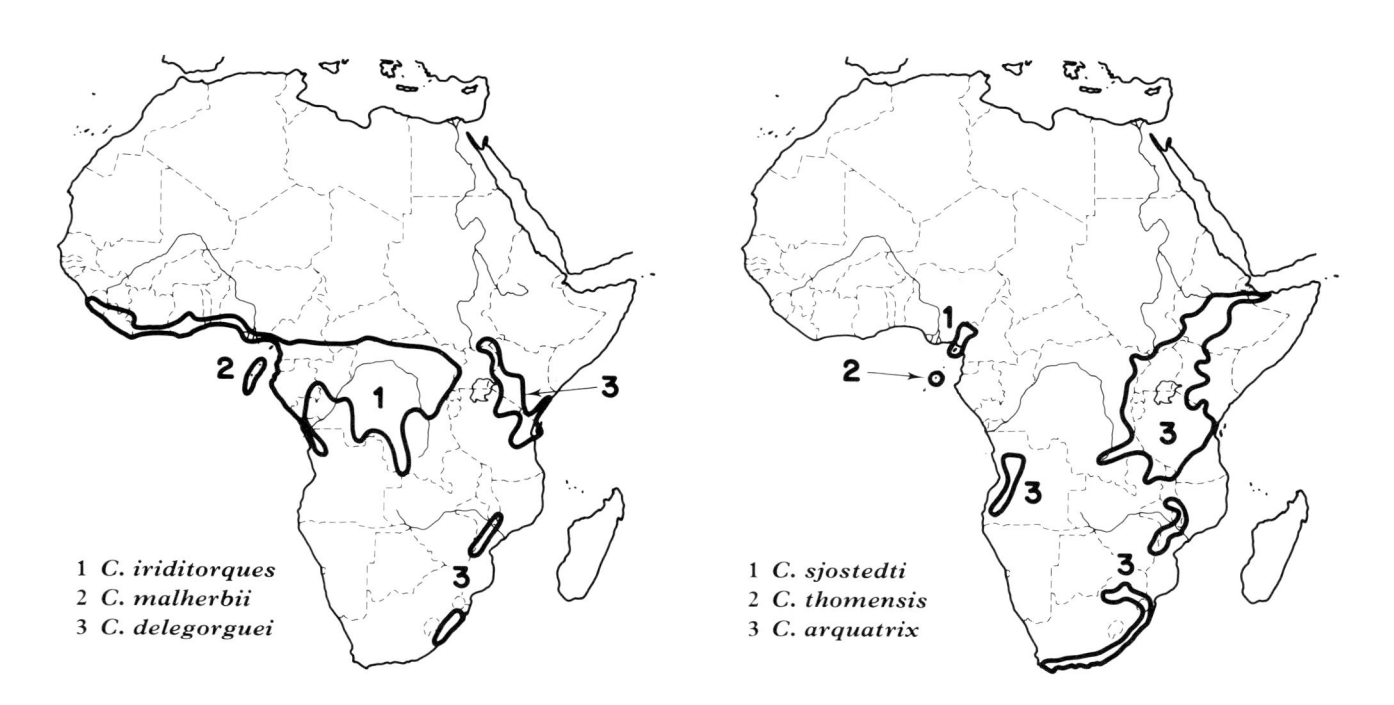

1 *C. iriditorques*
2 *C. malherbii*
3 *C. delegorguei*

1 *C. sjostedti*
2 *C. thomensis*
3 *C. arquatrix*

Columba iriditorques Cassin. Western Bronze-naped Pigeon. Pigeon à nuque bronzée.

Columba iriditorques Cassin, 1856. Proc. Acad. Nat. Sci. Philadelphia 8, p. 254: St Paul's R., Gabon.

Forms a superspecies with *C. delegorguei* and *C. malherbii*.

Range and Status. Endemic resident, lowland forests from Sierra Leone to W Ivory Coast, in S Ghana, SW Nigeria, NW Angola (Cuanza Norte and probably Cabinda and Lunda) and whole of Congo basin from S Cameroon and W Congo east to SW Uganda (Bwamba Forest at 900 m and Impenetrable Forest at 1500 m) and Rwanda. Not uncommon in Rwanda, Liberia (Mt Nimba, up to 900 m) and Nigeria; uncommon or rare in Cameroon and throughout Congo basin; but easily overlooked.

Description. ADULT ♂: forehead, crown, nape, face, sides of head and throat dark blue-grey (chin paler), rather sharply divided from hindneck and breast; hindcrown lightly glossed green, and nape and lower sides of neck strongly glossed green. Hindneck and upper mantle copper-bronze, strongly glossed pink; lower mantle feathers black with broad iridescent golden green, green and violet-blue fringes; back to uppertail-coverts and wings black with grey bloom and slight iridescence. P9 and P8 strongly emarginated. Middle tail feathers slate, broadly but obscurely tipped dark grey; outer rectrices with usually broad buff ends and mainly dark chestnut on inner webs; underside of tail dusky chestnut broadly tipped buff. Breast and belly dark mauve pink; undertail-coverts chestnut. Bill blue-grey, whitish at tip, cere dark red; eye pink-red, also said to be golden, pink, grey or greenish yellow to blue; orbital ring red; legs and feet light red. ADULT ♀: crown and nape rufous; hindneck and mantle glossy green or violet; face and chin pale brown, throat and breast rufous-brown, belly rufous, vent and undertail-coverts bright chestnut. SIZE: (15 ♂♂, 18 ♀♀) wing, ♂ av. 160 ± 5, ♀ av. 153 ± 4; tail, ♂ av. 90 ± 7·6, ♀ av. 83 ± 4·3; bill, ♂ av. 20·6 ± 1·0, ♀ av. 20·2 ± 1·2 WEIGHT ♂: (n = 15) av. 130 ± 8, ♀ (n = 18) av. 122 ± 10.

NESTLING: uniform brown, all contour feathers fringed rufous with narrow blackish subterminal band; throat feathered later than rest of body; belly with thick buffy down. Bill and eye brown; legs dark brown.

Field Characters. A small pigeon of forest canopy; ♂ with grey head, slaty back, wings, and (closed) tail, broad bronzy hindneck collar, dull mauve breast and belly, and chestnut undertail-coverts. Spread tail appears black with broad buff tip (central feathers all black). Seen from below tail is chestnut with broad buff end. ♀ has mainly rufous brown head and underparts, rich chestnut belly and undertail-coverts, slate black wings and closed tail, and iridescent green/violet hind collar. Spread tail like ♂'s but much duller. See also Eastern Bronze-naped Pigeon *C. delegorguei*.

Voice. Tape-recorded (44, C). Song starts with 3–4 low, rather quiet coos, followed suddenly by 4–5 loud, explosive coos, ending abruptly, but followed by 1–2 quiet, almost inaudible coos; total duration 4–5 s. Alighting call sounds like piece of cloth being ripped (Gee and Heigham 1977).

General Habits. Inhabits riparian evergreen forests and dense thickets away from water; primary lowland forests from sea level up to 1500 m; also old secondary growth, forested slopes and gallery forests. Solitary, in pairs or flocks of 3–4, usually well hidden in canopy of tall trees; sometimes occurs in low vegetation and even near forest floor. Mixes with Afep Pigeon *C. unicincta* when foraging in fruiting trees (A. Brosset, pers. comm.).

Food. Fruits of *Musanga* and *Eisterya* (Gabon: A. Brosset, pers. comm.) and *Haronga* (Cameroon); also seeds. Eats grit.

Breeding Habits. Monogamous.
NEST: 1 nest found in *Haronga* tree (Cameroon); only nest described (Zambia) was flimsy platform, 1·2 m above ground in dense thicket, with 1 egg.
EGG: smooth; slightly glossed; pale cream. SIZE: 30·2 × 22·3 (Benson 1959).
LAYING DATES: Liberia, Mar (and ♀♀ with large ovaries Apr, July and Sept); Zaïre, Dec–Mar; Zambia, Oct.

References
Benson, C. W. (1959).
Fry, C. H. *et al.* (1985).

460   COLUMBIDAE

Plate 30
(Opp. p. 449)

Columba malherbii Verreaux. São Tomé Bronze-naped Pigeon. Pigeon de Malherbe.

Columba malherbii J. & E. Verreaux, 1851. Rev. Mag. Zool. p. 514; Gabon in error for São Tomé.

Forms a superspecies with *C. iriditorques* and *C. delegorguei*.

Range and Status. Principe, São Tomé and Pagalu Is. (Gulf of Guinea); absent Bioko. In São Tomé, frequent to uncommon at low and middle altitudes; in Principe evidently uncommon 80 years ago but now very common at low level (R. de Naurois, pers. comm.); on Pagalu, very common in 1902 but uncommon in 1955 and 1959 (Fry 1961).

ochreous mainly on inner webs; tail below pale ochreous grey. Bill grey, with pale tip; eye pale grey; legs and feet bright red. ADULT ♀: like ♂ but underparts darker grey; lower breast and upper belly feathers finely freckled ochreous; lower belly, vent and undertail-coverts pale rufous, speckled with grey. SIZE: wing, ♂ (n = 7) 178–182, ♀ (n = 7) 165–176; tail, ♂♀ (n = 14) 94–115; bill, ♂♀ (n = 14) 16–18; tarsus, ♂♀ (n = 14) 22–23.

IMMATURE: green/violet hindneck collar less iridescent and less extensive than in adult; forehead pale grey, forecrown pale rufous; hindcrown like adult; underparts heavily freckled ochreous or rufous.

NESTLING: undescribed.

Field Characters. ♂ has pale grey face and throat to belly, slaty wings and tail, hindneck and mantle glossy green or violet, and rufous undertail-coverts. ♀ is similar but strongly suffused or speckled with rufous-buff below. São Tomé Lemon Dove *C. larvata* is very similar but has bronze-green iridescence on nape and mantle, brown upperparts, whitish belly and undertail-coverts.

Voice. Not tape-recorded. Like voice of *C. iriditorques*.

General Habits. Inhabits deep forests at 400–500 m on Pagalu, forests and plantations on São Tomé and Principe. Singly, or in flocks of up to 7. Feeds in middle stratum of trees (3–16 m above ground). Never seen to drink.

Description. ADULT ♂: differs from ♂ Western Bronze-naped Pigeon *C. iriditorques* as follows: hindneck and upper mantle strongly glossed green or pink according to direction of light, but on slaty (not copper-bronze) background; scapulars, wing-coverts, and mantle to uppertail-coverts blackish with oily green sheen. Breast and belly grey (concolorous with throat), vent and undertail-coverts pale rufous speckled with grey. Tail above very dark grey, outer feathers washed

Breeding Habits. Monogamous.

NEST: substantial platform built 5–12 m above ground in secondary forest tree, *Erythrina*, and plantation cocoa trees, often near trunk.

EGGS: 1–2 (n = 25, Principe). Oval; glossy; dirty white.

LAYING DATES: Principe, Nov–Jan; São Tomé, probably Feb.

Plate 30
(Opp. p. 449)

Columba delegorguei Delegorgue. Eastern Bronze-naped Pigeon; Delegorgue's Pigeon. Pigeon de Delegorgue.

Columba delegorguei Delegorgue, 1847. Voy. Afr. Austr. 2, p. 615; Durban.

Forms a superspecies with *C. iriditorques* and *C. malherbii*.

Range and Status. Endemic resident. Discontinuous distribution from SE Sudan to South Africa (Transkei). Sudan, common in Imatong, Dongotona and Didinga Mts, 1100–2800 m; Uganda, Mt Lonyili in NE and Mt Elgon in E; Kenya and Tanzania, widespread and locally common in highland forests from Mt Kenya to Ulugurus and Ngurus, also, uncommonly, at Sokoke,

Rabai, Shimba Hills, Mbulumbulu and Mt Hanang (Fuggles-Couchman 1984); and 1 record Zanzibar (sea level); Malaŵi, Mt Thyolo, at 1475 m; E Zimbabwe, common in Lusitu, Haroni and Makurupini R. forests, below 500 m, and adjacently in Mozambique to Mt Gorongoza; South Africa, forests of Transkei and E Griqualand to Natal and W Zululand.

Columba delegorguei

C. d. sharpei Salvadori: Sudan to Tanzania. ♂ lacks purple tinge on lower mantle which is blackish; green rather than mauve-amethyst reflections on neck; belly and flanks slate grey, not dull vinous. Some have white neck-patch reduced in size. ♀ has greener neck iridescence than ♀ *delegorguei*.

Field Characters. Bronze-naped pigeons (*C. delegorguei*, *C. iriditorques*, *C. malherbii*) are small, shy, dark pigeons of the forest canopy, often difficult to see. Similar-sized Lemon Dove *Columba larvata* also shy but lives on forest floor. ♂ Eastern Bronze-naped Pigeon distinctive with grey head and white hind collar, but ♀ with brown head and bronze hindneck resembles Lemon Dove; upperparts blacker, underparts grey (sympatric races of Lemon Dove have cinnamon underparts).

Voice. Tape-recorded (10, 14, 74, 75, C, F). Advertising call starts with 3–4 short, muffled, low-pitched coos, then 2–3 more emphatic, higher-pitched ones, ending with fast series of 5–6 coos in descending scale and *diminuendo*; whole sequence not exceeding 4 s.

General Habits. Inhabits canopy of tall evergreen lowland and adjacent riparian fringing forests (Zimbabwe); in Natal restricted to climax forests in mist belt. Also dense woodland and thick bush, sometimes gardens and mature pine plantations (South Africa). Occurs singly or in small parties (max. 30) in canopy of high trees. Walks easily between fallen branches and logs on forest floor; feeds mainly early morning and afternoon. May make local movements (Kenya).

Food. Fruits of *Podocarpus latifolius*, *Macaranga capensis*, *Phialodiscus zambesianus*, wild figs and berries of *Trema orientalis*; some seeds and insect larvae.

Breeding Habits. ♂ courtship described as 'walking backwards and forwards along a branch, filling out his chest and cooing like an ordinary tame Pigeon'.

NEST: platform of twigs, c. 7–10 m above ground in trees.

EGGS: 2. White. SIZE: 30 × 22.

LAYING DATES: E Africa: Region A, Mar; Region B, Mar–June, Dec; Zimbabwe, Jan.

Description. *C. d. delegorguei* Delegorgue: Malaŵi to South Africa. ADULT ♂: differs from ♂ Western Bronze-naped Pigeon *C. iriditorques* as follows: head darker grey, chin dark (not pale) grey; hindneck collar strongly glossed green or pink according to light, but on slaty (not copper-bronze) background, and separated from lower mantle by broad white crescent with pearly sheen, crescent fringed pink-violet posteriorly. Mantle, scapulars and wing-coverts black washed maroon, with some iridescent green fringes on lower mantle. Tail black with 4 mm very dark grey tip (below, tip is paler). Belly and undertail-coverts dark grey. ADULT ♀: differs from ♀ *C. iriditorques* in having hindneck and upper mantle glossy copper-green (rather than violet); underparts dark grey very finely peppered with ochre (looks tawny grey). SIZE: wing, ♂ (n = 6) 178–185 (182), ♀ (n = 6) 163–179 (173); tail, ♂♀ 100–110; bill, ♂ (n = 4) 19–22·5 (20·4), ♀ (n = 4) 19–22 (20·7); tarsus, ♂♀ 19–21. WEIGHT: ♂ (n = 6, E Africa) 136–175 (153), ♀ (n = 6) 133–154.

IMMATURE: no white collar; upperparts dark greyish brown; underparts dark rufous; some iridescence on nape.

NESTLING: not known.

Columba larvata Temminck. Lemon Dove; Cinnamon Dove. Tourterelle à masque blanc. **Plate 32**
(Opp. p. 465)

Columba larvata Temminck, 1810. In Knip, Les Pigeons, p. 71; Knysna, Cape Prov., South Africa.

Range and Status. Endemic resident. Patchily distributed, in forests from N Ethiopia to coastal South Africa, west to E Zaïre, NW Zambia and Cape Town. Also in W Africa: Liberia (Mt Nimba and Paiata), Cameroon, Gabon, Bioko, Principe, São Tomé and Pagalu Is. Single records on upper R. M'Bomu (N Zaïre/E Central African Republic), lower Congo R. (Congo/W Zaïre), and probably occurs in N Angola (Salazar; C. J. Vernon, pers. comm.). From sea level up to 2100 m in W and southern Africa, but from Limpopo R. to Ethiopia is more montane, up to 3200 m. Generally uncommon, but very elusive, and locally frequent in Gulf of Guinea islands and Ethiopia; density of 2–3 pairs in 12 ha forest (Zambia/Malaŵi: F. Dowsett-Lemaire, pers. comm.). Possibly endangered through destruction of forest habitat, but has adapted to plantations of exotic trees, cocoa (Principe Is.) and well-wooded gardens (SW Cape).

462   COLUMBIDAE

Columba larvata

Description. *C. l. larvata* Temminck: S Sudan to South Africa, west to Uganda (except Ruwenzoris), W Tanzania (Mt Mahari), Malaŵi and Cape. ADULT ♂: crown, forehead and forecrown white, hindcrown to upper mantle brown with strong bronze, green, violet or pink iridescence according to light; foreneck with some green iridescence, sides of neck shading to pink-brown; face, chin and throat white to pale greyish white; lower mantle, wing-coverts, tertials, back, rump and uppertail-coverts dark brown with faint mauve or green iridescence; tail with central pair of rectrices brown, others dark brown with 3 cm grey terminal band (grey palest on outermost rectrix). Breast pink-brown, feathers fringed iridescent green (strongly so at sides of breast), merging to rufous belly and cinnamon undertail-coverts. Wings dark brown; P9–P7 emarginated. Underwing brown. Bill black; naked orbital skin purple; eye variable, brown to crimson; legs and feet purplish red. Sexes alike. SIZE: wing, ♂ (n = 5 South Africa) 145–154 (151), ♀ (n = 5) 143–147 (145); tail, ♂♀ 91–117 (102); bill to skull, ♂♀ 20–22·5 (21·3); tarsus, ♂♀ 26·5–29 (27·6) (Rowan 1983). WEIGHT: (E Africa) ♂ (n = 7) 130–210 (151), ♀ (n = 11) 125–175 (146); (South Africa) ♂ (n = 7) 156–195 (169); (Zambia/Malaŵi) ♂♀ adult (n = 31) 123–191 (153), juvenile (n = 3) 121–146 (129) (R. J. Dowsett, pers. comm.).

IMMATURE: dark brown above; forehead and throat whitish; breast dark brown, belly rufous-buff. All brown contour feathers have rufous fringe and narrow black subterminal band; rufous fringes broad on breast which can look mainly rufous. Grey rectrix tips obscure.

NESTLING: covered with fairly sparse, golden yellow down.

C. l. bronzina Rüppell: Ethiopia, and Boma Hills (Sudan). Like *larvata* but smaller. Wing, ♂ (n = 12) 124–134 (130), ♀ (n = 8) 120–134 (127).

C. l. jacksoni (Sharpe): SW Uganda (Impenetrable Forest to Budongo, Mabira Forest), E Zaïre, W Tanzania (Mt Kungwe), NW Zambia. ♂ upperparts darker than *larvata*; breast grey or brownish grey, paler on ventral region; orbital skin grey. ♀ like ♀ *larvata*. Larger: wing, ♂♀ 147–163. WEIGHT: ♂ (n = 5) 133–160 (147), ♀ (n = 1) 134.

C. l. principalis (Hartlaub): Principe Is. ♂ below intermediate between ♂ *larvata* and ♂ *jacksoni*, greyish pink, pale pink towards vent, grey or pink undertail-coverts; above like ♂ *larvata*. ♀ like ♀ *larvata* but lower belly and vent pale buff, pale pink or whitish.

C. l. inornata (Reichenow): Liberia, Cameroon, Gabon, Bioko and Pagalu Is. ♂, above dark (like *jacksoni*), below clear grey. Throat whitish; lower belly, vent and undertail-coverts either clear grey or white. ♀ like ♀ *principalis*. Variable. Cameroon birds tend to be slightly darker (*'plumbescens'*) and Liberian ones slightly paler. Smaller than *larvata*; wing, ♂♀ 140–152. WEIGHT: ♂ (n = 3) 130–170 (155), ♀ (n = 2) 85–120 (103).

C. l. simplex (Hartlaub): São Tomé Is. ♂ and ♀, above like ♂ *larvata*, below like ♂ *inornata*. Variable, sometimes brownish.

Field Characters. A medium-sized, stocky, dark brown and grey or cinnamon dove of forest floor and low undergrowth; rather short-tailed and long-legged, walks with upright stance. Very wary; when disturbed seldom seen again; flight very noisy. Similar-sized bronze-naped pigeons (*C. delegorguei*, *C. iriditorques*, *C. malherbii*) live in forest canopy and have blackish grey rather than brown upperparts; for further distinctions, see under those species. Blue-headed Wood Dove *Turtur brehmeri* is smaller, slimmer, with proportionately longer tail, blue head contrasting with rufous plumage.

Voice. Tape-recorded (14, 44, 74, 75, F). A short, low-pitched 'hoo', slurring upward, repeated after pause 14 times in 20 s. At height of breeding ♀ also sings, duetting with ♂ (R. J. Dowsett, pers. comm.). Also a quiet rodent-like squeak every 3–4 s, perhaps a contact call (Dowsett 1971); a low gruff warning note and a hissing sound during attack on rival (Rowan 1983).

General Habits. Inhabits lowland and montane evergreen forest; also plantations, thickets and large wooded gardens. Keeps well down in forest and spends greater part of time on ground, in deep shade. Forages in pairs among dead leaves and forest litter; sometimes also singly or in flocks of up to 10. Walks swiftly and easily; reluctant to fly, and when flushed flies short distance and alights on low branch. Rises with clatter of wings, then flight is silent. Shy; but can become tame and glean crumbs at picnic sites. Drinks in early and late afternoon; visits bird baths; bathes regularly.

Sedentary, but concentrates seasonally at fruiting trees (Prigogine 1971) and minor altitudinal movements noted Zimbabwe.

Food. Seeds and fallen fruits on forest floor. Fond of *Calodendrum capense* fruit. Seeds of bamboo, *Podocarpus* and *Kiggelaria africana* (82 seeds in 1 crop); tubers of small ground orchid. Fruits or seeds of 17 spp. in Zambia and Malaŵi (F. Dowsett-Lemaire, pers. comm.). Some small snails; termites.

Breeding Habits. Monogamous, territorial, solitary. ♂♂ fight by wing-cuffing, accompanied by hissing sound.

NEST: a quite substantial platform of twigs and rootlets (1 entirely of pine needles), placed 1–9 m (usually 2·5 m) high in tangle of creepers, on debris in fork of tree, or in middle of horizontal bough; nearly always in deep shade. Built by ♂ and ♀, in 7–8 days. One 2 m high on frond of *Encephalarctos* was used for 3 successive summers.

EGGS: 1–3 laid at 1-day intervals; southern Africa (32 clutches): 4 clutches of 1 egg, 26 of 2, 2 of 3. Buff or cream. SIZE: (n = 14, southern Africa) 25·8–30·5 × 20·8–23·6 (28·7 × 22·0).

LAYING DATES: Cameroon, Jan, Aug, Oct; Ethiopia, Mar–Apr; Rwanda, probably every month; Malaŵi and Zambia, Sept–Dec; East Africa: Region B, Feb, Apr–May, July; Region D, Jan, Mar–May, July–Sept; Zimbabwe, Oct–Jan, Mar, July, Aug; South Africa (Cape Province) Mar, Sept–Dec, (Natal) Feb–Apr, July, Oct–Dec; in southern Africa mainly Nov–Dec.

INCUBATION: carried out by ♀ only (van Someren 1956). Period: 14–18 days.

DEVELOPMENT AND CARE OF YOUNG: at 1 nest hatchling was fed regularly by brooding parent every 10–20 min, by convulsive regurgitation; at another, both parents fed young 3 times in a day. 1 nestling may be much more advanced than other. Fledging period: 20 days. Young remain with parents for 2 months (van Someren 1956).

BREEDING SUCCESS/SURVIVAL: preyed upon by African Goshawk *Accipiter tachiro*.

References
Fry, C. H. *et al.* (1985).
Rowan, M. K. (1983).
van Someren, V. G. L. (1956).

Columba arquatrix Temminck. Olive Pigeon; Rameron Pigeon. Pigeon rameron.

Colomba (sic) arquatrix Temminck, 1809. In Knip, Les Pigeons, p. 11, pl. 5; Knysna, Cape Province.

Forms a superspecies with *C. sjostedti* and *C. thomensis*.

Plate 30
(Opp. p. 449)

Range and Status. Endemic, resident, somewhat nomadic. Discontinuously distributed in forests of eastern Africa from Ethiopia to Cape (see map) and in W Angola. Frequent to uncommon in Ethiopia; locally common in NW Somalia (Mt Wagar, Darass, Gadabursi hills, Birdeh); in E Africa, widespread and frequent or locally common, particularly above 1500 m, ranging down to 700 m; in Malaŵi locally common 760–1300 m; in Zimbabwe frequent 900–1500 m, flocks often flying high over lower land. Altitudinal limits *c*. 300–3200 m. In South Africa frequent to uncommon, commonest from Natal to Knysna in flocks of 10–200; but until about 50 years ago evidently much commoner, locally abundant, with flocks of up to 2500 in forests of ironwood *Olea capensis* in Knysna (Cape) (Phillips 1927). Decline in southern Africa (and Zaïre) attributed to excessive hunting and habitat destruction; now protected in South Africa and in Natal has increased markedly in last 20 years. Density varies according to that of main food plants, on Nyika Plateau (Zambia/Malaŵi) from 1 pair/20 ha to 4 pairs/4·7 ha (av. 1 pair/3 ha) (Dowsett-Lemaire 1983.)

Description. ADULT ♂: forecrown dark purplish grey, hindcrown and nape silvery grey; feathers of hindneck and upper mantle pointed, dark brown with pale mauve tips (brown part normally concealed); face dark vinous shading to mauve on chin and throat; sides of neck and foreneck dull mauve forming, with hindneck, a broad mauve collar; upper mantle rich dark maroon with some white spots at sides, lower mantle brown; back, rump and uppertail-coverts dark brown; tail slaty black. Upper breast dull mauve with silvery sheen; lower breast feathers dark maroon, each with silvery mauve tip; belly dark maroon, profusely spotted white (spots smaller and less distinct on lower belly, absent near vent). Remiges dark brown; scapulars dark vinous maroon, each feather with 2 white spots; greater primary and secondary coverts unspotted dark grey; median and lesser coverts dark vinous maroon spotted with white; underwing dark blue-grey. Bill, large area of bare skin around eye, legs and feet, yellow; eye greyish yellow to light brown. Sexes alike but ♀ duller. SIZE: (n = 5, South Africa) wing, ♂ 220–229 (224), ♀ 213–224 (222); tail, ♂♀ 125–145 (138); bill, ♂♀ 23·5–25 (24·1); tarsus, ♂♀ 25–27 (26·0). WEIGHT: (E Africa) ♂ (n = 13) 370–429 (405), ♀ (n = 10) 300–420 (352); (Zambia/Malaŵi) ♂ (n = 8) 269–425 (373), ♀ (n = 3) 288–425 (348) (R. J. Dowsett, pers. comm.).

IMMATURE: upperparts dark earth brown, forehead, scapulars and lesser wing-coverts fringed rusty, nape and median wing-coverts bluish grey; underparts warm rufescent brown with pale fringes to breast feathers and some small soft white spots on upper belly.

NESTLING: densely covered with long yellow down.

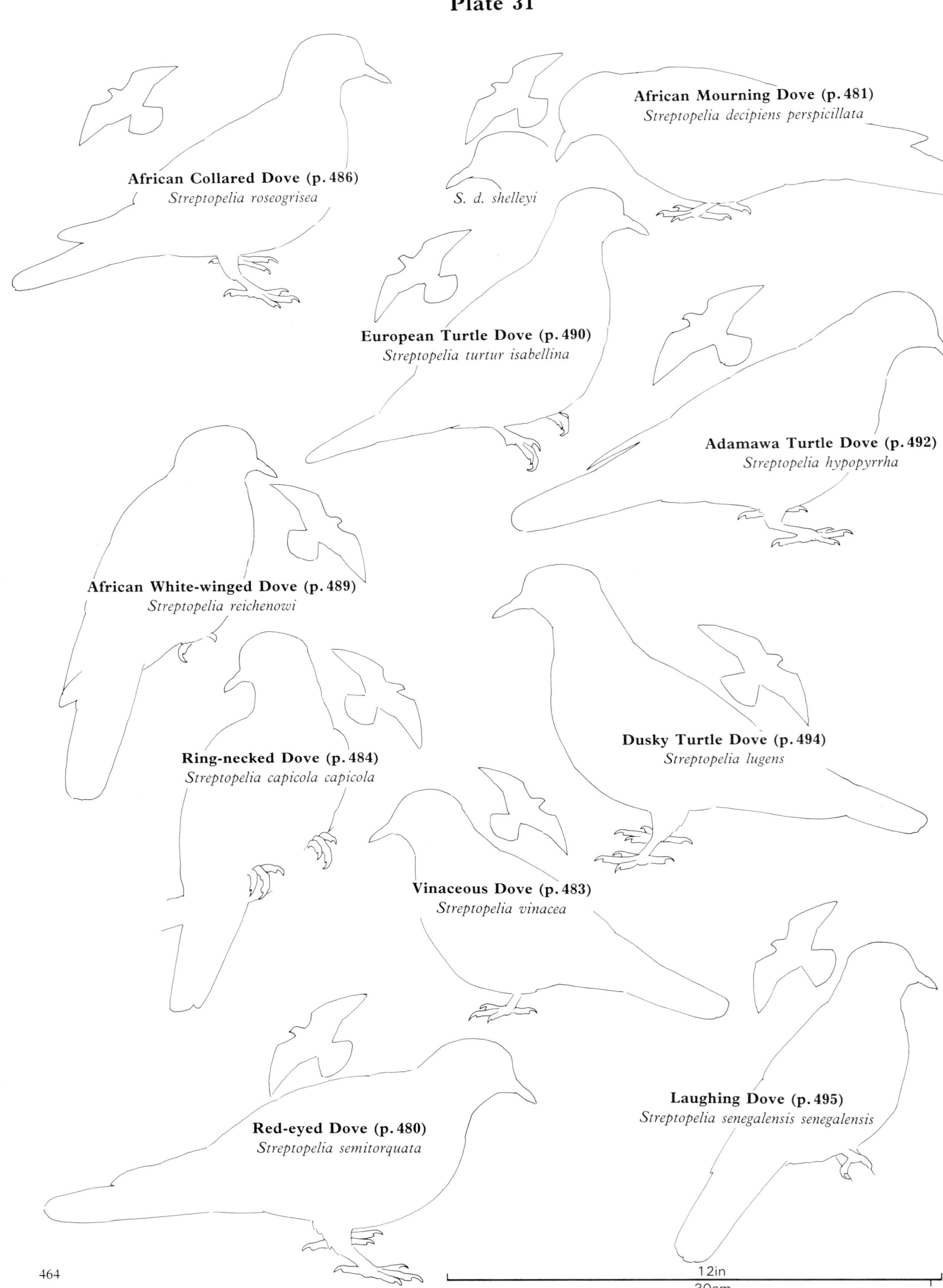

Plate 31

African Collared Dove (p. 486)
Streptopelia roseogrisea

African Mourning Dove (p. 481)
Streptopelia decipiens perspicillata

S. d. shelleyi

European Turtle Dove (p. 490)
Streptopelia turtur isabellina

Adamawa Turtle Dove (p. 492)
Streptopelia hypopyrrha

African White-winged Dove (p. 489)
Streptopelia reichenowi

Ring-necked Dove (p. 484)
Streptopelia capicola capicola

Dusky Turtle Dove (p. 494)
Streptopelia lugens

Vinaceous Dove (p. 483)
Streptopelia vinacea

Red-eyed Dove (p. 480)
Streptopelia semitorquata

Laughing Dove (p. 495)
Streptopelia senegalensis senegalensis

12in
30cm

MWW 1981

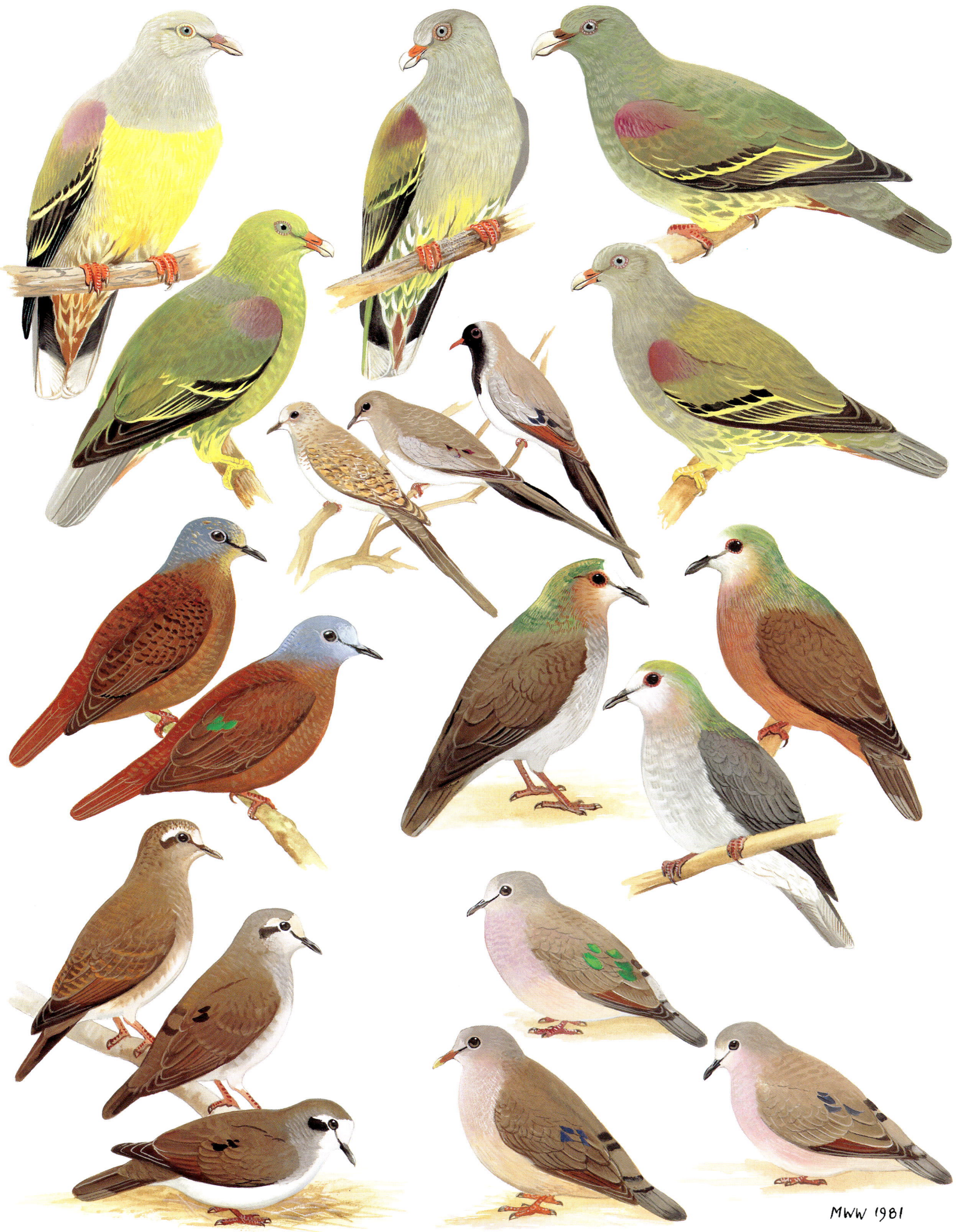
MWW 1981

Plate 32

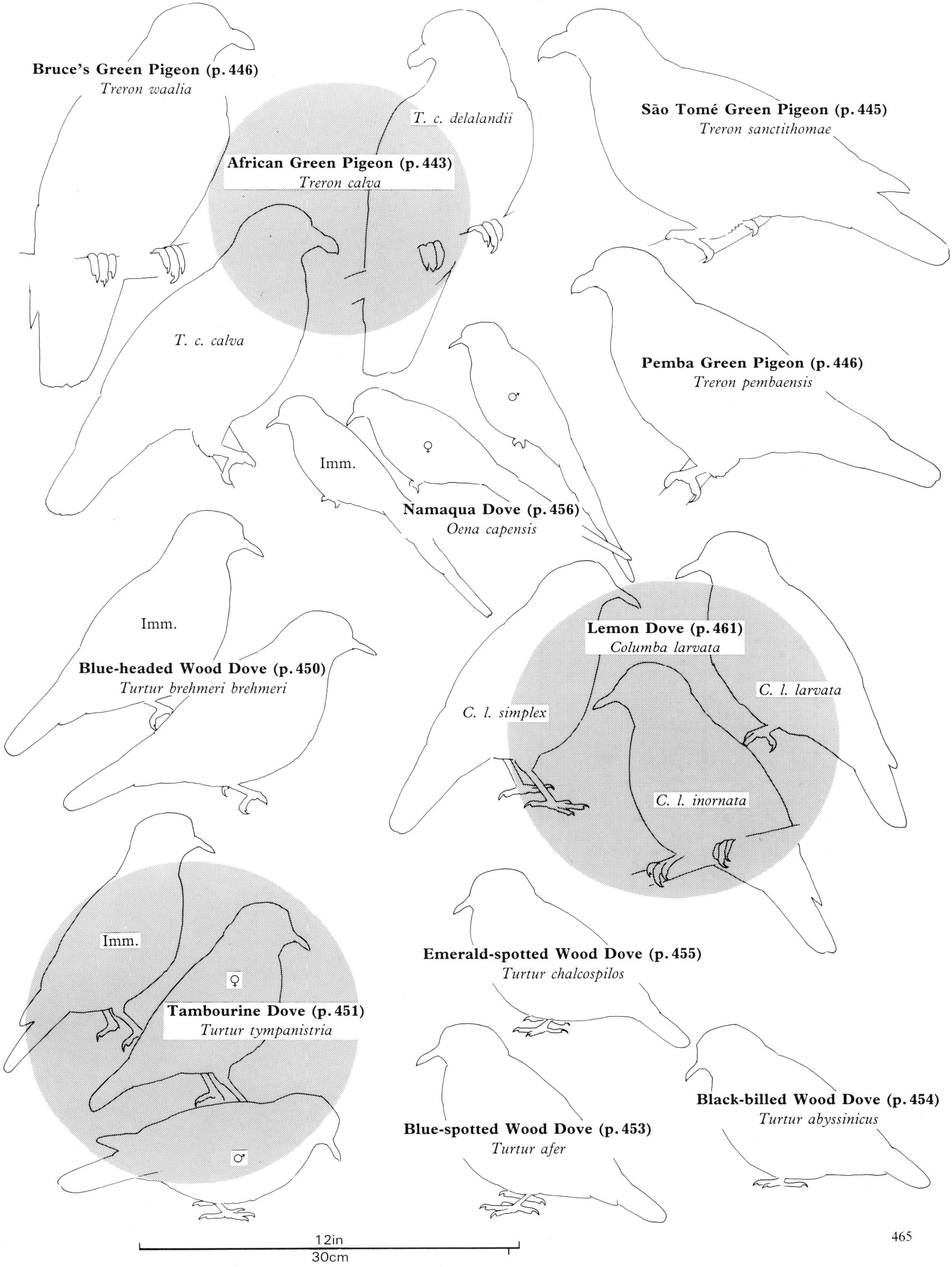

Bruce's Green Pigeon (p. 446)
Treron waalia

T. c. delalandii

African Green Pigeon (p. 443)
Treron calva

São Tomé Green Pigeon (p. 445)
Treron sanctithomae

T. c. calva

Pemba Green Pigeon (p. 446)
Treron pembaensis

Imm. ♀ ♂

Namaqua Dove (p. 456)
Oena capensis

Imm.

Blue-headed Wood Dove (p. 450)
Turtur brehmeri brehmeri

Lemon Dove (p. 461)
Columba larvata

C. l. simplex *C. l. larvata*

C. l. inornata

Imm. ♀

Tambourine Dove (p. 451)
Turtur tympanistria

♂

Emerald-spotted Wood Dove (p. 455)
Turtur chalcospilos

Blue-spotted Wood Dove (p. 453)
Turtur afer

Black-billed Wood Dove (p. 454)
Turtur abyssinicus

12in / 30cm

Field Characters. A large, dark, rather uniform pigeon, predominantly vinous with greyer head, whole bird looking blackish in poor light. Best field character is vivid yellow bill, eye area and legs, diagnostic of this pigeon and distinguishing it from otherwise rather similar Speckled Pigeon *C. guinea*. Eastern Bronze-naped Pigeon *C. delegorguei*, which can occur in same forest canopy, is very much smaller, unspotted, ♂ with white hind collar, ♀ with bronze-brown nape concolorous with head (nape contrastingly pale grey in Olive Pigeon).

Voice. Tape-recorded (74, C, F, 217, 244, 258, 296). Advertising call short, very deep, rumbling note followed by short series of monotonous, somewhat quavering coos; duration 5–6 s. During display flight a bleating 'meeeeh' (like excitement cry of Red-eyed Dove *Streptopelia semitorquata*).

General Habits. Occurs in canopy of evergreen forest, mainly over 1400 m; also in riparian and coastal forests, and in exotic pine, wattle and eucalyptus plantations. Flocks move freely and for quite long distances between forests, flying high up over intervening highland and lowland habitats.

Feeds mainly in canopy of large trees; also in midstratum. Climbs on lianas, and able to edge out along slender branches which its weight can break, often with flapping and wing-clatter when reaching out for fruits like a parrot. Feeds mainly in trees but also on ground; foraging with ponderous walk on seeds on ground and berries of low-growing weeds. Drinks only rarely; but observed to mix with White-naped Pigeon *C. albinucha* when drinking and feeding in trees. When disturbed, slips out of far side of tree canopy with clatter of wings audible 1 km away. Flies strongly, in flocks of 6–10. Said to make daily movements (Uganda (Ruwenzori Mts), South Africa) between highland forests, where it roosts, and deep valleys and lowland forests where it spends day foraging; but significance not understood, for not all individuals move, and same ripe fruits occur on mountain slopes and in valleys (Rowan 1983). On Nyika plateau (Zambia/Malaŵi) does *not* make such movements (R. J. Dowsett, pers. comm.). Descends singly or in pairs from high altitude very fast with wings and tail partly or fully folded.

Makes seasonal movements in South Africa, Zambia and Zimbabwe, perhaps related to fruiting of trees. Formerly large flocks appeared on Natal coast, June–Aug. At 1 locality in Drakensberg Mts recorded in winter only, but at another 50 km away in summer only. In parts of Natal, Stellenbosch (Cape) and elsewhere resident, numbers fluctuating with local wanderings.

Food. Olive-like fruits (drupes); other fruits and seeds and some insects, including many caterpillars. 30 native fruit spp. recorded eaten in South Africa (Rowan 1983) including *Podocarpus, Kiggelaria, Celtis, Sideroxylon, Trema, Calodendrum, Prunus, Ficus* and *Phytolacca*. 10 fruit spp. recorded in Zambia (F. Dowsett-Lemaire, pers. comm.), where eats particularly *Afrocrania volkensii* and *Olea capensis* fruits. Also eats cultivated mulberries, olives and grapes, and seeds of exotic *Pinus pinea, Acacia cyclops* and Black Wattle *A. mearnsii*. 1 bird ate daily 100–150 *Olea capensis* drupes (20–25 × 10–15 mm). In Natal now eats fruit of exotic bugweed *Solanum mauritianum* in preference to most native fruits, even in native forests; 95% of birds (n = 100) fed exclusively on it (Oatley 1984). On Nyika Plateau (Zambia/Malaŵi) *Jasminum* attracts birds to lower levels (Dowsett-Lemaire 1983).

Breeding Habits. Monogamous; territorial. In Zambia, density of nests depends on density of food trees *Afrocrania* and *Olea*, with nest 'territory' of 0·8–3·0 ha. Displaying ♂ makes short rising flight with noisy wing-clapping and glides down in wide circle around ♀; or ♂ makes almost vertical flight up from perch and then nose-dives toward nearby perch with wings outspread; during this flight utters bleating 'meeeh' call. Display-flights daily, whatever the weather. During bowing display ♂ bows very low with neck swollen, walks on branch between bows, then attempts to mount; not clear whether tail is depressed (Verheyen 1955).

NEST: platform of twigs taken from trees and not from ground (1 observation), lined with a few green leaves, diam. (n = 2) 13–19 cm. Built in shrubby evergreens and trees in forests and more open habitats. Height above ground: 1–6 m (n = 9, Zaïre) and 6–14 m (n = 7, South Africa).

EGGS: 1–2 (rarely 2). White; slightly glossy; rather cylindrical. SIZE: (n = 10, South Africa) 36·2–46·5 × 27·3–33·5.

LAYING DATES: Ethiopia, Jan–May, Nov; Sudan, Nov–Jan; S Zaïre, Apr–June; Zambia, Sept–Nov; E Africa: Region A, Jan, Mar–Apr, Aug–Sept, Dec (peak: wet months); Region B, Nov–Mar (dry months); Region D, Feb–Mar, June, Aug, Oct, Dec (peak: dry months); Malaŵi, Sept–Nov; Zimbabwe, Sept, Nov, Jan, Mar, May; South Africa: (Natal) Nov–Mar, May–June (mainly May, 11 of 21 records); (Cape) probably every month.

INCUBATION: 20 days (n = 1) and (n = 2, captivity) 17 days.

DEVELOPMENT AND CARE OF YOUNG: fledging period: *c.* 19 days.

BREEDING SUCCESS/SURVIVAL: preyed upon by Peregrine Falcon *Falco peregrinus* and sparrowhawks, including *Accipiter rufiventris* (?), *A. tachiro* and *A. melanoleucos*.

References
Oatley, T. B. (1984).
Phillips, J. F. V. (1927).
Rowan, M. K. (1983).
Verheyen, R. (1955).

Columba sjostedti Reichenow. Cameroon Olive Pigeon. Pigeon du Cameroun.

Columba sjostedti Reichenow, 1898. J. f. Orn., 46, p. 138; Cameroon Mt.

Forms a superspecies with *C. arquatrix* and *C. thomensis*.

Range and Status. Endemic resident, Cameroon (Mt Cameroon, Manenguba, Bambulu, Oku), E Nigeria (Obudu Plateau) and Bioko. Restricted to highland forests up to 2500 m where fairly common.

Description. ADULT ♂: differs from Olive Pigeon *C. arquatrix* as follows: entire head uniform dark bluish grey; lower breast profusely white-spotted; white spots on belly larger than in *C. arquatrix*; bill yellow, dark red at base; eye yellowish, no bare skin around eye; legs and feet dark purple, nails yellow. Sexes alike. SIZE: (3 ♂♂, 3 ♀♀) wing, ♂ 211–225, ♀ 209–212; tail, ♂ 130–138, ♀ 118–125; bill, ♂♀ 18–20; tarsus, ♂ 25–26, ♀ 22–24.

IMMATURE: differs from immature *C. arquatrix* only in having nape less grey and breast and upper belly more heavily spotted with soft white.

NESTLING: not known.

Field Characters. A large, dark maroon pigeon with grey head and rump, white spots on underparts and wing-coverts, yellow bill and eye. Western Bronze-naped Pigeon *C. iriditorques* smaller, with pale band at end of tail, brown underparts, ♂ with light brown hind collar, ♀ with brown head; smaller Lemon Dove *Columba larvata* of forest floor has whitish face, pale belly, bronze hindneck, plain brown upperparts.

Voice. Not known.

General Habits. Restricted to dense, misty montane forests and forested gullies; on Mt Cameroon at 1000–2500 m, where trees give give way to tree-ferns; on Obudu Plateau at *c.* 1800 m; in flocks of up to 10 travelling considerable distances in search of food. Shy and wary.

Food. Fruits with watery pulp and hard seeds.

Breeding Habits.

NEST: only nest found was loosely built platform, almost flat on top, diam. 18 cm, depth 5 cm; built of quite stout criss-crossed twigs, some mossy, in horizontal fork 8 m high near end of upper branch of soft-leaved tree in thicket in forest clearing.

EGG: 1. Pure white; smooth and glossy; elliptical with 1 end slightly blunt. SIZE: 37·9 × 27·2;

LAYING DATES: Cameroon (Buea, 1950 m) May (Serle 1965).

Columba thomensis Bocage. São Tomé Olive Pigeon. Pigeon de São Tomé.

Columba arquatrix var *thomensis* Bocage, 1888. Jorn. Sci. Lisboa, XII, p. 230; São Tomé Island.

Forms a superspecies with *C. sjostedti* and *C. arquatrix*.

Range and Status. Resident, endemic to São Tomé and Rollas I. (Gulf of Guinea); now extinct Rollas I. as result of heavy deforestation. From sea level to 2024 m, except on eastern part of São Tomé only above 1300 m (de Naurois, unpub.).

Description. ADULT ♂: entire head dark slate grey; nape feathers pointed, blackish at base and blue-grey with pearly sheen at tip; neck tinged dull purple; mantle rich deep maroon; back and rump slaty black; uppertail-coverts very dark brown. Underparts rich maroon with indistinct small whitish spots on belly and flanks. Wings and tail same as in Olive Pigeon *C. arquatrix*. Bill yellowish horn; eye olive-brown; legs and feet yellow. ADULT ♀: like ♂ but duller, with only slight tinge of maroon on breast and wing-coverts. SIZE: wing, ♂ (n = 3) 232–241 (236), ♀ (n = 3) 221–235 (227); tail, ♂♀ (n = 6) 148–173 (161); bill, ♂♀ 21·5; tarsus, ♂♀, 28. WEIGHT: ♂ 520, 530, ♀ 350.

IMMATURE: dark brown, most feathers edged deep chestnut, without any white spotting.
NESTLING: not known.

Field Characters. A large, dark maroon pigeon with dark grey head and black tail; a few small white spots on wing-coverts and belly. São Tomé Bronze-naped Pigeon *C. malherbii* and Lemon Dove *C. larvata* much smaller; former with grey underparts and face, brown undertail-coverts, violet and green iridescence on nape and hindneck; latter with whitish face and belly, bronzy hindneck and mantle, dark brown upperparts.

Voice. Not tape-recorded. A short rolling 'crrrrr'.

General Habits. Inhabits primary forests, usually above 900 m, but exceptionally at sea level. Very tame, sitting about on tree-tops and allowing close approach (Snow 1950). Attracted by bush fires lit by mountain people in order to catch them.

Food. Berries of *Scheffleria mannii* found in 3 crops.

Reference
Collar, N. J. and Stuart, S. N. (1985).

Plate 30
(Opp. p. 449)

Columba albinucha Sassi. **White-naped Pigeon. Pigeon à nuque blanche.**

Columba arquatrix albinucha Sassi, 1901. Orn. Monatsb; 19, p. 68; Moera, eastern Congo.

Range and Status. Endemic resident, dense lowland forests in E Zaïre (primary and transitional forests up to 1500 m) and around Ruwenzoris in W Uganda (Bwamba, 700–1300 m; Kibale, 1800 m); also W Cameroon (Rumpi Hills, 1100 m). Locally common in E Zaïre; few records in Uganda, and in Cameroon only 3 records at 1 locality.

Description. ADULT ♂: crown and lores purplish maroon; nape and hindneck (as far as eye) white; sides of neck and throat ash grey, face darker grey; chin whitish; feathers of upper mantle pointed, dark purple fringed pale grey, giving scaly appearance; mantle purple; back, rump and uppertail-coverts slate grey with indistinct scaly effect; tail with terminal third very pale grey (almost white on underside), basal part brown; central rectrices dark with lighter base. Breast ash grey; belly dark purple, each feather with 2 white spots; vent dark grey; undertail-coverts pale grey. Remiges slaty brown; wing-coverts and underwing dark slate. Bill with dark base and yellow tip; eye buff or yellow with outer ring orange-red; legs and feet red. ADULT ♀: like ♂ but nuchal patch pale grey, upper mantle greyish purple, breast vinous grey. SIZE: wing, ♂ (n = 5) 203–218 (209), ♀ (n = 5) 192–214 (208); bill to feathers, ♂ (n = 1) 17, to skull 25·5; tail (n = 1) 113; tarsus (n = 1) 24. WEIGHT: (n = 3) 280–290.
IMMATURE: upperparts dark earth brown, head greyer, mantle feathers narrowly fringed rusty, uppertail-coverts blue-grey, tail grey with pale grey end. Breast feathers dark brown with pale rufous fringes, belly feathers blue-grey with sandy or rufous fringes.
NESTLING: not known.

Field Characters. Similar to sympatric Olive Pigeon *C. arquatrix*, but ♂ distinguished by large white hind-crown patch (grey in ♀); both sexes have plain wing-coverts, pale ends to tail feathers, brownish red bill, feet and toes.

Voice. Not tape-recorded. A deep, rather quavering, deliberate 'tuu-uu' followed by 3 or 4 'tuu-tu-tu' in decreasing volume; like call of Olive Pigeon. In flight a bleating note very like 'meeeeh' of Olive Pigeon.

Columba albinucha

General Habits. Usually in canopy but often feeds at mid-height in trees, rarely on ground; inhabits thick palm forest; in parties up to 6. Essentially a bird of dense lowland forest and forested slopes, hence largely separated ecologically from Olive Pigeon, although at times the 2 species forage and drink together. Sits motionless in thick foliage. If mildly alarmed freezes in foliage; if more so, flies out of tree unobtrusively. Flight fast and noiseless.

Food. Tree fruits; seeds.

Breeding Habits. Bird shot in Cameroon (Rumpi Hills) Feb contained yolking egg. Nothing else known.

Columba unicincta Cassin. Afep Pigeon; Grey Wood Pigeon. Pigeon gris.

Columba unicincta Cassin, 1860. Proc. Acad. Sci. Philad. for 1859, p. 143; Ogoue River, Gabon.

Plate 30
(Opp. p. 449)

Range and Status. Endemic resident. Lowland forests of W and equatorial Africa, up to 1600 m. S Cameroon to SE Uganda, NW Tanzania (Bukoba), south in Zaïre to 7°S; evidently isolated populations in Liberia and Ivory Coast (Abidjan to Taï, Nimba, Sipilou and Lamto), Ghana, W Cameroon, N Uganda (Mt Morongole), NW Angola (Luachimo R., Junda, Roca Canzele, Quiculungo), and E Angola/S Zaïre/NW Zambia. Rare (or seldom observed) in W and E Africa but frequent and locally common in S Cameroon and Gabon; frequent in NW Angolan coffee forests; sometimes frequent in Zambia (Mwinilunga, where may be partially a migrant).

Description. ADULT ♂: crown, nape, hindneck, sides of neck and face pale grey; foreneck pale grey becoming pinkish on upper breast; chin and throat white; mantle grey, becoming slate grey on back and rump, each feather edged pale grey giving scaly appearance; uppertail-coverts dark grey; tail broadly tipped blackish, a broad white band across outer rectrices, grey across central ones. Breast pale vinous pink turning pale grey on sides of belly and flanks; belly and undertail-coverts white. Primaries, secondaries and tertials blackish, paler on outer webs; scapulars and wing-coverts slate, each feather narrowly edged pale grey. Bill slate blue with greyish tip; eye orange-red to crimson, naked area around eye crimson; legs and feet slaty blue-grey. ADULT ♀: like ♂ but breast duller pink. SIZE: wing, ♂ (n = 12) 205–220, ♀ (n = 10) 200–217; tail, ♂ (n = 4) 104–136 (118), ♀ (n = 3) 108–120; bill from feathers, ♂♀ 18–22; tarsus, ♂♀ (n = 22) 23–27. WEIGHT: ♂ (n = 5) 357–490 (423), ♀ (n = 3) 356–360 (358).

IMMATURE: markedly darker on upperparts, breast brownish vinous. Fledgling has crown feathers striped blackish and tipped rufous-brown, lesser wing-coverts broadly edged rufous, primary coverts edged white, remiges and rectrices tipped white.

NESTLING: undescribed.

Field Characters. A large, true forest pigeon, difficult to locate in canopy of trees, and often identified only by deep far-carrying coo. In flight appears a large, pale grey pigeon with blackish wings and a broad white subterminal band across black-tipped tail. At close range scaly grey upperparts, white belly and undertail-coverts (and large size) distinguish it from large sympatric forest pigeons, namely Cameroon Olive Pigeon *C. sjostedti*, White-naped Pigeon *C. albinucha* and Western Bronze-naped Pigeon *C. iriditorques*.

Voice. Tape-recorded (44, 244). A series of drawn-out, medium-pitched 'hoo' notes, delivered slowly and regularly in a monotone, each coo of *c.* 1 s duration (probably advertising coo). May give *c.* 20 coos in unbroken succession.

General Habits. Lives in canopy of tall trees in primary, secondary and montane forests; occasionally in clearings; canopy of moist evergreen and riparian forests, also *Marquesia* thickets away from water (Zambia); coffee forest (Angola).

A large pigeon with strong, fast flight; over forest strongly reminiscent of Wood Pigeon *C. palumbus*. Generally in small parties of 3–5 in trees, but in flocks of 15–30 picking fruit from low *Solanum torvum* bushes in clearings (when food scarce, Gabon: A. Brosset, pers. comm.). In Zambia (Salujinga, R. Lisombo, Sakeji) frequent Aug–Nov but scarce or absent Mar–May, so may move north in rains (Benson *et al.* 1971).

Food. Fruit of *Solanum torvum*, *Musanga*, *Coelocaryon*, *Eisterya* (Gabon: A. Brosset, pers. comm.); *Sapium mannianum* and various species of *Ficus*; and seeds (Zambia). 1 specimen (Zaïre) had many small winged termites.

Breeding Habits. Builds stick nest high in tree; clutch is 1 egg and observations of singing and newly fledged young indicate dry season nesting (June–Sept) in Gabon (A. Brosset, pers. comm.); moreover, during the short dry season (Jan–Feb) there it resumes singing. Breeds Zaïre July and probably Jan–Apr, and Uganda probably Mar–Apr.

470 COLUMBIDAE

Plate 30
(Opp. p. 449)

***Columba palumbus* Linnaeus. Wood Pigeon. Pigeon ramier.**

Columba palumbus Linnaeus, 1758. Syst. Nat. (10th ed.) i, p. 163; 'Europa, Asia', restricted locality: Sweden.

Columba palumbus

Range and Status. W Palearctic: NW Africa, Portugal, Ireland and Norway east to about 70°E in central Asia and 90°E in Himalayas.

Resident in and European visitor to NW Africa. Widely but sparsely distributed as breeding bird in N Morocco and N Algeria (Rif, Middle and High Atlas up to 2500 m; Anti-Atlas and Sous valley; Tlemcen and Saïda Mts; abundant in cedar forest up to 1900 m at Aurès); in winter locally common (and in NE Morocco abundant), with migratory movements of tens and occasionally thousands of birds seen most years, particularly near Moroccan coast. In Tunisia scarce resident in NW south to Le Kef, and irregular winter visitor Nov–Mar. Vagrant Mauritania (Nouakchott, Apr 1981).

Description. *C. p. excelsa* Bonaparte: NW Africa. ADULT ♂: entire head grey-blue; nape and hindneck metallic green with purplish reflections; sides of neck metallic green with conspicuous white patch; mantle grey-brown; back, rump and uppertail-coverts grey-blue; tail dark grey, with broad blackish end. Breast brown-purple merging to pale grey belly and undertail-coverts. Primaries dark brown, outer webs broadly edged white; secondaries brown washed grey; scapulars and tertials grey-brown; primary coverts dark brown; outermost wing-coverts white, forming broad white bar across wing. Underwing-coverts grey-blue. Bill yellow with horn tip and pink basal half; eyes yellowish white; legs and feet vinous pink. Sexes alike but ♀ usually has smaller white neck-patch and less pink breast. SIZE: wing, ♂ (n = 17) 251–268 (257), ♀ (n = 6) 244–259 (253). See *palumbus* below.

IMMATURE: greyer and paler than adult; neck lacks metallic colours and white patch.

NESTLING: skin lead blue covered with sparse tufts of straw-coloured down. Bill dark grey, pink inside, long, pliable, spatulate, with upper mandible fitting into much broader lower mandible; tip white with narrow black transverse line.

C. p. palumbus Linnaeus: Europe; winter visitor to NW Africa. Separable only by smaller av. size: wing, ♂ (n = 27) 240–258 (251), ♀ (n = 19) 238–258 (248); tail, ♂ (n = 13) 158–185 (169), ♀ (n = 9) 158–172 (164); bill, ♂ (n = 19) 20·0–23·5 (21·7), ♀ (n = 9) 19·7–21·3 (20·4); tarsus, ♂ (n = 13) 30–34 (31·5), ♀ (n = 9) 29–35 (31·8). WEIGHT: (Sweden, Aug–Sept) ♂ (n = 17) 465–613 (539), ♀ (n = 11) 420–600 (498); (Germany, Feb–Mar) ♂ (n = 12) 466–534 (498), ♀ (n = 9) 260–498 (471).

Field Characters. A large pigeon distinguished from other NW African pigeons by white crescent across wing and (adult only) white patch on side of neck. Distinctive song. Migrates in quite high-flying flocks.

Voice. Tape-recorded (62, 73). Advertising call a series of 2–4 phrases, each lasting 2–3 s and consisting of 5 notes: a short introductory 'cu', an emphatic, deep-throated 'cooo', and 3 softer coos, with gap before last 2: 'cu-cooo-cooo, coo-coo'. The series often ends with a single short 'cu'.

General Habits. Inhabits evergreen and deciduous oak, cedar, pine and thuja forests; riverine stands of poplar and ash; groves of olives and argan (*Argania sideroxylon*); thuja (*Terraclinus articulata*) in Anti-Atlas; and (Tunisia) plains with oak and Aleppo pines. Breeds (Morocco) in forests above 800 m with tall trees and at least 500 mm rainfall. Flies strongly, straight and fast with regular wing-beats; when disturbed at perch dashes off noisily. In Middle Atlas Mts flies down from mountain forests to forage by day in plains. Feeds in trees and on ground; surprisingly agile in trees where can cling to small branches, even upside down, in spite of weight. Wary. Solitary and in pairs, but after breeding, congregates in parties of hundreds, joined by European birds; winter flocks wander south in Atlas Mts (Morocco).

Food. Not studied in Africa; eats olives. In Europe, mainly clover *Trifolium* leaves, also cereals at spring sowings (prefers wheat to barley), weed seeds in May–June, tree leaf and flower buds, mainly peas and beans *Pisum* in July–Oct, and in autumn tree fruits, mainly of beech *Fagus* and oak *Quercus*, and stubble grain (also a few insects, worms and molluscs) (Murton *et al.* 1964).

Breeding Habits. Monogamous, territorial. In display flight ♂ rises steeply, claps wings loudly and then glides downwards with wings outstretched horizontally. In bowing display (very like that of Stock Dove *C. oenas*) neck is swollen and head bowed very low; tail is raised almost vertically, spread during raising and closed when it reaches highest position. Tail not depressed and spread when approaching another bird. In aggression, head held erect, plumage sleeked and wings part opened; ♂ and ♀ may fight standing side-by-side on nest, hitting each other with wings. Submissive posture is with head lowered, neck withdrawn, plumage fluffed, and bill pointing down. ♀ on nest nibbles ♂'s plumage. Common sexual behaviour is touching mate's bill and face ('billing'). Bouts of billing alternate with both birds making repeated little nods (Murton 1960). Bowing display frequently leads to billing and copulation.

NEST: platform of dry twigs, occasionally with lining of grasses and leaves, diam. 25–40 cm, depth 9–12 cm. Usually placed at least 2 m high in almost any tree (rarely on ground) (Europe).

EGGS: 2; elliptical-ovate; rather glossy; white. SIZE: (n = 100, Britain) 36·2–47·8 × 25–30 (41 × 30). WEIGHT: 16–22 (19).

LAYING DATES: Morocco, Apr–May.

INCUBATION: intermittent with 1st egg, intense with 2nd egg; ♂ incubates chiefly by day and ♀ at night. Period: 16–17 days (Europe).

DEVELOPMENT AND CARE OF YOUNG: fed on crop milk, then pulped seeds, by both parents; 2 young feed together, both plunging bill deep into parent's throat at same time. Growth rapid; fully feathered by c. 14 days, yellow down adhering to tips of juvenile feathers. Wing-feathers lose adhering down first, at stage when feathered head and neck still quite downy, and unfeathered belly and tail region thickly downy (Murton 1960). Fledging period: 28–29 (exceptionally 34) days (Europe), but young cannot fly until 35th day. Well-feathered nestlings still retain blackish, swollen bill, quite different in shape from adults'.

BREEDING SUCCESS/SURVIVAL: killed for sport and as pest. Predators (Europe) include Carrion Crow *Corvus corone*, Goshawk *Accipiter gentilis*, Sparrowhawk *A. nisus*, Red Kite *Milvus milvus* and Long-eared Owl *Asio otus*.

References
Cramp, S. (ed.) (1985).
Glutz, U. N. and Bauer, K. (1980).

Columba oenas Linnaeus. Stock Dove. Pigeon colombin.

Plate 30
(Opp. p. 449)

Columba oenas Linnaeus, 1758. Syst. Nat. (10th ed.) i, p. 162; Europe.

Range and Status. W Palearctic: NW Africa, Spain, Ireland and Norway east to about 85°E in central Asia.

Resident in and European visitor to NW Africa. Sparse resident in N Morocco from near coast inland to Rif and Middle and High Atlas Mts; locally common after breeding; occurs erratically up to 2300 m on central plateau but not resident there. Probably some European birds winter near Straits of Gibraltar (not proven by ringing). In Algeria, old records from Tilremt but only 2 recent observations, a (migrating?) flock and 1 singing in Babor hills. Status uncertain in Tunisia; 1 dubious breeding record; regular autumn and spring migrant at Cap Bon in small numbers; only 2 midwinter records (Jan). Vagrant in Libya (Tobruk, Benghazi), NE Niger (Dec, Fairon 1971), and Egypt. Population limited by availability of old hollow trees necessary for nesting.

Description. ADULT ♂: crown, nape and hindneck mid grey; sometimes a metallic green half-collar across hindneck; sides of neck shifting metallic green or purple; foreneck dull purplish grey; face, throat and chin mid grey; mantle and scapulars slate grey tinged brown; back, rump and uppertail-coverts blue-grey; tail dark blue-grey, with broad blackish tip. Underparts blue-grey. Primaries dark brown, inner webs browner, both webs narrowly bordered whitish, basal half of inner primaries blue-grey; secondaries blackish, blue-grey proximally and on outer webs; tertials brown-grey, the 3 innermost with subterminal black patch on outer web; primary-coverts dark brown, rest of wing-coverts grey-blue, innermost greater coverts with subterminal black patch and usually 2–3 inner median coverts with smaller black patch (the patches make short wide bars on inner part of wing). Under-wing blue-grey. Bill yellowish horn, vinous at base; eye dark brown, eyelids pink; legs and feet pinkish mauve. Sexes alike. SIZE: (12 ♂♂, 5 ♀♀) wing, ♂ 210–226 (214), ♀ 209–225 (215); tail, ♂ 102–115; bill, ♂ 19–21; tarsus, ♂ 28–32. WEIGHT: ♂ (n = 8, Germany) 303–365 (337), ♀ 286·5, 290.

IMMATURE: very like adult but browner; metallic patches on neck sides absent; black marks on some inner wing-feathers absent.

NESTLING: lead-blue skin sparsely covered with coarse, hairy, deep yellow down; underparts mainly naked; inside mouth pink.

Field Characters. A small, pale grey NW African pigeon with black tail-tip and blackish wing-tips, likely to be confused only with Wood Pigeon *C. palumbus* and Rock Dove *C. livia*. Smaller and rather darker than Wood Pigeon, and lacks its white neck and wing marks. Rock Dove is same size, with white rump, and much longer black bars on inner wing.

Voice. Tape-recorded (62, 73). Advertising call a double note 'coo-oo' about same pitch as in Wood Pigeon but with different rhythm; rather less guttural; 'coo' is emphatic and separated from shorter and less accentuated 'oo' by short, deep 'u', sounding at close quarters more like 'coo-u-oo'; call is repeated once per s for 8–10 s. Display coo a faint, buzzing 'oo', synchronized with bowing and interspersed with clicks, audible only at close range.

in succession, each preceded by ♂ raising head with iridescent neck side feathers erected (see **A**). Unlike *C. livia*, ♂ does not turn around or away from ♀. Display flight horizontal or nearly so, in circles above or near nest-site from which it usually starts. Claps wings in flight (rather louder and slower than in *C. livia*). After copulating ♂ usually performs bowing display.

NEST: in holes in old pines, cedar, oak and evergreen oak trees in high-altitude forests and in pistachio trees in subdesert; eggs sometimes laid on bare wood, but usually on thin platform of twigs, roots and straws. In E Morocco, the same old-cedar cavities used every year and eventually fill up, so that good nesting places are declining and may limit the population (Brosset 1961). Exceptionally nests on ground (Europe).

EGGS: usually 2. Blunt ovals; slightly glossy; white or sometimes creamy. SIZE: (n = 100, Britain) 34–43 × 26·5–31 (37·9 × 29). WEIGHT: (n = 6, Belgium) 14·7–17·5.

A

General Habits. In breeding season inhabits pine, oak and cedar forests with old trees, between 1000 and 2300 m; after breeding inhabits open woodlands and fields. Feeds essentially on ground. In Middle Atlas Mts flies down to valleys to forage; large parties congregate after breeding and move south to foothills of High Atlas. Extremely sedentary (in Britain).

Food. In Europe primarily weed seeds (*Sinapsis, Brassica, Stellaria, Polygonum, Chenopodium*), also some cereal grain, and 1–5% leaves, tree buds, fruits and animal matter; diet of nestlings same as that of adults (Murton *et al.* 1964).

Breeding Habits. Monogamous; nests solitarily in tree-holes. In bowing display ♂ bows very low near ♀, with tail almost vertical; tail spread whilst being raised but closed when fully elevated. Several bows may follow

LAYING DATES: Morocco, May–June.
INCUBATION: carried out by both sexes; begins with 1st egg. Period (Britain): 16–18 days.
DEVELOPMENT AND CARE OF YOUNG: fed with crop milk by ♂ and ♀. Leaves nest prematurely, staying nearby, making fledging period difficult to measure; has been given as 18–30 days (in Britain av. 26 days).
BREEDING SUCCESS/SURVIVAL: in Europe many killed by hard weather, predators (Goshawk *Accipiter gentilis*, Red Kite *Milvus milvus*, Peregrine *Falco peregrinus*) and by flying into power lines. Annual adult survival (Britain) 53–57% (O'Connor and Mead 1984).

References
Cramp, S. (ed.) (1985).
Murton, R. K. *et al.* (1964).
O'Connor, R. J. and Mead, C. J. (1984).

***Columba oliviae* Clarke. Somali Stock Dove; Somaliland Pigeon. Pigeon de Somalie.** Plate 30 (Opp. p. 449)

Columba oliviae Clarke, 1918. Bull Br. Orn. Club 38, p. 61; Dubbar, Somalia.

Range and Status. Endemic resident; restricted to arid hills near coasts of N and NE Somalia, up to 800 m, where local and uncommon.

Description. ADULT ♂: crown, nape and upper hindneck dull purplish or vinous grey; sides of neck, foreneck, face, chin and throat pale grey; lower hindneck and upper mantle coppery dark pink with slight greenish sheen, half-collar 1–2 cm broad; mantle, scapulars, tertials and wing-coverts isabelline grey; back and rump pale bluish grey; uppertail-coverts dark bluish grey; tail blue-grey with narrow blackish bar 4 cm from tip and broadly tipped (2 cm) black. Underparts pale grey; undertail-coverts grey. Primaries light greyish isabelline tipped brown; these tips *c*. 2 cm broad, ill-defined and almost concolorous on 3 outermost; secondaries like primaries but terminal bar darker and 3 cm broad; scapulars indistinctly shot bluish green; tertials concolorous with mantle; wing-coverts like mantle but some specimens show black spots; underwing pale grey. Bill black, cere powdery white; broad purple-red naked area around eye; eye yellowish; legs and feet pink. Sexes alike. SIZE: wing, ♂ (n = 9) 193–211 (201), ♀ (n = 5) 195–203 (201); tail, ♂ (n = 5) 86–95 (90·2), ♀ (n = 5) 90–96 (92·8); bill to feathers, ♂ (n = 5) all 22, ♀ (n = 5) 21·5–22 (21·9); tarsus, ♂ (n = 5) 26–27·5 (26·8), ♀ (n = 5) 25–27·5 (26·1).

IMMATURE AND NESTLING: not known.

Field Characters. A small, bluish-grey pigeon with isabelline-grey back and wings, black-tipped grey tail, dark brown trailing edge to wing, dull purplish grey crown, and shiny pink-rufous upper mantle. Red orbital skin visible at close quarters. On ground resembles a *Streptopelia* dove more closely than a *Columba* pigeon; but told from *Streptopelia* spp. by purplish grey crown.

Voice. Not tape-recorded. Display coo described as 'wuk-wuk-wuk-oh, wuk-ow', the 'wuks' being typical pigeon coos and the 'oh' a deep growl.

General Habits. Inhabits rocky and sandstone hills, mainly at 300–800 m altitude, almost waterless and with scant vegetation 'khansa', Solsolaceae, 'arman', Acacia). Forages on ground in small parties, at times in larger flocks when food abundant (e.g. grain spilled near camels). Not shy. Locally migrant, since evidently absent from Guban cliffs in hot weather (May–Sept); habitat between May and Sept still unknown but suspected to be in mountainous region extending eastwards to Ras Asayr (Cape Guardafui).

Food. Seeds, berries and cereals.

Breeding Habits. Displays not known.
NEST: in recess in roof of caves on hillsides, nest-site almost dark.
EGGS: 2 nests found, each with 1 egg (not known whether clutch complete).
LAYING DATES: May, Aug.

Reference
Archer, G. and Godman, E. M. (1937).

***Columba guinea* Linnaeus. Speckled Pigeon; Speckled Rock Pigeon. Pigeon de Guinée.** Plate 30 (Opp. p. 449)

Columba guinea Linnaeus, 1758. Syst. Nat. (10th ed.), p. 163; Senegal.

Range and Status. Endemic, resident. Ranges from Senegambia and S Mauritania (Rosso) to Ethiopia (to 16°N) and NW Somalia, south to derived savannas bordering rain forest zone and through Uganda and W Kenya to S Tanzania and N Malawi. Another race ranges from NE Zimbabwe and SW Angola to Cape. From sea level up to 3000 m. Frequent to abundant; flocks of several thousand occur in E–central Mali, and 83,500 estimated in 635 ha around Makalle (Ethiopia). An increasingly common urban bird throughout W and E Africa, roosting and nesting in high density on some buildings; several hundred on hospital roof in Nigeria (Zaria). Has recently spread dramatically into urban areas in E and S Somalia, including Mogadishu. Par-

Columba guinea

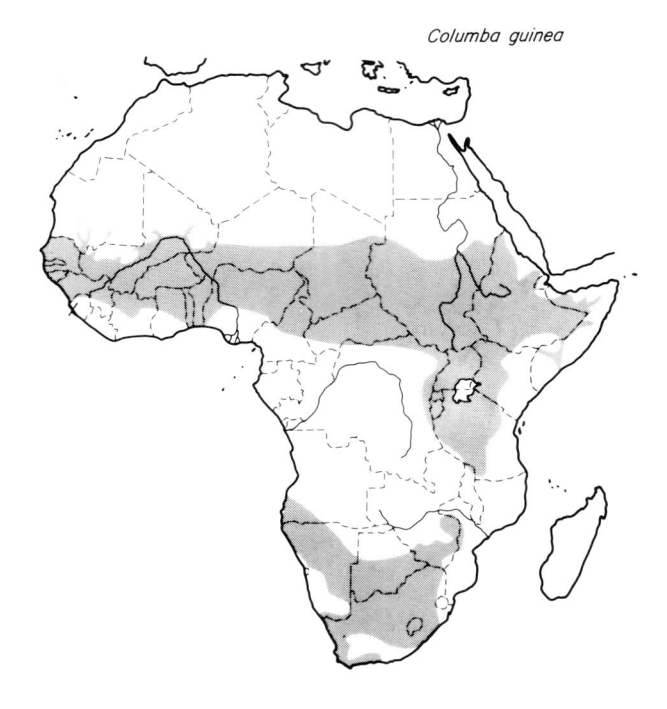

ticularly abundant in cereal-growing plains. However, distribution patchy: evidently absent from some areas which appear suitable. Rarely complained of as pest, for it forages mainly in harvested fields. Locally shot for sport.

In southern Africa Speckled Pigeon co-exists in towns and city suburbs with Rock Dove/Feral Pigeon *C. livia*. Elsewhere the 2 species evidently displace each other, and in many cities one is common and other scarce. In small region of geographical overlap near southern border of Sahara, Speckled Pigeon lives around buildings and palms and Rock Dove around cliffs and wells.

Description. *C. g. guinea* Linnaeus: W and E Africa south to Malaŵi. ADULT ♂: entire head and upper neck grey, darker on crown; neck, upper breast and scapular feathers stiff and bifurcate, chestnut with grey tips, faintly glossed greenish; mantle, scapulars and lesser wing-coverts vinous chestnut; rump and uppertail-coverts pale grey (paler tips giving scaly appearance); tail clear grey with broad black tip and indistinct dark grey subterminal bar; underparts grey; remiges brownish grey; greater coverts vinous chestnut with triangular white tips; underwing grey. Bill blackish, cere whitish; eye yellow to light brown, broadly surrounded by wine red naked skin as far as ear; legs and feet purplish pink. Sexes alike, ♀ slightly duller. SIZE: wing, ♂ (n = 53) 212–240 (221), ♀ (n = 54) 210–230 (221); tail, ♂ (n = 53) 110–125 (114), ♀ (n = 54) 105–125 (115); bill, ♂ (n = 9) 21·4–26·7 (24·1), ♀ (n = 5) 21·5–24·2 (22·9); tarsus, ♂ (n = 9) 30·5–37·0 (34·1), ♀ (n = 5) 31·6–35·4 (34·2). Ethiopian birds large: wing, ♂ av. 247, ♀ av. 235. WEIGHT: (Nigeria, decropped) ♂ (n = 53) 288–383 (337), ♀ (n = 54) 277–364 (320); (Ethiopia) ♂ (n = 9) 344–371 (359), ♀ (n = 5) 307–383 (340).

IMMATURE: brown where adult grey, new grey feathers showing among brown ones; neck feathers not bifurcate; naked area around eye chocolate brown.

NESTLING: skin reddish pink (later turns blackish,) well covered with yellow down; bill dark with blackish ring near white tip, egg-tooth yellow; claws white.

C. g. phaeonotus Gray: from 17° S (SW Angola and Zimbabwe) to Cape Province. Crown slate grey; mantle and wings chocolate brown (not vinous-chestnut), rump dark grey, underparts slate. Slightly larger: wing, ♂♀ 219–233 (226). WEIGHT: (South Africa: Cape Town) monthly means 315–340 (Jan–Mar) and 365–390 (May–July).

Field Characters. A large pale grey and vinous-chestnut pigeon of farmland and gardens; on buildings, in trees; feeds on ground. Much larger than the several species of *Streptopelia* doves which occur in towns; the only other urban pigeon is Rock Dove/Feral Pigeon *C. livia* in Sahara and cities throughout Africa. Speckled Pigeon readily distinguished from all of these by vinous chestnut plumage with heavily white-spotted wings, also large patch of bare red skin around eye, black bill and dark red legs. Olive Pigeon *C. arquatrix* is similarly spotted dark vinous, but inhabits forest and has yellow bill, eye-skin and legs. Head, rump and belly pale grey, rump in strong contrast with dark upperparts, particularly in flight. Often lives close to man, singing and courting on rooftop or windowsill, but flocks foraging in fields shy and hard to approach.

Voice. Tape-recorded (10, 20, 44, 74, 75). Advertising coo like that of Rock Dove/Feral Pigeon, but higher pitched and twice as fast, c. 120/min; song begins with faint muffled notes and quickly increases in emphasis and volume. Display coo is emphatic, somewhat explosive 'coo-co', followed by short muffled note, lower in pitch and inaudible at distance. Distress call, a plaintive 'o-orr'.

General Habits. Inhabits variety of open country: cereal farmland, pasture, savanna dotted with inselbergs (kopjies), gardens. Eastern and southern populations mainly associated with rocks and western populations with *Borassus* palms and Baobabs. In Ethiopia abundant in grassland at 300–2750 m and frequent to common in wooded country at 1200–2400 m; also occurs in deserts with annual grasses, rocky terrain, cliffs, gorges, and coastal sand dunes. Common in and around villages (particularly with 2-storey stone dwellings) and city suburbs.

A strong flier. Leaves roost at sunrise to forage up to 25 km away; urban birds generally forage in fields outside town. Forages on ground, in parties of up to 700 (av. flock size 4·05 birds, Ethiopia), often in company of other pigeons and doves. Has slow waddling walk, and makes short runs. Usually feeds in harvested cereal fields; also on other crops and under fruiting trees, and in Ethiopia in market places mixing with White-collared Pigeons *C. albitorques*. Roosts on cliffs, buildings, on arches under road bridges, and in tall trees; also recently, in southern Africa, on newly ploughed fields.

Food. Seeds and small fruits; wild foods include figs and *Tribulus* seeds; but most of diet is cash crops at planting and harvesting seasons: wheat, *Eragrostis tef*, lentils, barley, field beans, sorghum, maize (Ethiopia), ground nuts, guinea-corn and millet (Nigeria); bread and grain in markets and *Helianthus* seeds picked from standing flower heads (South Africa); snails. Eats coarse salt put out for cattle. In Nigeria, weight of crop content lowest in Mar–Apr (av. 7 g: wild seeds) and highest Nov–Jan (av. 21 g: almost exclusively ground-nuts and guinea-corn) (Shotter 1978).

Breeding Habits. Monogamous. Nests solitarily or quite gregariously. Display flight consists of 2–3 wing-claps then a short glide before clapping is repeated. Clapping flight may occur without interspersed glides. When pair flies from ground to nest, only 1 bird may clap. Immediate vicinity of nest is defended by ♂ and ♀; and sometimes a larger area. Performs bowing display (see **A**).

NEST: an untidy, substantial platform of 12 to several hundred dry twigs gathered within 100 m of nest. Twigs are 3–29 cm long and *c.* 5 mm diam.; in Senegambia, Ethiopia and South Africa uses twigs mainly of *Themeda triandra*, *Syringa*, *Melia azedarach* and *Cassia siamea*. Sometimes incorporates scraps of plastic refuse and bits of wire (total mass up to 360 g—Kok 1984). Nest has shallow cup of smaller twigs, also of the grass *Eleusine floccifolia*. Sited on buildings (windowsills, holes in masonry, gutters) and in trees (*Borassus* palms, Nigeria, Mali, and on old nests of Abdim's Stork *Ciconia abdimii*, Mali). Nest weighs 60–167 g; external diam. 30–37 cm, diam. of shallow cup 11–14 cm; depth 3–5 cm at centre and 5–10 cm at edge. One bird carries material in bill, while other stays by nest, receives material and builds nest. 26 twigs were brought in 60 min. Of 162 sites (South Africa) 30% were used for 2 nestings per season, 10% for 3 and 1% for 4 nestings. Time between 2 layings in same nest, (n = 33) 5–77 (17·8 ± 5·8) days.

EGGS: 1–3, laid about midday at 24 h intervals. Of 269 clutches (Cape Town), 266 were of 2 eggs, 2 of 1 egg and 1 of 3 eggs; most other records of clutches of 1 egg refer to addled egg or incomplete clutch (Elliott and Cooper 1980). Shape variable, usually ovoid; chalky or with slight sheen; white. SIZE: (n = 62, South Africa) 34·5–41·1 × 25·0–31·2 (36·4 × 27·6); (n = 76, Ethiopia) 38·3 ± 1·4 × 22·8 ± 0·9. WEIGHT: (n = 5) 14·9 ± 0·6; initial weight, Ethiopia, *c.* 20.

LAYING DATES: all months but mainly dry season. Senegambia, all months, peak Dec–Apr; Mali, Jan–June, Oct; Nigeria, mainly Sept–Apr; Sudan, Jan–May; Ethiopia, July–Apr, peak Oct–Mar; E Africa: Region A, Feb, Mar, Oct–Dec; Region B, Nov; Region C, Jan, Feb, Aug; Mau Narok (Kenya), all months; Zimbabwe, Feb–Nov, peak Mar–May; South Africa, (Orange Free State and Natal) all months, (Transvaal) peak Apr and July–Oct, (Cape) Sept–Feb with peak varying from Oct–Nov to Jan–Feb.

INCUBATION: carried out by both sexes, in 2 shifts per day: one bird incubates for 6 h (from about 10·30 h to 16·30 h) and the other for 18 h. Incubation only slight until 2nd egg laid. Incubating bird tame. Period: (n = 5, Cape) 14–16 (14·8) days, (Ethiopia) 17–18 days.

DEVELOPMENT AND CARE OF YOUNG: on 4th day eyes open (iris light blue), dark quills appear in feather tracts, skin turns blackish. Young brooded closely for 6 days; brooding ceases by 10th day. Both parents feed young. By 6th day still weak, but attempts to stand and snaps bill when disturbed. On 9th day some yellow down still present; disturbed bird puffs out breast and snaps bill vigorously. Plumage well developed by 14th day and fully developed by 20th day except for head; on 20th day primaries vaned for three-quarters of length. Young weighs *c.* 20 g on 2nd day and grows fast: *c.* 100 g on 7th, *c.* 200 g on 12th and *c.* 250 g on 22nd day (Wilson and Lewis 1977). Nestling period: (n = 5 nests) 20–25 (23·6) days; av. probably 25 days if undisturbed. Fledgling stays near nest until it can fly; some able to fly at only 20 days.

BREEDING SUCCESS/SURVIVAL: hatching success (n = 600 eggs) 66%; chick rearing success (n = 354 hatchlings) 83% (Cape Town: Elliott and Cooper 1980). 23% of nests are deserted shortly after laying. Nestlings killed by Black Magpie *Ptilostomus afer* (Senegambia).

References
Elliott, C. C. H. and Cooper, J. (1980).
Rowan, M. K. (1983).
Wilson, R. T. and Lewis, L. G. (1977).

476 COLUMBIDAE

Plate 30
(Opp. p. 449)

Columba albitorques Rüppell. **White-collared Pigeon. Pigeon à collier blanc.**

Columba albitorques Rüppell, 1837. N. Wirbelth., p. 63; Taranta Mts.

Range and Status. Endemic resident in rocky highlands of Ethiopia, in NE (Eritrea), W and SE, south to Jimma and Arussi Mts; from 1800 to over 4000 m. Also in cities (Addis Ababa). Abundant; in flocks of 50–200.

Description. ADULT ♂: crown and nape very dark grey; a white collar, c. 10 mm broad when neck retracted, at base of nape, running from ear to ear (and occasionally, very attenuated, onto throat); below white collar a broad collar of pointed hackles with faint greenish iridescence and silvery tips (except on sides of neck and foreneck); face and chin very dark grey; mantle slaty brown; back and rump dark bluish grey; uppertail-coverts browner than rump; tail blackish with broad subterminal dark grey band. Remaining underparts slaty grey; undertail-coverts slightly darker. Primaries grey-brown, with a very obscure narrow blackish bar 10 mm from tip of 4th to 7th; distal one-third of outer web of fresh 2nd–4th primaries sometimes white; proximal one-third of inner primaries sometimes pure white. Secondaries, scapulars and tertials grey-brown. Outer greater primary coverts dark grey, grading through pale grey middle ones to 2–5 pure white inner ones. Some median primary coverts sometimes white. Black spots on outer webs of median and greater secondary coverts (closed wing looks sparsely black-spotted); underwing dark grey. Bill black, cere powdery white; eye dark brown; legs and feet red. Sexes alike but white collar narrower in ♀. SIZE: wing, ♂ (n = 10) 215–230 (221), ♀ (n = 6) 210–228 (219·5); tail, ♂ (n = 10) 81–98 (88), ♀ (n = 7) 85–98 (90). WEIGHT: ♂ 292, ♀ 262.
IMMATURE: crown feathers have pale edges, collar feathers lack iridescence.
NESTLING: undescribed.

Field Characters. A large, sooty grey rock and town pigeon of highland Ethiopia, with sharply defined white hindneck collar reaching from ear to ear, and white patch in wing (mainly inner greater primary coverts). All lower neck feathers, below level of white patch, are lanceolate hackles. Irregular black spots on dark brown-grey folded wing, variable in size and number, usually c. 15. In flight wings make whistling or creaking noise.

Voice. Not tape-recorded. In captivity display coo is often 'hu ... hô-hô', audible only within 30 m. Advertising coo is long, soft, drawn-out 'coo-oo' or 'gu-hu-ho' or 'ooh, ooooh' or 'ooh, ooh, ooh, oooohooh'.

General Habits. Grassland, cultivation, woodland and other montane habitats above 1800 m; always near buildings, cliffs and gullies, rock formations or villages; common in Addis Ababa and other large towns.

Occurs singly, in pairs and small flocks, but at feeding and resting places often forms large mixed flocks with Speckled Pigeons *C. guinea*. White-collared Pigeon predominates above 2400 m. Forages in fallow and harvested fields up to 4400 m and in Addis Ababa market places and university campus grounds, in flocks of hundreds. Large flocks mixed with *C. guinea* do not break up when scared up from ground. Drinks on flat rocks and sandbars by rivers. Flight fast and strong; able to fly vertically up cliff and building faces (like Rock Pigeon *C. livia*, whose ecological niche White-collared Pigeon occupies, but unlike *C. guinea*). Momentarily hovers before landing on crowded ledge. Freely enters buildings, cliffs, gullies and caves, at about 2000 m altitude.

Daily movements from low roosting to high foraging stations are impressive. In mornings, about 08.00–09.00 h, flocks of 25–50 slowly ascend about 1500 m up escarpment of Simien Mts, in spiralling flight, resting at top before flying across plateau to foraging grounds. In evening flocks of 50–100 return to top of escarpment (e.g. Geech plateau, Sederek) and small groups make spectacular headlong descent at speed estimated at 120 km/h, travelling close to ground and, on reaching sheer precipice, flying almost vertically downward. Of 232 birds 33% descended singly or in pairs and 67% in flocks of 3–11 birds.

Food. Grain, bread. Feeds in barley fields and on newly sown wheat (June).

Breeding Habits. Solitary nester, presumably monogamous. Indulges in aerial chasing. Display flight is a few noisy flaps followed by brief level glide with wings stiffly raised in shallow V, and may end in upward swoop.
NEST: sited mainly on sheltered ledges of buildings, also inside buildings (attics, internal cornices); probably also in dark natural rock crevices and caves.
EGGS: 2. Creamy white with slight gloss. SIZE: (n = 2) 34·5 × 22–25; (n = 10, in captivity: Taibel 1954) 37·3 × 27·6.
LAYING DATES: every month.
INCUBATION: period c. 16 days (in captivity).
DEVELOPMENT AND CARE OF YOUNG: fledging period 27–31 days (in captivity).

References
Boswall, J. and Demment, M. (1970).
Taibel, A. M. (1954).

Columba livia Gmelin. Rock Dove; Feral Pigeon. Pigeon biset. Plate 30

Columba domestica β *livia* Gmelin, 1789. Syst. Nat., 1, ii, p. 769; no locality given. (Opp. p. 449)

Range and Status. Northern Africa and Mediterranean basin to Kazakhstan, W Mongolia, Kashmir, Nepal and India; sea coasts and islands from Faeroes and Britain to Madeira, Canary and Cape Verde Is. Resident.

Atlantic, Mediterranean and Red Sea coasts of N Africa (except N Tunisia and Gulf of Sirte, Libya) from Morocco (Essaouira Is. and probably Tarfaya) to Ethiopia (Asmara, Massawa, Assab and Madote Is.), extending inland up to 3000 m in Atlas Mts and Red Sea hills; in Sahara, widespread in Hoggar, Aïr, Tibesti, Ennedi, Darfur and Kordofan Mts (Algeria, Niger, Chad, Sudan), also an abundant lowland population in Mali between Mandingo Mts and Hombori. Coastal and inland records south to 10°N in Senegambia, Guinea (Loss Is.), Ghana (Gambaga) and Nigeria; and species is probably far more more widespread than shown in map. Locally abundant on N African coast, e.g. 150 pairs on Cala Iris Is. (Morocco), flocks of 800 at Ouarzazate and 3000 in High Atlas (Morocco); abundant in Mali where flocks of 10–20,000 assemble to drink; frequent to common in rest of range.

Wild Rock Dove is ancestor of Feral Pigeon of European cities, which now thrives also in numerous tropical African towns south at least to Angola and Mozambique. Feral Pigeons are highly variable in plumage but wild-type variants are frequent; in E Africa inhabit Kampala, Nairobi, Isiolo, Karatina, Malindi, Mombasa and Dar-es-Salaam.

Description. *C. l. livia* Gmelin: Morocco to Tripolitania. ADULT ♂: head and throat dark blue-grey; neck with metallic green or purple reflections; mantle light grey, upper mantle with some purple metallic reflections; rump white, lower rump and uppertail-coverts mid grey; tail slate grey or bluish grey, broadly tipped black; underparts slate grey, upper breast with purplish iridescence; primaries mid grey to grey-brown with darker tips; secondaries ash grey, outer ones and tertials tipped brown, inner ones with broad black subterminal bar on one or both webs; scapulars light ash grey; wing-coverts ash grey, greater coverts with broad subterminal black bar forming, with secondaries, 2 black bars across wing; underwing white. Bill lead grey, cere powdery white; eye deep orange with inner ring yellow; legs and feet red. Sexes alike but ♂ appreciably brighter than ♀. SIZE: wing, ♂ (n = 24) 209–231 (221), ♀ (n = 20) 207–224 (214); tail, ♂ (n = 14) 111–130 (120), ♀ (n = 15) 105–123 (114); bill, ♂ (n = 15) 17·7–20·2 (18·9), ♀ (n = 19) 16·7–20·3 (18·8); tarsus, ♂ (n = 14) 31·2–34·8 (32·5), ♀ 30·6–33·6 (31·4). WEIGHT: (Czechoslovakia) ♂ (n = 41) 290–535 (393), ♀ (n = 36) 304–433 (359).

IMMATURE: duller, black wing-bars less conspicuous, neck feathers with little or no iridescence.

NESTLING: covered with sparse, coarse yellowish down; bill lead grey with pinkish white tip, legs and feet greyish.

C. l. canariensis Bannerman: Essaouira Is., Morocco (part of population on these islands shows characters of Canary Is. race). Rump grey, not white.

C. l. gaddi Zarudny and Loudon: Cyrenaica, NW Egypt. Mantle, back, rump and forewing pale grey (2 black wing-bars standing out in strong contrast); rump sometimes white and always in sharp contrast with mid grey uppertail-coverts. Legs partly feathered. Larger: wing, ♂ (n = 11, Iran) 230–240 (234).

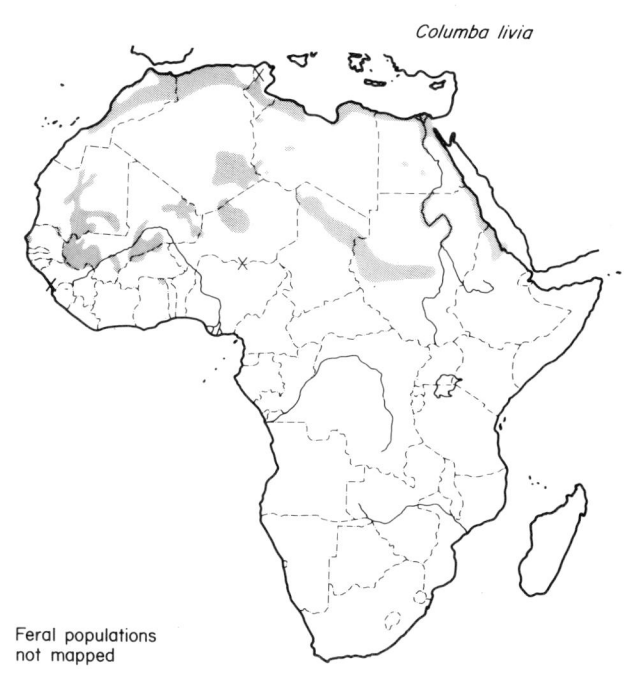

C. l. schimperi Bonaparte: Nile valley south to Khartoum; Red Sea hills from E Egypt to N Ethiopia. Like *gaddi* but smaller: wing, ♂ (n = 8) 194–217 (206).

C. l. dakhlae Meinertzhagen: Dakhla and Kharga Oases, south to central Egypt. Rump pure white; back, lower breast and belly very pale grey, almost white; 2 wing-bars dark grey (not black).

C. l. targia Geyr von Schweppenburg: central Sahara (Hoggar, Aïr, Tibesti, Ennedi Mts) to central Sudan. Like *gaddi* but rather darker and rump grey (never white); violet iridescence of breast less pronounced. Small: wing, ♂ (n = 12) 203–220 (212), ♀ (n = 12) 199–210 (203).

C. l. gymnocyclus G. R. Gray: Mauritania, Mali, Gambaga (Ghana), coasts of Senegambia and Guinea. Dark. Head and upperparts slate grey; neck iridescence more green, less violet than other races, and extends to throat and chin; mantle and wing-coverts dark blue or slate grey; rump pure white; conspicuous bare red skin around eye. WEIGHT: ♀ (n = 3, Senegambia) 255–290 (275).

C. l. 'domestica': the highly variable urban population; probably domesticated about 4500 BC (Géroudet 1983). Main morphs are blue-grey, pale cinnamon, and blackish. Melanistic birds predominate in E African cities.

Field Characters. A small *Columba* pigeon, plump and stocky with swift, dashing flight; blue-grey with white rump (grey in some populations) and 2 conspicuous black bars across inner wing (tertials and secondaries). Habitat characteristic: islands and sea-cliffs, bare rocky hills and buildings. In S Sahara readily told from Speckled Pigeon *C. guinea* (which has wine red back and white-spotted wing-coverts); in N Africa Stock Dove *C. oenas* is same size and colour but lacks white rump and black bars on secondaries, and Wood Pigeon *C. palumbus* is larger, with white flashes on neck and wrist.

Voice. Tape-recorded (44, 62, 72, 73, C). Wild and domestic birds sound alike. Advertising coo a long, drawling 'o-o-orr', repeated monotonously; display coo a hurried, rambling 'co-roo-ooo-ooo', the syllables more or less discrete according to intensity of display. Distress, a plaintive 'o-oor'.

General Habits. Inhabits sea-cliffs and rocky, offshore islands, escarpments, sandstone gullies, gorges, bare rocky hills, desert steppe; Roman ruins (Tunisia); shallow caves, mine shafts and ruins. Takes flight with clatter of wings; often flies low, over water or ground; glides along cliff faces, or in arc out from cliff. In pairs, small parties, or large flocks. Feeds on ground in rock-strewn terrain and in fields, sometimes mixing with other pigeons and doves. In E Morocco large flocks fly from hills to plains in morning to forage. Aggregates at oases and desert waterholes to drink, often in large numbers, with sandgrouse. Mali population is somewhat arboreal. Dependent upon water and not found further than 20 km from spring or well. Roosts on cliffs, and in Hoggar exceptionally in tamarisks. Sedentary.

Food. Seeds, especially of *Colocynthis* spp.; also of *Salvadora persica*, *Schowia purpurea*, and millet. Diet of nestlings (England) 62–94% barley Feb–May and 49–75% wheat Jun–Nov; also oats, *Polygonum*, *Stellaria*, *Brassica* and wild grass seeds and a few weed leaves, earthworms and molluscs (Murton and Clarke 1968); adult diet much the same (Murton and Westwood 1966).

Breeding Habits. Monogamous; barely territorial although nests usually well spaced; solitary, or can form loose colonies of up to 1000 birds. In courtship display flight, ♂ leaves perch with loud wing-clapping, flies with slower beats than usual, then glides in long arc with tail fanned and wings held high in steep V (shallow V when display not intense). After display flying ♂ may alight and display to ♀ by holding head up high with neck swollen, then bowing deeply, cooing, turning round in full or half circle, then stretching neck up again. Nods head quickly in pauses between bowing and circling. Turns away after 1 or several displays. On landing, excited bird may wing-lift (**A**). Bowing display used in courtship but also in aggression to another ♂; between bows, ♀ hops or runs towards ♀ or ♂ with tail spread and depressed. After copulating ♂ and ♀ may indulge in display flight.

NEST: small untidy pad of coarse, dry vegetable material, on wide bare rock or narrow ledge, in dark cavity: either at back of wide-mouthed cavern, or in small shaded rock fissure, or crevice among boulders. Feral Pigeons use ledges high on buildings, making nest in corner or in the angle between horizontal and vertical stones. Nest usually becomes heavily fouled with excreta during incubation period.

EGGS: 1–3, usually 2 (82 clutches, Morocco, were 21 clutches of 1 egg, 60 of 2 eggs and 1 of 3 eggs). Blunt ovals, white with slight gloss. SIZE: (n = 12, Morocco) 38–40 × 27·5–30·5. WEIGHT: (n = 30 Feral Pigeons) 13·8–19·8 (17·0).

LAYING DATES: N Morocco Nov–July with peak Apr–June; Sudan Mar; probably lays all months.

INCUBATION: by both sexes, mainly ♀; begins with 2nd egg. Period: 17–19 days (Europe). Pair has 2–3 broods each year.

DEVELOPMENT AND CARE OF YOUNG: young fed by both parents; young takes food from parent's throat, at first 'milk', later softened grain. In nest for 23–25 days; independent of parents at 30–35 days.

BREEDING SUCCESS/SURVIVAL: of 812 eggs 66% hatched (England), remainder being predated, deserted or infertile; of those hatched, 71% of nestlings fledged (47% of eggs laid). Nestling losses due to exposure and predation. Adults extensively preyed upon by Peregrine *Falco peregrinus* (e.g. on Moroccan coastal islands); other predators (Europe) are Merlin *F. columbarius* and Raven *Corvus corax*.

References
Cramp, S. (ed.) (1985).
Murton, R. K. and Clarke, S. P. (1968).

Genus *Streptopelia* Bonaparte

Turtle-doves, tending to be smaller and more lightly built than *Columba* spp.; wings usually less pointed, tail longer than in *Columba*, slightly graduated and tipped white. Plumage grey or light brown, monomorphic, usually with black half-collar. Bare parts dark red, or black. Tarsus bare, not concealed by plumage. Tail rounded. Gregarious, ubiquitous, most species abundant. Characteristic alighting calls. Use trees freely but feed on ground, on grain.

African species fall into 3 groups: collared doves *S. semitorquata*, *S. decipiens*, *S. vinacea/S. capicola* superspecies and *S. roseogrisea/S. decaocto/S. reichenowi* superspecies; rufous turtle-doves *S. turtur* and *S. hypopyrrha/S. lugens* superspecies; and the Laughing Dove *S. senegalensis*.

***Streptopelia vinacea* superspecies**

1 *S. vinacea*
2 *S. capicola*

***Streptopelia roseogrisea* superspecies**

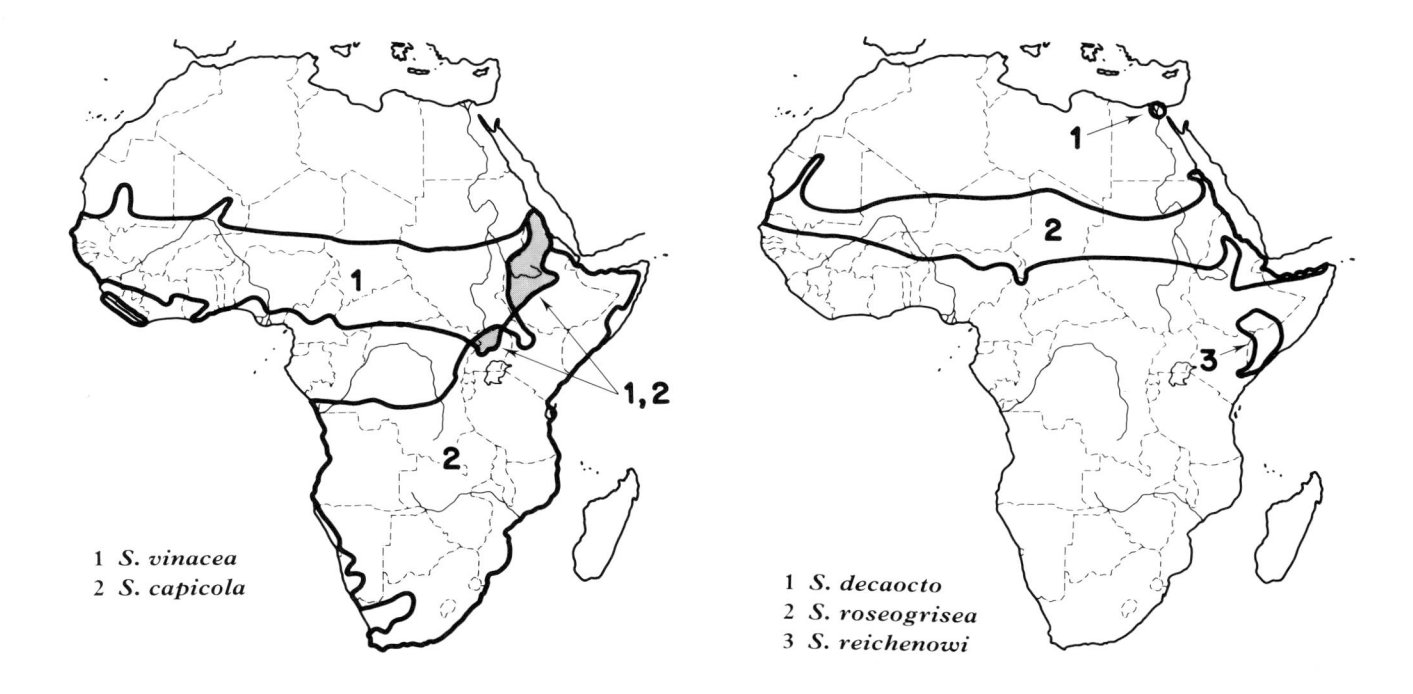

1 *S. decaocto*
2 *S. roseogrisea*
3 *S. reichenowi*

***Streptopelia lugens* superspecies**

1 *S. hypopyrrha*
2 *S. lugens*

480 COLUMBIDAE

Plate 31

(Opp. p. 464)

Streptopelia semitorquata (Rüppell). Red-eyed Dove. Tourterelle à collier.

Columba semitorquata Rüppell, 1837. Neue Wirbelth. p. 66; Taranta Mts, Eritrea.

Range and Status. Africa and SW Arabia. Dense woodlands from S Senegambia to Ethiopia and NW Somalia (north to 12–13°N in Mali, Nigeria and Chad and 9°N in Sudan), and south to N Namibia, N Botswana, Transvaal, Natal, Lesotho, SE and SW Cape Province (Brooke 1984). Up to 3000 m. Frequent to abundant. Scarce or absent in N Kenya, E and W Congo basin, S Ivory Coast, and most of Liberia. More widespread and numerous in southern Africa than 50 years ago.

Description. ADULT ♂: crown pale grey becoming whitish on forehead, nape grey; hindneck, sides of neck and forehead dull mauvish pink; black half-collar between hindneck and upper mantle; rest of upperparts (including inner wing-coverts) earth brown but rump washed grey; breast and belly dull mauvish pink shading to grey of vent, flanks and undertail-coverts. 2 middle rectrices earth brown with darker bases, others black above and below, with terminal quarter brown above and dirty white below. Primaries brown narrowly edged whitish, secondaries brown; outermost wing-coverts very dark brown giving bend of wing slight contrast; underwing-coverts grey-brown. Bill black; bare skin around eye vinous red, eye red, orange or yellow; legs and feet vinous red. Sexes alike. SIZE: (southern Africa) wing, ♂ (n = 16) 182–201 (192), ♀ (n = 19) 181–193 (186); tail, ♂♀ 119–133 (125); bill to skull, ♂♀ 21·5–22·5 (21·8); tarsus, ♂♀ 23–27 (24·6). Low-latitude birds brighter and smaller. (W Africa): wing, ♂ 172–186, ♀ 105–125; tail, ♂♀ 105–125. WEIGHT: (E Africa) ♂ (n = 15) 170–213 (196), ♀ (n = 8) 163–220 (190); (Zimbabwe) ♂ (n = 44) 208–326 (241), ♀ (n = 44) 211–261 (229); (SW Cape) ♂♀ (n = 337) 162–310 (252), heavier in winter and spring (av. 256) than in rest of year (av. 246) (Rowan 1983).

IMMATURE: mantle and wing-coverts with pale rufous edging, remiges tipped rufous; breast strongly washed with rufous; black collar poorly developed. Orbital skin grey.

NESTLING: skin greyish black or pinkish-brown with wiry buff or white down, denser on upper than underparts.

Field Characters. A large, dark dove (largest of the 'collared' doves) of well wooded habitats; forehead and throat whitish, upperparts brown rather than grey. Tail uniform brown from above except when alighting, when black base displayed; from below, grey outer half contrasts with black base. Distinguished from African Mourning Dove *S. decipiens* by dark wing-shoulder, lack of white in tail, very narrow dark red eye-ring only visible at close range (red eye-ring of Mourning Dove broad and conspicuous), different voice and habitat.

Voice. Tape-recorded (5, 22, 44, 61, 74, 75). Song or advertising call a series of 6 deep coos; the 1st and particularly the 2nd emphasized, then 4 softer, quieter and lower notes at slightly greater frequency: 'oo OO, oo-oo-oo-oo'. Song lasts 1·5–2·0 s and tends to be repeated every few s up to 40 times. Bout ends with 2 emphatic coos. Much variation, temporally by same bird and regionally, e.g. 'oo-oo-oo, OO oo' (Grahamstown, South Africa), 'OO OO oo oo' etc. (Orange Free State), 'oo OO, oo-oo-oo, oo-oo' (Zambia, E Zimbabwe) (Rowan 1983). Excitement call, a double nasal note, mainly on alighting. Call by both sexes selecting nest-site: 'krooooo oo oo'. During bowing, 3–4 harsh, deep, growling notes: 1st long, others shorter and higher; ♂ bows during 1st note and utters others in lowest position (A. Sala, pers. comm.).

Streptopelia semitorquata

General Habits. Inhabits well wooded country near permanent water, mangrove, gallery forests, forest edges and clearing with some tall trees, parks and gardens. Not in interior of rain forest. Forages (on ground), drinks and roosts solitarily or in small parties (up to 30). Sometimes forages in trees, eating berries. 2 foraging bouts, early morning and (mainly) afternoon; in hot weather rests at midday. ♂♂ sing in chorus from 45 min before until 1 h after sunrise, concealed in foliage, and for *c.* 1 h before sunset, perched on tree-tops.

Moves locally in Gambia, Nigeria and Cameroon at onset of wet season (thousands daily, June–July, in flocks of 40–200), possibly roosting/foraging movements (no evidence of distant migration).

Food. Seeds; also berries (*Lantana camara* etc.), rhizomes (*Cyperus esculentus*), flowers (*Cedrella toona*) and arthropods (termites, a millepede). 82 birds from Zimbabwean farmland contained (in decreasing abundance) *Cyperus* rhizomes, seeds of *Helianthus*, *Ricinus communis*, *Pennisetum typhoides*, grass and sorghum, and ground-nuts (Smithers 1965). Elsewhere, eats maize, croton, bamboo and cowpea seeds, ground-nuts in the shell, fruits of *Trema orientalis* and mulberry, and discarded fruit pulp. Usually eats only 1 food at a time: 1 crop held 120 *Cyperus* 'nuts', 1 held 70 *Ricinus* seeds (10 g), 1 held 38 maize seeds (20 g) and 1 held 1500 millet seeds (13 g) (Rowan 1983).

Breeding Habits. Monogamous, pairs for life. Territories continuously occupied by same pairs and well defended by ♂. In courtship flight ('towering'), bird flies steeply up then glides at low angle, wings and tail outspread, to perch, alighting silently or with excitement call after 1 or 2 loud flaps. Sometimes towers several times before alighting; glide may be punctuated by single flaps. In bowing display, ♂ inflates neck and bows deeply while calling 2 hoarse, throaty coos; directed at rival (see **A**). ♂ and ♀ preen each other. 2 ♂♂ fight in tree by cuffing with wings; 1 may stand on other to beat its head.

A

NEST: quite substantial platform of twigs, often with lining of grass or pine needles, 15–20 cm diam. and 10 cm deep. Built 30 cm to 18 m high in bushes and trees (av. of 137 nests 4 m, median 3 m); low nests are always in safe places, e.g. in *Ficus verruculosa* thicket over water (Okavango swamp). 40% of nests (n = 37) in vegetation beside or overhanging water, and 8% on old nests of pigeons, egrets, crows and thrushes. ♂ brings material collected from trees up to 90 m away, and from ground, at 2 min interval (taking 1·5 min to choose twig), mainly in early morning; ♀ stays by nest, receives then incorporates each twig. Building can continue during incubation (Rowan 1983). ♂ may take no part in building in Nigeria and Kenya.

EGGS: 1–2; 12% of clutches (n = 130) in SW Cape and Zimbabwe and 20–25% (n = 28) in E South Africa are of 1 egg. Laid at *c.* 24 h interval. Oval, smooth, slightly glossy, pure white, cream-white or sometimes light buff. SIZE: (n = 40): 28·7–34·0 × 22·2–25·2 (31·2 × 24·2).

LAYING DATES: Senegambia, Guinea-Bissau and Sierra Leone, Dec–Apr; Mali, Nigeria, every month; Ghana, S Congo, Mar–Aug; Cameroon, Feb; Gabon, Dec–Mar; Central African Republic, Nov–Dec; Ethiopia, Sept–June; East Africa: Region A, Jan–Mar, July–Sept; Region B, Jan–June, Aug, Oct–Dec; Region C, Jan, June–July, Nov–Dec; Region D, every month; Zimbabwe and South Africa, every month in summer rainfall areas and probably E Cape, Aug–Jan with peak Sept in winter rainfall areas.

INCUBATION: begins with 1st egg; ♂ incubates by day (from 07.15–08.55 to 13.45–16.25 h), ♀ for rest of 24 h. Incubating bird calls, particularly near relief time. Period: 14·5 days (and perhaps up to 17 days). Eggs hatch on same or successive days; parent drops shell up to 80 m away.

DEVELOPMENT AND CARE OF YOUNG: at 1 week head skin dark but body pink. Eyes fully open at 10 days. Well feathered at 12 days, when chick snaps bill and lunges at source of threat, flailing wings. Period: 14–17 days. Chick closely brooded at first; material added to nest even when young well grown; disturbed parent has 'broken wing' distraction display. *c.* 4 feeds per day, each lasting 3 min. Fledgling stays near nest for several days, then follows parents closely for 3 weeks, soliciting food with quivering wings, prodding parent's gape with bill.

BREEDING SUCCESS/SURVIVAL: *c.* 40% of nests (n = 33) survive to fledging. Flimsy nests destroyed by wind; some young die of cold. Squabs preyed on by African Goshawk *Accipiter tachiro* and Pied Crow *Corvus albus* (Rowan 1983).

References
Brooke, R. K. (1984).
Rowan, M. K. (1983).
van Someren, V. G. L. (1956).

Streptopelia decipiens (Hartlaub and Finsch). African Mourning Dove. Tourterelle pleureuse.

Plate 31
(Opp. p. 464)

Turtur decipiens Hartlaub and Finsch, 1870. *In* Finsch and Hartlaub, Vog. Ostafr., p. 544; Dongola, N Sudan.

Range and Status. Endemic resident, widespread in dry tropical Africa; absent from rain forest and adjacent wet savannas, and with markedly discontinuous distribution in southern Africa (e.g. Angola and Zimbabwe: see map). Strongly associated with dense riparian *Acacia* woods and wadis. Up to 1400 m in E Africa and 2000 m in Ethiopia. Sedentary, but marked local seasonal movements. North of about 12°N common to abundant, thousands congregating at drinking pools; abundant Sudan, Somalia; common to frequent in Kenya, Tanzania and Zambia.

Description. *S. d. shelleyi* (Salvadori): S Mauritania and Senegambia to S Niger and central Nigeria. ADULT ♂: crown and nape ash grey; upper hindneck, foreneck and sides of neck mauve-pink; black half collar with white upper margin separates hindneck from mantle; face ash grey, chin and throat pale grey to whitish; mantle, back, rump, uppertail-coverts, scapulars, tertials and inner wing-coverts earth brown (sides of rump with some grey). 2 central rectrices earth brown, next 2 paler with whitish tip, outermost grey-brown with broad whitish tip; below, basal two-thirds black, rest of tail white tinged grey. Breast dull mauve-pink shading to grey or rosy grey on belly; flanks grey; undertail-coverts grey tipped whitish; 3 outermost primaries brown thinly edged whitish,

Streptopelia decipiens

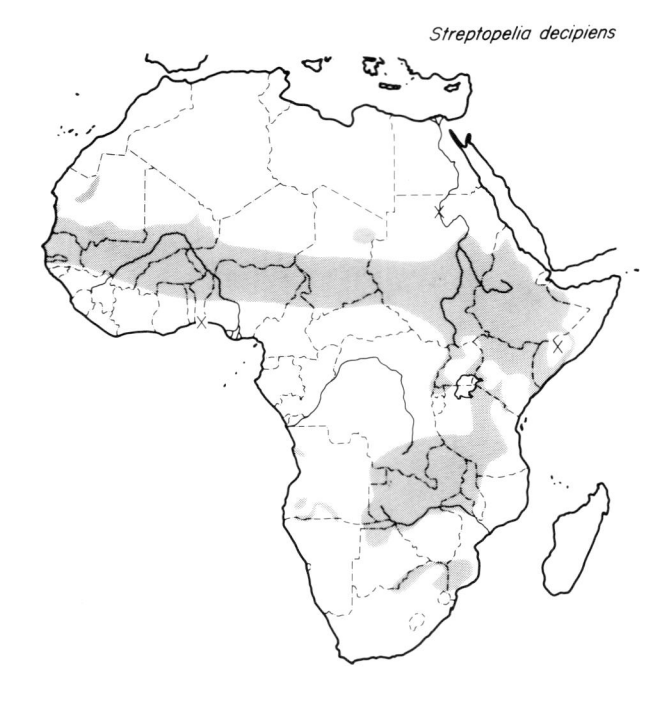

others with grey basal half and brown distal half; secondaries grey-brown; underwing-coverts grey. Bill black; skin round eye red; iris orange-red or red with narrow yellow inner ring; legs and feet wine red. SIZE: wing, ♂ (n = 9) 176–188 (182), ♀ (n = 5) 167–176 (172); tail, ♂ (n = 9) 127–142 (133), ♀ (n = 6) 125–131 (128); bill, ♂ (n = 9) 17–19 (17·9), ♀ (n = 6) 16–17 (16·7); tarsus, ♂ (n = 9) 22–27 (25·0), ♀ (n = 6) 26–29 (26·2). WEIGHT: (Senegambia) ♂ (n = 89) 156–230 (175), ♀ (n = 90) 140–200 (168); (Ethiopia) ♂♀ (n = 312) 113–199 (156).

IMMATURE: above like adult; crown and nape brown, breast light brown; remiges tipped russet; iris pale brown.

NESTLING: on hatching covered with down, longest over crop, deep buff, fawn on back.

S. d. logonensis (Reichenow): L. Chad basin east to S Sudan (L. No), NE and E Zaïre, W and N Uganda. Like *shelleyi* but breast vinous grey. Smaller: wing, ♂♀ 157–172. WEIGHT: (E Africa) ♂ (n = 10) 110–150 (134), ♀ (n = 8) 110–140 (124).

S. d. decipiens (Hartlaub and Finsch): Darfur (Sudan), Ethiopia except S, NW Somalia. Paler than *logonensis*, especially above.

S. d. perspicillata (Fischer and Reichenow): W Kenya, central Tanzania (south to 10°S). Like *decipiens* but belly, flanks and undertail-coverts white; eye pale yellow. Small: wing, 143–168.

S. d. elegans (Zedlitz): S Ethiopia, S Somalia, N and E Kenya. Paler than *perspicillata*, white of underparts extending onto lower breast.

S. d. ambigua (Bocage): Zambia, SE Zaïre, Malaŵi, Angola (Cuanza, Cunene, Okavango valleys), Zambezi and Limpopo valleys and adjacent lowlands. Like *decipiens* but centre of belly white; undertail-coverts grey, boldly edged white; iris yellow. Small: wing, ♂ (n = 22) 158–173 (168), ♀ (n = 11) 153–166 (160).

Field Characters. A large dark grey dove of green vegetation in open and arid country, with broad red eye-ring; iris red in W, yellow in E and S of range. Tail base black on outer feathers, grey-brown in centre; outer half white from above except for central feathers, all white from below. Paler and greyer than Red-eyed Dove *S. semitorquata*, with grey forehead, different voice and habitat; best mark is pale grey wing-shoulder which often looks bluish in the field (Red-eyed has dark shoulder). Vinaceous, Ring-necked and African Collared Doves *S. vinacea*, *S. capicola* and *S. roseogrisea* are much paler, have dark eyes without red eye-ring and have different voices.

Voice. Tape-recorded (10, 34, 35, 44, 58, 74, 75). Very characteristic song, lasting up to 13 s, is 1–2 fast falling trills (each of duration 1·3 s), followed by 3 coos repeated *c.* 7 times, and sometimes ending with 3–4 further trills alternating with short soft coo, 'wu'. Thus typical song sequence: 'krrrrrrrrrrow, oo-OO oo, oo-OO oo, oo-OO oo, oo-OO oo, oo-OO oo, krrrrrrrrrrow wu krrrrrrrrrrow wu krrrrrrrrrrow'. Trill is easily-imitable excited treble gargle, lacking soothing quality of purring song of turtle-doves (*S. hypopyrrha*, *S. lugens*, *S. turtur*). Of the 3 coos 'oo-OO oo', 2nd is emphasized and 3rd slightly lower-pitched. Trill also accompanies bowing display.

General Habits. Closely associated with thorny and evergreen riverside woods and thickets in arid savanna and with stands of *Acacia* within 10 km of water, particularly with tall trees such as *A. albida*; also in palms (*Borassus*, *Hyphaene*), and dense native and exotic trees in villages, farmlands and gardens. Forages on nearby open sandy or cultivated ground, alone or in parties of 25–30, often with other doves. Also eats grain on standing stalks (e.g. *Sorghum*). Rather tame; tends to perch high and in dense foliage; drinks usually twice a day, early morning and late afternoon, along riverbanks and ponds (never in tiny holes or by wells), sometimes from overhanging branches. In Senegambia roosts in hundreds on *Aeschynomene elaphroxylon* trees standing in water, flying to and fro in small flocks.

Not migratory, but makes daily movements, often in large flocks, between roosting, foraging and drinking places, and in wet season moves locally away from riverine woods.

Food. Seeds; also berries and termites. In Senegambia (n = 1147 birds) eats 49 seed and fruit species; in wet years, about same weight of grasses/cereals (mainly *Panicum laetum* and *Dactyloctenium aegyptium*), dicotyledonous plants (mainly *Tribulus terrestris* and *Gisekia pharnacioides*); in dry years, almost exclusively dicotyledons. Other important foods: *Colocynthis*, *Arachis*, *Vigna* (beans), and *Cocculus pendulus*, *Maytenus senegalensis*, *Boscia senegalensis*, *Salvadora persica* (fruits). Items weigh 0·25–892 mg; crop content is up to 18 g (dry weight) (av. 6 g, n = 45).

Breeding Habits. Monogamous. Nests solitarily or in scattered groups in trees and high bushes in wooded savanna and gardens. Courtship poorly known; 'towering' display flight and bowing display on ground much like those of Red-eyed Dove (*q. v.*).

NEST: platform of twigs and petioles with shallow central depression, lined with fine rootlets, diam. 15 cm, sometimes so thick that eggs invisible from below.

Often well hidden in foliage; in Senegambia uses mainly flooded *Acacia nilotica*, also *Balanites aegyptiaca* and *Acacia raddiana*. Av. height of nest (n = 54) above ground, 3·1 m; av. height in exotic trees 5·3 m.

EGGS: 1–2, usually 2. Oval; matt or with only slight gloss; white or creamy. SIZE: 27·0–35·5 × 21·5–28·0. WEIGHT: (n = 5) av. 8·5.

LAYING DATES: Senegambia every month; Sudan, Mar, May, Nov; Ethiopia, Jan, Apr; Zimbabwe, Sept, Nov, Dec; East Africa: Region A, Jan, Feb, Nov; Region B, Feb–July, Nov; Region C, May; Region D, Jan, July, Aug.

INCUBATION: period 13–14 days (in captivity).

DEVELOPMENT AND CARE OF YOUNG: fledging period 15–18 days.

Reference
Rowan, M. K. (1983).

Streptopelia vinacea **(Gmelin). Vinaceous Dove. Tourterelle vineuse.**

Columba vinacea Gmelin, 1789. Syst. Nat. 1, pt. 2, p. 782; Senegal.

Forms a superspecies with *S. capicola*.

Plate 31
(Opp. p. 464)

Range and Status. Endemic resident. Throughout northern tropics between rain forest zone and Sahara, but absent east of 40°E. Extends north to 18°N in Red Sea hills (Ethiopia) and 21°N in Mauritania; elsewhere to 15–16°N, and south to 1°N (Uganda). Sea level to 2000 m. Abundant. Recent increase in N Senegambia on large sugar cane scheme.

Description. ADULT ♂: crown pale pink, nape pink-grey, tinged brown; black half-collar on hindneck with greyish upper edge; sides of neck, foreneck and face pale vinous pink; forehead, chin and throat whitish; very narrow black line from gape to eye; mantle, scapulars, inner wing-coverts, tertials and uppertail-coverts pale earth brown; back and rump like mantle, slightly washed grey-blue. Central rectrices like mantle, next 2 paler, 3 outer ones with 25 mm white tip and black base (outermost narrowly edged white); below, tail black (brown in centre) with broad white tip. Breast pale vinous pink merging to whitish on belly; undertail-coverts white; flanks pale blue-grey; remiges dark brown, when fresh, outer primaries narrowly edged whitish; outer wing-coverts pale greyish blue visible on edge of closed wing; underwing pale grey. Bill black; eye dark brown; legs and feet vinous red. Sexes alike. SIZE: wing, ♂ (n = 11) 137–143, ♀ (n = 5) 134–141; tail, ♂ 90–97, ♀ 90–94; tarsus, ♂♀ 19–22. WEIGHT: ♂♀ (n = 18) 97–124 (110).

IMMATURE: like adult but pale edges to wing-coverts give scaly appearance.

NESTLING: skin dark purple; down pale buff and rather thick.

Field Characters. Over most of its range easily the most abundant dove; distinctive, far-carrying, monotonously repeated song is constantly heard everywhere. A rather small, whitish grey or very pale pinkish dove with light brown back and wings, darker flight feathers, and tail with black base and largely white at sides. Black hindneck collar. Bill black; legs dull carmine. Smaller, much paler, and in drier savannas than Red-eyed and African Mourning Doves *S. semitorquata* and *S. decipiens* and further distinguished from them by dark brown (not red-orange) eye and lack of red orbital ring. Overlaps in north of range with African Collared Dove *S. roseogrisea* which is very similar but lacks black at base of upperside of tail, and has remiges concolorous with (not darker than) rest of wing; also different voice. In Ethiopia and Uganda overlaps with extremely similar Ring-necked Dove *S. capicola*, which is distinguished by grey (not pink) forehead and dark grey (not pale grey) undersides of wings. Unlike Ring-necked Dove, Vinaceous Dove does not roll 2nd note of 3-note song.

Voice. Tape-recorded (44, 235, 253, 267, 296). Song a far-carrying, rapid, high-pitched, monotonous 'oo, oo-oo', with slight emphasis on 1st note, lasting *c*. 0·5 s and repeated at variable frequency up to 60/min. Sings all day and often at night. In display, a rolling 'krrroo' synchronized with bow. Creaking, nasal excitement cry on alighting.

General Habits. Inhabits dry woodland, bushy grassland and cultivation; but not a suburban or garden bird. Forages on open ground, roosts in trees, drinks twice a day in early morning and late afternoon; usually in pairs, but small parties commonplace and often congregates in large flocks.

In N Senegambia moves daily between roosting, foraging and drinking places, and seasonally between riparian habitats and adjacent bush savanna; probably disperses or migrates northward in rains. Huge movements in Mali (up to 300,000 birds) in Sahel and R. Niger inundation zone between Jan and Apr (Lamarche 1980); also in NE Zaïre.

Food. Seeds; also other vegetable matter, termites, snails and caterpillars. In N Senegambia eats 88 spp. of seeds and fruit (n = 852 birds); in wet years up to 80% of food are grass seeds (principally *Panicum laetum* and *Brachiaria*) and in dry years eats mainly legumes (*Zornia* and *Alysicarpus*). Also *Gisekia pharnacioides*, *Colocynthis*, *Arachis*; rhizomes of Cyperaceae, fruit stones (*Maytenus* and *Boscia*) and some *Tribulus*. Dried crop contents weigh up to 7 g (n = 10, Dec).

Breeding Habits. Monogamous, territorial. In 'towering' courtship flight, rises to tree-top height then descends in shallow glide with stiffly spread wings and tail, alighting on low bush with alighting call. In bowing display on ground ♂ pumps head up and down in shallow bows (*c.* 120/min), bill not coming close to ground; each bow is accompanied by throaty coo.

NEST: frail, thin platform of twigs, built at av. height of 2·6 m above ground (n = 111 nests, Senegambia), in *Balanites* and *Acacia raddiana* trees, also *Sclerocarya*, *Commiphora*, and 13 other tree spp.

EGGS: 2, laid at 1-day interval. Oval; white; glossy. SIZE: (n = 10) 23·6–26·5 × 18·9–20·5 (25·0 × 19·4). WEIGHT: (n = 28) av. 5·9 g.

LAYING DATES: Senegambia and Mali, every month; Ethiopia, Mar; E Africa: Region A, Aug; Region B, Aug–Sept.

INCUBATION: eggs hatch 1 day apart. Period: 14 days (in captivity).

DEVELOPMENT AND CARE OF YOUNG: in captivity young leave nest at 13–14 days, returning to it irregularly for 2–3 more days.

BREEDING SUCCESS/SURVIVAL: minimum interval between successful layings 34 days (in captivity).

References
Morel, G. and Morel, M.-Y. (1972).
Morel, M.-Y. (1980).

Plate 31
(Opp. p. 464)

Streptopelia capicola (Sundevall). **Ring-necked Dove; Cape Turtle Dove. Tourterelle du Cap.**

Columba vinacea var. *capicola* Sundevall, 1857. Kongl. Sv. Vet.-Akad. Handl. 2, art. 3, p. 54; Rondebosch, Cape Prov.

Forms a superspecies with *S. vinacea*.

Range and Status. Africa and Comoro Is. Resident savannas of southern and eastern Africa from Cape north to S Congo, S and E Zaïre, Uganda, Sudan (SE Equatoria), Ethiopia and W Somalia. Abundant, and probably increasing with agriculture and deforestation. Hundreds, sometimes thousands, congregate in harvested cereal fields (sorghum; Zimbabwe, Botswana). Common Zanzibar. Mainly below 2000 m; up to 3000m (Mau Narok, Kenya).

Description. *S. c. capicola* (Sundevall): W Cape Province, South Africa. ADULT ♂: crown bluish grey, forehead somewhat paler; nape, sides of neck and hindneck mauvish pink; black half-collar across hindneck; face light bluish grey, chin and throat whitish; very narrow black stripe from gape to eye; mantle, back, scapulars and inner wing-coverts dull earth brown; rump and uppertail-coverts dark bluish grey. Central rectrices like mantle, next 1 tinged grey, next 3 with dark base and increasing amounts of white at tip, outermost with distal half and all outer web white; tail below black (browner in centre) with distal half white (central feathers brown). Underparts pale mauve suffused grey, belly whitish, undertail-coverts white, flanks pale grey. Primaries dark brown, thinly edged whitish on outer web and with pale tips; secondaries dull greyish brown with outer web edged pale grey; tertials earth brown; outer wing-coverts pale grey-brown shading to pale bluish grey on outermost; underwing pale grey. Bill black; eye dark brown; legs and feet dark purple. Sexes alike but in ♀ grey parts tinged dull brown. SIZE: wing, ♂ (n = 5) 150–156 (154), ♀ (n = 6) 147–152 (149); tail, ♂♀ 102–121 (110); bill, ♂♀ 14–17 (16·0); tarsus, ♂♀ 21–27 (24·0). WEIGHT: (n = 1200, South Africa) 106–200 (153·5), monthly means from 141 g (Dec) to 161 g (June) (Siegfried 1971).

IMMATURE: like adult but feathers of back, wings and breast fringed with buff.

NESTLING: dark bluish grey skin covered with pale yellowish down. Bill black with white tip.

S. c. tropica (Reichenow): central Kenya to Angola (except SW), Zimbabwe (except SW) and South Africa (except W Cape). Crown paler than *capicola*, mantle warmer brown (less grey), breast, flanks and belly clearer pink (less grey). WEIGHT: ♂♀ (n = 7) 120–150 (129).

S. c. damarensis (Hartlaub and Finsch): Namibia and Botswana (except N) and SW Zimbabwe. Paler than *capicola*; wing-coverts and rump greyer, belly whiter. Large: wing, ♂♀ 155–170. WEIGHT: ♂ (n = 4) 130–140 (138), ♀ (n = 4) 95–140 (119).

S. c. onguati Macdonald: N Namibia and SW Angola. Much paler than *capicola*; greyer above; breast pale lilac-grey.

S. c. somalica (Erlanger): E Ethiopia, Somalia, N Kenya (south to Uaso Nyiro R.). Paler than *tropica*. Small: wing, ♂♀ 135–147.

S. c. electa (Madarasz): W Ethiopia. Darker than *tropica*, with greyer face and belly.

Field Characters. Far-carrying song of this ubiquitous, rather small pale grey dove forms constant aural background in the bush. Distinguished from Red-eyed and African Mourning Doves *S. semitorquata* and *S. decipiens* by dark brown (not red-orange) iris and lack of red orbital ring; bird is much paler, has broadly white-edged tail and different voice. Overlaps in Ethiopia and Uganda with Vinaceous Dove *S. vinacea* and almost identical with it; for distinctions, and other differences of both from congeners, see that species.

Voice. Tape-recorded (4, 8, 22, 61, 66, 69). Trisyllabic song, a most characteristic noise of African bush. 1st and 3rd notes a short, explosive 'puk', middle note a rolled 'crrrr', accented and longer: 'puk, crrrr puk'. Song is repeated without pause, at rate of 14 cycles per 20 s, for 10–60 s or more. Rhythm is even, so song can be heard as 'crrrr, puk puk, crrrr, puk puk .....'. Rather high-pitched. Sings monotonously all day and often at night, throughout year. ♂ and ♀ selecting nest-site give bisyllabic 'cuk cooo'. Display bowing accompanied by rolled middle note 'crrrr' of trisyllabic song. Excitement call a harsh, nasal 'eeerrr', usually given on alighting.

General Habits. Inhabits all woodlands and open tree savannas; occurs around villages, in plantations (wattles, conifers, gums) and farmland. Spends much time on ground, foraging for seeds, with easy walk and bobbing head, changing direction every few paces. Sometimes perches and feeds on standing cereal heads. Feeds in pairs, often close to its own and other kinds of doves, mainly in early morning and late afternoon; rather shy, and when disturbed all doves rise as one flock. Rises with clatter of wings; flight swift and strong, about 50–80 (65) km/h. Drinks at any time of day, mainly before 10.00 h and after 17.00 h, freely associating with other doves and congregating in large flocks. Sunbathes by squatting on ground with wings half spread and back feathers raised. Water-bathes by paddling in up to thighs, standing quietly, then splashing. At crowded waterhole may fly in, briefly hover with feet dipping in surface, or actually settle on water with wings half spread. Rain-bathes with 1 or both wings raised up. Does not dust-bathe. Rests during heat of day, and roosts at night, gregariously in isolated grove of trees. Competes with equally ubiquitous Laughing Dove *S. senegalensis*, sometimes threatening one with raised wings; the 2 species tend to replace each other numerically (Rowan 1983). Known also to threaten intruding shrikes.

More abundant May–Aug in W central Zambia and low-lying Zululand, and also Feb–Mar in Kruger National Park, than in rest of year; but highly sedentary in temperate South Africa (few recoveries beyond 8 km SW Cape) (Rowan 1983).

Food. Seeds of grasses, cereals (*Sorghum*, oats), *Euphorbia*, conifers, wattles, lupins; some large (*Quercus*, whole maize kernels); bulbs of sedge *Cyperus esculentus*; fruits of *Rhus*, *Lantana*, *Pyrecanthus*, seed stems of *Acacia cyclops*, succulent leaves of mesembryanthemums; nectar of *Aloe marlothii*. Some invertebrates (earthworms, termites, weevils, aphids, locust nymphs). Av. dry weight of crop contents (n = 28 crops), 3·4 g (Rowan 1983).

Breeding Habits. Monogamous; pairs for life. Territorial; territory size *c.* 0·5 ha; nests can be only 25 m apart; long tenure of established territories and slow shifting of boundaries with changes in individual occupants: in 1 ha wooded area 2 territories lasted for 8 years (Rowan 1983). Territorial ♂ has 'towering' display flight, rising with clapping wings to level above treetops, then gliding down in long shallow arc with stiffly outspread wings and tail, alighting with harsh call. Bowing display, on ground or large limb, is shallow quick bows with neck plumage inflated, accompanied by special coos. Paired birds preen each other, and ♂ courtship-feeds ♀, each a prelude to copulation.

NEST: frail structure of twigs, sometimes lined, diam. *c.* 13 cm, thicker (7–15 cm deep) with re-use; built by ♀ with material brought by ♂ for a few hours each day for 3–8 days. Nests in trees and bushes, sometimes (2%) over water, or re-uses other birds' nests including doves. Uses native trees (81% of 244 nests, Zimbabwe) and exotics (88% of 844 nests, SW Cape: mainly *Acacia* and *Eucalyptus*). Normally well hidden in foliage; sometimes uses bare tree or man-made structure. Height above ground 0·5–15 m, medians 3·0–3·5 m (Zimbabwe, SW Cape).

EGGS: 1–2, av. 1·91 (n = 178 clutches, SW Cape). Clutches of 3 (n = 6) and 4 (n = 3) probably each laid by 2 ♀♀. Eggs laid usually about 08·00 h, at interval of 24–36 h. Oval; pure white; slightly glossy. SIZE: (*S. c. tropica*, n = 88 eggs) 24·0–32·2 × 17·0–22·5 (27·0 × 21·1); (*S. c. capicola*, n = 61) 24·2–30·9 × 19·5–23·4 (28·6 × 22·0). WEIGHT: av. 3·05.

LAYING DATES: Ethiopia, Feb–Sept; E Africa: Region A, June, Aug, Dec; Region B, Jan–Feb, Apr–July, Sept, Nov–Dec; Region C, Jan–Apr, June–Nov; Region D, Jan, July–Aug; Zimbabwe, all months, of 556 clutches, 39% Sept–Oct, 10% Nov–Jan, 33% Feb–May, 8% June–July, 10% Aug; South Africa, all months, peak month in SW Cape varying from Sept to Dec: of 971 clutches, 24% Oct, 52% Sept, Nov–Dec, 13% Jan–Mar, 3% Apr–July, 7% Aug (Rowan 1983).

INCUBATION: begins with 1st egg; by ♀ from c. 16.30 h to c. 10.00 h and by ♂ for rest of time (relief times vary widely). Sits very tight. Relieving bird calls from distance and is answered by mate on nest. Period: 13–15 days (SW Cape), usually 14 ± 3 h.

DEVELOPMENT AND CARE OF YOUNG: brooded closely, day and night, for 6–12 days; disturbed brooding bird drops to ground and hurries off for 10–100 m dragging outspread wings. Down thickens as chick grows; eyes open and wing-quills show within 7 days; most feathers open by 10th day, tail-quills appear by 11th day. Chick defends itself by snapping bill and flailing wings. 2 chicks fed simultaneously by regurgitation. 1 young often seems twice size of other. Fledging period: 16–17 days. Remains near nest for several days after fledging, sometimes returning to nest to sleep for 1–6 nights. Becomes independent of parents for food within 12–18 days of first leaving nest.

BREEDING SUCCESS/SURVIVAL: of 731 eggs 53% hatched; of 333 young 71% fledged (38% of eggs laid, SW Cape). Pair can rear 9–10 broods a year. Annual mortality of juveniles 65% and of adults 35%. Exceptionally lives 10 years in wild and 32–35 years in captivity (Rowan 1983). Preyed upon by genet *Genetta* sp., cats, eagles (*Aquila wahlbergi*), falcons (*F. biarmicus*) and goshawks; particularly vulnerable when coming to drink, even Marabou Storks *Leptoptilos crumeniferus* gathering at water to kill doves (Brown 1963). In SW Cape the bird most commonly killed by road traffic; also commonly killed by flying into wires. Like other granivores prone to seed-dressing poisoning. Susceptible to non-lethal avian pox (10% of juveniles). See also Siegfried (1984).

Reference
Rowan, M. K. (1983).
Siegfried, W. R. (1984).

Plate 31
(Opp. p. 464)

Streptopelia roseogrisea (Sundevall). African Collared Dove; Pink-headed Dove. Tourterelle rieuse.

Columba roseogrisea Sundevall, 1857. Kongl. Sv. Vet.-Akad. Handl. (n.s.) 2, no. 1, p. 154; Nubia.

Forms a superspecies with *S. decaocto* and *S. reichenowi*.

Range and Status. Africa, SW Arabia.

Resident, sedentary and partially migratory. Sahel zone; from N Senegambia and S Mauritania to Red Sea coast of E Sudan, Jebel Elba (SE Egypt), N Ethiopia, and coastal lowlands of Somalia (Gulf of Aden, east to 49° E). Below 1000 m. Abundant; in N Senegambia density varies annually from 1·4 to 7·0 birds per 10 ha (Morel and Morel 1978).

Description. *S. r. roseogrisea* (Sundevall): Mauritania and Senegal east to Darfur, N Sudan and W Ethiopia. ADULT ♂: crown and nape pale rosy grey, forehead paler; black half-collar on hindneck edged above with white; sides of neck, foreneck, face and throat pale mauve-pink, chin whitish; mantle, back, rump, uppertail-coverts, scapulars and inner wing-coverts isabelline or pale earth brown. Central rectrices pale grey-brown, others greyer with increasingly broad white tip, and outermost with outer web edged white; tail below black with white distal half (central rectrices pale brown). Underparts pale mauve-pink, but belly whitish, vent white and flanks pale grey. Primaries brown narrowly edged buff, innermost mostly pale grey-brown with browner tip; secondaries pale grey tinged isabelline; tertials like mantle; wing-coverts a tone paler than mantle, outermost pale grey washed light brown. Bill black; eye dark red, narrow white orbital ring; legs and feet wine red. Sexes alike. SIZE: wing, ♂ (n = 88) 169 ± 5, ♀ (n = 91) 165 ± 4; tail, ♂ (n = 15) 105–123 (117), ♀ (n = 16) 95–111 (103); bill, ♂ (n = 15) 14–18 (16·1), ♀ (n = 13) 13–18 (14·8); tarsus, ♂ (n = 13) 24–30 (26·8), ♀ (n = 14) 23–30 (25·2). WEIGHT: (n = 600) 130–166 (149).

IMMATURE: like adult but collar not well defined; crown and breast isabelline; upperparts and most wing-coverts with pale margins. Eye pale-brown; legs and feet dull pink.

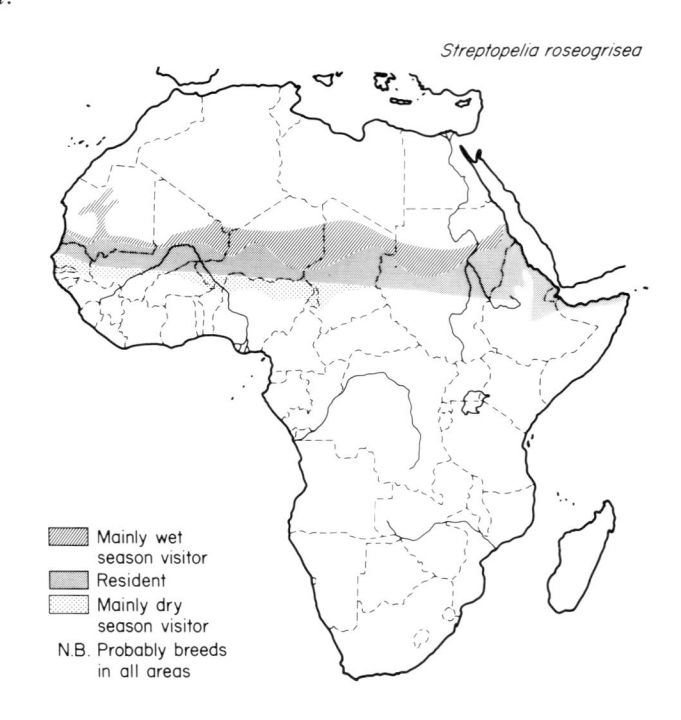

NESTLING: head skin purple sparsely covered with short yellow down; rest of body pink, covered with long dense yellow down.

S. r. arabica (Neumann): Sudan (Red Sea Province), Ethiopia (Eritrea, Awash valley) and Somalia. Darker; underwing coverts tinged grey.

Field Characters. A very pale sandy and pinkish-grey dove of sandy, subdesert habitats, with characteristic bisyllabic song. In size between African Mourning and Vinaceous Doves *S. decipiens* and *S. vinacea*. For distinctions from Eurasian Collared Dove *S. decaocto* (N Africa), see that species. Among sympatric doves most nearly resembles Vinaceous but paler than it, without dark primaries. White orbital ring emphasizes dark eye.

Voice. Tape-recorded (44). Song of 2 notes, a short 'coo' or 'hoo' and a lower and longer 'rrroo'. Each note followed by 'wa' made by strong air inspiration. Song duration *c.* 1·5 s. Excitement cry a high-pitched, whinnying or jeering 'he-he-he-he-he'.

General Habits. Inhabits thornbush, sandy riverbeds, open desert with annual grass, farmland within 25 km of water. Avoids riparian forests; exceptionally in cities (Nouakchott, Mauritania). Forages on ground in pairs or groups of up to 25, often with other doves; hundreds congregate at water and thousands fly to roost. Can eat unshed grain. In N Senegambia drinks early morning and late afternoon at riverbanks, ponds and by wells, never from small puddles; in morning hundreds perch in tree-tops near wells; during hottest hours rest in shade of foliage. Roosts alone or up to 10 in same tree or in large flocks in trees by river or lake.

Mainly sedentary, moving daily between foraging, drinking and roosting places. Migratory at north and south edges of range; in Chad extends north to 16°N Aug–Oct, and moves south into N sudanian zone (Nigeria) for dry season (Nov to mid–May), a few remaining to breed. Southern limit (Nigeria) 11°40'N and (Cameroon) 10°N.

Food. Seeds; also vegetable matter, insects and snails. In N Senegambia 1302 birds contained 68 spp. of seeds; in wet years, 75% of food by weight are monocotyledons (mainly *Panicum laetum*, also cereals and sedge nuts); in dry years eats mainly dicotyledons (up to 80% of *Tribulus terrestris*; also *Gisekia pharnacioides*, *Colocynthis*, *Arachis*, *Cocculus*, *Tinospora*, *Boscia*, *Salvadora* and *Commiphora*). Seeds weigh 0·24–892 mg; av. dry weight of stomach contents usually <5 g.

A

Breeding Habits. Monogamous; territorial. Courtship bowing (studied in Barbary Doves *S.* '*risoria*', the domestic derivative of African Collared Dove) very emphatic: ♂ draws itself up to full height, giving 1st note of song, and bows deeply and slowly (bow av. 2·4 s duration: Davies 1970), bill pointing vertically down (see **A**); 2nd note given when head in lowest position.

NEST: frail open cup on branches easy of access. In N Senegambia 219 nests were mainly in *Balanites aegyptiaca*, also *Acacia raddiana*, *Grewia bicolor*, *Commiphora africana* and 7 other tree species; height above ground 2·5 ± 0·8 m; uses old nests of other species.

EGGS: 1–2, usually 2; laying interval *c.* 36 h. Oval, white. SIZE: av. 29·2 × 23·0 mm. At least 50 days between successive clutches. WEIGHT: (n = 89) 7·5 ± 0·4 g.

LAYING DATES: Senegambia, Mali, every month; Chad, Sept–Oct; Sudan, Dec–June; Ethiopia, Dec–Feb.

INCUBATION: carried out by ♂ and ♀ with reliefs in morning and afternoon; continuous, but eggs can be left unattended at late afternoon relief period. Period: 15 days (in captivity, Senegambia).

DEVELOPMENT AND CARE OF YOUNG: fledging period *c.* 15 days (in captivity, Senegambia).

References
Cramp, S. (ed.) (1985).
Davies, S. J. J. F. (1970).
Morel, G. and Morel, M.-Y. (1972).

Streptopelia decaocto (Frivaldszky). Eurasian Collared Dove. Tourterelle turque.

Not illustrated in colour

Columba risoria L. var. *decaocto* Frivaldszky, 1838, K. magyar tudos Társaság Evkönyvi, 3, 3, p. 183; Turkey.

Forms a superspecies with *S. roseogrisea* and *S. reichenowi*.

Range and Status. Faeroes, Norway, Britain and NW Spain to Sri Lanka and Burma, also Kazakhstan and Mongolia; introduced China and Japan. Has bred Iceland. In W Palearctic natural range has greatly increased, in last 50 years from SE to NW Europe and in last 30 years from Israel to Sinai.

In Africa 1st recorded in 1979, 3 birds in Cairo (Egypt), probably escapes, and 26 near Suez, probably breeding in 1980; recorded in Cairo again in 1980 and at Alexandria in 1983; now at least 50 birds near Suez (Baha El Din, pers. comm.). 2 seen at Asilah (Morocco). Abundant outside Africa, e.g. 1·0–1·4 million W Germany.

488 COLUMBIDAE

Streptopelia decaocto

Voice. Tape-recorded. Advertising call a monotonously repeated trisyllable: 'cu-cooo-cu' or 'gu-guu-gu', accented on 2nd syllable, lacking resonant quality, slightly nasal and more treble than most congeners. Call lasts 1·2 s, repeated usually 3–12 times; 3rd syllable can be muffled by inhalation and tends to be omitted towards end of prolonged singing. Flight call, loud, swearing, very nasal 'kwurr'. Numerous other variant calls of adult and young known (Cramp 1985).

General Habits. In Egypt inhabits mature *Casuarina* plantations in desert and does not (yet) occur habitually in towns, which is also the case in SW Asian heartland of range; but in central and NW Europe a bird almost exclusively of suburbs, villages and farmyards. In parts of range noticeably commoner near coasts than inland. Forages on ground, rests in trees, perches freely on buildings and wires. Not markedly dependent on access to water.

In Europe keeps in pairs all year, juveniles flocking in autumn, later joined by adults; gregarious in winter, flocks usually <100 but up to 10,000 recorded. Single bird or pair flies freely, usually below roof or tree height and for short distance. Flight fast with frequent bursts of clipped beats and glides; 'shooting' acceleration when leaving cover. Mincing walk; can walk far when foraging undisturbed. Roosts usually communally, even when breeding, on outer branches of dense foliage (holly *Ilex*, hawthorn *Crataegus*, willow *Salix*), in conifers in cold weather, sometimes on buildings; sleeps with bill buried in breast feathers and feet in belly plumage. Loafs and sunbathes with plumage fluffed and eyes closed; bathes in rain and damp grass; preens often; drinks by sucking. Feeds mainly on ground; occasionally flies up to seize berry or takes seeds from sunflower *Helianthus* head or insects in tree.

Sedentary; some young birds disperse up to 200 km in spring following year of birth; adults seldom move >10 km. Somewhat of a migrant in mountains and in N Europe.

Food. Cereal grain, seeds and fruits of herbs and grasses; some green parts of plants, insects, breadcrumbs. Eats 11–26 g daily; c. 30 food plants known (Cramp 1985).

Breeding Habits. Breeding not yet proven in Africa. Monogamous, maintains pair-bond all year and probably perennially. Territorial, defending nest area strictly and surrounding area weakly; in Europe usually 1 pair per ha or several ha, but in parks can reach 10–15 pairs per ha with nest 6–50 m apart. ♂ guards territory with advertising calls and display flights, ascending steeply from high perch to c. 10 m high then glides with spread wings and tail back to perch. May wing-clap during ascent. ♀ occasionally displays similarly. ♂ threatens intruder by calling and flying towards it, then perching to perform bowing display, facing rival and alternating repeatedly between upright and crouched postures; with swollen neck and depressed (not fanned) tail, ♂ bows until bill low; 15 bows in c. 25 s.

Description. *S. d. decaocto*: only race in Africa. ADULT ♂: forehead and crown pale grey merging into vinous-buff nape and hindneck and sandy grey-brown mantle. Black collar across hindneck, narrowly edged whitish above and below. Mantle, back, scapulars, lesser wing-coverts and tertials sandy grey-brown; rump, uppertail-coverts and tail sandy grey-brown with grey feather-bases showing, outer tail pale grey (dark grey at base) merging to broadly greyish white tip. Lores and ear-coverts pale grey, merging into delicate vinous buff chin, throat, sides of neck and breast and into pale grey belly and flanks and mid grey undertail-coverts. Proximal half of undertail blackish grey, distal half greyish white. Alula, greater and primary coverts mainly pale grey; secondaries pale grey proximally merging to brownish grey tips, primaries dusky brown; underwing greyish white, remiges dusky. Bill black; eye deep red with cream orbital skin ring; legs and feet mauve-red. Sexes alike. SIZE: wing, ♂ (n = 30) 172–184 (179), ♀ (n = 18) 163–184 (174); tail ♂ (n = 30) 130–153 (141), ♀ (n = 17) 121–144 (134); bill to feathers, ♂ (n = 26) 15–18·5 (16·4), ♀ (n = 18) 15–18 (16·3); tarsus, ♂ (n = 25) 23–26 (24·1), ♀ (n = 18) 22–26 (23·8). WEIGHT: (Cairo) ♂ 133; (Germany, summer) ♂ (n = 29) 150–196 (172), ♀ (n = 18) 125–196 (166).

IMMATURE: lacks black half-collar, otherwise like adult but sandier, feathers of upperparts narrowly margined buff.

Field Characters. Medium-sized *Streptopelia*, pale sandy grey and palest vinous pink, much paler than any congener except African Collared Dove *S. roseogrisea*, distinguished from turtle-doves (*S. turtur*, *S. lugens*, *S. hypopyrrha*) by lack of rufous and by broader white undertail tip, and from most other congeners by white (not red) orbital ring. Slightly larger than African Collared Dove; undertail-coverts grey (not white) and wing-tips darker, but not safely distinguishable from it except by voice.

NEST: flimsy platform of *c.* 140 twigs, stems and roots, av. diam. (n = 25) 21 × 26 cm; built in 3–4 days mainly by ♂, with ♀ gathering material in single near or distant area; sited (n = 595) 1·9–16 m (6·8 m) above ground, in tree or hedge, rarely on building.

EGGS: 2. Oval; white; smooth, slightly glossy. SIZE: (n = 80) 28–34 × 22–25 (31 × 24); weight *c.* 9. 3–6 clutches per year.

LAYING DATES: Europe, all months, mainly Mar–Sept.

INCUBATION: begins with 2nd egg; by ♀ at night and ♂ for 8 h by day. Period: 14–16 (or 18) days; hatching interval 12–40 h.

DEVELOPMENT AND CARE OF YOUNG: brooded continuously when small; cared for by both ♂ and ♀; old enough to fledge at 14 days but stays in nest until 15–16 days when starts making short flights; may keep returning to nest until 19–20 days, then fed by parents out of nest for 1 week, after which independent. Breeds in 1st spring after year of birth; can breed at only 4 months.

BREEDING SUCCESS/SURVIVAL: of 588 nests (Germany) 1–2 eggs hatched in 65% and chicks fledged in 49%. Fledging success increases from Mar (32%) to Aug–Oct (70%). Egg losses mainly due to Magpie *Pica pica*.

References
Cramp, S. (ed.) (1985).
Glutz, U. N. and Bauer, K. M. (1980).
Sueur, F. (1982).

Streptopelia reichenowi (Erlanger). African White-winged Dove. Tourterelle à ailes blanches.

Plate 31 (Opp. p. 464)

Turtur reichenowi Erlanger, 1901. Orn. Monatsb. 9, p. 182; Salakle, Juba R., S Somalia.

Forms a superspecies with *S. decaocto* and *S. roseogrisea*.

Range and Status. Endemic resident, middle reaches of R. Webbi Shebelle (SE Ethiopia) and entire length of R. Juba (SW Somalia) and its Ethiopian tributaries (R. Webbi Gestro, R. Ganale Doria and R. Daua Parma which borders NE Kenya); also *c.* 30, Bura, N Kenya, 1982. Common to abundant; restricted to immediate environs of large rivers in Somalia but rather more widespread in Ethiopia (S Harare, S Bale, SE Sidamo Provinces).

Description. ADULT ♂: crown grey-brown, hindcrown browner; nape grey-brown; brown-black half-collar, 5–8 mm broad, on hindneck; sides of neck and face light grey-brown; chin and throat whitish; mantle (sharply demarcated from head by collar) earth brown; back light blue-grey; rump and uppertail-coverts brown. Central tail feathers earth brown, others with white tip increasing up to 2 cm wide on outermost; below, tail brown-black with distal half whitish, browner on middle feathers. Breast light grey tinged brown, merging to whitish on belly; flanks pale grey tinged brown. Primaries dark brown, secondaries paler; scapulars and tertials earth brown; outer greater and median coverts pale bluish grey with broad whitish edge to outer web forming whitish bar across opened wing; white edging of coverts visible on closed wing; underwing pale bluish grey. Bill brown-black; eye yellowish; legs and feet pink. Sexes similar but ♀ has grey parts browner. SIZE: wing, ♂ (n = 2) 149–150, ♀ (n = 2) 143–144; ♂♀ (n = 8) 134–149 (144). WEIGHT: ♂♀ (n = 14) 98–135 (119) (Ash *et al.* 1974).

IMMATURE: like adult but buff fringes to most feathers.
NESTLING: undescribed.

Field Characters. A pallid brown dove of riverine palms and trees in Somalia and SE Ethiopia. Appearance and habits very like African Collared Dove *S. roseogrisea, q.v.* Eye yellowish, surrounded by conspicuous white ring of tiny feathers. White crescent near bend of wing (outer webs of greater and median coverts) conspicuous in flight but not at rest. Distinctive voice.

Voice. Not tape-recorded. Song a deep-toned 'kok-kooorrr-kok-kooorrr' repeated rapidly *c.* 15 times; also low, crooning 'crooo-crooo-crooo-crooo' (Brown 1977).

General Habits. In Somalia confined to riparian woodland dominated by doum and fan palms (*Hyphaene* sp., *Borassus aethiopicum*) on banks and floodplains of large rivers, and avoids adjacent *Acacia* bush. In SE Ethiopia also inhabits broad-leaved trees within 2 km of rivers, and occurs on irrigation schemes with windbreaks in overgrazed former *Acacia* woodland.

490 COLUMBIDAE

May roost in reedbeds. Gregarious, often in groups of 20, sometimes with Ring-necked and African Mourning Doves *S. capicola* and *S. decipiens*. Forages mainly or entirely on ground. Drinks in morning.

Food. 4 crops were full of red berries.

Breeding Habits.

NEST: 1 was a flimsy semi-transparent platform of loosely-interlaced twigs and petioles, diam. 15 cm, sited 2·5 m high in *Parkinsonia* tree.

EGGS: 2. Oval; white.

INCUBATION: carried out by ♂ and ♀, sitting very tight (human approach to within 1 m); 1 (♀ ?) incubates all night, other by day from c. 08·40 h to 16.00–18.00 h. Relieving bird alights 20 m from nest and approaches by short flights from perch to perch; it calls and is answered by sitting bird, and (near to nest) raises and lowers tail showing pale outer rectrices.

LAYING DATES: Ethiopia, Jan–Feb.

References
Ash, J. S. *et al.* (1974).
Brown, L. H. (1977).

Plate 31 (Opp. p. 464)

Streptopelia turtur Linnaeus. European Turtle Dove. Tourterelle des bois.

Columba turtur Linnaeus, 1758. Syst. Nat. (10th ed.), i, p. 164; India, in error for England.

Range and Status. W Palearctic (Canary Is., N Africa and Britain east to Iran, Afghanistan and 85°E in central Asia), wintering in Africa (Saharan populations resident).

Resident and Palearctic winter visitor. Common and widespread breeding summer visitor in Morocco north of 28°N, N Algeria, Tunisia (south to Remada-Dehibat); less common and more local breeding visitor to coastal Libya and to Egypt (Nile valley, Dakhla oasis). Partially or wholly resident in Algerian oases and Saharan parts of Libya, Niger and Chad (Hoggar, Tassili, Aïr, Tibesti, Ennedi Mts). European and Asiatic birds abundant on autumn and spring passage through N Africa (a few rarely wintering there) and winter south of Sahara in sahelian zone, occurring locally in great abundance. From sea level up to 2750 m. In Mali c. 700,000 winter in Niger R. inundation zone, near Mopti and Niafunke, south to Segou. In Gambia 1 million at Kaur, Feb 1976 and hundreds of thousands at Jakhaly Feb–Mar many years. In Nigeria widespread and frequent south to Zaria (10°N) in flocks of tens or (rarely, in north) 1000. In N Cameroon c. 60,000 in Waza National Park Nov and Feb but uncommon Dec–Jan (Pettet 1976). Migrants common in Chad except south, but winter status there unclear. Common winter visitor in savannas throughout Ethiopia except east, but absent from Somalia and E Africa. Vagrant to Uganda (c. 7, 1983), Kenya (1 record), Somalia (4), Namibia (1) and South Africa (1). Midwinter distribution remains mysterious; probably species is widely but sparsely spread (see map) but has strong tendency to congregate in wooded wetland regions on migration and at times of drought.

Description. *S. t. turtur* Linnaeus: Europe, Asia; winters Senegambia to Ethiopia. ADULT ♂ (breeding): forehead pale grey; crown, nape and hindneck grey-blue; patch of white-tipped black feathers on side of neck which (when raised) appear as 4 diagonal white lines across oval black patch, nearly meeting on hindneck; mantle and back light brown faintly mottled black, each dusky feather broadly edged brown; sides of back blue-grey; rump greyish blue mottled light brown; uppertail-coverts dull grey-blue tipped pale rufous; central rectrices grey-brown, others dark bluish grey with broad white tips, outermost with whole outer web white. Face and chin pale pinkish buff, throat somewhat pinker; breast dull vinous pink, belly and undertail-coverts white. Primaries dark brown and secondaries dull grey-blue, outer webs narrowly margined whitish; scapulars and tertials blackish, broadly edged rufous; primary-coverts grey-black, bluish at base; all outer wing-coverts bluish grey; rest of wing-coverts blackish broadly edged pale rufous. Underwing grey. Undertail black with 25 mm broad white end, and white outer web of outermost feather. Bill blackish; iris yellow with black perimeter, bare patch (wider in front of and behind eye, than above and below it) dark pink. Legs and feet dark pink. ADULT ♂ (non-breeding): rufous feather edges wear, making black centres more conspicuous. ADULT ♀: like ♂ but with duller breast. SIZE: wing, ♂ (n = 8) 172–182 (177), ♀ (n = 4) 170–176 (173); tail, ♂ (n = 12) 96–112; bill, ♂ (n = 12) 17–19; tarsus, ♂ (n = 12) 23–24. WEIGHT: (Senegambia) ♂♀ (n = 18, May) 186 ± 13, ♂♀ (n = 7, July–Sept) 126 ± 15; (France) ♂♀ (n = 48, Apr–May) 100–156 (125); (Portugal) ♂♀ (n = 262, Sept) 85–170 (124).

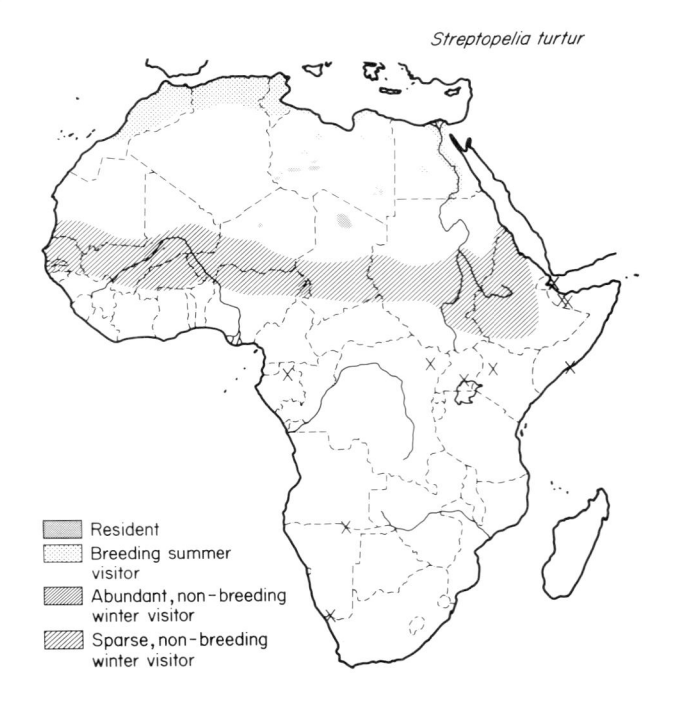

IMMATURE: paler than adult; lacks black and white neck patch until Oct; wing-coverts tipped whitish and tawny; breast pale isabelline, scaly.

NESTLING: skin lead blue with tufts of pale yellow hair-like down.

S. t. arenicola Hartert: NW Africa (Morocco to Fezzan, Libya) and SW Asia (Israel to Iran and Turkestan); winters Senegambia to Ethiopia. Considerably paler; slightly smaller: wing, ♂♀ 166–180 (173).

S. t. hoggara Geyr von Schweppenburg: Saharan mountains (Hoggar, Aïr, Tibesti, Ennedi); resident. General appearance sandy, in tone intermediate between *arenicola* and *isabellina*. Crown buffier, less grey than *S. t. turtur*, back more rufous, wing-coverts more broadly edged rufous.

S. t. isabellina Bonaparte: Libya (Kufra oasis), Egypt, N Sudan; winters in Sudan south to 11°N and in NW Ethiopia. Crown and back pale isabelline-brown; wing-coverts more broadly edged rufous than in *hoggara*; rump feathers and central rectrices tipped rufous. *S. t. rufescens* (Dakhla Oasis, Egypt; winters in Blue Nile Province, Sudan), is probably not separable from *isabellina* (D. Goodwin, pers. comm.).

Field Characters. A small, slender, highly gregarious dove readily distinguished from all but 2 other *Streptopelia* spp. by rufous shoulders, quite apparent at distance in perched and flying bird. 4 white and 3 black diagonal lines on side of neck form large oval patch, unique among African doves. Song is throaty purr, like Adamawa and Dusky Turtle Doves *S. hypopyrrha* and *S. lugens* (the other 2 'rufous' doves) and quite unlike remaining congeners. Inhabits stands of large separated trees in cereal-growing wetland areas. Habitat (and neck-patch) distinguish it from Dusky Turtle Dove, which overlaps in Ethiopia. Overlaps with Adamawa Turtle Dove in N Nigeria; for distinctions see that species.

Voice. Tape-recorded (44, 62, 73, C). Song a monotonous, purring 'rrrr rroorrrrrrrrr' with strongly rolled 'r' s dominating the 'oo'. Duration *c.* 2 s; song repeated 3–4 times with intervals of few s. Not far-carrying. In display a hurried 'rroorr' in time with rapid head-bobbing. Excitement cry, a short percussive note sometimes given on alighting.

General Habits. Inhabits open woodland, farmland, olive orchards and date-palm oases; specially common in open red juniper, evergreen and cork oak, *Thuya* forests interspersed with farms; along forested watercourses, in cultivated valleys, cypress windbreaks of citrus orchards, eucalyptus forest edges. Avoids cedar forests (Morocco: Brosset 1961, Beaubrun and Thévenot, in press). Winter habitat: *Acacia* and *Combretum* savannas, coastal *Suaeda* bush. Forages entirely on ground, often under fierce sun; in Mali in stubble and treeless plains, in Nigeria on wheat fields, on sandy ground and around marshes, always near woods. Roosts in tall trees in wooded dunes and riverine *Acacia* woods by flood water and trees standing in water (Mali). Sunbathes (Curry 1974). Drinks in late afternoon, often also early morning, flocking in thousands (Mali). Solitary or in pairs when breeding; highly gregarious on passage and in winter, in flocks of tens to thousands.

NW African breeding population arrives late Feb to early Apr (in 12 years av. date of 1st arrival, Morocco, 24 Mar) and departs Sept–Oct (in 10 years av. date of last departure, Morocco, 9 Oct); a few depart in Nov and a few (subspecies?) winter. Eurasian populations enter Africa mainly through Iberian Peninsula, on broad front across central Mediterranean, and from Balkan Peninsula to NE Africa. Small and large flocks migrate by day. They do not concentrate at Straits of Gibraltar but there are strong coasting movements along Moroccan Atlantic coast in autumn (Aug–Oct) and spring (Apr–May). Autumn passage stronger than spring passage in N Ethiopia, Chad, Mali and E Senegambia. 600 km further west, in Gambia and W Senegambia, autumn passage almost unknown but large concentrations occur from Feb to mid May (Morel and Roux 1966). Whether wintering birds move westward and have a 'loop migration' (Curry-Lindahl 1981) is questionable. Strong autumn movement across Mediterranean coast of Egypt; formerly (but not presently) strong spring migration through Nile delta (Woldhek 1980). Ringing recoveries southward: England to Mali; Belgium (5), Holland (2), Germany (2), France (2), England and Spain to N Morocco; Belgium to SW Morocco; Morocco to Senegambia. Northward: Senegambia (3) to Morocco and Spain; Morocco to Spain and France; Chad to Greece. Extent of mixing of various subspecies wintering in Africa is uncertain; *turtur*, *arenicola* and *isabellina* all occur in Ethiopia (*isabellina* only in Nov).

Food. Mainly seeds, in Senegambia of 28 spp. (281 crops), particularly *Panicum laetum* and *Tribulus terrestris*. Also wheat, rice, *Cyperus* rhizomes. In Tunisia eats mainly legumes even where wheat abundant. In Europe eats mainly fumitory *Fumaria* seeds; also leaves of sainfoin *Onybrychis*, seeds of *Plantago*, *Stellaria*, *Persicaria*, and a few small snails.

Breeding Habits. Monogamous, territorial. Rises steeply in display flight then glides in long arc, often back to same tree where resumes singing; tail spread during ascent and descent. ♂ bowing display consists of *c.* 6 rapid bows or bobs with breast pouted and bill pointing down, accompanied by 'rroorr' call, performed for 5–10 s before ♀ in tree or on ground.

NEST: very slight, flat platform of fine twigs; diam. 20–24 cm, depth 3–7 cm, built in Aleppo pines, olive trees, date palms, thorn trees, *Eucalyptus*, argan, carob and Phoenician Juniper.

EGGS: 2; laid at interval of 39–48 h. Oval; white. SIZE: (n = 100, Britain) 31·5–33·4 × 23·8–24·7 (30·7 × 22·98).

LAYING DATES: Morocco, May–June (perhaps earlier in south than north); Tunisia, Apr–July.

INCUBATION: begins on 2nd day; carried out by ♀ and ♂. Period: 13–16 (14) days.

DEVELOPMENT AND CARE OF YOUNG: fed on crop milk by both parents. Spends 18–23 days in nest and flies 2–3 days later.

COLUMBIDAE

BREEDING SUCCESS/SURVIVAL: European Turtle Doves are massively hunted on N African coast. They are one of hunters' main quarry species in Morocco, and in Egypt places are rented where 1 gun can shoot 100 birds daily in Sept. Formerly hunted in spring in Nile delta citrus groves (Woldhek 1980).

References
Beaubrun, P. and Thévenot, M. (in press).
Cramp, S. (ed.) (1985).
Curry, P. J. (1974).
Morel, M. Y. (1985).

Plate 31
(Opp. p. 464)

Streptopelia hypopyrrha (Reichenow). Adamawa Turtle Dove. Tourterelle à poitrine rose.

Turtur hypopyrrhus Reichenow, 1910. Orn. Monatsb. 18, p. 174; Benué River (Adamawa, Cameroon).

Forms a superspecies with *S. lugens*.

Range and Status. Endemic, resident and local migrant. N central Nigeria (Jos-Bauchi plateau, from Kaduna to Maiduguri), E Nigeria (Mambilla plateau), N Cameroon (Garoua, Ngaoundéré) and SW Chad (Fianga, Léré). Common and widespread in suburban and rural areas of Jos-Bauchi plateau, elsewhere locally common to rare. Somewhat nomadic.

Description. ADULT ♂: differs from Dusky Turtle Dove *S. lugens* as follows: forehead, face and throat silver grey, upper breast bluish grey, lower breast and belly dark salmon pink with purplish tinge. Tertials and upperwing-coverts more boldly fringed with rufous (like European Turtle Dove *S. turtur*). SIZE: wing, ♂ 175–183; tail, 123–125; bill, 17–18; tarsus, 25.
NESTLING: skin dark flesh-brown with wiry white and pale yellow down.

Field Characters. A medium-sized turtle dove of N Nigeria and N Cameroon, with rufous shoulders and greater wing-coverts which distinguish it from all other doves except Dusky Turtle (E Africa) and European Turtle. Overlaps with European Turtle Dove in Sept–Mar; differs in having solid black (not black and white) patch on side of neck, silver grey face and breast and pink belly; also larger and less gregarious. Distinctive purring voice (European Turtle does not often sing south of Sahara).

Voice. Tape-recorded (44). 3-note song of deep, purring, rolled coos, very like song of European Turtle Dove but rather richer and deeper: 'croorr crr-croor'. Sometimes with 4th syllable: 'croorr crr-crorr coor'. Also given in slower and heavy-toned version: 'croorr, croorr croo', reminiscent of Wood Pigeon *Columba palumbus*. Display coo like that of European Turtle Dove: 'crooa-crooa-crooa-crooa'. Harsh rattling call during Penguin Display.

General Habits. Inhabits edges of dense woods in broken or rolling upland country; wooded ravines, scrub-covered hillsides near water, mangoes and other native and exotic shade trees in parkland, cultivation and gardens. Forages, singly when breeding and in flocks of up to 60 at other times, almost entirely on ground, on bare soil, stubble and farmland, often far from trees. When feeding maintains slow forward walk, pecking small items from ground in front as it moves. Associates with Speckled Pigeons *C. guinea* and Vinaceous Doves *S. vinacea*, and when disturbed the 3 species fly off separately, Adamawa Turtle Doves in compact flock. Not shy.

Sedentary or probably moves only very locally on Jos Plateau; elsewhere in Nigeria reported at Maiduguri only in wet season and on Mambilla Plateau only in dry season.

Food. Very small seeds.

Breeding Habits. Monogamous, territorial. Breeding density of 3 pairs per 2–3 ha, Nigeria (Vom). Advertising coo (song) delivered seasonally from one of several

A

B

favoured perches, at edge of canopy near top of tall tree, or on tall boulder. Neighbouring territory-owners commonly fight, invariably in tree: 2 birds stand close together, parallel, each flicking open wing nearest opponent or holding it briefly aloft; they then strike each other with near wing, rotating through 180° while manoeuvring for position, fighting for up to 30 min; then 1 bird takes flight and is chased away. Does not call while fighting.

In display flight ♂ rises steeply, gives 1–2 wing-claps while ascending until clear of tree-tops, then descends in shallow glide on stiff wings with tail broadly spread, often returning to same perch. Flight lasts up to 30 s. Bowing display and copulation occur on traditional site, usually horizontal branch c. 3 m high, free of obstruction, allowing pair to stand across or along branch. ♂ initiates display by leaning forward to ♀, lowering body until crouched horizontally, raising neck feathers, then moving head rapidly up and down while uttering short 'crooa' notes (**A**). Result is rhythmic calling in time with bowing. ♀ begins to pace up and down in front of ♂, then raises head and approaches him closely; as they meet each assumes vertical stance with neck stretched, and they bill together with closed (sometimes opened) bills. Then ♀ invites copulation by crouching in front of ♂; ♂ mounts and as copulation is completed fluffs throat feathers and flaps wings vigorously, dismounting by unbalancing backward and flying to land next to ♀. Then both birds give Penguin Display, assuming vertical posture and pirouetting slightly until more or less facing each other; each raises head skywards and then back, with head laid far back into shoulders, breast feathers raised and bill tucked into breast (**B**). Penguin Display lasts 3–4 s and is followed by extended self- and allopreening.

NEST: thin, flat, loosely-woven platform of slender twigs; diam. 15 cm; built 4–5 m high in mango (n = 4). ♂ and ♀ collect material, but at any one time only 1 is engaged in doing so. Twigs collected from ground within 20 m and occasionally from nearby trees. Building spasmodic.

EGGS: 2. Elliptical; pure white. SIZE: (n = 1) 27·5 × 22.

LAYING DATES: Nigeria, Dec; but probably Aug–Mar.

Reference
Wood, B. (1975).

494 COLUMBIDAE

Plate 31
(Opp. p. 464)

Streptopelia lugens (Rüppell). Dusky Turtle Dove. Tourterelle cendrée.

Columba lugens Rüppell, 1837. Neue Wirbelth, p. 64; Taranta Mts, Acchele Guzai District, Eritrea.

Forms a superspecies with *S. hypopyrrha*.

Range and Status. Africa and SW Arabia.

Resident, eastern African montane forests, about 1800–3200 m alt. Frequent to abundant in W and SE highlands of Ethiopia. Occurs in Somalia (Alaüli, Golis Mts, Merodijeh, Hullia, Artu, Gildessa); S Sudan (Boma hills, Didinga and Imatong Mts) and N Uganda (Mts Morongole, Lonyili and Moroto); from E Uganda through highlands of Kenya to NE Tanzania (Marsabit, Moyale, Elgon, Nandi, Elgeyu, Mau, Mt Kenya, Machakos, Mbulu, Mt Kilimanjaro); and from SW Uganda down Rift Valley mountains to N Malawî (Kigezi, E Zaïre, Rwanda, Burundi, Mahari Mts, Njombe, Songea, Nyankhowa Mt, Mafinga Mts, Misuku hills) and E Zambia (W Nyika, Mwanda Mt). Frequent to common throughout E African range; in Ethiopia 100 seen in 1 week at Tertale and 550 in 5 days on Arussi plateau; prone to wander, even away from forest, giving rise to such records as one at Kalakol (N Kenya).

Description. ADULT ♂: crown, nape, hindneck and foreneck dark grey; black patch (not in sharp contrast) on side of neck; face light grey brown; chin and throat pale buff; mantle, scapulars and uppertail-coverts dark earth brown, each feather with indistinct paler fringe, scapulars appearing scaly; back and rump dark blue-grey; central tail-feathers brown, others brown-black broadly tipped pale grey. Underparts uniform dark grey, lower breast tinged with rufous. Remiges brown; tertials dark brown broadly edged rufous on outer webs; greater primary coverts, outermost lesser and median coverts dark brown with dark bluish grey edges; inner greater coverts broadly edged rufous; rest of wing-coverts dark earth brown, narrowly edged paler brown, appearing scaly. Underwing dark bluish grey; tail below with central feathers brown edged grey, others black with sharply contrasting broad pale grey tip. Bill blackish; eye orange; orbital ring purple; legs and feet purplish red. ADULT ♀: slightly paler and duller, eye sometimes yellow, black neck-patch a little smaller. SIZE: wing, ♂♀ (n = 10) 175–186 (181); tail, ♂ (n = 7) 107–125 (116), ♀ (n = 3) 100–115 (109). WEIGHT: ♂ (n = 10) 140–205 (173), ♀ (n = 6) 120–190 (154).

IMMATURE: paler and browner, most feathers edged rufous.
NESTLING: skin dark flesh-brown, covered with wiry white and pale yellow hairy down; bill large, soft, purple-brown (van Someren 1956).

Field Characters. A large, rather dark *Streptopelia* of E African montane forests, in size between African Mourning Dove *S. decipiens* and Red-eyed Dove *S. semitorquata* (which occur in same habitats). Distinguished from them by bold dark rufous tertials and greater coverts and from slighter, paler European Turtle Dove *S. turtur* by solid black (not pied) neck-patch. Resembles Adamawa Turtle Dove *S. hypopyrrha*, 2000 km to west. Usually in tree-top foliage and difficult to spot.

Voice. Tape-recorded (10, C, 244, 253, 264, 296). Song of 3–4 heavily rolled coos, 'orrr', slow and rather deep and raucous. Song repeated monotonously. Coo slightly bisyllabic, described as 'koo-or, koo-oor' and 'cu-oor, cu-oor'.

General Habits. Inhabits montane forests (*Juniperus* and *Eucalyptus*), bamboo, gardens, cultivation, pine plantations, and lower heath zone; also cedar and *Hagenia* (Tanzania, south to N Kilosa). Probably moves locally to lower grounds (200 m, Ethiopia, during rains). Very tame in vicinity of camps. In pairs when breeding; thereafter in flocks of up to 100. Forages gregariously on tilled ground, visiting ripening finger-millet gardens May–Aug around Livingstonia (Malawî). Hundreds gather in sunflower fields (Kenya). Occasionally takes standing grain. Drinks about midday.

Food. Weed seeds, cereals, sunflower seeds, small rhizomes, *Salvadora* berries; insects, molluscs.

Breeding Habits. Monogamous, territorial. Display flight like that of European Turtle Dove, ♂ often landing in nesting tree.

NEST: thin, transparent platform of criss-crossed twigs; sometimes lined with fine rootlets when eggs cannot be seen through nest. Sited 2–7 m above ground in tree or level spot among creepers; once in leafless thorn bush. Some nest material said to be collected by ♀ (van Someren 1956), from ground.

EGGS: 2. Oval; white. SIZE: 32 × 23.

LAYING DATES: Ethiopia, all months except July–Aug, mainly in dry season; E Africa: Region A, Jan–Feb, Apr, Nov–Dec; Region C, Mar; Region D, Feb–Mar, Mar–July, Sept–Dec.

INCUBATION: carried out by ♀ and ♂, to same schedule as in congeners, incubating at first very closely. Period: c. 20 days (van Someren 1956). ♂ feeds incubating ♀ by regurgitation (she also feeds for herself).

DEVELOPMENT AND CARE OF YOUNG: at first brooded by ♀ and ♂ very closely. Fed c. 4 times daily, about 08.00, 11.00, 16.00 and 18.00 h; feeding visit lasts 3–5 min; both chicks plunge bills into parent's mouth and food is pumped vigorously, then parent disengages and feeds each chick separately, afterwards picking up and swallowing any fallen seeds in floor of nest. Growth very rapid; chicks large and well feathered by c. 10 days.

Reference
van Someren, V. G. L. (1956).

Streptopelia senegalensis (Linnaeus). Laughing Dove; Palm Dove. Tourterelle maillée.

Plate 31
(Opp. p. 464)

Columba senegalensis Linnaeus, 1766. Syst. Nat. (12th ed.) i, p. 283; Senegal.

Range and Status. Africa, Arabia, Afghanistan, Turkestan and India. Introduced and well established in Turkey, Malta, Lebanon, Syria, Israel and W Australia. Vagrant S and N Europe.

Sedentary. In N Africa increasing in Algeria, partly as result of introductions (north to El Kantara and Algiers, where breeds, west to Bechar and Beni Abbès, south to Timimoun, Adrar, Ouargla, Aïn Amenas and Tamanrasset, Tazrout and Hirafok in Hoggar Mts); vagrant Morocco (Erfoud) 1979; common and increasing in Tunisia except NW and S; common near Libyan coast east to Tawarga, also at Ghadames in W and Serir in E, and several recently found in Fezzan (Mayaud 1984); common in entire Nile valley (Egypt, Sudan), also Dakhla oasis (central Egypt). Occurs in wooded hills and mountains in S Sahara: breeds W and central Mauritania; Adrar (NE Mali), Aïr (NW Niger), Tibesti and Ennedi (N and E Chad), and Red Sea hills (Gebel Elba, SE Egypt/NW Sudan). Throughout subsaharan Africa except for rain forest zone and Somali and SW African deserts; abundant, ubiquitous in villages and towns, widespread in 'bush' (but usually not the commonest dove). São Tomé and Principe Is. (introduced?), Mafia, Socotra I. From sea level to 2750 m (Ethiopia) and 3000 m (Kenya).

Description. *S. s. senegalensis* (Linnaeus): Africa south of about 24°N. ADULT ♂: face, forehead and chin pale vinous pink; rest of head and neck vinous brown or pink, with broad half-collar on throat and sides of neck, each feather black with forked, glossy rufous tip (collar appears rufous with black speckles, particularly when neck inflated); mantle, scapulars, tertials and inner wing-coverts light brown, each feather edged pale rufous; back and rump grey-blue; uppertail-coverts pale grey-brown. Central rectrices pale brown, next 1 grey edged brown, next 1 pale grey with ill-defined broad whitish tip, 3 outermost ones white with dark grey base; tail below grey with black base (inner rectrices greyish with blackish bases). Breast vinous pink, shading to whitish belly and white vent and undertail-coverts. Primaries grey-brown (brown when worn); secondaries dull dark grey; outermost wing-coverts blue-grey; underwing blue-grey. Bill dark brown; eye brown or very dark red; legs and feet dull purplish red. Sexes alike but ♀ has paler head and breast. SIZE: wing, ♂ (n = 16) 136–145 (140), ♀ (n = 13) 128–143 (134); tail, ♂♀ 104–118 (110); bill, ♂♀ 16–21 (18·7); tarsus, ♂♀ 20–25 (22·5).
WEIGHT: heavier in cooler, wetter regions: (N Senegambia, crop emptied) ♂ (n = 232) 81–105 (93), ♀ (n = 176) 77–104 (92); (E Africa) ♂ (n = 23) 60–123 (94), ♀ (n = 12) 78–103 (88); (Namib Desert) ♂♀ (n = 20) 64–103 (87); (SW Cape) ♂♀ (n = 592) 72–139 (103).

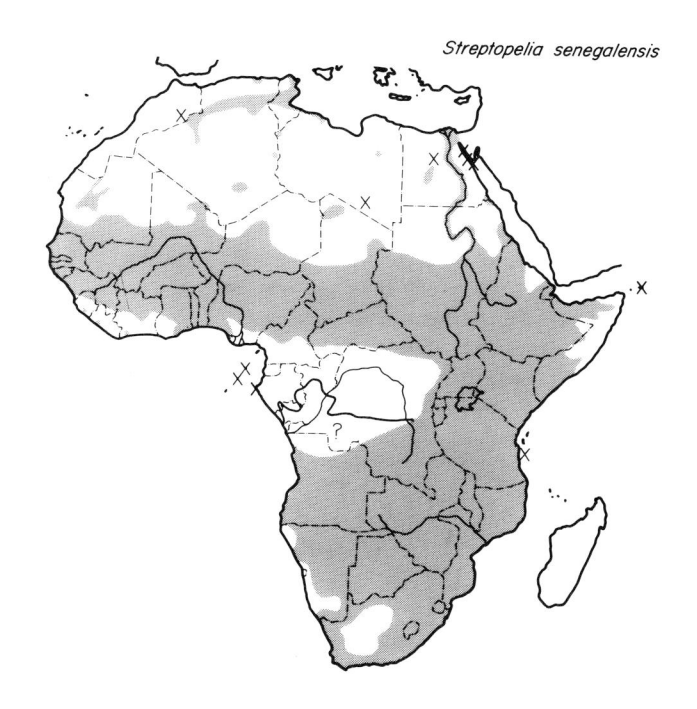

IMMATURE: like adult but head dull rosy brown, half-collar absent, mantle and wing-coverts pale brown washed rufous, breast strongly washed pale rufous, inner primaries broadly tipped rufous.

NESTLING: skin dark purplish grey, soon turning darker, covered with profuse yellow down. Bill blackish, tipped white.

S. s. phoenicophila Hartert: Algeria, Tunisia and Libya north of about 24°N. Larger, browner; wing, ♀ (n = 9) 138–144 (142).

S. s. aegyptiaca (Latham): Egypt. Darker; collar more rufous. Large; wing, ♂ (n = 9) 138–148 (143), ♀ (n = 4) 136–146 (143).

S. s. sokotrae Grant: Socotra Is. Very pale. Small; wing, ♂♀ 123–133.

Field Characters. Commonest dove of subsaharan villages, parks and gardens; small, slim, short-winged, unobtrusive; head and neck pinkish, with broad, black-speckled rufous half-collar at base of foreneck; back rufescent brown, rump grey, tail blackish brown with broad white tips to outer feathers; belly white; wing-coverts pale blue-grey, contrasting with blackish remiges and brown back and conspicuous in flight. Distinctive voice. Flights short, rather weak. Juvenile lacks half-collar.

Voice. Tape-recorded (5, 22, 44, 66, 74, 75). Commonly given song is quiet, soft, rather high-pitched, hurried phrase of 4–8 coos, the 1st muted and last slightly emphatic, rising slightly in scale; duration c. 1·5 s. Muted nasal bleating note after copulation and sometimes on alighting.

General Habits. Inhabits most types of wooded savanna and all kinds of human settlements: farmland, farmsteads, villages, abandoned country buildings, suburban gardens, and central city areas with trees and grass. Keeps within 10 km of water. Much commoner in *Acacia* and other dry biomes than in mesic *Isoberlinia*, *Brachystegia* and *Baikiaea* woods. Solitary or in pairs; less gregarious than other doves but often forms flocks of 3–4, and hundreds may assemble at water. Forages on ground, rather inconspicuously, keeping near to shrubs; rarely plucks fruits, sorghum and millet seeds from standing crop. Walks quickly and easily. Not shy; garden birds put to flight often alight within 2–5 m. Short flights look weak; longer ones swift and strong, 55–68 km/h. Drinks usually twice daily, about 08.00–10.00 h on sunny days or only in afternoon, about 16.30 h, on wet days; shy at water, remaining at small puddle for only few s (max. 23 s). Bathes by standing or splashing in water; sunbathes on limb or ground by squatting down with drooping wings, head up, and mantle feathers raised, sometimes spreading and lifting wings. Rises with wing-clatter. Sings all day, mainly early morning starting 1 h before dawn and late afternoon until 1 h after sunset; sometimes sings at night. During heat of day uses perches less than 4 m above ground, inside foliage or on shady ground.

Partially migratory, at least in southern Africa, where scarcity in 1 season (Barotseland, rains; SE Botswana, dry season; Fish R., Namibia, hot season) and greater abundance in another (Kruger Park, Transvaal, summer; Windhoek, Namibia, winter) suggest westerly movement about Mar–Apr and easterly return to E coast lowveld about Aug–Sept.

4000 birds moved west on Zambezi R. at Feira (Mozambique) on 12 mornings of 4 days in Apr (Tree 1963). 85% of recoveries are within 5 km of ringing place; max. distances 700 km (Mazoe to Beit Bridge, Zimbabwe) and 800 km (Maun, Botswana, to Gatooma, Zimbabwe). Mainly sedentary in tropics but numbers increase in Tsavo E National Park (E Africa) in short rains. Varying abundance at Serir, Libya, suggests migration (Bundy 1976).

Food. Primarily seeds; size mainly <2 mm but up to whole grains of maize and sunflower. Also seeds of wheat, oats, sorghum, millet, *Croton*, *Celosia*, *Amaranthus*, *Oxalis*, *Rhus*, *Polygala*, *Acacia* and legumes. Fond of *Poa annua*, eating seeds off panicle. At Barberspan (South Africa) and in Botswana eats mainly herb and grass seeds Aug–Sept and Dec–Jan, particularly *Eleusine indica*, and cereals for rest of year. Sorghum up to 70% of food (by weight, Botswana).

Also eats salt, nectar of *Aloe*, fruits of *Securinega*, *Atriplex*, *Salvadora*, *Ficus*, pawpaw, olive, succulent stems of *Acacia cyclops*; termites (particularly *Hodotermes mossambicus*), ants, fly larvae and pupae, and snails (Rowan 1983). In N Senegambia eats 96 spp. of seeds and fruit (n = 1214 birds), weighing 0·22–892 mg. Crop content weighs (dry) up to 9 g (av. 2·4 g: *Panicum laetum*). In good rainfall years eats 70–80% monocotyledons (mainly *P. laetum*); in poor years 70% dicotyledons (mainly *Gisekia pharnacioides* and *Tribulus terrestris*). Eats oily seeds (*Colocynthis*, *Arachis*) and protein-rich seeds (*Zornia glochidiata*, *Vigna*). In bad years eats spilled *Pennisetum* meal and cereals gleaned in stubbles or stacks (Morel and Morel 1972).

Breeding Habits. Monogamous; pairs probably for life; territorial. Av. of 1 nest per 0·19 ha (Dean 1980), with up to 21 nests in 4 ha and once 4 nests in 1 tree (some 3 m apart). ♂ defends territory, threatening other ♂♂ with bowing display or chasing and fighting them. In 'towering' display bird rises high with noisy wingbeats, then glides down with outstretched wings and tail, to perch in tree. Bowing display: bird pursues another, puffing neck out and bobbing forepart of body, accompanied by muffled coos. Bows quick and shallow, with rufous half-collar emphasized. Fighting with flailing wings may follow. ♂ and ♀ preen each other. Copulation preceded by displacement self-preening behind wing or by billing, ♀ placing her bill in ♂'s bill (food probably but not certainly passed).

NEST: frail transparent platform of roots, twigs and petioles; diam. 8–14 cm, depth 3–4 cm, but up to 12 cm thick with multiple re-use. 1 nest composed of 180 rootlets; 1 of 85 twigs and 6 grass bents. Built in bush or tree; in N Senegambia uses 19 tree spp., mainly (n = 258 nests) *Acacia raddiana*, *Balanites aegyptiaca* and *Zizyphus mauritiaca*; sited at up to 15 m high, av. 2·3 m. In SW Cape 70% of nests are in exotic trees, and 1% on wooden parts of building; in northern Africa such artificial sites may be commoner. Also builds on old bulbuls' and thrushes' nests. ♂ collects material, mainly loose on ground, sometimes growing, and is highly selective; brings 10–20 items per h. ♀ remains at nest, receives item from ♂, tucks it beneath her, and shuffles to shape nest. 1 nest was built in 2 days. 37% of nests used twice, 15% 3 times, 9% 4–5 times by same or different pair (Transvaal, n = 302 nests: Dean 1980).

EGGS: 1–2; of 619 clutches, 28 of 1 egg and 582 (94%) of 2 eggs; 8 of 3 and 1 of 4 eggs probably each laid by 2 ♀♀ (Dean 1980). Laid in morning of successive days. Oval; white; slightly glossy. SIZE: (n = 140) 24·5–29·5 × 18·0–22·8 (26·2 × 20·1).

LAYING DATES: Tunisia, June; Egypt, Feb; Senegambia, every month; São Tomé and Principe, Dec; E Africa: Region B, May, Sept–Dec; Region C, Jan–Sept (mainly Apr–July); Region D, Jan–Oct (mainly May) (throughout E Africa breeds just after rains and into dry season); Zimbabwe, every month (n = 327 nests), 53% in Aug–Oct; South Africa, every month (n = 1850 nests) with peaks in Sept (Transvaal highveld) or Oct (Cape, Transvaal lowveld). Data from other (tropical) countries suggest breeding in every month.

INCUBATION: by ♀ at night and ♂ by day, starting with 1st or 2nd egg or sitting before 1st egg (Rowan 1983). Relief times: 09.00–12.00 h and 14.00–17.00 h; nest-relief usually accompanied by cooing by both birds. Usually sits very tight and can be touched by person. Period: 12–13 days.

DEVELOPMENT AND CARE OF YOUNG: hatchling weighs 5 g. Wing-quills appear by day 4; all body feathers start breaking open on day 6; well feathered by day 10 (but development rate varies greatly; often 1 chick in nest twice as developed as other). Eyes open on days 4–6. Chick brooded for 6–9 days, parents using same relief schedule as with incubation. When alarmed, parent tumbles from nest to ground, where flutters along feebly for 20–30 m with flapping wings then flies strongly away. Chick defends itself by bill snapping and wing flailing. Leaves nest at 12–13 days when still flightless; stays nearby for 3–4 days, returning to nest to sleep or receive food. No longer returns to nest when able to fly. 2 chicks fed simultaneously, by regurgitation, 1–5 times/h at first, later 0–2 times/h. Each feed of 10 min duration.

BREEDING SUCCESS/SURVIVAL: of 548 eggs 61% hatched; of 340 chicks 75·5% fledged (46% of eggs laid). 3–13% of eggs predated: predators include shrikes, crows, coucals, squirrels and snakes (Rowan 1983); other clutches fail because of desertion or wind. Up to 8 attempted nestings, av. 5·5, per year; min. of 35 days between successful layings (in captivity, Senegambia); 1 pair raised 15 broods in 3 years. Major predator of fledglings is mongoose *Cynictis penicillata* (Transvaal). Other survivals, 53% and 64% of eggs laid in hot dry season and 33% and 46% in cool wet season. Estimated post-fledging survival: 33% in 1st year and 47% annually thereafter. Lives at least 8 years. Adults have numerous predators and are particularly at risk when coming to drink. Can contract fatal Newcastle disease. In SW Cape only 6 other bird species are more commonly killed by road traffic.

References

Cramp, S. (ed.) (1985).
Dean, W. R. J. (1977, 1979a, b, 1980).
Rowan, M. K. (1983).

BIBLIOGRAPHY

Our bibliography has been arranged in two parts: (1) a list of general and regional works and check lists concerning African birds which have been consulted for more or less every family. These have not then been repeated in (2), a list of references grouped under family headings containing all the works referred to by the authors for each section. Therefore, if a reference quoted in the text does not appear under the appropriate family heading in list 2, it will be found in list 1.

These two lists cover all the works consulted by the authors whether actually quoted in the text or not.

1. General and Regional References

Ali, S. and Ripley, S. D. (1969). 'Handbook of the Birds of India and Pakistan', Vol 3. Oxford University Press, Bombay.

Altenburg, W., Engelmoer, M., Mes, R. and Piersma, T. (1982). 'Wintering Waders on the Banc d'Arguin, Mauritania'. Report of Netherlands Orn. Mauritanian Exped. NOME 1980, Groningen.

Archer, G. and Godman, E. M. (1937). 'The Birds of British Somaliland and the Gulf of Aden'. Vol 2. Gurney and Jackson, London.

Ash, J. S. and Miskell, J. E. (1983). Birds of Somalia, their habitat, status and distribution. *Scopus*, Special Suppl. **1**, 1–97.

Bannerman, D. A. (1930–1951). 'The Birds of Tropical West Africa'. Vols 1–8. Crown Agents, London.

Basilio, A. (1963). 'Aves de la isla de Fernando Po'. Editorial Coculsa, Madrid.

Bates, G. L. (1930). 'Handbook of the Birds of West Africa'. John Bale, Sons and Danielsson Ltd, London.

Beaubrun, P. and Thévenot, M. (In press). Statut et repartition actuelle des Galliformes, Charadriiformes et Columbiformes nicheurs au Maroc. *Proc. Symp. Internat. Faune Sauvage Méditerraneénne* (Fès, March 1983).

Benson, C. W. and Benson, F. M. (1977). 'The Birds of Malaŵi'. Montfort Press, Limbe, Malaŵi.

Benson. C. W., Brooke, R. K., Dowsett, R. J. and Irwin, M. P. S. (1971). 'The Birds of Zambia'. Collins, London.

Berruti, A. (1980). Birds of Lake St Lucia. *Southern Birds* **8**, 1–60.

Bouet, G. (1955–1961). 'Oiseaux de l'Afrique Tropicale'. Off. Rech. Scient. Tech. Outre-Mer, Paris.

de Bournonville, D. (1967). Notes d' ornithologie Guinéenne. *Gerfaut* **57**, 145–158.

Britton, P. L. (Ed). (1980). 'Birds of East Africa'. East Africa Natural History Society, Nairobi.

Britton, P. L. and Zimmerman, D. A. (1979). The avifauna of Sokoke Forest, Kenya. *J. E. Afr. Nat. Hist. Soc.* **169**, 1–16.

Brosset, A. (1969). La vie sociale des oiseaux dans une forêt equatoriale du Gabon. *Biol. Gabon.* **5**, 26–69.

Brosset, A. and Erard, C. (1977). New faunistic records from Gabon. *Bull. Br. Orn. Club* **97**, 125–132.

Brown, L. H. and Britton P. L. (1980). 'The Breeding Seasons of East African Birds'. East Africa Natural History Society, Nairobi.

Browne, P. W. P. (1982). Palaearctic birds wintering in southwest Mauritania: species, distributions and population estimates. *Malimbus* **4**, 69–92.

Brunel, J. (1958). Observations sur les oiseaux du Bas-Dahomey. *Oiseau et R.F.O.* **28**, 1–38.

Brunel, J. and Thiollay, J. M. (1969). Liste préliminaire des oiseaux de Côte-d'Ivoire. *Alauda* **37**, 230–254.

Bundy, G. (1976). 'The Birds of Libya'. British Ornithologists' Union Check List No. 1. BOU, London.

Burton, P. J. K. (1974). 'Feeding and Feeding Apparatus in Waders. A Study of Anatomy and Adaptation in the Charadrii.' Brit. Museum (Natural History), London.

Cave, F. O. and Macdonald, J. D. (1955). 'Birds of the Sudan'. Oliver and Boyd, Edinburgh.

Chapin, J. P. (1932, 1939). The birds of the Belgian Congo. *Bull. Am. Mus. Nat. Hist.* 65, 75.

Chappuis, C. (1975). Illustration sonore de problèmes bioacoustiques posés par les oiseaux de la zone éthiopienne. *Alauda*, Suppl. Sonore, **43**, 427–474.

Cheke, R. A. and Walsh, J. F. (1984). Further bird records from the Republic of Togo. *Malimbus* **6**, 15–22.

Clancey, P. A. (1964). 'The Birds of Natal and Zululand'. Oliver and Boyd, Edinburgh.

Clancey, P. A. (1967). 'Gamebirds of Southern Africa'. Purnell, Cape Town.

Clancey, P. A. (1971). A handlist of the birds of southern Moçambique. Lourenço Marques: *Inst. Invest. Cientifica Moçambique*, Ser. A, **10**, 145–302; **11**, 1–167.

Clancey, P. A. (Ed.) (1980). 'Checklist of Southern African Birds', Southern African Ornithological Society, Johannesburg.

Collar, N. J. and Stuart, S. N. (1985). 'Threatened Birds of Africa and Its Islands'. ICBP Bird Red Data Book, 3rd ed., pt. 1. Int. Council for Bird Preserv., Cambridge.

Collet, J. (1982). Birds of the Cradock District. *Southern Birds* **9**, 1–65.

Colston, P. and Curry-Lindahl, K. (In press). The birds of the Mt Nimba region of Liberia. *Bull. Br. Mus. Nat. Hist. Zool.*

Cracraft, J. (1981). Toward a phylogenetic classification of Recent birds of the World (Class Aves). *Auk* **98**, 681–714.

Cramp, S. and Simmons, K. E. L. (Eds) (1980, 1983). 'The Birds of the Western Palearctic', Vols 2, 3. Oxford University Press, Oxford.

Cramp, S. (Ed.) (1985). 'The Birds of the Western Palearctic', Vol. 4. Oxford University Press, Oxford.

Curry-Lindahl, K. (1960). Ecological studies on mammals, birds, reptiles and amphibians in the eastern Belgian Congo. *Annl. Mus. r. Congo belge* **87**, 1–170.

Curry-Lindahl, K. (1981). 'Bird Migration in Africa', Vols 1, 2. Academic Press, London.

Cyrus, D. and Robson, N. (1980). 'Bird Atlas of Natal'. University of Natal Press, Pietermaritzburg, South Africa.

Dean, W. R. J. (1971). Breeding data for the birds of Natal and Zululand. *Durban Mus. Novit.* **9**(6), 59–61.

Dean, W. R. J. (1974). Breeding and distributional notes on some Angolan birds. *Durban Mus. Novit.* **10**, 109–125.

Dupuy, A. (1969). Catalogue ornithologique du Sahara Algérien. *Oiseau et R.F.O.* **39**, 140–160, 225–241.

Dupuy, A. (1984). Synthèse sur les oiseaux de mer observés au Sénégal. *Malimbus* **6**, 79–84.

Eisentraut, M. (1972). Die Wirbeltierfauna von Fernando Poo und Westkamerun. *Bonn. zool. Monogr.* no 3.

Elgood, J. H. (1982). 'The Birds of Nigeria'. British Ornithologists' Union Check List No. 4. BOU, London.

Elgood, J. H., Sharland, R. E. and Ward, P. (1966). Palaearctic migrants in Nigeria. *Ibis* **108**, 84–116.

Elgood, J. H., Fry, C. H. and Dowsett, R. J. (1973). African migrants in Nigeria. *Ibis* **115**, 1–45.

Etchécopar, R. D. and Hüe, F. (1967). 'The Birds of North Africa'. Oliver and Boyd, Edinburgh.

Fairon, J. (1975). Contribution à l'ornithologie de l'Aïr (Niger). *Gerfaut* **65**, 107–134.

Field, G. D. (1974). 'Birds of the Freetown Peninsula'. Fourah Bay College Bookshop, Freetown.

Fjeldså, J. (1977). 'Guide to the Young of European Precocial Birds'. Skarv Nature Publ., Strandgarden.

Frade, F. (1951). 'Catálogo das Aves de Moçambique'. Ministerio das Colonias, Lisbon.

Frade, F. and Bacelar, A. (1955). Catálogo das aves da Guiné Portuguesa. *Anml. Junta. Invest. Ultram.* **10**, 4.

Friedmann, H. (1930). Birds collected by the Childs Frick expedition to Ethiopia and Kenya Colony. *Bull. U.S. Nat. Mus.* **153** 1930, 1–516.

Friedmann, H. (1978). Results of the Lathrop Central African Republic Expedition 1976, ornithology. *Los Angeles Co. Mus. Contrib. Sci.* **287**, 1–22.

Fry, C. H. (1981–1985). Coded bibliography of African ornithology, 1980–1984. *Malimbus* Suppl. **2**, 1–14; **3**, 1–21; **4**, 1–15; **5**, 1–21; **6**, 1–17.

Gaugris, Y., Prigogine, A. and Vande weghe, J. P. (1981). Additions et corrections à l'avifaune du Burundi. *Gerfaut* **71**, 3–39.

Gee, J. P. (1984). The birds of Mauritania. *Malimbus* **6**, 31–66.

Ginn, P. J. (1976). Birds of Makgadikgadi: a preliminary report. *Wagtail* **15**, 21–96.

Glutz von Blotzheim, U. N. and Bauer, K. M. (1973–1980). 'Handbuch der Vögel Mitteleuropas', Vols 5–9. Akademische Verlagsgesellschaft, Wiesbaden.

Goodman, S. M. and Ames, P. L. (1983). A contribution to the ornithology of the Siwa Oasis and Qattara Depression, Egypt. *Sandgrouse* **5**, 82–96.

Goodman, S. M. and Storer, R. W. (In press). The seabirds of the Egyptian Red Sea and adjacent waters with notes on selected Ciconiiformes. *Gerfaut*.

Gore, M. E. J. (1981). 'The Birds of The Gambia'. British Ornithologists' Union Check List No. 3. BOU, London.

Green, A. A. (1983). The birds of Bamingui-Bangoran National Park, Central African Republic. *Malimbus* **5**, 17–30.

Green, A. A. and Sayer, J. A. (1979). The birds of Pendjari and Arli National Parks (Benin and Upper Volta). *Malimbus* **1**, 14–28.

Grimes, L. G. (1972). 'The Birds of the Accra Plains'. Unpubl. ms, Legon.

Hall, D. G. (1983). Birds of Mataffin, Eastern Transvaal. *Southern Birds* **10**, 1–55.

Harrison, C. (1975). 'A Field Guide to the Nests, Eggs and Nestlings of European Birds'. Collins, London.

Heim de Balsac, H. and Mayaud, N. (1962). 'Les Oiseaux du Nord-Ouest de l'Afrique'. Paul Lechevalier, Paris.

Hogg, P., Dare, P. J. and Rintoul, J. V. (1984). Palaearctic migrants in the central Sahara. *Ibis* **126**, 307–331.

Irwin, M. P. S. (1981). 'The Birds of Zimbabwe'. Quest Publ., Harare.

Jackson, F. J. and Sclater, W. L. (1938). 'The Birds of Kenya Colony and the Uganda Protectorate'. Gurney and Jackson, London.

Jehl, J. R., Jr (1968). Relationships in the Charadrii (shorebirds): a taxonomic study based on color patterns of the downy young. San Diego Society Nat. Hist., Memoir 3, 54 pp.

Jensen, J. V. and Kirkeby, J. (1980). 'The Birds of The Gambia'. Aros Nature Guides, Århus.

Johnsgard, P. A. (1981). 'The Plovers, Sandpipers, and Snipes of the World'. University of Nebraska Press, Lincoln.

Keith, G. S. (1971). 'Birds of the African Rain Forests'. Sounds of Nature #9, Fed. Ontario Naturalists and Amer. Mus. Nat. Hist., New York.

Keith, S., Twomey, A., Friedmann, H. and Williams, J. (1969). The avifauna of the Impenetrable Forest, Uganda. *Am. Mus. Novit.* 2389.

Lamarche, B. (1980). Liste commentée des oiseaux du Mali. *Malimbus* **2**, 121–158.

Lawson, P. C. and Edmonds, J. A. (1983). 'Birds of KaNgwane (Mswati District)'. *Southern Birds* **11**, 1–84.

Ledant, J. P., Jacob, J. P., Jacobs, P., Malher, F., Ochando, B. and Roche, J. (1981). Mise à jour de l'avifaune Algérienne. *Gerfaut* **71**, 295–398.

Lippens, L. and Wille, H. (1976). 'Les Oiseaux du Zaïre'. Lannoo, Tielt.

Louette, M. (1974–1975). Contribution to the ornithology of Liberia. *Rev. Zool. Bot. Afr.* **88**(4), 741–748; **92**(3), 639–643.

Louette, M. (1981). 'The Birds of Cameroon: an Annotated Check-List'. Paleis der Academiën, Brussels.

Macdonald, J. D. (1957). 'Contribution to the Ornithology of Western South Africa'. Trustees of the British Museum, London.

Mackworth-Praed, C. W. and Grant, C. H. B. (1957). 'Birds of Eastern and North Eastern Africa'. Vol. 1. Longmans, London.

Mackworth-Praed, C. W. and Grant, C. H. B. (1962). 'Birds of the Southern Third of Africa'. Vol. 1. Longmans, London.

Mackworth-Praed, C. W. and Grant, C. H. B. (1970). 'Birds of West Central and Western Africa'. Vol. 1. Longmans, London.

Maclean, G. L. (1984). 'Roberts' Birds of Southern Africa', 5th ed. Trustees of the John Voelcker Bird Book Fund, Cape Town.

Malbrant, R. (1952). 'Faune de Central Africain Français (Mammifères et Oiseaux)'. Encyclopédie Biol. 15, Lechevalier, Paris.

Malzy, P. (1962). La faune avienne du Mali. *Oiseau et R. F. O.* **32**, no. spécial, pp. 1–81.

Mayaud, M. (1982–1984). Les oiseaux du nord-ouest de l'Afrique. Notes complémentaires. *Alauda* **50**, 45–67, 116–145, 286–309; **51**, 271–301; **52**, 266–284.

McLachlan, G. R. and Liversidge, R. (1978). 'Roberts' Birds of Southern Africa', 4th ed. Trustees of the John Voelcker Bird Book Fund, Cape Town.

Meinertzhagen, R. (1930). 'Nicoll's Birds of Egypt'. Hugh Rees Ltd., London.

Meininger, P. L. and Mullié, W. C. (1981). Egyptian wetlands as threatened wintering areas for waterbirds. *Sandgrouse* **3**, 62–77.

Moltoni, E. and Ruscone, G. G. 1940–1944. 'Gli uccelli dell'Africa Orientale Italiana', 4 Vols, Milan.

Moreau, R. E. (1966). 'The Bird Faunas of Africa and its Islands'. Academic Press, London and New York.

Moreau, R. E. (1972). 'The Palaearctic-African Bird Migration Systems'. Academic Press, London and New York.

Morel, G. J. (1972). 'Liste commentée des oiseaux du Sénégal et de la Gambie'. Off. Rech. Scient. Tech. Outre-Mer, Dakar.

Morel, G. J. (1980). Supplement No. 1 to the 'Liste commentée des oiseaux du Sénégal et de la Gambie'.

(Mimeographed) Off. Rech. Scient. Tech. Outre-Mer, Dakar.

Morel, G. J. and Morel, M. Y. (1982). Dates de reproduction des oiseaux de Sénégambie. *Bonn. zool. Beitr.* **33**, 249–268.

Morel, G. J. and Roux, F. (1966). Les migrateurs paléarctiques au Sénégal. *Terre Vie* **113**, 19–72, 143–176.

Morel, G. J. and Roux, F. (1973). Les migrateurs paléarctiques au Sénégal: notes complémentaires. *Terre Vie* **27**, 523–550.

de Naurois, R. (1969). Peuplements et cycles de reproduction des oiseaux de la côte occidentale d'Afrique. *Mem. Mus. Nat. Hist. Nat.* **56**, 1–312.

Newby, J. E. (1979, 1980). The birds of the Ouadi Rime—Ouadi Achim Faunal Reserve: a contribution to the study of the Chadian avifauna. *Malimbus* **1**, 90–109; **2**, 29–50.

Newman, K. (1978). 'Bird Life in Southern Africa', 2nd ed. Macmillan, Johannesburg.

Newman, K. (1980). 'Birds of the Kruger National Park'. Macmillan, Johannesburg.

Newman, K. (1983). 'Newman's Birds of Southern Africa'. Macmillan South Africa, Johannesburg.

Nikolaus, G. (1984). Further notes of birds new or little known in Sudan. *Scopus* **8**, 36–38; 38–42.

NOME (1982). See Altenburg *et al.* (1982).

Olson, S. L. (1982). A critique of Cracraft's classification of birds. *Auk* **99**, 733–739.

Pakenham, R. H. W. (1979). 'The Birds of Zanzibar and Pemba'. British Ornithologists' Union Check List No. 2. BOU, London.

Peters, J. E. (1934, 1937). 'Check-List of Birds of the World'. Vols 2, 3. Museum of Comparative Zoology, Cambridge, Massachusetts.

Prigogine, A. (1971, 1978, 1984). Les oiseaux de l'Itombwe et de son hinterland. *Mus. Roy. Afr. Centr. Ann., Sér* 8, nos. 185, 223 and 243.

Rand, A. L. (1951). Birds from Liberia. *Fieldiana Zool.* **32**(9), 558–653.

Rand, A. L., Friedmann, H. and Traylor, M. A. (1959). Birds from Gabon and Moyen Congo. *Fieldiana Zool.* **41**, 221–411.

Reichenow, A. (1900–1901). 'Die Vögel Afrikas', Vols 1, 2. Neumann, Neudamm.

Ripley, S. D. and Bond, G. M. (1966). The birds of Socotra and Abd-el-Kuri. *Smithsonian Misc. Coll.* **151**, 1–37.

da Rosa Pinto, A. A. (1970). Um catálogo das aves do distrito da Huila [Angola]. *Mem. Trab. Inst. Invest. Ci. Angola* **6**. Luanda, Angola.

da Rosa Pinto, A. A. and Lamm, D. W. (1953–1960). Contribution to the study of the ornithology of Sul do Save (Mozambique), pts 1–4. *Mem. Mus. Alvaro de Castro* **2**, 65–85; **3**, 125–159; **4**, 107–167; **5**, 69–126.

Salvan, J. (1967–1969). Contribution à l'étude des oiseaux du Tchad. *Oiseau et R.F.O.* **37**, 255–284; **38**, 53–85, 127–150; **39**, 38–69.

Salvan, J. (1972). Notes ornithologiques du Congo-Brazzaville. *Oiseau et R.F.O.* **42**, 241–252.

Schmidl, D. (1982). 'The Birds of the Serengeti National Park, Tanzania'. British Ornithologists' Union Check List No. 5. BOU, London.

Schouteden, H. (1949). De vogels van belgish Congo en van Ruanda-Urundi, II. *Ann. Mus. Congo belge*, C. Zoologie (4), **2**, fasc. 2.

Sclater, W. L. (1924, 1930). 'Systema Avium Aethiopicarum', pt 1, 2. BOU, London.

Serle, W. (1939–1940). Field observations on some northern Nigerian birds. *Ibis* Ser. **14**(3), 654–699; Ser. **14**(4), 1–17.

Serle, W. (1943). Further field observations on Northern Nigerian birds. *Ibis* **85**, 264–300, 413–437.

Serle, W. (1948–1949). Notes on the birds of Sierra Leone. *Ostrich* **19**, 129–141, 187–199; **20**, 70–85.

Serle, W. (1957). A contribution to the ornithology of the eastern region of Nigeria. *Ibis* **99**, 371–418.

Serle, W. (1965). A third contribution to the ornithology of the British Cameroons. *Ibis* **107**, 60–94.

Serle, W. (1981). The breeding seasons of birds in the lowland rain forest and in the montane forest of West Cameroon. *Ibis* **123**, 62–74.

Short, L. L. and Horne, J. F. M. (1981). Bird observations along the upper Nile. *Sandgrouse* **3**, 43–61.

Sibley, C. G. and Ahlquist, J. E. (1972). A comparative study of the egg-white proteins of non-passerine birds. *Bull. Peabody Mus. Nat. Hist.* **39**, 1–276.

Simon, P. (1965). Synthèse de l'avifaune du massif montagneux du Tibesti. *Gerfaut* **55**, 26–71.

Sinclair, J. C. (1984). S.A.O.S. Rarities Committee report. *Bokmakierie* **36**, 64–68.

Smith, K. D. (1957). An annotated check list of the birds of Eritrea. *Ibis* **99**, 1–26, 307–337.

Smith, K. D. (1965). On the birds of Morocco. *Ibis* **107**, 493–526.

Smith, K. D. (1968). Spring migration through southeast Morocco. *Ibis* **110**, 452–492.

Smithers, R. H. N. (1964). 'A Check List of the Birds of the Bechuanaland Protectorate and the Caprivi Strip'. Trustees of the National Museum of Southern Rhodesia, Salisbury.

Snow, D. W. (Ed.) (1978). 'An Atlas of Speciation in African Non-Passerine Birds'. British Museum (Natural History), London.

Storer, R. W. (1971). Classification of birds. In 'Avian Biology'. Farner, D. S., King, J. R. and Parkes, K. C., Eds, Vol. 1. Academic Press, New York.

Strauch, J. G. (1978). The phylogeny of the Charadriiformes (Aves): a new estimate using the method of character compatibility analysis. *Trans. Zool. Soc. Lond.* **34**, 263–345.

Taylor, P. B. (1979). Palaearctic and intra-African migrant birds in Zambia: a report for the period May 1971 to December 1976. *Zambian Orn. Soc. Occ. Pap.* **1**, 1–103.

Thévenot, M., Beaubrun, P., Baouab, R. E. and Bergier, P. (1982). 'Compte-rendu d'ornithologie Marocaine, année 1981'. *Documents de l'Institut Scientifique* **7**, Univ. Mohammed V, Rabat.

Thiollay, J. M. (1971). L'avifaune de la région de Lamto (moyenne Côte-d'Ivoire). *Ann. Univ. Abidjan, sér. E. Ecol.*, **IV** (1), 5–135.

Thomsen, P. and Jacobsen, P. (1979). 'The Birds of Tunisia'. Odenae, Denmark.

Traylor, M. A. (1963). 'Check-list of Angolan birds'. Comp. Diam. Angola, Museo do Dundo, Lisbon.

Urban, E. K. and Brown, L. H. (1971). 'A Checklist of the Birds of Ethiopia'. Haile Sellassie I University Press, Addis Ababa.

van Someren, V. G. L. (1956). Days with birds. *Fieldiana Zool.* **38**, 1–520.

Vaurie, C. (1959, 1965). 'The Birds of the Palearctic Fauna'. 2 vols. Witherby, London.

Verschuren, J. (1978). Observations ornithologiques dans les parcs nationaux du Zaire, 1968–1974. *Gerfaut* **68**, 3–24.

Vielliard, J. (1971–1972). Données biogeographiques sur l'avifaune d'Afrique Centrale. *Alauda* **39**, 227–248; **40**, 63–92.

Voous, K. H. (1973). List of recent Holarctic bird species. Non-passerines. *Ibis* **115**, 612–638.

Welch, G. R. and Welch, H. J. (1984). 'Djibouti Expedition March 1984. A Preliminary Survey of *Francolinus ochropectus* and the Birdlife of the Country'. Publ. by the authors, Knottingley.

Wetmore, A. (1960). A classification for the birds of the world.

Smithsonian Misc. Coll. **139**(11), 1–37.
White, C. M. N. (1965). 'A Revised Check List of African Non-passerine Birds'. Government Printer, Lusaka.
Williams, J. G. and Arlott, N. (1980). 'A Field Guide to the Birds of East Africa'. Collins, London.

Winterbottom, J. M. (1971). 'A Preliminary Check List of the Birds of South West Africa'. S. W. Afr. Scient. Soc., Windhoek.
Witherby, H. F., Jourdain, F. C. R., Ticehurst, N. F. and Tucker, B. W. (1938–1941). 'The Handbook of British Birds'. Vols 4, 5. Witherby, London.

2 References for Each Family

Order GALLIFORMES

Family PHASIANIDAE

Subfamily NUMIDINAE: guineafowl

Aire, T. A., Makinde, M. O., Olowo-Okurun, M. O. and Ayeni, J. S. O. (1980). Visceral organ weights of male and female guinea fowl (*Numida meleagris*). *Afr. J. Ecol.* **18**, 259–264.

Angus, A. and Wilson, K. J. (1964). Observations on the diet of some game birds and Columbidae in Northern Rhodesia. 1. The Helmeted Guineafowl (*Numida meleagris*). *Puku* **2**, 1–9.

Appleton, C. C. (1983). Tetrameriasis in Crested Guineafowl from Natal. *Ostrich* **54**, 238–240.

Ayeni, J. S. O. (1979). A guineafowl research programme in Nigeria. *Malimbus* **1**, 32–35.

Ayeni, J. S. O. (1981). The biology of Helmeted Guineafowl (*Numida meleagris galeata* Pallas) in Nigeria. *World Pheasant Assoc. J.* **6**, 31–39.

Ayeni, J. S. O. (1983). Home range size, breeding behaviour, and activities of Helmeted Guineafowl *Numida meleagris* in Nigeria. *Malimbus* **5**, 37–43.

Basson, N. C. J. (1971). Effect of dieldrin and its photoisomerization product 'photodieldrin' on birds. *Phytophylactica* **3**, 115–124.

Bechinger, F. (1964). Beobachtungen am Weissbrust-Waldhuhn (*Agelastes meleagrides*) im Freileben und in der Gerfangenschaft. *Gefied. Welt* **88**, 61–62.

Benson, C. W. (1948). Geographical voice-variation in African birds. *Ibis* **90**, 48–71.

Benson, C. W. (1960). Breeding seasons of some game and protected birds in Northern Rhodesia. *Black Lechwe* **2**, 149–158.

Benson, C. W. (1961). Breeding season of some game birds. *Black Lechwe* **3**, 8–11.

Benson, C. W. (1963). Breeding seasons of game birds in the Federation of Rhodesia and Nyasaland. *Puku* **1**, 51–69.

Berry, P. S. M. (1972). Distribution of Crested Guineafowl. *Bull. Zamb. Orn. Soc.* **4**, 22.

Board, R. G. (1982). Properties of avian egg shells and their adaptive value. *Biol. Rev.* **57**, 1–28.

Board, R. G. and Perrott, H. R. (1982). The fine structure of the outer surface of the incubated eggshells of the Helmeted Guineafowl (*Numida meleagris*). *J. Zool., Lond.* **196**, 445–451.

Board, R. G., Tullett, S. G. and Perrott, H. R. (1977). An arbitrary classification of the pore systems in avian eggshells. *J. Zool., Lond.* **182**, 251–265.

Bock, W. J. (1969). Origin and radiation of birds. *Ann. N.Y. Acad. Sci.* **167**, 147–155.

Boetticher, H. von (1954). 'Die Perlhühner'. Neue Brehm Bucherei, Ziemsen, Wittenberg, Lutherstadt.

Bonke, B. A., Bonke, D. and Scheich, H. (1979). Connectivity of the auditory forebrain nuclei in the Guinea Fowl (*Numida meleagris*). *Cell Tissue Res.* **200**, 101–121.

Bonke, D., Scheich, H. and Langer, G. (1979). Responsiveness of units in the auditory neostriatum of the Guinea Fowl (*Numida meleagris*) to species-specific calls and synthetic stimuli. I. Tonotopy and functional zones of field L. *J. Comp. Physiol.* **132**, 243–255.

Crowe, T. M. (1977). Variation in intestinal helminth infestation of the Helmeted Guineafowl. *S. Afr. J. Wildl. Res.* **7**, 1–3.

Crowe, T. M. (1978a). The evolution of guinea-fowl (Galliformes, Phasianidae, Numidinae): taxonomy, phylogeny, speciation and biogeography. *Ann. S. Afr. Mus.* **76**, 43–136.

Crowe, T. M. (1978b). Limitation of population in the Helmeted Guineafowl. *S. Afr. J. Wildl. Res.* **8**, 117–126.

Crowe, T. M. (1979). Adaptive morphological variation in Helmeted Guineafowl *Numida meleagris* and Crested Guineafowl *Guttera pucherani*. *Ibis* **121**, 313–320.

Crowe, T. M. (in press). Anti-predator behaviour in Helmeted Guineafowl. *World Pheasant Assoc. J.*

Crowe, T. M. and Crowe, A. A. (1979). Anatomy of the vascular system of the head and neck of the Helmeted Guineafowl *Numida meleagris*. *J. Zool., Lond.* **188**, 221–233.

Crowe, T. M. and Siegfried, W. R. (1978). It's raining guineafowl in the northern Cape. *S. Afr. J. Sci.* **74**, 261–262.

Crowe, T. M. and Snow, D. W. (1978). Numididae. *In* 'An Atlas of Speciation in African Non-Passerine Birds' Snow, D. W. (Ed.), pp. 132–135. British Museum (Natural History), London.

Crowe, T. M. and Withers, P. C. (1979). Brain temperature regulation in Helmeted Guineafowl. *S. Afr. J. Sci.* **75**, 362–365.

DeJager, S. (1963). Breeding habits of guinea-fowl. *Afr. Wildl.* **17**, 300–302.

Dekker, D. (1971). Breeding Vulturine Guineafowl *Acryllium vulturinum*. *Int. Zoo. Yb.* **11**, 98.

Dixon, J. E. W. (1965). Parasites of the Crested Guineafowl *Guttera edouardi*. *Ostrich* **36**, 90.

Dixon, J. E. W. (1978). A preliminary investigation of some economic factors involved in the hunting of some gamebirds. *S. Afr. J. Wildl. Res.* **8**, 81–82.

Dzerzhinsky, F. Y. (1982). Adaptive features in the structure of maxillary system in some Anseriformes and probable ways of evolution of the order. *Zoologichesky Zhurnal* **61**, 1030–1041.

Elbin, S. B. (1979). 'Social organization in a group of free-ranging, Domestic Guinea Fowl'. Unpublished M. Sc. Dissertation, Pennsylvania State University.

Farkas, T. (1965). Interesting facts about the Crowned Guinea Fowl (*Numida meleagris*). *Fauna Flora*, Pretoria **16**, 23–28.

George, J. C. (1966). 'Avian Myology'. Academic Press, London and New York.

Ghigi, A. (1936). 'Galline di Faraone e Tacchini'. V. Hoepli, Milan.

Glenny, F. H. (1951). A systematic study of the main arteries in the region of the heart. Aves XII. Galliformes Pt. 1. *Ohio J. Sci.* **51**, 47–54.

Glenny, F. H. (1955). Modifications of pattern in the aortic arch system of birds and their phylogenetic significance. *Proc. U. S. Nat. Mus.* **104**, 525–621.

Grafton, R. N. (1971). Winter food of the Helmeted Guineafowl in Natal. *Ostrich* Suppl. **8**, 475–485.

Grahame, I. (1969). Breeding of the Vulturine Guinea-fowl (*Acryllium vulturinum*). *Avic. Mag.* **75**, 24–26.

Hall, B. P. (1961). The relationship of the guineafowls *Agelastes meleagrides* Bonaparte and *Phasidus niger* Cassin. *Bull. Br. Orn. Club* **81**, 132.

Happold, D. C. D. (1969). Crested Guinea Fowl in Gambari Forest Reserve. *Bull. Niger. Orn. Soc.* **6**, 60–61.

Hill, G. (1974). Observations on a relationship between Crested Guineafowl and Vervet Monkeys. *Bull. Br. Orn. Club* **94**, 68–69.

Horsburgh, B. R. (1978). 'The Gamebirds and Waterfowl of South Africa'. Winchester Press, Transvaal.

Hudson, G. E., Parker, R. A., Vanden Berge, J. and Lanzellotti, P. J. (1966). A numerical analysis of the modifications of appendicular muscles in various genera of gallinaceous birds. *Amer. Midl. Natur.* **76**, 1–73.

James H. W. (1970). 'Catalogue of birds' eggs in the collection of the National Museums of Rhodesia'. Queen Victoria Museum Trustees, Salisbury.

Kettle, P. R. (1981). The phylogenetic relationship within the Galliformes indicated by their lice (Insecta: Phthiraptera). *Notornis* **28**, 161–167.

Kitto, G. B. and Wilson, A. C. (1966). Evolution of malate dehydrogenase in birds. *Science* **153**, 1408–1410.

Maier, V. (1982). Acoustic communication in the Guinea Fowl (*Numida meleagris*): Structure and use of vocalizations, and the principles of message coding. *Z. Tierpsychol.* **59**, 29–83.

Mainardi, D. (1963). Immunological distance and Phylogenetic relationships in birds. *Proc. XIII Int. Orn. Congr.* pp. 103–114.

Meinertzhagen, R. (1940). Autumn in central Morocco. Part II. *Ibis* Ser. 14, **4**, 187–234.

Mentis, M. T., Poggenpoel, B. and Maguire, R. R. K. (1975). Food of Helmeted Guineafowl in highland Natal. *J. S. Afr. Wildl. Mgmt. Ass.* **5**, 23–25.

Mitchell, P. C. (1901). On the intestinal tract of birds; with remarks on the valuation and nomenclature of zoological characters. *Trans. Linn. Soc., Lond.* 2nd Ser. *Zool.* **8**, 173–275.

Olson, S. L. (1974). *Telecrex* restudied: A small Eocene guineafowl. *Wilson Bull.* **86**, 246–250.

Olson, S. L. and Feduccia, A. (1980). *Presbyornis* and the origin of the Anseriformes (Aves: Charadriomorphae). *Smithsonian Contrib. Zool.* **323**, 1–24.

Oosthuizen, J. H. and Markus, M. B. (1967). The haematozoa of South African birds. *Ibis* **109**, 115–117.

Ortlepp, R. J. (1963). Observations on the cestode parasites of guineafowl from southern Africa. *Onderstepoort J. Vet. Res.* **30**, 95–118.

Pfeffer, P. (1961). Etude d'une collection d'oiseaux de Côte d'Ivoire. *Bull. Mus. Hist. Nat.* **33**, 357–368.

Prager, E. M. and Wilson, A. C. (1976). Congruency of phylogenies derived from different proteins. *J. Mol. Evol.* **9**, 45–47.

Prager, E. M. and Wilson, A. C. (1981). Phylogenetic relationships and rates of evolution in birds. *Proc. XVII Int. Orn. Congr.* pp. 1209–1214.

Scheich, H., Bonke, B. A., Bonke, D. and Langer, G. (1979). Functional organization of some auditory nuclei in the Guinea Fowl (*Numida meleagris*) demonstrated by the 2-deoxyglucose technique. *Cell Tissue Res.* **204**, 17–27.

Scheich, H., Langner, G. and Bonke, D. (1979). Responsiveness of units in the auditory neostriatum of the Guinea Fowl (*Numida meleagris*) to species-specific calls and synthetic stimuli. II. Discrimination of iambus-like calls. *J. Comp. Physiol.* **132**, 257–276.

Siegfried, W. R. (1965a). Guineafowl using telephone poles as roosts. *Bokmakierie* **17**, 17.

Siegfried, W. R. (1965b). Fiscal Shrike attacking young guineafowl. *Ostrich* **36**, 224.

Siegfried, W. R. (1966). Growth, plumage development and moult in the Crowned Guineafowl *Numida meleagris coronata* Gurney. *Dept. Nature Conserv. Investig. Rep.* **8**, 1–52.

Simonetta, A. M. (1963). Cinesi e morphlgia del cranio negli uccelli non passeriformi: Studio su varie tendenze evolutive, parte 1. *Arch. Zool. Ital.* **48**, 53–135.

Skead, C. J. (1962). A study of the Crowned Guinea Fowl *Numida meleagris coronata* Gurney. *Ostrich* **33**, 51–65.

Stock, A. D. and Bunch, T. D. (1982). The evolutionary implications of chromosome banding pattern homologies in the bird order Galliformes. *Cytogenet. Cell Genet.* **34**, 136–148.

Stokes, A. W. and Williams, H. W. (1971). Courtship feeding in gallinaceous birds. *Auk* **88**, 543–559.

Steyn, P. and Tredgold, D. (1967). Crop contents of Crowned Guineafowl *Numida meleagris*. *Ostrich* **38**, 286.

Swank, W. G. (1977). Food of three upland game birds in Selengei Area, Kajiado District, Kenya. *E. Afr. Wildl. J.* **15**, 99–105.

Van Niekerk, J. H. (1979). Social and breeding behaviour of Crowned Guineafowl in the Krugersdorp Game Reserve. *Ostrich* **50**, 188–189.

Verheyen, R. (1956). Contribution à l'anatomie et à la systématique des Galliformes. *Bull. Inst. Sci. Nat. Belg.* **32**, 1–24.

Wilson, K. J. (1965). A note on the crop contents of two Crested Guinea-fowl *Guttera edouardi* (Hartlaub). *Ostrich* **36**, 103–106.

Withers, P. C. and Crowe, T. M. (1980). Brain temperature fluctuations in Helmeted Guineafowl under semi-natural conditions. *Condor* **82**, 99–100.

Wolff, S. (1976). Egg-dumping by the Helmeted Guineafowl. *Bokmakierie* **28**, 97–98.

Wolff, S. W. and Milstein, P. le S. (1976). 'A Guide to the Terrestrial Gamebirds of the Transvaal'. Transvaal Provincial Administration, Nature Conservation Division, Pretoria.

Wolstenholme, P. (1956). Meet my guinea fowl. *Afr. Wildl.* **10**, 153–155.

Subfamily PAVONINAE: Congo Peacock

and

Subfamily GALLINAE: pheasants, quail, partridges, francolins

Alers, I. E. (1943). A 'shot' at a favourite gamebird. *Ostrich* **14**, 31–34.

Appelman, F. J. (1961). The Congo Peacock. *Avic. Mag.* **67**, 41–42.

Arias de Reyna, L. and Alvarez, F. (1974). Comportamiento de la Perdiz Moruna (*Alectoris barbara*) en cautividad. *Doñana Acta Vet.* **1**, 69–82.

Ash, J. S. (1978). The undescribed female of Harwood's Francolin *Francolinus harwoodi* and other observations on the species. *Bull. Br. Orn. Club* **98**, 50–55.

Ash, J. S. (1979). A new species of serin from Ethiopia. *Ibis* **121**, 1–7.

Bannerman, D. A. (1935). On the adult, young and eggs of the Ahanta Francolin (*Francolinus ahantensis ahantensis*). *Bull. Br. Orn. Club* **55**, 132–134.

Beasley, A. J. (1977). Some notes on the birds of the Chimanimani Mountains. *Honeyguide* **90**, 16–21.

Benson, C. W. (1948). Geographical voice-variation in African birds. *Ibis* **90**, 48–71.

Benson, C. W. (1960). Breeding seasons of some game and protected birds in Northern Rhodesia. *Black Lechwe* **2**, 149–158.

Benson, C. W. (1961). Breeding season of some game birds. *Black Lechwe* **3**, 8–11.

Benson, C. W. (1963). Breeding seasons of game birds in the Federation of Rhodesia and Nyasaland. *Puku* **1**, 51–69.

Benson, C. W. and Irwin, M. P. S. (1966). The Common Quail *Coturnix coturnix* in the Ethiopian and Malagasy regions. *Arnoldia (Rhod.)* **2**, 1–14.

Bernstein, M. H. (1973). Development of thermoregulation in the Painted Quail *Excalfactoria chinensis*. *Comp. Biochem. Physiol. (A)*. **44**, 355–366.

Bourquin, O. (1980). 'Biology of the Quail (*Coturnix coturnix* Linnaeus)'. Unpublished Ph. D. dissertation, University of Natal–Pietermaritzburg.

Bowen, W. W. (1930). The relationships and distribution of the bare-throated francolins (*Pternistis*). *Proc. Acad. Nat. Sci. Phil.* **82**, 149–164.

Braine, J. and Braine, S. G. (1970). Two francolin species laying in one nest. *Ostrich* **41**, 263.

Brooke, R. K. (1971). Breeding seasons of Rhodesian francolins. *Honeyguide* **67**, 18–19.

Brosset, A. (1974). La nidification des oiseaux en forêt Gabonaise: Architecture, situation des nids et predation. *Terre Vie* **28**, 579–610.

Button, J. A. (1965). Harlequin Quail at Ilaro and Ibadan. *Bull. Niger. Orn. Soc.* **6**, 53.

Cardwell, P. (1971). The mewing note of Swainson's Francolin. *Wits. Bird Club News Sheet* **76**, 2.

Cave, F. O. (1947). Note on Heuglin's Banded Francolin. *Bull. Br. Orn. Club.* **68**, 3–7.

Cave, F. O. (1949). Some notes on the Banded Francolin, *Francolinus schlegelii* Heuglin. *Bull. Br. Orn. Club* **69**, 103–104.

Chapin, J. P. (1936). A new peacock-like bird from the Belgian Congo. *Rev. Zool. Bot. Afr.* **29**, 1–6.

Chapin, J. P. (1938). The Congo Peacock. *Proc. IX Int. Orn. Congr.* pp. 101–109.

Chapin, J. P. (1946). The range of *Francolinus finschi* extended northward. *Auk* **63**, 434–435.

Chenaux–Repond, R. (1983). Avian prey of the Spotted Eagle Owl. *Honeyguide* **116**, 18.

Clancey, P. A. (1976). Miscellaneous taxonomic notes on African birds 46. On the Quail *Coturnix coturnix* (Linnaeus) in the South African Sub-region. *Durban Mus. Novit.* **11**, 163–176.

Clancey, P. A. (1977). Miscellaneous taxonomic notes on African birds 50. Revised characters and ranges for the South African races of *Francolinus shelleyi* Ogilvie-Grant, 1890. *Durban Mus. Novit.* **11**, 248–251.

Collier, F. S. (1935). The Ahanta Francolin in Ibadan Province, Nigeria. *Ibis* **5**, 665–666.

Courtney–Latimer, M. and Clancey, P. A. (1960). The occurrence of migratory Palaearctic quail in South Africa. *Ostrich* **31**, 169–172.

Crowe, T. M. and Crowe, A. A. (1985). The genus *Francolinus* as a model for avian evolution and biogeography in Africa.: I. Relationships among species. In 'African Vertebrates: Systematics, Phylogeny and Evolutionary Ecology' (Schuchmann, K. L. Ed.) pp. 207–240. Selbstverlag, Bonn.

De Boer, L. E. M. and van Bocxstaele, R. (1981). Somatic chromosomes of the Congo Peafowl (*Afropavo congensis*) and their bearing on the species' affinities. *Condor* **83**, 204–208.

Degen, A. A., Pinshow, B. and Alkon, P. U. (1983). Summer water turnover rates in free-living Chukars and Sand Partridges in the Negev Desert. *Condor* **85**, 333–337.

Delacour, J. (1977). 'The Pheasants of the World'. World Pheasant Association, Surrey.

Dixon, J. E. W. (1978). A preliminary investigation of some economic factors involved in the hunting of some gamebirds. *S. Afr. J. Wildl. Res.* **8**, 81–82.

Dorst, J. and Jouanin, C. (1954). Précisions sur la position systématique et l'habitat de *Francolinus ochropectus*. *Oiseau et R.F.O.* **24**, 161–170.

Dudley, E. P. V. (1971). Development of chicks of the Harlequin Quail. *Ostrich* **42**, 79–80.

Durrer, H. and Villiger, W. (1975). Schillerstruktur des Kongopfaus (*Afropavo congensis* Chapin, 1936) im Elektronenmikroskop. *J. Orn.* **116**, 94–102.

Fitzgerald, T. C. (1969). 'The Coturnix Quail'. The Iowa State University Press, Ames, Iowa.

Friedmann, H. (1962). The Machris Expedition to Tchad, Africa Birds. Los Angeles Co. Mus. *Contrib. Sci.* **59**, 1–27.

Friedmann, H. and Williams, J. G. (1969). The birds of Sango Bay forests, Buddu County, Masaka District, Uganda. Los Angeles Co. Mus. *Contrib. Sci.* **162**, 1–48.

Frost, P. G. H. (1975). The systematic position of the Madagascar Partridge *Margaroperdix madagascariensis*. *Bull. Br. Orn. Club* **95**, 64–58.

Ghigi, A. (1949). Sulla posizione sistematica di *Afropavo congensis* Chapin. *Mem. R. Accad. Sci. Inst. Bologna*, **102**. Sci. Fis. Ser. 10, 3–7.

Grant, C. H. B. and Mackworth-Praed, C. W. (1935). On the Handsome Francolin (*Francolinus nobilis* Reichenow). *Ibis* Ser. 10, **5**, 582–584.

Gysels, H. and Rabaey, M. (1963). Taxonomic relationship of *Afropavo congensis* Chapin 1936 by means of biochemical techniques. *Bull. Soc. Roy. Zool. Anvers* **26**, 25–61.

Haagner, A. K. (1913). The nidification of the Crowned Francolin (*Francolinus sephaena*) in captivity. *J. S. Afr. Orn. Union* **9**, 63–64.

Hachisuka, M. (1937). Note on *Afropavo congensis* Chapin. *Bull. Br. Orn. Club* **57**, 124.

Hall, B. P. (1960). The ecology and taxonomy of some Angola birds. *Br. Mus. Nat. Hist. (Zool).* **6**, 367–453.

Hall, B. P. (1961). The faunistic importance of the Scarp of Angola. *Ibis* **102**, 420–442.

Hall, B. P. (1963). The francolins, a study in speciation. *Bull. Br. Mus. Nat. Hist. (Zool).* **10**, 105–204.

Hall, B. P. and Moreau, R. E. (1962). A study of the rare birds of Africa. *Br. Mus. Nat. Hist. (Zool).* **8**, 315–378.

Harrap, K. S. (1964). Breeding the Natal Francolin in captivity. *Avic. Mag.* **70**, 146–147.

Harrison, C. J. O. (1975). The pair-bond in *Excalfactoria*. *Bull. Br. Orn. Club* **95**, 128.

Harrison, C. (1975). 'A Field Guide to the Nests, Eggs and Nestlings of European Birds'. Collins, London.

Heinze, H. T. and Krott, N. (1979). Zur vogelwelt Marokkos (2). *Vogelwelt* **100**, 225–227.

Holman, F. C. (1947). Birds of the Gold Coast. *Ibis* **89**, 623–650.

Hopkinson, E. (1923). The gamebirds and pigeons of the Gambia. *Avic. Mag.* **1**, 125–131.

Hudson, G. E., Parker, R. A., Vanden Berge, J. and Lanzelotti, P. J. (1966). A numerical analysis of the modifications of appendicular muscles in various genera of gallinaceous birds. *Amer. Midl. Natur.* **76**, 1–73.

Hulselmans, J. L. J. (1963). The comparative myology of the pelvic limb of *Afropavo congensis* Chapin 1936. *Bull. Soc. Roy. Zool. Anvers* **26**, 25–61.

Irwin, M. P. S. (1971). The Red-necked and Swainson's

Francolin in Rhodesia. *Honeyguide* **66**, 29–33.
James, H. W. (1970). 'Catalogue of birds, eggs in the collection of the National Museums of Rhodesia.' Queen Victoria Museum Trustees, Salisbury.
Jeggo, D. (1972). Courtship displays of the Congo Peacock *Afropavo congensis*. *Report Jersey Wildl. Preserv. Trust.* **9**, 43–49.
Kinahan, J. (1975). Preliminary report on a study of *Francolinus hartlaubi*. Unpub. ms.
Kruger, F. J. (1981). Die Seisoenale ekologiese vereistes van die Bosveldfisant *Francolinus* Swainsonnii (Smith 1836). Unpublished report. Department of Nature Conservation Transvaal Provincial Administration. Pretoria, South Africa.
Lovel, T. W. I. (1976). The present status of the Congo Peacock. *J. World Pheasant Assoc.* **1**, 48–57.
Lowe, P. R. (1938). Some preliminary notes on the anatomy and systematic position of *Afropavo congensis* Chapin. *Proc. IX Int. Orn. Congr.* pp. 219–230.
Lynn-Allen, G. (1951). 'Shotgun and Sunlight: The Game Birds of East Africa.' The Batchworth Press, London.
Mackworth-Praed, C. W. and Grant, C. H. B. (1953). On the status of *Pternistis cooperi* Roberts. *Ostrich* **24**, 123.
Mainardi, D. (1963). Immunological distance and phylogenetic relationships in birds. *Proc. XIII Int. Orn. Congr.* pp. 103–114.
Mallet, J. (1964). Breeding of the Red-billed Francolin. *Avic. Mag.* **70**, 72–73.
Marchant, S. (1951). The Scaly Francolin. *Nigerian Field* **16**, 164–166.
Mentis, M. T. (1970). Swainson's Francolin. *Farmer's Weekly* April, p. 75.
Mentis, M. T. (1973). A comparative ecological study of Greywing and Redwing Francolins in the Natal Drakensberg. Unpublished M. Sc. Thesis, University of Stellenbosch.
Mentis, M. T. (1978). Principal foods of African Quail in southern and western Natal. *Lammergeyer* **24**, 1–4.
Mentis, M. T. and Bigalke, R. C. (1973). Management for Greywing and Redwing Francolins in Natal. *J. S. Afr. Wildl. Mgmt. Ass.* **3**, 41–47.
Mentis, M. T. and Bigalke, R. C. (1979). Some effects of fire on two grassland francolins in the Natal Drakensberg. *S. Afr. J. Wildl. Res.* **9**, 1–8.
Mentis, M. T. and Bigalke, R. C. (1980). Breeding, social behaviour and management of Greywing and Redwing Francolins. *S. Afr. J. Wildl. Res.* **10**, 133–139.
Mentis, M. T. and Bigalke. R. C. (1981a). Ecological isolation in Greywing and Redwing Francolins. *Ostrich* **52**, 84–97.
Mentis, M. T. and Bigalke. R. C. (1981b). The effect of scale of burn on the densities of grassland francolins in the Natal Drakensberg. *Biol. Conserv.* **21**, 247–261.
Mentis, M. T. and Bigalke. R. C. (1985). Counting francolins in grassland. *S. Afr. J. Wildl. Res.*
Meyer, H. F. (1971a). Shelley's Francolin. *Honeyguide* **66**, 27–28.
Meyer, H. F. (1971b). Notes on Coqui Francolin. *Honeyguide* **65**, 29–30.
de Naurois, R. (1981). Les Phasianidae de l'île de São Tomé. *Cyanopica* **2** (3) 29–36.
Newby-Varty, B. V. (1946). The Blue Quail. *Ostrich* **17**, 371–372.
Orcutt, F. S. and Orcutt, A. B. (1976). Nesting and parental behaviour in domestic Common Quail. *Auk* **93**, 135–141.
Ogilvie-Grant, W. R. (1896). 'A Handbook to the Gamebirds'. Edward Lloyd, Limited, London.
Oosthuizen, J. H. and Markus, M. B. (1967). The haematozoa of South African birds. *Ibis* **109**, 115–117.
Parker, S. A. (1963). Eggs of *Francolinus africanus lorti* and *Creatophora cinerea* from Somaliland. *Ool. Rec.* **37**, 41–42.

Peek, J. R. (1972). 'The Francolin of Kyle National Park, with special reference to Red-necked and Swainson's Francolin.' Unpublished Dissertation for a Certificate in Field Ecology, University of Rhodesia (Zimbabwe).
Pitman, C. R. S. (1948). The occurrence of *Francolinus n. nobilis* in south-western Uganda. *Ibis* **90**, 129–130.
da Rosa Pinto, A. A. (1966). Notas sobre as coleccoes ornithologicas recolhidas em Angola nas expedicoes efectuadas pelo Instituto de Investigação Cientifica de Angola de 1959 a 1961. *Sep. Biol. Inst. Invest. Cient. Angola*, **3**, 149–236.
da Rosa Pinto, A. A. (1983). 'Ornithologia de Angola', Vol. I. Institutes de Investigação Cientifica Tropical, Lisbon.
Redhead, R. E. (1972). Armyworm (*Spodoptera exempta*) predation by Yellow-necked Spurfowl (*Pternistis leucoscepus*) in the Longido Game controlled Area. *E. Afr. Wildl. J.* **10**, 65–66.
Roberts, A. (1947). A new *Pternistis* from Salisbury, S. Rhodesia. *Ostrich* **18**, 197.
Roles, D. G. (1973). Breeding the Red-necked Francolin, *Pternistes afer* at the Jersey Wildlife Preservation Trust. *Avic. Mag.* **79**, 204–207.
Seftel, S. M. (1974). Note to editor. *Bokmakierie* **26**, 52–53.
Serle, W. (1962). The Cameroon Mountain Francolin. *Nigerian Field* **27**, 34–36.
Sessions, P. J. B. (1967). Notes on the birds of Lengetia Farm, Mau Narok. *J. E. Afr. Nat. Hist. Soc.* **26**, 18–48.
Siegfried, W. R. (1971). Chukar Partridge on Robben Island. *Ostrich* **42**, 158.
Stock, A. D. and Bunch, T. D. (1982). The evolutionary implications of chromosome banding pattern homologies in the bird order Galliformes. *Cytogenet. Cell Genet.* **34**, 136–148.
Stokes, A. W. and Williams, H. W. (1971). Courtship feeding in gallinaceous birds. *Auk* **88**, 543–559.
Stronach, B. (1966). The feeding habits of the Yellow-necked Spurfowl (*Pternistis leucoscepus*) in northern Tanzania. *E. Afr. Wildl. J.* **4**, 76–81.
Swank, W. G. (1977). Food of three upland game birds in Selengi Area, Kajiado District, Kenya. *E. Afr. Wildl. J.* **15**, 99–105.
Taibel, A. M. (1961). Analogie fisio-etologische nel settore riproduttivo tra *Afropavo* Chapin e Penelope Merrem. *Natura, Milano* **52**, 57–64.
Thiollay, J.-M. (1970). Recherches écologiques dans la savane de Lamto (Côte d'Ivoire): le peuplement avien. Essai d'étude quantitative. *Terre Vie* **1**, 108–144.
Thiollay, J.-M. (1973). Place des oiseaux dans les chaînes trophiques d'une zone préforestière en Côte d'Ivoire. *Alauda* **61**, 273–300.
Traylor, M. A. (1960a). Notes on the birds of Angola, non-passeres. *Publ. cult. Co. Diam. Ang., Lisboa* **51**, 129–186.
Traylor, M. A. (1960b). *Francolinus schlegelii* Heuglin in Cameroon. *Bull. Br. Orn. Club* **80**, 86–88.
Trollope, J. (1966). Some observations on the Harlequin Quail (*Coturnix delegorguei*). *Avic. Mag.* **72**, 5–6.
Turner, D. A. (1977). Status and distribution of East African endemic species. *Scopus* **1**, 2–11.
Van Bemmel, A. C. (1961). L'élévage en captivité du Paon congolais (*Afropavo congensis* Chapin) à l'initiative de notre Société. *Zoo, Antwerp* **26**, 94–99.
Van den Bergh, W. (1975). Breeding the Congo Peacock at the Royal Society of Antwerp. *In* 'Breeding Endangered Species in Captivity' (Martin, R. D. Ed.), pp. 75–86. Academic Press, London and New York.
Van den Bergh, L., Chardome, M. and Peel, E. (1963). Les parasites du Paon Congolais, *Afropavo congensis* Chapin. 1. *Haemoproteus chapini*; 2. *Microfilaria chapini*. *Rev. Zool. Bot. Afr.* **67**, 74–80.
Van Niekerk, J. H. (1983). Observations on courtship in

Swainson's Francolin. *Bokmakierie* **35**, 90–92.
van Someren, V. G. L. (1916). A list of birds collected in Uganda and British East Africa, with notes on their nesting and other habits. *Ibis* Ser. 10, **4**, 193–252.
Verheyen, R. (1953). 'Exploration du Parc National de l'Upemba. Oiseaux.' (Mission G. F. de Witte), **19**, 1–687.
Verheyen, R. (1956). Contribution à l'anatomie et à la systématique des Galliformes. *Bull. Inst. Sci. nat. Belg.* **32**, 1–24.
Verheyen, W. (1962). The Congo Peacock *Afropavo congensis* Chapin 1936 at Antwerp Zoo. *Int. Zoo Yb.* **4**, 87–91.
Verheyen, W. N. (1962). Monographie du Paon congolais *Afropavo congensis* Chapin 1936. *Bull. Soc. Roy. Zool. Anvers* **26**, 1–98.
Verheyen, W. N. (1965). 'Der Kongopfau'. Ziemsen, Wittenberg, Lutherstadt.
Verschuren, J. and Mankarika, M. B. (1982). Note sur les oiseaux "apparents" du Congo et remarques sur l'adaptabilité aux facteurs anthropiques. *Gerfaut* **72**, 307–323.

Vince, M. A. (1970). Communication between quail embryos and the synchronisation of hatching. *Proc. XV Int. Orn. Congr.* pp. 357–362.
Vincent, A. W. (1945). On the breeding habits of some African birds. *Ibis* **87**, 203–216.
Vincent, J. (1934). The birds of northern Portuguese East Africa. Part III. *Ibis* Ser. 13, **4**, 305–340.
Walls, E. S. (1933). Field-notes on the Sierra Leone Bushfowl (*Francolinus bicalcaratus thornei*). *Ibis* Ser. 13, **3**, 129–132.
White, C. M. N. (1952). On the genus *Pternistis*. *Ibis* **94**, 306–309.
Woldhek, S. (1980). 'Bird Killing in the Mediterranean' European Committee for the Prevention of Mass Destruction of Migratory Birds. Zeist, Netherlands.
Wolff, S. (1978). The use of tape-recorded calls to locate and census Orange River Partridges. *S. Afr. J. Wildl. Res.* **8**, 135–136.
Wolff, S. W. (1978). Secretary Bird swallowing egg. *Bokmakierie* **30**, 53.

Order GRUIFORMES

Family TURNICIDAE: button-quail

Aspinwall, D. R. (1978). Striped Crakes and other dambo birds near Lusaka. *Bull. Zamb. Orn. Soc.* **10**(2), 52–56.
Britton, P. L. (1970). Birds of the Balovale District of Zambia. *Ostrich* **41**(3), 145–190.
Butler, A. G. (1905). On breeding *Turnix lepurana* in German birdrooms. *Avic. Mag.* **3**, 195–203.
Clancey, P. A. (Ed) (1965). Ninth report of the S. A. O. S. List Committee. *Ostrich* **36**, 51–58.
Clancey, P. A. (1967). The South African races of the hemipode *Turnix sylvatica* (Desfontaines). *Bull. Br. Orn. Club* **87**, 114–117.
Clancey, P. A. (1978). Miscellaneous taxonomic notes on African birds 52. On *Ortygis lepurana* Smith, 1836. *Durban Mus. Novit.* **11**, 311–312.
Elgood, J. H. (1964). Provisional checklist of the birds of Nigeria. *Bull. Niger. Orn. Soc.* **1**(1), 13–25.
Gee, J. and Higham, J. (1977). Birds of Lagos, Nigeria. *Bull. Niger. Orn. Soc.* **13**, 43–51, 103–132.
Gray, H. H. (1965). Some notes on the birds of Tivland. *Bull. Niger. Orn. Soc.* **2**, 66–68.
Hendrickson, H. T. (1969). A comparative study of the egg white proteins of some species of the avian order Gruiformes. *Ibis* **111**, 80–91.
Hoesch, W. (1959). Zur Biologie des sudafrikanischen Laufhunchens *Turnix sylvatica lepurana*. *J. Orn.* **100**, 341–349.
Hoesch, W. (1960). Zum Brutverhalten des Laufhunchens *Turnix sylvatica lepurana*. *J. Orn.* **101**, 265–275.
Horsbrugh, B. (1912). 'The Game-birds and Water-fowl of South Africa'. Witherby, London.
Jackson, F. J. (1926). 'Game-birds of Kenya and Uganda'. Williams and Norgate, London.
Lack, P. C. (1975). Range expansion of the Quail Plover. *Bull. E. Afr. Nat. Hist. Soc.* **1975**, 110.
Lack, P. C., Leuthold, W. and Smeenk, C. (1980). Check-list of the birds of Tsavo East National Park. *J. E. Afr. Nat. Hist. Soc. and Nat. Mus.* **170**, 1–24.
Lamm, D. W. and Horwood, M. T. (1958). Species recently added to the list of Ghana birds. *Ibis* **100**, 175–178.
Lawson, W. J. (1970). The downy plumage of the Kurrichane Button-quail. *Ostrich* **41**, 253–254.
Lowe, P. R. (1923). Notes on the systematic position of *Ortyxelos* together with some remarks on the relationships of the Turnicomorphs and the position of the Seed-Snipe (Thinocoridae) and Sand-Grouse. *Ibis* **5**, 276–299.
Lynes, H. (1925). On the birds of North and Central Darfur, with notes on the West-Central Kordofan and North Nuba Provinces of British Sudan. (Part V). *Ibis* Ser. 12, 541–590.
McInnis, D. (1933). Nesting habits of some East African birds. *J. E. Afr. and Uganda Nat. Hist. Soc.* **48**, 128–135.
Masterson, A. N. B. (1969). Flufftails and button quails. *Honeyguide* **59**, 15–17.
Masterson, A. N. B. (1973). Notes on the Hottentot Buttonquail. *Honeyguide* **74**, 12–16.
Paget-Wilkes, A. H. (1938). Notes on the breeding of some species in north-east Uganda. *Ibis* **2**, 118–129.
Pakenham, R. H. W. (1936). Field notes on the birds of Zanzibar and Pemba. *Ibis* **6**, 249–272.
Ripley, S. D. and Heinrich, G. (1960). Additions to the avifauna of northern Angola. *Postilla* **47**, 1–7.
Roberts, A. (1935). Birds in: Scientific results of the Vernay-Lang Kalahari expedition, March to September, 1930. *Ann. Trans. Mus.* **16**, 1–185.
Sharland, R. E. and Wilkinson, R. (1981). The birds of Kano State, Nigeria. *Malimbus* **3**(1), 7–30.
Steyn, P. (1980). Observations on the prey and breeding success of Wahlberg's Eagle. *Ostrich* **51**, 56–59.
Stoneham, H. F. (1931). A race of the Lesser Button-Quail *Turnix nana* Sund., new to science. *Bateleur* **3**, 79–80.
Stoneham, H. F. (1932). The breeding of the East African Lesser Button Quail (*Turnix nana luciana*) in Kenya Colony. *Ool. Rec.* **12**, 14–16.
Trollope, J. (1970). Behaviour notes on the Barred and Andalusian Hemipodes (*Turnix suscitator* and *Turnix sylvatica*). *Avic. Mag.* **76**, 219–227.
van Someren, V. G. L. (1926). The birds of Kenya and Uganda, Part 4. *J. E. Afr. and Uganda Nat. Hist. Soc.* **27**, 197–212.
Wintle, C. C. (1975). Notes on the breeding habits of the Kurrichane Buttonquail. *Honeyguide* **82**, 27–30.

Family RALLIDAE: rails, flufftails, crakes, gallinules, moorhens, coots

Alley, R. and Boyd, H. (1950). Parent-young recognition in the Coot *Fulica atra*. *Ibis* **92**, 46–51.

Anon. (1983). Another new record for Tanzania. *Museum Avifauna News*, National Museum of Kenya, Dept. of Ornithology, May, p. 21.

Ash, J. S. (1978). *Sarothrura* crakes in Ethiopia. *Bull. Br. Orn. Club* **98**, 26–29.

Aspinwall, D. R. (1978). Striped Crakes and some other dambobirds near Lusaka. *Bull. Zamb. Orn. Soc.* **10**, 52–56.

Astley Maberly, C. T. (1935a). Notes on *Sarothrura elegans*. *Ostrich* **6**, 39–42.

Astley Maberly, C. T. (1935b). Further notes upon *Sarothrura elegans*. *Ostrich* **6**, 101–104.

Astley Maberly, C. T. (1961). *Sarothrura rufa*—Red-chested Flufftail Rail. *Ostrich* **32**, 137.

Bates, G. L. (1927). Notes on some birds of Cameroon and the Lake Chad Region: their status and breeding times. *Ibis* Ser. 12. **2**(1), 1–64.

Baur, S. (1980). A breeding record of the Baillon's Crake. *Bokmakierie* **32**, 108.

Benson, C. W. (1957). Migrants at the south end of Lake Tanganyika. *Bull. Br. Orn. Club* **77**, 88.

Benson, C. W. (1964). Some intra-African migratory birds. *Puku* **2**, 53–56.

Benson, C. W. and Holliday, C. S. (1964). *Sarothrura affinis* and some other species on the Nyika Plateau. *Bull. Br. Orn. Club* **84**, 131–132.

Benson, C. W. and Irwin, M. P. S. (1965). Some intra-African migratory birds II. *Puku* **3**, 45–55.

Benson, C. W. and Irwin, M. P. S. (1971). A South African male of *Sarothrura ayresii* and other specimens of the genus in the Leiden Museum. *Ostrich* **42**, 227–228.

Benson, C. W. and Pitman, C. R. S. (1984). Further breeding records from Northern Rhodesia (No. 4). *Bull. Br. Orn. Club* **84**, 54–60.

Benson, C. W. and Pitman, C. R. S. (1966a). Further breeding records from Zambia (formerly Northern Rhodesia) (No. 5). *Bull. Br. Orn. Club* **86**, 21–33.

Benson, C. W. and Pitman, C. R. S. (1966b). On the breeding of Baillon's Crake *Porzana pusilla* (Pallas) in Africa and Madagascar. *Bull. Br. Orn. Club* **86**, 141–143.

Benson, C. W. and Winterbottom, J. M. (1968). The relationship of the Striped Crake *Crecopsis egregia* (Peters) and the White-throated Crake *Porzana albicollis* (Vieillot). *Ostrich* **39**, 177–179.

Boyd, H. J. and Alley, R. (1948). The function of the head-colouration of the nestling Coot and other nestling Rallidae. *Ibis* **90**, 582–593.

Brass, D. (1981). Back Yard Birds: Bandera Lane, Nairobi. *Bull. E. Afr. Nat. Hist. Soc.* **1981**, 98–99.

Broekhuysen, G. J., LeStrange, G. K. and Myburgh, N. (1964). The nest of the Red-chested Flufftail (*Sarothrura rufa*) (Vieillot). *Ostrich* **35**, 117–120.

Brooke, R. K. (1968). On the distribution, movements and breeding of the Lesser Reedhen *Porphyrio alleni* in southern Africa. *Ostrich* **39**, 259–262.

Brooke, R. K. (1974). The Spotted Crake *Porzana porzana* (Aves: Rallidae) in south-central and southern Africa. *Durban Mus. Novit.* **10**(3), 43–52.

Brooke, R. K. (1975) Cooperative breeding, duetting, allopreening and swimming in the Black Crake. *Ostrich* **46**, 190–191.

Brosset, A. (1968). Localisation écologique des oiseaux migrateurs dans la forêt équatoriale du Gabon. *Biol. Gabon.* **4**(3), 211–226.

Brown, L. H. (1966). A report on the National Geographic Society/World Wildlife Fund Expedition to study the Mountain Nyala *Tragelaphus buxtoni*. Mimeo, typed, 69 pp.

Cheesman, R. E. (1935). On a collection of birds from north-western Abyssinia. *Ibis* Ser. 8, **5**, 297–329.

Cooper, J. (1970). Nest of the African Crake. *Honeyguide* **62**, 34.

Cottrell, C. B. (1949). Notes on some nesting habits of the Buff-spotted Pigmy Rail. *Ostrich* **20**, 168–170.

Cowan, P. J. (1982). Birds in west central Libya, 1980–81. *Bull. Br. Orn. Club* **102**, 32–35.

Curry-Lindahl, K. (1956). Ecological studies on mammals, birds, reptiles and amphibians in the eastern Belgian Congo. *Annl. Mus. r. Congo belge* Ser. 8, **42**.

Dahm, A. G. (1969). A Corn-crake, *Crex crex* L., trapped in Kumasi, Ghana. *Bull. Br. Orn. Club* **89**, 76–78.

Dean, W. R. J. (1976). Breeding records of *Crex egregia*, *Myrmecocichla nigra* and *Cichladusa ruficauda* from Angola. *Bull. Br. Orn. Club* **96**, 48–49.

Dean, W. R. J. (1980). Brood division by Red-knobbed Coot. *Ostrich* **51**, 125–126.

Dean, W. R. J. and MacDonald, I. A. W. (1981). A review of African birds feeding in association with mammals. *Ostrich* **52**, 135–155.

Dean, W. R. J. and Skead, D. M. (1978). Problems in sexing Red-knobbed Coots. *Safring News* **7**, 9–11.

Dean, W. R. J. and Skead, D. M. (1979). Moult and mass of the Red-knobbed Coot. *Ostrich* **50**, 199–202.

Deetjen, H. (1969). Zur feldornithologischen Kennzeichnung von *Fulica atra* und *F. cristata*. *J. Orn.* **110**, 107.

Dorst, J. and Roux, F. (1973). L'avifaune des forêts de *Podocarpus* de la province de l'Arussi, Ethiopie. *Oiseau et R.F.O.* **43**, 269–304.

Douthwaite, R. J. (1978). Geese and Red-knobbed Coot on the Kafue Flats in Zambia, 1970–1974. *E. Afr. Wildl. J.* **16**, 29–47.

Dupuy, A. (1969). Catalogue ornithologique du Sahara Algérien. *Oiseau et R.F.O.* **39**, 140–160.

Erard, C. and Larigauderie, F. (1972). Migration prénuptiale dans l'ouest de la Libye. *Oiseau et R.F.O.* **42**, 81–169, 253–284.

Erard, C. and Vielliard, J. (1977). *Sarothrura rufa* (Vieillot) au Togo. *Oiseau et R.F.O.* **47**, 309–310.

Fagan, M. J., Schmitt, M. B. and Whitehouse, P. J. (1976). Moult of Purple Gallinule and Moorhen in the southern Transvaal. *Ostrich* **47**, 226–227.

Friedmann, H. (1962). The Machris expedition to Tchad, Africa. Birds. *Los Angeles Co. Mus. Contrib. Sci.* **59**.

Fry, C. H. (1966). On the feeding of Allen's Gallinule. *Bull. Niger. Orn. Soc.* **3**, 97.

Gaugris, Y. (1979). Les oiseaux aquatiques de la plaine de la basse Rusizi (Burundi) (1973–1978). *Oiseau et R.F.O.* **49**, 133–153.

Gillard, L. (1976). The secretive flufftails. *Bokmakierie* **28**, 98–99.

Guichard, K. M. (1947–48). Notes on *Sarothrura ayresii* and three birds new to Abyssinia. *Bull. Br. Orn. Club* **68**, 102–104.

Hamling, H. H. (1949). King Reed-Hen or Purple Gallinule. *Porphyrio madagascariensis*. *Ostrich* **20**, 91–94.

Harrison, C. J. O. and Parker, S. A. (1967). The eggs of Woodford's Rail, Rouget's Rail, and the Malayan Banded Crake. *Bull. Br. Orn. Club* **87**, 14–16.

Harvey, W. G. (1971). A note on Lesser Moorhen *Gallinula angulata* in Tanzania. *Bull. E. Afr. Nat. Hist. Soc.* **1971**, 94–95.

Holyoak, D. T. (1970). The behaviour of captive Purple Gallinules (*Porphyrio porphyrio*). *Avic. Mag.* **76**, 98–109.

Hopkinson, G. and Masterson, A. N. B. (1975). Notes on the Striped Crake. *Honeyguide* **84**, 12–21.

Hopkinson, G. and Masterson, A. N. B. (1977). On the occurrence near Salisbury of the Whitewinged Flufftail. *Honeyguide* **91**, 25–28.

Hudson, R. (1974). Allen's Gallinule in Britain and the Palearctic. *British Birds* **67**, 405–413.

Keith, S. (1973). The voice of *Sarothrura insularis*, with

further notes on members of the genus. *Bull. Br. Orn. Club* **93**, 130–136.

Keith, S., Benson, C. W. and Irwin, M. P. S. (1970). The genus *Sarothrura* (Aves, Rallidae). *Bull. Am. Mus. Nat. Hist.* **143**, Art 1.

Lambert, K. (1969). Amerikanisches Sultanshuhn (*Porphyrula martinica*) vor der westafrikanischen Kuste. *J. Orn.* **110**, 219–220.

Langley, C. H. (1979). Lesser Gallinule from the Cape Peninsula. *Ostrich* **50**, 62.

Liversidge, R. (1968). The first plumage of *Sarothrura rufa*. *Ostrich* **39**, 200.

Macleay, K. N. G. (1960). 'The Birds of Khartoum Province'. *Univ. Khartoum Nat. Hist. Mus. Bull* **1**.

Madden, S. T. and Schmitt, M. B. (1976). Scavenging by some Rallidae in the Transvaal. *Ostrich* **47**, 68.

Malbrant, R. (1957). La Foulque noire au Tchad. *Oiseau et R.F.O.* **27**, 303–304.

Marshal, H. W. (1958). Observations on two Purple Gallinules *Porphyrio alba* (White) and their single chick. *Ostrich* **29**, 10–13.

Mendelsohn, J. M., Sinclair, J. C. and Tarboton, W. R. (1983). Flushing flufftails out of vleis. *Bokmakierie* **35**, 9–11.

Milstein, P. le S. (1975). The biology of Barberspan, with special reference to the avifauna. *Ostrich* Suppl. 10.

Neuby Varty, B. V. (1953). Reichenow Striped Flufftail. *Bokmakierie* **5**, 52.

Neumann, O. (1904). Vögel von Schoa und Sud Athiopien. *J. Orn.* **52**, 321–410.

Oatley, T. B. and Pinnell, N. R. (1968). The birds of Winterskloof, Natal. *S. Afr. Av. Ser.* **358**, 20 pp.

Olson, S. (1973). A classification of the Rallidae. *Wilson Bull.* **85**, 381–416.

Pakenham, R. H. W. (1943). Field notes on the birds of Zanzibar and Pemba. *Ibis* **85**, 165–189.

Parnell, G. W. (1967). Spotted Crake at Marandellas. *Honeyguide* **50**, 10.

Parrot, J. (1979). Kaffir Rail *Rallus caerulescens* in West Africa. *Malimbus* **1**, 145–146.

Petrie, M. (1983). Female Moorhens compete for small fat males. *Science* **220**, 413–415.

Pitman, C. R. S. (1929). Notes on the breeding habits and nesting of *Limnocorax flavirostra*—The Black Rail. *Ool. Rec.* **9**, 37–41.

Pitman, C. R. S. (1965). The nest and eggs of the Striped Crake, *Porzana marginalis* Hartlaub. *Bull. Br. Orn. Club* **85**, 32–40.

Plumb, W. J. (1978). Cooperative feeding of young by adult and sub-adult Red-knobbed Coots. *Bull. E. Afr. Nat. Hist. Soc.* **1978**, 88.

Porter, R. N. (1970). Nesting of the Buff-spotted Flufftail. *Ostrich* **41**, 219.

Pye-Smith, G. (1950). The nest and eggs of the Uganda White-spotted Crake (*Sarothrura pulchra centralis*). *Ool. Rec.* **24**, 48–49.

Read, M. E. (1982). Observations on a farm. *Bokmakierie* **34**, 24.

Ripley, S. D. (1977). 'Rails of the World'. David R. Godine, Boston, Mass.

Ripley, S. D. and Heinrich, G. (1966). Additions to the avifauna of northern Angola II. *Postilla* **95**.

Roux, F. and Benson, C. W. (1969a). A note on *Sarothrura lugens*. *Bull. Br. Orn. Club* **89**, 67–68.

Roux, F. and Benson, C. W. (1969b). The Buff-spotted Flufftail *Sarothrura elegans* in Ethiopia. *Bull. Br. Orn. Club* **89**, 119–120.

Roux, F. and Benson, C. W. (1970). The Red-chested Flufftail *Sarothrura rufa* in the Central African Republic. *Bull. Br. Orn. Club* **90**, 117.

Schmitt, M. B. (1975). Observations on the Black Crake in the southern Transvaal. *Ostrich* **46**, 129–138.

Schmitt, M. B. (1976). Observations on the Cape Rail in the southern Transvaal. *Ostrich* **47**, 16–26.

Sclater, W. L. and Moreau, R. E. (1932). Taxonomic and field notes on some birds of northeastern Tanganyika Territory. *Ibis* Ser.13, **2**, 487–522.

Serle, W. (1939). Observations on the breeding habits of some Nigerian Rallidae. *Ool. Rec.* **19**, 61–70.

Shewell, E. L. (1959). The waterfowl of Barberspan. *Ostrich* Suppl. **3**, 160–179.

Short, L. L. and Horne, J. F. M. (1981). Bird observations along the Egyptian Nile. *Sandgrouse* **3**, 43–60.

Shuel, R. (1938). Notes on the breeding habits of birds near Zaria, N. Nigeria, with descriptions of their nests and eggs. *Ibis* Ser. 14, **2**, 230–244.

Siegfried, W. R. and Frost, P. G. H. (1973). Regular occurrence of *Porphyrula martinica* in South Africa. *Bull. Br. Orn. Club* **93**, 36–37.

Siegfried, W. R. and Frost, P. G. H. (1975). Continuous breeding and associated behaviour in the Moorhen *Gallinula chloropus*. *Ibis* **117**, 102–109.

Silbernagl, H. P. (1982). Seasonal and spatial distribution of the American Purple Gallinule in South Africa. *Ostrich* **53**, 236–240.

Skead, D. M. (1981) Recovery distribution of Red-knobbed Coots ringed at Barberspan. *Ostrich* **52**, 126–128.

Skead, D. M. and Dean, W. R. J. (1977). Status of the Barberspan avifauna, 1971–1975. *Ostrich* Suppl. **12**, 3–42.

Skinner, K. L. (Ed) (1925). Eggs of Allen's Gallinule, '*Porphyrio alleni* Thompson'. *Ool. Rec.* **5**, 67.

Strijbos, J. P. (1955). The plundering gallinule. *Bokmakierie* **7**, 9–10.

Taylor, P. B. (1980). Little Crake *Porzana parva* at Ndola, Zambia. *Scopus* **4**, 93–94.

Taylor, P. B. (1985). Field studies of the African Crake *Crex egregia* in Zambia and Kenya. *Ostrich* **56**, 117–185.

Tree, A. J. (1970). Lesser Gallinule in Luderitz District, S. W. Africa. *Ostrich* **41**, 214.

Van Heerden, J. (1972). Observations on a coot. *Bokmakierie* **24**, 88–89.

Verheyen, R. (1953). 'Exploration du Parc National de l'Upemba. Oiseaux'. Institut des Parcs Nationaux du Congo Belge, Bruxelles.

Vincent, A. W. (1945). On the breeding habits of some African birds. *Ibis* **87**, 345–365.

Visser, J. (1974). The post-embryonic development of the Coot *Fulica atra*. *Ardea* **62**, 172–189.

Von Erlanger, C. (1905). Beitrage zur Vogelfauna Nordostafrikas. *J. Orn.* **53**, 42–158.

Von Heuglin, T. (1869–1873). 'Ornithologie Nordost-Afrikas'. Kassel.

Watson, R. M., Singh, T. and Parker, I. S. C. (1970). The diet of duck and coot on Lake Naivasha. *E. Afr. Wildl. J.* **8**, 131–144.

Wilkinson, R., Beecroft, R. and Aidley, D. J. (1982). Nigeria, a new wintering area for the Little Crake, *Porzana parva*. *Bull. Br. Orn. Club* **102**, 139–140.

Winterbottom, J. M. (1966). Some notes on the Red-knobbed Coot *Fulica cristata* Gmelin in South Africa. *Ostrich* **37**, 92–94.

Wolff, S. W. and Milstein, P. le S. (1976). Rediscovery of the Whitewinged Flufftail in South Africa. *Bokmakierie* **28**, 33–36.

Wood, N. A. (1974). The breeding behaviour and biology of the Moorhen. *British Birds* **67**, 104–115, 137–158.

Wood, N. A. (1975). Habitat preference and behaviour of Crested Coots in winter. *British Birds* **68**, 116–118.

Wood, N. A. (1977). A feeding technique of Allen's Gallinule *Porphyrio alleni*. *Ostrich* **48**, 120–121.

Wood, R. C. (1935). Another note on *Sarothrura elegans*. *Ostrich* **6**, 105–106.
Yealland, J. J. (1952). Notes on some birds of the British Cameroons forest. *Avic. Mag.* **58**, 48–59.
Zipp, W. C. H. (1939). The African Moorhen. *Ostrich* **10**, 54–56.

Family GRUIDAE: cranes

Alerstam, T. (1975). Crane *Grus grus* migration over sea and land. *Ibis* **117**, 489–495.
Anon. (1967). Wattle Crane family. *Black Lechwe* **6**(3), 23–25.
Anon. (1983). Gregarious Blue Cranes. *The Bee-eater* **34**(2), 19.
Archibald, G. W. (1976). Crane taxonomy as revealed by the unison call. *Proc. Int. Crane Workshop* **1**, 225–251.
Archibald, G. W. (1981). Aerial censusing as a valuable technique in crane conservation. *Bokmakierie* **33**(3), 60–61.
Archibald, G. W., Shigeta, Y., Matsumoto, K. and Momose, K. (1981). Endangered Cranes. *In* 'Crane Research Around the World' (Lewis, J. C. and Masatomi, H. Eds), pp. 1–12. Robinson Press, Ft. Collins, Colorado.
Ash, J. S., Ferguson-Lees, I. J. and Fry, C. H. (1967). B.O.U. expedition to Lake Chad, northern Nigeria, March–April 1967. *Ibis* **109**, 478–486.
Bannerman, D. A. and Bannerman, W. M. (1958). 'Birds of Cyprus'. Oliver and Boyd, Edinburgh.
Benson, C. W. and Pitman, C. (1964). Further breeding records from Northern Rhodesia (No. 4). *Bull. Br. Orn. Club* **84**, 54–68.
Berg, B. (1930). 'To Africa with the Migrating Birds'. G. P. Putnams Sons, London.
Berger, G. (1972). Beitrag zur Brutbiologie der Kronenkraniche (*Balearica pavonina* L. und *Balearica gibbericeps* Rchw.). *Der Zoologische Garten N. F. Leipzig* **42**, 308–324.
Bernis, F. (1960). About wintering and migration of the Common Crane (*Grus grus*) in Spain. *Proc. XII Inter. Orn. Congr.*, Helsinki **1**, 110–117.
Boulton, R., Brown, D. and Morris, A. (1982). The species survey—Crowned and Wattled Cranes. *Honeyguide* **109**, 10–13.
Brelsford, V. (1947). Notes on the birds of the Lake Bangweulu area in northern Rhodesia. *Ibis* **89**, 57–77.
Britton, H. A. (1977). E.A.N.H.S. nest record scheme: July 1976–December 1977. *Scopus* **1**(5), 132–146.
Britton, H. A. (1980). E.A.N.H.S. nest record scheme: 1980. *Scopus* **3**(5), 123–133.
Britton, P. L. (1970). Some non-passerine bird weights from East Africa. *Bull. Br. Orn. Club* **90**, 142–144, 152–154.
Brooke, R. K. (1963). Birds round Salisbury, then and now. *S. Afr. Av. Ser.* **9**, 1–67.
Brooke, R. K. (1964). Avian observations on a journey across central Africa and additional information on some of the species seen. *Ostrich* **35**, 277–291.
Brown, L. H. (1965). 'Africa.' Hamish Hamilton, London.
Browne, P. W. P. (1979). Bird observations in southwest Mauritania during 1978 and 1979. Unpub. ms, 9 pp.
Burke, V. E. M. (1965). A count of Crowned Cranes (*Balearica regulorum* (Bennett)) in the Kisii district, Kenya. *J. E. Afr. and Uganda Nat. Hist. Soc.* **25**(2)(111), 162–163.
Campbell, N. A. and Miles, H. M. (1956). Bird counts in a highveld dam in southern Rhodesia. *Ostrich* **27**, 56–69.
Carthew, W. R. (1966). Breeding of Grey-necked Crowned Cranes (*Balearica regulorum*). *Avic. Mag.* **72**(1), 1–3.
Cawkell, E. M. and Moreau, R. E. (1963). Notes on birds in The Gambia. *Ibis* **105**, 156–178.
Cheesman, R. E. and Sclater, W. L. (1935). On a collection of birds from North-western Abyssinia. *Ibis* Ser. 13, **5**, 151–191, 297–329.
Cheke, R. A. and Walsh, J. F. (1980). Bird records from the Republic of Togo. *Malimbus* **2**, 112–120.
Clarke, H. W. and Amadei, L. (1969). Breeding in captivity of the Black-necked Crowned Crane (*Balearica pavonina*). *Avic. Mag.* **75**(2), 37–39.
Conway, W. and Hamer, A. (1977). A 36-year laying record of a Wattled Crane at the New York Zoological Park. *Auk* **94**(4), 786–787.
Cooper, J. (1969). Observations at a nest of the Wattled Crane. *Honeyguide* **60**, 17–20.
Cowan, P. J. (1983). Birds in the Brak and Sabha regions of central Libya, 1981–82. *Bull. Br. Orn. Club* **103**(2), 44–47.
Cusack, E. B. (1943). The Crowned Crane (*Balearica regulorum*). *Ostrich* **13**(4), 212–218.
Day, D. H. (1980). The crane study group, 1980. *Bokmakierie* **32**(3), 90–92.
Day, D. H. (1981). Some thoughts on our cranes. *Bokmakierie* **33**(3), 58–60.
Day, D. H. (1982). Cranes, cranes, cranes. *Honeyguide* **109**, 7–9.
D'Eath, J. O. (1972). Some notes on breeding the Stanley Crane (*Anthropoides paradisea*). *Avic. Mag.* **78**(5), 165–169.
Deetjen, H. (1968). Notes du Moyen Atlas. *Alauda* **36**, 287.
Douthwaite, R. J. (1974). An endangered population of Wattled Cranes (*Grus carunculatus*). *Biol. Conserv.* **6**(2), 134–142.
Dowsett, R. J. and de Vos, A. (1965). The ecology and numbers of aquatic birds on the Kafue Flats, Zambia. *Wildfowl Trust, 16th Annual Report*, 67–73.
Elliott, C. C. H. (1983). Unusual breeding records made from a helicopter in Tanzania. *Scopus* **7**(2), 33–36.
Fairon, J. (1972). Analyse du contenus stomachaux d'oiseaux provenant du Kaouor, Niger. *Gerfaut* **62**, 325–330.
Farkas, T. (1962). Contributions to the bird fauna of Barberspan. *Ostrich* Suppl. 4, 39
Feduccia, A. (1980). 'The Age of Birds'. Harvard University Press, Cambridge, Mass.
Field, D. (1978). First aerial survey for Wattled Cranes. *Bokmakierie* **30**(1), 18–20.
Frame, G. W. (1982). East African Crowned Crane *Balearica regulorum gibbericeps* ecology and behaviour in Tanzania. *Scopus* **6**, 60–69.
Fry, C. H. (1981). West African Crowned Crane status. *In* 'Crane Research Around the World' (Lewis, J. C. and Masatomi, H. Eds), pp. 251–253. Robinson Press, Ft. Collins, Colorado.
Fry, C. H. (in press). New data on West African status of Crowned Cranes. *Proc. Int. Crane Workshop*, Bharatpur.
Gamble, K. (1980). Letter to editor: Report on the crane study group for 1978/79—*Bokmakierie*, **31**(3). *Bokmakierie* **32**(1), 24.
Ginn, P. J. (1968). Birds of the Marandellas District. *S. Afr. Av. Ser.* **61**, 1–18.
Golding, F. D. (1934). Notes on some birds of the Lake Chad area. *Ibis* **75**, 738–757.
Guichard, K. M. (1956). Observations on wintering birds near Tripoli, Libya. *Ibis* **98**, 311–316.
Hanmer, D. B. (1976). Birds of the lower Zambezi. *Southern Birds* **2**, 2–3, 6–11, 24–27.
Harwin, R. M. (1976). The pattern of occurrence of some migrant species at Rainham Farm. *Honeyguide* **86**, 21–23.
Holmes, D. A. (1972). Bird notes from the plains south of Lake Chad. Winter 1971–1972. Pt. 2. *Bull. Niger. Orn. Soc.* **9**, 76–84.
Ingold, J. L., Guttman, S. I. and Osborne, D. R. (in press)

Biochemical systematics of the cranes: Crowned cranes (*Balearica*). *Proc. Int. Crane Workshop*, Bharatpur.

Ingold, J. L., Guttman, S. I. and Osborne, D. R. (1983). Endangered species of cranes: electrophoretic determination of relationships. Unpub. MS, 12 pp.

Johnsgard, P. (1983). 'Cranes of the World.' Indiana University Press, Bloomington.

Keith, S. (1968). Notes on birds of East Africa, including additions to the avifauna. *Am. Mus. Novit.*, 2321.

Konrad, P. M. (1981). Status and ecology of Wattled Crane in Africa. In 'Crane Research Around the World' (Lewis, J. C. and Masatomi, H. Eds), pp. 220–237. Robinson Press, Fort Collins, Colorado.

Konrad, P. M. (in press). Man, wetlands, and cranes in southern Africa. *Proc. Int. Crane Workshop*, Bharatpur.

Konrad, P. M. (in press). Wet season ecology of Grey Crowned Cranes in the Luangwa Valley, Zambia. *Proc. Int. Crane Workshop*, Bharatpur.

Koslova, E. V. (1978). Birds of the 'Zonal Steppes' and wastelands of Central Asia. Translated by E. Anderson. Unpub. MS.

Kuhk, R. and Schüz, E. (1971). Ungewöhnliche Vogelmarkierungen. *Vogelwarte* 26(2), 197–202.

Latta, S. C. and Archibald, G. W. (1980). A research proposal concerning the conservation of the Demoiselle Crane in Morocco. Unpub. MS. 15 pp.

Lees, S. G. (1977). Crowned crane nesting in a tree. *Honeyguide* 92, 49.

Lötter, W. J. (1975). Is Kraanvoëls ware landlewende voëls? *Bokmakierie* 27(4), 75–78.

Lynes, H. (1924). On the birds of North and Central Darfur, with notes on the west-central Kordofan and north Nuba Provinces of British Sudan. *Ibis* 1924, 399–446.

Macartney, P. (1968). Wattled Cranes in Zambia. *Bokmakierie* 20(2), 38–41.

Macleay, K. N. G. (1960). The birds of Khartoum Province. *Univ. Khartoum Nat. Hist. Mus. Bull.* 1, 33 pp.

Malbrant, R. (1937). Sur le passage de la Grue de Numidie au Tchad. *Oiseau et R.F.O.* 7, 378.

Malbrant, R. and Maclatchy, A. (1958). A propos de l'occurrence de deux oiseaux d'Afrique australe au Gabon: Le Manchot du Cap *Spheniscus demersus* Linné et la Grue Couronnée, *Balearica regulorum* Bennett. *Oiseau et R.F.O.* 28(1), 84–86.

Marchant, S. (1941). Notes on the birds of the Gulf of Suez. *Ibis* 5, 265–295.

Mathiasson, S. (1963). Visible diurnal migration in the Sudan. *Proc. XIII Int. Orn. Congr.*, pp. 430–435.

Meinertzhagen, R. (1949). Notes on Saudi Arabian birds. *Ibis* 91, 465–482.

Meinertzhagen, R. (1955). The speed and altitude of bird flight (with notes on other animals). *Ibis* 97(1), 81–117.

Moreau, R. E. (1953). Migration in the Mediterranean area. *Ibis* 95, 329–364.

Moreau, R. E. (1967). Water birds over the Sahara. *Ibis* 109(2), 232–259.

Morris, A. (in press). Cranes in Zimbabwe. *Proc. Int. Crane Workshop*, Bharatpur.

Mweya, A. N. (1973). Ornithological notes from southeast of Lake Bangweulu. *Puku* 7, 151–161.

Oatley, T. B. (1969). Unusual breeding behaviour of Blue Cranes, *Tetrapteryx paradisea*. *Lammergeyer* 10, 87–88.

Owre, O. T. (1966). The Crowned Crane at Lake Rudolf. *Bull. Br. Orn. Club* 86(3), 54–56.

Peters, J. L. (1934). 'Check-list of Birds of the World', Vol. II. Harvard University Press, Cambridge.

Pettet, A. (1976). The avifauna of Waza National Park. *Bull Niger. Orn. Soc.* 12, 18–24.

Pomeroy, D. E. (1980a). Aspects of the ecology of Crowned Cranes *Balearica regulorum*. *Scopus* 4, 29–35.

Pomeroy, D. E. (1980b). Growth and plumage changes of the Grey Crowned Crane *Balearica regulorum gibbericeps*. *Bull. Br. Orn. Club*. 100(4), 219–223.

Quintin, W. H. (1903). On breeding the Demoiselle Crane (*Anthropoides virgo*) in captivity. *Avic. Mag.* 1 (2), 390–393.

Robinson, J., Robinson, St. C. and Winterbottom, J. M. (1957). Notes on the birds of the Cape Agulhas Region. *Ostrich* 28, 147–163.

Schmitt, M. B. and Baur, S. (1976). An unusual sighting of Blue Cranes. *Bokmakierie* 28(2), 47.

Siegfried, W. R., Frost, P. G. H., Cooper, J. and Kemp, A. C. (1976). 'South Africa Red Data Book—Aves'. S. Afr. National Sci. Prog. Report No. 7. National Scientific Programmes, Pretoria.

Sievi, J. and Manson, A. (1974). Letter to Editor. *Honeyguide* 77, 48.

Smith, V. W. (1962). Some birds which breed near Vom, Northern Nigeria. *Nigerian Field* 27, 4–34.

Steyn, P. and Ellman-Brown, P. (1974). Crowned Crane nesting on a tree. *Ostrich* 45(1), 40–42.

Steyn, P. and Tredgold, D. (1977). Crowned Cranes covering eggshells. *Bokmakierie* 29, 82–83.

Storer, R. W. (1971). Classification of birds. In 'Avian Biology' (Farner, D. S., King, J. R. and Parkes, K. C. Eds) Vol. 1. Academic Press, New York.

Tarboton, W. (1984). The status and conservation of the Wattled Crane in the Transvaal. *Proc. V Pan Afr. Orn. Congr.*, 665–678.

Tree, A. J. (1973). Crowned Crane *Balearica pavonina*. *Ostrich* 44(2), 128.

Trott, A. C. (1947). Notes on birds seen and collected at Jedda and in Arabia during 1937, 1938, 1939, and 1940. *Ibis* 89, 77–98.

Ullman, M. (1982). High numbers of cranes in Tunisia in December 1979. *Dutch Birding* 4(3), 87.

Urban, E. K. (1981). The Sudan Crowned Crane. In 'Crane Research Around the World' (Lewis, J. C. and Masatomi, H. Eds), p. 254. Robinson Press, Ft. Collins, Colorado.

Urban, E. K. (in press). The cranes of Africa: an overview. *Proc. Int. Crane Workshop*, Bharatpur.

Urban, E. K. and Walkinshaw, L. H. (1967). The Sudan Crowned Crane in Ethiopia. *Ibis* 109(3), 431–433.

Urban, E. K. and Walkinshaw, L. H. (1967). The Wattled Crane in Ethiopia. *Auk* 84(2), 263–264.

Van der Ven, J. A. (1981). Common Cranes in Europe. In 'Crane Research Around the World' (Lewis, J. C. and Masatomi, H. Eds), pp. 181–183. Robinson Press, Ft. Collins, Colorado.

Van Ee, C. A. (1966). Notes on the breeding behaviour of the Blue Crane *Tetrapteryx paradisea*. *Ostrich* 37(1), 23–29.

Van Ee, C. A. (1981). Status of the Blue Crane in South and Southwest Africa. In 'Crane Research Around the World' (Lewis, J. C. and Masatomi, H. Eds), p. 259. Robinson Press, Ft. Collins, Colorado.

Verschuren, J. (1967). Note sur les oiseaux da la région du Parc National de Serengeti (Tanzania). *Gerfaut* 57, 77–81.

Vincent, A. W. (1945). On the breeding habits of some African birds. *Ibis* 87, 345–365.

Walkinshaw, L. H. (1963). Some life history studies of the Stanley Crane. *Proc. XIII Int. Orn. Congr.*, pp. 344–353.

Walkinshaw, L. H. (1964). The African Crowned Cranes. *Wilson Bull.* 76(4), 355–377.

Walkinshaw, L. H. (1965). Attentiveness of cranes at their nests. *Auk* 82, 465–476.

Walkinshaw, L. H. (1965). Territories of cranes. Papers Michigan Acad. Science, Arts and Letters 50, 75–88.

Walkinshaw, L. H. (1965). The Wattled Crane *Bugeranus*

carunculatus (Gmelin). *Ostrich* **36**(2), 73–81.
Walkinshaw, L. H. (1966). The West African Crowned Crane on the Jos Plateau, Northern Nigeria. *Bull. Niger. Orn. Soc.* **3**(9), 6–10.
Walkinshaw, L. H. (1973). 'Cranes of the World.' Winchester Press, New York.
West, O. (1956). Nestling plumage, Wattled Crane. *Ostrich* **27**(1), 41.
West, O. (1963). Notes on the Wattled Crane *Bugeranus carunculatus* (Gmelin). *Ostrich* **34**(2), 63–77.
West, O. (1976). Notes on the distribution and status of the southern population of Wattled Cranes in Africa. Unpub. MS.
Wetmore, A. (1960). A classification for the birds of the world. *Smithsonian Misc. Coll.* **139**(11), 1–37.
Wilson, R. T. (1982). Environmental changes in western Darfur, Sudan, over half a century and their effects on selected bird species. *Malimbus* **4**, 15–26.
Wright, P. J. (1966). A note on the Wattled Crane (*Grus carunculatus*). *Puku* **4**, 196–198.
Wright, P. J. (1971). Distraction display in Wattled Cranes. *Honeyguide* **65**, 37.

Family HELIORNITHIDAE: finfoots

Beddard, F. E. (1890). On the anatomy of *Podica senegalensis*. *Proc. Zool. Soc. Lond.* **1890**, 425–443.
Brooke, R. K. (1984). Taxonomic subdivisions within the Heliornithidae. *Ostrich* **55**, 171–173.
Carver, J. and Carver, J. (1979). Peter's Finfoot nest. *Bull. E. Afr. Nat. Hist. Soc.* **1979**, 113–114.
Erard, C. and Benson, C. W. (1975). The race of the African Finfoot *Podica senegalensis* (Vieillot) in Ethiopia. *Bull. Br. Orn. Club* **95**, 147–148.
Ginn, P. J. (1977). The elusive African Finfoot. *Fauna and Flora* **28**, 12–14.
Jubb, R. A. (1982). Peter's Finfoot: a rather shy bird. *Naturalist* **26**, 7–9.
Mitchell, S. (1977). Apparently atypical habitat and behaviour of Peters' Finfoot. *Honeyguide* **92**, 43.
Percy, Lord W. (1963). Further notes on the African Finfoot, *Podica senegalensis* (Vieillot). *Bull. Br. Orn. Club* **83**, 127–132 (includes note by C. R. S. Pitman).
Pitman, C. R. S. (1962). Notes on the African Finfoot, *Podica senegalensis* (Vieillot) with particular reference to Uganda. *Bull. Br. Orn. Club* **62**, 156–160.
Skead, C. J. (1962). Peters' Finfoot *Podica senegalensis* (Vieillot) at the nest. *Ostrich* **33**, 31–33.
del Toro, M. A. (1971). On the biology of the American Finfoot in southern Mexico. *Living Bird* **10**, 79–88.
Traylor, M. A. and Hart, R. C. (1965). Some interesting birds from Barotseland. *Puku* **3**, 133–141.
Vernon, C. (1983). Display of the Finfoot. *The Bee-eater*, Suppl. **10**, 12.
Whateley, A. (1982). Anting in the African Finfoot. *Ostrich* **53**, 177.

Family OTIDIDAE: bustards

Alamargot, J. (1978). Observations ornithologiques dans le sud-ouest de l'Ethiopie (Province du Guémou-Gofa) en 1973–74. Institut d'Elèvage et de Médecine Vétérinaire des Pays Tropicaux. Maisons-Alfort, France.
Allen, P. M. and Clifton, M. P. (1972). Aggressive behaviour of Kori Bustard *Otis kori*. *Bull. E. Afr. Nat. Hist. Soc.* **1972**, 188–189.
Ash, J. S. (1977). Four species of birds new to Ethiopia and other notes. *Bull. Br. Orn. Club* **97**, 4–9.
Aspinwall, D. R. (1979). Bird notes from Zambezi District, North-Western Province. *Zambian Orn. Soc. Occ. Pap.* **2**.
Astley-Maberly, C. T. (1937). Notes on birds from north-eastern Transvaal. *Ostrich* **8**, 10–19.
Astley-Maberly, C. T. (1967). An unusual display of the Red-crested Korhaan. *Bokmakierie* **19**, 41.
Bell, J. (1970). The white-quilled Black Bustard. *Animal Kingdom* **73**(6), 25–28.
Benson, C. W. and Irwin, M. P. S. (1964). Some additions and corrections to a 'Check List of the Birds of Northern Rhodesia'. No. 5. *Occ. Pap. Nat. Mus. Sth. Rhod.* **27**B, 106–127.
Benson, C. W. and Irwin, M. P. S. (1967). A contribution to the ornithology of Zambia. *Zambia Mus. Pap.* **1**.
Benson, C. W. and Irwin, M. P. S. (1968). Taxonomic notes on Congolese birds. *Rev. Zool. Bot. Afr.* **77**, 44–49.
Benson, C. W. and Irwin, M. P. S. (1972). Variation in tarsal and other measurements in *Otis denhami*, with some distributional notes. *Bull. Br. Orn. Club* **92**, 70–77.
Blancou, L. (1933). Contribution à l'étude des oiseaux de l'Oubangui-Chari. *Oiseau et R.F.O.* **3**, 8–58.
Blancou, L. (1938). Contribution à l'étude des oiseaux de l'Oubangui-Chari occidental. *Oiseau et R.F.O.* **8**, 642–649.
Blancou, L. (1959). A propos de la nidification de l'Outarde de Denham. *Oiseau et R.F.O.* **29**, 66–68.
Bonde, K. (1981). 'An Annotated Checklist of the Birds of Lesotho'. Prelim. ed., Maseru.
Brooke, R. K. (1984). 'South African Red Data Book: Birds'. Foundt. Res. Dev. CSIR, Pretoria.
Butler, A. L. (1905). A contribution to the ornithology of the Egyptian Soudan. *Ibis* **5**(8), 301–401.
Butler, A. L. (1908). A second contribution to the ornithology of the Egyptian Soudan. *Ibis* **3**(9), 205–263.
Cassels, K. A. H. and Elliott, H. F. I. (1975). An undescribed display of the Red-crested Korhaan. *Bull. Br. Orn. Club* **95**, 116–117.
Chappuis, C., Erard, C. and Morel, G. J. (1979). Données comparatives sur la morphologie et les vocalisations des diverses formes d'*Eupodotis ruficrista* (Smith). *Malimbus* **1**, 74–89.
Cheesman, R. E. and Sclater, W. L. (1935). On a collection of birds from north-western Abyssinia, pt. 2. *Ibis Ser.* **13**, 5, 297–329.
Clancey, P. A. (1972–1973). The magnificent bustards. A preliminary assessment. *Bokmakierie* **24**, 74–79; **25**, 10–14.
Clancey, P. A. (1977). Miscellaneous taxonomic notes on African birds, 49. *Durban Mus. Novit.* **11**, 223–238.
Collar, N. J. (1980). The world status of the Houbara: a preliminary review. Symposium papers on the Great Bustard *Otis tarda* (Sofia, Bulgaria, May 26 1978) and the Houbara Bustard *Chlamydotis undulata* (Athens, Greece, May 24, 1979). Fondation Internationale pour la Sauvegarde du Gibier/Conseil Internationale de la Chasse/Game Conservancy, 12 pp.
Collar, N. J. and Goriup, P. D. (1980). Problems and progress in the captive breeding of Great Bustards *Otis tarda* in quasi-natural conditions. *Avic. Mag.* **86**, 131–140.
Collar, N. J. and Goriup, P. D. (Eds) (1983). Results of the ICBP Fuerteventura Houbara Expedition, 1979. *Bustard Studies* **1**.
Collar, N. J. (in press). The world status of the Great Bustard. *Bustard Studies* **2**.
Collins, D. R. (in prep.). A study of the Canarian Houbara (*Chlamydotis undulata fuertaventurae*), with special reference to its behaviour and ecology.
Daly, M. and Daly, S. (1975). Oiseaux observés à Beni-Abbès, Sahara algérien. *Oiseau et R.F.O.* **45**, 337–340.

Ena, V., Lucio, A. and Purroy, F. J. (in press). The Great Bustard in Leon, Spain. *Bustard Studies* 2.
von Frisch, O. (1976). Zur Biologie der Zwergtrappe (*Tetrax tetrax*). *Bonn. zool. Beitr.* 27, 21–38.
Gewalt, W. (1959). 'Die Grosstrappe'. Neue Brehm-Bücherei. Wittenburg-Lutherstadt.
Gewalt, W. and Gewalt, I. (1966). Über Haltung und Zucht der Grosstrappe *Otis tarda* L. *Zool. Garten* (NF) 32, 265–322.
Hoesch, W. (1938). Zur Balz von *Choriotis kori*. *Orn. Monatsber.* 46, 110–111.
Hopkins, G. H. E. (1942). The Mallophaga as an aid to the classification of birds. *Ibis* 6(14), 94–106.
Hollom, P. A. D. (1980). Observations of nesting Houbara Bustards. *British Birds* 73, 410–411.
Horseburgh, B. R. (1907). The Blue Korhaan. *Avic. Mag.* (NS) 5, 139–142.
Howells, W. W. and Finn, K. J. (1979). The occurrence of Denham's Bustard at Wankie National Park and in north west Rhodesia with notes on movements and behaviour. *Honeyguide* 97, 4–12.
Irwin, M. P. S. and Benson, C. W. (1967). Notes on the birds of Zambia, 3. *Arnoldia (Rhod.)* 3(4).
Jackson, F. J. (1926). 'Notes on the Game Birds of Kenya and Uganda.' Williams and Norgate, London.
Jackson, H. D. (1969). Notes on a collection of birds from the Khwae River in Botswana. *Arnoldia (Rhod.)* 4(24).
Johst, E. (1972). Die Haltung und künstliche Aufzucht der Senegal-Trappe (*Eupodotis senegalensis* Vieillot). *Gefied. Welt* 96, 61–64.
Jubb, R. A. (1981). The stately Kori Bustard. *Naturalist* 25(3), 9–11.
Kemp, A. C. and Tarboton, W. R. (1976). Small South African bustards. *Bokmakierie* 28, 40–43.
Kostin, Y. V. (1978). Is the Little Bustard doomed? *Bull. Moscow Nat. Soc. Biol. Sect.* 83(3), 67–71.
Little, J. de V. (1964). Observations on the call of the Black-bellied Korhaan. *Bokmakierie* 16, 15.
Lynes, H. (1920). Ornithology of the Maroccan 'Middle-Atlas'. *Ibis* 2, (11) 260–301.
Lynes, H. (1925). On the birds of North Central Darfur, with notes on the West-Central Kordofan and North Nuba Provinces of British Sudan. (Part V). *Ibis* 1, (12), 541–590.
Lynn-Allen, B. G. (1951). 'Shot-gun and Sunlight. The Game Birds of East Africa.' Batchworth Press, London.
Macdonald, J. D. and Hall, B. P. (1957). Ornithological results of the Bernard Carp/Transvaal Museum Expedition to the Kaokoveld, 1951. *Ann. Transv. Mus.* 23, 1–39.
Maclean, G. L., Maclean, C. M., Geldenhuys, J. N. and Allan, D. G. (1983). Group size in the Blue Korhaan. *Ostrich* 54, 244–245.
Malbrant, R. (1940). Les oiseaux gibiers de la région du Pool. *Bull. Soc. Zool. Bot. Congolaises* 3(4), 1–12.
Mendelsohn, H., Marder, U. and Stavy, M. (1979). Captive breeding of the Houbara *Chlamydotis undulata macqueenii* and a description of its display. *XIII ICBP Bulletin*, 134–149.
Morgan-Davies, A. M. (1965). On the Kori Bustard, *Ardeotis kori* (Burchell), in north-western Tanzania. *Bull. Br. Orn. Club* 85, 145–147.
Niethammer, G. (1940). Beobachtungen über die Balz und Untersuchungen über den Oesophagus südafrikanischer Trappen. *Orn. Monatsber.* 48, 29–33.
Nikolaus, G. (in prep.). 'Birds of South Sudan'.
Otero, C. (in press). A key for sex differentiation of Little Bustards *Tetrax tetrax* based on wing and age. *Bustard Studies* 2.
Pineau, J. and Giraud-Audine, M. (1977). Notes sur les oiseaux nicheurs de l'extrême nord-ouest du Maroc: reproduction et mouvements. *Alauda* 45, 75–104.
Pitman, C. R. S. (1957). Uganda's bustards. *Uganda Wildl. Sport* 1(2).
Quinton, W. F. (1948). The Karroo Korhaan (*Eupodotis vigorsii vigorsii*). *Ostrich* 19, 235–236.
Reynolds, J. F. (1973). Some observations on bustards in Nairobi National Park. *Bull. E. Afr. Nat. Hist. Soc.* 1973, 98–99.
Roberts, A. (1936). Some unpublished field notes made by Dr (Sir) Andrew Smith. *Ann. Transv. Mus.* 18, 271–323.
Rockingham-Gill, D. V. (1982). Kori Bustard enquiry. Newsletter 2, Unpub.
Rockingham-Gill, D. V. (1983). On the distribution of the Kori Bustard (*Otis kori*) in Zimbabwe. *In* 'Bustards in Decline' (Goriup, P. D. and Vardhan, H. Eds), pp. 97–103. Tourism and Wildlife Society of India, Jaipur.
Sarro, A., Pons Oliveras, J. R. and Gutierrez Pagés, L. (1968). Observaciónes ornitológicas en Mauritania. *Ardeola* 14, 175–183.
Schaller, G. B. (1973). 'Golden Shadows, Flying Hooves'. A. A. Knopf, New York.
Schulz, H. (1980). Zur Bruthabitatwahl der Zwergtrappe *Tetrax t. tetrax* in der Crau (Südfrankreich). *Braunschw. Naturk. Schr.* 1, 141–160.
Schulz, H. (in press). On the social behavior of the Little Bustard *Tetrax tetrax*: a preliminary report. *Bustard Studies* 2.
Skead, C. J. (1967a). Ecology of birds in the Eastern Cape Province. *Ostrich* Suppl. 7.
Skead, C. J. (1967b). A pet Stanley Bustard. *Bokmakierie* 19, 16–17.
Skead, D. M. (1977). Weights of birds handled at Barberspan. *Ostrich* Suppl. 12, 117–131.
Smith, V. W. (1966). Breeding records for the Plateau province over 3000 ft., 1957–1966. *Bull. Niger. Orn. Soc.* 3, 78–91.
Tarboton, W. R. (in press). Breeding behaviour of Denham's Bustard. *Bustard Studies*.
Thomas, D. K. (1960). Birds – notes on breeding in Tanganyika: 1958–1959. *Tanganyika Notes and Records* 55, 225–243.
Turner, D. A. (1982). The status and distribution of the Arabian Bustard *Otis arabs* in northeastern Africa and its possible occurrence in northern Kenya. *Scopus* 6, 20–21.
Urban, E. K., Brown, L. H., Brown, B. and Newman, K. B. (1978). Kori Bustard eating gum. *Bokmakierie* 30, 105.
Uys, C. J. (1963). Some observations on the Stanley Bustard (*Neotis denhami*) at the nest. *Bokmakierie* 15, 2–4.
Vernon, C. (1983). Notes from the border–11. Blue Korhaan group size. *The Bee-eater* Suppl. 10, 1–2.
Viljoen, P. J. (1983). Distribution, numbers and group size of the Karoo Korhaan in Kaokoland, South West Africa. *Ostrich* 54, 50–51.
Wilson, V. J. (1972). Notes on *Otis denhami jacksoni* from the Nyika Plateau. *Bull. Br. Orn. Club* 92, 77–81.
Winterbottom, J. M. (1966). Results of Percy FitzPatrick Institute–Windhoek State Museum joint ornithological expeditions. 5. The birds of the Kaokoveld and Cunene River. *Cimbebasia* 19.

Family JACANIDAE: jacanas

Benson, C. W. (1961). Jacanas and other birds perching on hippo. *Bull. Br. Orn. Club* 81, 85–87.
Cunningham-van Someren, G. R. and Robinson, C. (1962).

Notes on the African Lilytrotter *Actophilornis africanus* (Gmelin). *Bull. Br. Orn. Club* **82**, 67–72.

Dowsett, R. J. and de Vos, A. (1964). The ecology and numbers of aquatic birds on the Kafue Flats, Zambia. The Wildfowl Trust, *6th Annual Report*, 67–73.

Forbes, W. A. (1881). Notes on the anatomy and systematic position of the jaçanás (Parridae). *Proc. Zool. Soc. Lond.* **1881**, 639–647.

Fry, C. H. (1983a) The jacanid radius and *Microparra*, a neotenic genus. *Gerfaut* **73**, 173–184.

Fry, C. H. (1983b). Incubation, brooding, and a structural character of the African Jacana. *Ostrich* **54**, 175–176.

Hopcraft, J. B. D. (1968). Some notes on the chick-carrying behavior in the African Jacana. *Living Bird* **7**, 85–88.

James, G. L. (1948). Notes on a jacana, shrike and coly. *Ostrich* **19**, 167–170.

Lowe, P. R. (1925). On the systematic position of the Jacanidae (Jaçanás), with some notes on a hitherto unconsidered anatomical character of apparent taxonomic value. *Ibis* **67**, 132–144.

Macdonald, M. A. (1979). Breeding data for birds in Ghana. *Malimbus* **1**, 36–42.

Maclean, G. L. (1972). Waders of waterside vegetation. *Afr. Wildl.* **26**, 163–167.

Masterson, A. (1969). Distraction behaviour in certain birds. *Honeyguide* **58**, 27–28.

Miller, W. T. (1951). The bird that walks on water. *Afr. Wildl.* **5**, 283–289.

Phelan, P. I. (1970). Egg covering. *Lammergeyer* **11**, 82–83.

Pitman, C. R. S. (1960). A note on the African Jaçana, *Actophilornis africanus* (Gmelin). *Bull. Br. Orn. Club* **80**, 103–105.

Postage, A. (1984). The behaviour of breeding African Jacanas. *Bokmakierie* **36**, 12–14.

Simpson, C. D. (1961). The African Jacana, *Actophilornis africanus* (Gmelin). *Bull. Br. Orn. Club.* **81**, 82–85.

Steyn, P. (1973). African Jacana at last. *Bokmakierie* **25**, 34–37.

Tarboton, W. (1976). Notes on South African jacanas. *Fauna and Flora* **27**, 5–7.

Tarboton, W. R. and Fry, C. H. (in press). Breeding and other behaviour of the Lesser Jacana. *Ostrich*.

Vernon, C. J. (1973). Polyandrous *Actophilornis africanus*. *Ostrich* **44**, 85.

Wilson, G. (1974). Incubating behaviour of the African Jacana. *Ostrich* **45**, 185–188.

Winterbottom, J. M. (1961). African Jacana in the Richtersveld. *Ostrich* **32**, 99.

Family ROSTRATULIDAE: painted-snipe

Baird, D. A. (1979). Twenty-eight additions to Archer & Godman's 'Birds of British Somaliland and the Gulf of Aden'. *Bull. Br. Orn. Club* **99**(1), 6–9.

Benson, C. W. (1982). Migrants in the Afrotropical Region south of the equator. *Ostrich* **53**, 31–49.

Dowsett, R. J. (1965). Weights of some Zambian birds. *Bull. Br. Orn. Club.* **85**(8), 150–152.

Elgood, J. H. and Donald, R. G. (1962). Breeding of the Painted Snipe *Rostratula benghalensis* in southwest Nigeria. *Ibis* **104**, 253–256.

Komeda, Shigemoto. (1983). Nest attendance of parent birds in the Painted Snipe (*Rostratula benghalensis*). *Auk* **100**, 48–55.

Schmidt, R. K. (1961). Incubation period of the Painted Snipe *Rostratula benghalensis*. *Ostrich* **32**(4), 183–184.

Skead, D. M. (1977). Weights of birds handled at Barberspan. *Ostrich*, Suppl. **12**, 117–131.

Summers, R. W. and Waltner, M. (1978). Seasonal variation in the mass of waders in southern Africa, with special reference to migration. *Ostrich* **50**, 21–37.

Family DROMADIDAE: Crab Plover

Feare, C. J. (1979). Ecology of Bird Island, Seychelles. *Atoll Res. Bull.* **226**, 1–29.

Moore, R. J. and Balzarotti, M. A. (1977). Report of the 1976 expedition to the Suakim Archipelago (Sudanese Red Sea). Results of Marine Turtle Survey and notes on marine and bird life. Mimeographed, iv + 27 pp.

Moore, R. J. and Balzarotti, M. A. (1983). Observations of sea birds nesting on islands of the Sudanese Red Sea. *Bull. Br. Orn. Club* **103**(2), 65–71.

Ticehurst, C. B. (1926). On the down plumages of some Indian birds. *J. Bombay Nat. Hist. Soc.* **31**, 368–378.

Urban, E. K. and Boswall, J. (1969). Bird observations from the Dahlak Archipelago, Ethiopia. *Bull. Br. Orn. Club* **89**, 121–129.

Family HAEMATOPODIDAE: oystercatchers

Baker, A. J. and Hockey, P. A. R. (1984). Behavioural and vocal affinities of the African Black Oystercatcher (*Haematopus moquini*). *Wilson Bull.* **96**, 656–671.

Dowsett, R. J. (1980). The migration of coastal waders from the Palaearctic across Africa. *Gerfaut* **70**, 3–35.

Hall, K. R. L. (1959). Observations on the nest-sites and nesting behaviour of the Black Oystercatcher *Haematopus moquini* on the Cape Peninsula. *Ostrich* **30**, 143–154.

Hockey, P. A. R. (1981a). Morphometrics and sexing of the African Black Oystercatcher. *Ostrich* **52**, 244–247.

Hockey, P. A. R. (1981b). Feeding techniques of the African Black Oystercatcher *Haematopus moquini*. *In* 'Proceedings of the Symposium of Birds of the Sea and Shore, 1979' (Cooper, J. Ed) pp. 99–115. African Seabird Group, Cape Town.

Hockey, P. A. R. (1982a). Adaptiveness of nest site selection and egg coloration in the African Black Oystercatcher *Haematopus moquini*. *Behav. Ecol. Sociobiol.* **11**, 117–123.

Hockey, P. A. R. (1982b). The taxonomic status of the Canary Islands Oystercatcher, *Haematopus meadewaldoi*. *Bull. Br. Orn. Club.* **102**, 77–84.

Hockey, P. A. R. (1983a). The distribution, population size, movements and conservation of the African Black Oystercatcher *Haematopus moquini*. *Biol. Conserv.* **25**, 233–262.

Hockey, P. A. R. (1983b). Aspects of the breeding biology of the African Black Oystercatcher. *Ostrich* **54**, 26–35.

Hockey, P. A. R. (1984a). Growth and energetics of the African Black Oystercatcher *Haematopus moquini*. *Ardea* **72**, 111–117.

Hockey, P. A. R. (1984b). Behaviour patterns of non-breeding African Black Oystercatchers at offshore islands. *Proc. V Pan Afr. Orn. Congr.*, 707–728.

Hockey, P. A. R. (1985). Observations on the communal roosting of the African Black Oystercatcher. *Ostrich* **56**, 52–57.

Hockey, P. A. R. and Branch, G. M. (1983). Do oystercatchers influence limpet shell shape? *Veliger* **26**, 139–141.

Hockey, P. A. R. and Branch, G. M. (1984). Oystercatchers and limpets: impact and implications. A preliminary assessment. *Ardea* **72**, 199–206.

Hockey, P. A. R. and Cooper, J. (1980). Paralytic shellfish poisoning: a controlling factor in Black Oystercatcher populations? *Ostrich* **51**, 188–190.

Hockey, P. A. R. and Cooper, J. (1982). Occurrence of the European Oystercatcher *Haematopus ostralegus* in southern Africa. *Ardea* **70**, 55–58.

Hockey, P. A. R. and Underhill, L. G. (1984). Diet of the African Black Oystercatcher on rocky shores: spatial, temporal and sex-related variation. *S. Afr. J. Zool.* **19**, 1–11.

Prater, A. J., Marchant, J. H. and Vuorinen, J. (1977). Guide to the identification and ageing of Holarctic Waders. British Trust for Ornithology Guide **17**.

Randall, R. M. and Randall, B. M. (1982). The hard-shelled diet of African Black Oystercatcher chicks at St. Croix Island, South Africa. *Ostrich* **53**, 157–163.

Summers, R. W. and Cooper, J. (1977). The population, ecology and conservation of the Black Oystercatcher *Haematopus moquini*. *Ostrich* **48**, 28–40.

Family RECURVIROSTRIDAE: stilts, avocets

Adret, A. (1982). The sound signals of the adult Avocet *Recurvirostra avosetta* during the perinatal phase. *Ibis* **124**, 275–287.

Ash, J. S. (1972). Charadriiform birds in the Ethiopian Rift Valley. *Walia* **4**, 14–18.

Ash, J. A. (1980). Migrational status of Palaearctic birds in Ethiopia. *Proc. IV Pan Afr. Orn. Congr.*, pp. 199–208.

Ash, J. S., Ferguson-Lees, I. J. and Fry, C. H. (1967). B.O.U. expedition to Lake Chad, northern Nigeria, March–April 1967. *Ibis* **109**, 478–486.

Berruti, A. (1980). Status and review of waterbirds breeding at Lake St. Lucia. *Lammergeyer* **28**, 1–19.

Berry, H. H. and Berry, C. U. (1975). A check list and notes on the birds of Sandvis, South West Africa. *Madoqua* **9**(2), 5–18.

Betham, R. M. (1929). Some observations on the nesting birds in the vicinity of Cape Town, South Africa. *Ibis* **5**, 71–104.

Bie, de S. and Zijlstra, M. (1979). Some remarks on the behaviour of the Avocet (*Recurvirostra avosetta* L.) in relation to difficult breeding places. *Ardea* **67**, 68–69.

Blondel, J. and Blondel, C. (1964). Remarques sur l'hivernage des limicoles et autres oiseaux aquatiques au Maroc (janvier 1964). *Alauda* **32**(4), 250–279.

Bologna, G. and Petretti, F. (1978). Sur Cavaliera d'Italia (*Himantopus himantopus*). *Riv. ital. Orn.* **48**, 179–180.

Boyd, H. (1962). Mortality and fertility of European Charadrii. *Ibis* **104**, 368–387.

Britton, H. A. (1978). E.A.N.H.S. nest record scheme: July 1976–December 1977. *Scopus* **1**(5), 132–141.

Britton, H. A. (1979). E.A.N.H.S. nest record scheme, 1978. *Scopus* **2**(5), 126–132.

Broekhuysen, G. J. and Macleod, J. G. R. (1948). Avocets (*Recurvirostra avosetta*) breeding in the vicinity of Cape Town. *Ostrich* **19**(2), 148–149.

Brooke, R. K. (1964). Avian observations on a journey across central Africa and additional information on some species seen. *Ostrich* **35**(4), 277–292.

Brown, P. E. (1950). Avocets in England. *Occ. Publ. RSPB* **14**.

Browne, P. W. P. (1979). Bird observations in southwestern Mauritania during 1978 and 1979. Typed MS, 9 pp.

Burger, J. (1980). Age differences in foraging Black-necked Stilts in Texas. *Auk* **97**, 633–636.

Burnier, E. (1979). Notes sur l'ornithologie Algérienne. *Alauda* **47**(2), 93–102.

Cackett, K. (1974). News from the Lowveld. *Honeyguide* **78**, 21–23.

Cadbury, C. J. and Olney, P. J. S. (1978). Avocet population dynamics in England. *British Birds* **71**, 102–121.

Cooper, J. (1971). Avocet and Glossy Ibis at Lake McIlwaine. *Honeyguide* **67**, 34.

Cooper, J., Summers, R. W. and Pringle, J. S. (1976). Conservation of coastal habitats of waders in the South-Western Cape, South Africa. *Biol. Conserv.* **10**, 239–247.

Cowan, P. J. (1982). Birds in west central Libya, 1980–81. *Bull. Br. Orn. Club* **102**(1), 32–35.

Cunningham-van Someren, G. R. (1975). A note on the Lake Magadi Avocets and other birds. *Bull. E. Afr. Nat. Hist. Soc.* Aug/Sep, 90–92.

Dawson, R. (1975). Marsh Sandpipers associating with feeding avocets and other species. *British Birds* **68**(7), 294–295.

Din, N. A. (1979). Notes on bird counts from Hippo Point in Rwenzori National Park, Uganda. *Pakistan J. Zool.* **11**(1), 149–162.

Dowsett, R. J. (1969). B.O.U. supported research at Lake Chad in 1968. *Ibis* **111**, 449–552.

Dupuy, A. (1968). La migration des Laro-Limicoles au Sahara Algérien. *Alauda* **36**(1–2), 27–35.

Dupuy, A. R. and Verschuren, J. (1978). Note sur les oiseaux, principalement aquatiques, de la région du Parc National du Delta du Saloum (Sénégal). *Gerfaut* **68**, 321–345.

Every, B. (1974). Abnormal clutch size for the Blackwinged Stilt. *Ostrich* **45**(4), 260.

Gore, G. and Gore, P. (1966). The Avocet *Recurvirostra avosetta* in Senegambia. *Ibis* **108**(2), 281.

Goriup, P. D. (1982). Behaviour of Black-winged Stilts. *British Birds* **75**, 12–24.

Goss-Custard, J. D. (1970). Feeding dispersion in some overwintering wading birds. In 'Social Behaviour in Birds and Mammals' (Crook, J. H. Ed) pp. 3–35. Academic Press, London and New York.

Goutner, V. (1984). Belly-soaking in the Avocet. *Ostrich* **55**(3), 167–168.

Gowthorpe, P. (1980). Sur la reproduction d'*Ardea cinerea* et d'*Himantopus himantopus* au Sénégal. *Oiseau et R.F.O.* **50**(3–4), 345.

Hall, P. (1976). The status of Cape Wigeon *Anas capensis*, Three-banded Plover *Charadrius tricollaris* and Avocet *Recurvirostra avosetta* in Nigeria. *Bull. Niger. Orn. Soc.* **12**(41), 43.

Hamilton, R. B. (1975). Comparative behavior of the American Avocet and the Black-necked Stilt (Recurvirostridae). *Orn. Monogr.* **17**.

Harvey, W. G. (1971). A breeding record of the Black-winged Stilt in coastal Tanzania. *Bull. E. Afr. Nat. Hist. Soc.* Nov **1971**, 176–177.

Harvey, W. G. (1974). The occurrence of waders in the Dar es Salaam area of Tanzania, Pt. 3. *Bull. E. Afr. Nat. Hist. Soc.* **1974**, 90–92.

Jacob, J-P. and Jacob, A. (1980). Nouvelles données sur l'avifaune du Lac de Boughzoul (Algérie). *Alauda* **48**(4), 209–219.

Jacobs, W. N. (1973). Avocet feeding behaviour. *Lammergeyer* **19**, 33.

Joubert, C. S. W. (1974). Black-winged Stilts breeding on Lake Kariba. *Honeyguide* **80**, 28–31.

Kallas, J. (1974). Nesting ecology of the Avocet in Käira Bay. In 'Estonian Wetlands and their Life' (Kumari, E. Ed) pp. 119–138. Tallinin.

Lack, P. C., Leuthold, W. and Smeenk, C. (1980). Check-list of the birds of Tsavo East National Park, Kenya. *J. E. Afr. Nat. Hist. Soc. Nat. Mus.* **170**, 25 pp.

Line, L. J. (1942). Two rare visitors to the Dargle district.

Ostrich **13**(2), 93–95.
Liversidge, R. (1963). *Himantopus himantopus meridionalis* (Brehm). *Ostrich* **34**(3), 167.
Macdonald, M. A. (1978). Seasonal changes in numbers of waders at Cape Coast, Ghana. *Bull. Niger. Orn. Soc.* **14**(45), 28–35.
Macdonald, M. A. (1979). Breeding data for birds in Ghana. *Malimbus* **1**, 36–42.
Maclean, G. L. (1975). Belly-soaking in the Charadriiformes. *J. Bombay Nat. Hist. Soc.* **72**, 74–82.
Makkink, G. F. (1936). An attempt at an ethogram of the European Avocet (*Recurvirostra avosetta* L.), with ethological and psychological remarks. *Ardea* **25**(1–2), 1–62.
Mayr, E. and Short, L. L. (1970). Species taxa of North American birds. *Publ. Nuttall Orn. Club*, **9**.
McLachlan, A., Wooldridge, T., Schramm, M. and Kuhn, M. (1980). Seasonal abundance, biomass and feeding of shorebirds on sandy beaches in the eastern Cape, South Africa. *Ostrich* **51**(1), 44–52.
Moreau, R. E. (1967). Water-birds over the Sahara. *Ibis* **109**, 232–259.
Morgan-Davies, A. M. (1960). A nesting colony of Avocets at Lake Manyara, Tanganyika. *J. E. Afr. and Uganda Nat. Hist. Soc.* **23**(6), 241–243.
Moser, M. R. (Ed) (1981). Shorebird studies in North West Morocco. Durham University Sidi Moussa Expedition, Durham, UK, 100 pp.
Olney, P. J. S. (1970). Studies of Avocet behaviour. *British Birds* **63**, 206–209.
Olson, S. R. and Feduccia, A. (1980a). *Presbyornis* and the origin of the Anseriformes (Aves: Charadriomorphae). *Smithsonian Contrib. Zool.* **323**, 24 pp.
Olson, S. L. and Feduccia, A. (1980b). Relationships and evolution of flamingoes (Aves: Phoenicopteridae). *Smithsonian Contrib. Zool.* **316**, 73.
Pienkowski, M. W. (1975). Studies on coastal birds and wetlands in Morocco 1972. Univ. East Anglia Expeditions to Morocco 1971–72, Norwich, England.
Pienkowski, M. W. and Dick, W. J. A. (1976). Some biases in cannon netted and mist netted samples of wader populations. *Ringing Migr.* **1**(2), 105–107.
Pineau, H. and Giraud-Audine, M. (1977). Notes sur les oiseaux nicheurs de l'extrême nord-ouest du Maroc: reproduction et mouvements. *Alauda* **45**(1), 75–104.
Plowes, D. C. H. (1946). Additional notes on the birds of Bloemhof District. *Ostrich* **17**(3), 131–144.
Poulsen, H. (1965). Breeding stilts *Himantopus* spp. at Copenhagen Zoo. *Int. Zoo Yb.* **5**, 130–131.
Prater, A. J. (1976). The distribution of coastal waders in Europe and North Africa. *Proc. Int. Conf. Conservation Wetlands Waterfowl*, Heiligenhafen, 2–6 December 1974. Int. Wtrfwl Res. Bur., Slimbridge, pp. 255–271.
Pringle, J. S. and Cooper, J. (1977). Wader populations (Charadrii) of the marine littoral of the Cape Province, South Africa. *Ostrich* **48**, 98–105.
Reynolds, J. F. (1975). Chestnut-banded Sand Plover responding to a supernormal stimulus. *Bull. E. Afr. Nat. Hist. Soc.* **1975**, 130.
Reynolds, J. F. (1977). Thermo-regulatory problems of birds nesting in areas in East Africa: a review. *Scopus* **1**(3), 57–68.
Robertson, H. G. (1981). Annual summer and winter fluctuations of Palaearctic and resident waders (Charadrii) at Langebaan Lagoon, South Africa, 1915–1979. In 'Proceedings of the Symposium on Birds of the Sea and Shore, 1979' (Cooper, J. Ed), pp. 335–345. African Seabird Group, Cape Town.
Robin, P. (1968). L'avifaune de l'Iriki (Sud-Marocain). *Alauda* **36**, 237–253.
Robinson, G. P. (1975). Breeding records for the Black-winged Stilt at Mufulira. *Bull. Zamb. Orn. Soc.* **7**(1), 32–33.
Roux, F. (1959). Quelques données sur les Anatidés et Charadriidés paléarctiques hivernants dans la basse Vallée du Sénégal et sur leur écologie. *Terre Vie* **106**, 315–321.
Roux, F. (1976). The status of wetlands in the West African Sahel: their value for waterfowl in their future. *Proc. Int. Conf. Conserv. Wetlands and Waterfowl*, Heiligenhafen. Int. Wtrfwl Res. Bur., Slimbridge, pp. 272–287.
Rydzewski, W. (1978). The longevity of ringed birds. *The Ring* **96–97**, 218–262.
Schnell, G. D. and Hellack, J. J. (1978). Flight speeds of Brown Pelicans, Chimney Swifts and other birds. *Bird Banding* **49**(2), 108–112.
Seebohm, G. D. (1886). A review of the species of the genus *Himantopus*. *Ibis* **4**(3), 224–237.
Sharland, R. E. and Wilkinson, R. (1981). The birds of Kano State, Nigeria. *Malimbus* **3**, 7–30.
Simmons, K. E. L. and Crowe, R. W. (1953). Displacement-sleeping in the Avocet and Oystercatcher as a reaction to predators. *British Birds* **46**, 405–410.
Skead, D. M. (1977). Weights of birds handled at Barberspan. *Ostrich* Suppl. **12**, 117–131.
Skinner, N. J. (1969). Avocet *Recurvirostra avosetta* at Zaria. *Bull. Niger. Orn. Soc.* **6**(21), 34.
de Smet, K. and Van Gompel, J. (1980). Observations sur la côte sénégalaise en décembre et janvier. *Malimbus* **2**, 56–70.
Smith, K. D. (1950). The birds of Chirundu, middle Zambesi. *Ostrich* **21**, 62–71.
Spitz, F. (1969). Present extent of knowledge of the number of waders wintering in Europe, the Mediterranean basin and North Africa. *Int. Wtrfwl Res. Bur. Bull.* **27/28**, 12–14.
Summers, R. W. (1977). Distribution, abundance and energy relationships of waders (Aves: Charadrii) at Langebaan Lagoon. *Trans. Roy. Soc. S. Afr.* **53**, 483–495.
Summers, R. W., Cooper, J. and Pringle, J. S. (1977). Distribution and numbers of coastal waders (Charadrii) in the Southwestern Cape, South Africa, summer 1975–76. *Ostrich* **48**, 85–97.
Summers, R. W. and Waltner, M. (1978). Seasonal variations in the mass of waders in southern Africa, with special reference to migration. *Ostrich* **50**, 21–37.
Szlivka, L. (1957). Der Stelzenläufer, *Himantopus himantopus*, in Mali Idjos. *Larus* **9/10**, 225.
Taylor, J. S. (1957). Notes on the birds of inland waters in eastern Cape Province with special reference to the Karoo. *Ostrich* **28**, 1–80.
Tree, A. J. (1969). The status of Ethiopian waders in Zambia. *Puku* **5**, 181–205.
Tree, A. J. (1974). Waders in the Salisbury area 1972/74. *Honeyguide* **80**, 13–27.
Tree, A. J. (1978). The roosting habits of Charadriiformes in Rhodesia. *Safring News* **7**(1), 2–7.
Tree, A. J. (1979). Recent interesting observations in Mashonaland. *Honeyguide* **97**, 18–24.
Tree, A. J. (1980). Migration as an ecological adaptation in Central African Charadrii. *Honeyguide* **102**, 16–25.
Trembsky, A. and Trembsky, J. (1978). Observations ornithologiques effectuées au Maroc au cours des mois de juillet 1974 et 1975. *Aves* **15**(1), 1–16.
Underhill, L. G., Cooper, J. and Waltner, M. (1980). The status of waders (Charadrii) and other birds in the coastal region of the southern and eastern Cape, summer 1978/79. Western Cape Wader Study Group, Pinelands, South Africa.
Vernon, C. (1982). Notes from the border. *The Bee-eater* **33**, 48.
Vincent, A. W. (1945). On the breeding habits of some

African birds. *Ibis* **87**, 345–365.
Walters, J. (1980). Onset of moult in Avocet. *Bird Study* **27**, 108.
Waltner, M. (1979). Waders and wader-ringing. *Safring News* **8**(1), 15–18.
Westernhagen, W. von (1968). Limicolen-Vorkommen an der Westafrikanischen Küste auf der Banc d'Arguin (Mauritania). *J. Orn.* **109**, 185–205.
Whitelaw, D. A., Underhill, L. G., Cooper, J. and Clinning, C. F. (1978). Waders (Charadrii) and other birds on the Namib coast: counts and conservation priorities. *Madoqua* **2**(2), 137–150.
Winterbottom, J. M. (1962). Systematic notes on birds of the Cape Province. XXI. *Himantopus himantopus* (L.). *Ostrich* **33**(2), 74–75.
Winterbottom, J. M. (1972). The ecological distribution of birds in Southern Africa. Monograph Percy Fitzpatrick Institute African Ornithology, **1**, 81 pp.
Wyndham, C. (1942). Nest and eggs of the Avocet (*Recurvirostra avosetta*). *Ostrich* **13**(2), 70–74.
Zwarts, L. (1972). Bird counts in Merja Zerga, Morocco, December 1970. *Ardea* **60**, 120–123.

Family BURHINIDAE: thick-knees, stone curlews, dikkops

Anon. (1981). Recent Egyptian records. *Bull. Orn. Soc. Middle East* **6**, 12–13.
Bates, G. L. (1934). Birds of the southern Sahara and adjoining countries in French West Africa—Part 2. *Ibis* Ser. 13, **4**, 61–79.
Bigalke, R. (1933). Observations on the breeding habits of the Cape Thick-knee, *Burhinops capensis capensis* (Lcht.), in captivity. *Ostrich* **4**, 41–48.
Bird, G. (1933). Some habits of the Stone-curlew. *British Birds* **27**, 114–116.
Britton, P. L. (1970). Some non-passerine bird weights from East Africa. *Bull. Br. Orn. Club.* **90**, 142–144.
Broekhuysen, G. J. (1964). A description and discussion of threat- and anxiety behaviour of *Burhinus capensis* (Lichtenstein) during incubation. *Zool. Meded.* **39**, 240–248.
Browne, P. W. P. (1979). Bird observations in southwestern Mauritania during 1978 and 1979. Unpub. ms, 9 pp.
Friedmann, H. and Northern, J. R. (1975). Results of the Taylor South West African Expedition 1972: Ornithology. Los Angeles Co. Mus. *Contrib. Sci.*, **266**.
Heinzel, H., Fitter, R. and Parslow, J. (1972). 'The Birds of Britain and Europe with North Africa and the Middle East'. Collins, London.
Kemp, A. C. (1971). Observations on birds of the Eastern Caprivi Strip. *Bull. Transv. Mus.*, July 1971.
Loos, K. (1910). Beobachtungen über den Triel aus der Umgebung von Liboch. *Orn. Monatsber.* **35**, 369–381.
Maclean, G. L. (1966). Studies on the behaviour of a young Cape Dikkop *Burhinus capensis* (Lichtenstein) reared in captivity. *Ostrich* Suppl. **6**, 155–170.
Pearson, D. J. (1983). East African bird report 1981. *Scopus* **5**(5), 131–153.
Pitman, C. R. S. (1928). The breeding of *Burhinus vermiculatus büttikoferi*—the West African Water Dikkop, on the shores and islands of Victoria Nyanza. *Ool. Rec.* **7**(4), 102–105.
Pitman, C. R. S. (1931). Notes from Uganda on the breeding of *Burhinus s. senegalensis* (Swainson). *Ool. Rec.* **11**, 62–64.
Prater, A. J., Marchant, J. H. and Vuorinen, J. (1977). 'Guide to the Identification and Ageing of Holarctic Waders'. Brit. Trust Ornithol., Tring.

Prigogine, A. (1976). Additions à l'avifaune du Zaïre. *Gerfaut* **66**, 307–308.
Tuck, G. S. (1964). Reports of land birds at sea. *Sea Swallow* **17**, 40–51.
Wadewitz, O. (1955). Zur Brutbiologie des Triels, *Burhinus oedicnemus* (L.). *Beitr. Vogelkde.* **4**, 86–107.
Watt, J. S. (1934). Note on the Cape Thickknee. *Ostrich* **5**(2), 77–78.
Westwood, N. J. (1983). Breeding of Stone-curlews at Weeting Heath, Norfolk. *British Birds* **76**, 291–304.

Family GLAREOLIDAE: Egyptian Plover, coursers, pratincoles

Benson, C. W. (1982). Migrants in the Afrotropical region south of the equator. *Ostrich* **53**, 31–49.
Benson, C. W. and Pitman, C. R. S. (1964). Further breeding records from Northern Rhodesia (No. 4). *Bull. Br. Orn. Club.* **84**, 54–60.
Brehm, A. E. (1879). Thierleben, 2nd ed. Vögel, Vol. 6. Bibliographisches Institut, Leipzig.
Britton, P. L. (1970). Some non-passerine bird weights from East Africa. *Bull. Br. Orn. Club* **90**, 142–144.
Britton, P. L. (1977). The Madagascar Pratincole in Africa. *Scopus* **1**, 94–97.
Brosset, A. (1957). Les oiseaux de la Steppe de Berguent. Remarques particulières sur leurs migrations. *Alauda* **25**, 196–208.
Brosset, A. (1979). Le cycle de reproduction de la glaréole *Glareola nuchalis*; ses déterminants écologiques et comportementaux. *Terre et Vie* **33**, 95–108.
Browne, P. W. P. (1979). Bird observations in southwest Mauritania during 1978 and 1979. Unpub. ms, 9 pp.
Browne, P. W. P. (1981). Breeding of six Palaearctic birds in southwest Mauritania. *Bull. Br. Orn. Club.* **101**(2), 306–310.
Browne, P. W. P. (1981). New bird species in Mauritania. *Malimbus* **3**, 62–72.
Butler, A. L. (1931). The chicks of the Egyptian Plover. *Ibis* Ser. 13, **1**, 345–347.
Cassidy, R., Salinger, T. and Wolhuter, D. (1984). The Threebanded Courser in South Africa. *Ostrich* **55**(1), 34.
Cheke, R. A. (1980). A small breeding colony of the Rock Pratincole *Glareola nuchalis* in Togo. *Bull. Br. Orn. Club* **100**, 175–178.
Cheke, R. A. (1982). Additional information on the Rock Pratincole *Glareola nuchalis* in Togo. *Bull. Br. Orn. Club* **102**, 116–117.
Clancey, P. A. (1979). Miscellaneous taxonomic notes on African birds. *Durban Mus. Novit.* **12**, 49–50.
Clancey, P. A. (1981). Miscellaneous taxonomic notes on African birds. *Durban Mus. Novit.* **12**, 224–227.
Cowan, P. J. (1983). Birds in the Brak and Sabha regions of central Libya, 1981–82. *Bull. Br. Orn. Club* **103**(2), 44–47.
Cramp, S. and Reynolds, J. F. (1972). Studies of less familiar birds. 168. Cream-coloured Courser. *British Birds* **65**, 120–124.
Curry, P. J. and Sayer, J. A. (1979). The inundation zone of the Niger as an environment for Palaearctic migrants. *Ibis* **121**, 20–40.
Dixon, J. E. W. (1975). Nasal salt secretion from Burchell's Courser. *Madoqua* **9**, 63–64.
Fehl, H. (1976). Les oiseaux de l'île de Kembé (RCA). *Alauda* **44**(2), 153–167.
Friedmann, H. and Northern J. R. (1975). Results of the Taylor South West African Expedition 1972: Ornithology. Los Angeles Co. Mus. *Contrib. Sci.* **266**.

Fry, C. H. (1970). Bird distribution on west/central African great rivers at high water. *Bull. Niger. Orn. Soc.* **7**, 7–23.

Howell, T. R. (1979). Breeding biology of the Egyptian Plover *Pluvianus aegyptius*. *Univ. Calif. Publ. Zool.* **113**.

Kemp, A. C. (1974). 'The Distribution and Status of the Birds of the Kruger National Park'. National Parks Board of Trustees, Pretoria.

Kemp, A. C. and Maclean, G. L. (1973a). Neonatal plumage patterns of Three-banded and Temminck's Coursers and their bearing on courser genera. *Ostrich* **44**, 80–81.

Kemp, A. C. and Maclean, G. L. (1973b). Nesting of the Three-banded Courser. *Ostrich* **44**, 82–83.

Koenig, A. (1926). Ein weiterer Teilbeitrag zur Avifauna Aegyptiaca ... der Wat-oder Sumpfvögel (*Grallatores*). *J. Orn.* **74**, Sonderheft.

Lambert, F. R. (1985). Large-scale movements of Common Pratincole *Glareola pratincola* at Juba, Sudan. *Malimbus* **7**, 136.

Maclean, G. L. (1967). The breeding biology and behaviour of the Double-banded Courser *Rhinoptilus africanus* (Temminck). *Ibis* **109**, 556–569.

Maclean, G. L. (1970). The neonatal plumage pattern of the Double-banded Courser. *Ostrich* **41**, 215–216.

Maclean, G. L. (1978). Pratincoles and coursers. In 'Bird Families of the World' (Harrison, C. J. O. Ed) pp. 105–107. Elsevier-Phaidon, Oxford.

Malbrant, R. and Maclatchy, A. (1949). Faune de l'équateur Africain français. In 'Oiseau', Vol. 1. Lechevalier, Paris.

Meinertzhagen, R. (1959). 'Pirates and Predators'. Oliver and Boyd, Edinburgh and London.

Milon, P., Petter, J. J. and Randrianasolo, G. (1973). Faune de Madagascar. XXXV. Oiseaux. ORSTOM, Tananarive, CNRS, Paris.

Nikolaus, G. and Backhurst, G. C. (1982). First ringing report for the Sudan. *Scopus* **6**(4), 77–90.

Payn, W. H. (1948). Notes from Tunisia and eastern Algeria: February 1943–April 1944. *Ibis* **90**, 1–21.

Pearson, D. J. (1983). East African bird report 1981. *Scopus* **5**(5), 131–153.

Penry, E. H. (1979). The Rock Pratincole *Glareola nuchalis* at Greystone, Kitwe, and a review of its migratory movements. *Bull. Zamb. Orn. Soc.* **11**(2), 2–32.

Reynolds, J. F. (1968). The nestling plumage of four East African coursers. *E. Afr. Wildl. J.* **6**, 141.

Reynolds, J. F. (1975). Coursers of the dry savannas. *Wildlife* **17**, 274–278.

Reynolds, J. F. (1977). Thermo-regulatory problems of birds nesting in arid areas in East Africa: a review. *Scopus* **1**(3), 57–68.

Steyn, P. (1965). Temminck's Courser. *Afr. Wildl.* **19**(1), 29–32.

Thiollay, J–M. (1973). Arrivée de migrateurs paléarctiques au sud du Sahara. *Nos Oiseaux* **32**(6), 168–171.

Vernon, C. (1982). Coursers in the Eastern Cape. *The Bee-eater* **33**, 31–34.

Family CHARADRIIDAE

Subfamily CHARADRIINAE: plovers

Al-Hussaini, A. H. (1939). Further notes on the birds of Ghardaqa (Hurghada), Red Sea. *Ibis* **3**(14), 343–347.

Anon. (1964). Two interesting plateau birds: the Red-capped Lark *Calandrella cinerea* and the Three-banded Plover *Afroxyechus tricollaris*. *Bull. Niger. Orn. Soc.* **1**(2), 4.

Anon. (1976). Recoveries of interest. *Safring News* **5**(2), 16.

Anon. (1981a). East African bird report 1980. *Scopus* **4**(5), 103–120.

Anon. (1981b). Recent Egyptian reports. *Bull. Orn. Soc. Middle East*, **6**, 12–13.

Armstrong, E. A. (1952). The distraction displays of the Little Ringed Plover and territorial competition with the Ringed Plover. *British Birds* **45**, 55–59.

Ash, J. S. (1969). Spring weights of trans-Sahara migrants in Morocco. *Ibis* **111**(1), 1–10.

Ash, J. S. (1972). Charadriiform birds in the Ethiopian Rift Valley. *Walia* **4**, 14–18.

Ash, J. S. (1980a). Migrational status of Palaearctic birds in Ethiopia. *Proc. IV Pan Afr. Orn. Congr.*, pp. 199–208.

Ash, J. S. (1980b). The Lesser Golden Plover *Pluvialis dominica* in northeast Africa and the southern Red Sea. *Scopus* **4**, 64–66.

Ash, J. S. (1981). Spring passage of Whimbrel *Numenius phaeopus* and other waders off the coast of Somalia. *Scopus* **5**, 71–76.

Ash, J. S., Ferguson-Lees, I. J. and Fry, C. H. (1967). B.O.U. expedition to Lake Chad, northern Nigeria, March–April 1967. *Ibis* **109**, 478–486.

Aspinwall, D. R. (1973). Sight record of Great Sandplover *Charadrius leschenaultii*: a species new to Zambia. *Bull. Zamb. Orn. Soc.* **5**(1), 29.

Aspinwall, D. R. (1975). Great Sandplover (*Charadrius leschenaultii*) and Eastern Golden Plovers (*Pluvialis dominica*) at Lusaka. *Bull. Zamb. Orn. Soc.* **7**(2), 101–102.

Backhurst, G. C. (1981). Eastern African bird ringing report 1977–1981. *J. E. Afr. Nat. Hist. Soc. and Nat. Mus.* **174**, 19 pp.

Bannerman, D. A. (1961). 'The Birds of the British Isles', Vol. 10. Oliver and Boyd, London.

Bates, G. L. (1932a). *Charadrius marginatus hesperius*, nom. nov. *Bull. Br. Orn. Club* **53**, 11.

Bates, G. L. (1932b). *Charadrius marginatus migirius*, nom. nov. *Bull. Br. Orn. Club* **53**, 75–76.

Becker, P. (1965). Ornithologische Beobachtungen in Lüderitzbucht. *Mitt. Orn. Arbeitsgr. S. W. Afr. Wiss. Ges.* **1**(7/9), 2–5.

Benson, C. W. (1968). Kittlitz's Sandplover *Charadrius pecuarius* Temminck, in Angola. *Ostrich* **39**(1), 41.

Berruti, A. (1980). Status and review of waterbirds breeding at Lake St. Lucia. *Lammergeyer* **28**, 1–19.

Berry, H. H. and Berry, C. U. (1975). A check list and notes on the birds of Sandvis, South West Africa. *Madoqua* **9**(2), 5–18.

Betham, R. M. (1929). Some observations on the nesting of birds in the vicinity of Cape Town, South Africa. *Ibis* **5**, 71–104.

Beven, G. and Chiazzari, W. L. (1943). Waders, dikkops, coursers and common larks at Oudtshoorn. *Ostrich* **14**, 139–151.

Blaker, D. (1966). Notes on the sandplovers *Charadrius* in southern Africa. *Ostrich* **37**, 95–102.

Blondel, J. and Blondel, C. (1964). Remarques sur l'hivernage des limicoles et autres oiseaux aquatiques au Maroc (janvier 1964). *Alauda* **32**(4), 250–279.

Bock, W. J. (1958). A generic review of the plovers (Charadriinae, Aves). *Bull. Mus. Comp. Zool.* **118**(2), 27–97.

Boyd, H. (1962). Mortality and fertility of European Charadrii. *Ibis* **104**, 368–387.

Boyd, R. L. (1972). Breeding biology of the Snowy Plover at Cheyenne Bottoms Wildlife Management area, Barton County, Kansas. MS thesis, Kansas State Teachers College, Emporia.

Brinkman, J. (1980) Greater Sand Plover *Charadrius leschenaultii* near Korba, Tunisia. *Dutch Birding* **1**(4), 106.

Britton, H. A. (1978). E.A.N.H.S. nest record scheme: July 1976–December 1977. *Scopus* **1**(5), 132–141.

Britton, H. A. (1979). E.A.N.H.S. nest record scheme, 1978. *Scopus* **2**(5), 126–132.

Britton, P. L. (1982). Identification of sand plovers. *British*

Birds 75, 94–95.
Britton, P. L. and Britton, H. (1966). Some interesting records from Mashonaland. *Honeyguide* 48, 18–19.
Britton, P. L. and Dowsett, R. J. (1969). More bird weights from Zambia. *Ostrich* 40, 55–60.
Broekhuysen, G. J. (1955). Occurrence and movement of migratory species in Rhodesia and southern Africa during the period 1950–1953 (Part I). *Ostrich* 26, 99–114.
Broekhuysen, G. J. (1971). Third report on bird migration in southern Africa. *Ostrich* 42, 41–64, 211–225, 235–250.
Broekhuysen, G. J. and Stanford, W. P. (1955). Display in Ringed Plover (*Charadrius hiaticula*) while in the winter quarter. *Ostrich* 26, 41–43.
Brooke, R. K. (1959). Avian highlights of a journey across South Africa. *Ostrich* 30, 82–83.
Brooke, R. K. (1963). Birds round Salisbury, then and now. *Sth. Afr. Av. Ser.* 9, 1–67.
Brooke, R. K. (1964). Avian observations on a journey across central Africa and additional information on some species seen. *Ostrich* 35(4), 277–292.
Brooks, W. S. (1967). Organisms consumed by various migrating shorebirds. *Auk* 84, 128–130.
Brown, L. H. (1948). Notes on birds of Kabba, Ilorin and N. Benin Provinces of Nigeria. *Ibis* 90, 525–538.
Browne, P. W. P. (1979). Bird observations in southwest Mauritania during 1978 and 1979. Unpub. ms, 9 pp.
Browne, P. W. P. (1981). Breeding of six Palaearctic birds in southwest Mauritania. *Bull. Br. Orn. Club* 101, 306–310.
Browne, P. W. P. (1981). New bird species in Mauritania. *Malimbus* 3, 62–72.
Bundy, G. and Morgan, J. H. (1969). Notes on Tripolitanian birds. *Bull. Br. Orn. Club.* 89, 139–144, 151–159.
Burnier, E. (1977). Sur l'hivernage de Pluvier Guignard en Algérie. *Nos Oiseaux* 34, 74.
Burnier, E. (1979). Notes sur l'ornithologie Algérienne. *Alauda* 47(2), 93–102.
Cabanis, J. (1884). Bericht über die Mai-Sitzung. *J. Orn.* 32, 437–438.
Campbell, L. (1972). Caspian Plover *Charadrius asiaticus* at the coast. *Bull. E. Afr. Nat. Hist. Soc.* 1972, 65.
Campbell, N. A. (1970). Identification of the sandplovers in Rhodesia. *Honeyguide* 62, 17–25.
Campbell, N. A. and Miles, H. M. (1956). Bird counts in a highveld dam in Southern Rhodesia. *Ostrich* 27, 56–66.
Carlisle, J. S. (1929). African breeding Charadriidae. *Ool. Rec.* 9(41), 74–80.
Clancey, P. A. (1962). Chestnut-banded Sandplover *Charadrius pallidus pallidus* Strickland in southern Mozambique. *Ostrich* 33(1), 26.
Clancey, P. A. (1971a). Miscellaneous taxonomic notes on African birds, 33. The southern African races of the White-fronted Sandplover *Charadrius marginatus* Vieillot. *Durban Mus. Novit.* 9(9), 113–118.
Clancey, P. A. (1971b). Miscellaneous taxonomic notes on African birds, 33. Variation in Kittlitz's Sandplover *Charadrius pecuarius* Temminck. *Durban Mus. Novit.* 9(9), 109–112.
Clancey, P. A. (1975). Miscellaneous taxonomic notes on African birds, 43. On the relationship of *Charadrius marginatus* Vieillot of Ethiopian Africa to *Charadrius alexandrinus* Linnaeus. *Durban Mus. Novit.* 11(1), 1–9.
Clancey, P. A. (1979). Miscellaneous taxonomic notes on African birds, 53. *Durban Mus. Novit.* 12(1), 1–17.
Clancey, P. A. (1980). On birds from the mid-Okavango Valley on the South West Africa/Angola border. *Durban Mus. Novit.* 12(9), 87–127.
Clancey, P. A. (1981). Miscellaneous taxonomic notes on African birds, 60. Variation in the Afrotropical populations of *Charadrius tricollaris* Vieillot. *Durban Mus. Novit.* 13(1), 1–6.
Clancey, P. A. (1982a). Miscellaneous taxonomic notes on African birds, 63. On the Ringed Plover *Charadrius hiaticula* Linnaeus in southern Africa. *Durban Mus. Novit.* 13(10), 127–130.
Clancey, P. A. (1982b). Miscellaneous taxonomic notes on African birds, 63. The Great Sandplover *Charadrius leschenaultii* Lesson in southern and eastern Africa. *Durban Mus. Novit.* 13(10), 131–132.
Clark, A. (1982a). Some observations on the breeding behaviour of Kittlitz's Sand-Plover. *Ostrich* 53, 120–122.
Clark, A. (1982b). Some observations on the behaviour of the Threebanded Plover. *Ostrich* 53(4), 222–227.
Colston, P. R. (1971). Additional non-passerine bird weights from East Africa. *Bull. Br. Orn. Club.* 91(4), 110–111.
Connors, P. G. (1983). Taxonomy, distribution, and evolution of Golden Plovers (*Pluvialis dominica* and *Pluvialis fulva*). *Auk* 100, 607–620.
Conway, W. G. and Bell, J. (1968). Observations on the behavior of Kittlitz's Sandplover at the New York Zoological Park. *Living Bird* 7, 57–70.
Cooper, J. (1972). A checklist of the birds of the Zambezi Valley from Kariba to Zumbo. *Sth. Afr. Av. Ser.* 85.
Cooper, J., Summers, R. W. and Pringle, J. S. (1976). Conservation of coastal habitats of waders in the South-Western Cape, South Africa. *Biol. Conserv.* 10, 239–247.
Cowan, P. J. (1983). Birds in the Brak and Sabha regions of central Libya, 1981–82. *Bull. Br. Orn. Club.* 103(2), 44–47.
Cramp, S. and Conder, P. J. (1970). A visit to the oasis of Kufra, spring 1969. *Ibis* 112, 261–263.
Cunningham-van Someren, G. R. (1971). On the nesting of the 'Chestnut-banded' or 'Magadi' Plover *Charadrius pallidus venustus* Fischer and Reichenow, at Lake Nakuru. *Bull. E. Afr. Nat. Hist. Soc.* 1971, 130–131.
Curry, P. J. and Sayer, J. A. (1979). The inundation zone of the Niger as an environment for Palaearctic migrants. *Ibis* 121, 20–40.
Curry-Lindahl, K. (1961). Exploration du Parc National Albert et du Parc National de la Kagera, II. Mission Kai Curry-Lindahl. Bruxelles, Institut des Parcs Nationaux du Congo et du Ruanda-Urundi, fasc. 1.
Dathe, H. (1953). 'Der Flussregenpfeifer'. Leipzig.
Day, D. H. (1975). Some bird weights for the Transvaal and Botswana. *Ostrich* 46, 192–194.
Dhondt, A. A. (1975). Note sur les échassiers (Charadrii) de Madagascar. *Oiseau et R.F.O.* 45, 73–82.
Dick, W. J. A. (1975). Oxford and Cambridge Mauritanian Expedition, 1973, Report. Cambridge.
Dick, W. J. A. and Pienkowski, M. W. (1979). Autumn and early winter weights of waders in north-west Africa. *Orn. Scand.* 10, 117–123.
Dowsett, R. J. (1965). Weights of some Zambian birds. *Bull. Br. Orn. Club* 85(8), 150–152.
Dowsett, R. J. (1969a). B.O.U. supported research at Lake Chad in 1968. *Ibis* 111, 449–452.
Dowsett, R. J. (1969b). Greater Sandplovers *Charadrius leschenaultii* Lesson at Lake Chad. *Bull. Br. Orn. Club* 89(3), 73–74.
Dowsett, R. J. (1977). The distribution of some falcons and plovers in East Africa. *Scopus* 1(3), 73–78.
Dowsett, R. J. (1979). Recent additions to the Zambian list. *Bull. Br. Orn. Club* 99, 94–98.
Dowsett, R. J. (1980). The migration of coastal waders from the Palearctic across Africa. *Gerfaut* 70(1), 3–36.
Dowsett, R. J. and Lemaire, F. (1976). A second Zambian record of Mongolian Sandplover. *Bull. Zamb. Orn. Soc.* 8(2), 68–69.

Drury, W. H. (1961). Breeding biology of shorebirds. *Auk* **78**, 176–219.

Dugan, P. J., Evans, P. R., Goodyer, L. R. and Davidson, N. C. (1981). Winter fat reserves in shorebirds: disturbance of regulated levels by severe weather conditions. *Ibis* **123**, 359–363.

Dupuy, A. R. (1968). Le migration des Laro-Limicoles au Sahara Algérien. *Alauda* **36**(1–2), 27–43.

Dupuy, A. R. (1974). Observations de deux pluviers dorés (*Pluvialis apricaria*) au Sénégal. *Bull. IFAN* 36, ser. A, **3**, 754–755.

EANHS Ornithological Subcommittee. (1977). Some recent records of Palaearctic birds in Kenya and Tanzania. *Scopus* **1**(2), 39–43.

Edwards, P. J. (1982). Plumage variation, territoriality and breeding displays of the Golden Plover *Pluvialis apricaria* in southwest Scotland. *Ibis* **124**, 88–96.

Elliott, H. F. I. (1956). Some field-notes on the Caspian Plover. *British Birds* **49**, 282–283.

Ellis, M. (1971). Chestnut-banded Plover at Baringo. *Bull. E. Afr. Nat. Hist. Soc.* **1971**, 82.

Erard, C. (1963). Sur le comportement de diversion du Gravelot à collier interrompu *Charadrius alexandrinus* L. à l'égard de l'Homme. *Alauda* **31**(4), 262–284.

Evans, S. and Campbell, K. (1977). Ringed Plover taken by Fish Eagle. *Bull. E. Afr. Nat. Hist. Soc.* **1977**, 63.

Farmer, R. (1979). Checklist of birds of the Ile-Ife area, Nigeria. *Malimbus* **1**, 56–64.

Field, G. D. (1974). Nearctic waders in Sierra Leone: Lesser Golden Plover and Buff-breasted Sandpiper. *Bull. Br. Orn. Club* **94**(2), 76–78.

Fraser, W. (1969). Chestnut-banded Sandplover *Charadrius pallidus*. *Ostrich* **40**(3), 131.

Freeman, R. J. (1970). Feeding behaviour of a Treble-banded Plover. *Ostrich* **41**(4), 263–264.

Friedmann, H. and Williams, J. G. (1971). The birds of the lowlands of Bwamba, Toro Province, Uganda. Los Angeles Co. Mus. *Contrib. Sci.* **211**, 70 pp.

Fry, C. H. and Horne, J. (1972). Kentish Plover at Lake Rudolf. *Bull. E. Afr. Nat. Hist. Soc.* **1972**, 139.

Gaul, W. D. (1973). Breeding biology of the Mountain Plover. *Wilson Bull.* **85**, 60–70.

Gee, J. and Heigham, J. (1977.) Birds of Lagos, Nigeria. *Bull. Nig. Orn. Soc.* **13**(43), 43–51; **13**(44), 103–132.

Ginn, P. J. (1968). Birds of the Marandellas District. *Sth. Afr. Av. Ser.* **61**.

Goodman, S. M. and Atta, G. A. M. (in press). The birds of southeastern Egypt. *Gerfaut*.

Goodman, S. M. and Watson, G. E. (1983). bird specimen records of some uncommon or previously unrecorded forms in Egypt. *Bull. Br. Orn. Club* **103**(3), 101–107.

Greig-Smith, P. W. (1977). Breeding data of birds in Mole National Park, Ghana. *Bull. Niger. Orn. Soc.* **13**(44), 89–93.

Greve, K. (1958). Polygamie beim Seeregenpfeifer (*Charadrius alexandrinus*). *Vogelwelt* **79**, 184.

Greve, K. (1969). Zur Siedlungsdichte einiger Brütvogel auf der Nordseeinsel Neuwerk. *Orn. Mitteil.* **21**, 169.

Grimes, L. G. (1974). Radar tracks of Palaearctic waders departing from the coast of Ghana in spring. *Ibis* **116**, 165–171.

Grimes, L. G. and Vanderstichelen, G. (1974). Initial departure directions of waders and other water birds in spring at Accra. *Bull. Niger. Orn. Soc.* **10**(38), 62–63.

Hall, K. R. L. (1958). Observations on the nesting sites and nesting behaviour of the Kittlitz's Sandplover *Charadrius pecuarius*. *Ostrich* **29**, 113–125.

Hall, K. R. L. (1959). Nest records and additional behaviour notes for Kittlitz's Sandplover *Charadrius pecuarius* in the S. W. Cape Province. *Ostrich* **30**, 33–38.

Hall, K. R. L. (1960). Egg-covering by the White-fronted Sandplover, *Charadrius marginatus*. *Ibis* **102**(4), 545–553.

Hall, K. R. L. (1965). Nest records and behaviour notes for three species of plovers in Uganda. *Ostrich* **36**, 107–108.

Hall, P. (1976). The status of Cape Wigeon *Anas capensis*, Three-banded Plover *Charadrius tricollaris* and Avocet *Recurvirostra avosetta* in Nigeria. *Bull. Niger. Orn. Soc.* **12**(41), 43.

Hanmer, D. B. (1976). Birds of the lower Zambezi. *Southern Birds* **2**, 1–66.

Harrison, C. J. O. (1963). Eggs of the Great Sand Plover, *Charadrius leschenaultii* Lesson, from Somaliland. *Bull. Br. Orn. Club* **83**(9), 158–159.

Harrison, J. M. (1938). Remarks on *Charadrius hiaticula major*. *Bull. Br. Orn. Club* **59**(1), 17–18.

Harvey, W. G. (1972). Caspian Plover *Charadrius asiaticus* at the coast. *Bull. E. Afr. Nat. Hist. Soc.* **1972**, 175.

Harvey, W. G (1973). Lesser Golden Plovers *Pluvialis dominica* near Dar-es-Salaam in Tanzania. *Bull. E. Afr. Nat Hist. Soc.* **1973**, 84–85.

Heaton, A. M. and Heaton, A. E. (1980). The birds of Obudu, Cross River State, Nigeria. *Malimbus* **2**, 16–24.

Hockey, P. A. R. (1982). Waders (Charadrii) and other coastal birds in the Lüderitz region of South West Africa. *Madoqua* **13**, 27–33.

Hodgson, C. J. and Hodgson, L. K. (1966). White-fronted Sandplovers (*Charadrius marginatus*) on the Mashonaland Plateau. *Honeyguide* **48**, 14.

Höhn, E. O. (1957). Display in Arctic birds. *Auk* **74**, 203–214.

Hopson, A. J. (1974). A further record of the Kentish Plover at Lake Rudolf. *Bull. E. Afr. Nat. Hist. Soc.* **1974**, 18.

Hussell, D. J. T. and Page, G. W. (1976). Observations on the breeding biology of the Black-bellied Plover on Devon Is., NWT, Canada *Wilson Bull.* **88**(4), 632–653.

Irwin, M. P. S. and Benson, C. W. (1966). Notes on the birds of Zambia, Part II. *Arnoldia (Rhod.)* **2**(37).

Jacob, J-P. and Jacob, A. (1980). Nouvelles données sur l'avifaune du Lac de Boughzoul (Algérie). *Alauda* **48**(4), 209–219.

James, H. W. (1922). Notes on the breeding habits of South African sandpipers. *Ool. Rec.* **2**(1), 1–7.

Jeffery, R. G. and Liversidge, R. (1951). Notes on the Chestnut-banded Sandplover, *Charadrius pallidus pallidus*. *Ostrich* **22**(2), 68–76.

Jehl, H. (1974). Quelques migrateurs paléarctiques en Republique Centrafricaine. *Alauda* **42**(4), 397–406.

Johnston, D. W. and McFarland, R. W. (1967). Migration and bioenergetics of flight in the Pacific Golden Plover. *Condor* **69**, 156–168.

Kemp, A. C. (1971). Observations on birds of the Eastern Caprivi Strip. *Bull. Transv. Mus.* July 1971, 3 pp.

Kemp, A. C. (1974). The distribution and status of the birds of the Kruger National Park. *Koedoe Monogr.* **2**, 1–352.

Kieser, J. A. (1981). Field identification of the Lesser Golden Plover. *Bokmakierie* **33**(3), 62–64.

Kieser, J. A. and Liversidge, R. (1981). The identification of White-fronted Sand Plover *Charadrius marginatus*. *Dutch Birding* **3**(3), 81–84.

Kitson, A. R., Marr, B. A. E. and Porter, R. F. (1980). Greater Sand Plover new to Britain and Ireland. *British Birds* **73**, 568–573.

Knight, P. J. and Dick, W. J. A. (1975). Recensement de limicoles au Banc d'Arguin (Mauritanie). *Alauda* **43**, 363–385.

Kutilek, M. J. (1974). Notes on Kittlitz's Sandplover and Blacksmith Plover at Lake Nakuru. *E. Afr. Wildl. J.* **12**, 87–91.

Lack, P. C., Leuthold, W. and Smeenk, C. (1980). Check-list of the birds of Tsavo East National Park, Kenya. *J. E. Afr. Nat. Hist. Soc. Nat. Mus.* **170**, 25 pp.

Laferrère, M. (1974). *Charadrius tricollaris* Vieillot dans le

Sud-Est du Mali. *Oiseau et R.F.O.* **44**, 346–347.
Little, J. de V. (1966). Sight record of the Chestnut-banded Sandplover *Charadrius pallidus* Strickland at Lake Chrissie, Eastern Transvaal. *Ostrich* **37**(4), 238.
Liversidge, R. (1965). Egg covering in *Charadrius marginatus*. *Ostrich* **36**(2), 59–61.
Liversidge, R. (1968). Bird weights. *Ostrich* **39**, 223–227.
Liversidge, R., Broekhuysen, G. J. and Thesen, A. R. (1958). The birds of Langebaan Lagoon. *Ostrich* **29**(3), 95–106.
Low, C. (1939). Remarks on the races of *Charadrius hiaticula hiaticula*. *Bull. Br. Orn. Club* **59**(3), 48–50.
Macdonald, M. A. (1978a). Lesser Golden Plover in Ghana. *Bull. Niger. Orn. Soc.* **14**(45), 47–48.
Macdonald, M. A. (1978b). Seasonal changes in numbers of waders at Cape Coast, Ghana. *Bull. Niger. Orn. Soc.* **14**(45), 28–35.
Macdonald, M. A. (1979). Evidence for migration and other movements of African birds in Ghana. *Rev. Zool. Afr.* **93**(2), 413–424.
Mackworth-Praed, C. W. and Grant, C. H. B. (1941). Systematic notes on East African birds. Part XXVIII. On the occurrence of the Golden Plover in eastern Africa. *Ibis* (4), 617–618.
Maclean, G. L. (1965). Nest transfer in the Three-banded Sandplover *Charadrius tricollaris* Vieillot. *Ostrich* **36**(2), 62–63.
Maclean, G. L. (1972a). Clutch size and evolution in the Charadrii. *Auk* **89**, 299–324.
Maclean, G. L. (1972b). Problems of display postures in the Charadrii (Aves: Charadriiformes). *Zool. Afr.* **7**, 57–74.
Maclean, G. L. (1973). Waders of the beach. *Afr. Wildl.* **27**, 132–135.
Maclean, G. L. (1974). Egg-covering in the Charadrii. *Ostrich* **45**, 167–174.
Maclean, G. L. (1975). Belly-soaking in the Charadriiforms. *J. Bombay Nat. Hist. Soc.* **72**, 74–82.
Maclean, G. L. and Moran, V. C. (1965). The choice of nest site in the White-fronted Sandplover *Charadrius marginatus* Vieillot. *Ostrich* **36**(2), 63–72.
Macleay, K. N. G. (1960). The birds of Khartoum Province. *Univ. Khartoum Nat. Hist. Mus. Bull.* **1**, 33 pp.
Madge, S. G. (1972). Eastern Golden Plover (*Pluvialis dominica*) at Ndola: a species new to Zambia. *Bull. Zamb. Orn. Soc.* **4**(2), 59.
Mansfield, D. (1965). Some breeding records from Nyasaland. *Ostrich* **36**, 37.
Marchant, S. (1941). Notes on the birds of the Gulf of Suez. *Ibis* **5**(14), 265–295, 378–396.
Martin, J. (1972). Nesting habits of our three resident sandplovers. *Bokmakierie* **24**(2), 40–41.
McLachlan, G. R. and Jeffery, R. G. (1949). Nesting of the Chestnut-banded Sandplover. *Ostrich* **20**(1), 36–37.
McLachlan, A., Wooldridge, M., Schramm, M. and Kühn, M. (1980). Seasonal abundance, biomass and feeding of shore birds on sandy beaches in the eastern Cape, South Africa. *Ostrich* **51**, 44–52.
Meise, W. (1961). *Pipus heteroclitus* (A. Lichtenstein), der erste Sandregenpfeifer von Ghana. *Vogelwarte* **31**(2), 144–146.
Moreau, R. E. (1944). Some weights of African and of wintering Palaearctic birds. *Ibis* **86**, 16–29.
Moreau, R. E. (1946). Coition of Three-banded Plover, *Charadrius tricollaris tricollaris* Vieillot. *Ibis* **1946**, 524–525.
Moreau, R. E. (1967). Water-birds over the Sahara. *Ibis* **109**, 232–259.
Morel G. J. (1984). Liste commentée des Oiseaux de Sénégambie. Mimeo. ORSTOM, Dakar.
Morgan-Davies, A. M. (1960). The Chestnut-banded Plover at Lake Manyara, Northern Tanganyika. *J. E. Afr. and Uganda Nat. Hist. Soc.* **23**(7), 299.
Moser, M. E. (Ed) (1981). Shorebird studies in North West Morocco. Durham University Sidi Moussa Expedition. Durham, UK, 100 pp.
Mullié, C. and Meininger, P. L. (1981). Numbers, measurements and stomach contents of Dunlins, Little Stints and Kentish Plovers from Lake Manzala, Egypt. *Wader Study Group Bull.* **32**, 25–29.
Nethersole-Thompson, D. (1973). 'The Dotterel'. Collins, London.
Neumann, O. (1929). Ueber den Formenkreis des *Charadrius alexandrinus*. *Novitates Zoologicae* **35**, 212–216.
Nielsen, B. P. (1971). Migration and relationships of four Asiatic plovers Charadriinae. *Orn. Scand.* **2**(2), 137–142.
Nielsen, B. P. (1975). Affinities of *Eudromias morinellus* (L.) to the genus *Charadrius* L. *Orn. Scand.* **6**, 65–82.
Nikolaus, G. and Backhurst, G. C. (1982). First ringing report for the Sudan. *Scopus* **6**, 77–90.
Pearson, D. J. (1979). East African bird report 1978. *Scopus* **2** (5), 105–125.
Pearson, D. J. (1983). East African bird report 1981. *Scopus* **5**(5), 131–153.
Pearson, D. J. and Britton, P. L. (1980). Arrival and departure times of Palaearctic waders on the Kenya coast. *Scopus* **4**, 84–88.
Pearson, D. J. and Stevenson, T. (1980). A survey of wintering palaearctic waders in the southern part of the Kenyan Rift Valley. *Scopus* **4**, 59–63.
Pearson, R. G. and Parker, G. A. (1973). Sequential activities in the feeding behaviour of some Charadriiformes. *J. Nat. Hist.* **7**, 573–589.
Penry, E. H. (1975). Forbes's Plover (*Charadrius forbesi*) in Kitwe. *Bull. Zamb. Orn. Soc.* **7**(1), 30–31.
Perry, I. A. (1975). Chestnut-banded Sandplover at Aiselby, Bulawayo. *Honeyguide* **84**, 44–45.
Petetin, M. and Trotignon, J. (1972). Prospection hivernale au Banc d'Arguin (Mauritanie). *Alauda* **40**(3), 195–243.
Phillips, R. E. (1980). Behaviour and systematics of New Zealand plovers. *Emu* **80**, 177–197.
Pienkowski, M. W. (1975). Studies on coastal birds and wetlands in Morocco 1972. Univ. East Anglia Expeditions to Morocco 1971–72. Norwich, England.
Pienkowski, M. W. and Dick, W. J. A. (1976). Some biases in cannon netted and mist netted samples of wader populations. *Ringing Migr.* **1**(2), 105–107.
Pienkowski, M. W. and Knight, P. J. (1977). Le migration post-nuptial des Limicoles sur la côte Atlantique du Maroc. *Alauda* **45**, 165–170.
Pienkowski, M. W., Knight, P. J., Stanyard, D. J. and Argyle, F. B. (1976). The primary moult of waders on the Atlantic coast of Morocco. *Ibis* **118**, 347–365.
Pineau, J. and Giraud-Audine, M. (1977). Notes sur les oiseaux nicheurs de l'extrême nord-ouest du Maroc: réproduction et mouvements. *Alauda* **45**(1), 75–104.
Pitman, C. R. S. (1956). A strange injury to a Ringed Plover (*Charadrius hiaticula tundrae* (Lowe)). *Bull. Br. Orn. Club*. **76**(3), 52.
Pitman, C. R. S. (1965). The eggs and nesting habits of the St. Helena Sand-Plover or Wirebird, *Charadrius pecuarius sanctae-helenae* (Hartig). *Bull. Br. Orn. Club* **85**, 121–129.
Plumb, W. J. (1978). The Lesser Golden Plover in Kenya. *Scopus* **2** (3), 72–73.
Pollard, C. J. (1980). A visual record of Chestnut-banded Sandplover at the Victoria Falls. *Honeyguide* **102**, 37.
Prater, A. J. (1976). The distribution of coastal waders in Europe and North Africa. *Proc. Int. Conf. Conservation Wetlands Waterfowl*, Heiligenhafen 2–6 December 1974. Int. Wtrfwl Res. Bur., Slimbridge, pp. 255–271.
Prater, A. J., Marchant, J. H. and Vuorinen, J. (1977). Guide to the identification and ageing of Holarctic waders.

British Trust for Ornithology Guide 17.
Pringle, J. S. and Cooper, J. (1975). The Palaearctic wader population of Langebaan Lagoon. *Ostrich* **46**, 213–218.
Pringle, J. S. and Cooper, J. (1977). Wader populations (Charadrii) of the marine littoral of the Cape Province, South Africa. *Ostrich* **48**, 98–105.
Pym, A. (1982). Identification of Lesser Golden Plover and status in Britain and Ireland. *British Birds* **75**, 112–124.
Recher, A. G. (1966). Some aspects of ecology of migrant shorebirds. *Ecology* **47**, 393–407.
Reynolds, J. F. (1972). Photographs of immature Caspian Plovers. *British Birds* **65**, 124–125.
Reynolds, J. F. (1975). Chestnut-banded Sand Plover responding to a supernormal stimulus. *Bull. E. Afr. Nat. Hist. Soc.* **1975**, 130.
Reynolds, J. F. (1977). Thermo-regulatory problems of birds nesting in areas in East Africa: a review. *Scopus* **1**(3), 57–68.
Reynolds, J. F. (1984). Response of a Blacksmith Plover *Vanellus armatus* to ants attacking a hatching egg. *Scopus* **8**, 79.
Richards, D. K. (1974). Eastern Golden Plover (*Pluvialis dominica*) at Ndola. *Bull. Zamb. Orn. Soc.* **6**(1), 25.
Richards, D. K. (1980). Distribution of the Chestnut-banded Sandplover *Charadrius pallidus* in Tanzania. *Scopus* **4**(1), 24.
Rittinghaus, H. (1956). Untersuchungen am Seeregenpfeifer (*Charadrius alexandrinus* L.) auf der Insel Oldeoog. *J. Orn.* **97**(2), 117–155.
Rittinghaus, H. (1961). 'Der Seeregenpfeifer'. Neue Brehm-Bucherei 282. A. Ziemsen Verlag, Wittenberg.
Roberts, M. G. (1977). Belly-soaking in the Whitefronted Plover. *Ostrich* **48**, 111–112.
Robertson, H. G. (1981). Annual summer and winter fluctuations of Palaearctic and resident waders (Charadrii) at Langebaan Lagoon, South Africa, 1975–1979. *In* 'Proceedings of the Symposium on Birds of the Sea and Shore, 1979' (Cooper, J. Ed) pp. 335–345. African Seabird Group, Cape Town.
Robin, P. (1968). L'avifaune de l'Iriki (Sud-Marocain). *Alauda* **36**, 237–253.
Robinson, G. P. (1973). Forbes's Plover near Ndola. *Bull. Zamb. Orn. Soc.* **5**(1), 33.
Robinson, J. R. and Robinson, C. S. M. (1951). Some notes on the Chestnut-banded Sandplover *Charadrius venustus*. *Ostrich* **22** (3), 198.
Rudebeck, G. (1963). Studies on some Palearctic and Arctic birds in their winter quarters in South Africa. *S. Afr. Animal Life* **9**, 414–453.
Rydzewski, W. (1978). The longevity of ringed birds. *The Ring* **96–97**, 218–262.
Sauer, E. G. F. (1962). Ethology and ecology of Golden Plovers on St. Lawrence Island, Bering Sea. *Psychol. Forsch.* **26**, 399–470.
Serle, W. (1956). Exhibition of eggs of *Charadrius forbesi* (Shelley.) *Bull. Br. Orn. Club* **76**(5), 79.
Sharland, R. E. and Wilkinson, R. (1981). The birds of Kano State, Nigeria. *Malimbus* **3**, 7–30.
Shewell, E. L. (1951). Notes on the nesting of the White-fronted Sandplover *Charadrius marginatus* at Gautoos River mouth in 1950. *Ostrich* **22**(2), 117–119.
Simmons, K. E. L. (1951). Distraction-display in the Kentish Plover. *British Birds* **44**, 181–187.
Simmons, K. E. L. (1953a). Some aspects of the aggressive behaviour of three closely related plovers (*Charadrius*). *Ibis* **95**, 115–127.
Simmons, K. E. L. (1953b). Some studies on the Little Ringed Plover. *Avic. Mag.* **59**(6), 190–207.
Simmons, K. E. L. (1956). Territory in the Little Ringed Plover *Charadrius dubius*. *Ibis* **98**, 390–397.
Simmons, K. E. L. (1961). Foot-movements in plovers and other birds. *British Birds* **54**, 34–39.
Sinclair, J. C. and Nicholls, G. H. (1980). Winter identification of Greater and Lesser Sand Plovers. *British Birds* **73**, 206–213.
Skead, D. M. (1977). Weights of birds handled at Barberspan. *Ostrich*, Suppl. **12**, 117–131.
Slight, D. A. (1966). Mating behaviour of Kittlitz's Sandplover. *Ostrich* **37**(4), 277.
Sluiters, J. E. (1954). Waarnemingen over de drie bij Amsterdam broedende pluviersoorten (*Leucopolius a. alexandrinus*, *Charadrius dubius curonicus* en *Ch. h. hiaticula*). *Limosa* **27**, 71–86.
de Smet, K. and Van Gompel, J. (1980). Observations sur la côte sénégalaise en décembre et janvier. *Malimbus* **2**, 56–70.
Smith, K. D. (1943). Notes on Rhodesian birds. *Ostrich* **14**, 114–116.
Smith, K. D. (1950). The birds of Chirundu, middle Zambesi. *Ostrich* **21**, 62–71.
Smith, N. G. (1969). Polymorphism in ringed plovers. *Ibis* **111**(2), 177–188.
Smithers, R. H. N. (1956). Some interesting Rhodesian and Bechuanaland records, III. *Ostrich* **27**, 14–17.
Spitz, R. (1969). Present extent of knowledge of the number of waders wintering in Europe, the Mediterranean basin and North Africa. *Int. Wtrfwl Res. Bur. Bull.* **27/28**, 12–14.
Steyn, P. (1966). White-fronted Sandplover. *Bokmakierie* **18**, 75.
Summers, R. W. (1971). Distribution, abundance and energy relationships of waders (Aves: Charadrii) at Langebaan Lagoon. *Trans. Royal Soc. S. Afr.* **53**, 483–495.
Summers, R. W. and Hockey, P. A. R. (1980). Breeding biology of the White-fronted Plover (*Charadrius marginatus*) in the south-western Cape, South Africa. *J. Nat. Hist.* **14**, 433–445.
Summers, R. W. and Waltner, M. (1979). Seasonal variation in the mass of waders in southern Africa, with special reference to migration. *Ostrich* **50**, 21–37.
Summers, R. W., Cooper, J. and Pringle, J. S. (1977). Distribution and number of coastal waders (Charadrii) in the Southwestern Cape, South Africa, summer 1975–76. *Ostrich* **48**, 85–97.
Tarboton, W. A. (1974). Birds of the Mosdene Nature Reserve, Naboomspruit. *Sth. Afr. Avifauna Ser.*, P.F.I.A.O., **78**, 1–51.
Taylor, J. S. (1957). Notes on the birds of inland waters in eastern Cape Province with special reference to the Karoo. *Ostrich* **28**, 1–80.
Taylor, P. B. (1977). Great Sandplover (*Charadrius leschenaultii*) at Ndola. *Bull. Zamb. Orn. Soc.* **9**(2), 67–68.
Taylor, P. B. (1978). Mongolian Plover (*Charadrius mongolus*) at Ndola. *Bull. Zamb. Orn. Soc.* **10**(2), 79–80.
Taylor, P. B. (1980). Little Ringed Plover *Charadrius dubius* at Lunshya, Zambia. *Scopus* **4**(3), 69.
Taylor, P. B. (1980). Migration of the Ringed Plover *Charadrius hiaticula*. *Orn. Scand.* **11**, 30–42.
Taylor, P. B. (1982). First Zambian records of Chestnut-banded Sandplover *Charadrius pallidus* and observations of White-fronted Sandplovers *C. marginatus* and Cape Teal *Anas capensis* at the same locality. *Bull. Br. Orn. Club.* **102**(1), 5–7.
Thielcke, G. (1951). Beobachtungen und Feststellungen am Seeregenpfeifer. *Vogelwelt* **79**, 184.
Ticehurst, C. B. (1929). Palaearctic waders in Africa. *Bateleur*, Nairobi, **1**, 111–114.
Took, J. M. E. (1967). Mating behaviour of Kittlitz's Sandplover. *Ostrich* **38**(3), 199.
Tree, A. J. (1964). Forbes' Plover *Charadrius forbesi* in the Central Province. *Puku* **2**, 129–130.
Tree, A. J. (1969). The status of Ethiopian waders in Zambia.

Puku 5, 181–197.
Tree, A. J. (1973a). Ageing the Kittlitz Plover. *Safring News* 2(2), 23–25.
Tree, A. J. (1973b). Golden Plover *Pluvialis dominica*. *Ostrich* 44(2), 128.
Tree, A. J. (1974) A comparative ecological study of the Kittlitz Plover and Treble-banded Plover at Lake McIlwaine. Unpub. ms thesis.
Tree, A. J. (1976). Waders in central Mashonaland 1974/75. *Honeyguide* 85, 17–27.
Tree, A. J. (1977a). Notes on the Ringed Plover *Charadrius hiaticula* in southern Africa. *Safring News* 6(1), 25–29.
Tree, A. J. (1977b). Waders in central Mashonaland 1975/77. *Honeyguide* 92, 31–41.
Tree, A. J. (1978). The roosting habits of Charadriiformes in Rhodesia. *Safring News* 7(1), 2–7.
Tree, A. J. (1979). Occurrence of the Ringed Plover in Zimbabwe-Rhodesia. *Honeyguide* 100, 14–19.
Tree, A. J. (1980a). Migration as an ecological adaptation in Central African Charadrii. *Honeyguide* 102, 16–25.
Tree, A. J. (1980b). Small plover studies in southern Africa. *Safring News* 9, 3–9.
Trembsky, A. and Trembsky, J. (1978). Observations ornithologiques effectuées au Maroc au cours de mois de juillet 1974 et 1975. *Aves* 15(1), 1–16.
Trotignon, J. (1976). The international ornithological importance of the Banc d'Arguin. *Proc. Int. Conf. Conserv. Wetlands and Waterfowl*, Heiligenhafen. Int. Wtrfwl Res. Bur. Slimbridge, pp. 130–133.
Trotignon, E., Trotignon, J., Baillou, M., Dejonghe, J-F., Duhautois, L. and Lecomte, M. (1980). Récensement hivernal des limicoles et autres oiseaux aquatiques sur le Banc d'Arguin (Mauritanie) (hiver 1978/79). *Oiseau et R.F.O.* 50, 323–343.
Tyler, S. (1978). Observations on the nesting of the Three-banded Plover *Charadrius tricollaris*. *Scopus* 2(2), 39–41.
Underhill, L. G., Cooper, J. and Waltner, M. (1980). The status of waders (Charadrii) and other birds in the coastal region of the southern and eastern Cape, summer 1978/79. Western Cape Wader Study Group, Pinelands, South Africa. Special Mimeog. report, 248 pp.
Underhill, L. G. and Whitelaw, D. A. (1977). An ornithological expedition to the Namib coast, summer 1976/77. Western Cape Wader Study Group, 106 pp.
Vander weghe, J. P. (1981). Additions à l'avifaune du Rwanda. *Gerfaut* 71, 175–184.
Vaurie, C. (1964). Systematic notes on Palaearctic birds. no. 53. Charadriidae: The genera *Charadrius* and *Pluvialis*. *Am. Mus. Novit.* 2177, 1–22.
Vernon, C. J. (1971). Grey Plover at Lake Kyle. *Honeyguide* 67, 34.
Vincent, A. W. (1945). On the breeding habits of some African birds. *Ibis* 87, 345–365.
Vrydagh, J-H. (1953). Comportement de quelques oiseaux Eurasiens dans leurs quartiers d'hiver au Congo Belge. *Gerfaut* 43, 301–327.
Walters, J. (1954). Der Brutanteil der Geschlechter beim Seeregenpfeifer *Leucopolius alexandrinus* (L.) *Limosa* 27, 19–24.
Walters, J. (1959a). Regenpfeifer-Notizen aus dem 'Seewinkel' in Burgenland (Oesterreich). *Vogelwelt* 80, 33–42.
Walters, J. (1959b). Waarnemingen bij twee Strandplevierenbroedsels, *Charadrius alexandrinus*, op Texel. *Ardea* 47 (1/2), 48–67.
Walters, J. (1962). Zum Thema: Bolzflug und Ruf des Seeregenpfeifers (*Charadrius alexandrinus*). *Vogelwelt* 83, 139–142.
Waltner, M. (1979). Waders and wader ringing. *Safring News* 8 (1), 15–18.
Waltner, M. (1981). Unusual waders at Langebaan. *Safring News* 10, 9–11.
Westernhagen, W. von. (1968). Limicolen-Vorkommen an der westafrikanischen Küste auf der Banc d'Arguin (Mauritanien). *J. Orn.* 109, 185–205.
Whitelaw, D. A., Underhill, L. G., Cooper, J. and Clinning, C. F. (1978). Waders (Charadrii) and other birds on the Namib coast: counts and conservation priorities. *Madoqua* 2(2), 137–150.
Wilson, J. (1981). Observations of waders at Mogobane Dam, South-east Botswana. *Botswana Bird Club Publ.* 1, 8–11.
Winterbottom, J. M. (1963). Comments on the ecology and breeding of sandplovers *Charadrius* in Southern Africa. *Rev. Zool. Bot. Afr.* 62(1–2), 11–16.
Winterbottom, J. M. (1972). The ecological distribution of birds in Southern Africa. Monograph Percy Fitzpatrick Institute African Ornithology, 1, 81 pp.
Zwarts, L. (1972). Bird counts in Merga Zerga, Morocco. *Ardea* 60, 120–123.

Subfamily VANELLINAE: lapwings

Ade, B. (1976). The Crowned Plover: an open space nester. *Honeyguide* 86, 32–34.
Ade, B. (1979). Some observations on the breeding of Crowned Plovers (1977). *Bokmakierie* 31, 9–16.
Bainbridge, W. R. (1965). Nesting behaviour of the White-headed Wattled Plover. *Puku* 3, 171–173.
Baker, N. E. (1982). Notes on some Libyan birds. *Orn. Soc. Middle East Bulletin* 8, 4.
Begg, G. W. and Maclean, G. L. (1976). Belly-soaking in the White-crowned Plover. *Ostrich* 47, 65.
Benson, C. W. (1982). Migrants in the Afrotropical Region south of the equator. *Ostrich* 53, 31–49.
Bock, W. J. (1958). A generic review of the plovers (Charadriinae, Aves). *Bull. Mus. Comp. Zool.* 118, No. 2.
Bowen, P. St. J. (1979). Brown-chested Wattled Plover *Vanellus superciliosus* at Mwinilunga: a new species for Zambia. *Bull. Zamb. Orn. Soc.* 11, 33–35.
Broekhuysen, G. J. (1948). Breeding record of the Blacksmith Plover *Hoplopterus armatus* for the neighbourhood of Capetown. *Ostrich* 19, 237–239.
Brooke, R. K. (1967). On the food of the Senegal Wattled Plover. *Ostrich* 38, 202–203.
Brown, L. H. (1972). Partial burying of eggs by a Blacksmith Plover. *Ostrich* 43, 130.
Clarke, J. D. (1936). The Brown-chested Wattled Plover, *Anomalophrys superciliosus*. *Nigerian Field* 5(2), 72–73.
Clifton, M. (1972). Unusual behaviour of Spur-winged Plover. *Bull. E. Afr. Nat. Hist. Soc.* 1972, 190.
Cowan, P. J. (1983). Birds in the Brak and Sabha regions of Central Libya, 1981–2. *Bull. Br. Orn. Club* 103, 44–47.
Cyrus, D. P. (1982). Blackwinged Plovers nesting on the coastal plain of Zululand. *Ostrich* 53, 248.
Dean, A. R., Fortey, J. E. and Phillips, E. G. (1977). White-tailed Plover: new to Britain and Ireland. *British Birds* 70, 65–471.
Dean, W. R. J. (1978). An analysis of avian stomach contents from southern Africa. *Bull. Br. Orn. Club* 98, 10–13.
Dejonghe, J. F. and Czajkowski, M. A. (1983). Sur la longévité des oiseaux bagués en France métropolitaine, dans les départements d'outre-mer et dans les pays d'influence française. *Alauda* 51, 27–47.
Dowsett, R. J. (1977). The distribution of some falcons and plovers in East Africa. *Scopus* 1(3), 73–78.
Dragesco, J. (1960). Notes biologiques sur quelques oiseaux d'Afrique Equatoriale. *Alauda* 28, 262–273.
Etchécopar, R. D. and Morel, G. (1960). Notes on the clutch

size of the Black-headed Plover. *Ool. Rec.* **34**, 6–8.

Gaugris, Y. (1979). Les oiseaux aquatiques de la plaine de la basse Rusizi (Burundi) (1973–1978). *Oiseau et R.F.O.* **49**, 133–153.

Girard-Audine, M. and Pineau, J. (1972). *Emberiza striolata* et *Vanellus gregarius* dans le Tangerois. *Alauda* **41**, 317.

Greenham, R. and Greenham, L. M. (1975). Blackhead Plover nesting. *Bull. E. Afr. Nat. Hist. Soc.* **1975**, 66–67.

Grisser, P. (1983). Observation d'un Vanneau sociable *Chettusia gregaria* (Pall.), en Dordogne. *Alauda* **51**, 231–233.

Guichard, K. M. (1947). Birds of the inundation zone of the River Niger, French Soudan. *Ibis* **89**, 450–489.

Hall, K. R. L. (1959). A study of the Blacksmith Plover *Hoplopterus armatus* in the Cape Town area: I. Distribution and breeding data. *Ostrich* **30**, 117–126.

Hall, K. R. L. (1964). A study of the Blacksmith Plover *Hoplopterus armatus* in the Cape Town area: II. Behaviour. *Ostrich* **35**, 3–16.

Hall, K. R. L. (1965). Nest records and behaviour notes for three species of plover in Uganda. *Ostrich* **36**, 107–108.

Hanley, A. W. D. (1983). The 1982/1983 breeding season of the Humewood Kiewietjes. *The Bee-eater* **34**, 9–11.

Heslop, I. R. P. (1937). *Anomalophrys superciliosus* in Nigeria. *Ibis* 14, **1**, 174–175.

Howells, W. W. (1977). Lesser Blackwinged Plovers in Wankie National Park. *Honeyguide* **92**, 48.

Hume, R. A. (1983). Swan Hellenic/R. S. P. B. Nile Cruise 28 Sept–14 Oct 1983: ornithological report. R. S. P. B.

Imboden, C. (1974). Zug, Fremdansiedlung und Brutperiode des Kiebitz *Vanellus vanellus* in Europa. *Orn. Beob.* **71**, 5–134.

Irwin, M. P. S. (1976). Some little known and inadequately documented Rhodesian birds. *Honeyguide* **90**, 9–13.

Johnson, A. R. and Biber, O. (1974). Winter waterfowl counts along the Atlantic coast of Morocco in January 1974. *Int. Wtrfwl Res. Bur. Bull.* **37**, 76–81.

Kutilek, M. J. (1974). Notes on Kittlitz's Sandplover and the Blacksmith Plover at Lake Nakuru. *E. Afr. Wildl. J.* **12**, 87–91.

Lack, P. C., Leuthold, W. and Smeenk, C. (1980). Check-list of the birds of Tsavo East National Park, Kenya. *J.E. Afr. Nat. Hist. Soc.* **170**.

Lack, P. C. and Quicke, D. L. J. (1978). Dietary notes on some Kenyan birds. *Scopus* **2**, 86–91.

Lewis, A. D. (1981). Birds of the Kaisut Desert, northern Kenya. *Bull. E. Afr. Nat. Hist. Soc.* **1981**, 60–64.

Little, J. de V. (1967). Some aspects of the behaviour of the Wattled Plover *Afribyx senegallus* (Linnaeus). *Ostrich* **38**, 259–280.

Lombard, A. L. (1965). Notes sur les oiseaux de Tunisie. *Alauda* **33**, 1–33.

Lorber, P. (1970). Lesser Black-winged Plover (?) in the Wankie National Park. *Honeyguide* **63**, 25.

McInnis, D. (1933). Nesting habits of some East African birds. *J. E. Afr. and Uganda Nat. Hist. Soc.* **11**, 128–135.

Maclean, G. L. (1982). Opportunistic breeding of plovers under manipulated conditions. *Ostrich* **53**, 52–54.

Milstein, P. le S. (1975). The biology of Barberspan, with special reference to the avifauna. *Ostrich* Suppl. **10**.

Moore, R. and Vernon, C. (1973). Crowned Plover nesting in loose colonies. *Ostrich* **44**, 262.

Mundy, P. J. and Cook, A. W. (1972). The birds of Sokoto, Part I. *Bull. Niger. Orn. Soc.* **9**, 26–47.

North, M. E. W. (1936). Breeding habits of the Crested Wattled Plover (*Sarciophorus tectus latifrons*). *J. E. Afr. and Uganda Nat. Hist. Soc.* **13**, 132–145.

Olson, C. (1976). Summary of field observations of birds from Begemder and Simien Province. *Walia* **7**, 16–27.

Parnell, G. W. (1968). Behaviour and breeding of the Senegal Wattled Plover. *Honeyguide* **55**, 28.

Penry, E. H. (1980). Blacksmith Plover *Vanellus armatus* breeding on the Copperbelt. *Bull. Zamb. Orn. Soc.* **12**, 10–12.

Pettet, A. (1982). Feeding behaviour of White-tailed Plover. *British Birds* **75**, 182.

Quinn, S. (1982). Bird observations along Egyptian Nile made during The American Museum of Natural History winter 1982 Nile Discovery Cruise. American Museum of Natural History, New York.

Quinn, S. (1983). Bird observations along the Egyptian Nile made during The American Museum of Natural History winter 1983 Nile Discovery Cruise. American Museum of Natural History, New York.

Redfern, C. P. F. (1983). Aspects of the growth and development of Lapwings *Vanellus vanellus*. *Ibis* **125**, 266–272.

Reynolds, J. F. (1968). Observations on the White-headed Plover. *E. Afr. Wildl. J.* **6**, 142–144.

Reynolds, J. F. (1980). Sympatry of Blacksmith and Spur-winged Plovers *Vanellus armatus* and *spinosus* at Amboseli. *Scopus* **4**, 43.

Riley, J. W. and Rooke, K. B. (1962). Sociable Plover in Dorset. *British Birds* **55**, 233–235.

Ruwet, J.-C. (1965) Les Oiseaux des plaines et du lac-barrage de la Lufira supérieure (Katange Méridional). Université de Liège, Editions F.U.L.R.E.A.C.

Saunders, C. R. (1970). Observations on breeding of the Long-toed or White-winged Plover *Hemiparra crassirostris leucoptera* (Reichenow). *Honeyguide* **62**, 27–29.

Serle, W. (1956). Notes on *Anomalophrys superciliosus* (Reichenow) in West Africa with special reference to its nidification. *Bull. Brit. Orn. Club* **76**, 101–104.

Sessions, P. H. B. (1967). Notes on the birds of Lengetia Farm, Mau Narok. *J. E. Afr. Nat. Hist. Soc.* **26**, (1) (113), 18–48.

Short, L. L. (1977). Wild life along the Nile, 28 October to 17 November, 1977. Report on The American Museum of Natural History's Nile Tour to Egypt. American Museum of Natural History, New York.

Skead, C. J. (1955). A study of the Crowned Plover *Stephanibyx coronatus coronatus* (Boddaert). *Ostrich* **26**, 88–98.

Spronk, W. (1965). Wattled Plover. *Bokmakierie* **17**, 42.

Symmes, T. C. L. (1952). Some observations on the breeding of the Crowned Plover *Stephanibyx coronatus* Bodd. *Ostrich* **23**, 85–87.

Tarboton, W. R. and Nel, F. (1980). On the occurrence of the White-crowned Plover in the Kruger National Park. *Bokmakierie* **32**, 19–21.

Thomas, D. H. (1983). Aposematic behaviour in the Blacksmith Plover. *Ostrich* **54**, 51.

Tree, A. J. (1969). The status of Ethiopian waders in Zambia. *Puku* **5**, 181–205.

Tree, A. J. (1978). Little known plovers in Rhodesia. *Honeyguide* **96**, 23.

Tree, A. J. (1981). Age at which Crowned Plover may breed. *Ostrich* **52**, 64.

Turner, M. I. M. (1964). Some observations of bird behaviour made from an aircraft in the Serengeti National Park. *Bull. Br. Orn. Club* **84**, 65–67.

Urban, E. K., Brown, L. H., Buer, C. E. and Plage, G. D. (1970). Four descriptions of nesting, previously undescribed, from Ethiopia. *Bull. Br. Orn. Club* **90**, 162–164.

Urban, E. K. (1978). 'Ethiopia's Endemic Birds.' Ethiopian Tourist Organisation, Addis Ababa.

Vande weghe, J.-P. and Monfort-Braham, N. (1975). Quelques aspects de la séparation écologique des vanneaux du Parc National de l'Akagera. *Alauda* **43**, 143–166.

Verheyen, R. (1953). 'Exploration du Parc National de

l'Upemba'. Institut des Parcs Nationaux du Congo Belge, Bruxelles.

Vesey-Fitzgerald, D. and Beesley, J. S. S. (1960). An annotated list of the birds of the Rukwa Valley. *Tanganyika Notes and Records* **54**, 91–110.

Vincent, J. (1934). Birds of northern Portuguese East Africa. *Ibis* Ser. 13, 495–527.

Vincent, A. W. (1945). On the breeding habits of some African birds. *Ibis* **87**, 345–365.

Von Heuglin, T. (1869–1873). 'Ornithologie Nordost-Afrikas.' Kassel.

Walker, R. B. (1965). Two Sudan savanna birds at Zaria. *Bull. Niger. Orn. Soc.* **2**(5), 22–23.

Walters, J. R. (1979). Interspecific aggressive behaviour by Long-toed Lapwings (*Vanellus crassirostris*). *Anim. Behav.* **27**, 969–981.

Walters, J. R. (1982). Parental behavior in Lapwings (Charadriidae) and its relationships with clutch sizes and mating systems. *Evolution* **36**, 1030–1040.

Welch, G. (1983). Swan Hellenic/R.S.P.B. Nile Cruise 9–25 March 1983; ornithological report. R.S.P.B.

Winterbottom, J. M. (1972). The ecological distribution of birds in southern Africa. Percy Fitzpatrick Inst. Afr. Orn. *Monogr.* **1**.

Wright, P. J. (1963). Nesting behaviour of the Wattled Plover. *Puku* **1**, 218.

Family SCOLOPACIDAE: sandpipers, snipe and phalaropes

Anon (1982). Recent reports. *Bull. Orn. Soc. Middle East* **9**, 8.

Ash, J. S. (1978). Inland and coastal occurrences of Broad-billed Sandpipers *Limicola falcinellus* in Ethiopia and Djibouti. *Bull. Br. Orn. Club* **98**, 24–26.

Ash, J. S. (1980). Migrational status of Palaearctic birds in Ethiopia. *Proc. IV. Pan Afr. Orn. Congr.* pp. 199–208.

Ash, J. S. (1981). Spring passage of Whimbrel *Numenius phaeopus* and other waders off the coast of Somalia. *Scopus* **5**, 71–76.

Ash, J. S. (1983). Over fifty additions to the Somalia list including two hybrids, together with notes from Ethiopia and Kenya. *Scopus* **7**, 54–79.

Ash, J. S. and Ashford, O. M. (1977.) Great Black-headed Gulls *Larus ichthyaetus* and Red-necked Phalaropes *Phalaropus lobatus* inland in Ethiopia. *J. E. Afr. Nat. Hist. Soc. Nat. Mus.* **31**(162), 1–3.

Ash, J. S. and Miskell, J. E. (1981). The Dunlin *Calidris alpina* in Ethiopia and Somalia. *Scopus* **5**, 32.

Backhurst, G. C. (1969). A record of *Gallinago stenura* from Kenya. *Bull. Br. Orn. Club.* **89**, 95–96.

Backhurst, G. C. and Britton, P. L. (1969). A record of *Calidris subminuta* from Kenya. *Bull. Br. Orn. Club* **89**, 121.

Backhurst, G. C., Britton, P. L. and Mann, C. F. (1973). The less common Palaearctic migrant birds of Kenya and Tanzania. *J. E. Afr. Nat. Hist. Soc. Nat. Mus.* **27**(140), 1–38.

Baird, D. A. (1979.) Twenty-eight additions to Archer & Godman's 'Birds of British Somaliland and the Gulf of Aden'. *Bull. Br. Orn. Club* **99**(1), 6–9.

Baker, N. E. (1982). Notes on some Libyan birds. *Bull. Orn. Soc. Middle East* **8**, 4.

Becker, P. (1974). Beobachtungen am paläarktischen Zugvögeln in ihren Winterquartier Südwestafrika. Wissenschaftliche Forschung in Südwestafrika, 12. Folge. Verlag und Schriftleitung, Windhoek.

Becker, P. (1977). Ornithologische Notizen von der Küste Südwestafrikas. *J. S.W.A. Wiss. Gesellschaft* **31**, 65–82.

Becker, P., Jensen, R. A. C. and Berry, H. H. (1974). Broad-billed Sandpiper *Limicola falcinellus* in South West Africa. *Madoqua* **1**(8), 67–71.

Beere, W. (1981). Camouflage of crouching Jack Snipe. *British Birds* **74**, 440–441.

Berry, H. H. and Berry, C. U. (1975). A check list and notes on the birds of Sandvis, South West Africa. *Madoqua* **9**(2), 5–18.

Blanchet, A. (1957). Oiseaux de Tunisie. *Mém. Soc. Sci. Nat. Tunisie* **1**(1–2), 1–216.

Blondel, J. and Blondel, C. (1964). Remarques sur l'hivernage des limicoles et autres oiseaux aquatiques au Maroc. *Alauda* **32**, 250–279.

Bourne, W. R. P. (1964). Observations of seabirds. *Sea Swallow* **16**, 9–39.

Bourne, W. R. P. (1965). Observations of sea birds. *Sea Swallow* **17**, 10–39.

Bourne, W. R. P. (1973). Observations of sea birds 1967–1969. *Sea Swallow* **22**, 29–60.

Bourne, W. R. P. and Dixon, T. J. (1975). Observations of seabirds 1970–1972. *Sea Swallow* **24**, 65–87.

Bourne, W. R. P. and Radford, M. C. (1962). Observations of seabirds. *Sea Swallow* **15**, 7–27.

Branson, N. J. B. A., Ponting, E. D. and Minton, C. D. T. (1978). Turnstone migrations in Britain and Europe. *Bird Study* **25**, 181–187.

Britton, P. L. (1974). Broad-billed Sandpipers and Herring Gulls wintering on the North Kenya coast. *Bull. E. Afr. Nat. Hist. Soc.* **1974**, 112–113.

Britton, P. L. and Britton, H. A. (1979.) Phalaropes in coastal Kenya. *Scopus* **3**(2), 58.

Broberg, L. (1967). A Long-toed Stint in Ethiopia. *Ibis* **109**, 440.

Broekhuysen, G. J. (1971). Third report on migration in southern Africa. *Ostrich* **42**(1), 41–64, 211–225.

Brogger-Jensen, S. (1977). First record of Temminck's Stint in Zambia. *Bull. Zamb. Orn. Soc.* **9**, 23.

Brosset, A. (1956). Les oiseaux du Maroc oriental de la Mediterranée à Berguent. *Alauda* **24**, 161–205.

Brown, R. G. B. (1979). Seabirds of the Senegal upwelling and adjacent waters. *Ibis* **121**(3), 283–292.

Browne, P. W. P. (1950). Notes on Broad-billed and Terek Sandpipers at Aden. *British Birds* **42**, 333–334.

Button, E. L. (1973). The Common Snipe *Gallinago gallinago* in Zambia. *Bull. Br. Orn. Club* **93**, 174.

Carswell, M. J. (1977). The Black-tailed Godwit *Limosa limosa* in Uganda. *Scopus* **1**, 49.

Castan, R. (1964). Capture d'un Becasseau Rousset *Tryngites subruficollis* (Vieillot) dans le Sud Tunisien. *Alauda* **32**, 129–132.

Cawkell, F. O. and Moreau, R. E. (1963). Notes on the birds in The Gambia. *Ibis* **105**, 156–178.

Cheeseman, R. E. and Sclater, W. L. (1935). A collection of birds in Northwestern Abyssinia, Part II. *Ibis* **13**(5), 297–334.

Clancey, P. A. (1964a). On the races of the Whimbrel *Numenius phaeopus* wintering in southeastern Africa. *Bull. Br. Orn. Club* **84**, 138–140.

Clancey, P. A. (1964b). The first records of the Red-necked Stint *Calidris ruficollis* from the Ethiopian region. *Ibis* **106**, 254–255.

Clancey, P. A. (1965). The Knot *Calidris canutus* in Natal. *Ostrich* **36**, 143–144.

Clancey, P. A. (1974). Subspeciation studies in some Rhodesian birds. *Arnoldia* **6**(28), 1–43.

Clarke, A., Madden, S. T. and Milstein, P le S. (1974). Black-tailed Godwit influx. *Bokmakierie* **26**, 101.

Cooper, J., Robertson, H. G. and Shaughnessy, P. D. (1980). Waders (Charadrii) and other coastal birds of the Diamond coast and the islands off Southwest Africa. *Madoqua* **12**, 51–58.

Curry, P. J. and Sayer, J. A. (1979) The inundation zone of the Niger as an environment for Palaearctic migrants. *Ibis* **121**, 20–40.

Dean, A. R. (1977). Moult of the Little Stint in South Africa.

Ardea 65, 73–79.

Dick, W. J. A. (1975). Oxford and Cambridge Mauritanian Expedition 1973 Report. Cambridge.

Dick, W. J. A. (1978). Ringing recoveries from the Oxford and Cambridge Mauritanian Expedition 1973. *Wader Study Group Bull.* 24, 31–33.

Dick, W. J. A. (1979). Results of the WSG project on the spring migration of Siberian Knot *Calidris canutus* 1979. *Wader Study Group Bull.* 27, 8–13.

Dick, W. J. A. and Pienkowski, M. W. (1979). Autumn and early winter weights of waders in north-west Africa. *Orn. Scand.* 10, 117–123.

Dick, W. J. A., Pienkowski, M. W., Waltner, M. and Minton, C. D. T. (1976). Distribution and geographical origins of Knot *Calidris canutus* in Europe and Africa. *Ardea* 64, 22–47.

Dowsett, R. J. (1969). Migrants at Mallam' fatori, Lake Chad, autumn 1968. *Bull. Niger. Orn. Soc.* 6, 39–45.

Dowsett, R. J. (1980). The migration of coastal waders from the Palaearctic across Africa. *Gerfaut* 70, 3–35.

Dubois, P. and Duhautois, L. (1977). Notes sur l'ornithologie marocaine. *Alauda* 45, 285–291.

Dupuy, A. (1968). La migration des laro-limicoles au Sahara Algérien. *Alauda* 36, 27–35.

E. A. N. H. S. Ornithological Sub-committee. (1977). Some recent records of Palaearctic birds in Kenya and Tanzania. Part I. *Scopus* 1, 39–43.

Elliott, C. C. H., Waltner, M., Underhill, L. G., Pringle, J. S. and Dick, W. J. A. (1976). The migration system of the Curlew Sandpiper *Calidris ferruginea* in Africa. *Ostrich* 47, 191–213.

Erard, C. and Etchécopar, R. D. (1970). Some notes on the birds of Angola. *Bull. Br. Orn. Club* 90, 158–161.

Ferguson-Lees, I. J. (1955). Some photographic studies of Wilson's Phalarope and Stilt Sandpiper. *British Birds* 48, 30–33.

Field, G. D. (1974). Nearctic waders in Sierra Leone—Lesser Golden Plover and Buff-breasted Sandpiper. *Bull. Br. Orn. Club* 94, 76–78.

Fogden, M. P. L. (1963). Early autumn migrants in coastal Kenya. *Ibis* 105, 112–113.

Fournier, St. F. and Dick, W. J. A. (1981). Preliminary survey of the Archipel des Bijagos, Guinea Bissau. *Wader Study Group Bull.* 31, 24–25.

François, J. (1975). L'avifaune annuelle du Lac de Boughzoul (Algérie). *Alauda* 43, 125–133.

Gaugris, Y. (1979). Les oiseaux aquatiques de la pleine de la basse Rusizi (Burundi) (1973–1978). *Oiseau et R.F.O.* 49, 133–153.

Germain, M., Dragesco, J., Roux, F. and Garcin, H. (1973). Contribution à l'ornithologie du Sud-Cameroun. *Oiseau et R.F.O.* 40, 69–81.

Goss-Custard, J. D. (1969). The winter feeding ecology of the Redshank. *Ibis* 111, 338–356.

Grimes, L. G. (1969). The Spotted Redshank *Tringa erythropus* in Ghana. *Ibis* 111, 246–251.

Grimes, L. G. (1974). Radar tracks of Palaearctic waders departing from the coast of Ghana in spring. *Ibis* 116, 165–171.

Grosskopf, G. (1958–59). Zur Biologie des Rotschenkels. *J. Orn.* 99, 1–17; 100, 210–236.

Hale, W. G. (1971). A review of the taxonomy of the Redshank. *Zool. J. Linn. Soc.* 50, 199–268.

Harvey, W. G. (1971a). A Tanzanian record of the Knot. *Bull. E. Afr. Nat. Hist. Soc.* 1971, 75.

Harvey, W. G. (1971b). The second Broad-billed Sandpiper for Tanzania. *Bull. E. Afr. Nat. Hist. Soc.* 1971, 161.

Hopson, A. J. and Hopson, J. (1974). A Buff-breasted Sandpiper *Tryngites subruficollis* at Kerio Bay, Lake Rudolf. *Bull. E. Afr. Nat. Hist. Soc.* 1974, 17–18.

Hopson, J. and Hopson, A. J. (1972). Broad-billed Sandpiper at Lake Rudolf. *Bull. E. Afr. Nat. Hist. Soc.* 1972, 170–171.

Hopson, J. and Hopson, A. J. (1975). Preliminary notes on the birds of the Lake Turkana area. (Cyclostyled). Kitale.

Hockey, P. A. R., Cooper, J., Tree, A. J. and Sinclair, J. C. (in prep). The occurrence of rare and vagrant Palaearctic waders in southern Africa.

Höhn, E. O. (1971). Observations on the breeding behaviour of Grey and Red-necked Phalaropes. *Ibis* 113(3), 335–348.

Hume, R. A. (1978). Posture of Jack Snipe whilst feeding. *British Birds* 71, 79.

Keith, S. (1968). Notes on birds of East Africa, including additions to the avifauna. *Am. Mus. Novit.*, 2321.

Kitson, A. R. (1978). Identification of Long-toed Stint, Pintail Snipe and Asiatic Dowitcher. *British Birds* 71, 558.

Knight, P. and Dick, W. J. A. (1975). Recensement des limicoles au Banc d'Arguin (Mauritanie). *Ardea* 64, 22–47.

Lambert, K. (1971). Seevogelbeobachtungen auf zwei Reisen im östlichen Atlantik mit besonderer Berücksichtigung des Seegebietes von Südwestafrika. *Beitr. Vogelk.* 17, 1–32.

Lister, S. M. (1981). Le Grand Maubeche *Calidris tenuirostris* nouveau pour l'ouest du Paléarctique. *Alauda* 49, 227–228.

Macdonald, M. A. (1977). Short-billed Dowitcher in Ghana. *Bull. Niger. Orn. Soc.* 13, 148.

Macdonald, M. A. (1978). Seasonal changes in numbers of waders at Cape Coast, Ghana. *Bull. Niger. Orn. Soc.* 14, 28–35.

Meadows, B. S. (1977). Exceptional numbers of Black-tailed Godwits *Limosa limosa* wintering at Lake Naivasha during 1976/77. *Scopus* 1, 49–50.

Meeth, P. (1981). Baird's Sandpiper *Calidris bairdii* in Senegal in Dec 1965. *Dutch Birding* 3, 51.

Meinertzhagen, R. (1919). A preliminary study of the relationship between geographical distribution and migration with special reference to the Palaearctic Region. *Ibis* Ser. 11, 1, 279–292.

Middlemiss, E. (1961). Biological aspects of *Calidris minuta* while wintering in Southwest Cape. *Ostrich* 32, 107–121.

Moreau, R. E. (1967). Water birds over the Sahara. *Ibis* 109, 232–259.

Mundy, P. J. and Cook, A. W. (1972). The birds of Sokoto, Part I. *Bull. Niger. Orn. Soc.* 9, 26–47.

Nikolaus, G. (in prep.). A checklist of the birds of the South Sudan.

Orsini, P. (1980). Première observation hivernale du Phalarope à bec étroit *Phalaropus lobatus* dans le Moyen-Atlas (Maroc). *Alauda* 48, 258–259.

Paige, J. P. (1965). Field identification and winter range of the Asiatic Dowitcher *Limnodromus semipalmatus*. *Ibis* 107, 95–97.

Pakenham, R. H. W. (1939). Field notes on the birds of Zanzibar and Pemba. *Ibis* 14(3), 322–354.

Paton, D. C. and Wykes, B. J. (1978). Reappraisal of moult in Red-necked Stints in southern Australia. *Emu* 78, 54–60.

Pearson, D. J. (1974). The timing of wing moult in some Palaearctic waders wintering in East Africa. *Wader Study Group Bull.* 12, 6–12.

Pearson, D. J. (1977). The first year moult of the Common Sandpiper *Tringa hypoleucos* in Kenya. *Scopus* 1, 89–93.

Pearson, D. J. (1981). The wintering and moult of Ruffs *Philomachus pugnax* in the Kenyan rift valley. *Ibis* 123, 158–182.

Pearson, D. J. (1983). East African bird report 1981. *Scopus* 5(5), 131–153.

Pearson, D. J. (1984a). A Pintail Snipe *Gallinago stenura* at Naivasha. *Scopus* **8**, 45–46.

Pearson, D. J. (1984b). Some counts of wintering waders on the Kenya coast. *Scopus* **8**, 93–95.

Pearson, D. J. (1984c). The moult of the Little Stint *Calidris minuta* in the Kenyan rift valley. *Ibis* **126**, 1–15.

Pearson, D. J. and Britton, P. L. (1980). Arrival and departure times of Palaearctic waders on the Kenya coast. *Scopus* **4**, 84–88.

Pearson, D. J., Phillips, J. H. and Backhurst, G. C. (1970). Weights of some Palaearctic waders wintering in Kenya. *Ibis* **112**, 199–208.

Pearson, D. J. and Stevenson, G. (1980). A survey of wintering waders in the southern part of the Kenya rift valley. *Scopus* **4**, 59–63.

Pienkowski, M. W. (Ed.) (1975). Studies of coastal birds and wetlands in Morocco, 1972. Joint Report of the East Anglian Expedition to Tarfaya Province, Morocco 1972 and the Cambridge Sidi Moussa Expedition 1972. Norwich.

Pienkowski, M. W. (1978). Ringing recoveries from the Moroccan wader expeditions. *Wader Study Group Bull.* **24**, 28–30.

Pienkowski, M. W. and Dick, W. J. A. (1975). The wintering and migration of Dunlin *Calidris alpina* in North-west Africa. *Orn. Scand.* **6**, 151–167.

Pienkowski, M. W. and Knight, P. (1977). La migration post-nuptiale des limicoles sur la côte Atlantique du Maroc. *Alauda* **45**, 165–190.

Piersma, T., Engelmoer, M., Altenburg, W. and Mes, R. (1980). A wader expedition to Mauritania. *Wader Study Group Bull.* **29**, 14.

Pitelka, F. A. (1948). Problematical relationships of the Asiatic shorebird *Limnodromus semipalmatus*. *Condor* **50**, 259–269.

Pocock, T. N. (1962). Red-necked Phalarope *Phalaropus lobatus* and other birds at Iscor Dams, Vanderbijl Park. *Ostrich* **33**, 41–44.

Prater, A. J., Marchant, J. H. and Vuorinen, J. (1977). Guide to the identification and ageing of Holarctic waders. British Trust for Ornithology Guide **17**.

Pringle, J. S. and Cooper, J. (1975). The Palaearctic wader population of Langebaan lagoon. *Ostrich* **46**, 213–218.

Pringle, J. S. and Cooper, J. (1977). Wader populations (Charadrii) of the marine littoral of the Cape Peninsula, South Africa. *Ostrich* **48**, 98–105.

Pringle, J. S. and Pringle, W. S. (1971). Red-necked Phalarope *Phalaropus lobatus* at Strandfontein. *Ostrich* **42**(3), 228–229.

Prozeshy, O. P. M. (1964). Notes on grey phalarope (*Phalaropus fulicarius* Linnaeus) found at Vootrekkerhoogte, Pretoria, Transvaal. *Ostrich* **35**(b) (1) 70.

Puttick, G. M. (1978). Diet of the Curlew Sandpiper at Langebaan lagoon, South Africa. *Ostrich* **49**, 158–167.

Puttick, G. M. (1979). Foraging behaviour and activity budgets of Curlew Sandpipers. *Ardea* **67**, 111–122.

Reynolds, J. F. (1965). On the occurrence of Palaearctic waders in western Tanzania. *E. Afr. Wildl. J.* **3**, 131–132.

Richards, D. K. (1982). The birds of Conakry and Kakulima, Democratic Republic of Guinée. *Malimbus* **4**, 93–103.

Robinson, N. (1972). Bird notes from Republic of Togo. *Bull. Niger. Orn. Soc.* **9**, 85–89.

Roux, F. (1959). Quelques données sur les Anatidés et Charadriidés Paléarctiques hivernant dans la basse vallée du Sénégal et sur leur écologie. *Terre Vie* **106**, 315–321.

Roux, F. (1973). Censuses of Anatidae in the central delta of the Niger and the Senegal delta—January 1972. *Wildfowl* **24**, 63–80.

Scott, D. A. (1981). A preliminary inventory of wetlands of international importance in West Europe and Northwest Africa. Int. Wtrfwl Res. Bur. special publication 2.

Schmitt, M. B. and Waterhouse, P. J. (1976). Moult and mensural data of Ruff on the Witwatersrand. *Ostrich* **47**, 179–190.

Simmons, K. E. L. (1951). Behaviour of Common Sandpiper in winter quarters. *British Birds* **44**, 415–416.

Sinclair, J. C. (1977). The phalaropes. *Bokmakierie* **29**(4), 90–91.

Sinclair, J. C. (1984). S.A.O.S. Rarities Committee report. *Bokmakierie* **36**(3), 64–68.

Sinclair, J. C. and Hockey, P. A. R. (1980.) Wilson's Phalarope in South Africa. *Bokmakierie* **32**, 114–115.

Sinclair, J. C. and Nicolls, H. (1976). Red-necked Stint identification. *Bokmakierie* **28**, 59–60.

Sinclair, J. C., Robson, N. F. and Bull, G. (1974). New birds in Natal. *Bokmakierie* **26**, 68–69.

Smalley, M. E. (1979). Dowitcher in The Gambia. *Malimbus* **1**, 68–69.

Smart, J. B. and Forbes-Watson, A. (1971). Occurrence of the Asiatic Dowitcher in Kenya. *Bull. E. Afr. Nat Hist. Soc.***1971**, 74–75.

de Smet, K. and Van Gompel, J. (1979). Wader counts at the Senegalese coast in winter, 1978–1979. *Wader Study Group Bull.* **27**, 30.

Smith, K. D. (1964). Nearctic waders in Morocco. *Ibis* **106**, 530–531.

Smith, P. A. (1966). Palaearctic waders on the Niger Delta. *Bull. Niger. Orn. Soc.* **3**, 2–6.

Spitz, F. (1969). Present extent of knowledge of the numbers of waders wintering in Europe, the Mediterranean basin and North Africa. *Int. Wtrfwl Res. Bur. Bull.* **27/28**, 12–14.

Stanford, W. P. (1953). Winter distribution of the Grey Phalarope *Phalaropus fulicarius*. *Ibis* **95**, 483–491.

Stanford, W. P. (1954). Note on sexual behaviour of Greenshank (*Tringa nebularia*) in winter quarters. *Ostrich* **25**, 100.

Sudbury, A. and Field, G. D. (1972). An African record of the Hudsonian Curlew *Numenius phaeopus hudsonicus* Latham. *Bull. Br. Orn. Club* **92**, 148–149.

Summers, R. W., Cooper, J. and Pringle, J. S. (1977). Distribution and numbers of coastal waders (Charadrii) in the southwestern Cape, South Africa, summer 1975–76. *Ostrich* **48**, 85–97.

Summers, R. W. and Waltner, M. (1979). Seasonal variations in the mass of waders in southern Africa, with special reference to migration. *Ostrich* **50**, 21–37.

Taylor, P. B. (1978). Two records of Knot from the Kenya coast. *Scopus* **2**, 97–98.

Taylor, P. B. (1980a). The field separation of Common, Ethiopian and Great Snipe (*Gallinago gallinago, nigripennis* and *media*). *Scopus* **4**, 1–5.

Taylor, P. B. (1980b). Pectoral Sandpiper *Calidris melanotos* and Lesser Yellowlegs *Tringa flavipes* in Zambia. *Bull. Br. Orn. Club* **100**, 233–235.

Taylor, P. B. (1981). First East African record of the Red-necked Stint *Calidris ruficollis*. *Scopus* **5**, 126–127.

Taylor, P. B. (1982). Pectoral Sandpiper *Calidris melanotos* at Mombasa. *Scopus* **6**, 21.

Taylor, P. B. (1984). Field identification of Pintail Snipe and recent records in Kenya. *Dutch Birding* **6**(3), 77–90.

Thévenot, M. (1973). Compte rendu d'activité de la Station de baguage de l'Institut Chérifien, année 1971. *Bull. Soc. sci. nat. Maroc.* **53**, 199–245.

Tree, A. J. (1971). Notes on Palaearctic migrants in the eastern Cape. *Ostrich* **42**, 198–208.

Tree, A. J. (1974a). The use of primary moult in ageing the 6–15 month age class of some Palaearctic waders. *Safring News* **3**, 21–24.

Tree, A. J. (1974b). Waders in the Salisbury area, 1972/74.

Honeyguide **80**, 17–27.
Tree, A. J. (1977). Waders in Central Mashonaland, 1975/77. *Honeyguide* **92**, 31–41.
Tree, A. J. (1979). Biology of the Greenshank in southern Africa. *Ostrich* **50**, 240–251.
Tree, A. J. (1981). Greater and Lesser Yellowlegs in southern Africa. *Bokmakierie* **33**, 44–46.
Tree, A. J. and Kieser, J. A. (1982). Field separation of Lesser Yellowlegs and Wood Sandpiper. *Honeyguide* **2**, 40–41.
Trotignon, E. (1979). Parc National du Banc d'Arguin, Mauritanie. Cyclostyled.
Trotignon, E., Trotignon, J., Baillon, M., Dejonghe, J.-F., Duhautois, L. and Lecompte, M. (1980). Récensement hivernal des limicoles et autres oiseaux aquatiques sur le Banc d'Arguin (Mauritanie) (hiver 1978/79). *Oiseau et R.F.O.* **50**, 323–343.
van Someren, R. A. L. and van Someren, V. G. L. (1911). 'Studies of Birdlife in Uganda'. John Bale and Danielson, London.
Wallace, D. I. M. (1969). Lesser Yellowlegs at Lagos: a species new to Nigeria. *Bull. Niger. Orn. Soc.* **6**, 58.
Walsh, F. (1971). Early Palaearctic waders at Kainji and New Bussa in 1969. *Bull. Niger. Orn. Soc.* **8**, 32–34.
Waltner, M. and Sinclair, J. C. (1981). Distribution, biometrics and moult of the Terek Sandpiper *Xenus cinereus* in southern Africa. *In* 'Proc. Symp. Birds of the Sea and Shore, 1979' (Cooper, J. Ed.), pp. 233–266. African Seabird Group, Cape Town.
Westernhagen, N. von (1968). Limicolen-Verkommen an der westafrikanischen Küste auf der Banc d'Arguin (Mauritanien). *J. Orn.* **109**, 185–205.
Whitelaw, D. A., Underhill, L. G., Cooper, J. and Clinning, C. F. (1978). Waders (Charadrii) and other birds on the Namib coast: counts and conservation priorities. *Madoqua* **2**(2), 137–150.
Wilds, C. (1982). Separating the Yellowlegs. *Birding* **14**, 172–188.
Wilson, J. R. (1981). Sightings of Pectoral Sandpiper, Knot, Redshank and Bar-tailed Godwit in Botswana. *Ostrich* **52**, 255.
Wilson, J. R., Czajkowski, M. A. and Pienkowski, M. W. (1980). The migration through Europe and wintering in West Africa of the Curlew Sandpiper. *Wildfowl* **31**, 107–122.
Winkler, H. (1980). Zum Nahrungserwerb des Terek wasserlanges (*Xenus cinereus*) in Winterquartier. *Egretta* **23**, 56–59.

Family STERCORARIIDAE: skuas

Andrew, J. O. (1969). Notes on some Palearctic migrants in The Gambia, spring 1969. *Bull Niger. Orn. Soc.* **6**, 95.
Ash, J. S. (1975). Six species of birds new to Ethiopia. *Bull. Br. Orn. Club* **93**, 3–6.
Ash, J. S. (1983). Over fifty additions of birds to the Somalia list including two hybrids, together with notes from Ethiopia and Kenya. *Scopus* **7**, 54–79.
Backhurst, G. C. (1971). A sight record of a new bird for Kenya. *Bull. E. Afr. Nat. Hist. Soc.* March, **1971**, 40.
Backhurst, G. C. and Backhurst, D. E. G. (1970). 'A Preliminary Checklist of East African Birds.' Kabete, Nairobi.
Backhurst, G. C., Britton, P. L. and Mann, C. F. (1973). The less common Palearctic migrant birds of Kenya and Tanzania. *J. E. Afr. Nat. Hist. Soc. Nat. Mus.* **140**, 1–38.
Becker, P. (1974). Beobachtungen am paläarktischen Zugvögeln in ihren Winterquartier Südwestafrika. Wissenschaftliche Forschung in Südwestafrika, 12. Folge. Verlag und Schriftleitung, Windhoek.

Becker, P. (1977). Ornithologische Notizen von der Küste Südwestafrikas. *J. S. W. Afr. Wiss. Gesellschaft* **31**, 65–82.
Benson, C. W. (1957). Migrants at the south end of Lake Tanganyika. *Bull. Br. Orn. Club* **77**, 88.
Britton, P. L. and Britton, H. A. (1974). Pomarine Skua on the Kenya coast. *Bull. E. Afr. Nat. His. Soc.* Jan. **1974**, 4–5.
Brooke, R. K. (1977). The South Polar Skua is a probable visitor. *Cormorant* **3**, 14.
Brooke, R. K. (1978). The *Catharacta* skuas (Aves: Laridae) occurring in South African waters. *Durban Mus. Novit.* **11**(18), 295–308.
Brooke, R. K. (1981a). The seabirds of Moçâmedes province, Angola. *Gerfaut* **71**, 209–225.
Brooke, R. K. (1981b). What is *Stercorarius madagascariensis* Bonaparte? *Ardea* **69**(1), 144.
Brown, R. G. B. (1979). Seabirds of the Senegal upwelling and adjacent waters. *Ibis* **121**, 283–292.
Devillers, P. (1977). The skuas of the North American Pacific coast. *Auk* **94**, 417–429.
Devillers, P. (1978). Distribution and relationships of South American skuas. *Gerfaut* **68**, 374–417.
Furness, B. L. (1983). The feeding behaviour of Arctic Skuas *Stercorarius parasiticus* wintering off South Africa. *Ibis* **125**(2), 245–251.
Furness, R. W. (1982). Population, breeding biology and diets of seabirds in Foula in 1980. *Seabird Report* **6**, 5–12.
Griffiths, A. M. and Sinclair, J. C. (1982). The occurrence of Holarctic seabirds in the African sector of the Southern Ocean. *Cormorant* **10**, 35–44.
Harrison, P. (1983). 'Seabirds'. Houghton Mifflin, Boston.
Heinze, von J., Krott, N. and Mittendorf, H. (1978). Zur Vogelwelt Marokkos. *Die Vogelwelt* **99**, 132–137.
Herroelen, P. (1983). Un Labbe parasite *Stercorarius parasiticus* au Katanga: une nouvelle espèce pour le Zaïre. *Malimbus* **5**, 78.
Joubert, S. C. J. (1977). Avian marine vagrants to the Kruger National Park. *Koedoe* **20**, 185.
Lambert, R. (1971). Seevogelbeobachtungen auf zwei Reisen im östlichen Atlantik mit besonderer Berücksichtigung des Seegebietes von Südwestafrika. *Beitr. Vogelk.* **17**, 1–32.
Lambert, R. (1980). Ein Überwinterungsgebiet der Falkenraubmöwe *Stercorarius longicaudus* Vieill. 1819, von Südwest und Südafrika entdeckt. *Beitr. Vogelk.* **26**, 199–212.
Liversidge, R. (1959). The place of South Africa in the distribution and migration of ocean birds. *Ostrich* Suppl. **3**, 47–67.
Mann, C. F. (1974). A second Pomarine Skua on the Kenya coast. *Bull. Afr. Nat. Hist. Soc.* Mar. **1974**, 31.
Moore, R. D. (1981). Long-tailed Skuas *Stercorarius longicaudus* at the Kenya coast. *Scopus* **5**, 79.
Murphy, R. C. (1936). 'Oceanic Birds of South America'. American Museum of Natural History, New York.
O'Donald, P. (1983). 'The Arctic Skua: an Account of the Ecology, Genetics and Sociobiology of a Polymorphic Seabird'. Cambridge University Press, Cambridge.
Robertson, P., Hopson, J. and Hopson, T. (1974). Skuas at Lake Rudolf. *Bull. E. Afr. Nat. Hist. Soc.* Mar. **1974**, 31–32.
de Roo, A. and van Damme, G. (1970). A first record of the Long-tailed Skua, *Stercorarius longicaudus* Vieillot, from the Ethiopian Region (Aves: Stercorariidae). *Rev. Zool. Bot. Afr.* **82** (1/2), 157–162.
Sinclair, J. C. (1980). Subantarctic Skua *Catharacta antarctica* predation techniques on land and at sea. *Cormorant* **8**, 3–6.
Smith, K. D. (1972). *Stercorarius longicaudus* in tropical African waters. *Bull. Br. Orn. Club* **92**, 102.
Spearpoint, J. A. (1981). A Long-tailed Skua *Stercorarius*

longicaudus in the Kalahari-Gemsbok National Park, South Africa. *Cormorant* **9**, 45.

Summerhayes, C. P., Hofmeyer, P. K. and Rioux, R. H. (1974). Seabirds off the south-western coast of Africa. *Ostrich* **45**, 83–109.

Taylor, I. R. (1979). The kleptoparasitic behaviour of the Arctic Skua *Stercorarius parasiticus* with three species of tern. *Ibis* **121**, 274–282.

Tree, A. J. (1973). Antarctic Skua *Stercorarius skua*. *Ostrich* **44**(2), 129.

Wallace, D. I. M. (1973). Sea-birds at Lagos and in the Gulf of Guinea. *Ibis* **115**, 559–571.

Watson, G. E. (1966). Seabirds of the tropical Atlantic Ocean. *Smithsonian Publication* **4680**.

Watson, G. E. (1975). 'Birds of the Antarctic and Sub-Antarctic'. American Geophysical Union, Washington D. C.

Wells, D. R. (1966). A Pomarine Skua inland. *Bull. Niger. Orn. Soc.* **3**, 97.

Family LARIDAE: gulls

Ash, J. S. and Ashford, O. M. (1977). Great Black-headed Gulls *Larus ichthyaetus* and Red-necked Phalaropes *Phalaropus lobatus* inland in Ethiopia. *J. E. Afr. Nat. Hist. Soc. Nat. Mus.* **31**(162), 1–3.

Bailey, R. M. (1971). Sea-bird observations off Somalia. *Ibis* **113**, 29–41.

Beaubrun, P. C. (1983). Le Goéland d'Audouin (*Larus audouinii* Payr.) sur les côtes du Maroc. *Oiseau et R.F.O* **53**, 209–226.

Bridgeford, P. A. (1982). New breeding locality data for Southern African seabirds. *Cormorant* **10**, 125–126.

Britton, H. A. and Britton, P. L. (1968). Seafowl observed on a voyage from Cape Town to Southampton, 24th January to 5th February, 1968. *Bull. Br. Orn. Club* **88**, 93–96.

Britton, P. L. and Brown, L. H. (1974). The status and breeding behaviour of East African Lari. *Ostrich* **45**, 63–82.

Broekhuysen, G. J. and Elliott, C. C. H. (1974). Hartlaub's Gulls breeding on the roof of a building. *Bokmakierie* **26**, 66–69.

Brooke, R. K. and Cooper, J. (1979a). The distinctiveness of southern African *Larus dominicanus* (Aves: Laridae). *Durban Mus. Novit.* **12**, 27–37.

Brooke, R. K. and Cooper, J. (1979b). What is the feeding niche of the Kelp Gull in South Africa? *Cormorant* **7**, 27–29.

Brooke, R. K., Avery, G. and Brown, P. C. (1982). First specimen of the nominate race of the Kelp Gull *Larus dominicanus* in Africa. *Cormorant* **10**, 117.

Burger, J. and Gochfeld, M. (1981a). Colony and habitat selection of six Kelp Gull *Larus dominicanus* colonies in South Africa. *Ibis* **123**, 298–310.

Burger, J. and Gochfeld, M. (1981b). Nest site selection by Kelp Gulls in southern Africa. *Condor* **83**, 243–251.

Burger, J. and Shisler, J. (1980). The process of colony formation among Herring Gulls *Larus argentatus* nesting in New Jersey. *Ibis* **122**, 15–26.

Christy, P. (1983). La Mouette rieuse *Larus ridibundus* au Gabon. *Oiseau et R.F.O* **53**, 293.

Clapham, C. S. (1964). The birds of the Dahlac Archipelago. *Ibis* **106**, 376–388.

Crawford, R. J. M., Cooper, J. and Shelton, P. A. (1982). Distribution, population size, breeding and conservation of the Kelp Gull in southern Africa. *Ostrich* **53**, 164–177.

Dillingham, I. H. (1972). A rare colour aberration in the Southern Black-backed Gull, *Larus dominicanus*. *Bull. Br. Orn. Club* **92**, 101.

Donnelly, B. G. (1974). The Lesser Black-backed Gull *Larus fuscus* in southern and central Africa. *Bull. Br. Orn. Club* **94**, 63–68.

Dragesco, J. (1961). Observations éthologiques sur les oiseaux du Banc d'Arguin. *Alauda* **2**, 81–98.

Dupuy, A.-R. (1975). Laridés dans les deltas du Siné-Saloum et du fleuve Sénégal en juin 1974. *Oiseau et R.F.O.* **45**, 313–317.

Dupuy, A.-R. (1983). Reproduction de la Mouette rieuse *Larus ridibundus* au Sénégal. *Oiseau et R.F.O.* **53**, 294.

Dupuy, A.-R. (1984). Quelques données nouvelles sur l'avifaune du Sénégal ainsi que sur celle des Isles de la Madeleine. *Alauda* **52**(3), 177–183.

Dwight, J. (1925). The gulls (Laridae) of the world; their plumages, molts, variations, relationships and distribution. *Bull. Am. Mus. Nat. Hist.* **52**, 63–336.

Erard, C., Guillou, J. J. and Mayaud, N. (1984). Sur l'identité spécifique de certains Laridés nicheurs au Sénégal. *Alauda* **52**(3), 184–188.

Fogden, M. P. L. (1964). The reproductive behaviour and taxonomy of Hemprich's Gull *Larus hemprichii*. *Ibis* **106**, 299–320.

Fordham, R. A. (1963). Individual and social behaviour of the Southern Black-backed Gull. *Notornis* **10**, 206–222.

Fry, C. H. (1965). Sabine's Gull *Xema sabini* (Sabine) off West Africa. *Bull. Niger. Orn. Soc.* **2**(6), 47–48.

Furness, B. L. and Furness, R. W. (1982). Biometrics of Sabine's Gulls *Larus sabini* in the South Atlantic during the northern winter. *Cormorant* **10**, 31–34.

Gowthorpe, P. (1979). Reproduction de Laridés et d'Ardéidès dans la delta du Siné-Saloum (Sénégal). *Oiseau et R.F.O.* **49**, 105–112.

Grant, P. J. (1982). 'Gulls: a Guide to Identification.' Poyser, Calton.

Harris, M. P. (1964). Aspects of the breeding biology of the gulls *Larus argentatus*, *L. fuscus* and *L. marinus*. *Ibis* **106**, 432–456.

Hockey, P. A. R. (1980). Kleptoparasitism by Kelp Gulls *Larus dominicanus* of African Black Oystercatchers *Haematopus moquini*. *Cormorant* **8**, 97–98.

Hoogendoorn, W. (1982). Ring-billed Gull in Morocco in August 1982. *Dutch Birding* **4**(3), 91–92.

Isenmann, P. (1976). Contribution à l'étude de la biologie de la reproduction et de l'étho-écologie du Goéland railleur *Larus genei*. *Ardea* **64**, 48–61.

Jacob, J.-P. (1979). Resultats d'un recensement hivernal de Laridés en Algérie. *Gerfaut* **69**, 425–436.

Jacob, J.-P. and Courbet, B. (1980). Oiseaux de mer nicheurs sur la côte Algérienne. *Gerfaut* **70**, 385–401.

Johnstone, R. E. (1982). Distribution, status and variation of the Silver Gull *Larus novaehollandiae* Stephens, with notes on the *Larus cirrocephalus* species-group. *Rec. West. Aust. Mus.* **10**(2), 133–165.

de Juana, E., Bueno, J. M., Carbonell, M., Mellado, V. P. and Varela, J. (1979). Aspectos de la alimentación y biología de reproducción de *Larus audouinii* Payr en su gran colonia de cria de las Islas Chafarinas (año 1976). *Bol. Estación Central de Ecologia* **8**(16), 53–65.

de Juana, E., Varela, J. and Witt, H.-H. (in press). On the conservation of the Chafarinas Islands. *Proc. ICBP Conference 1981*.

Kennerley, P. R. (1979). Goéland atricille *Larus atricilla* au Maroc. *Alauda* **47**, 214–215.

Kilpi, M. and Saurola, P. (1984). Migration and wintering strategies of juvenile and adult *Larus marinus*, *L. argentatus* and *L. fuscus* from Finland. *Ornis Fennica* **61**(1), 1–8.

Lambert, K. (1975). Beobachtungen, Schwalbenmöwen *Xema sabini* (Sabine 1819), im Südsommer 1972/73 im

Südafrikanischen Winterquartier. *Beitr. Vogelkd.* **21**, 410–415.

Latour, M. (1973). Nidification de cinq espèces de Laridés au voisinage de l'embouchure du fleuve Sénégal. *Oiseau et R.F.O.* **43**, 89–96.

Miller, W. T. (1951). The Grey-headed Gull. *Bokmakierie* **3**, 30–31.

Moynihan, M. (1962). Hostile and sexual behaviour patterns of South American and Pacific Laridae. *Behaviour* Suppl. **8**.

Nikolaus, G. (1984a). Distinct status changes of certain Palaeartic migrants in the Sudan. *Scopus* **8**(2), 36–38.

Nikolaus, G. (1984b). Further notes of birds new or little known in Sudan. *Scopus* **8**(2), 38–42.

Oreel, G. J. (1975). Letter on 'Slender-billed Gull in Botswana?'. *Bokmakierie* **27**, 64.

Robertson, H. G. and Wooller, R. D. (1981). Seasonal decrease in the clutch size of Hartlaub's Gulls *Larus hartlaubii* at Strandfontein in 1978. *Cormorant* **9**, 23–26.

Robin, P. (1968). L'avifaune de l'Iriki (Sud-Marocain). *Alauda* **36**, 226–236.

Shaughnessy, P. D. (1980). Food of Kelp Gulls *Larus dominicanus* at Square Point, South West Africa. *Cormorant* **8**, 99–100.

Siegfried, W. R. (1977). Mussel-dropping behaviour of Kelp Gulls. *S. Afr. J. Sci.* **79**, 337–341.

Sinclair, J. C. (1977). Interbreeding of Grey-headed and Hartlaub's Gulls. *Bokmakierie* **29**, 70–71.

Sinclair, J. C. (1984). S. A. O. S. Rarities Committee report. *Bokmakierie* **36**(3), 64–68.

Steyn, P. (1975). Slender-billed Gull in Botswana? *Bokmakierie* **27**, 10–11.

Tinbergen, N. (1953). 'The Herring Gull's World'. Collins, London.

Tinbergen, N. (1957). Treading motion in feeding Hartlaub's Gull. *Ostrich* **28**, 171–172.

Tinbergen, N. and Broekhuysen, G. J. (1954). On the threat and courtship behaviour of Hartlaub's Gull, *Hydrocoloeus novaehollandiae hartlaubi* (Bruch). *Ostrich* **25**, 50–61.

Trotignon, J. (1976). La nidification sur le Banc d'Arguin (Mauritanie) au printemps 1974. *Alauda* **44**, 119–133.

Wallace, D. I. M. (1964). Studies of less familiar birds. 128. Slender-billed Gull. *British Birds* **57**, 242–247.

Wallace, D. I. M. (1973). Sea-birds at Lagos and in the Gulf of Guinea. *Ibis* **115**, 559–571.

Walter, C. B. (1984). Fish prey remains in Swift Tern and Hartlaub's Gull pellets at Possession Island, off Namibia. *Ostrich* **55**, 166–167.

Williams, A. J. (1977a). Midwinter diurnal activity and energetics of Kelp Gulls. *Cormorant* **3**, 4–9.

Williams, A. J. (1977b). Notes on the breeding biology of the Hartlaub's Gull. *Cormorant* **3**, 15.

Williams, A. J., Cooper, J. and Hockey, P. A. R. (1984). Aspects of breeding biology of the Kelp Gull at Marion Island and in South Africa. *Ostrich* **55**, 147–154.

Witt, H. (1977a). Zur Biologie der Korallenmöwe *Larus audouinii*—Brut und Ernährung. *J. Orn.* **118**, 134–155.

Witt, H. (1977b). Zur Verhaltensbiologie der Korallenmöwe *Larus audouinii*. *Z. Tierpsychol.* **43**, 46–67.

Witt, H., Crespo, J., de Juana, E. and Varela, J. (1981). Comparative feeding ecology of Audouin's Gull *Larus audouinii* and the Herring Gull *L. argentatus* in the Mediterranean. *Ibis* **123**, 519–526.

Zoutendyk, P. (1968). The occurrence of Sabine's Gull *Xema sabini* off the Cape Peninsula. *Ostrich* **34**, 9–11.

Family STERNIDAE: terns

Ash, J. S. (1980). Common and Lesser Noddy *Anous stolidus* and *A. tenuirostris* in Somalia. *Scopus* **4**, 6–9.

Ash, J. S. and Karani, A. A. (1981). Roseate and Sooty Terns *Sterna dougallii* and *S. fuscata* breeding on islets in southern Somalia. *Scopus* **5**, 22–27.

Ashmole, N. P. (1962). The Black Noddy *Anous tenuirostris* on Ascension Island. *Ibis* **103**b, 235–273.

Ashmole, N. P. and Ashmole, M. J. (1967). Comparative feeding ecology of sea birds of a tropical oceanic island. *Bull. Peabody Mus. Nat. Hist.* **24**, 1–131.

Becker, P. (1976). Raubseeschwalbe (*Hydroprogne caspia*) als Brutvogel an der Südwestafrikanischen Küste. *Namib und Meer* **7**, 21–23.

Begg, G. W. (1973). The feeding habits of the Whitewinged Black Tern on Lake Kariba. *Ostrich* **44**, 149–153.

Bergman, G. (1953). Verhalten und Biologie der Raubseeschwalbe (*Hydroprogne tschegrava*). *Acta Zool. Fenn.* **77**, 1–50.

Britton, P. L. (1977). First African records of two Malagasy seabirds. *Bull. Br. Orn. Club* **97**, 54–56.

Britton, P. L. and Brown, L. H. (1974). The status and breeding behaviour of East African Lari. *Ostrich* **45**, 63–82.

Bourne, W. R. P., Bogan, J. A., Bullock, D., Diamond, A. W. and Feare, C. J. (1977). Abnormal terns, sick sea and shore birds, organochlorines and arboviruses in the Indian Ocean. *Mar. Pollut. Bull.* **8**, 155–158.

Brooke, R. K. and Sinclair, J. C. (1978). Preliminary list of southern African seabirds. *Cormorant* **4**, 10–17.

Brown, R. G. B. (1979). Seabirds of the Senegal upwelling and adjacent waters. *Ibis* **121**, 283–292.

Brown, W. Y. (1976). Prolonged parental care in the Sooty Tern and Brown Noddy. *Condor* **78**, 128–129.

Browne, P. W. P. (1980). Birds observed near Lomé, Togo in 1976 and 1977. *Malimbus* **2**, 51–55.

Buckley, F. G. and Buckley, P. A. (1972). The breeding ecology of Royal Terns *Sterna* (*Thalasseus*) *maxima maxima*. *Ibis* **114**, 344–359.

Buckley, P. A. and Buckley, F. G. (1969). Juvenile Royal Tern killing a downy conspecific. *Ardea* **58**, 95–96.

Campredon, P. (1978). Reproduction de la Sterne caugek, *Thalasseus sandvicensis* Lath., sur le Banc d'Arguin (Gironde). Aperçu de sa distribution hivernale. *Oiseau et R.F.O.* **48**, 263–280.

Castan, R. (1961). Nouvelles recherches sur l'avifaune des îlots de la côte sud-est de Tunisie. *Alauda* **29**, 31–52.

Cawkell, E. M. and Moreau, R. E. (1963). Birds in The Gambia. *Ibis* **105**, 156–178.

Chaniot, G. E. (1970). Notes on color variation in downy Caspian Terns. *Condor* **72**, 460–465.

Clancey, P.A. (1971). Miscellaneous taxonomic notes on African birds 33. Comments on southern African Caspian Terns *Hydroprogne caspia* (Pallas). *Durban Mus. Novit.* **9**, 118–120.

Clancey, P. A. (1975). *Sterna bergii* and *Sterna maxima* in the South African Sub-region, with observations on their relationship. *Durban Mus. Novit.* **10**, 191–206.

Clancey, P. A. (1976). Further on the characters and status of *Sterna hirundo tibetana* Saunders in southern Africa. *Ostrich* **47**, 228.

Clancey, P. A. (1977). Data from Sooty Terns from Natal and Zululand. *Ostrich* **48**, 43–44.

Clancey, P. A. (1982). The Little Tern in southern Africa. *Ostrich* **53**, 102–106.

Clancey, P. A. and Wooldridge, T. (1975). The Whitecapped Noddy *Anous minutus* Boie in South African waters. *Durban Mus. Novit.* **10**, 227–230.

Clapham, C. S. (1964). The birds of the Dahlac Archipelago. *Ibis* **106**, 376–388.

Clinning, C. F. (1978a). The biology and conservation of the Damara Tern in South West Africa. *Madoqua* **11**, 31–39.

Clinning, C. F. (1978b). Breeding of the Caspian Tern in South West Africa. *Cormorant* **5**, 15–16.

Cooper, J. (1976). Seasonal and spatial distribution of the Antarctic Tern in South Africa. *S. Afr. J. Antarctic Res.* **6**, 30–32.

Craig, A. (1974). Whiskered Terns feeding on Arum Frogs. *Ostrich* **45**(2), 142.

Cullen, J. M. (1960). The aerial display of the Arctic Tern and other species. *Ardea* **48**, 1–37.

Diamond, A. W. (1976). Subannual breeding and moult cycles in the Bridled Tern *Sterna anaethetus* in the Seychelles. *Ibis* **118**, 414–419.

Dragesco, J. (1961). Observations éthologiques sur les oiseaux du Banc d'Arguin. *Alauda* **2**, 81–98.

Dunn, E. K. (1972a). Effect of age on the fishing ability of Sandwich Terns *Sterna sandvicensis*. *Ibis* **114**, 360–366.

Dunn, E. K. (1972b). Studies on terns with particular reference to feeding ecology. Unpublished Ph.D. thesis, Durham Univ., Durham.

Dunn, E. K. (1973). Robbing behaviour of Roseate Terns. *Auk* **90**, 641–651.

Dupuy, A.-R. (1975). Laridés dans les deltas du Siné-Saloum et du fleuve Sénégal en juin 1974. *Oiseau et R.F.O.* **45**, 313–317.

Dupuy, A.-R. (1979). Reproduction de *Sterna fuscata* et de *S. albifrons* dans la delta du Sénégal. *Oiseau et R.F.O.* **49**, 324.

Elliott, C. C. H. (1971). Analysis of the ringing and recoveries of three migrant terns. *Ostrich* Suppl. **9**, 71–82.

Erard, C. and Etchécopar, R. D. (1970). Some notes on the birds of Angola. *Bull. Br. Orn. Club* **90**, 158–161.

Feare, C. J. (1975). Post-fledging parental care in Crested and Sooty Terns. *Condor* **77**, 368–370.

Feare, C. J. (1976a). The breeding of the Sooty Tern *Sterna fuscata* in Seychelles and the effects of experimental removal of its eggs. *J. Zool. Lond.* **179**, 317–360.

Feare, C. J. (1976b). Desertion and abnormal development in a colony of Sooty Terns infested by virus-infected ticks. *Ibis* **118**, 112–115.

Feare, C. J. (1981). Breeding schedules and feeding strategies of Seychelles seabirds. *Ostrich* **52**, 179–185.

Frazier, J., Salas, S. and Abbas, M. (1984). Ornithological observations along the Egyptian Red Sea coast, spring 1982: with notes on migratory and breeding species. *Courser* **1**, 17–27.

Frost, P. G. H. and Shaughnessy, G. (1976). Breeding adaptations of the Damara Tern *Sterna balaenarum*. *Madoqua* **9**, 33–39.

Fry, C. H. (1961). Notes on the birds of Annobon and other islands in the Gulf of Guinea. *Ibis* **103a**, 267–276.

Fuggles-Couchman, N. R. (1962). Nesting of Whiskered Tern *Chlidonias hybrida sclateri* in Tanganyika. *Ibis* **104**, 563–564.

Furness, B. L. (1983). The feeding behaviour of Arctic Skuas *Stercorarius parasiticus* wintering off South Africa. *Ibis* **125**, 245–251.

Géroudet, P. (1965). 'Water-birds with Webbed Feet'. English language edition translated by P. Barclay-Smith. Blandford Press, London.

Griffiths, A. M. (1982). Observations of pelagic seabirds feeding in the African sector of the Southern Ocean. *Cormorant* **10**, 9–14.

Grimes, L. G. (1978). Occurrence of Javan Little Tern *Sterna albifrons sinensis* in West Africa. *Bull. Br. Orn. Club* **98**, 114.

Harrison, C. J. O. (1983). The occurrence of Saunder's Little Tern in the Upper Arabian Gulf. *Sandgrouse* **5**, 100–101.

Hockey, P. A. R. and Hockey, C. T. (1980). Notes on Caspian Terns *Sterna caspia* breeding near the Berg River, Southwestern Cape. *Cormorant* **8**, 7–10.

Jacob, J.-P. and Courbet, B. (1980). Oiseaux de mer nicheurs sur la côte Algérienne. *Gerfaut* **70**, 385–401.

Kasparek, M. (1982). Zur Zuggeschwindigkeit der Fluzseeschwalbe *Sterna hirundo*. *J. Orn.* **123**, 297–305.

Kilpi, M. and Saurola, P. (1984). Migration and survival areas of Caspian Terns *Sterna caspia* from the Finnish coast. *Ornis Fennica* **61**(1), 24–29.

Langham, N. P. E. (1971). Seasonal movements of British terns in the Atlantic Ocean. *Bird Study* **18**, 155–175.

Latour, M. (1973). Nidification de cinq espèces de Laridés au voisinage de l'embouchure du fleuve Sénégal. *Oiseau et R.F.O.* **43**, 89–96.

Lind, H. (1963). The reproductive behaviour of the Gull-billed Tern, *Sterna nilotica* Gmelin. *Vidensk. Medd. dansk. naturh. Foren.* **125**, 407–448.

Lomont, H. (1945). Les conditions de la nidification en Camargue de la Guifette moustac, *Chlidonias hybrida* (Pallas). *Bull. Mus. Hist. Nat. Marseille* **5**, 106–110.

Mees, G. F. (1977). The subspecies of *Chlidonias hybridus* (Pallas), their breeding distribution and migrations (Aves, Laridae, Sterninae). *Zool. Verh.* **157**, 1–64.

Moore, R. F. (1984). Notes on birds nesting in the Suakin Archipelago, Red Sea (Sudan). ms.

Moore, R. F. and Balzarotti, M. A. (1983). Observations of sea birds nesting on islands of the Sudanese Red Sea. *Bull. Br. Orn. Club* **103**, 65–71.

Moreau, R. E. (1967). Water-birds over the Sahara. *Ibis* **109**, 232–259.

de Naurois, R. and Roux, F. (1974). Précisions concernant la morphologie, les affinités et la position systématique de quelques oiseaux du Banc d'Arguin (Mauritanie). *Oiseau et R.F.O.* **44**, 72–84.

Nicholls, G. H. (1977). Studies of less familiar birds. Bridled Tern. *Bokmakierie* **29**, 20–23.

Nikolaus, G. (1984a). Distinct status changes of certain Palaeartic migrants in the Sudan. *Scopus* **8**(2), 36–38.

Nikolaus, G. (1984b). Further notes of birds new or little known in Sudan. *Scopus* **8**(2), 38–42.

Nisbet, I. C. T. (1981). Biological characteristics of the Roseate Tern *Sterna dougallii*. U.S. Dept. of the Interior, Fish and Wildlife Service, Office of Endangered Species.

North, M. E. W. (1945). Notes on the sea-birds of Brava. *J. E. Afr. Nat. Hist. Soc.* **18**(1/2), 32–40.

Palmer, R. S. (1941). A behaviour study of the Common Tern (*Sterna hirundo hirundo* L.). *Proc. Boston Soc. Nat. Hist.* **42**, 1–119.

Pitman, C. R. S. (1967). Seafowl observed on a voyage, Capetown to London, 23rd January to 8th February 1967. *Bull. Br. Orn. Club* **87**, 117–120.

Randall, R. M. and McLachlan, A. (1982). Damara Terns breeding in the Eastern Cape, South Africa. *Ostrich* **53**, 50–51.

Randall, R. M. and Randall, B. M. (1978). Diet of Roseate Tern during the breeding season at St Croix Island, Algoa Bay. *Cormorant* **5**, 4–10.

Randall, R. M. and Randall, B. M. (1981). Roseate Tern breeding biology and factors responsible for low chick production in Algoa Bay, South Africa. *Ostrich* **52**, 17–24.

Robertson, W. B. (1969). Transatlantic migration of juvenile Sooty Terns. *Nature (Lond.)* **222**, 632–634.

Robin, P. (1968). L'avifaune de l'Iriki (Sud-Marocain). *Alauda* **36**, 226–236.

Rowan, M. K. (1962). Mass mortality among European Common Terns in South Africa in April-May 1961. *British Birds* **55**, 103–114.

Schmitt, M. B., Milstein, P. le S., Hunter, H. C. and

Hopcraft, C. J. (1973). Black Terns in the Transvaal. *Bokmakierie* **25**, 91–92.

Sears, H. F. (1978). Nesting behaviour of the Gull-billed Tern. *Bird-Banding* **49**, 1–16.

Serventy, D. L., Serventy, V. and Warham, J. (1971). 'The Handbook of Australian Sea-birds'. Reed, Sydney.

Sinclair, J. C. (1977). Black-naped Tern in Natal. *Bokmakierie* **29**, 18–19.

Sinclair, J. C. (1982). Common Terns *Sterna hirundo* roosting at sea. *Cormorant* **10**, 49.

Sinclair, J. C. (Ed) (1983). S. A. O. S. Rarities Committee report. *Bokmakierie* **35**, 35–40.

Smith, K. D. (1951). Notes on the Bridled Tern in the Red Sea. *British Birds* **44**, 325–326.

Steyn, P. (1960). Nesting of Whiskered Tern in Southern Cape. *Bokmakierie* **12**, 35–36.

Steyn, P. (1966). Whiskered Terns. *Bokmakierie* **18**, 83–85.

Sutton, R. W. W. (1970). Bird records from Ghana in 1967 and 1968/9. *Bull. Niger. Orn. Soc.* **7**(27), 54.

Swift, J. J. (1960). Notes on the behaviour of Whiskered Terns. *British Birds* **53**, 559–572.

Tarboton, W. R., Clinning, C. F. and Grond, M. (1975). Whiskered Terns breeding in the Transvaal. *Ostrich* **46**, 188.

Thomas, D. K. and Elliott, H. F. I. (1973). Nesting of the Roseate Tern (*Sterna dougallii*) near Dar-es-Salaam. *Bull. Br. Orn. Club* **93**, 21–23.

Trotignon, J. (1976). La nidification sur le Banc d'Arguin (Mauritanie) au printemps 1974. *Alauda* **44**, 119–133.

Tye, A. (1983). Caspian Tern feeding young in winter quarters. *Malimbus* **5**, 91.

Wallace, D. I. M. (1973). Sea-birds at Lagos and in the Gulf of Guinea. *Ibis* **115**, 559–571.

Walter, C. B. (1984). Fish prey remains in Swift Tern and Hartlaub's Gull pellets at Possession Island, off Namibia. *Ostrich* **55**, 166–167.

Warham, J. (1958). Photographic studies of some less familiar birds 91. Bridled Tern. *British Birds* **51**, 303–308.

Whitfield, A. K. and Blaber, S. J. M. (1978). Feeding ecology of piscivorous birds at Lake St Lucia. Part 1: diving birds. *Ostrich* **49**, 185–198.

Woodward, P. W. (1972). The natural history of Kure Atoll, North-western Hawaii Islands. *Atoll Res. Bull.* **164**, 1–318.

Family RYNCHOPIDAE: skimmers

Attwell, R. I. G. (1959). The African Skimmer *Rhynchops flavirostris*: population counts and breeding in the Nsefu Game Reserve. *Ostrich* **30**, 69–72.

Beven, G. (1944). Nesting of African Skimmer. *Ostrich* **15**, 138–139.

Britton, P. L. and Brown, L. H. (1974). The status and breeding of East African Lari. *Ostrich* **45**, 63–82.

Burger, J. (1981). Sexual differences in parental activities of breeding Black Skimmers. *Am. Nat.* **117**, 975–984.

Chubb, E. C. (1943). Record of nesting of Skimmer at St. Lucia. *Ostrich* **14**, 111–112.

Dowsett, R. J. (1975). How does the Skimmer wet its eggs? *Bull. E. Afr. Nat. Hist. Soc.* **1975**, 13.

Gardiner, N. (1975). How does the Skimmer wet its eggs? *Bull. E. Afr. Nat. Hist. Soc.* **1975**, 96–97.

Garland, I. (1944). Skimmers again nesting and other birds observed at Lake St. Lucia. *Ostrich* **15**, 75–76.

Hanmer, D. B. (1982). First record of the African Skimmer breeding in Malaŵi. *Ostrich* **53**, 189.

Harrison, P. (1983). 'Seabirds, an Identification Guide'. Croom Helm, Beckenham, UK.

Harvey, W. G. (1973). A recent breeding record of Skimmers *Rynchops flavirostris*—Tanzania. *Bull. E. Afr. Nat. Hist. Soc.* **1973**, 139.

Hawksley, G. (1974). African Skimmer on the highveld. *Honeyguide* **79**, 44–45.

Modha, M. L. and Coe, M. J. (1969). Notes on the breeding of the African Skimmer *Rynchops flavirostris* on Central Island, Lake Rudolf. *Ibis* **111**, 593–608.

Moynihan, M. (1959). A revision of the family Laridae (Aves). *Am. Mus. Novit.* **1928**.

Pitman, C. R. S. (1932). Notes on the breeding habits and eggs of *Rhynchops flavirostris* (Vieill.)—African Skimmer or Scissor-bill. *Ool. Rec.* **12**, 51–54.

Roberts, M. G. (1976). Belly-soaking and chick transport in the African Skimmer. *Ostrich* **47**, 126.

Schildmacher, H. (1931). Ueber das 'Wasserpflügen' der Scherenschnäbel (*Rynchops*). *Orn. Monatsber.* **39**, 37–41.

Schnell, G. D. (1970). A phenetic study of the suborder Lari (Aves). 2. Phenograms, discussion and conclusions. *Syst. Zool.* **19**, 264–302.

Sears, H. F., Moseley, L. J. and Mueller, H. C. (1976). Behavioral evidence on skimmers' evolutionary relationships. *Auk* **93**, 170–174.

Terres, J. K. (1980). 'The Audubon Society Encyclopedia of North American Birds'. Alfred A. Knopf, New York.

Turner, D. A. and Gerhart, J. (1971). 'Foot-wetting' by incubating African Skimmers *Rynchops flavirostris*. *Ibis* **113**, 244.

Wetmore, A. (1919). A note on the eye of the Black Skimmer (*Rynchops nigra*). *Proc. Biol. Soc. Wash.* **32**, 195.

Zusi, R. L. (1962). Structural adaptations of the head and neck in the Black Skimmer *Rynchops nigra* Linnaeus. *Publ. Nuttall Orn. Club* **3**, 1–101.

Zusi, R. L. and Bridge, D. (1981). On the slit pupil of the Black Skimmer (*Rynchops niger*). *J. Field Orn.* **52**, 338–340.

Family ALCIDAE: auks

Bannerman, D. A. (1961). 'The Birds of the British Isles', Vol. 12. Oliver and Boyd, London and Edinburgh.

Bedard, J. (1969). Histoire naturelle du Gode, *Alca torda* L., dans le golfe Saint-Laurent, province de Quebec, Canada. Étude du service Canadien de la Faune, 7.

Birkhead, T. R. (1978). Behavioural adaptations to high density nesting in the Common Guillemot *Uria aalge*. *Anim. Behav.* **26**, 321–331.

Conder, P. J. (1950). On the courtship and social displays of three species of auks. *British Birds* **43**(3), 65–69.

Evans, P. G. H. (1981). Ecology and behaviour of the Little Auk *Alle alle* in West Greenland. *Ibis* **123**(2), 1–18.

Harris, M. P. (1980). Breeding performance of puffins *Fratercula arctica* in relation to nest density, laying date and year. *Ibis* **122**(2), 193–209.

Kozlova, E. V. (1957). Fauna of USSR. 'Birds'. Vol II (3). Zoological Institute of the Academy of Sciences of the USSR, New Series. No. 65. Moscow. Israel Program for Scientific Translations, Jerusalem (1961).

Lockley, R. M. (1962). 'Puffins'. Doubleday & Co., New York.

Spring, L. (1971). A comparison of functional and morphological adaptations in the Common Murre (*Uria aalge*) and Thick-billed Murre (*Uria lomvia*). *Condor* **73**, 1–27.

Tuck, L. M. (1960). 'The Murres'. Canadian Wildlife Service, Queen's Printer, Ottawa.

Family PTEROCLIDAE: sandgrouse

Brooke, R. K. (1968). On the status of the Yellow-throated Sandgrouse south of the Zambezi. *Ostrich* **39**, 33–34.

Cade, T. J. (1965). Relations between raptors and columbiform birds at a desert water hole. *Wilson Bull.* **77**, 340–345.

Cade, T. J. and Maclean, G. L. (1967). Transport of water by adult sandgrouse to their young. *Condor* **69**, 323–343.

Cade, T. J., Willoughby, E. J. and Maclean, G. L. (1966). Drinking behavior of sandgrouse in the Namib and Kalahari Deserts, Africa. *Auk* **83**, 124–126.

Casado, M. A., Levassor, C. and Parra, F. (1983). Régime alimentaire estival du Ganga cata *Pterocles alchata* (L.) dans le centre de l'Espagne. *Alauda* **51**, 203–209.

Christensen, G. C., Bohl, W. H. and Bump, G. (1964). A study and review of the Common Indian Sandgrouse and the Imperial Sandgrouse. Spec. Sci. Rep. Wildlife No. 84, U.S. Fish and Wildlife Service, Washington, D.C.

Clancey, P. A. (1967). Systematic notes on austral African sandgrouse. *Bull. Br. Orn. Club* **87**, 102–111.

Clancey, P. A. (1979). On *Pterocles namaqua* (Gmelin) in South West Africa. *Madoqua* **11**, 261–265.

Dixon, J. E. W. (1976). A record of the Namaqua Sandgrouse *Pterocles namaqua* from Rhodesia. *Madoqua* **9**, 55.

Dixon, J. E. W. (1977). Miscellaneous notes on South West African birds. *Madoqua* **10**, 149–151.

Dixon, J. E. W. (1978). Animal remains recovered from sandgrouse (Aves, Pteroclidae) crops in the Etosha National Park. *Madoqua* **11**, 75–76.

Dixon, J. and Louw, G. (1978). Seasonal effects on nutrition, reproduction and aspects of thermoregulation in the Namaqua Sandgrouse (*Pterocles namaqua*). *Madoqua* **11**, 19–29.

Fjeldså, J. (1976). The systematic affinities of sandgrouse, Pteroclididae. *Vidensk. Meddr. dansk naturh. Foren.* **139**, 179–243.

Fuggles-Couchman, N. R. (1984). The distribution of, and other notes on, some birds of Tanzania. *Scopus* **8**, 1–17.

George, U. (1969). Über das Tränken der Jungen andere Lebensäusserungen des Senegal-Flughuhns, *Pterocles senegallus*, in Marokko. *J. Orn.* **110**, 181–191.

George, U. (1970). Beobachtungen an *Pterocles senegallus* und *Pterocles coronatus* in der Nordwest-Sahara. *J. Orn.* **111**, 175–188.

George, U. (1976). 'In den Wüsten dieser Erde.' Hoffmann und Campe, Hamburg.

Ginn, P. J. (1977). Sandgrouse drinking behaviour in Botswana. *Honeyguide* **90**, 23–27.

Goodman, S. M. and Watson, G. E. (1983). Bird specimen records of some uncommon or previously unrecorded forms in Egypt. *Bull. Br. Orn. Club* **103**, 101–106.

Guichard, K. M. (1955). The birds of Fezzan and Tibesti. *Ibis* **97**, 393–424.

Hüe, F. and Etchécopar, R. D. (1957). Les ptéroclididés. *Oiseau et R.F.O.* **27**, 35–58.

Hüe, F. and Etchécopar, R. D. (1970). 'Les Oiseaux du Proche et du Moyen Orient'. N. Boubée & Cie, Paris.

Joubert, C. S. W. and Maclean, G. L. (1973). The structure of the water-holding feathers of the Namaqua Sandgrouse. *Zool. Afr.* **8**, 141–152.

Kalchreuter, H. (1979). Zur Mauser der äquatorialen Flughühner *Pterocles exustus* und *P. decoratus*. *Bonn. zool. Beitr.* **30**, 102–116.

Kalchreuter, H. (1980). The breeding season of the Chestnut-bellied Sandgrouse *Pterocles exustus* and the Black-faced Sandgrouse *P. decoratus* in northern Tanzania and its relation to rainfall. *Proc. IV Pan Afr. Orn. Congr.*, pp. 277–282.

Kemp, A. C. (1974). The distribution and status of the birds of the Kruger National Park. *Koedoe Monogr.* **2**, 1–352.

Lewis, A. D., Loefler, I. J. P. and Pearson, D. J. (1984). Four-banded Sandgrouse *Pterocles quadricinctus* in northwest Kenya. *Scopus* **8**, 46–48.

Lunais, B. (1984). Données sur l'avifaune terrestre résidente du Parc National du Banc d'Arguin (Mauritanie). *Alauda* **52**, 256–265.

Maclean, G. L. (1967). Die systematische Stellung der Flughühner (Pteroclididae). *J. Orn.* **108**, 203–217.

Maclean, G. L. (1968). Field studies on the sandgrouse of the Kalahari Desert. *Living Bird* **7**, 209–235.

Maclean, G. L. (1976). Adaptations of sandgrouse for life in arid lands. *Proc. XVI Int. Orn. Congr.* 1974, pp. 502–516.

Maclean, G. L. (1983) Water transport in sandgrouse. *Bioscience* **33**, 365–369.

Maclean, G. L. (1984). Evolutionary trends in the sandgrouse (Pteroclidae). *Malimbus* **6**, 75–78.

Marchant, S. (1961). Observations on the breeding of the sandgrouse *Pterocles alchata* and *senegallus*. *Bull. Br. Orn. Club* **81**, 134–141.

Marchant, S. (1962). Watering of young in *Pterocles alchata*. *Bull. Br. Orn. Club* **82**, 123–124.

McLachlan, G. R. (1985). The breeding season of the Namaqua Sandgrouse. *Ostrich* **56**, 210–212.

Meinertzhagen, R. (1934). The biogeographical status of the Ahaggar Plateau in the central Sahara, with special reference to birds. *Ibis* **13**(4), 528–571.

Morel, G. and Morel, M.-Y. (1970) Adaptations écologiques de la reproduction chez les oiseaux granivores de la savane sahélienne. *Ostrich* Suppl. **8**, 323–331.

Mungure, S. A. (1974). A brief interesting observation on sandgrouse at Seronera River pool. *Bull. E. Afr. Nat. Hist. Soc.* **1974**, 52–53.

Parra, F. and Levassor, C. (1981). Winter food of the sandgrouse, *Pterocles alchata*, in the Mancha region, Spain. *Bol. Est. Cent. Ecol.* **19**, 99–108.

Pineau, J. and Giraud-Audine, M. (1974). Notes sur les migrateurs traversant l'extrême nord-ouest du Maroc. *Alauda* **42**, 159–188.

Pitman, C. R. S. (1928). The nesting of *Eremialector quadricinctus lowei*—the Eastern Four-banded Sandgrouse. *Ool. Rec.* **8**, 79–81.

de Smet, K. and Van Gompel, J. (1980). Observations sur la côte sénégalaise en décembre et janvier. *Malimbus* **2**, 56–70.

Thomas, D. H. (1984a). Sandgrouse as models of avian adaptations to deserts. *S. Afr. J. Zool.* **19**, 113–120.

Thomas, D. H. (1984b). Adaptations of desert birds: Sandgrouse (Pteroclididae) as highly successful inhabitants of Afro-Asian arid lands. *J. Arid Environm.* **7**, 157–181.

Thomas, D. H. and Maclean, G. L. (1981). Comparison of physiological and behavioural thermoregulation and osmoregulation in two sympatric sandgrouse species (Aves: Pteroclididae). *J. Arid Environm.* **4**, 335–358.

Thomas, D. H. and Robin, A. P. (1977). Comparative studies of thermoregulatory and osmoregulatory behaviour and physiology of five species of sandgrouse (Aves: Pteroclididae) in Morocco. *J. Zool. Lond.* **183**, 229–249.

Thomas, D. H. and Robin, A. P. (1983). Description of the downy young of Lichtenstein's Sandgrouse *Pterocles lichtensteinii* and the significance of 'unpatterned' downy young in the Pteroclididae. *Bull. Br. Orn. Club* **103**, 40–43.

Thomas, D. H., Maclean, G. L. and Clinning, C. F. (1981). Daily patterns of behaviour compared between two sandgrouse species (Aves: Pteroclididae) in captivity. *Madoqua* **12**, 187–198.

Vincent, J. (1944). On the occurrence of Namaqua Sandgrouse in Natal. *Ostrich* **15**, 235–236.

Von Frisch, O. (1968). Vögel, die man säugen muss;

Spiessflughühner—ganz aussergewöhnliche Vögel. *Vogel-Kosmos* 5, 364–368.
Von Frisch, O. (1969). Zur Jugendentwicklung und Ethologie des Spiessflughuhns (*Pterocles alchata*). *Bonn. zool. Beitr.* 20, 130–144.
Von Frisch, O. (1970). Zur Brutbiologie und Zucht des Spiessflughuhns (*Pterocles alchata*) in Gefangenschaft. *J. Orn.* 111, 189–195.

Family COLUMBIDAE: pigeons and doves

Ash, J. S., Erard, C. and Prevost, J. (1974). Statut et distribution de *Streptopelia reichenowi* en Ethiopie. *Oiseau et R.F.O.* 44, 340–345.
Benson, C. W. (1959). *Turturoena iriditorques* in the Mwinilunga district, Northern Rhodesia. *Ibis* 101, 240.
Benson, C. W. and Irwin, M. P. S. (1966). The Bronze-naped Pigeon *Columba delegorguei* (Delegorgue) in Rhodesia. *Arnoldia (Rhod.)* 2(23), 1–4.
Boswall, J. & Demment, M. (1970). The daily altitudinal movement of the White-collared Pigeon *Columba albitorques* in the High Simien, Ethiopia. *Bull. Br. Orn. Club* 90, 105–107.
Brooke, R. K. (1981). The feral pigeon – a 'new' bird for the South African list. *Bokmakierie* 33, 37–40.
Brooke, R. K. (1984). A history of the Redeyed Dove in the southwestern Cape Province, South Africa. *Ostrich* 55, 12–16.
Brosset, A. (1956). Les oiseaux au Maroc oriental. *Alauda* 24, 161–205.
Brosset, A (1961). Ecologie des oiseaux du Maroc oriental. *Travaux Inst. Sci. Chérifien sér. biol.* 22.
Brosset, A. (1971). L'"imprinting" chez les Columbidés—Etude des modifications comportementales au cours du vieillissement. *Z. Tierpsychol.* 29, 279–300.
Brosset, A. (1976). 'La Vie dans la Forêt Equatoriale', 126 pp. F. Nathan, Paris.
Brown, L. H. (1977). The White-winged Dove *Streptopelia reichenowi* in S. E. Ethiopia, comparisons with other species, and a field key for identification. *Scopus* 1, 107–109.
Cawkell, E. M. and Moreau, R. E. (1963). Notes on birds in The Gambia. *Ibis* 105, 156–178.
Cooper, J. (1975). Primary moult, weight and breeding cycles of the Rock Pigeon on Dassen Island. *Ostrich* 45, 154–156.
Corbin, K. W. (1967). Evolutionary relationships in the avian genus *Columba* as indicated by ovalbumin tryptic peptides. *Evolution* 21, 355–368.
Corbin, K. W. (1968). Taxonomic relationships of some *Columba* species. *Condor* 70, 1–13.
Cornwallis, L. and Porter, R. F. (1982). Spring observations of the birds of North Yemen. *Sandgrouse* 4, 1–36.
Curry, P. J. (1974). The occurrence and behaviour of Turtle Doves in the inundation zone of the Niger, Mali. *Bristol Orn.* 7, 67–71.
Davies, S. J. J. F. (1970). Patterns of inheritance in the bowing display and associated behaviour of some hybrid *Streptopelia* doves. *Behaviour* 36, 187–214.
Davies, S. J. J. F. (1974). Studies of the three coo-calls of the male Barbary Dove. *Emu* 74, 18–26.
Dean, W. R. J. (1977). Population, diet and the annual cycle of the Laughing Dove at Barberspan. I. Life expectancy and survival estimates. *Ostrich* Suppl. 12, 102–107.
Dean, W. R. J. (1979a). Population, diet and the annual cycle of the Laughing Dove at Barberspan. II. Diet. *Ostrich* 50, 215–19.
Dean, W. R. J. (1979b). Population, diet and the annual cycle of the Laughing Dove at Barberspan. III. The annual cycle. *Ostrich* 50, 234–239.
Dean, W. R. J. (1980). Population, diet and the annual cycle of the Laughing Dove at Barberspan. IV. Breeding data and population estimates. *Ostrich* 51, 80–91.
Dorst, J. and Roux, F. (1972). Esquisse écologique sur l'avifaune des Monts du Balé, Ethiopie. *Oiseau et R.F.O.* 42, 203–240.
Dorst, J. and Roux, F. (1973). L'avifaune des forêts de *Podocarpus* de la province de l'Arussi, Ethiopie. *Oiseau et R.F.O.* 43, 269–304.
Dowsett, R. J. (1971). A call of the Lemon Dove *Aplopelia larvata*. *Ostrich* 42, 296.
Dowsett-Lemaire, F. (1983). Ecological and territorial requirements of montane forest birds on the Nyika Plateau, south-central Africa. *Gerfaut* 73, 345–378.
Dunbar, R. P. (1974). Mammals and birds of the Simien Mountains National Parks. *Walia* 5, 4–5.
Dupuy, A. (1966). Liste des oiseaux rencontrés en hiver au cours d'une mission dans le Sahara algérien. *Oiseau et R.F.O.* 36, 131–144.
Elliott, C. C. H. and Cooper, J. (1980). The breeding biology of an urban population of Rock Pigeons, *Columba guinea*. *Ostrich* 51, 198–203.
Fairon, J. (1971). Exploration ornithologique au Kahouar (hiver 1970). *Gerfaut* 61, 141–161.
Fry, C. H. (1961). Notes on the birds of Annobon and other islands in the Gulf of Guinea. *Ibis* 103, 267–276.
Fry, C. H., Keith, S. and Urban, E. K. (1985). Evolutionary expositions from 'The Birds of Africa': *Halcyon* song phylogeny; cuckoo host partitioning; systematics of *Aplopelia* and *Bostrychia*. *Proc. Int. Symp. Afr. Vert.* (Bonn, 1984) (Schuchmann, K.-L., Ed.), pp. 163–180.
Fuggles-Couchman, N. R. (1984). The distribution of, and other notes on, some birds of Tanzania, pt. 2. *Scopus* 8, 73–78; 81–92.
Gee, J. and Heigham, J. (1977). Birds of Lagos, Nigeria. *Bull. Niger. Orn. Soc.* 13, 103–132.
Germain, M., Dragesco, J., Roux, F. and Garcin, H. (1973). Contribution à l'ornithologie du sud-Cameroun. I. Non-Passériformes. *Oiseau et R.F.O.* 43, 119–183.
Géroudet, P. (1965). Notes sur les oiseaux du Maroc. *Alauda* 33, 294–308.
Géroudet, P. (1983). 'Limicoles, Gangas et Pigeons d'Europe', Vol. 2. Delachaux & Niestlé Neuchâtel, Lausanne, Paris.
Goodman, S. M. and Atta, G. A. M. (in press). The birds of southeastern Egypt. *Gerfaut*.
Goodman, S. M. and Houlihan, P. F. (1981). The Collared Turtle Dove *Streptopelia decaocto* in Egypt. *Bull. Br. Orn. Club* 101, 334–336.
Goodwin, D. (1956). Observations on the voice and some displays of certain pigeons. *Avic. Mag.* 62, 17–33, 63–70.
Goodwin, D. (1959). Taxonomy of the genus *Columba*. *Bull. Mus. Nat. Hist. Zool.* 6, 1–23.
Goodwin, D. (1960). Sexual dimorphism in pigeons. *Bull. Br. Orn. Club.* 80, 45–52.
Goodwin, D. (1983). 'Pigeons and Doves of the World', 3rd ed. Brit. Mus. Nat. Hist., London.
Greig-Smith, P. W. and Davidson, N. C. (1977). Weights of West African savanna birds. *Bull. Br. Orn. Club* 97, 96–99.
Guichard, K. M. (1955). The birds of Fezzan and Tibesti. *Ibis* 97, 393–424.
Hall, P. (1976). The birds of Mambilla Plateau. *Bull. Niger. Orn. Soc.* 12, 67–72.
Hall, P. (1977). The birds of Maiduguri. *Bull. Niger. Orn. Soc.* 13, 15–36.
Harrison, C. J. O. (1967). Apparent zoogeographical dispersal patterns in two avian families. *Bull. Br. Orn. Club.* 87, 49–56.

Hoffman, K. (1969). Zum Tagesrhythmus der Brutablösung beim Kaptäubchen (*Oena capensis* L.) und bei anderen Tauben. *J. Orn.* **110**, 448–464.

Husain, K. Z. (1958). Subdivisions and zoogeography of the genus *Treron* (green fruit-pigeons). *Ibis* **100**, 334–398.

de Juana, E. and Santos, T. (1981). Observations sur l'hivernage des oiseaux dans le Haut-Atlas (Maroc). *Alauda* **49**, 1–12.

Kok, O. B. and Kok A. C. (1984) Ongewone nests van Kransduiwe (*Columba guinea*). *Ostrich* **55**, 168–170.

Lilyestrom, W. E. (1974). Birds of the Simien highlands. *Walia* **5**, 2–3.

Meininger, P. L., Mullié, W. C. and Goodman, S. M. (1983). Atlas of breeding birds in Egypt: second progress report. *Orn. Soc. Middle East Bull.* **11**, 1–7.

Morel, G. and Morel, M.-Y. (1972). Etude comparative du régime alimentaire de cinq espèces de tourterelles dans une savane semi-aride du Sénégal. Premiers résultats. *In* Proc. General Meeting Working Group Graniv. Birds (Kendeigh, S. C. F. and Pinowski, J. Eds) Warszawa; pp. 351–355. IBP, PT section, The Hague 1970.

Morel, G. and Morel, M.-Y. (1978). Recherches écologiques sur une savane sahélienne du Ferlo septentrional, Sénégal. Etude d'une communauté avienne. *Cah. ORSTOM (sér. biol)* **43**(4), 347–358.

Morel, M.-Y. (1975). Comportement de sept espèces de tourterelles aux points d'eau naturels et artificiels dans une savane sahélienne du Ferlo septentrional, Sénégal. *Oiseau et R.F.O.* **45**, 97–125.

Morel, M.-Y. (1980). Coexistence of seven species of doves in a semi-arid tropical savanna of northern Senegal. *Proc. IV Pan Afr. Orn. Congr.*, pp. 283–290.

Morel, M.-Y. (1983). La mue de *Streptopelia rosegrisea* dans une région tropicale semi-aride (Nord Sénégal). *Alauda* **51**, 179–202.

Morel. M.-Y. (1985). La Tourterelle des bois, *Streptopelia turtur*, en Sénégambie: évolution de la population au cours de l'année et identification des races. *Alauda* **53**, 100–110.

Mountfort, G. (1981). Diurnal migration of Turtle Doves [Gambia]. *British Birds* **74**, 265–266.

Murton, R. K. (1960). Some photographs of Woodpigeon behaviour and feeding. *British Birds* **53**, 321–324.

Murton, R. K. (1968). Breeding, migration and survival of Turtle Doves. *British Birds* **61**, 193–212.

Murton, R. K. and Clark, S. P. (1968). Breeding biology of Rock Doves. *British Birds* **61**, 429–448.

Murton, R. K. and Isaacson, A. J. (1964). The feeding habits of the Wood-pigeon *Columba palumbus*, Stock Dove *C. oenas* and Turtle Dove *Streptopelia turtur*. *Ibis* **106**, 174–188.

Murton, R. K. and Westwood, N. J. (1966). The foods of the Rock Dove and the Feral Pigeon. *Bird Study* **13**, 130–146.

Murton, R. K., Thearle, J. P. and Thompson, J. (1972). Ecological studies of the Feral Pigeon *Columba livia* var. *J. Appl. Ecol.* **9**, 835–874.

Naether, C. (1975). The White-collared Pigeon, *Columba albitorques*. *Avic. Mag.* **8**, 228–229.

North, M. E. W. and Simms, E. (1979). 'Witherby's Sound Guide to British Birds'. London.

Oatley, T. B. (1984). Exploitation of a new niche by the Rameron Pigeon *Columba arquatrix* in Natal. *Proc. V Pan-Afr. Orn. Congr.*, pp. 323–330.

O'Connor, R. J. and Mead, C. J. (1984). The Stock Dove in Britain, 1930–80. *British Birds* **77**, 181–201.

Peirce, M. A. (1984). Weights of birds from Balmoral, Zambia. *Bull. Br. Orn. Club* **104**, 84–85.

Pettet, A. (1976). The avifauna of Waza National Park, Cameroun, in December. *Bull. Nigerian Orn. Soc.* **12**, 18–24.

Phillips, J. F. V. (1927). The role of the 'Bushdove' *Columba arquatrix* T. & K. in fruit-dispersal in the Knysna forests. *S. Afr. J. Sci* **24**, 435–440.

Pineau, J. and Giraud-Audine, M. (1976). Notes sur les oiseaux hivernant dans l'extrême nord-ouest du Maroc et sur leurs mouements. *Alauda* **44**, 47–75.

Pitwell, L. R. and Goodwin, D. (1964). Some observations on pigeons in Addis Ababa. *Bull. Br. Orn. Club* **84**, 41–45.

Prigogine, A. (1965). Le Pigeon à nuque blanche pour la première fois dans un jardin zoologique. *Zoo* **3**.

Rand, A. L. (1949). The races of the African Wood Dove *Turtur afer*. *Fieldiana Zool.* **31**, 307–312.

Robinson, C. (1956). Observations on the nesting of a pair of Laughing Doves (*Stigmatopelia senegalensis*). *Ostrich* **27**, 70–75.

Rowan, M. K. (1983). 'The Doves, Parrots, Loeries and Cuckoos of Southern Africa'. David Philip, Cape Town.

Sclater, W. L. and Moreau, R. E. (1932). Taxonomic and field notes on some birds of north-eastern Tanganyika Territory. Pt. I. *Ibis* Ser. 13, **2**, 487–522.

Serle, W. (1959). The West African races of the Lemon Dove *Aplopelia larvata* (Temm. and Knip). *Bull. Br. Orn. Club* **79**, 38–41.

Shotter, R. A. (1978). Aspects of the biology and parasitology of the Speckled Pigeon *Columba guinea* L. from Ahmadu Bello University Campus, Zaria, North Central State, Nigeria. *Zool. J. Linn. Soc.* **62**, 193–203.

Siegfried, W. R. (1971). Weights of three species of *Streptopelia* doves. *Ostrich* **42**, 155–147.

Siegfried, W. R. (1984). Group size of Cape Turtle Doves at desert water holes. *Proc. V Pan-Afr. Orn. Congr.*, 745–752.

Skead, D. M. (1971). A study of the Rock Pigeon, *Columba guinea*. *Ostrich* **42**, 65–69.

Smith, K. D. (1965). A note on *Streptopelia reichenowi*. *Ibis* **107**, 544–545.

Smithers, R. H. N. (1965). Notes on the feeding habits of the Red-eyed Dove, *Streptopelia semitorquata* (Rüppell) in a peri-urban area in Rhodesia. *Arnoldia (Rhod.)* **1**, 1–8.

Snow, D. W. (1950). The birds of São Tomé and Principe in the Gulf of Guinea. *Ibis* **92**, 579–595.

Stresemann, E. (1927–34). Aves *in* 'Handbuch der Zoologie' (Kükenthal, W. and Krumbach, T. Eds) Vol. 7, 2. W. de Gruyter, Berlin.

Stresemann, E. and Stresemann, V. (1966). Die Mauser der Vögel. *J. Orn.* **107**, 1–447.

Sueur, F. (1982). Notes sur la Tourterelle Turque *Streptopelia decaocto* en Picardie. *Alauda* **50**, 250–259.

Taibel, A. M. (1954). Notizie sulla riproduzione in cattività del Colombo dal collare bianco (*Columba albitorques* Rüppell). *Riv. Ital. Orn.* Ser. 2, 195–203.

Tarboton, W. R. and Vernon, C. J. (1971). Notes on the breeding of the Green Pigeon *Treron australis*. *Ostrich* **42**, 190–192.

Taylor, I. R. and Macdonald, M. A. (1978). The status of some northern Guinea savanna birds in Mole National Park, Ghana. *Bull. Niger. Orn. Soc.* **14**, 4–8.

Tree, A. J. (1963). Laughing Dove *Streptopelia senegalensis* passage on the Zambezi at Feira. *Ostrich* **34**, 180.

Uys, C. J. (1967). Breeding of Rameron Pigeon *Columba arquatrix* Temm. & Knip in the De Hoop region. *Ostrich* **38**, 200–202.

Verheyen, R. (1955). Le Pigeon bleu (*Columba arquatrix* Temm.) du Ruwenzori. *Gerfaut* **2**, 127–145.

Vaurie, C. (1961). Systematic notes on palearctic birds, no. 49. Columbidae: the Genus *Streptopelia*. *Am. Mus. Novit*, **1854**, pp. 1–25.

Walsh, J. F. (1980). Inter-sibling conflict in the Laughing Dove. *Ostrich* **51**, 191.

Walsh, J. F. (1981). Rotating behaviour of the incubating

Yellow-bellied Fruit Pigeon, *Treron waalia*. *Bull. Br. Orn. Club.* **101**, 311.

Wilson, R. T. and Lewis, L. G. (1977). Observations on the Speckled Pigeon *Columba guinea* in Tigrai, Ethiopia. *Ibis* **119**, 195–198.

Woldhek, S. (1980). 'Bird Killing in the Mediterranean'. Eur. Comm. Prevent. Mass Destruct. Migr. Birds, Utrecht. pp. 62.

Wood, B. (1975). Observations on the Adamawa Turtle Dove. *Bull. Br. Orn. Club.* **95**, 68–73.

3. Acoustic References

Section A: Discs and Cassettes

2. Queeny, E. M. (1951). Songs of East African Birds. American Museum of Natural History, New York. Three 12-inch, 78 r.p.m. discs; approximately 48 species. Some errors on the labels and misidentifications were noted by Boswall and North (1967).

3. Queeny, E. M. (1951). Birds of Lake Nyibor. American Museum of Natural History, New York. One 12-inch, 78 r.p.m. disc; 14 species.

4. Cowles, R. B. (1956). Sounds of a South African Homestead. Folkways Records and Service Corp., 117 W 46 St., New York 10036. One 12-inch, 33⅓ r.p.m. disc, No. FPX 151. 30 species.

5. North, M. E. W. (1958). Voices of African Birds. Cornell University Press. 159 Sapsucker Woods Road, Ithaca, N. Y. 14850. One 12-inch, 33⅓ r.p.m. disc. 42 species. The first African record concerned mainly with identification. Species are presented in systematic order, grouped on separate bands, and details given of circumstances, place and date of recording.

7. Haagner, C. H. (1961). Birds of the Kruger National Park. International Library of African Music, P.O. Box 138, Roodeport, near Johannesburg, South Africa. Two 7-inch, 45 r.p.m. discs, Nos XTR 17044 and XTR 27045. 31 species in systematic order following Roberts (1957. 'Birds of South Africa', Trustees of the John Voelcker Bird Book Fund, Cape Town.) and with the Roberts number; each on a separate band.

8. Seed, F. (1963). Safari: the Story of a Night in the East African Bush. Michael Orme, High Fidelity Productions Ltd., Nairobi, Kenya. Two 12-inch, 33⅓ r.p.m. discs, nos. ZB 8058–59. Mainly concerned with atmosphere and ambience, with much commentary and background of animal voices; 16 species.

9. Haagner, C. H. (1964). Birds of Zululand. Same publisher as No. 7. Two 7-inch, 45 r.p.m. discs, Nos. XTR 4 7094 and XTR 5 7095. 27 species.

10. North, M.E.W. and McChesney, D. S. (1964). More Voices of African Birds. Houghton Mifflin Co., Boston, U. S. A. One 12-inch, 33⅓ r.p.m. disc. 90 species. Details of recordings are given in an accompanying booklet. These 2 discs (Nos 5 and 10) together contain the voices of 132 species, and are the first major reference work for African bird voices.

11. Stannard, J. and Niven, P. (1966). Bird Songs of Amanzi. Percy Fitzpatrick Institute of African Ornithology, University of Capetown, Rondebosch 7700, South Africa. One 12-inch 33⅓ r.p.m. disc, No. ACP 524; No 1 in 'Bird Song Series'. 37 species. On one side the birds are heard in their natural surroundings, the emphasis being on atmosphere or ambience; on the other side they are singled out and identified.

12. Pooley, A. C. (1966). Wildlife Calls of Africa. Percy Fitzpatrick Institute (address under No.11). One 12-inch, 33⅓ r.p.m. disc. 24 species.

13. Hayes, C. and Hayes, J. (1966). East African Birdsong; No.2 in *Heartbeat of Africa*, Series 1. Sapra Studios, Box 5882, Kimathi and York Streets, Nairobi, Kenya. One 7-inch, 45 r.p.m. disc. 25 species.

14. Stannard, J. and Niven, P. (1967). Bird Song of the Forest. Percy Fitzpatrick Institute (address under No.11). One 12-inch 33⅓ r.p.m. disc. GALP 1559. 32 species. Same format as No.11. Some species appear on both records.

15. Walker, A. (1967). Bird Song of Southern Africa. African Music Society and International Library of African Music, Roodeport, South Africa. One 12-inch 33⅓ r.p.m. disc, GALP 1501. 33 species.

17. Reucassel, R. and Pooley, A. C. (1967). Calls of the Bushveld. Published by the authors and obtainable from the Wildlife Society of South Africa. One 12-inch 33⅓ r.p.m. disc, WL2; also available as a cassette. 28 species.

20. Henley, A. and Pooley, A. C. (1970). Birds of the Drakensberg. Published by the authors and obtainable from Wildlife Society of South Africa. One 12-inch 33⅓ r.p.m. stereo disc, BD 100. 41 species.

21. Reucassel, R. and Adendroff, A. (1970). Nature's Melody. Published by the authors; obtainable from Wildlife Society of South Africa. One 12-inch 33⅓ r.p.m. stereo disc, SWL 3. 53 species.

22. Walker, A. (1970). Garden Birds of Southern Africa. Gallo (Africa) Ltd., Johannesburg; obtainable from Wildlife Society of South Africa. One 12-inch 33⅓ r.p.m. stereo disc, SGALP 1598. 40 species.

25. Dangerfield, G. (1970). Sounds of the Serengeti. Music for Pleasure Ltd., Astronaut House, Hounslow Road, Feltham, Middlesex, England. One 12-inch 33⅓ r.p.m. stereo disc, MFP 1371. 24 species.

27. Hayes, J. (c. 1970). Bird Song of Africa. *Heartbeat of Africa* series 2. Sapra Studios (address under No.13). One 7-inch, 45 r.p.m. disc. 12 species.

30. Roché, J.-C. (1971). Birds of Kenya. *Birds and Wild Beasts of Africa*, No. 1. L'Oiseau Musicien, France. One 12-inch, 33⅓ r.p.m. stereo disc, G.07. 32 species.

32. Keith, G. S. and Gunn, W. W. H. (1971). Birds of the African Rain Forests. *Sounds of Nature* No.9. Federation of Ontario Naturalists, 1262 Don Mills Road, Don Mills, Ontario M3B 2WB, Canada, and American Museum of Natural History, New York. Two 12-inch, 33⅓ r.p.m. discs. 95 species. The most important reference work since the records of North (Nos. 5 and 10) and the first specializing in forest birds, many of which are here published for the first time. Most species are from East Africa, some from central Africa. Species are arranged in systematic order and grouped in bands; a simple announcement of the name accompanies each species, but a lot of information is provided on the jacket.

33. Stannard, J. (1971). Bird Sounds and Songs. Fitzpatrick Institute (address under No.11). Issued in conjunction with *Ostrich* Supplement 9. One 7-inch, 45 r.p.m. disc, NV1. 20 species.

34. Chappuis, C. (1971). Ambiances des plaines et savanes d'Afrique orientale. *Afrique Sauvage* No.1. One 12-inch, 33⅓ r.p.m. disc, JAC 9. Edition *Jacana*, 32 rue St. Marc, 75002 Paris. 44 species.

35. Martin, R. B. (1971). Journey Across Africa. Parlophone PCSJ (D) 12.79. Obtainable from Wildlife Society of South Africa. One 12-inch, 33⅓ r.p.m. disc. 34 species.

36. Ker, A. (1972). Safari 99. Equator Sound Studios Ltd., P.O. Box 30068, Nairobi, Kenya. One 12-inch, 33⅓ r.p.m. disc, ESS 1001. 63 species.
38. Keibel, W. D. (1972). Wildlife of South West Africa. Wildlife Society of South Africa, P.O. Box 3508, Windhoek, Namibia. 1 cassette, 48 species.
39. Roché, J.-C. (1973a). Birds of South Africa. *Birds and Wild Beasts of Africa*, No.2. L'Oiseau Musicien, France. One 12-inch, 33⅓ r.p.m. stereo disc, G. 08. About 65 species. 7 environments are presented without commentary, created by 3 or 4 birds singing simultaneously.
40. Roché, J.-C. (1973b). Birds of West Africa—Senegal. *Birds and Wild Beasts of Africa* No. 3. L'Oiseau Musicien, France. One 12-inch, 33⅓ r.p.m. disc, G. 09. 26 species.
42. Worman, D. (1974). African Birds. Soundpics Enterprises (Pty) Ltd, P.O. Box 61055, Marshalltown 2107, South Africa. One 10-inch 33⅓ r.p.m. disc, SP 002, and 16 colour slides. 16 species.
43. Chappuis, C. (1974a). Le Niokolo-Koba. National Parks of Senegal, Tambacounda, Senegal. One 7-inch, 45 r.p.m. disc, with 9 colour photographs and booklet. 6 species.
44. Chappuis, C. (1974b). *Les Oiseaux de l'Ouest Africain*, Disc 1; Columbidae and Cuculidae. *Alauda*, Sound Supplement No.1; accompanying commentary in *Alauda* 42, 197–222. Société d'études ornithologiques, 46 Rue d'Ulm, 75230 Paris. One 12-inch, 33⅓ r.p.m. disc, ALA1. 44 species. This record is the first of a series whose aim is to present all known recordings for species of a particular region, including different forms of songs and calls and geographical variation. Details of the recordings are provided in the accompanying article in *Alauda*, of which reprints may be requested when ordering the record.

    This record represents a landmark in the history of African voice-recording. It is the beginning of a lengthy series covering large numbers of species in great detail, and the accompanying commentaries in *Alauda* are of considerable scientific value.
46. Anon. (1966). A Night at Treetops. Sapra Studios (address under No.13). *Heartbeat of Africa*, Series 1, No. 3. One 17-cm 45 r.p.m. disc. About 10 species.
47. Jones, B. (1969). The Rhino Story. Wildlife Society of South Africa, P.O. Box 44189, Linden 2104, South Africa. One 17-cm 45 r.p.m. disc. 12 species.
48. Chappuis, C. (1970). Pages sonores d'un monde inconnu. Jacana, 30, Rue St. Marc, 75002 Paris. One 17-cm, 45 r.p.m. flimsy disc. 12 birds, of which 7 Afrotropical. Contains the only known recording of Forest Francolin *Francolinus lathami*.
49. Reucassel, D. (1975). Calls of the Wild. Published by the author and available from Wildlife Society of South Africa. One 30-cm, 33⅓ r.p.m. disc, AV1; also available as a cassette. Comes with 32 colour slides. 17 species.
50. Hart, S. (1975). Listen to the Wild—in the Bush. EMI/Brigadiers (Pty) Ltd., South Africa. 30-cm, 33⅓ r.p.m. stereo disc, Brigadiers Music LTW(W)1. 17 species.
51. Hart, S. (1975). Listen to the Wild—Among the Rocks. See No. 50. 16 species.
52. Hart, S. and Bannister, A. (1975) Listen to the Wild—with the Insects. See No. 50. 9 species.
53. Chappuis, C. (Ed.). (1975). *Les Oiseaux de l'Ouest Africain*; disc 4; Phoenicopteridae, Anatidae, Rallidae, Heliornithidae, Podicipedidae, Jacanidae. *Alauda* sound supplement; commentary in *Alauda* 43, 427–441. One 30-cm, 33⅓ r.p.m. disc, ALA 7/8. 43 species. See No.44. On this disc Stuart Keith provided recordings of 6 species of *Sarothrura*, and the voices of 4 rallid species are here published for the first time.
54. Petersen, H. (1975). Belauschte Welt der Tiere. Penny, Postfach 1, 6000 Frankfurt 1, West Germany. 1 30-cm, 33⅓ r.p.m. stereo disc, Penny S 1475/10. 9 species.
58. Natal Bird Club (c. 1978). Bird Calls, Vols 1 and 2. Natal Bird Club, P.O. Box 10909, Marine Parade, Durban 4056, South Africa. 2 cassettes. 136 species presented in random order. Lengthy and numerous cuts are provided for each species.
60. Roché, J. C. (1968). Guide sonore des oiseaux d'Europe, Tome II: Maghreb. Edwards Records, 58, Rue du Docteur Calmette, 59320 Sequedin, France. Five 17-cm, 33⅓ r.p.m. discs. Disc 1 contains the voice of Double-spurred Francolin *Francolinus bicalcaratus*.
61. Slater, A. and Slater, D. (1970). Bird Chorus on the Limpopo. Published by A. V. Slater, 30, 9th St., Parkhurst 2193, South Africa. One 30-cm, 33⅓ r.p.m. disc, BS 001. 12 species.
62. Palmer, S. and Boswall, J. (1969–1972). A Field Guide to the Bird Songs of Britain and Europe. SR Records, Swedish Broadcasting Corp., 105 10 Stockholm, Sweden. Twelve 12-inch, 33⅓ r.p.m. discs, RFLP 5001–5012. 530 species, nesting or accidental in Europe, mostly wintering in Africa. Presented in systematic order, on separate bands, announced by scientific name. The most important reference work for Palearctic birds wintering in Africa.
63. Palmer, S. and Boswall, J. (1973). A sequel to No. 62. 2 discs, RFLP 5013 and 5014. Includes 23 African species.
64. Hayes, J. and Allan, J. O. Wild Africa. Andrew Crawford Productions, P.O. Box 42004, Nairobi, Kenya. One 12-inch 33⅓ r.p.m. disc, ACP1001. 8 species.
66. Kabaya, T. (1978). Birds of the World. I: Africa. King Records Co., Japan. One 30-cm, 33⅓ r.p.m. stereo disc, King Records SKS (H) 2007. 20 species.
69. Walker, A. (1980). Sounds of the Zimbabwe Bush. Available from the author at 1 Northmoor Road, Oxford OXZ 6UW, England, or Queen Victoria Museum, Harare, Zimbabwe. One stereo cassette. 27 species.
70. Palmer, S. and Boswall, J. (1980). A Field Guide to the Bird Songs of Britain and Europe. A sequel to No. 63; one disc, RFLP 5015.
72. Audio Three (1981). Bird Calls. See No. 58. 3 cassettes, of which the first 2 are the same as those of No. 58; the third contains additional species.
73. Palmer, S. and Boswall, J. (1981). A Field Guide to the Bird Songs of Britain and Europe. 16 cassettes, RFLP 5021–5036. An updated edition of Nos. 62, 63 and 70. 612 species, in boxes of 4 cassettes with commentary and list of species in each box. A first class reference collection.
74. Gillard, L. and Gibbon, G. (1982). A Field Guide to the Bird Calls of southern Africa. Published by the authors; P.O. Box 394, Greenside 2034, Johannesburg, or P.O. Box 10123, Ashwood, 3600 Pinetown, South Africa. 2 cassettes, about 420 species. The large number of species makes this one of the most important and comprehensive collections of African bird voices so far published. Several types of songs or calls are often given per species. Species are grouped by environment, and in systematic order within the environment.
75. Audio Three. Bird Calls: Bird Families, Vol IV. 2 cassettes, 171 species. Many of these species already appear on No.72, but here all are in systematic order.
76. Chappuis, C. (1984). Oiseaux migrateurs et gibier d'eau en hiver—waterfowl and waders in winter. Obtainable from the author, Les Chardonnerets, Vallon du Fer à Cheval, La Bouille, 76530 Grand Couronne, France. One cassette. A revised and enlarged version of the 1966

publication *Oiseaux de France*, Vol 1. Gives only flight and contact calls, not songs, of Palearctic birds; useful because these are the vocalizations typically made in Africa by migrants.

Section B: Most Important Discs and Cassettes by Region

East Africa: Nos 5, 10, 32, 36
West Africa: Nos 44, 53
Southern Africa: Nos 58, 72, 74, 75.
Palearctic migrants: Nos 62, 63, 70, 73, 76

Section C: Institutions with Sound Libraries

A. Audio Three, 6, Larch Road, Durban, South Africa.
B. British Library of Wildlife Sounds (BLOWS). The British Library, National Sound Archive, 29 Exhibition Road, London SW7 2AS.
C. Cornell University, Library of Natural Sounds, Laboratory of Ornithology, 159 Sapsucker Woods Road, Ithaca N.Y. 14850.
F. Fitzpatrick Bird Communication Library, Bird Department, Transvaal Museum, P.O. Box 413, Pretoria 0001.
N. Natal Bird Club, P.O. Box 10909, Marine Parade, Durban 4056.
S. South African Broadcasting Corporation. Library of Wildlife Sounds, P.O. Box 4559, Johannesburg 2000.

Section D: Individual Recordists

(Recordists whose names are followed by an institution have deposited copies of their tapes in that institution.)

200. Adendorff, G.
201. Allan, J. O.
202. Anon.
203. Armstrong, E. A.
204. Aspinwall, D. R.
205. Attenborough, D. (BBC), BLOWS
206. Beamish, H. H.
207. Bell, Fairfax, BLOWS
208. Berry, H. H.
209. Bird, L. (BBC) BLOWS
210. Blencowe, E. J.
211. Boston (BBC), BLOWS
212. Boulton, R., Cornell
213. Bourguignon, C.
214. Broekhuysen, G. J.
215. Brunel, J.
216. Carnochan, J.
217. Chappuis, C.
218. Cowles, R. B.
219. Crook, J. H.
220. Dangerfield, G.
221. Despin, B.
222. Downey, S. P.
223. Duval, C. T., BLOWS
224. Elders, D.
225. Elgood, J. H.
226. Erard, C.
227. Farkas, T.
228. Fisher, J. (BBC), BLOWS
229. Forbes-Watson, A.
230. Foster, B.
231. Gibbon, G.
232. Gill, F.
233. Gillard, L.
234. Gregory, A. R.
235. Grimes, L., BLOWS
236. Gunn, W. W. H.
237. Guttinger, H. R.
238. Haagner, C. H.
239. Hart, S.
240. Hayes, C.
241. Hayes, J. (BBC), BLOWS
242. Helb, H. W.
243. Henley, T.
244. Horne, J.
245. Howell, T. R.
246. Johnson, E. D. H., BLOWS
247. Jones, B.
248. Jouventin, P.
249. Kabaya, T.
250. Kaestner, J., Cornell
251. Kaestner, P., Cornell
252. Keibel, W. D.
253. Keith, G. S., Cornell
254. Ker, A.
255. Koch, L.
255a. Komen, J.
256. König, C.
257. Le Maho, Y.
258. Lemaire, F.
259. Lernoud, J. M.
260. Liversidge, R., Cornell
261. Low, G. C., BLOWS
262. Lutgens, H., BLOWS
263. McChesney, D., Cornell
264. McVicker, R.
265. Martin, R. B.
266. Martin Gauntlet, F., BLOWS
267. Mees, V., BLOWS
268. Morel, G.
269. Neal, E. (BBC), BLOWS
270. Nicolai, J.
271. Nightingale, T.
272. Niven, P.
273. North, M. E. W., BLOWS, Cornell
274. Oatley, T. B.
275. Parelius, D.
276. Parker, T., Cornell
277. Payne, R. B.
278. Petersen, H.
279. Pooley, T.
280. Queeny, E. M.
281. Reucassel, D.
282. Roché, J. C.
283. Root, A.
284. Rose, M. E.
285. Seed, F.
286. Sessions, P. H. B.
287. Short, L.
288. Simms, E.
289. Slater, A.
290. Slater, D.
291. Smith (BBC), BLOWS
292. Smithers, R. H. W.
293. Snow, D.
294. Stafford Smith, T.
295. Stannard, J.
296. Stjernstedt, R., BLOWS
297. Strinati, P.
298. Swales, M. K.
299. Thorpe, W. H.

300. Tibbles, M. (BBC), BLOWS
301. Tollu, B.
302. Turner, D.
303. Vielliard, J.
304. Voisin, J. F.
305. Walker, A.
306. Watts, D. E.
307. Worman, D.
308. Zimmermann, D. and Zimmerman, M.
309. Zino, A.

ERRATA, VOLUME I

Over 40 reviews of Volume I have appeared in print and most have greeted it with unqualified acclaim. Some have criticized one or another aspect of the work in general terms, such as inaccuracies in mapping or difficulties in 'reading' the Plates, and we are attempting to remedy these failings in this and remaining volumes. Other critics have listed omissions, perceived omissions, and a few errors of commission. We are grateful for all such comments and suggestions. Errors and omissions in Volume I are as follows:

Mapping inaccuracies

White-headed Petrel, Hadada, African Pygmy Goose, Yellow-billed Duck, Vulturine Fish Eagle, Rufous-chested Sparrowhawk and Long-legged Buzzard in southern Africa (see P. A. Clancey, *Honeyguide* 30, 1984, 41–47); Black-necked Grebe in Namibia (see M. K. Rowan, *Ostrich* 55, 1984, 39–41); Cape Shoveler in Zaïre (see D. R. Aspinwall, *Black Lechwe* 5, 1983, 35); Malagasy Pond Heron, Cape Teal, African Hobby and Taita Falcon in Botswana (see N. D. Hunter, *Babbler* 7, 1984, 50–51); Ostrich, Mountain Buzzard and Grey Kestrel in Zambia (see D. L. Berkvens, *Bull. Zambian Orn. Soc.* 13–15, 1983, 139–141); Olive Ibis, Mountain Buzzard and Verreaux's Eagle in Zaïre (see A. Prigogine, *Gerfaut* 73, 1983, 215–217); Black Heron, Saddle-billed Stork, Glossy Ibis, White-backed Duck, African Pygmy Goose, Cape Teal, Osprey, African Fish Eagle, Bat Hawk, African Swallow-tailed Kite, Hooded Vulture, Booted Eagle, Cassin's Hawk Eagle and African Marsh Harrier in East Africa, Bean Goose in Mali, and Red Kite in South Africa (see D. A. Turner, *Scopus* 7, 1983, 100–101); Rufous-bellied Heron, African Swallow-tailed Kite, Bat Hawk, Secretary Bird and Dickinson's Kestrel in East Africa and Imperial Eagle in Cameroon (see D. A. Zimmerman, *Auk* 100, 1983, 1005–1009).

Colour corrections

Plate 2, Figures 33 and 34 (see Zimmerman 1983); Plate 9, Figure 4 (see Zimmerman 1983); Plate 11, Figures 13 and 17 (see Prigogine 1983); Plate 22, Figures 8 and 21 (see Zimmerman 1983); Plate 25, Figure 8, cere should be grey or black; Plate 27, Figure 13, eye should be ivory white (see Zimmerman 1983); Figures 23 and 25 (see Zimmerman 1983).

Other Errata, 1st Printing

p. 10, Figure 8, IV, left column, line 2: add comma after 'reduced'.

p. 31, top left, figure should read: 'greater coverts', not 'greater wing-coverts', and in figure of bird's head (above hawk) 'eye-stripe' (by bill) should read 'lores'.

p. 35, amend Figure 26 to read: '*D. chlororhynchos bassi*'.

p. 50: 'Atlantic Petrel' should precede 'Schlegel's Petrel'.

p. 96, Red-tailed Tropic-bird, Voice, after 'Recorded': add (2a).

p. 164, amend mean egg size from '(60 × 73)' to '(60 × 43)'.

p. 215, References: Berry's name should precede Brown.

p. 230, Figure illustrates Trumpeter Swan, not Mute Swan.

p. 266, General Habits, line 17: for 'Santa da Bandeira' read 'Sa da Bordeira'.

p. 277, Figure A illustrates Goldeneye, not Southern Pochard.

p. 284, Figure A illustrates Ring-necked Duck; replace it with Figure 2 on p. 277.

p. 286, right column, 5 lines from foot: insert 'underparts and' before secondary patches.

p. 388, right column, line 26: read 'forest greenbuls', not 'Forest Greenbuls'.

p. 400, *Buteo auguralis*, second English name should be 'African Red-tailed Buzzard'.

p. 436, left column, 3 lines from foot: Darfur, not Daifur.

p. 481, left column, line 7: for Boon read Bonn.

p. 507, reference to Tinbergen (1932) is out of alphabetical sequence.

p. 517, Duck, Comb: delete 273.

p. 517, Duck, Marbled: **273**, not 273.

INDEXES

Bold page numbers indicate the main account of an entry in the index; italic, the relevant plate illustration.

Scientific Names

A

aalge, *Colymbus* 415
aalge, *Uria* 353, 409, **415**, 419
abdimii, *Ciconia* 475
abyssinica, *Chalcopelia* 454
abyssinicus, *Turtur* 450, **454**, *465*
Accipiter gentilis 471, 472
  melanoleucos 466
  nisus 471
  rufiventris 466
  tachiro 463, 466
Acryllium **6**
  vulturinum 7, *96*
Actitis **325**
  hypoleucos 187, 289, 304, 323, 325, *326*, *352*
Actodromas 291
Actophilornis **181**
  africana 121, *177*, **181**, 184, 280
adamauae, *Francolinus* 54
adansonii, *Coturnix c.* **16**, *32*
adspersus, *Francolinus* 33, **59**
Aegialtis 234
aegyptiaca, *Streptopelia s.* **495**
aegyptius, *Charadrius* 206
aegyptius, *Pluvianus* xii, **206**, *255*, *265*, *320*, *321*
Aenigmatolimnas **110**
  marginalis 93, 99, 107, 109, **110**, *112*
aequatorialis, *Gallinago n.* **302**
aeruginosus, *Circus* 130
aethiopica, *Threskiornis* 280, 344, 356, 382, 393, 415
afer, *Columba* 453
afer, *Francolinus* 27, *33*, **70**
afer, *Francolinus a.* *33*, **71**
afer, *Ptilostomus* 475
afer, *Tetrao* 70
afer, *Turtur* **453**, *465*
affinis, *Crex* 94
affinis, *Sarothrura* **94**, *113*
affinis, *Sarothrura a.* **94**, *113*
afra, *Eupodotis* 168, *176*
afra, *Eupodotis a.* **169**
afra, *Otis* 168
afraoides, *Eupodotis a.* **168**, *176*
africana, *Actophilornis* 121, *177*, **181**, 184, 280
africana, *Coturnix c.* 14
africana, *Parra* 181
africanus, *Bubo* 53
africanus, *Cursorius* 213, *320*, *321*
africanus, *Cursorius a.* **213**, *320*
africanus, *Francolinus* 25, **36**, *49*
africanus, *Phalacrocorax* 357, 365, 394
Afropavo **11**
  congensis **11**, *96*
agaze, *Neotis n.* 154
Agelastes **1**
  meleagrides **2**, 6, *96*

niger 3, 4, 6, 11, *96*
ahantensis, *Francolinus* 26, **45**, *48*
alba, *Calidris* 238, 244, **286**, *305*, *328*, *352*
alba, *Gygis* 374, 385
alba *Tringa* 286
alba, *Tyto* 22
albiceps, *Vanellus* 241, *255*, *321*
albicollis, *Porzana* 98
albidorsalis, *Sterna m.* 379, 401
albifrons, *Sterna* 377, **398**, 401, *417*
albifrons, *Sterna a.* **398**
albinucha, *Columba* 449, **468**
albinucha, *Columba a.* **468**
albionis, *Uria a.* **415**
albitorques, *Columba* 449, **476**
alboaxillaris, *Numenius* 311
albogularis, *Francolinus* 25, **31**, *32*
albogularis, *Francolinus a.* **31**, *32*
albus, *Corvus* 209, 264, 276, 428
Alca **418**
  torda 353, **418**
Alcae **415**
alchata, *Pterocles* **422**, *448*
alchata, *Tetrao* 448
Alcidae 415
Alectoris 15, **21**
  barbara 20, **21**, *32*, 54
  chukar 21, *32*, 67
  graeca 22
  rufa 22
alexandrinus, *Charadrius* xii, 225, **235**, *256*, *265*, *352*
alexandrinus, *Charadrius a.* **235**, *256*
Alle **419**
  alle 353, **419**
alle, *Alca* 419
alle, *Alle* 353, **419**
alle, *Alle a.* **419**
alleni, *Porphyrio* **116**, 122, *160*
alleni, *Turnix s.* 79
alpina, *Calidris* **294**, 296, *305*, *352*
alpina, *Calidris a.* **294**, *305*
alpina, *Tringa* 294
Amaurornis **114**
  flavirostris 105, **114**, *160*, 280
  phoenicurus 114
ambigua, *Streptopelia d.* **482**
Ammoperdix 15, **20**
  heyi **20**, *32*
anaethetus, *Sterna* 393, **400**, *417*
Anas undulata 105
angolensis, *Gallinago n.* **302**
angolensis, *Francolinus* 29
angulata, *Gallinula* **125**, *160*
Anhimidae 1
Anhinga melanogaster 147
Anous 386, **408**
  minutus **400**, **408**, *417*
  stolidus 395, **400**, **410**, *417*
  tenuirostris **400**, **409**
Anseriformes 1, 180

ansorgei, *Numida* 9
ansorgei, *Pterocles b.* **437**
ansorgei, *Treron c.* **444**
antarctica, *Catharacta s.* 338
Anthropoides **136**
  paradisea 97, **138**, 275
  virgo 97, 132, **137**
antonii, *Sarothrura a.* **94**
Aplopelia 458
apricaria, *Charadrius* 248
apricaria, *Pluvialis* **248**, *257*, *321*
aquaticus, *Rallus* **103**, *112*
aquaticus, *Rallus a.* **103**, *112*
Aquila rapax 131, 168
  wahlbergi 81, 486
arabica, *Streptopelia r.* **486**
arabs, *Ardeotis* xii, 152, 154, **158**, *161*
arabs, *Ardeotis a.* **159**
arabs, *Otis* 158
Aramidae 149
archeri, *Francolinus* 42
arctica, *Alca* 420
arctica, *Calidris a.* **295**
arctica, *Fratercula* 353, **420**
Ardea cinera 131, 378
  goliath 415
  melanocephala 100
Ardeola ralloides 182, 233, 280
ardeola, *Dromas* **188**, *240*, *321*
Ardeotis 151, **157**
  arabs xii, 152, 154, **158**, *161*
  kori xii, 158, **159**, *161*
arenaceus, *Charadrius m.* 237
Arenaria **327**
  interpres 257, 286, **327**, *352*
Arenariinae 327
arenicola, *Streptopelia t.* **491**
argentatus, *Larus* 343, 367, **369**, 416
argentatus, *Larus a.* **370**
Argus 11
armatus, *Charadrius* 262
armatus, *Vanellus* 207, 241, **262**, *321*
arquata, *Numenius* 240, **313**, *321*, *350*
arquata, *Numenius a.* **313**
arquata, *Scolopax* 313
arquatrix, *Columba* 449, **458**, *463*
asiatica, *Zenaida* xii
asiaticus, *Charadrius* **245**, *257*, *278*, *352*
Asio capensis 114
  otus 471
assimilis, *Puffinus* 339
atlanticus, *Anous m.* **400**, **408**
atlantis, *Larus a.* 370
atra, *Fulica* **126**, *127*, *160*, 370
atra, *Fulica a.* **127**, *160*
atricilla, *Larus* 349
atrifrons, *Francolinus c.* **64**
audouinii, *Larus* 361, **368**, 416
auguralis, *Buteo* 261

539

auritus, Nettapus 117
australis, Treron 442
avosetta, Recurvirostra 194, **196**, 236, *240, 321*
ayesha, Francolinus b. **54**
ayresii, Coturnicops 95
ayresii, Sarothrura **95**, *113*

B

bairdii, Actodromas 291
bairdii, Calidris **291**, **305**, *352*
balaenarum, Sterna **396**, *401, 417*
Balearica 141
  pavonina xii, 97, 137, **141**
  regulorum xii, 97, 135, 139, **141**, *143*
Balearicinae 140
bangsi, Sterna d. **386**
barbara, Alectoris 20, **21**, *32, 54*
barbara, Alectoris b. **21**, *32*
barbara, Perdix 21
barbata, Alectoris b. **22**
barbata, Guttera p. **5**
barrowii, Eupodotis s. 170, **174**, *176*
bassana, Sula 361
batesi, Canirallus o. **86**
batesi, Sarothrura p. **87**
bellicosus, Polemaetus 163
bengalensis, Sterna 346, 357, 375, **382**, *401, 417*
benghalensis, Rallus 186
benghalensis, Rostratula 177, **186**, *352*
benghalensis, Rostratula b. **177**, **186**
bergii, Sterna 344, 355, **381**, *401, 417*
bergii, Sterna b. **381**, *401*
biarmicus, Falco 264, 276, 428, 432, 433, 486
bicalcaratus, Francolinus 22, 26, *32*, **54**
bicalcaratus, Francolinus 22, 26, *32*, **54**
bicalcaratus, Tetrao 54
bicinctus, Pterocles **437**, *448*
bicinctus, Pterocles b. **437**, *448*
bilineatus, Pogoniulus 88
bisignatus, Cursorius a. **214**, *320*
blancoui, Numida 9
bodalyae, Numida 9
boehmi, Sarothrura 17, 83, **93**, *113*
bonapartei, Sarothrura r. **91**
bottegi, Francolinus 64
brehmeri, Chalcopelia 450
brehmeri, Turtur **450**, **462**, *465*
brehmeri, Turtur b. **450**, *465*
brehmi, Ptilopachus 23
brevicera, Treron c. **444**
brevirostris, Pterodroma 339
bronzina, Columba l. **462**
Bubo africanus 53
Bubulcus ibis 120, 144
buckleyi, Francolinus a. **31**
Bucorvus leadbeateri 70
Bugeranus 133
  carunculatus 97, **133**, 137, 139, 144
burchelli, Pterocles **440**, *448*
Burhinidae *198*
Burhinus 199

capensis 177, **204**, *321*
  oedicnemus 177, **199**, *321*
  senegalensis 177, **201**, *321, 360*
  vermiculatus 177, **203**, *321*
Buteo auguralis 261
butleri, Ardeotis a. **159**
butleri, Ptilopachus 23
buttikoferi, Burhinus v. **203**

C

cabanisi, Lanius 280
cachinnans, Larus a. **370**
caerulescens, Eupodotis **172**, *176*
caerulescens, Otis 172
caerulescens, Rallus **104**, **112**, 115
cailliautii, Campethera xiii
Calidridinae 283
Calidris 230, 242, **283**, 298, 330, 365
  alba 238, 244, **286**, *305, 328, 352*
  alpina **294**, 296, *305, 352*
  bairdii **291**, *305, 352*
  canutus **284**, *305, 352*
  ferruginea 238, 244, **292**, *305, 325, 352*
  fuscicollis 290
  maritima **294**, *305, 352*
  melanotos **292**, *305, 352*
  minuta 265, **288**, 296, *305, 352*
  ruficollis **287**, *305*
  subminuta **290**, *305*
  temminckii **289**, *305, 352*
  tenuirostris 283
callewaerti, Numida 9
Calonectris diomedea 339, 361
calva, Columba 443
calva, Treron **443**, *465*
calva, Treron c. **444**, *465*
camerunensis, Francolinus 27, 48, **61**
camerunensis, Podica s. **146**
campbelli, Francolinus 29
Campethera cailliautii xiii
  permista xiii
canariensis, Columba l. **477**
canicollis, Eupodotis s. **174**, *176*
canidorsalis, Francolinus 39
Canirallus 86
  oculeus **86**, *113*
canorus, Melierax 168
canus, Larus **363**, 368, 373, *416*
canus, Larus c. **363**
canutus, Calidris **284**, *305, 352*
canutus, Calidris c. **284**
canutus, Tringa 284
capensis, Asio 114
capensis, Burhinus 117, **204**, *321*
capensis, Burhinus c. **177**, **204**
capensis, Columba 456
capensis, Corvus 173
capensis, Francolinus 33, **58**
capensis, Microparra 177, **184**
capensis, Oedicnemus 204
capensis, Oena **456**, *465*
capensis, Parra 184
capensis, Phalocrocorax 365, 397
capensis, Sula 365
capensis, Tetrao 58
capicola, Columba v. 484
capicola, Streptopelia xii, 258, **464**, *479*, **484**
capicola, Streptopelia c. **464**, **484**

carbo, Phalacrocorax 365
carunculata, Ardea 133
carunculatus, Bugeranus 97, **133**, 137, 139, 144
caspia, Sterna **377**, *401, 417*
castaneicollis, Francolinus 27, 48, **64**
castaneicollis, Francolinus c. 48, **64**
castaneiventer, Francolinus a. 33, **70**
Catharacta 338
  chilensis 338
  maccormicki **339** *353*
  skua 333, **338**, *353*
caudacutus, Pterocles a. **423**, *448*
cayanus, Hoploxypterus 250
ceciliae, Balearica p. **142**
centralis, Sarothrura p. **87**
Centropus superciliosus 280
Ceryle rudis 110
Chalcopelia 450, 454
chalcopterus, Cursorius **217**, *320, 321*
chalcospilos, Columba 455
chalcospilos, Turtur 450, **455**, *465*
chapini, Francolinus 62
chapini, Guttera 5
Charadrii 180
Charadiidae 224
Charadriiformes 76, **180**
Charadriinae 225
Charadrioidea **190**
Charadrius 225
  alexandrinus xii, 225, **235**, 256, *265, 352*
  asiaticus **245**, 257, 278, *352*
  dubius **226**, *256, 352*
  forbesi 225, **234**, *256, 352*
  hiaticula **228**, *256, 265, 352*
  leschenaultii **243**, *256, 286, 293, 321, 325*
  marginatus 225, **237**, *256, 352, 397*
  mongolus **242**, *256, 286, 293, 321*
  morinellus **246**, *257, 352*
  pallidus **239**, *256, 352*
  pecuarius **229**, *256, 265, 352*
  tricollaris 225, **231**, *256, 258, 352*
Chettusia 250, 279
chilensis, Catharacta 338
chinensis, Coturnix **16**, *32, 83, 93*
chinensis, Coturnix c. **16**
Chlamydotis 156
  undulata 150, 154, **156**, *161, 163*
Chlidonias 399, **403**
  hybridus 392, 400, **403**, *417*
  leucopterus 400, **406**, *417*
  niger 400, **405**, *417*
chloropus, Fullica 122
chloropus, Gallinula xv, 105, 117, **122**, *147, 160, 182*
chloropus, Gallinula c. **122**
chobiensis, Francolinus 68
cholmleyi, Ammoperdix h. **20**
chukar, Alectoris 21, *32, 67*
Ciconia abdimii 475
  episcopus 134
Cinclus 102
cinctus, Cursorius **216**, *320, 321*
cinctus, Cursorius c. **216**, *320*
cinctus, Hemerodromus 216
cinerea, Ardea 131, 378
cinerea, Creatophora 280

INDEXES: SCIENTIFIC NAMES 541

cinerea, Glareola xii, **223**, 258, *320*
cinerea, Scolopax 324
cinereus, Xenus **304**, **324**, *352*
Circus 280
    aeruginosus 130
    ranivorus 105, 106, 116
cirrocephalus, Larus 198, 343, **356**, *369*, 380, 404, 415, *416*
Cladorhynchus leucocephalus 193
clappertoni, Francolinus 26, **32**, **56**
clappertoni, Francolinus c. **32**
colchicus, Phasianus 13
Columba **458**
    albinucha 449, **468**
    albitorques 449, **476**
    arquatrix 449, 458, **463**
    delegorguei 449, 458, **460**
    guinea 449, **473**, 492
    iriditorques 449, 458, **459**
    larvata 452, **461**, 465
    livia 449, **477**
    malherbii 449, 458, **460**
    oenas 449, **471**
    oliviae 449, **473**
    palumbus 449, **470**, 492
    risoria 487
    sjostedti 449, 458, **467**
    thomensis 449, 458, **467**
    unicincta 449, **469**
columbarius, Falco 478
Columbidae **442**
Columbiformes 180, **442**
Columbinae **447**
columbinus, Charadrius 1. **244**, 256
Colymbus 415
congensis, Afropavo **11**, *96*
coprotheres, Gyps 365
coqui, Francolinus 25, **29**, *33*
coqui, Francolinus c. **29**, *33*
coqui, Perdix 29
Coracias naevia 174
corax, Corvus 347, 411
coronata, Numida m. **9**
coronatus, Charadrius 272
coronatus, Pterocles **432**, **448**
coronatus, Pterocles c. **432**, **448**
coronatus, Vanellus 217, **241**, **272**, *321*
coronatus, Vanellus c. **241**, **272**
corone, Corvus 471
Corvus albus 209, 264, 276, 428
    capensis 173
    corax 347, 411
    corone 471
    rhipidurus 415
    ruficollis 388
Coturnicops 87, **95**
Coturnix **13**, **77**
    chinensis **16**, *32*, 83, *93*
    coturnix **14**, 20, *32*, *79*, *83*
    delegorguei **18**, *32*, *79*, *83*
coturnix, Coturnix **14**, 20, *32*, *79*, *83*
coturnix, Coturnix c. **14**, *32*
coturnix, Tetrao 14
Cracidae 1, **11**
cranchii, Francolinus a. *33*, **71**
crassirostris, Charadrius l. **244**
crassirostris, Chettusia 279
crassirostris, Vanellus **241**, *279*, *321*
crassirostris, Vanellus c. **241**, **279**
crawshayi, Francolinus 37
Creatophora cinerea 280

Crecopsis 98
Crex **98**
    crex 100, 111, *112*
    egregia **98**, 109, 111, *112*
crex, Rallus 100
cristata, Fulica xii, 126, **128**, *160*, 182, 404
crumeniferus, Leptoptilos 267, 486
cunenesis, Francolinus 42
curonicus, Charadrius d. **226**, 256
cursor, Charadrius 210
cursor, Cursorius **210**, *320*, *321*
cursor, Cursorius c. **210**
Cursoriinae **206**
Cursorius 77, **209**
    africanus **213**, *320*, *321*
    chalcopterus **217**, *320*, *321*
    cinctus **216**, *320*, *321*
    cursor **210**, *320*, *321*
    rufus **211**, *320*
    temminckii **212**, *320*, *321*

D

dactylatra, Sula 382
dakhlae, Columba l. **477**
damarensis, Burhinus c. **204**
damarensis, Francolinus 68
damarensis, Numida m. **9**
damarensis, Streptopelia c. **485**
decaocto, Streptopelia 479, **487**
decaocto, Streptopelia d. **488**
decaocto, Streptopelia r. 487
decipiens, Streptopelia 464, **481**
decipiens, Streptopelia d. **482**
decipiens, Turtur 481
decoratus, Pterocles **433**, **448**
decoratus, Pterocles d. **433**, **448**
delalandii, Chlidonias h. **400**, *404*
delalandii, Treron c. **444**, *465*
delawarensis, Larus 362
delegorguei, Columba 449, 458, **460**
delegorguei, Columba d. 449, **461**
delegorguei, Coturnix **18**, *32*, *79*, *83*
delegorguei, Coturnix d. **18**, *32*
demersus, Spheniscus 365, 387, 391
demissus, Vanellus c. **273**
denhami, Neotis **151**, 159, *161*, 162
denhami, Neotis d. **151**, *161*
denhami, Otis 151
dewittei, Francolinus a. **31**, *32*
diomedea, Calonectris 339, 361
dodsoni, Burhinus c. **204**
domestica, Columba l. **477**
dominica, Pluvialis **247**, *257*, *321*
dominica, Pluvialis d. **247**
dominicanus, Larus 343, **364**, *369*, *379*, *382*, 388, *397*, *416*
dominicanus, Larus d. **364**
dominicus, Charadrius 247
dougallii, Sterna **386**, *401*, 411, *417*
dougallii, Sterna d. **386**
Dromadidae **188**
Dromadoidea **188**
Dromas **188**
    ardeola **188** *240*, *321*
dubius, Charadrius **226**, 256, *352*
duvaucelii, Vanellus 250

E

ecaudatus, Terathopius 258

edouardi, Guttera 5
edouardi, Guttera p. **5**, *96*
egregia, Crex **98**, 109, 111, *112*
egregia, Ortygometra 98
Egretta garzetta 360
    intermedia 120, 393
electa, Streptopelia c. **485**
elegans, Gallinula 88
elegans, Sarothrura **88**, *113*
elegans, Sarothrura e. **88**
elegans, Streptopelia d. **482**
elgonensis, Francolinus p. **41**, 48
elizabethae, Sarothrura r. **91**
ellenbecki, Francolinus p. 40
ellenbecki, Pterocles **434**
ellioti, Pterocles e. **425**
emini, Cursorius c. **216**
emini, Ptilopachus 23
enigma, Sterna b. **381**
episcopus, Ciconia 134
erckelii, Francolinus 27, 48, **66**
erckelii, Perdix 66
erlangeri, Coturnix c. **14**, *32*
erlangeri, Eupodotis s. **174**
erythropus, Scolopax 314
erythropus, Tringa **304**, **314**, *352*
Esacus 199
etoschae, Eupodotis a . **169**
Eupodotis **165**
    afra **168**, *176*
    caerulescens **172**, *176*
    hartlaubii 166, *176*, **179**
    humilis **171**, *176*
    melanogaster 166, **175**, *176*
    rueppellii 166, **170**, *176*
    ruficrista **166**, *176*
    senegalensis 154, 163, **173**, *176*
    vigorsii 166, **169**, *176*
Excalfactoria 13
excelsa, Columba p. 449, **470**
exustus, Pterocles **424**, **448**
exustus, Pterocles e. **425**, **448**

F

falcinellus, Limicola 295, **296**, 305, *352*
falcinellus, Scolopax 296
Falco biarmicus 264, 276, 428, 432, 433, 486
    columbarius 478
    peregrinus 432, 466, 471, 478
    rupicoloides 428
ferruginea, Calidris 238, 244, **292**, 305, 325, *352*
ferruginea, Tringa 292
Ficedula hypoleuca 362
finschi, Francolinus 25, **38**, *49*
fitzsimonsi, Eupodotis r. **170**
flavipes, Scolopax 322
flavipes, Tringa **322**, *352*
flavirostra, Gallinula 114
flavirostris, Amaurornis 105, **114**, *160*, 280
flavirostris, Limnocorax 114
flavirostris, Rynchops 258, **400**, **412**, *417*
florentiae, Ptilopachus 23
floweri, Pterocles e. **425**
forbesi, Aegialtis 234
forbesi, Charadrius 225, **234**, 256, *352*

Francolinus 11, 15, 20, **24**
  adspersus 33, **59**
  afer 27, 33, **70**
  africanus 25, **36**, *49*
  ahantensis 26, **45**, *48*
  albogularis 25, **31**, *32*
  bicalcaratus 22, 26, **32**, *54*
    camerunensis 27, *48*, **61**
  capensis 33, **58**
  castaneicollis 27, *48*, **64**
  clappertoni 26, **32**, *56*
  coqui 25, **29**, *33*
  erckelii 27, *48*, **66**
  finschi 25, **38**, *49*
  griseostriatus 26, **46**, *48*
  hartlaubi **49**, *50*
  harwoodi 26, **49**, *57*
  hildebrandti 26, **49**, *51*
  icterorhynchus 26, **32**, *55*
  jacksoni 27, *48*, **63**
  lathami **28**, *48*
  leucoscepus 27, 33, **73**
  levaillantii 25, **37**, *49*
  levaillantoides 25, 33, **41**
  nahani **47**, *48*
  natalensis 26, 33, **52**
  nobilis 27, *48*, **62**
  ochropectus xii, 27, *48*, **65**
  pondicerianus 67
  psilolaemus xii, 25, **40**, *48*
  rufopictus 27, 33, **72**
  schlegelii 25, **34**, *49*
  sephaena 23, 33, **42**
  shelleyi 25, 33, **39**
  squamatus 26, **44**, *48*, 85
  streptophorus 25, **35**, *49*
  swainsonii 27, 33, **68**
  swierstrai 27, *48*, **60**
Fratercula **420**
  arctica 353, **420**
frommi, Numida 9
fuertaventurae, Chlamydotis 157
Fulica 122, **126**
  atra 126, **127**, *160*, 370
  cristata xii, 126, **128**, *160*, 182, 404
fulicaria, Tringa 331
fulicarius, Phalaropus 305, **331**, *352*, 354
fulva, Pluvialis d. **247**, *257*
fuscata, Sterna 334, **395**, *400*, 408, 410, *417*
fuscata, Sterna f. 395
fuscicollis, Calidris **290**
fuscicollis, Tringa 290
fuscus, Larus **366**, *369*, 415, *416*
fuscus, Larus f. **366**, *369*

G

gaddi, Columba l. **477**
galeata, Numida m. **9**, *96*
Galliformes 1, 76
Gallinae 13
Gallinagininae 299
Gallinago **300**
  gallinago *177*, **300**, 324
  media *177*, **303**
  nigripennis *177*, **302**, *352*
  stenura *177*, **306**
gallinago, Gallinago *177*, **300**, 324

gallinago, Gallinago g. *177*, **300**
gallinago, Scolopax 300
Gallini 13
Gallinula **121**
  angulata **125**, *160*
  chloropus xv, 105, 117, **122**, *147*, *160*, 182
Gallus 2, 13
  gallus 19, 60, 69, 72
gallus, Gallus 19, 60, 69, 72
gambensis, Plectopterus 135, 279
garzetta, Egretta 360
Gaviiformes 180
gedgii, Francolinus c. 32
Gelochelidon **374**
  nilotica 360, **375**, *378*, 384, 396, *400*, *417*
genei, Larus **359**, *368*, *416*
gentilis, Accipiter 471, 472
georgiae, Sterna v. **391**
gibbericeps, Balearica r. *142*, **143**
gibberifrons, Treron c. **444**
gilli, Francolinus 68
gindiana, Eupodotis r. **167**
Glareola **218**
  cinerea xii, **223**, *258*, *320*
  lactea xii
  nordmanni **220**, *320*
  nuchalis **222**, *320*
  ocularis **221**, *320*
  pratincola **218**, *320*
glareola, Tringa 110, 233, *304*, **323**, *329*, *352*
Glareolidae **206**
Glareolinae **218**
glaucoides, Larus 371
gofanus, Francolinus 64
goliath, Ardea 415
grabae, Fratercula a. **420**
gracilis, Cursorius a. **214**
graeca, Alectoris 22
graellsii, Larus f. **366**, *369*
granti, Cursorius a. **214**
granti, Francolinus s. **43**
granti, Treron c. **444**
gravis, Puffinus 339
gregarius, Charadrius 276
gregarius, Vanellus 257, **276**, *321*
griseostriatus, Francolinus 26, **46**, *48*
griseus, Limnodromus 308
Grues **76**
Gruidae **131**
Gruiformes **76**, 180, 185
Gruinae **131**
Grus **131**
  grus 97, **132**, 137
  leucogeranus 133
grus, Ardea 132
grus, Grus 97, **132**, 137
grus, Grus g. 97, **132**
guinea, Columba 449, **473**, 492
guinea, Columba g. 449, **474**
guinea, Sterna a. **399**
Guttera 2, 3, 50
  plumifera 3, **4**, *96*
  pucherani 2, **5**, *7*, *96*
gutturalis, Francolinus l. **42**
gutturalis, Pterocles **438**, *448*
gutturalis, Pterocles g. **438**, *448*
Gygis alba 374, 385
gymnocyclus, Columba l. 449, **477**

H

Haematopodidae **190**
Haematopus **190**, 255
  moquini **191**, 239, *240*, *321*, 365
  ostralegus **190**, *240*, *321*
haematopus, Himantornis **84**, *86*, *113*
Haliaeetus vocifer 131, 258
haliaetus, Pandion 365
hamiltoni, Catharacta s. 338
harterti, Francolinus a. **71**
hartingi, Cursorius a. **214**
hartlaubi, Francolinus **49**, *50*
hartlaubii, Eupodotis 166, *176*, **179**
hartlaubii, Larus 335, 339, 343, **355**, *369*, *382*, *416*
hartlaubii, Larus n. 355
hartlaubii, Otis 179
harwoodi, Francolinus 26, **49**, *57*
Heliopais 146
Heliornis 145
Heliornithidae **145**
Hemerodromus 216
Hemiparra 250
hemprichii, Larus **344**, *369*, *382*, 394, 411, *416*
heuglini, Larus a. *369*, **370**
heuglinii, Neotis **155**, *161*
heuglinii, Otis 155
heyi, Ammoperdix **20**, *32*
heyi, Perdix 20
hiaticula, Charadrius **228**, *256*, 265, *352*
hiaticula, Charadrius h. **228**
hildebrandti, Francolinus 26, **49**, *51*
hildebrandti, Scleroptera 51
Himantopus **193**
  himantopus **193**, *197*, *240*, 265, *278*, *321*
himantopus, Charadrius 193
himantopus, Himantopus **193**, *197*, *240*, 265, *278*, *321*
himantopus, Himantopus h. **194**, *240*
Himantornis **84**
  haematopus **84**, *86*, *113*
Himantornithinae **84**
Hirundo 218
hirundo, Sterna 335, 337, 365, **388**, *401*, 406, *417*
hirundo, Sterna h. **388**, *401*
histrionica, Coturnix d. **18**
hoeschianus, Francolinus 29
hoggara, Streptopelia t. **491**
hopkinsoni, Francolinus 46
Hoploxypterus cayanus 250
hottentotta, Turnix **82**, *112*
hottentotta, Turnix h. **82**, *112*
hottentottus, Turnix 82
Houbaropsis 165
hubbardi, Francolinus c. **30**
hudsonicus, Numenius p. **312**
humilis, Eupodotis **171**, *176*
humilis, Sypheotides 171
hybrida, Sterna 403
hybrida, Vanellus 279
hybridus, Chlidonias 392, *400*, **403**, *417*
hybridus, Chlidonias h. **404**
Hydrophasianus 181
Hydroprogne 376

hyperboreus, Larus 353, **371**, *416*
hyperboreus, Larus h. **371**
hypoleuca, Ficedula 362
hypoleucos, Actitis 187, 289, *304*, 323, 325, **326**, *352*
hypoleucos, Tringa 326
hypopyrrha, Streptopelia 464, 479, **492**
hypopyrrhus, Turtur 492

I

ibis, Bubulcus 120, 144
ibis, Mycteria 393
ichthyaetus, Larus 347, **353**, *416*
icterorhynchus, Francolinus 26, 32, **55**
incerta, Pterodroma 339
indicus, Pterocles 422
inermis, Numida 8
infelix, Turtur b. **451**
inornata, Columba l. **462**, 465
inornatus, Burhinus s. 102
insolata, Turnix h. 82
intercedens, Francolinus 71
intermedia, Egretta 120, 393
intermedia, Numida 8
intermedia, Porzana p. **108**, *112*
interpres, Arenaria 257, 286, **327**, *352*
interpres, Arenaria i. 257, **327**
interpres, Tringa 327
iriditorques, Columba 449, 458, **459**
isabella, Stiltia 206, 218
isabellina, Streptopelia t. 464, **491**
islandica, Alca t. **418**

J

Jacana 180
Jacanidae 180
Jacanoidea 180
jacksoni, Columba l. **462**
jacksoni, Francolinus 27, *48*, **63**
jacksoni, Neotis d. **151**, *161*
johnstoni, Francolinus 52, 53
jugularis, Francolinus l. **42**

K

kaffanus, Francolinus 48, 64
kalaharica, Eupodotis a. **169**
kalaharica, Francolinus 41
kasaicus, Francolinus 29
kathleenae, Guttera 5
kikuyuensis, Francolinus l. **37**, *49*
koenigi, Alectoris b. **22**
konigseggi, Francolinus 57
kori, Ardeotis xii, 158, **159**, *161*
kori, Ardeotis k. **162**
kori, Otis 159
korustes, Sterna d. **401**

L

lactea, Glareola xii
Lanius cabanisi 280
 senator 362
lapponica, Limosa 240, 307, **310**, 312, *321*
lapponica, Limosa l. **310**

lapponica, Scolopax 310
Lari 332
Laridae 340
Larus **343**
 argentatus 343, **367**, 369, *416*
 atricilla **349**
 audouinii **361**, *368*, *416*
 canus **363**, *368*, 373, *416*
 cirrocephalus 198, 343, **356**, *369*, 380, 404, 415, *416*
 delawarensis **362**
 dominicanus 343, **364**, *369*, 379, 382, 388, 397, *416*
 fuscus **366**, *369*, 415, *416*
 genei **359**, *368*, *416*
 glaucoides 371
 hartlaubii 335, 339, 343, **355**, *369*, 382, *416*
 hemprichii **344**, *369*, 382, 388, 394, 411, *416*
 hyperboreus 353, **371**, *416*
 ichthyaetus 347, **353**, *416*
 leucophthalmus **345**, *369*
 marinus 353, **372**, *416*
 melanocephalus **348**, *368*, *416*
 minutus **351**, *368*, *416*
 novaehollandiae 355
 pipixcan **350**, *368*, *416*
 ridibundus **358**, *368*, *416*
 sabini 334, 335, 337, **354**, *368*, 373, *416*
larvata, Columba 452, **461**, *465*
larvata, Columba l. **462**, 465
lateralis, Vanellus s. **252**
lathami, Francolinus **28**, *48*
lathami, Francolinus l. **28**
latifrons, Vanellus t. **260**
leadbeateri, Bucorvus 70
lehmanni, Francolinus 70
Leptoptilos crumeniferus 267, 486
lepurana, Turnix s. **79**, *112*
leschenaultii, Charadrius **243**, *256*, 286, 293, *321*, 325
Lestris 332
leucocephalus, Cladorhynchus 193
leucogeranus, Grus 133
leucoparaeus, Francolinus a. **71**
leucophthalmus, Larus **345**, *369*
leucoptera, Sterna 406
leucoptera, Vanellus c. **279**
leucopterus, Chlidonias 400, **406**, *417*
leucoscepus, Francolinus 27, *33*, **73**
leucurus, Charadrius 277
leucurus, Vanellus 257, **277**, *321*
levaillantii, Francolinus 25, **37**, *49*
levaillantii, Francolinus l. **37**, *49*
levaillantii, Perdix 37
levaillantoides, Francolinus 25, *33*, **41**
levaillantoides, Francolinus l. *33*, **41**
levaillantoides, Perdix 41
liberiae, Glareola n. **222**
lichtensteinii, Pterocles **434**, *448*
lichtensteinii, Pterocles l. **435**
Limicola **296**
 falcinellus 295, **296**, *305*, *352*
Limnocorax flavirostris 114
Limnodromus **307**
 griseus 308
 scolopaceus 308

 semipalmatus **307**
Limosa **309**
 lapponica 240, 307, **310**, 312, *321*
 limosa 240, **309**, *321*
limosa, Limosa 240, **309**, *321*
limosa, Limosa l. 240, **309**
limosa, Scolopax 309
limpopoensis, Numida 9
littoralis, Cursorius c. **210**
livia, Columba 449, **477**
livia, Columba d. 477
livia, Columba l. 477
lividicollis, Guttera 5
loangwae, Francolinus 71
lobata, Tringa 329
lobatus, Phalaropus 305, **329**, *352*
Lobivanellus 259
logonensis, Streptopelia d. **482**
longicaudus, Stercorarius **336**, *353*
longipes, Haematopus o. **191**
lonnbergi, Catharacta s. 338
lorti, Francolinus l. *33*, **42**
loveridgei, Pterocles d. **434**
luciana, Turnix h. 82
ludwigii, Neotis **153**, *161*
ludwigii, Otis 153
lugens, Columba 494
lugens, Crex 92
lugens, Sarothrura **92**, *113*
lugens, Sarothrura l. **92**, *113*
lugens, Streptopelia 464, 479, **494**
lugubris, Charadrius 268
lugubris, Vanellus 241, 251, **268**, *321*
lundazi, Francolinus s. **68**
Lymnocryptes **299**
 minimus *177*, **299**, 301
lynesi, Ardeotis a. **159**
lynesi, Sarothrura l. **92**, *113*

M

maccormicki, Catharacta **339**, *353*
maccormicki, Stercorarius 339
mackenziei, Eupodotis s. **174**
macqueenii, Chlamydotis u. **156**
macroceras, Numida 8
Macrorhamphus 307
maculosus, Burhinus c. **204**
madagascariensis, Catharacta s. 338
madagascariensis, Porphyrio p. **119**, *160*
maharao, Francolinus c. **30**, *33*
major, Numida 8
major, Ptilopachus p. **23**
major, Vanellus 252
malherbii, Columba 449, 458, **460**
Mallophaga 149
marchei, Numida 9
marginalis, Aenigmatolimnas 93, 99, 107, 109, **110**, *112*
marginalis, Porzana 110
marginatus, Charadrius 225, **237**, 256, *352*, 397
marginatus, Charadrius m. **237**, 256
marinus, Larus 353, **372**, *416*
maritima, Calidris **294**, *305*, *352*
maritima, Tringa 294
martinica, Fulica 118
martinica, Porphyrio xii, **118**, *122*, *160*

marungensis, Numida m. **9**
mathewsi, Sterna s. 385
maxima, Numida 9
maxima, Sterna 379, **401**, *417*
mechowi, Charadrius m. 237
media, Gallinago 177, **303**
media, Scolopax 303
Megapodioidea 1
meiffrenii, Ortyxelos 77, *112*
meinertzhageni, Turnix 77
meinertzhageni, Francolinus 31
melanocephala, Ardea 100
melanocephalus, Larus 348, 368, *416*
melanocephalus, Lobivanellus 259
melanocephalus, Vanellus 241, **259**, *321*
melanogaster, Anhinga 147
melanogaster, Eupodotis 166, **175**, *176*
melanogaster, Eupodotis m. **175**, *176*
melanogaster, Francolinus a. 33, **71**
melanogaster, Otis 175
melanoleuca, Scolopax 319
melanoleuca, Tringa **319**
melanoleucos, Accipiter 466
melanopterus, Charadrius 270
melanopterus, Vanellus 241, 251, **270**
melanopterus, Vanellus m. **271**
melanotos, Calidris **292**, *305*, *352*
melanotos, Tringa 292
meleagrides, Agelastes 2, **6**, 96
Meleagrididae 1, 13
Meleagris 12
meleagris, Numida 6, 7, **8**, 96
meleagris, Numida m. **8**, 96
meleagris, Phasianus 8
Melierax canorus 168
meridionalis, Gallinula c. **122**, *160*
Merops orientalis 182
mexicanus, Himantopus h. 194
michahellis, Larus a. 367, **369**
Microparra **184**
  capensis 177, **184**
migrans, Milvus 121, 209
Milvus migrans 121, 209
  milvus 471, 472
minima, Scolopax 299
minimus, Lymnocryptes 177, **299**, 301
minor, Vanellus m. **241**, *270*
minuta, Calidris 265, **288**, *296*, *305*, *352*
minuta, Tringa 288
minutus, Anous 400, **408**, *417*
minutus, Larus 351, **368**, *416*
Mirafra 77
mitrata, Numida m. **9**, 96
mollis, Pterodroma 339
mongolus, Charadrius **242**, *256*, *286*, *293*, *321*
moquini, Haematopus **191**, *239*, *240*, *321*, *365*
morinellus, Charadrius **246**, *257*, *352*
Motacilla 265
multicolor, Pterocles b. **437**
Mycteria ibis 393

N

naevia, Coracias 174
nahani, Francolinus **47**, *48*

namaqua, Eupodotis v. **170**
namaqua, Pterocles **426**, *448*
namaqua, Tetrao 426
nana, Turnix h. 82, *112*
natalensis, Francolinus 26, **33**, **52**
neavei, Francolinus 53
nebularia, Tringa 304, **318**, *352*
nebularis, Scolopax 318
neglectus, Phalacrocorax 365
Neotis **151**
  denhami **151**, *159*, *161*, *162*
  heuglinii **155**, *161*
  ludwigii **153**, *161*
  nuba **154**, *161*
Nettapus auritus 117
neumanni, Numida 8
nicolli, Ammoperdix h. **20**, *32*
niger, Agelastes 3, 4, 6, **11**, *96*
niger, Chlidonias **400**, **405**, *417*
niger, Chlidonias n. **400**, **405**
niger, Phasidus 3
niger, Sterna 405
nigripennis, Gallinago 177, **302**, *352*
nigripennis, Gallinago n. **302**
nigrosquamatus, Francolinus 57
nilotica, Gelochelidon 360, **375**, *378*, *384*, *396*, **400**, *417*
nilotica, Gelochelidon n. **375**, *400*
nilotica, Sterna 375
nisus, Accipiter 471
nobilis, Francolinus 27, **48**, **62**
nordmanni, Glareola **220**, *320*
notatus, Francolinus 70
notophila, Eupodotis m. **175**
Notornis 116
novaehollandiae, Larus 355
nuba, Neotis **154**, *161*
nuba, Neotis n. **154**, *161*
nuba, Otis 154
nuchalis, Glareola **222**, *320*
nuchalis, Glareola n. **222**
nudirostris, Treron c. **443**
Numenius **311**
  arquata 240, **313**, *321*, *350*
  phaeopus 240, **311**, *321*
  tenuirostris 240, **312**, *321*
Numida **8**
  meleagris 6, 7, **8**, 96
Numidinae 1
Nycticryphes 185
Nyctiperdix 422

O

ochropectus, Francolinus xii, 27, **48**, **65**
ochropus, Tringa 304, **322**, *352*
ocularis, Glareola **221**, *320*
oculea, Gallinula 86
oculeus, Canirallus **86**, *113*
Odontophorini 13
Oedicnemus 201, 203, 204
oedicnemus, Burhinus 177, **199**, *321*
oedicnemus, Burhinus o. **199**
oedicnemus, Charadrius 199
Oena **456**
  capensis **456**, *465*
oenas, Columba 449, **471**
ogilvie-granti, Francolinus 54
ogoensis, Francolinus 64
olivascens, Pterocles e. **425**

oliviae, Columba 449, **473**
omoensis, Numida 8
onguati, Streptopelia c. **485**
orientalis, Merops 182
orientalis, Numenius a. **240**, **313**
orientalis, Pterocles **429**, *448*
orientalis, Pterocles o. **429**, *448*
orientalis, Tetrax t. **150**
Ortyxelos 76
  meiffrenii **77**, *112*
ostralegus, Haematopus **190**, *240*, *321*
ostralegus, Haematopus o. **190**, *240*
Otides **148**
Otididae **148**
Otis **164**
  tarda *161*, *163*, **164**
otus, Asio 471

P

Pachyptila vittata 339
pallasi, Guttera 5
pallidior, Francolinus **33**, *41*
pallidus, Charadrius **239**, *256*, *352*
pallidus, Charadrius p. **239**, *256*
palumbus, Columba 449, **470**, *492*
palumbus, Columba, p. **470**
pamirensis, Charadrius m. **234**, *256*
Pandion haliaetus 365
papillosa, Numida 9
paradisaea, Sterna 335, 337, **390**, *401*, *417*
paradisea, Anthropoides 97, **138**, *275*
paradisea, Ardea 138
parasiticus, Larus 334
parasiticus, Stercorarius **334**, *353*, *354*, *355*, *365*, *389*
Parra 181, 184, 251
parva, Porzana 103, **106**, *111*, *112*
parvus, Rallus 106
Pavo 11
pavonina, Ardea 141
pavonina, Balearica xii, 97, **137**, **141**
pavonina, Balearica p. **97**, **142**
Pavoninae 11, 13
pecuarius, Charadrius **229**, *256*, *265*, *352*
pecuarius, Charadrius p. **229**, *256*
pelegrinoides, Falco p. **432**
pembaensis, Treron 443, **446**, *465*
Penelope 11
pentoni, Francolinus e. 67
Perdicini 13
Perdix 20, 21, 29, 37, 41, 42, 66, 68
peregrinus, Falco **432**, *466*, *471*, *478*
permista, Campethera xiii
perspicillata, Streptopelia d. 464, **482**
petersii, Podica s. **147**
petiti, Himantornis 85
petrosus, Ptilopachus **23**, *32*
petrosus, Ptilopachus p. **23**, *32*
petrosus, Tetrao 23
phaeonotus, Columba g. **474**
phaeopus, Numenius 240, **311**, *321*
phaeopus, Numenius p. 240, **311**
phaeopus, Scolopax 311
Phalacrocorax africanus 357, 365, 394

capensis 365, 397
carbo 365
neglectus 365
Phalaropodinae **328**
Phalaropus **328**
  fulicarius 305, **331**, *352*, 354
  lobatus 305, **329**, *352*
  tricolor **328**
Phasianidae **1**
Phasianinae 1, 13
Phasianoidea 1
Phasianus 11, 24
  colchicus 13
Phasidus 3
Philomachus **297**
  pugnax 257, 293, **297**, 316, 318, 324, *352*
phoenicophila, Streptopelia s. **495**
phoenicoptera, Treron 442
Phoenicopteridae 180
phoenicurus, Amaurornis 114
Pica pica 489
pica, Pica 489
pipixcan, Larus 350, 368, *416*
Plectropterus gambensis 135, 279
plumbeigularis, Anous s. **411**
plumbescens, Columba 462
plumifera, Guttera 3, **4**, 96
plumifera, Guttera p. **4**, 96
plumifera, Numida 4
Pluvialis 245, 246, **247**
  apricaria **248**, 257, *321*
  dominica **247**, 257, *321*
  squatarola 244, **249**, 257, 312, *321*
Pluvianidae 206
Pluvianus **206**
  aegyptius xii, **206**, 255, 265, *320*, *321*
Podica **145**
  senegalensis **146**, *160*
Pogoniulus bilineatus 88
Poicephalus 444
poicephalus, Larus c. **356**, *369*
Polemaetus bellicosus 163
pollenorum, Francolinus 163
Polyboroides typus 223
Polyplectron 11
pomarinus, Lestris 332
pomarinus, Stercorarius **332**, *353*
pondicerianus, Francolinus 67
Porphyrio **116**
  alleni **116**, 122, *160*
  martinica xii, **118**, 122, *160*
  porphyrio xii, 116, **119**, *160*
porphyrio, Fulica 119
porphyrio, Porphyrio xii, 116, **119**, *160*
porphyrio, Porphyrio p. **120**, *160*
Porphyriops 121
Porphyriornis 121
portlandica, Sterna 381
Porzana **106**
  albicollis 98
  parva 103, **106**, 111, *112*
  porzana 99, 103, **109**, 111, *112*, 125
  pusilla 103, **107**, 111, *112*
  porzana, Porzana 99, 103, **109**, 111, *112*, 125
porzana, Rallus 109

pratincola, Glareola **218**, *320*
pratincola, Hirundo 218
principalis, Columba l. **462**
psilolaemus, Francolinus xii, 25, **40**, 48
psilolaemus, Francolinus p. **40**, 48
Psittaciformes 442
Psophia 84, 156
Psophiidae 84, 149
Pternistis 25, 72
Pterocles **422**
  alchata **422**, *448*
  bicinctus **437**, *448*
  burchelli **440**, *448*
  coronatus **432**, *448*
  decoratus **433**, *448*
  exustus **424**, *448*
  gutturalis **438**, *448*
  indicus 422
  lichtensteinii **434**, *448*
  namaqua **426**, *448*
  orientalis **429**, *448*
  quadricinctus **435**, *448*
  senegallus **430**, *448*
Pteroclidae 209, **422**
Pterocliformes 180, **422**
Pterodroma brevirostris 339
  incerta 339
  mollis 339
Ptilinopus 442
Ptilopachus **23**
  petrosus **23**, *32*
Ptilostomus afer 475
pucherani, Guttera 2, **5**, 7, 96
pucherani, Guttera p. **5**, 96
pucherani, Numida 5
Puffinus assimilis 339
  gravis 339
  puffinus 348
puffinus, Puffinus 348
pugnax, Philomachus 257, 293, **297**, 316, 318, 324, *352*
pugnax, Tringa 297
pulchra, Crex 87
pulchra, Sarothrura **87**, *113*
pulchra, Sarothrura p. **87**
pusilla, Porzana 103, **107**, 111, *112*
pusilla, Porzana p. **108**
pusillus, Rallus 107

Q

quadricinctus, Pterocles **435**, *448*

R

raffertyi, Cursorius a. **214**
Rallicula 86, 87
Rallidae **84**, 185
Rallinae 85
ralloides, Ardeola 182, 233, 280
Rallus 102
  aquaticus **103**, *112*
  caerulescens **104**, *112*, 115
ranivorus, Circus 105, 106, 116
rapax, Aquila 131, 168
Recurvirostra **196**
  avosetta 194, **196**, 236, *240, 321*
Recurvirostridae **193**
regulorum, Anthropoides 143
regulorum, Balearica xii, **97**, 135, 139, 141, **143**
regulorum, Balearica r. **97**, **143**
reichenovi, Sarothrura e. **89**
reichenowi, Numida m. **9**, 96
reichenowi, Streptopelia xii, *464*, 479, **489**
reichenowi, Turtur 489
repressa, Sterna **392**, *401*, 404, *417*
Rheinartia 11
rhipidurus, Corvus 415
ridibundus, Larus **358**, 368, *416*
rikwae, Numida 9
risoria, Columba 487
Rissa 373
  tridactyla 351, 368, **373**, *416*
robusta, Tringa t. 316
roseogrisea, Columba 486
roseogrisea, Streptopelia *464*, 479, **486**
roseogrisea, Streptopelia r. **486**
Rostratula **185**
  benghalensis 177, **186**, *352*
Rostratulidae **185**
rougetii, Rallus 101
rougetii, Rougetius **101**, 105, *113*
Rougetius **101**
  rougetii **101**, 105, *113*
rovuma, Francolinus s. **33**, *43*
ruandae, Francolinus 29
rudis, Ceryle 110
rueppellii, Eupodotis 166, **170**, *176*
rueppellii, Eupodotis r. **170**, *176*
rueppellii, Otis 170
rufa, Alectoris 22
rufa, Sarothrura **90**, *113*
rufa, Sarothrura r. **90**, *113*
rufescens, Streptopelia t. 491
ruficollis, Calidris **287**, 305
ruficollis, Corvus 388
ruficollis, Trynga 287
ruficrista, Eupodotis **166**, *176*
ruficrista, Eupodotis r. **166**, *176*
ruficrista, Otis 166
rufipictus, Francolinus 27, **33**, 72
rufiventris, Accipiter 466
rufopictus, Pternistis 72
rufus, Cursorius **211**, *320*
rufus, Rallus 90
rupicoloides, Falco 428
rusticola, Scolopax 177, **308**, *352*
Rynchopidae **412**
Rynchops **412**
  flavirostris 258, *400*, **412**, *417*

S

sabini, Larus 334, 335, 337, **354**, 368, 373, *416*
sabyi, Numida m. **9**, 96
Sagittarius serpentarius 144, 276
saharae, Burhinus o. 177, **199**
sanctithomae, Treron 443, **445**, *465*
sandvicensis, Sterna 335, 375, **383**, *401*, *417*
sandvicensis, Sterna s. **384**, *401*
Sarothrura **87**
  affinis **94**, *113*
  ayresii **95**, *113*
  boehmi 17, 83, **93**, *113*
  elegans **88**, *113*
  lugens **92**, *113*

pulchra **87**, *113*
rufa **90**, *113*
saturatior, *Pterocles g.* **439**
saturatior, *Ptilopachus* 23
saundersi, *Sterna* 377, *401*, **402**, 411, *417*
saundersi, *Sterna s.* **402**
savilei, *Eupodotis r.* **167**
schalowi, *Treron c.* **444**
schimpera, *Columba l.* **477**
schinzii, *Calidris a.* **295**
schlegelii, *Francolinus* 25, **34**, *49*
schoutedeni, *Guttera* 5
schubotzi, *Francolinus l.* **28**
schubotzi, *Guttera p.* **4**
sclateri, *Guttera p.* **5**, *96*
Scleroptera 51
scolopaceus, *Limnodromus* 308
Scolopacidae 283
Scolopacinae 308
Scolopax **308**
  rusticola 177, **308**, *352*
Scopus umbretta 276
seebohmi, *Cursorius c.* **216**, *320*
semipalmatus, *Limnodromus* **307**
seimpalmatus, *Macrorhamphus* 307
semitorquata, *Columba* 480
semitorquata, *Streptopelia* **464**, 466, **480**
senator, *Lanius* 362
senegalensis, *Burhinus* 177, **201**, *321*, 360
senegalensis, *Burhinus s.* **202**
senegalensis, *Columba* 495
senegalensis, *Eupodotis* 154, 163, **173**, *176*
senegalensis, *Eupodotis s.* **173**, *176*
senegalensis, *Heliopais* 146
senegalensis, *Oedicnemus* 201
senegalensis, *Otis* 173
senegalensis, *Podica* **146**, *160*
senegalensis, *Podica s.* **146**, *160*
senegalensis, *Streptopelia* 452, **464**, **495**
senegalensis, *Streptopelia s.* **464**, **495**
senegalla, *Parra* 251
senegallus, *Pterocles* **430**, *448*
senegallus, *Tetrao* 430
senegallus, *Vanellus* 241, **251**, *321*
senegallus, *Vanellus s.* **241**, **252**
sephaena, *Francolinus* 23, **33**, **42**
sephaena, *Francolinus s.* **33**, **43**
sephaena, *Perdix* 42
sequestris, *Francolinus* 39
serpentarius, *Sagittarius* 144, 276
seth-smithi, *Guttera* 5
sharpei, *Columba d.* **461**
sharpei, *Cursorius a.* **214**
sharpei, *Treron c.* **444**
sharpii, *Francolinus* **32**, 57
shelleyi, *Francolinus* 25, **33**, **39**
shelleyi, *Francolinus s.* **33**, **39**
shelleyi, *Streptopelia d.* **481**
simplex, *Columba l.* **462**, *465*
sinensis, *Sterna a.* **399**
sjostedti, *Columba* 449, 458, **467**
skua, *Catharacta* 333, **338**, *353*
skua, *Catharacta s.* **338**
sokotrae, *Streptopelia s.* **495**
solitaneus, *Vanellus* 252
solitaria, *Tringa* **304**, **322**, *352*

somalica, *Streptopelia c.* **485**
somaliensis, *Cursorius c.* **210**
somaliensis, *Numida m.* **9**, *96*
somereni, *Podica s.* **147**
spatzi, *Alectoris b.* **22**
Spheniscus demersus 365, 387, 391
spilogaster, *Francolinus s.* **43**
spinetorum, *Francolinus c.* **30**
spinosus, *Charadrius* 264
spinosus, *Vanellus* 241, **264**, *321*
squamatus, *Vanellus* 241, **264**, *321*
squamatus, *Francolinus* 26, **44**, *48*, 85
squamatus, *Francolinus s.* **48**
squatarola, *Pluvialis* 244, **249**, *257*, *312*, *321*
squatarola, *Tringa* 249
stagnatilis, *Tringa* 265, 278, **304**, *317*, *329*, *352*
stagnatilis, *Totanus* 317
stanleyi, *Neotis d.* **151**
Steganopus 328
stenura, *Gallinago* 177, **306**
stenura, *Scolopax* 306
Stephanibyx 251
Stercorariidae 332
Stercorarius **332**, 408, 410, 411
  longicaudus **336**, *353*
  parasiticus **334**, *353*, *354*, *355*, 365, 389
  pomarinus **332**, *353*
Sterna **376**, 407
  albifrons 377, **398**, *401*, *417*
  anaethetus **393**, **400**, *417*
  balaenarum **396**, *401*, *417*
  bengalensis 346, 357, 375, **382**, *401*, *417*
  bergii 344, 355, **381**, *401*, *417*
  caspia 377, *401*, *417*
  dougallii **386**, *401*, 411, *417*
  fuscata 334, **395**, **400**, 408, 410, *417*
  hirundo 335, 337, 365, **388**, *401*, *406*, *417*
  maxima **379**, *401*, *417*
  paradisaea 335, 337, **390**, *401*, *417*
  repressa **392**, *401*, *404*, *417*
  sandvicensis 335, 375, **383**, *401*, *417*
  saundersi 377, *401*, **402**, 411, *417*
  sumatrana **385**, *401*, *417*
  vittata **391**, *401*, *417*
Sternidae 374
stieberi, *Ardeotis a.* **158**, *161*
Stiltia isabella 206, 218
stolida, *Sterna* 410
stolidus, *Anous* 395, **400**, **410**, *417*
stolidus, *Anous s.* **410**
Streptopelia **473**, *474*, *478*
  capicola xii, 258, **464**, 479, **484**
  decaocto 479, **487**
  decipiens **464**, **481**
  hypopyrrha **464**, 479, **492**
  lugens **464**, 479, **494**
  reichenowi xii, **464**, 479, **489**
  roseogrisea **464**, 479, **486**
  semitorquata **464**, 466, **480**
  senegalensis 452, **464**, **495**
  turtur **464**, **490**
  vinacea xii, **464**, 479, **483**

streptophorus, *Francolinus* 25, **35**, *49*
struthiunculus, *Ardeotis k.* *161*, **162**
subminuta, *Calidris* **290**, *305*
subminuta, *Tringa* 290
subruficollis, *Tringa* 297
subruficollis, *Tryngites* **297**, *305*, *352*
sukensis, *Pterocles l.* **435**
Sula bassana 361
  capensis 365
  dactylatra 382
sumatrana, *Sterna* **385**, *401*, *417*
superciliosus, *Centropus* 280
superciliosus, *Lobivanellus* 267
superciliosus, *Vanellus* 241, **267**, *321*
suscitator, *Turnix* 78
swainsonii, *Francolinus* 27, **33**, **68**
swainsonii, *Francolinus s.* **33**, **68**
swainsonii, *Perdix* 68
swierstrai, *Francolinus* 27, *48*, **60**
swynnertoni, *Francolinus a.* **71**
sylvatica, *Turnix* 18, **78**, *112*
sylvatica, *Turnix s.* **79**
sylvaticus, *Tetrao* 78
symonsi, *Guttera* 5
Sypheotides 149, 171
Syrrhaptes tibetanus 422

T

tachiro, *Accipiter* 463, 466
taimyrensis, *Larus a.* 370
tanki, *Turnix* 76
tarda, *Otis* 161, *163*, **164**
tarda, *Otis t.* *161*, **164**
targia, *Columba l.* **449**, *477*
targius, *Pterocles l.* **434**, *448*
tectus, *Charadrius* 260
tectus, *Vanellus* 241, **260**, *321*
tectus, *Vanellus t.* 241, **260**
Telophorus zeylonus 239
temminckii, *Calidris* **289**, *305*, *352*
temminckii, *Cursorius* **212**, *320*, *321*
temminckii, *Tringa* 289
tenuirostris, *Anous* **400**, **409**
tenuirostris, *Anous t.* **400**, **409**
tenuirostris, *Calidris* 283
tenuirostris, *Numenius* 240, **312**, *321*
tenuirostris, *Sterna* 409
tenuirostris, *Totanus* 283
Terathopius ecaudatus 258
Tetrao 14, 23, 54, 58, 70, 78, 422, 426, 430
Tetraoninae 1, 13
Tetrax **149**
  tetrax **149**, 156, *161*, 163
tetrax, *Otis* 149
tetrax, *Tetrax* **149**, 156, *161*, 163
tetrax, *Tetrax t.* **149**, *161*
thalassina, *Sterna b.* **381**
theresae, *Francolinus* 41
thikae, *Francolinus* 30
thomae, *Columba s.* 445
thomensis, *Columba* 449, 458, **467**
thomensis, *Columba a.* 467
thornei, *Francolinus* 54
Threskiornis aethiopica 280, 344, 356, 382, 393, 415
tibetana, *Sterna h.* **388**
tibetanus, *Syrrhaptes* 422
torda, *Alca* *353*, **418**
toruensis, *Numida* 8

Totanus 317
totanus, Scolopax 315
totanus, Tringa 304, **315**, *352*
totanus, Tringa t. **304**, **315**
transvaalensis, Numida 9
traylori, Cursorius a. **214**
Treron **442**
   australis 442
   calva **443**, *465*
   pembaensis 443, **446**, *465*
   phoenicoptera 442
   sanctithomae 443, **445**, *465*
   waalia **446**, *465*
Treroninae **442**
Tribonyx 121
tricollaris, Charadrius 225, **231**, 256, *258*, *352*
tricollaris, Charadrius t. **231**, *256*
tricolor, Phalaropus **328**
tricolor, Steganopus 328
tridactyla, Larus 373
tridactyla, Rissa 351, **368**, **373**, *416*
Tringa 307, **314**, 326
   erythropus 304, **314**, *352*
   flavipes **322**, *352*
   glareola 110, 233, 304, **323**, 329, *352*
   melanoleuca **319**
   nebularia 304, **318**, *352*
   ochropus, 304, **322**, *352*
   solitaria 304, **322**, *352*
   stagnatilis 265, 278, 304, **317**, 329, *352*
   totanus 304, **315**, *352*
Tringinae **309**
tristanesis, Sterna v. **391**
tropica, Streptopelia c. **485**
Tryngites **297**
   subruficollis **297**, **305**, *352*
tundrae, Charadrius h. **228**, *256*
Turnicidae **76**
Turnix 14, 77, **78**
   hottentotta **82**, *112*
   suscitator 78
   sylvatica 18, **78**, *112*
   tanki 76
Turtur **450**, 457
   abyssinicus 450, **454**, *465*
   afer **453**, *465*

brehmeri **450**, **462**, *465*
chalcospilos 450, **455**, *465*
tympanistria **451**, *465*
turtur, Columba 490
turtur, Streptopelia 464, **490**
turtur, Streptopelia t. **490**
tympanistria, Columba 451
tympanistria, Turtur **451**, *465*
typus, Polyboroides 223
Tyto alba 22

U

uellensis, Treron c. **444**
uhehensis, Numida 8
uluensis, Francolinus 39
umbretta, Scopus 276
undulata, Anas 105
undulata, Chlamydotis 150, 154, **156**, *161*, 163
undulata, Chlamydotis u. **156**, *161*
undulata, Psophia 156
unicincta, Columba 449, **469**
Uria **415**
   aalge 353, 409, **415**, 419
usambarae, Francolinus 45, **48**
ussuriensis, Tringa t. **316**

V

Vanellinae **250**
Vanellus xii, **250**
   albiceps 241, **255**, *321*
   armatus 207, 241, **262**, *321*
   coronatus 217, 241, **272**, *321*
   crassirostris 241, **279**, *321*
   duvaucelii 250
   gregarius 257, **276**, *321*
   leucurus 257, **277**, *321*
   lugubris 241, 251, **268**, *321*
   melanocephalus 241, **259**, *321*
   melanopterus 241, 251, **270**
   senegallus 241, **251**, *321*
   spinosus 241, **264**, *321*
   superciliosus 241, **267**, *321*
   tectus 241, **260**, *321*
   vanellus 241, **280**, *321*
vanellus, Tringa 280
vanellus, Vanellus 241, **280**, *321*

variegatus, Numenius p. **312**
velox, Sterna b. **381**, *401*
venustus, Charadrius p. **239**
vermiculatus, Burhinus 177, **203**, *321*
vermiculatus, Burhinus v. **177**, 203
vermiculatus, Oedicnemus 203
vernayi, Francolinus 29
verreauxi, Guttera p. **5**
vetula, Larus d. **364**, *369*
vigorsi, Otis 169
vigorsii, Eupodotis 166, **169**, *176*
vigorsii, Eupodotis v. **169**, *171*, *176*
vinacea, Columba 483
vinacea, Streptopelia xii, **464**, 479, **483**
virescens, Treron c. **444**
virgo, Anthropoides 97, 132, **137**
virgo, Ardea 137
vittata, Pachytila 339
vittata, Sterna **391**, *401*, *417*
vittata, Sterna v. **391**
vocifer, Haliaeetus 131, 258
vulturina, Numida 7
vulturinum, Acryllium **7**, 96
vylderi, Treron c. **444**

W

waalia, Treron **446**, *465*
wahlbergi, Aquila 81, 486
wakefieldii, Treron c. **444**
whitesidei, Himantornis 85
whytei, Francolinus s. **39**

X

Xenus **324**
   cinereus 304, **324**, *352*
xerophilus, Vanellus c. **273**

Z

zambesiae, Francolinus s. **43**
Zenaida asiatica xii
zenkeri, Sarothrura p. **87**
zeylonus, Telophorus 239

English Names

A

Auk, Little *353*, **419**
Avocet, Eurasian 194, **196**, 236, *240*, *321*

B

Bateleur 258
Bee-eater, Little Green 182
Bokmakierie 239
Booby, Masked 382
Bustard, Arabian 154, **158**, *161*
   Black 168
   Black-bellied **175**, *176*
   Blue 172
   Crested **166**, *176*
   Denham's **151**, 159, *161*, *162*

   Great *161*, 163, **164**
   Hartlaub's *176*, **179**
   Heuglin's **155**, *161*
   Houbara 156
   Kori **159**, *161*
   Little **149**, 156, *161*, 163
   Little Brown **171**, *176*
   Ludwig's **153**, *161*
   Nubian **154**, *161*
   Rüppell's 170
   Stanley 151
   Vigors' 169
   White-bellied 154, 163, **173**, *176*
Button-Quail, Black-rumped **82**, *112*
   Hottentot 82
   Kurrichane 78
   Little xv, 18, **78**, *112*

Buzzard, Red-necked 261

C

Chukar 21, *32*, 67
Coot xii
   Crested xii, 128
   Eurasian xii, **127**, *160*, 370
   Red-knobbed xii, **128**, *160*, 182, 404
Cormorant, Bank 365
   Cape 365, 397
   Great 365
   Long-tailed 357, 365, 394
Corncrake **100**, 111, *112*
Coucal, White-browed 280
Courser, Australian 206
   Bronze-winged **217**, *320*, *321*

Burchell's **211**, *320*
Cream-coloured **210**, *320*, *321*
Double-banded **213**, *320*, *321*
Heuglin's 216
Temminck's **212**, *320*, *321*
Three-banded **216**, *320*, *321*
Violet-tipped 217
Crab Plover **188**, *240*, *321*
Crake, African **98**, 109, 111, *112*
African Black 114
Baillon's 103, **107**, 111, *112*
Black 105, **114**, *160*, 280
Buff-spotted 88
Chestnut-tailed 94
Little 103, **106**, 111, *112*
Spotted 99, 103, **109**, 111, *112*, 125
Streaky-breasted Pygmy 93
Striped 93, 99, 107, 109, **110**, *112*
White-spotted 87
White-winged 95
Crane, Black Crowned xii, **97**, 137, **141**
Blue **97**, **138**, 275
Common **97**, **132**, 137
Demoiselle *97*, 132, **137**
Grey Crowned xii, **97**, 135, 139, **143**
Siberian 133
Stanley 138
Wattled **97**, **133**, 137, 139, 144
Crow, Carrion 471
Pied 209, 264, 276, 428
Curlew, Eurasian *240*, **313**, *321*, 350
Slender-billed *240*, **312**, *321*
Stone 199

D

Darter 147
Dikkop, Spotted 204
Water 203
Dotterel, Eurasian **246**, *257*, *352*
Red-capped xii
Dove, Adamawa Turtle **464**, **492**
African Collared *464*, **486**
African Mourning *464*, **481**
African White-winged xii, *464*, **489**
Black-billed Wood **454**, *465*
Blue-headed Wood **450**, **462**, *465*
Blue-spotted Wood **453**, *465*
Cape Turtle 484
Cinnamon 461
Dusky Turtle *464*, **494**
Emerald-spotted Wood **455**, *465*
Eurasian Collared 487
European Turtle *464*, **490**
Laughing 452, *464*, **495**
Lemon 452, **461**, *465*
Long-tailed 456
Namaqua **456**, *465*
Palm 495
Pink-headed 486
Red-billed Wood 453
Red-eyed *464*, 466, **480**
Ring-necked 258, *464*, **484**
Rock *449*, 477
Somali Stock *449*, **473**

Stock *449*, **471**
Tambourine **451**, *465*
Vinaceous *464*, **483**
White-winged xii
Dovekie 419
Dowitcher, Asiatic **307**
Long-billed 308
Short-billed 308
Duck, Yellow-billed 105
Dunlin **294**, 296, **305**, *352*

E

Eagle, African Fish **131**, 258
Martial 163
Tawny 131, 168
Wahlberg's 81
Egret, Cattle 120, 144
Little 360
Yellow-billed 120, 393
Egyptian Plover xii, **206**, 255, 265, *320*, *321*

F

Falcon, Barbary 432
Lanner 264, 276, 428, 432, 433
Peregrine 466, 471, 472, 478
Finfoot, African **146**, *160*
Peter's 146
Fiscal, Long-tailed 280
Flufftail, Böhm's 17, 83, **93**, *113*
Buff-spotted **88**, *113*
Chestnut-headed **92**, *113*
Red-chested **90**, *113*
Striped **94**, *113*
White-spotted **87**, *113*
White-winged **95**, *113*
Flycatcher, Pied 362
Francolin, Ahanta 45, **48**
Cameroon Mountain **48**, **61**
Cape *33*, **58**
Chestnut-naped *48*, **64**
Clapperton's *32*, **56**
Coqui **29**, *33*
Crested 23, *33*, **42**
Djibouti xii, *48*, **65**
Double-spurred 22, *32*, **54**
Erckel's *48*, **66**
Finsch's **38**, *49*
Grey 67
Grey-breasted *33*, 72
Grey striped **46**, *48*
Grey-wing **36**, *49*
Handsome *48*, **62**
Hartlaub's **49**, 50
Harwood's *49*, **57**
Heuglin's *32*, **55**
Hildebrandt's *49*, **51**
Jackson's *48*, **63**
Latham's Forest **28**, *48*
Moorland xiii, **40**, *48*
Nahan's **47**, *48*
Natal *33*, **52**
Orange River *33*, **41**
Red-billed *33*, **59**
Red-necked *33*, **70**
Red-wing **37**, *49*
Ring-necked **35**, *49*
Scaly **44**, *48*, 85
Schlegel's **34**, *49*

Shelley's *33*, **39**
Swainson's *33*, **68**
Swierstra's *48*, **60**
White-throated **31**, *32*
Yellow-necked *33*, **73**

G

Gallinule, Allen's **116**, 122, *160*
American Purple 118
Common 122
Purple xii, **118**, 119, 122, *160*
Gannet, Cape 365
Northern 361
Godwit, Bar-tailed *240*, 307, **310**, 312, *321*
Black-tailed *240*, **309**, *321*
Goose, African Pygmy 117
Spur-winged 135, 279
Goshawk, African 463
Northern 471, 472
Pale Chanting 168
Greenshank 318
Common *304*, **318**, *352*
Guillemot *353*, 409, **415**, 419
Guineafowl, Black **3**, 4, 6, 11, *96*
Crested 2, **5**, 7, **50**, *96*
Helmeted 6, 7, **8**, *96*
Plumed 3, **4**, *96*
Vulturine 7, *96*
White-breasted **2**, 6, *96*
Gull, Audouin's **361**, *368*, *416*
Black-headed **358**, *368*, *416*
Common 363
Common Black-headed 358
Dominican 364
Franklin's **350**, *368*, *416*
Glaucous *353*, **371**, *416*
Great Black-backed *353*, **372**, *416*
Great Black-headed **347**, *353*, *416*
Grey-headed 198, **356**, *369*, 380, 404, *416*
Grey-hooded 356
Hartlaub's 335, 339, **355**, *369*, 382, *416*
Hemprich's **344**, *369*, 382, 388, 394, 411, *416*
Herring **367**, *369*, *416*
Iceland 371
Kelp **364**, *369*, 379, 382, 388, 397, *416*
Laughing 349
Lesser Black-backed **366**, *369*, *416*
Little **351**, *368*, *416*
Mediterranean **348**, *368*, *416*
Mew **363**, *368*, 373, *416*
Ring-billed 362
Sabine's 334, 335, 337, **354**, *368*, 373, *416*
Slender-billed **359**, *368*, *416*
Sooty 344
White-eyed **345**, *369*, *416*

H

Hammerkop 276
Harrier, African Marsh 105, 106, 116
Marsh 130

INDEXES: ENGLISH NAMES

Hawk, African Harrier 223
Heron, Black-headed 100
  Grey 131, 378
  Squacco 182, 233, 280
Hornbill, Ground 70
Houbara 150, 154, **156**, *161*, 163

I

Ibis, Sacred 280, 344, 356, 382, 393, 415

J

Jacana, African 121, *177*, **181**, 184, 280
  Lesser *177*, **184**
Jaeger, Long-tailed 336
  Parasitic 334
  Pomarine 332
Junglefowl 19, 60, 69, 72

K

Kestrel, Greater 428
Kingfisher, Pied 110
Kite, Black 121, 209
  Red 471, 472
Kittiwake, Black-legged 351, *368*, **373**, *416*
Knot xii
  Great **283**
  Red xii, **284**, *305*, *352*
Korhaan, Black **168**, *176*
  Black-bellied 175
  Blue **172**, *176*
  Crested 166
  Karoo **169**, *176*
  Rüppell's **170**, *176*
  White-bellied 173

L

Lapwing, African Wattled *241*, **251**, *321*
  Black-headed *241*, **260**, *321*
  Blacksmith *241*, **262**, *321*
  Black-winged *241*, **270**, *321*
  Brown-chested *241*, **267**, *321*
  Crowned 217, *241*, **272**, *321*
  Lesser Black-winged *241*, **268**, *321*
  Long-toed *241*, **279**, *321*
  Northern *241*, **280**, *321*
  Sociable *257*, **276**, *321*
  Spot-breasted *241*, **259**, *321*
  Spur-winged *241*, **264**, *321*
  White-headed *241*, **255**, *321*
  White-tailed *257*, **277**, *321*
  White-winged 279
Lark Quail 77
Lily-trotter 181
  Lesser 184

M

Magpie 489
  Black 475
Marabou 267
Merlin 478
Moorhen, Common xv, 105, 117, 122, 147, *160*, 182
  Lesser **125**, *160*
Murre, Common 415

N

Noddy, Black *400*, **408**, *417*
  Brown 395, *400*, **410**, *417*
  Common 410
  Lesser *400*, **409**
  White-capped 408

O

Osprey 365
Owl, African Marsh 114
  Barn 22
  Long-eared 471
  Spotted Eagle 53
Oystercatcher, African Black **191**, 239, *240*, *321*, 365
  Eurasian **190**, *240*, *321*

P

Painted-Snipe, Greater *177*, **186**, *352*
Partridge, Barbary 20, **21**, *32*, 54
  Red-legged 22
  Rock 22
  Sand 20, *32*
  Stone **23**, *32*
Peacock, Congo **11**, *96*
Penguin, Jackass 365, 387, 391
Petrel, Atlantic 339
  Kerguelen 339
  Soft-plumaged 339
Phalarope, Grey *305*, **331**, *352*, 354
  Northern 329
  Red 331
  Red-necked *305*, **329**, *352*
  Wilson's **328**
Pheasant, Ring-necked 13
Pigeon, Afep *449*, **469**
  African Green *443*, *465*
  Bruce's Green **446**, *465*
  Cameroon Olive *449*, **467**
  Delegorgue's 460
  Eastern Bronze-naped *449*, **460**
  Feral 477
  Grey Wood *449*, **469**
  Olive *449*, **463**
  Pemba Green **446**, *465*
  Rameron 463
  São Tomé Bronze-naped *449*, **460**
  São Tomé Green **445**, *465*
  São Tomé Olive *449*, **467**
  Somaliland 473
  Speckled *449*, **473**, 492
  Speckled Rock 473
  Western Bronze-naped *449*, **459**
  White-collared *449*, **476**
  White-naped *449*, **468**
  Wood *449*, **470**, 492
Plover, Black-bellied 249
  Blackhead 260
  Blacksmith 262
  Black-winged 270
  Brown-chested Wattled 267
  Caspian **245**, *257*, **278**, *352*
  Chestnut-banded **239**, *256*, *352*
  Chestnut-banded Sand- 239
  Common Ringed 228
  Crab 188
  Crowned 272
  Egyptian 206
  Forbes' **234**, *256*, *352*
  Great Sand- **243**, *256*, 286, 293, *321*
  Greater Golden **248**, *257*, *321*
  Greater Sand- 243
  Grey 244, **249**, *257*, 312, *321*
  Kentish xii, **235**, *256*, 265, *352*
  Kittlitz's 229, *256*, 265, *352*
  Kittlitz's Sand- 229
  Lesser Golden **247**, *257*, *321*
  Lesser Sand- 242
  Little Ringed **226**, *256*, *352*
  Mongolian **242**, *256*, 286, 293, *321*
  Ringed **228**, *256*, 265, *352*
  Senegal 268
  Snowy xii, 235
  Sociable 276
  Spot-breasted 259
  Spur-winged 264
  Three-banded **231**, *256*, 258, *352*
  Treble-banded 231
  Wattled 251
  White-crowned 255
  White-fronted **237**, *256*, *352*, 397
  White-fronted Sand- 237
  White-tailed 277
Pratincole, Australian 206
  Black-winged **220**, *320*
  Collared **218**, *320*
  Common 218
  Grey xii, **223**, *258*, *320*
  Little xii
  Madagascar **221**, *320*
  Rock **222**, *320*
  White-collared 222
Prion, Broad-billed 339
Puffin, Atlantic *353*, **420**

Q

Quail, Blue **16**, *32*, 83, 93
  Common **14**, *20*, *32*, 79, 83
  Harlequin **18**, *32*, 79, 83
Quail Plover 77, *112*

R

Rail, African Water 104
  Grey-throated **86**, *113*
  Kaffir **104**, *112*, 115
  Nkulengu **84**, 86, *113*
  Rouget's **101**, 105, *113*
  Water **103**, *112*
Raven, Brown-necked 388
  Common 347, 411
Razorbill *353*, **418**
Redshank 315
  Common *304*, **315**, *352*
  Spotted *304*, **314**, *352*
Roller, Roufous-crowned 174
Rook, Cape 173
Ruff *257*, 293, **297**, 316, 318, 324, *352*

S

Sanderling 237, 244, **286**, *305*, 328, *352*
Sandgrouse, Black-bellied **429**, *448*
  Black-faced **433**, *448*
  Burchell's **440**, *448*
  Chestnut-bellied **424**, *448*
  Coronetted 432
  Crowned **432**, *448*
  Double-banded **437**, *448*
  Four-banded **435**, *448*
  Lichtenstein's **434**, *448*
  Namaqua **426**, *448*
  Pin-tailed **422**, *448*
  Spotted **430**, **440**, *448*
  Yellow-throated **438**, *448*
Sandpiper, Baird's **291**, *305*, *352*
  Broad-billed **295**, **296**, *305*, *352*
  Buff-breasted **297**, *305*, *352*
  Common 187, **289**, *304*, 323, 325, **326**, *352*
  Curlew 238, 244, **292**, *305*, 325, *352*
  Green *304*, **322**, *352*
  Marsh 265, 278, *304*, **317**, 329, *352*
  Pectoral **292**, *305*, *352*
  Purple **294**, *305*, *352*
  Solitary *304*, **322**, *352*
  Terek *304*, **324**, *352*
  White-rumped **290**
  Wood 110, 233, *304*, **323**, 329, *352*
Sand-Plover, Chestnut-banded 239
  Great **243**, 256, 286, **293**, *321*, 325
  Greater 243
  Kittlitz's 229
  Lesser 242
  White-fronted 237
Secretary Bird 144, 276
Shearwater, Cory's 339, 361
  Great 339
  Little 339
  Manx 348
Shrike, Woodchat 362

Skimmer, African 258, *400*, **412**, *417*
Skua, Arctic **334**, *353*, 354, 355, 365, 389
  Brown 338
  Great 333, **338**, *353*
  Long-tailed **336**, *353*
  McCormick's 339
  Pomarine **332**, *353*
  South Polar **339**, 353
Snipe, African 177, **302**, *352*
  Common 177, **300**, 324
  Great 177, **303**
  Jack 177, **299**, 301
  Pintail 177, **306**
Sparrowhawk, European 471
Spurfowl, Grey-breasted 72
  Red-necked 70
  Yellow-necked 73
Starling, Wattled 280
Stilt, Australian Banded 193
  Black-winged 193
  Common **193**, 197, *240*, 265, 278, *321*
Stint, Little 265, **288**, 296, *305*, *352*
  Long-toed **290**, *305*
  Red-necked 287
  Rufous-necked **287**, *305*
  Temminck's **289**, *305*, *352*
Stone Curlew 177, **199**, *321*
Stork, Abdim's 475
  Marabou 486
  Woolly-necked 134
  Yellow-billed 393
Swamphen, Purple xii, 116, **119**, 160

T

Tern, Antarctic **391**, *401*, *417*
  Arctic 335, 337, **390**, *401*, *417*
  Black *400*, **405**, *417*
  Black-naped **385**, *401*, *417*
  Bridled **393**, *400*, *417*
  Brown-winged 393
  Caspian **377**, *401*, *417*

  Common 335, 337, 365, **388**, *401*, 406, *417*
  Damara **396**, *401*, *417*
  Greater Crested 381
  Gull-billed 360, **375**, 378, 384, 396, *400*, *417*
  Lesser Crested 346, 357, 375, **382**, *401*, *417*
  Little **398**, *401*, *417*
  Roseate **386**, *401*, 411, *417*
  Royal **379**, *401*, *417*
  Sandwich 335, 375, **383**, *401*, *417*
  Saunders' *401*, **402**, 411, *417*
  Sooty 334, **395**, *400*, 408, 410, *417*
  Swift 344, 355, **381**, *401*, *417*
  Whiskered **392**, *400*, **403**, *417*
  White 374, 385
  White-cheeked **392**, *401*, **404**, *417*
  White-winged 406
  White-winged Black *400*, **406**, *417*
Thick-knee, Eurasian 199
  Senegal 177, **201**, *321*, 360
  Spotted 177, **204**, *321*
  Water 177, **203**, *321*
Tinkerbird, Golden-rumped 88
Turnstone xii, 327
  Ruddy xii, *257*, 286, **327**, *352*

V

Vulture, Cape 365

W

Whimbrel *240*, **311**, *321*
Woodcock, Eurasian 177, **308**, *352*
Woodpecker, Green-backed xiii
  Little Spotted xiii
  Speckle-breasted xii
  Uganda Spotted xii
Wood Pigeon *449*, **470**, 492
  Grey *449*, **469**
Yellowlegs, Greater **319**
  Lesser **322**, *352*

French Names

A

Avocette élégante **196**

B

Barge à queue noire **309**
  rousse **310**
Bargette de Terek **324**
Bécasse des bois **308**
Bécasseau de Baird **291**
  de Bonaparte **290**
  cocorli **292**
  à col roux **287**
  falcinelle **296**
  Grand maubéche **283**
  à longs doigts **290**
  maubèche **284**
  minute **288**

  rousset **297**
  sanderling **286**
  tachète **292**
  de Temminck **289**
  variable **294**
  violet **294**
Bécassine africaine **302**
  double **303**
  des marais **300**
  à queue peinte **306**
  sourde **299**
Bec-en-ciseaux d'Afrique **412**

C

Caille arlequine **18**
  des blés **14**
  bleue **16**
Chevalier aboyeur **318**

  arlequin **314**
  combattant **297**
  culblanc **322**
  gambette **315**
  Grand à pattes jaunes **319**
  guignetté **326**
  Petit à pattes jaunes **322**
  solitaire **322**
    stagnatile **317**
    sylvain **323**
  Courlis à bec grêle **312**
    cendré **313**
    corlieu **311**
  Courvite à ailes violettes **217**
    de Burchell **211**
    à double bande **213**
    isabelle **210**
    de Temminck **212**
    à triple collier **216**

D

Drome ardéole **188**

E

Échasse blanche **193**

F

Foulque à crête **128**
   macroule **127**
Francolin d'Ahanta **45**
   à ailes grises **36**
   d'Archer **41**
   à bandes grises **46**
   à bec jaune **55**
   à bec rouge **59**
   de Clapperton **56**
   à collier **35**
   coqui **29**
   à cou jaune **73**
   à cou roux **64**
   criard **58**
   à double éperon **54**
   écaillé **44**
   d'Erckel **66**
   de Finsch **38**
   à gorge blanche **31**
   à gorge rouge **70**
   de Hartlaub **50**
   de Harwood **57**
   de Hildebrandt **51**
   huppé **42**
   de Jackson **63**
   de Latham **28**
   de Levaillant **37**
   montagnard **40**
   du Mont Cameroun **61**
   de Nahan **47**
   du Natal **52**
   noble **62**
   à poitrine grise **72**
   de Schlegel **34**
   de Shelley **39**
   des Somalis **65**
   de Swainson **68**
   de Swierstra **60**

G

Gallinule africaine **125**
   poule d'eau **122**
Ganga bibande **437**
   de Burchell **440**
   cata **422**
   couronné **432**
   à face noire **433**
   à gorge jaune **438**
   de Lichtenstein **434**
   namaqua **426**
   quadribande **435**
   tacheté **430**
   unibande **429**
   à ventre brun **424**
Glaréole à ailes noires **220**
   aureolée **222**
   à collier **218**
   grise **223**
   malgache **221**
Goéland argenté **367**
   atricille **349**
   de Audouin **361**
   à bec cerclé **362**
   bourgmestre **371**
   brun **366**
   cendré **363**
   dominicain **364**
   d'Hemprich **344**
   ichthyaète **347**
   à iris blanc **345**
   marin **372**
   railleur **359**
Gravelot, Grand **228**
   Petit **226**
Grébifoulque du Sénégal **146**
Grue caronculée **133**
   cendrée **132**
   couronnée **141**
   demoiselle **137**
   de paradis **138**
   royal **143**
Guifette leucoptère **406**
   moustac **403**
   noire **405**
Guillemot de Troïl **415**

H

Huîtrier de Moquin **191**
   pie **190**

J

Jacana nain **184**
   à poitrine dorée **181**

L

Labbe, Grand **338**
   à longue queue **336**
   parasite **334**
   pomarine **332**
Limnodrome simipalmé **307**

M

Macareux moine **420**
Marouette de Baillon **107**
   noire **114**
   ponctuée **109**
   poussin **106**
   rayée **110**
Mergule nain **419**
Mouette de Franklin **350**
   de Hartlaub **355**
   mélanocéphale **348**
   pygmée **351**
   rieuse **358**
   de Sabine **354**
   à tête grise **356**
   tridactyle **373**

N

Noddi brun **410**
   marianne **409**
   noir **408**

O

Oedicnème criard **199**
   du Sénégal **201**
   du tachard **204**
   vermiculé **203**
Outarde arabe **158**
   canepetière **149**
   de Denham **151**
   Grande **164**
   de Hartlaub **179**
   de Heuglin **155**
   houbara **156**
   houpette **166**
   korhaan **168**
   kori **159**
   de Ludwig **153**
   nubienne **154**
   plombée **172**
   de Rüppell **170**
   du Sénégal **173**
   somalienne **171**
   à ventre noir **175**
   de Vigors **169**

P

Paon du Congo **11**
Perdrix gambra **21**
   de Hay **20**
Phalarope à bec étroit **329**
   à bec large **331**
   de Wilson **328**
Pigeon biset **477**
   du Cameroun **467**
   à collier blanc **476**
   colombin **471**
   de Delegorgue **460**
   gris **469**
   de Guinée **473**
   de Malherbe **460**
   à nuque blanche **468**
   à nuque bronzée **459**
   rameron **463**
   ramier **470**
   de São Tomé **467**
   de Somalie **473**
   vert à front nu **443**
   vert de Pemba **446**
   vert de São Tomé **445**
   vert waalia **446**
Pingouin, Petit **418**
Pintade noire **3**
   plumifère **4**
   à poitrine blanche **2**
   de Pucheran **5**
   sauvage **8**
   vulturine **7**
Pluvian d'Egypte **206**
Pluvier argenté **249**
   asiatique **245**
   à collier interrompu **235**
   doré **248**
   élégant **239**
   fauve **247**
   de Forbes **234**
   à front blanc **237**
   guignard **246**
   de Leschenault **243**
   mongol **242**
   pâtre **229**
   à triple collier **231**
Poule de rocher **23**

R

Râle bleuâtre 104
  de Böhm 93
  à camail 90
  d'eau 103
  des genets 100
  à gorge grise 86
  à miroir 95
  perlé 87
  à pieds rouges 84
  ponctué 88
  des prés 98
  à queue rousse 94
  de Rouget 101
  à tête rousse 92
Rhynchée peinte 186

S

Skua de McCormick 339
Sterne arctique 390
  des baleiniers 396
  bridée 393
  caspienne 377
  caugek 383
  couronnée 391
  diamante 385
  de Dougall 386
  fuligineuse 395
  hansel 375
  huppée 381
  à joues blanches 392
  naine 398
  pierregarin 388
  royal 379
  de Saunders 402
  voyageuse 382

T

Talève d'Allen 116
  poulesultane 119
  pourprée 118
Tournepierre à collier 327
Tourtelette d'Abyssinie 454
  améthystine 453
  demoiselle 450
  à masque de fer 456
  tambourette 451
Tourterelle à ailes blanches 489
  des bois 490
  du Cap 484
  cendrée 494
  à collier 480
  émeraudine 455
  maillée 495
  à masque blanc 461
  pleureuse 481
  à poitrine rose 492
  rieuse 486
  turque 487
  vineuse 483
Turnix d'Afrique 78
  de Meiffren 77
  nain 82

V

Vanneau d'Abyssinie 259
  à ailes blanches 279
  à ailes noires 270
  armé 262
  caronculé 267
  coiffé 260
  couronné 272
  demi-deuil 268
  éperonné 264
  huppé 280
  à queue blanche 277
  du Sénégal 251
  sociable 276
  à tête blanche 255